Remote Sensing Handbook, Volume III (Six Volume Set)

Volume III of the Six Volume *Remote Sensing Handbook*, Second Edition, is focused on agriculture; food security; vegetation; phenology; rangelands; soils; and global biomass modeling, mapping, and monitoring using multi-sensor remote sensing. It discusses the application of remote sensing in agriculture systems analysis, phenology, cropland mapping and modeling, terrestrial vegetation studies, physically based models, food and water security, precision farming, crop residues, global view of rangelands, and soils. This thoroughly revised and updated volume draws on the expertise of a diverse array of leading international authorities in remote sensing and provides an essential resource for researchers at all levels interested in using remote sensing. It integrates discussions of remote sensing principles, data, methods, development, applications, and scientific and social context.

FEATURES

- Provides the most up-to-date comprehensive coverage of remote sensing science in agriculture, vegetation, and soil studies.
- Discusses and analyzes data from old and new generations of satellites and sensors spread across 60 years.
- Provides comprehensive assessment of modeling, mapping, and monitoring agricultural crops, vegetation, and soils from a wide array of sensors, methods, and techniques.
- Includes numerous case studies on advances and applications at local, regional, and global scales.
- Introduces advanced methods in remote sensing such as machine learning, cloud computing, and AI.
- Highlights scientific achievements over the last decade and provides guidance for future developments.

This volume is an excellent resource for the entire remote sensing and GIS community. Academics, researchers, undergraduate and graduate students, as well as practitioners, decision makers, and policymakers, will benefit from the expertise of the professionals featured in this book, and their extensive knowledge of new and emerging trends.

Contents

Foreword

Satellite remote sensing has progressed tremendously since the first Landsat was launched on June 23, 1972. Since the 1970s, satellite remote sensing and associated airborne and in situ measurements have resulted in geophysical observations for understanding our planet through time. These observations have also led to improvements in numerical simulation models of the coupled atmosphere-land-ocean systems at increasing accuracies and predictive capabilities. This was made possible by data assimilation of satellite geophysical variables into simulation models, to update model variables with more current information. The same observations document the Earth's climate and have driven consensus that *Homo sapiens* are changing our climate through greenhouse gas emissions.

These accomplishments are the work of many scientists from a host of countries and a dedicated cadre of engineers who build and operate the instruments and satellites that collect geophysical observation data from satellites, all working toward the goal of improving our understanding of the Earth. This edition of *Remote Sensing Handbook* (Second Edition, Volumes I–VI) is a compendium of information for many research areas of the Earth system that have contributed to our substantial progress since the 1970s. The remote sensing community is now using multiple sources of satellite and in situ data to advance our studies of planet Earth. In the following paragraphs, I will illustrate how valuable and pivotal satellite remote sensing has been in climate system study since the 1970s. The chapters in *Remote Sensing Handbook* provide other specific studies on land, water, and other applications using Earth observation data of the past 60+ years.

The Landsat system of Earth-observing satellites led the way in pioneering sustained observations of our planet. From 1972 to the present, at least one and frequently two Landsat satellites have been in operation (Wulder et al. 2022; Irons et al. 2012). Starting with the launch of the first NOAA-NASA Polar Orbiting Environmental Satellites NOAA-6 in 1978, improved imaging of land, clouds, and oceans and atmospheric soundings of temperature were accomplished. The NOAA system of polar-orbiting meteorological satellites has continued uninterrupted since that time, providing vital observations for numerical weather prediction. These same satellites are also responsible for the remarkable records of sea surface temperature and land vegetation index from the Advanced Very High-Resolution Radiometers (AVHRR) that now span more than 46 years as of 2024, although no one anticipated valuable climate records from these instruments before the launch of NOAA-6 in 1978 (Cracknell 2001). AVHRR instruments are expected to remain in operation on the European MetOps satellites into 2026 and possibly beyond.

The successes of data from the AVHRR led to the MODerate resolution Imaging Spectrometer (MODIS) instruments on NASA's Earth Observing System of satellite platforms that improved substantially upon the AVHRR. The first of the EOS platforms, Terra, was launched in 2000, and the second of these platforms, Aqua, was launched in 2002. Both of these platforms are nearing their operational end-of-life and many of the climate data records from MODIS will be continued with the Visible Infrared Imaging Suite (VIIRS) instrument on the Joint Polar Satellite System (JPSS) meteorological satellites of NOAA. The first of these missions, the NPOES Preparation Project, was launched in 2012 with the first VIIRS instrument that is operating currently along with similar instruments on JPSS-1 (launched in 2017) and JPSS-2 (launched in 2022). However, unlike the morning/afternoon overpasses of MODIS, the VIIRS instruments are all in an afternoon overpass orbit. One of the strengths of the MODIS observations was morning and afternoon data from identical instruments.

Continuity of observations is crucial for advancing our understanding of the Earth's climate system. Many scientists feel the crucial climate observations provided by remote sensing satellites are among the most important satellite measurements because they contribute to documenting the current state of our climate and how it is evolving. These key satellite observations of our climate are second in importance only to the polar orbiting and geostationary satellites needed for numerical weather prediction that provide natural disaster alerts.

The current state of the art for remote sensing is to combine different satellite observations in a complementary fashion for what is being studied. Climate study is an example of using disparate observations from multiple satellites coupled with in situ data to determine if climate change is occurring and where it is occurring, and to identify the various component processes responsible.

1. **Planet warming quantified by satellite radar altimetry**. Remotely sensed climate observations provide the data to understand our planet and to identify what forces influence our climate. The primary sea level climate observations come from radar altimetry that started in late 1992 with TOPEX-Poseidon and has been continued by Jason-1, Jason-2, Jason-3, and Sentinel-6 to provide an uninterrupted record of global sea level. Changes in global sea level provide unequivocal evidence that our planet is warming, cooling, or staying at the same temperature. Radar altimetry from 1992 to date has shown global sea level increases of ~3.5 mm/y; hence our planet is warming (Figure 0.1). Sea level rise has two components, ocean thermal expansion and ice melt from the ice sheets of Greenland and Antarctica, and to a lesser extent for glacier concentrations in places like the Gulf of Alaska and Patagonia. The combination of GRACE and GRACE Follow-On gravity measurements quantifies the ice mass losses of Greenland and Antarctica to a high degree of accuracy. Combining the gravity data with the flotilla of almost 4,000 Argo floats provides the temperature data with the depth necessary to quantify ocean temperatures and isolate the thermal component of sea level rise.
2. **Our Sun is remarkedly stable in total solar irradiance**. Observations of total solar irradiance have been made from satellites since 1979 and show total solar irradiance has varied only ±1 part in 500 over the past 35 years, establishing that our Sun is not to blame for global warming (Figure 0.2).
3. **Determining ice sheet contributions to sea level rise**. Since 2002 gravity observations from the Gravity Recovery and Climate Experiment Satellite, or GRACE, mission and the

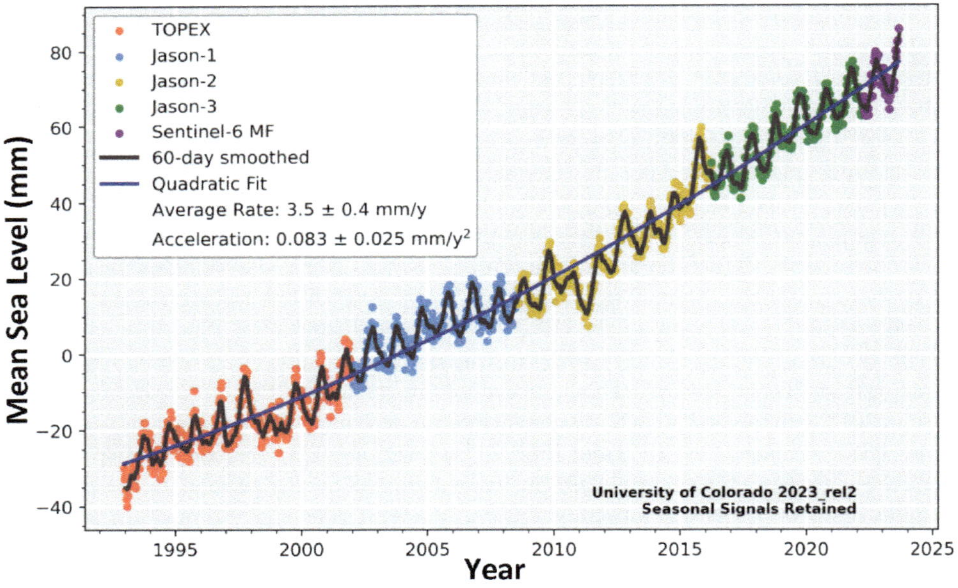

FIGURE 0.1 Seasonal sea level from five satellite radar altimeters from later 1992 to the present. Sea level is the unequivocal indicator of the Earth's climate—when sea level rises, the planet is warming; when sea level falls, the planet is cooling. (Nerem et al. 2018, updated to 2023; https://sealevel.colorado.edu/data/total-sea-level-change.)

GRACE Follow-On mission have been measured. GRACE data quantify ice mass changes from the Antarctic and Greenland ice sheets that constitute 98% of the ice mass on land (Luthcke et al. 2013). GRACE data are truly remarkable—their retrieval of variations in the Earth's gravity field is quantitatively and directly linked to mass variations. With GRACE data we are able for the first time to determine the mass balance with time of the Antarctic and Greenland ice sheets and concentrations of glaciers on land. GRACE data show sea level rise is 60% explained by ice sheet mass loss (Figure 0.3). GRACE data have many other uses, such as changes in groundwater storage. See www.csr.utexas.edu/grace/.

4. **Forty percent sea level rise explained by thermal expansion in the planet's oceans measured by in situ ~ 3,700 Argo drifting floats**. The other contributor to sea level rise is the thermal expansion or "steric" component of our planet's oceans. To document this necessitates using diving and drifting floats or buoys in the Argo network to record temperature with depth (Romerich et al. 2009 and Figure 0.4). Argo floats are deployed from ships; they then submerge and descend slowly to 1,000 m depth, recording temperature,

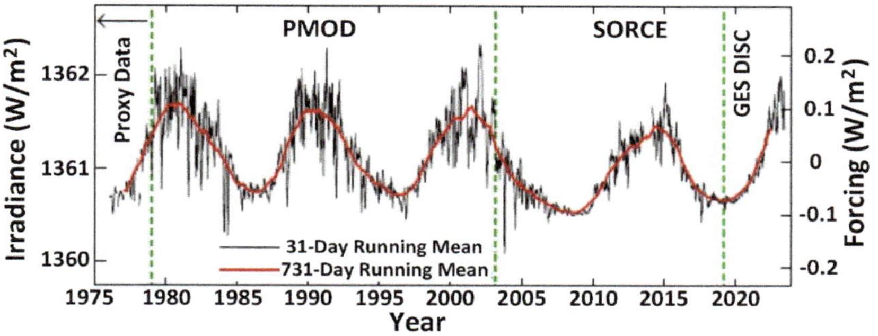

FIGURE 0.2 The Sun is not to blame for global warming, based on total solar irradiance observations from satellites. The few Watts/m^2 solar irradiance variations covary with the sunspot cycle. The luminosity of the Sun varies 0.2% over the course of the 11-year solar and sunspot cycle. The SORCE TSI dataset continues these important observations with improved accuracy on the order of ±0.035. (Kopp et al., 2024, and from https://lasp.colorado.edu/sorce/data/tsi-data/.)

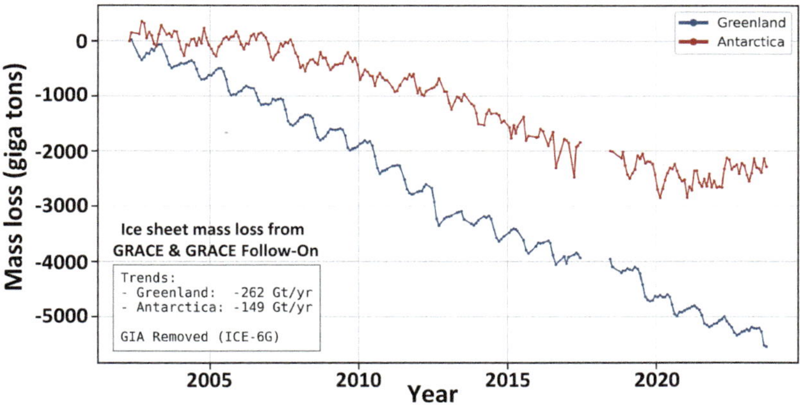

FIGURE 0.3 Sixty percent sea level rise is explained by mass balance of melting of ice measured by GRACE and GRACE Follow-On satellites. Ice mass variations are from 2003 to 2023 for the Antarctica and the Greenland ice sheets using gravity data (Croteau et al. 2021, updated to 2023). The Antarctic and Greenland Ice Sheets constitute 98% of the Earth's land ice.

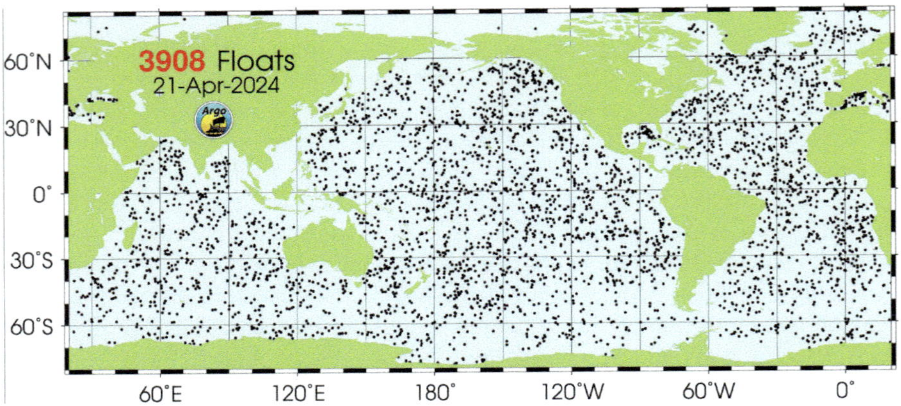

FIGURE 0.4 Forty percent sea level rise explained by thermal expansion in the planet's oceans measured in situ by ~3,908 drifting floats that were in operation on April 21, 2024. These floats provide the data needed to document thermal expansion of the oceans. (Roemmich & The Argo Float Team 2009, updated to 2024 and www.argo.ucsd.edu/.)

pressure, and salinity as they descend. At 1,000 m depth, they drift for ten days, continuing their measurements of temperature and salinity. After ten days, they slowly descend to 3,000 m and then ascend to the surface, all the time recording their measurements. At the surface, each float transmits all the data collected on the most recent excursion to a geostationary satellite and then descends again to repeat this process.

Argo temperature data show that 40% of sea level rise results from warming and thermal expansion of our oceans. Combining radar altimeter data, GRACE and GRACE Follow-On data, and Argo data provide confirmation of sea level rise and show what is responsible for it and in what proportions. With total solar irradiance being near constant, what is driving global warming can be determined. Analysis of surface in situ air temperature coupled with lower tropospheric air temperature and stratospheric temperature data from remote sensing infrared and microwave sounders show the surface and near-surface is warming while the stratosphere is cooling. This is an unequivocal confirmation that greenhouse gases are warming the planet.

Combining sea level radar altimetry, GRACE and GRACE Follow On gravity data to quantify ice sheet mass loses, and Argo floats to measure ocean temperatures with depth enables reconciliation of sea level increases with mass loss of ice sheets and ocean thermal expansion. The ice and steric expansion explain 95% of sea level rise (Figure 0.5).

5. **The global carbon cycles**. Many scientists are actively working to study the Earth's carbon cycle and there are several chapters in this *Remote Sensing Handbook* (Volumes I–VI) on various components under study (Figure 0.6).

Carbon cycles through reservoirs on the Earth's surface in plants and soils, exists in the atmosphere as gases such as carbon dioxide (CO_2), and exists in ocean water in phytoplankton and in marine sediments. CO_2 is released to the atmosphere from the combustion of fossil fuels, by land cover changes on the Earth's surface, by respiration of green plants, and by decomposition of carbon in dead vegetation and in soils, including carbon in permafrost.

Land gross primary production has been a MODIS product that is extended into the VIIRS era (Running et al. 2004; Román et al. 2024). MODIS data also provide burned area and CO2 emissions from wildfire (Giglio et al. 2016). Oceanic gross primary production will be provided by the Plankton, Aerosol, Cloud, and ocean Ecosystem, or PACE, satellite that was launched in early 2024

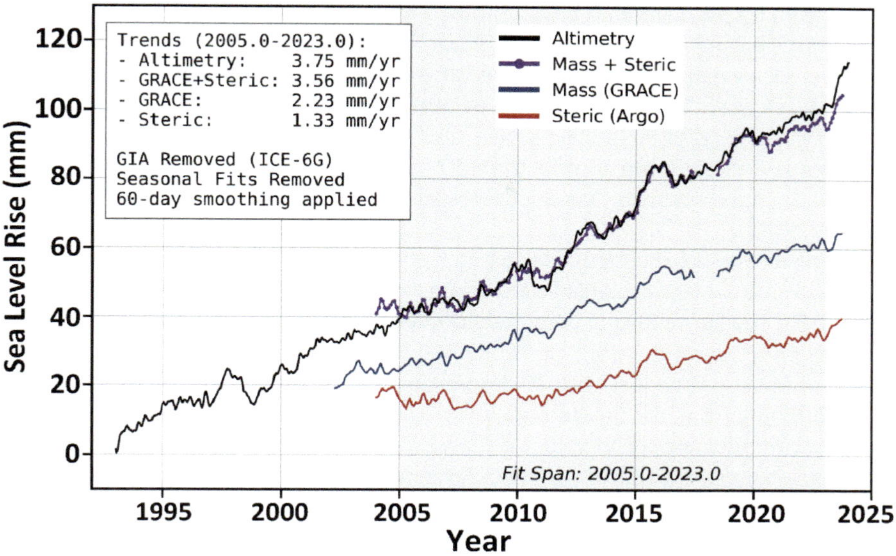

FIGURE 0.5 Sea level rise with the gravity tide mass loss and the Argo thermal expansion quantities added to the plot of global mean sea level. The GRACE and GRACE Follow On ice sheet gravity term and Argo thermal expansion term together explain 95% of sea level rise. (Croteau et al. 2021, updated to 2023.)

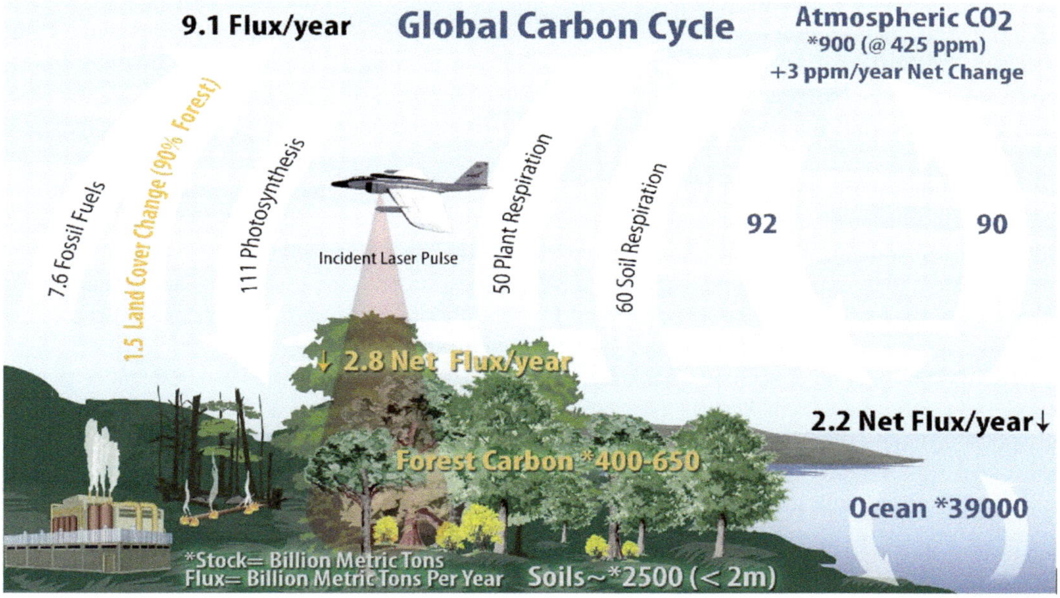

FIGURE 0.6 Global carbon cycle measurements from a multitude of satellite sensors. A representation of the global carbon cycle showing our best estimates of carbon fluxes and carbon reservoirs as of 2024. A series of satellite observations are needed simultaneously to understand the carbon cycle and its role in the Earth's climate system (Ciais et al. 2014, updated to 2023). The major unknowns in the global carbon cycle are fluxes between different reservoirs, oceanic gross primary production, carbon in soils, and the carbon in woody vegetation.

(Gorman et al. 2019). This complements the GPP land portion of the carbon cycle and will enable global gross primary production to be determined by MODIS-VIIRS and PACE.

Furthermore, Harmonized Landsat-8, Landsat-9, and Sentinel-2 30 m data (HLS) provide multispectral time series data at 30 m with a revisit frequency of three days at the equator (Crawford et al. 2023; Masek et al. 2018). This will enable time series improvements in spatial detail to 30 m from the 250 m scale of MODIS. The revisit time of Sentinel-2 with 10 m data is five days at the equator, which is a major improvement from 30 m. Multispectral time series observations are the basis for providing gross primary production estimates on land that are also used for food security (Claverie et al. 2018).

Refinements in satellite multispectral spatial resolution to the 50 cm to 3–4 m scale provided by commercial satellite data have enabled tree carbon to be determined from large areas of trees outside of forests. NASA has started using commercial satellite data to complement MODIS, Landsat, and other observations. One of the uses for Planet 3–4 m and Maxar <1 m data has been for mapping trees outside of forests (Brandt et al. 2020, Reiner et al. 2023, Tucker et al. 2023). Tucker et al. (2023) mapped ten billion trees at the 50-cm scale over 10 million km^2 and converted them into carbon at the tree level with allometry. The value of Planet and Maxar (formerly Digital Globe) data allows carbon studies to be extended into areas with discrete trees and Huang et al. (2024) has successfully mapped one tree species across the entire Sahelian and Sudanian Zones of Africa.

The height of trees is an important measurement to determine their carbon content. For areas of contiguous tree crowns, GEDI and ICESat laser altimetry (Magruder et al. 2024) coupled with Landsat and Sentinel-2 observations, enable improved estimates of carbon in these forests (Claverie et al. 2018).

The key to closing several uncertainties in the carbon cycle is to quantify fluxes among the various components. Passive CO_2 retrieval methods from the Greenhouse gases Observing SATellite (GOSAT) (Noël et al. 2021) and the Orbiting Carbon Observatory-2 (OCO-2) (Jacobs et al. 2024) are inadequate to provide this. Passive methods are not possible at night, in all seasons, and require specific Sun-target-sensor viewing perspectives and conditions. A recent development of the Aerosol and Carbon dioxide Detection Lidar (ACDL) instrument (Dai et al. 2023) by our Chinese colleagues offers a tenfold coverage improvement in CO_2 retrievals over those provided by OCO-2, and 20-fold coverage improvement over GOSAT. The reported uncertainty of ACDL is on the order of ±0.6 ppm.

Understanding the carbon cycle requires a "full court press" of satellite and in situ observations, because all of these observations must be made at the same time. Many of these measurements have been made over the past 30 to 40 years but new measurements are needed to quantify carbon storage in vegetation, to quantify CO_2 fluxes, to quantify land respiration, and to improve numerical carbon models. Similar work needs to be performed for the role of clouds and aerosols in climate and to improve our understanding of the global hydrological cycle.

The remote sensing community has made tremendous progress over the last six decades as captured in various chapters of *Remote Sensing Handbook* (Second Edition, Volumes I–VI). Handbook chapters provide comprehensive understanding of land and water studies through detailed methods, approaches, algorithms, synthesis, and key references. Every type of remote sensing data obtained from systems such as optical, radar, LiDAR, hyperspectral, and hyperspatial are presented and discussed in different chapters. Chapters in this volume address remote sensing data characteristics, within and between sensor calibrations, classification methods, and accuracies by taking data from over last five decades from a wide array of remote sensing sensors and platforms. Volume I also brings in new remote sensing technologies such as radio occultation and reflectometry from the global navigation satellite system, or GPS, satellites, crowdsourcing, drones, cloud computing, artificial intelligence, machine learning, hyperspectral, radar, and remote sensing law. The chapters in *Remote Sensing Handbook* are written by leading remote sensing scientists of the world and ably edited by Dr. Prasad S. Thenkabail, senior scientist (ST) at U.S. Geological Survey (USGS) in Flagstaff, Arizona. The importance and the value of *Remote Sensing Handbook* is clearly demonstrated by the need for a second edition. The *Remote Sensing Handbook* (First Edition, Volumes I–III) was published in 2014, and now after ten years, *Remote Sensing Handbook* (Second Edition, Volumes I–VI) with 91 chapters and nearly 3,500 pages will be published. It is certainly monumental work in remote sensing science, and for this I want to compliment Dr. Prasad S. Thenkabail. Remote sensing is now important to many scientific disciplines beyond our community, and I recommend *Remote Sensing Handbook* (Second Edition, six volumes) to not only remote sensers but also to the entire scientific community.

We can look forward in the coming decades to improving our quantitative understanding of the global carbon cycle, understanding the interaction of clouds and aerosols in our radiation budget, and understanding the global hydrological cycle.

by Compton J. Tucker
Satellite Remote Sensing Beyond 2025
NASA/Goddard Space Flight Center
Earth Science Division
Greenbelt, Maryland 20771 USA

REFERENCES

Brandt, M., Tucker, C.J., Kariryaa, A., et al. 2020. An unexpectedly large count of trees in the West African Sahara and Sahel. *Nature* 587:78–82. doi: 10.1038/s41586-020-2824-5.

Ciais, P., et al. 2014. Current systematic carbon-cycle observations and the need for implementing a policy-relevant carbon observing system. *Biogeosciences* 11(13):3547–3602.

Claverie, M., Ju, J., Masek, J.G., Dungan, J.L., Vermote, E.F., Roger, J.-C., Skakun, S.V., et al. 2018. The Harmonized Landsat and Sentinel-2 surface reflectance data set. *Remote Sensing of Environment* 219:145–161. doi: 10.1016/j.rse.2018.09.002.

Cracknell, A. 2001. The exciting and totally unanticipated success of the AVHRR in applications for which it was never intended. *Advances in Space Research* 28:233–240. doi: 10.1016/S0273-1177(01)00349-0.

Crawford, C.J., Roy, D.P., Arab, S., Barnes, C., Vermote, E., Hulley, G., et al. 2023. The 50-year Landsat collection 2 archive. *Science of Remote Sensing* 8:100103. ISSN 2666-0172. doi: 10.1016/j.srs.2023.100103. (www.sciencedirect.com/science/article/pii/S2666017223000287).

Croteau, M.J., Sabaka, T.J., and Loomis, B.D. 2021. GRACE fast mascons from spherical harmonics and a regularization design trade study. *Journal of Geophysical Research:Solid Earth* 126:e2021JB022113. doi: 10.1029/2021JB022113.10.1029/2021JB022113.

Dai, G., Wu, S., Sun, K., Long, W., Liu, J., and Chen, W. 2023. Aerosol and Carbon dioxide Detection Lidar (ACDL) overview. Presentation at ESA-JAXA EarthCare Workshop, November.

Giglio, L., Schroeder, W., and Justice, C.O. 2016. The collection 6 MODIS active fire detection algorithm and fire products. *Remote Sensing of Environment* 178:31–41. doi: 10.1016/j.rse.2016.02.054.

Gorman, E.T., Kubalak, D.A., Patel, D., Dress, A., Mott, D.B., Meister, G., and Werdell, P.J. 2019. The NASA Plankton, Aerosol, Cloud, ocean Ecosystem (PACE) mission: An emerging era of global, hyperspectral Earth system remote sensing. *Sensors, Systems, and Next Generation Satellites* 23:11151. doi: 10.1117/12.2537146.

Huang, K., et al. 2024. Mapping every adult baobab (Adansonia digitata L.) across the Sahel to uncover the co-existence with rural livelihoods. *Nature Ecology and Evolution* doi: 10.21203/rs.3.rs-3243009/v1.

Irons, J.R., Dwyer, J.L., and Barsi, J.A. 2012. The next Landsat satellite: The Landsat Data Continuity Mission. *Remote Sensing of Environment* 122:11–21. doi: 10.1016/j.rse.2011.08.026.

Jacobs, N., et al. 2024. The importance of digital elevation model accuracy in X_{CO2} retrievals: Improving the Orbiting Carbon Observatory-2 Atmospheric Carbon Observations from Space version 11 retrieval product. *Atmospheric Measurement Techniques* 17(5):1375–1401. doi: 10.5194/amt-17-1375-2024.

Kopp, G., Nèmec, N.E., and Shapiro, A. 2024. Correlations between total and spectral solar irradiance variations. *Astrophysical Journal* 964(1). doi: 10.3847/1538-4357/ad24e5.

Magruder, L.A., Farrell, S.L., Neuenschwander, A., Duncanson, L., Csatho, B., Kacimi, S., et al. 2024. Monitoring Earth's climate variables with satellite laser altimetry. *Nature Reviews of Earth and Environment* 5(2):120–136. doi: 10.1038/s43017-023-00508-8.

Masek, J., Ju, J., Roger, J.-C., Skakun, S., Claverie, M., and Dungan, J. 2018. Harmonized Landsat/Sentinel-2 products for land monitoring, IGARSS 2018-2018. *IEEE International Geoscience and Remote Sensing Symposium*, Valencia, Spain, pp. 8163–8165. doi: 10.1109/IGARSS.2018.8517760.

Nerem, R.S, Beckley, B.D., Fasullo, J.T., and Mitchum, G.T. 2018. Climate-change—driven accelerated sea-level rise detected in the altimeter era. *Proceeding of the National Academy of Sciences* 115(9):2022–2025. doi: 10.1073/pnas.1717312115.

Noël, S., et al. 2021. XCO_2 retrieval for GOSAT and GOSAT-2 based on the FOCAL algorithm. *Atmospheric Measurement Techniques* 14(5):3837–3869. doi: 10.5194/amt-14-3837-2021.

Reiner, F., et al. 2023. More than one quarter of Africa's tree cover is found outside areas previously classified as forest. *Nature Communications* doi: 10.1038/s41467-023-37880-4.

Roemmich, D., and The Argo Steering Team. 2009. Argo—the challenge of continuing 10 years of progress. *Oceanography* 22(3):46–55.

Román, M., et al. 2024. Continuity between NASA MODIS Collection 6.1 and VIIRS Collection 2 land products. *Remote Sensing of Environment* 302. doi: 10.1016/j.rse.2023.113963.

Running, S.W., Nemani, R.R., Heinsch, F.A., Zhao, M.S., Reeves, M., and Hashimoto, H. 2004. A continuous satellite-derived measure of global terrestrial primary production. *Bioscience* 54(6):547–560. doi: 10.1641/0006-3568.

Tucker, C., Brandt, M., Hiernaux, P., Kariryaa, A., et al. 2023. Sub-continental-scale carbon stocks of individual trees in African drylands. *Nature* 615:80–86. doi: 10.1038/s41586-022-05653-6.

Wulder, M.A., Roy, D.P., Radeloff, V.C., Loveland, T.R., Anderson, M.C., Johnson, D.M., et al. 2022. Fifty years of Landsat science and impacts. *Remote Sensing of Environment* 280:113195. ISSN 0034-4257. doi: 10.1016/j.rse.2022.113195. (www.sciencedirect.com/science/article/pii/S0034425722003054).

Preface

The overarching goal of this six-volume, 91-chapter, about 3,500-page *Remote Sensing Handbook* (Second Edition, Volumes I–VI) was to capture and provide the most comprehensive state of the art of remote sensing science and technology development and advancement in the last 60+ years, by clearly demonstrating the (1) scientific advances, (2) methodological advances, and (3) societal benefits achieved during this period, as well as to provide a vision of what is to come in years ahead. The book volumes are, to date and to my best knowledge, the most comprehensive documentation of the scientific and methodological advances that have taken place in understanding remote sensing data, methods, and a wide array of applications. Written by 300+ leading global experts in the area, each chapter: (1) focuses on specific topic (e.g., data, methods, and specific set of applications), (2) reviews existing state-of-the-art knowledge, (3) highlights the advances made, and (4) provides guidance for areas requiring future development. Chapters in the book cover a wide array of subject matter concerning remote sensing applications. *Remote Sensing Handbook* (Second Edition, Volumes I–VI) is planned as reference material for a broad spectrum of remote sensing scientists to understand the fundamentals as well as the latest advances, and a wide array of applications such for land and water resource practitioners, natural and environmental practitioners, professors, students, and decision makers.

Special features of the six-volume *Remote Sensing Handbook* (Second Edition) include the following:

1. Participation of an outstanding group of remote sensing experts, an unparalleled team of writers for such a book project
2. Exhaustive coverage of a wide array of remote sensing science: data, methods, applications
3. Each chapter being led by a luminary and most chapters written by writing teams that further enriched the chapters
4. Broadening the scope of the book to make it ideal for expert practitioners as well as students
5. Global team of writers, global geographic coverage of study areas, and wide array of satellites and sensors
6. Plenty of color illustrations

Chapters in the book cover the following aspects of remote sensing:

State of the art on satellites, sensors, science, technology, and applications
Methods and techniques
Wide array of applications such as land and water applications, natural resources management, and environmental issues
Scientific achievements and advancements over last 60+ years
Societal benefits
Knowledge gaps
Future possibilities in the 21st century

Great advances have taken place over the last 60+ years in the study of planet Earth from remote sensing, especially using data gathered from the multitude of Earth Observation (EO) satellites launched by various governments as well as private entities. A large part of the initial remote sensing technology was developed and tested during the two world wars. In the 1950s, remote sensing slowly began its foray into civilian applications. But, during the years of cold war remote sensing applications both in civilian and military increased swiftly. But, it was also an age when remote sensing was the domain of very few top experts, often having multiple skills in engineering, science, and computer technology. From the 1960s onward, there have been many governmental agencies that have initiated civilian remote sensing. The

National Aeronautics and Space Administration (NASA) of the United States has been at the forefront of many of these efforts. Others who have provided leadership in civilian remote sensing include, but are not limited to, European Space Agency (ESA) of European Union; Indian Space Research Organization (ISRO) of India; the Centre national d'études spatiales (CNES) of France; the Canadian Space Agency (CSA), Canada; the Japan Aerospace Exploration Agency (JAXA), Japan; the German Aerospace Center (DLR), Germany; the China National Space Administration (CNSA), China; the United Kingdom Space Agency (UKSA), UK; and Instituto Nacional de Pesquisas Espaciais (INPE), Brazil. Many private entities such as the Planet Labs PBC have launched and operate satellites. These government and private agencies and enterprises have launched, and continue to launch and operate, a wide array of satellites and sensors that capture data of the planet Earth in various regions of electromagnetic spectrum and in various spatial, radiometric, and temporal resolutions, routinely and repeatedly. However, the real thrust for remote sensing advancement came during the last decade of the 20th century and the beginning of the 21st century. These initiatives included the launch of series of new generation EO satellites to gather data more frequently and routinely, release of pathfinder datasets, web-enabling the data for free by many agencies (e.g., USGS release of the entire Landsat archives as well as real-time acquisitions of the world for free by making them web accessible), and providing processed data ready to users (e.g., the Harmonized Landsat and Sentinel-2 or HLS data, surface reflectance products of MODIS). Other efforts like the Google Earth made remote sensing more popular and brought in a new platform for easy visualization and navigation of remote sensing data. Advances in computer hardware and software made it possible to handle big data. Crowdsourcing, web access, cloud computing such as in Google Earth Engine (GEE) platform, machine learning, deep learning, coding, artificial intelligence, mobile apps, and mobile platforms (e.g., drones) added new dimension to how remote sensing data is used. Integration with global positioning systems (GPS) and global navigation satellite systems (GNSS), and inclusion of digital secondary data (e.g., digital elevation, precipitation, temperature) in analysis has made remote sensing much more powerful. Collectively, these initiatives provided new vision in making remote sensing data more popular, more widely understood, and increasingly used for diverse applications, hitherto considered difficult. Freely available archival data when combined with more recent acquisitions has also enabled quantitative studies of change over space and time. ***Remote Sensing Handbook* (Volumes I–VI) is targeted to capture these vast advances in data, methods, and applications, so a remote sensing student, scientist, or a professional practitioner will have the most comprehensive, all-encompassing reference material in one place**.

Modern-day remote sensing technology, science, and applications are growing exponentially. This growth is as a result of combination of factors that include (1) advances and innovations in data capture, access, processing, computing, and delivery (e.g., big data analytics, harmonized and normalized data, inter-sensor relationships, web enabling of data, cloud computing such as in Google Earth Engine (GEE), crowdsourcing, mobile apps, machine learning, deep learning, coding in Python and Java Script, and artificial intelligence); (2) an increasing number of satellites and sensors gathering data of the planet, repeatedly and routinely, in various portions of the electromagnetic spectrum as well as in array of spatial, radiometric, and temporal resolutions; (3) efforts at integrating data from multiple satellites and sensors (e.g., Sentinels with Landsat); (4) advances in data normalization, standardization, and harmonization (e.g., delivery of data in surface reflectance, inter-sensor calibration); (5) methods and techniques for handling very large data volumes (e.g., global mosaics), (6) quantum leap in computer hardware and software capabilities (e.g., ability to process several terabytes of data); (7) innovation in methods, approaches, and techniques leading to sophisticated algorithms (e.g., spectral matching techniques, neural network perceptron); and (8) development of new spectral indices to quantify and study specific land and water parameters (e.g., hyperspectral vegetation indices or HVIs). As a result of these all-round developments, remote sensing science is today very mature and is widely used in virtually every discipline of Earth sciences for quantifying, mapping, modeling, and monitoring our planet Earth. Such rapid advances are captured in a number of remote sensing and Earth science journals. However, students, scientists, and practitioners of remote sensing science and applications have significant difficulty in gathering a complete understanding of the various developments and advances that have taken place because of their vastness spread across the last 60+ years. Thereby, the chapters in *Remote Sensing Handbook* are designed to give a whole picture of scientific and technological advances of the last 60+ years.

Today, the science, art, and technology of remote sensing is truly ubiquitous and increasingly part of everyone's everyday life, often without the user even knowing it. Whether looking at your own home or farm (e.g., Figure 0.7), helping you navigate when you drive, visualizing a phenomenon occurring in a distant part of the world (e.g., Figure 0.7), monitoring events such as droughts and floods, reporting weather, detecting and monitoring troop movements or nuclear sites, studying deforestation, assessing biomass carbon, addressing disasters like earthquakes or tsunamis, and a host of other applications (e.g., precision farming, crop productivity, water productivity, deforestation, desertification, water resources management), remote sensing plays a key role. Already, many new innovations are taking place. Companies such as the Planet Labs PBC and Skybox are capturing very high spatial resolution imagery and even videos from space using a large number of microsatellite (CubeSat) constellations. Planet Labs also will soon launch hyperspectral satellites called Tanager. There are others (e.g., Pixxel, India) who have launched and continue to launch constellations of hyperspectral or other sensors. China is constantly putting a wide array of satellites into orbit. Just as the smartphone and social media connected the world, remote sensing is making the world our backyard (e.g., Figure 0.7). No place goes unobserved, and no event gets reported without an image. True liberation for any technology and science comes when it is widely used by common people who often have no idea on how it all comes together, but understand the information provided intuitively. That is already happening (e.g., how we use smartphones is significantly driven by satellite data-driven maps and GPS-driven locations). These developments make it clear that not only do we need to understand the state of the art but also

FIGURE 0.7 Google Earth can be used to seamlessly navigate and precisely locate any place on Earth, often with very high spatial resolution data (VHRI; sub-meter to 5 m) from satellites such as IKONOS, Quickbird, and Geoeye (Note: this image is from one of the VHRI). Here, the editor in chief (EiC) of this *Remote Sensing Handbook* (Volumes I–VI) (Thenkabail) located his village home and surroundings, which has land cover such as secondary rainforests, lowland paddy farms, areca nut plantations, coconut plantations, minor roads, walking routes, open grazing lands, and minor streams (typically, first and second order) (Note: land cover based on ground knowledge of the EiC). The first primary school attended by the EiC is located precisely. Precise coordinates (13 45 39.22 Northern latitude, 75 06 56.03 Eastern longitude) of Thenkabail's village house on the planet are located, and the date of image acquisition (March 1, 2014) is noted. Google Earth Images are used for visualization as well as for numerous science applications such as accuracy assessment, reconnaissance, determining land cover, establishing land use, and for various ground surveys.

have a vision on where the future of remote sensing is headed. Thereby, in a nutshell, the goal of *Remote Sensing Handbook* (Volumes I–VI) is to cover the developments and advancement of six distinct eras (listed here) in terms of data characterization and processing as well as myriad land and water applications:

Pre-civilian remote sensing era of pre-1950s: World War I and II when remote sensing was a military tool.

Technology demonstration era of 1950s and 1960s: Sputnik-I and NOAA AVHRR era of the 1950s and 1960s.

Landsat era of 1970s: when first truly operational land remote sensing satellite (Earth Resources Technology Satellite or ERTS, later renamed Landsat) was launched and operated.

Earth observation era of 1980s and 1990s: when a number of space agencies began launching and operating satellites (e.g., Landsat 4, 5 by USA, SPOT-1, 2 by France; IRS-1a, 1b by India).

Earth observation and new millennium era of 2000s: when data dissemination to users became as important as launching, operating, and capturing data (e.g., MODIS terra/acqua, Landsat-8, Resourcesat).

Twenty-first century era starting 2010s: when new generation micro/Nano satellites or CubeSats (e.g., Planet Labs PBC, Skybox), hyperspectral satellite sensors (e.g., Tanager-1, DESIS, PRISMA, EnMAP, upcoming NASA SBG) add to increasing constellation of multi-agency sensors (e.g., Sentinels, Landsat-8, 9, upcoming Landsat-Next).

Motivation to take up editing the six-volume *Remote Sensing Handbook* (Second Edition) wasn't easy. It is a daunting work and requires an extraordinary commitment over two to three years. After repeated requests from Ms. Irma Shagla-Britton, manager and leader for Remote Sensing and GIS books of Taylor and Francis/CRC Press, and considerable thought, I finally agreed to take the challenge in 2022. Having earlier edited the three-volume *Remote Sensing Handbook*, published in 2014, I was pleased that the books were of considerable demand for a second edition. This was enough motivation. Further, I wanted to do something significant at this stage of my career that will make a considerable contribution to the global remote sensing community. When I edited the first edition during the 2012–2014 period, I was still recovering post colon cancer surgery and chemotherapy. But this second edition is a celebration of my complete recovery from the dreaded disease. I have not only fully recovered but never felt so completely full of health and vigor. This, naturally, gave me the sufficient energy and enthusiasm required to back my motivation to edit this monumental six-volume *Remote Sensing Handbook*. At least for me, this is the *magnum opus* that I feel proud to have accomplished and feel confident of the immense value for students, scientists, and professional practitioners of remote sensing who are interested in a standard reference on the subject. They will find these six volumes of *Remote Sensing Handbook*: "Complete and comprehensive coverage of the state-of-the-art remote sensing, capturing the advances that have taken place over last 60+ years, which will set the stage for a vision for the future."

Above all, I am indebted to some 300+ authors and co-authors of the chapters who have spent so much of their creative energy to work on the chapters, deliver them on time, and patiently address all edits and comments. These are amongst the very best remote sensing scientists from around the world. Extremely busy people, making time for the book project and making outstanding contributions. I went back to everyone who contributed to *Remote Sensing Handbook* (First Edition, three volumes) published in 2014 and requested them to revise their chapters. Most of the lead authors of the chapters agreed to revise, which was reassuring. However, some were not available, due to retirement or for other reasons. In such cases, I adopted two strategies: (1) invite a few new chapter authors to make up for this gap, and (2) update the chapters myself in other cases. I am convinced this strategy worked very well to ensure capturing the latest information and to maintain the integrity of every chapter. What was also important was to ensure that the latest advances in remote sensing science were adequately covered. The authors of the chapters amazed me by their commitment and attention to detail. First, the quality of each of the chapters was of the highest standards. Second, with very few exceptions, chapters were delivered on time. Third, edited chapters were revised thoroughly and returned on time. Fourth, all my requests on various formatting and quality enhancements were addressed. My heartfelt gratitude to these great authors for their dedication to quality science. It has been my great honor and privilege to work with these dedicated legends. Indeed, I call them my "heroes" in the true sense. These are highly accomplished, renowned, pioneering scientists of the highest merit in remote sensing science, and I am ever grateful to have their time, effort, enthusiasm, and outstanding intellectual contributions. I am indebted to their kindness and generosity. In the end, we had 300+ authors writing 91 chapters.

Overall, the ***Remote Sensing Handbook*** **(Volumes I–VI)** took about two years, from the time book chapters and authors were identified to final publication of the book. The six volumes of *Remote Sensing Handbook* were designed in such a way that a reader can have all six volumes as standard reference or have individual volumes to study specific subject areas. The six volumes are:

Remote Sensing Handbook, Second Edition, Vol. I
Volume I: *Sensors, Data Normalization, Harmonization, Cloud Computing, and Accuracies—9781032890951*

Remote Sensing Handbook, Second Edition, Vol. II
Volume II: *Image Processing, Change Detection, GIS, and Spatial Data Analysis—9781032890975*

Remote Sensing Handbook, Second Edition, Vol. III
Volume III: *Agriculture, Food Security, Rangelands, Vegetation, Phenology, and Soils—9781032891019*

Remote Sensing Handbook, Second Edition, Vol. IV
Volume IV: *Forests, Biodiversity, Ecology, LULC and Carbon—9781032891033*

Remote Sensing Handbook; Second Edition, Vol. V
Volume V: *Water, Hydrology, Floods, Snow and Ice, Wetlands, and Water Productivity—9781032891453*

Remote Sensing Handbook; Second Edition, Vol. VI
Volume VI: *Droughts, Disasters, Pollution, and Urban Mapping—9781032891484*

There are 18, 17, 17, 12, 13, and 14 chapters, respectively, in the six volumes.

A wide array of topics covered in the six volumes.

The topics covered in the **chapters of Volume I** include (1) satellites and sensors, (2) global navigation satellite systems (GNSS), (3) remote sensing fundamentals, (4) data normalization, harmonization, and standardization, (5) vegetation indices and their within and across sensor calibration, (6) crowdsourcing, (7) cloud computing, (8) Google Earth Engine supported remote sensing, (9) accuracy assessments, and (10) remote sensing law.

The topics covered in the **chapters of Volume II** include (1) digital image processing fundamentals and advances; (2) digital image classifications for applications such as urban, land use, and land cover; (3) hyperspectral image processing methods and approaches; (4) thermal infrared image processing principles and practices; (5) image segmentation; (6) object-oriented image analysis (OBIA), including geospatial data integration techniques in OBIA; (7) image segmentation in specific applications like land use and land cover; (8) LiDAR digital image processing; (9) change detection; and (10) integrating geographic information systems (GIS) with remote sensing in geoprocessing workflows, democratization of GIS data and tools, fronters of GIScience, and GIS and remote sensing policies.

The topics covered in the **chapters of Volume III** include (1) vegetation and biomass, (2) agricultural croplands, (3) rangelands, (4) phenology and food security, and (5) soils.

The topics covered in the **chapters of Volume IV** include (1) forests, (2) biodiversity, (3) ecology, (4) land use and land cover, and (5) carbon. Under each of the preceding broad topics, there are one or more chapters.

The **chapters of Volume V** has focus on hydrology, water resources, ice, wetlands, and crop water productivity. The chapters are broadly classified into (1) geomorphology, (2) hydrology and water resources, (3) floods, (4) wetlands, (5) crop water use and crop water productivity, and (6) snow and ice.

The **chapters of Volume VI** has focus on water resources, disasters, and urban remote sensing. The chapters are broadly classified into (1) droughts and drylands, (2) disasters, (3) volcanoes, (4) fires, and (5) nightlights.

There are many ways to use *Remote Sensing Handbook* (Second Edition, six volumes). A lot of thought went into organizing the volumes and chapters. So, you will see a "flow" from chapter to chapter and volume to volume. As you read through the chapters, you will see how they are interconnected and how reading all of them provides you with greater in-depth understanding. You will also realize, as someone deeply interested in one of the topics, that you will have greater interest in one volume. Having all six volumes as reference material is ideal for any remote sensing expert, practitioner, or student. However, you can also refer to individual volumes based on your interest. We have also made great attempts to ensure chapters are self-contained. That way, you can focus on a chapter and read it through, without having to be overly dependent on other chapters. Taking this perspective, a small amount of material (~5 to 10%) may be repeated across chapters. This is done deliberately. For example, when you are reading a chapter on LiDAR or Radar, you don't want to go all the way back to another chapter to understand characteristics of these data. Similarly, certain indices (e.g., vegetation condition index (VCI) or temperature condition index (TCI)) that are defined in one chapter (e.g., on drought) may be repeated in another chapter (also on drought). Such minor overlaps help the reader avoid going back to another chapter to understand a phenomenon or an index or a characteristic of a sensor. However, if you want a lot of details of these sensors or indices or phenomenon, then you will have to read the appropriate chapter where there is in-depth coverage of the topic.

Each volume has a summary chapter (the last chapter of each volume). The summary chapter can be read two ways: (1) either as last chapter to recapture the main points of each of the chapters, or (2) as an initial overview to get the first feeling for what is in the volume, before diving into to read each chapter in detail. I suggest the readers do it both ways: read it first before reading chapters in detail to gather an idea on what to expect in each chapter and then read it at the end to recapture what is being read in each of the chapters.

It has been a great honor as well as humbling experience to edit the *Remote Sensing Handbook* (Volumes I–VI). I truly enjoyed the effort, albeit felt overwhelmed at times with never-ending work. What an honor to work with luminaries in your field of expertise. I learned a lot from them and am very grateful for their support, encouragement, and deep insights. Also, it has been a pleasure working with the outstanding professionals of Taylor and Francis Inc./ CRC Press. There is no joy greater than being immersed in pursuit of excellence, knowledge gain, and knowledge capture. At the same time, I am happy it is over. If there will be a third edition in a decade or so from now, it will be taken up by someone else (individually or as a team) and certainly not me!

I expect the book to be a standard reference of immense value to any student, scientist, professional, and practical practitioners of remote sensing. Any book that has the privilege of 300+ of the best brains of truly outstanding and dedicated remote sensing scientists ought to be a *magnum opus* deserving to be standard reference on the subject.

Dr. Prasad S. Thenkabail, PhD
Editor in Chief (EiC)
Remote Sensing Handbook (Second Edition, Volumes I–VI)

Volume I: Sensors, Data Normalization, Harmonization, Cloud Computing, and Accuracies
Volume II: Image Processing, Change Detection, GIS, and Spatial Data Analysis
Volume III: Agriculture, Food Security, Rangelands, Vegetation, Phenology, and Soils
Volume IV: Forests, Biodiversity, Ecology, LULC and Carbon
Volume V: Water, Hydrology, Floods, Snow and Ice, Wetlands, and Water Productivity
Volume VI: Droughts, Disasters, Pollution, and Urban Mapping

About the Editor

Dr. Prasad S. Thenkabail, PhD, is a senior scientist with the U.S. Geological Survey (USGS), specializing in remote sensing science for agriculture, water, and food security. He is a world-recognized expert in remote sensing science with multiple major contributions in the field sustained for 40+ years. Dr. Thenkabail has conducted pioneering research in hyperspectral remote sensing of vegetation, global croplands mapping for water and food security, and crop water productivity. His work on hyperspectral remote sensing of agriculture and vegetation are widely cited. His papers on hyperspectral remote sensing are first of its kind and, collectively, they have (1) determined optimal hyperspectral narrowbands (OHNBs) in study of agricultural crops; (2) established hyperspectral vegetation indices (HVIs) to model and map crop biophysical and biochemical quantities; (3) created framework and sample data for the global hyperspectral imaging spectral libraries of crops (GHISA); (4) developed methods and techniques of overcoming Hughes' phenomenon; (5) demonstrated the strengths of hyperspectral narrowband (HNB) data in advancing classification accuracies relative to multispectral broadband (MBB) data; (6) showed advances one can make in modeling crop biophysical and biochemical quantities using HNB and HVI data relative to MBB data; and (7) created a body of work in understanding, processing, and utilizing HNB and HVI data in agricultural cropland studies. This body of work has become a widely referred reference worldwide. In studies of global croplands for food and water security, he has led the release of the world's first 30-m Landsat Satellite-derived global cropland extent product at 30 m (GCEP30; https://www.usgs.gov/apps/croplands/app/map) (Thenkabail et al., 2021; https://lpdaac.usgs.gov/news/release-of-gfsad-30-meter-cropland-extent-products/) and Landsat-derived global rainfed and irrigated area product at 30 m (LGRIP30) (Teluguntla et al., 2023; https://lpdaac.usgs.gov/products/lgrip30v001/). Earlier, he led producing the world's first global irrigated area map (GIAM; https://lpdaac.usgs.gov/products/lgrip30v001/; https://lpdaac.usgs.gov/products/gfsad1kcdv001/) product at nominal 1-km spatial resolution. The global cropland datasets using satellite remote sensing demonstrates a "paradigm shift" in global cropland mapping using remote sensing through big data analytics, machine learning, and petabyte-scale cloud computing on the Google Earth Engine (GEE). The LGRIP30 and GCEP30 products are released through NASA's LP DAAC and published in USGS professional paper 1868 (Thenkabail et al., 2021). He has been principal investigator of many projects over the years including the NASA-funded global food security support analysis data in the 30-m (GFSAD) project (www.usgs.gov/wgsc/gfsad30).

His career scientific achievements can be gauged by successfully making the list of the world's top 1% of scientists as per the Stanford study ranking world's scientists from across 22 scientific fields and 176 subfields based on deep analysis evaluating about ten million scientists based on SCOPUS data from Elsevier from 1996 to 2023 (Ioannidis, 2023; Ioannidis et al., 2020). Dr. Thenkabail was recognized as Fellow of the American Society of Photogrammetry and Remote Sensing (ASPRS) in 2023. Dr. Thenkabail has published more than 150 peer-reviewed scientific papers and edited 15 books. His scientific papers have won several awards over the years, demonstrating world-class highest quality research. These include: 2023 Talbert Abrams Grand Award, the highest scientific paper award of the ASPRS (with Itiya Aneece); 2015 ASPRS ERDAS award for best scientific paper in remote sensing (with Michael Marshall); 2008 John I. Davidson ASPRS President's Award for Practical Papers (with Pardhasaradhi Teluguntla); and 1994 Autometric Award for the outstanding paper in remote sensing (with Dr. Andy Ward).

Dr. Thenkabail' s contributions to series of leading edited books places him as a world leader in remote sensing science. There are three seminal book-sets with a total of 13 volumes that he edited that have demonstrated his major contributions as an internationally acclaimed remote sensing scientist. These are (1) *Remote Sensing Handbook* (Second Edition, six-volume book-set, 2024) with 91 chapters and nearly 3,000 pages and for which he is the sole editor; (2) *Remote Sensing Handbook* (First Edition, three-volume book-set, 2015) with 82 chapters and 2,304 pages and for which he was the sole editor; and (3) *Hyperspectral Remote Sensing of Vegetation* (four-volume book-set, 2018) with 50 chapters and 1,632 pages that he edited as the chief editor (co-editors: Prof. John Lyon and Prof. Alfredo Huete).

Dr. Thenkabail is at the center of rendering scientific service to the world's remote sensing community over long periods of service. This includes serving as editor in chief (2011–present) of *Remote Sensing Open Access Journal*; associate editor (2017–present) of *Photogrammetric Engineering and Remote Sensing* (PE&RS); Editorial Advisory Board (2016–present) of the International Society of Photogrammetry and Remote Sensing (ISPRS); and Editorial Board Member (2007–2017) of the Remote Sensing of Environment.

The USGS and NASA selected him as one of the three international members on the Landsat Science Team (2006–2011). He is an Advisory Board member of the online library collection to support the United Nations' Sustainable Development Goals (UN SDGs), and currently scientist for the NASA and ISRO (Indian Space Research Organization) Professional Engineer and Scientist Exchange Program (PESEP) program for 2022–2024. He was the chair, International Society of Photogrammetry and Remote Sensing (ISPRS) Working Group WG VIII/7 (land cover and its dynamics) from 2013–2016; played a vital role for USGS as global coordinator, Agricultural Societal Beneficial Area (SBA), Committee for Earth Observation (CEOS) (2010–2013) during which he co-wrote the global food security case study for the CEOS for the *Earth Observation Handbook* (EOS), Special Edition for the UN Conference on Sustainable Development, presented in Rio de Janeiro, Brazil; was the co-lead (2007–2011) of IEEE "Water for the World" initiative, a nonprofit effort funded by IEEE that worked in coordination with the Group on Earth Observations (GEO) in its GEO Water and GEO Agriculture initiatives.

Dr. Thenkabail worked as a postdoctoral researcher and research faculty at the Center for Earth Observation (YCEO), Yale University (1997–2003), and led remote sensing programs in three international organizations including the following:

- International Water Management Institute (IWMI), 2003–2008
- International Center for Integrated Mountain Development (ICIMOD), 1995–1997
- International Institute of Tropical Agriculture (IITA), 1992–1995

He began his scientific career as a scientist (1986–1988) working for the National Remote Sensing Agency (NRSA) (now renamed National Remote Sensing Center, or NRSC), Indian Space Research Organization (ISRO), Department of State, government of India.

Dr. Thenkabail's work experience spans over 25 countries including East Asia (China), South-East Asia (Cambodia, Indonesia, Myanmar, Thailand, Vietnam), Middle East (Israel, Syria), North America (United States, Canada), South America (Brazil), Central Asia (Uzbekistan), South Asia (Bangladesh, India, Nepal, and Sri Lanka), West Africa (Republic of Benin, Burkina Faso, Cameroon, Central African Republic, Cote d'Ivoire, Gambia, Ghana, Mali, Nigeria, Senegal, and Togo), and Southern Africa (Mozambique, South Africa). Dr. Thenkabail is regularly invited as keynote speaker or invited speaker at major international conferences and at other important national and international forums every year.

Dr. Thenkabail obtained his PhD in agricultural engineering from The Ohio State University, USA, in 1992. He has a master's degree in hydraulics and water resources engineering, and bachelor's degree in civil engineering (both from India). He has 168 publications including 15 books; 175+ peer-reviewed journal articles, book chapters, and professional papers/monographs; and 15+ significant major global and regional data releases.

REFERENCES

Journals

Ioannidis, J.P.A. (2023). October 2023 data-update for "updated science-wide author databases of standardized citation indicators". Elsevier Data Repository, V6. https://doi.org/10.17632/btchxktzyw.6

Ioannidis, J.P.A., Boyack, K.W., & Baas, J. (2020). Updated science-wide author databases of standardized citation indicators. PLoS Biology, 18(10): e3000918. https://doi.org/10.1371/journal.pbio.3000918

Scientific Papers

https://scholar.google.com/citations?user=9IO5Y7YAAAAJ&hl=en

USGS Professional Paper, Data and Product Gateways, Interactive Viewers

Thenkabail, P.S., Teluguntla, P.G., Xiong, J., Oliphant, A., Congalton, R.G., Ozdogan, M., Gumma, M.K., Tilton, J.C., Giri, C., Milesi, C., Phalke, A., Massey, R., Yadav, K., Sankey, T., Zhong, Y., Aneece, I., and Foley, D. (2021). Global cropland-extent product at 30-m resolution (GCEP30) derived from Landsat satellite time-series data for the year 2015 using multiple machine-learning algorithms on Google Earth Engine cloud: U.S. Geological Survey Professional Paper 1868, 63 p., https://doi.org/10.3133/pp1868 (research paper) https://lpdaac.usgs.gov/news/release-of-gfsad-30-meter-cropland-extent-products/ (download data, documents) www.usgs.gov/apps/croplands/app/map (view data interactively)

P. Teluguntla, P. Thenkabail, A. Oliphant, M. Gumma, I. Aneece, D. Foley and R. McCormick (2023a). Landsat-derived Global Rainfed and Irrigated-Cropland Product @ 30-m (LGRIP30) of the World (GFSADLGRIP30WORLD). The Land Processes Distributed Active Archive Center (LP DAAC) of NASA and USGS. Pp. 103. https://lpdaac.usgs.gov/news/release-of-lgrip30-data-product/ (download data, documents)

Books

Remote Sensing Handbook (Second Edition, Six Volumes, 2024)

Thenkabail, Prasad. 2024. *Remote Sensing Handbook (Second Edition, Six Volume Book-set), Volume I: Sensors, Data Normalization, Harmonization, Cloud Computing, and Accuracies*. Taylor and Francis Inc./CRC Press, Boca Raton, London, New York. 978-1-032-89095-1—CAT# T132478. Print ISBN: 9781032890951. eBook ISBN: 9781003541141. Pp. 581.

Thenkabail, Prasad. 2024. *Remote Sensing Handbook (Second Edition, Six Volume Book-set), Volume II: Image Processing, Change Detection, GIS, and Spatial Data Analysis*. Taylor and Francis Inc./CRC Press, Boca Raton, London, New York. 978-1-032-89097-5—CAT# T133208. Print ISBN: 9781032890975. eBook ISBN: 9781003541158. Pp. 464.

Thenkabail, Prasad. 2024. *Remote Sensing Handbook (Second Edition, Six Volume Book-set), Volume III: Agriculture, Food Security, Rangelands, Vegetation, Phenology, and Soils*. Taylor and Francis Inc./CRC Press, Boca Raton, London, New York. 978-1-032-89101-9—CAT# T133213. Print ISBN: 9781032891019; eBook ISBN: 9781003541165. Pp. 788.

Thenkabail, Prasad. 2024. *Remote Sensing Handbook (Second Edition, Six Volume Book-set), Volume IV: Forests, Biodiversity, Ecology, LULC, and Carbon*. Taylor and Francis Inc./CRC Press, Boca Raton, London, New York. 978-1-032-89103-3—CAT# T133215. Print ISBN: 9781032891033. eBook ISBN: 9781003541172. Pp. 501.

Thenkabail, Prasad. 2024. *Remote Sensing Handbook (Second Edition, Six Volume Book-set), Volume V: Water, Hydrology, Floods, Snow and Ice, Wetlands, and Water Productivity*. Taylor and Francis Inc./CRC Press, Boca Raton, London, New York. 978-1-032-89145-3—CAT# T133261. Print ISBN: 9781032891453. eBook ISBN: 9781003541400. Pp. 516.

Thenkabail, Prasad. *Remote Sensing Handbook (Second Edition, Six Volume Book-set), Volume VI: Droughts, Disasters, Pollution, and Urban Mapping*. Taylor and Francis Inc./CRC Press, Boca Raton, London, New York. 978-1-032-89148-4 — CAT# T133267. Print ISBN: 9781032891484; eBook ISBN: 9781003541417. Pp. 467.

Hyperspectral Remote Sensing of Vegetation (Second Edition, Four Volumes, 2018)

Thenkabail, P.S., Lyon, G.J., and Huete, A. (Editors) 2018. *Hyperspectral Remote Sensing of Vegetation* (Second Edition, four-volume set).

Volume I: *Fundamentals, Sensor Systems, Spectral Libraries, and Data Mining for Vegetation*. CRC Press-Taylor and Francis Group, Boca Raton, London, New York. p. 449, Hardback ID: 9781138058545; eBook ID: 9781315164151.

Volume II: *Hyperspectral Indices and Image Classifications for Agriculture and Vegetation*. CRC Press-Taylor and Francis Group, Boca Raton, London, New York. p. 296. Hardback ID: 9781138066038; eBook ID: 9781315159331.

Volume III: *Biophysical and Biochemical Characterization and Plant Species Studies*. CRC Press-Taylor and Francis Group, Boca Raton, London, New York. p. 348. Hardback: 9781138364714; eBook ID: 9780429431180.

Volume IV: *Advanced Applications in Remote Sensing of Agricultural Crops and Natural Vegetation*. CRC Press-Taylor and Francis Group, Boca Raton, London, New York. p. 386. Hardback: 9781138364769; eBook ID: 9780429431166.

Remote Sensing Handbook (First Edition, Three Volumes, 2015)

Thenkabail, P.S., (Editor-in-Chief), 2015. "Remote Sensing Handbook"

Volume I: *Remotely Sensed Data Characterization, Classification, and Accuracies*. Taylor and Francis Inc./CRC Press, Boca Raton, London, New York. ISBN 9781482217865—CAT# K22125. Print ISBN: 978-1-4822-1786-5; eBook ISBN: 978-1-4822-1787-2. p. 678.

Volume II: *Land Resources Monitoring, Modeling, and Mapping with Remote Sensing*. Taylor and Francis Inc./CRC Press, Boca Raton, London, New York. ISBN 9781482217957—CAT# K22130. p. 849.

Volume III: *Remote Sensing of Water Resources, Disasters, and Urban Studies*. Taylor and Francis Inc./CRC Press, Boca Raton, London, New York. ISBN 9781482217919—CAT# K22128. p. 673.

Hyperspectral Remote Sensing of Vegetation (First Edition, Single Volume, 2013)

Thenkabail, P.S., Lyon, G.J., and Huete, A. (Editors) 2012. *Hyperspectral Remote Sensing of Vegetation*. CRC PressTaylor and Francis Group, Boca Raton, London, New York. p. 781 (80+ pages in color). www.crcpress.com/product/isbn/9781439845370

Remote Sensing of Global Croplands for Food Security (First Edition, Single Volume, 2009)

Thenkabail, P., Lyon, G.J., Turral, H., and Biradar, C.M. (Editors) 2009. *Remote Sensing of Global Croplands for Food Security*. (CRC Press-Taylor and Francis Group, Boca Raton, London, New York. p. 556 (48 pages in color). Published in June 2009.

FIGURE Snap shots of the Editor-in-Chief's work and life.

Contributors

Jay Angerer
Livestock and Range Research Laboratory
Miles City, MT, USA

Damien Arvor
Research Geographer
Montpellier, France

Clement Atzberger
Mantle-Labs Ltd.
UK
and
University of Natural Resources and Life Sciences
Institute of Geomatics
Vienna

Frédéric Baret
Research Director
INRA-EMMAH/CAPTE
Avignon, France

Agnès Bégué
Research Scientist
CIRAD, UMR TETIS
34093 Montpellier, France

Henrique Bellinaso
Agronomist, Planning Assistant and Rural Extensionist
Remote Sensing Applied to Soils and Soil Conservation
São Paulo State Government, Secretariat of Agriculture and Supply
Coordination of Integrate Technical Assistance—CATI, Piracicaba Regional
Piracicaba, São Paulo, Brazil

Juliane Bendig
Institute of Bio-and Geosciences
Plant Sciences (IBG-2)
Forschungszentrum Jülich GmbH
Jülich, Germany

Eyal Ben-Dor
Department of Geography and Human Environment
Tel Aviv University
Israel

Trenton Benedict
U.S. Geological Survey EROS Center
Sioux Falls, South Dakota

D. Terrance Booth
Rangeland Resources Research Unit
Cheyenne, Wyoming

Jesslyn F. Brown
U.S. Geological Survey
Earth Resources Observation and Science (EROS) Center
Sioux Falls, South Dakota

Molly E. Brown
Research Scientist
Department of Geographical Sciences
University of Maryland College Park
College Park, Maryland

Indrė Bručienė
Department of Agricultural Engineering and Safety
Faculty of Engineering, Agriculture Academy
Vytautas Magnus University
Kaunas, Lithuania

James B. Campbell
Department of Geography
Virginia Tech
Blacksburg, Virginia

Sabine Chabrillat
Senior Scientist
Helmholtz Centre Potsdam
GFZ German Research Centre for Geosciences
Potsdam, Germany

Russell G. Congalton
University of New Hampshire
Durham, New Hampshire

Samuel E. Cox
Wyoming State Office
Bureau of Land Management
Cheyenne, Wyoming

Craig S.T. Daughtry
USDA-ARS Hydrology and Remote Sensing Laboratory
Beltsville, Maryland

Kirsten de Beurs
Professor/Chair
Laboratory of Geo-information Science and Remote Sensing
Wageningen University
Netherlands

José A.M. Demattê
Professor and Researcher Agronomist
Soil Scientist, Remote Sensing Applied to Soils
University of São Paulo—"Luiz de Queiroz" College of Agriculture
Department of Soil Science
Piracicaba, São Paulo, Brazil

Venkateswarlu Dheeravath
United Nations World Food Program
Erbil, Iraq

Marston H.D. Franceschini
Agronomist, Researcher in Remote Sensing
University of São Paulo—"Luiz de Queiroz" College of Agriculture
Department of Soil Science
Piracicaba, São Paulo, Brazil

Chandra Giri
U.S. Environmental Protection Agency
Research Triangle Park, North Carolina

Kathryn Grace
Professor
University of Minnesota
Department of Geography, Environment and Society
Minneapolis, Minnesota

Murali Krishna Gumma
International Crops Research Institute for the Semi-Arid Tropics (ICRISAT)
Patancheru, Hyderabad, India

Joanne V. Hall
Department of Geographical Sciences
College Park, University of Maryland
College Park, Maryland

Lea Hallik
Tartu Observatory
Faculty of Science and Technology
University of Tartu
Tartu County, Estonia

Alexander J. Hernandez
Department of Wildland Resources
Utah State University
Logan, Utah

David Herrera
Institute of Bio- and Geosciences
Plant Sciences (IBG-2)
Forschungszentrum Jülich GmbH
Jülich, Germany

Yanbo Huang
Research Agricultural Engineer
Genetics and Sustainable Agriculture Research Unit
USDA-Agricultural Research Services
Mississippi State, Mississippi

Alfredo Huete
Professor
School of Life Sciences
University of Technology Sydney
Ultimo, NSW, Australia

E. Raymond Hunt Jr.
USDA-ARS Beltsville Agricultural Research Center
Hydrology and Remote Sensing Laboratory
Beltsville, Maryland

Markus Immitzer
Mantle-Labs Ltd.
United Kingdom
and
University of Natural Resources and Life Sciences
Institute of Geomatics
Vienna

Ruchita Ingle
Water Systems and Global Change Group
Wageningen University
Wageningen, The Netherlands

Christoph Jedmowski
Institute of Bio-and Geosciences
Plant Sciences (IBG-2)

Forschungszentrum Jülich GmbH
Jülich, Germany

Laura Verena Junker-Frohn
Institute of Bio- and Geosciences
Plant Sciences (IBG-2)
Forschungszentrum Jülich GmbH
Jülich, Germany

Justin George Kalambukattu
Agriculture and Soils Department
Indian Institute of Remote Sensing
Indian Space Research Organisation
Dehradun, Uttarakhand, India

Wasantha Kulawardhana
Department of Biological and Environmental Sciences
Alabama A&M University
Huntsville, Alabama

Lalit Kumar
School of Environmental and Rural Science
University of New England
Armidale, NSW, Australia

Suresh Kumar
Agriculture, Forestry and Ecology Group
Indian Institute of Remote Sensing
Indian Space Research Organisation
Dehradun, Uttarakhand, India

Kristina Lekavičienė
Department of Agricultural Engineering and Safety, Faculty of Engineering, Agriculture Academy
Vytautas Magnus University
Kaunas, Lithuania

Camille Lelong
Research Scientist
CIRAD, UMR TETIS
Montpellier, France

Tatiana Loboda
Department of Geographical Sciences
University of Maryland
College Park, Maryland

Thomas Loveland
U.S. Geological Survey EROS Center
Sioux Falls, South Dakota

Xuanlong Ma
Professor
MOE Key Laboratory of Western China's Environmental Systems
College of Earth and Environmental Sciences
Lanzhou University

Richard Massey
The Aerospace Corporation
Colorado Springs, Colorado

Heather McNairn
Science and Technology Branch
Agriculture and Agri-Food Canada
Ottawa, Ontario, Canada

Arjan J.H. Meddens
School of the Environment
Washington State University
Pullman, Washington

Graciela Metternicht
Institute of Environmental Studies
University of New South Wales Australia
Sydney, NSW, Australia

Yuxin Miao
Director of Precision Agriculture Center
Department of Soil, Water and Climate
University of Minnesota
St. Paul, Minnesota

Cristina Milesi
NASA Ames Research Center
Moffett Field, California

Pedro Millikan
U.S.D.A. Forest Service, Geospatial Technology and Applications Center (GTAC)
Rocky Mountain Research Station
Vallejo, California

Cristine L.S. Morgan
Chief Scientific Officer
Soil Health Institute
Morrisville, North Carolina

David J. Mulla
W.E. Larson Endowed Chair of Soil and Water Sciences
Department of Soil, Water and Climate

University of Minnesota
Saint Paul, Minnesota

Vilma Naujokienė
Department of Agricultural Engineering and Safety, Faculty of Engineering, Agriculture Academy
Vytautas Magnus University
Kaunas, Lithuania

Adam Oliphant
U.S. Geological Survey (USGS)
Flagstaff, Arizona

Mutlu Ozdogan
University of Wisconsin
Madison, Wisconsin

Anna Pacheco
Science and Technology Branch
Agriculture and Agri-Food Canada
Ottawa, Ontario, Canada

Aparna Phalke
The University of Alabama
Huntsville, Alabama

Guillermo Ponce-Campos
School of Natural Resources and the Environment
The University of Arizona
Tucson, Arizona

Juan Quiros-Vargas
Institute of Bio- and Geosciences
Plant Sciences (IBG-2)
Forschungszentrum Jülich GmbH
Jülich, Germany

R. Douglas Ramsey
Department of Wildland Resources
Utah State University Logan
Logan, Utah

Uwe Rascher
Institute of Bio- and Geosciences
Plant Sciences (IBG-2)
Forschungszentrum Jülich GmbH
Jülich, Germany

Matthew Reeves
Research Ecologist
Rocky Mountain Research Station
Missoula, Montana

Natalia Restrepo-Coupe
Designated Campus Colleague
Ecology and Evolutionary Biology
The University of Arizona
Tucson, Arizona

Angus Retallack
Department of Ecology and Evolutionary Biology
The University of Adelaide
Adelaide, SA, Australia

Matthew Rigge
USGS Earth Resources Observation and Science (EROS) Center
Sioux Falls, South Dakota

Rodnei Rizzo
Agronomist, Soil Scientist, PhD in Applied Ecology
Center of Nuclear Energy in Agriculture (CENA)
University of São Paulo, Luiz de Queiroz
Piracicaba, São Paulo, Brasil

Temuulen Tsagaan Sankey
School of Informatics, Computing, and Cyber Systems
Northern Arizona University
Flagstaff, Arizona

Egidijus Šarauskis
Department of Agricultural Engineering and Safety
Faculty of Engineering, Agriculture Academy
Vytautas Magnus University
Kaunas, Lithuania

Guy Serbin
Spatial Analysis Unit
Rural Economy and Development Programme
Teagasc Food Research Centre
Ashtown, Dublin, Ireland

Asfa Siddiqui
Scientist/Engineer "SE"

Urban and Regional Studies Department
Indian Institute of Remote Sensing
Indian Space Research Organisation
Dehradun, Uttarakhand, India

Bastian Siegmann
Institute of Bio- and Geosciences
Plant Sciences (IBG-2)
Forschungszentrum Jülich GmbH
Jülich, Germany

Margareth Simões
Research Engineer
Embrapa—Brazilian Agricultural Research Corporation
UERJ—Rio de Janeiro State University
Rio de Janeiro, Brazil

Priyakant Sinha
Ecosystem Management
School of Environmental and Rural Science
University of New England
Armidale, NSW, Australia

William K. Smith
School of Natural Resources and the Environment
University of Arizona
Tucson, Arizona

Terry Sohl
Geological Survey EROS Center
Sioux Falls, South Dakota

Carson A. Stam
Department of Wildland Resources
Utah State University
Logan, Utah

Pardhasaradhi Teluguntla
U.S. Geological Survey (USGS)
Flagstaff, Arizona
and
Bay Area Environmental Research Institute (BAERI)
NASA Ames Research Park
Moffett Field, California

Fabrício da S. Terra
Agronomist, Professor/Researcher Agronomist
Soil Scientist, Spectral Sensing (Proximal and Remote) Applied to Soils
Institute of Agricultural Science (ICA)
Federal University of Jequitinhonha and Mucuri Valleys (UFVJM)
Campus Unaí
Unaí, Minas Gerais, Brazil

Prasad S. Thenkabail
U.S. Geological Survey (USGS)
Flagstaff, Arizona

James Tilton
NASA Goddard Space Flight Center (GSFC)
Greenbelt, Maryland

Richa Upadhyay
Scientist/Engineer "SE"
Geo-Sciences Department
Indian Institute of Remote Sensing
Indian Space Research Organisation
Dehradun, Uttarakhand, India

Gustavo M. Vasques
Soil Researcher
Embrapa Soils
Rio de Janeiro, RJ, Brazil

Elodie Vintrou
Research Scientist
CIRAD, UPR AIDA
France

Letícia G. Vogel
Agronomist
Researcher in Soil Science, Remote Sensing Applied to Soils
University of São Paulo—"Luiz de Queiroz" College of Agriculture
Department of Soil Science
São Paulo, Brazil

Cuizhen Wang
Department of Geography
University of South Carolina
Columbia, South Carolina

Robert Washington-Allen
Agriculture, Veterinary and Rangeland Sciences
University of Nevada
Reno, Nevada

Johanna Wetterlind
Research Assistant
Swedish University of Agricultural Sciences
Skara, Sweden

Jun Xiong
The Climate Corporation
San Francisco, California

Kamini Yadav
U.S. Forest Service
Corvallis, Oregon

Wen Zhang
School of Natural Resources and the Environment
University of Arizona
Tucson, Arizona

Yongguang Zhang
International Institute for Earth System Sciences
Nanjing University, Nanjing
Jiangsu, China

Baojuan Zheng
School of Geographical Sciences and Urban Planning
Arizona State University
Tempe, Arizona

Acknowledgments

Remote Sensing Handbook (Second Edition, Volumes I–VI) brought together a galaxy of highly accomplished, renowned remote sensing scientists, professionals, and legends from around the world. The lead authors were chosen by me after careful review of their accomplishments and sustained publication record over the years. The chapters in the second edition were written/revised over a period of two years. All chapters were edited, and revised.

Gathering such a galaxy of authors was the biggest challenge. These are all extremely busy people, and committing to a book project that requires substantial work is never easy. However, almost all those whom I requested agreed to write a chapter specific to their area of specialization, and only a few I had to convince to make time. The quality of the chapters should convince readers why these authors are such highly rated professionals and why they are so successful and accomplished in their fields of expertise. They not only wrote very high-quality chapters but also delivered them on time, addressed any editorial comments in a timely manner without complaints, and were extremely humble and helpful. Their commitment for quality science is what makes them special. I am truly honored to have worked with such great professionals.

I would like to mention the names of everyone who contributed and made ***Remote Sensing Handbook*** **(Second Edition, Volumes I–VI) possible**. In the end, we had 91 chapters, a little over 3,000 pages, and a little more than 400 authors. My gratitude goes to each one of them. These are well-known **"who is who"** in remote sensing science in the world. A list of all authors is provided here. The names of the authors are organized chronologically for each volume and the chapters. Each lead author of the chapter is in bold type. **The names of the 400+ authors who contributed to six volumes are as follows:**

Volume I: Sensors, Data Normalization, Harmonization, Cloud Computing, and Accuracies—18 chapters written by 53 authors (editor-in-chief: Prasad S. Thenkabail):

Drs. Sudhanshu S. Panda, Mahesh N. Rao, Prasad S. Thenkabail, Debasmita Misra, and James P. Fitzgerald; **Mohinder S. Grewal**; **Kegen Yu**, Chris Rizos, and Andrew Dempster; **D. Myszor**, O. Antemijczuk, M. Grygierek, M. Wierzchanowski, K.A. Cyran; **Natascha Oppelt** and Arnab Muhuri; **Philippe M. Teillet**; **Philippe M. Teillet** and Gyanesh Chander; **Rudiger Gens** and Jordi Cristóbal Rosselló; **Aolin Jia** and Dongdong Wang; **Tomoaki Miura**, Kenta Obata, Hiroki Yoshioka, and Alfredo Huete; **Michael D. Steven**, Timothy J. Malthus, and Frédéric Baret; **Fabio Dell'Acqua** and Silvio Dell'Acqua; **Ramanathan Sugumaran**, James W. Hegeman, Vivek B. Sardeshmukh and Marc P. Armstrong; **Lizhe Wang**, Jining Yan, Yan Ma, Xiaohui Huang, Jiabao Li, Sheng Wang, Haixu He, Ao Long, and Xiaohan Zhang; **John E. Bailey** and Josh Williams; **Russell G. Congalton**; **P.J. Blount**; **Prasad S. Thenkabail**.

Volume II: Image Processing, Change Detection, GIS and Spatial Data Analysis—17 chapters written by 64 authors (editor-in-chief: Prasad S. Thenkabail):

Sunil Narumalani and Paul Merani; **Mutlu Ozdogan**; **Soe W. Myint**, Victor Mesev, Dale Quattrochi, and Elizabeth A. Wentz; **Jun Li**, Paolo Gamba, and Antonio Plaza; **Qian Du**, Chiranjibi Shah, Hongjun Su, and Wei Li; **Claudia Kuenzer**, Philipp Reiners, Jianzhong Zhang, Stefan Dech; **Mohammad D. Hossain** and Dongmei Chen; **Thomas Blaschke**, Maggi Kelly, Helena Merschdorf; **Stefan Lang** and Dirk Tiede; **James Tilton**, Selim Aksoy, and Yuliya Tarabalka; **Shih-Hong Chio**, Tzu-Yi Chuang, Pai-Hui Hsu, Jen-Jer Jaw, Shih-Yuan Lin, Yu-Ching Lin, Tee-Ann Teo, Fuan Tsai, Yi-Hsing Tseng, Cheng-Kai Wang, Chi-Kuei Wang, Miao Wang, and Ming-Der Yang; **Guiying Li**, Mingxing Zhou, Ming Zhang, Dengsheng Lu; **Jason A. Tullis**, David P. Lanter, Aryabrata Basu, Jackson D. Cothren, Xuan Shi, W. Fredrick Limp, Rachel F. Linck, Sean G. Young, Jason Davis, and

Tareefa S. Alsumaiti; **Gaurav Sinha**, Barry J. Kronenfeld, and Jeffrey C. Brunskill; **May Yuan**; **Stefan Lang**, Stefan Kienberger, Michael Hagenlocher, and Lena Pernkopf; **Prasad S. Thenkabail**.

Volume III: Agriculture, Food Security, Rangelands, Vegetation, Phenology, and Soils—17 chapters written by 110 authors (editor-in-chief: Prasad S. Thenkabail):

Alfredo Huete, Guillermo Ponce-Campos, Yongguang Zhang, Natalia Restrepo-Coupe, and Xuanlong Ma; **Juan Quiros-Vargas,** Bastian Siegmann, Juliane Bendig, Laura Verena Junker-Frohn, Christoph Jedmowski, David Herrera, Uwe Rascher; **Frédéric Baret**; **Lea Hallik**, Egidijus Šarauskis, Ruchita Ingle, Indrė Bručienė, Vilma Naujokienė, and Kristina Lekavičienė; **Clement Atzberger** and Markus Immitzer; **Agnès Bégué**, Damien Arvor, Camille Lelong, Elodie Vintrou, and Margareth Simões; **Pardhasaradhi Teluguntla,** Prasad S. Thenkabail, Jun Xiong, Murali Krishna Gumma, Chandra Giri, Cristina Milesi, Mutlu Ozdogan, Russell G. Congalton, James Tilton, Temuulen Tsagaan Sankey, Richard Massey, Aparna Phalke, and Kamini Yadav; **Yuxin Miao**, David J. Mulla, and Yanbo Huang; **Baojuan Zheng**, James B. Campbell, Guy Serbin, Craig S.T. Daughtry, Heather McNairn, and Anna Pacheco; **Prasad S. Thenkabail**, Itiya Aneece, Pardhasaradhi Teluguntla, Richa Upadhyay, Asfa Siddiqui, Justin George Kalambukattu, Suresh Kumar, Murali Krishna Gumma, Venkateswarlu Dheeravath; **Matthew Reeves,** Robert Washington-Allen, Jay Angerer, E. Raymond Hunt Jr., Wasantha Kulawardhana, Lalit Kumar, Tatiana Loboda, Thomas Loveland, Graciela Metternicht, R. Douglas Ramsey, Joanne V. Hall, Trenton Benedict, Pedro Millikan, Angus Retallack, Arjan J.H. Meddens, William K. Smith, and Wen Zhang; **E. Raymond Hunt Jr.,** Cuizhen Wang, D. Terrance Booth, Samuel E. Cox, Lalit Kumar, and Matthew Reeves; **Lalit Kumar**, Priyakant Sinha, Jesslyn F Brown, R. Douglas Ramsey, Matthew Rigge, Carson A. Stam, Alexander J. Hernandez, E. Raymond Hunt, Jr., and Matthew Reeves; **Molly E Brown**, Kirsten de Beurs, and Kathryn Grace; **José A.M. Demattê**, Cristine L.S. Morgan, Sabine Chabrillat, Rodnei Rizzo, Marston H.D. Franceschini, Fabrício da S. Terra, Gustavo M. Vasques, Johanna Wetterlind, Henrique Bellinaso, and Letícia G. Vogel; **E. Ben-Dor**, José Alexandre M. Demattê; **Prasad S. Thenkabail**.

Volume IV: Forests, Biodiversity, Ecology, LULC and Carbon: 12 chapters written by 71 authors (editor-in-chief: Prasad S. Thenkabail):

E.H. Helmer, Nicholas R. Goodwin, Valéry Gond, Carlos M. Souza Jr., and Gregory P. Asner; **Juha Hyyppä**, Xiaowei Yu, Mika Karjalainen, Xinlian Liang, Anttoni Jaakkola, Mike Wulder, Markus Hollaus, Joanne C. White, Mikko Vastaranta, Jiri Pyörälä, Tuomas Yrttimaa, Ninni Saarinen, Josef Taher, Juho-Pekka Virtanen, Leena Matikainen, Yunsheng Wang, Eetu Puttonen, Mariana Campos, Matti Hyyppä, Kirsi Karila, Harri Kaartinen, Matti Vaaja, Ville Kankare, Antero Kukko, Markus Holopainen, Hannu Hyyppä, Masato Katoh, Eric Hyyppä; **Gregory P. Asner**, Susan L. Ustin, Philip A. Townsend, and Roberta E. Martin; **Sylvie Durrieu**, Cédric Véga, Marc Bouvier, Frédéric Gosselin, Jean-Pierre Renaud, Laurent Saint-André; **Thomas W. Gillespie**, Morgan Rogers, Chelsea Robinson, Duccio Rocchini; **Stefan LANG**, Christina CORBANE, Palma BLONDA, Kyle PIPKINS, Michael FÖRSTER; **Conghe Song**, Jing Ming Chen, Taehee Hwang, Alemu Gonsamo, Holly Croft, Quanfa Zhang, Matthew Dannenberg, Yulong Zhang, Christopher Hakkenberg, Juxiang Li; **John Rogan** and Nathan Mietkiewicz; **Zhixin Qi**, Anthony Gar-On Yeh, Xia Li, Qianwen Lv; **R.A. Houghton; Wenge Ni-Meister; Prasad S. Thenkabail**.

Volume V: Water Resources: Hydrology, Floods, Snow and Ice, Wetlands, and Water Productivity—13 chapters written by 60 authors (editor-in-chief: Prasad S. Thenkabail):

James B. Campbell and Lynn M. Resler; **Sadiq I. Khan**, Ni-Bin Chang, Yang Hong, Xianwu Xue, Yu Zhang; **Santhosh Kumar Seelan; Allan S. Arnesen**, Frederico T. Genofre, Marcelo P. Curtarelli, and Matheus Z. Francisco; **Allan S. Arnesen**, Frederico T. Genofre, Marcelo P. Curtarelli, and Matheus Z. Francisco; **Sandro Martinis**, Claudia Kuenzer, and André Twele; **Le Wang**, Jing Miao, Ying Lu; **Chandra Giri; D. R. Mishra**, X. Yan, S. Ghosh, C. Hladik, J. L. O'Connell, H. J.

Cho; **Murali Krishna Gumma**, Prasad S. Thenkabail, Pranay Panjala, Pardhasaradhi Teluguntla, Birhanu Zemadim Birhanu, Mangi Lal Jat; **Trent W. Biggs**, Pamela Nagler, Anderson Ruhoff, Triantafyllia Petsini, Michael Marshall, George P. Petropoulos, Camila Abe, Edward P. Glenn; **Antônio Teixeira**, Janice Leivas; Celina Takemura, Edson Patto, Edlene Garçon, Inajá Sousa, André Quintão, Prasad S. Thenkabail, and Ana Azevedo; **Hongjie Xie,** Tiangang Liang, Xianwei Wang, Guoqing Zhang, Xiaodong Huang, and Xiongxin Xiao; **Prasad. S. Thenkabail**.

Volume VI: Droughts, Disasters, Pollution, and Urban Mapping—14 chapters written by 53 authors (editor-in-chief: Prasad S. Thenkabail)

Felix Kogan and Wei Guo; **F. Rembold**, M. Meroni, O. Rojas, C. Atzberger, F. Ham and E. Fillol; **Brian D.Wardlow**, Martha A. Anderson, Tsegaye Tadesse, Mark S. Svoboda, Brian Fuchs, Chris R. Hain, Wade T. Crow, and Matt Rodell; **Jinyoung Rhee**, Jungho Im, and Seonyoung Park; **Marion Stellmes**, Ruth Sonnenschein, Achim Röder, Thomas Udelhoven, Gabriel del Barrio, and Joachim Hill; **Norman Kerle; Stefan LANG**, Petra FÜREDER, Olaf KRANZ, Brittany CARD, Shadrock ROBERTS, Andreas PAPP; **Robert Wright; Krishna Prasad Vadrevu** and Kristofer Lasko; **Anupma Prakash**, Claudia Kuenzer, Santosh K. Panda, Anushree Badola, Christine F. Waigl; **Hasi Bagana**, Chaomin Chena, and Yoshiki Yamagata; **Yoshiki Yamagata**, Daisuke Murakami, Hajime Seya, and Takahiro Yoshida; **Qingling Zhang**, Noam Levin, Christos Chalkias, Husi Letu and Di Liu; **Prasad S. Thenkabail**.

The authors not only delivered excellent chapters, but they also provided valuable insights and inputs for me in many ways throughout the book project.

I was delighted when **Dr. Compton J. Tucker**, senior Earth scientist, Earth Sciences Division, Science and Exploration Directorate, NASA Goddard Space Flight Center (GSFC) agreed to write the foreword for the book. For anyone practicing remote sensing, Dr. Tucker needs no introduction. He has been a "godfather" of remote sensing and has inspired a generation of remote sensing scientists. I have been a student of his without ever really being one. I mean, I have not been his student in a classroom, but have followed his legendary work throughout my career. I remember reading his highly cited paper (now with citations nearing 7,700!):

- Tucker, C.J. (1979) "Red and Photographic Infrared Linear Combinations for Monitoring Vegetation," *Remote Sensing of Environment,* **8(2)**,127–150.

I first read this paper in 1986 when I had just joined the National Remote Sensing Agency (NRSA; now NRSC), Indian Space Research Organization (ISRO). Ever since, Dr. Tucker's pioneering works have been a guiding light for me. After getting his PhD from the Colorado State University in 1975, Dr. Tucker joined NASA GSFC as a postdoctoral fellow in 1975 and became a full-time NASA employee in 1977. Ever since, he has conducted several path-finding research studies. He has used NOAA AVHRR, MODIS, SPOT Vegetation, and Landsat satellite data for studying deforestation, habitat fragmentation, desert boundary determination, ecologically coupled diseases, terrestrial primary production, glacier extent, and how climate affects global vegetation. He has authored or coauthored more than 280 journal articles that have been **cited more than 93,000 times**, is an adjunct professor at the University of Maryland, is a consulting scholar at the University of Pennsylvania's Museum of Archaeology and Anthropology and has appeared in more than 20 radio and TV programs. He is a Fellow of the American Geophysical Union and has been awarded several medals and honors, including NASA's Exceptional Scientific Achievement Medal, the Pecora Award from the US Geological Survey, the National Air and Space Museum Trophy, the Henry Shaw Medal from the Missouri Botanical Garden, the Galathea Medal from the Royal Danish Geographical Society, and the Vega Medal from the Swedish Society of Anthropology and Geography. He was the NASA representative to the U.S. Global Change Research Program from 2006 to 2009. He is instrumental in releasing the AVHRR 33-year (1982–2014) **Global Inventory Modeling and Mapping Studies** (GIMMS) data. **I strongly recommend that everyone read his excellent foreword before**

reading the book. In the foreword, Dr. Tucker demonstrates the importance of data from Earth Observation (EO) sensors from orbiting satellites to maintaining a reliable and consistent climate record. Dr. Tucker further highlights the importance of continued measurements of these variables of our planet in the new millennium through new, improved, and innovative EO sensors from Sun-synchronous and/or geostationary satellites.

I want to acknowledge with thanks for the encouragement and support received by my U.S. Geological Survey (USGS) colleagues. I would like to mention the late Mr. Edwin Pfeifer, Dr. Susan Benjamin (my director at the Western Geographic Science Center), Dr. Dennis Dye, Mr. Larry Gaffney, Mr. David F. Penisten, Ms. Emily A. Yamamoto, Mr. Dario D. Garcia, Mr. Miguel Velasco, Dr. Chandra Giri, Dr. Terrance Slonecker, Dr. Jonathan Smith, Timothy Newman, and Zhouting Wu. Ofcouse, my dear colleagues at USGS, Dr. Pardhasaradhi Teluguntla, Dr. Itiya Aneece, Mr. Adam Oliphant, and Mr. Daniel Foley, have helped me in numerous ways. I am ever grateful for their support and significant contributions to my growth and this body of work. Throughout my career, there have been many postdoctoral level scientists who have worked with me closely and contributed in my scientific growth in different ways. They include Dr. Murali Krishna Gumma, head of Remote Sensing at the International Crops Research Institute for the Semi-Arid Tropics; Dr. Jun Xiong, Geo ML ≠ ML with GeoData, Climate Corp., Dr. Michael Marshall, associate professor, University of Twente, Netherlands; Dr. Isabella Mariotto, former USGS postdoctoral researcher; Dr. Chandrashekar Biradar, country director, India for World Agroforestry; and numerous others. I am thankful for their contributions. I know I am missing many names: too numerous to mention them all, but my gratitude for them is the same as the names I have mentioned here.

There is a very special person I am very thankful for: the late Dr. Thomas Loveland. I first met Dr. Loveland at USGS, Sioux Falls, for an interview to work for him as a scientist in the late 1990s when I was still at Yale University. But even though I was selected, I was not able to join him as I was not a U.S. citizen at that time and working for USGS required that. He has been my mentor and pillar of strength over two decades, particularly during my Landsat Science Team days (2006–2011) and later once I joined USGS in 2008. I have watched him conduct Landsat Science Team meetings with great professionalism, insights and creativity. I remember him telling my PhD advisor on me being hired at USGS: "We don't make mistakes!" During my USGS days, he was someone I could ask for guidance and seek advice and he would always be there to respond with kindness and understanding. And, above all, share his helpful insights. It is too sad that we lost him too early. I pray for his soul. Thank you, Tom, for your kindness and generosity.

Over the years, there are numerous people who have come into my professional life who have helped me grow. It is a tribute to their guidance, insights, and blessings that I am here today. In this regard, I need to mention a few names as gratitude: (1) Prof. G. Ranganna, my master's thesis advisor in India at the National Institute of Technology (NIT), Surathkal, Karnataka, India. Prof. Ranganna is 92 years old (2024). I met him few months back and to this day, he remains my guiding light on how to conduct oneself with fairness and dignity in professional and personal conduct. Prof. Ranganna's trait of selflessly caring for his students throughout his life is something that influenced me to follow. (2) Prof. E.J. James, former director of the Center for Water Resources Development and Management (CWRDM), Calicut, Kerala, India. Prof. James was my master's thesis advisor in India, whose dynamic personality in professional and personal matters had an influence on me. Dr. James' always went out of his way to help his students in spite of his busy schedules. (3) The late Dr. Andrew Ward, my PhD advisor at The Ohio State University in Columbus, Ohio. He funded my PhD studies in the U.S. through grants. Through him I learned how to write scientific papers and how to become a thorough professional. He was a tough task master, your worst critic (to help you grow), but also a perfectionist who helped you grow as a peerless professional, and above all a very kind human being at the core. He would write you long memos on the flaws in your research, but then help you out of it by making you work double the time! To make you work harder, he would tell you, "You won't get my sympathy." Then when you accomplished the task, he would tell you, "You have paid back for your scholarship many times over!!" (4) Dr. John G. Lyon, also my PhD advisor at The Ohio State University. He was a peerless motivator, encouraged you to believe in yourself. (5) Dr. Thiruvengadachari, a scientist at the National Remote Sensing Agency (NRSA), which is now the National Remote Sensing Center (NRSC), India. He was my first boss at the Indian Space Research Organization (ISRO) and through him I learned the initial steps in remote sensing science. I was just 25 years of age then and had joined NRSA after my Master of Engineering (hydraulics and water resources) and Bachelor of Engineering (civil engineering) degrees. The first day in office Dr. Thiruvengadachari asked me how much remote sensing I knew. I told him "Zero" and instantly thought, he will ask me to leave the room. But his response was "very good!" and he gave me a manual on remote sensing from Laboratory for Applications of Remote Sensing (LARS), Purdue University, to study. Those were the days when there was no formal training in remote sensing in universities. So, my remote sensing lessons began working practically on projects and one of our first projects was "drought monitoring for India using NOAA AVHRR data." This was an intense period of learning the fundamentals of remote sensing science for me by practicing on a daily basis. Data came in 9 mm tapes, data was read on massive computing systems, image processing was done mostly working on night shifts by booking time on centralized computing, fieldwork was conducted using false color composite (FCC) outputs and topographic maps (there was no global positioning systems or GPS), geographic information system (GIS) was in its infancy, and a lot of calculations were done using calculators, as we had just started working in IBM 286 computers with floppy disks. So, when I decided to resign my NRSA job and go to the United States to do my PhD, Dr. Thiruvengadachari told me, "Prasad, I am losing my right hand, but you can't miss this opportunity." Those initial wonderful days of learning from Dr. Thiruvengadachari will remain etched in my memory. I am also thankful to my very good old friend Shri C.J. Jagadeesha, who was my colleague at NRSA/NRSC, ISRO. He was a friend who encouraged me to grow as a remote sensing scientist through our endless rambling discussions over tea in Iranian restaurants outside NRSA those days and elsewhere.

I am ever grateful to my former professors at The Ohio State University, Columbus, Ohio, USA: the late Prof. Carolyn Merry, Dr. Duane Marble, and Dr. Michael Demers. They have taught, and/or encouraged, and/or inspired, and/or gave me opportunities at the right time. The opportunity to work for six years at the Center for Earth Observation of the Yale University (YCEO) was incredibly important. I am thankful to Prof. Ronald G. Smith, director of YCEO for the opportunity, guidance, and kindness. At YCEO, I learned and advanced myself as a remote sensing scientist. The opportunities I got working for the International Institute of Tropical Agriculture (IITA) based in Nigeria and the International Water Management Institute (IWMI) based in Sri Lanka, where I worked on remote sensing science pertaining to a number of applications such as agriculture, water, wetlands, food security, sustainability, climate, natural resources management, environmental issues, droughts, and biodiversity water, were extremely important in my growth as a remote sensing scientist—especially from the point of view of understanding the real issues on the ground in real-life situations. Finding solutions and applying one's theoretical understanding to practical problems and seeing them work has its own nirvana.

As it is clear from the preceding, it is of great importance to have guiding pillars of light at crucial stages of your education. That is where you become what you become in the end, grow, and make your own contributions. I am so blessed to have had these wonderful guiding lights come into my professional life at right time of my career (which also influenced me positively in my personal life). From that firm foundation, I could build on from what I learned and through the confidence of knowledge and accomplishments pursue my passion for science and do several significant pioneering research projects throughout my career.

I mention all of the preceding as a gratitude for my ability today to edit such a monumental *Remote Sensing Handbook* (Second Edition, Volumes I–VI).

I am very thankful to Ms. Irma Shagla-Britton, manager and leader for Remote Sensing and GIS books at Taylor and Francis/CRC Press. Without her consistent encouragement to take on this responsibility of editing *Remote Sensing Handbook*, especially in trusting me to accomplish this momentous work over so many other renowned experts, I would never have gotten to work on this in the first place. Thank you, Irma. Sometimes you need to ask several times, before one can say yes to something!

I am very grateful to my wife (Ms. Sharmila Prasad), daughter (Dr. Spandana Thenkabail), and son-in-law (Mr. Tejas Mayekar) for their usual unconditional understanding, love, and support. My wife and daughter have always been pillars of my life, now joined by my equally loving son-in-law. I learned the values of hard work and dedication from my revered parents. This work wouldn't come through without their life of sacrifices to educate their children and their silent blessings. My father's vision in putting emphasis on education and sending me to the best of places to study despite our family's very modest income and my mother's endless hard work are my guiding light and inspiration. Ofcouse, there are many, many others to be thankful for, but too many to mention here. Finally, it must be noted that the work of this magnitude, editing monumental *Remote Sensing Handbook* (Second Edition, Volumes I–VI) continuing from the three-volume first edition, requires blessings of almighty. I firmly believe nothing happens without the powers of the universe blessing you and providing needed energy, strength, health, and intelligence. To that infinite power my humble submission of everlasting gratefulness.

It has been my deep honor and great privilege to have edited the *Remote Sensing Handbook* (Second Edition, Volumes I–VI) after having edited the three-volume first edition that was published in 2014. Now, after ten years, we will have a six-volume second edition in the year 2024. A huge thanks to all the authors, publisher, family, friends, and everyone who made this huge task possible.

Dr. Prasad S. Thenkabail, PhD
Editor in Chief
Remote Sensing Handbook (Second Edition, Volumes I–VI)

Volume I: Sensors, Data Normalization, Harmonization, Cloud Computing, and Accuracies
Volume II: Image Processing, Change Detection, GIS, and Spatial Data Analysis
Volume III: Agriculture, Food Security, Rangelands, Vegetation, Phenology, and Soils
Volume IV: Forests, Biodiversity, Ecology, LULC and Carbon
Volume V: Water, Hydrology, Floods, Snow and Ice, Wetlands, and Water Productivity
Volume VI: Droughts, Disasters, Pollution, and Urban Mapping

Part I

Vegetation and Biomass

1 Measuring Terrestrial Productivity from Space

Alfredo Huete, Guillermo Ponce-Campos, Yongguang Zhang, Natalia Restrepo-Coupe, and Xuanlong Ma

ACRONYMS AND DEFINITIONS

AGB	Aboveground Biomass
APAR	Absorbed Photosynthetically Active Radiation
AVHRR	Advanced Very-High-Resolution Radiometer
EO	Earth Observation
EVI	Enhanced Vegetation Index
FLUXNET	A network of micrometeorological tower sites to measure carbon dioxide, water, and energy balance between terrestrial systems and the atmosphere
GPP	Gross Primary Productivity
LAI	Leaf Area Index
LiDAR	Light Detection and Ranging
LST	Land Surface Temperature
LUE	Light Use Efficiency
MERIS	Medium Resolution Imaging Spectrometer
MODIS	Moderate-Resolution Imaging Spectroradiometer
NASA	National Aeronautics and Space Administration
NDVI	Normalized Difference Vegetation Index
NIR	Near-Infrared
NOAA	National Oceanic and Atmospheric Administration
NPP	Net Primary Productivity
PAR	Photosynthetically Active Radiation
PRI	Photochemical Reflectance Index
SOC	Soil Organic Carbon
SPOT	Satellite Pour l'Observation de la Terre, French Earth Observing Satellites
TOA	Top of Atmosphere
TROPOMI	TROPOspheric monitoring instrument
VI	Vegetation Index

1.1 INTRODUCTION

Terrestrial productivity is defined as the process by which plants use sunlight to produce organic matter from carbon dioxide through photosynthesis. In the process of photosynthesis, plants absorb CO_2 from the atmosphere and use sunlight to combine it with water to form sugar molecules (CH_2O) and oxygen (O_2). The chemical reaction is:

$$CO_2 + H_2O + \text{light energy} = CH_2O + O_2 \tag{1.1}$$

DOI: 10.1201/9781003541165-2

The organic compounds produced by plants are assimilated for building plant structures (leaves, stems, wood branches, trunks, and roots) and stored in biomass (Figure 1.1). Vegetation productivity forms the basis of terrestrial biosphere functioning and carbon, energy, and water budgets. Accurate estimates of vegetation productivity across space and time are necessary for quantifying carbon balances at regional to global scales (Schimel, 1998).

Part of the carbon fixed in photosynthesis by plants is returned to the atmosphere through the process of respiration, in which oxygen combines with sugar to release water, CO_2, and energy. Through autotrophic respiration, plants break down the sugar to get the energy needed to grow.

The basic chemical reaction in respiration is:

$$CH_2O + O_2 = CO_2 + H_2O + \text{energy} \tag{1.2}$$

1.1.1 Key Productivity Definitions

There are various ways to define productivity, including:

1. The rate of carbon fixation by the terrestrial biosphere, or total plant organic carbon produced per unit of time and over a defined area is termed photosynthesis, or gross primary productivity (GPP). GPP is the basis for food, fiber, and wood production, and

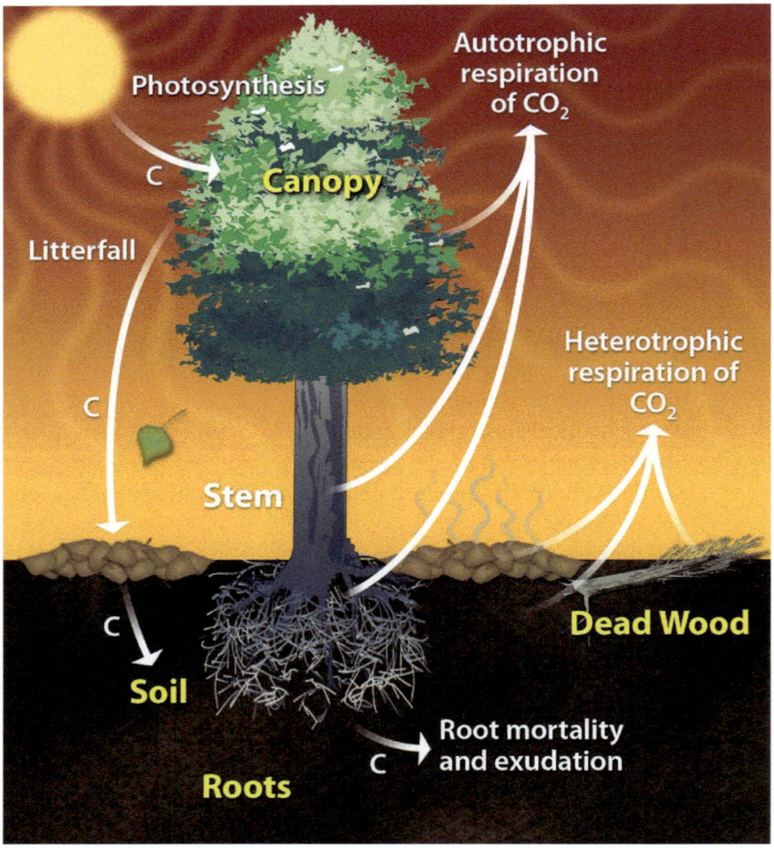

FIGURE 1.1 Terrestrial carbon cycle. (Source: https://public.ornl.gov/site/gallery/.)

has important implications for human welfare. It is the largest carbon flux between the terrestrial biosphere and the atmosphere and is a key measure of ecosystem metabolism (Figure 1.2).

2. Terrestrial net primary productivity (NPP) is the rate of carbon fixed by plants and accumulated as biomass (Cramer et al., 1999; Zhao and Running, 2010). As part of the carbon fixed in photosynthesis by plants is returned to the atmosphere through autotrophic respiration (R_a), NPP is the net balance between GPP and R_a:

$$\text{NPP} = \text{GPP}—R_a \quad (1.3)$$

3. The accumulated plant biomass may be consumed by other organisms, along food chains, which further break down the plant sugar to get energy (heterotrophic respiration, or R_h). Net ecosystem productivity (NEP) is defined as GPP minus total ecosystem respiration (ER), which is the sum of autotrophic and heterotrophic respiration, including respiration from soils, microbes, plants, and litter:

$$\text{NEP} = \text{NPP}—R_h \quad (1.4)$$

4. Net biome productivity (NBP) is defined as the overall net ecosystem carbon balance, and includes other processes such as deforestation, harvest, fire, disease, and disturbance that can also lead to the loss of, and changes in, carbon (Figure 1.2):

$$\text{NBP} = \text{NEP minus disturbance carbon losses} \quad (1.5)$$

An area will be designated a land carbon sink when GPP exceeds ecosystem respiration and disturbance carbon losses, that is, positive NBP. When the reverse occurs, an area is considered as a land carbon source.

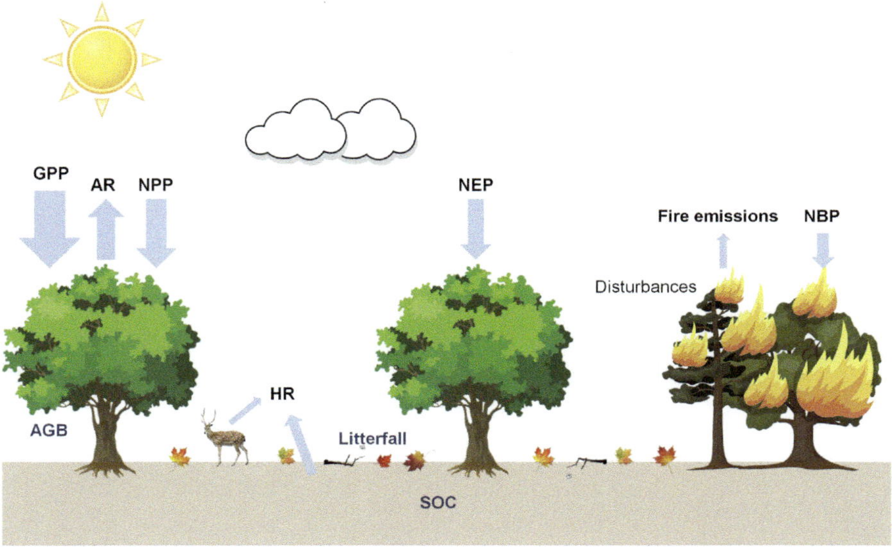

FIGURE 1.2 Terrestrial carbon fluxes: GPP—Gross Primary Productivity; AR—Autotrophic Respiration; NPP—Net Primary Productivity; HR—Heterotrophic Respiration; NEP—Net Ecosystem Productivity; NBP—Net Biome Productivity. Carbon stocks include AGB—Above Ground Biomass; SOC—Soil Organic Carbon. (Source, Xiao et al., 2019, Figure 1.1.)

1.1.2 Seasonal Biosphere Productivity

Vegetation productivity will vary at seasonal to annual time scales, as plants grow, develop, and senesce through their growing cycle phenophases. There is a tightly coupled relationship between the atmosphere and biosphere that can be observed with CO_2 fluctuations in the atmosphere with the changing seasons. When the large land masses of the Northern Hemisphere green up during spring and summer, atmosphere CO_2 decreases as carbon is removed from the atmosphere though increased photosynthesis. In the autumn and winter, many plants senesce and decay with GPP declining while ecosystem respiration increases, thus returning CO_2 back to the atmosphere.

1.1.3 Constraints on Productivity

Vegetation productivity is also spatially variable, limited or co-limited by spatially and temporally varying resource constraints (e.g., nutrients, light, water, temperature; Nemani et al., 2003) (Figure 1.3). GPP and microbial respiration will be primarily limited by water availability in arid regions, by cold temperatures at high latitudes and elevations, and by solar radiation in humid regions. This results in spatially distributed high biomass, productive ecosystems as well as low biomass areas (e.g., deserts) with sparse production over the planet in response to climatic, geologic (nutrients), and topographic variations. Improved knowledge of these resource constraints and main drivers of plant productivity is needed for predictable assessments of climate change. GPP is essentially an integrator of resource availability and disturbances, and according to the resource optimization theory (Field et al., 1995), ecological processes tend to adjust plant characteristics over time periods of weeks or months to match the capacity of the environment to support photosynthesis and maximize growth.

1.2 MEASURES OF PRODUCTIVITY

Accurate estimates of photosynthesis and productivity in the biosphere are necessary for quantifying carbon balances across space and time, and at regional to global scales, but are quite challenging to accomplish (Ryu et al., 2019: Baldocchi et al., 2001; Schimel, 1998). The determination of productivity across diverse landscapes is traditionally carried out in various, often inconsistent ways,

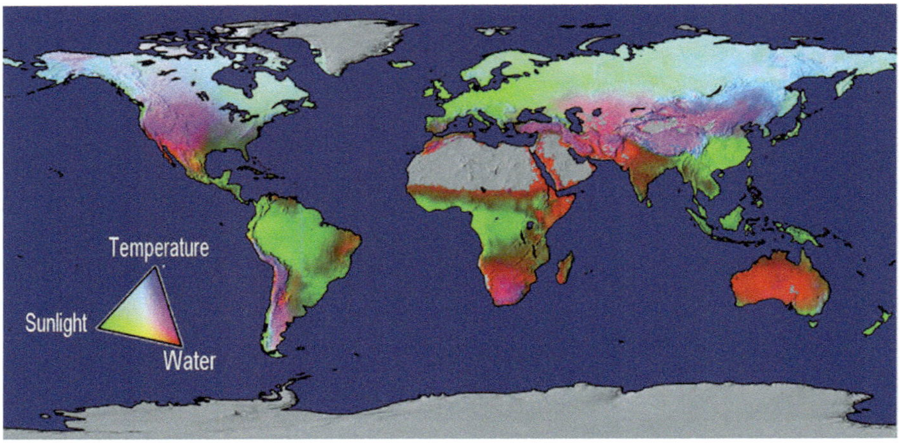

FIGURE 1.3 Potential limits and co-limitations to vegetation net primary production based on fundamental physiological limits of solar radiation, water balance, and temperature. Greener colors depict biomes increasingly limited by radiation, while red colors are water-limited and blue colors are temperature-limited. Many regions are limited by more than one factor. (Adapted from Nemani et al., 2003.)

that include field and plot scale biomass measurements, plant harvests, micrometeorological flux measurements, Earth Observations (EO), and empirical to process-based models.

1.2.1 Field Methods

In situ measures of productivity and carbon stocks involve sampling methods that will vary with biome type, and can include tree inventories, litter traps, grassland forage estimates, destructive sampling, cropland harvests, and market statistics. Traditionally, national-scale carbon monitoring has been accomplished with networks of field inventory plots, which provide direct carbon measurements of only very small areas of forest, and are further difficult to install, monitor, and maintain over time (Chambers et al., 2009).

Established long-term experimental plots further enable cross-site production comparisons. Plot-level methods generally measure aboveground net primary production (ANPP) from which gross primary production (GPP) can be estimated by correcting for respiratory losses (Field et al., 1995). Agricultural yield statistics combined with maps of cropland areas provide large-scale ANPP estimates from local to national level census statistics (Monfreda et al., 2008; Guanter et al., 2014). These methods are amenable to many uncertainties due to differences in site-based procedures, and in some cases, inconsistent sampling methods over time at a given site (Sala et al., 1988; Biondini et al., 1991; Moran et al., 2014). For example, established procedures for biomass clipping of grass sampling plots vary in timing from either the peak of the growing season or the end of the growing season (Moran et al., 2014).

1.2.2 Eddy Covariance Flux Towers

A global network of "regional networks" encompassing over 1,000 active and historic flux micrometeorological tower sites, known as FLUXNET, provide continuous measurements of carbon, water, and energy exchanges between ecosystems and the atmosphere (Running et al., 1999; Baldocchi, 2008). The eddy covariance method is used to directly measure fluxes at a spatial scale of hundreds of meters, by computing the covariance between the vertical velocity and target scalar mixing ratios at each tower site. The carbon gas flux measured is the net amount resulting from ecosystem respiration (ER) and photosynthesis (GPP), and is termed net ecosystem exchange (NEE):

$$\text{NEE} = \text{GPP} + \text{ER} \tag{1.6}$$

This yields information on diurnal, daily, and seasonal dynamics plus inter-annual variations of net ecosystem exchange (NEE) of carbon dioxide between the land surface and the atmosphere (Baldocchi et al., 2001; Verma et al., 2005). GPP is the residual calculated from the direct measurements of NEE and estimates of respiration. This generates valuable in situ information for validation and independent assessments of EO-based productivity products and carbon models (Xiao et al., 2019).

1.2.3 Earth Observation Data

For five decades, EO data has played an increasingly important role in global carbon cycle studies (Xiao et al., 2019; Ryu et al., 2019). EO sensors offer synoptic-scale observations of ecosystem states, land cover, landscape dynamics, disturbances, and fire, and are seen as invaluable tools to help fill the large spatial gaps and restrictive coverage afforded by situ measurements, as well as to better constrain and improve the accuracies of terrestrial ecosystem models. EO provides much needed data and information of under-sampled critical regions (e.g., tropical, arctic/boreal, and arid environments) and facilitates broad-scale patterns of ecosystem functioning. A strategic combination of EO and in situ data, can provide the dense sampling in space and time required to characterize the heterogeneity of ecosystem structure and function.

A wide variety of satellite sensors have been used to quantify carbon fluxes (Running et al., 2004; Xiao et al., 2019) and biomass stocks (Saatchi et al., 2011) at various spatial and temporal scales. These include sensors in the optical, thermal, and microwave regions of the electromagnetic (EM) spectrum, and include measures of vegetation chlorophyll content, leaf area, biomass, non-photosynthetic vegetation (NPV), solar-induced chlorophyll fluorescence (SIF), canopy height and structure, and atmosphere carbon gases (Figure 1.4). Recent satellites with high resolution spectrometers, which can retrieve concentrations of key atmospheric gases (such as the Orbiting Carbon Observatory-2; OCO-2), are providing valuable insights into global variations in, and sources of, greenhouse gases. This combination renders EO a powerful tool for studying vegetation productivity at local, regional, and global scales (Gitelson et al., 2006; Schimel et al., 2015). A summary of the various EO-based estimates of GPP is shown in Table 1.1.

1.3 EO PRODUCTIVITY STUDIES

There are many empirical, diagnostic, and process-based models that have been developed during the EO era to quantify land productivity, with many of these methods integrating independently derived carbon flux measurements from satellite data, EC tower fluxes, and field measurements. Vegetation productivity is directly related to the interaction of solar radiation with the plant canopy. As plants grow, the productivity gain resulting from their conversion of CO_2 into biomass through photosynthesis, can be quantified with EO approaches involving vegetation indices (VIs), solar-induced chlorophyll fluorescence (SIF), and light use efficiency (LUE) models.

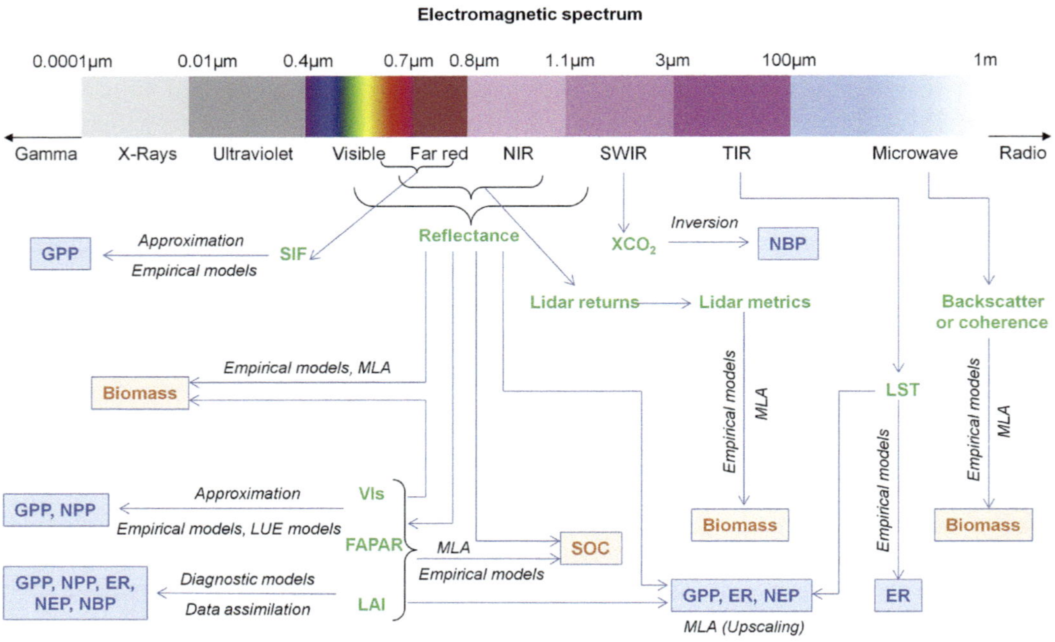

FIGURE 1.4 Earth Observations (EO) of terrestrial ecosystem productivity and carbon cycle studies. (Source: Xiao et al., 2019, Figure 1.2.)

TABLE 1.1

Examples of Remote Sensing Methods of Deriving Gross Primary Productivity, GPP, with Some References

Gross Primary Productivity Measurement	Biome/Location	Satellite Products used	Other Non-Satellite Drivers	Method/Approach	Equation	R2	Reference
BIOME-BGC (Biogeochemical Cycles) MODIS GPP/NPP, where GPP is Gross Primary Productivity and NPP is Net Primary Productivity	Continental	MODIS FPAR (MOD15); photosynthetic active radiation (PAR) as 0.45 × SWrad (shortwave downward solar radiation)	Maximum light use efficiency (ε_{max}) from a biome-properties look-up table and maximum daily vapor pressure deficit (VPD) and minimum daily air temperature (Tmin) from forcing meteorology.	See equation	$GPP = \varepsilon_{max} \times 0.45 \times SWrad \times fPAR \times f(VPD) \times f(Ta)$	NA	Running et al., 2004
GPP and light use efficiency (LUE)	North American ecosystems from evergreen needleleaf and deciduous forest to grassland to savanna	MODIS EVI	NA	Linear regression	GPP = m × EVI + bLUE = m × EVI + b	0.76 MODIS GPP - GPP0.92 EVI-GPP0.76 EVI - LUE0.62 MODIS LUE -LUE	Sims et al., 2006
GPP	Tropical forests and converted pastures at the Amazon basin	MODIS EVI	NA	Linear regression	GPP = m × EVI + b	0.5	Huete et al., 2006
GPP and maximum Net Ecosystem Exchange (NEE_{max})	Northern Europe ecosystems from evergreen needleleaf and deciduous forest to grasslands	MODIS EVI	NA	Linear regression	GPP = m × EVI + bNEE_{max} = m × EVI + b	GPP-EVI:0.81 deciduous, 0.69, coniferous forests NEE_{max}-EVI:0.83 deciduous. 0.72, coniferous forests	Olofsson et al., 2008
GPP	Dry to humid tropical forest sites in Southeast Asia	MODIS EVI	NA	Linear regression	GPP = 8282 × EVI + 2118, GPP [kgC ha^{-1} mo^{-1}]	0.74	Huete et al., 2008

(Continued)

TABLE 1.1 *(Continued)*

Examples of Remote Sensing Methods of Deriving Gross Primary Productivity, GPP, with Some References

Gross Primary Productivity Measurement	Biome/Location	Satellite Products used	Other Non-Satellite Drivers	Method/Approach	Equation	R2	Reference
GPP	African tropical savanna ecosystems including shrubland, woodlands, crops and grasslands	MODIS EVI	NA	Linear regression	GPP = m × EVI + b	NA	Sjöström et al., 2011
GPP	Northern Australian mesic and xeric tropical savannas	MODIS EVI and MODIS GPP product	Eddy covariance measured water availability index (EF) and PAR	Linear regression	GPP = m × EVI + bGPP = b + m(MODIS GPP) GPP = b + m(EVI × PAR) GPP = b + m(EVI × PAR × EF)	Linear regression EVI-GPP ranges from 0.89 (woodland) to 0.52 (wooded grassland)	Ma et al., 2013
Temperature and Greenness Model (T-G)	North American ecosystems from evergreen needleleaf and deciduous forest to grassland to savanna	MODIS daytime land surface temperature (LST) and EVI	NA	See equation	GPP = ($EVI_{scaled \times}$ LST_{scaled}) × *m* LST_{scaled} = min[(LST/30); (2.5 – (0.05 × LST))] EVI_{scaled} = EVI–0.10	NA	Sims et al., 2008
Greenness and Radiation (G-R) model	Crops, including soybean and maize–soybean rotation	MODIS NDVI and a chlorophyll-related spectral index (VI_{chl}): EVI or wide dynamic range vegetation index (WDRVI)	PAR_{toc} is the top of canopy measured PAR	See equation	GPP = VI_{chl} × PAR_{toc} GPP = NDVI × PAR_{toc}	0.84 GPP = EVI2 × PAR_{toc} 0.87 GPP = Red edge NDVI × PAR_{toc} 0.9 to 0.9 GPP = EVI × PAR_{toc}	Peng and Gitelson, 2012
Greenness and Radiation (G-R) model	Northern Australian mesic and xeric tropical savannas	MODIS EVI	PAR_{toa} is the top of atmosphere PAR	See equation	GPP = EVI × PAR_{toa}	NA	Ma et al., 2014

Temperature and Greenness Model (T-G) and Greenness and Radiation (G-R) model	Temperate and boreal forest ecosystems in North America	MODIS EVI	NA	Linear regression	GPP = m × EVI + bGPP = ScaledEVI × ScaledLST	T-G model GPP:0.27 to 0.91 at non-forests~0.9 at deciduous forests 0.28 to 0.91 evergreen forests	Wu et al., 2011
Vegetation Photosynthesis Model (VPM)	Single temperate deciduous broadleaf ecosystem forest	MODIS EVI, NDVI, LSWI, water (W_{scalar}), leaf phenology (P_{scalar})	Temperature (air) and leaf phenology information T_{scalar}, P_{scalar}, respectively	See equation	GPP = ε × fAPARchl × PARtocε = εmax × Tscalar × Wscalar × PscalarGPP = −74.4 + 179.4 × NDVIGPP = −68.3 + 299.7 × EVI, GPP [gC m^{-2} 10-day]	GPP-NDVI, 0.64GPP-EVI, 0.84GPP - VPM GPP, 0.92	Xiao et al., 2004
Vegetation Photosynthesis and Respiration Model (VPRM) Net Ecosystem Exchange, NEE = GPP - ecosystem respiration	Nine vegetation classes, including evergreen and deciduous forests, grasslands, and shrub sites in North America	MODIS EVI, and LSWI	Incoming solar radiation and air temperature	Model	NA	Monthly NEE - VPRM NEE ranges from 0.96 (deciduous temperate forest) to > 1 at grasses and agricultural areas	Mahadevan et al., 2008
Light use efficiency, LUE	Crops: sunflower	Photochemical reflectance index (PRI)	NA	Linear regression	LUE = m × PRI + b	NA	Gamon et al., 1992;
GPP	Cropland and grassland ecosystems	Solar-induced chlorophyll fluorescence (SIF)	NA	Linear regression	US croplands: GPP = −0.88 + 3.55 × SIFEurope grasslands: GPP = 0.35 + 3.71 × SIFAll sites: GPP = −0.17 + 3.48 × SIF, GPP [gC m^{-2} d^{-1}]	0.92, US croplands0.79, Europe grasslands0.87, All sites	Guanter et al., 2014

1.3.1 EO Measures of GPP

EO-based estimates of GPP have been implemented at global scales, based on the LUE equation that defines the amount of carbon fixed through photosynthesis as proportional to the solar energy absorbed by green vegetation multiplied by the efficiency with which the absorbed light is used in carbon fixation (Pei et al., 2022; Ryu et al., 2019):

$$GPP = \varepsilon_{par} \times APAR = \varepsilon_{par} \times fAPAR \times PAR \tag{1.7}$$

where

- ε_{par}, or LUE, is the efficiency of conversion of absorbed light into aboveground biomass;
- PAR is the photosynthetically active radiation;
- APAR is the absorbed PAR integrated over a time period; and
- *f*APAR is the fraction of APAR to the PAR available.

Monteith (1972) suggested that productivity of stress-free annual crops would be linearly related to vegetation absorbed PAR.

EO-based modeling approaches focus on the fAPAR term, which is derived through spectral vegetation index (VI) relationships (Asrar et al., 1984; Sellers, 1985; Goward and Huemmrich, 1992). VIs exploit the chlorophyll-absorbing red band relative to the non-absorbing and high scattering near-infrared (NIR) band to assess the amount and vigor of a vegetated pixel. The normalized difference vegetation index (NDVI, Tucker, 1979) is written as,

$$NDVI = (\rho_{NIR} - \rho_{red}) / (\rho_{NIR} + \rho_{red}) \tag{1.8}$$

where ρ_{NIR} and ρ_{red} are spectral reflectance values (unitless) for NIR and red band, respectively.

Asrar et al. (1984) showed the NDVI was linearly related with APAR and thereby related to productivity through the potential capacity of vegetation to absorb light for photosynthesis (Figure 1.5). The linear relationship between NDVI and *f*APAR has been documented through field measurements (Fensholt et al., 2004) and theoretical analyses (Sellers, 1985; Goward and Huemmrich, 1992; Myneni and Williams, 1994), although these relationships were not universal, and appeared to be unique to vegetation type, structure, and soil optics.

1.3.2 VI Relationships with Flux Tower GPP

Monteith and Unsworth (1990) noted that VIs can legitimately be used to estimate the rate of processes that depend on absorbed light, such as photosynthesis and transpiration. The integration of tower measured carbon fluxes with VIs has thus been the focus of many investigations across many ecosystems (Zeng et al., 2022; Sjöström et al., 2011; Glenn et al., 2008; Sims et al., 2006; Gitelson et al., 2006; Rahman et al., 2005). As top-of canopy measurements, flux towers do not require knowledge of LAI or details of canopy architecture to estimate fluxes facilitating their comparisons with satellite spectral index measures that similarly involve community properties to estimate greenness, and thereby minimizing the need for meteorological and LUE information. With the measurement footprint of flux towers at least partially overlapping the pixel size of daily return satellites (e.g., 250 m for MODIS), VIs can simplify the upscaling of carbon fluxes, such as photosynthesis, from the network of flux towers to larger landscape units and to regional scales (Schimel et al., 2015).

Several studies reported that the Enhanced Vegetation Index (EVI) provided reasonably accurate estimates of GPP across a wide range of ecosystems (including dense forests) in North America (Rahman et al., 2005), Northern Europe (Olofsson et al., 2008), African tropical savanna ecosystems (Sjöström et al., 2011), mesic and xeric tropical savannas in northern Australia (Ma et al.,

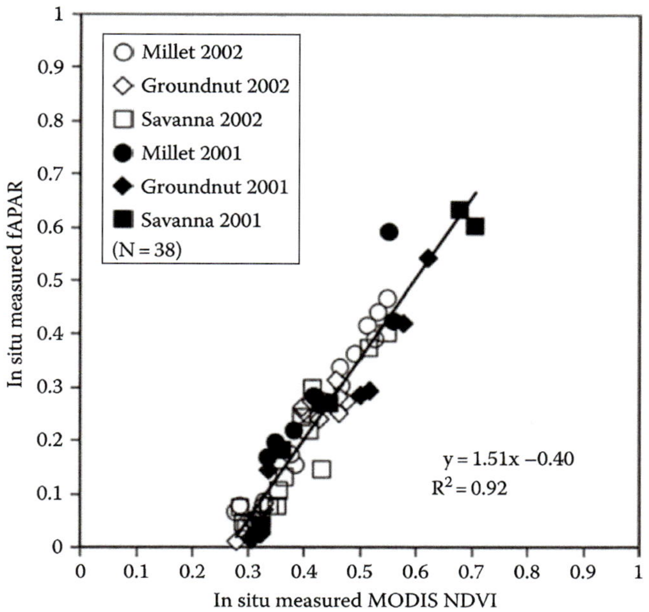

FIGURE 1.5 Linear relationship between in situ NDVI and field measured *f*APAR across multiple cropland and biome sites in Africa. (Source: Fensholt et al., 2004.)

2013), and in dry to humid tropical forest sites in Southeast Asia and the Amazon (Xiao et al., 2005; Huete et al., 2006, 2008). EVI is expressed as:

$$EVI = G \cdot \frac{\rho_{NIR} - \rho_{red}}{\rho_{NIR} + L + C_1 \cdot \rho_{red} - C_2 \cdot \rho_{blue}} \tag{1.9}$$

where ρ_{NIR}, ρ_{red}, and ρ_{blue} are atmospherically corrected spectral reflectances; G is a gain factor; and C_1 and C_2 are aerosol resistance coefficients; and L functions as the soil-adjustment factor, with all terms dimensionless (Huete et al., 2002). In the MODIS EVI product, C_1 and C_2 are 6.0 and 7.5, respectively, and G is 2.5.

The direct relationships between EVI and tower GPP may be partly a result of fairly good correlations between LUE and EVI that make an independent estimate of LUE less necessary. Sims et al. (2006) reported that LUE derived from nine flux towers in North America was well-correlated with EVI ($r^2 = 0.76$), while Wu et al. (2011) reported moderate correlation between EVI and tower LUE in temperate and boreal forest ecosystems in North America. Further, the 16-day averaging period may also remove the influences of short-term fluctuations in solar radiation and other environmental parameters, thereby minimizing the need for climatic drivers.

On the hand, the strength of the linear relationship between EVI and tower GPP can vary significantly with biome type, being greater in seasonally contrasting biomes (e.g., deciduous forests) compared with more aseasonal evergreen forests (Rahman et al., 2005; Sims et al., 2006). Sims et al. (2006) further noted that when data from the winter period of inactive photosynthesis was excluded by use of land surface temperature (LST) observations below 0°C, the EVI-tower GPP relationship improved.

Similarly, distinct differences in VI relationships with tower GPP were found between phenology-driven and meteorologically driven Australian ecosystems. Restrepo-Coupe et al. (2016) found that in primarily meteorologically driven (e.g., PAR, temperature, and/or precipitation) and relatively aseasonal ecosystems, there were no statistically significant relationships between GPP and MODIS vegetation products (LAI, fAPAR, VIs). The poor correlations were observed where meteorology and phenology were asynchronous (e.g., Mediterranean ecosystems) and VI greenness values may hold constant even when temperatures inhibit photosynthesis or when warming temperatures induce high rates of photosynthesis with no corresponding change in greenness. On the other hand, in phenology-driven ecosystems, changes in the vegetation status can be well-represented by VIs, and highly correlated VI-GPP relationships can be found in locations where key meteorological variables and vegetation phenology were synchronous.

1.3.3 Temperature-Greenness (T-G) Model

Sims et al. (2008) introduced the temperature and greenness (T-G) model, using combined daytime LST and EVI MODIS products to take into account the restrictive role of non-optimal temperatures on vegetation productivity. They found the T-G model substantially improved the correlation between predicted and measured GPP at 11 EC flux tower sites across North American biomes compared with the MODIS GPP product or MODIS EVI alone, while keeping the model based entirely based on remotely sensed variables without any ground-based meteorological inputs (Sims et al., 2008). The T-G model may be described as follows,

$$GPP = (EVI_{scaled} \times LST_{scaled}) \times m \tag{1.10}$$

$$LST_{scaled} = \min[(LST/30); (2.5-(0.05 \times LST))] \tag{1.11}$$

$$EVI_{scaled} = EVI-0.10 \tag{1.12}$$

where LST_{scaled} sets GPP to zero when LST is less than zero, and defines the inactive winter period; EVI_{scaled} adjusts EVI values to a zero baseline value in which GPP is known to be zero; m is a scalar that varies between deciduous and evergreen sites, with units of mol C m^{-2} day^{-1}; and LST_{scaled} also accounts for low temperature limitations to photosynthesis when LST is between 0°C and 30°C, and accounts for high temperature and high VPD stress in sites that exceed LST values of 30°C (Sims et al., 2008; Figure 1.6).

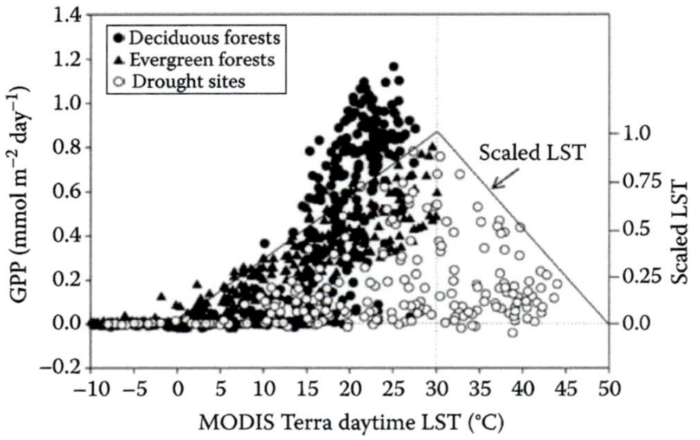

FIGURE 1.6 Gross primary productivity (GPP) measured at the eddy covariance flux towers as a function of daytime land surface temperature (LST) measured by the MODIS satellite. Solid line represents scaled LST from the Temperature-Greenness (T-G) model. GPP is enhanced by increasing temperatures, but only to approximately 30°C before being negatively influenced (Source: Sims et al., 2008.)

LST is closely related to VPD and thus can provide a measure of drought stress, consistent with the BIOME-BGC model, where temperature and VPD are used as scalars directly modifying LUE (Running et al., 2004). LST is a useful measure of physiological activity of the upper canopy leaves, provided that leaf cover is high enough that LST is not significantly affected by soil surface temperature. Thus, the T-G model has been found less useful in sparsely vegetated ecosystems (e.g., shrublands) where soil surface temperatures significantly influence derived LST values, rendering them less useful as indicators of plant physiology. As an example, Ma et al. (2014) found coupling EVI with LST showed no improvements in predicting savanna GPP compared with using EVI alone over the relatively open tropical savannas in northern Australia, with appreciable soil exposure. This may also be due to temperature not being a limiting factor or significant driver of photosynthesis in tropical savannas (Leuning et al., 2005; Cleverly et al., 2013; Kanniah et al., 2013a).

1.3.4 Greenness and Radiation (G-R) LUE Models

Chlorophyll spectral indices can be coupled with measures of light energy, PAR, to provide robust estimates of GPP. Canopy level chlorophyll represents a community property that is most relevant in quantifying the amount of absorbed radiation used for productivity. The fraction of PAR specifically absorbed by chlorophyll throughout the canopy (*f*APAR*chl*) can lead to more accurate GPP estimates than through more general *f*APAR products that may include non-photosynthetic PAR absorption (Zhang et al., 2014b). Gitelson et al. (2006) showed that for the same LAI amount, the chlorophyll content during the green-up stage might be more than two times higher than the chlorophyll content in leaves in the reproductive and senescence stages. In the Greenness and Radiation (G-R) model, both *f*APAR and LUE are driven by total chlorophyll content with strong correlations between GPP/PAR and canopy chlorophyll content (Gitelson et al., 2006; Peng et al., 2011),

$$\mathrm{GPP} \propto \mathrm{VI}chl \times \mathrm{PAR}toc \tag{1.13}$$

where VI*chl* is a chlorophyll-related spectral index and PAR*toc* is the top-of-canopy measured PAR (MJ m^{-2} day^{-1}).

Chlorophyll spectral indices may include (1) commonly used VIs, such as EVI and the Wide Dynamic Range Vegetation Index (WDRVI), which indirectly indicate total chlorophyll content through "greenness" estimates; and (2) narrow-band chlorophyll spectral indices, such as the MERIS Terrestrial Chlorophyll Index (MTCI), which more directly represent canopy scale chlorophyll content (Peng et al., 2013). The WDRVI equation is:

$$\mathrm{WDRVI} = (a * \rho_{\mathrm{NIR}} - \rho_{\mathrm{red}}) / (a * \rho_{\mathrm{NIR}} + \rho_{\mathrm{red}}) \tag{1.14}$$

where a is a weighing coefficient with value between 0.1 and 0.2 (Gitelson, 2004; Gitelson et al., 2006).

MTCI is the ratio of the difference in reflectance between an NIR and red edge band and the difference in reflectance between red edge and red band as:

$$\mathrm{MTCI} = \left(\rho_{753.75} - \rho_{708.75}\right) / \left(\rho_{708.75} - \rho_{681.25}\right) \tag{1.15}$$

where $\rho_{753.75}$, $\rho_{708.75}$, and $\rho_{681.25}$ are reflectances in the center wavelengths of the MERIS narrow-band channel settings (Dash and Curran, 2004). The red edge band has been used extensively to retrieve leaf chlorophyll contents (Zhang et al., 2022). The G-R model has been successfully applied in estimating GPP in natural ecosystems (Sjöström et al., 2011; Wu et al., 2011, 2014) and croplands, including maize, soybeans, and wheat (Wu et al., 2010; Peng et al., 2011).

The LUE term in G-R model may be defined as either the ratio of GPP to APAR, or the ratio of GPP to PAR (Gower et al., 1999), with the latter sometimes referred to as ecosystem-LUE, or *e*LUE,

$$\varepsilon = \text{GPP/APAR}, \tag{1.16}$$

$$e\text{LUE} = \text{GPP / PAR} = f\text{APAR} \times \varepsilon \tag{1.17}$$

An advantage of using chlorophyll-based VIs in G-R models is that the biological drivers of photosynthesis, fAPAR and ε, resulting from environmental stress and leaf age phenology, are combined into eLUE, thereby simplifying EO-based productivity estimates.

Other measures of PAR that have been used include "potential' PAR, or maximal clear-sky PAR (PAR*potential*; Peng et al., 2013; Rossini et al., 2014) and top-of-atmosphere PAR (PAR*toa*). PAR*potential* can be calibrated from long term PAR*toc* measurements or modeled using an atmosphere radiative transfer code (Kotchenova and Vermote, 2007). Gitelson et al. (2012) found an improved performance of PAR*potential* relative to PAR*toc* noting that decreases in PAR*toc* during the day do not always imply a decrease in GPP. Further, Kanniah et al. (2013b) showed that the negative forcings of wet season cloud cover over Australian tropical savannas were partly compensated by enhanced LUE resulting from a greater proportion of diffuse radiation. Ma et al. (2014) found that coupling of EVI with PAR*toa* better predicted GPP than coupling EVI with PAR*toc* and attributed this to tower sensor-based measurement uncertainties of PAR*toc*, as well as better approximations of meteorological controls on GPP by PAR*toa*.

1.3.5 Solar-Induced Chlorophyll Fluorescence (SIF) as Proxy for GPP

More recent EO capabilities to measure solar-induced chlorophyll fluorescence (SIF) provide a more direct approach for estimating GPP with more accurate projections of terrestrial GPP (Mohammed et al., 2019; He et al., 2019: Frankenberg et al., 2011). The PAR absorbed by chlorophyll in leaves is used to drive photosynthesis, but some radiation is dissipated as heat or emitted back into the atmosphere at longer wavelengths (650–850 nm). This re-emitted red and near infrared (NIR) light from illuminated plants as a by-product of photosynthesis, has been found to more strongly correlate with GPP compared with VIs (Baker, 2008; Meroni et al., 2009; Frankenberg et al., 2011; Guanter et al., 2012). Chlorophyll fluorescence may be conceptualized as:

$$\text{SIF} = \varepsilon_f \times \text{PAR} \times f\text{APAR} \tag{1.18}$$

where ε_f is the *f*APAR photons that are re-emitted from the canopy as SIF photons, or yield of fluorescence photons.

SIF data provide information on both the light absorbed and the efficiency with which it is being used for photosynthesis. It is an independent measurement, linked to chlorophyll absorption, theoretically less sensitive to soil background, providing unique information on photosynthesis relative to VIs. The SIF expression can be combined with the GPP-based LUE equation to yield:

$$\text{GPP} = (\varepsilon_{par}/\varepsilon_f) \times \text{SIF} \tag{1.19}$$

Empirical studies at the leaf and canopy scale indicate that the two LUE terms tend to covary under the conditions of EO measurement (Flexas et al., 2002). SIF is seen as one way to increase the effective remotely sensed temporal resolution of vegetation photosynthesis, with near real time capabilities. SIF has been found to be more dynamic than greenness measures, and respond more quickly to environmental stress, though both change in stress-induced LUE and canopy light absorption (Porcar-Castell et al., 2014: Schimel et al., 2015).

As SIF responds to both incoming radiation and fluorescence yield, one may also normalize SIF by PAR (SIF/PAR) to better relate SIF signals to canopy properties (Shen et al., 2020). Shen et al. (2020) also combined SIF and EVI, as measure of SIF per unit greenness. Zhang et al., 2016 used the SCOPE model to better understand the relationship between SIF and GPP for EO applications. They reported the SIF-GPP relationship was primarily driven by APAR, especially at the seasonal scale.

Current sensors with capabilities to measure SIF include, NASA's Orbiting Carbon Observatory-2/3 (OCO-2/3) satellite (Frankenberg et al., 2014), the Global Ozone Monitoring Instrument (GOME-2), and the more recent TROPOspheric Monitoring Instrument (TROPOMI) on board the Copernicus Sentinel-5, and the Chinese TanSat. The Sentinel-5 Precursor Tropospheric Monitoring Instrument (TROPOMI) (Veefkind et al., 2012) satellite mission now provides advance spectrometer and fluorescence data with significantly finer spatial resolution. Preparatory studies are also underway for the future European FLuorescence EXplorer (FLEX) satellite mission (Meroni et al., 2009) that will provide measurements characterizing the spectral shape of fluorescence emission and enable estimates of photosynthesis rates under different vegetation stress conditions.

1.4 EO LIGHT USE EFFICIENCY PRODUCTIVITY MODELS

The simple LUE-based productivity equation comprises a large amount of biological complexity, resulting in numerous productivity modeling approaches through knowledge of the two conversion coefficients, *f*APAR and LUE (ε). The LUE concept has been widely adopted by the EO community to assess and extrapolate carbon processes and has been one of the most important methods to map GPP and NPP regionally or globally (Pei et al., 2022; Running et al., 2004).

LUE is very difficult to measure as it dynamically varies with plant functional type, vegetation phenophase, and different environmental stress conditions (Pei et al., 2022; Jenkins et al., 2007; Sims et al., 2006; Turner et al., 2004; Ruimy et al., 1995). LUE models assume the actual LUE (ε_g) is downregulated from its theoretical maximum LUE (ε_0) by environmental conditions, such as temperature or water stress. There are scarce measurements of LUE available, particularly at the landscape scale, and maximum LUE values have only been specified for a limited set of biome types, with these values down-regulated by environmental stress scalars derived from meteorological inputs (Zhao et al., 2005; Heinsch et al., 2006). EO measures also provide water stress and incident radiation (Zhang et al., 2014a), as well as spatially explicit information on land cover type (Friedl et al., 2010) that determines maximum LUE and other model parameters.

The general form of the LUE-based productivity model can be expressed as:

$$\varepsilon_g = \varepsilon_0 \times f(\mathrm{T}, \mathrm{W}, \ldots) \tag{1.20}$$

where $f(\mathrm{T}, \mathrm{W}, \ldots)$ represents environmental conditions, such as air temperature $f(\mathrm{Ta})$ and soil water $f(\mathrm{SW})$.

1.4.1 BIOME-BGC Model

The BIOME-BGC (BioGeochemical Cycles) model calculates daily GPP as a function of incoming solar radiation, conversion coefficients, and environmental stresses (Running et al., 2004). This was implemented as the first operational standard satellite product for MODIS (MOD17), providing global estimates of GPP (Figure 1.7), expressed as follows,

$$\mathrm{GPP} = \varepsilon_{\max} \times 0.45 \text{ x } \mathrm{SW}_{\mathrm{rad}} \times f\,\mathrm{PAR} \times f(\mathrm{VPD}) \times f(\mathrm{T}_{\min}) \tag{1.21}$$

where

- ε_{max} is the maximum light-use-efficiency (g C MJ^{-1}) obtained from a biome-properties look-up table (BPLUT);
- SW_{rad} is short-wave downward solar radiation (MJ^{-1} d^{-1}), of which 45% is assumed to be PAR;
- f(VPD) and $f(T_{min})$ are vapor pressure deficit and minimum air temperature reduction scalars for the biome specific ε_{max} values, and
- fAPAR is directly input from the MODIS FPAR (MOD15) product (Running et al., 2004; Zhao et al., 2005).

MODIS FPAR retrievals are physically based and use biome specific lookup tables (LUT) generated using a three-dimensional radiative transfer model (Myneni et al., 2002). The reduction scalars encompass LUE variability resulting from water stress (high daily VPD) and low temperatures (low daily minimum temperature T_{min}) (Running et al., 2004). The MODIS GPP product is directly linked to remote sensing and weather forecast products and can provide near-real time information on productivity and the influence of anomalies such as droughts. A consistent forcing meteorology is based upon the NCEP/NCAR (National Centres for Environmental Prediction/National Centre for Atmospheric Research) Reanalysis II datasets (Kanamitsu et al., 2002) (Figure 1.7).

Ichii et al. (2007) used the BIOME-BGC model to simulate seasonal variations in GPP for different rooting depths, from 1m to 10m, over Amazon forests and determine which rooting depths best estimated GPP consistent with satellite-based EVI. They were subsequently able to map rooting depths at regional scales and improve the assessments of carbon, water, and energy cycles in tropical forests.

The utility and accuracy of MODIS GPP/NPP products have been validated in various FLUXNET studies, which have also demonstrated the value of independent tower flux measures to better understand the satellite based GPP/NPP products (Leuning et al., 2005; Zhao et al., 2005; Turner et al., 2006). These studies highlight the capabilities of MODIS GPP to correctly predict observed fluxes at tower

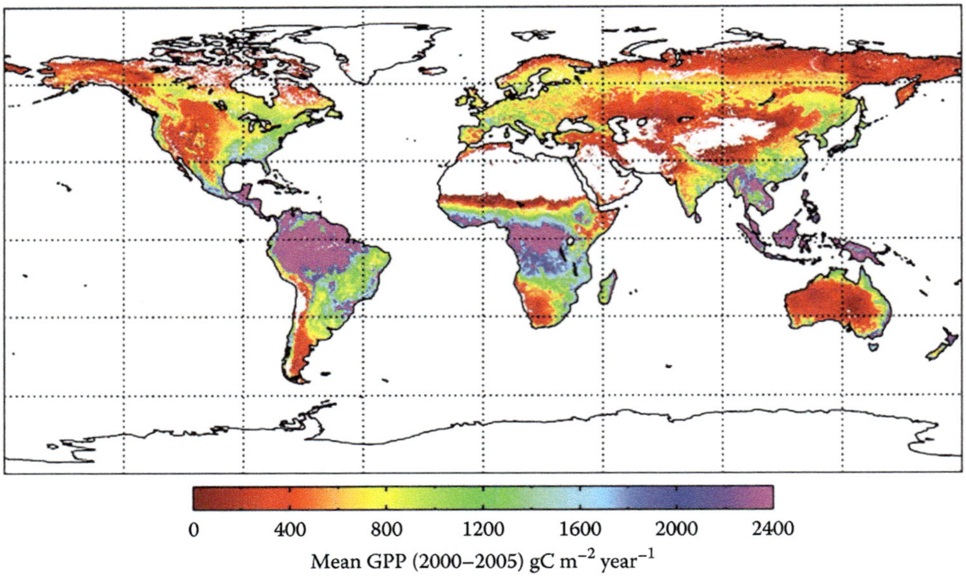

FIGURE 1.7 MODIS Gross Primary Production EO product (MOD17), displaying the mean GPP across years 2000–2005 for the global terrestrial surface. The highest GPP rates are seen in the tropical forests of southeast Asia, the Amazon basin, and equatorial Africa. The lowest rates are seen in Australia, South Africa, western North America, the Sahel, and Atacama desert. (Courtesy of Numerical Terradynamic Simulation Group, University of Montana, www.ntsg.umt.edu/project/mod17.)

sites, but also draw attention to some of the uncertainties associated with use of coarse resolution and interpolated meteorology inputs, uncertainties with the LUT-based values, noise, and uncertainties in the satellite *f*APAR inputs, and difficulties in constraining the light-use-efficiency term (Zhao et al., 2005; Heinsch et al., 2006; Yuan et al., 2010; Sjöström et al., 2013). Since meteorological inputs are often not available at sufficiently detailed temporal and spatial scales, they can introduce large errors into the carbon exchange estimates.

Turner et al. (2006) concluded that although the MODIS NPP/GPP products are generally responsive to spatial-temporal trends associated with climate, land cover, and land use, they tend to overestimate GPP at low productivity sites and underestimate GPP at high productivity sites. Similarly, Sjöström et al. (2013) found that although MODIS-GPP described seasonality at 12 African flux tower sites quite well, it tended to underestimate tower GPP at the dry sites in the Sahel region due to uncertainties in the meteorological and *f*APAR input data and the underestimation of ε_{max}. Jin et al. (2013) reported the MODIS GPP product to substantially underestimate tower GPP during the green-up phase at a woodland savanna site in Botswana, while overestimating tower-GPP during the brown-down phase.

Some studies have found that when properly parameterized with site-level meteorological measurements, MODIS GPP becomes more closely aligned with flux tower derived GPP (Turner et al., 2003; Sjöström et al., 2013; Kanniah et al., 2009). Kanniah et al. (2011), however, found that utilizing site-based meteorology could only improve GPP estimates during the wet season over northern Australian savannas, and suggested the MODIS GPP product has a systematic limitation in the estimation of savanna GPP in arid and semi-arid areas due to the lack of the representation of soil moisture. Sjöström et al. (2013) also found soil moisture information to be quite important for accurate GPP estimates in drier African savannas.

1.4.2 Vegetation Photosynthesis Model (VPM)

Xiao et al. (2004) developed a mostly satellite-based Vegetation Photosynthesis Model (VPM) that estimates GPP using satellite inputs of EVI and the land surface water index (LSWI),

$$GPP = \varepsilon \times fAPAR_{chl} \times PAR_{toc} \quad (1.22)$$

$$\varepsilon = \varepsilon_{max} \times T_{scalar} \times W_{scalar} \times P_{scalar} \quad (1.23)$$

where

- $fAPAR_{chl}$ is estimated as a linear function of EVI;
- PAR_{toc} is measured at the site; and
- T_{scalar}, W_{scalar}, P_{scalar} are scalars for the effects of temperature, water and leaf phenology on vegetation, respectively (Figure 1.8).
- T_{scalar} is based on air temperature and uses minimum, maximum, and optimum temperature for photosynthesis at each time step; the
- W_{scalar} is based on satellite-derived LSWI that accounts for the effect of water stress on photosynthesis;

$$W_{scalar} = (1 + LSWI)/(1 + LSWI_{max}), \quad (1.24)$$

$$LSWI = (\rho_{nir} - \rho_{swir})/(\rho_{nir} + \rho_{swir}), \quad (1.25)$$

where

- ρ_{swir} is the reflectance in a broadband shortwave infrared band (e.g., MODIS, 1580–1750 nm) and
- $LSWI_{max}$ is the maximum value for the growing season.

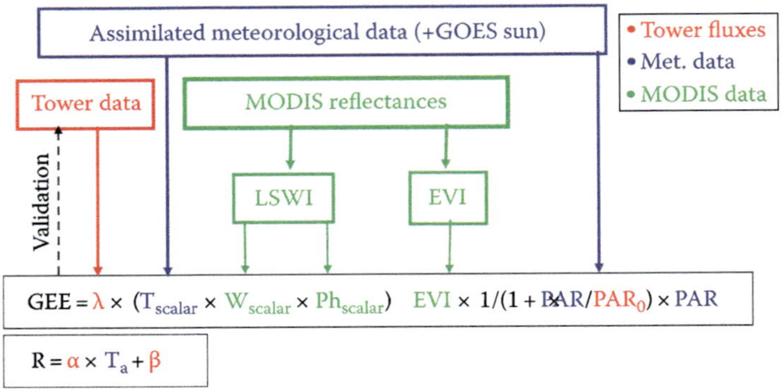

FIGURE 1.8 Schematic diagram of the Vegetation Photosynthesis and Respiration Model (VPRM) utilizing EVI, LSWI, and scalars for temperature, leaf phenology, and canopy water content, T_{scalar}, P_{scalar}, and W_{scalar}, respectively. The VPM model uses primarily remote sensing data along with air temperatures, while the VPRM model additionally assimilates tower flux and meteorological information. (Source: Mahadevan et al., 2008.)

The P_{scalar} accounts for the effect of leaf age on photosynthesis and is dependent on the growing season life expectancy of the leaves (Wilson et al., 2001). P_{scalar} is calculated over two phenophases as,

$$P_{scalar} = (1 + LSWI)/2 \tag{1.26}$$

from bud burst to full leaf expansion, and $P_{scalar} = 1$, after full expansion (Xiao et al., 2004).

The VPM model has been applied to both MODIS and SPOT-4 VEGETATION sensor data to produce tower-calibrated estimates of GPP across a wide range of biomes, including evergreen and deciduous forests, grasslands, and shrub sites in temperate North America and in seasonally moist tropical evergreen forests in the Amazon (Mahadevan et al., 2008; Xiao et al., 2005; Jin et al., 2013).

1.5 EO MEASURES OF NET PRIMARY PRODUCTIVITY (NPP)

There are many empirical, diagnostic and process-based models that have been developed over the past decades to monitor and assess vegetation productivity, with many of these methods employing EO data in conjunction with in situ measurements to varying extents.

1.5.1 Annual Integrated VI Estimates of Productivity

Vegetation indices (VIs) may be more appropriate measures of net primary productivity, NPP (see Equation 1.3), the balance between GPP and plant autotrophic respiration, that is, VIs may be more closely related to the vegetation remaining after respiration. Several studies have suggested that ecosystem NPP can be captured by the annual VI integral, that integrates growing season production through VI relationships with *f* APAR. Goward et al. (1985) found good relationships between aboveground NPP (ANPP) and integrated NDVI from AVHRR, over annual growing periods of North American biomes (Figure 1.9). Wang et al. (2004) found that the NDVI integral over the early growing season was strongly correlated to in situ forest measurements of diameter increase and tree ring width in the U.S. central Great Plains.

More recently, the integrated Enhanced Vegetation Index (iEVI) was also found to be a good proxy of ANPP. Ponce Campos et al. (2013) compiled in situ field measures of ANPP across ten sites in the United States ranging from arid grassland to forest and directly compared annual integrated values of the MODIS enhanced vegetation index (iEVI) (Figure 1.10):

$$ANPP = 51.42 \times iEVI^{1.15} \tag{1.27}$$

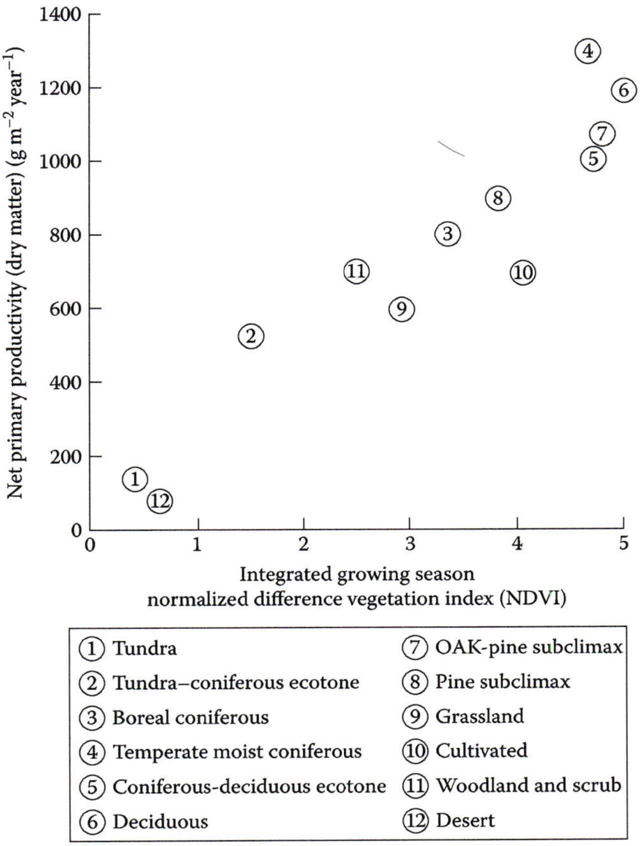

FIGURE 1.9 Relationship between biome averaged integrated NDVI from NOAA-AVHRR sensors and net primary productivity rates. (Source: Goward et al., 1985.)

where a log-log relation accounted for the uneven distribution of ANPP estimates over time.

The annual integrated VI offers a robust approximation of vegetation productivity because, in general, VIs provide both a measure of the capacity to absorb light energy, as well as reflect recent environmental stress acting on the canopy, with stress forcings showing up as reductions in NDVI expressed as either less chlorophyll or less foliage (Running et al., 2004).

Moran et al. (2014) found plot-scale measurements of ANPP at arid and mesic grassland sites were significantly related to MODIS iEVI over a decadal time period in a log-log relation ($r^2 = 0.71$, $P < 0.01$). Zhang et al. (2013) studied the ecological impacts of rainfall intensification on vegetation productivity through use of iEVI as a surrogate measure of ANPP. They found extreme precipitation patterns, associated with heavy rainfall events followed by longer dry periods, caused higher water stress conditions which resulted in strong negative influences on ANPP across biomes and reduced rainfall use efficiencies (20% on average).

Generally, in situ measures of productivity are made at discrete times within the growing season and there is a need to synchronize the satellite measurements with scheduled or variable destructive or harvesting sampling dates to reduce in situ ANPP-VI uncertainties. Often it is difficult to predict and sample at peak productivity and greenness periods. Continuous VI growing season productivity profiles allow one to better synchronize VI temporal values with actual in situ sampling periods. For example, Moran et al. (2014) found significant improvements in productivity–iEVI relationships across a range of grassland sites, when the EVI was only partially integrated from the beginning

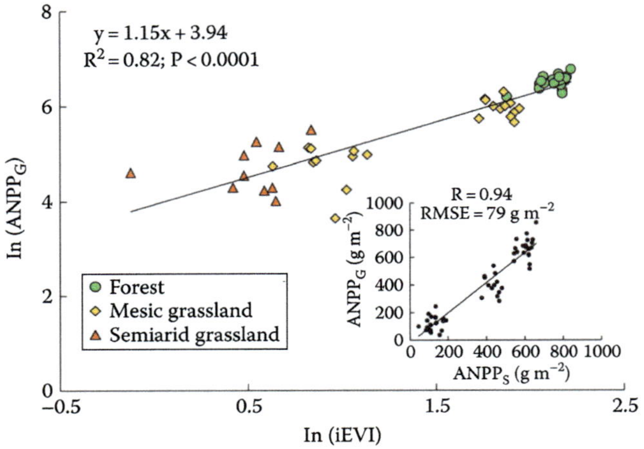

FIGURE 1.10 Relationship between in situ estimates of aboveground net primary production (ANPPg) and annual integrated EVI (iEVI) derived from MODIS data (2000–2009) for ten sites across several biomes. The solid line represents the linear regression used to estimate ANPP from iEVI (ANPPs). (Source: Ponce Campos et al., 2013.)

to the peak of the growing season period (rather than the full season). This was due to the synchronization of time periods to peak biomass periods when grassland ANPP destructive sampling is typically conducted. In such cases, EO data provides better temporal stability and opportunities to reduce productivity uncertainties. In the case of the VPM model, phenological factors such as leaf age and life expectancy, which play important roles in productivity, were explicitly incorporated (Xiao et al., 2004).

1.5.2 EO NPP Model Products

Estimates of annual NPP are now routinely produced operationally over the global terrestrial surface at 1 km spatial resolution through production efficiency models with near real-time satellite data inputs from the Moderate Resolution Imaging Spectroradiometer (MODIS) (Turner et al., 2006) (Figure 1.11). MODIS NPP is determined by first computing daily net photosynthesis values, which is the difference of GPP and maintenance respiration. The NPP product is a cumulative composite of GPP values composited over a year and at global scale, NPP is estimated to be about half of the GPP (see Section 4.1, BIOME-BGC model.) and synthesized NPP. Using the MODIS NPP products, Zhao and Running (2010) found that global NPP declined slightly by 0.55 petagram carbon (Pg C, with Pg = 10^{15} grams) due to the early 21st-century global drought from 2000 to 2009.

Another popular EO-based LUE model for quantifying NPP or GPP is the Carnegie Ames Stanford Approach (CASA) Biosphere model (Potter et al., 1993). CASA has been widely used to simulate carbon dynamics at regional to global scales using VI inputs. Liu et al., (2019) used the CASA model to investigate global urban expansions on terrestrial NPP and found urbanization-induced decrease in NPP offset 30% of the climate-driven increase over the period (2000–2010). Further, Jay et al. (2016) fused high resolution Landsat land cover data with MODIS EVI values to generate a finer version of the CASA ecosystem model. Their monthly estimates of NPP performed very high against tower flux data from Ameriflux with R^2 of 0.82. This demonstrated the potential to improve LUE models and estimates of NPP though more accurate land cover parameter inputs. Low correlations between CASA and tower derived NPP mainly occurred in disturbed areas.

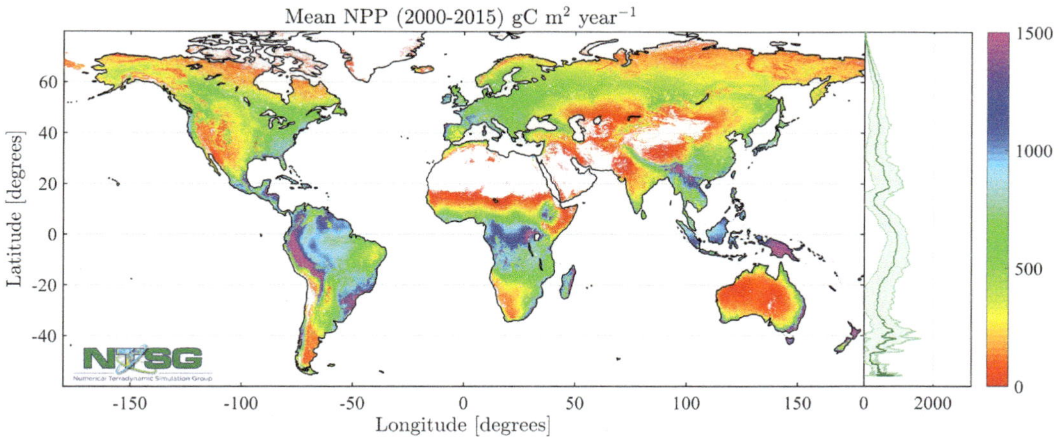

FIGURE 1.11 MODIS Net Primary Production satellite product (MOD17). Example showing the mean NPP across years 2000–2005 for the global terrestrial surface. The highest production is seen across the equatorial zone encompassing southeast Asia, the Amazon basin, and equatorial Africa. The least productive regions appear in Australia and the Sahelian region. (Courtesy of Numerical Terradynamic Simulation Group, University of Montana, www.ntsg.umt.edu/project/mod17.)

1.5.3 Photochemical Reflectance Index (PRI)

There has been much interest in reducing the uncertainties in GPP/NPP models through direct remote sensing assessments of LUE. The photochemical reflectance index (PRI) is a hyperspectral index that provides a scaled LUE measure, or photosynthetic efficiency, based on light absorption processes by carotenoids (Ryu et al., 2019; Gamon et al., 1992; Garbulsky et al., 2011),

$$PRI = (\rho_{531nm} - \rho_{570nm})/(\rho_{531nm} + \rho_{570nm}) \tag{1.28}$$

Spectral variations at 531 nm are closely associated with the dissipation of excess light energy by xanthophyll pigments (a major carotenoid group of yellow pigments) in order to protect the photosynthetic leaf apparatus (Ripullone et al., 2011). Carotenoids function in processes of light absorption in plants as well as protecting plants from the harmful effects of high light conditions, hence, lower carotenoid/chlorophyll ratios, indicate lower physiological stress (Goerner et al., 2009; Peñuelas et al., 1995; Guo and Trotter, 2004).

Several studies have shown the linear relationship between PRI and LUE over different vegetation types (e.g., Nichol et al., 2000). Rahman et al. (2004) produced a "continuous field" retrieval of LUE from EO data, using the PRI as a proxy of LUE, without the need of look-up tables or predetermined biome specific LUE values. However, Barton and North (2001) showed that PRI was most sensitive to changes in leaf area index (LAI) and Gitelson et al. (2006) noted that in order to use PRI to predict LUE, one would need an independent estimate of LAI.

1.5.4 Net Ecosystem Productivity and Net Biome Productivity

Net ecosystem productivity (NEP) and net biome productivity (NBP) (Figure 1.2) both estimate net ecosystem carbon uptake/release. The fluxes contributing to NEP are GPP and autotrophic (Ra) and heterotrophic (Rh) respiration, while NBP is mainly determined by NEP minus the loss of carbon by processes such as fire and harvest:

$$NEP = GPP - Ra - Rh \quad (1.29)$$

$$NBP = NEP - fire - harvest \quad (1.30)$$

Mahadevan et al. (2008) further developed the Vegetation Photosynthesis and Respiration Model (VPRM), a satellite-based assimilation scheme that estimates hourly values of NEE using EVI, LSWI, and high-resolution meteorology observations of sunlight and air temperature (Figure 1.8). NEE represents the difference between uptake (photosynthesis) and loss (respiration) processes that vary over a wide range of timescales (Goulden et al., 1996; Katul et al., 2001). The VPRM model provides fine-grained fields of surface CO_2 fluxes for application in inverse models at continental and smaller scales (Mahadevan et al., 2008). This capability is presently limited by the number of vegetation classes for which NEE can be constrained using eddy covariance tower flux data.

The recent availability of column CO_2 concentration retrievals from newly launched satellites has made it feasible to quantify NEP and NBP from satellite observations. These satellites use high resolution spectrometers to measure the intensity of sunlight, at different wavelengths, for retrievals of columnar CO_2 and CH_4 concentrations. For example, Japanese GOSAT (launched in 2009), NASA's OCO-2 (launched in 2014), and Chinese TanSat (launched in 2016) provide comprehensive, global measurements of CO_2 in the atmosphere, including seasonal fluctuations of the greenhouse gas and their spatial sources and sinks. This enables a better understanding of how ecosystems absorb and release CO_2, both seasonally and across years in response to interannual climate variability. Basu et al. (2013) estimated the global distribution of CO_2 fluxes using column CO_2 measurements from the GOSAT instrument. Inversions of satellite CO_2 observations provide useful constraints on terrestrial carbon sinks and sources.

Ma et al. (2016) investigated carbon fluxes during an unusual, extremely wet year across Australia in 2011, in which an exceptionally large land carbon sink anomaly was recorded, of which more than half was attributed to Australia (Figure 1.12). They amassed multiple EO carbon measures

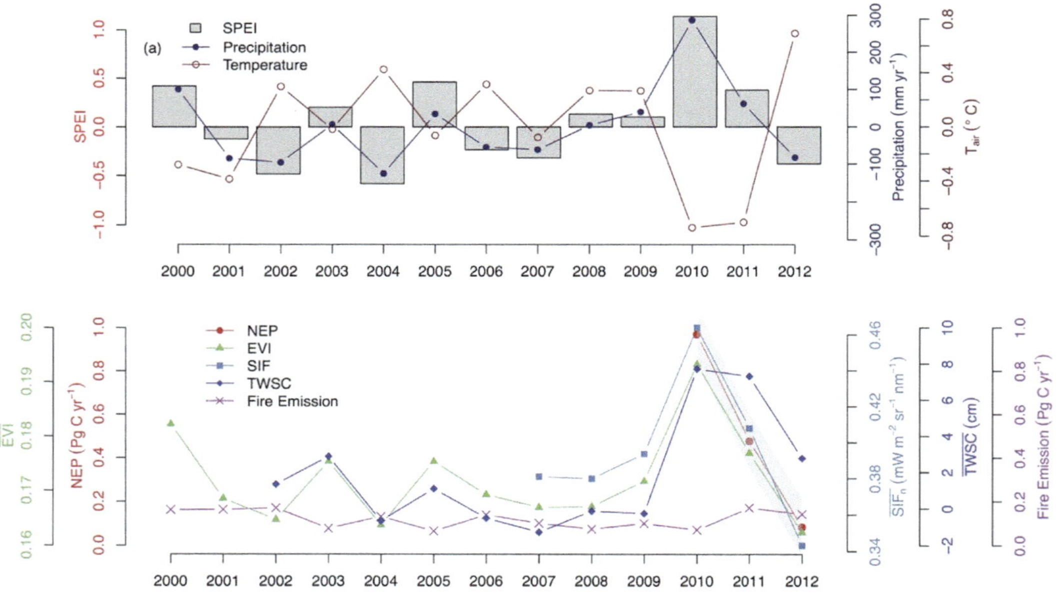

FIGURE 1.12 Australian continental averages of annual integrated EVI, SIFn (PAR normalized SIF), Total Water Storage Change (TWSC, cm), continental total fire carbon emission (Pg C yr^{-1}), and interannual variations of NEP (Pg C yr^{-1}). Interannual variations in NEP is derived from GOSAT atmospheric inversion modeling. The gray shaded area represents the Bayesian uncertainty range of inverted NEP ($\pm 1\sigma$). (Source: Ma et al. (2016), Figure 1.1.)

including atmospheric CO_2 from GOSAT, EVI from MODIS, and SIF from GOME-2, and found semi-arid Australian net CO_2 uptake was highly transient and rapidly dissipated by subsequent drought. They showed the accuracies of EO-based CO_2 retrievals to be approaching the levels needed to constrain CO_2 fluxes for accurate determinations of net biome productivity (NBP) at regional to continental scales.

1.6 DISCUSSION AND FUTURE

Modeling of carbon, water, and energy fluxes between terrestrial surfaces and the atmosphere are increasingly important for hydrological and climate studies. Ecophysiological information relating to photosynthetic activity, biomass, productivity, water content, phenology, soil moisture, and nutrient status may be acquired and analyzed consistently and repeatedly over large areas.

Space-based earth observations (EO) strongly complement in situ observations in providing quantitative ecosystem production assessments and information globally. While EO spaceborne measurements have uncertainty and bias errors of their own, they can help in reducing bias errors associated with relatively sparse in situ systems and benchmarking land surface models. Given the challenges of long term, in situ observations at regional scales, EO measurements have made increasingly important contributions, complementing the detailed information available in situ by providing broad spatial and temporal coverage. To have comprehensive knowledge of carbon stocks and fluxes everywhere and over time involves the coordination of in situ and remote observations. It still remains a challenge to develop such a coherent set of terrestrial ecosystem observations of the carbon cycle, and be able to develop early warning and prediction capabilities of carbon cycle feedbacks.

New space-based observations, including fluorescence, hyperspectral, thermal, microwave and LiDAR are greatly expanding the number of ecosystem properties that can be quantified from space. New hyperspectral missions will provide future data fusion opportunities for the scaling and extension of leaf physiologic processes and phenology from species and ecosystem to regional and global scales. This will enable improvements in chlorophyll indices and use of PRI in LUE productivity models. Microwave remote sensing, specifically using metrics like Vegetation Optical Depth, that describes the vegetation's attenuation of radiation by biomass and water content and in the microwave domain, are increasingly being used for carbon flux studies, including GPP (Dou et al., 2023). VOD measurements can improve productivity monitoring through observations under cloud cover, enabling consistent, uninterrupted monitoring. VODCA2GPP is a recent long-term (1988–2020) GPP dataset derived from microwave remote sensing (Wild et al., 2022).

Advances made in global weather forecasting accuracies and the development of next generation geostationary satellites with sub-daily (10- to 30-minute) spectral and thermal observations also facilitate the acquisition of cloud-free observations as well as reveal water and temperature stress impacts on diurnal vegetation patterns. Daily MODIS VI-PAR-based GPP products at 250 m were recently developed with the NIRv index (NIRv = NDVI*NIR) over the U.S. (Jiang et al., 2021).

Machine learning is increasingly being used for data fusion and extrapolating of in situ data with EO data, for example, Zhan et al. (2022) applied machine learning to improve flux tower NEE partitioning into respiration and GPP by establishing the relationships between SIF and GPP. Through combinations of the techniques and observations described earlier will enable a more thorough coupling of the environmental conditions that plants experience with improved characterization of their biophysical states, and with better monitoring capabilities to track plant responses to environmental changes. These advances are providing a better understanding of the dynamics of terrestrial productivity and the use of EO data to drive productivity models of the land surface. This has the potential for routine estimates of the key carbon fluxes and stocks shown in Figures 1.1, 1.2, and 1.12 using EO instruments and techniques. These observations will have the common goal of helping us better understand our planet's vegetation and understand how it is changing in regard to our food, health, and economic security needs.

REFERENCES

Asrar, G., Fuchs, M., Kanemasu, E.T. & Hatfield, J.L. 1984. Estimating absorbed photosynthetic radiation and leaf area index from spectral reflectance in wheat. Agronomy Journal, 76, 2, 300–306.

Baker, N.R. 2008. Chlorophyll fluorescence: A probe of photosynthesis in vivo. Annual Review of Plant Biology, 59, 1, 89–113.

Baldocchi, D. 2008. 'Breathing' of the terrestrial biosphere: Lessons learned from a global network of carbon dioxide flux measurement systems. Australian Journal of Botany, 56, 1–26. https://doi.org/10.1071/BT07151.

Baldocchi, D., Falge, E., Lianhong, G., Olson, R., Hollinger, D., Running, S., Anthoni, P., Bernhofer, C., Davis, K., Evans, R., Fuentes, J., Goldstein, A., Katul, G., Law, B., Xuhui, L., Malhi, Y., Meyers, T., Munger, W., Oechel, W. & Paw, U.K.T. 2001. FLUXNET: A new tool to study the temporal and spatial variability of ecosystem-scale carbon dioxide, water vapor, and energy flux densities. Bulletin of the American Meteorological Society, 82, 11, 2415.

Barton, C.V.M. & North, P.R.J. 2001. Remote sensing of canopy light use efficiency using the photochemical reflectance index: Model and sensitivity analysis. Remote Sensing of Environment, 78, 3, 264–273.

Basu, S., Guerlet, S., Butz, A., Houweling, S., Hasekamp, O., Aben, I., Krummel, P., Steele, P., Langenfelds, R., Torn, M., Biraud, S., Stephens, B., Andrews, A. & Worthy, D. 2013. Global CO_2 fluxes estimated from GOSAT retrievals of total column CO_2. Atmospheric Chemistry and Physics, 13, 17, 8695–8717. https://doi.org/10.5194/acp-13-8695-2013.

Biondini, M.E., Lauenroth, W.K. & Sala, O.E. 1991. Correcting estimates of net primary production: Are we overestimating plant production in rangelands? Journal of Range Management, 194–198.

Chambers, J.Q., Negrón-Juárez, R.I., Hurtt, G.C., Marra, D.M. & Higuchi, N. 2009. Lack of intermediate-scale disturbance data prevents robust extrapolation of plot-level tree mortality rates for old-growth tropical forests. Ecology Letters, 12, 12, E22–E25.

Cleverly, J., Boulain, N., Villalobos-Vega, R., Grant, N., Faux, R., Wood, C., Cook, P.G., Yu, Q., Leigh, A. & Eamus, D. 2013. Dynamics of component carbon fluxes in a semi-arid Acacia woodland, central Australia. Journal of Geophysical Research: Biogeosciences, 118, 3, 1168–1185.

Cramer, W., Kicklighter, D., Bondeau, A., Iii, B.M., Churkina, G., Nemry, B., Ruimy, A., Schloss, A., Intercomparison, T. & Model, P.O.T.P.N. 1999. Comparing global models of terrestrial net primary productivity (NPP): Overview and key results. Global Change Biology, 5, S1, 1–15.

Dash, J. & Curran, P.J. 2004. The MERIS terrestrial chlorophyll index. International Journal of Remote Sensing, 25, 23, 5403–5413.

Dou, Y., Tian, F., Wigneron, J.-P., Tagesson, T., Du, J., Brandt, M., Liu, Y., Zou, L., Kimball, J. S. & Fensholt, R. 2023. Reliability of using vegetation optical depth for estimating decadal and interannual carbon dynamics. Remote Sensing of Environment, 285, 113390. https://doi.org/10.1016/j.rse.2022.113390.

Fensholt, R., Sandholt, I. & Rasmussen, M.S. 2004. Evaluation of MODIS LAI, fAPAR and the relation between fAPAR and NDVI in a semi-arid environment using in situ measurements. Remote Sensing of Environment, 91, 3–4, 490–507.

Field, C.B., Randerson, J.T. & Malmström, C.M. 1995. Global net primary production: Combining ecology and remote sensing. Remote Sensing of Environment, 51, 1, 74–88.

Flexas, J., Bota, J., Escalona, J.M., Sampol, B. & Medrano, H. 2002, April. Effects of drought on photosynthesis in grapevines under field conditions: An evaluation of stomatal and mesophyll limitations. Functional Plant Biology, 29, 4, 461–471. https://doi.org/10.1071/PP01119. PMID: 32689491.

Frankenberg, C., Butz, A. & Toon, G.C. 2011. Disentangling chlorophyll fluorescence from atmospheric scattering effects in O2 A-band spectra of reflected sun-light. Geophysical Research Letters, 38, 3, L03801.

Frankenberg, C., O'Dell, C., Berry, J., Guanter, L., Joiner, J., Köhler, P., Pollock, R. & Taylor, T.E. 2014. Prospects for chlorophyll fluorescence remote sensing from the orbiting carbon observatory-2. Remote Sensing of Environment, 147, 0, 1–12.

Friedl, M.A., Sulla-Menashe, D., Tan, B., Schneider, A., Ramankutty, N., Sibley, A. & Huang, X. 2010. MODIS collection 5 global land cover: Algorithm refinements and characterization of new datasets. Remote Sensing of Environment, 114, 1, 168–182. https://doi.org/10.1016/j.rse.2009.08.016.

Gamon, J.A., Peñuelas, J. & Field, C.B. 1992. A narrow-waveband spectral index that tracks diurnal changes in photosynthetic efficiency. Remote Sensing of Environment, 41, 1, 35–44.

Garbulsky, M.F., Peñuelas, J., Gamon, J., Inoue, Y. & Filella, I. 2011. The photochemical reflectance index (PRI) and the remote sensing of leaf, canopy and ecosystem radiation use efficiencies: A review and meta-analysis. Remote Sensing of Environment, 115, 281–297.

Gitelson, A.A. 2004. Wide dynamic range vegetation index for remote quantification of biophysical characteristics of vegetation. Journal of Plant Physiology, 161, 2, 165–173.

Gitelson, A.A., Keydan, G.P. & Merzlyak, M.N. 2006. Three-band model for noninvasive estimation of chlorophyll, carotenoids, and anthocyanin contents in higher plant leaves. Geophysical Research Letters, 33, 11, L11402.

Gitelson, A.A., Peng, Y., Masek, J.G., Rundquist, D.C., Verma, S., Suyker, A., Baker, J.M., Hatfield, J.L. & Meyers, T. 2012. Remote estimation of crop gross primary production with Landsat data. Remote Sensing of Environment, 121, 0, 404–414.

Glenn, E.P., Huete, A.R., Nagler, P.L. & Nelson, S.G. 2008. Relationship between remotely-sensed vegetation indices, canopy attributes and plant physiological processes: What vegetation indices can and cannot tell us about the landscape. Sensors, 8, 4, 2136–2160.

Goerner, A., Reichstein, M. & Rambal, S. 2009. Tracking seasonal drought effects on ecosystem light use efficiency with satellite-based PRI in a Mediterranean forest. Remote Sensing of Environment, 113, 5, 1101–1111.

Goulden, M.L., Munger, J.W., Fan, S.-M., Daube, B.C. & Wofsy, S.C. 1996. Measurements of carbon sequestration by long-term eddy covariance: Methods and a critical evaluation of accuracy. Global Change Biology, 2, 3, 169–182.

Goward, S., Tucker, C. & Dye, D. 1985. North American vegetation patterns observed with the NOAA-7 advanced very high resolution radiometer. Vegetatio, 64, 1, 3–14.

Goward, S.N. & Huemmrich, K.F. 1992. Vegetation canopy PAR absorptance and the normalized difference vegetation index: An assessment using the SAIL model. Remote Sensing of Environment, 39, 2, 119–140.

Gower, S.T., Kucharik, C.J. & Norman, J.M. 1999. Direct and indirect estimation of leaf area index, fAPAR, and net primary production of terrestrial ecosystems. Remote Sensing of Environment, 70, 1, 29–51.

Guanter, L., Frankenberg, C., Dudhia, A., Lewis, P.E., Gómez-Dans, J., Kuze, A., Suto, H. & Grainger, R.G. 2012. Retrieval and global assessment of terrestrial chlorophyll fluorescence from GOSAT space measurements. Remote Sensing of Environment, 121, 0, 236–251.

Guanter, L., Zhang, Y., Jung, M., Joiner, J., Voigt, M., Berry, J.A., Frankenberg, C., Huete, A.R., Zarco-Tejada, P., Lee, J.-E., Moran, M.S., Ponce-Campos, G., Beer, C., Camps-Valls, G., Buchmann, N., Gianelle, D., Klumpp, K., Cescatti, A., Baker, J.M. & Griffis, T.J. 2014. Global and time-resolved monitoring of crop photosynthesis with chlorophyll fluorescence. Proceedings of the National Academy of Sciences, 111, 14, E1327–E1333.

Guo, J. & Trotter, C.M. 2004. Estimating photosynthetic light-use efficiency using the photochemical reflectance index: Variations among species. Functional Plant Biology, 31, 3, 255–265.

He, L., Chen, J. M., Liu, J., Zheng, T., Wang, R., Joiner, J., Chou, S., Chen, B., Liu, Y., Liu, R. & Rogers, C. 2019. Diverse photosynthetic capacity of global ecosystems mapped by satellite chlorophyll fluorescence measurements. Remote Sensing of Environment, 232, 111344. https://doi.org/10.1016/j.rse.2019.111344.

Heinsch, F.A., Maosheng, Z., Running, S.W., Kimball, J.S., Nemani, R.R., Davis, K.J., Bolstad, P.V., Cook, B.D., Desai, A.R., Ricciuto, D.M., Law, B.E., Oechel, W.C., Hyojung, K., Hongyan, L., Wofsy, S.C., Dunn, A.L., Munger, J.W., Baldocchi, D.D., Liukang, X., Hollinger, D.Y., Richardson, A.D., Stoy, P.C., Siqueira, M.B.S., Monson, R.K., Burns, S.P. & Flanagan, L.B. 2006. Evaluation of remote sensing based terrestrial productivity from MODIS using regional tower eddy flux network observations. <cite lang="en">IEEE Transactions on Geoscience and Remote Sensing,</cite> 44, 7, 1908–1925, July. https://doi.org/10.1109/TGRS.2005.853936.

Huete, A., Didan, K., Miura, T., Rodriguez, E.P., Gao, X. & Ferreira, L.G. 2002. Overview of the radiometric and biophysical performance of the MODIS vegetation indices. Remote Sensing of Environment, 83, 1–2, 195–213.

Huete, A.R., Didan, K., Shimabukuro, Y.E., Ratana, P., Saleska, S.R., Hutyra, L.R., Yang, W., Nemani, R.R. & Myneni, R. 2006. Amazon rainforests green-up with sunlight in dry season. Geophysical Research Letters, 33, 6, 612. https://doi.org/10.1029/2005gl025583.

Huete, A.R., Restrepo-Coupe, N., Ratana, P., Didan, K., Saleska, S.R., Ichii, K., Panuthai, S. & Gamo, M. 2008. Multiple site tower flux and remote sensing comparisons of tropical forest dynamics in Monsoon Asia. Agricultural and Forest Meteorology, 148, 5, 748–760.

Ichii, K., Hashimoto, H., White, M.A., Potter, C., Hutyra, L.R., Huete, A.R., Myneni, R.B. & Nemani, R.R. 2007. Constraining rooting depths in tropical rainforests using satellite data and ecosystem modeling for accurate simulation of gross primary production seasonality. Global Change Biology, 13, 1, 67–77.

Jay, S., Potter, C., Crabtree, R., Genovese, V., Weiss, D. J. & Kraft, M. 2016. Evaluation of modelled net primary production using MODIS and landsat satellite data fusion. Carbon Balance and Management, 11, 1, 8. https://doi.org/10.1186/s13021-016-0049-6.

Jenkins, J.P., Richardson, A.D., Braswell, B.H., Ollinger, S.V., Hollinger, D.Y. & Smith, M.L. 2007. Refining light-use efficiency calculations for a deciduous forest canopy using simultaneous tower-based carbon flux and radiometric measurements. Agricultural and Forest Meteorology, 143, 1–2, 64–79.

Jiang, C., Guan, K., Wu, G., Peng, B. & Wang, S. 2021. A daily, 250 m and real-time gross primary productivity product (2000—present) covering the contiguous United States. Earth System Science Data, 13, 2, 281–298. https://doi.org/10.5194/essd-13-281-2021.

Jin, C., Xiao, X., Merbold, L., Arneth, A., Veenendaal, E. & Kutsch, W.L. 2013. Phenology and gross primary production of two dominant savanna woodland ecosystems in Southern Africa. Remote Sensing of Environment, 135, 0, 189–201.

Kanamitsu, M., Ebisuzaki, W., Woollen, J., Yang, S.-K., Hnilo, J.J., Fiorino, M. & Potter, G.L. 2002. NCEP—DOE AMIP-II Reanalysis (R-2). Bulletin of the American Meteorological Society, 83, 11, 1631–1643.

Kanniah, K.D., Beringer, J. & Hutley, L.B. 2011. Environmental controls on the spatial variability of savanna productivity in the Northern Territory, Australia. Agricultural and Forest Meteorology, 151, 11, 1429–1439.

Kanniah, K.D., Beringer, J. & Hutley, L.B. 2013a. Response of savanna gross primary productivity to interannual variability in rainfall: Results of a remote sensing based light use efficiency model. Progress in Physical Geography, 37, 5, 642–663.

Kanniah, K.D., Beringer, J. & Hutley, L.B. 2013b. Exploring the link between clouds, radiation, and canopy productivity of tropical savannas. Agricultural and Forest Meteorology, 182–183, 0, 304–313.

Kanniah, K.D., Beringer, J., Hutley, L.B., Tapper, N.J. & Zhu, X. 2009. Evaluation of collections 4 and 5 of the MODIS gross primary productivity product and algorithm improvement at a tropical savanna site in northern Australia. Remote Sensing of Environment, 113, 9, 1808–1822.

Katul, G.G., Leuning, R., Kim, J., Denmead, O.T., Miyata, A. & Harazono, Y. 2001. Estimating CO_2 source/sink distributions within a rice canopy using higher-order closure model. Boundary-Layer Meteorology, 98, 1, 103–125.

Kotchenova, S.Y. & Vermote, E.F. 2007. Validation of a vector version of the 6S radiative transfer code for atmospheric correction of satellite data. Part II: Homogeneous Lambertian and anisotropic surfaces. Applied Optics, 46, 20, 4455–4464.

Leuning, R., Cleugh, H.A., Zegelin, S.J. & Hughes, D. 2005. Carbon and water fluxes over a temperate Eucalyptus forest and a tropical wet/dry savanna in Australia: Measurements and comparison with MODIS remote sensing estimates. Agricultural and Forest Meteorology, 129, 3–4, 151–173.

Liu, X., Pei, F., Wen, Y., Li, X., Wang, S., Wu, C., Cai, Y., Wu, J., Chen, J., Feng, K., Liu, J., Hubacek, K., Davis, S. J., Yuan, W., Yu, L. & Liu, Z. 2019. Global urban expansion offsets climate-driven increases in terrestrial net primary productivity. Nature Communications, 10, 1, 5558. https://doi.org/10.1038/s41467-019-13462-1.

Ma, X., Huete, A., Cleverly, J., Eamus, D., Chevallier, F., Joiner, J., Poulter, B., Zhang, Y., Guanter, L., Meyer, W., Xie, Z. & Ponce-Campos, G. 2016. Drought rapidly diminishes the large net CO2 uptake in 2011 over semi-arid Australia. Scientific Reports, 6, 37747. https://doi.org/10.1038/srep37747.

Ma, X., Huete, A., Yu, Q., Coupe, N.R., Davies, K., Broich, M., Ratana, P., Beringer, J., Hutley, L.B., Cleverly, J., Boulain, N. & Eamus, D. 2013. Spatial patterns and temporal dynamics in savanna vegetation phenology across the North Australian tropical transect. Remote Sensing of Environment, 139, 0, 97–115.

Ma, X., Huete, A., Yu, Q., Restrepo-Coupe, N., Beringer, J., Hutley, L.B., Kanniah, K.D., Cleverly, J. & Eamus, D. 2014. Parameterization of an ecosystem light-use-efficiency model for predicting savanna GPP using MODIS EVI. Remote Sensing of Environment, 154, 0, 253–271.

Mahadevan, P., Wofsy, S.C., Matross, D.M., Xiao, X., Dunn, A.L., Lin, J.C., Gerbig, C., Munger, J.W., Chow, V.Y. & Gottlieb, E.W. 2008. A satellite-based biosphere parameterization for net ecosystem CO2 exchange: Vegetation photosynthesis and respiration model (VPRM). Global Biogeochemical Cycles, 22, 2, GB2005. https://doi.org/10.1029/2006GB002735.

Meroni, M., Rossini, M., Guanter, L., Alonso, L., Rascher, U., Colombo, R. & Moreno, J. 2009. Remote sensing of solar-induced chlorophyll fluorescence: Review of methods and applications. Remote Sensing of Environment, 113, 10, 2037–2051.

Mohammed, G. H., Colombo, R., Middleton, E. M., Rascher, U., Van Der Tol, C., Nedbal, L., Goulas, Y., Pérez-Priego, O., Damm, A., Meroni, M., Joiner, J., Cogliati, S., Verhoef, W., Malenovský, Z., Gastellu-Etchegorry, J.-P., Miller, J. R., Guanter, L., Moreno, J., Moya, I., . . . Zarco-Tejada, P. J. 2019. Remote sensing of solar-induced chlorophyll fluorescence (SIF) in vegetation: 50 years of progress. Remote Sensing of Environment, 231, 111177. https://doi.org/10.1016/j.rse.2019.04.030.

Monfreda, C., Ramankutty, N. & Foley, J.A. 2008. Farming the planet: 2. Geographic distribution of crop areas, yields, physiological types, and net primary production in the year 2000. Global Biogeochemical Cycles, 22, 1, GB1022.

Monteith, J.L. 1972. Solar radiation and productivity in tropical ecosystems. Journal of Applied Ecology, 9, 3, 747.

Monteith, J.L. & Unsworth, M.H. 1990. Principles of Environmental Physics. 2nd ed., Antony Rowe Ltd., Eastbourne, UK.

Moran, M.S., Ponce-Campos, G.E., Huete, A., McClaran, M.P., Zhang, Y., Hamerlynck, E.P., Augustine, D.J., Gunter, S.A., Kitchen, S.G., Peters, D.P.C., Starks, P.J. & Hernandez, M. 2014. Functional response of U.S. grasslands to the early 21st-century drought. Ecology, 95, 8, 2121–2133.

Myneni, R.B., Hoffman, S., Knyazikhin, Y., Privette, J.L., Glassy, J., Tian, Y., Wang, Y., Song, X., Zhang, Y., Smith, G.R., Lotsch, A., Friedl, M., Morisette, J.T., Votava, P., Nemani, R.R. & Running, S.W. 2002. Global products of vegetation leaf area and fraction absorbed PAR from year one of MODIS data. Remote Sensing of Environment, 83, 1–2, 214–231.

Myneni, R.B. & Williams, D.L. 1994. On the relationship between FAPAR and NDVI. Remote Sensing of Environment, 49, 3, 200–211.

Nemani, R.R., Keeling, C.D., Hashimoto, H., Jolly, W.M., Piper, S.C., Tucker, C.J., Myneni, R.B. & Running, S.W. 2003. Climate-driven increases in global terrestrial net primary production from 1982 to 1999. Science, 300, 5625, 1560–1563.

Nichol, C.J., Huemmrich, K.F., Black, T.A., Jarvis, P.G., Walthall, C.L., Grace, J. & Hall, F.G. 2000. Remote sensing of photosynthetic-light-use efficiency of boreal forest. Agricultural and Forest Meteorology, 101, 2–3, 131–142.

Olofsson, P., Lagergren, F., Lindroth, A., Lindström, J., Klemedtsson, L., Kutsch, W. & Eklundh, L. 2008. Towards operational remote sensing of forest carbon balance across Northern Europe. Biogeosciences, 5, 3, 817–832.

Pei, Y., Dong, J., Zhang, Y., Yuan, W., Doughty, R., Yang, J., Zhou, D., Zhang, L. & Xiao, X. 2022. Evolution of light use efficiency models: Improvement, uncertainties, and implications. Agricultural and Forest Meteorology, 317, 108905. https://doi.org/10.1016/j.agrformet.2022.108905.

Peng, Y. & Gitelson, A.A. 2012. Remote estimation of gross primary productivity in soybean and maize based on total crop chlorophyll content. Remote Sensing of Environment, 117, 440–448.

Peng, Y., Gitelson, A.A., Keydan, G., Rundquist, D.C. & Moses, W. 2011. Remote estimation of gross primary production in maize and support for a new paradigm based on total crop chlorophyll content. Remote Sensing of Environment, 115, 4, 978–989.

Peng, Y., Gitelson, A.A. & Sakamoto, T. 2013. Remote estimation of gross primary productivity in crops using MODIS 250m data. Remote Sensing of Environment, 128, 0, 186–196.

Peñuelas, J., Filella, I. & Gamon, J.A. 1995. Assessment of photosynthetic radiation-use efficiency with spectral reflectance. New Phytologist, 131, 3, 291–296.

Ponce Campos, G.E., Moran, M.S., Huete, A., Zhang, Y., Bresloff, C., Huxman, T.E., Eamus, D., Bosch, D.D., Buda, A.R., Gunter, S.A., Scalley, T.H., Kitchen, S.G., McClaran, M.P., McNab, W.H., Montoya, D.S., Morgan, J.A., Peters, D.P.C., Sadler, E.J., Seyfried, M.S. & Starks, P.J. 2013. Ecosystem resilience despite large-scale altered hydroclimatic conditions. Nature, 494, 7437, 349–352.

Porcar-Castell, A., Tyystjärvi, E., Atherton, J., van der Tol, C., Flexas, J., Pfündel, E.E., Moreno, J., Frankenberg, C. & Berry, J.A. 2014. Linking chlorophyll a fluorescence to photosynthesis for remote sensing applications: Mechanisms and challenges. Journal of Experimental Botany, 65, 15, 4065–4095.

Potter, C.S., Randerson, J.T., Field, C.B., Matson, P.A., Vitousek, P.M., Mooney, H.A. & Klooster, S.A. 1993. Terrestrial ecosystem production: A process model based on global satellite and surface data. Global Biogeochemical Cycles, 7, 4, 811–841.

Rahman, A.F., Cordova, V.D., Gamon, J.A., Schmid, H.P. & Sims, D.A. 2004. Potential of MODIS ocean bands for estimating CO_2 flux from terrestrial vegetation: A novel approach. Geophysical Research Letters, 31, 10, L10503.

Rahman, A.F., Sims, D.A., Cordova, V.D. & El-Masri, B.Z. 2005. Potential of MODIS EVI and surface temperature for directly estimating per-pixel ecosystem C fluxes. Geophysical Research Letters, 32, 19, L19404.

Restrepo-Coupe, N., Huete, A., Davies, K., Cleverly, J., Beringer, J., Eamus, D., van Gorsel, E., Hutley, L.B. & Meyer, W.S. 2016. MODIS vegetation products as proxies of photosynthetic potential along a gradient of meteorologically and biologically driven ecosystem productivity. Biogeosciences, 13, 19, 5587–5608.

Ripullone, F., Rivelli, A.R., Baraldi, R., Guarini, R., Guerrieri, R., Magnani, F., Peñuelas, J., Raddi, S. & Borghetti, M. 2011. Effectiveness of the photochemical reflectance index to track photosynthetic activity over a range of forest tree species and plant water statuses. Functional Plant Biology, 38, 3, 177–186.

Rossini, M., Migliavacca, M., Galvagno, M., Meroni, M., Cogliati, S., Cremonese, E., Fava, F., Gitelson, A., Julitta, T., Morra di Cella, U., Siniscalco, C. & Colombo, R. 2014. Remote estimation of grassland gross primary production during extreme meteorological seasons. International Journal of Applied Earth Observation and Geoinformation, 29, 0, 1–10.

Ruimy, A., Jarvis, P., Baldocchi, D. & Saugier, B. 1995. CO_2 fluxes over plant canopies and solar radiation: A review. Advances in Ecological Research, 26, 1–68.

Running, S.W., Baldocchi, D.D., Turner, D.P., Gower, S.T., Bakwin, P.S. & Hibbard, K.A. 1999. A global terrestrial monitoring network integrating tower fluxes, flask sampling, ecosystem modeling and EOS satellite data. Remote Sensing of Environment, 70, 1, 108–127.

Running, S.W., Heinsch, F.A., Zhao, M., Reeves, M., Hashimoto, H. & Nemani, R.R. 2004. A continuous satellite-derived measure of global terrestrial primary production. BioScience, 54, 6, 547–560.

Ryu, Y., Berry, J.A. & Baldocchi, D.D. 2019. What is global photosynthesis? History, uncertainties and opportunities. Remote Sensing of Environment, 223, 95–114. https://doi.org/10.1016/j.rse.2019.01.016.

Saatchi, S.S., Harris, N.L., Brown, S., Lefsky, M., Mitchard, E.T.A., Salas, W., Zutta, B.R., Buermann, W., Lewis, S.L., Hagen, S., Petrova, S., White, L., Silman, M. & Morel, A. 2011. Benchmark map of forest carbon stocks in tropical regions across three continents. Proceedings of the National Academy of Science U.S.A., 108, 9899–9904.

Sala, O.E., Parton, W.J., Joyce, L.A. & Lauenroth, W.K. 1988. Primary production of the central grassland region of the United States. Ecology, 69, 1, 40–45.

Schimel, D., Pavlick, R., Fisher, J.B., Asner, G.P., Saatchi, S., Townsend, P., Miller, C., Frankenberg, C., Hibbard, K. & Cox, P. 2015. Observing terrestrial ecosystems and the carbon cycle from space. Global Change Biology, 21, 1762–1776. https://doi.org/10.1111/gcb.12822.

Schimel, D.S. 1998. Climate change: The carbon equation. Nature, 393, 6682, 208–209.

Sellers, P.J. 1985. Canopy reflectance, photosynthesis and transpiration. International Journal of Remote Sensing, 6, 8, 1335–1372.

Shen, J., Huete, A., Ma, X., Tran, N.N., Joiner, J., Beringer, J., Eamus, D. & Yu, Q. 2020. Spatial pattern and seasonal dynamics of the photosynthesis activity across Australian rainfed croplands. Ecological Indicators, 108, 105669. https://doi.org/10.1016/j.ecolind.2019.105669.

Sims, D.A., Rahman, A.F., Cordova, V.D., El-Masri, B.Z. & Baldocchi, D.D. 2006. On the use of MODIS EVI to assess gross primary productivity of North American ecosystems. Journal of Geophysical Research, 111.

Sims, D.A., Rahman, A.F., Cordova, V.D., El-Masri, B.Z., Baldocchi, D.D., Bolstad, P.V., Flanagan, L.B., Goldstein, A.H., Hollinger, D.Y., Misson, L., Monson, R.K., Oechel, W.C., Schmid, H.P., Wofsy, S.C. & Xu, L. 2008. A new model of gross primary productivity for North American ecosystems based solely on the enhanced vegetation index and land surface temperature from MODIS. Remote Sensing of Environment, 112, 4, 1633–1646.

Sjöström, M., Ardö, J., Arneth, A., Boulain, N., Cappelaere, B., Eklundh, L., de Grandcourt, A., Kutsch, W.L., Merbold, L., Nouvellon, Y., Scholes, R.J., Schubert, P., Seaquist, J. & Veenendaal, E.M. 2011. Exploring the potential of MODIS EVI for modeling gross primary production across African ecosystems. Remote Sensing of Environment, 115, 4, 1081–1089.

Sjöström, M., Zhao, M., Archibald, S., Arneth, A., Cappelaere, B., Falk, U., de Grandcourt, A., Hanan, N., Kergoat, L., Kutsch, W., Merbold, L., Mougin, E., Nickless, A., Nouvellon, Y., Scholes, R.J., Veenendaal, E.M. & Ardö, J. 2013. Evaluation of MODIS gross primary productivity for Africa using eddy covariance data. Remote Sensing of Environment, 131, 0, 275–286.

Tucker, C.J. 1979. Red and photographic infrared linear combinations for monitoring vegetation. Remote Sensing of Environment, 8, 2, 127–150.

Turner, D.P., Ollinger, S.V. & Kimball, J.S. 2004. Integrating remote sensing and ecosystem process models for landscape- to regional-scale analysis of the carbon cycle. BioScience, 54, 6, 573–584.

Turner, D.P., Ritts, W.D., Cohen, W.B., Gower, S.T., Running, S.W., Zhao, M., Costa, M.H., Kirschbaum, A.A., Ham, J.M., Saleska, S.R. & Ahl, D.E. 2006. Evaluation of MODIS NPP and GPP products across multiple biomes. Remote Sensing of Environment, 102, 3–4, 282–292.

Turner, D.P., Urbanski, S., Bremer, D., Wofsy, S.C., Meyers, T., Gower, S.T. & Gregory, M. 2003. A cross-biome comparison of daily light use efficiency for gross primary production. Global Change Biology, 9, 3, 383–395.

Veefkind, J.P., Aben, I., McMullan, K., Förster, H., de Vries, J., Otter, G., Claas, J., Eskes, H.J., de Haan, J.F., Kleipool, Q., van Weele, M., Hasekamp, O., Hoogeveen, R., Landgraf, J., Snel, R., Tol, P., Ingmann, P., Voors, R., Kruizinga, B., Vink, R., Visser, H. & Levelt, P.F. 2012. TROPOMI on the ESA Sentinel-5 precursor: A GMES mission for global observations of the atmospheric composition for climate, air quality and ozone layer applications. Remote Sensing of Environment, 120, 0, 70–83.

Verma, S.B., Dobermann, A., Cassman, K.G., Walters, D.T., Knops, J.M., Arkebauer, T.J., Suyker, A.E., Burba, G.G., Amos, B., Yang, H., Ginting, D., Hubbard, K.G., Gitelson, A.A. & Walter-Shea, E.A. 2005. Annual carbon dioxide exchange in irrigated and rainfed maize-based agroecosystems. Agricultural and Forest Meteorology, 131, 1–2, 77–96.

Wang, J., Rich, P.M., Price, K.P. & Kettle, W.D. 2004. Relations between NDVI and tree productivity in the central Great Plains. International Journal of Remote Sensing, 25, 16, 3127–3138.

Wild, B., Teubner, I., Moesinger, L., Zotta, R.-M., Forkel, M., Van Der Schalie, R., Sitch, S. & Dorigo, W. 2022. VODCA2GPP—a new, global, long-term (1988–2020) gross primary production dataset from microwave remote sensing. Earth System Science Data, 14, 3, 1063–1085. https://doi.org/10.5194/essd-14-1063-2022.

Wilson, K.B., Baldocchi, D.D. & Hanson, P.J. 2001. Leaf age affects the seasonal pattern of photosynthetic capacityand net ecosystem exchange of carbon in a deciduous forest. Plant, Cell & Environment, 24, 6, 571–583.

Wu, C., Chen, J.M. & Huang, N. 2011. Predicting gross primary production from the enhanced vegetation index and photosynthetically active radiation: Evaluation and calibration. Remote Sensing of Environment, 115, 12, 3424–3435.

Wu, C., Gonsamo, A., Zhang, F. & Chen, J.M. 2014. The potential of the greenness and radiation (GR) model to interpret 8-day gross primary production of vegetation. ISPRS Journal of Photogrammetry and Remote Sensing, 88, 0, 69–79.

Wu, C., Niu, Z. & Gao, S. 2010. Gross primary production estimation from MODIS data with vegetation index and photosynthetically active radiation in maize. Journal of Geophysical Research. Atmospheres, 115, 12.

Xiao, J., Chevallier, F., Gomez, C., Guanter, L., Hicke, J.A., Huete, A.R., Ichii, K., Ni, W., Pang, Y., Rahman, A.F., Sun, G., Yuan, W., Zhang, L. & Zhang, X. 2019. Remote sensing of the terrestrial carbon cycle: A review of advances over 50 years. Remote Sensing of Environment, 233, 111383. https://doi.org/10.1016/j.rse.2019.111383.

Xiao, X., Zhang, Q., Braswell, B., Urbanski, S., Boles, S., Wofsy, S., Moore III, B. & Ojima, D. 2004. Modeling gross primary production of temperate deciduous broadleaf forest using satellite images and climate data. Remote Sensing of Environment, 91, 2, 256–270.

Xiao, X., Zhang, Q., Saleska, S., Hutyra, L., De Camargo, P., Wofsy, S., Frolking, S., Boles, S., Keller, M. & Moore III, B. 2005. Satellite-based modeling of gross primary production in a seasonally moist tropical evergreen forest. Remote Sensing of Environment, 94, 1, 105–122.

Yuan, W., Liu, S., Yu, G., Bonnefond, J.-M., Chen, J., Davis, K., Desai, A.R., Goldstein, A.H., Gianelle, D., Rossi, F., Suyker, A.E. & Verma, S.B. 2010. Global estimates of evapotranspiration and gross primary production based on MODIS and global meteorology data. Remote Sensing of Environment, 114, 7, 1416–1431.

Zeng, Y., Hao, D., Huete, A., Dechant, B., Berry, J., Chen, J., Joiner, J., Frankenberg, C., Bond-Lamberty, B., Ryu, Y., Xiao, J., Asrar, G. & Chen, M. 2022. Optical vegetation indices for monitoring terrestrial ecosystems globally. Nature Reviews Earth and Environment, 3, 7, 477–493. https://doi.org/10.1038/s43017-022-00298-5.

Zhan, W., Yang, X., Ryu, Y., Dechant, B., Huang, Y., Goulas, Y., Kang, M. & Gentine, P. 2022. Two for one: Partitioning CO2 fluxes and understanding the relationship between solar-induced chlorophyll fluorescence and gross primary productivity using machine learning. Agricultural and Forest Meteorology, 321, 108980. https://doi.org/10.1016/j.agrformet.2022.108980.

Zhang, H., Li, J., Liu, Q., Lin, S., Huete, A., Liu, L., Croft, H., Clevers, J., Zeng, Y., Wang, X., Gu, C., Zhang, Z., Zhao, J., Dong, Y., Mumtaz, F. & Yu, W. 2022. A novel red-edge spectral index for retrieving the leaf chlorophyll content. Methods in Ecology and Evolution, 13, 12, 2771–2787. https://doi.org/10.1111/2041-210X.13994.

Zhang, Y., Guanter, L., Berry, J.A., Joiner, J., van der Tol, C., Huete, A., Gitelson, A., Voigt, M. & Köhler, P. 2014a. Estimation of vegetation photosynthetic capacity from space-based measurements of chlorophyll fluorescence for terrestrial biosphere models. Global Change Biology, 20, 3727–3742. https://doi.org/10.1111/gcb.12664.

Zhang, Y., Guanter, L., Berry, J.A., Van Der Tol, C., Yang, X., Tang, J. & Zhang, F. 2016. Model-based analysis of the relationship between sun-induced chlorophyll fluorescence and gross primary production for remote sensing applications. Remote Sensing of Environment, 187, 145–155. https://doi.org/10.1016/j.rse.2016.10.016.

Zhang, Y., Susan Moran, M., Nearing, M.A., Ponce Campos, G.E., Huete, A.R., Buda, A.R., Bosch, D.D., Gunter, S.A., Kitchen, S.G., Henry McNab, W., Morgan, J.A., McClaran, M.P., Montoya, D.S., Peters, D.P.C. & Starks, P.J. 2013. Extreme precipitation patterns and reductions of terrestrial ecosystem production across biomes. Journal of Geophysical Research: Biogeosciences, 118, 1, 148–157.

Zhang, Y., Yu, G., Yang, J., Wimberly, M.C., Zhang, X., Tao, J., Jiang, Y. & Zhu, J. 2014b. Climate-driven global changes in carbon use efficiency. Global Ecology and Biogeography, 23, 2, 144–155.

Zhao, M., Heinsch, F.A., Nemani, R.R. & Running, S.W. 2005. Improvements of the MODIS terrestrial gross and net primary production global data set. Remote Sensing of Environment, 95, 2, 164–176.

Zhao, M. & Running, S.W. 2010. Drought-induced reduction in global terrestrial net primary production from 2000 through 2009. Science, 329, 5994, 940–943.

2 Remote Sensing of Solar-Induced Chlorophyll Fluorescence

Juan Quiros-Vargas, Bastian Siegmann, Juliane Bendig, Laura Verena Junker-Frohn, Christoph Jedmowski, David Herrera, and Uwe Rascher

ACRONYMS AND DEFINITIONS

ESA	European Space Agency
GPP	Gross Primary Productivity
LST	Land Surface Temperature
LUE	Light Use Efficiency
MERIS	Medium Resolution Imaging Spectrometer
MODIS	Moderate Resolution Imaging Spectroradiometer
NIR	Near-Infrared
TROPOMI	TROPOspheric Monitoring Instrument
UAV	Unmanned Aerial Vehicles

2.1 INTRODUCTION: WHAT IS SOLAR-INDUCED CHLOROPHYLL FLUORESCENCE (SIF)?

The solar radiation that interacts with vegetation surfaces on earth can follow three possible paths: be reflected, transmitted, or absorbed (Figure 2.1a). The absorbed solar energy is captured by chlorophyll molecules, pigments located in the thylakoid membrane of chloroplasts (cellular organelles also composed by outer and inner membranes, stroma, granum, thylakoids, and lamellae). The energy of the photons absorbed by chlorophyll molecules is transferred to the reaction center of photosystem II (Figure 2.1b), where a series of physiological reactions (electron transport chain) is triggered leading what is called photochemical quenching (PQ; herewith understood as carbon fixation or photosynthesis). However, not all of the absorbed energy can be photochemically quenched, and a fraction of it has to be dissipated either as heat (non-photochemical quenching, -NPQ-) or emitted as solar-induced chlorophyll fluorescence (SIF); which can thus be defined as a low intensity red to far-red signal continuously emitted from ~600–800 nm, with two peaks located at 690 nm (SIF_{Red}) and 740 nm ($SIF_{Far\text{-}red}$; Mohammed et al. 2019).

Due to its close relation with photosynthesis, the use of the fluorescence signal from vegetation has a great potential to be used as a source of information to analyze the physiological status of vegetation. It is worth mentioning that such information can also be obtained from active sensing techniques like the pulse amplitude modulation, which measures the fluorescence signal generated by saturating (blue or red) light pulses; yet, the focus of this chapter is on the use of the information contained on the passively sensed SIF signal. In particular, the use of SIF for early stress detection has gained great interest in the last years, since its emission depends not only on factors like the observed species, phenology, and the diurnal (Siegmann et al. 2021) and seasonal contexts

DOI: 10.1201/9781003541165-3

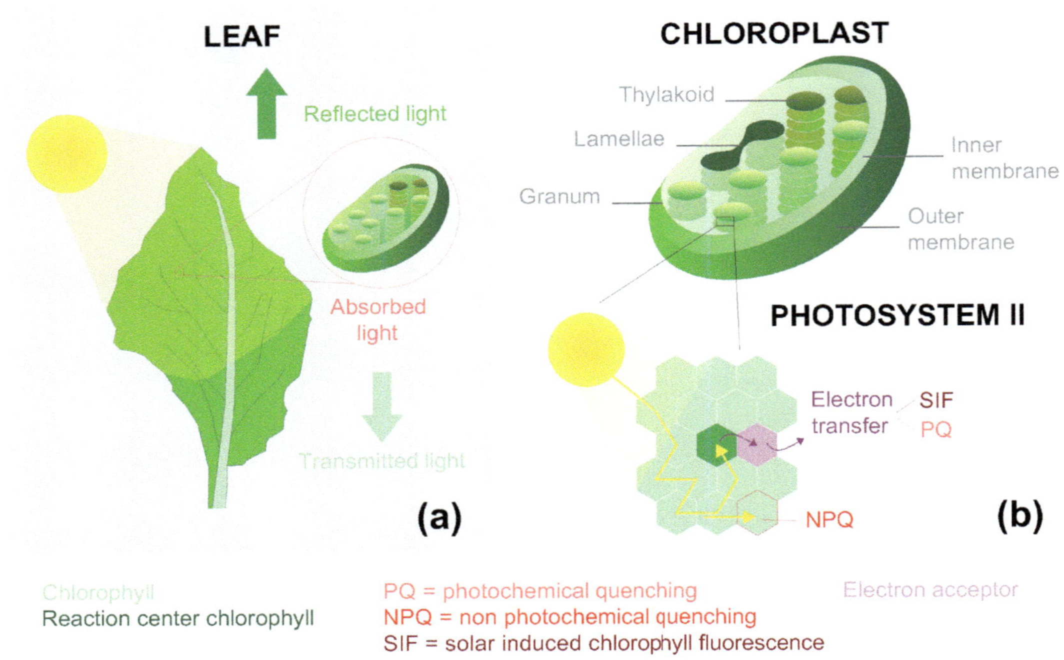

FIGURE 2.1 Interaction of the sunlight with vegetation at leaf (a), and chloroplast and reaction center (b) levels.

(Mengistu et al. 2021) when the data acquisition takes place, but also on the stress/health status of plants (Zeng et al. 2022). Moreover, when estimated from remote sensing platforms, the SIF data can provide additional information on the spatiotemporal dynamics of physiological changes caused by biotic (Zarco-Tejada et al. 2018) or abiotic stressors (Panigada et al. 2014, Ač et al. 2015).

Despite of its relevance for understanding the spatiotemporal dynamics of physiological status of open field vegetation surfaces, the remote sensing of SIF has to cope with multiple challenges: (1) the connection between SIF and variations in the photosynthetic activity is already weak; (2) the SIF signal is strongly affected by canopy structure (by scattering and reabsorption); and (3) the low intensity of the emitted SIF (representing only 1–5% of the total reflected radiation in the near infrared; Meroni et al. 2009) makes it difficult to disentangle that signal from the environmental light and the reflected radiation by plants. Additionally, the intensity of the SIF signal is distinctly decreasing with increasing distance from the molecular (photosystem) level, due to scattering and re-absorption processes (Figure 2.2a), and at the canopy level it can be reabsorbed by other canopy constituents (Porcar-Castell et al. 2014). Moreover, the SIF signal leaving the canopy is weakened by the scattering caused by atmosphere molecules (Figure 2.2b). Despite of all these challenges, advances in remote sensing technology and methods in the last decade, nowadays, make it possible to retrieve SIF from aerial and orbital platforms.

SIF retrieval methods attempt to overcome those challenges through the use of what is called absorption features in the electromagnetic spectrum. Solar absorption features are also called Fraunhofer lines, which are dark lines in the spectrum caused by the absorption properties of chemical elements at specific wavelengths in the Sun photosphere. Assuming constancy in the reflectance and fluorescence signals, Fraunhofer line depth (FLD) methods were developed to retrieve SIF by calculating the ratio between the incoming solar radiance inside and outside a Fraunhofer line (within the chlorophyll fluorescence emission spectrum), and the apparent reflectance (including the contribution of SIF) inside and outside the same Fraunhofer line (Theisen 2002, and Plascyk 1975; cited by Meroni et al. 2009). The result of these calculation show a slightly higher apparent reflectance inside the FL, which is proportional to the emitted SIF signal. Besides, other retrieval methods exploit the absorption of solar light by elements in the Earth atmosphere through the so-called telluric bands. In particular, for the retrieval of SIF, oxygen (O_2) telluric bands are used since they absorb the solar light at ~759 nm (O_2-A) and ~687 nm (O_2-B; Tubuxin et al. 2015; Figure 2.2b) from where $SIF_{Far\text{-}red}$ (hereinafter referred to as SIF) and SIF_{Red} can be measured, respectively.

Novel spectral fitting methods (SFMs) were developed to retrieve SIF either from the O_2 absorption bands or from the full chlorophyll fluorescence spectrum (Cogliati et al. 2015; Cogliati et al. 2019) by using high resolution spectral data from contiguous bands within the spectral region of interest. SFM approaches offer some advantages over the FLD-based approaches, since they overcome the assumption of a constant reflectance and fluorescence signal (Meroni et al. 2010), they are based on the principles of the radiative transfer theory, and they allow for the correction of interferences of the atmosphere (e.g., aerosols, surface pressure, water vapor, etc.). A detailed revision of approaches to retrieve SIF can be found in Bandopadhyay et al. (2020).

Imaging and non-imaging (point) spectrometers are employed to measure the high spectral resolution reflectance at sub-nanometer scale used to retrieve SIF. At ground scale, point spectrometers like the fluorescence box (Flox; JB Hyperspectral) are used to monitor SIF with high temporal resolution, but low spatial coverage. On the other hand, a higher spatial coverage, but low temporal resolution, can be provided by imaging spectrometers mounted on airborne platforms, for example, the imaging fluorimeter (IBIS; Gamon et al. 2018), the chlorophyll fluorescence imaging spectrometer (CFIS; Frankenberg et al. 2018) and the high-performance airborne imaging spectrometer (HyPlant; Rascher et al. 2015; Siegmann et al. 2019).

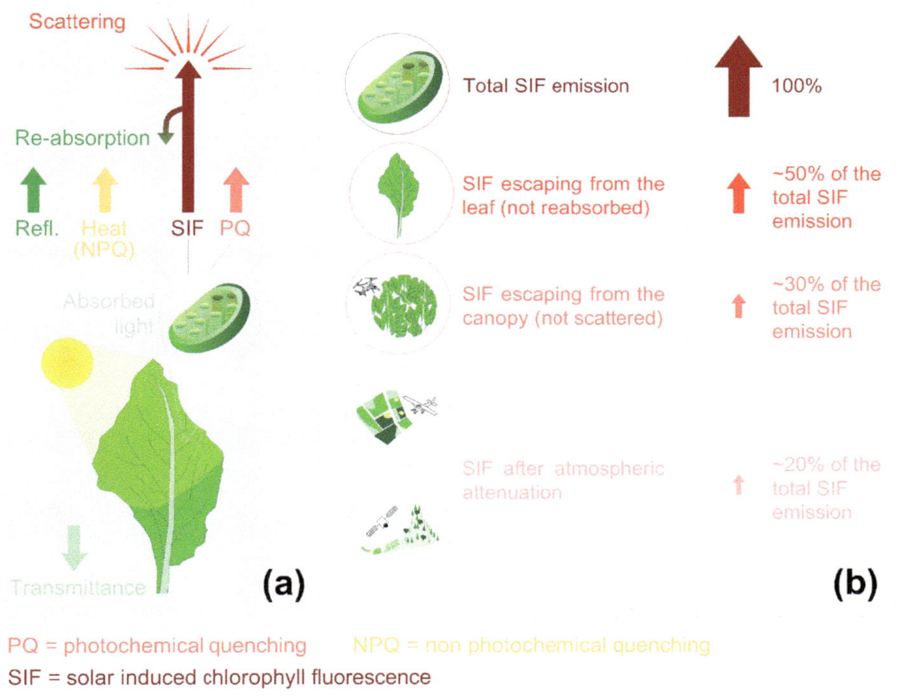

FIGURE 2.2 Possible ways followed by the absorbed light (a), and the attenuation of the solar-induced chlorophyll fluorescence (SIF) signal from the molecular- to the ground- and aerial- (unmanned aerial vehicle (UAV) and airborne) and satellite-scales (b). For a more detailed graphical representation about how SIF is attenuated across spatial scales, readers are referred to Porcar-Castell et al. (2021).

Despite the great advances achieved during the last decades, the retrieval of SIF carries instrument- and retrieval-associated uncertainties, which are yet a matter of investigation to be solved. Specific sensor characteristics like the used detector and fore optics, as well as the type of calibration used for one specific device, makes it difficult to standardize and compare SIF measurements across spatiotemporal scales and sensors. Moreover, the major source of uncertainty associated to the retrieval comes from the atmospheric interference, since it is sensitive to variations in factors like water vapor, terrain elevation, aerosol optical thickness, and surface pressure. Thus, the method used (either empirical corrections or those based on radiative transfer models) to correct for the interference of atmosphere will have a strong impact on the output data (European Space Agency, ESA 2022a; Photoproxy, report No. 3). Here, numeric inaccuracies in the atmospheric correction algorithms, and the use of atmospheric transfer models with an erroneous parameterization are probably the main sources of uncertainty (ESA 2022b, and ESA 2022c; corresponding to FLEXSense campaigns final reports 2018 and 2019, respectively).

It is worth mentioning that remote sensing studies often require the integration of data from several methods (parametric, non-parametric, radiative transfer, etc.; Gerhards et al. 2019). This is due to the complexity of the vegetation processes being measured, which can be caused by several factors, which at the same time produce different responses in the plant. For instance, the several functions water has in plant physiology (turgor, thermal regulation, photolysis, etc.) can cause many responses at the root, leaf, and canopy structure levels (Jonard et al. 2020). While Vis can be used to monitor changes in the canopy pigments composition, thermal data and SIF can be used to analyze alterations in canopy temperature and photosynthetic activity, respectively.

2.2 SIF ASSESSMENTS AT UNMANNED AERIAL VEHICLE (UAV) SCALE

Vegetation in open fields is always heterogeneous due to edaphic and micro-climatic variations. The assessment of spatial variability across vegetation surfaces is necessary to improve management practices in agriculture, and to understand ecosystem dynamics. Unmanned aerial vehicles (UAVs) are the most cost-effective technology fur such tasks in areas of ≤20 ha for example in viticulture (Matese et al. 2015), forestry (Torresan et al. 2017), environmental monitoring (Manfreda et al. 2018), and in several branches of agronomy, for example, crop protection (Psirofonia et al. 2017), site specific management, and field phenotyping (Sankaran et al. 2015).

UAV platforms, namely multirotor UAVs, fixed-wing gliders and more recently hybrid platforms (Norasma et al. 2019; Ziliani et al. 2018, Oldeland et al. 2021) have been used for vegetation monitoring for more than a decade (Aber et al. 2010, Colomina and Molina 2014). After the first attempts to install lightweight spectroradiometers on drones were successful (Burkart et al. 2014), systems dedicated to SIF retrieval followed. Since, the retrieval of SIF requires high (sub-nm) spectral resolution (full width at half maximum, FWHM) and a high dynamic range detector (at least 16-bit) for obtaining high signal-to-noise ratio SIF measurements (Cendrero-Mateo et al. 2019), the challenge was to find the optimum configuration that meets the space and weight limitations of UAVs.

A simple system that combined a spectroradiometer and RGB camera was developed (Garzonio et al. 2017), with the drawback of lacking the option to directly measure downwelling irradiance. A similarly simple system was the Floxplane, a fixed wing platform aimed to characterize the interference of the (multiple km) atmosphere column on the retrieval of SIF (Quiros-Vargas et al. 2020a). Consequently, systems with bifurcated fibers and fiber multiplexers were developed than measure the upwelling radiance from vegetation with a bare fiber, or a field of view restricting fore optic, and the downwelling radiance from the Sun with a diffuser of 180° field of view. These systems were either comprising one visible-near infrared spectroradiometer (Bendig et al. 2020, Wang et al. 2021) or two spectroradiometers, of which one covers only the spectral region of SIF emission but with higher spectral resolution (650–800 nm) (Mac Arthur et al. 2014). An alternative system was developed by Chang et al. (2020), where instead of a bifurcated fiber, a mechanical arm was pointing the fiber with the diffuser at the canopy and the Sun in an alternating manner, to avoid light loss

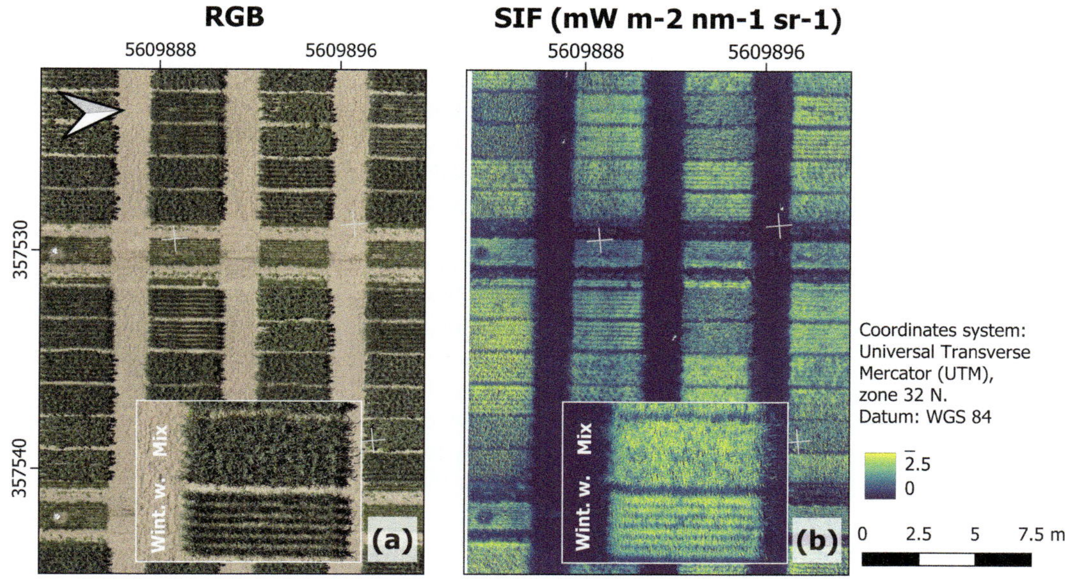

FIGURE 2.3 Red-green-blue (RGB) and solar-induced chlorophyll fluorescence (SIF) images collected from an unmanned aerial vehicle (UAV) in winter wheat and mixed (winter wheat and bean) breeding plots in West Germany. The SIF image was obtained with the SIFcam (Kneer et al. 2023).

at the fiber entrance. Compared to the other systems, this system had a hemispherical field of view of 180° for vegetation observations.

All systems have the challenge in common that point spectrometers add complexity to data collection and analysis. One of the major difficulties is the location and characterization of the sensor footprint, which was solved either by a very detailed payload characterization in combination with inertial measurement units (Gautam et al. 2020), or by combining laser range finders and RGB cameras for field of view characterization (Wang et al. 2021). As most studies conducted involved agricultural crops, the applicability of these systems in small scale field experiments was limited by the accuracy of the footprint characterization and the navigation accuracy of the UAV (Bendig et al. 2021).

Ultimately, a simple imaging system was developed for SIF measurements at 760 nm, as snapshot dual camera array which enables centimeter-level spatial resolution (Kneer et al. 2023). The system called SIFcam enables insights in the spatial variation of SIF in the canopy, for example, looking into differences between shaded and sunlit canopy parts (Bendig et al. 2023). An example a UAV-based SIF image is presented in Figure 2.3, next to a RGB image for visual comparison. Both the simple imaging system as well as the spectrally more versatile non-imaging systems have justifications in SIF research. For example, Wang et al. (2021, 2022a, 2022b) utilized the FluorSpec system to investigate water stress in sugar beet and performed a comparison the airborne imaging spectrometer HyPlant. Although being a highly specialized research field, it is expected that more studies will be conducted considering the potential for UAVs to assist in the calibration and validation operations of the upcoming Fluorescence Explorer (FLEX) satellite mission (Rossini et al. 2022).

2.3 SIF ASSESSMENTS AT AIRBORNE SCALE

The UAV-based systems mentioned earlier typically offer insights limited to the intra-field scale. In contrast, airborne-based remote sensing imagery allows for the examination of vegetation surfaces on a larger scale. For instance, since the first airborne SIF maps were published (Rascher et al.

2015), data from the HyPlant imaging spectrometer (Siegmann et al. 2019) has pushed forward our understanding of larger scale vegetation processes. There are numerous studies using HyPlant data since then: for example, Wieneke et al. (2016) analyzed HyPlant SIF to improve GPP estimates, Tagliabue et al. (2020) proposed the use of HyPlant SIF for the assessment of the functional diversity as a reference of ecosystems functioning and stability, Hornero et al. (2021) investigated the contributions of tree crowns, understory and soil to the SIF signal measured by HyPlant, and Zeng et al. (2022) and Damm et al. (2022) showed the potential of HyPlant SIF data for the early detection of vegetation stress.

The correct physiological interpretation of retrieved canopy SIF is affected by different confounding effects. In particular, a better understanding of re-absorption and scattering processes of SIF within the canopy has been a topic of great research interest in recent years, since it is essential for comparing SIF observations recorded at different scales. Canopy-scale SIF is different from SIF measured at the leaf or photosystem scale and thus cannot be directly used to quantitatively detect variations in plant physiology. For this reason, different vegetation indices have been developed that allow for the downscaling of far-red SIF from the canopy to the leaf scale, that is, the fluorescence correction vegetation index (FCVI) (Yang et al. 2020), the NIR reflectance of vegetation index (NIRv) (Zeng et al. 2019), and the further developed and improved versions of NIRv (NIRvH1 and NIRvH2) (Zeng et al. 2021). Those indices can be derived from multi- or hyperspectral image data that are collected in parallel to the SIF imagery and further assist the quantification and interpretation of plant physiological changes reflected in SIF measurements. While NIRv was predominantly applied to canopy SIF measured with point spectrometers mounted on towers (e.g., Dechant et al. 2020) or satellite image data (e.g., Zhang et al. 2020a), Siegmann et al. (2021) showed the potential of FCVI applied to a diurnal HyPlant SIF dataset to downscale canopy SIF to the leaf level and additionally calculated SIF emission efficiency (Ef). *Ef* also referred as fluorescence quantum yield can be regarded as SIF measured at leaf scale normalized for incoming radiation and canopy structure. The authors point out that information about SIF downscaled from the canopy to the leaf scale is of particular importance when environmental constraints limit photosynthetic processes and when ecosystem models require a better physiology-based parameterization.

The larger spatial coverage enabled by airborne sensors becomes important for the assessment of water stress, since the soil water content is strongly determined by the topography and soil physical characteristics of a whole landscape. The use of SIF for water stress assessment is related to two of the main emerging SIF applications mentioned by Porcar-Castell et al. (2021): (1) the pre-visual stress detection and (2) the water cycle studies. Besides, the understanding of the SIF response to varying soil water contents can lead to applications in climate modeling, since it can help to elucidate the mechanistic basis of SIF towards better constraining transpiration and photosynthetic dynamics.

The state-of-the-art of airborne-based SIF measurements are, in general, in a "proof of concepts" stage. Namely, airborne SIF information (in most of the cases from HyPlant) is being used to prove research concepts over large vegetation surfaces (Siegmann et al. 2021). In particular, investigations about the use of airborne SIF data for water stress assessment are currently gaining relevance. In this context, Damm et al. (2022) recently published the first study showing how SIF data detected an early physiological plant response to drought effects so called "double SIF response" (a SIF increase in the first about two days after a water limitation was initiated in maize experimental plots, followed by a strong decrease in the signal), which previously was just theoretically known. In addition to the Damm et al. (2022) study, in the last years several studies have been published aiming to elucidate how the soil water availability can be related to the emission of SIF (De Cannière et al. 2022; Berger et al. 2022). In a similar direction, Quiros-Vargas et al. (2020b) found HyPlant SIF to be more sensitive than Vis to the effect of heat in lower soil water retention capacity areas. In the same study, the authors reported a significant match (that was not observed with NDVI) between the spatial patterns of SIF and soil homogeneous units similarly as it was reported by von Hebel et al. (2018).

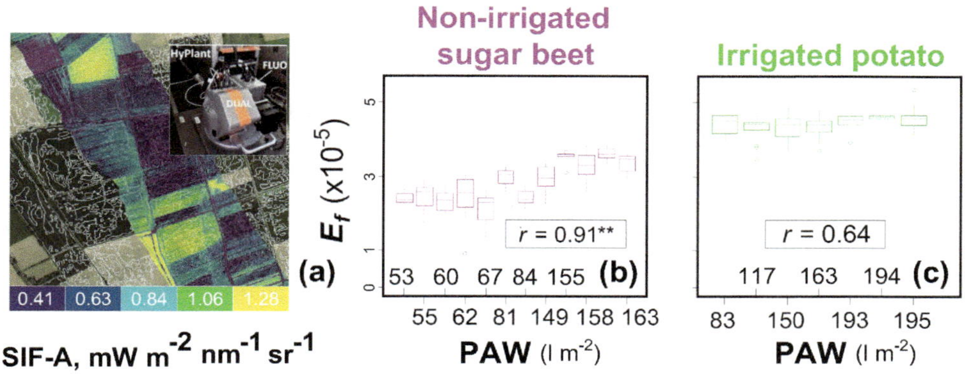

FIGURE 2.4 The high-performance airborne imaging spectrometer (HyPlant) and an example of the solar-induced chlorophyll fluorescence (SIF) imagery used in the study case (a). The main results of the relation between the SIF emission efficiency (E_f) and the estimated plant available water (PAW) in the non-irrigated sugar beet and irrigated potato fields are presented in (b) and (c), respectively.

To complement this section of the chapter, we developed a study case regarding the spatial relation between airborne-based SIF (HyPlant; Figure 2.4a) and the plant available water (PAW; derived from the map published by Brogi et al., 2019) in the soil over several irrigated potato, and non-irrigated sugar beet fields. Our results show a strong SIF-PAW correlation in non-irrigated sugar beet fields (Figure 2.4b), that was not present in the irrigated potato case (Figure 2.4c); potentially due to the fact that for that crop the water requirements were fully supplied and thus not reflected in the SIF signal. For interpretation of these results, it is essential to take into account that the SIF-PAW relation can vary, for instance according to the stress severity and the spatiotemporal scale of data. Therefore, in the following subsection we provide an insight about the SIF-soil water availability up-scaled to the satellite level.

Future studies assessing the effect of water shortage on plants should use actual soil water content. Moreover, alongside SIF information, reflectance-based vegetation indices (e.g., PRI, MTCI and NDVI) can provide complementary qualitatively data about vegetation traits and the effects of a water deficit (Damm et al. 2018). Indeed, the integration of multiple-sensor data (especially SIF- and thermal-based) has been reported as the most effective way to understand the effects of stressors in vegetation (Zarco-Tejada et al. 2018).

2.4 SIF ASSESSMENTS AT SATELLITE SCALE

SIF retrieved at satellite scale has been based on data from missions that were originally planned for atmospheric chemistry applications. Such spaceborne-based SIF data provides key information for vegetation functioning studies from regional to global scale with high resolution time series. The medium resolution imaging spectrometer (MERIS, from the European space agency—ESA) and the moderate resolution imaging spectroradiometer (MODIS; from the national aeronautics and space administration—NASA) were the firsts sensors allowing space born SIF retrieval based on the so-called "fluorescence line height algorithm" (Gower et al. 1999; Gower et al. 2004). Nevertheless, those studies were focused on the analysis of phytoplankton fluorescence and thus had low impact for the remote sensing of vegetation community. Further, those early studies were strongly affected by atmospheric conditions like the aerosol optical thickness (Bandopadhyay et al. 2020).

The first spaceborne-SIF retrieval of MERIS data from vegetation was performed using the FLD principle by Guanter (2007), who validated the satellite estimations with ground and airborne measurements. The subsequent launch of the greenhouse gases observing satellite (GOSAT) by the Japan aerospace exploration agency (JAXA) motivated further research on the retrieval of the seasonal SIF dynamics at global scale using the equipped thermal and near-infrared sensor for carbon observation—fourier transform spectrometer (TANSO-FTS) (Joiner et al. 2011). The low spatiotemporal resolution and poor signal to noise ratio (SNR) of GOSAT's data were partly overcome with the launch of the global ozone monitoring experiment 2 (GOME-2; Joiner et al. 2013) instrument on board the MetOp-A satellite by the European organization for the exploitation of meteorological satellites (EUMETSAT) and ESA. SIF derived from GOME-2 data was used as an indicator of crop photosynthetic capacity (Zhang et al. 2014).

Besides the impact of the aforementioned platforms and sensors, satellite-based SIF retrievals were greatly improved with the launch of the orbiting carbon observatory 2 (OCO-2; Sun et al. 2017) satellite. Among other applications, OCO-2 data was used to analyze the relation of SIF and GPP in numerous studies (Bandopadhyay et al. 2020). Later, the launch of the tropospheric monitoring instrument (TROPOMI) onboard the Sentinel-5 Precursor satellite raised expectations to retrieve SIF with a quality similar to the OCO-2, but with higher spatiotemporal resolution (Guanter et al. 2015). This was confirmed by Köhler et al. (2018) who published a time series of global SIF dataset (TROPOMI-SIF) with a spatial resolution of 7x3.5 km pixel^{-1} providing daily information over several years. A similar SIF product was recently also released by ESA in the frame of the TROPOSIF project (Guanter et al. 2021). The carrying satellite and a global SIF map example from TROPOMI are shown in Figure 2.5a and 2.5b. In addition, based on NASA's public information, Figure 2.5c shows the SIF data availability from some of the North American missions since 1995 to the present. Readers are directed to Sun et al. (2023), for an overview of the past, current and future orbital platforms from which SIF can be retrieved. The unprecedented high spatiotemporal resolution of the TROPOMI-based SIF products encouraged novel studies addressing, for example, in more detail the SIF-GPP relation (Li and Xiao 2022), as well as variations of SIF in the dry season of tropical forests (Doughty et al. 2019).

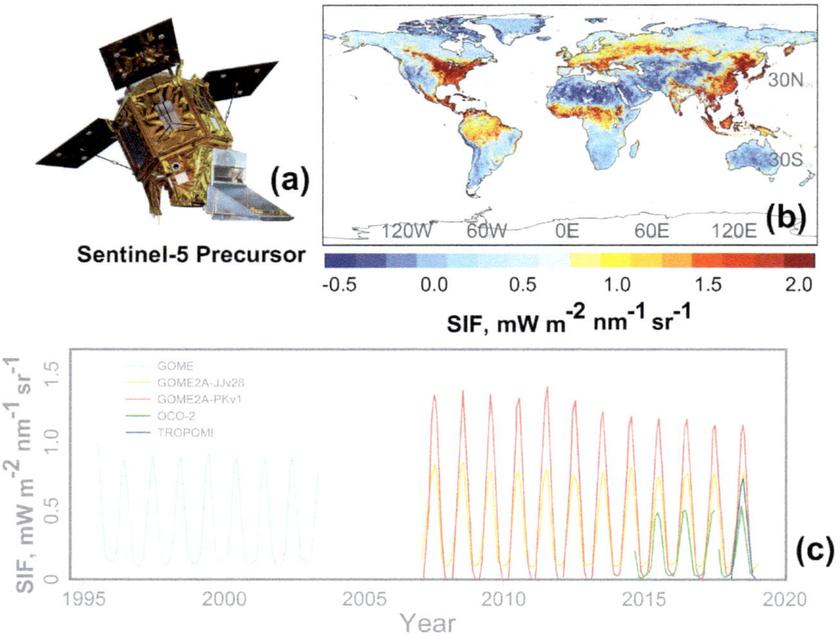

FIGURE 2.5 The Sentinel-5 precursor satellite (a) holding the tropospheric monitoring instrument (TROPOMI) sensor from which worldwide solar-induced chlorophyll fluorescence (SIF) data can be retrieved. (Source: https://sentinel.esa.int/web/sentinel/missions/sentinel-5p.) An example of a global SIF map, recorded in July 2018, is presented in panel (b). Global SIF data since 1992 from GOME, OCO-2, and TROPOMI platforms (c) (adapted from https://climatesciences.jpl.nasa.gov/sif/download-data/level-2/).

The assessment of SIF from spaceborne sensors is to date in a "proof of specific research concepts" stage. Investigations on the satellite scale address regional to continental and global scale vegetation functioning assessment. For instance, the use of satellite SIF data to analyze the effect of water limitations was recently addressed by Jonard et al. (2022). The authors reported nonlinear relations of photosynthesis with light and water at global scale. Another topic of utmost interest within the SIF research community is about the SIF-GPP relation during water scarcity periods. This has been addressed in some studies with ground data (Martini et al. 2021), yet, it is still unknown how it may behave in a regional to continental and global scale.

Using European scale TROPOMI-SIF information integrated with satellite based soil moisture, information from NASA's soil moisture active/passive (SMAP) mission, and the GPP data from ESA's TerrA-P project (Figure 2.6a), Quiros-Vargas et al. (2022b) analyzed how the SIF-GPP relation is influenced by soil moisture at the European scale during a heatwave in summer 2018. This was done in order to understand the continental scale response of vegetation to abnormal high temperatures. The authors report a strong positive SIF-soil moisture relation ($r = 0.91, p < 0.01$; Figure 2.6b) and a lower SIF but more heat sensitive SIF pattern across time in the lower soil moisture classes (Figure 2.6c). Moreover, the results suggest that the positive SIF-GPP relation observed under normal conditions becomes negative under abnormal high temperature conditions during a heat wave (as recently reported by Martini et al. 2021) in regions with soil moisture below 15 l m^{-2}, but remains positive in areas with higher soil water content (Figure 2.6d).

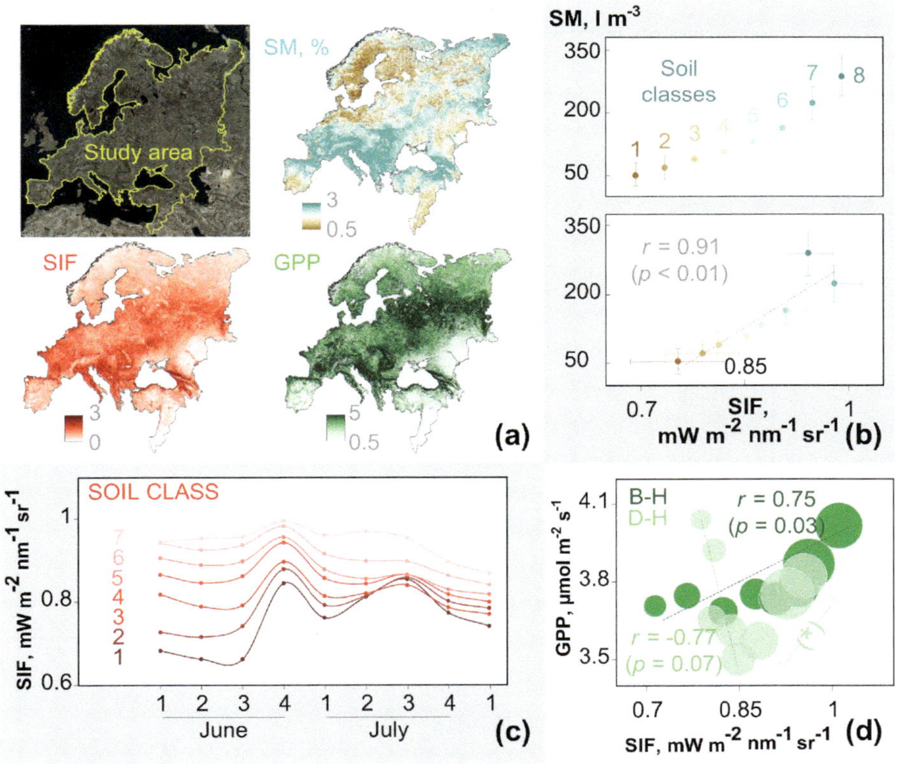

FIGURE 2.6 Average soil moisture (SM), solar-induced chlorophyll fluorescence (SIF), and gross primary productivity (GPP) maps computed over the study area (yellow boundary) across the nine time points analyzed (a). Values of the soil moisture classes and their correlation with SIF (b). Temporal variation of SIF for each soil moisture class before (B-H, June) and during (D-H, July) the peak of the heatwave (c). B-H (dark green) and D-H (light green) relation of GPP, SIF, and soil moisture (circles size; d). (*) Positive GPP-SIF relation kept D-H in regions with higher soil moisture.

2.5 THE SIF-SCALING ISSUE: IMPORTANCE OF DOWNSCALING

The spatiotemporal resolution of remote sensing data determines the amount and quality of information that can further be used to assess vegetation functioning. While the spatial resolution is mainly determined by the sensor characteristics and the sensor surface distance, the temporal resolution is driven by the amount of data collected across time. In general, proximal-, aerial-, and satellite-based information are more suitable to detect changes in the regulatory processes, canopy reflectance and structure and plant morphology, respectively (Figure 2.7, based on Gamon et al. 2019). That is, fine-scale remote sensing information (in centimeters and minutes-hours) can capture a wide range of plant responses to stress, from regulatory processes which can potentially be useful for the early (asymptomatic) stress detection, to changes in the pigments composition and leaf angles; whereas lower spatiotemporal resolution information (in meters and days) might not be able to sense subtle changes in physiological process, but still can track changes in the canopy structure and color (reflectance), useful for the assessment of vegetation productivity. Coarser remote sensing information (in the order of kilometers and months to years) can only capture strong alterations on vegetation morphology over regional to global scales, which is generally used to quantify the impact of severe stress, or to study energy exchanges between the surface and atmosphere on the biome level.

SIF information, compared with reflectance- and thermal-based data, is more affected by the scaling issue in the spatial domain. The spatial resolution of SIF products is particularly hindered

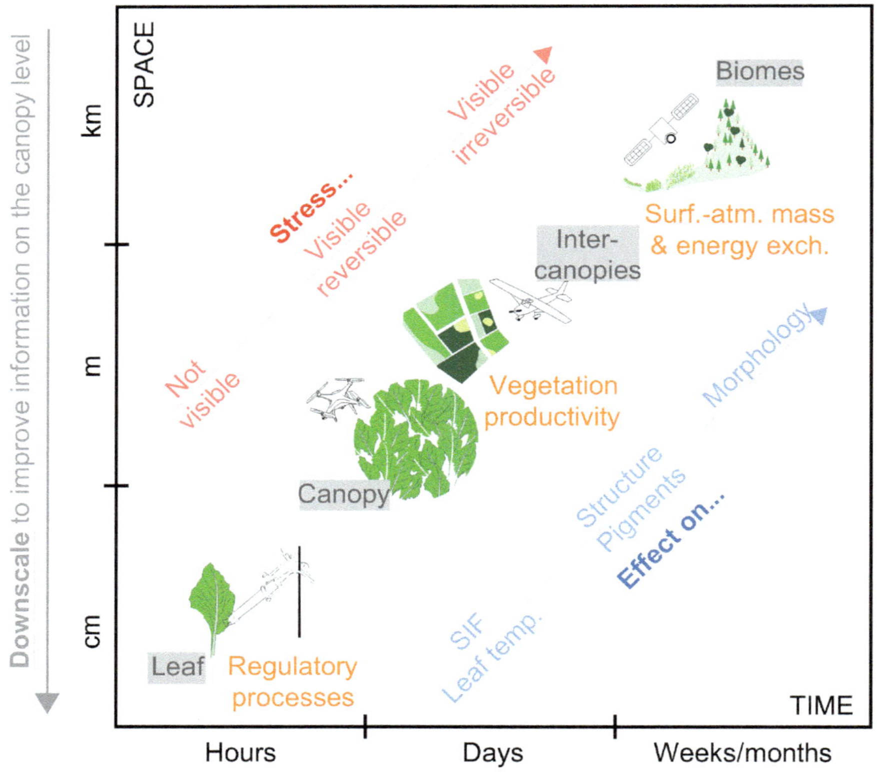

FIGURE 2.7 Spatial (*y*-axis, cm to km) and temporal (*x*-axis, hours to months) scales of ground-, unmanned aerial vehicles (UAV)-, airborne-, and satellite-based information, and the vegetation processes (in orange) for which each scale is particularly more suitable. The importance of downscaling, as well as the types of stress and their effect are presented in gray, red, and blue, respectively. The figure is based on Gamon et al. (2019).

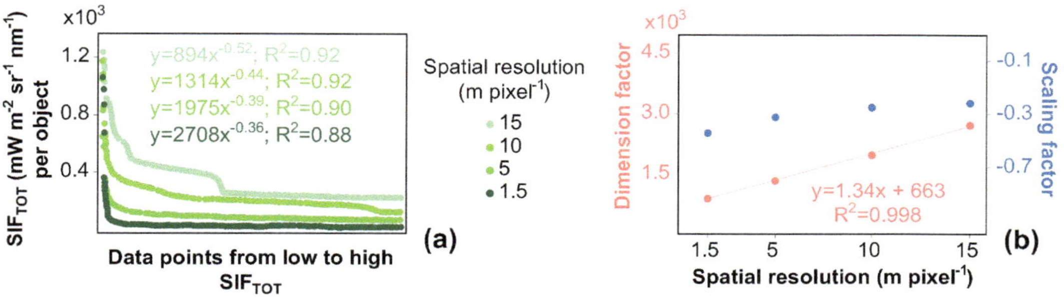

FIGURE 2.8 Total solar-induced chlorophyll fluorescence (SIF_{TOT}) power law (PL) distributions observed at 1.5, 5, 10, and 15 m pixel^{-1} spatial scales (a), and the respective dimension and scaling factors (b).

by technical limitations; for example, the low intensity of the SIF signal makes it necessary to sense larger areas in order to integrate a signal with high SNR. Consequently the SIF downscaling, herewith understood as the increase in the spatial resolution, is nowadays of utmost importance within the SIF research community targeting the improvement of the amount and quality of information of SIF imagery. Currently, the SIF downscaling research is generally focused on the use of linear relations between SIF and explanatory variables like the light use efficiency (LUE; Duveiller et al. 2020), land surface temperature (LST) and VIs (Zhang et al. 2020b), which are derived from remote sensing at higher spatial resolution. Yet, besides the aforementioned efforts, more flexible SIF-downscaling approaches have to be investigated trying to meet the dynamism of SIF in diverse ecosystems.

In Quiros-Vargas et al. (2022e) we propose to consider the use of the fractal theory for SIF-downscaling, whose developer said: "the success of fractals depend on people being familiar with the basic ideas and pushing them in different directions with more specialized topics" (Mandelbrot 2012). The theory states that natural phenomena can be described as repetition of patterns (fractal geometry) across spatiotemporal scales. The presence of fractal geometry can be recognized through different mathematical approaches, for example, based on power laws (PL's; Nagajothi et al. 2021) as addressed in Quiros-Vargas et al. (2022e), where we found that the total SIF (SIF_{TOT}) of vegetation objects within a 60 ha soybean field followed a PL distribution across spatial scales (1.5, 5, 10, and 15 m pixel^{-1}; Figure 2.8a). According to the fractal theory this indicates the presence of a fractal geometry composed by patterns where few incidences of high SIF_{TOT} values contrast with abundant occurrences of small values. We furthermore observed a linear increase and a nearly steady behavior of the dimension and scaling factors of the PLs across scales (Figure 2.8b), which can be interpreted as evidence of the scale invariant property of fractals.

Future SIF downscaling efforts in this direction might aim to find explanatory variable(s) that can describe the SIF_{TOT} distribution through bi-variate SIF PL's. A potential variable could be related to the object geometry properties (like area and perimeter). Special interest has to be paid to the use of the object size as it was found to strongly determine the spatial dependency of NIRv (Badgley et al. 2017; Quiros-Vargas et al. 2022c). Finally, in preliminarily analyses we observed that the direct linear relation between SIF_{TOT} and the average object size becomes a nearly perfect PL if the second variable is inverted (Quiros-Vargas et al. 2022d), a point which will be addressed in future studies.

2.6 CONCLUSIONS

The use of SIF for the assessment of the physiological status of vegetation has gained great interest in the last years, and such interest is reflected in the rapid increase on the number of remote sensing platforms and sensors, and their associated methods, available to retrieve SIF at multiple scales.

Such information, integrated with other types (e.g., reflectance-based) of remote sensing products, provides a rich data pool to increase our understanding about the effect of varying environmental conditions on the vegetation functioning over large scales.

Even though the panorama for the use of remotely sensed SIF for vegetation monitoring is promising, it is still in the research and development stage, where efforts are being focused on topics like: (1) the downscaling of the SIF signal to the leaf level, as well as (2) the spatial downscale (increase in the spatial resolution), (3) the understanding of the SIF response to different (e.g., water) stress conditions, and (4) the relation of the signal with the GPP. In addition, another field of active research is (5) the development of new sensors (especially at UAV and satellite scales). Thus, the progress on the remote sensing of SIF research towards advancing the understanding of the signal and its link with physiological processes, will determine the delivery of operational SIF-based tools to support agriculture and forestry sectors in the future.

REFERENCES

J.S. Aber, I. Marzolff, J. Ries, *Small-Format Aerial Photography: Principles, Techniques and Geoscience Applications* (Elsevier Science, Países Bajos, 2010).

A. Ač, Z. Malenovský, J. Olejníčková, A. Gallé, U. Rascher, G. Mohammed, Meta-analysis assessing potential of steady-state chlorophyll fluorescence for remote sensing detection of plant water, temperature and nitrogen stress. *Remote Sens. Environ.* 168, 420–436 (2015).

C. Brogi, et al., Large-scale soil mapping using multi-configuration EMI and supervised image classification. *Geoderma* 335, 133–148 (2019). https://doi.org/10.1016/j.geoderma.2018.08.001

G. Badgley, C.B. Field, J.A. Berry, Canopy near-infrared reflectance and terrestrial photosynthesis. *Sci. Adv.* 3, e1602244 (2017).

S. Bandopadhyay, A. Rastogi, R. Juszczak, Review of top-of-canopy sun-induced fluorescence (SIF) studies from Ground, UAV, airborne to spaceborne observations. *Sensors* 20(4), 1144 (2020).

J. Bendig, et al., "Imaging spatial heterogeneity of solar-induced chlorophyll fluorescence (SIF) with very high spatial resolution drone imagery" IGARSS 2023–2023 in *IEEE International Geoscience and Remote Sensing Symposium* (Pasadena, CA, USA, 2023), pp. 4654–4657.

J. Bendig, et al., Solar-induced chlorophyll fluorescence measured from an unmanned aircraft system: Sensor etaloning and platform motion correction. *IEEE Trans. Geosci. Remote Sens.* 58(5), 3437–3444 (2020).

J. Bendig, C.Y. Chang, N. Wang, J. Atherton, Z. Malenovský, U. Rascher, "Measuring solar-induced fluorescence from unmanned aircraft systems for operational use in plant phenotyping and precision farming" in *Proceedings of the 38th IEEE International Geoscience and Remote Sensing Symposium* (IGARSS, Brussels, Belgium, 2021), pp. 1921–1924.

K. Berger, M. Kycko, M. Celesti, J. Verrelst, M. Machwitz, U. Rascher, I. Herrmann, V.S. Paz, S. Fahrner, R. Pieruschka, E.T. Gormus, M. Gerhards, A. Halabuk, C. Atzberger, C. van der Tol, M. Rossini, M. Foerster, B. Siegmann, G. Tagliabue, T. Hank, H. Aasen, I. Pocas, M. Garcia, S.C. Kefauver, S. Bandopadhyay, A. Damm, E. Tomelleri, O. Rozenstein, L. Filchev, G. Stancile, M. Schlerf, Stress detection in agriculture with focus on the synergistic use of different optical domains: A review. *Remote Sens. Environ.* 280, 113198 (2022).

A. Burkart, S. Cogliati, A. Schickling, U. Rascher, A novel UAV-based ultra-light weight spectrometer for field spectroscopy. *IEEE Sens. J.* 14(1), 62–67 (2014).

M.P. Cendrero-Mateo, Sun-induced chlorophyll fluorescence III: Benchmarking retrieval methods and sensor characteristics for proximal sensing. *Remote Sens.* 11, 962 (2019).

C.Y. Chang, R. Zhou, O. Kira, S. Marri, J. Skovira, L. Gu, Y. Sun, An unmanned aerial system (UAS) for concurrent measurements of solar-induced chlorophyll fluorescence and hyperspectral reflectance toward improving crop monitoring. *Agric. For. Meteorol.* 294, 108145 (2020).

S. Cogliati, M. Celesti, I. Cesana, F. Miglietta, L. Genesio, T. Julitta, D. Schuettemeyer, M. Drusch, U. Rascher, P. Jurado, R. Colombo, A spectral fitting algorithm to retrieve the fluorescence spectrum from canopy radiance. *Remote Sens.* 11, 1840 (2019).

S. Cogliati, W. Verhoef, S. Kraft, N. Sabater, L. Alonso, J. Vicent, J. Moreno, M. Drusch, R. Colombo, Retrieval of sun-induced fluorescence using advanced spectral fitting methods. *Remote Sens. Environ.* 169, 344–357 (2015).

I. Colomina, P. Molina, Unmanned aerial systems for photogrammetry and remote sensing: A review. *ISPRS J. Photogramm. Remote Sens.* 92, 79–97 (2014).

A. Damm, S. Cogliati, R. Colombo, L. Fritsche, A. Genangeli, L. Genesio, J. Hanus, A. Peressotti, P. Rademske, U. Rascher, D. Schuettemeyer, B. Siegmann, J. Sturm, F. Miglietta, Response times of remote sensing measured sun-induced chlorophyll fluorescence, surface temperature and vegetation indices to evolving soil water limitation in a crop canopy. *Remote Sens. Environ.* 273, 112957 (2022).

A. Damm, E. Paul-Limoges, E. Haghighi, C. Simmer, F. Morsdorf, F.D. Schneider, C. van der Tol, M. Migliavacca, U. Rascher, Remote sensing of plant-water relations: An overview and future perspectives. *J. Plant Physiol.* 277, 3–19 (2018).

S. De Cannière, H. Vereecken, P. Defourny, F. Jonard, Remote sensing of instantaneous drought stress at canopy level using sun-induced chlorophyll fluorescence and canopy reflectance. *Remote Sens.* 14, 2642 (2022).

B. Dechant, et al., Canopy structure explains the relationship between photosynthesis and sun-induced chlorophyll fluorescence in crops. *Remote Sens. Environ.* 241, 111733 (2020).

R. Doughty, P. Köhler, C. Frankenberg, T.S. Magney, X. Xiao, Y. Qin, X. Wu, B. Moore, TROPOMI reveals dry-season increase of solar-induced chlorophyll fluorescence in the Amazon forest. *Proc. Natl. Acad. Sci. USA(PNAS)* 116(44), 22393–22398 (2019).

G. Duveiller, F. Filipponi, S. Walther, P. Köhler, C. Frankenberg, L. Guanter, A. Cescatti, A spatially downscaled sun-induced fluorescence global product for enhanced monitoring of vegetation productivity. *Earth Syst. Sci. Data* 12, 1101–1116 (2020).

European Space Agency, "Photosynthesis report" ESA. Photosynthetic-Proxy (Photoproxy, report No. 3, 2022a).

European Space Agency, "Flexsense campaign 2018" ESA. Flexsense (final report, 2022b).

European Space Agency, "Flexsense campaign 2019" ESA. Flexsense (final report, 2022c).

C. Frankenberg, P. Köhler, T.S. Magney, S. Geier, P. Lawson, M. Schwochert, J. McDuffie, D.T. Drewry, R. Pavlick, A. Kuhnert, The chlorophyll fluorescence imaging spectrometer (CFIS), mapping far red fluorescence from aircraft. *Remote Sens. Environ.* 217, 523–536 (2018).

J. Gamon, G. Hmimina, G. Miao, K. Guan, K. Springer, R. Wang, R. Yu, H. Gholizadeh, R. Moore, E. Walter-Shea, T. Arkebauer, A. Suyker, T. Franz, B. Wardlow, D. Wedin., "Imaging spectrometry and fluorometry in support of flex: What can we learn from multi-scale experiments?" in *Proceedings of the 38th IEEE International Geoscience and Remote Sensing Symposium* (IGARSS, Valencia, Spain, 2018), pp. 3931–3934.

J.A. Gamon, B. Somers, Z. Malenovský, E.M. Middleton, U. Rascher, M.E. Schaepman, "Assessing vegetation function with imaging spectroscopy" in *Surveys in Geophysics*, M. Rycroft, Ed. (Springer, 2019), pp. 489–513.

R. Garzonio, et al., Surface reflectance and sun-induced fluorescence spectroscopy measurements using a small hyperspectral UAS. *Remote Sens.* 9, 472 (2017).

D. Gautam, A. Lucieer, J. Bendig, Z. Malenovský, Footprint determination of a spectroradiometer mounted on an unmanned aircraft system. *IEEE Trans. Geosci. Remote Sens.* 58(5), 3085–3096 (2020).

M. Gerhards, M. Schlerf, K. Mallick, T. Udelhoven, Challenges and future perspectives of multi-/hyperspectral thermal infrared remote sensing for crop water-stress detection: A review. *Remote Sens.* 11(10), 1240 (2019).

J.F.R. Gower, L. Brown, G.A. Borstad, Observation of chlorophyll fluorescence in west coast waters of Canada using the MODIS satellite sensor. *Can. J. Remote Sens.* 30, 17–25 (2004).

J.F.R. Gower, R. Doerer, G.A. Borstad, Interpretation of the 685 nm peak in water-leaving radiance spectra in terms of fluorescence, absorption and scattering, and its observation by MERIS. *Int. J. Remote Sens.* 20, 1771–1786 (1999).

L. Guanter, "New algorithms for atmospheric correction and retrieval of biophysical parameters in earth observation." PhD Thesis, Universitat de València, Valencia, Spain, 2007.

L. Guanter, I. Aben, P. Tol, J.M. Krijger, A. Hollstein, P. Köhler, A. Damm, J. Joiner, C. Frankenberg, J. Landgraf, Potential of the TROPOspheric monitoring instrument (TROPOMI) onboard the Sentinel-5 precursor for the monitoring of terrestrial chlorophyll fluorescence. *Atmos. Meas. Tech.* 8, 1337–1352 (2015).

L. Guanter, C. Bacour, A. Schneider, I. Aben, T.A. van Kempen, F. Maignan, C. Retscher, P. Köhler, C. Frankenberg, J. Joiner, Y. Zhang, The TROPOSIF global sun-induced fluorescence dataset from the Sentinel-5P TROPOMI mission. *Earth Syst. Sci. Data* 13, 5423–5440 (2021).

A. Hornero, et al., Assessing the contribution of understory sun-induced chlorophyll fluorescence through 3-D radiative transfer modelling and field data. *Remote Sens. Environ.* 253, 112195 (2021).

JB Hyperspectral Devices. Available at https://www.jb-hyperspectral.com/ (accessed January 18, 2022).

J. Joiner, L. Guanter, R. Lindstrot, M. Voigt, A.P. Vasilkov, E.M. Middleton, K.F. Huemmrich, Y. Yoshida, C. Frankenberg, Global monitoring of terrestrial chlorophyll fluorescence from moderate-spectral-resolution near-infrared satellite measurements: Methodology, simulations, and application to GOME-2. *Atmos. Meas. Tech.* 6, 2803–2823 (2013).

J. Joiner, Y. Yoshida, A.P. Vasilkov, Y. Yoshida, L.A. Corp, E.M. Middleton, First observations of global and seasonal terrestrial chlorophyll fluorescence from space. *Biogeosciences* 8, 637–651 (2011).

F. Jonard, S. De Cannière, N. Brüggemann, P. Gentine, D.J.S. Gianotti, G. Lobet, D.G. Miralles, C. Montzka, B.R. Pagán, U. Rascher, H. Vereecken, Value of sun-induced chlorophyll fluorescence for quantifying hydrological states and fluxes: Current status and challenges. *Agric. For. Meteorol.* 291, 108088 (2020).

F. Jonard, A.F. Feldman, D.J.S. Gianotti, D. Entekhabi, Observed water- and light-limitation across global ecosystems. *European Geophysical Union (EGU)* [Preprint] (2022). https://doi.org/10.5194/bg-2022-25 (accessed July 28, 2022).

C. Kneer, et al., A snapshot imaging system for the measurement of solar-induced chlorophyll fluorescence—addressing the challenges of high-performance spectral imaging. *IEEE Sens. J.* 23(19), 2023.

P. Köhler, C. Frankenberg, T.S. Magney, L. Guanter, J. Joiner, J. Landgraf, Global retrievals of solar-induced chlorophyll fluorescence with TROPOMI: First results and intersensor comparison to OCO-2. *Geophys. Res. Lett.* 45, 10–456 (2018).

A. Mac Arthur, "A dual-field-of-view spectrometer system for reflectance and fluorescence measurements (Piccolo Doppio) and correction of etaloning" in *Proceedings of the 5th International Workshop on Remote Sensing of Vegetation Fluorescence* (Paris, France, 2014).

B. Mandelbrot, *The Fractalist: Memoir of a Maverick Scientist* (Vintage Books, New York, 2012), pp. 264–278.

S. Manfreda, M.F. McCabe, P.E. Miller, R. Lucas, V. Pajuelo Madrigal, et al., On the use of unmanned aerial systems for environmental monitoring. *Remote Sens.* 10, 641 (2018).

D. Martini, K. Sakowska, G. Wohlfahrt, J. Pacheco-Labrador, C. van der Tol, A. Porcar-Castell, T.S. Magney, A. Carrara, R. Colombo, T.S. El-Madany, R. Gonzalez-Cascon, M. Pilar Martín, T. Julitta, G. Moreno, U. Rascher, M. Reichstein, M. Rossini, M. Migliavacca, Heatwave breaks down the linearity between sun-induced fluorescence and gross primary production. *New Phytol.* 233, 2415–2428 (2021).

A. Matese, P. Toscano, S.F. Di Gennaro, L. Genesio, F.P. Vaccari, J. Primicerio, C. Belli, A. Zaldei, R. Bianconi, B. Gioli, Intercomparison of UAV, aircraft and satellite remote sensing platforms for precision viticulture. *Remote Sens.* 7, 2971–2990 (2015).

A.G. Mengistu, G.M. Tsidu, G. Koren, M.L. Kooreman, K.F. Boersma, T. Tagesson, J. Ardö, Y. Nouvellon, W. Peters, Sun-induced fluorescence and near-infrared reflectance of vegetation track the seasonal dynamics of gross primary production over Africa. *Biogeosciences* 18, 2843–2857 (2021).

M. Meroni, L. Busetto, R. Colombo, L. Guanter, J. Moreno, W. Verhoef, Performance of spectral fitting methods for vegetation fluorescence quantification. *Remote Sens. Environ.* 114(2), 363–374 (2010).

M. Meroni, M. Rossini, L. Guanter, L. Alonso, U. Rascher, R. Colombo, J. Moreno, Remote sensing of solar-induced chlorophyll fluorescence: Review of methods and applications. *Remote Sens. Environ.* 113, 2037–2051 (2009).

G.H. Mohammed, R. Colombo, E.M. Middleton, U. Rascher, C. van der Tol, L. Nedbal, Y. Goulas, O. Pérez-Priego, A. Damm, M. Meroni, J. Joiner, S. Cogliati, W. Verhoeaf, Z. Malenovský, J.-P. Gastellu-Etchegorry, J.R. Miller, L. Guanter, J. Moreno, I. Moya, J.A. Berry, C. Frankenberg, P.J. Zarco-Tejada, Remote sensing of solar-induced chlorophyll fluorescence (SIF) in vegetation: 50 years of progress. *Remote Sens. Environ.* 231, 111177 (2019).

K. Nagajothi, H.M. Rajashekara, B.S.D. Sagar, Universal fractal scaling laws for surface water bodies and their zones of influence. *IEEE Geosci. Remote Sens. Lett.* 18(5), 781–785 (2021). https://doi.org/10.1109/LGRS.2020.2988119.

C.Y.N. Norasma, M.A. Fadzilah, N.A. Rolsin, Z.W.N. Zanariah, Z. Tarmidi, F.S. Candra, "Unmanned aerial vehicle applications in agriculture" in *Proceedings of the 1*st *South Aceh International Conference on Engineering and Technology* (SAICOET, Kabupaten Aceh Selatan, Indonesia, 2019), p. 012063.

J. Oldeland, et al., New tools for old problems—comparing drone- and field-based assessments of a problematic plant species. *Environ. Monit. Assess.* 193, 90 (2021).

C. Panigada, et al., Fluorescence, PRI and canopy temperature for water stress detection in cereal crops. *Int. J. Appl. Earth Obs. Geoinf.* 30, 167–178 (2014).

J.A. Plascyk, The MK II Fraunhofer line discriminator (FLD-II) for airborne and orbital remote sensing of solar-stimulated luminescence. *Opt. Eng.* 14, 339–346 (1975).

A. Porcar-Castell, et al., Chlorophyll a fluorescence illuminates a path connecting plant molecular biology to earth-system science. *Nat. Plants* 7, 998–1009 (2021).

A. Porcar-Castell, E. Tyystjärvi, J. Atherton, C. van der Tol, J. Flexas, E.E. Pfündel, J. Moreno, C. Frankenberg, J.A. Berry, Linking chlorophyll a fluorescence to photosynthesis for remote sensing applications: Mechanisms and challenges. *J. Exp. Bot.* 65(15), 4065–4095 (2014).

P. Psirofonia, V. Samaritakis, P. Eliopoulos, Use of unmanned aerial vehicles for agricultural applications with emphasis on crop protection: Three novel case-studies. *J. Agric. Sci.Technol.* 5(1), 30–39 (2017).

J. Quiros-Vargas, J. Bendig, A. Mac Arthur, A. Burkart, T. Julitta, K. Maseyk, R. Thomas, B. Siegmann, M. Rossini, M. Celesti, D. Schüttemeyer, T. Kraska, O. Muller, U. Rascher, Unmanned Aerial Systems (UAS)-based methods for Solar Induced Chlorophyll Fluorescence (SIF) retrieval with non-imaging spectrometers: State-of-the-art. *Remote Sens.* 12, 1624 (2020a).

J. Quiros-Vargas, C. Brogi, V. Krieger, B. Siegmann, M. Celesti, M. Rossini, S. Cogliati, L. Weihermüller, U. Rascher, "Solar induced chlorophyll fluorescence and vegetation indices for heat stress assessment in three crops at different geophysics-derived soil units" in *Proceedings of the American Geophysical Union* (AGU, online, 2020b).

J. Quiros-Vargas, R.D. Caldeira, N. Zendonadi dos Santos, L. Zimmermann, B. Siegmann, T. Kraska, M.W. Vasconcelos, U. Rascher, O. Muller, "Response of bean (Phaseolus vulgaris L.) to elevated CO2 in Yield, biomass and chlorophyll fluorescence" in *Proceedings of the IEEE International Geoscience and Remote Sensing Symposium* (IGARSS, Brussels, Belgium, 2021), pp. 5861–5864.

J. Quiros-Vargas, L.R. Khot, Potential of low altitude multispectral imaging for in-field apple tree nursery inventory mapping. *IFAC-PapersOnLine* 49(16), 421–425 (2016).

J. Quiros-Vargas, B. Siegmann, A. Damm, C. Brogi, P. Köhler, R. Van Hoolst, D. Martini, O. Muller, U. Rascher, "Solar-Induced Chlorophyll Fluorescence (SIF) relation with Soil Moisture (SM) and Gross Primary Productivity (GPP) at European scale in a heat wave" in *Proceedings of the European Association of Remote Sensing Laboratories* (EARSeL, Potsdam, Germany, 2022b).

J. Quiros-Vargas, B. Siegmann, A. Damm, V. Krieger, O. Muller, U. Rascher, "Spatial dependency of Solar-Induced Chlorophyll Fluorescence (SIF)-emitting objects in the Footprint of a Fluorescence Explorer (FLEX) pixel: A SIF-downscaling perspective" in *Proceedings of the European Geophysical Union* (EGU, Vienna, Austria, 2022c).

J. Quiros-Vargas, B. Siegmann, A. Damm, R. Wang, J. Gamon, V. Krieger, B.S.D. Sagar, O. Muller, U. Rascher, "Sun-Induced Chlorophyll Fluorescence (SIF)-downscaling from the fractal geometry perspective" in *Proceedings of the Living Planet Symposium* (LPS, Bonn, Germany, 2022d).

J. Quiros-Vargas, B. Siegmann, A. Damm, R. Wang, J. Gamon, V. Krieger, B.S.D. Sagar, O. Muller, U. Rascher, "Fractal geometry and the downscaling of sun-induced chlorophyll fluorescence imagery" in *Encyclopedia of Mathematical Geosciences*, B.S. Daya Sagar, Q. Cheng, J. McKinley, F. Agterberg, Eds. (Springer Nature, Switzerland, 2022e).

M. Rossini, et al., Evaluation of the spatial representativeness of in situ SIF observations for the validation of medium-resolution satellite SIF products. *Remote Sens.* 14(20), 5107 (2022).

U. Rascher, L. Alonso, A. Burkart, C. Cilia, S. Cogliati, R. Colombo, A. Damm, M. Drusch, L. Guanter, J. Hanus, T. Hyvärinen, T. Julitta, J. Jussila, K. Kataja, P. Kokkalis, S. Kraft, T. Kraska, M. Matveeva, J. Moreno, O. Muller, C. Panigada, M. Pikl, F. Pinto, L. Prey, R. Pude, M. Rossini, A. Schickling, U. Schurr, D. Schüttemeyer, J. Verrelst, et al., Sun-induced fluorescence—a new probe of photosynthesis: First maps from the imaging spectrometer HyPlant. *Glob. Change Biol.* 21(12), 4673–4684 (2015).

S. Sankaran, L.R. Khot, C.Z. Espinoza, S. Jarolmasjed, V.R. Sthuvalli, G.J. Vandemark, P.H. Miklas, A.H. Carter, M.O. Pumphrey, N.R. Knowels, M.J. Pavek, Low-altitude, high-resolution aerial imaging systems for row and field crop phenotyping: A review. <cite lang="fr">*Eur. J. Agron.* 70, 112–123 (2015).

B. Siegmann, L. Alonso, M. Celesti, S. Cogliati, R. Colombo, A. Damm, S. Douglas, L. Guanter, J. Hanuš, K. Kataja, T. Kraska, M. Matveeva, J. Moreno, O. Muller, M. Pikl, F. Pinto, J. Quirós Vargas, P. Rademske, F. Rodriguez-Morene, N. Sabater, A. Schickling, D. Schüttemeyer, F. Zemek, U. Rascher, The high-performance airborne imaging spectrometer HyPlant—from raw images to top-of-canopy reflectance and fluorescence products: Introduction of an automatized processing Chain. *Remote Sens.* 11, 2760 (2019).

B. Siegmann, M.-P. Cendrero-Mateo, S. Cogliati, A. Damm, J. Gamon, D. Herrera, C. Jedmowski, Laura Verena Junker-Frohn, T. Kraska, O. Muller, P. Rademske, C. van der Tol, J. Quiros-Vargas-Vargas, P. Yang, U. Rascher, Downscaling of far-red solar-induced chlorophyll fluorescence of different crops from canopy to leaf level using a diurnal dataset acquired by the airborne imaging spectrometer HyPlant. *Remote Sens. Environ.* 264, 112609 (2021).

Y. Sun, et al., From remotely-sensed solar-induced chlorophyll fluorescence to ecosystem structure, function, and service: Part II—Harnessing data. *Glob. Change Biol.* 29, 2893–2925 (2023).

Y. Sun, C. Frankenberg, J.D. Wood, D.S. Schimel, M. Jung, L. Guanter, D.T. Drewry, M. Verma, A. Porcar-Castell, et al. OCO-2 advances photosynthesis observation from space via solar-induced chlorophyll fluorescence. *Science* 358, eaam5747 (2017).

G. Tagliabue, C. Panigada, M. Celesti, S. Cogliati, R. Colombo, M. Migliavacca, U. Rascher, D. Rocchini, D. Schüttemeyer, M. Rossini, Sun—induced fluorescence heterogeneity as a measure of functional diversity. *Remote Sens. Environ.* 247, 111934 (2020).

A.F. Theisen, "Detecting chlorophyll fluorescence from Orbit: The Fraunhofer line depth model" in *From Laboratory Spectroscopy to Remotely Sensed Spectra of Terrestrial Ecosystems*, R.S. Muttiah, Ed. (Kluwer Academic Publishers, Dordrecht, 2002), pp. 203–232.

C. Torresan, A. Berton, F. Carotenuto, S.F. Di Gennaro, B. Gioli, A. Matese, F. Miglietta, C. Vagnoli, A. Zaldei, L. Wallace, Forestry applications of UAVs in Europe: A review. *Int. J. Remote Sens.* 38(8), 2427–2447 (2017).

B. Tubuxin, P. Rahimzadeh-Bajgiran, Y. Ginnan, F. Hosoi, K. Omasa, Estimating chlorophyll content and photochemical yield of photosystem II (ΦPSII) using solar-induced chlorophyll fluorescence measurements at different growing stages of attached leaves. *J. Exp. Bot.* 66(18), 5595–5603 (2015).

C. von Hebel, M. Matveeva, E. Verweij, P. Rademske, M.S. Kaufmann, C. Brogi, H. Vereecken, U. Rascher, J. Van der Kur, Understanding soil and plant interaction by combining ground-based quantitative electromagnetic induction and airborne hyperspectral data. *Geophys. Res. Lett.* 45, 3 (2018).

N. Wang, et al., Comparison of a UAV- and an airborne-based system to acquire far-red sun-induced chlorophyll fluorescence measurements over structurally different crops. *Agric. For. Meteorol.* 323, 109081 (2022a).

N. Wang, et al., Potential of UAV-based sun-induced chlorophyll fluorescence to detect water stress in sugar beet. *Agric. For. Meteorol.* 323, 109033 (2022b).

N. Wang, J. Suomalainen, H. Bartholomeus, L. Kooistra, D. Masiliūnas, J.G.P.W. Clevers, Diurnal variation of sun-induced chlorophyll fluorescence of agricultural crops observed from a point-based spectrometer on a UAV. *Int. J. Appl. Earth Obs. Geoinf.* 96, 102276 (2021).

S. Wieneke, H. Ahrends, A. Damm, F. Pinto, A. Stadler, M. Rossini, U. Rascher, Airborne based spectroscopy of red and far-red sun-induced chlorophyll fluorescence: Implications for improved estimates of gross primary productivity. *Remote Sens. Environ.* 184, 654–667 (2016).

P. Yang, C. van der Tol, P.-K.-E. Campbell, E.-M. Middleton, Fluorescence correction vegetation index (FCVI): A physically based reflectance index to separate physiological and non-physiological information in far-red sun-induced chlorophyll fluorescence. *Remote Sens. Environ.* 240, 111676 (2020).

P.J. Zarco-Tejada, C. Camino, P.S.A. Beck, R. Calderon, A. Hornero, et al., Previsual symptoms of Xylella fastidiosa infection revealed in spectral plant-trait alterations. *Nat. Plants.* 4, 432–439 (2018).

Y. Zeng, et al., Estimating near-infrared reflectance of vegetation from hyperspectral data. *Remote Sens. Environ.* 267, 112723 (2021).

Y. Zeng, M. Chen, D. Hao, A. Damm, G. Badgley, U. Rascher, J.E. Johnson, B. Dechant, B. Siegmann, Y. Ryu, H. Qiu, V. Krieger, C. Panigada, M. Celesti, F. Miglietta, X. Yang, J.A. Berry, Combining near-infrared radiance of vegetation and fluorescence spectroscopy to detect effects of abiotic changes and stresses. *Remote Sens. Environ.* 270, 112856 (2022).

Z. Zhang, et al., Reduction of structural impacts and distinction of photosynthetic pathways in a global estimation of GPP from space-borne solar-induced chlorophyll fluorescence. *Remote Sens. Environ.* 240, 111722 (2020a).

Y. Zhang, L. Guanter, J.A. Berry, J. Joiner, C. Van der Tol, A. Huete, A. Gitelson, M. Voigt, P. Köhler, Estimation of vegetation photosynthetic capacity from space-based measurements of chlorophyll fluorescence for terrestrial biosphere models. *Glob. Chang. Biol.* 20, 3727–3742 (2014).

Z. Zhang, W. Xu, Q. Qin, Z. Long, Downscaling solar-induced chlorophyll fluorescence based on convolutional neural network method to monitor agricultural drought. *IEEE Trans. Geosci. Remote Sens.* 59(2), 1012–1028 (2020b).

Y. Zeng, et al., A practical approach for estimating the escape ratio of near-infrared solar-induced chlorophyll fluorescence. *Remote Sens. Environ.* 232, 111209 (2019).

M.G. Ziliani, S.D. Parkes, I. Hoteit, M.F. McCabe, Intra-season crop height variability at commercial farm scales using a fixed-wing UAV. *Remote Sens.* 10(12), 2007 (2018).

3 Canopy Biophysical Variables Retrieval from the Inversion of Reflectance Models

Frédéric Baret

ACRONYMS AND DEFINITIONS

AVHRR	Advanced Very-High-Resolution Radiometer
CHRIS	Compact High Resolution Imaging Spectrometer
GIMMS	Global Inventory Modeling and Mapping Studies
GLAI	Green Leaf Area Index
JRC	Joint Research Center
LAI	Leaf Area Index
MERIS	Medium Resolution Imaging Spectrometer
MODIS	Moderate-Resolution Imaging Spectroradiometer
PAR	Photosynthetically Active Radiation
PROSPECT	Radiative transfer model to measure leaf optical properties spectra
SVM	Support Vector Machines
VI	Vegetation Index

3.1 INTRODUCTION

Estimates of canopy biophysical characteristics are required for a wide range of agricultural, ecological, hydrological, and meteorological applications (Wan et al. 2024; Liu et al. 2023; Zhu et al. 2023; Martínez-Ferrer et al. 2022; Sun et al. 2022; Chaabouni et al. 2021; Schiefer et al. 2021; Abdelbaki et al. 2019; Verrelst et al. 2019a, 2019b; Xie et al. 2019; Sun et al. 2018; Wocher et al. 2018; Zeng et al. 2018; Danner et al. 2017). These should cover exhaustively large spatial domains at several scales: from the very local one corresponding to precision agriculture where cultural practices are adapted to the within field variability, through environmental management generally approached at the landscape scale, up to biogeochemical cycling and vegetation dynamics investigated at national, continental, and global scales. Remote sensing observations answer well these requirements with spatial resolution spanning from kilometric down to decametric resolution observations according to the nomenclature proposed by (Morisette 2010). Further, remote sensing from satellites brings the unique capacity to monitor the dynamics required to access the functioning of the vegetation.

Few biophysical variables have been recognized as essential climate variables (ECV) for their key role played in the main vegetation canopy processes such as photosynthesis and evapotranspiration (GCOS 2011). These ECVs include the Leaf Area Index (LAI) and the Fraction of Absorbed Photosynthetically Active Radiation (FAPAR). Since the 1980s, considerable improvement in the quality of terrestrial estimates of LAI and FAPAR derived from satellite or airborne systems have been achieved due to the advances of measurement capability of satellite instruments and to our understanding of the radiation regime within vegetation canopies (Liang 2004). However, remote sensing observations sample the radiation field reflected or emitted by the surface, and thus do not

DOI: 10.1201/9781003541165-4

provide directly LAI or FAPAR estimates. It is therefore necessary to transform the radiance values recorded by the sensor into LAI or FAPAR values. The retrieval algorithms used should ideally be accurate, precise, and computationally efficient. Most importantly, they should require minimal calibration since they are supposed to be applied over diverse locations, seasons, and conditions (Walthall et al. 2004).

Many methods have been proposed to retrieve land surface characteristics from remote sensing observations (Schiefer et al. 2021; Verrelst et al. 2019a, 2019b; Baret and Buis 2007; Goel 1989; Houborg and Boegh 2008; Kimes et al. 2000; Laurent et al. 2013; Myneni et al. 1988; Pinty and Verstraete 1991a; Verger et al. 2011a). They include empirical methods with calibration over experimental datasets. These simple methods are limited by the size and diversity of the calibration dataset as well as by the uncertainties attached to the ground measurements. More complex ones based on the use of radiative transfer models have been proposed where no in situ calibration dataset is required. Radiative transfer models describe the physical processes involved in the photon transport within vegetation canopies. They simulate the radiation field reflected by the surface for a given observational configuration, once the vegetation and the background are known. Retrieving canopy characteristics from the radiation field as sampled by the sensor aboard satellite needs to "invert" the radiative transfer model, that is, to estimate some input variables from the measurement of the outputs of the model.

This book chapter aims at reviewing how canopy biophysical variables may be derived both from kilometric and decametric resolution remote sensing observations. It will be illustrated by LAI and FAPAR variables that will first be defined before describing the principles of the radiative transfer model inversion used to retrieve them. Then the theoretical performances of LAI and FAPAR will be investigated. Several techniques improve the retrievals will be discussed in detail. Finally the possible combinations of methods, products, and sensors will be presented. A conclusion will highlight the main issues to tackle, suggesting future research avenues.

3.2 THE SEVERAL DEFINITIONS OF LAI AND FAPAR

3.2.1 Leaf Area Index: LAI, GLAI, PAI, GAI, Effective and Apparent Values

LAI is defined as half of the total developed area of green vegetation elements per unit ground area (m^2/m^2) (Chen and Black 1992; Stenberg 2006). It is a structural variable which describes the size of the interface for exchange of energy and mass between the canopy and the atmosphere. It governs photosynthesis, transpiration, and rain interception processes. For photosynthesis and transpiration, the LAI definition should be restricted to the green active area leading to the GLAI definition (Green Leaf Area Index). Further, the area of other organs such as stems, branches or fruits should be accounted for if they are green, leading to the GAI (Green Area Index) definition. LAI, GLAI, and GAI may be measured using destructive techniques. However, this is tedious and time-consuming and indirect methods based on canopy gap fraction (Po) measurements have been developed (Jonckheere et al. 2004; Weiss et al. 2004). Since no distinction is made by these devices between green and non-green elements, neither between leaves and the other elements, the actual quantity measured is PAI (Plant Area Index). However, directional photos taken from the top of the canopy may be also used to compute the green fraction, GF, defined as the fraction of green area seen in the considered direction. Assuming that the green leaves are mostly at the top of the canopy, which is generally the case, such technique provides an estimate of the GAI (Baret et al. 2010). Similarly, remote sensing observations are mainly sensitive to the green elements of the canopy, and thus are mostly related to the GAI (Zhu et al. 2023; Schiefer et al. 2021; Sun et al. 2018; Duveiller et al. 2011; Raymaekers et al. 2014). Table 3.1 clearly shows that indirect methods are mainly accessing GAI and PAI depending on the capacity to distinguish green from non-green elements.

The derivation of PAI or GAI from indirect measurements requires some assumptions on canopy architecture. The turbid medium assumption is the most commonly used, considering that leaves

TABLE 3.1
Definitions of LAI, GLAI, GAI, and PAI and the associated indirect measurement methods. All quantities are expressed in m²·m⁻².

		Only Green	Green + Non-green	Only Leaves	All Elements	Indirect Measurement Method
LAI	Leaf Area Index		✓	✓		Only destructive methods
GLAI	Green Leaf Area Index	✓		✓		Only destructive methods
GAI	Green Area Index	✓			✓	GF from the top, remote sensing
PAI	Plant Area Index		✓		✓	Po measurements

have infinitesimal size and are randomly distributed in the canopy volume. However, this simple assumption is not always verified by actual canopies, leaves having a finite dimension and being clumped at several scales including the shoot (leaves grouped in shoots), plant (shoots grouped in plants), stand (plants grouped in stands) to landscape (stands distributed in the landscape). This creates artefacts in the estimation of the corresponding PAI (Walter et al. 2003) from gap fraction measurements or GAI from reflectance measurements (Chen et al. 2005). Therefore, "effective" and "apparent" quantities need to be introduced to complement the actual "true" PAI or GAI definitions. The effective PAI or GAI is the quantity that can be derived from the directional gap fraction or green fraction based on Miller's formula (Miller 1967) that assumes leaves randomly distributed in the canopy volume (Ryu et al. 2010). However, the application of Miller's formula requires the measurement of Po or GF in all the directions of the hemisphere, which is rarely possible. We therefore estimate an "apparent" PAI or GAI value, which depends on the directional sampling used. Similarly, estimates of GAI from remote sensing are "apparent" values (Martonchik 1994) that will depend on the observational configuration used, the inverse technique employed including the assumptions on canopy architecture embedded in the radiative transfer model considered as we will see in the following sections (Liu et al. 2023; Abdelbaki et al. 2019; Wocher et al. 2018; Zeng et al. 2018; Danner et al. 2017).

3.2.2 FAPAR: Illumination Conditions and Green/Non-Green Elements

FAPAR is defined as the fraction of the photosynthetically active radiation (PAR, solar radiation in the 400–700 nm spectral domain) absorbed by a vegetation canopy (Mõttus et al. 2011). FAPAR is widely used as input into a number of primary productivity models (McCallum et al. 2009). It is therefore necessary to consider only the green photosynthetically active elements, that is, the green parts of the canopy. Similarly to what was presented for the LAI definition, FAPAR measurements can be computed from the radiation balance in the 400–700 nm PAR spectral domain (Mõttus et al. 2011). The FAPAR value can be also approximated by the fraction of intercepted radiation, FIPAR, that is, the complement to unity of the gap faction (Liu et al. 2023; Xie et al. 2019; Sun et al. 2018; Danner et al. 2017; Asrar 1989; Begué et al. 1991; Gobron et al. 2006; Russel et al. 1989). However, it is not possible to distinguish the absorption or interception of the light by the green elements from that of the non-green elements with these measurements techniques. Conversely, measurements of the green fraction, GF, from the top of the canopy in the illumination direction provide a direct estimate of the FIPAR.

FAPAR and FIPAR variables are not intrinsic properties of the vegetation, but result from the interaction of the light with the canopy. FAPAR and FIPAR will thus depend on the illumination conditions. Similarly to albedo (Martonchik 1994), the illumination conditions could be described by a component coming only from the Sun direction, the black sky FAPAR or FIPAR, and a diffuse component coming from the sky hemisphere, the white sky FAPAR or FIPAR. The black sky FAPAR or FIPAR values depend on the Sun direction. Most FAPAR products are defined as the black sky values corresponding to the Sun position at the time of the satellite overpass (Weiss et al. 2014), that is, around 10:30 solar time. Note that the black sky FAPAR or FIPAR values at 10:00 have been demonstrated to be a good estimation of the daily integrated value of FAPAR or FIPAR (Baret et al. 2004).

3.3 RADIATIVE TRANSFER MODEL INVERSION METHODS

The light reflected by the canopy results from the radiative transfer processes within the vegetation. It depends on canopy state variables as well as on the illumination conditions and the observational configuration that defines the sampling of the reflectance field: wavebands, view direction(s), frequency of observations, and spatial resolution. State variables characterizing the canopy structure and the optical properties of the vegetation elements include therefore some of the variables of interest for the applications such as LAI (Figure 3.1). Other variables such as FAPAR can also be computed from the knowledge of the canopy state variables and the illumination configuration considered using the same radiative transfer model.

The causal relationship between the variables of interest and remote sensing data corresponds to the forward (or direct) problem. They could be either described through empirical relationships calibrated over experiments or using radiative transfer models based on a more or less close approximation of the actual physical processes, canopy architecture, and optical properties of the elements including the background. Conversely, retrieving the variables of interest from remote sensing measurements corresponds to the inverse problem, that is, developing algorithms to estimate the variables of interest from remote sensing data as observed in a given configuration. Prior information on the type of surface and on the distribution of the variables of interest can also be included in the retrieval process to improve the performances as we will see later. Note that the estimation of FAPAR could be achieved in two steps: first, the canopy state variables are retrieved by inverting a radiative transfer model. Then, the FAPAR is computed under specific illumination conditions using the same radiative transfer model and the estimates of canopy state variables.

The retrieval techniques can be split into two main approaches depending whether the emphasis is put on the inputs (the canopy biophysical variables driven approach) or the outputs (radiometric data driven approach) of the radiative transfer model (Figure 3.2).

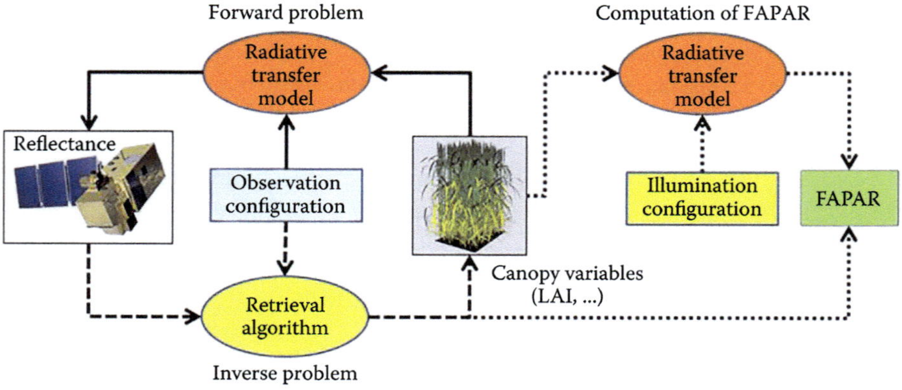

FIGURE 3.1 Forward (solid lines) and inverse (dashed lines) problems in remote sensing. The computation of FAPAR from the retrieved canopy variables is also illustrated.

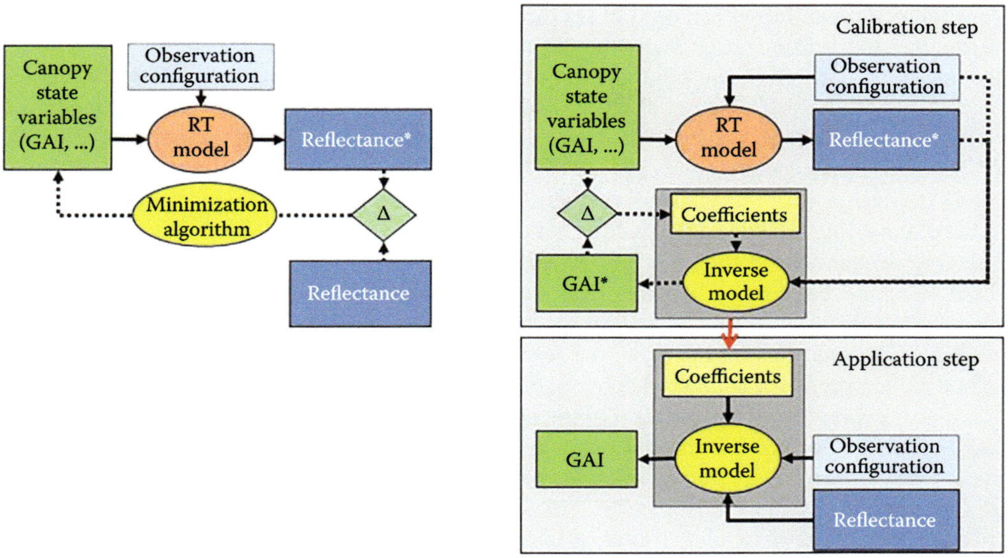

FIGURE 3.2 The two main approaches used to estimate canopy characteristics from remote sensing data for LAI estimation. On the left side, the approach focuses on radiometric data showing the solution search process leading to the estimated *LAI* value, *LAI**. On the right side the approach focuses on the biophysical variables showing the calibration of the inverse model (top) and the application using the inverse model with its calibrated coefficients (bottom). "Δ" represents the cost function to be minimized over the biophysical variables (right) or over the radiometric data (left).

3.3.1 Radiometric Data-Driven Approach: Minimizing the Distance between Observed and Simulated Reflectance

The radiometric data driven approach focuses on the outputs of the radiative transfer model: it aims at finding the best match between the measured reflectance values and those simulated by a radiative transfer model (Figure 3.2, right). The misfit is quantified by a cost function (*J*) that should account for measurements and model uncertainties. It can be theoretically derived from the maximum likelihood (Tarentola 1987) assuming that uncertainties associated to each configuration used are independent and Gaussian distributed:

$$J = (R - \hat{R})^t \times W^{-1} \times (R - \hat{R}) \quad (3.1)$$

Where/and/are, respectively, the vectors of observed and estimated reflectances and *W* is the covariance matrix of uncertainties. One main limitation in applying this formalism is the difficulty to get the covariance matrix *W*. In most cases, just the diagonal terms corresponding to the variance associated to the uncertainties (σ^2) are known. In these conditions, Equation 3.1 simplifies into the normalized Euclidian distance:

$$J = \sum_{n=1}^{N} \frac{(R_n - \hat{R}_n)^2}{\sigma_n^2} \quad (3.2)$$

Where *N* is the number of configurations used (bands, directions, etc.). More sophisticated cost functions have been proposed to include a regularization term that prevents the solution to be too far away from its prior expectation. This will be reviewed later in a separate section.

Several techniques have been used to get the solution corresponding to the minimum of the cost function: iterative minimization including the simplex algorithm (Nelder and Mead 1965), gradient descent based algorithms (Zhu et al. 2023; Sun et al. 2022; Chaabouni et al. 2021; Schiefer et al. 2021; Wocher et al. 2018; Zeng et al. 2018; Danner et al. 2017; Bacour et al. 2002a; Bicheron and Leroy 1999; Combal et al. 2002; Combal et al. 2000; Goel and Deering 1985; Goel et al. 1984; Goel and Thompson 1984; Jacquemoud et al. 1995; Kuusk 1991a, 1991b; Lauvernet et al. 2008; Pinty et al. 1990; Privette et al. 1996; Voßbeck et al. 2010), Monte Carlo Markov Chains (Zhang et al. 2005), simulated annealing (Bacour 2001) and genetic algorithms (Fang et al. 2003; Renders and Flasse 1996). One of the major difficulties associated with these techniques is the possibility to get sub-optimal solutions that corresponds to a local minimum of the cost function. This can be mitigated by using several initial guesses spread over the space of canopy realization as well as allowing some flexibility and randomness along the search path towards the solution. This is unfortunately achieved at the expense of additional computation time. However, the process can be speed up using an analytical expression of the gradient of the cost function, that is, the adjoint model (Lauvernet et al. 2008). Further, to increase the computation speed, the actual radiative transfer model could be emulated into a meta-model that additionally eases the derivation of the adjoint model (Jamet et al. 2005). Note that the meta-model can be considered as an interpolation between a set of simulated cases that can be used to populate a look-up table as described later. To limit the problem of possible local minimum when iteratively minimizing the cost function, a regularization term could be added based on the knowledge of the prior distribution of the input variables (Tarentola 1987) or integrating some constraints. This will be more detailed in a following section.

The look-up tables (LUT) working on pre-computed simulations containing the input canopy variables and the corresponding simulated reflectance values have also been used directly without interpolation (Darvishzadeh et al. 2008; Ganguly et al. 2012; Knyazikhin et al. 1998; Vohland et al. 2010; Weiss et al. 2000). This technique is more tractable in terms of computation requirements and limits the possibility to get trapped in a local minimum of the cost function since this cost function is evaluated systematically over each case of the LUT. To populate the LUT, the space of canopy realization has to be sampled to represent the surface response, that is, with better sampling where the sensitivity of reflectance to canopy characteristics is the higher (Combal et al. 2002; Weiss et al. 2000). This is different from the sampling of the training data base required in canopy biophysical variables driven approaches as explained earlier. The cases in the LUT are sorted according to the cost function value (J). Then the solution may be considered as the one corresponding to the best match obtained with the minimum value of J, similarly to what is done with the iterative minimization techniques. It can be also defined as a fraction of the initial population of cases such as in (Combal et al. 2002; Weiss et al. 2000) or using a threshold defined by measurements and model uncertainties (Knyazikhin et al. 1998). A more rigorous way of exploiting the solutions would be to weigh each case according to its likelihood as done in the GLUE method (Beven and Binsley 1992; Makowski et al. 2002).

3.3.2 Canopy Biophysical Variables Driven Approach: Machine Learning

This approach belongs to the machine learning type of algorithm that requires first to calibrate an inverse parametric model (Figure 3.2, right). The calibration mainly consists in adjusting the coefficients of the inverse model to minimize the distance between the estimated variable of interest (GAI in this example) and the ones populating the calibration dataset. For FAPAR, the RT model is used a second time to simulate the corresponding FAPAR values for a given illumination condition. The inverse model is then calibrated by minimizing the distance between the estimated FAPAR value and the one simulated in the calibration dataset. Once calibrated, the parametric inverse model can be used in the forward mode to compute the variables of interest from the observed reflectance values. The learning dataset can be generated either using simulations of radiative transfer models, or based on concurrent experimental measurements of the variables of interest and reflectance data.

The inverse model may be calibrated both over experimental or synthetic datasets (Wan et al. 2024; Martínez-Ferrer et al. 2022; Xie et al. 2019; Asrar et al. 1984; Chen et al. 2002; Deng et al. 2006; Huete 1988; Richardson et al. 1992; Verrelst et al. 2012; Wiegand et al. 1990; Wiegand et al. 1992). However, the use of experimental datasets may be limited by its representativeness regarding the possible conditions encountered over the targeted surfaces, that is, combinations of geometrical configurations, type of vegetation and state including variability in development stage, stress level and type and background (bare soil, understory) and state (roughness, moisture). Measurement errors associated both to the variables of interest and to the reflectance values may also propagate to uncertainties and biases in the algorithm and should be explicitly accounted for (Fernandes and Leblanc 2005; Huang et al. 2006). Further, since ground measurements having a footprint ranging from few meters to few decameters, specific sampling designs should be developed to represent the sensor pixel (Weiss et al. 2007). This task is obviously more difficult for medium and coarse resolution sensors (Camacho et al. 2013; Morisette et al. 2006; Weiss et al. 2007). Radiative transfer models could be used efficiently to generate a calibration dataset covering a wide range of situations and configurations (Bacour et al. 2006; Banari et al. 1996; Baret and Guyot 1991; Baret et al. 2007; Ganguly et al. 2012; Gobron et al. 2000; Huete et al. 1997; Knyazikhin et al. 1999; Leprieur et al. 1994; Rondeaux et al. 1996; Sellers 1985; Verstraete and Pinty 1996).

3.3.2.1 Vegetation Index (VI) Based Approaches

The simplest methods are based on the calibration of linear or polynomial multiple regression functions where the dependent variable is the biophysical variable of interest. The independent variables are either the top of canopy reflectance in few bands, or a transform and/or a combination of these reflectances resulting into a vegetation index (VI). VIs are designed to minimize the influence of confounding factors such as soil reflectance (Baret and Guyot 1991; Richardson and Wiegand 1977) or atmospheric effects (Huete and Lui 1994). The strong nonlinearity between reflectances and canopy variables is reduced using these reflectance transforms or band combination allowing using linear statistical models. Based on these principles, operational algorithms developed for medium resolution sensors are currently used: MGVI for MERIS further extended to other sensors (Gobron et al. 2008), MODIS back-up algorithm based on NDVI (Myneni et al. 2002), POLDER algorithm based on DVI computed from bidirectional reflectance factor (BRF) (Roujean and Lacaze 2002). Nevertheless, although quite often effective, VIs are intrinsically limited by the empiricism of their design and the small number of bands concurrently used (generally two to three).

3.3.2.2 Machine Learning Approaches

Alternatively, more sophisticated machine learning methods have been proposed since the beginning of the nineties. Neural networks have been used intensively (Wan et al. 2024; Sun et al. 2022; Wocher et al. 2018; Zeng et al. 2018; Danner et al. 2017; Abuelgasim et al. 1998; Atkinson and Tatnall 1997; Danson et al. 2003; Gong et al. 1999; Kimes et al. 1998; Smith 1992; 1993). Baret et al. (1995) and Verger et al. (2011a) demonstrated that neural networks used with individual bands were performing better than classical approaches based on VIs. Fang and Liang (2005) found that neural networks were performing as well as the projection pursuit multiple regression. It was applied over MERIS (Bacour et al. 2006) and VEGETATION (Baret et al. 2007) kilometer spatial resolution data. The principles have been also applied at decametric resolution over airborne POLDER (Weiss et al. 2002a), LANDSAT (Fang and Liang 2003), CHRIS (Verger et al. 2011a) and FORMOSAT (Claverie et al. 2013) sensors. Although neural networks are becoming very popular, Verrelst et al. (2012) investigated alternative machine learning methods including support vector regression (SVM) and Gaussian Process Regression (GPR). They demonstrated the potentials of GPR when the training was achieved over experimental datasets. However, when applied to a large number of simulated cases, GPR is limited by the computation capacity (Mackay 2003). Further, one advantage of the GPR is the possibility to get an estimation of the associated uncertainties when applied to experimental data. In the case of model simulations, the uncertainties attached to the reflectance measurements need to be specified, which is not an easy task.

The training dataset is obviously a major component of the machine learning methods. It should represent the distributions and co-distributions of the input canopy biophysical variables. This is where the prior information is mainly embedded in machine learning methods that can be considered as a Bayesian approach. The density of cases that populate the space of canopy realization may rapidly decrease as a function of its dimensionality defined by the number of required canopy variables. Experimental plans may be conveniently used to limit local sparseness of the training dataset (Bacour et al. 2002b). Machine learning systems can be also considered as smoothers. They thus mainly "interpolate" between cases in the training dataset. Extrapolation outside the definition domain (corresponding to the convex hull of the input reflectance of the training dataset) is likely to provide unrealistic estimates. Further, cases that are simulated but never observed may be discarded to get a more compact training dataset and efficient learning process (Baret and Buis 2007). However, it requires compiling a large data base of reflectance measurements that should be representative of all the possible situations available.

3.3.3 Pros and Cons Associated with the Retrieval Approaches

The several approaches just briefly reviewed will be discussed regarding several aspects listed here:

- ***Computation requirements.*** Machine learning approaches, once calibrated, are obviously very little demanding in terms of computation. The inverse model is generally relatively simple and could be run very quickly. However, the calibration (or learning or training) process could require large computer resources, particularly for complex parametric model with a significant number of coefficients to be tuned and when the training dataset is large. The implementation of a LUT technique in algorithmic operational chains is very efficient because the radiative transfer model is run offline. Conversely, iterative minimization methods require large computer resources because of its iterative nature. Improvements are possible using a meta-model. Further, automatic segmentation or discretization of the reflectance space (Pinty et al. 2011) will also reduce the number of inversions to be completed over a whole set of images.
- ***Flexibility of the observational configuration.*** Iterative optimization methods allow retrieving canopy characteristics from several observational configurations. It is even possible to invert radiative transfer models concurrently over several pixels. This opens great potentials for exploiting additional temporal or spatial constraints as we will see later. LUT could theoretically cope with variable configurations at the expense of the dimensionality and thus the size of the tables, making them more difficult to manipulate. Conversely, machine learning methods require a fixed number of inputs. The characteristics of the configuration need thus to be used as inputs of the inverse parametric model as illustrated in Figure 3.3 where the illumination and view directions are explicitly used. However, this increases the dimensionality of the system, making the calibration step more demanding and more difficult. One alternative is to calibrate several parametric models for each individual configuration and then select the proper calibrated inverse model.
- ***Integration of prior information.*** The radiometric data driven approaches integrate the prior information directly in the cost function within the regularization term (see Equation 3.3). However, in the case of LUTs, it is also possible to restrict the simulations to the range of situations to be encountered as is done within the MODIS LAI and FAPAR algorithm that depends on the biome type considered (Shabanov et al. 2005). For the machine learning approaches, the prior information is introduced through the distributions and

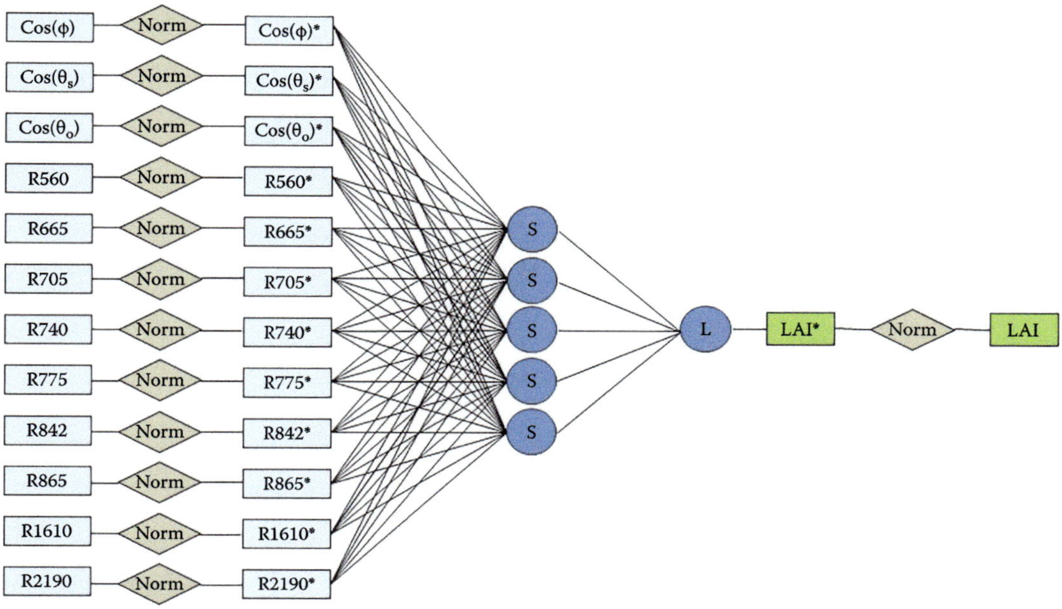

FIGURE 3.3 An example of a neural network used to estimate LAI from Sentinel 2 top of canopy reflectance. “”represents the normalization of the inputs or output (LAI). “S” and “L” represent tangent-sigmoid and linear transfer functions associated to each neurone. ϕ, θsθo represent, respectively, the relative azimuth between Sun and view directions, Sun and view zenith angles. “R560” to “R2190” represent the top of canopy reflectance in the several Sentinel 2 bands. From (Baret et al. 2009).

co-distributions of the inputs of the radiative transfer model: when LAI and FAPAR have to be estimated under situations where the type of canopies and their stage of development are known, it is more efficient to calibrate a specific inverse model for each individual situation. Note that (Qu et al. 2008) proposed to use Bayesian networks where model simulations could be exploited along with a description of the distribution of the variables that may depend on growth stages or canopy types.

- ***Associated uncertainties***. The radiometric data driven approaches allow getting some estimates of the uncertainties associated to the solution by propagating the uncertainties associated to the measurements and to the model using the partial derivatives of the cost function with regards to the measurements (Lauvernet 2005). When using LUTs, uncertainties could be estimated by Monte-Carlo methods or approximated by the standard deviation of the ensemble of solutions defined by the uncertainties in the measurements (Knyazikhin et al. 1998). For machine learning methods, the error on the measurements may be assessed in different ways as proposed by Aires et al. (2004), which may also include the errors associated to the retrieval process itself. A more simple alternative solution is also proposed by Baret et al. (2013) based on the training dataset. Although the estimation of uncertainties of the retrievals is possible, it is generally limited by the poor knowledge on the input uncertainties associated to the reflectance measurements and radiative transfer models used. Knyazikhin et al. (1998) used a 20% relative uncertainty applied to MODIS top of canopy reflectance for LAI and FPR retrieval. Baret et al. (2007) proposes to use an additive uncertainty around 0.05 and a multiplicative uncertainty around 3%. This example shows that the uncertainties attached to each band is poorly known. Further, the structure of the uncertainties may also play an important role and is unfortunately very difficult to describe.

- ***Robustness of the retrieval and quality assessment***. A quality index needs to be associated to the retrieved values to inform about the status of the inversion process. For iterative optimization techniques, it could be criterions relative to the stop of the iterations (Gilbert 2002). As a matter of facts, the algorithm may sometimes encounter numerical problems occurring generally with very small values of *J*. No numerical problems are expected for LUT and machine learning approaches, and the quality index should mainly indicate whether the input reflectances were inside the definition domain and if the output solution is in the expected range of variation (Baret et al. 2013). The performances of the approach will both depend on the minimization algorithm itself and on the level of illposedness of the inverse problem as a function of measurement configuration and model and measurement uncertainties.

3.4 THEORETICAL PERFORMANCES OF BIOPHYSICAL VARIABLES ESTIMATION

Several biophysical variables are potentially accessible as reviewed previously. However, depending on the assumptions on canopy structure and the observational configuration considered, the "apparent" values retrieved from remote sensing observations will be associated with contrasted performances. Further, the several possible definitions for GAI and FAPAR need also to be discussed in terms of the associated uncertainties. The theoretical estimation performances were thus investigated using a simple synthetic experiment. The SLC radiative transfer model (Verhoef and Bach 2007) coupled with the PROSPECT model (Jacquemoud and Baret 1990) was used to simulate the canopy reflectance in the Sentinel 2 (Malenovský et al. 2012) bands for a large set of combination of canopy characteristics (Figure 3.5) covering the expected range of variation of each of the canopy, leaf, and soil input variables. The seven bands considered (560 670 705 740 865 1610 2190 nm) were chosen to sample well the main absorption features of chlorophyll and water. The SLC model allows simulating leaf clumping at the plant scale: plants are randomly sown and are represented by ellipsoidal envelopes filled with randomly distributed leaves. The leaf clumping is mainly driven here by the crown fraction, that is, the fraction of ground area casted by the crowns in the vertical direction. Therefore, SLC allows also simulating turbid medium canopies when the crowns cover fully the background (crown cover = 1.0). Three typical Sun positions and five view directions were considered. The black sky FAPAR ($FAPAR_{bs}$) and white sky FAPAR ($FAPAR_{ws}$), the green fraction (GF) and the effective GAI, GAI_{eff} were simulated in addition to the input GAI, GAI_{true} . The simulated dataset was used as a LUT to retrieve the five variables of interest: $FAPAR_{bs}, FAPAR_{ws}, GF, GAI_{eff}$ *and* GAI_{true}. A subsample of the simulated cases was used as the test dataset. The corresponding reflectances were contaminated with realistic measurement uncertainties. The solution is finally selected as the case in the LUT that corresponds to the minimum of the cost function presented in Equation 3.2, where σ^2 is the variance of the reflectance of the test dataset computed from the measurement uncertainties introduced. Note that no constraints or prior information were used in the cost function. The retrieval was achieved over turbid medium or clumped test cases using LUT based either on turbid medium or clumped canopy structure assumption. More details can be found in (Kandasamy et al. 2010).

Results presented in Figure 3.4 show that GF and $FAPAR_{ws}$ are the best estimated variables. Further, the good performances are relatively independent from the assumptions on canopy structure. The black sky, $FAPAR_{bs}$, is still well estimated, with however a significant degradation of the retrieval performances when the test cases correspond to clumped canopies. GAI values are much more difficult to estimate, particularly the actual GAI_{true} value for the clumped test cases. Conversely, the effective GAI, GAI_{eff}, provides relatively stable performances independently from the assumptions on canopy structure. Note that the turbid medium test cases retrieved with a LUT made of clumped canopies provides poorer estimates as compared to those derived from the turbid

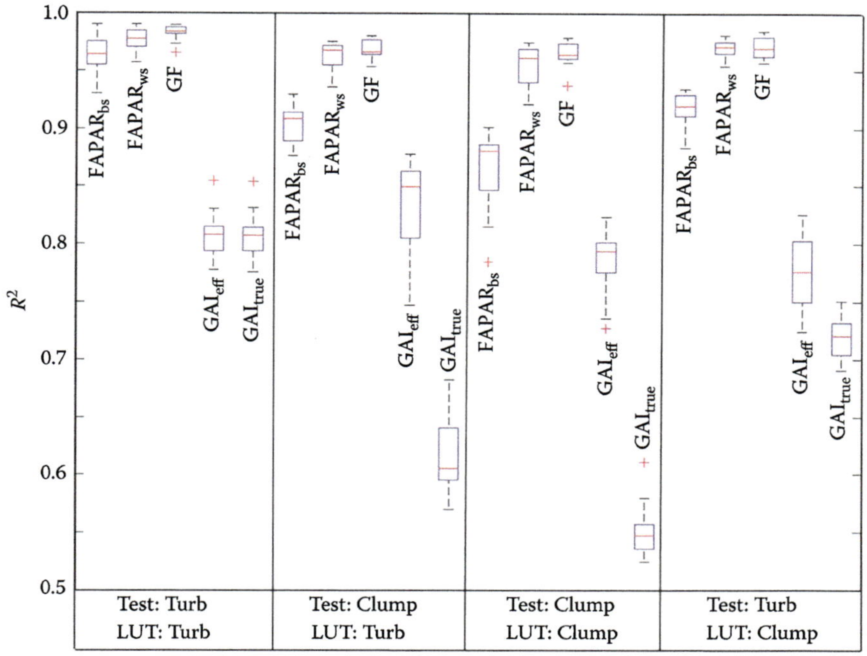

FIGURE 3.4 Theoretical performances (R^2 of the regression between the reference variable and the estimated value) of [$FAPAR_{bs}$, $FAPAR_{ws}$, GF, GAI_{eff}, GAI_{true}] estimation depending on the test cases considered (turbid or clumped) and the assumptions on canopy structure used in the LUT (turbid or clumped). Box plot representation of the results for the 15 observational configurations (three Sun position and five view directions). The median value is the red line in the boxplot that contains 50% of the data. Whiskers extend to the extreme values except if they are considered as outliers that are represented by a red "+."

medium LUT. Although turbid medium cases are included in the LUT made with clumped canopies, the degradation of performances is explained by the smaller number of turbid medium cases contained in the LUT populated with clumped canopies (less cases with crown cover close to 1.0). Further, possible ambiguities between turbid medium and clumped cases providing very similar reflectance values may be encountered. The variability of retrieval performances depending on the observations configuration is larger for the $FAPAR_{bs}$ and GAI_{eff} and more particularly for GAI_{true}, in agreement with the overall performances associated to the retrieval of these variables.

This simple numerical experiment demonstrates that the retrieval of the true GAI from mono-directional reflectance measurements is likely to be relatively inaccurate, particularly in the case of clumped canopies. The use of a clumped canopy model in the inversion process does not help the retrieval: more constraints or prior information is needed to compensate for the additional unknown variables required to describe canopy clumping as discussed in the next section.

3.5 MITIGATING THE UNDERDETERMINED AND ILL-POSED NATURE OF THE INVERSE PROBLEM

3.5.1 Underdetermination and Ill-Posedness of the Inverse Problem

Canopy reflectance models will depend on a set of input variables characterizing the several components: soil, leaf and canopy structure (Figure 3.5). Several models have been proposed to describe the soil reflectance. They are either physically sound ones mostly focusing on the bidirectional variability

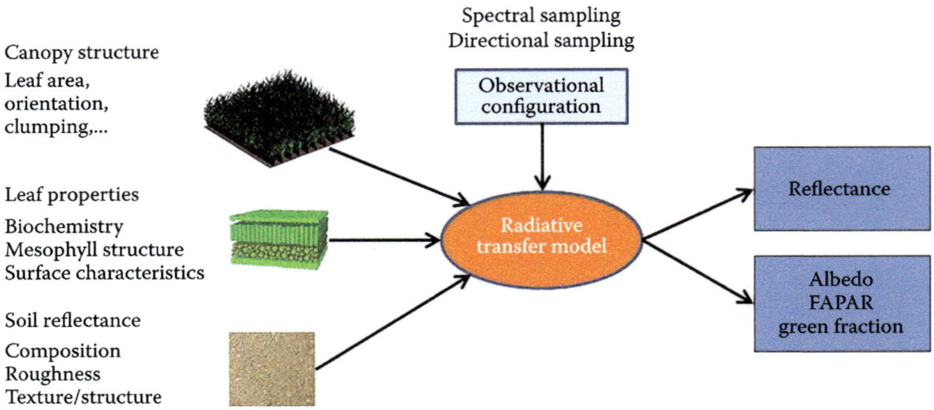

FIGURE 3.5 The radiative transfer models used to simulate canopy reflectance and additional canopy properties as a function of the leaf, soil, and canopy characteristics.

of the reflectance (Chaabouni et al. 2021; Schiefer et al. 2021; Abdelbaki et al. 2019; Verrelst et al. 2019a; Cierniewski et al. 2002; Hapke 1981; Jacquemoud et al. 1992; Liang and Townshend 1996) or more empirical ones describing the spectral variability (Bach and Mauser 1994; Liu et al. 2002; Price 1990). At least 6 parameters are required to describe both the directional and spectral variation of soil properties. Leaf reflectance and transmittance may be simulated from the knowledge of its composition in main absorbers (chlorophyll, water and dry matter), the mesophyll structure and the surface features (Dawson et al. 1998; Jacquemoud and Baret 1990; Jacquemoud et al. 2009). At least four parameters are required here. The simplest description of canopy structure could be achieved with two parameters: leaf area index and the orientation of the leaves. Therefore, the whole spectral and directional reflectance field of the canopy could be simulated with at least 12 parameters that are mainly unknown to be estimated through radiative transfer model inversion. This should be solved with at least the same amount of independent reflectance measurements provided by the observational configuration, that is, combination of bands and view or illumination conditions.

The actual dimensionality of remote sensing measurements has been evaluated in different ways, generally by considering independently the spectral and directional dimensions. Several studies report that the bidirectional reflectance distribution function could be decomposed using empirical or semi-empirical orthogonal functions with generally 2 to 4 kernels (Bréon et al. 2002; Lucht 1998; Weiss et al. 2002b) (Weiss et al. 2002b; Zhang et al. 2002a; Zhang et al. 2002b). Other studies report a high level redundancy between bands (Wan et al. 2024; Liu et al. 2023; Abdelbaki et al. 2019; Wocher et al. 2018; Zeng et al. 2018; Danner et al. 2017; Green and Boardman 2001; Liu et al. 2002; Price 1994, 1990; Thenkabail et al. 2004) with a dimensionality varying between 5 to 60 depending on the data considered and the method used to quantify the dimensionality. More recently, (Laurent et al. 2011) found a dimensionality of three to four using a singular decomposition method applied to CHRIS images having a high spectral resolution and several view directions. These finding confirmed those of Settle (2004) and Simic and Chen (2008), showing a high degree of redundancy between bands and directions. It is therefore clear that in most situations, the radiative transfer model inversion is an underdetermined problem since the number of unknown variables to be estimated is larger than the actual dimensionality of the observations.

Because of its underdetermination and uncertainties attached to models and measurements, the inverse problem is generally ill posed: the solution is not unique and does not depend continuously on the observations (Garabedian 1964). In these conditions, very similar reflectance spectra simulated by a radiative transfer model (Figure 3.2, left) may correspond to a wide range of solutions. This may be due to two main factors:

- ***Lack of sensitivity of canopy reflectance to a given variable.*** This is the case for large LAI values because of the well-known saturation problem: a small variation in the measurements may correspond to a very large variation in the retrieved LAI value. Under the same high LAI conditions, the retrieved soil reflectance will be very poor, since the measured reflectance will be no more sensitive to soil background reflectance.
- ***Compensation between variables.*** This is obviously the case when some variables appear combined together always the same way in the model, such as in the form of a product: it is thus impossible to estimate separately each variable in this situation. However, this is also observed for other variables that are not formally appearing as products in the model as reported by several authors (Chaabouni et al. 2021; Schiefer et al. 2021; Abdelbaki et al. 2019; Baret and Buis 2007; Baret et al. 1999; Shoshany 1991; Teillet et al. 1997; Weiss et al. 2000).

The ill-posedness of radiative transfer model inversion should be mitigated by exploiting additional information (Baret et al. 2000; Combal et al. 2001; Knyazikhin et al. 1999). This could be achieved both by using prior information on the distribution of the variables, and by exploiting some constraints on the variables. Further, reducing the model uncertainties when possible by a proper selection of the radiative transfer model will also improve the accuracy of the retrieval. These issues will be investigated separately in the following.

3.5.2 Reducing Model Uncertainties

The realism of the radiative transfer model impacts largely the retrieval performances. The model should be physically sound and the embedded assumptions on canopy architecture and leaf and soil optical properties should be consistent with the actual canopy considered. The soil is relatively well described mainly by empirical models as reviewed in a previous section. The leaf optical properties are also quite well described by the PROPSECT (Jacquemoud et al. 2009) or LIBERTY (Dawson et al. 1998; Moorthy et al. 2008) models, at least if the directional effects are not considered (Comar et al. 2014). The canopy architecture is therefore recognized as the main limiting factor in the modeling of vegetation reflectance. To account for particular architectural features of a given canopy, prior knowledge on the type of vegetation viewed is therefore mandatory. Depending on the spatial resolution of the observation and the heterogeneity of the scene, this information is not always accessible. Observations at kilometric spatial resolution are often corresponding to a mix of different vegetation types making the use of specific radiative transfer models challenging. Conversely, at decametric spatial resolution, pixels are more likely to be "pure" and the type of vegetation may be more easily identified. In these conditions, the inversion using a radiative transfer model for which the architecture is described in a more realistic way will reduce the error associated to the radiative transfer model and contribute to improve the retrieval performances. Lopez-Lozano (2008) compared the inversion of a turbid medium reflectance model where leaves are assumed randomly distributed within the canopy volume and of infinitesimal size to that of a 3D model adapted to maize and vineyard crops (Figure 3.6). The results showed clearly that GAI estimation is much improved with a 3D description of canopy architecture for the vineyard case, where the turbid medium assumption is very far from reality as compared to the maize case. The estimated GAI using a turbid medium shows a systematic underestimation due to the leaf clumping: when the assumptions on canopy architecture are not verified by the canopy observed, the retrieved GAI value will be termed "apparent." This apparent GAI value is the one which is accessible from the measurement and the interpretation pipeline. It will thus depend on the inverse technique and on the radiative transfer model used. In addition, the apparent value may also strongly depend on the observational configuration used as in the case of the vineyard canopy where the row orientation has to be accounted for. Using more realistic canopy architecture implies more characteristics to describe

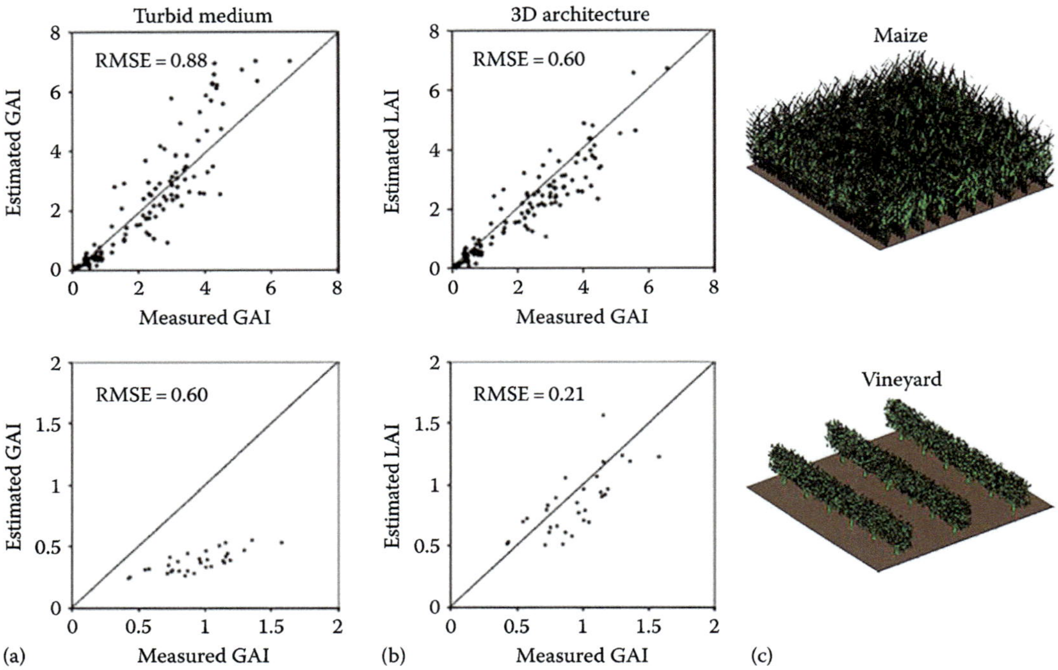

FIGURE 3.6 Comparison of GAI retrieval performances when using a turbid medium radiative transfer model (a) and 3D realistic canopy architecture (b). (c) Examples over maize (top) and vineyard (bottom). RMSE is the Root Mean Square Error. (From Lopez-Lozano, 2008.)

the vertical and horizontal distribution of the green area density. The gain in realism obtained at the expense of additional unknown canopy variables should be counterbalanced by prior information on the distribution of these additional canopy structure variables.

3.5.3 Using Prior Information

The prior information characterizes the knowledge available on the distribution and co-distribution of the input variables of the radiative transfer models. It is used directly in machine learning approaches to generate a calibration dataset that reflects this knowledge. For LUTs and iterative optimization methods, the prior information is introduced in the cost function through a regularization term:

$$J = \underbrace{(R-\hat{R})^t \times W^{-1} \times (R-\hat{R})}_{\text{Radiometric information}} + \underbrace{(\hat{V}-V_p)^t \times C^{-1} \times (\hat{V}-V_p)}_{\text{Prior information}} \quad (3.3)$$

where $\hat{V}$ and V_p are, respectively, the vectors of the estimated and prior values of the input biophysical variables and C is the covariance matrix characterizing the prior information. Note that the first part of this equation corresponds to Equation 3.2. The second part of Equation 3.3 corresponds to the distance between the values of the estimated variables and those of the prior information. The theory behind this equation derives from Bayes' theorem (Bayes and Price 1763) that was extensively used in parameter estimation (Tarantola 2005). However, if the theory is well known, it is not yet largely used in the community (Combal et al. 2002; Lauvernet et al. 2008; Lewis et al. 2012; Pinty et al. 2011).

Implementing the cost function as expressed by Equation 3.3 requires some reasonable estimates of covariance matrices W and C as well as of the prior values V_p. The terms of W should reflect both measurement and radiative transfer model uncertainties. While some rough estimates of the measurement uncertainties could be derived from the sensor specification, model uncertainties are far more difficult to estimate. Further, they may depend significantly on the situation considered, such as low or high vegetation amount and the discrepancy between the canopy structure description embedded in the radiative transfer model and that of the observed canopy. Even more difficult to estimate, are the covariance terms in W: measurement and model uncertainties may have important structure that translates into high covariance terms which are however very poorly known. When using simultaneously a large number of configurations as in the case of hyperspectral observations, these covariance terms will allow weighing properly the several configurations used. It thus accounts for the large redundancy exhibited between spectral bands. The difficulty to estimate the covariance terms in W explains why a small number of configurations is often selected when a larger number is available as in the case of hyperspectral and/or multidirectional observations. Parsimony, thus dimensionality reduction of the observations is highly desired in most retrieval problems (Tenenbaum et al. 2000). For machine learning methods, a reduced dimensionality is also beneficial since the number of coefficients of the inverse parametric model will grow with the number of observations used as inputs, making the calibration process more difficult and instable.

Introducing prior information in the inversion process improves the precision by reducing the variability of the posterior distribution of the estimated variables. However, this is achieved at the expense of a loss of accuracy: the solution is biased towards the prior information value as observed in Figure 3.7.

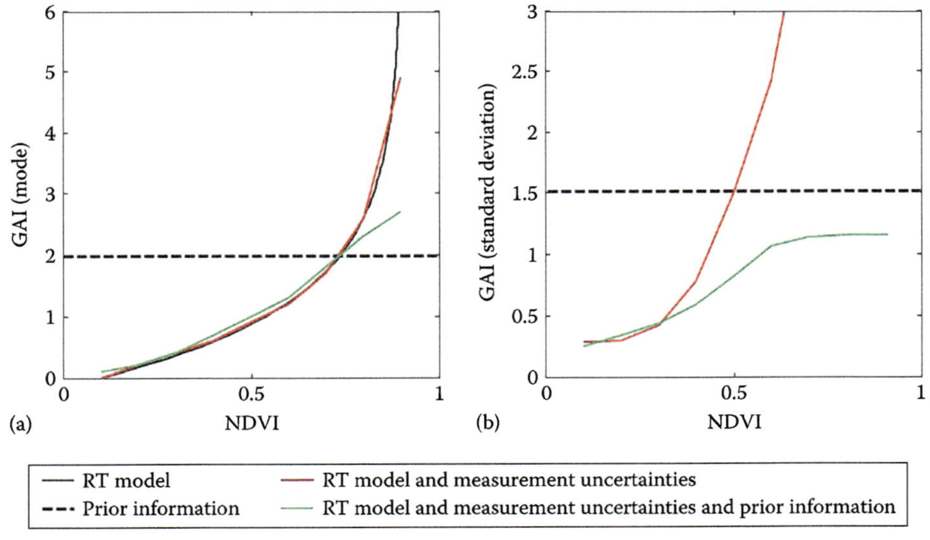

FIGURE 3.7 Mode (a) of the distribution of the solution (GAI) of the inverse problem as a function of the measured value. In this case, GAI is estimated using a simple empirical RT model (NDVI = RT(GAI)). The mode corresponds to the maximum PDF value, that is, the maximum likelihood. Four estimates are displayed: (1) using only prior information (i.e., no measurements are used) (Prior information); (2) using RT model (GAI = RT^{-1}(NDVI)) assumed to be perfect with perfect measurements (no uncertainties accounted for); (3) using RT model and measurements with their associated uncertainties; and (4) using RT model and measurements with their associated uncertainties and prior information. (b) The standard deviation of the distribution of the solution is also displayed for the several cases. The case with perfect RT model and measurements is not displayed here because its standard deviation is null by definition. (From Baret and Buis, 2007.)

3.5.4 Using Additional Constraints

3.5.4.1 Temporal Constraints

The dynamics of canopy structure and leaf optical properties results from incremental processes under the control of climate, soil and the genetic characteristics of the plants. Very brutal and chaotic time course are therefore not expected, at the exception of accidents such as fire, flooding, harvesting, or lodging. The smooth character of the dynamics of canopy variables may be exploited as additional constraints in the retrieval process as proposed by (Lewis et al. 2012). The use of a models describing the time course of some of the variables was proposed by Kötz et al. (2005) to improve remote sensing estimates of GAI in maize crops: results show a significant improvement of estimates, particularly for the larger GAI values where saturation of reflectance is known to be a problem. The semi-empirical nature of the model with parameters having some biological meaning, prior information on them could be accumulated and efficiently exploited. However, the results show that the improvement in GAI retrieval is mainly coming from the "smoothing" effect of the model: fitting the GAI dynamics model over the instantaneous estimates corresponding to each individual date of observation provides similar performances (Kötz et al. 2005). This explains why compositing techniques applied to kilometric resolution observations are very popular: very little prior information is available on the dynamics of the surface except the expected smoothness of the temporal profiles (Atkinson et al. 2012; Chen et al. 2004; Kandasamy et al. 2013; Lewis et al. 2012; Refslund et al. 2013; Verbesselt et al. 2010; Zhu et al. 2013). The usual sigmoidal shape of vegetation growth and senescence curves (Jönsson and Eklundh 2004; Zhang et al. 2003) and the possible use of the climatology, (Samain et al. 2007; Verger et al. 2012a) have been also exploited.

3.5.4.2 Spatial constraints

Most of the algorithms are currently applied to independent pixels, neglecting the possible use of spatial structure as observed on most images. However, some authors attempted to exploit these very obvious patterns at high spatial resolution. The "object retrieval" approach proposed by Atzberger (2004) is based on the use of covariance between variables as observed over a limited cluster of pixels representing the same class of object such as an agricultural field. Results show quite significant improvement of the retrieval performances for GAI, chlorophyll and water contents, presumably because of a better handling of possible compensations between GAI and leaf inclination in the retrieval process as suggested by Atzberger (2004). The principles were further extended later using simple heuristics that could apply at the field scale for agriculture applications (Atzberger and Richter 2010). This implies that objects, sometimes called "patches" are first identified which is now becoming a very common approach in remote sensing image segmentation techniques (Blaschke 2010; Peña-Barragán et al. 2011; Vieira et al. 2012). The objects need then to be classified to exploit some features shared by the pixels of a single patch.

3.5.4.3 Holistic Retrieval Over Coupled Models: From Inversion to Assimilation

Retrieval of characteristics of some element of the observational system without solving the whole system at once will be sub-optimal: each element of the system imposes constraints on the other elements through the radiative transfer physical processes, temporal or spatial constraints as seen previously. This is clearly demonstrated in the case of the radiative coupling between the leaves and the canopy: when estimating structural canopy characteristics from bottom of the atmosphere reflectance measurements in several bands and directions, the inversion process may be split into several parallel and independent inversions for each band. The leaf characteristics, that is, reflectance and transmittance, sometimes grouped into the single scattering albedo, need to be estimated (Pinty and Verstraete 1991b) for each of the bands considered. This may lead to inconsistent estimates of the structure characteristics derived from the inversion applied independently on each band. Further, it may lead to spectrally inconsistent leaf optical properties estimates, since no spectral constraints coming from a leaf optical properties model are imposed. Solving the whole system at once using coupled leaf and canopy radiative transfer models will therefore improve the consistency of the estimates by imposing the spectral constraints coming from the leaf radiative transfer model. The interest of such holistic approach was recently highlighted by Laurent et al. (2011) when using coupled canopy and atmosphere radiative transfer models.

Lauvernet et al. (2008) proposed a "multitemporal patch" inversion scheme to account both for spatial and temporal constraints. Reflectance data are here considered observed from the top of the atmosphere. Atmosphere/canopy/leaf/soil radiative transfer models are thus coupled to simulate top of the atmosphere reflectance from the set of input variables of each sub-model. Spatial and temporal constraints are based on the assumption that the atmosphere is stable over a limited area (typically few kilometers) but varies from date to date, and that surface characteristics vary only marginally over a limited temporal window (typically ±7 days) but may strongly change from pixel to pixel (Hagolle et al. 2008). This has obviously important consequences on the underdetermined nature of the inverse problem since atmospheric characteristics will be shared between the pixels of a patch while vegetation characteristics will be shared during a limited time period. Results on the performances achieved demonstrate the interest of the approach for the estimation of most of the variables, particularly for the aerosol characteristics and for canopy characteristics such as GAI.

However, the improvement of retrieval performances based on such holistic approach is gained at the expense of additional complexity in terms of the number of unknown variables to be estimated and of the computational resources required to run the coupled models. Machine learning approaches may reach their limits in such conditions. Iterative optimization efficiently implemented using the adjoint model (Lauvernet et al. 2008; Lewis et al. 2012; Voßbeck et al. 2010) provides a convenient solution. This could be used ultimately to couple the radiative transfer model to a functional-structural plant model as proposed by (Weiss et al. 2001). However, considerable efforts are still needed to describe the dynamics of the canopy structure consistently with both the radiative transfer modeling and with the canopy functioning.

3.6 COMBINATION OF METHODS AND SENSORS TO IMPROVE THE RETRIEVALS

3.6.1 Hybrid Methods and Ensemble Products

Verger et al. (2008) demonstrated that neural networks could be used efficiently to replace the actual MODIS algorithm (Shabanov et al. 2005) which is based on a LUT method: neural networks were calibrated over an empirical training dataset containing the MODIS top of canopy BRF values and the corresponding MODIS LAI products. This approach is therefore different from calibrating a machine learning algorithm directly on radiative transfer model simulations as done by Bacour et al. (2006) or Baret et al. (2007). It is termed hybrid because a canopy biophysical driven method is calibrated over the outputs of a radiometric data driven approach. This principle was later used to relate the long-time series of AVHRR NDVI vegetation index (Tucker et al. 2005) to LAI and FAPAR MODIS products during an overlapping period between both sensors (2000–2009) (Zhu et al. 2013).

With the compilation of results derived from several initiatives dedicated to the validation of global remote sensing products, the performances of products started to be quantified in a more representative way (Garrigues et al. 2008). This allowed to select the more consistent available products and to eventually combine them and propose a new "ensemble" product that capitalizes over past development efforts (Figure 3.8): a training dataset is first built that contains a globally representative sample of MODIS and CYCLOPES products along with reflectance as measured by a sensor from which the "ensemble" product is generated (Baret et al. 2013; Verger et al. 2014; Xiao et al. 2014). The original biophysical products in the training database need to share the same spatial and temporal support to be consistently combined. This is achieved by applying interpolation

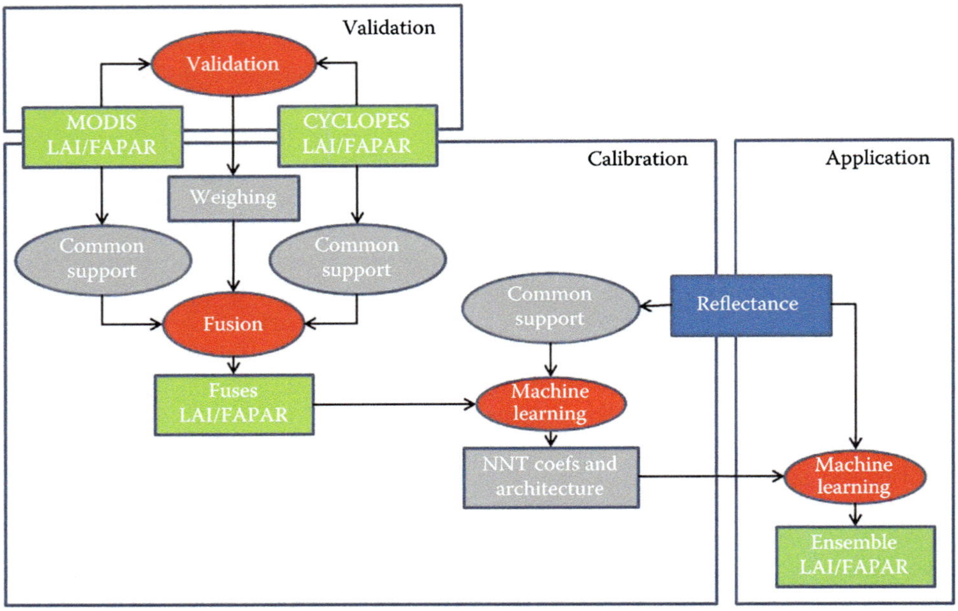

FIGURE 3.8 Principle of the GEOV1 (Baret et al. 2013), GEOV2 (Verger et al. 2014), and GLASS (Xiao et al. 2014) algorithms to generate "ensemble" products.

methods that will further smooth possible spatial or temporal discrepancies. A weighed average of the original products is then computed to get the fused products (Figure 3.8). The weights are derived from the results of the validation of the original products using either the associated uncertainties (Xiao et al. 2014) or heuristic arguments (Baret et al. 2013). The fused products and the corresponding reflectance values are then used to calibrate a machine learning algorithm. The calibrated machine learning algorithm is finally used to transform the reflectance values into the corresponding "ensemble" product (Figure 3.8). In the case of GEOV1 (Baret et al. 2013), the transformation is applied using a back propagation neural network over each individual observation to get instantaneous fused LAI and FAPAR values. A smoothing and gap filling algorithm is then applied over the time series of fused products (Verger et al. 2011b). In the case of GLASS products (Xiao et al. 2014), a whole year of reflectance observations in the red and near-infrared is used to get the corresponding yearly time series of LAI products using a Generalized Regression Neural Network (Specht 1991). Results show that these ensemble products are generally over-performing the original products (Wan et al. 2024; Zhu et al. 2023; Verrelst et al. 2019a, 2019b, Camacho et al. 2013; Fang et al. 2013; Xiao et al. 2014).

3.6.2 Combining Sensors to Build Long, Dense, and Consistent Time Series

Monitoring the dynamics at the seasonal or multi-annual scale allows to better characterize the canopy functioning including the phenology (Ganguly et al. 2010; Jönsson and Eklundh 2004) and detect anomalies (Bessemoulin et al. 2004; Ciais et al. 2005), breaks (Verbesselt et al.) or trends (Wan et al. 2024; Sun et al. 2022; Xie et al. 2019; Alcaraz-Segura et al. 2010; de Jong et al. 2012; Fensholt et al. 2012; Herrmann et al. 2005) across long time series of consistent observations. The revisit frequency, consistency, and length of the period when observations are accumulated are the main limiting factors when exploiting the time series. For global scale applications, observations are currently provided by kilometric spatial resolution sensors on polar orbit (Figure 3.10). They have a relatively large swath allowing to map the whole Earth within 1 day. However, this potential daily observation frequency is reduced because of cloud occurrence. The combination of observations by different sensors will provide only marginal gain in terms of the number of cloud-free dates of observations because of the strong spatiotemporal correlation of the distribution of clouds: Yang et al. (2006) reported no improvement when combining MODIS products derived from AQUA and TERRA. However, Hagolle et al. (2005) reported an improvement of both the completeness and the precision of top of canopy reflectance when compositing the two VEGETATION instruments as compared to the use of a single one. Similarly, Verger et al. (2011b) fused MODIS and VEGETATION data which resulted both in a significant reduction of the fraction of missing products as well as an improvement of the accuracy and precision of LAI estimates. These contrasting results are explained by the very different compositing algorithms used in these studies, highlighting the importance of the compositing process that mainly consists in smoothing and eventually gap filling the time series (Kandasamy et al. 2013).

The interest of fusing the data coming from several sensors is obvious when considering the decametric spatial resolution observations for which several days are needed to map the whole Earth. However, except in the case of the Rapid-eye and DMC constellation of satellites (Sun et al. 2002), very little attention has been carried out on the fusion between different decametric resolution satellites. Although the satellites currently orbiting provide great potentials for seasonal monitoring of the vegetation at decametric spatial resolution, this has not been exploited because of the difficulty and cost associated to the images of these sensors that are used commercially. However, the development of the fusion between different decametric satellites does not pose great technical difficulties as illustrated by Figure 3.9: a very good temporal consistency of estimates derived from different sensors using the same algorithm is generally observed. This confirms the results of Verger et al. (2008) and Gobron et al. (2008) who demonstrated that applying a single algorithm to different sensors provides generally consistent products if the differences in observational configurations are carefully accounted for.

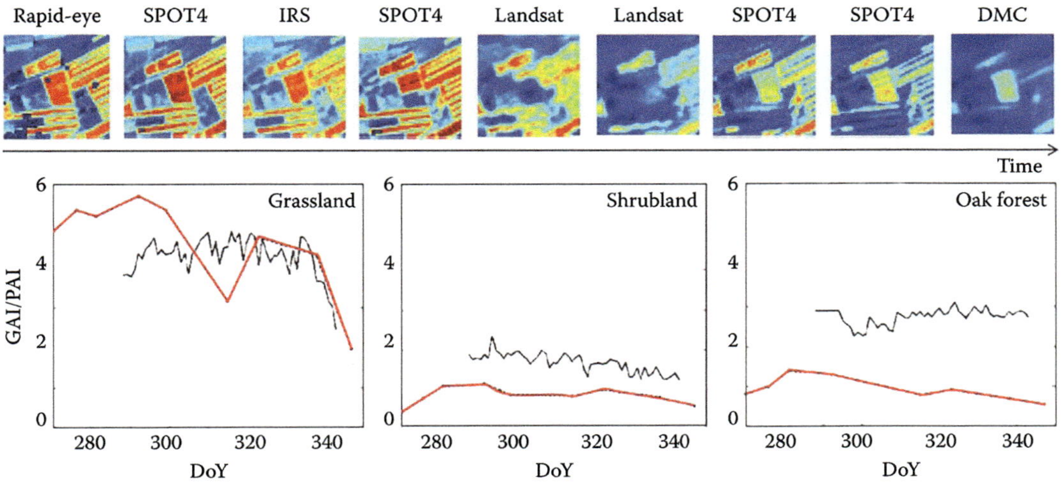

FIGURE 3.9 Exploitation of an heterogeneous constellation of satellites to derive seasonal variation of GAI. On top, the GAI images derived from each individual sensor along the growing season. On bottom, the seasonal variation over three sites. The red line corresponds to GAI estimates. The black line corresponds to reference ground measurements of PAI (Plant Area Index). Unpublished results obtained on the Crau site (43.5° latitude, 4.9° longitude). The algorithm used for all the sensors is similar to that described by Verger et al. (2011a) with no compositing applied to the data. The difference between satellite estimates and ground measurements is mainly explained by the difference in the definition of the variable accessed from remote sensing (GAI) and that measured on the ground (PAI).

The fusion between decametric resolution images and daily kilometric resolution data is very appealing because it potentially provides daily decametric products. However, this combination has been rarely investigated for deriving decametric dynamics of biophysical variables. It has mainly been applied for classification (Karkee et al. 2009), for reflectances (Faivre and Fischer 1997) including pan-sharpening (Fasbender et al. 2008) and for vegetation indices (Cardot et al. 2008; Gao et al. 2006). More studies should be directed towards the development of the fusion between biophysical variables obtained from decametric and kilometric spatial resolution sensors.

The succession of several kilometric sensors allows building long time series of global observations since 1981 (Figure 3.10). However, the consistency between the several sensors used to build the time series has to be very high in order to identify possible trends that may be very small (Beck et al. 2011). This is currently achieved by applying a single algorithm to the succession of sensors available. Zhu et al. (2013) transformed the long time series of NDVI derived from the several AVHRR sensors (Figure 3.10) into LAI and FAPAR by calibrating a neural network on MODIS products during and overlapping period between AVHRR and MODIS. The consistency and the compositing is here achieved at the NDVI level, based on the GIMMS products (Tucker et al. 2005). Verger et al. (2012b) built also a long time series of observations based on AVHRR up to 2000 and then using VEGETATION data. The input reflectance values were carefully processed according to (Nagol et al. 2009). Then neural networks were calibrated over an overlapping period between AVHRR, VEGETATION, and MODIS. The LAI and FAPAR products from MODIS and VEGETATION were fused to be used as target products. Finally, a compositing algorithm was applied to eliminate outliers, smooth out the resulting data, and fill possible gaps Verger et al. (2012a).

3.7 CONCLUSION

This review of retrieval techniques for canopy biophysical variables shows that great advancement in the maturity of the algorithms has been achieved these last years (Wan et al. 2024; Liu et al. 2023;

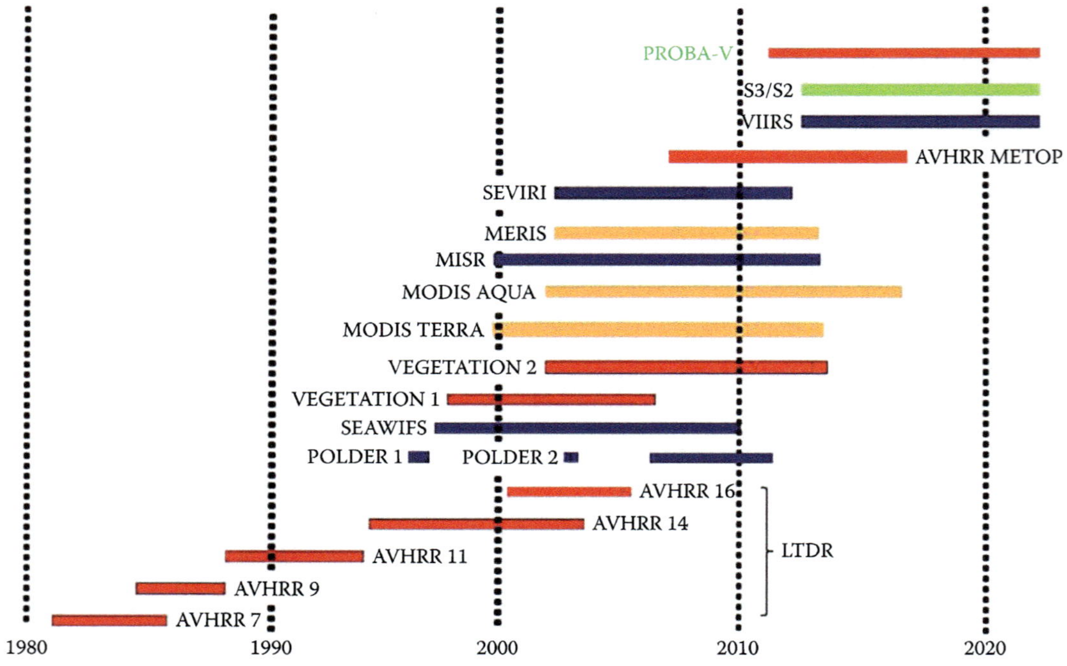

FIGURE 3.10 The series of kilometric spatial resolution sensors available from 1980 up to now.

Zhu et al. 2023; Martínez-Ferrer et al. 2022; Sun et al. 2022; Chaabouni et al. 2021; Schiefer et al. 2021; Abdelbaki et al. 2019; Verrelst et al. 2019a, 2019b; Xie et al. 2019; Sun et al. 2018; Wocher et al. 2018; Zeng et al. 2018; Danner et al. 2017). Several products were released to the wider community, mainly derived from kilometric resolution sensors as illustrated by Table 3.2. The multiplicity of products allows building enough confidence from the consistency observed between some of them as well as with ground measurements. The validation exercise is therefore mandatory to identify possible problems, improve the products and finally quantify the associated uncertainties. The RMSE values associated to FAPAR are on the order of 0.10–0.15 in absolute value (Weiss et al. 2014) while LAI are estimated within a RMSE slightly smaller than 1.0. However, the currently limited number of available ground measurements at the kilometric resolution limits the evaluation of the accuracy of remote sensing products.

When very few information or constraints are available as it is the case for the kilometric resolution observations, the variables that are the better estimated are the green fraction and the black and white sky FAPAR or FIPAR ones (Figure 3.11). Conversely, the "apparent" leaf area index derived

TABLE 3.2
The Several LAI and FAPAR Global Products Currently Available

Products	Sensors	LAI	FAPAR	Spatial Resolution	Time Sampling (days)	Time Period	Reference
MODIS C5	MODIS	✓	✓	1 km	8	2000–	Myneni et al. 2002
CYCLOPES V3	VEGETATION	✓	✓	0.009°	10	1999–2007	Baret et al. 2007
GLOBCARBON	VEGETATION	✓	✓	0.009°	30	1999–2007	Deng et al. 2006
JRC-FAPAR	SEAWIFS		✓	2 km	1	1997–2006	Gobron et al. 2006
JRC-TIP	MODIS	✓	✓	0.01°	16	2000–	Pinty et al. 2011
GIMMS_3g	AVHRR	✓	✓	8 km	30	1981–2013	Ganguly et al. 2010
GLASS	MODIS/AVHRR	✓	✓	1 km	10	1981–2014	Xiao et al. 2012
GEOV1_VEG	VEGETATION	✓	✓	0.009°	10	1999–	Baret et al. 2013
GEOV2_VEG	VEGETATION	✓	✓	0.009°	10	1999–	Verger et al. 2014

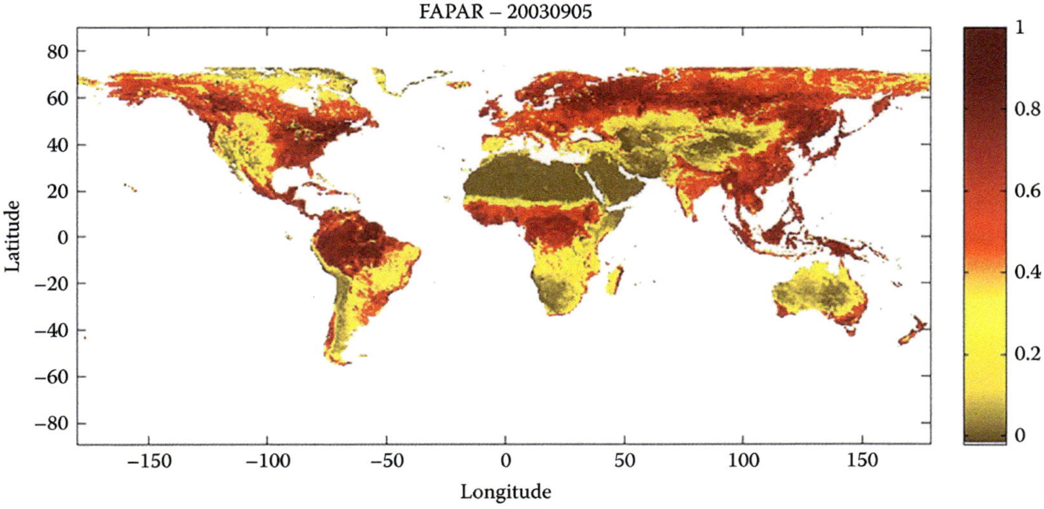

FIGURE 3.11 A global map of GEOV1 FAPAR product for September 5, 2003.

from the reflectance observations is more closely linked with the effective green area index, GAI_{eff}, while the true leaf area index is poorly estimated with uncertainties that are very dependent on the observational configuration. This finding should be much better reflected to the users of current LAI products derived from remote sensing, although some attempts were proposed to correct for this effect (Xiao et al. 2012). However, focusing on the GF variable will allow reaching more easily a better consistency with the canopy functioning models that have their own specific description of the canopy architecture.

Two main types of canopy biophysical variables retrieval approaches were identified. Canopy biophysical variables driven approaches when trained over empirical datasets with ground measurements of the canopy variables defined in a consistent way with the variables accessible from remote sensing observations would be ideal: they implicitly integrate the measurement uncertainties while no model uncertainties have to be included since no radiative transfer model is used. Further, machine learning approaches are very computer efficient once trained, allowing easy implementation within operational processing chains. However, because of the difficulty of getting a representative sampling of cases to populate the training dataset, training over a data base made of radiative transfer model simulations is often preferred. The radiative transfer models need to be well adapted to the type of canopy they target. Unfortunately, using a more realistic description of the canopy architecture requiring more variables may create problems in the inversion process if no additional prior information or constraints are exploited. Radiometric data driven approaches such as iterative minimization appear to be very appealing to handle a wide range of constraints and prior information that may be available at the decametric resolution. This may ultimately lead to the assimilation of calibrated radiances into structural-functional vegetation models coupled with atmospheric models which is currently in the infancy stage of development. The expected increasing accessibility of frequent decametric observations will certainly push investigations in such a direction, exploiting explicitly the whole set of available information and knowledge on physical and biological processes.

REFERENCES

Abdelbaki, A., Schlerf, M., Verhoef, W., & Udelhoven, T. (2019). Introduction of variable correlation for the improved retrieval of crop traits using canopy reflectance model inversion. *Remote Sensing*, *11*(22), 2681. https://doi.org/10.3390/rs11222681

Abuelgasim, A.A., Gopal, S., & Strahler, A.H. (1998). Forward and inverse modelling of canopy directional reflectance using a neural network. *International Journal of Remote Sensing*, *19*, 453–471.

Aires, F., Prigent, C., & Rossow, W.B. (2004). Neural network uncertainty assessment using Bayesian statistics: A remote sensing application. *Neural Computation*, *16*, 2415–2458.

Alcaraz-Segura, D., Liras, E., Tabik, S., Paruelo, J., & Cabello, J. (2010). Evaluating the consistency of the 1982–1999 NDVI trends in the Iberian Peninsula across four time-series derived from the AVHRR sensor: LTDR, GIMMS, FASIR, and PAL-II. *Sensors*, *10*, 1291–1314.

Asrar, G. (1989). *Theory and Applications of Optical Remote Sensing*. New-York: Wiley.

Asrar, G., Fuchs, M., Kanemasu, E.T., & Hatfield, J.L. (1984). Estimating absorbed photosynthetic radiation and leaf area index from spectral reflectance in wheat. *Agronomy Journal*, *76*, 300–306.

Atkinson, P.M., Jeganathan, C., Dash, J., & Atzberger, C. (2012). Inter-comparison of four models for smoothing satellite sensor time-series data to estimate vegetation phenology. *Remote Sensing of Environment*, *123*, 400–417.

Atkinson, P.M., & Tatnall, A.R.L. (1997). Neural network in remote sensing. *International Journal of Remote Sensing*, *18*, 699–709.

Atzberger, C. (2004). Object-based retrieval of biophysical canopy variables using artificial neural nets and radiative transfer models. *Remote Sensing of Environment*, *93*, 53–67.

Atzberger, C., & Richter, K. (2010). Vegetation biophysical variable retrieval using object-based inversion of hyperspectral CHRIS/PROBA data.

Bach, H., & Mauser, W. (1994). Modelling and model verification of the spectral reflectance of soils under varying moisture conditions. In *IGARSS'94* (pp. 2354–2356). Pasadena: IEEE.

Bacour, C. (2001). Contribtion à la détermination des paramètres biophysiques des couverts végétaux par inversion de modèles de réflectance: analyse de sensibilité et configurations optimales. In *Méthodes physiqes en télédétection* (p. 206). Paris: Université Paris 7—Denis Diderot.

Bacour, C., Baret, F., Béal, D., Weiss, M., & Pavageau, K. (2006). Neural network estimation of LAI, fAPAR, fCover and LAIxCab, from top of canopy MERIS reflectance data: Principles and validation. *Remote Sensing of Environment*, *105*, 313–325.

Bacour, C., Jacquemoud, S., Leroy, M., Hautecoeur, O., Weiss, M., Prévot, L., Bruguier, N., & Chauki, H. (2002a). Reliability of the estimation of vegetation characteristics by inversion of three canopy reflectance models on airborne POLDER data. *Agronomie*, *22*, 555–566.

Bacour, C., Jacquemoud, S., Tourbier, Y., Dechambre, M., & Frangi, J.P. (2002b). Design and Analysis of numerical experiments to compare four canopy reflectance models. *Remote Sensing of Environment*, *79*, 72–83.

Banari, A., Huete, A.R., Morin, D., & Zagolski, F. (1996). Effets de la couleur et de la brillance du sol sur les indices de végétation. *International Journal of Remote Sensing*, *17*, 1885–1906.

Baret, F., Bacour, C., Weiss, M., Pavageau, K., Béal, D., Bruniquel, V., Regner, P., Moreno, J., Gonzalez, C., & Chen, J. (2004). Canopy biphysical variables estimation from MERIS observations based on neural networks and radiative transfer modelling: Principles and validation. In ESA (Ed.), *ENVISAT conference*, Salzburg: ESA.

Baret, F., & Buis, S. (2007). Estimating canopy characteristics from remote sensing observations. Review of methods and associated problems. In S. Liang (Ed.), *Advances in Land Remote Sensing: System, Modeling, Inversion and Application* (pp. 171–200). New York: Springer.

Baret, F., Clevers, J.G.P.W., & Steven, M.D. (1995). The robustness of canopy gap fraction estimates from red and near infrared reflectances: A comparison of approaches. *Remote Sensing of the Environment*, *54*, 141–151.

Baret, F., De Solan, B., Lopez-Lozano, R., Ma, K., & Weiss, M. (2010). GAI estimates of row crops from downward looking digital photos taken perpendicular to rows at 57.5° zenith angle. Theoretical considerations based on 3D architecture models and application to wheat crops. *Agricultural and Forest Meteorology*, *150*, 1393–1401.

Baret, F., & Guyot, G. (1991). Potentials and limits of vegetation indices for LAI and APAR assessment. *Remote Sensing of the Environment*, *35*, 161–173.

Baret, F., Hagolle, O., Geiger, B., Bicheron, P., Miras, B., Huc, M., Berthelot, B., Nino, F., Weiss, M., Samain, O., Roujean, J.L., & Leroy, M. (2007). LAI, fAPAR and fCover CYCLOPES global products derived from VEGETATION: Part 1: Principles of the algorithm. *Remote Sensing of Environment*, *110*, 275–286.

Baret, F., Hagolle, O., Geiger, B., Bicheron, P., Miras, B., Huc, M., Berthelot, B., Weiss, M., Samain, O., Roujean, J.L., & Leroy, M. (2007). LAI, fAPAR and fCover CYCLOPES global products derived from VEGETATION. Part 1: Principles of the algorithm. *Remote Sensing of Environment*, *110*, 275–286.

Baret, F., Knyazikhin, J., Weiss, M., Myneni, R., & Pragnère, A. (1999). Overview of canopy biophysical variables retrieval techniques. In *ALPS'99*. Meribel.

Baret, F., Weiss, M., Bicheron, P., & Bethelot, B. (2009). S2PAD—Sentinel-2 MSI—Level 2B Products Algorithm Theoretical Basis Document. In Vega, GMBH.

Baret, F., Weiss, M., Lacaze, R., Camacho, F., Makhmara, H., Pacholcyzk, P., & Smets, B. (2013). GEOV1: LAI and FAPAR essential climate variables and FCOVER global time series capitalizing over existing products. Part1: Principles of development and production. *Remote Sensing of Environment*, *137*, 299–309.

Baret, F., Weiss, M., Troufleau, D., Prévot, L., & Combal, B. (2000). Maximum information exploitation for canopy characterisation by remote sensing. *Remote Sensing in Agriculture* (pp. 71–82), Association of Applied Biologists.

Bayes, T., & Price, R. (1763). An essay towards solving a problem in the doctrine of chance. By the late Rev. Mr. Bayes, communicated by Mr. Price, in a letter to John Canton, A. M. F. R. S. *Philosophical Transactions of the Royal Society of London*, *53*, 370–418.

Beck, H.E., McVicar, T.R., van Dijk, A.I., Schellekens, J., de Jeu, R.A., & Bruijnzeel, L.A. (2011). Global evaluation of four AVHRR—NDVI data sets: Intercomparison and assessment against Landsat imagery. *Remote Sensing of Environment*, *115*, 2547–2563.

Begué, A., Desprat, J.F., Imbernon, J., & Baret, F. (1991). Radiation use efficiency of pearl millet in the Sahelian zone. *Agricultural and Forest Meteorology*, *56*, 93–110.

Bessemoulin, P., Bourdette, Shabanov, N., Courtier, P., & Manach, J. (2004). La canicule d'aout 2003 en France et en Europe. *La météorologie*, *46*, 25–33.

Beven, K.J., & Binsley, A.M. (1992). The future of distributed models: Model calibration and uncertainty predictions. *Hydrological Processes*, *6*, 279–298.

Bicheron, P., & Leroy, M. (1999). A method of biophysical parameter retrieval at global scale by inversion of a vegetation reflectance model. *Remote Sensing of Environment*, *67*, 251–266.

Blaschke, T. (2010). Object based image analysis for remote sensing. *ISPRS Journal of Photogrammetry and Remote Sensing*, *65*, 2–16.

Bréon, F.M., Maignan, F., Leroy, M., & Grant, I. (2002). Analysis of hot spot directional signatures measured from space. *Journal of Geophysical Research*, *107*, AAC 11–15.

Camacho, F., Cernicharo, J., Lacaze, R., Baret, F., & Weiss, M. (2013). GEOV1: LAI, FAPAR essential climate variables and FCOVER global time series capitalizing over existing products. Part 2: Validation and intercomparison with reference products. *Remote Sensing of Environment*, *137*, 310–329.

Cardot, H., Maisongrande, P., & Faivre, R. (2008). Varying-time random effects models for longitudinal data: Unmixing and temporal interpolation of remote sensing data. *Journal of Applied Statistics*, *35*, 827–846.

Chaabouni, S., Kallel, A., & Houborg, R. 2021. Improving retrieval of crop biophysical properties in dryland areas using a multi-scale variational RTM inversion approach. *International Journal of Applied Earth Observation and Geoinformation*, *94*, 102220, ISSN 1569–8432, https://doi.org/10.1016/j.jag.2020.102220. (www.sciencedirect.com/science/article/pii/S0303243420308631)

Chen, J., Jönsson, P., Tamura, M., Gu, Z., Matsushita, B., & Eklundh, L. (2004). A simple method for reconstructing a high quality NDVI time series data set based on the Savitzky-Golay filter. *Remote Sensing of Environment*, *91*, 332–344.

Chen, J.M., & Black, T.A. (1992). Defining leaf area index for non-flat leaves. *Plant, Cell and Environment*, *15*, 421–429.

Chen, J.M., Menges, C.H., & Leblanc, S.G. (2005). Global mapping of foliage clumping index using multi-angular satellite data. *Remote Sensing of Environment*, *97*, 447–457.

Chen, J.M., Pavlic, G., Brown, L., Cihlar, J., Leblanc, S., White, H.P., Hall, R.J., Peddle, D.R., King, D.J., Trofymov, J.A., Swift, E., Van der Sanden, J., & Pellika, P.K.E. (2002). Derivation and validation of Canada wide coarse resolution leaf area index maps using high resolution satellite imagery and ground measurements. *Remote Sensing of Environment*, *80*, 165–184.

Ciais, P., Reichstein, M., Viovy, N., Granier, A., Ogée, J., Allard, V., Aubinet, M., Buchmann, N., Bernhofer, C., Carrara, A., Chevallier, F., De Noblet, N., Friend, A.D., Friedlingstein, P., Grünwald, T., Heinesch, B., Keronen, P., Knohl, A., Krinner, G., Loustau, D., Manca, G., Matteucci, G., Miglietta, F., Ourcival, J.M., Papale, D., Pilegaard, K., Rambal, S., Seufert, G., Soussana, J.F., Sanz, M.J., Schulze, E.D., Vesala, T., & Valentini, R. (2005). Europe wide reduction in primary productivity caused by the heat and draught in 2003. *Nature*, *437*, 529–533.

Cierniewski, J., Verbrugghe, M., & Marlewski, A. (2002). Effects of farming works on soil surface bidirectional reflectance measurements and modelling. *International Journal of Remote Sensing*, *23*, 1075–1094.

Claverie, M., Vermote, E., Weiss, M., Baret, F., Hagolle, O., & Demarez, V. (2013). Validation of coarse spatial resolution LAI and FAPAR time series over cropland in southwest France. *Remote Sensing of Environment*, *139*, 216–230.

Comar, A., Baret, F., Obein, G., Simonot, L., Meneveaux, D., Vienot, F., & de Solan, B. (2014). ACT: A leaf BRDF model taking into account the azimuthal anisotropy of monocotyledonous leaf surface. *Remote Sensing of Environment*, *143*, 112–121.

Combal, B., Baret, F., & Weiss, M. (2001). Improving canopy variables estimation from remote sensing data by exploiting ancillary information. Case study on sugar beet canopies. *Agronomie*, *22*, 2–15.

Combal, B., Baret, F., Weiss, M., Trubuil, A., Macé, D., Pragnère, A., Myneni, R., Knyazikhin, Y., & Wang, L. (2002). Retrieval of canopy biophysical variables from bi-directional reflectance data. Using prior information to solve the ill-posed inverse problem. *Remote Sensing of Environment*, *84*, 1–15.

Combal, B., Ochshepkov, S.L., Sinyuk, A., & Isaka, H. (2000). Statistical framework of the inverse problem in the retrieval of vegetation parameters. *Agronomie*, *20*, 65–77.

Danner, M., Berger, K., Wocher, M., Mauser, W., and Hank, T. 2017. Retrieval of biophysical crop variables from multi-angular canopy spectroscopy. *Remote Sensing*, *9*(7), 726. https://doi.org/10.3390/rs9070726

Danson, F.M., Rowland, C.S., & Baret, F. (2003). Training a neural network with a canopy reflectance model to estimate crop leaf area index. *International Journal of Remote Sensing*, *24*, 4891–4905.

Darvishzadeh, R., Skidmore, A., Schlerf, M., & Atzberger, C. (2008). Inversion of a radiative transfer model for estimating vegetation LAI and chlorophyll in a heterogeneous grassland. *Remote Sensing of Environment*, *112*, 2592.

Dawson, T.P., Curran, P.J., & Plummer, S.E. (1998). LIBERTY-Modeling the effects of leaf biochemical concentration on reflectance spectra. *Remote Sensing of Environment*, *65*, 50–60.

de Jong, R., Verbesselt, J., Schaepman, M., & De Bruin, S. (2012). Trend changes in global greening and browning: Contribution of short-term trends to longer-term change. *Global Change Biology*, *18*, 642–655.

Deng, F., Chen, J.M., Chen, M., & Pisek, J. (2006). Algorithm for global leaf area index retrieval using satellite imagery. *IEEE Transactions on Geoscience and Remote Sensing*, *44*, 2219–2229.

Duveiller, G., Weiss, M., Baret, F., & Defourny, P. (2011). Retrieving wheat Green Area Index during the growing season from optical time series measurements based on neural network radiative transfer inversion. *Remote Sensing of Environment*, *115*, 887.

Faivre, R., & Fischer, A. (1997). Predicting crop reflectances using satellite data observing mixed pixels. *Journal of Agricultural,Biological and Environmental Statistics*, *2*, 87–107.

Fang, H., Jiang, C., Li, W., Wei, S., Baret, F., Chen, J.M., Garcia-Haro, J., Liang, S., Liu, R., Myneni, R.B., Pinty, B., Xiao, Z., & Zhu, Z. (2013). Characterization and intercomparison of global moderate resolution leaf area index (LAI) products: Analysis of climatologies and theoretical uncertainties. *Journal of Geophysical Research: Biogeosciences*, *118*, 529–548.

Fang, H., & Liang, S. (2003). Retrieving leaf area index with a neural network method: Simulation and validation. *IEEE Transactions on Geoscience and Remote Sensing*, *41*, 2052–2062.

Fang, H., & Liang, S. (2005). A hybrid inversion method for mapping leaf area index from MODIS data: Experiments and application to broadleaf and needleleaf canopies. *Remote Sensing of Environment*, *94*, 405–424.

Fang, H., Liang, S., & Kuusk, A. (2003). Retrieving leaf area index using a genetic algorithm with a canopy radiative transfer model. *Remote Sensing of Environment*, *85*, 257–270.

Fasbender, D., Radoux, J., & Bogaert, P. (2008). Bayesian data fusion for adaptable image pansharpening. *IEEE Transactions on Geoscience and Remote Sensing*, *46*, 1847–1857.

Fensholt, R., Langanke, T., Rasmussen, K., Reenberg, A., Prince, S.D., Tucker, C., Scholes, R.J., Le e, Q.B., Bondeau, A., Eastman, R., Epstein, H., Gaughan, A.E., Hellden, U., Mbow, C., Olsson, L., Paruelo, J., Schweitzer, C., Seaquist, J., & Wessels, K.J. (2012). Greenness in semi-arid areas across the globe 1981–2007—an Earth Observing Satellite based analysis of trends and drivers. *Remote Sensing of Environment*, *212*, 144–158.

Fernandes, R., & Leblanc, S.G. (2005). Parametric (modified least squares) and non-parametric (Theil-Sen) linear regressions for predicting biophysical parameters in the presence of measurements errors. *Remote Sensing of Environment*, *95*, 303–316.

Ganguly, S., Friedl, M.A., Tan, B., Zhang, X., & Verma, M. (2010). Land surface phenology from MODIS: Characterization of the Collection 5 global land cover dynamics product. *Remote Sensing of Environment*, *114*, 1805.

Ganguly, S., Nemani, R.R., Zhang, G., Hashimoto, H., Milesi, C., Michaelis, A., Wang, W., Votava, P., Samanta, A., Melton, F., Dungan, J.L., Vermote, E., Gao, F., Knyazikhin, Y., & Myneni, R.B. (2012). Generating global leaf area index from Landsat: Algorithm formulation and demonstration. *Remote Sensing of Environment*, *122*, 185–202.

Gao, F., Masek, J.G., Schwaller, M., & Forrest, H. (2006). On the blending of the Landsat and MODIS surface reflectance. *IEEE Transactions on Geoscience and Remote Sensing*, *44*, 2207–2219.

Garabedian, P. (1964). *Partial Differential Equations*. New York: Wiley.

Garrigues, S., Lacaze, R., Baret, F., Morisette, J., Weiss, M., Nickeson, J., Fernandes, R., Plummer, S., Shabanov, N.V., Myneni, R., & Yang, W. (2008). Validation and intercomparison of global leaf area index products derived from remote sensing data. *Journal of Geophysical Research*, *113*.

GCOS (2011). Global Climate Observing System—Systematic Observation Requirements for Satellite-Based Products for Climate—2011 Update, Supplemental Details to the Satellite Based Component of the Implementation Plan for the Global Observing System for Climate in Support of the UNFCCC (2010 Update). In (p. 138). Geneva, Switzerland: World Meteorological Organization.

Gilbert, J.C. (2002). *Optimisation différentiable: Théorie et algorithmes*.

Gobron, N., Pinty, B., Aussedat, O., Chen, J.M., Cohen, W.B., Fensholt, R., Gond, V., Huemmrich, K.F., Lavergne, T., Mélin, F., Privette, J.L., Sandholt, I., Taberner, M., Turner, D.P., Verstraete, M.M., & Widlowski, J.L. (2006). Evaluation of fraction of absorbed photosynthetically active radiation products for different canopy radiation transfer regimes: Methodology and results using Joint Research Center products derived from SeaWiFS against ground-based estimations. *Journal of Geophysical Research*, *111*. https://doi.org/10.1029/2005JD006511

Gobron, N., Pinty, B., Aussedat, O., Taberner, M., Faber, O., Melin, F., Lavergne, T., Robustelli, M., & Snoeij, P. (2008). Uncertainty estimates for the FAPAR operational products derived from MERIS -Impact of top-of-atmosphere radiance uncertainties and validation with field data. *Remote Sensing of Environment, 112*, 1871–1883.

Gobron, N., Pinty, B., Verstraete, M., & Widlowski, J.L. (2000). Advanced vegetation indices optimized for up-coming sensors: Design, performances and applications. *IEEE Transactions on Geoscience and Remote Sensing, 38*, 2489–2505.

Goel, N.S. (1989). Inversion of canopy reflectance models for estimation of biophysical parameters from reflectance data. In G. Asrar (Ed.), *Theory and Applications of Optical Remote Sensing* (pp. 205–251). Hoboken, NJ: Wiley Interscience.

Goel, N.S., & Deering, D.W. (1985). Evaluation of a canopy reflectance model for LAI estimation through its inversion. *IEEE Transactions on Geoscience and Remote Sensing, GE-23*, 674–684.

Goel, N.S., Strebel, D.E., & Thompson, R.L. (1984). Inversion of vegetation canopy reflectance models for estimating agronomic variables. II. Use of angle transforms and error analysis as illustrated by the SUIT model. *Remote Sensing of Environment, 14*, 77–111.

Goel, N.S., & Thompson, R.L. (1984). Inversion of vegetation canopy reflectance models for estimating agronomic variables. IV. Total inversion of the SAIL model. *Remote Sensing of Environment, 15*, 237–253.

Gong, P., Wang, D.X., & Liang, S. (1999). Inverting a canopy reflectance model using a neural network. *International Journal of Remote Sensing, 20*, 111–122.

Green, R.O., & Boardman, J. (2001). Exploration of the relationship between information content and signal to noise ratio and spatial resolution in AVIRIS spectral data. In R.O. Gree (Ed.), *AVIRIS Workshop*. Pasadena, CA: Jet Propulsion Laboratory.

Hagolle, O., Dedieu, G., Mougenot, B., Debaecker, V., Duchemin, B., & Meygret, A. (2008). Correction of aerosol effects on multi-temporal images acquired with constant viewing angles: Application to Formosat-2 images. *Remote Sensing of Environment, 112*, 1689.

Hagolle, O., Lobo, A., Maisongrande, P., Cabot, F., Duchemin, B., & De Pereyra, A. (2005). Quality assessment and improvement of temporally composited products of remotely sensed imagery by combination of VEGETATION 1 and 2 images. *Remote Sensing of Environment, 94*, 172–186.

Hapke, B. (1981). Bidirectional reflectance spectroscopy. 1. Theory. *Journal of Geophysical Research, 86*, 3039–3054.

Herrmann, S.M., Anyamba, A., & Tucker, C.J. (2005). Recent trends in vegetation dynamics in the African Sahel and their relationship to climate. *Global Environmental Change, 15*, 394–404.

Houborg, R., & Boegh, E. (2008). Mapping leaf chlorophyll and leaf area index using inverse and forward canopy reflectance modeling and SPOT reflectance data. *Remote Sensing of Environment, 112*, 186.

Huang, D., Yang, W., Tan, B., Rautiainen, M., Zhang, P., Hu, J., Shabanov, N., Linder, S., Knyazikhin, Y., & Myneni, R. (2006). The importance of measurement error for deriving accurate reference leaf area index maps and validation of the MODIS LAI products. *IEEE Transactions on Geoscience and Remote Sensing, 44*, 1866–1871.

Huete, A.R. (1988). A soil adjusted vegetation index (SAVI). *Remote Sensing of the Environment, 25*, 295–309.

Huete, A.R., Liu, H.Q., Batchily, K., & van Leeuwen, W. (1997). A comparison of vegetation indices over a global set of TM images for EOS-MODIS. *Remote Sensing of Environment, 59*, 440–451.

Huete, A.R., & Lui, H.Q. (1994). An error and sensitivity analysis of the atmospheric and soil correcting variants of the NDVI for the MODIS-EOS. *IEEE Transactions on Geoscience and Remote Sensing, 32*, 897–905.

Jacquemoud, S., & Baret, F. (1990). PROSPECT: A model of leaf optical properties spectra. *Remote Sensing of Environment, 34*, 75–91.

Jacquemoud, S., Baret, F., Andrieu, B., Danson, M., & Jaggard, K. (1995). Extraction of vegetation biophysical parameters by inversion of the PROSPECT+SAIL model on sugar beet canopy reflectance data. Application to TM and AVIRIS sensors. *Remote Sensing of the Environment, 52*, 163–172.

Jacquemoud, S., Baret, F., & Hanocq, J.F. (1992). Modeling spectral and directional soil reflectance. *Remote Sensing of the Environment, 41*, 123–132.

Jacquemoud, S., Verhoef, W., Baret, F., Bacour, C., Zarco-Tejada, P.J., Asner, G.P., François, C., & Ustin, S.L. (2009). PROSPECT + SAIL models: A review of use for vegetation characterization. *Remote Sensing of Environment, 113*, S56–S66.

Jamet, C., Thiria, S., Moulin, C., & Crepon, M. (2005). Use of a neurovariational inversion for retrieving oceanic and atmospheric constituents from ocean color imagery: A feasibility study. *Journal of Atmospheric and Oceanic Technology*, *22*, 460–475.

Jonckheere, I., Fleck, S., Nackaerts, K., Muys, B., Coppin, P., Weiss, M., & Baret, F. (2004). Review of methods for in situ leaf area index determination. Part I: Theories, sensors and hemispherical photography. *Agricultural and Forest Meteorology*, *121*, 19–35.

Jönsson, P., & Eklundh, L. (2004). TIMESAT—a program for analyzing time series of satellite sensor data. *Computers and Geosciences*, *30*, 833–845.

Kandasamy, S., Baret, F., Verger, A., Neveux, P., & Weiss, M. (2013). A comparison of methods for smoothing and gap filling time series of remote sensing observations: Application to MODIS LAI products. *Biogeosciences*, *10*, 4055–4071.

Kandasamy, S., Lopez-Lozano, R., Baret, F., & Rochdi, N. (2010). The effective nature of LAI as measured from remote sensing observations. In *Proceedings of the 2010IEEE International Geoscience and Remote Sensing Symposium* (pp. 789–792). Honolulu.

Karkee, M., Steward, B.L., Tang, L., & Aziz, S.A. (2009). Quantifying sub-pixel signature of paddy rice field using an artificial neural network. *Computers and Electronics in Agriculture*, *65*, 65–76.

Kimes, D.S., Knyazikhin, Y., Privette, J.L., Abuelgasim, A.A., & Gao, F. (2000). Inversion methods for physically-based models. *Remote Sensing Reviews*, *18*, 381–439.

Kimes, D.S., Nelson, R.F., Manry, M.T., & Fung, A.K. (1998). Attributes of neural networks for extracting continuous vegetation variables from optical and radar measurements. *International Journal of Remote Sensing*, *19*, 2639–2663.

Knyazikhin, Y., Glassy, J., Privette, J.L., Tian, Y., Lotsch, A., Zhang, Y., Wang, Y., Morissette, J.T., Votava, P., Myneni, R.B., Nemani, R.R., & Running, S.W. (1999). MODIS Leaf area index (LAi) and fraction of photosynthetically active radiation absorbed by vegetation (FPAR) product (MOD15) algorithm theoretical basis document. In: http://eospso.gsfc.nasa.gov/atbd/modistables.html

Knyazikhin, Y., Martonchik, J.V., Myneni, R.B., Diner, D.J., & Running, S.W. (1998). Synergistic algorithm for estimating vegetation canopy leaf area index and fraction of absorbed photosynthetically active radiation from MODIS and MISR data. *Journal of Geophysical Research*, *103*, 32257–32275.

Kötz, B., Baret, F., Poilvé, H., & Hill, J. (2005). Use of coupled canopy structure dynamic and radiative transfer models to estimate biophysical canopy characteristics. *Remote Sensing of Environment*, *95*, 115–124.

Kuusk, A. (1991a). Determination of vegetation canopy parameters from optical measurements. *Remote Sensing of the Environment*, *37*, 207–218.

Kuusk, A. (1991b). The inversion of the Nilson-Kuusk canopy reflectance model, a test case. In *Int. Geosci. and Remote Sens. Symp. (IGARSS'91)* (pp. 1547–1550). Helsinki (Finland).

Laurent, V.C., Verhoef, W., Damm, A., Schaepman, M.E., & Clevers, J.G. (2013). A Bayesian object-based approach for estimating vegetation biophysical and biochemical variables from APEX at-sensor radiance data. *Remote Sensing of Environment*, *139*, 6–17.

Laurent, V.C.E., Verhoef, W., Clevers, J.G.P.W., & Schaepman, M.E. (2011). Inversion of a coupled canopy-atmosphere model using multi-angular top-of-atmosphere radiance data: A forest case study. *Remote Sensing of Environment*, *115*, 2603–2612.

Lauvernet, C. (2005). Assimilation variationnelle d'observations de télédétection dans les modèles de fonctionnement de la végétation: utilisation du modèle adjoiint et prise en compte de contraintes spatiales. In *Mathématiques appliquées* (p. 205). Grenoble: Université Joseph Fourier.

Lauvernet, C., Baret, F., Hascoët, L., Buis, S., & Le Dimet, F.X. (2008). Multitemporal-patch ensemble inversion of coupled surface-atmosphere radiative transfer models for land surface characterization. *Remote Sensing of Environment*, *112*, 851–861.

Leprieur, C., Verstraete, M.M., & Pinty, B. (1994). Evaluation of the performance of various vegetation indices to retrieve vegetation cover from AVHRR data. *Remote Sensing Reviews*, *10*, 265–284.

Lewis, P., Gómez-Dans, J., Kaminski, T., Settle, J., Quaife, T., Gobron, N., Styles, J., & Berger, M. (2012). An Earth Observation Land Data Assimilation System (EO-LDAS). *Remote Sensing of Environment*, *120*, 219–235.

Liang, S. (2004). *Quantitative Remote Sensing of Land Surfaces*.

Liang, S., & Townshend, J.R.G. (1996). A parametric soil BRDF model: A four stream approximation for multiple scattering. *International Journal of Remote Sensing*, *17*, 1303–1315.

Liu, L., Li, S., Yang, W., Wang, X., Luo, X., Ran, P., & Zhang, H. 2023. Forest canopy water content monitoring using radiative transfer models and machine learning. *Forests*, *14*(7), 1418. https://doi.org/10.3390/f14071418

Liu, W., Baret, F., Gu, X., Tong, Q., Zheng, L., & Zhang, B. (2002). Relating soil surface moisture to reflectance. *Remote Sensing of Environment*, *81*, 238–246.

Lopez-Lozano, R. (2008). Tecnologias de informacion geografica en la cartografia de parametros biofisicos de parcelas de maiz y vina para agricultura de precision. In *Geografia y ordenacion del territorio* (p. 211). Zaragoza: Universidad de Zaragoza.

Lucht, W. (1998). Expected retrieval accuracies of bidirectional reflectance and albedo from EOS-MODIS and MISR angular sampling. *Journal of Geophysical Research*, *103*, 8763–8778.

Mackay, D.J.C. (Ed.) (2003). *Information Theory, Inference and Learning Algorithms*. Cambridge, UK: Cambridge University Press.

Makowski, D., Wallach, D., & Tremblay, M. (2002). Using Bayesian approach to parameter estimation; comparison of the GLUE and MCMC methods. *Agronomie*, *22*, 191–203.

Malenovský, Z., Rott, H., Cihlar, J., Schaepman, M.E., García-Santos, G., Fernandes, R., & Berger, M. (2012). Sentinels for science: Potential of Sentinel-1, -2, and -3 missions for scientific observations of ocean, cryosphere, and land. *Remote Sensing of Environment*, *120*, 91–101.

Martínez-Ferrer, M., Moreno-Martínez, A., Campos-Taberner, M., García-Haro, J., Muñoz-Marí, F.J., Running, S.W., Kimball, J., Clinton, N., Camps-Valls, G. 2022. Quantifying uncertainty in high resolution biophysical variable retrieval with machine learning. *Remote Sensing of Environment*, *280*, 113199, ISSN 0034–4257, https://doi.org/10.1016/j.rse.2022.113199. (www.sciencedirect.com/science/article/pii/S0034425722003091)

Martonchik, J.V. (1994). Retrieval of surface directional reflectance properties using ground level multiangle measurements. *Remote Sensing of Environment*, *50*, 303–316.

McCallum, I., Wagner, W., Schmullius, C., Shvidenko, A., Obersteiner, M., Fritz, S., & Nilsson, S. (2009). Satellite-based terrestrial production efficiency modeling. *Carbon Balance and Management*, *4*(8).

Miller, J.B. (1967). A formula for average foliage density. *Australian Journal of Botany*, *15*, 141–144.

Moorthy, I., Miller, J.R., & Noland, T.L. (2008). Estimating chlorophyll concentration in conifer needles with hyperspectral data: An assessment at the needle and canopy level. *Remote Sensing of Environment*, *112*, 2824–2838.

Morisette, J., Baret, F., Privette, J.L., Myneni, R.B., Nickeson, J., Garrigues, S., Shabanov, N., Weiss, M., Fernandes, R., Leblanc, S., Kalacska, M., Sanchez-Azofeifa, G.A., Chubey, M., Rivard, B., Stenberg, P., Rautiainen, M., Voipio, P., Manninen, T., Pilant, D., Lewis, T., Iiames, J., Colombo, R., Meroni, M., Busetto, L., Cohen, W., Turner, D., Warner, D., Petersen, G.W., Seufert, G., & Cook, R. (2006). Validation of global moderate resolution LAI Products: A framework proposed within the CEOS Land Product Validation subgroup. *IEEE Transactions on Geoscience and Remote Sensing*, *44*, 1804–1817.

Morisette, J.T. (2010). Toward a standard nomenclature for imagery spatial resolution. *International Journal of Remote Sensing*, *31*, 2347–2349.

Mõttus, M., Sulev, M., Baret, F., Reinart, A., & Lopez, R. (2011). Photosynthetically active radiation: Measurement and modeling. In R. Meyers (Ed.), *Encyclopedia of Sustainability Science and Technology* (pp. 7902–7932). New York: Springer.

Myneni, R.B., Gutschick, V.P., Asrar, G., & Kanemasu, E.T. (1988). Photon transport in vegetation canopies with anisotropic scattering part II. Discrete-ordinates/exact-kernel technique for one-angle photon transport in slab geometry. *Agricultural and Forest Meteorology*, *42*, 17–40.

Myneni, R.B., Hoffman, S., Knyazikhin, Y., Privette, J.L., Glassy, J., Tian, Y., Wang, Y., Song, X., Zhang, Y., Smith, G.R., Lotsch, A., Friedl, M., Morisette, J.T., Votava, P., Nemani, R.R., & Running, S.W. (2002). Global products of vegetation leaf area and absorbed PAR from year one of MODIS data. *Remote Sensing of Environment*, *83*, 214–231.

Nagol, J.R., Vermote, E.F., & Prince, S.D. (2009). Effects of atmospheric variation on AVHRR NDVI data. *Remote Sensing of Environment*, *113*, 392–397.

Nelder, J.A., & Mead, R.A. (1965). A simplex method for function optimization. *Computer Journal*, *7*, 308–313.

Peña-Barragán, J.M., Ngugi, M.K., Plant, R.E., & Six, J. (2011). Object-based crop identification using multiple vegetation indices, textural features and crop phenology. *Remote Sensing of Environment*, *115*, 1301–1316.

Pinty, B., Jung, M., Kaminski, T., Lavergne, T., Mund, M., Plummer, S., Thomas, E., & Widlowski, J.L. (2011). Evaluation of the JRC-TIP 0.01° products over a mid-latitude deciduous forest site. *Remote Sensing of Environment*, *115*, 3567–3581.

Pinty, B., & Verstraete, M.M. (1991a). Extracting information on surface properties from bidirectional reflectance measurements. *Journal of Geophysical Research*, *96*, 2865–2874.

Pinty, B., & Verstraete, M.M. (1991b). Extracting information on surface properties from bidirectional reflectance measurements. *Journal of Geophysical Research*, *96*, 2865–2874.

Pinty, B., Verstraete, M.M., & Dickinson, R.E. (1990). A physical model of the bidirectional reflectance of vegetation canopies. 2. Inversion and validation. *Journal of Geophysical Research*, *95*, 11767–11775.

Price, J.C. (1990). On the information content of soil reflectance spectra. *Remote Sensing of Environment*, *33*, 113–121.

Price, J.C. (1994). How unique are spectral signatures? *Remote Sensing of the Environment*, *49*, 181–186.

Privette, J.L., Emery, W.J., & Schimel, D.S. (1996). Inversion of a vegetation reflectance model with NOAA AVHRR data. *Remote Sensing of Environment*, *58*, 187–200.

Qu, Y., Wang, J., Wan, H., Li, X., & Zhou, G. (2008). A Bayesian network algorithm for retrieving the characterization of land surface vegetation. *Remote Sensing of Environment*, *112*, 613–622.

Raymaekers, D., Garcia, A., Di Bella, C., Beget, M.E., Llavallol, C., Oricchio, P., Straschnoy, J., Weiss, M., & Baret, F. (2014). SPOT-VEGETATION GEOV1 biophysical parameters in semi-arid agro-ecosystems. *International Journal of Remote Sensing*, *35*, 2534–2547.

Refslund, J., Dellwik, E., Hahmann, A., Barlage, M., & Boegh, E. (2013). Development of satellite green vegetation fraction time series for use in mesoscale modeling: Application to the European heat wave 2006. *Theoretical and Applied Climatology*, 1–16.

Renders, J.-M., & Flasse, S.P. (1996). Hybrid methods using genetic algorithms for global optimization. *IEEE Transactions on Systems, Man, and Cybernetics*, *26*, 243–258.

Richardson, A.J., & Wiegand, C.L. (1977). Distinguishing vegetation from soil background information. *Photogrammetric Engineering and Remote Sensing*, *43*, 1541–1552.

Richardson, A.J., Wiegand, C.L., Wanjura, D.F., Dusek, D., & Steiner, J.L. (1992). Multisite analyses of spectral-biophysical data for sorghum. *RSE*, *41*(71), 82.

Rondeaux, G., Steven, M.D., & Baret, F. (1996). Optimization of soil adjusted vegetation indices. *Remote Sensing of the Environment*, *55*, 95–107.

Roujean, J.L., & Lacaze, R. (2002). Global mapping of vegetation parameters from POLDER multiangular measurements for studies of surface-atmosphere interactions: A pragmatic method and validation. *Journal of Geophysical Research*, *107*, ACL 6 1–14.

Russel, G., Jarvis, P.G., & Monteith, J.L. (1989). Absorption of radiation by canopies and stand growth. In G. Russel, B. Marshall, & P.G. Jarvis (Eds.), *Plant Canopies: Their Growth, Form and Function* (pp. 21–39). New York: Cambridge University Press.

Ryu, Y., Nilson, T., Kobayashi, H., Sonnentag, O., Law, B.E., & Baldocchi, D.D. (2010). On the correct estimation of effective leaf area index: Does it reveal information on clumping effects? *Agricultural and Forest Meteorology*.

Samain, O., Roujean, J.L., & Geiger, B. (2007). Use of Kalman filter for the retrieval of surface BRDF coefficients with time-evolving model based on ECOCLIMAP land cover classification. *Remote Sensing of Enviroment, submitted.*

Schiefer, F., Schmidtlein, S., & Kattenborn, T. 2021. The retrieval of plant functional traits from canopy spectra through RTM-inversions and statistical models are both critically affected by plant phenology. *Ecological Indicators*, 121, 107062, ISSN 1470–160X, https://doi.org/10.1016/j.ecolind.2020.107062. (www.sciencedirect.com/science/article/pii/S1470160X20310013)

Sellers, P.J. (1985). Canopy reflectance photosynthesis and transpiration. *International Journal of Remote Sensing*, *3*, 1335–1372.

Settle, J. (2004). On the dimensionality of multi-view hyperspectral measurements of vegetation. *Remote Sensing of Enviroment*, *90*, 235–242.

Shabanov, N.V., Huang, D., Yang, W., Tan, B., Knyazikhin, Y., Myneni, R.B., Ahl, D.E., Gower, S.T., & Huete, A.R. (2005). Analysis and optimization of the MODIS leaf area index algorithm retrievals over broadleaf forests. *IEEE Transactions on Geoscience and Remote Sensing*, *43*, 1855–1865.

Shoshany, M. (1991). The equifinality of bidirectional distribution functions of various microstructures. *International Journal of Remote Sensing*, *12*, 2267–2281.

Simic, A., & Chen, J.M. (2008). Refining a hyperspectral and multiangle measurement concept for vegetation structure assessment. *Canadian Journal of Remote Sensing*, *34*, 174–171.

Smith, J.A. (1992). LAI inversion from optical reflectance using neural network trained with multiple scattering model. In W. R. (Ed.), *IGARS'92* (pp. 757–759). South Shore Harbour Resort and Conference Center, Houston, Texas, USA.

Smith, J.A. (1993). LAI inversion using backpropagation neural network trained with multiple scattering model. *IEEE Transactions on Geoscience and Remote Sensing*, *31*, 1102–1106.

Specht, D.F. (1991). A general regression neural network. *IEEE Transactions on Neural Networks*, *2*, 568–576.

Stenberg, P. (2006). A note on the G-function for needle leaf canopies. *Agricultural and Forest Meteorology*, *136*, 76.

Sun, W., Stephens, J., & Sweeting, M. (2002). Micro-mini-satellites for affordable EO constellations: Rapid-eye & DMC. *Photogrammetrie Fernerkundung Geoinformation*, 31–36.

Sun, J., Wang, L., Shi, S., Li, Z., Yang, J., Gong, W., Wang, S., & Tagesson, T. 2022. Leaf pigment retrieval using the PROSAIL model: Influence of uncertainty in prior canopy-structure information. *The Crop Journal*, *10*(5), 1251–1263, ISSN 2214–5141, https://doi.org/10.1016/j.cj.2022.04.003. (www.sciencedirect.com/science/article/pii/S2214514122000873)

Sun, S., Shi, S., Yang, J., Du, L., Gong, W., Chen, B., & Song, S. 2018. Analyzing the performance of PROSPECT model inversion based on different spectral information for leaf biochemical properties retrieval. *ISPRS Journal of Photogrammetry and Remote Sensing*, 135, 74–83, ISSN 0924–2716, https://doi.org/10.1016/j.isprsjprs.2017.11.010. (www.sciencedirect.com/science/article/pii/S0924271617303465)

Tarantola, A. (2005). *Inverse Problem Theory Problem and Methods for Model Parameter Estimation*. Society for Industrial and Applied Mathematics.

Tarentola, A. (1987). *Inverse Problem Theory:Methods for Data Fitting and Model Parameter Estimation*. Amsterdam, The Netherlands: Elsevier Science Publisher B.V.

Teillet, P.M., Gauthier, R.P., Staenz, K., & Fournier, R.A. (1997). BRDF equifinality studies in the context of forest canopies. In G. Guyot & T. Phulpin (Eds.), *Physical Measurements and Signatures in Remote Sensing* (pp. 163–170). Courchevel: Balkema, Rotterdam.

Tenenbaum, J.B., de Silva, V., & Langford, J.C. (2000). A global geometric framework for nonlinear dimensionality reduction. *Science*, *290*, 2319–2323.

Thenkabail, P.S., Enclona, E.A., Ashton, M.S., & Van Der Meer, B. (2004). Accuracy assessments of hyperspectral waveband performance for vegetation analysis applications. *Remote Sensing of Enviroment*, *91*, 354–376.

Tucker, C.J., Pinzón, J.E., Brown, M.E., Slayback, D.A., Pak, E.W., Mahoney, R., Vermote, E., & El Saleous, N. (2005). An extended AVHRR 8-km NDVI dataset compatible with MODIS and SPOT vegetation NDVI data. *International Journal of Remote Sensing*, *26*, 4485–4498.

Verbesselt, J., Hyndman, R., Zeileis, A., & Culvenor, D. (2010). Phenological change detection while accounting for abrupt and gradual trends in satellite image time series. *Remote Sensing of Environment*, *114*, 2970.

Verger, A., Baret, F., & Camacho de Coca, F. (2011a). Optimal modalities for radiative transfer-neural network estimation of canopy biophysical characteristics: Evaluation over an agricultural area with CHRIS/PROBA observations. *Remote Sensing of Environment*, *115*, 415–426.

Verger, A., Baret, F., & Weiss, M. (2008). Performances of neural networks for deriving LAI estimates from existing CYCLOPES and MODIS products. *Remote Sensing of Environment*, *112*, 2789–2803.

Verger, A., Baret, F., & Weiss, M. (2011b). A multisensor fusion approach to improve LAI time series *Remote Sensing of Enviroment*, *115*, 2460–2470.

Verger, A., Baret, F., & Weiss, M. (2014). Near real-time vegetation monitoring at global scale. *JSTARS,IEEE Journal of Selected Topics in Applied Earth Observations and Remote Sensing*, *99*, 1–9.

Verger, A., Baret, F., Weiss, M., Kandasamy, S., & Vermote, E. (2012a). The CACAO method for smoothing, gap filling and characterizing anomalies in satellite time series. *IEEE Transactions on Geoscience and Remote Sensing*, *51*, 1963–1972.

Verger, A., Baret, F., Weiss, M., Lacaze, R., Makhmara, H., & Vermote, E. (2012b). Long term consistent global GEOV1 AVHRR biophysical products. In *1st EARSeL Workshop on Temporal Analysis of Satellite Images* (pp. 1–6). Mykonos.

Verhoef, W., & Bach, H. (2007). Coupled soil-leaf-canopy and atmosphere radiative transfer modeling to simulate hyperspectral multi-angular surface reflectance and TOA radiance data. *Remote Sensing of Enviroment*, *109*, 166–182.

Verrelst, J., Malenovský, Z., Van der Tol, C., et al. (2019a). Quantifying vegetation biophysical variables from imaging spectroscopy data: A review on retrieval methods. *Surveys in Geophysics*, *40*, 589–629. https://doi.org/10.1007/s10712-018-9478-y

Verrelst, J., Muñoz, J., Alonso, L., Delegido, J., Rivera, J.P., Camps-Valls, G., & Moreno, J. (2012). Machine learning regression algorithms for biophysical parameter retrieval: Opportunities for Sentinel-2 and -3. *Remote Sensing of Environment*, *118*, 127–139.

Verrelst, J., Vicent, J., Rivera-Caicedo, J.P., Lumbierres, M., Morcillo-Pallarés, P., & Moreno, J. (2019b). Global sensitivity analysis of leaf-canopy-atmosphere RTMs: Implications for biophysical variables retrieval from top-of-atmosphere radiance data. *Remote Sensing*, *11*(16), 1923. https://doi.org/10.3390/rs11161923

Verstraete, M., & Pinty, B. (1996). Designing optimal spectral indexes for remote sensing applications. *IEEE Transactions on Geoscience and Remote Sensing*, *34*, 1254–1265.

Vieira, M.A., Formaggio, A.R., Rennó, C.D., Atzberger, C., Aguiar, D.A., & Mello, M.P. (2012). Object based image analysis and data mining applied to a remotely sensed Landsat time-series to map sugarcane over large areas. *Remote Sensing of Environment*, *123*, 553–562.

Voßbeck, M., Clerici, M., Kaminski, T., Lavergne, T., Pinty, B., & Giering, R. (2010). An inverse radiative transfer model of the vegetation canopy based on automatic differentiation. *Inverse Problems*, *26*, 15.

Vohland, M., Mader, S., & Dorigo, W. (2010). Applying different inversion techniques to retrieve stand variables of summer barley with PROSPECT SAIL. *International Journal of Applied Earth Observation and Geoinformation*, *12*, 71–80.

Walter, J.M.N., Fournier, R.A., Soudani, K., & Meyer, E. (2003). Integrating clumping effects in forest canopy structure: An assessment through hemispherical photographs. *Canadian Journal of Remote Sensing*, *29*, 388–410.

Walthall, C.L., Dulaney, W.P., Anderson, M.C., Norman, J.M., Fang, H., & Liang, S. (2004). A comparison of empirical and neural network approaches for estimating corn and soybean leaf area index from Landsat ETM+ imagery. *Remote Sensing of Enviroment*, *92*, 465–474.

Wan, L., Ryu, Y., Dechant, B., Lee, J., Zhong, Z., & Feng, H. 2024. Improving retrieval of leaf chlorophyll content from Sentinel-2 and Landsat-7/8 imagery by correcting for canopy structural effects. *Remote Sensing of Environment*, *304*, 114048, ISSN 0034–4257, https://doi.org/10.1016/j.rse.2024.114048. (www.sciencedirect.com/science/article/pii/S0034425724000592)

Weiss, M., Baret, F., Block, T., Koetz, B., Burini, A., Scholze, B., Lecharpentier, P., Brockmann, C., Fernandes, R., Plummer, S., Myneni, R., Gobron, N., Nightingale, J., Schaepman-Strub, G., Camacho, F., & Sanchez-Azofeifa, A. (2014). On Line Validation Exercise (OLIVE): A web based service for the validation of medium resolution land products: Application to FAPAR products. *Remote Sensing*, *6*, 4190–4216.

Weiss, M., Baret, F., Garrigues, S., Lacaze, R., & Bicheron, P. (2007). LAI, fAPAR and fCover CYCLOPES global products derived from VEGETATION. part 2: Validation and comparison with MODIS Collection 4 products. *Remote Sensing of Environment*, *110*, 317–331.

Weiss, M., Baret, F., Leroy, M., Hautecoeur, O., Bacour, C., Prévot, L., & Bruguier, N. (2002a). Validation of neural net techniques to estimate canopy biophysical variables from remote sensing data. *Agronomie*, *22*, 547–554.

Weiss, M., Baret, F., Myneni, R., Pragnère, A., & Knyazikhin, Y. (2000). Investigation of a model inversion technique for the estimation of crop characteristics from spectral and directional reflectance data. *Agronomie*, *20*, 3–22.

Weiss, M., Baret, F., Smith, G.J., Jonckheered, I., & Coppin, P. (2004). Review of methods for in situ leaf area index determination, part II: Estimation of LAI, errors and sampling. *Agricultural and Forest Meteorology*, *121*, 37–53.

Weiss, M., Jacob, F., Baret, F., Pragnère, A., Bruchou, C., Leroy, M., Hautecoeur, O., Prévot, L., & Bruguier, N. (2002b). Evaluation of kernel-driven BRDF models for the normalization of Alpilles/ReSeDA POLDER data. *Agronomie*, *22*, 531–536.

Weiss, M., Troufleau, D., Baret, F., Chauki, H., Prévot, L., Olioso, A., Bruguier, N., & Brisson, N. (2001). Coupling canopy functioning and canopy radiative transfer models for remote sensing data assimilation. *Agricultural and Forest Meteorology*, *108*, 113–128.

Wiegand, C.L., Gerberman, A.H., Gallo, K.P., Blad, B.L., & Dusek, D. (1990). Multisite analyses of spectral-biophysical data for corn. *Remote Sensing of the Environment*, *33*, 1–16.

Wiegand, C.L., Maas, S.J., Aase, J.K., Hatfield, J.L., Pinter, P.J.J., Jackson, R.D., Kanemasu, E.T., & Lapitan, R.L. (1992). Multisite analysis of spectral biophysical data for wheat. *Remote Sensing of the Environment*, *42*, 1–22.

Wocher, M., Berger, K., Danner, M., Mauser, W., & Hank, T. 2018. Physically-based retrieval of canopy equivalent water thickness using hyperspectral data. *Remote Sensing*, *10*(12), 1924. https://doi.org/10.3390/rs10121924

Xiao, Z., Liang, S., Wang, J., Chen, P., Yin, X., Zhang, L., & Song, J. (2014). Use of general regression neural networks for generating the GLASS leaf area index product from time-series MODIS surface reflectance. *IEEE Transactions on Geoscience and Remote Sensing*, *52*, 209–223.

Xiao, Z., Liang, S., Wang, J., Yin, X., Xiang, Y., Song, J., & Ma, H. (2012). GLASS leaf area index product derived from MODIS time series remote sensing data. In *IGARSS* Munich: IEEE.

Xie, Q., Dash, J., Huete, A., Jiang, A., Yin, G., Ding, Y., Peng, D., Hall, C.C., Brown, L., Shi, Y., Ye, H., Dong, Y., & Huang, W. (2019). Retrieval of crop biophysical parameters from Sentinel-2 remote sensing imagery. *International Journal of Applied Earth Observation and Geoinformation*, *80*, 187–195, ISSN 1569–8432, https://doi.org/10.1016/j.jag.2019.04.019. (www.sciencedirect.com/science/article/pii/S0303243419301199)

Yang, W., Shabanov, N.V., Huang, D., Wang, W., Dickinson, R.E., Nemani, R.R., Knyazikhin, Y., & Myneni, R.B. (2006). Analysis of leaf area index products from combination of MODIS Terra and Aqua data. *Remote Sensing of Environment*, *104*, 297–312.

Zeng, Y., Xu, B., Yin, G., Wu, S., Hu, G., Yan, K., Yang, B., Song, W., & Li, J. 2018. Spectral invariant provides a practical modeling approach for future biophysical variable estimations. *Remote Sensing*, *10*(10), 1508. https://doi.org/10.3390/rs10101508

Zhang, Q., Xiao, X., Braswell, B., Linder, E., Baret, F., & Moore Iii, B. (2005). Estimating light absorption by chlorophyll, leaf and canopy in a deciduous broadleaf forest using MODIS data and a radiative transfer model. *Remote Sensing of Environment*, *99*, 357.

Zhang, X., Friedl, M.A., Schaaf, C.B., Strahler, A.H., Hodges, J.C.F., Gao, F., Reed, B.C., & Huete, A.R. (2003). Monitoring vegetation phenology using MODIS. *Remote Sensing of Environment*, *84*, 471–475.

Zhang, Y., Shabanov, N., Knyazikhin, Y., & Myneni, R.B. (2002a). Assessing the information content of multiangle satellite data for mapping biomes I. Theory. *Remote Sensing of Environment*, *80*, 435–446.

Zhang, Y., Tian, Y., Myneni, R.B., Knyazikhin, Y., & Woodcock, C.E. (2002b). Assessing the information content of multiangle satellite data for mapping biomes I. Statistical analysis. *Remote Sensing of Environment*, *80*, 418–434.

Zhu, J., Lu, J., Li, W., Wang, Y., Jiang, J., Cheng, T., Zhu, Y., Cao, W., & Yao, X. 2023. Estimation of canopy water content for wheat through combining radiative transfer model and machine learning. *Field Crops Research*, *302*, 109077, ISSN 0378–4290, https://doi.org/10.1016/j.fcr.2023.109077. (www.sciencedirect.com/science/article/pii/S0378429023002708)

Zhu, Z., Bi, J., Pan, Y., Ganguly, S., Anav, A., Xu, L., Samanta, A., Piao, S., Nemani, R.R., & Myneni, R.B. (2013). Global data sets of vegetation Leaf Area Index (LAI)3g and Fraction of Photosynthetically Active Radiation (FPAR)3g derived from Global Inventory Modeling and Mapping Studies (GIMMS) Normalized Difference Vegetation Index (NDVI3g) for the period 1981 to 2011. *Remote Sensing*, *5*, 927–948.

Part II

Agricultural Croplands

4 Agricultural Crop Biophysical and Biochemical Quantity Retrievals Using Remote Sensing with Multi-Sensor Synergy, Machine Learning, and Radiative Transfer Models

Lea Hallik, Egidijus Šarauskis, Ruchita Ingle, Indrė Bručienė, Vilma Naujokienė, and Kristina Lekavičienė

ACRONYMS AND DEFINITIONS

ANNs	Artificial Neural Networks
CEOS	Committee on Earth Observation Satellites
EnMAP	Environmental Mapping and Analysis Program
EO	Earth Observation
ET	Evapotranspiration
LAI	Leaf Area Index
LiDAR	Light Detection and Ranging
MODIS	Moderate-Resolution Imaging Spectroradiometer
PROSAIL	Combination of PROSPECT and SAIL, the two nondestructive physically based models to measure biophysical and biochemical properties
PROSPECT	Radiative transfer model to measure leaf optical properties spectra
SAIL	Scattering by Arbitrary Inclined Leaves
SAR	Synthetic Aperture Radar
TIR	Thermal Infrared
UAV	Unmanned Aerial Vehicles

4.1 ADVANCING EARTH OBSERVATION SCIENCE FOR SUSTAINABLE AGRICULTURE

The field of Earth Observation (EO) science is undergoing a transformative era with the advent of new satellite missions such as hyperspectral imaging spectrometer platforms. These sophisticated instruments promise to provide information-rich retrievals that can be used to better understand plant and ecosystem traits, function, and diversity (Berger et al., 2020b; Verrelst et al., 2021; Ranghetti et al., 2022). Hyperspectral data provides a comprehensive view of the electromagnetic spectrum, offering a rich source of information about the Earth's surface. Unlike multispectral sensors, which capture data in a few discrete bands, hyperspectral sensors measure hundreds or even thousands of narrow spectral bands, allowing for a much finer-grained analysis of surface composition. While hyperspectral data

DOI: 10.1201/9781003541165-6

holds immense potential, its full utilization is currently limited by data availability. However, a few satellite missions, such as PRISMA (launched on March 22, 2019) and EnMAP (launched on April 1, 2022), are already providing complementary hyperspectral time-series data. Optical remote sensing in general has witnessed remarkable advancements in recent years, with agricultural applications of proximal and spaceborne RS approaching Technology Readiness Levels (TRL) 8–9. These advancements, driven by improvements in unmanned aerial vehicles (UAVs) and satellite technologies, have significantly enhanced our ability to accurately model plant ecophysiological processes, inform agricultural breeding programs, and optimize agricultural management decisions (Berger et al., 2022; Ustin & Middleton, 2021; Weiss et al., 2020; Wu et al., 2015). Remote sensing of reflected (visible, near infrared, shortwave infrared) and emitted radiance (solar-induced fluorescence, thermal infrared) allows to capture distinct plant responses occurring under short-, medium-term or severe chronic stress exposure.

The continuous advancements in satellite observations, characterized by enhanced spatial, spectral, and temporal resolutions, coupled with the introduction of novel missions, have enabled the generation of vast datasets for the estimation of pertinent variables for agricultural applications. A prime example is the Sentinel-1 and Sentinel-2 missions, which provide high-resolution observations of optical and microwave radiation, generating petabytes of processed and unprocessed data for estimating soil and vegetation properties. In research fields that demand detailed local environmental information, these repositories serve as a transformative resource by offering global coverage of high-resolution data. However, the utilization of these satellite missions faces a multitude of technical and scientific challenges. A primary challenge lies in obtaining internally consistent retrievals of multiple land surface variables from diverse satellite types, as required for large-scale modeling (Bauer et al., 2021; Purcell & Neubauer, 2023).

The convergence of new proximal and remote sensing technologies, and vast satellite datasets presents a unique opportunity to revolutionize our understanding of plant and ecosystem processes and inform sustainable agricultural practices. Multi-sensor scalable remote sensing approaches have demonstrated their effectiveness in accurately diagnosing specific plant stressors in agricultural contexts (Berger et al., 2022). Leveraging the quantitative capabilities of remote sensing technologies for plant biophysical trait retrievals and environmental assessments in a multidisciplinary manner, enables the parameterization of crop growth models (CGMs), optimization of resource use, and enhanced agricultural resilience in the face of global challenges (Huang et al., 2019a). Increasing usage of satellite and unmanned aerial vehicle data in parallel with a shift in methods from simpler parametric approaches towards more advanced physically based and hybrid models has been highlighted as a current trend in agricultural remote sensing (Berger et al., 2022).

4.2 HARNESSING THE POWER OF REMOTE SENSING FOR DIGITAL TWINS

Digital Twins, employing simulators that digitally mimic reality with (near) real-time data, empower scientists to analyze the effects of various management or mitigation scenarios. By harnessing big data techniques, Digital Twins facilitate advanced data-driven modeling and simulation, generating novel insights unattainable through traditional observation models (Barricelli et al., 2019; Blair, 2021; El Saddik, 2018). Due to the spatially contiguous and repetitive nature of satellite observations, remote sensing is poised to play a crucial role in such Digital Twins for agriculture (Berger et al., 2022; Wu et al., 2023). To ensure the high-accuracy performance of Digital Twins, interactive and dynamic simulations of the Earth system, guided by comprehensive observational datasets, demand consistent satellite observations of multiple essential variables (e.g., soil moisture, land surface temperature, ecosystem structure, and density) at unprecedented levels of detail (Bauer et al., 2021). Current satellite remote sensing data products, providing information on key environmental structure, processes, and parameters, are not directly suitable for such simulations. This is attributed to their independent estimation and generally coarse resolutions. For instance, predicting the impact of drought requires the integration of multiple key land surface parameters while discrepancies between estimated parameters and overly coarse spatiotemporal resolutions can lead to significant underestimations of these impacts (Fisher et al., 2017).

4.3 RADIATIVE TRANSFER MODELS (RTMS)

Remote sensing research typically involves scientists acquiring suitable data, pre-processing it, and ultimately retrieving relevant state variables. To obtain multiple outputs from remote sensing observations, mathematical functions known as forward operators are employed to connect the Earth system's state variables (often referred to as biophysical parameters) to the observations made by remote sensing instruments. Both statistically and physically based methods can provide the capability to produce multiple outputs. Among these, only radiative transfer model (RTM) retrieval allows for the incorporation of relationships between dependent variables (Verrelst et al., 2015). RTMs link satellite observations with vegetation and soil biophysical parameters and such physically based models exists in both optical and microwave domain. Utilizing a single RTM model to generate multiple outputs (rather than employing multiple single-output models) ensures consistency. Additionally, compatibility between optical and microwave RTMs enables the utilization of cross-relationships, enhancing the retrieval of specific variables (Lauvernet et al., 2008; Mousivand et al., 2014).

4.3.1 Common RTMs Used for Agricultural Crops

PROSAIL is a combined model that integrates the PROSPECT leaf radiative transfer model (Féret et al., 2021; Feret et al., 2008; Jacquemoud & Baret, 1990) with the SAIL canopy radiative transfer model (Verhoef, 1984) to simulate optical radiation processes within canopies and soils (Jacquemoud et al., 2006). Within the PROSAIL model, the reflectances and transmittances of the isolated canopy layer are determined by leaf optical properties calculated by PROSPECT, the leaf inclination angle distribution function, and the observational geometry. PROSAIL's accuracy for continuous canopies, such as crop fields, has been extensively evaluated through multiple studies, demonstrating its effectiveness (Pinty et al., 2001; Widlowski et al., 2007; Jiang et al., 2018).

A two-layer canopy reflectance model (ACRM) utilizes also PROSPECT leaf optical model. Within ACRM vegetation canopy is supposed to consist of a main homogeneous layer of vegetation, and a geometrically thin layer of vegetation on ground surface (for example weed layer under main crop). Both vegetation layers are characterized by a similar set of biophysical parameters: leaf area index (LAI), leaf angle distribution (LAD) parameters, leaf size, and leaf biochemical parameters of PROSPECT model which control the optical properties of leaves (Kuusk, 2001).

SCOPE model (Soil Canopy Observation, Photochemistry and Energy fluxes) is an integrated radiative transfer and energy balance model, which allows to link visible and thermal infrared (0.4 to 50 μm) radiance spectra observed above the canopy to the fluxes of water, heat and carbon dioxide, considering vegetation structure and the vertical profiles of temperature. In this model photosynthesis rate and chlorophyll fluorescence are calculated at the leaf level as a function of net radiation and leaf temperature. SCOPE model can help to design the algorithms for the retrieval of evapotranspiration from optical and thermal Earth Observation data (van der Tol et al., 2009).

In the microwave domain, the Water Cloud Model (WCM) is a widely used tool for retrieving soil moisture from radar observations. WMC is a semi-empirical model preferred in retrieving soil moisture and modeling of the scattering of vegetated areas mainly due to its simplicity. This model has demonstrated promising results in various studies, particularly for retrieving soil moisture under mature fully developed crops like winter wheat (Baghdadi et al., 2017; Gherboudj et al., 2011; Li & Wang, 2018). WCM-like models have consistently shown acceptable accuracy in soil moisture retrieval over vegetated areas (Quaife et al., 2022; Choker et al., 2017).

4.3.2 Emulators

Radiative transfer models (RTMs) are generally computationally expensive. While this may not pose a significant limitation for remote sensing imagery with a low number of pixels, in large-scale applications, the computational cost of running RTMs directly can become prohibitive. Additionally, the complexity of RTMs and their challenging interfacing with other codes can rapidly become limiting factors in their direct application (Verrelst et al., 2019). To address these challenges, statistical emulators can be employed as a promising alternative (Caicedo et al., 2014; Gómez-Dans et al., 2016; Verrelst et al., 2017). Emulators are machine learning models that approximate the complex relationships between biophysical parameters and corresponding reflectance or backscatter signals. Emulators not only offer a significant computational advantage over the original models but also provide access to the Jacobian and Hessian matrices of the underlying models, enabling faster inferences and uncertainty quantification.

4.4 TOWARDS CONSISTENT AND SCALABLE BIOPHYSICAL PARAMETER RETRIEVAL: A MULTI-SENSOR APPROACH LEVERAGING CLOUD COMPUTING ENVIRONMENTS

Inconsistencies in remote sensing variables, such as the failure to adequately incorporate vegetation dynamics into soil moisture estimations, data gaps caused by cloud/snow cover or sensor malfunctions, and the absence of quantified uncertainties hinder their combined use in biophysical models, leading to potentially large errors. This issue has been recognized in various remote sensing applications, including evapotranspiration estimation, crop health monitoring, and vegetation dynamics assessment (Fisher et al., 2017; Wen et al., 2021; Zhang et al., 2021). Data gaps are typically addressed using interpolation and filtering techniques applied to satellite-derived variables (Pan et al., 2017; Wang et al., 2019; Yuan et al., 2011). However, this approach introduces inconsistencies between simultaneously derived variables due to the disregard for interdependencies between variables. To address the substantial inconsistencies between variables derived from different remote sensing products (Fisher et al., 2017), multi-sensor approaches offer a promising solution by exploiting the covariance relationships between multiple variables (Salcedo-Sanz et al., 2020; Verrelst et al., 2015). The synergistic combination of optical and fluorescence observations has demonstrated significant improvements in biophysical parameter retrieval (Verrelst et al., 2012). Similarly, the complementarity of synthetic aperture radar (SAR) and optical data in joint retrieval of vegetation properties has enhanced the retrieval of biophysical parameters compared to using only one spectral domain (Quaife et al., 2022). These findings underscore the potential of multi-sensor approaches to overcome the limitations of single-domain approaches. However, the current landscape of tools poses limitations for the further development of such multi-sensor frameworks. Two primary challenges, which need to be addressed, are computational scalability and flexibility. While proprietary tools like ArcGIS, the Sentinel Application Platform (SNAP), and the Google Earth Engine (GEE) platform offer substantial scalability, they often lack flexibility, which significantly hinders scientific progress in multivariate retrievals (Pérez-Cutillas et al., 2023). For example, despite GEE's ability to perform physical retrieval of vegetation properties over large spatial extents (Estévez et al., 2022), its implementation often requires extensive workarounds (Pipia et al., 2019), limiting its adaptability for multivariate retrieval. Conversely, scientific tools like the ARTMO framework (Verrelst et al., 2011) and the EOLDAS framework (Lewis et al., 2012) prioritize flexibility in multivariate retrieval but provide limited support (in terms of workload distribution in cloud-computing environments) for upscaling workflows to broader spatial extents (Quaife et al., 2022). MULTIPLY framework (written in Python and available at https://github.com/multiply-org) has been developed to address these issues. MULTIPLY is an open source, user-extendable high performance framework for merging different types of satellite observations, including MODIS, Sentinel-1, and Sentinel-2 for synergistically retrieving multiple land surface parameters. It contains mature components to download the satellite raw data, to pre-process optical and microwave observations and merge them together in a Bayesian assimilation approach with a priori information to derive gap-filled vegetation and soil variables and their respective uncertainties (Figure 4.1). MULTIPLY framework provides a high level of scalability and can be used in large-scale computational infrastructures to allow researchers to test satellite synergies in actual applications over larger spatial and temporal extents to retrieve multiple biophysical traits of soil and vegetation

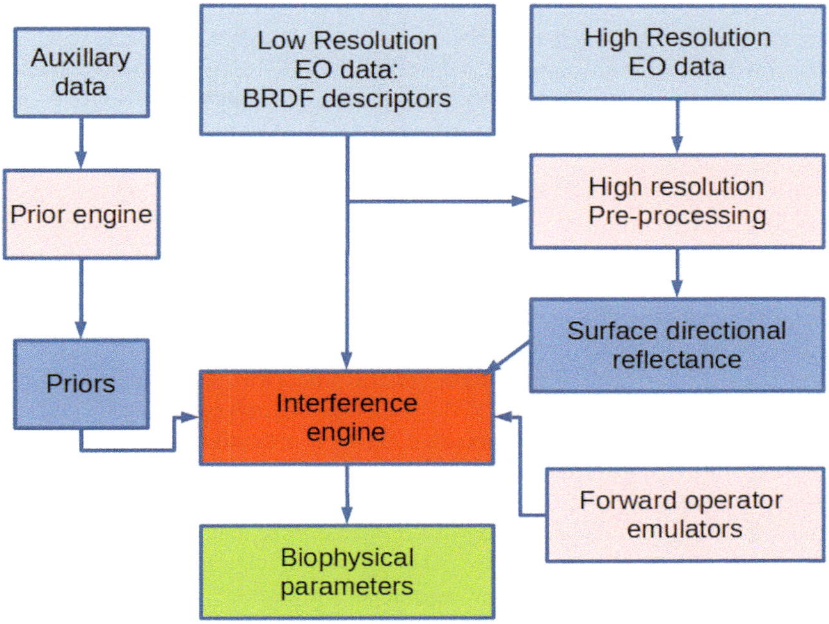

FIGURE 4.1 MULTIscale SENTINEL land surface information retrieval Platform MULTIPLY (available at https://github.com/multiply-org).

synergistically. MULTIPLY employs statistical emulators for radiation transfer models (RTMs) as forward operators, utilizing Gaussian process emulators and neural networks. While emulators offer substantial computational advantages over their physical RT counterparts, a crucial requirement for MULTIPLY is access to the Jacobian and Hessian matrices of the underlying models, enabling faster inferences and uncertainty quantification. The MULTIPLY prior engine comprises two modules: one for vegetation priors to assist PROSAIL and another to deliver soil priors for WCM. Specific priors include information on soil brightness and wetness, leaf chlorophyll, carotenoid, dry matter and water content, hotspot parameter, leaf area index, soil moisture, roughness, and vegetation water content. Current default implementation of MULTIPLY uses PROSAIL as an optical model, which is suitable for continuous crop canopies but has limited use in more complex landscapes.

4.5 UNCERTAINTY BUDGETS

A major obstacle for fully realizing the potential of applied remote sensing technology in agriculture is the accuracy assessment of the measured data and derived products. Uncertainty estimates should be traceable from the satellite sensor, through levels of data processing, to final products describing the geophysical state of the Earth (Lewis et al., 2012). Any quantitative information should be accompanied by an uncertainty budget, including remote sensing observations, retrieval algorithms, and crop model parameterizations (Huang et al., 2019a). In metrology, this is achieved through three fundamental principles: traceability to International System of Units (SI), uncertainty analysis and comparison. Adhering to these three concepts is crucial for generating reliable datasets, which form the foundation for informed decision making.

In recent years, the recognition of the importance of uncertainty assessment has gained traction among space agencies, data providers, and both the scientific and data user communities. Initiatives such as Quality Assurance for Earth Observation (QA4EO, https://qa4eo.org/) and Metrology for Earth Observation and Climate (MetEOC, www.meteoc.org/) have emerged to establish guidelines for uncertainty assessment. These guidelines and best practice documents serve as a foundation for uncertainty assessment procedures, tailored to specific measurement and data processing scenarios. Although methods and techniques for in situ, airborne, and satellite data are being developed (Mittaz et al., 2019), they have yet to be fully integrated into the remote sensing community's practices. Consequently, there is a need for training among both RS scientists and data end-users to adopt the processes established by the metrology community and integrate uncertainty budgets operationally into sensor-to-product chains.

For field-related measurements, there is no overarching global governing body responsible for accrediting field teams. While best practice guides for various field spectroscopy measurements have been available for some time, continuous training of field personnel would be essential to ensure these practices are effectively implemented. Building, reinforcing, and maintaining awareness of uncertainty sources within the community requires training, comparison exercises, and ideally, software tools that can identify data collection issues. For novel measurement approaches, such as UAV-based sensors, standardization efforts are still underway, as exemplified by the Fiducial Reference Measurements for Vegetation (FRM4VEG, https://frm4veg.org/) project funded by ESA.

Quantitative measurements, whether obtained through proximal or remote sensing methods, must be accompanied by propagated uncertainty estimates to ensure their reliability and accuracy (Mittaz et al., 2019). Measurements and derived products must be traceable to internationally recognized standards to support the generation of actionable information for effective decision making (Kaminski et al., 2017). To address these challenges effectively, there is significant potential in fostering collaboration and knowledge exchange between the remote sensing, plant phenotyping, precision agriculture, and metrology communities.

4.6 SYNERGY BETWEEN REMOTE SENSING AND CROP GROWTH MODELS: DIGITAL TWINS FOR AGRICULTURE

4.6.1 Bridging Remote Sensing and Crop Growth Models

Remote sensing enables quantitative evaluation of crop biophysical traits. Assimilation of RS data into Crop Growth Models (CGMs) hold immense potential for agriculture, enabling informed decision making and improved yield predictions (Ines et al., 2013; Wu et al., 2023). CGMs are dynamic, process-driven models that encapsulate the current knowledge of plant physiologists regarding the interplay between climatic drivers, soil, and the plant under specific management practices. However, for a comprehensive and realistic representation of even the major processes, these CGMs typically involve a substantial number of parameters and variables that need to be specified for model execution (Rosenzweig et al., 2013). Data assimilation frameworks such as EOLDAS and MULTIPLY provide a way of combining remotely sensed observations with crop growth models, merging the real-time monitoring capabilities of EO data with the predictive and explanatory power of crop growth models. Integration of Crop Growth Models (CGM) with radiative transfer modeling (RTM) allows for the retrieval of vegetation biophysical features from remote or proximal sensing data (Huang et al., 2019a). Diverse observations from a range of sensor modalities could offer suitable constraints to parameterize/calibrate CGMs (Berger et al., 2022). Integration of remote sensing data into crop models such as EPIC and WOFOST, enhances yield predictions by providing valuable insights into vegetation traits and soil conditions (Kasampalis et al., 2018; Wang et al., 2022). Precision and regenerative agriculture can benefit from the assimilation of RS data in CGMs to deduce crop status information and support decision making. To robustly infer the crop's physiological status, it is crucial to calibrate the CGM at (sub)field scale. The continuous depiction of crop growth and development using mechanistic models offers the possibility to assess processes that would otherwise be unobservable (Berger et al., 2022). While RS can detect vegetation anomalies/stress, subsequent integration with CGMs can facilitate the identification of stressors (e.g., water, nutrients, temperature, etc.), and even predict this stress in advance before it occurs. Thus, an integrated RS-CGM can form digital twins of agroecosystems (Figure 4.2) that inform

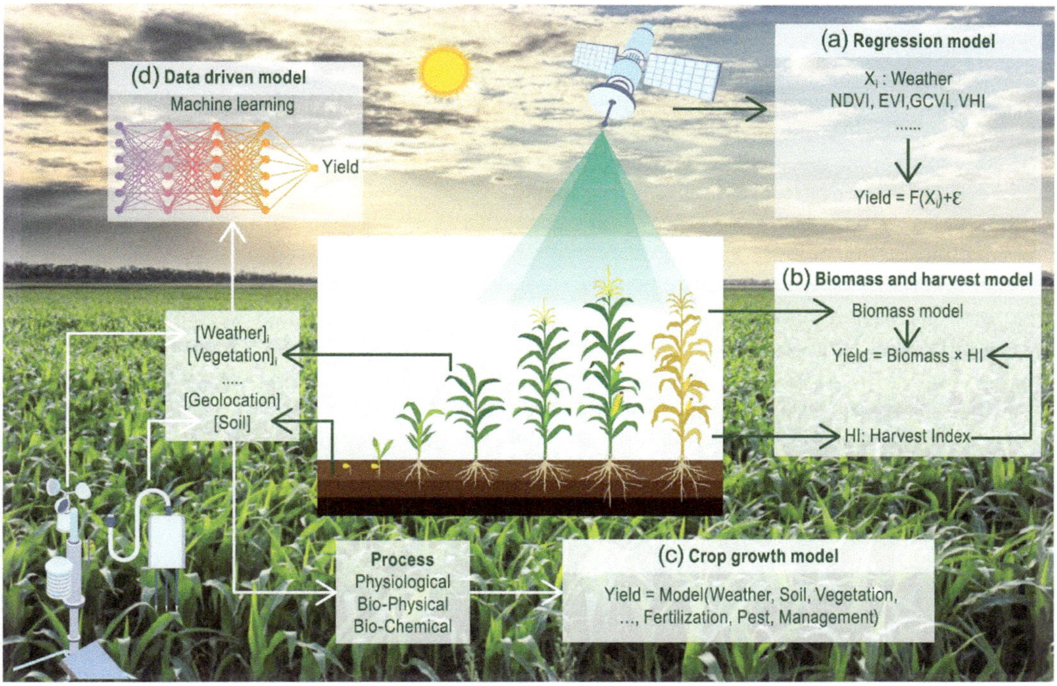

FIGURE 4.2 RS-driven digital twins of agroecosystems. (Wu et al., 2023.)

farmers' decisions in a manner that supports risk management, yield optimization, and sustainable production (Berger et al., 2022). The expansive nature of an increasing number of satellite constellations provides continuous coverage, making CGM and digital twin approaches more operational and cost-effective. Satellite-driven methods allow predicting crop yields weeks or months ahead of harvest, and current food security programs already operationally use such approaches that combine satellite data with agroclimate indices, which are calibrated and then transformed into a final crop yield to predict food production (Wu et al., 2023).

4.6.2 Scaling Up and Down Crop Growth Models

In precision agriculture, UAVs may also be integrated into the Internet of Things (IoT) and other interconnectable proximal sensing technologies for environmental monitoring to feed into CGMs. While most CGMs were developed and tested under homogeneous field conditions, they are increasingly being applied beyond the field scale. Inadequate representation of the spatial variability at a larger scale can introduce significant errors in the models' predictions, yet attention to this topic is lacking. The selection of optimal crop models and their inputs when moving from the field to a regional scale must be carefully executed using strict guidelines while considering uncertainty propagation. A successful example of data assimilation (DA) implementation has been demonstrated by Kang and Özdoğan (2019), where a coarse-scale calibration of a crop model was initially performed at the county level, followed by refinement at the Landsat pixel (30 m) resolution through the assimilation of Landsat observations. Achieving accurate yield estimates in fragmented landscapes through DA at high or medium resolutions continues to be a challenge (Huang et al., 2019b).

4.7 SENSOR SYNERGIES FOR BIOPHYSICAL TRAIT RETRIEVALS: BRIDGING THE GAP BETWEEN REMOTE SENSING AND PLANT PHENOTYPING

Traditional RS field measurements of plant physiological properties often rely on single-sensor instruments, limiting the range of information gathered. Recent advancements in sensor technology and unmanned aerial vehicle (UAV) integration have enabled the simultaneous collection of data from multi-sensor payloads, mirroring the developments in satellite platforms used for large-scale information retrieval (Maes & Steppe, 2019). Hyperspectral, multispectral, and thermal imaging sensor combinations provide synergistic information, offering more comprehensive insights into nutrient, biotic, and abiotic stress assessments across scales (Berger et al., 2022). UAV mounted LiDAR can be used to estimate canopy density parameters for crops (Bates et al., 2021). Furthermore, RS technologies provide transferability in time and space, meaning that advancements achieved from proximal platform trials can be seamlessly scaled to spaceborne sensors that provide continuous global coverage. Despite these promising advances, the full potential of multiple remote data sources remains untapped, necessitating a synergistic approach to address gaps in scalable and transferable RS technologies, knowledge, and methodologies for stress detection in agriculture (Berger et al., 2022; Machwitz et al., 2021; Wu et al., 2023).

The development of novel RS and field phenotyping synergies in technology, data exchange, and data tools (artificial intelligence, machine learning) is crucial for optimizing vegetation trait monitoring. By bridging the gap between RS and plant phenotyping (PP), we can contribute to resilient agriculture by enabling the monitoring and modeling of stress symptoms and the development of stress-resilient breeding lines. Collaboration and exchange between the two communities are expected to advance research and create innovations on both sides (Machwitz et al., 2021). RS-based observation methods and analysis tools for field trials can enhance phenotyping-related analysis to effectively monitor experimental progress. RS-based High Throughput Phenotyping Platforms (HTPPs) hold promise for rapidly developing disease-tolerant and weather-resistant crop varieties. These platforms may incorporate high-resolution visible (RGB), multispectral, thermal (TIR), hyperspectral, and solar-induced fluorescence (SIF) imaging (Pieruschka & Schurr, 2019; Camino et al., 2019; Mohammed et al., 2019). While RGB cameras offer cost-effectiveness, their full potential in combination with other sensors remains to be explored. For instance, RGB+TIR cameras are commercially available but underutilized. RGB imagery has demonstrated its effectiveness in estimating chlorophyll content, leaf area index (LAI), and vegetation cover (Araus & Kefauver, 2018). Crop variables estimated from RGB or multispectral sensors combined with thermal (TIR) camera images are sufficient for estimating crop evapotranspiration using surface energy balance approaches (Gómez-Candón et al., 2021).

4.8 GROUND VALIDATION OF LAND SURFACE PRODUCTS

The CEOS Land Product Validation (LPV) subgroup is a working group within the Committee on Earth Observation Satellites (CEOS) that coordinates and promotes the quantitative validation of satellite-derived land surface products. The LPV subgroup's mission is to ensure that land surface products from different sources are consistent, accurate, and reliable for a wide range of applications. They develop and recommend best practices for validation, establish and maintain a standardized validation hierarchy for land products, and provide the platform for the exchange of validation data and results. These coordination and standardization efforts are essential for ensuring the consistency, robustness, and comparability of land surface products from various sources. The LPV subgroup contributes to the development of new validation methodologies, algorithms, and tools to enhance the accuracy and reliability of land surface products. They identify and address emerging validation challenges, such as the use of multi-platform and multi-sensor data, the integration of in situ data, and the assessment of uncertainty. The LPV subgroup consists of ten Focus Areas: biophysical—LAI (Leaf Area Index) and fAPAR (fraction of Absorbed Photosynthetic Active

Radiation), land surface phenology, vegetation indices, land cover, snow cover, albedo, land surface temperature and emissivity, soil moisture, and active fire and burned area. Recommended best practice protocols for ground measurements to parameterize and/or validate satellite-derived products can be found at https://lpvs.gsfc.nasa.gov/.

4.9 COMMONLY ESTIMATED CROP BIOPHYSICAL TRAITS

4.9.1 Leaf Area Index (LAI)

The concept of the leaf area index (LAI) was first proposed by Watson (1947), who defined LAI as the total one-sided green leaf area per unit ground surface area. LAI is a dimensionless quantity that is widely used to characterize plant canopies and it is recognized as an Essential Climate Variable (ECV) by the Global Climate Observing System (GCOS). A higher LAI indicates a denser and more developed canopy, leading to increased light interception and subsequently, enhanced photosynthetic activity. LAI is a direct indicator of canopy development, providing insights into crops health and growth. It can serve as a reference for management practices to maintain optimal crop condition at each growth stage (Patil et al., 2018). The LAI indicates the condition of the vegetation, productivity, and the ability of agricultural plants to absorb sunlight energy, which is important for maximizing the yield potential of plants (Pearce et al., 1965; Fang et al., 2019; Bates et al., 2021). LAI is also a crucial parameter in crop modeling and yield prediction. It serves as a key input in models that simulate the growth and development of crops under different environmental conditions. By incorporating remotely sensed LAI data, these models can provide valuable insights into crop growth trajectories, allowing farmers to anticipate yield potential and make informed decisions regarding crop management practices (Novelli et al., 2019; Dong et al., 2020; Zhuo et al., 2022). LAI is also closely linked to transpiration and water use efficiency as crop canopies with higher LAI values tend to transpire more water influencing the overall water demand and irrigation requirements. From an agronomic point of view, LAI is also an important indicator for weed control, plant-weed competition, plant water use capacity and soil erosion (Wicks et al., 2004; Bavec et al., 2007). In addition, LAI is considered as a criterion of crop nitrogen status and helps to decide on the appropriate fertilization to keep the crop in optimal condition and increase yield (Addai & Alimiyawo, 2015; Bates et al., 2021).

Optical estimation methods allow LAI to be divided into two subcategories: the green LAI, which consists of green photosynthetically active leaves, and the brown LAI, where brown senescent leaves have lost photosynthetic function (Delegido et al., 2015; Yan et al., 2019). When crops (e.g., winter wheat and other cereal crops) start to turn brown, their spectral reflectance becomes more similar to that of the soil. Although the green LAI is the most important index for predicting crop yields, it is useful to know the timing, locations, and extent of crop damage due to drought, pests, or other damages before harvesting (Fang et al., 2019). Therefore, green and brown LAI assessments separately are valuable to obtain a more complete picture of crop management (Bates et al., 2021)

LAI measurement consists of direct and indirect methods (Breda, 2003; Bates et al., 2021). The direct method measures leaf area using a leaf area meter or scanner, and leaf area can also be calculated from a specific leaf area (Fang et al., 2019). The direct method is the most accurate, but is time-consuming, destructive, and lacks spatial representativeness (Hicks & Lascano, 1995). Direct measurement methods determine true LAI values and are often used as a reference for indirect measurement methods (Fang et al., 2019). Indirect measurements are based on statistical methods and analysis of light transmission through the canopy (Ross, 1981; De Bei et al., 2016). Indirect methods still need to be calibrated, so direct and indirect methods are complementary (Breda, 2003). Given the importance of LAI, new methods of assessment are proliferating and new objectives for crop monitoring are emerging. Remote sensing methods monitor crops over a large area of the land surface, which has complemented the limited measurement capabilities of direct and indirect methods. Remote sensing is carried out from the air and from space, providing a more complete picture of the LAI values and changes under study compared to ground-based measurements (Ilniyaz et al., 2023; Bates et al., 2021). However, remote LAI estimates need to be calibrated against ground-based measurements. Two-step calibration method by coupling Sentinel-2 and UAV observations has been shown to reduce errors by more than 50% (Revill et al., 2020). Global LAI products such as CYCLOPES, GLASS, MODIS, and GEOV1 have been released in recent decades (Ilniyaz et al., 2023).

In recent decades, the use of indirect non-contact LAI measurement methods based on light transmission through canopies has led to the development of new LAI measurement methods. Combined with remote sensing techniques, indirect non-contact methods have provided a timely and efficient solution for LAI assessment over large spatial and long-time scales (Zhang et al., 2012; Brown et al., 2020). For the assessment of LAI, when remote sensing techniques are used, there is a choice between physically based and statistical regression empirical models. For the estimation of LAI, statistical regression empirical models are commonly used due to their simplicity (e.g., Kross et al., 2015; Nguy-Robertson et al., 2012). The flexibility of these models is an important factor in combining data from several sources. In a statistical approach, the regression method relates the observed LAI to predictor variables. Parametric regression methods are often used, defined by mathematically linear or nonlinear functions. However, in the face of multiple forecasts, these methods may suffer from poor accuracy. Meanwhile, non-parametric regression methods, such as Artificial Neural Networks (ANNs) models, are used to discover relationships between remotely sensed data as independent variables and vegetation characteristics as target variables (Kimes et al., 1998). ANNs are a class of machine learning algorithms that are inspired by the structure and function of the human brain. They consist of interconnected layers of artificial neurons, which are mathematical functions that process and transmit information. ANNs can learn from data. Deep learning is a subfield of machine learning that focuses on the use of ANNs with multiple hidden layers. These deep neural networks (DNNs) are able to learn more complex patterns in data than shallow ANNs. This makes deep learning particularly well-suited for tasks that require a high level of abstraction or generalization such as classification and feature approximation. Machine learning methods are suitable for many agricultural applications and they are also used to analyze remote sensing information on crops. Deep convolutional neural networks (CNNs) are able to extract a large number of features and perform accurate prediction tasks and are computationally efficient and are particularly successful for image classification and regression applications, outperforming classical machine learning methods and are often used in precision agriculture practices (Ilniyaz et al., 2023; Pandey & Jain, 2022).

LAI is essentially a proxy for biomass (Dong et al., 2020). Biomass is accumulated dry weight of plant material and is a measure of crop productivity. It can be assessed by both active and passive remote sensing technologies. Simple RGB images obtained by UAVs equipped with digital cameras can produce cost-effective 3D point cloud data through SfM technology, which represents a low-cost remote sensing alternative to airborne and terrestrial LiDAR (Yang et al., 2023). Combination of spectral and texture features can improve LAI estimation from multispectral images (Yang et al., 2021). An interesting technological alternative is laser rangefinder sensors (Ehlert et al., 2009). Accurate estimation of biomass with remote sensing tools is essential for assessing the overall performance of crops and determining their potential economic value. It serves as a valuable indicator of yield potential, enabling farmers to make decisions regarding harvest timing and resource allocation. Remote sensing assessment of aboveground dry biomass of crops and harvest index allows yield prediction (Wu et al., 2015; Schauberger et al., 2020; Basso & Liu, 2019). Additionally, remote sensing based biomass estimation is crucial for monitoring the effectiveness of agronomic practices, such as fertilization and irrigation (Wu et al., 2020). By tracking changes in biomass over time, farmers can evaluate the impact of different management strategies on crop productivity and make adjustments accordingly. This information is invaluable for optimizing resource use efficiency and minimizing input costs.

4.9.2 Chlorophylls

In the visible wavelength band (400–700 nm), absorption by leaf pigments is the most important process, resulting in low reflectance and transmittance values. The main light-absorbing pigments are chlorophylls a and b (Chl a and Chl b), carotenoids, xanthophylls and polyphenols, and all pigments have overlapping absorption properties. Chlorophyll a shows maximum absorption in the 410–430 nm and 600–690 nm regions, while Chlorophyll b shows maximum absorption in the 450–470 nm range. These strong absorption bands cause a reflection peak in the green region around 550 nm. Carotenoids absorb most efficiently between 440 and 480 nm. In the near-infrared (NIR) wavelength band (700–1,300 nm), leaf pigments and cellulose are almost transparent, so absorption is very low, and reflectance and transmittance reach their highest values. This is caused by internal scattering at air, cell and water interfaces in leaves.

Chlorophyll is the primary pigment involved in photosynthesis, the process by which plants convert light energy into chemical energy. Chlorophyll is responsible for absorbing light energy from the Sun and transferring it to the chemical reactions of photosynthesis. It is a valuable indicator of plant health and vitality. As a result, chlorophyll measurements can be used to assess plant health and monitor the impact of environmental factors such as nutrient deficiencies, drought, or pests (Eitel et al., 2007; Delloye et al., 2018; Berger et al., 2020b; Croft et al., 2020) Much more than a hundred vegetation indices (VI) have been created to estimate the amount of chlorophylls (Hallik et al., 2017). Common tools such as SPAD mechanistically assess leaf chlorophyll content.

Canopy chlorophyll content (CCC) is a measure of the total amount of chlorophyll a and b pigments in a contiguous group of plants per unit ground area. It is the product of chlorophyll content per unit leaf area (g m^{-2}) and leaf area index (LAI) (m^2 m^{-2})), which describes the distribution of chlorophyll pigments in three-dimensional space on the crop surface. CCC can be estimated using various methods, both direct and indirect. Direct methods involve measuring the chlorophyll content of individual leaves or small vegetation patches. Indirect methods rely on remote sensing techniques, such as satellite imagery or airborne sensors, to derive estimates of CCC based on the spectral reflectance of vegetation canopies (Clevers & Gitelson, 2013; Singhal et al., 2019). CCC is a key indicator of vegetation productivity and is closely related to factors such as leaf area index (LAI), photosynthetic rates, and carbon assimilation (Delloye et al., 2018). Higher CCC values generally indicate healthier and more productive vegetation, while lower values suggest stress or senescence.

4.9.3 Nitrogen

Nitrogen (N) is an essential macronutrient for plants. Nitrogen deficiency in soil and plants generally limits root development and growth, inhibits lateral root development, increase the carbon/nitrogen (C:N) ratio of the plant, reduce photosynthesis and lead to premature leaf senescence (Féret et al., 2021). Thus, N strongly influences crop growth and quality. In addition, the N content of plants can give an indication of yield potential, while excessive levels have adverse effects on both crop quality and the environment. Therefore, the analysis of plant biophysical traits such as height, leaf area and color from remotely sensed data can provide a lot of useful information on the nitrogen content of plants, in order to properly select and manage the application of nitrogen fertilizers according to the field conditions (Bavec et al., 2007; Yu et al., 2021; Verrelst et al., 2021; Ranghetti et al., 2022). Nitrogen deficiency in plants can usually be seen in the lightening of leaves and stems, when photosynthesis and plant metabolism are impaired due to a reduction in chlorophyll content. In addition to the assessment of plant structural variables and chlorophyll pigments, the use of hyperspectral imaging spectroscopy allows the quantification of nutrients such as nitrogen (Berger et al., 2020b; Zheng et al., 2022).

The N status of a crop can be expressed in terms of an area-based measure—"nitrogen content" (N_{area}), or in terms of a mass-based measure—"nitrogen concentration" (N%). The leaf nitrogen content (N_{area}, g/cm²) is commonly calculated as the product of leaf nitrogen concentration and leaf mass per area (LMA). Canopy nitrogen content is a product of leaf nitrogen content and LAI (Berger et al., 2020b). The main indicators that characterize N in plants are: leaf nitrogen content (LNC), canopy nitrogen content (CNC), leaf chlorophyll content (LCC), canopy chlorophyll content (CCC), and leaf protein content (Cp). Nitrogen Nutrition Index (NNI) is defined as the ratio of the actual crop N concentration to the critical N concentration for indicating whether the crop N concentration is at an optimum level (Delloye et al., 2018; Zheng et al., 2022). The assessment of the N status of crops can be carried out using traditional vegetation indices (VI) such as Nitrogen Reflectance Index, Water Resistance Nitrogen Index, Nitrogen Planar Domain Index, Double-peak Canopy Nitrogen Index, Canopy Chlorophyll Content Index and Modified Chlorophyll Absorption Ratio Index (Clevers & Gitelson, 2013; Eitel et al., 2007; Zheng et al., 2022). However, compared to physical models such as PROSAIL inversion, estimation of canopy N with simple Vis has shown much higher validation errors when tested on the independent dataset (Bossung et al., 2022).

Nitrogen deficiency often manifests as chlorosis, a condition characterized by yellowing leaves due to a decline in chlorophyll content. Chlorophyll content can be measured easily enough and is therefore commonly used as a proxy for N identification. Most N estimation methods rely on close relationship between chlorophyll content and N (Eitel et al., 2007; Herrmann et al., 2010; Clevers & Gitelson, 2013; Delloye et al., 2018; Berger et al., 2020b; Croft et al., 2020). However, chlorophyll, the pigment present in plants, contains only a small proportion (less than 2%) of N (Verrelst et al., 2021). Meanwhile, nitrogen is an important and major component of proteins, and the relationship between N and proteins is stronger. Therefore, protein-nitrogen correlation has recently been increasingly taken into account in hyperspectral data validation models. It was advances in optical modeling, where the spectral decomposition of leaf dry matter content into nitrogen-based proteins and other carbon-based constituents has been parameterized, that have led to the successful development of models for CNC retrieval from hyperspectral data (Berger et al., 2020a; Féret et al., 2021; Verrelst et al., 2021). The PROSPECT-PRO model is often used to quantify LNC from Cp and the carbon-to-nitrogen ratio has been successfully estimated in leaves from the LOPEX dataset (Féret et al., 2021). PROSAIL inversion can be also successfully used to estimate canopy N from multispectral data such as Sentinel-2 images (Bossung et al., 2022).

Recently, machine learning (ML) methods have received a great deal of attention, due to their ability to process large amounts of input data and to perform nonlinear tasks (often from several different sources) (Chlingaryan et al., 2018; Yu et al., 2021; Zheng et al., 2022). Random Forest (RF), Gaussian Process Regression (GPR), and support vector regression (SVR) are among the most popular machine learning algorithms used for regression tasks, which involve predicting a continuous numerical value based on a set of input features. They are good at handling high-dimensional data and nonlinear relationships. RF is a supervised learning ensemble method that combines multiple decision trees to make predictions. It helps to reduce overfitting and makes RF also very robust to noise and outliers. GPR is a non-parametric regression algorithm that models the relationship between input features and target variables using a Gaussian process. GPR is simple to train and works well with a relatively small dataset. The goal of SVR is to find a hyperplane that best separates the support vectors. This hyperplane is then used to predict new data points. SVR is particularly good at handling nonlinear relationships between the features and the target variable. Vegetation indices (VI) are commonly used as input data for ML models in combination with satellite-derived climatic data or ground measurements from environmental loggers such as temperature and soil moisture (Yu et al., 2021; Zhang et al., 2021b). Hybrid methods, combining physically based models with machine learning algorithms, have been recognized as the most promising direction for N retrieval using hyperspectral (Berger et al., 2020b; Verrelst et al., 2021; Ranghetti et al., 2022) as well as multispectral remote sensing data (Dehghan-Shoar et al., 2023).

4.9.4 Canopy Temperature for Water Stress Management

One of the key factor in precision agriculture is assessment of crop water stress for sustainable water management. Direct field measurements such as pressure chamber and porometer are expensive and time consuming with limited spatiotemporal coverage. Plant closes stomata as a conservative response of to save water which leads to increase in canopy temperatures via reduced transpiration (Blonquist et al., 2009). Canopy temperature is an important driver of water stress in plants as it shows plants ability to optimize water uptake and release. Elevated canopy temperatures have been used as a reliable indicator of real time assessment of water stress in crops and is globally used to get insight into severity and duration of the stress for implementation of optimal irrigation strategies to alleviate the impacts of water scarcity (Jackson et al., 1981). Canopy temperature is a direct indicator of plants physiological response to the water deficit and it can be mapped explicitly by capturing thermal infrared radiation emitted by crops (Blonquist et al., 2009; Monteith, 1965). Thermal infrared cameras are able to record thermal radiation emitted by objects without direct contact at various spatial scales from handheld cameras to spaceborne satellite sensors. By capturing the thermal radiation emitted by plant canopies, these sensors provide a direct indicator of the plant's physiological response to water availability. Thermal images using FLIR handheld camera for example allow successful detection of irrigated areas due to their lower temperatures (Taghvaeian et al., 2013). Additionally, thermal infrared cameras can provide continuous monitoring of canopy temperature through the growing season allowing for timely interventions. With recent developments in UAVs, thermal images with high spatial and temporal resolutions have become available at a low cost (Khanal et al., 2017). Satellite data provides insight into crops health and water stress over large agriculture area by satellite derived canopy temperatures map. Some of the recent developments are spaceborne and airborne land-surface temperature, optical multispectral, hyperspectral, and LiDAR data for estimation of water stress in crops (Safdar et al., 2023).

Majority of studies have used crop water stress index (CWSI) to detect water stress in plant (Jackson et al., 1981). Another common approach is to use water deficit index (WDI) along with VI and composite surface temperature to estimate water stress for moderately vegetated fields (Moran et al., 1994). Available remotely sensed data products now represent almost all aspects of drought propagation being either meteorological, agricultural or hydrological drought and the relationships between them (West et al., 2019). Landsat 8 derived crop water stress index (CWSI) was used to estimate water stress and irrigation scheduling in sugarcane fields in Iran. The results suggested that the CWSI based approach to monitor water stress allows irrigation scheduling in sugarcane fields using satellite imagery without any need for ground ancillary data (Veysi et al., 2017).

UAVs equipped with thermal sensors can capture canopy temperature from an aerial perspective (Gómez-Candón et al., 2021; Zhang et al., 2019). This allows for medium-scale monitoring and provides a broader view of temperature variations across fields. Satellites equipped with thermal sensors can capture temperature data at a large scale. While spaceborne sensors may have coarser resolution compared to other methods (Fisher et al., 2017), they provide valuable information for regional or global assessments. Aircraft equipped with thermal sensors can capture high-resolution canopy temperature data over larger areas compared to UAVs which is particularly useful for research and large-scale agricultural assessments. Each method of water stress estimation has its own pros and cons and the choice of method depends on multiple factors such as the scale of measurement, spatial resolution required, accessibility of the area, and budget considerations. Integrating multiple methods may provide a comprehensive understanding of canopy temperature dynamics in different environments and under varying management practices.

Leaf or canopy temperature is a widely measured variable that has long been used as an empirical estimate of plant water stress. However, with a few supplemental measurements such as an estimate of plant height and environmental data available on automated weather stations, and application of biophysical principles, infrared measurement of canopy temperature can be used to calculate canopy stomatal conductance, a physiological variable derived from the energy balance for a plant canopy (Blonquist et al., 2009). Evapotranspiration (ET), that is, the loss of water to the atmosphere, can be also estimated from local to global scale using remotely sensed optical (VNIR) and thermal infrared (TIR) observations (Allen et al., 2011; Anderson et al., 2011; Bastiaanssen et al., 2005; Fisher et al., 2017; Gómez-Candón et al., 2021). ET is a crucial variable that links the water cycle (evaporation), energy cycle (latent heat flux), and carbon cycle (transpiration-photosynthesis trade-off) (Monteith, 1965; Fisher, 2013). It is the predominant variable needed for water management in agriculture to control irrigation levels so that applied water approximates atmospheric demand for ET (Allen et al., 1998; Anderson et al., 2011). Reliable estimation of ET requires a combination of accurate information from sensors of TIR (especially in field scales), VNIR (for vegetation characteristics), and meteorology. Incorporating complementary soil moisture and carbon cycle observations of vegetation response, such as chlorophyll, carotenoids, and SIF can also aid in better discriminating coupled water and carbon responses in ET assessment (Fisher et al., 2017). To facilitate adequate decision making it has been suggested that remote sensing based ET measurements should be acquired at high spatial resolutions (10–100 m) and high temporal resolutions (daily, diurnal) for agricultural applications (Allen et al., 2007; Anderson et al., 2011; Fisher et al., 2017).

4.10 CONCLUSIONS

Technological advancements from new satellite missions (hyperspectral, SIF) to proximal sensors in High Throughput Phenotyping Platforms (HTPP) open a new era in agricultural remote sensing. Faced with climate change and more frequent and intense stress events, there is a need for both phenotyping resilient breeding lines and providing early stress detection and precision farming recommendations. Driven by the need for new concepts in sustainable agriculture, the increased use of remote sensing approaches in field phenotyping and precision agriculture applications have been geared towards informing the adaptive management of agriculture to meet new global challenges.

Current trends in monitoring crop stress include increasing usage of satellite and unmanned aerial vehicle data in parallel with a shift in methods from simpler parametric approaches and vegetation indices towards more advanced physically based and hybrid models to estimate biophysical variables. Combined acquisition of data from multiple sensors and simultaneous retrieval of multiple plant traits combining multi-domain radiative transfer models and machine learning methods can be recognized as a way forward. Quantitative measurements are the basis of proximal and remote sensing approaches and should be accompanied by propagated uncertainty estimates. Assimilation of remote sensing estimated plant traits into integrated crop growth models is an emerging new trend. As a future outlook, frameworks such as MULTIPLY can facilitate combining multiple remote sensing data streams into crop model assimilation schemes to build up Digital Twins of agroecosystems, which may provide the most efficient way to detect various stresses and enable timely intervention with management decisions.

ACKNOWLEDGMENTS

The authors would like to thank ERA-NET for their financial support under the ReLive project. This work is funded through the partners of the Joint Call of the Co-fund ERA-Nets SusCrop (Grant N° 771134), FACCE ERA-GAS (Grant N° 696356), ICT-AGRI-FOOD (Grant N° 862665) and SusAn (Grant N° 696231).

REFERENCES

Addai, I.K., & Alimiyawo, M. (2015). Graphical determination of leaf area index and its relationship with growth and yield parameters of sorghum (Sorghum bicolor L. Moench) as affected by fertilizer application. Journal of Agronomy, 14, 272–278.

Allen, R.G., Pereira, L.S., Raes, D., & Smith, M. (1998). Crop evapotranspiration: Guidelines for computing crop water requirements (FAO irrigation and drainage paper), 328 pp., FAO—Food and Agric. Organ. Of the U. N., Rome.

Allen, R.G., Pereira, L.S., Howell, T.A., & Jensen, M.E. (2011). Evapotranspiration information reporting: I. Factors governing measurement accuracy. Agricultural Water Management, 98(6), 899–920.

Allen, R.G., Tasumi, M., & Trezza, R. (2007). Satellite-based energy balance for mapping evapotranspiration with internalized calibration (METRIC)-model. Journal of Irrigation and Drainage Engineering, 133, 380–394.

Anderson, M.C., Kustas, W.P., Norman, J.M., Hain, C.R., Mecikalski, J.R., Schultz, L., González-Dugo, M.P., Cammalleri, C., d'Urso, G., Pimstein, A., & Gao, F. (2011) Mapping daily evapotranspiration at field to continental scales using geostationary and polar orbiting satellite imagery. Hydrology and Earth System Sciences, 15, 223–239. https://doi.org/10.5194/hess-15-223-2011

Araus, J.L., & Kefauver, S.C. (2018). Breeding to adapt agriculture to climate change: Affordable phenotyping solutions. Current Opinion in Plant Biology, 45, 237–247.

Baghdadi, N., Hajj, M. El, Zribi, M., & Bousbih, S. (2017). Calibration of the water cloud model at C-band for winter crop fields and grasslands. Remote Sens (Basel), 9. https://doi.org/10.3390/rs9090969

Barricelli, B.R., Casiraghi, E., & Fogli, D., 2019. A survey on digital twin: Definitions, characteristics, applications, and design implications. IEEE Access. https://doi.org/10.1109/ACCESS.2019.2953499

Basso, B., & Liu, L. (2019) Chapter Four—seasonal crop yield forecast: Methods, applications, and accuracies. Advances in Agronomy, 154, 201–255. https://doi.org/10.1016/bs.agron.2018.11.002

Bastiaanssen, W.G.M., Noordman, E., Pelgrum, H., Davids, G., Thoreson, B., & Allen, R. (2005). SEBAL model with remotely sensed data to improve water-resources management under actual field conditions. Journal of Irrigation & Drainage Engineering, 131(1), 85–93.

Bates, J.S., Montzka, C., Schmidt, M., & Jonard, F. (2021). Estimating canopy density parameters time-series for winter wheat using UAS Mounted LiDAR. Remote Sensing, 13(4), 710.

Bauer, P., Stevens, B., & Hazeleger, W., 2021a. A digital twin of Earth for the green transition. Nature Climate Change. https://doi.org/10.1038/s41558-021-00986-y

Bavec, M., Vukovič, K., Grobelnik, S., Rozman, C., & Bavec, F. (2007) Leaf area in winter wheats: Response on seed rate and nitrogen application by different varieties. Journal of Central European Agriculture, 8, 337–342.

Berger, K., Machwitz, M., Kycko, M., Kefauver, S.C., Van Wittenberghe, S., Gerhards, M., Verrelst, J., Atzberger, C., van der Tol, C., Damm, A., Rascher, U., Herrmann, I., Paz, V.S., Fahrner, S., Pieruschka, R., Prikaziuk, E., Buchaillot, M.L., Halabuk, A., Celesti, M., Koren, G., Gormus, E.T., Rossini, M., Foerster, M., Siegmann, B., Abdelbaki, A., Tagliabue, G., Hank, T., Darvishzadeh, R., Aasen, H., Garcia, M., Pôças, I., Bandopadhyay, S., Sulis, M., Tomelleri, E., Rozenstein, O., Filchev, L., Stancile, G., & Schlerf, M. (2022). Multi-sensor spectral synergies for crop stress detection and monitoring in the optical domain: A review. Remote Sensing of Environment, 280: 113198. https://doi.org/10.1016/j.rse.2022.113198.

Berger, K., Verrelst, J., Féret, J.B., Hank, T., Wocher, M., Mauser, W., & Camps-Valls, G. (2020a). Retrieval of aboveground crop nitrogen content with a hybrid machine learning method. International Journal of Applied Earth Observation and Geoinformation, 92, 102174.

Berger, K., Verrelst, J., Féret, J. B., Wang, Z., Wocher, M., Strathmann, M., . . . & Hank, T. (2020b). Crop nitrogen monitoring: Recent progress and principal developments in the context of imaging spectroscopy missions. Remote Sensing of Environment, 242, 111758.

Blair, G.S., 2021. Digital twins of the natural environment. Patterns, 2. https://doi.org/10.1016/j.patter.2021.100359

Blonquist Jr, J.M., Norman, J.M., & Bugbee, B. (2009). Automated measurement of canopy stomatal conductance based on infrared temperature. Agricultural and Forest Meteorology, 149(11), 1931–1945.

Bossung, C., Schlerf, M., & Machwitz, M. (2022). Estimation of canopy nitrogen content in winter wheat from Sentinel-2 images for operational agricultural monitoring. Precision Agriculture, 23(6), 2229–2252.

Breda, N.J.J., 2003. Ground-based measurements of leaf area index: A review of methods, instruments and current controversies. Journal of Experimental Botany, 54, 2403–2417.

Brown, L.A., Meier, C., Morris, H., Pastor-Guzman, J., Bai, G., Lerebourg, C., Gobron, N., Lanconelli, C., Clerici, M., & Dash, J. (2020) Evaluation of global leaf area index and fraction of absorbed photosynthetically active radiation products over North America using Copernicus ground based observations for validation data. Remote Sensing of Environment, 247, 111935.

Caicedo, J.P.R., Verrelst, J., Munoz-Mari, J., Moreno, J., & Camps-Valls, G. (2014). Toward a semiautomatic machine learning retrieval of biophysical parameters. IEEE Journal of Selected Topics in Applied Earth Observations and Remote Sensing, 7, 1249–1259. https://doi.org/10.1109/JSTARS.2014.2298752

Camino, C., Gonzalez-Dugo, V., Hernandez, P., & Zarco-Tejada, P.J. (2019). Radiative transfer Vcmax estimation from hyperspectral imagery and SIF retrievals to assess photosynthetic performance in rainfed and irrigated plant phenotyping trials. Remote Sensing of Environment, 231, 111186.

Chlingaryan, A., Sukkarieh, S., & Whelan, B. (2018). Machine learning approaches for crop yield prediction and nitrogen status estimation in precision agriculture: A review. Computers and Electronics in Agriculture, 151, 61–69.

Choker, M., Baghdadi, N., Zribi, M., El Hajj, M., Paloscia, S., Verhoest, N.E.C., Lievens, H., & Mattia, F., 2017. Evaluation of the Oh, Dubois and IEM backscatter models using a large dataset of SAR data and experimental soil measurements. Water (Switzerland), 9. https://doi.org/10.3390/w9010038

Clevers, J.G., & Gitelson, A.A. (2013). Remote estimation of crop and grass chlorophyll and nitrogen content using red-edge bands on Sentinel-2 and-3. International Journal of Applied Earth Observation and Geoinformation, 23, 344–351.

Croft, H., Arabian, J., Chen, J.M., Shang, J., & Liu, J. (2020). Mapping within-field leaf chlorophyll content in agricultural crops for nitrogen management using Landsat-8 imagery. Precision Agriculture, 21, 856–880.

De Bei, R., Fuentes, S., Gilliham, M., Tyerman, S., Edwards, E., Bianchini, N., Smith, J., & Collins, C. (2016). VitiCanopy: A free computer app to estimate canopy vigor and porosity for grapevine. Sensors, 16(4), 585.

Dehghan-Shoar, M.H., Pullanagari, R.R., Kereszturi, G., Orsi, A.A., Yule, I.J., & Hanly, J. (2023). A unified physically based method for monitoring grassland nitrogen concentration with Landsat 7, Landsat 8, and Sentinel-2 satellite data. Remote Sensing, 15(10), 2491.

Delegido, J., Verrelst, J., Rivera, J.P., Ruiz-Verdú, A., & Moreno, J. (2015). Brown and green LAI mapping through spectral indices. International Journal Applied Earth Observation and Geoinformation, 35(Part B), 350–358.

Delloye, C., Weiss, M., & Defourny, P. (2018). Retrieval of the canopy chlorophyll content from Sentinel-2 spectral bands to estimate nitrogen uptake in intensive winter wheat cropping systems. Remote Sensing of Environment, 216, 245–261.

Dong, T., Liu, J., Qian, B., He, L., Liu, J., Wang, R., Jing, Q., Champagne, C., McNairn, H., Powers, J., Shi, Y., Chen, J.M., & Shang, J. (2020). Estimating crop biomass using leaf area index derived from Landsat 8 and Sentinel-2 data. ISPRS Journal of Photogrammetry and Remote Sensing, 168, 236–250. https://doi.org/10.1016/j.isprsjprs.2020.08.003

Ehlert, D., Adamek, R., & Horn, H.J. (2009) Laser rangefinder-based measuring of crop biomass under field conditions. Precision Agriculture, 10, 395–408. https://doi.org/10.1007/s11119-009-9114-4

Eitel, J.U.H., Long, D.S., Gessler, P.E., & Smith, A.M.S. (2007). Using in-situ measurements to evaluate the new RapidEye™ satellite series for prediction of wheat nitrogen status. International Journal of Remote Sensing, 28(18), 4183–4190.

El Saddik, A. (2018). Digital twins: The convergence of multimedia technologies. IEEE Multimedia, 25. https://doi.org/10.1109/MMUL.2018.023121167

Estévez, J., Salinero-Delgado, M., Berger, K., Pipia, L., Rivera-Caicedo, J.P., Wocher, M., Reyes-Muñoz, P., Tagliabue, G., Boschetti, M., & Verrelst, J. (2022). Gaussian processes retrieval of crop traits in Google Earth Engine based on Sentinel-2 top-of-atmosphere data. Remote Sensing Environment, 273. https://doi.org/10.1016/j.rse.2022.112958

Fang, H., Baret, F., Plummer, S., & Schaepman-Strub, G. (2019). An overview of global Leaf Area Index (LAI): Methods, products, validation, and applications. Reviews of Geophysics, 57, 739–799.

Féret, J.B., Berger, K., De Boissieu, F., & Malenovský, Z. (2021). PROSPECT-PRO for estimating content of nitrogen-containing leaf proteins and other carbon-based constituents. Remote Sensing of Environment, 252, 112173. https://doi.org/10.1016/j.rse.2020.112173

Feret, J.B., François, C., Asner, G.P., Gitelson, A.A., Martin, R.E., Bidel, L.P.R., Ustin, S.L., le Maire, G., & Jacquemoud, S. (2008). PROSPECT-4 and 5: Advances in the leaf optical properties model separating photosynthetic pigments. Remote Sensing of Environment, 112, 3030–3043. https://doi.org/10.1016/j.rse.2008.02.012

Fisher, J.B. (2013). Land-atmosphere interactions: Evapotranspiration. In Encyclopedia of Remote Sensing, edited by E. Njoku, pp. 1–5, Springer, Berlin.

Fisher, J.B., Melton, F., Middleton, E., Hain, C., Anderson, M., Allen, R., McCabe, M.F., Hook, S., Baldocchi, D., Townsend, P.A., Kilic, A., Tu, K., Miralles, D.D., Perret, J., Lagouarde, J.P., Waliser, D., Purdy, A.J., French, A., Schimel, D., Famiglietti, J.S., Stephens, G., & Wood, E.F. (2017). The future of evapotranspiration: Global requirements for ecosystem functioning, carbon and climate feedbacks, agricultural management, and water resources. Water Resources Research, 53, 2618–2626. https://doi.org/10.1002/2016WR020175

Gherboudj, I., Maga gi, R., Berg, A.A., & Toth, B. 2011. Soil moisture retrieval over agriculture fields from multi-polarized and multi-angular Radarsat-2 SAR data. Remote Sensing of Environment, 115, 33–43.

Gómez-Candón, D., Bellvert, J., & Royo, C. (2021) Performance of the Two-Source Energy Balance (TSEB) model as a tool for monitoring the response of durum wheat to drought by high-throughput field phenotyping. Frontiers in Plant Science, 12: 658357. https://doi.org/10.3389/fpls.2021.658357.

Gómez-Dans, J.L., Lewis, P.E., & Disney, M. (2016). Efficient emulation of radiative transfer codes using gaussian processes and application to land surface parameter inferences. Remote Sensing (Basel), 8, e119. https://doi.org/10.3390/rs8020119

Hallik, L., Kazantsev, T., Kuusk, A., Galmés, J., Tomás, M., & Niinemets, Ü. (2017) Generality of relationships between leaf pigment contents and spectral vegetation indices in Mallorca (Spain). Regional Environmental Change, 17, 2097–2109. https://doi.org/10.1007/s10113-017-1202-9

Herrmann, I., Karnieli, A., Bonfil, D.J., Cohen, Y., & Alchanatis, V. (2010). SWIR-based spectral indices for assessing nitrogen content in potato fields. International Journal of Remote Sensing, 31(19), 5127–5143.

Hicks, S.K., & Lascano, R.J. (1995). Estimation of leaf area index for cotton canopies using the LI-COR LAI-2000 plant canopy analyzer. Agronomy Journal, 87(3), 458–464.

Huang, J., Gómez-Dans, J.L., Huang, H., Ma, H., Wu, Q., Lewis, P.E., Liang, S., Chen, Z., Xue, J.H., Wu, Y., Zhao, F., Wang, J., & Xie, X. (2019a). Assimilation of remote sensing into crop growth models: Current status and perspectives. Agricultural and Forest Meteorology, 276, 107609.

Huang, J., Ma, H., Sedano, F., Lewis, P., Liang, S., Wu, Q., Su, W., Zhang, X., & Zhu, D. (2019b). Evaluation of regional estimates of winter wheat yield by assimilating three remotely sensed reflectance datasets into the coupled WOFOST-prosail model. European Journal of Agronomy, 102, 1–13.

Ilniyaz, O., Du, Q., Shen, H., He, W., Feng, L., Azadi, H., . . . & Chen, X. (2023). Leaf area index estimation of pergola-trained vineyards in arid regions using classical and deep learning methods based on UAV-based RGB images. Computers and Electronics in Agriculture, 207, 107723.

Ines, A.V.M., Das, N.N., Hansen, J.W., & Njoku, E.G. (2013). Assimilation of remotely sensed soil moisture and vegetation with a crop simulation model for maize yield prediction Remote Sensing of Environment, 138, 149–164, 10.1016/j.rse.2013.07.018

Jackson, R.D., Idso, S.B., Reginato, R.J., & Pinter, P.J. (1981). Canopy temperature as a crop water stress indicator. Water Resources Research, 17(4), 1133–1138. https://doi.org/10.1029/WR017i004p01133

Jacquemoud, S., & Baret, F. (1990). PROSPECT: A model of leaf optical properties spectra. Remote Sensing of Environment, 34, 75–91. https://doi.org/10.1016/0034-4257(90)90100-Z

Jacquemoud, S., Zarco-Tejada, P.J., Verhoef, W., Asner, G.P., Ustin, S.L., Baret, F., & François, C. (2006). PROSPECT+SAIL: 15 Years of use for land surface characterization. International Geoscience and Remote Sensing Symposium (IGARSS), 00, 1992–1995. https://doi.org/10.1109/IGARSS.2006.516

Jiang, J., Comar, A., Burger, P., Bancal, P., Weiss, M., & Baret, F. (2018). Estimation of leaf traits from reflectance measurements: Comparison between methods based on vegetation indices and several versions of the PROSPECT model. Plant Methods, 14. https://doi.org/10.1186/s13007-018-0291-x

Kaminski, T., Pinty, B., Voßbeck, M., Lopatka, M., Gobron, N., & Robustelli, M. 2017. Consistent retrieval of land surface radiation products from EO, including traceable uncertainty estimates. Biogeosciences, 14, 2527–2541, https://doi.org/10.5194/bg-14-2527-2017

Kang, Y., & Özdoğan, M. (2019). Field-level crop yield mapping with Landsat using a hierarchical data assimilation approach. Remote Sensing of Environment, 228, 144–163.

Kasampalis, D.A., Alexandridis, T.K., Deva, C., Challinor, A., Moshou, D., & Zalidis, G., (2018). Contribution of remote sensing on crop models: A review. Journal of Imaging. https://doi.org/10.3390/jimaging4040052

Khanal, S., Fulton, J., & Shearer, S. (2017). An overview of current and potential applications of thermal remote sensing in precision agriculture. Computers and Electronics in Agriculture, 139, 22–32. 10.1016/j.compag.2017.05.001

Kimes, D.S., Nelson, R.F., Manry, M.T., & Fung, A.K. (1998). Attributes of neural networks for extracting continuous vegetation variables from optical and radar measurements. International Journal of Remote Sensing, 19(14), 2639–2663.

Kross, A., McNairn, H., Lapen, D., Sunohara, M., & Champagne, C. (2015). Assessment of RapidEye vegetation indices for estimation of leaf area index and biomass in corn and soybean crops. International Journal of Applied Earth Observation and Geoinformation, 34(1), 235–248. https://doi.org/10.1016/j.jag.2014.08.002

Kuusk, A. (2001). A two-layer canopy reflectance model. Journal of Quantitative Spectroscopy and Radiative Transfer, 71(1), 1–9. https://doi.org/10.1016/S0022-4073(01)00007-3

Lauvernet, C., Baret, F., Hascoët, L., Buis, S., & Le Dimet, F.X. (2008). Multitemporal-patch ensemble inversion of coupled surface-atmosphere radiative transfer models for land surface characterization. Remote Sensing of Environment, 112. https://doi.org/10.1016/j.rse.2007.06.027

Lewis, P., Gómez-Dans, J., Kaminski, T., Settle, J., Quaife, T., Gobron, N., Styles, J., & Berger, M. (2012). An Earth Observation Land Data Assimilation System (EO-LDAS). Remote Sensing of Environment, 120, 219–235. https://doi.org/10.1016/j.rse.2011.12.027

Li, J., & Wang, S. (2018). Using SAR-derived vegetation descriptors in a water cloud model to improve soil moisture retrieval. Remote Sensing, 10, 1370.

Machwitz, M., et al. (2021). Bridging the gap between remote sensing and plant phenotyping— challenges and opportunities for the next generation of sustainable agriculture. Frontiers in Plant Science, 2334.

Maes, W.H., & Steppe, K. (2019). Perspectives for remote sensing with unmanned aerial vehicles in precision agriculture. Trends in Plant Science, 24(2), 152–164.

Mittaz, J., Merchant, C., & Woolliams, E.R. (2019). Applying principles of metrology to historical Earth observations from satellites. Metrologia, 56(3), https://doi.org/10.1088/1681-7575/ab1705

Mohammed, G.H., et al. (2019). Remote sensing of solar-induced chlorophyll fluorescence (SIF) in vegetation: 50 years of progress. Remote Sensing of Environment, 231, 111177.

Monteith, J.L. (1965). Evaporation and the environment. Symposia of the Society for Experimental Biology, 19, 205–234.

Moran, M.S., Clarke, T.R., Inoue, Y., & Vidal, A. (1994). Estimating crop water deficit using the relation between surface-air temperature and spectral vegetation index. Remote Sensing of Environment, 49(3), 246–263. https://doi.org/10.1016/0034-4257(94)90020-5

Mousivand, A., Menenti, M., Gorte, B., & Verhoef, W. (2014). Global sensitivity analysis of the spectral radiance of a soil-vegetation system. Remote Sensing Environmental, 145. https://doi.org/10.1016/j.rse.2014.01.023

Nguy-Robertson, A., Gitelson, A., Peng, Y., Viña, A., Arkebauer, T., & Rundquist, D. (2012). Green leaf area index estimation in maize and soybean: Combining vegetation indices to achieve maximal sensitivity. Agronomy Journal, 104(5), 1336–1347. https://doi.org/10.2134/agronj2012.0065

Novelli, F., Spiegel, H., Sandén, T., & Vuolo, F. (2019). Assimilation of Sentinel-2 leaf area index data into a physically-based crop growth model for yield estimation. Agronomy, 9, 255. https://doi.org/10.3390/agronomy9050255

Pan, Z., Hu, Y., & Cao, B. (2017). Construction of smooth daily remote sensing time series data: A higher spatiotemporal resolution perspective. Open Geospatial Data, Software and Standards, 2. https://doi.org/10.1186/s40965-017-0038-z

Pandey, A., & Jain, K. (2022). An intelligent system for crop identification and classification from UAV images using conjugated dense convolutional neural network. Computers and Electronics in Agriculture, 192, 106543.

Patil, P.B., Biradar, P., Bhagawathi, A.U., & Hejjegar, I. (2018). A review on leaf area index of horticulture crops and its importance. International Journal of Current Microbiology Applied Sciences, 7, 505–513.

Pearce, R.B., Brown, R.H., & Blaser, R.E. (1965). Relationships between leaf area index, light interception and net photosynthesis in orchardgrass. Crop Science, 5, 553–556.

Pérez-Cutillas, P., Pérez-Navarro, A., Conesa-García, C., Zema, D.A., & Amado-Álvarez, J.P. (2023). What is going on within google earth engine? A systematic review and meta-analysis. Remote Sensing Applications, 29, 100907. https://doi.org/10.1016/J.RSASE.2022.100907

Pieruschka, R., & Schurr, U. (2019). Plant phenotyping: Past, present, and future. Plant Phenomics, 2019, 7507131. https://doi.org/10.34133/2019/7507131

Pinty, B., Gobron, N., Widlowski, J.L., Gerstl, S.A.W., Verstraete, M.M., Antunes, M., Bacour, C., Gascon, F., Gastellu, J.P., Goel, N., Jacquemoud, S., North, P., Qin, W., & Thompson, R. (2001). Radiation transfer model intercomparison (RAMI) exercise. Journal of Geophysical Research Atmospheres, 106, 11937–11956. https://doi.org/10.1029/2000JD900493

Pipia, L., Muñoz-Marí, J., Amin, E., Belda, S., Camps-Valls, G., & Verrelst, J. (2019). Fusing optical and SAR time series for LAI gap filling with multioutput Gaussian processes. Remote Sensing Environment, 235, 111452. https://doi.org/10.1016/J.RSE.2019.111452

Purcell, W., & Neubauer, T. (2023). Digital twins in agriculture: A state-of-the-art review. Smart Agricultural Technology, 3, 100094. https://doi.org/10.1016/J.ATECH.2022.100094

Quaife, T., Pinnington, E.M., Marzahn, P., Kaminski, T., Vossbeck, M., Timmermans, J., Isola, C., Rommen, B., & Loew, A. (2022). Synergistic retrievals of leaf area index and soil moisture from Sentinel-1 and Sentinel-2. International Journal of Image and Data Fusion, 1–18. https://doi.org/10.1080/19479832.2022.2149629

Ranghetti, M., Boschetti, M., Ranghetti, L., Tagliabue, G., Panigada, C., Gianinetto, M., . . . & Candiani, G. (2022). Assessment of maize nitrogen uptake from PRISMA hyperspectral data through hybrid modelling. European Journal of Remote Sensing, 1–17.

Revill, A., Florence, A., MacArthur, A., Hoad, S., Rees, R., & Williams, M. (2020). Quantifying uncertainty and bridging the scaling gap in the retrieval of leaf area index by coupling Sentinel-2 and UAV observations. Remote Sensing, 12(11), 1843.

Rosenzweig, C., Jones, J.W., Hatfield, J.L., Ruane, A.C., Boote, K.J., Thorburn, P., Antle, J.M., Nelson, G.C., Porter, C., Janssen, S., Asseng, S., Basso, B., Ewert, F., Wallach, D., Baigorria, G., & Winter, J.M. (2013). The Agricultural Model Intercomparison and Improvement Project (AgMIP): Protocols and pilot studies. Agricultural and Forest Meteorology, 170, 166–182.

Ross, J. (1981). The Radiation Regime and Architecture of Plant Stands, 1st Edition, p. 391, Dr W. Junk Publishers, The Hague, The Netherlands. https://doi.org/10.1007/978-94-009-8647-3

Safdar, M., Shahid, M.A., Sarwar, A., Rasul, F., Majeed, M.D., & Sabir, R.M. (2023) Crop water stress detection using remote sensing techniques. Environmental Science Proceedings, 25, 20. https://doi.org/10.3390/ECWS-7-14198

Salcedo-Sanz, S., Ghamisi, P., Piles, M., Werner, M., Cuadra, L., Moreno-Martínez, A., Izquierdo-Verdiguier, E., Muñoz-Marí, J., Mosavi, A., & Camps-Valls, G. (2020). Machine learning information fusion in Earth observation: A comprehensive review of methods, applications and data sources. Information Fusion, 63, 256–272. https://doi.org/10.1016/J.INFFUS.2020.07.004

Schauberger, B., Jägermeyr, J., & Gornott, C. (2020). A systematic review of local to regional yield forecasting approaches and frequently used data resources. European Journal of Agronomy, 120, 126153. https://doi.org/10.1016/j.eja.2020.126153

Singhal, G., Bansod, B., Mathew, L., Goswami, J., Choudhury, B.U., & Raju, P.L.N. (2019). Estimation of leaf chlorophyll concentration in turmeric (*Curcuma longa*) using high-resolution unmanned aerial vehicle imagery based on kernel ridge regression. Journal of the Indian Society of Remote Sensing, 47, 1111–1122. https://doi.org/10.1007/s12524-019-00969-9

Taghvaeian, S., Luis Chávez, J., Trout, T.J., Dejonge, K.C., Chávez, J.L., Altenhofen, J., Trout, T., & Dejonge, K. (2013). Remote sensing for evaluating crop water stress at field scale using infra-red thermography: Potential and limitations. www.researchgate.net/publication/290484352

Ustin, S.L., & Middleton, E.M. (2021). Current and near-term advances in Earth observation for ecological applications. Ecological Processes, 10(1), 1–57.

Van der Tol, C., Verhoef, W., Timmermans, J., Verhoef, A., & Su, Z. (2009). An integrated model of soil-canopy spectral radiances, photosynthesis, fluorescence, temperature and energy balance. Biogeosciences, 6, 3109–3129. https://doi.org/10.5194/bg-6-3109-2009

Verhoef, W. (1984). Light scattering by leaf layers with application to canopy reflectance modeling: The SAIL model. Remote Sensing Environment, 16, 125–141. https://doi.org/10.1016/0034-4257(84)90057-9

Verrelst, J., Camps-Valls, G., Muñoz-Marí, J., Rivera, J.P., Veroustraete, F., Clevers, J.G.P.W., & Moreno, J. (2015). Optical remote sensing and the retrieval of terrestrial vegetation bio-geophysical properties—A review. ISPRS Journal of Photogrammetry and Remote Sensing, 108, 273–290. https://doi.org/10.1016/j.isprsjprs.2015.05.005

Verrelst, J., Malenovský, Z., Van der Tol, C., Camps-Valls, G., Gastellu-Etchegorry, J.P., Lewis, P., North, P., & Moreno, J. (2019). Quantifying vegetation biophysical variables from imaging spectroscopy data: A review on retrieval methods. Surveys in Geophysics, 40, 589–629. https://doi.org/10.1007/s10712-018-9478-y

Verrelst, J., Rivera, J.P., Alonso, L., Lindstrot, R., & Moreno, J. (2012). Potential retrieval of biophysical parameters from FLORIS, S3-OLCI and its synergy. In International Geoscience and Remote Sensing Symposium (IGARSS). https://doi.org/10.1109/IGARSS.2012.6352021

Verrelst, J., Rivera, J.P., Alonso, L., & Moreno, J. (2011). ARTMO: An automated radiative transfer models operator toolbox for automated retrieval of biophysical parameters through model inversion. In EARSeL 7th SIG-Imaging Spectroscopy Workshop 2011.

Verrelst, J., Rivera Caicedo, J.P., Muñoz-Marí, J., Camps-Valls, G., & Moreno, J. (2017). SCOPE-based emulators for fast generation of synthetic canopy reflectance and sun-induced fluorescence Spectra. Remote Sensing (Basel), 9. https://doi.org/10.3390/rs9090927

Verrelst, J., Rivera-Caicedo, J.P., Reyes-Muñoz, P., Morata, M., Amin, E., Tagliabue, G., . . . & Berger, K. (2021). Mapping landscape canopy nitrogen content from space using PRISMA data. ISPRS Journal of Photogrammetry and Remote Sensing, 178, 382–395.

Veysi, S., Naseri, A.A., Hamzeh, S., & Bartholomeus, H. (2017). A satellite based crop water stress index for irrigation scheduling in sugarcane fields. Agricultural Water Management, 189, 70–86. https://doi.org/10.1016/j.agwat.2017.04.016

Wang, L., Qi, F., Shen, X., & Huang, J. (2019). Monitoring multiple cropping index of Henan Province, China Based on MODIS-EVI time series data and savitzky-golay filtering algorithm. CMES—Computer Modeling in Engineering and Sciences, 119. https://doi.org/10.32604/cmes.2019.04268

Wang, Z., Ye, L., Jiang, J., Fan, Y., & Zhang, X. (2022). Review of application of EPIC crop growth model. Ecological Modelling. https://doi.org/10.1016/j.ecolmodel.2022.109952

Watson, D.J. (1947). Comparative physiological studies on the growth of field crops: I. Variation in net assimilation rate and leaf area between species and varieties, and within and between years. Annals of Botany, 11, 41–76.

Weiss, M., Jacob, F., & Duveiller, G. (2020). Remote sensing for agricultural applications: A meta-review. Remote Sensing of Environment, 236. https://doi.org/10.1016/j.rse.2019.111402

Wen, W., Timmermans, J., Chen, Q., & van Bodegom, P.M. (2021). A review of remote sensing challenges for food security with respect to salinity and drought threats. Remote Sensing (Basel). https://doi.org/10.3390/rs13010006

West, H., Quinn, N., & Horswell, M. (2019). Remote sensing for drought monitoring & impact assessment: Progress, past challenges and future opportunities. Remote Sensing Environment, 232, 111291. https://doi.org/10.1016/j.rse.2019.111291

Wicks, G., Nordquist, P., Baenziger, P., Klein, R., Hammons, R., & Watkins, J. (2004). Winter wheat cultivar characteristics affect annual weed suppression. Weed Technology, 18(4), 988–998. https://doi.org/10.1614/WT-03-158R1

Widlowski, J.L., Taberner, M., Pinty, B., Bruniquel-Pinel, V., Disney, M., Fernandes, R., Gastellu-Etchegorry, J.P., Gobron, N., Kuusk, A., Lavergne, T., Leblanc, S., Lewis, P.E., Martin, E., Mõttus, M., North, P.R.J., Qin, W., Robustelli, M., Rochdi, N., Ruiloba, R., Soler, C., Thompson, R., Verhoef, W., Verstraete, M.M., & Xie, D. (2007). Third Radiation Transfer Model Intercomparison (RAMI) exercise: Documenting progress in canopy reflectance models. Journal of Geophysical Research Atmospheres, 112. https://doi.org/10.1029/2006JD007821

Wu, B., Gommes, R., Zhang, M., Zeng, H., Yan, N., Zou, W., Zheng, Y., Zhang, N., Chang, S., Xing, Q., et al. (2015). Global crop monitoring: A satellite-based hierarchical approach. Remote Sensing, 7, 3907–3933. https://doi.org/10.3390/rs70403907

Wu, B., Ma, Z., & Yan, N. (2020). Agricultural drought mitigating indices derived from the changes in drought characteristics. Remote Sensing Environment, 244, 111813. https://doi.org/10.1016/j.rse.2020.111813

Wu, B., Zhang, M., Zeng, H., Tian, F., Potgieter, A.B., Qin, X., Yan, N., Chang, S., Zhao, Y., Dong, Q., Boken, V., Plotnikov, D., Guo, H., Wu, F., Zhao, H., Deronde, B., Tits, L., & Loupian, E. (2023). Challenges and opportunities in remote sensing-based crop monitoring: A review. National Science Review, 10(4), nwac290, https://doi.org/10.1093/nsr/nwac290

Yan, G., Hu, R., Luo, J., Weiss, M., Jiang, H., Mu, X., Xie, D., & Zhang, W. (2019). Review of indirect optical measurements of leaf area index: Recent advances, challenges, and perspectives. Agricultural and Forest Meteorology, 265, 390–411.

Yang, J., Xing, M., Tan, Q., Shang, J., Song, Y., Ni, X., . . . & Xu, M. (2023). Estimating effective leaf area index of winter wheat based on UAV point cloud data. Drones, 7(5), 299.

Yang, K., Gong, Y., Fang, S., Duan, B., Yuan, N., Peng, Y., Wu, X., & Zhu, R. (2021). Combining spectral and texture features of UAV images for the remote estimation of rice LAI throughout the entire growing season. Remote Sensing, 13, 3001.

Yu, J., Wang, J., Leblon, B., & Song, Y. (2021). Nitrogen estimation for wheat using UAV-based and satellite multispectral imagery, topographic metrics, leaf area index, plant height, soil moisture, and machine learning methods. Nitrogen, 3(1), 1–25.

Yuan, H., Dai, Y., Xiao, Z., Ji, D., & Shangguan, W. (2011). Reprocessing the MODIS Leaf Area Index products for land surface and climate modelling. Remote Sensing Environment, 115, 1171–1187. https://doi.org/10.1016/j.rse.2011.01.001

Zhang, D., Geng, X., Chen, W., Fang, L., Yao, R., Wang, X., & Zhou, X. (2021). Inconsistency of global vegetation dynamics driven by climate change: Evidences from spatial regression. Remote Sensing (Basel), 13. https://doi.org/10.3390/rs13173442

Zhang, L., Niu, Y., Zhang, H., Han, W., Li, G., Tang, J., & Peng, X. (2019). Maize canopy temperature extracted from UAV thermal and RGB imagery and its application in water stress monitoring. Frontiers in Plant Science, 10. https://doi.org/10.3389/fpls.2019.01270

Zhang, L., Zhang, Z., Luo, Y., Cao, J., Xie, R., & Li, S. (2021b). Integrating satellite-derived climatic and vegetation indices to predict smallholder maize yield using deep learning. Agricultural and Forest Meteorology, 311. https://doi.org/10.1016/j.agrformet.2021.108666

Zhang, Y., Qu, Y., Wang, J., Liang, S., & Liu, Y. (2012). Estimating leaf area index from MODIS and surface meteorological data using a dynamic bayesian network. Remote Sensing Environment, 127, 30–43.

Zheng, J., Song, X., Yang, G., Du, X., Mei, X., & Yang, X. (2022). Remote sensing monitoring of rice and wheat canopy nitrogen: A review. Remote Sensing, 14(22), 5712.

Zhuo, W., Fang, S., Gao, X., Wang, L., Wu, D., Fu, S., Wu, Q., & Huang, J. (2022). Crop yield prediction using MODIS LAI, TIGGE weather forecasts and WOFOST model: A case study for winter wheat in Hebei, China during 2009–2013. International Journal of Applied Earth Observation and Geoinformation, 106. https://doi.org/10.1016/j.jag.2021.102668

5 Agriculture

Clement Atzberger and Markus Immitzer

ACRONYMS AND DEFINITIONS

EO	Earth Observation
GHG	Greenhouse Gas Emissions
GOES	Geostationary Operational Environmental Satellite
LAI	Leaf Area Index
LiDAR	Light Detection and Ranging
LSP	Land Surface Phenology
LST	Land Surface Temperature
LUE	Light Use Efficiency
LULC	Land Use, Land Cover
MARS	Monitoring Agricultural Resources action of the European Commission
MODIS	Moderate-Resolution Imaging Spectroradiometer
NDVI	Normalized Difference Vegetation Index
NOAA	National Oceanic and Atmospheric Administration
NPP	Net Primary Productivity
OBIA	Object Oriented Image Analysis
PAR	Photosynthetically Active Radiation
SAR	Synthetic Aperture Radar
VCI	Vegetation Condition Index
VI	Vegetation Index

5.1 INTRODUCTION

The development of remote sensing in the 1970s was originally driven by agricultural information needs (Becker-Reshef et al., 2010; Wulder et al., 2022). Since that time, the importance of Earth Observation (EO) for the agricultural sector has continued to increase and non-agricultural applications were added as abandonly exemplified in this compendium.

The utility of EO in agriculture can be explained by the specific challenges the agricultural sector faces and the fact that agricultural activities significantly contribute to the current planetary crises (Box 5.1).

5.1.1 What's Unique about Agricultural Activities?

Agricultural support and monitoring applications at different cadences (Table 5.1) are strongly needed as several external drivers require a quick and widespread adaptation of agriculture practices:

- The environmental impacts of agriculture must be minimized (Zaks and Kucharik, 2011) while increasing the carbon stored in the soil and improving soil health (Lal, 2009; Tiessen et al., 1994).
- More food calories must be produced on less land so that land can be spared for reforestation and restauration activities (Paul and Knoke, 2015; Rayden et al., 2023), while feeding the nine-billion people predicted by mid-century (Foley et al., 2011)—this should also

DOI: 10.1201/9781003541165-7

BOX 5.1. SPECIFIC ISSUES WITHIN THE AGRICULTURAL VALUE CHAIN THAT FAVOR THE USE OF EARTH OBSERVATION (EO) TECHNIQUES

Challenges favoring the use of EO data	Impacts pushing for an EO-based monitoring
Growing conditions are highly variable in time and space	Release of GHG (mainly from the livestock sector)
Agricultural products and management practices are highly variable	Main direct/indirect driver for deforestation
	Main global freshwater user
Agricultural production follows strong seasonal patterns	Groundwater pollution
Pests, diseases, and other unfavorable growing conditions can quickly affect the agricultural productivity	Land degradation
	Soil health impacts
Many agricultural items are perishable	Impacts on biodiversity
Many countries are still not food secure but subject to large price volatilities	

As a result of these particularities:

- agricultural activities must be monitored to track its status and impacts;
- decision support is needed to optimize agricultural activities.

The necessary information should be provided in a cost-efficient, repetitive, and transparent manner, while lag times should be minimized. Depending on the information need, reasonably detailed pixel sizes (GSD) are required, usually paired with a high revisit frequency (Bruinsma, 2003). Only remote sensing can offer this capacity at local to global scales, while offering a coherent historical look back in time (Hunt et al., 2019b).

The unique capabilities of EO explain the large range of agricultural applications developed till date (Table 5.1)—with this chapter focusing on yield and production estimates. Most probably, additional application will emerge in the future, as the observational capacities of the various platforms (satellites, aircraft, drones, etc.) continue to improve, while additional sensor modalities are developed (see Chapter 1 of Volume 1 of the *Remote Sensing Handbook*) and improved analytics become available.

be accompanied by shifts in diets (Garnett, 2013; Searchinger et al., 2019) and efforts to minimize food waste (Kummu et al., 2012).

- To mitigate climate change, low productivity farms (for example grazing on degraded pastures) must be converted into either forests or agroforestry systems (Knoke et al., 2014) while avoiding leakage problems, planting of non-native species, and conversion of non-forest biomes (da Conceição Bispo et al., 2024; Jose and Bardhan, 2012; Lorenz and Lal, 2014).
- As food and farming are responsible for more biodiversity loss than any other sector (BirdLife International, 2022; Soulé, 1985; Tilman et al., 2017), the conversion of forest, etc., to arable land must stop, as well as the use of agricultural land for biofuel production and urban expansion (Demirbas and Balat, 2006; Grau et al., 2013; Searchinger et al., 2022).

EO-based agricultural monitoring systems are excellent tools to monitor the progress of the necessary adaptations (Section 5.5). With its historical data record, policy makers and stakeholders cannot only be informed about the current state of the agricultural sector, but also the pathway that led to it. This delivers critical feedback to decision makers regarding the actual impact of their policies and investments. Reliable information also facilitates risk reduction. Information derived across a range of scales also improves statistical analyses, enabling a timely and accurate national to regional agricultural statistical reporting. The landowners and farmer themselves can strongly benefit from improved information with respect to various decision-making tasks, as outlined in Chapter 8 of this volume.

TABLE 5.1

Major EO-Based Agricultural Support and Monitoring Applications Together with Appropriate Scales, Typical Spatial Coverages, and Cadences. For each application, one recommended reading is provided: (a) applications with seasonal to annual updating cadence; (b) applications with weekly to monthly updating cadence, and (c) applications with daily to sub-daily updating cadence.

	Scale						Cover			Cadence						
(a)	**centimeter**	**decimeter**	**meter**	**decameter**	**hectometer**	**kilometer**	**local**	**regional**	**continental**	**hourly**	**daily**	**weekly**	**monthly**	**seasonally**	**Annually**	**Reference**
Crop rotation monitoring				x				x	x						x	Blickensdörfer et al. (2022)
Yield gap quantification			x	x	x	x	x	x	x					x	x	Lobell (2013)
Mapping of water infiltration capacity				x	x		x	x						x	x	Wassenaar et al. (2005)
Assessment of soil water holding capacity				x	x		x	x						x	x	Ferrant et al. (2016)
Monitoring land use intensity					x	x		x	x						x	Kuemmerle et al. (2013)
Identification of cropping patterns					x	x		x	x						x	Mahlayeye et al. (2022)
Identification of agricultural land use systems					x	x		x	x						x	Bellón et al. (2017)
Monitoring of agricultural expansion					x	x		x	x						x	Arvor et al. (2013)
Monitoring of agricultural abandonment					x	x		x	x						x	Löw et al. (2015)
Support for selective harvesting			x	x	x		x							x		Johnson et al. (2001)
Delineation of SMZ			x	x			x								x	Reyes et al. (2023)
Mapping of soil types				x	x		x	x	x						x	Hahn and Gloaguen (2008)
Assessment of soil health status			x	x	x		x	x							x	Rinot et al. (2019)
Mapping of Soil organic carbon (SOC)			x	x			x	x							x	Stevens et al. (2010)
Mapping of soil texture				x	x		x	x							x	Barnes et al. (2000)
Mapping of soil properties			x	x	x		x	x							x	Forkuor et al. (2017)
Salinization effects			x	x			x	x						x	x	Wen et al. (2023)
Cropping practices			x	x	x			x	x					x	x	Bégué et al. (2018)
Cover crop adoption				x	x			x	x					x	x	Seifert et al. (2018)
Assessment of GHG budgets			x	x	x	x	x	x						x	x	IPCC (2022)

(Continued)

TABLE 5.1 (*Continued*)

Major EO-Based Agricultural Support and Monitoring Applications Together with Appropriate Scales, Typical Spatial Coverages, and Cadences. For each application, one recommended reading is provided: (a) applications with seasonal to annual updating cadence; (b) applications with weekly to monthly updating cadence, and (c) applications with daily to sub-daily updating cadence.

	Scale						Cover			Cadence						
(b)	centimeter	decimeter	meter	decameter	hectometer	kilometer	local	regional	continental	hourly	daily	weekly	monthly	seasonally	Annually	**Reference**
Crop acreage estimates				x	x			x	x				x	x	x	Inglada et al. (2015)
Crop yield estimates				x	x			x	x				x	x	x	Battude et al. (2016)
Crop production estimates				x	x			x	x				x	x	x	Weiss et al. (2020)
Crop status & condition monitoring			x	x	x	x	x	x	x			x	x			Defourny et al. (2019)
Agricultural water consumption				x	x	x		x				x	x	x	x	Demarez et al. (2019)
Assessment of nutrient leaching				x	x	x		x					x	x	x	Hively et al. (2009)
Monitoring of crop phenology				x	x	x		x	x			x	x			Zeng et al. (2020)
Monitoring of crop establishment & harvest				x	x	x		x	x				x	x		Gao et al. (2020)
Assessment of disastrous climatic events					x	x		x	x			x	x	x		Zaitchik et al. (2006)
Pasture quality assessment				x	x		x	x	x			x	x	x		Mutanga et al. (2005)
Support for index-insurance					x	x		x	x			x	x	x		Jensen et al. (2019)
Monitoring of soil nutrients			x	x			x	x				x	x			Udelhoven et al. (2003)
Monitoring salinization levels				x	x			x					x	x	x	Metternicht and Zinck (2003)
Quantification of crop residues				x	x		x	x				x	x	x	x	Biard and Baret (1997)
Monitoring of tillage practice				x	x		x	x					x	x		Daughtry et al. (2006)
Monitoring of residue burning				x	x	x	x	x				x	x	x		McCarty (2011)
Verification of legal requirements (CAP)			x	x			x	x					x	x		Boix-Fayo and de Vente (2023)
MRV for voluntary carbon markets			x	x	x		x	x					x	x		COWI et al. (2021)
Practice monitoring for in- and off-setting			x	x	x		x	x					x	x		Costa et al. (2020)

(*Continued*)

TABLE 5.1 (*Continued*)

Major EO-Based Agricultural Support and Monitoring Applications Together with Appropriate Scales, Typical Spatial Coverages, and Cadences. For each application, one recommended reading is provided: (a) applications with seasonal to annual updating cadence; (b) applications with weekly to monthly updating cadence, and (c) applications with daily to sub-daily updating cadence.

	Scale						Cover			Cadence						
(c)	centimeter	decimeter	meter	decameter	hectometer	kilometer	local	regional	continental	hourly	daily	weekly	monthly	seasonally	Annually	Reference
Crop nutrient requirements			x	x	x		x	x			x	x	x			Hatfield et al. (2008)
Crop water requirements			x	x	x		x	x			x	x	x			Courault et al. (2010)
Irrigation monitoring				x	x		x	x			x	x	x			Fieuzal et al. (2011)
Soil Moisture monitoring				x	x	x	x	x	x		x	x	x			El Hajj et al. (2019)
Nutrient and water stress assessment				x	x		x	x			x	x	x			Tilling et al. (2007)
Water stress				x	x	x	x	x		x	x	x				Suárez et al. (2008)
Heat stress				x	x	x	x	x			x	x				Dobrowski et al. (2005)
Identification of disturbances & stresses		x	x	x			x	x			x	x	x			Thenkabail et al. (2014)
Pest management			x	x			x	x			x	x				Pinter et al. (2003)
Infestation detection		x	x	x			x	x			x	x				Zarco-Tejada et al. (2018)
Monitoring of vegetation anomalies					x	x		x	x		x	x	x			Meroni et al. (2019)
Support for precision farming		x	x	x			x				x	x	x			Moran et al. (1997)
Support for phenotyping	x	x	x				x			x	x	x				White et al. (2012)

5.1.2 Pathways for Increased Efficiency and Sustainability

Natural resources are under strong pressure. The main drivers are increasing consumption of calorie- and meat-intensive diets (FAO, 2009; Foley et al., 2011; Hill et al., 2006; Pelletier and Tyedmers, 2010; Ranganathan et al., 2016; Sy et al., 2019). On a global scale, the diets themselves are closely related to per capita gross domestic product (GDP), while substantial amounts of food are wasted along the value chain (Parfitt et al., 2010).

The resulting agricultural expansion and intensification have strong negative impacts (Foley et al., 2011; Tilman et al., 2011):

- Deforestation is in large parts driven by agricultural expansion (Rudel et al., 2009)
- Greenhouse gas (GHG) emissions from land clearing, crop production and fertilization contribute already to one-third of global GHG emissions (Burney et al., 2010)
- Biodiversity is threatened by land clearing and habitat fragmentation (Dirzo and Raven, 2003)
- Global nitrogen and phosphorus cycles have been disrupted, with impacts on water quality, aquatic ecosystems, and marine fisheries (Canfield et al., 2010; Vitousek et al., 1997)
- Freshwater resources are depleted, as nearly 80% of freshwater currently used by humans is for irrigation (Postel et al., 1996; Thenkabail, 2010; Thenkabail et al., 2009)

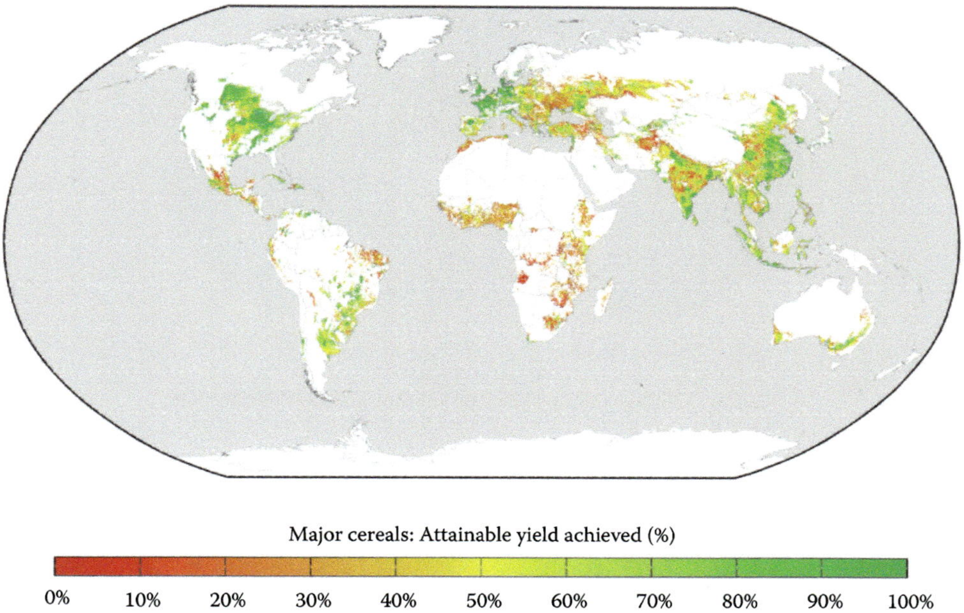

FIGURE 5.1 Global distribution of the yield gap of major cereals. The yield gap is here defined as the difference between the realized productivity and the best that can be achieved using current plant material for a given set of environmental (growing) conditions. (From Mueller, N.D. et al., Nature, 490, 254, 2012.)

Assuming that the GDP and global population will continue to increase in the near future, the past trend of strongly increasing food demand is expected to last for three to four decades. Tilman et al. (2011), for example, project that per capita demand for crops will double between 2005 and 2050.

Food production must therefore grow substantially for meeting the world's future food security and sustainability needs. At the same time, agriculture's environmental footprint must shrink dramatically (Baulcombe et al., 2009; Foley et al., 2011; Godfray et al., 2010; Searchinger et al., 2019). Future food demands must be met without undermining further the integrity of the Earth's environmental systems (Mueller et al., 2012; Searchinger et al., 2015). The necessary transformation will have to take place in times of climate change, adding supplementary difficulties (Griscom et al., 2017; Jones and Thornton, 2003; Trnka et al., 2014). The agricultural transition phase should be monitored at various temporal and spatial scales.

The environmental impacts of an increased global crop production will depend on how this increase is pursued (Foley et al., 2011; Tilman et al., 2011). As extensification implies clearing or adapting additional land for crop production, this path is counterproductive. The "land sparing trajectory," on the other hand, achieves higher yields through increased inputs, improved agronomic practices (e.g., drop irrigation), improved crop varieties and other innovations (Folberth et al., 2020). Increasing global production with small environmental footprint is the preferred solution, as closing the yield gap minimizes both land clearing pressure and GHG emissions, compared to the current situation, that is, an extensification in the poorer countries (Tilman et al., 2011). The large existing yield gaps are exemplified in Figure 5.1 for cereals.

This view on intensification is also shared by Foley et al. (2011). Their analysis showed how many calories could be produced by closing existing yield gaps (Figure 5.2a). Additional calories could be produced by allocating a higher fraction of the cropland to growing food crops (crops that are directly consumed by people) instead of using this land for animal feed, bioenergy crops and fibers, etc. (Figure 5.2b and c).

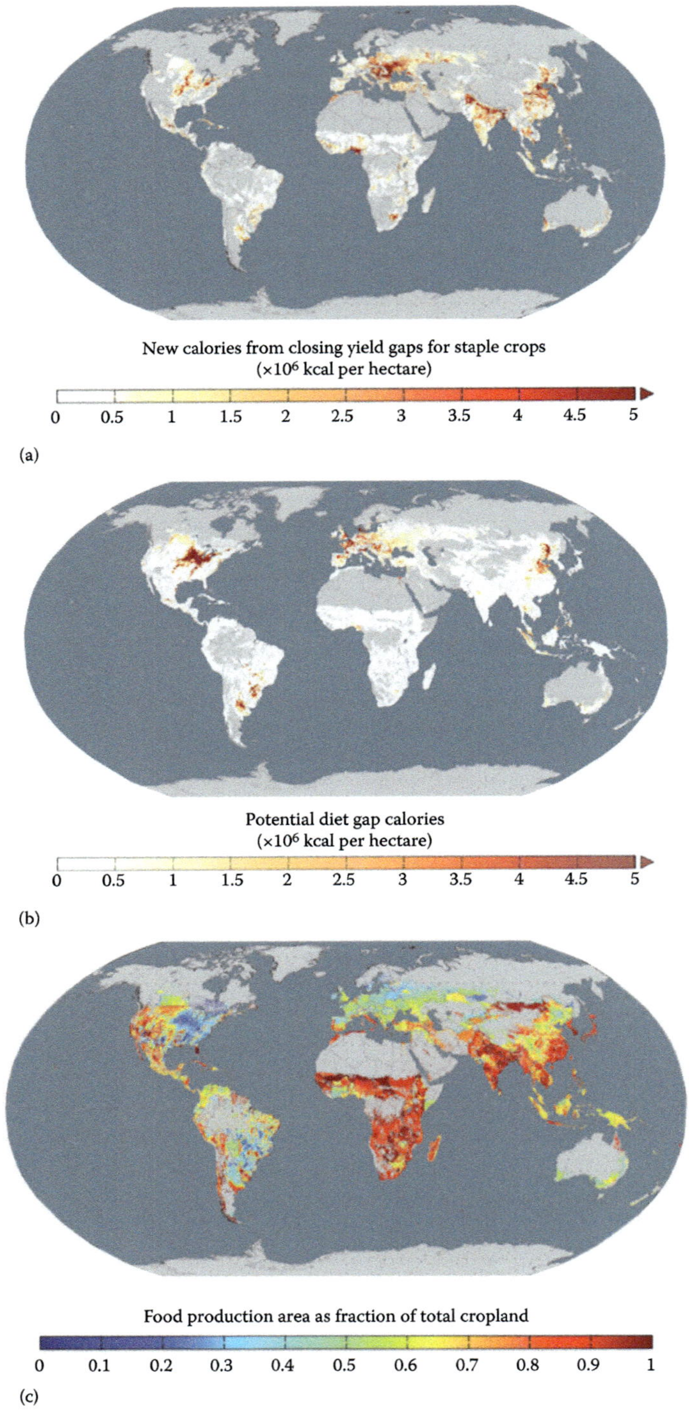

FIGURE 5.2 Pathways for increasing agricultural production to reduce pressure on forests and natural resources: (a) additional calories that could be produced by closing current yield gaps of crops, (b) increased food supply (in calories) by shifting crops to 100% human food and away from current mix of uses, and (c) fraction of cropland that is allocated in 2000 to growing food crops (crops that are directly consumed by people) versus all other crop uses, including animal feed and bioenergy crops. (From Foley, J.A. et al., Nature, 478, 337, 2011.)

Foley et al. (2011) recommends adopting five strategies:

- Closing yield gaps on underperforming lands
- Increasing cropping efficiency
- Shifting diets away from meat
- Reducing waste within the agricultural production chain
- Halting agricultural expansion

Together, these five strategies could double food production, while greatly reducing the environmental impacts of agriculture. Similar conclusions are drawn by Godfray et al. (2010). They promote a multifaceted and linked global strategy to ensure sustainable and equitable food security. To provide the right incentives, these pathways should ideally be embedded in the voluntary carbon market and accompanied by regulations that enforce traceability as well as in- and off-setting mechanisms (Ahonen et al., 2022; COWI et al., 2021; Ebersold et al., 2023; Michaelowa et al., 2023). Obviously, additional regulations and legislations are needed to prevent further deforestation (Bager et al., 2020), while creating favorable conditions for solving the biodiversity crisis (Singh, 2002).

5.2 THE MONITORING OF CROP STATUS AND DEVELOPMENT

Crops develop from sowing to harvest as a function of meteorological drivers (e.g., temperature, sunlight, and precipitation). The growth and development processes are modified by soil and plant characteristics (e.g., genetics) as well as farming practices (e.g., planting dates, nutrient management). As changes in crop vigor, density, health, and productivity affect canopy optical properties, crop development and growth can be monitored remotely (Jones and Vaughan, 2010) (Figure 5.3).

5.2.1 Complementarity of Different Spatial Resolutions

Crop monitoring can be attempted at a large range of scales using sensors with spatial resolutions (ground sampling distance; GSD) ranging from centimeters to kilometers. As important sensor characteristics are often intrinsically linked, trade-offs exist between spatial resolution and spectral resolution, signal-to-noise-ratio (SNR), temporal revisit frequency, synoptic spatial coverage, BRDF

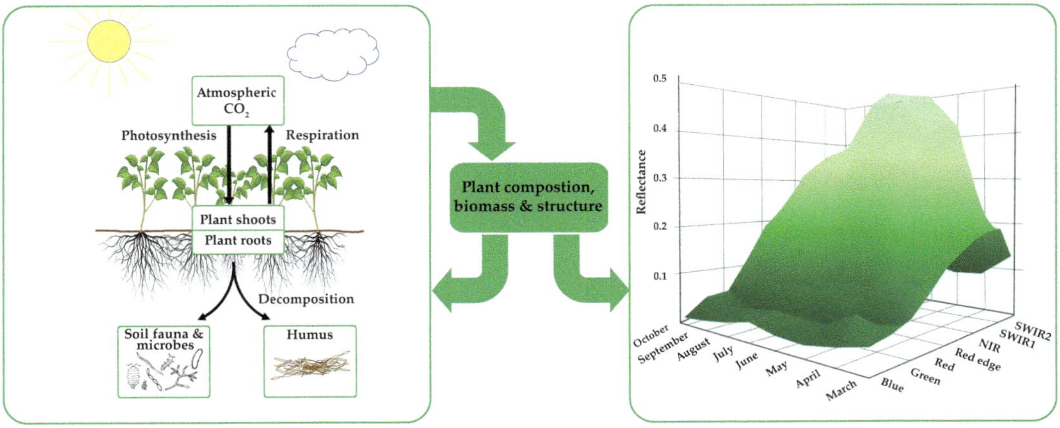

FIGURE 5.3 Crop functioning and crop optical properties. The illustration shows the close relation between growth processes, composition, and structure of crops (left), which in turn drive the spectral-temporal signature of the crop (right). This makes EO techniques well suited within the agricultural value chain for (1) mapping and monitoring of plant composition, structure, and biomass, as well as (2) for constraining dynamic process models.

TABLE 5.2
Complementarity of Different Optical EO Sensors and Platforms for Agricultural Applications. Eight typical sensor types—with typical examples—are distinguished. (1) spectrometer: ASD Field Spec 4; (2) drone: Altum PT; (3) aerial photography: Microsoft Ultracam Eagle; (4) VHR-satellite: WorldView-3; (5) HR-satellite (commercial): Planet Scope; (6) HR-satellite (scientific): Sentinel-2; (7) LR-satellite: MODIS-Aqua; (8) geostationary: GOES-16.

	Trade-offs								
Sensor & Platform Type	Spatial resolution	Spectral resolution	SNR & radiometric resolution	Revisit frequency	Synoptic spatial coverage	BRDF effects	Geometric stability	Costs (USD/km²)	Insights for Agriculture
Spectro-meter	++	++	++	−−	−−	++	++	−−	Physical foundation
	point								
Drone	++	+	−	−−	−−	−−	−−	−−	Characterization of field plots and phenotyping
	mm—cm								
Aerial photography	++	−	++	−−	−	−	++	−	Regional enhancement for fine-scale inventories
	cm—dm								
VHR-Satellite	++	++	+	−	−	++	−	+	Regional enhancement for fine-scale inventories
	dm—m								
HR-Satellite (commercial)	+	+	−	++	+	+	+	+	Global backbone for agricultural monitoring
	dam								
HR-Satellite (scientific)	+	++	++	+	+	+	++	++	Global backbone for agricultural monitoring
	dam								
LR-Satellite	−	++	++	++	++	−	++	++	Large scale crop growth anomalies
	hm—km								
GEO-Satellite	−−	+	++	++	++	−	++	++	Large scale diurnal land surface variability
	km								

effects, geometric stability, and unit costs (km^{-2}) (Table 5.2). The potential of EO for different use cases is defined by all sensor characteristics together. This implies that aiming for a higher spatial resolution is not always productive.

5.2.1.1 Point Sensors

The relationship between the spectral properties of crops and their biomass/yield has been recognized since the very first spectrometric field experiments (Bauer et al., 1986; Tucker, 1979; Tucker et al., 1980). Those early insights have since been complemented by many other field and laboratory studies (Atzberger et al., 2010; Peñuelas et al., 1995), and enabled a sound theoretical understanding and description of the interactions between electromagnetic radiation and canopy elements (Baret et al., 1987; Ustin et al., 2009; Verhoef, 1985).

5.2.1.2 Milli- to Decimetric Sensors

For very local analyses drones can provide interesting data, in particular if equipped with multiple sensors such as hyperspectral sensors and LiDAR. This is for example useful for phenotyping studies (Blancon et al., 2019; Jin et al., 2022), and in general, to characterize field sampling sites for later upscaling to larger areas (Daryaei et al., 2020; Som-ard et al., 2024, 2021).

To monitor areas larger than a square kilometer, drones are excessively expensive. In such cases, classical aerial remote sensing is favored as data in similar spatial resolution can be acquired, but at much lower cost per unit area (Table 5.2).

As both, drones and aircraft are usually flown with large instantaneous field of view (IFOV) and with 60–90% overlap, data from both sensors can be analyzed through photogrammetry, that is, for deriving 3D information (e.g., canopy height) (Murakami et al., 2012; Sofonia et al., 2019). This adds useful and complementary information.

5.2.1.3 Decimetric to Metric Sensors

VHR satellite sensors offer interesting and complementary data in decimetric to metric resolution for many agricultural applications (Amorós López et al., 2011). For study regions in the size range of 10 to 100 km^2, satellite data can be tasked at reasonable costs (ca 10 USD/km^2) while providing decimetric to metric resolution with scientific grade optical instruments and stereo-capability.

5.2.1.4 Decametric Sensors

Whereas drones provide useful data for small-scale studies—and spectrometric studies are useful for enhancing our physical understanding of the radiative transfer in canopies at plot scale—only satellite imagery permits to fully leverage the potential of remote sensing in agriculture, that is, the repeated quantification and analysis of radiative properties for the characterization and monitoring of the crops, respectively, production systems.

The outstanding potential of satellite sensors (from decimetric to decametric spatial resolution) can for example be seen in the applications listed in Table 5.1, which are mostly based on this sensor category. With the U.S. Landsat program (since 1972), respectively, the European Copernicus program (since 2014), plus imagery provided by commercial data provider such as Planet and Maxar, high-quality multispectral time series with a daily global coverage offer rich intra-field information while covering large areas in synoptic view (Belward and Skøien, 2015; Blickensdörfer et al., 2022; Hunt et al., 2019a; Veloso et al., 2017).

The combined analysis of different sensor modalities offers additional benefits (Berger et al., 2022; Ghamisi et al., 2019; Guanter et al., 2019; Potočnik Buhvald et al., 2022). However, the combination of data from different sensors is still challenging as images are acquired from various orbits and in different spatial and temporal resolutions, with similar albeit not perfectly matching spectral channels. Approaches based on joint embeddings are an active area of research (Balestriero et al., 2023).

5.2.1.5 Hecto- to Kilometric Sensors

Thanks to their large swath width, low resolution systems have a much better synoptic view and temporal revisit frequency compared to high spatial resolution sensors with narrow swaths (Table 5.2). Obviously, the intrinsic drawback of these sensors is related to their low spatial resolution (hecto- to kilometric), typically far above average field sizes. Consequently, recorded spectral radiances are weighted averages of several surface types (Cracknell, 1998).

Specific unmixing approaches have been developed to derive sub-pixel information (Atkinson et al., 1997; Atzberger and Rembold, 2013; Busetto et al., 2008; Foody, 2004; Foody and Cox, 1994). However, this seriously complicates the interpretation (and validation) of the signal, as well as the reliability of the derived information products.

Without any spatial down-scaling, low resolution imagery from sensors such as MODIS offer excellent, large-scale indicators of agricultural productivity, in particular in water-limited ecosystems (Henricksen and Durkin, 1986; Johnson et al., 1987; Maselli et al., 1993). Most of the current global food security monitoring systems still heavily rely on such data (Becker-Reshef et al., 2010;

Duveiller et al., 2013; Franch et al., 2019), often complemented by (gridded) weather parameter (Duveiller et al., 2017) and market information.

5.2.2 Features Derived from EO Sensor Data

To analyze multi- to hyper-spectral EO data, a large range of tools have been developed. Useful examples and overviews are for example given in the following publications: Camps-Valls and Bruzzone (2009); Dumeur et al. (2024); Hosseiny et al. (2024); Justice et al. (2002); Macdonald and Hall (1980); Plaza et al. (2009); Song et al. (2014); Verstraete et al., (1996) and Wu et al. (2022).

With respect to the explored features, most approaches analyze the recorded spectral and/or temporal signatures (e.g., Badhwar et al., 1982; Lobell and Asner, 2004; Mello et al., 2013c; Rußwurm and Körner, 2020; Udelhoven et al., 2009; Vuolo and Atzberger, 2012; Wardlow et al., 2007). The predominance of these features is the related to the fact that spectral-temporal signatures implicitly track the (co)evolution of important biophysical traits such as Leaf Area Index (LAI), leaf pigmentation and water content. These traits not only permit to quantify (directly or indirectly) important crop functions and statuses, but also to identify crop types.

Accessible through imaging spectrometer data, fluorescence information has a great potential for agricultural applications (Damm et al., 2015; Guanter et al., 2014; Meroni et al., 2009). Amongst all features which can potentially be acquired by EO sensor data, fluorescence has probably the strongest potential as it is casually linked to a plant's photosynthetic status (Lichtenthaler, 1988; Moya et al., 1992). However, as the signal represents a flux, it is subject to rapid temporal fluctuations, at least similar to changes in land surface temperature (LST).

Further useful information can be retrieved from textural, respectively, structural information (Blaschke, 2010; Vintrou et al., 2012). Whereas texture has a direct link to crop biophysical properties and their spatial covariation (e.g., Atzberger, 2004; Atzberger and Richter, 2012), structural information reflects planting and management pattern and usually requires that the sensor's spatial resolution is significantly smaller than the object size.

Less often used are approaches analyzing the directional reflectance properties of vegetation (e.g., Barnsley et al., 1997; Clevers et al., 1994; Gobron et al., 2002; Koukal and Atzberger, 2012; Roujean and Lacaze, 2002; Schlerf and Atzberger, 2012; Vuolo et al., 2008), polarization properties (Maignan et al., 2009; Rondeaux and Herman, 1991) or fluorescence signals (Damm et al., 2015; Guanter et al., 2014; Meroni et al., 2009).

5.3 REMOTE SENSING FOR ASSESSING YIELD AND BIOMASS

Remote sensing techniques are well suited for assessing yield and/or crop biomass in a cost-efficient and transparent way. Excellent examples are given in several studies (Bernardes et al., 2012; Doraiswamy et al., 2005; Duveiller et al., 2013; Kooistra et al., 2024; Mello et al., 2013a; Meroni et al., 2013b; Mulianga et al., 2013; Rembold et al., 2013; Zhang et al., 2005).

For mapping and monitoring of crop biomass and yield, four main techniques can be distinguished (Table 5.3). The four groups also summarize the evolution from comparative to more quantitative and process-based approaches and hence—in some way—the history of agricultural remote sensing for yield and production assessments:

- Anomaly detection
- Regression modeling
- Application of Monteith's (light use) efficiency equation
- Assimilation of remote sensing data into mechanistic and dynamic crop growth models

To obtain total production, the yield indicator must obviously be combined with acreage information, which we will cover in a separate chapter (Section 5.4).

TABLE 5.3
Summary of Four Different Approaches for Estimating Crop Yield Using Remote Sensing Techniques

Method	Description	Formulation	Reference
Anomaly-based	For each pixel location, a spectral indicator known to be proportional to yield is compared against its historical (multi-year) minimum and maximum value at the same location and time of year.	$Y_{rel} = (VI - VI_{min})/(VI_{max} - VI_{min})$	Kogan (1990)
Regression-based	Based on a labeled dataset, a regression model is built between spectral indicator(s) and the reference yield (dependent variable). Once calibrated, the model is applied to the entire region of interest.	$Y = a_0 + a_1 \times VI$	Rudorff and Batista (1990)
LUE-based	The total amount of energy absorbed by the canopy during its life cycle is derived using a spectrally determined light harvesting fraction (fAPAR) for each time step between emergence and harvest which is convolved with the incoming PAR. The accumulated energy is converted into biomass and yield using the dry matter conversion rate (ε_b) and the harvest index (HI).	$Y = HI \times \varepsilon_b \times \Sigma[PAR_t \times fAPAR_t]$ $fAPAR = f(VI)$	Song et al. (2013)
CGM & EO data assimilation	Starting from emergence, the amount of absorbed light energy is calculated at daily time steps based on the actual—dynamically updated—status of the light harvesting system (i.e., LAI_t). The model converts the harvested energy into new leaf area, considering the actual energy conversion into assimilates and their allocation into leaf, root, and reproduction systems (which depend on phenology and growth conditions), whereas the finite life span of leaves is taken into account. Spectral data is used to parameterize and/or initialize the model for each pixel location in the region of interest.	$Y = f_{fb}$ (weather, soil, plant, management)	Delécolle et al. (1992) Huang et al. (2019)

5.3.1 Comparative Crop Monitoring and Anomaly Detection

Anomaly detection methods are based on the comparison of the actual crop status to the "normal" situation at the same time and location. Detected divergences (or "anomalies") are then used to draw conclusions on possible yield limitations. The crop status is usually quantified using spectral indices such as the NDVI.

The usefulness of arithmetic combinations of reflectance in different spectral bands (so called "vegetation indices") was established in the early 80s by Tucker, Deering and co-workers (Deering, 1978; Tucker, 1979; Tucker et al., 1980). Thanks to its simplicity, the proposed NDVI became subsequently the most popular indicator for studying vegetation health and crop production.

The usefulness of NDVI is underpinned by the fact that it is relatively closely related to the canopy LAI and fAPAR (fraction of Absorbed Photosynthetically Active Radiation) (Baret and Guyot, 1991; Prince, 1991). Indeed, due to its almost linear relation with fAPAR, the NDVI can also be seen as an indirect measure of primary productivity (Box et al., 1989; Myneni et al., 1992).

A well-established index for quantifying such anomalies is the VCI (vegetation condition index) (Figure 5.4). The VCI scales the observed vegetation index (VI) of a given month between the historical maximum (VI_{max}) and minimum (VI_{min}) of that same period in the year at the same location (Kogan, 1990):

$$VCI = 100 \times \frac{VI - VI_{min}}{VI_{max} - VI_{min}} \tag{5.1}$$

Hecto- to kilometric resolution satellites with a long history (e.g., MODIS) are best suited for such approaches, as they offer a high temporal revisit frequency with an extended geographical coverage at low data costs per unit area.

To quantify the actual and "typical" crop status (Equation 5.1), respectively, many spectral indices have been used, as well as biophysical products derived from the optical data. Especially for

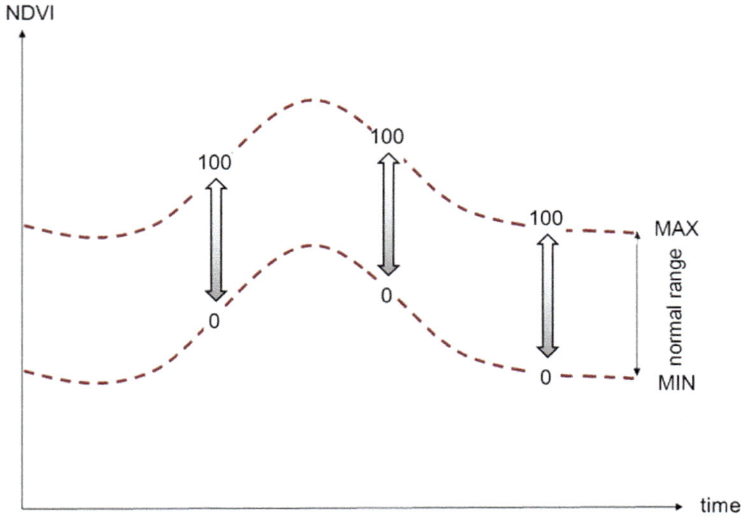

FIGURE 5.4 Principle for calculating the vegetation condition index (VCI) to detect crop growth anomalies from EO time series. Multiple years of historical time series of vegetation indices (here NDVI) are used to define for each time step (e.g., each month) the respective historical minimum and maximum values. Once the upper and lower bounds are established (broken lines), any observed NDVI can be linearly scaled between these two extremes. Negative growth conditions indicative of agricultural drought can be readily seen by VCI values approaching zero.

near-real-time applications, sophisticated filtering approaches are necessary to remove unwanted artifacts related to poor atmospheric conditions and (undetected) clouds (Atzberger and Eilers, 2011; Meroni et al., 2019).

Crop monitoring systems making use of "anomaly maps" are particularly useful in arid and semiarid countries, where temporal and geographic rainfall variability leads to high inter-annual fluctuations in primary production and to a large risk of famines (Hutchinson, 1991; Klisch and Atzberger, 2016). These environmental situations, along with the wide extent of the areas to monitor and the generally poor availability of efficient agricultural data collection systems, represent a scenario where anomaly based approaches can produce valid information for releasing early warnings about possible crop stress. Such systems are for example been used by the MARS Crop Yield Forecasting System of the European Commission (Seguini et al., 2019). The anomaly indices themselves can be forecasted using autoregressive models and deep-learning techniques such as LSTM (Lees et al., 2022), model ensembles of neural nets of support vectors (Adede et al., 2019a) or mixed models of neural nets (Adede et al., 2019b), amongst other techniques.

5.3.2 Crop Yield Predictions Using Regression Analysis

Regression based approaches quantify the expected yield in absolute units (e.g., in t/ha), instead of solely providing relative comparisons (e.g., "anomalies") of spectral indices described in (Section 5.3.1). Three regression methods will be covered here in more detail to exemplify different approaches:

- Spectral models
- Spectral bio-climatic models
- Yield correlation masking

In contrast to the anomaly based approaches, the regression approaches must be calibrated using appropriate reference information. In most cases, agricultural statistics on crop yields are used as reference (e.g., dependent variable). Obviously, this limits the applicability of the approach.

Importantly, whatever functional form is used for regression, it must be noted that the correlation between crop yield and spectral measurements varies during the growing season, and regression coefficients show strong temporal variations (Rudorff and Batista, 1990). Indeed, the relation between crop yield (Y) and spectral data (e.g., a VI) is only indirect—and strongly dependent on three additional factors (HI, Lr, and SLA) which are not directly accessible through remote sensing and moreover change over time and space (Equation 5.**2**):

$$Y = \frac{Y}{DM} \times \frac{DM}{LDM} \times \frac{LDM}{LAI} \times LAI = HI \times \frac{1}{Lr} \times \frac{1}{SLA} \times \frac{1}{f(VI)} \tag{5.2}$$

where

Y	: yield [kg·m^{-2}]
DM	: total aboveground dry mass at harvest [kg·m^{-2}]
LDM	: leaf dry mass [kg·m^{-2}]
HI	: harvest index [fraction] = yield [kg·m^{-2}]/total aboveground dry mass at harvest [kg·m^{-2}]
Lr	: leaf mass ratio [fraction] = leaf dry mass [kg·m^{-2}]/total aboveground dry mass at harvest [kg·m^{-2}]
SLA	: specific leaf area [m^2·kg^{-1}] = leaf area [m^2]/leaf dry mass [kg]
f(VI)	: a function relating a remotely sensed indicator (e.g., a vegetation index) to the crop LAI [m^2·m^{-2}]

The model in Equation 5.2 (F Baret, pers. comm.) highlights that the spectral model f(VI) only links the observed spectral data to LAI and not directly to the final yield (Y). How much this LAI represents in terms of leaf dry matter depends on the specific leaf area (SLA) of the crop. As the proportion of dry matter (DM) in the leaves compared to the total DM evolves during the growing season, the leaf mass ratio (Lr) needs also to be considered, as well as the harvest index (HI) which indicates the part of the DM in the reproductive organs. Only in cases where the three scaling factors HI, SLA and Lr, as well as f(VI), are either known or stable over space and time, one can expect robust yield estimates using empirical regression approaches. Established relationships are therefore, to some degree, "good fortune" and usually time and site specific (Baret et al., 1989).

5.3.2.1 Spectral Models

The well-established relationship between spectral indices/fAPAR and (leaf) biomass enables the estimation of crop yield, since the yield of many crops is mainly determined by the photosynthetic activity of agricultural plants in certain periods prior to harvest (Baret et al., 1989; Benedetti and Rossini, 1993).

The basic assumption of this method is that sufficiently long and consistent time series of both remote sensing images and agricultural statistics are available. Using multi-year datasets enhances the generalization of the developed models. Examples of NDVI/yield regressions for cereals at national level are shown in Figure 5.5.

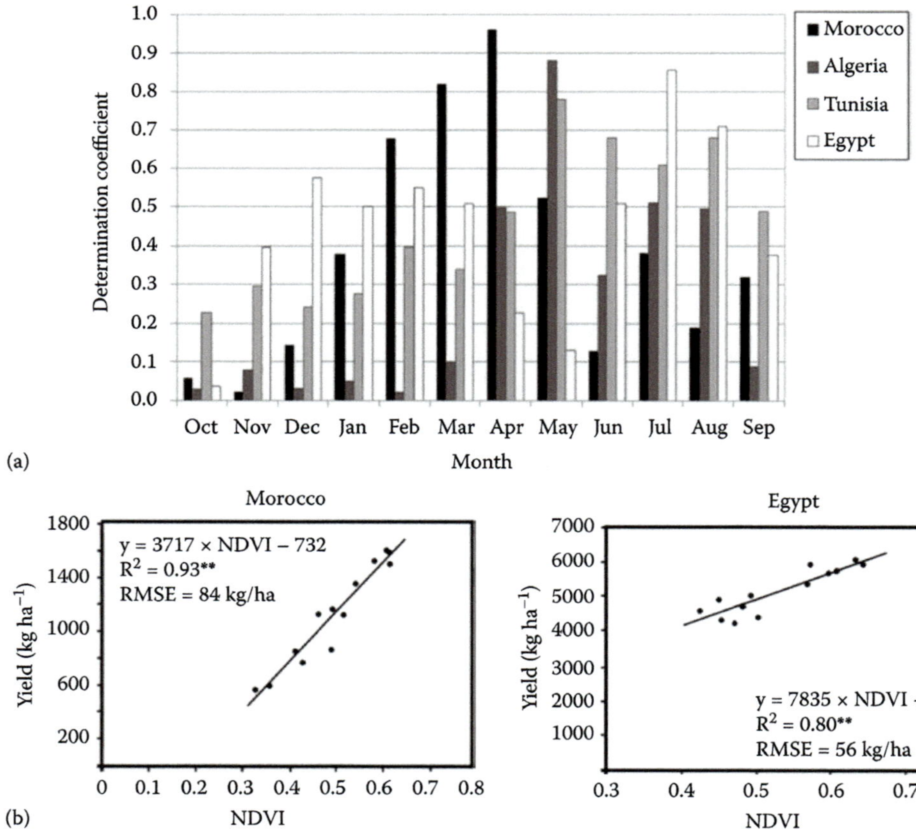

FIGURE 5.5 NDVI/yield linear regressions for cereals in North Africa. (a) Evolution of the coefficient of determination (R2) between NDVI and yield over time. (b) Scatter plots between NDVI and cereal yield for Morocco (left) and Egypt (right). Each dot corresponds to the annual yield for agricultural areas at national level and to the monthly NDVI best correlated to yield. (Modified from Maselli, F. and Rembold, F., *Photogramm. Eng. Remote Sens.*, 67, 593, 2001.)

Many studies reported useful statistical relationships using NDVI values at the peak of the growing season and final crop yield. The different empirical techniques appear to be relatively accurate for crops with low final production because biomass is the limiting factor to yield and the relationship between LAI and the vegetation response (NDVI) is below the range of saturation (Delécolle et al., 1992). Empirical relationships also appear to be relatively accurate for grass crops, where dry matter is the harvestable yield.

Linear regression models relating NDVI to crop yield have for example been developed by Rasmussen (1992) and Groten (1993) for Burkina Faso and by Maselli et al. (1993) for Niger. The same and other investigations showed that yields can often be best forecasted using NDVI data of specific periods; the "optimum" time window depends on the eco-climatic conditions and the types of crops grown (Hayes and Decker, 1996; Lewis et al., 1998; Maselli et al., 2000).

Besides classical (multiple) linear regression, other statistical techniques such as partial least squares regression (PLSR) may be more appropriate to model the relation between the sought variable(s) and the spectral reflectance (Atzberger et al., 2010; Hansen and Schjoerring, 2003; Nguyen et al., 2006). Obviously, nonlinear machine learning models such as random forest (RF) or neural nets (NN) can be used as well (dela Torre et al., 2021; Som-ard et al., 2024).

With high dimensional EO datasets a panoply of features can possibly be used to estimate crop yield. The choice of the deployed predictor variable(s) is therefore important. As crops develop over time, it can be expected that accumulated radiometric data are more closely related to crop production than instantaneous measurements. For example, Pinter et al. (1981) argued that the accumulation of radiometric data was similar to a measure of the duration of green leaf area. They consequently related yield of wheat and barley to an accumulated NDVI index. While good results were achieved, the performance was strongly linked to identifying the optimum integration period. When the optimum dates could not be specified accurately, predictions were less accurate. In a similar manner, Meroni et al. (2013b) estimated winter wheat yield at the national level of Tunisia. Instead of using a fixed integration period, the integral was computed between the start of the growing period and the beginning of the descending phase. The two dates were computed for each pixel and each crop season separately.

Several other choices of temporal NDVI analysis can be found, reaching from the simple selection of the maximum NDVI value of the season (which in itself is related to the average NDVI over the season), to the average of the peak values (plateau), to the sum of the total NDVI values of the total crop cycle.

A less used but interesting time series technique involves the concept of aging or senescence, first developed by Idso et al. (1980). Idso and co-workers found that yield of wheat could be estimated by an evaluation of the rate of senescence as measured by a ratio index following heading. The lower the rate of senescence the larger the yield, as stressed plants begin to senesce sooner. The same technique was later applied by Baret and Guyot (1986).

5.3.2.2 Spectral Bio-Climatic Models

One important limitation of simple spectral regression approaches is that the developed spectral models are linked to the specific environmental characteristics of the area of interest. This implies that locally calibrated spectral forecasting methods usually cannot be transferred to larger regions and/or different growing seasons. At least part of the variability can be compensated for by adding independent meteorological (or bio-climatic) variables into the regression models.

Several bio-climatic variables have proven to be highly correlated with yield for certain crops in specific areas (Lewis et al., 1998; Rasmussen, 1998; Reynolds et al., 2000). These variables can be either measured directly (e.g., rainfall from synoptic weather stations), derived from satellites (e.g., gridded rainfall estimates), or can be the outputs of other models (e.g., actual evaporation or soil moisture).

Potdar et al. (1999), for example, observed that the spatiotemporal rainfall distribution could be successfully incorporated into crop yield models (in addition to vegetation indices), to predict crop yield of different cereal crops grown in rain-fed conditions. Such hybrid models often show higher correlation and predictive capability compared to models using solely remote sensing indicators (Balaghi et al., 2008; Manjunath et al., 2002) as the input variables complement each other.

Indeed, the bio-climatic variables introduce information about the environmental drivers of vegetation growth (e.g., solar radiation, temperature, air humidity, and soil water availability) whereas the spectral component introduces information about the actual growth outcome of such drivers, thus indirectly taking into account crop management, varieties and other stresses not directly considered by the agro-meteorological models (Rudorff and Batista, 1990).

Although linear regression modeling is likely the most common method to produce yield predictions by using remote sensing-derived indicators together with bio-climatic information, this is not the only one. Numerous other methods have been developed that include, for instance, similarity analysis and (nonlinear) neural networks (Stathakis et al., 2006). In all cases, however, the covariation of the different predictor variables should be considered and corrected when integrating bio-climatic and spectral indicators into multiple regression models.

5.3.2.3 Yield Correlation Masking

One obstacle to successful empirical modeling and prediction of crop yields using remotely sensed imagery is the identification of suitable crop masks, as crops often rotate (Kastens et al., 2005). In areas with crop rotation, for each growing season, new crop masks need to be established to identify the locations of the crop(s) of interest. To overcome the shortcomings related to cropland masking and crop-specific masking, Kastens et al. (2005) proposed a new masking technique, called yield-correlation masking. The main idea behind this concept is that all vegetated pixel in a region (i.e., crops and natural vegetation) integrate the season's cumulative growing conditions in some fashion. Hence, in the yield-masking approach, all pixels are considered for use in crop yield prediction.

In practical terms, yield-correlation masking generates a unique mask for each spectral predictor variables such as NDVI (e.g., each time step at which NDVI is available) and each combined pair of crop type x region. The technique is initiated by correlating each of the historical, pixel-level NDVI variable values with the region's yield history. The highest correlating pixels are retained for further processing and evaluation of the (NDVI) variable at hand. A diagram outlining this process for a single NDVI variable is shown in Figure 5.6.

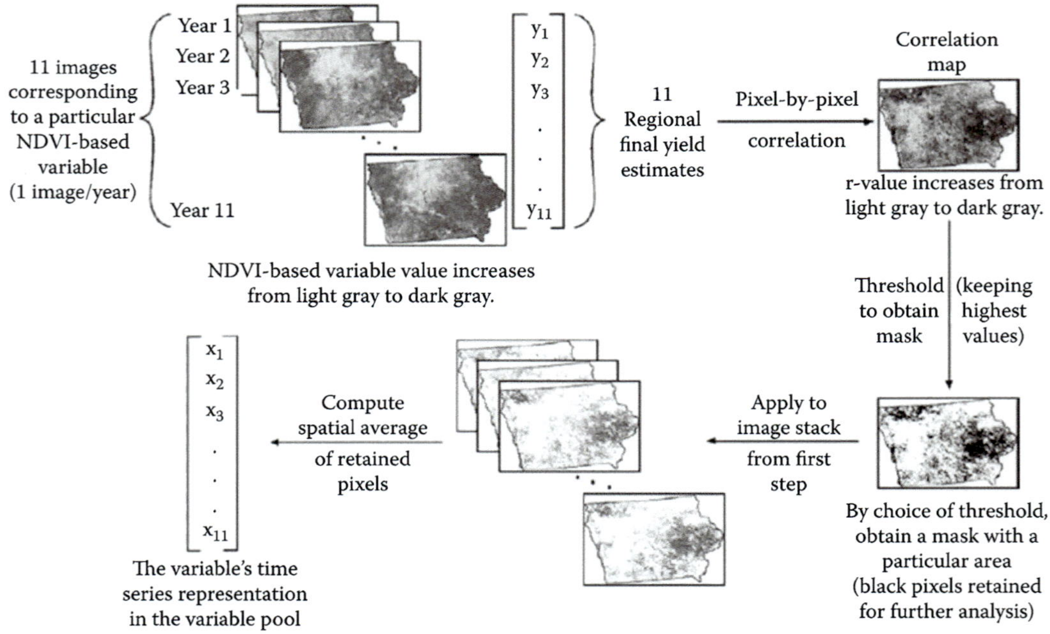

FIGURE 5.6 Illustration of the yield-masking approach involving a dataset of 11 years. (From Kastens, J.H. et al., *Remote Sens. Environ.*, 99, 341, 2005.)

Unlike other empirical approaches, yield-correlation masking can be readily applied to low-producing regions and regions possessing sparse crop distribution. Also, since yield-correlation masks are not constrained to include pixels dominated by cropland, they are not necessarily hindered by the weak and insensitive NDVI responses exhibited by crops early in their respective growing seasons. Furthermore, once the issue of identifying optimal mask size (i.e., determining how many pixels should be included in the masks) is addressed, the entire modeling procedure becomes completely automated. The most important appeal of yield-correlation masking is that no crop-type maps are required to implement the procedure, while the procedure results in forecasts of comparable accuracy to those obtained when using cropland masking or crop-specific masks (Kastens et al., 2005). The procedure only requires an adequate time series of imagery and a corresponding record of the region's crop yields.

Problems regarding this approach can be expected when dramatic land cover changes happen on the locations of the selected yield proxies. In addition, the procedure used to select the pixels to be retained in the mask increases the parameterization of the final yield forecast model, so that its predictive power must be carefully scrutinized.

5.3.3 Use of Monteith's Light Use Efficiency Equation

Remotely sensed images were first proposed in the 1980s for assessing and mapping of the crop's assimilation potential. One of the first steps in this direction was the introduction of the so-called light use efficiency (LUE) equation by Monteith (Monteith, 1977, 1972) (Equation 5.3):

$$\Delta DM = PAR \times fAPAR \times \varepsilon_b \tag{5.3}$$

where:

ΔDM : increase in dry matter (DM) or net primary production (NPP) ($g \cdot m^{-2} \cdot d^{-1}$)
PAR : incident photosynthetically active solar radiation ($MJ \cdot m^{-2} \cdot d^{-1}$)
fAPAR : fraction of intercepted and absorbed PAR calculated from spectral data (dimensionless)
ε_b : photosynthetic efficiency ($g \cdot MJ^{-1}$)

Experimental data confirms the linear relation between accumulated absorbed PAR (fAPAR × PAR) and biomass (Figure 5.7), where the slope (here 1.743) equals ε_b. Operational NPP products based on Monteith's LUE model are nowadays readily available (Morisette et al., 2006; Xiao et al., 2019). Excellent reviews of LUE models, their prospects, and shortcomings are provided by Song et al. (2013) and Field et al. (1995).

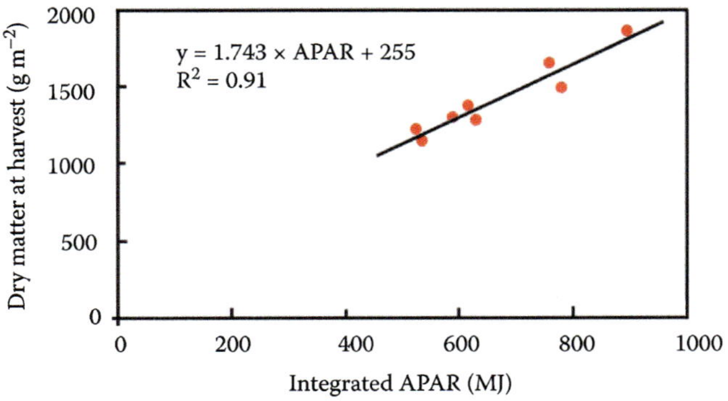

FIGURE 5.7 Linear relation between the seasonally integrated absorbed PAR (from sowing to harvest) and dry matter at harvest (g·m–2). Each point corresponds to one commercial winter wheat plot (n = 9). (From Atzberger, C., *Estimates of Winter Wheat Production through Remote Sensing and Crop Growth Modelling: A Case Study on the Camargue Region,* Verlag für Wissenschaft und Forschung, Berlin, Germany, 1997.)

To calculate final yield (Y) from Equation 5.3, the fraction of the total biomass which is located in the reproductive organs (HI) has to be measured or estimated as this value not only depends on crop type and/or variety but also on the environmental conditions during the growth period (Equation 5.4):

$$Y = HI \times \sum_{\text{sowing}}^{\text{harvest}} \Delta DM \tag{5.4}$$

Monteith's LUT thus assumes that the biomass production (ΔDM) during a given time interval can be described as the simple multiplication of three variables: the incident photosynthetically active radiation (PAR, 400–700 nm), the PAR fraction which is absorbed by the vegetation layer (fAPAR), and finally e_b, the energy to dry matter conversion factor.

The amount of radiation available to the photosynthetic process is the absorbed solar radiation (APAR) and is a function of the incoming PAR and the crop's PAR interception capacity, fAPAR:

$$\text{fAPAR} = \text{APAR/PAR} \tag{5.5}$$

fAPAR depends mainly (but not solely) on the leaf area of the canopy (Baret et al., 1989). Generally, an exponential relation between LAI and fAPAR is admitted:

$$\text{fAPAR} = \text{fAPARmax}\left(1 - \exp\left(-\text{k} \times \text{LAI}\right)\right) \tag{5.6}$$

with fAPARmax between 0.93 and 0.97 and extinction coefficient k between 0.6 and 2.2 (Baret et al., 1989), mainly depending on the leaf angle distribution and the leaf chlorophyll content.

As both, NDVI and fAPAR increase in an exponential manner with LAI, a close link between NDVI and fAPAR exists (Myneni and Williams, 1994). Often, a quasi-linear relation between NDVI and fAPAR is assumed:

$$fAPAR = a + b \times NDVI \tag{5.7}$$

Typical values for the slope (*b*) are between 1.2 and 1.4 and for the intercept (*a*) between −0.2 and −0.4. The negative intercept reflects the fact that the NDVI of bare soils (i.e., fAPAR = 0) is often between 0.2 and 0.4.

The relation between fAPAR and NDVI is not surprising because PAR interception and canopy reflectance/NDVI are functionally interdependent as they both depend on the same factors (Baret, 1988; Baret et al., 1989). The main factors determining PAR interception and canopy reflectance/NDVI are (in order of decreasing importance): (1) leaf area index, (2) leaf optical properties (especially leaf pigment concentration), (3) leaf angle distribution, (4) soil optical properties, and (5) the Sun-target-sensor geometry.

Given the close link between spectral variables and fAPAR (e.g., Equation 5.7), the temporal profile of fAPAR over the course of the growing season can be deducted—at daily time steps—from remotely sensed time series, for example using curve fitting or (linear) interpolation techniques (Belda et al., 2020; Kandasamy et al., 2013; Moreno et al., 2014). Incident PAR can be readily obtained at daily time steps from meteorological stations or gridded weather products, so that the incremental gain in dry matter (ΔDM) can be calculated over the growing season for an assumed dry matter conversion factor (ε_b) (Equation 5.3; Figure 5.7).

When calculated over the entire growth cycle—and in the absence of growth stresses—the light-use efficiency (ε_b) is relatively constant for crops like winter wheat (with a value of about 2.0 g·MJ^{-1}) (Baret et al., 1989). Over shorter time intervals, however, the light-use efficiency is not constant (Leblon et al., 1991; Steinmetz et al., 1990). The short-term variability of the light-use efficiency is a result of temperature, nutrient, and water conditions that eventually can lead to plant stress (Field et al., 1995). Attempts to model these dependencies are for example described in Eerens et al. (2004).

The seasonal integration of radiometric measurements theoretically improves the capability of estimating biomass compared to mono-temporal approaches, since the LUE model is based on sound physical and biological theory, whereas the relationship between instantaneous measurements of canopy reflectance and biomass is mainly empirical (Baret et al., 1989).

The main disadvantage of Monteith's LUE model relates to the need for a complete time series of fAPAR values from sowing to harvest. In excessively cloudy regions even the dense time series provided by Sentinel-2 might occasionally not be sufficient to enable a reliable reconstruction of the seasonal fAPAR profile. Second, both the harvest index (HI) and the dry matter conversion factor (ε_b) are associated with high uncertainties. Together, this sometimes introduces substantial errors in the derived products. To tackle these issues, more process-oriented models were developed (Section 5.3.4), where (1) any number of (cloud-free) observations can be accommodated, and (2) where growth-limiting factors are made explicit, while (3) feedbacks between the different components of the crop/soil system are considered.

5.3.4 Remote Sensing Data Assimilation into Dynamic Crop Growth Models

The approaches described in the previous sections aimed either to assess vegetation vigor by comparing observed vegetation greenness against the "normal" situation (Section 5.3.1) or to quantitatively estimate the crop yield using semi-empirical regression techniques (Section 5.3.2), respectively, Monteith's LUE equation (Section 5.3.3). In this section, we will present a group of techniques involving modeling of crop physiology. Such approaches are also known as crop growth modeling (CGM), SVAT (Soil Vegetation Atmosphere) modeling, or agro-meteorological (AgMet) models. Compared to the previously described LUE model (§ 5.3.3), crop growth models make growth-limiting factors explicit and also involve feedbacks between the different components of the crop/soil system.

5.3.4.1 Formalization of Growth and Development

As defined by Delécolle et al. (1992), crop growth modeling involves the use of mathematical simulation models formalizing the analytical knowledge previously gained by plant physiologists. The models describe the primary physiological mechanisms of crop growth (e.g., phenological development, photosynthesis, dry matter portioning, and organogenesis), as well as their interactions with the underlying environmental driving variables (e.g., air temperature, soil moisture, nutrient availability) using mechanistic equations (Delécolle et al., 1992). Importantly, state variables (such as phenological development stage, biomass, leaf area index, soil water content, etc.) are updated in a computational loop that is usually performed daily (Guérif and Delécolle, 1993) (Figure 5.8).

In the computational loop (Figure 5.8), model state variables such as development stage, organ dry mass and LAI are linked to environmental driving variables such as temperature and precipitation, which are usually provided with a daily time step (Delécolle et al., 1992; Huang et al., 2019). Soil and plant parameters are used to mimic the plant's reaction to these driving variables. Whereas model state variables are constantly updated within the computational loop, model parameters remain unchanged during the simulation run (e.g., soil texture information). All state variables must be initialized at the beginning of the simulation run.

5.3.4.2 Distinct Characteristics of Crop Growth Models (CGM)

CGMs are excellent analytical tools because they exhibit three distinct characteristics that distinguish them from the previously described approaches (Delécolle et al., 1992):

- They are dynamic, in that they operate on a time step for ordering input data and updating state variables.
- They contain parameters that allow a general scheme of equations to be adopted to the specific growth behavior of different crop species.
- They include a strategy for describing phenological development of a crop to order organ appearance, organ death, and portioning/division of photosynthetic products.

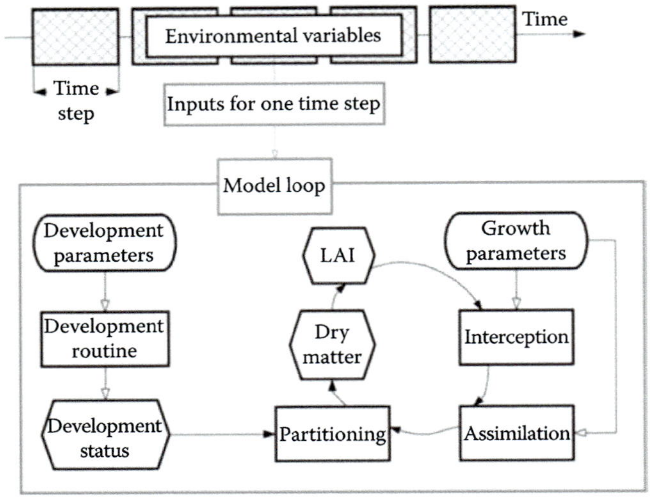

FIGURE 5.8 Simplified scheme of a crop process model. Model state variables such as development phase, organ dry mass, or leaf area index are linked to input variables, including weather, geographic, and management variables. (Modified from Delécolle, R. et al., *ISPRS J. Photogramm.*, 47, 145, 1992.)

5.3.4.3 Prominent Examples and Application Challenges

The first CGMs were developed by the end of World War II (Sinclair and Seligman, 1996). In subsequent decades, they became both more complex and potentially more useful (Boote et al., 1996). Deterministic CGMs have been validated for cereals, as well as for potato, sugar beet, oilseed, rice, canola, and sunflower. Most of these models include water and energy balance modules and run on a daily time basis over the whole life cycle of a crop. Prominent models are, for example, CERES (Jones and Kiniry, 1986), WOFOST (Supit et al., 1994), OILCROPSUN (Villalobos et al., 1996), CROPSYST (Stöckle et al., 2003), and STICS (Brisson et al., 1998). Some simpler models (without water and energy balance) such as SAFY (Claverie et al., 2012; Duchemin et al., 2008) and GRAMI (Maas, 1992) also exist. More sophisticated models attempt to integrate numerous factors that affect crop growth and development, such as plant available soil water, temperature, wind, genetics, management choices, and pest infestations (Hank et al., 2015).

The strength of CGMs as research tools resides in their ability to capture the soil-environment-plant interactions, but their initialization and parameterization for operational purposes generally requires several physiological and pedological parameters that are not easily available (J. Huang et al., 2019). In addition, given the high number of model parameters, careful validation strategies must be employed for obtaining the required predictive power (Bellocchi et al., 2010).

5.3.4.4 Complementary between CGM and Remote Sensing

Crop growth models are covered here in some detail because CGM and remote sensing nicely complement each other: crop growth models provide a continuous estimate of crop growth over time, whilst remote sensing provides temporally discontinuous but spatially detailed pictures of crop actual status (e.g., LAI, soil water content, emergence date) within a given area (Clevers and Van Leeuwen, 1996; Doraiswamy et al., 2004, 2003; Guérif and Duke, 2000; Padilla et al., 2012). Remotely sensed images are particularly useful in spatially distributed modeling frameworks (Moulin et al., 1998; Weiss et al., 2001). As remote sensing provides gridded (spatial) status maps, the use of remotely sensed information makes the CGM more robust (Doraiswamy et al., 2003; Guérif and Duke, 2000; J. Huang et al., 2019; Moulin et al., 1998). This complementary nature of remote sensing and crop growth modeling was first recognized by S. Maas who described routines for using satellite-derived information in mechanistic crop models (Maas, 1988a).

Spatialized information is readily available concerning many meteorological drivers (e.g., from global circulation models). However, other parameters and initial conditions required by CGMs may not be available spatially, for instance: (1) soil, plant and management parameters, and (2) initial values of all crop state variables (Doraiswamy et al., 2003).

5.3.4.5 EO Data Assimilation

To minimize field data collection and to ensure applicability of CGM over large areas, different approaches for assimilating EO data into dynamic process models have been developed (Table 5.4). All ideas are extracted from the outstanding paper of Delécolle et al. (1992). Useful overviews are also given in Clevers et al. (1994), Moulin et al. (1998), Bach and Mauser (2003), and Dorigo et al. (2007). In Huang et al. (2019) the interested reader finds a comprehensive, and mathematically detailed description of data assimilation (DA) approaches.

5.3.4.6 Parameterization and Initialization

In the most straightforward way, remote sensing may be used to parameterize and/or initialize crop growth models. The term "parameterization" refers here to the provision of model parameters required by crop growth models, for example, soil texture information, photosynthetic pathway information, crop type, sowing date, etc. The term "initialization" refers to the provision of model state variables at the start of the simulation (e.g., the soil water content at sowing).

To illustrate initialization/parameterization one can imagine in the simplest case that remotely sensed data is used to provide information about crop type. With known crop type, plant specific parameter settings within the CGM can be assigned, for example related to thermal or photothermal crop development. Likewise, optical imagery of bare soil conditions may be used to map soil organic matter content, soil texture and soil albedo (Ben-Dor, 2002; Viscarra Rossel et al., 2009). These three model parameters influence nutrient release, water capacity and radiation budget (Ungaro et al., 2005) and often need to be specified within CGM. Microwave imagery may be used to provide an estimate of soil water content at the beginning of the simulation run, that is, at sowing (Wagner et al., 2007). This will result in a model initialization, as the state variable "soil water content" has been attributed a value for the start of the simulation.

5.3.4.7 Re-Parameterization

In the "re-parameterization" approach (also sometimes called "re-calibration"), one assumes that some parameters of the crop growth model are inaccurately parameterized (or calibrated), although the model—as a whole—is formally adequate (Delécolle et al., 1992). By providing "reference" observations of some key vegetation properties (e.g., remotely derived LAI), some crop model parameters which are internally linked to LAI can be calibrated (Figure 5.9). This is usually

TABLE 5.4

Techniques for Assimilating EO Data in Dynamic Crop Growth Models. The reader is referred to Huang et al. (2019) for a more comprehensive overview of data assimilation techniques.

Assimilation Technique	Example
Updating	Bach and Mauser (2003), Pellenq and Boulet (2004)
Forcing	Maas (1988a, 1988b), Bouman (1995), Clevers et al. (2002)
Re-initialization	Bach and Mauser (2003), Nouvellon et al. (2001), Guérif and Duke (1998), Doraiswamy et al. (2003)
Re-parameterization	Maas (1988a, 1988b), Bouman (1995, 1992), Clevers and van Leeuwen (1996), Launay and Guerif (2005), Clevers et al. (1994), Guérif and Duke (1998)

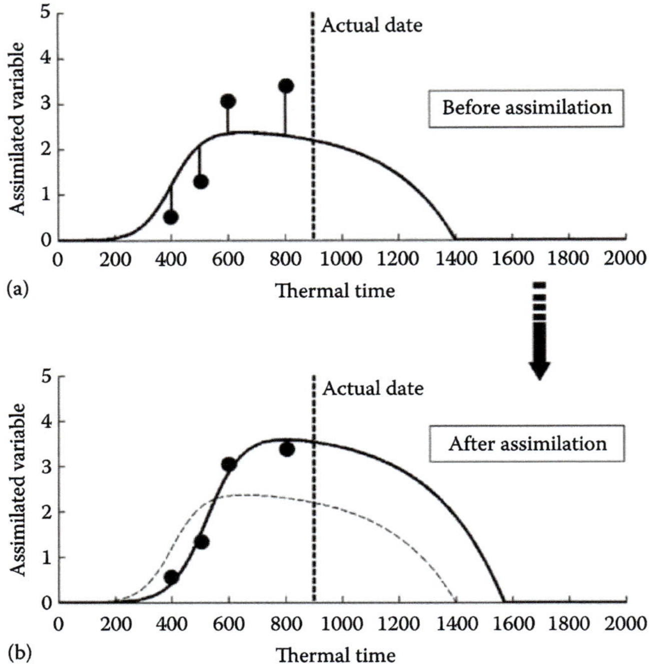

FIGURE 5.9 Schematic description of the recalibration method using remotely sensed state variables as inputs (here the assimilated variable corresponds to LAI). The crop growth model simulates the leaf development (LAI) over time. (a) Without assimilation, the simulated LAI is far from the four LAI observations. (b) After assimilation (e.g., after the nonlinear minimization procedure), new model coefficients are assigned to the crop growth model such that the residues between observed and simulated LAI are minimized. (Modified from Houlés, V., Mise au point d'un outile de modulation intra-parcellaire de la fertilization azotee du ble d'hiver base sur la teledetection et un modele de culture, Thèse de Doctorat, Institut National Agronomique Paris-Grignon, France, 2004.)

achieved by (iteratively) adjusting the model parameters until measured and simulated temporal profiles of the selected variable (here: LAI values) match each other (Doraiswamy et al., 2003)—a so-called "variational" approach (Huang et al., 2019). In spatially distributed modeling this re-calibration has of course to be done pixel by pixel. Contrary to Monteith's LUE (Section 5.3.3), the reference observations do not necessarily need to cover the entire growing season—already a single observation can be assimilated in the CGM.

5.3.4.8 Re-Initialization

The "re-initialization" of crop growth models works in a very similar way; however, instead of adjusting model parameters, one simply tunes the initial values of state variables until a good match between observed and simulated state variables is obtained. In both cases, the remote sensing derived variables are considered as an absolute reference for the model simulation. Again, the exact timing of the remotely sensed observations is of minor importance which is helpful for large scale applications. Already as few as one reference observations are useful (Atzberger, 1997; Baret et al., 2007; Launay and Guerif, 2005). Obviously, the more satellite observations are available, and the better they are distributed across the growing season, the more/better model parameters can be calibrated and/or initialized (Doraiswamy et al., 2003). Ideally, data from several sensor modalities are combined (e.g., optical, microwaves and LiDAR) to provide complementary status variables, sensitive to different sets of model parameters.

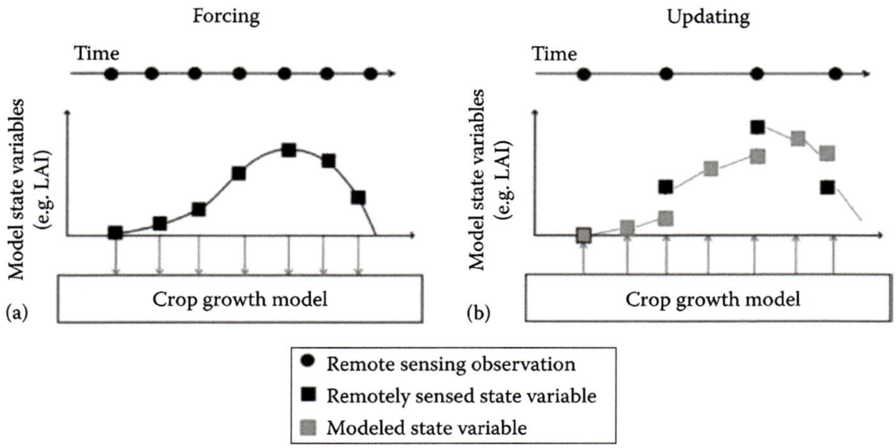

FIGURE 5.10 Schematic description of "forcing" and "updating" methods. (a) In the forcing approach, the complete time profile of a crop state variable (here: LAI) is reconstructed from remote sensing data and introduced (e.g., "forced") into the dynamic crop growth model at each time step in the simulation. (b) In the updating approach, the crop growth model is run (with standard parameter settings) until a remotely sensed state variable is available (in black). When new observations are available, the simulated state variable at this point is replaced by the remotely observed state variable and the crop growth simulation is continued without changing the parameter setting until another (if any) new observation becomes available. (Modified from Dorigo, W.A. et al., *Int. J. Appl. Earth Observ. Geoinform.*, 9(2), 165, 2007.)

5.3.4.9 Forcing

In the "forcing" approach, important remotely mapped state variables such as LAI are directly ingested for each time step into the model, thus "forcing" the model to follow the remotely sensed information (Figure 5.10a). Such a simplification makes crop growth models very similar to the Monteith's LUE model (Equations 5.3–5.4), by breaking the computational loop shown in Figure 5.8. As the model does no longer determine the values of that variable by itself, inconsistent model states may result (Delécolle et al., 1992).

5.3.4.10 Updating

The updating approach resembles a kind of Kalman Filter (KF) (Figure 5.10b): simulated values of crop state variables are replaced by their corresponding remotely sensed values each time these are available. The computations then continue with these updated values until new (remote sensing) inputs are provided. As for the "forcing" method, the replacement of simulated by observed state variables may result in inconsistent model states. Importantly, the approach does not correct the errors in the model calibration, which are causing the differences between simulated and observed state variables.

5.4 CROP ACREAGE ESTIMATION

Crop acreage is a core metric for production estimates (Table 5.1) and the acreage of the different crops must be known for each growing season for accurate production estimates (Baruth et al., 2008; Gallego, 2005).

Besides its direct use for production estimates, the "planting arrangement in time and space" has several other use cases:

- It informs about crop succession and cropping pattern (Bégué et al., 2018).
- Crop types, cropping patterns, and changes in LULC are indicative of land use intensity (Bégué et al., 2015; Kuemmerle et al., 2013; Waldner et al., 2015).

- Information about crop extent is necessary to better understand the role and response of regional cropping practices in relation to various environmental issues (e.g., climate change, groundwater depletion, soil erosion) (Galford et al., 2008).
- Monitoring the time and location of land-cover changes is important for establishing links between policy decisions, regulatory actions, and subsequent land-use activities, as outlined by Galford et al. (2008) and Som-ard et al. (2022).
- Determining the physical and temporal patterns of agricultural extensification or expansion and intensification is the first step in understanding their implications, for example, for long-term crop production and environmental, agricultural, and economic sustainability (Galford et al., 2008), respectively, the drivers of these changes (Hostert et al., 2011; Kuemmerle et al., 2011).

Together these information needs motivate the detailed regional-scale mapping of cropping patterns on a repetitive basis (Atzberger and Rembold, 2012; Bégué et al., 2018; Erb et al., 2013; Som-ard et al., 2022; Vieira et al., 2012; Wardlow et al., 2007).

The review paper of Olofsson et al. (2014) offers a guidance for accurate area estimation and change in land use, whereas the excellent work of Wardlow et al. (2007) provides a valuable discussion of different approaches focused on crop mapping, from which the following summary is derived. The interested reader is referred to Gallego (2004) for a detailed discussion on the limits of so-called "pixel counting" approaches and how to best integrate field sampling into statistically sound crop acreage estimates.

5.4.1 Crop Mapping Using Decametric Satellite Data

Considerable progress has been made classifying LULC patterns using multispectral, decametric resolution Landsat TM data as a primary input (Inglada et al., 2015; Vogelmann et al., 2001). With the launch of the Sentinel-2 twin satellites in 2016, an even finer spatial resolution can be achieved (Defourny et al., 2019; Immitzer et al., 2016; Vuolo et al., 2018). To remain concise, we only cover five of the major (supervised) approaches (Gómez et al., 2016):

- Classification of mono-temporal data
- Classification of multi-temporal data
- Time series analysis
- Object-based classification
- Semantic segmentation and deep learning approaches

5.4.1.1 Supervised Pixel-Wise Classification

In most cases, crop maps are generated by supervised classification of individual pixels (Beltran and Belmonte, 2001; Congalton et al., 1998). EO images for classification are generally acquired at key phenological stages for optimizing class separability. These approaches are labor and cost intensive and require amounts of cloud-free high spatial resolution imagery. This impedes an operational implementation over large areas and in multiple years (Lobell and Asner, 2004). In very cloudy areas data availability can be insufficient, particularly if specific crop stages need to be imaged (Perdigão and Annoni, 1997). Although generally feasible, the problems mentioned have historically limited the possibility of automatically updating land cover maps over large areas at regular (annual) intervals (Chang et al., 2007). With the advent of Sentinel-2 and commercial imagery since 2016, the data availability has greatly improved so that nowadays decametric data is available at daily time steps (see Table 5.2).

5.4.1.2 Classifications Using Multi-Temporal Data

In many parts of the world, relatively dense stacks of cloud-free images can be obtained thanks to the high revisit frequency of decametric sensors such as Sentinel-2 (typically between 10 and 30 cloud-free images per year). If the region of interest is not overly large (e.g., in the order of one or two tiles), this permits to simply stack the multispectral raster together to obtain highly reliable crop type maps (Ghassemi et al., 2022; Vuolo et al., 2018), without any further pre-processing into analysis-ready-data (ARD). Obviously, if regions become larger, the spatially varying acquisition and cloudiness pattern require some pre-processing to generate gap-free (and more or less consistent) spectral feature vectors (typically using compositing techniques). This permits to generate continental-scale crop type maps (Ghassemi et al., 2022).

Two main factors explain the additional information within multi-temporal data:

- Different crops have (more or less) different phenologies that are represented in the data leading to an enhanced discrimination.
- More images along the time axis enhance the chance of identifying crop-specific spectral differences between otherwise difficult to separate crop types.

The improved data availability in decametric resolution makes it now possible to update thematically rich LULC maps at annual time steps (Inglada et al., 2017). Depending on the environmental setting, the quality of the resulting maps is however subject to large variations (Inglada et al., 2015). Early crop type and cropland identification is also feasible (Matton et al., 2015), in particular when blending different sensor modalities (Blaes et al., 2005). For classes where texture and context are pertinent, convolutional NN can outperform traditional (e.g., RF-based) classifiers (Stoian et al., 2019).

5.4.1.3 Classifications Using Time Series Analysis

As shown in the early work of Badhwar (1984) and others (Guérif et al., 1996), a proper time series analysis involves more than stacking multi-temporal images together. Indeed, in time series analysis, a deterministic (parametric) model is fitted to the data (for example a double-logistic). Subsequently, the extracted model parameters are used for the classification. The advantage of such curve fitting approaches is that the exact dates of cloud free observations can be different for each pixel. If the system is not underdetermined, the absolute number of available observations can also be variable. This removes the need for data pre-processing (e.g., compositing) which is known to be sub-optimum (Verger et al., 2013). Note that instead of parametric functions, one can also use non-parametric approaches such as Fast Fourier Transform (FFT) and wavelets etc.—alternative approaches and good overviews and discussions are found in Pelletier et al. (2019) Pelletier and Valero (2021) and Rußwurm and Körner (2020).

5.4.1.4 Object-Based Classification

Conventional pixel-based classifications occasionally reveal difficulties regarding the automatic pattern recognition, mainly because of the phenological variability of crops, different cropping systems, sensor noise and non-uniform measurement conditions (e.g., atmospheric disturbances) (Vieira et al., 2012). In such a context, Object Based Image Analysis (OBIA) using mono- or multi-temporal (satellite) imagery appears promising (De Wit and Clevers, 2004; Peña et al., 2014; Whiteside and Ahmad, 2005).

The most common approach used to generate image objects is image segmentation (Benz et al., 2004; Pal and Pal, 1993). The segmentation process subdivides an image into homogeneous regions through the grouping of pixels in accordance with determined criteria of homogeneity and heterogeneity (Comaniciu and Meer, 2002; Haralick and Shapiro, 1985). For each resulting object, spectral, textural, morphic and contextual attributes can be generated and exploited for the purpose of classification (Blaschke, 2010). Textural and shape information can be particularly well retrieved from decimetric to metric (VHR) data but is obviously less relevant in decametric or coarser data (Peña-Barragán et al., 2011).

As an example, Vieira et al. (2012) combined OBIA and decision trees (DT) to map harvest-ready sugarcane in Brazil. To derive the binary map indicating the area of harvest-ready sugarcane, four Landsat images acquired between September 2000 and March 2001 were used. An overview of the methodology and processing steps are outlined in Figure 5.11. The images were automatically segmented to retrieve precise field boundaries (Figure 5.12). Many attributes were afterwards extracted for each polygon (object). The attributes included spectral, spatial, and textural features as described in Blaschke (2010).

Interestingly, only a small set of features was selected by the DT for obtaining an overall accuracy of 94% on independent test samples (i.e., NDVI, spectral signatures and one textural feature). As expected, multi-temporal information was necessary to differentiate between harvest-ready sugarcane and the other land uses. Textural attributes were relevant where and when areas with high-biomass sugarcane were confounded with other high-biomass areas (e.g., forests). Spatial attributes

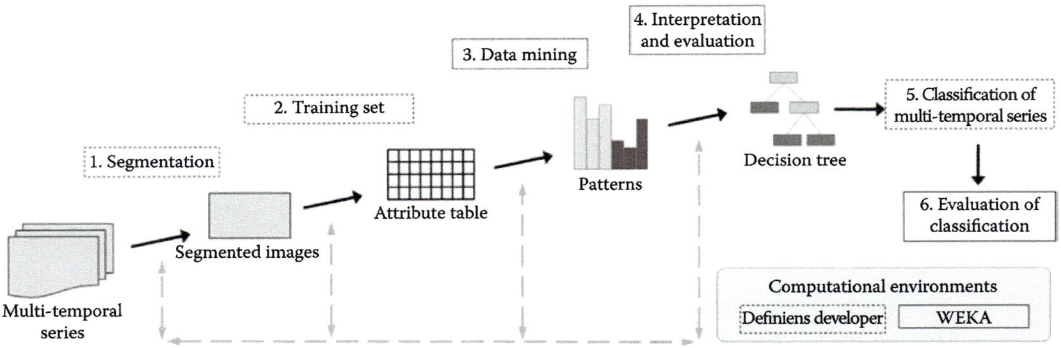

FIGURE 5.11 Flowchart illustrating the main stages that are part of the OBIA+ DM approach proposed to classify sugarcane areas (RH) from Landsat time-series images. Hachures illustrate the different computational environments used in each methodological stage. Broken-lined arrows indicate iteration possibilities. (From Vieira, M.A. et al., *Remote Sens. Environ.*, 123, 553, 2012.)

FIGURE 5.12 Example of the segmentation result used by Vieira et al. (2012) for mapping harvest-ready sugarcane. The underlying RGB composite consists of a TM image taken in the month of February 2011, with composition R(4) G(5) B(3). (From Vieira, M.A. et al., *Remote Sens. Environ.*, 123, 553, 2012.)

(e.g., shape, dimension), on the other hand, were not selected in this study since most agricultural fields, both sugarcane and the other crops, had similar geometric characteristics.

5.4.1.5 Semantic Segmentation

In "semantic segmentation" and other deep learning techniques, the two steps of segmentation and classification are executed simultaneously using neural nets (López et al., 2020; Xu et al., 2021). Recent overviews of these techniques are provided in Yuan et al. (2021) and Ofori-Ampofo et al. (2021), whereas Rustowicz et al. (2019) provide additional details and datasets. As the field is evolving quickly, the reader is referred to the preceding references for more information.

5.4.2 Crop Mapping Using Hecto- to Kilometric Resolution Satellite Time Series

At national to global scales, kilometric resolution imagery from VGT and AVHRR instruments played historically an important role in crop mapping (Loveland et al., 2000). However, with the improved availability of decametric time series at weekly revisit frequency (e.g., Sentinel-2), and the concurrent increase in compute power to handle big data, the importance of coarse resolution data has decreased over time.

To provide the reader a quick overview, we distinguish three main techniques:

- Hard classification
- Fuzzy classification and unmixing
- Bayesian inference

5.4.2.1 Hard Classifications: Analysis of Temporal Signatures.

The high temporal resolution provided by hecto- to kilometric resolution sensors allows land cover types to be discriminated based on their unique phenological (seasonal) characteristics (Fries et al., 1998; Vuolo and Atzberger, 2014, 2012). Due to the mixed nature of coarse resolution pixels, however, few of these mapping efforts have classified detailed, crop-related LULC patterns (Wardlow et al., 2007).

For areas with relatively large field sizes, MODIS provides higher resolution (250 m) compared to the mentioned kilometric resolution sensors. This offers good opportunities for a more detailed, large-area LULC characterization by providing global coverage with daily revisit frequency and hectometric spatial resolution (Justice et al., 2002). Several studies have already successfully demonstrated the potential of these data for detailed LULC characterization in an agricultural setting (Lobell and Asner, 2004; Lunetta et al., 2010; Wardlow et al., 2007).

Wardlow et al. (2007), for example, found that MODIS time-series at 250 m ground resolution had sufficient temporal and radiometric resolution to discriminate major crop types and crop-related land use practices in Kansas, US. For each crop, a unique multi-temporal VI profile consistent with the known crop phenology was detected. Most crop classes were separable at some point during the growing season based on their phenology-driven differences expressed in the VI temporal trajectory. Even regional intra-class variations were detected, reflecting the climate and planting date gradient in the study area. They also found that MODIS's 250 m spatial resolution was an appropriate scale at which to map the general cropping patterns of the US Central Great Plains.

Lunetta et al. (2010) used MODIS 16-day NDVI composite data to successfully develop annual cropland and crop-specific map products (corn, soybeans, and wheat) for the Laurentian Great Lakes Basin. The crop area distributions and changes in crop rotations were characterized by comparing annual crop map products for 2005, 2006, and 2007. Obviously, for regions with (much) smaller field sizes, the resolution of MODIS can be much less adequate.

5.4.2.2 Fuzzy Classification and Unmixing Approaches

The use of hecto- to kilometric resolution data is effective because it offers numerous advantages: global coverage and low cost, high temporal frequency, easy processing at a regional to continental scale, availability of long-term records (e.g., from 1980 thanks to AVHRR instruments onboard of NOAA satellites), and finally continuity of data provision as ensured by relatively well coordinated satellite programs. However, because of sub-pixel heterogeneity, the application of traditional hard classification approaches faces intrinsic methodological limitations and may result in significant errors in the estimated crop areas (Chang et al., 2007; DeFries et al., 1995). Indeed, as pointed out by Foody (1996), the conventional allocation of each image pixel to a land cover class is inappropriate for mixed pixel, and fuzzy approaches are required.

To address sub-pixel heterogeneity resulting from fragmented landscapes, Quarmby et al. (1992) used linear mixture model (LMM) techniques. Other authors investigated so-called unmixing approaches combining kilometric time series with decametric resolution images to improve sub-pixel crop monitoring capabilities (Doraiswamy et al., 2004; Maselli et al., 1998). While generally successful, insufficient contrast between endmembers often leads to an ill-posed inverse problem with unstable solutions, resulting in inaccurate fraction images with high uncertainties (Lobell and Asner, 2004). On the other hand, too few endmembers fail to correctly represent the input signature, as evident in high reconstruction errors.

A probabilistic linear unmixing approach with MODIS spectral/temporal data was developed and tested by Lobell and Asner (2004). The approach estimates sub-pixel fractions of crop area based on the temporal reflectance signatures throughout the growing season. In this approach, endmember sets are constructed using Landsat data to identify pure pixels, mainly located within large fields. Rather than defining endmembers with a single spectrum, endmembers are defined as a set of spectra which represent the full range of potential variability. The uncertainty in endmember fractions arising from endmember variability can then be quantified using Monte Carlo techniques.

To quantify inter-annual crop area changes based on kilometric NDVI time series, other studies relied on SAM (Spectral Angle Mapping) where pixels are matched to reference feature vectors using an n-dimensional angle (Rembold and Maselli, 2006, 2004). A reference period of seven years was required to obtain valuable results.

Regression tree analysis was used by Chang et al. (2007) for the percentage of the corn and soybean area mapping using 500 m MODIS time series dataset. Numerous phenological measures and data transformations were used as potentially discriminative features crop-type discrimination.

Neural networks (NN) fed with 1 km monthly NDVI composite images were used by Verbeiren et al. (2008) to model the area fraction images (AFI) of eight classes. Relatively good results were obtained, especially if the initial (pixel-based) results were aggregated to coarser regional levels. The portability across growing seasons was investigated in an accompanying paper on the same dataset by Bossyns et al. (2007). The NNs were trained on data of one growing season and then applied to left-out years. Compared to the R^2 of the training data, the rest data accuracy decreased by approximately 0.45 units demonstrating a lack of portability.

To better cope with the natural year-to-year variability of NDVI profiles of vegetated surfaces, Atzberger and Rembold (2013) trained networks with AVHRR time series. The target variable represented the sub-pixel winter crop fractional coverage. To permit the NN to distinguish for various proportions of non-arable land within the mixed pixels (e.g., forested areas, urban land, etc.), CORINE land cover information was used as additional input. A positive impact was demonstrated regarding the concurrent use of ancillary information. In-season predictions improved compared to the work of Rembold and Maselli (2006, 2004) using the same dataset and linear prediction models. On average (median), 79% of the spatial variability of the (sub-pixel) winter crop abundances were explained by the NN approach. For the individual years, the cross-validated R^2 ranged between 0.70–0.82. The cross-validated RMSE values were around ~10% (relative to the winter crop area). The same approach was tested by Atzberger and Rembold (2012) for its portability across years and its usefulness to derive regional statistics. Data from three years between 1988 and 2001 were used to train the NN. The trained net was then applied to the left-out period 2002–2009. Even though two years of the validation dataset had (extreme) conditions not previously seen by the NN (e.g., with exceptionally high and low winter wheat areas, respectively), the NN performed relatively well.

Atkinson et al. (1997) showed how NN can be used for unmixing single date (five wavebands) AVHRR imagery to map sub-pixel proportional land cover. The use of NN for estimating sub-pixel land cover from temporal signatures was investigated by Karkee et al. (2009). Braswell et al. (2003) demonstrated that NN-based approaches offer significant improvement relative to linear unmixing for estimation of sub-pixel land cover fractions in the heterogeneous disturbed areas of Brazilian Amazonia. The improvement was related to the fact that linear unmixing assumes the existence of pure sub-pixel classes (endmembers) with fixed reflectance signatures.

Shang et al. (2022) proposed GAN (generative adversarial networks) to directly (end-to-end) generate high-resolution land cover maps from the intermediate output of soft classification. This contrast with traditional soft classification approaches which "only" estimate fractional values of classes within low-resolution pixels. To solve the task, in the approach of Shang et al. (2022), pairs of low-resolution class fraction image and corresponding high-resolution land cover maps are required. The two steps of (1) deduction, and (2) conversion of fraction images to (high-resolution) categorical maps are afterwards fully integrated to avoid solving the problem in two separate steps.

5.4.2.3 Bayesian Inference Using Additional Context Variables.

The combined use of remotely sensed data and ancillary information was presented by Mello et al. (2013b) for the case of soybean mapping in Mato Grosso State, Brazil. The approach is based on Bayesian Networks (BN). These networks can incorporate experts' knowledge in complex classification tasks and therefore help to characterize phenomena through plausible reasoning inferences based on evidence.

Mato Grosso State (total size of about 900,000 km^2) was selected as area of interest as the state is the largest Brazilian soybean producer (about 30% of the total domestic production) and an important global hub for tropical agricultural production. For Mato Grosso, tabulated agricultural statistics at municipality level exist. These statistics are however only released with a delay of about two years. This delay hinders an effective and timely monitoring of the possible spread of soybean cropping into new, sometimes environmentally sensitive, areas. As such, there is demand for the use of remote sensing images as an accurate, efficient, timely, and cost-effective way to monitor agricultural crops (Rudorff et al., 2010).

The Bayes' theorem, which is used in Bayesian Networks, updates the knowledge (prior probability) of a specific event in the light of additional evidence (conditional probabilities), allowing one to have a plausible reasoning based on a degree of belief (posteriori probability) (McGrayne, 2011). Observations made upon variables that are related to a particular phenomenon may be used to develop plausible reasoning about the phenomenon, its causes, and consequences (Jaynes, 2003). When the number of variables increases or even when the complexity of the interactions among the variables involved in a specific phenomenon rises, the Bayesian Network is a representation suited to model and handle such tasks (Nielsen and Jensen, 2007; Pearl, 1988). The joint probability of any instantiation (sometimes called realization) of all the variables in a BN can be computed as the product of n probabilities. This ability to compute posterior probabilities given some evidence is called inference.

To apply this concept to the problem at hand, Mello et al. (2013b) named "target variable" the variable which represents the phenomenon, and "context variables" the variables that are somehow related to the phenomenon. Besides hectometric remotely sensed spectral and temporal information from the MODIS sensor, several other context variables related with soybean occurrence were chosen (e.g., soil type and distance to roads and other infrastructure facilities) (Garrett et al., 2013). The resulting probability image (PI) is shown in Figure 5.13.

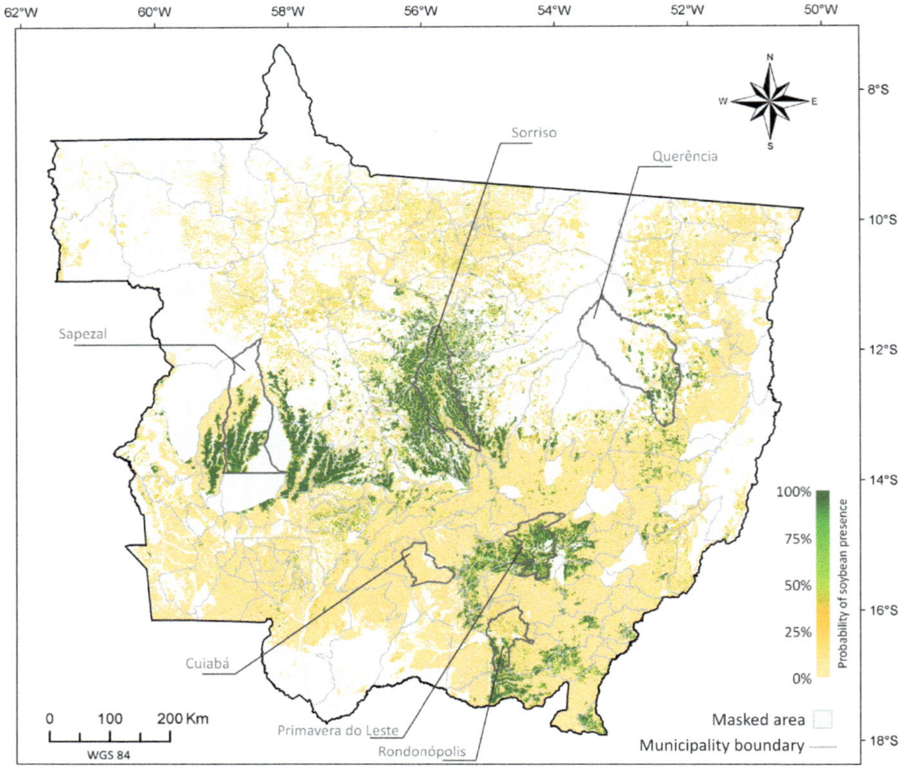

FIGURE 5.13 Probability image (PI) of soybean presence for Mato Grosso State, Brazil. Main soybean producer centers and the capital, Cuiabá, are highlighted. The color indicates the calculated probability of soybean presence in 2005/2006 given the observations made for the context variables. (From Mello, M.P. et al., Spatial statistic to assess remote sensing acreage estimates: An analysis of sugarcane in São Paulo State, Brazil, *Remote Sens.*, 5(11), 5999, 2013b.)

The map shows the spatial distribution of the probability of soybean crops throughout Mato Grosso territory in crop year 2005/2006. Mello et al. (2013b) found a high agreement of the mapped soybean acreage with (independent) official statistics. A quantification of the influence of each context variable found that the remotely sensed data were the most important variables used to infer about soybean occurrence followed by soil aptitude. Interestingly, the distance to the next road was only of minor importance, explained by the fact that soybean fields in Mato Grosso are usually very large, so that even very high transportation costs do not hinder soybean cultivation.

5.4.3 Accuracy Considerations

Olofsson et al. (2014) provides guidance, and good practice recommendations, on how to assess map accuracy in a consistent and transparent manner. We can only underline their statement that

> a key strength of remote sensing is that it enables spatially exhaustive, wall-to-wall coverage of the area of interest. However, as might be expected with any mapping process, the results are rarely perfect. Placing spatially and categorically continuous conditions into discrete classes may result in confusion at the categorical transitions. Error can also result from the mapping process, the data used, and analyst biases.
>
> *(Foody, 2010)*

As map users are acutely interested in understanding the quality of the provided maps, we invite interested readers to consult the work of Olofsson et al. (2014), as well as work published by Foody (2002), Strahler et al. (2006), Foody (2010) and Gallego (2012).

Several real world challenges to accuracy assessment were addressed in a recent work of Foody (2023). Especially if no explicit account is made for the effects of prevalence and reference data error, he urges to interpret classical map statistics with care. Using numerical experiments, he demonstrates that the direction and magnitude of accuracy metric mis-estimation were a function of prevalence (i.e., class imbalances) and the size and nature of the imperfections in the reference standard (i.e., partly incorrect reference samples). We invite the interested reader to consult this work to better understand the challenges of accuracy assessment and to find guidance in how to address both class imbalances and (unavoidable) errors in the reference data. To better comprehend the impacts of "ignored classes" (i.e., cases where the defined set of classes is not exhaustive), we recommend the additional work of Foody (2021).

5.5 CROP DEVELOPMENT AND PHENOLOGY

Mapping of a crop's phenological development is important as the phenology is often closely related to biomass production and crop yield (Meroni et al., 2014a) (see "partitioning" control in Figure 5.8). For example, cool summers may result in delayed heading and thus decreased yields. The end of the grain filling period needs to be known for optimum harvest results. Reliable information about development stages is also a relevant input in dynamic (mechanistic) crop growth models, for example to trigger the grain filling period, and more generally to define the start of season and planting dates, etc. Besides use for agriculture, multi-annual observations of phenology are also excellent proxies for climate change (Brown et al., 2012; de Jong et al., 2011). Several prominent applications for phenology are listed in Table 5.5, and the relation between phenology and vegetation productivity is reviewed in Kooistra et al. (2024).

TABLE 5.5
Applications of Land Surface Phenology (LSP) Metrices in Agriculture and for Environmental Monitoring

Application	Example
Mapping and monitoring of crop development	Sakamoto et al. (2011)
	Kawamura et al. (2005)
Parameterization of crop growth models	Moulin et al. (1998)
	Boschetti et al. (2009)
Yield forecasting	Xin et al. (2002)
	Meroni et al. (2013b)
	Bolton and Friedl (2013)
Cropland identification	Galford et al. (2008)
	Atkinson et al. (2012)
	Arvor et al. (2011)
Identification of crop types	Badhwar et al. (1982)
	Zhang et al. (2003)
	Guérif et al. (1996)
	Galford et al. (2008)
	Peña-Barragán et al. (2011)
Identification of cropping patterns	Mahlayeye et al. (2024)
	Liu et al. (2018)
Monitoring of LULC change	Galford et al. (2010)
	de Beurs and Henebry (2004)
Multi-annual proxy for climate change	Udelhoven et al. (2009)
	de Jong et al. (2011)

5.5.1 Phenology and Land Surface Phenology

Using skilled observers, the phenological development of plants can be characterized in situ at very detailed levels (e.g., the BBCH-scale). However, as the phenological development of crops is spatially and temporally variable—mainly driven by the inter- and intra-annual dynamics of the Earth's climatic and hydrologic regimes—this variability can only partly be traced using ground observations (Zhang et al., 2003).

To track rapid changes in plant development, a high revisit frequency of the employed sensor is mandatory. Historical this prerequisite was only met by hecto- to kilometric sensors. Since the launch of Sentinel-2 and the (commercial) Planet constellation, however, such dense time series are now also available at decametric resolution. This strongly reduces mixed-pixel effects (Section 5.4.2). Both, deca- and kilometric EO sensors now provide:

- Synoptic global coverage at high resolution
- Frequent temporal sampling (e.g., daily) at short leap times
- Free data (for noncommercial sensors) and easy access

However, it must be noted that EO does not really track the "real" phenological development of plants (e.g., flowering events, organ appearance). Instead, what is recorded is usually a proxy for growth conditions: for example, changes in (green biomass) (Figure 5.14) (Bolton et al., 2020). For this reason, one distinguishes between "phenology" in the proper sense of the meaning (i.e., referring to a plant's development stages; Figure 5.8), and "land surface phenology (LSP)"; the latter being a way to characterize the intra-annual growth profile of a (more or less) small piece of land, irrespective of its plant species composition.

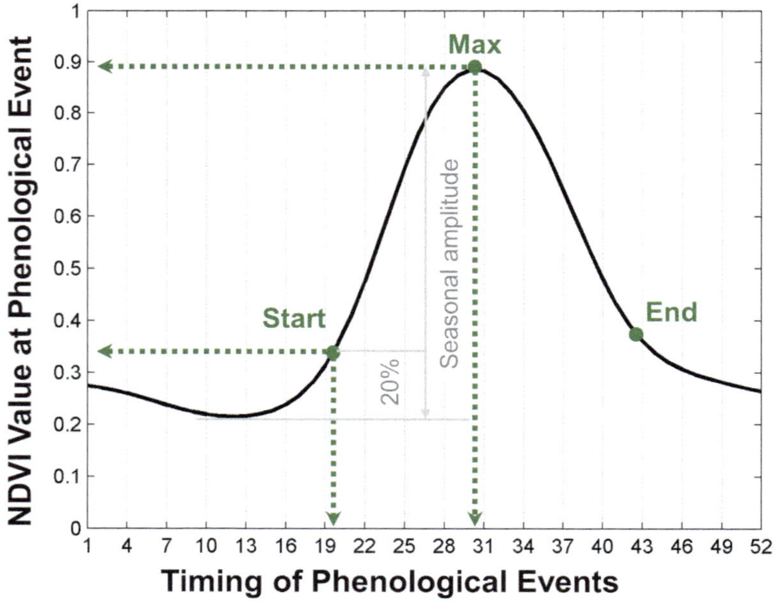

FIGURE 5.14 Basic phenological metrics that can be extracted from NDVI time series: start of season, maximum/peak of season, and end of season. For extracting the start of season (SOS), the relative threshold approach (20% of seasonal amplitude) is used and illustrated with gray lines. (From Atzberger, C. and Klisch, A., 2014.)

5.5.2 LSP Mapping and Monitoring

To map LSP metrices, usually spectral (vegetation) indices are used. Dense time series of NDVI and other spectral indices are useful to assess basic LSP metrices such as start, peak (maximum) and end of season, as they track the seasonal evolution of the aboveground (green) biomass (Figures 5.15 and 5.16).

To determine the timing of vegetation green-up and senescence from remotely sensed VI time series, several different approaches have been developed (Table 5.6). Following Beck et al. (2006), the methods can be grouped in two categories:

1. Methods estimating independently the timing of single phenological events (Badeck et al., 2004; Reed et al., 1994; White et al., 1997)
2. Methods modeling the entire time series using a mathematical function and where the parameters refer to different LSP events (Beck et al., 2006; Jönsson and Eklundh, 2002; Stöckli and Vidale, 2004)

5.5.3 Time Series Modeling

Modeling VI time series as a (parametric) function has the advantage of conserving a maximum amount of information in the VI data, while reducing the dimensionality of the data (Jönsson and

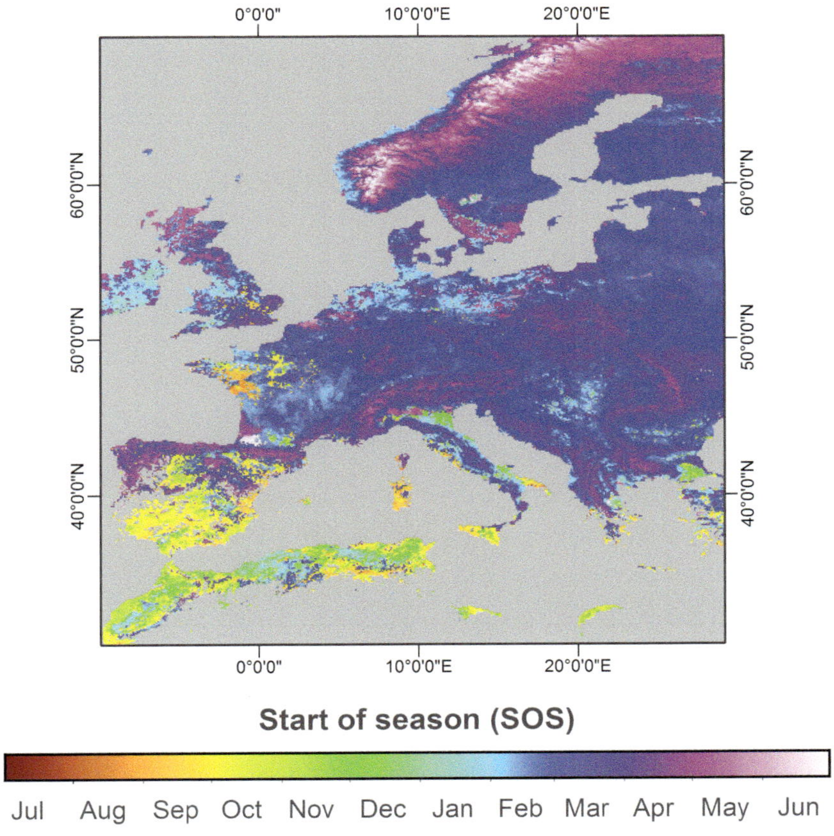

FIGURE 5.15 Spatial distribution of start of season (SOS) derived from MODIS time series in 2007 over Europe/Maghreb. Land pixels without vegetation and water surfaces are masked out (in gray). (From Atzberger et al., 2014.)

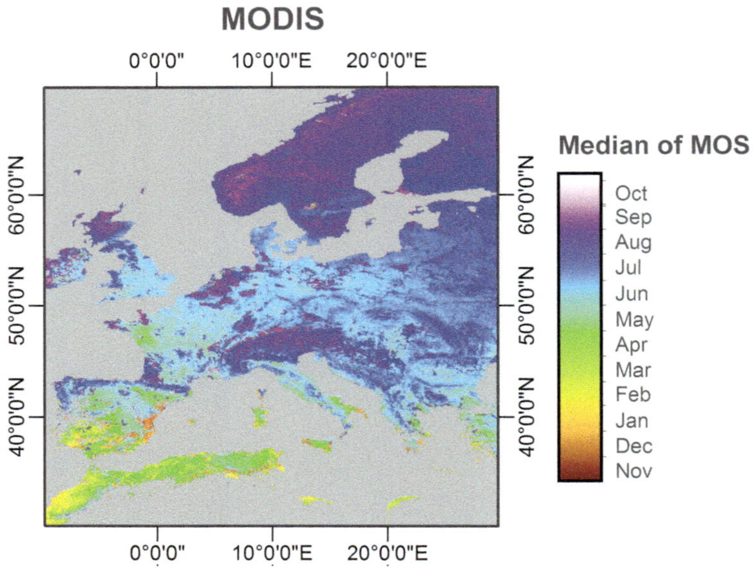

FIGURE 5.16 Spatial distribution of maximum of season (MOS) derived from MODIS time series: averaged MOS (median) over all years (2003–2011). Land pixels without vegetation and water surfaces are masked out (in gray). (From Atzberger et al., 2014.)

TABLE 5.6

Methods for Determining Land Surface Phenological (LSP) Events from EO Time Series (e.g., Green-Up Date)

Methods timing individual phenological events	
Use of specific (VI) thresholds	White et al. (1997), Lloyd (1990), Atzberger et al. (2014)
Detection of inflection points	Kaduk and Heimann (1996)
Local curve fitting	Zhang et al. (2003)
Backward-looking moving averages	Reed et al. (1994)
Methods modeling the entire time series	
Principle component analysis	Hirosawa et al. (1996)
Fourier and harmonic analysis	Atkinson et al. (2012), Azzali and Menenti (2000), Jakubauskas et al. (2001)
Wavelet decomposition	Anyamba and Eastman (1996), Sakamoto et al. (2005)
Global curve fitting	Zhang et al. (2003), Beck et al. (2006), Jönsson and Eklundh (2002), Meroni et al. (2014a)

Eklundh, 2004) (see also Section 5.4.2). Therefore, in addition to the phenological dates, other parameters can be readily retrieved from the fitted parameters (Beck et al., 2006).

On a downside, parametric functions are usually not easy to apply for large regions and generally do not apply well for ecosystems characterized by multiple growth cycles (e.g., double- or triple-cropping systems, semiarid systems with multiple rainy seasons, etc.). This was demonstrated, for example, by Atkinson et al. (2012) over India.

The traditional Fourier transform, for example, expects periodicity in the data not always given (e.g., in the case of land use change). Additionally, application of Fourier transforms often reveals spurious oscillations (Hermance, 2007). This happens frequently when many harmonics must be combined for fitting non-trivial temporal patterns (e.g., related to double/triple cropping).

Nonstationary data with irregular temporal shapes are better handled by the wavelet transform (Galford et al., 2008). In an agricultural application, wavelet-smoothed time series were successfully used to identify the start of the growing season and the time of harvest with relatively low errors (±2 weeks) (Sakamoto et al., 2005). Wavelet analysis can handle the range of agricultural patterns that occur through time, as well as the spatial heterogeneity of fields that result from precipitation and management decisions, because the transform is localized in time and frequency.

Curve fitting using predefined functions such as double logistic or gaussians is one of the most widely used approach for modeling the entire time series (Badhwar et al., 1982; Beck et al., 2006; Guérif et al., 1996; Meroni et al., 2014b). A fitted curve simplifies the parameterization necessary for identification of LSP metrics, such as start of season. In addition, data gaps are easily handled. A drawback of curve-fitting approaches is that *a priori* information is necessary to inform the algorithm about the number of cropping seasons within a 12-month period and the probable location of vegetation peaks (Jönsson and Eklundh, 2004).

Most of the mentioned issues are handled in a good manner by software packages such as TimeSat (Eklundh and Jönsson, 2015), TimeStats (Udelhoven, 2011) or SPIRITS (Rembold et al., 2015). Therefore, such fitting approaches are generally preferred over (threshold-based) event detection approaches. All approaches can be done at scales ranging from decameters to kilometers and at continental scale (Bolton et al., 2020).

5.5.4 Use of Thresholds

The alternative to curve fitting for the detection of LSP events is the application of simple thresholds to time series of spectral (vegetation) indices.

Threshold-based approaches are often employed for detecting the start of season (SOS) after cut events in managed grasslands (Dujakovic et al., 2024; Watzig et al., 2023), to map and monitor LSP over large regions (Huang et al., 2019; Ma et al., 2022), to detect cropping pattern (Liu et al., 2018), or to model and predict crop yields (Liu et al., 2020), amongst many other applications (Table 5.5).

While the required thresholds can be readily fine-tuned to local/regional conditions, the approaches generally suffer from a lack of model transferability. Without proper noise removal and gap-filling, such approaches are also prone to false detections (both false positives and false negatives) as only individual observations are thresholded, without modeling the full growing cycle.

5.6 CONCLUSIONS AND RECOMMENDATIONS

The chapter has illustrated the potential of remote sensing for solving important issues within the agricultural sector. The number of applications is huge, with a predominance of applications focusing on yield and area estimation, as well as change detection and other monitoring applications.

The use of remote sensing techniques is expected to further increase in the future as agricultural information is regularly needed for various decision makers. Indeed, as a better inventory of natural resources supports an informed decision making within the agricultural value chain, one can foresee an increased use of EO for example with respect to:

- Environmental compliance monitoring
- Deforestation and land conversion monitoring
- Food security applications
- Information needs of commodity markets
- Precision farming requirements
- Monitoring-Reporting-Verification (MRV) requirements of voluntary carbon markets and other in- and offsetting mechanisms

While only small parts of possible agricultural applications could be covered within the limited space of this book chapter, we hope that the reader realizes the tremendous power of remote sensing techniques for monitoring the entire agricultural value chain. With improved data availability and increased computational power, areas of any size can be monitored in a cost-effective and transparent/objective way. Although we mainly described approaches using globally available (deca- to kilometric) datasets, the reader should bear in mind that additional information can be derived from (very) high spatial resolution data with metric to sub-metric spatial resolution as well as from the plethora of supplementary sensor modalities (e.g., LiDAR, SAR).

Largely in line with the objectives of the GEOGLAM initiative (Soares et al., 2011), the following recommendations are drawn (Atzberger, 2013):

- For being useful (and used), applications should be scalable. This implies a mix of physical-based approaches (e.g., radiative transfer models and mechanistic process models) with modern approaches for self-supervised learning.
- The issue of timeliness should be dealt with as for most agricultural applications information is worth little if it comes too late.
- Product developers have only limited access to ground truth information to evaluate their products under various environmental settings. International efforts are needed to establish such networks of validation sites (Baret et al., 2006; Justice et al., 2000; Morisette et al., 2006; Olofsson et al., 2014, 2012; Stehman et al., 2012).
- Multi-sensor approaches that leverage synergies from data derived from different sensors and sensor modalities offer a more comprehensive view on plant status and functioning (Berger et al., 2022). This requires more efforts for sensor inter-calibration (Meroni et al., 2013a) but also ways for handling jointly different sensor modalities (e.g., optical and microwaves) acquired at different temporal resolution—ideally sensor integration should be handled within physical based radiative transfer models and also leverage the temporal domain.
- For potential users, the wide variety of products and promises can be confusing. The confusion is further amplified by the explosive growth of publications with insufficient scrutiny.
- When possible, the integration of crowd sourced information should be pursued (Foody et al., 2013; Foody and Boyd, 2013).
- The challenging accuracy assessment of map statistics should be better addressed, as outlined in the recent work of Foody (2023). This requires in particular an explicit account for the effects of prevalence and reference data error.

REFERENCES

Adede, C., Oboko, R., Wagacha, P.W., Atzberger, C., 2019a. Model ensembles of artificial neural networks and support vector regression for improved accuracy in the prediction of vegetation conditions and droughts in four Northern Kenya counties. ISPRS Int. J. Geo-Inf. 8, 562. https://doi.org/10.3390/ijgi8120562

Adede, C., Oboko, R., Wagacha, P.W., Atzberger, C., 2019b. A mixed model approach to vegetation condition prediction using Artificial Neural Networks (ANN): Case of Kenya's operational drought monitoring. Remote Sens. 11, 1099. https://doi.org/10.3390/rs11091099

Ahonen, H.-M., Kessler, J., Michaelowa, A., Espelage, A., Hoch, S., 2022. Governance of fragmented compliance and voluntary carbon markets under the Paris agreement. Polit. Gov. 10, 4759. https://doi.org/10.17645/pag.v10i1.4759

Amorós López, J., Izquierdo Verdiguier, E., Gómez Chova, L., Muñoz Marí, J., Rodríguez Barreiro, J.Z., Camps Valls, G., Calpe Maravilla, J., 2011. Land cover classification of VHR airborne images for citrus grove identification. ISPRS J. Photogramm. Remote Sens. 66, 115–123. https://doi.org/10.1016/j.isprsjprs.2010.09.008

Anyamba, A., Eastman, J.R., 1996. Interannual variability of NDVI over Africa and its relation to El Niño/Southern Oscillation. Int. J. Remote Sens. 17, 2533–2548. https://doi.org/10.1080/01431169608949091

Arvor, D., Dubreuil, V., Simões, M., Bégué, A., 2013. Mapping and spatial analysis of the soybean agricultural frontier in Mato Grosso, Brazil, using remote sensing data. GeoJournal. 78, 833–850. https://doi.org/10.1007/s10708-012-9469-3

Arvor, D., Jonathan, M., Meirelles, M.S.P., Dubreuil, V., Durieux, L., 2011. Classification of MODIS EVI time series for crop mapping in the state of Mato Grosso, Brazil. Int. J. Remote Sens. 32, 7847–7871. https://doi.org/10.1080/01431161.2010.531783

Atkinson, P.M., Cutler, M.E.J., Lewis, H., 1997. Mapping sub-pixel proportional land cover with AVHRR imagery. Int. J. Remote Sens. 18, 917–935.

Atkinson, P.M., Jeganathan, C., Dash, J., Atzberger, C., 2012. Inter-comparison of four models for smoothing satellite sensor time-series data to estimate vegetation phenology. Remote Sens. Environ. 123, 400–417. https://doi.org/10.1016/j.rse.2012.04.001

Atzberger, C., 1997. Estimates of winter wheat production through remote sensing and crop growth modelling: A case study on the Carmargue region, 1. Aufl. ed. Verlag für Wissenschaft und Forschung.

Atzberger, C., 2004. Object-based retrieval of biophysical canopy variables using artificial neural nets and radiative transfer models. Remote Sens. Environ. 93, 53–67. https://doi.org/10.1016/j.rse.2004.06.016

Atzberger, C., 2013. Advances in remote sensing of agriculture: Context description, existing operational monitoring systems and major information needs. Remote Sens. 5, 949–981. https://doi.org/10.3390/rs5020949

Atzberger, C., Eilers, P.H.C., 2011. A time series for monitoring vegetation activity and phenology at 10-daily time steps covering large parts of South America. Int. J. Digit. Earth. 4, 365–386.

Atzberger, C., Guérif, M., Baret, F., Werner, W., 2010. Comparative analysis of three chemometric techniques for the spectroradiometric assessment of canopy chlorophyll content in winter wheat. Comput. Electron. Agric. 73, 165–173. https://doi.org/10.1016/j.compag.2010.05.006

Atzberger, C., Klisch, A., Mattiuzzi, M., Vuolo, F., 2014. Phenological metrics derived over the European continent from NDVI3g data and MODIS time series. Remote Sens. 6, 257–284. https://doi.org/10.3390/rs6010257

Atzberger, C., Rembold, F., 2012. Portability of neural nets modelling regional winter crop acreages using AVHRR time series. Eur. J. Remote Sens. 45, 371–392. https://doi.org/10.5721/EuJRS20124532

Atzberger, C., Rembold, F., 2013. Mapping the spatial distribution of winter crops at sub-pixel level using AVHRR NDVI time series and neural nets. Remote Sens. 5, 1335–1354. https://doi.org/10.3390/rs5031335

Atzberger, C., Richter, K., 2012. Spatially constrained inversion of radiative transfer models for improved LAI mapping from future Sentinel-2 imagery. Remote Sens. Environ. 120, 208–218. https://doi.org/10.1016/j.rse.2011.10.035

Azzali, S., Menenti, M., 2000. Mapping vegetation-soil-climate complexes in southern Africa using temporal Fourier analysis of NOAA-AVHRR NDVI data. Int. J. Remote Sens. 21, 973–996. https://doi.org/10.1080/014311600210380

Bach, H., Mauser, W., 2003. Methods and examples for remote sensing data assimilation in land surface process modeling. IEEE Trans. Geosci. Remote Sens. 41, 1629–1637. https://doi.org/10.1109/TGRS.2003.813270

Badeck, F.-W., Bondeau, A., Böttcher, K., Doktor, D., Lucht, W., Schaber, J., Sitch, S., 2004. Responses of spring phenology to climate change. New Phytol. 162, 295–309. https://doi.org/10.1111/j.1469-8137.2004.01059.x

Badhwar, G.D., 1984. Classification of corn and soybeans using multitemporal thematic mapper data. Remote Sens. Environ. 16, 175–181. https://doi.org/10.1016/0034-4257(84)90061-0

Badhwar, G.D., Austin, W.W., Carnes, J.G., 1982. A semi-automatic technique for multitemporal classification of a given crop within a Landsat scene. Pattern Recognit. 15, 217–230. https://doi.org/10.1016/0031-3203(82)90073-5

Bager, S., Persson, M., Reis, T., 2020. Reducing commodity-driven tropical deforestation: Political feasibility and 'theories of change' for EU policy options. https://doi.org/10.2139/ssrn.3624073

Balaghi, R., Tychon, B., Eerens, H., Jlibene, M., 2008. Empirical regression models using NDVI, rainfall and temperature data for the early prediction of wheat grain yields in Morocco. Int. J. Appl. Earth Obs. Geoinformation. 10, 438–452.

Balestriero, R., Ibrahim, M., Sobal, V., Morcos, A., Shekhar, S., Goldstein, T., Bordes, F., Bardes, A., Mialon, G., Tian, Y., Schwarzschild, A., Wilson, A.G., Geiping, J., Garrido, Q., Fernandez, P., Bar, A., Pirsiavash, H., LeCun, Y., Goldblum, M., 2023. A cookbook of self-supervised learning. https://doi.org/10.48550/arXiv.2304.12210

Baret, F., 1988. Un modele simplifie de reflectance et d'absorptance d'un couvert vegetal, in: Guyenne, T.D., Hunt, J.J. (Eds.), Proceedings of the Conference Spectral Signatures of Objects in Remote Sensing, ESA SP-287. European Space Agency, Aussois.

Baret, F., Champion, I., Guyot, G., Podaire, A., 1987. Monitoring wheat canopies with a high spectral resolution radiometer. Remote Sens. Environ. 22, 367–378. https://doi.org/10.1016/0034-4257(87)90089-7

Baret, F., Guyot, G., 1986. Monitoring of the ripening period of wheat canopies using visible and near infra red radiometry [reflectance, vegetation index, senescence rate, water plateau]. Agronomie. 6, 509–516.

Baret, F., Guyot, G., 1991. Potentials and limits of vegetation indices for LAI and APAR assessment. Remote Sens. Environ. 35, 161–173. https://doi.org/10.1016/0034-4257(91)90009-U

Baret, F., Guyot, G., Major, D.J., 1989. Crop biomass evaluation using radiometric measurements. Photogrammetria. 43, 241–256.

Baret, F., Houlès, V., Guérif, M., 2007. Quantification of plant stress using remote sensing observations and crop models: The case of nitrogen management. J. Exp. Bot. 58, 869–880. https://doi.org/10.1093/jxb/erl231

Baret, F., Morissette, J.T., Fernandes, R.A., Champeaux, J.L., Myneni, R.B., Chen, J., Plummer, S., Weiss, M., Bacour, C., Garrigues, S., Nickeson, J.E., 2006. Evaluation of the representativeness of networks of sites for the global validation and intercomparison of land biophysical products: Proposition of the CEOS-BELMANIP. IEEE Trans. Geosci. Remote Sens. 44, 1794–1803. https://doi.org/10.1109/TGRS.2006.876030

Barnes, E.M.E., Clarke, T.R.T., Richards, S.E.S., Colaizzi, P.D.D., Haberland, J., Kostrzewski, M., Waller, P., Choi, C., Riley, E., Thompson, T., Lascano, R.J., Li, H., Moran, M.S.S., Robert, P.C., Rust, R.H., Larson, W.E., 2000. Coincident detection of crop water stress, nitrogen status and canopy density using ground-based multispectral data, in: Robert, P.C., Rust, R.H., Larson, W.E. (Eds.), Proceedings of the Fifth International Conference on Precision Agriculture. American Society of Agronomy.

Barnsley, M.J., Allison, D., Lewis, P., 1997. On the information content of multiple view angle (MVA) images. Int. J. Remote Sens. 18, 1937–1960. https://doi.org/10.1080/014311697217963

Baruth, B., Royer, A., Klisch, A., Genovese, G., 2008. The use of remote sensing within the MARS crop yield monitoring system of the European Commission. Int. Arch. Photogramm Remote Sens. Spat. Inf. Sci. 36, 935–940.

Battude, M., Al Bitar, A., Morin, D., Cros, J., Huc, M., Marais Sicre, C., Le Dantec, V., Demarez, V., 2016. Estimating maize biomass and yield over large areas using high spatial and temporal resolution Sentinel-2 like remote sensing data. Remote Sens. Environ. 184, 668–681. https://doi.org/10.1016/j.rse.2016.07.030

Bauer, M.E., Daughtry, C.S.T., Biehl, L.L., Kanemasu, E.T., Hall, F.G., 1986. Field Spectroscopy of agricultural crops. IEEE Trans. Geosci. Remote Sens. GE-24, 65–75. https://doi.org/10.1109/TGRS.1986.289589

Baulcombe, D., Crute, I., Davies, B., Dunwell, J., Gale, M., Jones, J., Pretty, J., Sutherland, W., Toulmin, C., 2009. Reaping the Benefits: Science and the Sustainable Intensification of Global Agriculture, RS Policy Document. The Royal Society, London.

Beck, P.S.A., Atzberger, C., Høgda, K.A., Johansen, B., Skidmore, A.K., 2006. Improved monitoring of vegetation dynamics at very high latitudes: A new method using MODIS NDVI. Remote Sens. Environ. 100, 321–334.

Becker-Reshef, I., Justice, C., Sullivan, M., Vermote, E., Tucker, C., Anyamba, A., Small, J., Pak, E., Masuoka, E., Schmaltz, J., Hansen, M., Pittman, K., Birkett, C., Williams, D., Reynolds, C., Doorn, B., 2010. Monitoring global croplands with coarse resolution earth observations: The Global Agriculture Monitoring (GLAM) project. Remote Sens. 2, 1589–1609. https://doi.org/10.3390/rs2061589

Bégué, A., Arvor, D., Bellon, B., Betbeder, J., De Abelleyra, D., Ferraz, P.D., Lebourgeois, R., Lelong, V., Simões, C., Verón, R., 2018. Remote sensing and cropping practices: A review. Remote Sens. 10, 99. https://doi.org/10.3390/rs10010099

Bégué, A., Arvor, D., Lelong, C., Vintrou, E., Simões, M., 2015. Agricultural systems studies using remote sensing, in: Land Resources Monitoring, Modeling, and Mapping with Remote Sensing, Remote Sensing Handbook. CRC Press, Taylor & Francis Group, Boca Raton, pp. 113–130.

Belda, S., Pipia, L., Morcillo-Pallarés, P., Rivera-Caicedo, J.P., Amin, E., De Grave, C., Verrelst, J., 2020. DATimeS: A machine learning time series GUI toolbox for gap-filling and vegetation phenology trends detection. Environ. Model. Softw. 127, 104666. https://doi.org/10.1016/j.envsoft.2020.104666

Bellocchi, G., Rivington, M., Donatelli, M., Matthews, K., 2010. Validation of biophysical models: Issues and methodologies. A review. Agron. Sustain. Dev. 30, 109–130.

Bellón, B., Bégué, A., Lo Seen, D., De Almeida, C.A., Simões, M., 2017. A remote sensing approach for regional-scale mapping of agricultural land-use systems based on NDVI time series. Remote Sens. 9, 600. https://doi.org/10.3390/rs9060600
Beltran, C.M., Belmonte, A.C., 2001. Irrigated crop area estimation using landsat TM imagery in La Mancha, Spain. Photogramm. Eng. Remote Sens. 67, 1177–1184.
Belward, A.S., Skøien, J.O., 2015. Who launched what, when and why; trends in global land-cover observation capacity from civilian earth observation satellites. ISPRS J. Photogramm. Remote Sens. 103, 115–128. https://doi.org/10.1016/j.isprsjprs.2014.03.009
Ben-Dor, E., 2002. Quantitative remote sensing of soil properties. Adv. Agron. 75, 173–243. https://doi.org/10.1016/S0065-2113(02)75005-0
Benedetti, R., Rossini, P., 1993. On the use of NDVI profiles as a tool for agricultural statistics: The case study of wheat yield estimate and forecast in Emilia Romagna. Remote Sens. Environ. 45, 311–326.
Benz, U.C., Hofmann, P., Willhauck, G., Lingenfelder, I., Heynen, M., 2004. Multi-resolution, object-oriented fuzzy analysis of remote sensing data for GIS-ready information. ISPRS J. Photogramm. Remote Sens., Integration of Geodata and Imagery for Automated Refinement and Update of Spatial Databases. 58, 239–258. https://doi.org/10.1016/j.isprsjprs.2003.10.002
Berger, K., Machwitz, M., Kycko, M., Kefauver, S.C., Van Wittenberghe, S., Gerhards, M., Verrelst, J., Atzberger, C., van der Tol, C., Damm, A., Rascher, U., Herrmann, I., Paz, V.S., Fahrner, S., Pieruschka, R., Prikaziuk, E., Buchaillot, M.L., Halabuk, A., Celesti, M., Koren, G., Gormus, E.T., Rossini, M., Foerster, M., Siegmann, B., Abdelbaki, A., Tagliabue, G., Hank, T., Darvishzadeh, R., Aasen, H., Garcia, M., Pôças, I., Bandopadhyay, S., Sulis, M., Tomelleri, E., Rozenstein, O., Filchev, L., Stancile, G., Schlerf, M., 2022. Multi-sensor spectral synergies for crop stress detection and monitoring in the optical domain: A review. Remote Sens. Environ. 280, 113198. https://doi.org/10.1016/j.rse.2022.113198
Bernardes, T., Moreira, M.A., Adami, M., Giarolla, A., Rudorff, B.F.T., 2012. Monitoring biennial bearing effect on coffee yield using MODIS remote sensing imagery. Remote Sens. 4, 2492–2509. https://doi.org/10.3390/rs4092492
Biard, F., Baret, F., 1997. Crop residue estimation using multiband reflectance. Remote Sens. Environ. 59, 530–536. https://doi.org/10.1016/S0034-4257(96)00125-3
BirdLife International, 2022. State of the World's Birds 2022: Insights and solutions for the biodiversity crisis-BirdLife International. BirdLife International, Cambridge, UK.
Blaes, X., Vanhalle, L., Defourny, P., 2005. Efficiency of crop identification based on optical and SAR image time series. Remote Sens. Environ. 96, 352–365. https://doi.org/10.1016/j.rse.2005.03.010
Blancon, J., Dutartre, D., Tixier, M.-H., Weiss, M., Comar, A., Praud, S., Baret, F., 2019. A high-throughput model-assisted method for phenotyping maize green leaf area index dynamics using unmanned aerial vehicle imagery. Front. Plant Sci. 10.
Blaschke, T., 2010. Object based image analysis for remote sensing. ISPRS J. Photogramm. Remote Sens. 65, 2–16. https://doi.org/10.1016/j.isprsjprs.2009.06.004
Blickensdörfer, L., Schwieder, M., Pflugmacher, D., Nendel, C., Erasmi, S., Hostert, P., 2022. Mapping of crop types and crop sequences with combined time series of Sentinel-1, Sentinel-2 and Landsat 8 data for Germany. Remote Sens. Environ. 269, 112831. https://doi.org/10.1016/j.rse.2021.112831
Boix-Fayos, C., de Vente, J., 2023. Challenges and potential pathways towards sustainable agriculture within the European Green Deal. Agric. Syst. 207, 103634. https://doi.org/10.1016/j.agsy.2023.103634
Bolton, D.K., Friedl, M.A., 2013. Forecasting crop yield using remotely sensed vegetation indices and crop phenology metrics. Agric. For. Meteorol. 173, 74–84. https://doi.org/10.1016/j.agrformet.2013.01.007
Bolton, D.K., Gray, J.M., Melaas, E.K., Moon, M., Eklundh, L., Friedl, M.A., 2020. Continental-scale land surface phenology from harmonized Landsat 8 and Sentinel-2 imagery. Remote Sens. Environ. 240, 111685. https://doi.org/10.1016/j.rse.2020.111685
Boote, K.J., Jones, J.W., Pickering, N.B., 1996. Potential uses and limitations of crop models. Agron. J. 88, 704–716.
Boschetti, M., Stroppiana, D., Brivio, P.A., Bocchi, S., 2009. Multi-year monitoring of rice crop phenology through time series analysis of MODIS images. Int. J. Remote Sens. 30, 4643–4662. https://doi.org/10.1080/01431160802632249
Bossyns, B., Eerens, H., van Orshoven, J., 2007. Crop area assessment using sub-pixel classification with a neural network trained for a reference year, in: 2007 International Workshop on the Analysis of Multi-Temporal Remote Sensing Images. Presented at the 2007 International Workshop on the Analysis of Multi-temporal Remote Sensing Images, pp. 1–8. https://doi.org/10.1109/MULTITEMP.2007.4293038

Bouman, B.A.M., 1992. Linking physical remote sensing models with crop growth simulation models, applied for sugar beet. Int. J. Remote Sens. 13, 2565–2581. https://doi.org/10.1080/01431169208904064
Bouman, B.A.M., 1995. Crop modelling and remote sensing for yield prediction. Neth. J. Agric. Sci. 43, 143–161. https://doi.org/10.18174/njas.v43i2.573
Box, E.O., Holben, B.N., Kalb, V., 1989. Accuracy of the AVHRR vegetation index as a predictor of biomass, primary productivity and net CO2 flux. Vegetatio. 80, 71–89. https://doi.org/10.1007/BF00048034
Braswell, B.H., Hagen, S.C., Frolking, S.E., Salas, W.A., 2003. A multivariable approach for mapping sub-pixel land cover distributions using MISR and MODIS: Application in the Brazilian Amazon region. Remote Sens. Environ. 87, 243–256. https://doi.org/10.1016/j.rse.2003.06.002
Brisson, N., Mary, B., Ripoche, D., Jeuffroy, M.H., Ruget, F., Nicoullaud, B., Gate, P., Devienne-Barret, F., Antonioletti, R., Durr, C., Richard, G., Beaudoin, N., Recous, S., Tayot, X., Plenet, D., Cellier, P., Machet, J.-M., Meynard, J.M., Delécolle, R., 1998. STICS: A generic model for the simulation of crops and their water and nitrogen balances. I. Theory and parameterization applied to wheat and corn. Agronomie. 18, 311–346.
Brown, M.E., de Beurs, K.M., Marshall, M., 2012. Global phenological response to climate change in crop areas using satellite remote sensing of vegetation, humidity and temperature over 26 years. Remote Sens. Environ. 126, 174–183. https://doi.org/10.1016/j.rse.2012.08.009
Bruinsma, J., 2003. World Agriculture: Towards 2015/2030 An FAO Perspective. Earthscan Publications, London.
Burney, J.A., Davis, S.J., Lobell, D.B., 2010. Greenhouse gas mitigation by agricultural intensification. Proc. Natl. Acad. Sci. 107, 12052–12057. https://doi.org/10.1073/pnas.0914216107
Busetto, L., Meroni, M., Colombo, R., 2008. Combining medium and coarse spatial resolution satellite data to improve the estimation of sub-pixel NDVI time series. Remote Sens. Environ. 112, 118–131.
Camps-Valls, G., Bruzzone, L., 2009. Kernel Methods for Remote Sensing Data Analysis. John Wiley & Sons. https://onlinelibrary.wiley.com/doi/book/10.1002/9780470748992
Canfield, D.E., Glazer, A.N., Falkowski, P.G., 2010. The evolution and future of earth's nitrogen cycle. Science. 330, 192–196. https://doi.org/10.1126/science.1186120
Chang, J., Hansen, M.C., Pittman, K., Carroll, M., DiMiceli, C., 2007. Corn and Soybean mapping in the United States using MODIS time-series data sets. Agron. J. 99, 1654–1664. https://doi.org/10.2134/agronj2007.0170
Claverie, M., Demarez, V., Duchemin, B., Hagolle, O., Ducrot, D., Marais-Sicre, C., Dejoux, J.-F., Huc, M., Keravec, P., Béziat, P., Fieuzal, R., Ceschia, E., Dedieu, G., 2012. Maize and sunflower biomass estimation in southwest France using high spatial and temporal resolution remote sensing data. Remote Sens. Environ. 124, 844–857. https://doi.org/10.1016/j.rse.2012.04.005
Clevers, J., Vonder, O., Jongschaap, R., Desprats, J.-F., King, C., Prévot, L., Bruguier, N., 2002. Using SPOT data for calibrating a wheat growth model under mediterranean conditions. Agronomie. 22, 687–694. https://doi.org/10.1051/agro:2002038
Clevers, J.G.P.W., Büker, C., van Leeuwen, H.J.C., Bouman, B.A.M., 1994. A framework for monitoring crop growth by combining directional and spectral remote sensing information. Remote Sens. Environ. 50, 161–170. https://doi.org/10.1016/0034-4257(94)90042-6
Clevers, J.G.P.W., Van Leeuwen, H.J.C., 1996. Combined use of optical and microwave remote sensing data for crop growth monitoring. Remote Sens. Environ. 56, 42–51.
Comaniciu, D., Meer, P., 2002. Mean shift: A robust approach toward feature space analysis. IEEE Trans. Pattern Anal. Mach. Intell. 24, 603–619. https://doi.org/10.1109/34.1000236
Congalton, R.G., Balogh, M., Bell, C., Green, K., Milliken, J.A., Ottman, R., 1998. Mapping and monitoring agricultural crops and other land cover in the lower Colorado River Basin. Photogramm. Eng. Remote Sens. 64, 1107–1113.
Costa, C., Dittmer, K.M., Shelton, S.W., Bossio, D.A., Zinyengere, N., Luu, P., Heinz, S., Egenolf, K., Rowland, B., Zuluaga, A., Klemme, J., Mealey, T., Smith, M., Wollenberg, E.K., 2020. How soil carbon accounting can improve to support investment-oriented actions promoting soil carbon storage.
Courault, D., Hadria, R., Ruget, F., Olioso, A., Duchemin, B., Hagolle, O., Dedieu, G., 2010. Combined use of FORMOSAT-2 images with a crop model for biomass and water monitoring of permanent grassland in Mediterranean region. Hydrol. Earth Syst. Sci. 14, 1731–1744. https://doi.org/10.5194/hess-14-1731-2010
COWI, Ecologic Institute, IEEP, 2021. Technical guidance handbook—Setting up and implementing result-based carbon farming mechanisms in the EU, Report to the European Commission. COWI, Kongens Lyngby.

Cracknell, A.P., 1998. Review article Synergy in remote sensing-what's in a pixel? Int. J. Remote Sens. 19, 2025–2047. https://doi.org/10.1080/014311698214848

da Conceição Bispo, P., Picoli, M.C.A., Marimon, B.S., Marimon Junior, B.H., Peres, C.A., Menor, I.O., Silva, D.E., de Figueiredo Machado, F., Alencar, A.A.C., de Almeida, C.A., Anderson, L.O., Aragão, L.E.O.C., Breunig, F.M., Bustamante, M., Dalagnol, R., Diniz-Filho, J.A.F., Ferreira, L.G., Ferreira, M.E., Fisch, G., Galvão, L.S., Giarolla, A., Gomes, A.R., de Marco Junior, P., Kuck, T.N., Lehmann, C.E.R., Lemes, M.R., Liesenberg, V., Loyola, R., Macedo, M.N., de Souza Mendes, F., do Couto de Miranda, S., Morton, D.C., Moura, Y.M., Oldekop, J.A., Ramos-Neto, M.B., Rosan, T.M., Saatchi, S., Sano, E.E., Segura-Garcia, C., Shimbo, J.Z., Silva, T.S.F., Trevisan, D.P., Zimbres, B., Wiederkehr, N.C., Silva-Junior, C.H.L., 2024. Overlooking vegetation loss outside forests imperils the Brazilian Cerrado and other non-forest biomes. Nat. Ecol. Evol. 8, 12–13. https://doi.org/10.1038/s41559-023-02256-w

Damm, A., Guanter, L., Paul-Limoges, E., van der Tol, C., Hueni, A., Buchmann, N., Eugster, W., Ammann, C., Schaepman, M.E., 2015. Far-red sun-induced chlorophyll fluorescence shows ecosystem-specific relationships to gross primary production: An assessment based on observational and modeling approaches. Remote Sens. Environ. 166, 91–105. https://doi.org/10.1016/j.rse.2015.06.004

Daryaei, A., Sohrabi, H., Atzberger, C., Immitzer, M., 2020. Fine-scale detection of vegetation in semi-arid mountainous areas with focus on riparian landscapes using Sentinel-2 and UAV data. Comput. Electron. Agric. 177, 105686. https://doi.org/10.1016/j.compag.2020.105686

Daughtry, C.S.T., Doraiswamy, P.C., Hunt, E.R., Stern, A.J., McMurtrey, J.E., Prueger, J.H., 2006. Remote sensing of crop residue cover and soil tillage intensity. Soil Tillage Res. 91, 101–108. https://doi.org/10.1016/j.still.2005.11.013

de Beurs, K.M., Henebry, G.M., 2004. Land surface phenology, climatic variation, and institutional change: Analyzing agricultural land cover change in Kazakhstan. Remote Sens. Environ. 89, 497–509. https://doi.org/10.1016/j.rse.2003.11.006

Deering, D.W., 1978. Rangeland reflectance characteristics measured by aircraft and spacecraft sensors (PhD thesis). A & M University, Texas.

Defourny, P., Bontemps, S., Bellemans, N., Cara, C., Dedieu, G., Guzzonato, E., Hagolle, O., Inglada, J., Nicola, L., Rabaute, T., Savinaud, M., Udroiu, C., Valero, S., Bégué, A., Dejoux, J.-F., El Harti, A., Ezzahar, J., Kussul, N., Labbassi, K., Lebourgeois, V., Miao, Z., Newby, T., Nyamugama, A., Salh, N., Shelestov, A., Simonneaux, V., Traore, P.S., Traore, S.S., Koetz, B., 2019. Near real-time agriculture monitoring at national scale at parcel resolution: Performance assessment of the Sen2-Agri automated system in various cropping systems around the world. Remote Sens. Environ. 221, 551–568. https://doi.org/10.1016/j.rse.2018.11.007

DeFries, R.S., Field, C.B., Fung, I., Justice, C.O., Los, S., Matson, P.A., Matthews, E., Mooney, H.A., Potter, C.S., Prentice, K., Sellers, P.J., Townshend, J.R.G., Tucker, C.J., Ustin, S.L., Vitousek, P.M., 1995. Mapping the land surface for global atmosphere-biosphere models: Toward continuous distributions of vegetation's functional properties. J. Geophys. Res. Atmospheres. 100, 20867–20882. https://doi.org/10.1029/95JD01536

de Jong, R., de Bruin, S., de Wit, A., Schaepman, M.E., Dent, D.L., 2011. Analysis of monotonic greening and browning trends from global NDVI time-series. Remote Sens. Environ. 115, 692–702. https://doi.org/10.1016/j.rse.2010.10.011

dela Torre, D.M.G., Gao, J., Macinnis-Ng, C., 2021. Remote sensing-based estimation of rice yields using various models: A critical review. Geo-Spat. Inf. Sci. 24, 580–603. https://doi.org/10.1080/10095020.2021.1936656

Delécolle, R., Maas, S.J., Guérif, M., Baret, F., 1992. Remote sensing and crop production models: Present trends. ISPRS J. Photogramm. Remote Sens. 47, 145–161.

Demarez, V., Helen, F., Marais-Sicre, C., Baup, F., 2019. In-season mapping of irrigated crops using Landsat 8 and Sentinel-1 time series. Remote Sens. 11, 118. https://doi.org/10.3390/rs11020118

Demirbas, M.F., Balat, M., 2006. Recent advances on the production and utilization trends of bio-fuels: A global perspective. Energy Convers. Manag. 47, 2371–2381. https://doi.org/10.1016/j.enconman.2005.11.014

De Wit, A.J.W., Clevers, J.G.P.W., 2004. Efficiency and accuracy of per-field classification for operational crop mapping. Int. J. Remote Sens. 25, 4091–4112. https://doi.org/10.1080/01431160310001619580

Dirzo, R., Raven, P.H., 2003. Global state of biodiversity and Loss. Annu. Rev. Environ. Resour. 28, 137–167. https://doi.org/10.1146/annurev.energy.28.050302.105532

Dobrowski, S.Z., Pushnik, J.C., Zarco-Tejada, P.J., Ustin, S.L., 2005. Simple reflectance indices track heat and water stress-induced changes in steady-state chlorophyll fluorescence at the canopy scale. Remote Sens. Environ. 97, 403–414. https://doi.org/10.1016/j.rse.2005.05.006

Doraiswamy, P.C., Hatfield, J.L., Jackson, T.J., Akhmedov, B., Prueger, J., Stern, A., 2004. Crop condition and yield simulations using Landsat and MODIS. Remote Sens. Environ., 2002 Soil Moisture Experiment (SMEX02) 92, 548–559. https://doi.org/10.1016/j.rse.2004.05.017

Doraiswamy, P.C., Moulin, S., Cook, P.W., Stern, A., 2003. Crop yield assessment from remote sensing. Photogramm. Eng. Remote Sens. 69, 665–674. https://doi.org/10.14358/PERS.69.6.665

Doraiswamy, P.C., Sinclair, T.R., Hollinger, S., Akhmedov, B., Stern, A., Prueger, J., 2005. Application of MODIS derived parameters for regional crop yield assessment. Remote Sens. Environ. 97, 192–202. https://doi.org/10.1016/j.rse.2005.03.015

Dorigo, W.A., Zurita-Milla, R., de Wit, A.J.W., Brazile, J., Singh, R., Schaepman, M.E., 2007. A review on reflective remote sensing and data assimilation techniques for enhanced agroecosystem modeling. Int. J. Appl. Earth Obs. Geoinformation, Advances in airborne electromagnetics and remote sensing of agro-ecosystems 9, 165–193. https://doi.org/10.1016/j.jag.2006.05.003

Duchemin, B., Maisongrande, P., Boulet, G., Benhadj, I., 2008. A simple algorithm for yield estimates: Evaluation for semi-arid irrigated winter wheat monitored with green leaf area index. Environ. Model. Softw. 23, 876–892. https://doi.org/10.1016/j.envsoft.2007.10.003

Dujakovic, A., Schaumberger, A., Klingler, A., Mayer, K., Atzberger, C., Klisch, A., Vuolo, F., 2024. Growth unveiled: Decoding the start of grassland seasons in Austria. Eur. J. Remote Sens. 0, 2323633. https://doi.org/10.1080/22797254.2024.2323633

Dumeur, I., Valero, S., Inglada, J., 2024. Self-Supervised spatio-temporal representation learning of satellite image time series. IEEE J. Sel. Top. Appl. Earth Obs. Remote Sens. 17, 4350–4367. https://doi.org/10.1109/JSTARS.2024.3358066

Duveiller, G., Donatelli, M., Fumagalli, D., Zucchini, A., Nelson, R., Baruth, B., 2017. A dataset of future daily weather data for crop modelling over Europe derived from climate change scenarios. Theor. Appl. Climatol. 127, 573–585. https://doi.org/10.1007/s00704-015-1650-4

Duveiller, G., López-Lozano, R., Baruth, B., 2013. Enhanced processing of 1-km spatial resolution fAPAR time series for sugarcane yield forecasting and monitoring. Remote Sens. 5, 1091–1116. https://doi.org/10.3390/rs5031091

Ebersold, F., Hechelmann, R.-H., Holzapfel, P., Meschede, H., 2023. Carbon insetting as a measure to raise supply chain energy efficiency potentials: Opportunities and challenges. Energy Convers. Manag. X 20, 100504. https://doi.org/10.1016/j.ecmx.2023.100504

Eerens, H., Piccard, I., Royer, A., Oralndi, S., 2004. Methodology of the MARS crop yield forecasting system. Vol. 3: Remote sensing information, data processing and analysis (No. EUR 21291 EN/3). Joint Research Centre European Commission.

Eklundh, L., Jönsson, P., 2015. TIMESAT: A software package for time-series processing and assessment of vegetation dynamics, in: Kuenzer, C., Dech, S., Wagner, W. (Eds.), Remote Sensing Time Series: Revealing Land Surface Dynamics, Remote Sensing and Digital Image Processing. Springer International Publishing, Cham, pp. 141–158. https://doi.org/10.1007/978-3-319-15967-6_7

El Hajj, M., Baghdadi, N., Zribi, M., 2019. Comparative analysis of the accuracy of surface soil moisture estimation from the C- and L-bands. Int. J. Appl. Earth Obs. Geoinformation. 82, 101888. https://doi.org/10.1016/j.jag.2019.05.021

Erb, K.-H., Haberl, H., Jepsen, M.R., Kuemmerle, T., Lindner, M., Müller, D., Verburg, P.H., Reenberg, A., 2013. A conceptual framework for analysing and measuring land-use intensity. Curr. Opin. Environ. Sustain., Human Settlements and Industrial Systems. 5, 464–470. https://doi.org/10.1016/j.cosust.2013.07.010

FAO, 2009. The State of Food Insecurity in the World 2009: Economic Crises—Impacts and Lessons Learned. The State of Food Security and Nutrition in the World (SOFI), Rome, Italy.

Ferrant, S., Bustillo, V., Burel, E., Salmon-Monviola, J., Claverie, M., Jarosz, N., Yin, T., Rivalland, V., Dedieu, G., Demarez, V., Ceschia, E., Probst, A., Al-Bitar, A., Kerr, Y., Probst, J.-L., Durand, P., Gascoin, S., 2016. Extracting soil water holding capacity parameters of a distributed agro-hydrological model from high resolution optical satellite observations series. Remote Sens. 8, 154. https://doi.org/10.3390/rs8020154

Field, C.B., Randerson, J.T., Malmström, C.M., 1995. Global net primary production: Combining ecology and remote sensing. Remote Sens. Environ., Remote Sensing of Land Surface for Studies of Global Chage. 51, 74–88. https://doi.org/10.1016/0034-4257(94)00066-V

Fieuzal, R., Duchemin, B., Jarlan, L., Zribi, M., Baup, F., Merlin, O., Hagolle, O., Garatuza-Payan, J., 2011. Combined use of optical and radar satellite data for the monitoring of irrigation and soil moisture of wheat crops. Hydrol. Earth Syst. Sci. 15, 1117–1129. https://doi.org/10.5194/hess-15-1117-2011

Folberth, C., Khabarov, N., Balkovič, J., Skalský, R., Visconti, P., Ciais, P., Janssens, I.A., Peñuelas, J., Obersteiner, M., 2020. The global cropland-sparing potential of high-yield farming. Nat. Sustain. 3, 281–289. https://doi.org/10.1038/s41893-020-0505-x

Foley, J.A., Ramankutty, N., Brauman, K.A., Cassidy, E.S., Gerber, J.S., Johnston, M., Mueller, N.D., O'Connell, C., Ray, D.K., West, P.C., Balzer, C., Bennett, E.M., Carpenter, S.R., Hill, J., Monfreda, C., Polasky, S., Rockström, J., Sheehan, J., Siebert, S., Tilman, D., Zaks, D.P.M., 2011. Solutions for a cultivated planet. Nature. 478, 337–342. https://doi.org/10.1038/nature10452

Foody, G.M., 1996. Approaches for the production and evaluation of fuzzy land cover classifications from remotely-sensed data. Int. J. Remote Sens. 17, 1317–1340. https://doi.org/10.1080/01431169608948706

Foody, G.M., 2002. Status of land cover classification accuracy assessment. Remote Sens. Environ. 80, 185–201. https://doi.org/10.1016/S0034-4257(01)00295-4

Foody, G.M., 2004. Sub-pixel methods in remote sensing, in: Jong, S.M.D., Meer, F.D.V. der (Eds.), Remote Sensing Image Analysis: Including the Spatial Domain. Springer, The Netherlands, Dordrecht, pp. 37–49. https://doi.org/10.1007/978-1-4020-2560-0_3

Foody, G.M., 2010. Assessing the accuracy of land cover change with imperfect ground reference data. Remote Sens. Environ. 114, 2271–2285. https://doi.org/10.1016/j.rse.2010.05.003

Foody, G.M., 2021. Impacts of ignorance on the accuracy of image classification and thematic mapping. Remote Sens. Environ. 259, 112367. https://doi.org/10.1016/j.rse.2021.112367

Foody, G.M., 2023. Challenges in the real world use of classification accuracy metrics: From recall and precision to the Matthews correlation coefficient. PLoS One. 18, e0291908. https://doi.org/10.1371/journal.pone.0291908

Foody, G.M., Boyd, D.S., 2013. Using volunteered data in land cover map validation: Mapping West African forests. IEEE J. Sel. Top. Appl. Earth Obs. Remote Sens. 6, 1305–1312. https://doi.org/10.1109/JSTARS.2013.2250257

Foody, G.M., Cox, D.P., 1994. Sub-pixel land cover composition estimation using a linear mixture model and fuzzy membership functions. Int. J. Remote Sens. 15, 619–631.

Foody, G.M., See, L., Fritz, S., Van der Velde, M., Perger, C., Schill, C., Boyd, D.S., 2013. Assessing the accuracy of volunteered geographic information arising from multiple contributors to an internet based collaborative project. Trans. GIS 17, 847–860. https://doi.org/10.1111/tgis.12033

Forkuor, G., Hounkpatin, O.K.L., Welp, G., Thiel, M., 2017. High resolution mapping of soil properties using remote sensing variables in South-Western Burkina Faso: A comparison of machine learning and multiple linear regression models. PLoS One. 12, e0170478. https://doi.org/10.1371/journal.pone.0170478

Franch, B., Vermote, E.F., Skakun, S., Roger, J.C., Becker-Reshef, I., Murphy, E., Justice, C., 2019. Remote sensing based yield monitoring: Application to winter wheat in United States and Ukraine. Int. J. Appl. Earth Obs. Geoinformation. 76, 112–127. https://doi.org/10.1016/j.jag.2018.11.012

Fries, R.S.D., Hansen, M., Townshend, J.R.G., Sohlberg, R., 1998. Global land cover classifications at 8 km spatial resolution: The use of training data derived from Landsat imagery in decision tree classifiers. Int. J. Remote Sens. 19, 3141–3168. https://doi.org/10.1080/014311698214235

Galford, G.L., Melillo, J., Mustard, J.F., Cerri, C.E.P., Cerri, C.C., 2010. The Amazon frontier of land-use change: Croplands and consequences for greenhouse gas emissions. Earth Interact. 14, 1–24. https://doi.org/10.1175/2010EI327.1

Galford, G.L., Mustard, J.F., Melillo, J., Gendrin, A., Cerri, C.C., Cerri, C.E.P., 2008. Wavelet analysis of MODIS time series to detect expansion and intensification of row-crop agriculture in Brazil. Remote Sens. Environ., Soil Moisture Experiments 2004 (SMEX04) Special Issue. 112, 576–587. https://doi.org/10.1016/j.rse.2007.05.017

Gallego, F.J., 2004. Remote sensing and land cover area estimation. Int. J. Remote Sens. 25, 3019–3047. https://doi.org/10.1080/01431160310001619607

Gallego, F.J., 2005. Stratified sampling of satellite images with a systematic grid of points. ISPRS J. Photogramm. Remote Sens. 59, 369–376. https://doi.org/10.1016/j.isprsjprs.2005.10.001

Gallego, F.J., 2012. The efficiency of sampling very high resolution images for area estimation in the European Union. Int. J. Remote Sens. 33, 1868–1880. https://doi.org/10.1080/01431161.2011.602993

Gao, F., Anderson, M., Daughtry, C., Karnieli, A., Hively, D., Kustas, W., 2020. A within-season approach for detecting early growth stages in corn and soybean using high temporal and spatial resolution imagery. Remote Sens. Environ. 242, 111752. https://doi.org/10.1016/j.rse.2020.111752

Garnett, T., 2013. Food sustainability: Problems, perspectives and solutions. Proc. Nutr. Soc. 72, 29–39. https://doi.org/10.1017/S0029665112002947

Garrett, R.D., Lambin, E.F., Naylor, R.L., 2013. Land institutions and supply chain configurations as determinants of soybean planted area and yields in Brazil. Land Use Policy, Themed Issue 1-Guest Editor Romy GreinerThemed Issue 2- Guest Editor Davide Viaggi. 31, 385–396. https://doi.org/10.1016/j.landusepol.2012.08.002

Ghamisi, P., Rasti, B., Yokoya, N., Wang, Q., Hofle, B., Bruzzone, L., Bovolo, F., Chi, M., Anders, K., Gloaguen, R., Atkinson, P.M., Benediktsson, J.A., 2019. Multisource and multitemporal data fusion in remote sensing: A comprehensive review of the state of the art. IEEE Geosci. Remote Sens. Mag. 7, 6–39. https://doi.org/10.1109/MGRS.2018.2890023

Ghassemi, B., Dujakovic, A., Żółtak, M., Immitzer, M., Atzberger, C., Vuolo, F., 2022. Designing a European-wide crop type mapping approach based on machine learning algorithms using LUCAS field survey and Sentinel-2 data. Remote Sens. 14, 541. https://doi.org/10.3390/rs14030541

Gobron, N., Pinty, B., Verstraete, M.M., Widlowski, J.-L., Diner, D.J., 2002. Uniqueness of multiangular measurements—Part II: Joint retrieval of vegetation structure and photosynthetic activity from MISR. IEEE Trans. Geosci. Remote Sens. 40, 1574–1592.

Godfray, H.C.J., Beddington, J.R., Crute, I.R., Haddad, L., Lawrence, D., Muir, J.F., Pretty, J., Robinson, S., Thomas, S.M., Toulmin, C., 2010. Food security: The challenge of feeding 9 billion people. Science. 327, 812–818. https://doi.org/10.1126/science.1185383

Gómez, C., White, J.C., Wulder, M.A., 2016. Optical remotely sensed time series data for land cover classification: A review. ISPRS J. Photogramm. Remote Sens. 116, 55–72. https://doi.org/10.1016/j.isprsjprs.2016.03.008

Grau, R., Kuemmerle, T., Macchi, L., 2013. Beyond 'land sparing versus land sharing': Environmental heterogeneity, globalization and the balance between agricultural production and nature conservation. Curr. Opin. Environ. Sustain., Human Settlements and Industrial Systems. 5, 477–483. https://doi.org/10.1016/j.cosust.2013.06.001

Griscom, B.W., Adams, J., Ellis, P.W., Houghton, R.A., Lomax, G., Miteva, D.A., Schlesinger, W.H., Shoch, D., Siikamäki, J.V., Smith, P., Woodbury, P., Zganjar, C., Blackman, A., Campari, J., Conant, R.T., Delgado, C., Elias, P., Gopalakrishna, T., Hamsik, M.R., Herrero, M., Kiesecker, J., Landis, E., Laestadius, L., Leavitt, S.M., Minnemeyer, S., Polasky, S., Potapov, P., Putz, F.E., Sanderman, J., Silvius, M., Wollenberg, E., Fargione, J., 2017. Natural climate solutions. Proc. Natl. Acad. Sci. 114, 11645–11650. https://doi.org/10.1073/pnas.1710465114

Groten, S.M.E., 1993. NDVI—crop monitoring and early yield assessment of Burkina Faso. Int. J. Remote Sens. 14, 1495–1515.

Guanter, L., Brell, M., Chan, J.C.-W., Giardino, C., Gomez-Dans, J., Mielke, C., Morsdorf, F., Segl, K., Yokoya, N., 2019. Synergies of spaceborne imaging spectroscopy with other remote sensing approaches. Surv. Geophys. 40, 657–687. https://doi.org/10.1007/s10712-018-9485-z

Guanter, L., Zhang, Y., Jung, M., Joiner, J., Voigt, M., Berry, J.A., Frankenberg, C., Huete, A.R., Zarco-Tejada, P., Lee, J.-E., Moran, M.S., Ponce-Campos, G., Beer, C., Camps-Valls, G., Buchmann, N., Gianelle, D., Klumpp, K., Cescatti, A., Baker, J.M., Griffis, T.J., 2014. Global and time-resolved monitoring of crop photosynthesis with chlorophyll fluorescence. Proc. Natl. Acad. Sci. 111, E1327–E1333. https://doi.org/10.1073/pnas.1320008111

Guérif, M., Blöser, B., Atzberger, C., Clastre, P., Guinot, J.-P., Delecolle, R., 1996. Identification de parcelles agricoles à partir de la forme de leur évolution radiométrique au cours de la saison de culture. Identif. Parcel. Agric. À Partir Forme Leur Évolution Radiométrique Au Cours Saison Cult. 34, 12–22.

Guérif, M., Delécolle, R., 1993. Introducing remote sensed estimates of canopy structure into plant models, in: Canopy Structure and Light Microclimate. Characterization and Applications. INRA, Paris, pp. 479–490.

Guérif, M., Duke, C., 1998. Calibration of the SUCROS emergence and early growth module for sugar beet using optical remote sensing data assimilation. Eur. J. Agron. 9, 127–136. https://doi.org/10.1016/S1161-0301(98)00031-8

Guérif, M., Duke, C.L., 2000. Adjustment procedures of a crop model to the site specific characteristics of soil and crop using remote sensing data assimilation. Agric. Ecosyst. Environ. 81, 57–69.

Hahn, C., Gloaguen, R., 2008. Estimation of soil types by non linear analysis of remote sensing data. Nonlinear Process. Geophys. 15, 115–126.

Hank, T.B., Bach, H., Mauser, W., 2015. Using a remote sensing-supported hydro-agroecological model for field-scale simulation of heterogeneous crop growth and yield: Application for wheat in Central Europe. Remote Sens. 7, 3934–3965. https://doi.org/10.3390/rs70403934

Hansen, P.M., Schjoerring, J.K., 2003. Reflectance measurement of canopy biomass and nitrogen status in wheat crops using normalized difference vegetation indices and partial least squares regression. Remote Sens. Environ. 86, 542–553. https://doi.org/10.1016/S0034-4257(03)00131-7

Haralick, R.M., Shapiro, L.G., 1985. Image segmentation techniques. Comput. Vis. Graph. Image Process. 29, 100–132. https://doi.org/10.1016/S0734-189X(85)90153-7

Hatfield, J.L., Gitelson, A.A., Schepers, J.S., Walthall, C.L., 2008. Application of spectral remote sensing for agronomic decisions. Agron. J. 100, S-117–S-131. https://doi.org/10.2134/agronj2006.0370c

Hayes, M.J., Decker, W.L., 1996. Using NOAA AVHRR data to estimate maize production in the United States Corn Belt. Int. J. Remote Sens. 17, 3189–3200. https://doi.org/10.1080/01431169608949138

Henricksen, B.L., Durkin, J.W., 1986. Growing period and drought early warning in Africa using satellite data. Int. J. Remote Sens. 7, 1583–1608.

Hermance, J.F., 2007. Stabilizing high-order, non-classical harmonic analysis of NDVI data for average annual models by damping model roughness. Int. J. Remote Sens. 28, 2801–2819. https://doi.org/10.1080/01431160600967128

Hill, J., Nelson, E., Tilman, D., Polasky, S., Tiffany, D., 2006. Environmental, economic, and energetic costs and benefits of biodiesel and ethanol biofuels. Proc. Natl. Acad. Sci. 103, 11206–11210. https://doi.org/10.1073/pnas.0604600103

Hirosawa, Y., Marsh, S.E., Kliman, D.H., 1996. Application of standardized principal component analysis to land-cover characterization using multitemporal AVHRR data. Remote Sens. Environ. 58, 267–281. https://doi.org/10.1016/S0034-4257(96)00068-5

Hively, W.D., Lang, M., McCarty, G.W., Keppler, J., Sadeghi, A., McConnell, L.L., 2009. Using satellite remote sensing to estimate winter cover crop nutrient uptake efficiency. J. Soil Water Conserv. 64, 303–313. https://doi.org/10.2489/jswc.64.5.303

Hosseiny, B., Mahdianpari, M., Hemati, M., Radman, A., Mohammadimanesh, F., Chanussot, J., 2024. Beyond supervised learning in remote sensing: A systematic review of deep learning approaches. IEEE J. Sel. Top. Appl. Earth Obs. Remote Sens. 17, 1035–1052. https://doi.org/10.1109/JSTARS.2023.3316733

Hostert, P., Kuemmerle, T., Prishchepov, A., Sieber, A., Lambin, E.F., Radeloff, V.C., 2011. Rapid land use change after socio-economic disturbances: The collapse of the Soviet Union versus Chernobyl. Environ. Res. Lett. 6, 045201. https://doi.org/10.1088/1748-9326/6/4/045201

Huang, J., Gómez-Dans, J.L., Huang, H., Ma, H., Wu, Q., Lewis, P.E., Liang, S., Chen, Z., Xue, J.-H., Wu, Y., Zhao, F., Wang, J., Xie, X., 2019. Assimilation of remote sensing into crop growth models: Current status and perspectives. Agric. For. Meteorol. 276–277, 107609. https://doi.org/10.1016/j.agrformet.2019.06.008

Huang, X., Liu, J., Zhu, W., Atzberger, C., Liu, Q., 2019. The optimal threshold and vegetation index time series for retrieving crop phenology based on a modified dynamic threshold method. Remote Sens. 11, 2725. https://doi.org/10.3390/rs11232725

Hunt, M.L., Blackburn, G.A., Carrasco, L., Redhead, J.W., Rowland, C.S., 2019a. High resolution wheat yield mapping using Sentinel-2. Remote Sens. Environ. 233, 111410. https://doi.org/10.1016/j.rse.2019.111410

Hunt, M.L., Blackburn, G.A., Rowland, C.S., 2019b. Monitoring the sustainable intensification of arable agriculture: The potential role of earth observation. Int. J. Appl. Earth Obs. Geoinformation. 81, 125–136. https://doi.org/10.1016/j.jag.2019.05.013

Hutchinson, C.F., 1991. Uses of satellite data for famine early warning in sub-Saharan Africa. Int. J. Remote Sens. 12, 1405–1421.

Idso, S.B., Pinter Jr., P.J., Jackson, R.D., Reginato, R.J., 1980. Estimation of grain yields by remote sensing of crop senescence rates. Remote Sens. Environ. 9, 87–91.

Immitzer, M., Vuolo, F., Atzberger, C., 2016. First experience with Sentinel-2 data for crop and tree species classifications in Central Europe. Remote Sens. 8, 1–27. https://doi.org/10.3390/rs8030166

Inglada, J., Arias, M., Tardy, B., Hagolle, O., Valero, S., Morin, D., Dedieu, G., Sepulcre, G., Bontemps, S., Defourny, P., Koetz, B., 2015. Assessment of an operational system for crop type map production using high temporal and spatial resolution satellite optical imagery. Remote Sens. 7, 12356–12379. https://doi.org/10.3390/rs70912356

Inglada, J., Vincent, A., Arias, M., Tardy, B., Morin, D., Rodes, I., 2017. Operational high resolution land cover map production at the country scale using satellite image time series. Remote Sens. 9, 95. https://doi.org/10.3390/rs9010095

IPCC, 2022. Climate Change 2022—Impacts, Adaptation and Vulnerability: Working Group II Contribution to the Sixth Assessment Report of the Intergovernmental Panel on Climate Change, 1st ed. Cambridge University Press, Cambridge, UK & New York, NY USA. https://doi.org/10.1017/9781009325844

Jakubauskas, M.E., Legates, D.R., Katens, J.H., 2001. Harmonic analysis of time-series AVHRR NDVI data. Photogramm. Eng. Remote Sens. 67, 461–470.

Jaynes, E.T., 2003. Probability Theory: The Logic of Science. Cambridge University Press, Cambridge.

Jensen, N., Stoeffler, Q., Fava, F., Vrieling, A., Atzberger, C., Meroni, M., Mude, A., Carter, M., 2019. Does the design matter? Comparing satellite-based indices for insuring pastoralists against drought. Ecol. Econ. 162, 59–73. https://doi.org/10.1016/j.ecolecon.2019.04.014

Jin, X., Yang, W., Doonan, J.H., Atzberger, C., 2022. Crop phenotyping studies with application to crop monitoring. Crop J., Crop Phenotyping Studies with Application to Crop Monitoring. 10, 1221–1223. https://doi.org/10.1016/j.cj.2022.09.001

Johnson, G.E., van Dijk, A., Sakamoto, C.M., 1987. The use of AVHRR data in operational agricultural assessment in Africa. Geocarto Int. 2, 41–60. https://doi.org/10.1080/10106048709354080

Johnson, L.F., Bosch, D.F., Williams, D.C., Lobitz, B.M., 2001. Remote sensing of vineyard management zones: Implications for wine quality. Appl. Eng. Agric. 17. https://doi.org/10.13031/2013.6454

Jones, C.A., Kiniry, J.R., 1986. Ceres-Maize: A Simulation Model of Maize Growth and Development, 1st ed. A&M University Press, College Station.

Jones, H.G., Vaughan, R.A., 2010. Remote sensing of vegetation: Principles, techniques, and applications. Oxford University Press, New York, NY.

Jones, P.G., Thornton, P.K., 2003. The potential impacts of climate change on maize production in Africa and Latin America in 2055. Glob. Environ. Change. 13, 51–59. https://doi.org/10.1016/S0959-3780(02)00090-0

Jönsson, P., Eklundh, L., 2002. Seasonality extraction by function fitting to time-series of satellite sensor data. IEEE Trans. Geosci. Remote Sens. 40, 1824–1832. https://doi.org/10.1109/TGRS.2002.802519

Jönsson, P., Eklundh, L., 2004. TIMESAT—a program for analyzing time-series of satellite sensor data. Comput. Geosci. 30, 833–845. https://doi.org/10.1016/j.cageo.2004.05.006

Jose, S., Bardhan, S., 2012. Agroforestry for biomass production and carbon sequestration: An overview. Agrofor. Syst. 86, 105–111. https://doi.org/10.1007/s10457-012-9573-x

Justice, C., Belward, A., Morisette, J., Lewis, P., Privette, J., Baret, F., 2000. Developments in the "validation" of satellite sensor products for the study of the land surface. Int. J. Remote Sens. 21, 3383–3390. https://doi.org/10.1080/014311600750020000

Justice, C.O., Townshend, J.R.G., Vermote, E.F., Masuoka, E., Wolfe, R.E., Saleous, N., Roy, D.P., Morisette, J.T., 2002. An overview of MODIS Land data processing and product status. Remote Sens. Environ., The Moderate Resolution Imaging Spectroradiometer (MODIS): A New Generation of Land Surface Monitoring. 83, 3–15. https://doi.org/10.1016/S0034-4257(02)00084-6

Kaduk, J., Heimann, M., 1996. A prognostic phenology scheme for global terrestrial carbon cycle models. Clim. Res. 6, 1–19. https://doi.org/10.3354/cr006001

Kandasamy, S., Baret, F., Verger, A., Neveux, P., Weiss, M., 2013. A comparison of methods for smoothing and gap filling time series of remote sensing observations—application to MODIS LAI products. Biogeosciences. 10, 4055–4071. https://doi.org/10.5194/bg-10-4055-2013

Karkee, M., Steward, B.L., Tang, L., Aziz, S.A., 2009. Quantifying sub-pixel signature of paddy rice field using an artificial neural network. Comput. Electron. Agric. 65, 65–76. https://doi.org/10.1016/j.compag.2008.07.009

Kastens, J.H., Kastens, T.L., Kastens, D.L.A., Price, K.P., Martinko, E.A., Lee, R.-Y., 2005. Image masking for crop yield forecasting using AVHRR NDVI time series imagery. Remote Sens. Environ. 99, 341–356.

Kawamura, K., Akiyama, T., Yokota, H., Tsutsumi, M., Yasuda, T., Watanabe, O., Wang, G., Wang, S., 2005. Monitoring of forage conditions with MODIS imagery in the Xilingol steppe, Inner Mongolia. Int. J. Remote Sens. 26, 1423–1436. https://doi.org/10.1080/01431160512331326783

Klisch, A., Atzberger, C., 2016. Operational drought monitoring in Kenya using MODIS NDVI time series. Remote Sens. 8, 267. https://doi.org/10.3390/rs8040267

Knoke, T., Bendix, J., Pohle, P., Hamer, U., Hildebrandt, P., Roos, K., Gerique, A., Sandoval, M.L., Breuer, L., Tischer, A., Silva, B., Calvas, B., Aguirre, N., Castro, L.M., Windhorst, D., Weber, M., Stimm, B., Günter, S., Palomeque, X., Mora, J., Mosandl, R., Beck, E., 2014. Afforestation or intense pasturing improve the ecological and economic value of abandoned tropical farmlands. Nat. Commun. 5, 5612. https://doi.org/10.1038/ncomms6612

Kogan, F.N., 1990. Remote sensing of weather impacts on vegetation in non-homogeneous areas. Int. J. Remote Sens. 11, 1405–1419. https://doi.org/10.1080/01431169008955102

Kooistra, L., Berger, K., Brede, B., Graf, L.V., Aasen, H., Roujean, J.-L., Machwitz, M., Schlerf, M., Atzberger, C., Prikaziuk, E., Ganeva, D., Tomelleri, E., Croft, H., Reyes Muñoz, P., Garcia Millan, V., Darvishzadeh, R., Koren, G., Herrmann, I., Rozenstein, O., Belda, S., Rautiainen, M., Rune Karlsen,

S., Figueira Silva, C., Cerasoli, S., Pierre, J., Tanır Kayıkçı, E., Halabuk, A., Tunc Gormus, E., Fluit, F., Cai, Z., Kycko, M., Udelhoven, T., Verrelst, J., 2024. Reviews and syntheses: Remotely sensed optical time series for monitoring vegetation productivity. Biogeosciences. 21, 473–511. https://doi.org/10.5194/bg-21-473-2024

Koukal, T., Atzberger, C., 2012. Potential of multi-angular data derived from a digital aerial frame camera for forest classification. IEEE J. Sel. Top. Appl. Earth Obs. Remote Sens. 5, 30–43. https://doi.org/10.1109/JSTARS.2012.2184527

Kuemmerle, T., Erb, K., Meyfroidt, P., Müller, D., Verburg, P.H., Estel, S., Haberl, H., Hostert, P., Jepsen, M.R., Kastner, T., Levers, C., Lindner, M., Plutzar, C., Verkerk, P.J., van der Zanden, E.H., Reenberg, A., 2013. Challenges and opportunities in mapping land use intensity globally. Curr. Opin. Environ. Sustain. 5, 484–493. https://doi.org/10.1016/j.cosust.2013.06.002

Kuemmerle, T., Olofsson, P., Chaskovskyy, O., Baumann, M., Ostapowicz, K., Woodcock, C.E., Houghton, R.A., Hostert, P., Keeton, W.S., Radeloff, V.C., 2011. Post-Soviet farmland abandonment, forest recovery, and carbon sequestration in western Ukraine. Glob. Change Biol. 17, 1335–1349. https://doi.org/10.1111/j.1365-2486.2010.02333.x

Kummu, M., de Moel, H., Porkka, M., Siebert, S., Varis, O., Ward, P.J., 2012. Lost food, wasted resources: Global food supply chain losses and their impacts on freshwater, cropland, and fertiliser use. Sci. Total Environ. 438, 477–489. https://doi.org/10.1016/j.scitotenv.2012.08.092

Lal, R., 2009. Challenges and opportunities in soil organic matter research. Eur. J. Soil Sci. 60, 158–169. https://doi.org/10.1111/j.1365-2389.2008.01114.x

Launay, M., Guerif, M., 2005. Assimilating remote sensing data into a crop model to improve predictive performance for spatial applications. Agric. Ecosyst. Environ. 111, 321–339.

Leblon, B., Guerif, M., Baret, F., 1991. The use of remotely sensed data in estimation of PAR use efficiency and biomass production of flooded rice. Remote Sens. Environ. 38, 147–158. https://doi.org/10.1016/0034-4257(91)90076-I

Lees, T., Tseng, G., Atzberger, C., Reece, S., Dadson, S., 2022. Deep learning for vegetation health forecasting: A case study in Kenya. Remote Sens. 14, 698. https://doi.org/10.3390/rs14030698

Lewis, J.E., Rowland, J., Nadeau, A., 1998. Estimating maize production in Kenya using NDVI: Some statistical considerations. Int. J. Remote Sens. 19, 2609–2617.

Lichtenthaler, H.K. (Ed.), 1988. Applications of chlorophyll fluorescence in photosynthesis research, stress physiology, hydrobiology and remote sensing, Proceedings of the first International Chlorophyll Fluorescence Symposium held in the Physikzentrum, Bad Honnef FRG, 6–8 June 1988. Dordrecht [etc.]: Kluwer Academic Publishers, Karlsruhe.

Liu, J., Atzberger, C., Huang, X., Shen, K., Liu, Y., Wang, L., 2020. Modeling grass yields in Qinghai Province, China, based on MODIS NDVI data—an empirical comparison. Front. Earth Sci. 14, 413–429. https://doi.org/10.1007/s11707-019-0780-x

Liu, J., Zhu, W., Atzberger, C., Zhao, A., Pan, Y., Huang, X., 2018. A phenology-based method to map cropping patterns under a wheat-maize rotation using remotely sensed time-series data. Remote Sens. 10, 1203. https://doi.org/10.3390/rs10081203

Lloyd, D., 1990. A phenological classification of terrestrial vegetation cover using shortwave vegetation index imagery. Int. J. Remote Sens. 11, 2269–2279. https://doi.org/10.1080/01431169008955174

Lobell, D.B., 2013. The use of satellite data for crop yield gap analysis. Field Crops Res., Crop Yield Gap Analysis—Rationale, Methods and Applications. 143, 56–64. https://doi.org/10.1016/j.fcr.2012.08.008

Lobell, D.B., Asner, G.P., 2004. Cropland distributions from temporal unmixing of MODIS data. Remote Sens. Environ. 93, 412–422. https://doi.org/10.1016/j.rse.2004.08.002

López, J., Torres, D., Santos, S., Atzberger, C., 2020. Spectral imagery tensor decomposition for semantic segmentation of remote sensing data through fully convolutional networks. Remote Sens. 12, 517. https://doi.org/10.3390/rs12030517

Lorenz, K., Lal, R., 2014. Soil organic carbon sequestration in agroforestry systems: A review. Agron. Sustain. Dev. 34, 443–454. https://doi.org/10.1007/s13593-014-0212-y

Loveland, T.R., Reed, B.C., Brown, J.F., Ohlen, D.O., Zhu, Z., Yang, L., Merchant, J.W., 2000. Development of a global land cover characteristics database and IGBP DISCover from 1 km AVHRR data. Int. J. Remote Sens. 21, 1303–1330. https://doi.org/10.1080/014311600210191

Löw, F., Fliemann, E., Abdullaev, I., Conrad, C., Lamers, J.P.A., 2015. Mapping abandoned agricultural land in Kyzyl-Orda, Kazakhstan using satellite remote sensing. Appl. Geogr. 62, 377–390. https://doi.org/10.1016/j.apgeog.2015.05.009

Lunetta, R.S., Shao, Y., Ediriwickrema, J., Lyon, J.G., 2010. Monitoring agricultural cropping patterns across the Laurentian Great Lakes Basin using MODIS-NDVI data. Int. J. Appl. Earth Obs. Geoinformation 12, 81–88. https://doi.org/10.1016/j.jag.2009.11.005

Ma, M., Liu, J., Liu, M., Zhu, W., Atzberger, C., Lv, X., Dong, Z., 2022. Quantitative assessment of the spatial scale effects of the vegetation phenology in the Qinling Mountains. Remote Sens. 14, 5749. https://doi.org/10.3390/rs14225749

Maas, S.J., 1988a. Use of remotely-sensed information in agricultural crop growth models. Ecol. Model. 41, 247–268.

Maas, S.J., 1988b. Using satellite data to improve model estimates of crop yield. Agron. J. 80, 655–662. https://doi.org/10.2134/agronj1988.00021962008000040021x

Maas, S.J., 1992. GRAMI: A Crop Growth Model That Can Use Remotely Sensed Information ARS 91. U.S. Department of Agriculture, Washington DC.

Macdonald, R.B., Hall, F.G., 1980. Global crop forecasting. Science. 208, 670–679.

Mahlayeye, M., Darvishzadeh, R., Nelson, A., 2022. Cropping patterns of annual crops: A remote sensing review. Remote Sens. 14, 2404. https://doi.org/10.3390/rs14102404

Mahlayeye, M., Darvishzadeh, R., Nelson, A., 2024. Characterising maize and intercropped maize spectral signatures for cropping pattern classification. Int. J. Appl. Earth Obs. Geoinformation. 128, 103699. https://doi.org/10.1016/j.jag.2024.103699

Maignan, F., Bréon, F.-M., Fédèle, E., Bouvier, M., 2009. Polarized reflectances of natural surfaces: Spaceborne measurements and analytical modeling. Remote Sens. Environ. 113, 2642–2650. https://doi.org/10.1016/j.rse.2009.07.022

Manjunath, K.R., Potdar, M.B., Purohit, N.L., 2002. Large area operational wheat yield model development and validation based on spectral and meteorological data. Int. J. Remote Sens. 23, 3023–3038.

Maselli, F., Conese, C., Petkov, L., Gilabert, M.A., 1993. Environmental monitoring and crop forecasting in the Sahel through the use of NOAA NDVI data: A case study: Niger 1986–89. Int. J. Remote Sens. 14, 3471–3487.

Maselli, F., Gilabert, M.A., Conese, C., 1998. Integration of high and low resolution NDVI data for monitoring vegetation in Mediterranean environments. Remote Sens. Environ. 63, 208–218. https://doi.org/10.1016/S0034-4257(97)00131-4

Maselli, F., Romanelli, S., Bottai, L., Maracchi, G., 2000. Processing of GAC NDVI data for yield forecasting in the Sahelian region. Int. J. Remote Sens. 21, 3509–3523.

Matton, N., Canto, G.S., Waldner, F., Valero, S., Morin, D., Inglada, J., Arias, M., Bontemps, S., Koetz, B., Defourny, P., 2015. An automated method for annual cropland mapping along the season for various globally-distributed agrosystems using high spatial and temporal resolution time series. Remote Sens. 7, 13208–13232. https://doi.org/10.3390/rs71013208

McCarty, J.L., 2011. Remote sensing-based estimates of annual and seasonal emissions from crop residue burning in the contiguous United States. J. Air Waste Manag. Assoc. 61, 22–34. https://doi.org/10.3155/1047-3289.61.1.22

McGrayne, S.B., 2011. The Theory That Would Not Die: How Bayes' Rule Cracked the Enigma Code, Hunted Down Russian Submarines, & Emerged Triumphant from Two Centuries of C. Yale University Press.

Mello, M.P., Aguiar, D.A., Rudorff, B.F.T., Pebesma, E., Jones, J., Santos, N.C.P., 2013a. Spatial statistic to assess remote sensing acreage estimates: An analysis of sugarcane in São Paulo State, Brazil, in: 2013 IEEE International Geoscience and Remote Sensing Symposium—IGARSS. Presented at the 2013 IEEE International Geoscience and Remote Sensing Symposium—IGARSS, pp. 4233–4236. https://doi.org/10.1109/IGARSS.2013.6723768

Mello, M.P., Risso, J., Atzberger, C., Aplin, P., Pebesma, E., Vieira, C.A.O., Rudorff, B.F.T., 2013b. Bayesian Networks for Raster Data (BayNeRD): Plausible reasoning from observations. Remote Sens. 5, 5999–6025. https://doi.org/10.3390/rs5115999

Mello, M.P., Vieira, C.A.O., Rudorff, B.F.T., Aplin, P., Santos, R.D.C., Aguiar, D.A., 2013c. STARS: A new method for multitemporal remote sensing. IEEE Trans. Geosci. Remote Sens. 51, 1897–1913. https://doi.org/10.1109/TGRS.2012.2215332

Meroni, M., Atzberger, C., Vancutsem, C., Gobron, N., Baret, F., Lacaze, R., Eerens, H., Leo, O., 2013a. Evaluation of agreement between space remote sensing SPOT-VEGETATION fAPAR time series. IEEE Trans. Geosci. Remote Sens. 51, 1951–1962. https://doi.org/10.1109/TGRS.2012.2212447

Meroni, M., Fasbender, D., Rembold, F., Atzberger, C., Klisch, A., 2019. Near real-time vegetation anomaly detection with MODIS NDVI: Timeliness vs. accuracy and effect of anomaly computation options. Remote Sens. Environ. 221, 508–521. https://doi.org/10.1016/j.rse.2018.11.041

Meroni, M., Marinho, E., Sghaier, N., Verstrate, M.M., Leo, O., 2013b. Remote sensing based yield estimation in a stochastic framework—Case study of Durum wheat in Tunisia. Remote Sens. 5, 539–557. https://doi.org/10.3390/rs5020539

Meroni, M., Rembold, F., Verstraete, M.M., Gommes, R., Schucknecht, A., Beye, G., 2014a. Investigating the relationship between the inter-annual variability of satellite-derived vegetation phenology and a proxy of biomass production in the Sahel. Remote Sens. 6, 5868–5884. https://doi.org/10.3390/rs6065868

Meroni, M., Rossini, M., Guanter, L., Alonso, L., Rascher, U., Colombo, R., Moreno, J., 2009. Remote sensing of solar-induced chlorophyll fluorescence: Review of methods and applications. Remote Sens. Environ. 113, 2037–2051. https://doi.org/10.1016/j.rse.2009.05.003

Meroni, M., Verstraete, M.M., Rembold, F., Urbano, F., Kayitakire, F., 2014b. A phenology-based method to derive biomass production anomalies for food security monitoring in the Horn of Africa. Int. J. Remote Sens. 35, 2472–2492. https://doi.org/10.1080/01431161.2014.883090

Metternicht, G.I., Zinck, J.A., 2003. Remote sensing of soil salinity: Potentials and constraints. Remote Sens. Environ. 85, 1–20. https://doi.org/10.1016/S0034-4257(02)00188-8

Michaelowa, A., Honegger, M., Poralla, M., Winkler, M., Dalfiume, S., Nayak, A., 2023. International carbon markets for carbon dioxide removal. PLoS Clim. 2, e0000118. https://doi.org/10.1371/journal.pclm.0000118

Monteith, J.L., 1972. Solar radiation and productivity in tropical ecosystems. J. Appl. Ecol. 9, 747. https://doi.org/10.2307/2401901

Monteith, J.L., 1977. Climate and the efficiency of crop production in Britain. Philos. Trans. R. Soc. B Biol. Sci. 281, 277–294. https://doi.org/10.1098/rstb.1977.0140

Moran, M.S., Inoue, Y., Barnes, E.M., 1997. Opportunities and limitations for image-based remote sensing in precision crop management. Remote Sens. Environ. 61, 319–346. https://doi.org/10.1016/S0034-4257(97)00045-X

Moreno, Á., García-Haro, F.J., Martínez, B., Gilabert, M.A., 2014. Noise reduction and gap filling of fAPAR time series using an adapted local regression filter. Remote Sens. 6, 8238–8260. https://doi.org/10.3390/rs6098238

Morisette, J.T., Baret, F., Privette, J.L., Myneni, R.B., Nickeson, J.E., Garrigues, S., Shabanov, N.V., Weiss, M., Fernandes, R.A., Leblanc, S.G., Kalacska, M., Sanchez-Azofeifa, G.A., Chubey, M., Rivard, B., Stenberg, P., Rautiainen, M., Voipio, P., Manninen, T., Pilant, A.N., Lewis, T.E., Iiames, J.S., Colombo, R., Meroni, M., Busetto, L., Cohen, W.B., Turner, D.P., Warner, E.D., Petersen, G.W., Seufert, G., Cook, R., 2006. Validation of global moderate-resolution LAI products: A framework proposed within the CEOS land product validation subgroup. IEEE Trans. Geosci. Remote Sens. 44, 1804–1817. https://doi.org/10.1109/TGRS.2006.872529

Moulin, S., Bondeau, A., Delécolle, R., 1998. Combining agricultural crop models and satellite observations: From field to regional scales. Int. J. Remote Sens. 19, 1021–1036.

Moya, I., Guyot, G., Goulas, Y., 1992. Remotely sensed blue and red fluorescence emission for monitoring vegetation. ISPRS J. Photogramm. Remote Sens. 47, 205–231. https://doi.org/10.1016/0924-2716(92)90033-6

Mueller, N.D., Gerber, J.S., Johnston, M., Ray, D.K., Ramankutty, N., Foley, J.A., 2012. Closing yield gaps through nutrient and water management. Nature. 490, 254–257. https://doi.org/10.1038/nature11420

Mulianga, B., Bégué, A., Simões, M., Todoroff, P., 2013. Forecasting regional sugarcane yield based on time integral and spatial aggregation of MODIS NDVI. Remote Sens. 5, 2184–2199. https://doi.org/10.3390/rs5052184

Murakami, T., Yui, M., Amaha, K., 2012. Canopy height measurement by photogrammetric analysis of aerial images: Application to buckwheat (*Fagopyrum esculentum* Moench) lodging evaluation. Comput. Electron. Agric. 89, 70–75. https://doi.org/10.1016/j.compag.2012.08.003

Mutanga, O., Skidmore, A.K., Kumar, L., Ferwerda, J., 2005. Estimating tropical pasture quality at canopy level using band depth analysis with continuum removal in the visible domain. Int. J. Remote Sens. 26, 1093–1108. https://doi.org/10.1080/01431160512331326738

Myneni, R.B., Ganapol, B.D., Asrar, G., 1992. Remote sensing of vegetation canopy photosynthetic and stomatal conductance efficiencies. Remote Sens. Environ. 42, 217–238. https://doi.org/10.1016/0034-4257(92)90103-Q

Myneni, R.B., Williams, D.L., 1994. On the relationship between FAPAR and NDVI. Remote Sens. Environ. 49, 200–211.

Nguyen, H.T., Kim, J.H., Nguyen, A.T., Nguyen, L.T., Shin, J.C., Lee, B.-W., 2006. Using canopy reflectance and partial least squares regression to calculate within-field statistical variation in crop growth and nitrogen status of rice. Precis. Agric. 7, 249–264. https://doi.org/10.1007/s11119-006-9010-0

Nielsen, T.D., Jensen, F.V., 2007. Bayesian Networks and Decision Graphs, 2nd ed. Springer, New York.

Nouvellon, Y., Moran, M.S., Seen, D.L., Bryant, R., Rambal, S., Ni, W., Bégué, A., Chehbouni, A., Emmerich, W.E., Heilman, P., Qi, J., 2001. Coupling a grassland ecosystem model with Landsat imagery for a 10-year simulation of carbon and water budgets. Remote Sens. Environ., Landsat. 7(78), 131–149. https://doi.org/10.1016/S0034-4257(01)00255-3

Ofori-Ampofo, S., Pelletier, C., Lang, S., 2021. Crop type mapping from optical and radar time series using attention-based deep learning. Remote Sens. 13, 4668. https://doi.org/10.3390/rs13224668

Olofsson, P., Foody, G.M., Herold, M., Stehman, S.V., Woodcock, C.E., Wulder, M.A., 2014. Good practices for estimating area and assessing accuracy of land change. Remote Sens. Environ. 148, 42–57. https://doi.org/10.1016/j.rse.2014.02.015

Olofsson, P., Stehman, S.V., Woodcock, C.E., Sulla-Menashe, D., Sibley, A.M., Newell, J.D., Friedl, M.A., Herold, M., 2012. A global land-cover validation data set, part I: Fundamental design principles. Int. J. Remote Sens. 33, 5768–5788. https://doi.org/10.1080/01431161.2012.674230

Padilla, F.L.M., Maas, S.J., González-Dugo, M.P., Mansilla, F., Rajan, N., Gavilán, P., Domínguez, J., 2012. Monitoring regional wheat yield in Southern Spain using the GRAMI model and satellite imagery. Field Crops Res. 130, 145–154. https://doi.org/10.1016/j.fcr.2012.02.025

Pal, N.R., Pal, S.K., 1993. A review on image segmentation techniques. Pattern Recognit. 26, 1277–1294. https://doi.org/10.1016/0031-3203(93)90135-J

Parfitt, J., Barthel, M., Macnaughton, S., 2010. Food waste within food supply chains: Quantification and potential for change to 2050. Philos. Trans. R. Soc. B Biol. Sci. 365, 3065–3081. https://doi.org/10.1098/rstb.2010.0126

Paul, C., Knoke, T., 2015. Between land sharing and land sparing—what role remains for forest management and conservation? Int. For. Rev. 17, 210–230. https://doi.org/10.1505/146554815815500624

Pearl, J., 1988. Probabilistic Reasoning in Intelligent Systems: Networks of Plausible Inference. Morgan Kaufmann. https://www.sciencedirect.com/book/9780080514895/probabilistic-reasoning-in-intelligent-systems

Pellenq, J., Boulet, G., 2004. A methodology to test the pertinence of remote-sensing data assimilation into vegetation models for water and energy exchange at the land surface. Agronomie. 24, 197–204. https://doi.org/10.1051/agro:2004017

Pelletier, C., Valero, S., 2021. Pixel-based classification techniques for satellite image time series, in: Atto, A.M., Bovolo, F., Bruzzone, L. (Eds.), Change Detection and Image Time Series Analysis 2: Supervised Methods. John Wiley & Sons, London.

Pelletier, C., Webb, G.I., Petitjean, F., 2019. Temporal convolutional neural network for the classification of satellite image time series. Remote Sens. 11, 523. https://doi.org/10.3390/rs11050523

Pelletier, N., Tyedmers, P., 2010. Forecasting potential global environmental costs of livestock production 2000–2050. Proc. Natl. Acad. Sci. 107, 18371–18374. https://doi.org/10.1073/pnas.1004659107

Peña, J.M., Gutiérrez, P.A., Hervás-Martínez, C., Six, J., Plant, R.E., López-Granados, F., 2014. Object-based image classification of summer crops with machine learning methods. Remote Sens. 6, 5019–5041. https://doi.org/10.3390/rs6065019

Peña-Barragán, J.M., Ngugi, M.K., Plant, R.E., Six, J., 2011. Object-based crop identification using multiple vegetation indices, textural features and crop phenology. Remote Sens. Environ. 115, 1301–1316. https://doi.org/10.1016/j.rse.2011.01.009

Peñuelas, J., Baret, F., Filella, I., 1995. Semi-empirical indices to assess carotenoids/chlorophyll a ratio from leaf spectral reflectance. Photosynthetica. 31, 221–230.

Perdigão, V., Annoni, A., 1997. Technical and methodological guide for updating CORINE Land Cover data base (No. EUR 17288EN). European Commission, Ispra.

Pinter, Jr., P.J., Hatfield, J.L., Schepers, J.S., Barnes, E.M., Moran, M.S., Daughtry, C.S.T., Upchurch, D.R., 2003. Remote sensing for crop management. Photogramm. Eng. Remote Sens. 69, 647–664. https://doi.org/10.14358/PERS.69.6.647

Pinter Jr, P.J., Jackson, R.D., Idso, S.B., Reginato, R.J., 1981. Multidate spectral reflectance as predictors of yield in water stressed wheat and barley. Int. J. Remote Sens. 2, 43–48.

Plaza, A., Benediktsson, J.A., Boardman, J.W., Brazile, J., Bruzzone, L., Camps-Valls, G., Chanussot, J., Fauvel, M., Gamba, P., Gualtieri, A., 2009. Recent advances in techniques for hyperspectral image processing. Remote Sens. Environ. 113, S110–S122.

Postel, S.L., Daily, G.C., Ehrlich, P.R., 1996. Human appropriation of renewable fresh water. Science. 271, 785–788. https://doi.org/10.1126/science.271.5250.785

Potdar, M.B., Manjunath, K.R., Purohit, N.L., 1999. Multi-season atmospheric normalization of NOAA AVHRR derived NDVI for crop yield modeling. Geocarto Int. 14, 52–57. https://doi.org/10.1080/10106049908542128

Potočnik Buhvald, A., Račič, M., Immitzer, M., Oštir, K., Veljanovski, T., 2022. Grassland use intensity classification using intra-annual Sentinel-1 and -2 time series and environmental variables. Remote Sens. 14, 3387. https://doi.org/10.3390/rs14143387

Prince, S.D., 1991. A model of regional primary production for use with coarse resolution satellite data. Int. J. Remote Sens. 12, 1313–1330. https://doi.org/10.1080/01431169108929728

Quarmby, N.A., Townshend, J.R.G., Settele, J.J., White, K.H., Milnes, M., Hindle, T.L., Silleos, N., 1992. Linear mixture modelling applied to AVHRR data for crop area estimation. Int. J. Remote Sens. 13, 415–425. https://doi.org/10.1080/01431169208904046

Ranganathan, J., Vennard, D., Waite, R., Searchinger, T., Dumas, P., Lipinski, B., 2016. Shifting diets: Toward a sustainable food future.

Rasmussen, M.S., 1992. Assessment of millet yields and production in northern Burkina Faso using integrated NDVI from the AVHRR. Int. J. Remote Sens. 13, 3431–3442.

Rasmussen, M.S., 1998. Developing simple, operational, consistent NDVI-vegetation models by applying environmental and climatic information. Part II: Crop yield assessment. Int. J. Remote Sens. 19, 119–139.

Rayden, T., Jones, K.R., Austin, K., Radachowsky, J., 2023. Improving climate and biodiversity outcomes through restoration of forest integrity. Conserv. Biol. 37, e14163. https://doi.org/10.1111/cobi.14163

Reed, B.C., Brown, J.F., VanderZee, D., Loveland, T.R., Merchant, J.W., Ohlen, D.O., 1994. Measuring phenological variability from satellite imagery. J. Veg. Sci. 5, 703–714. https://doi.org/10.2307/3235884

Rembold, F., Atzberger, C., Savin, I., Rojas, O., 2013. Using low resolution satellite imagery for yield prediction and yield anomaly detection. Remote Sens. 5, 1704–1733. https://doi.org/10.3390/rs5041704

Rembold, F., Maselli, F., 2004. Estimating inter-annual crop area variation using multi-resolution satellite sensor images. Int. J. Remote Sens. 25, 2641–2647. https://doi.org/10.1080/01431160310001657614

Rembold, F., Maselli, F., 2006. Estimation of inter-annual crop area variation by the application of spectral angle mapping to low resolution multitemporal NDVI images. Photogramm. Eng. Remote Sens. 72, 55–62. https://doi.org/10.14358/PERS.72.1.55

Rembold, F., Meroni, M., Urbano, F., Royer, A., Atzberger, C., Lemoine, G., Eerens, H., Haesen, D., 2015. Remote sensing time series analysis for crop monitoring with the SPIRITS software: New functionalities and use examples. Front. Environ. Sci. 3. https://doi.org/10.3389/fenvs.2015.00046

Reyes, F., Casa, R., Tolomio, M., Dalponte, M., Mzid, N., 2023. Soil properties zoning of agricultural fields based on a climate-driven spatial clustering of remote sensing time series data. Eur. J. Agron. 150, 126930. https://doi.org/10.1016/j.eja.2023.126930

Reynolds, C.A., Yitayew, M., Slack, D.C., Hutchinson, C.F., Huetes, A., Petersen, M.S., 2000. Estimating crop yields and production by integrating the FAO crop specific water balance model with real-time satellite data and ground-based ancillary data. Int. J. Remote Sens. 21, 3487–3508.

Rinot, O., Levy, G.J., Steinberger, Y., Svoray, T., Eshel, G., 2019. Soil health assessment: A critical review of current methodologies and a proposed new approach. Sci. Total Environ. 648, 1484–1491. https://doi.org/10.1016/j.scitotenv.2018.08.259

Rondeaux, G., Herman, M., 1991. Polarization of light reflected by crop canopies. Remote Sens. Environ. 38, 63–75. https://doi.org/10.1016/0034-4257(91)90072-E

Roujean, J.-L., Lacaze, R., 2002. Global mapping of vegetation parameters from POLDER multiangular measurements for studies of surface-atmosphere interactions: A pragmatic method and its validation. J. Geophys. Res. Atmospheres. 107, ACL 6-1–ACL 6-14. https://doi.org/10.1029/2001JD000751

Rußwurm, M., Körner, M., 2020. Self-attention for raw optical satellite time series classification. ISPRS J. Photogramm. Remote Sens. 169, 421–435. https://doi.org/10.1016/j.isprsjprs.2020.06.006

Rudel, T.K., Defries, R., Asner, G.P., Laurance, W.F., 2009. Changing drivers of deforestation and new opportunities for conservation. Conserv. Biol. 23, 1396–1405. https://doi.org/10.1111/j.1523-1739.2009.01332.x

Rudorff, B.F.T., Aguiar, D.A., Silva, W.F., Sugawara, L.M., Adami, M., Moreira, M.A., 2010. Studies on the rapid expansion of sugarcane for ethanol production in São Paulo state (Brazil) using Landsat data. Remote Sens. 2, 1057–1076. https://doi.org/10.3390/rs2041057

Rudorff, B.F.T., Batista, G.T., 1990. Yield estimation of sugarcane based on agrometeorological-spectral models. Remote Sens. Environ. 33, 183–192. https://doi.org/10.1016/0034-4257(90)90029-L

Rustowicz, R.M., Cheong, R., Wang, L., Ermon, S., Burke, M., Lobell, D., 2019. Semantic segmentation of crop type in Africa: A novel dataset and analysis of deep learning methods, Proceedings of the IEEE/CVF Conference on Computer Vision and Pattern Recognition Workshops. pp. 75–82.

Sakamoto, T., Wardlow, B.D., Gitelson, A.A., 2011. Detecting spatiotemporal changes of corn developmental stages in the U.S. corn belt using MODIS WDRVI data. IEEE Trans. Geosci. Remote Sens. 49, 1926–1936. https://doi.org/10.1109/TGRS.2010.2095462

Sakamoto, T., Yokozawa, M., Toritani, H., Shibayama, M., Ishitsuka, N., Ohno, H., 2005. A crop phenology detection method using time-series MODIS data. Remote Sens. Environ. 96, 366–374. https://doi.org/10.1016/j.rse.2005.03.008

Schlerf, M., Atzberger, C., 2012. Vegetation structure retrieval in Beech and Spruce forests using spectrodirectional satellite data. IEEE J. Sel. Top. Appl. Earth Obs. Remote Sens. 5, 8–17. https://doi.org/10.1109/JSTARS.2012.2184268

Searchinger, T., James, O., Dumas, P., Kastner, T., Wirsenius, S., 2022. EU climate plan sacrifices carbon storage and biodiversity for bioenergy. Nature. 612, 27–30. https://doi.org/10.1038/d41586-022-04133-1

Searchinger, T., Waite, R., Hanson, C., Ranganathan, J., Dumas, P., 2019. Creating a sustainable food future: A menu of solutions to feed nearly 10 billion people by 2050—final report (Report). World Resources Institute, Washington DC.

Searchinger, T.D., Estes, L., Thornton, P.K., Beringer, T., Notenbaert, A., Rubenstein, D., Heimlich, R., Licker, R., Herrero, M., 2015. High carbon and biodiversity costs from converting Africa's wet savannahs to cropland. Nat. Clim. Change. 5, 481–486. https://doi.org/10.1038/nclimate2584

Seguini, L., Bussay, A., Baruth, B., 2019. From extreme weather to impacts: The role of the areas of concern maps in the JRC MARS bulletin. Agric. Syst. 168, 213–223. https://doi.org/10.1016/j.agsy.2018.07.003

Seifert, C.A., Azzari, G., Lobell, D.B., 2018. Satellite detection of cover crops and their effects on crop yield in the Midwestern United States. Environ. Res. Lett. 13, 064033. https://doi.org/10.1088/1748-9326/aac4c8

Shang, C., Li, X., Foody, G.M., Du, Y., Ling, F., 2022. Superresolution land cover mapping using a generative adversarial network. IEEE Geosci. Remote Sens. Lett. 19, 1–5. https://doi.org/10.1109/LGRS.2020.3020395

Sinclair, T.R., Seligman, N.G., 1996. Crop modeling: From infancy to maturity. Agron. J. 88, 698–704.

Singh, J.S., 2002. The biodiversity crisis: A multifaceted review. Curr. Sci. 82, 638–647.

Soares, J., Williams, M., Jarvis, I., Bingfang, W., Leo, O., Fabre, P., Huynh, F., Kosuth, P., Lepoutre, D., Parihar, J.S., Scarascia-Mugnozza, G., Justice, C., Williams, D., 2011. The G20 Global Agricultural Monitoring initiative (GEO-GLAM) (Technical Report).

Sofonia, J., Shendryk, Y., Phinn, S., Roelfsema, C., Kendoul, F., Skocaj, D., 2019. Monitoring sugarcane growth response to varying nitrogen application rates: A comparison of UAV SLAM LiDAR and photogrammetry. Int. J. Appl. Earth Obs. Geoinformation. 82, 101878. https://doi.org/10.1016/j.jag.2019.05.011

Som-ard, J., Atzberger, C., Izquierdo-Verdiguier, E., Vuolo, F., Immitzer, M., 2021. Remote sensing applications in sugarcane cultivation: A review. Remote Sens. 13, 4040. https://doi.org/10.3390/rs13204040

Som-ard, J., Immitzer, M., Vuolo, F., Atzberger, C., 2024. Sugarcane yield estimation in Thailand at multiple scales using the integration of UAV and Sentinel-2 imagery. Precis. Agric. https://doi.org/10.1007/s11119-024-10124-1

Som-ard, J., Immitzer, M., Vuolo, F., Ninsawat, S., Atzberger, C., 2022. Mapping of crop types in 1989, 1999, 2009 and 2019 to assess major land cover trends of the Udon Thani Province, Thailand. Comput. Electron. Agric. 198, 107083. https://doi.org/10.1016/j.compag.2022.107083

Song, B., Li, J., Dalla Mura, M., Li, P., Plaza, A., Bioucas-Dias, J.M., Benediktsson, J.A., Chanussot, J., 2014. Remotely sensed image classification using sparse representations of morphological attribute profiles. IEEE Trans. Geosci. Remote Sens. 52, 5122–5136. https://doi.org/10.1109/TGRS.2013.2286953

Song, C., Dannenberg, M.P., Hwang, T., 2013. Optical remote sensing of terrestrial ecosystem primary productivity. Prog. Phys. Geogr. Earth Environ. 37, 834–854. https://doi.org/10.1177/0309133313507944

Soulé, M.E., 1985. What is conservation biology? BioScience. 35, 727–734. https://doi.org/10.2307/1310054

Stathakis, D., Savin, I.Y., Nègre, T., 2006. Neuro-fuzzy modeling for crop yield prediction, Proceedings of the ISPRS Commission VII Symposium, Remote Sensing: From Pixels to Processes, ISPRS. Enschede.

Stehman, S.V., Olofsson, P., Woodcock, C.E., Herold, M., Friedl, M.A., 2012. A global land-cover validation data set, II: Augmenting a stratified sampling design to estimate accuracy by region and land-cover class. Int. J. Remote Sens. 33, 6975–6993. https://doi.org/10.1080/01431161.2012.695092

Steinmetz, S., Guerif, M., Delecolle, R., Baret, F., 1990. Spectral estimates of the absorbed photosynthetically active radiation and light-use efficiency of a winter wheat crop subjected to nitrogen and water deficiencies. Int. J. Remote Sens. 11, 1797–1808.

Stevens, A., Udelhoven, T., Denis, A., Tychon, B., Lioy, R., Hoffmann, L., van Wesemael, B., 2010. Measuring soil organic carbon in croplands at regional scale using airborne imaging spectroscopy. Geoderma, Diffuse Reflectance Spectroscopy in Soil Science and Land Resource Assessment. 158, 32–45. https://doi.org/10.1016/j.geoderma.2009.11.032

Stöckle, C.O., Donatelli, M., Nelson, R., 2003. CropSyst, a cropping systems simulation model. Eur. J. Agron. 18, 289–307.

Stöckli, R., Vidale, P.L., 2004. European plant phenology and climate as seen in a 20-year AVHRR land-surface parameter dataset. Int. J. Remote Sens. 25, 3303–3330. https://doi.org/10.1080/0143116031000161814 9

Stoian, A., Poulain, V., Inglada, J., Poughon, V., Derksen, D., 2019. Land cover maps production with high resolution satellite image time series and convolutional neural networks: Adaptations and limits for operational systems. Remote Sens. 11, 1986. https://doi.org/10.3390/rs11171986

Strahler, A.H., Boschetti, L., Foody, G.M., Friedl, M.A., Hansen, M.C., Herold, M., Mayaux, P., Morisette, J.T., Stehman, S.V., Woodcock, C.E., 2006. Global land cover validation: Recommendations for evaluation and accuracy assessment of global land cover maps (GOFC-GOLD Report No. 25), EUR 22156. European Communities, Luxembourg.

Suárez, L., Zarco-Tejada, P.J., Sepulcre-Cantó, G., Pérez-Priego, O., Miller, J.R., Jiménez-Muñoz, J.C., Sobrino, J., 2008. Assessing canopy PRI for water stress detection with diurnal airborne imagery. Remote Sens. Environ., Soil Moisture Experiments. 2004 (SMEX04) Special Issue 112, 560–575. https://doi.org/10.1016/j.rse.2007.05.009

Supit, I., Hooijer, A.A., Diepen, C.A. van, 1994. System Description of the WOFOST 6.0 Crop Simulation Model Implemented in CGMS. European Commission.

Sy, V.D., Herold, M., Achard, F., Avitabile, V., Baccini, A., Carter, S., Clevers, J.G.P.W., Lindquist, E., Pereira, M., Verchot, L., 2019. Tropical deforestation drivers and associated carbon emission factors derived from remote sensing data. Environ. Res. Lett. 14, 094022. https://doi.org/10.1088/1748-9326/ab3dc6

Thenkabail, P.S., 2010. Global croplands and their importance for water and food security in the twenty-first century: Towards an ever green revolution that combines a second green revolution with a blue revolution. Remote Sens. 2, 2305–2312. https://doi.org/10.3390/rs2092305

Thenkabail, P.S., Biradar, C.M., Noojipady, P., Dheeravath, V., Li, Y., Velpuri, M., Gumma, M., Gangalakunta, O.R.P., Turral, H., Cai, X., Vithanage, J., Schull, M.A., Dutta, R., 2009. Global irrigated area map (GIAM), derived from remote sensing, for the end of the last millennium. Int. J. Remote Sens. 30, 3679–3733. https://doi.org/10.1080/01431160802698919

Thenkabail, P.S., Gumma, M.K., Teluguntla, P., Mohammed, I.A., 2014. Hyperspectral remote sensing of vegetation and agricultural crops. Photogramm. Eng. Remote Sens. 80, 697–723.

Tiessen, H., Cuevas, E., Chacon, P., 1994. The role of soil organic matter in sustaining soil fertility. Nature. 371, 783–785. https://doi.org/10.1038/371783a0

Tilling, A.K., O'Leary, G.J., Ferwerda, J.G., Jones, S.D., Fitzgerald, G.J., Rodriguez, D., Belford, R., 2007. Remote sensing of nitrogen and water stress in wheat. Field Crops Res., 'Ground—breaking Stuff'-Proceedings of the 13th Australian Society of Agronomy Conference, 10–14 September 2006, Perth, Western Australia 104, 77–85. https://doi.org/10.1016/j.fcr.2007.03.023

Tilman, D., Balzer, C., Hill, J., Befort, B.L., 2011. Global food demand and the sustainable intensification of agriculture. Proc. Natl. Acad. Sci. 108, 20260–20264. https://doi.org/10.1073/pnas.1116437108

Tilman, D., Clark, M., Williams, D.R., Kimmel, K., Polasky, S., Packer, C., 2017. Future threats to biodiversity and pathways to their prevention. Nature. 546, 73–81. https://doi.org/10.1038/nature22900

Trnka, M., Rötter, R.P., Ruiz-Ramos, M., Kersebaum, K.C., Olesen, J.E., Žalud, Z., Semenov, M.A., 2014. Adverse weather conditions for European wheat production will become more frequent with climate change. Nat. Clim. Change. 4, 637–643. https://doi.org/10.1038/nclimate2242

Tucker, C.J., 1979. Red and photographic infrared linear combinations for monitoring vegetation. Remote Sens. Environ. 8, 127–150.

Tucker, C.J., Holben, B.N., Elgin Jr, J.H., McMurtrey III, J.E., 1980. Relationship of spectral data to grain yield variation. Photogramm. Eng. Remote Sens. 46, 657–666.

Udelhoven, T., 2011. TimeStats: A software tool for the retrieval of temporal patterns from global satellite archives. IEEE J. Sel. Top. Appl. Earth Obs. Remote Sens. 4, 310–317. https://doi.org/10.1109/JSTARS.2010.2051942

Udelhoven, T., Emmerling, C., Jarmer, T., 2003. Quantitative analysis of soil chemical properties with diffuse reflectance spectrometry and partial least-square regression: A feasibility study. Plant Soil. 251, 319–329. https://doi.org/10.1023/A:1023008322682

Udelhoven, T., Stellmes, M., del Barrio, G., Hill, J., 2009. Assessment of rainfall and NDVI anomalies in Spain (1989–1999) using distributed lag models. Int. J. Remote Sens. 30, 1961–1976. https://doi.org/10.1080/01431160802546829

Ungaro, F., Calzolari, C., Busoni, E., 2005. Development of pedotransfer functions using a group method of data handling for the soil of the Pianura Padano-Veneta region of North Italy: Water retention properties. Geoderma. 124, 293–317.

Ustin, S.L., Gitelson, A.A., Jacquemoud, S., Schaepman, M., Asner, G.P., Gamon, J.A., Zarco-Tejada, P., 2009. Retrieval of foliar information about plant pigment systems from high resolution spectroscopy. Remote Sens. Environ., Imaging Spectroscopy Special Issue 113, S67–S77. https://doi.org/10.1016/j.rse.2008.10.019

Veloso, A., Mermoz, S., Bouvet, A., Le Toan, T., Planells, M., Dejoux, J.-F., Ceschia, E., 2017. Understanding the temporal behavior of crops using Sentinel-1 and Sentinel-2-like data for agricultural applications. Remote Sens. Environ. 199, 415–426. https://doi.org/10.1016/j.rse.2017.07.015

Verbeiren, S., Eerens, H., Piccard, I., Bauwens, I., Van Orshoven, J., 2008. Sub-pixel classification of SPOT-VEGETATION time series for the assessment of regional crop areas in Belgium. Int. J. Appl. Earth Obs. Geoinformation, Modern Methods in Crop Yield Forecasting and Crop Area Estimation. 10, 486–497. https://doi.org/10.1016/j.jag.2006.12.003

Verger, A., Baret, F., Weiss, M., Kandasamy, S., Vermote, E., 2013. The CACAO method for smoothing, gap filling, and characterizing seasonal anomalies in satellite time series. IEEE Trans. Geosci. Remote Sens. 51, 1963–1972. https://doi.org/10.1109/TGRS.2012.2228653

Verhoef, W., 1985. Earth observation modeling based on layer scattering matrices. Remote Sens. Environ. 17, 165–178. https://doi.org/10.1016/0034-4257(85)90072-0

Verstraete, M.M., Pinty, B., Myneni, R.B., 1996. Potential and limitations of information extraction on the terrestrial biosphere from satellite remote sensing. Remote Sens. Environ. 58, 201–214.

Vieira, M.A., Formaggio, A.R., Rennó, C.D., Atzberger, C., Aguiar, D.A., Mello, M.P., 2012. Object based image analysis and data mining applied to a remotely sensed Landsat time-series to map sugarcane over large areas. Remote Sens. Environ. 123, 553–562. https://doi.org/10.1016/j.rse.2012.04.011

Villalobos, F.J., Hall, A.J., Ritchie, J.T., Orgaz, F., 1996. OILCROP-SUN: A development, growth, and yield model of the sunflower crop. Agron. J. 88, 403–415.

Vintrou, E., Soumaré, M., Bernard, S., Bégué, A., Baron, C., Lo Seen, D., 2012. Mapping fragmented agricultural systems in the Sudano-Sahelian environments of Africa using random forest and ensemble metrics of coarse resolution MODIS imagery. Photogramm. Eng. Remote Sens. 78, 839–848. https://doi.org/10.14358/PERS.78.8.839

Viscarra Rossel, R.A., Cattle, S.R., Ortega, A., Fouad, Y., 2009. In situ measurements of soil colour, mineral composition and clay content by vis-NIR spectroscopy. Geoderma. 150, 253–266.

Vitousek, P.M., Aber, J.D., Howarth, R.W., Likens, G.E., Matson, P.A., Schindler, D.W., Schlesinger, W.H., Tilman, D.G., 1997. Human alteration of the global nitrogen cycle: Sources and consequences. Ecol. Appl. 7, 737–750. https://doi.org/10.1890/1051-0761(1997)007[0737:HAOTGN]2.0.CO;2

Vogelmann, J.E., Howard, S.M., Yang, L., Larson, C.R., Wylie, B.K., Van Driel, N., 2001. Completion of the 1990s national land cover data set for the conterminous United States from Landsat thematic mapper data and ancillary data sources. Photogramm. Eng. Remote Sens. 67.

Vuolo, F., Atzberger, C., 2014. Improving land cover maps in areas of disagreement of existing products using NDVI time series of MODIS—Example for Europe. Photogramm.—Fernerkund.—Geoinformation 393–407. https://doi.org/10.1127/1432-8364/2014/0232

Vuolo, F., Atzberger, C., 2012. Exploiting the classification performance of support vector machines with multi-temporal Moderate-Resolution Imaging Spectroradiometer (MODIS) data in areas of agreement and disagreement of existing land cover products. Remote Sens. 4, 3143–3167. https://doi.org/10.3390/rs4103143

Vuolo, F., Dini, L., D'Urso, G., 2008. Retrieval of leaf area index from CHRIS/PROBA data: An analysis of the directional and spectral information content. Int. J. Remote Sens. 29, 5063–5072. https://doi.org/10.1080/01431160802036490

Vuolo, F., Neuwirth, M., Immitzer, M., Atzberger, C., Ng, W.-T., 2018. How much does multi-temporal Sentinel-2 data improve crop type classification? Int. J. Appl. Earth Obs. Geoinformation. 72, 122–130. https://doi.org/10.1016/j.jag.2018.06.007

Wagner, W., Naeimi, V., Scipal, K., Jeu, R., Martínez-Fernández, J., 2007. Soil moisture from operational meteorological satellites. Hydrogeol. J. 15, 121–131.

Waldner, F., Lambert, M.-J., Li, W., Weiss, M., Demarez, V., Morin, D., Marais-Sicre, C., Hagolle, O., Baret, F., Defourny, P., 2015. Land cover and crop type classification along the season based on biophysical variables retrieved from multi-sensor high-resolution time series. Remote Sens. 7, 10400–10424. https://doi.org/10.3390/rs70810400

Wardlow, B.D., Egbert, S.L., Kastens, J.H., 2007. Analysis of time-series MODIS 250 m vegetation index data for crop classification in the U.S. Central Great Plains. Remote Sens. Environ. 108, 290–310. https://doi.org/10.1016/j.rse.2006.11.021

Wassenaar, T., Andrieux, P., Baret, F., Robbez-Masson, J.M., 2005. Soil surface infiltration capacity classification based on the bi-directional reflectance distribution function sampled by aerial photographs. The case of vineyards in a Mediterranean area. CATENA, Surface Characterisation for Soil Erosion Forecasting. 62, 94–110. https://doi.org/10.1016/j.catena.2005.05.004

Watzig, C., Schaumberger, A., Klingler, A., Dujakovic, A., Atzberger, C., Vuolo, F., 2023. Grassland cut detection based on Sentinel-2 time series to respond to the environmental and technical challenges of the Austrian fodder production for livestock feeding. Remote Sens. Environ. 292, 113577. https://doi.org/10.1016/j.rse.2023.113577

Weiss, M., Jacob, F., Duveiller, G., 2020. Remote sensing for agricultural applications: A meta-review. Remote Sens. Environ. 236, 111402. https://doi.org/10.1016/j.rse.2019.111402

Weiss, M., Troufleau, D., Baret, F., Chauki, H., Prévot, L., Olioso, A., Bruguier, N., Brisson, N., 2001. Coupling canopy functioning and radiative transfer models for remote sensing data assimilation. Agric. For. Meteorol. 108, 113–128.

Wen, W., Timmermans, J., Chen, Q., van Bodegom, P.M., 2023. Evaluating crop-specific responses to salinity and drought stress from remote sensing. Int. J. Appl. Earth Obs. Geoinformation. 122, 103438. https://doi.org/10.1016/j.jag.2023.103438

White, J.W., Andrade-Sanchez, P., Gore, M.A., Bronson, K.F., Coffelt, T.A., Conley, M.M., Feldmann, K.A., French, A.N., Heun, J.T., Hunsaker, D.J., Jenks, M.A., Kimball, B.A., Roth, R.L., Strand, R.J., Thorp, K.R., Wall, G.W., Wang, G., 2012. Field-based phenomics for plant genetics research. Field Crops Res. 133, 101–112. https://doi.org/10.1016/j.fcr.2012.04.003

White, M.A., Thornton, P.E., Running, S.W., 1997. A continental phenology model for monitoring vegetation responses to interannual climatic variability. Glob. Biogeochem. Cycles. 11, 217–234. https://doi.org/10.1029/97GB00330

Whiteside, T., Ahmad, W., 2005. A comparison of object-oriented and pixel-based classification methods for mapping land cover in North Australia, Proceedings of Spatial Science Institute Biennial Conference SSC2005. Spatial Intelligence, Innovation and Praxis. Presented at the The national biennial Conference of the Spatial Science Institute, Spacial Sciences Institute, Melbourne, pp. 1225–1231.

Wu, X., Hong, D., Chanussot, J., 2022. Convolutional neural networks for multimodal remote sensing data classification. IEEE Trans. Geosci. Remote Sens. 60, 1–10. https://doi.org/10.1109/TGRS.2021.3124913

Wulder, M.A., Roy, D.P., Radeloff, V.C., Loveland, T.R., Anderson, M.C., Johnson, D.M., Healey, S., Zhu, Z., Scambos, T.A., Pahlevan, N., Hansen, M., Gorelick, N., Crawford, C.J., Masek, J.G., Hermosilla, T., White, J.C., Belward, A.S., Schaaf, C., Woodcock, C.E., Huntington, J.L., Lymburner, L., Hostert, P., Gao, F., Lyapustin, A., Pekel, J.-F., Strobl, P., Cook, B.D., 2022. Fifty years of Landsat science and impacts. Remote Sens. Environ. 280, 113195. https://doi.org/10.1016/j.rse.2022.113195

Xiao, J., Chevallier, F., Gomez, C., Guanter, L., Hicke, J.A., Huete, A.R., Ichii, K., Ni, W., Pang, Y., Rahman, A.F., Sun, G., Yuan, W., Zhang, L., Zhang, X., 2019. Remote sensing of the terrestrial carbon cycle: A review of advances over 50 years. Remote Sens. Environ. 233, 111383. https://doi.org/10.1016/j.rse.2019.111383

Xin, J., Yu, Z., van Leeuwen, L., Driessen, P.M., 2002. Mapping crop key phenological stages in the North China plain using NOAA time series images. Int. J. Appl. Earth Obs. Geoinformation. 4, 109–117. https://doi.org/10.1016/S0303-2434(02)00007-7

Xu, J., Yang, J., Xiong, X., Li, H., Huang, J., Ting, K.C., Ying, Y., Lin, T., 2021. Towards interpreting multi-temporal deep learning models in crop mapping. Remote Sens. Environ. 264, 112599. https://doi.org/10.1016/j.rse.2021.112599

Yuan, X., Shi, J., Gu, L., 2021. A review of deep learning methods for semantic segmentation of remote sensing imagery. Expert Syst. Appl. 169, 114417. https://doi.org/10.1016/j.eswa.2020.114417

Zaitchik, B.F., Macalady, A.K., Bonneau, L.R., Smith, R.B., 2006. Europe's 2003 heat wave: A satellite view of impacts and land—atmosphere feedbacks. Int. J. Climatol. 26, 743–769. https://doi.org/10.1002/joc.1280

Zaks, D.P.M., Kucharik, C.J., 2011. Data and monitoring needs for a more ecological agriculture. Environ. Res. Lett. 6, 014017. https://doi.org/10.1088/1748-9326/6/1/014017

Zarco-Tejada, P.J., Camino, C., Beck, P.S.A., Calderon, R., Hornero, A., Hernández-Clemente, R., Kattenborn, T., Montes-Borrego, M., Susca, L., Morelli, M., Gonzalez-Dugo, V., North, P.R.J., Landa, B.B., Boscia, D., Saponari, M., Navas-Cortes, J.A., 2018. Previsual symptoms of Xylella fastidiosa infection revealed in spectral plant-trait alterations. Nat. Plants. 4, 432–439. https://doi.org/10.1038/s41477-018-0189-7

Zeng, L., Wardlow, B.D., Xiang, D., Hu, S., Li, D., 2020. A review of vegetation phenological metrics extraction using time-series, multispectral satellite data. Remote Sens. Environ. 237, 111511. https://doi.org/10.1016/j.rse.2019.111511

Zhang, P., Anderson, B., Tan, B., Huang, D., Myneni, R., 2005. Potential monitoring of crop production using a satellite-based Climate-Variability Impact Index. Agric. For. Meteorol. 132, 344–358. https://doi.org/10.1016/j.agrformet.2005.09.004

Zhang, X., Friedl, M.A., Schaaf, C.B., Strahler, A.H., Hodges, J.C.F., Gao, F., Reed, B.C., Huete, A., 2003. Monitoring vegetation phenology using MODIS. Remote Sens. Environ. 84, 471–475.

6 Agricultural Systems Studies Using Remote Sensing

Agnès Bégué, Damien Arvor, Camille Lelong, Elodie Vintrou, and Margareth Simões

ACRONYMS AND DEFINITIONS

ASAR	Advanced Synthetic Aperture Radar
FAO	Food and Agriculture Organization of the United Nations
GEOBIA	GEographic Object-Based Image Analysis
LULC	Land Use, Land Cover
MODIS	Moderate-Resolution Imaging Spectroradiometer
NDVI	Normalized Difference Vegetation Index
NIR	Near-Infrared
SPOT	Satellite Pour l'Observation de la Terre, French Earth Observing Satellites
USAID	United States Agency for International Development

6.1 INTRODUCTION

Agricultural systems of the world are going through massive transformations as a result of advances in smart technologies, increasing urbanization, demographic changes involving movement of population from rural to urban, changes in cropland areas as well as irrigation, water availability for agriculture, and the changing climate (Owusu et al., 2024, Rozenstein et al., 2024, Victor et al., 2024, Omia et al., 2023, Teluguntla et al., 2023, Kumar et al., 2022, Mahlayeye et al., 2022, Jung et al., 2021, Thenkabail et al., 2021, Karthikeyan et al., 2020, Segarra et al., 2020, Sishodia et al., 2020, Weiss et al., 2020, Hatab et al., 2019, Khanal et al., 2020, Tantalaki et al., 2019, Huang et al., 2018, Raymond et al., 2018, Jones et al., 2017, Steele-Dunne et al., 2017). The world population is expected to reach 9.3 billion in 2050 (UN, 2010). To feed this population, the Food and Agriculture Organization last global projection exercise forecasted that the world's agricultural production will need to increase by approximately 70% by 2050, compared with the 2005 production levels (FAO, 2011). Approximately 80% of the increased agricultural production will need to come from yield increases, and higher cropping intensities such as increased multiple cropping and/or shortening of fallow periods.

Such evolutions must cope with climate change (characterized by changing rainfall patterns and an increasing number of extreme weather events) and its consequences (changing distributions of plant and vector-borne diseases, and increased crop yield variability), more competition for land (increased competition between food and bioenergy production), and the associated increased environmental pressures (e.g., overexploitation of groundwater resources, water quality degradation, and soil degradation). As a consequence, in addition to the need to increase crop production, another major agricultural challenge is the task of improving the management of natural resources, especially through the adoption of more environmental-friendly practices, such as ecological intensification or conservation agriculture. Major agricultural powers such as Europe and Brazil have launched ambitious programs, e.g., the GAP (Good Agricultural Practice) guidelines and the ABC

DOI: 10.1201/9781003541165-8

Program (Brazilian Low Carbon Agriculture Program), respectively. These programs give a special role to multifunctional landscapes to establish sustainable agriculture. Landscapes must be considered a whole land use system at the heart of human-nature relationships that need to be efficiently managed to preserve and restore ecosystem services (DeFries and Rosenzweig, 2010), and to contribute to sustainable solutions, especially regarding food security challenges (Thenkabail et al., 2021, Verburg et al., 2013). In view of these global challenges, there is an urgent need to better characterize agricultural systems at the regional and global scales, with a particular emphasis on the various pathways towards agricultural intensification. Those systems are the key to understanding land use sustainability in agricultural territories.

Although everyone agrees on the need to qualify agricultural systems at the regional scale, few examples exist in the literature (Owusu et al., 2024, Teluguntla et al., 2023, Thenkabail et al., 2021, Steele-Dunne et al., 2017). Leenhardt et al. (2010) reviewed cropping system descriptions and locations at the regional scale, and concluded that both remain highly unclear for most world regions. The FAO continental farming system maps (Dixon et al., 2001) and the U.S. Agency for International Development (USAID) Famine Early Warning Systems Network (FEWS NET) national livelihood maps for Africa (USAID, 2009) are produced at very broad scales. More detailed, the regional maps of rice areas in Southeast Asia (Bridhikitti and Overcamp, 2012) or sugarcane areas in Brazil (Adami et al., 2012) have recently been produced using remote sensing data only. But these simple approaches, based on the dominant crop type with limited consideration of land management, are insufficient to draw a complete picture of coupled human-environment systems (Verburg et al., 2009).

So, evolving from traditional remote sensing land cover mapping to land use system mapping is not straightforward and requires processing new data, implementing new methods and, above all, an enhanced integration between land science research disciplines (Rozenstein et al., 2024, Victor et al., 2024, Omia et al., 2023, Karthikeyan et al., 2020, Tantalaki et al., 2019, Huang et al., 2018, Verburg et al., 2009, Koschke et al., 2013). Vaclavik et al. (2013) derived a global representation of land use systems using land use intensity datasets, environmental conditions and socio-economic indicators. Land use intensity was derived from satellite-based land cover maps and sub-national statistics. The authors noted that the scope of the study was limited because the quality of the statistical datasets they used was geographically distributed unevenly worldwide. Kuemmerle et al. (2013) proposed a review of the current input (crop type, cropping frequency, capital and labor intensity, etc.) and output (yields and carbon stock, etc.) land intensity metrics that could be provided directly or indirectly by satellite remote sensing. They concluded that satellite-based approaches are still experimental in that domain and cannot readily be applied across large areas. Despite these issues, new opportunities are arising.

The objective of the present study is to give an overview of remote sensing-based approaches for regional mapping of agricultural systems and to illustrate the diversity of these approaches through case studies (Victor et al., 2024, Teluguntla et al., 2023, Thenkabail et al., 2021, Raymond et al., 2018, Jones et al., 2017, Steele-Dunne et al., 2017, Bellon et al., 2018). To do this, we propose and introduce a general framework, including satellite data and land mapping approaches, to characterize agricultural systems at different scales. These approaches are illustrated by three case studies representing a wide diversity of agricultural systems across the tropical world. Based on these case studies and a literature review, the opportunities and challenges for agricultural systems mapping at regional and global scales are discussed, and further research is proposed.

6.2 ROLES OF REMOTE SENSING IN THE ASSESSMENT OF AGRICULTURAL SYSTEMS

6.2.1 Diversity of the Agricultural Systems in the World

To our knowledge, the most complete global agricultural map is the map produced by the Food and Agriculture Organization (FAO) and the World Bank (Dixon et al., 2001) which covers the six main regions of the developing world. This map represents 72 farming systems (Figure 6.1a) that were defined according to (1) the available natural resource base (water, land, climate, altitude, etc.); (2) the dominant pattern of farm activities and household livelihoods, including relationship to markets; and (3) the intensity of production activities. These detailed farming systems are grouped in eight broad categories (Figure 6.1b; Table 6.1). It is interesting to note that seven out of the eight broad farming systems categories are based on smallholder producers (less than 2 ha of land, according to FAO).

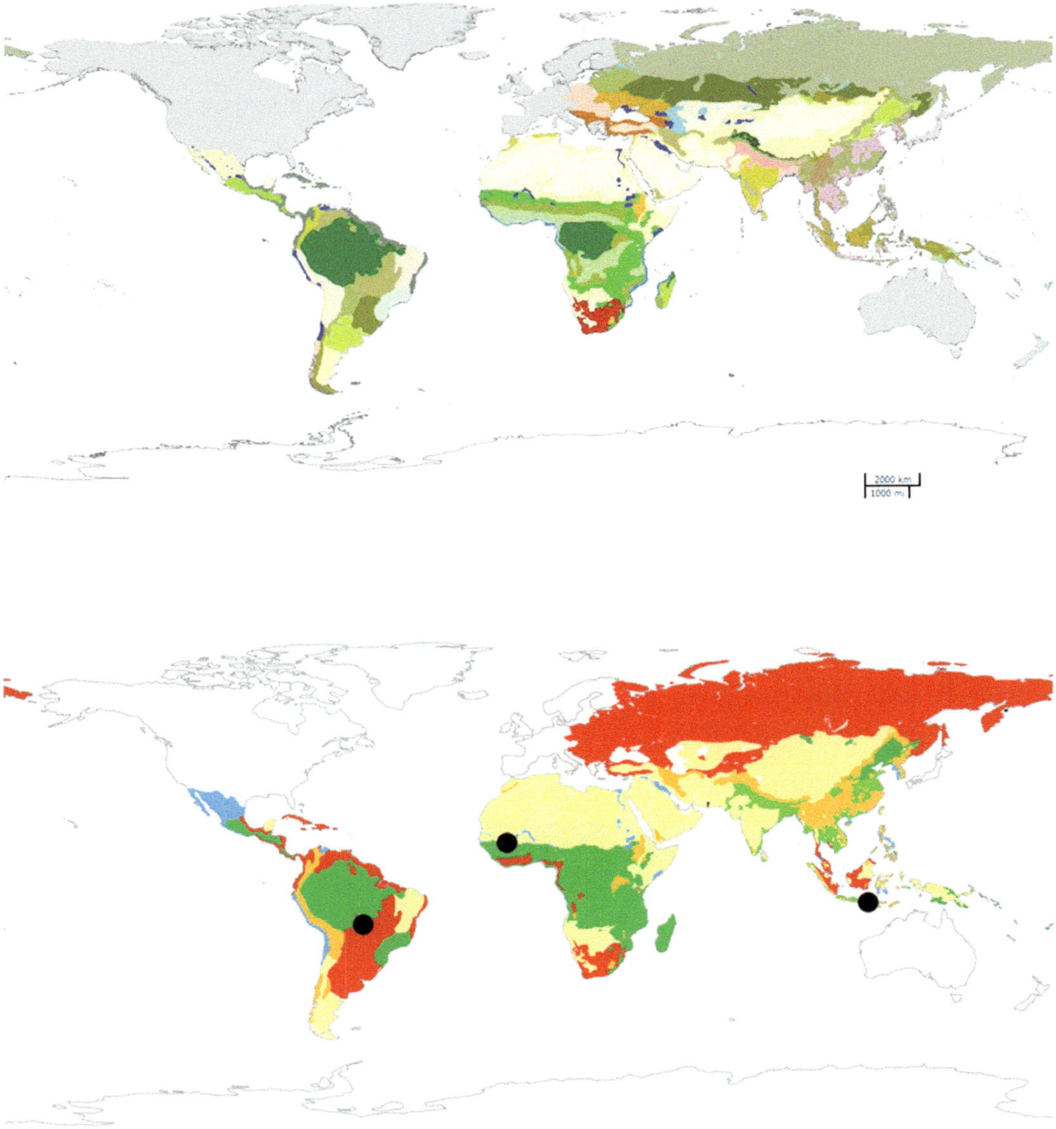

FIGURE 6.1 Farming system maps of the developing regions of the world (Dixon et al., 2001): (a) the original FAO 72-class map (see Dixon et al., 2001 for legend) and (b) the FAO eight broad categories (see Table 6.1 for legend). Black dots in (b) correspond to the location of the three case studies.

TABLE 6.1
Broad Category of Farming Systems (Dixon et al., 2001)

#		Farming System Name	Characteristics
1	●	Irrigated farming systems	Dominated by smallholder producers
2	●	Wetland rice based	Dominated by smallholder producers, dependent upon seasonal rains supplemented by irrigation
3	●	Rainfed farming systems in humid (and subhumid) areas	Dominated by smallholder producers, characterized by specific dominant crops or mixed crop-livestock systems
4	●	Rainfed farming systems in steep and highland areas	Dominated by smallholder producers, often mixed crop-livestock systems
5	●	Rainfed farming systems in dry or cold areas	Dominated by smallholder producers, with mixed crop-livestock and pastoral systems merging into systems with very low current productivity
6	●	Mixed large commercial and small holder	Dualistic, across a variety of ecologies and with diverse production patterns
7	●	Coastal artisanal fishing mixed	Dominated by smallholder producers, incorporates mixed farming elements
8	●	Urban based	Dominated by smallholder producers, typically focused on horticultural and livestock production

6.2.2 A Conceptual Framework Based on Land Mapping Issues

Remote sensing-based information can play different roles in the assessment of agricultural systems (Owusu et al., 2024, Teluguntla et al., 2023, Jung et al., 2021, Thenkabail et al., 2021). Figure 6.2 illustrates how satellite images can help derive "land maps" (land cover, land use, and land use system maps; ① in Figure 6.2) using various processing approaches (② in Figure 6.2). In the case of agriculture-dominated landscapes, these "land maps" can be interpreted as "agricultural system" maps (cropland, cropping system, and farming system; ③ in Figure 6.2).

Based on this framework, monitoring and mapping agricultural systems using remote sensing require clearly defined concepts and objects, that is, which "land maps" to monitor which "agricultural systems"? In the proposed conceptual framework (Figure 6.2), we tried to build bridges between the land maps (land cover, land use, and land use system), that can be obtained with the contribution of remote sensing data, and the agricultural systems (cropland, cropping system, and farming system, respectively) that are addressed in this chapter. These bridges are based on a set of definitions and hypotheses that are presented hereafter.

- Land cover addresses the description of the land surface in terms of soil and vegetation layers, including natural vegetation, crops, and human structures (Burley, 1961). Land use refers to the purpose for which humans exploit the land cover (Lambin et al., 2006), including land management techniques (Verburg et al., 2009). In remote sensing-derived maps, mixed LULC (land use/land cover) legends are often used because concepts concerning land cover and land use activities are closely related and, in many cases, can be used interchangeably (Anderson et al., 1976). Cropping systems are defined, at least, by the dominant crop type (Thenkabail et al., 2021, Weiss et al., 2020, Hatab et al., 2019, Khanal et al., 2020). Crop types, or at least crop groups (e.g., winter and summer crops; Atzberger and Rembold, 2013), are often represented in these satellite-derived LULC maps. More recently, information on the intensification mode, such as the use of irrigation (e.g., Thenkabail et al., 2010) or the adoption of multiple cropping (e.g., Arvor et al., 2011), appears in the LULC maps, improving the characterization of the cropping systems using remote sensing data.

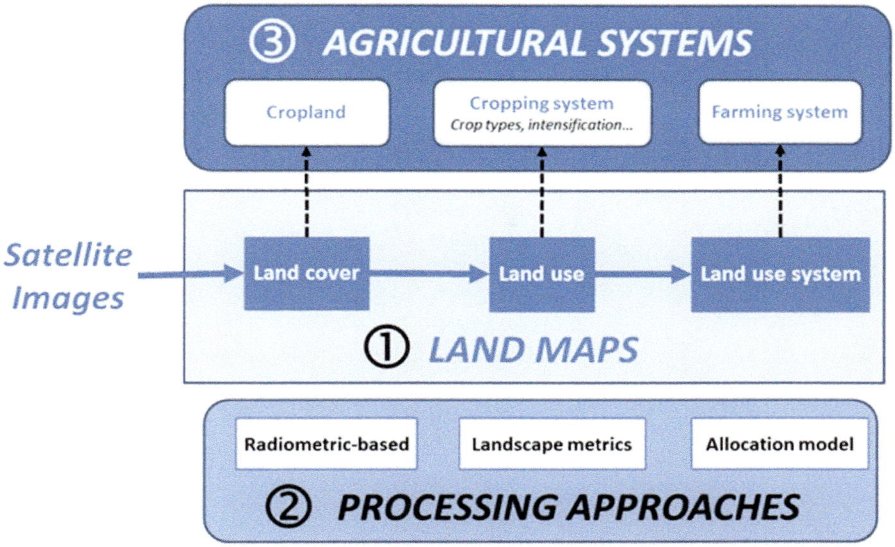

FIGURE 6.2 The conceptual framework used in this study.

- Land use system can be defined as a coupled human-environment system. It describes how land, as an essential resource, is being used and managed by. Remote sensing data do not record human activities and thus cannot be directly used for land use system mapping. Photo-interpreters historically used patterns, tones, textures, shapes, and site associations to derive initial land cover information into land use information (Anderson et al., 1976). This approach is consistent with Verburg et al. (2009) who proposed obtaining land use system maps from land cover maps supplemented by observations, inferred from landscape structures. Farming systems, defined by most experts as a combination of biophysical, socio-economic and human elements of a farm, can be seen as the land use system version for agriculture.

To conclude, LCLU mapping can be obtained by classifying satellite images, while land use system mapping needs a larger view and must be approached on a larger scale (landscape scale).

6.2.3 Processing Approaches

A large panel of methods and tools to produce agricultural system maps from remote sensing data are described in the literature (Owusu et al., 2024, Rozenstein et al., 2024, Victor et al., 2024, Omia et al., 2023, Teluguntla et al., 2023, Kumar et al., 2022, Mahlayeye et al., 2022, Jung et al., 2021, Thenkabail et al., 2021, Karthikeyan et al., 2020, Segarra et al., 2020, Sishodia et al., 2020, Weiss et al., 2020, Hatab et al., 2019, Khanal et al., 2020, Tantalaki et al., 2019, Huang et al., 2018, Raymond et al., 2018, Jones et al., 2017, Steele-Dunne et al., 2017). The methods can be grouped in three types: radiometric-based method, landscape approach and allocation models.

6.2.3.1 Radiometric-Based Methods

Radiometric-based methods are largely used for cropland and crop type mapping. Most of the publications report pixel or object based classifications, and photo-interpretation methods. Examples are discussed in Chapter 4, and this topic will not be further discussed in this paper.

Beyond crop type, many examples concerning remote sensing and cropping practices are found in the literature (see review by Begue et al., 2018). Most of the methods are based on statistical

TABLE 6.2
Literature Examples of Use of Remote Sensing for Mapping Cropping Practices

Cropping Practice	Crop (Sensor)	Example of Studies
Crop variety	Sugarcane (Hyperion)	Galvao et al., 2005
	Sugarcane (Landsat)	Fortes and Dematte, 2006
Double-cropping	Soybean & others (MODIS)	Arvor et al., 2011
	Cereals (MODIS)	Qiu et al., 2014
Harvest date	Sugarcane (SPOT)	Lebourgeois et al., 2007
	Sugarcane (SPOT)	El Hajj et al., 2007
Sowing date	Soybean (MODIS)	Maatoug et al., 2012
Harvest mode	Sugarcane (Landsat, DMC)	Aguiar et al., 2011
	Sugarcane (Landsat, CBERS)	Goltz et al., 2009
Irrigation	Various crops (MODIS)	Gumma et al., 2011
	Wheat (FORMOSAT, ASAR)	Hadria et al., 2009
	Review	**Ozdogan et al., 2010**
Crop residue	Various crops (Landsat)	Pacheco et al., 2006
	Review	**Zhang et al., 2011**
Tillage	Wheat (FORMOSAT, ASAR)	Hadria et al., 2009
	Various crops (Landsat)	Sullivan et al., 2008
Row orientation and width	Vineyard (aerial photos)	Delenne et al., 2008
	Olive groves (QuickBird)	Amoruso et al., 2009
	Orchards (Ikonos)	Aksoy et al., 2012
	Vineyard, cereals (aerial photos)	Lefebvre et al., 2011

Note: References in bold are review papers.

relationships between surface variables and image variables (reflectance, spectral index, texture index, etc.), while others use signal processing techniques. The examples listed in Table 6.2 show that there is a strong link between the type of cropping practice and the sensor. High resolution image primarily identifies inter- and mixed-cropping, and agroforestry composition and structure. High image acquisition frequency usually helps to identify double cropping practices, crop types or groups of crop types, and sowing/harvest dates, while spectral richness is used to distinguish cultivars. Irrigation, crop residues, and tillage practices are mainly obtained through multispectral image analyses conducted at different scales depending on the structure of the fields.

A detailed analysis of the publications on cropping practices and remote sensing shows that, even if the proportion of publications addressing this issue is increasing (4% of the total remote sensing and agriculture publications in the 1990s, and 9% currently), these publications primarily concern only one cropping practice at a time, and the analyses are generally conducted at local scale. Literature on the cropping system itself is still limited in terms of the number of publications (2% of the total published remote sensing and agriculture papers), and does not progress significantly.

6.2.3.2 Landscape Approach

Cropland and crop type maps can be viewed as a mosaic of patches, where the patches are the landscape elements (Owusu et al., 2024, Teluguntla et al., 2023, Thenkabail et al., 2021, Karthikeyan et al., 2020, Segarra et al., 2020, Sishodia et al., 2020, Weiss et al., 2020, Hatab et al., 2019, Khanal et al., 2020). In that case, landscape metrics can be used to characterize the agricultural system. The term "landscape metrics" refers to indices developed for categorical map patterns (McGarigal, 2014). Landscape metrics exist at the patch, class (patch type), and landscape level. At the class and landscape level, some of the metrics quantify the landscape composition (the relative abundance of crop patch types, for instance), while others quantify the landscape configuration (the position, connectivity or the edge-to-area ratios of the cropland, for example).

Although very few articles use landscape metrics to characterize agrosystems compared to ecosystems (see review by Uuemaa et al., 2013), some of them use crop class metrics as an input for ecological studies (e.g., Pocas et al., 2011), and a few use landscape research for agricultural perspectives. The aim of these latter is generally to evaluate different policies on agricultural landscapes or to assess the sustainability of the agricultural systems. For example, Plexida et al. (2014) discussed the role of modern cultivation methods on the simplification of landscape patterns in central Greece. They showed that the landscape in the agricultural lowlands was characterized by connectedness (high values of Patch Cohesion Index) and simple geometries (low values of fractal dimension index), whereas the landscape pattern of the pastoral uplands was found to be highly diverse (high Shannon Diversity index). Panigrahy et al. (2005) and Panigrahy et al. (2011) used landscape composition metrics to assess and evaluate the efficiency and sustainability of the agricultural systems in India. They proposed and calculated three indices, namely the Multiple Cropping Index (MCI), Area Diversity Index (ADI), and Cultivated Land Utilization Index (CLUI), using three satellite-derived seasonal land cover maps. The MCI measures the cropping intensity as the number of crops grown temporally in a particular area over a period of one year, the ADI measures the multiplicity of crops or farm products planted in a single year, and the CLUI measures how efficient the available land area has been used over the year (see Panigrahy et al., 2005) for formula). The indices were categorized as high, medium, and low to evaluate the cropping system performance in each of the districts.

An example of landscape metrics based on the spatial configuration of the classes is given in Colson et al. (2011). They used eight landscape metrics to quantify and investigate the spatial patterns of cattle pasture and cropland throughout the states of Pará, Mato Grosso, Rondônia, and Amazonas, and concluded that these metrics showed evidence of a possible measure for discerning the patterns of agriculture attached to a certain state.

6.2.3.3 Spatial Allocation Modeling

Global cropping system maps (crop type and irrigation) are emerging at coarse resolution (see Anderson et al. (2014) for the description and comparison of these products) (Owusu et al., 2024, Teluguntla et al., 2023, Thenkabail et al., 2021). They are based on statistical data downscaled at the administrative level into grid-cell specific values. An illustrative example of spatial allocation is the SPAM model (SPatial Allocation Model), developed at the meso-scale by You and Wood (2006) and You et al. (2009), to spatially disaggregate crop production data (acreage and yield) within geopolitical units (e.g., countries or sub-national provinces and districts), using a cross-entropy approach. The pixel-scale allocations are performed by compiling and merging relevant spatially explicit data, including production statistics, satellite-derived land cover data, biophysical crop suitability assessments, and population density. In such models, remote sensing is mainly used to locate cropland at regional scales as an input for the allocation models (to spatially disaggregate statistics data for instance), while the crop determining factors are generally established by expertise or statistical analyses (Leenhardt et al., 2010). Recent examples showed that satellite images can also be used to understand and model the environmental drivers of cropping systems. For example, Jasinski et al. (2005) used a multiple logistic regression to model the role of environmental variables (vegetation type, soil type, altitude, slope, rainfall) on the Southern Amazonian cropland dynamics previously assessed using remote sensing data. More recently, Arvor et al. (2014) showed that the adoption of intensive double-cropping practices was related to the spatial variability of rainfall regimes and favored by a high annual rainfall, a long rainy season and a low variability of the onset date.

However, a major drawback of the spatial allocation models approach is that it is not always possible to obtain deterministic relations between easily accessible factors (climate, soil, etc.) and cropping system elements, especially in "intensive systems" compared to "traditional systems," which are more dependent on environmental factors (Figure 6.3). According to Jouve (2006), in southern

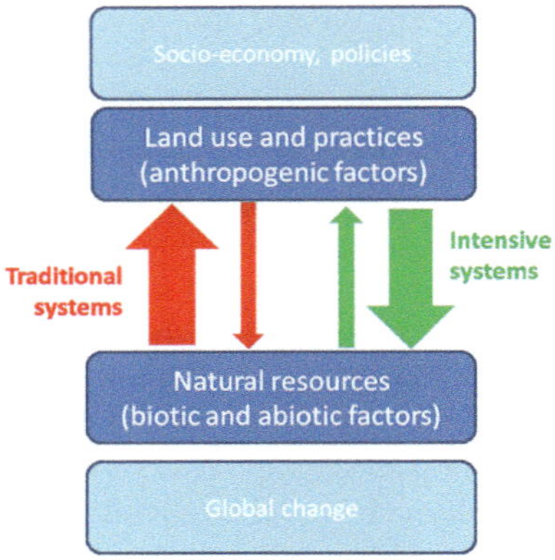

FIGURE 6.3 Relative weights of the determining factors in the traditional and intensive agricultural systems.

countries where traditional systems are important and make little use of modern means of production (mechanization, fertilization), the farmers capacity to artificialize their environment and get rid of the environmental constraints is limited. In those cases, the relationship between the cropping systems and environmental conditions is strong, and the spatial distribution of the cropping systems reflects more the environmental differences than the farming differences. Additionally, the relationship can be identified at the rural community scale. Inversely, in intensive systems, the determining factors approach is more difficult to set up and the spatial allocation models can be more difficult to implement.

6.3 EXAMPLES OF AGRICULTURAL SYSTEM STUDIES USING REMOTE SENSING

Three case studies—agroforestry in Bali (Indonesia), double cropping in Southern Amazon (Brazil), and traditional rainfed agriculture in Mali—were selected to illustrate the use of remote sensing for mapping agricultural systems. Two of them, Bali and Mali, are characterized by smallholder agriculture, while the Brazilian case is characterized by commercial agriculture (Figure 6.1b). These case studies are far from representing all of the possible uses of remote sensing, but they illustrate a diversity of technical and scientific approaches, while addressing some worldwide agricultural issues (geographic certification, agricultural system sustainability, food security, etc.) (Owusu et al., 2024, Rozenstein et al., 2024, Victor et al., 2024, Omia et al., 2023, Teluguntla et al., 2023, Kumar et al., 2022, Thenkabail et al., 2021, Karthikeyan et al., 2020, Lemettais et al., 2024).

6.3.1 Presentation of the Case Studies

6.3.1.1 Agro-Forestry in Bali

In tropical regions, small stakeholders' agroforestry is the most common traditional cropping system. It associates different crops inside a single plot, with multifunctional trees to produce fruits, cash-crops, wood, medicines, shading, or to conserve biodiversity in various proportions and organizations. This system allows a relative sustainability in food diversification, but not in incomes, which depends on the trading market fluctuations. Agroforestry is promoted by agronomists for environmental and livelihood quality, and is questioned by socio-economists because of the cash-crop vulnerability. This emphasizes the need for evaluating the actual environmental, social and economic benefits of such cropping system. Remote sensing studies now propose new tools to objectively characterize the agroforestry systems at the intra-plot scale (Victor et al., 2024, Omia et al., 2023, Teluguntla et al., 2023, Kumar et al., 2022, Mahlayeye et al., 2022, Jung et al., 2021, Thenkabail et al., 2021, Karthikeyan et al., 2020, Segarra et al., 2020, Sishodia et al., 2020, Coltri et al., 2013, Guillen-Climent et al., 2014, Peña-Barragán et al., 2004, Aksoy et al., 2012, Ursani et al., 2012, Mougel and Lelong, 2008), at the farm level (distribution among neighbors), and to replace it in the landscape matrix (Wästfelt et al., 2012, Lei et al., 2012). This allows associating different environmental, agricultural, and socioeconomic conditions in integrated analyses to understand the drivers of agricultural choices and resilience (Fox et al., 1994, Gobin et al., 2001, Kunwar, 2010), and the level of productivity and quality of the production.

The case study presented in this chapter is situated in Bali, an active volcanic island of Indonesia. Coffee is cropped almost everywhere in the central highlands. The study focused on a 220 km² area located in Kintamani county, which is famous for its coffee crops. The landscape is shaped by the local topography, which ranges from 300 m to 1,800 m (Figure 6.4). This work aims at producing a cropping system map in order to understand coffee quality drivers, and helps delimitating an area labeled by the distinction of the Protected Geographical Indication on Arabica coffee.

6.3.1.2 Double Cropping in Southern Amazon

For nearly 40 years, the Southeastern Amazon in Brazil experiences severe agricultural dynamics. Cropland expanded dramatically to support commercial cultivation of important commodities such as soybean, maize, and cotton. The severity of the agricultural dynamics explains the abundance of large-scale monitoring studies using remote sensing. To date, most remote sensing based studies were carried out with MODIS data for three reasons: (1) monitoring such a large area requires a huge number of high remote sensing data to be processed, (2) high cloud cover rates during the rainy season prevents the acquisition of good quality high resolution images during the cropping period, and (3) the mean field area is about 180 ha so that even 250 m medium resolution images are valid for crop type mapping. Consequently, most MODIS-based approaches to date were based on the interpretation of vegetation index (NDVI or EVI) time series. Such time series have long been successfully used to estimate cropland areas thus evidencing the rapid agricultural expansion during the 2000s (Owusu et al., 2024, Teluguntla et al., 2023, Thenkabail et al., 2021, Anderson et al., 2003, Morton et al., 2006).

In Mato Grosso state, Arvor et al. (2012) estimated that net cropped areas increased by 43% between 2000 and 2007, reaching an area of 55,988 km². In the same time, farmers adopted new agricultural management practices to intensify the production process. The cultivation of two successive crops, as soybean and cotton, benefits from a long rainy season (Arvor et al., 2014) and regular rainfall from mid-September to late May. In this context, the Mato Grosso case study aims at producing a cropping system map showing the main crop type and the intensification practices in relation to the rainfall, and a land use system map to analyze the agricultural transition in Mato Grosso.

6.3.2.3 Rainfed Agriculture in Mali

In the Sudano-Sahelian region, farming is the main source of income for many people, where millet and sorghum are the main food crops. The vast majority of the population (80%) consists of subsistence farmers. A few larger farms produce crops for sale (cash crops), mainly cotton and peanuts. In the Sudano-Sahelian zone, the strong dependence on rainfed agriculture implies exposure to climate variability in addition to the impacts population growth have on food security. Key deliverables of food security systems for crop monitoring consist of early estimates of cultivated area and crop-type distribution, cropping practices, detection of growth anomalies, and crop yield estimates. Unfortunately, the national statistics can be deficient in insecure countries, and remote sensing has an important role to play in delivering information for crop monitoring (e.g., Owusu

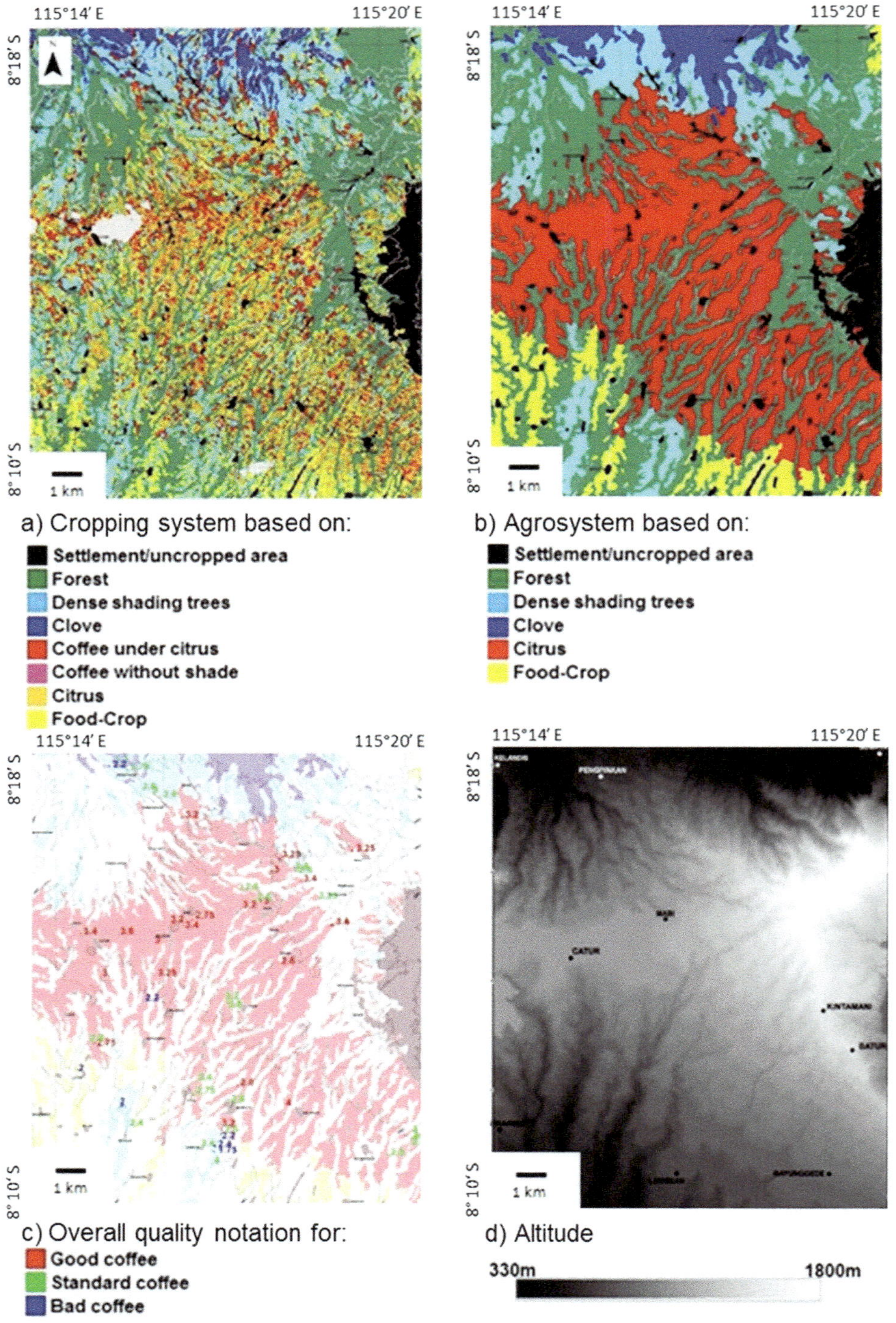

FIGURE 6.4 Map products in Central Bali: (a) main cropping systems map derived from QuickBird image visual interpretation, (b) agrosystems map derived from spatial analysis of the cropping system map, (c) location map of 40 sampled coffees and quality notation rate for each type of aromatic value, and (d) Digital Terrain Model derived from topographic maps.

et al., 2024, Teluguntla et al., 2023, Thenkabail et al., 2021, Hutchinson, 1991, Thenkabail et al., 2009, Begue et al., 2020). Remote sensing techniques face numerous challenges for crop mapping in regions where the cropland is fragmented, made of small, highly heterogeneous fields covered with many trees. In Mali, Vintrou et al. (2011) showed that 20% to 40% of cropland classification errors using MODIS is inherent to the structure of the landscape.

Southern Mali case study aims at producing farming system map (food-producing, intensive, and mixed agricultures) in support to food security analyses (USAID, 2009). Because local factors, such as climate, soil, water availability, access to markets, and fertilizers influence the agricultural systems, mapping these systems can help determine which region and which population may be vulnerable to different hazards. Additionally, the cropping system map can be used for spatialized agro-meteorological modeling and forecasting at regional scales (see example in Vintrou et al. (2014).

6.3.2 Remote Sensing Data and Methods

The data (remote sensing images, ancillary data) and methods used to produce agricultural maps (Owusu et al., 2024, Teluguntla et al., 2023, Thenkabail et al., 2021) are presented in Table 6.3 for the three case studies.

In Bali, a multispectral QuickBird image at 0.6 m resolution was photo-interpreted to delineate the field limits and identify six cropping systems based on the field survey: citrus monocrop, coffee monocrop without shade, coffee associated with light shadow (citrus), coffee under dense shadow (erythrina, albizias, leucaenas, etc.), clove crops associated or not with coffee, and food-crops. An agrosystem map was then obtained by applying a majority filter (1 ha square corresponding to a dozen of crop plots) on the cropping system map, and was defined by its upper vegetation layer in four classes: citrus, clove, dense shading trees, and food crops. The term *agrosystem* is preferred here to the term *farming system* whose definition goes beyond what is studied in this case.

In Mato Grosso, MOD13Q1 EVI products acquired from 2005 to 2008 period, were used to produce a cropping system map showing the main crop types (soybean, corn and cotton), and

TABLE 6.3
Typology, Data, and Methods Used to Produce Agricultural System Maps for the Three Case Studies

Case Study (Area)	Agriculture Type	Satellite Data (Acquisition Year)	Other Data	Method	Map Products
Bali island (224 km²)	Smallholder agriculture	QuickBird bundle (2003)	DEM 760 ground survey points	Photo interpretation Spatial analysis (majority filter; 1 ha window)	Cropping system Farming system (agrosystem)
Mato Grosso (906 000 km²)	Commercial agriculture	MOD13Q1 EVI product (2005–2008)		Pixel-based supervised classification Landscape analysis (land cover and land use classes metrics; 770 km² window)	Crop type Cropping system Farming system
Southern Mali (165 790 km²)	Smallholder agriculture	MOD13Q1 NDVI product (2007) MCD12Q2 phenology product (2007)	100 villages field survey (2001–2004) Cropland map at 250 m resolution Climate type, DEM, and population 4000 villages location	Texture analysis (MODIS NDVI) Landscape analysis (land cover classes metrics; 100 km² window) Random Forest classification	Farming system

their intensification practices (mono and double-cropping). Arvor et al. (2013a) used a landscape approach to better characterize the land use system across the state. The strategy consisted of applying a regular grid where each cell represented an approximation of a district territory (a district was considered as an administrative sub-level, below the municipality level). There were 1175 districts in Mato Grosso, a total of 906 000 km², and the grid cell was fixed at 25.75 × 27.75 km², approximating an area of 770 km². A set of landscape indices was then computed for each cell based on MODIS-based land use classifications and deforestation maps. Those indices referred to the proportion of wilderness areas, the proportion of cropped areas in deforested areas, and the proportion of intensive practices observed in cropped areas. Some thresholds were applied to identify different land use systems, such as pre-settlement area, non-cropland occupation, cropland occupation, non-cropland consolidation, cropland consolidation, non-cropland intensifying, cropland intensifying and intensive cropland.

In Mali, the field size and MODIS spatial resolution prevent from producing a crop type map. We then mapped directly the farming system map using a 3-class typology. This typology was defined at the village scale, and based on a field survey carried out in 100 villages in Southern Mali (Soumare, 2008). The typology was created using expert knowledge, and considering the main crop types cultivated in the village and the intensification of production (use of fertilizers, equipment, livestock, etc.): the "food-producing agriculture" class groups the millet and sorghum-based agricultural systems, the "intensive agriculture" class includes farms with maize and cotton, and the "mixed agriculture" class encompasses farms where both coarse grain (sorghum) and a cash crop (cotton) are found (Vintrou et al., 2012). A random forest algorithm (Breiman, 2001) was trained on the 100-village dataset, and on a set of 30 variables composed of four spectral metrics (annual maximum, annual mean, annual amplitude, and seasonal mean from May through November; MOD13Q1 product), 12 texture indices (maximum and mean of the variance and skewness indices, calculated with a pattern size of seven MODIS pixels for March, June, and September; MOD13Q1 product), seven phenology metrics (MCD12Q2 product), three spatial metrics (the fraction of cropped area, number of cultivated patches, and the mean cultivated patch size inside a 10 × 10 km² area centered on the village; MCD12Q1 product), three environmental indices (climate type, maximum and mean of elevation), and one population index. All of the indices were extracted for cropland only. The random forest model trained on the 100-village ground survey, was applied to the 4,000 villages in South Mali.

6.3.3 Results

6.3.3.1 Agroforestry in Bali

In Bali, the cropping system map is presented in Figure 6.4a. Photointerpretation performed on the ground-truth plots showed that confusion between citrus and coffee under citrus is less than 10%, whereas other class errors lie below 2%. The analysis of the distribution of each cropping system showed that the most frequent are the citrus-based crops (18%) and those shaded by large trees (15%), followed by the food-crops (12%), and the associated coffee and citrus crops (10%). The mean size of a plot is approximately 0.7 ha, but the clove plots are generally bigger (1.2 ha) and the food-crops are smaller (0.3 ha).

The agrosystem map is presented in Figure 6.4b. The citrus-based agrosystem is largely dominant. Coffee, as being cropped below the dominant trees, does not appear in the map legend.

At first glance, the cropping system and agrosystem spatial distribution looks complex because of a number of factors, such as a North/South contrast, altitude, and local geographic characteristics, such as river network density, slope, exposition to wind, and the presence of lava-flows and forests. The cropping and agrosystem maps were then used to analyze the distribution of each agricultural system, in relation to altimetry because of the strong relationship between coffee quality and altitude (Florinsky, 1998, Wintgens, 2004, Montagnon, 2006). The area covered by all of the different

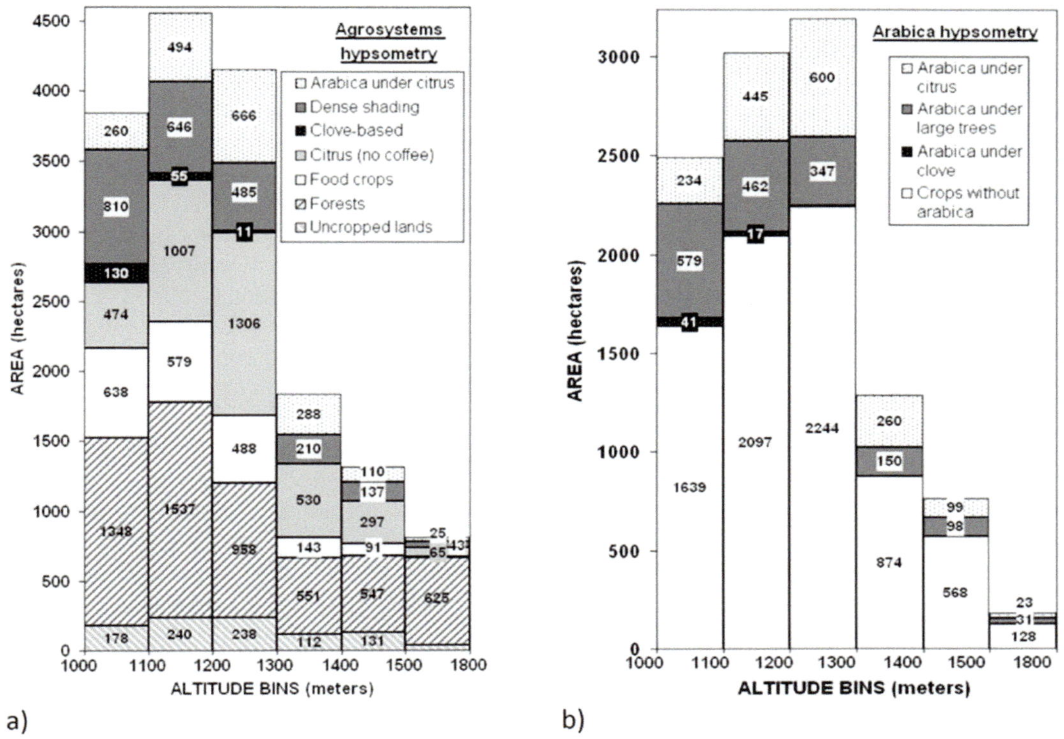

FIGURE 6.5 Areal altimetry distribution per bin of 100 m between 1,000 m and 1,800 m in Kintamani territory in Bali: (a) per cropping system class, and (b) per coffee-based cropping system class (Arabica monocrop is not displayed because it covers less than 2 ha).

cropping systems is plotted for each 100 m-altitude bin, between 1000 m and 1800 m in Figure 6.5a, while Figure 6.5b represents the altitude distribution for the area covered by the coffee-based cropping systems alone. The two principal coffee-based cropping systems were found to be those dominated by citrus or dense shading trees. The former is most common at high altitudes (64% from 1,200 m to 1,400 m), while the latter dominates coffee crops at lower altitudes (68% below 1,100 m). The third coffee-based cropping system, dominated by clove shading, covers a small acreage and is spatially restricted. It is present at the lowest altitudes, mainly below 1,100 m (68%) and 1,200 m (28%). The unshaded coffee monoculture is not typical in this territory.

The coffee samples location and sensorial quality rates were plotted in both the cropping and agrosystem maps to understand the spatial distribution of the coffee characteristics at the two scales (Figure 6.4c). A landscape analysis provided spatial and topographic distribution information about the three coffee quality classes, and helped to identify the relationships between quality of coffee beans and the local and regional environment. This integrated analysis suggests that good coffee is only found in the citrus-dominated agrosystem, even if it is not cultivated in association with citrus at the plot level, and cropped above 1200 m. This area was validated by both the coffee farmers and the traders, and accepted by the Indonesian government as the official limits of the labeled territory.

6.3.3.2 Double Cropping in Southern Amazon

Time series of vegetation indices were used to detect crop types and cropping practices using an analysis of agricultural calendars. The producers undertake two successive harvests per rainy season: they cultivate soybean from late September to early February, and then cultivate maize or cotton until June or July. The double cropping systems show very different patterns in their vegetation index time series and can be easily discriminated (Arvor et al., 2011). The user's and producer's accuracies of the cropland were higher than 95%. Main crop types were also correctly detected (Figure 6.6) with good Kappa index (0.68) and overall accuracy (74%). Once the double cropping classes are grouped (i.e., the "soybean + corn" and "soybean + cotton" classes; Figure 6.7a), the user's and producer's accuracies increased up to of 95% and 86%, respectively. The main uncertainties to be considered in these maps refer to sorghum or millet that is sometimes sown after the soybean harvest (to prevent soil erosion from intense rainfall) and can thus be confused with maize. Such issue highlights a main limitation of EVI time series-based classification (different crops with similar agricultural calendars may be confused) that could be overcome with a better spatial and radiometric resolution (since only Blue, Red and NIR bands are used to compute the Enhanced Vegetation Index used in that work).

Beyond such limitations, those results are in agreement with results obtained by different authors (Owusu et al., 2024, Jung et al., 2021, Thenkabail et al., 2021, Khanal et al., 2020, Tantalaki et al., 2019, Galford et al., 2008, Arvor et al., 2011, Brown et al., 2013) who successfully mapped double cropping systems in Mato Grosso and confirmed the generalization of such intensive practices. Arvor et al. (2012) estimated that the proportion of croplands permanently covered by double-cropping vegetation during the rainy season increased from 35% to 62% between 2000 and 2007. This trend raises a major issue regarding the sustainability of cropland systems in Mato Grosso. Fu et al. (2013) proved that the length of the rainy season is decreasing in the southern Amazon, which

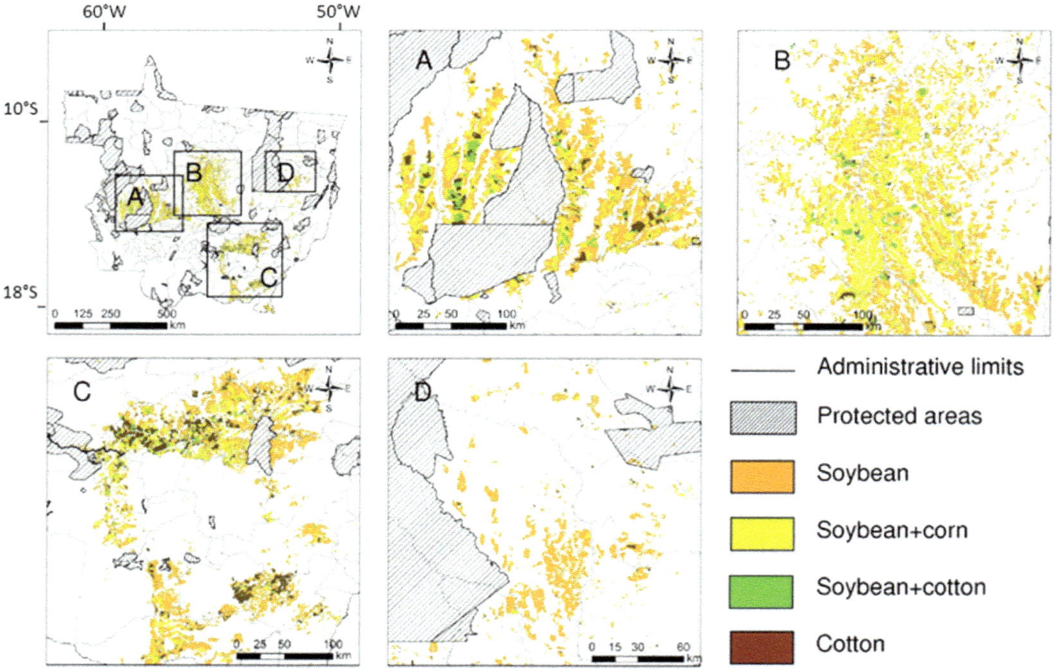

FIGURE 6.6 Maps of the three main crop types (soybean, corn, and cotton) for the 2006–2007 harvest for the four main agricultural regions in Mato Grosso: (a) Parecis plateau, (b) along the BR163 highway, (c) southeastern region, (d) eastern region (Arvor et al., 2011). Maps were obtained through supervised classification of MODIS vegetation index (EVI) time series.

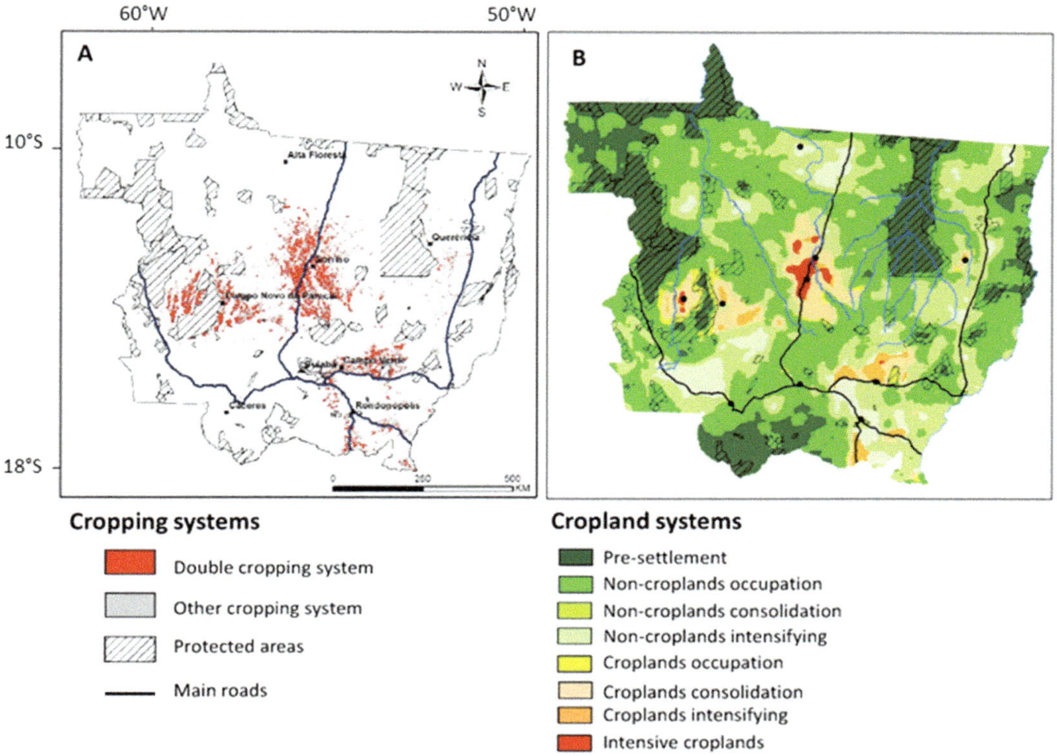

FIGURE 6.7 Maps of (a) cropping systems and (b) land use systems obtained from MODIS vegetation index time series and landscape analysis for the 2006–2007 harvest. (From Arvor et al., 2012, 2013a.)

leads to the question of whether the adoption of double cropping practices would still be viable in the changing climate. Even if intensive practices are a relevant strategy to contain deforestation, it raises new issues regarding agricultural sustainability in that region.

The land use system map shows a good overview of the soybean agricultural frontier in the southeastern Amazon (Figure 6.7b). It demonstrates the efficiency of public policies to simultaneously contend deforestation (through the creation of protected areas) and encourage crop expansion (through the construction of important infrastructures, such as the Transamazonian roads).

6.3.3.3 Rainfed Agriculture in Mali

The Random Forest model classified the agricultural systems with an estimated overall accuracy of 60% calculated from Out-of-Bag observations (Figure 6.8). The "food-producing agriculture" class was dominant in the Sudano-Sahelian part of the area. Sorghum and millet are well adapted to this zone because they are resistant, and have a short growth cycle of about 90 days. In the traditional cotton basin, the dominant system is agro-forestry/pastoral agriculture mainly with rainfed crops. Agriculture is focused on cotton, the main cash crop, and corresponds to the class "intensive agriculture." The Sudanian zone part of the area is also a cotton-based system zone, but is more diversified, with the simultaneous presence of "intensive agriculture" and "mixed agriculture" systems. The length of the rainy season in this region makes it possible to grow a wide range of species. Farmers usually cultivate different species and varieties to ensure a certain degree of production stability.

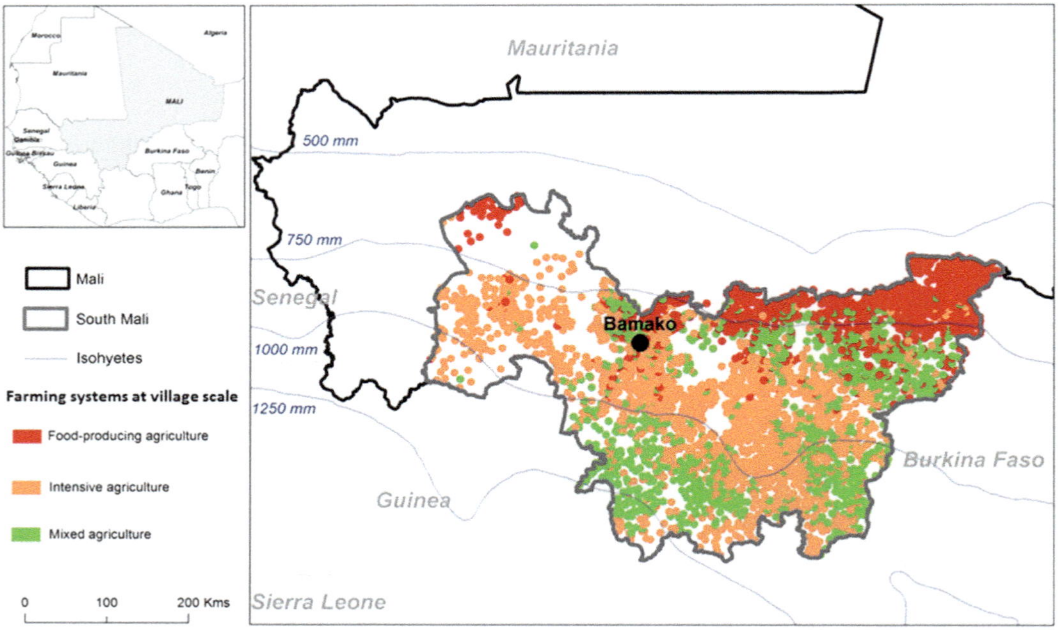

FIGURE 6.8 Village-based farming systems in South Mali predicted by the Random Forest model (Vintrou et al., 2012). The model was based on 100 village samples, and 30 MODIS-derived and socio-environmental metrics calculated on agricultural areas.

Class errors ranged from 30% to 50%. Globally, producer's and user's accuracies were reasonably balanced for each class (less than 10% difference), meaning that the village agricultural systems were estimated correctly. Misclassifications can be explained by three main factors: (1) the small size of the crop patches compared to the 250-m spatial resolution of MODIS sensor, and the natural and crop vegetation seasonal synchronization due to a short rainy season; (2) the size of the training dataset (100 villages); and (3) the definition of the classes (a rough proportion of different crop types, and crop intensification variables) that is expert-dependent and includes variables that cannot be directly related to landscape features.

The analysis of the contribution of the different metrics (Figure 6.9) shows the role of the texture of the MODIS images in the classification of the cropland, even if the fields are not visible at the MODIS resolution. The field crop information is hidden in these broad images, but can be identified with landscape metrics, such as image texture indices. This indirect analysis was confirmed by Bisquert et al. (2014) who showed that the texture of broad-scale images is an important variable for land stratification in relation to land cover, even if the land cover units are not detectable.

6.4 DISCUSSION

While remote sensing approaches have proven to be efficient for cropland (land cover) mapping, they still remain ill-suited for cropping system (land use) monitoring at the regional and global scales because of their inability to distinguish crop types and the associated practices (Monfreda et al., 2008). In this section, we consider the main present limitations of remote sensing studies for regional mapping of cropping systems, and introduce some emerging research areas to overcome such limitations (Victor et al., 2024, Omia et al., 2023, Thenkabail et al., 2021, Hatab et al., 2019, Khanal et al., 2020, Tantalaki et al., 2019, Raymond et al., 2018, Jones et al., 2017, Steele-Dunne et al., 2017, Lemettais et al., 2024). We then discuss the opportunity to work on an extended landscape agronomy approach.

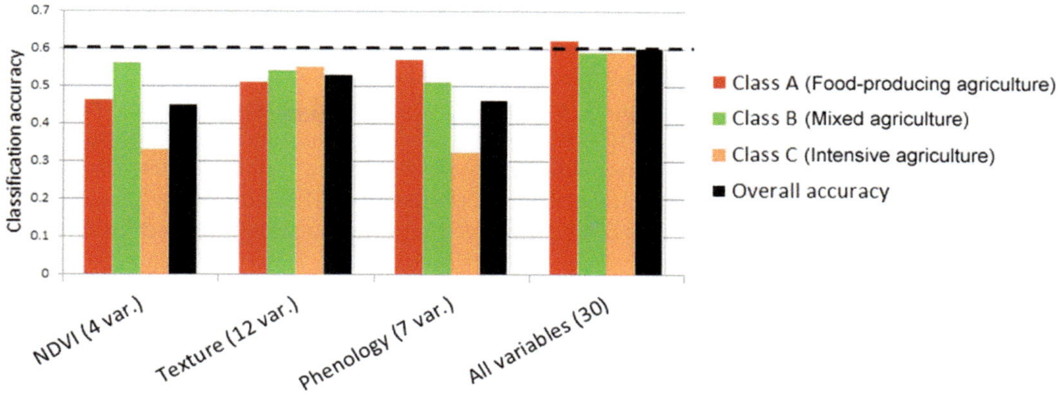

FIGURE 6.9 Accuracy of class and overall classification of Random Forest run with different sets of metrics (NDVI, texture, and phenology metrics). The dotted line corresponds to the overall accuracy obtained with all the metrics.

6.4.1 Difficulties of Mapping the Cropping Systems at Regional Scales

Remote sensing-based land use maps suffer from uncertainties related to the spatial and temporal resolutions of the observing system, and to the landscape structure.

The spatial resolution issue is particularly true for smallholders agriculture (Figure 6.1b), for which remote sensing data are unable to resolve individual fields (Kumar et al., 2022, Jung et al., 2021, Karthikeyan et al., 2020, Khanal et al., 2020, Raymond et al., 2018, Jones et al., 2017, Ozdogan, 2010). Rather than a sensor resolution issue, it should actually be considered as a scale issue to be addressed through the concept of H-resolution and L-resolution (Strahler et al., 1986, Blaschke et al., 2014). H- and L-resolution terms are different from high and low spatial resolution images as generally mentioned in remote sensing studies. In the latter, the resolution refers to the sensor spatial resolution independently of the geographic objects concerned. H-resolution model is valid when scene objects are much larger than the image spatial resolution, thus several pixels may represent a single object (a field, a tree, etc.). Meanwhile, L-resolution model is when objects are much smaller than the image spatial resolution. An image may contain both H- and L-resolution information (Hay et al., 2001). Marceau et al. (1994) place the limit between H- and L- when the dimension of the resolution cells is ½ to ¾ the size of the objects of interest in the scene. This threshold should be a guide for assessing whether the analysis should be performed at H- or L-resolution.

- For an H-resolution situation—agricultural fields in Mato Grosso using MODIS sensor—a cropping system can be assessed directly by characterizing crop types and their associated cropping practices using inner field information (derived from relatively pure pixels).
- For an L-resolution situation—cropped trees in Bali using QuickBird sensor or cropped fields in Mali using MODIS sensor—pixels correspond to a mixture of different crop (or trees) types and other landscape elements (natural vegetation, water bodies, buildings, roads, etc.).

The temporal resolution issue in crop mapping is highly dependent on the environmental and agronomic conditions. For example, in tropical dry areas where rainfall is the main driver of vegetation growth (e.g., the Sahelian part of Mali), natural and cultivated vegetation are difficult to separate using phenology. In equatorial areas (e.g., Bali) characterized by a low seasonality, it is difficult to discriminate crops due to fluctuating crop calendars. However, even in regions with contrasted seasons (e.g., Mato-Grosso), different cropping systems with similar agricultural calendars cannot be separated using MODIS EVI time series. A better temporal resolution (less than 16 days) would surely improve crop discrimination in most of the agricultural systems.

The quality of the land maps produced by image pixel-based classification, is usually evaluated using a set of indices (producer's accuracy, user's accuracy, overall accuracy and Kappa index) which are commonly calculated from an error matrix (or confusion matrix; see Congalton and Green, 1999). While such accuracy metrics have been widely accepted by the scientific community for a long time, they have also been regularly criticized (Pontius and Millones, 2011). These metrics tell nothing about the source of error that can be linked to the performance of the classification algorithm, or to the resolution of the remotely sensed data (Boschetti et al., 2004). For instance, Vintrou et al. (2011) using the Pareto boundary method showed that in Mali, 20% to 40% of cropland classification errors using MODIS data is inherent to the landscape structure. In this context, new processing and evaluation approaches are required to better consider landscape properties in order to overcome these limitations and allow an efficient monitoring of farming systems at regional scale.

6.4.2 Emerging Remote Sensing Research

There was a challenge in land cover mapping in the 2000s, and today, there is a challenge in land use system mapping (Owusu et al., 2024, Teluguntla et al., 2023, Thenkabail et al., 2021, Tantalaki et al., 2019, Huang et al., 2018, Raymond et al., 2018). It is an emerging area for the remote sensing community that needs to focus on land use and land function (Verburg et al., 2009). It requires developing new data, methods, and a further integration of the disciplines involved in land science research. These developments are presented hereafter according to the resolution situation (the direct and direct cases).

When the landscape elements are larger than the pixel size (H-resolution situation), many examples in the literature showed that cropping practices can be directly assessed (Table 6.2). Except for rare examples of mapping crop type and cropping intensity in regions where the size of the plots is compatible with broad scale sensors (Mato Grosso case study), the research was mainly developed at local scale, and for one practice at a time. To further characterize regional scale cropping practices, research needs to focus on developing automatic or semi-automatic crop type classification procedures, and on the combination of different sensors to catch different practices in the same area. Another way to work at broader scales is to properly translate local findings to larger regions by using case study results from specific land functions (Verburg et al., 2009). This approach needs to define the spatial extent and function for the local studies representative of a region. Land stratification into homogeneous landscape units could be a way to reach this objective (Lemettais et al., 2024). Bisquert et al. (2014) showed that processing broad-scale remote sensing data with spectral and textural segmentation techniques permit to delineate radiometrically homogeneous landscape that were consistent in terms of land cover.

When the landscape elements are smaller than the pixel size (L-resolution situation), research needs to focus on the role of landscape as a mean to characterize the cropping systems. Research on landscape metrics for agricultural systems characterization must be pursued and enhanced. Furthermore, given the multidimensional nature of agricultural systems, focusing on multiple metrics within a systems perspective is needed (Kuemmerle et al., 2013). As the current approaches based on the remote sensing data are not sufficient to develop a comprehensive understanding of situational changes for multiple land functions, remote sensing-based metrics should be completed by other types of metrics, such as socio-economic descriptors (demography, ethnic spatialized data, etc.). To merge heterogeneous information, new data processing tools, such as fuzzy logic and data mining tools (Vintrou et al., 2013, Korting et al., 2013), must be tested to characterize and map agricultural systems and processes.

To implement both approaches (direct and indirect), the scientific community should benefit from recent promising advances in remote sensing such as GEographic Object-Based Image Analysis (GEOBIA) and ontologies. GEOBIA is based on the hypothesis that partitioning an image into objects is related to the way humans conceptually organize the landscape to comprehend it (Hay and Castilla, 2008). It is actually based on two main components. First, a segmentation delineates regions (objects) of the image that have common attributes. Second, the approach incorporates the user (expert) knowledge in the image processing operation to produce reliable maps. However, to date, GEOBIA is still limited by important issues related to product evaluation and knowledge management. Indeed, it is still unclear how to assess a segmentation quality (actually considered as an ill-posed problem), although Clinton et al. (2010) proposed interesting metrics to assess GEOBIA segmentation goodness through vector-based measures. Although the integration of knowledge expertise in the image interpretation process is a main strength of GEOBIA, it can also be considered as a main limitation as long as two experts do not share a consensual knowledge (Belgiu et al., 2014). In such a context, it is likely that knowledge representation techniques such as ontologies can play a pivotal role (Arvor et al., 2013b). This point is especially meaningful in the case of agricultural system mapping where expert knowledge is crucial and often difficult to formalize. In case of land cover products, Comber et al. (2005) investigated the semantic and ontological meanings of land cover classes and concluded that current paradigms for reporting data quality do not adequately communicate the producer's knowledge. In case of land use and land use system products, the ontological meaning of the classes is even more difficult to formalize. For example, agricultural practices such as double cropping or no-tillage have been studied in various regions of the world although they might correspond to different practices on the ground (different types of crop, different levels of soil management). In conclusion, ontologies might play an important role to allow comparison of complex and heterogeneous land maps.

6.4.3 Towards an Extended Landscape Agronomy Approach

Landscape and agronomy have long been considered as closely associated (Kumar et al., 2022, Karthikeyan et al., 2020, Segarra et al., 2020, Weiss et al., 2020, Hatab et al., 2019, Tantalaki et al., 2019, Huang et al., 2018, Raymond et al., 2018, Jones et al., 2017, Steele-Dunne et al., 2017). The first references on the relationship between agricultural landscapes and field management appeared in the 1990s (e.g., Baudry, 1993, Deffontaines et al., 1995) and addressed how farming activities produce agricultural landscapes, that is, explained the spatial distribution of patches (fields and associated boundaries). Since then, very few studies were published on the relationship between agricultural practices and landscape properties (e.g., Galli et al., 2010, Herzog et al., 2006). Most of the research focused on the characterization and understanding of landscape patterns to relate them to ecological issues (e.g., Baudry, 1993, Herzog et al., 2006). Benoit et al. (2012) argued why and how agronomy can contribute to landscape research with a conceptual model. He suggested a new perspective on farming practices as a crucial driver in the landscape pattern-agricultural process relationship. He proposed to develop a new research area called Landscape Agronomy (see also Rizzo et al., 2013) defined as "the relations among farming practices, natural resources and landscape patterns, which are involved in the dynamics of agricultural landscapes."

We previously mentioned that few landscape studies related to agricultural issues use remote sensing. Although it is now widely understood that cropping practices adopted in agricultural systems shape rural landscapes, we believe it is time to use landscape agronomy and quantitative remote sensing sciences. Applying concepts of landscape ecology to agricultural systems monitoring and mapping is a major idea. The case studies from Bali and Mato Grosso illustrate this new trend in landscape agronomy research and show that, thanks to its ability to identify spatial land cover patterns at local (Bali) and regional (Mato-Grosso) scales, remote sensing has become an essential source of information to identify agricultural systems.

However, landscape agronomy research will have to face the same limitations as landscape research. These limitations concern the numerous sources of error or uncertainty with producing land cover/land use maps from remote sensing imagery, and on the choice of the landscape metrics which need to show a close association with the processes to be detected (Newton et al., 2009, Hurni et al., 2013). Another source of limitation is the simplistic approach of thematic mapping and the derivation of two-dimensional pattern metrics in landscape ecology (Newton et al., 2009), while remote sensing data have the potential to provide a three-dimensional characterization of landscapes and their component parts (as seen in Bali study case) and quantitative surface variables (as seen in Mali study case) that could be directly integrated in the landscape analysis. We showed through the Mali study case that the agricultural landscapes could be indirectly characterized by using a set of satellite-derived metrics (spectral, textural, temporal metrics) without going through a thematic map of the crop types. This approach is essential when the ratio between the field size and the sensor spatial resolution is low (L-resolution), meaning land use maps cannot be produced, but it can also be used in H-resolution situation.

6.5 CONCLUSIONS

It is widely recognized that accurate, updated, and spatially explicit information on cropping systems (and thus cropping intensity) is urgently needed at the global and regional scales to provide insight into the direction and magnitude of world agricultural production in terms of crop type acreage and yield (Owusu et al., 2024, Rozenstein et al., 2024, Victor et al., 2024, Omia et al., 2023, Teluguntla et al., 2023, Kumar et al., 2022, Mahlayeye et al., 2022, Jung et al., 2021, Thenkabail et al., 2021, Karthikeyan et al., 2020, Segarra et al., 2020, Sishodia et al., 2020, Weiss et al., 2020, Hatab et al., 2019, Khanal et al., 2020, Tantalaki et al., 2019, Huang et al., 2018, Raymond et al., 2018, Jones et al., 2017, Steele-Dunne et al., 2017, Lobell and Field, 2007), and in terms of agricultural impacts on natural environments (Galford et al., 2008) and water resources (Thenkabail et al., 2021, 2010). Additionally, information is needed locally to monitor resources, preserve cultural landscapes and for land certification (Jouve, 2006). This information is not yet included in the regional land cover datasets, and remote sensing entirely overlooks the actual practice of agriculture (what is grown, how it is grown, what inputs are used) at this scale (Monfreda et al., 2008, Begue et al., 2018).

In this paper, we showed how the current generation of Earth observation systems can contribute to the characterization of agricultural systems locally and regionally, through bibliographic studies and three case studies. We showed that the remote sensing ability to describe cropping systems is mainly related to the ratio between the spatial resolution of the sensor and the size of the landscape elements. This ratio determines if the fields (or the trees) can be identified by the observation system, or if the remote sensing data offers only a view of the cropland in its environment. This latter case leads to the development of new tools and methods to indirectly connect the spatial patterns of the agricultural landscape to the cropping management practices over large territories.

This bibliographic overview shows that the research community is now at a turning point where landscape research is not devoted to ecological issues only, but has started to embrace agricultural matters also. We believe that landscape agronomy is on the right track, and that the current and future Earth observing systems (such as Sentinel-2 or Landsat8) will have an important role to play in this new research area.

REFERENCES

Adami, M., Mello, M.P., Aguiar, D.A., Rudorff, B.F.T., Souza, A.F., 2012. A web platform development to perform thematic accuracy assessment of sugarcane mapping in South-Central Brazil. Remote Sensing, 4, 3201–3214.

Aksoy, S., Yalniz, I.Z., Tasdemir, K., 2012. Automatic detection and segmentation of orchards using very high resolution imagery. Ieee Transactions on Geoscience and Remote Sensing, 50, 3117–3131.

Anderson, J.R., Hardy, E.E., Roach, J.T., Witmer, R.E., 1976. A Land Use and Land Cover Classification System for Use with Remote Sensor Data. United States Government Printing Office, Washington, US, 41.

Anderson, L.O., Rojas, E., Shimabukuro, Y., 2003. Avanço da soja sobre os ecossistemas cerrado e floresta no Estado do Mato Grosso. XI Simpósio Brasileiro de Sensoriamento Remoto, Belo Horizonte, Brésil, 19–25.

Anderson, W., You, L.Z., Wood, S., Wood-Sichra, U., Wu, W., 2014. A Comparative Analysis of Global Cropping Systems Models and Maps. IFPRI, Washington DC (US), 33.

Arvor, D., Dubreuil, V., Ronchail, J., Simões, M., Funatsu, B.M., 2014. Spatial patterns of rainfall regimes related to levels of double cropping agriculture systems in Mato Grosso (Brazil). International Journal of Climatology, 34, 2622–2633.

Arvor, D., Dubreuil, V., Simões, M., Bégué, A., 2013a. Mapping and spatial analysis of the soybean agricultural frontier in Mato Grosso, Brazil, using remote sensing data. GeoJournal, 78, 833–860.

Arvor, D., Durieux, L., Andres, S., Laporte, M.A., 2013b. Advances in Geographic Object-Based Image Analysis with ontologies: A review of main contributions and limitations from a remote sensing perspective. ISPRS Journal of Photogrammetry and Remote Sensing, 82, 125–137.

Arvor, D., Jonathan, M., Meirelles, M.S.P., Dubreuil, V., Durieux, L., 2011. Classification of MODIS EVI time series for crop mapping in the state of Mato Grosso, Brazil. International Journal of Remote Sensing, 32, 7847–7871.

Arvor, D., Meirelles, M., Dubreuil, V., Begue, A., Shimabukuro, Y.E., 2012. Analyzing the agricultural transition in Mato Grosso, Brazil, using satellite-derived indices. Applied Geography, 32, 702–713.

Atzberger, C., Rembold, F., 2013. Mapping the spatial distribution of winter crops at sub-pixel level using AVHRR NDVI time series and neural nets. Remote Sensing, 5, 1335–1354.

Baudry, J., 1993. Landscape Dynamics and Farming Systems: Problems of Relating Patternes and Predicting Ecological Changes. Lewis Publishers Inc, Boca Raton.

Begue, A., Arvor, D., Bellon, B., Betbeder, J., de Abelleyra, D., Ferraz, R.P.D., Lebourgeois, V., Lelong, C., Simões, M., Veron, S., 2018. Remote sensing and cropping practices: A review. Remote Sensing, 10(1), 99, https://doi.org/10.3390/rs10010099.

Begue, A., Leroux, L., Soumaré, M., Faure, J.-F., Diouf, A.A., Augusseau, X., Touré, L., Tonneau, J.-P., 2020. Remote sensing products and services in support of agricultural public policies in Africa: Overview and challenges. Frontiers in Sustainable Food Systems, 4, 58, https://doi.org/10.3389/fsufs.2020.00058.

Belgiu, M., Hofer, B., Hofmann, P., 2014. Coupling formalized knowledge bases with object-based image analysis. Remote Sensing Letters, 5, 530–538.

Bellon, B., Begue, A., Lo Seen, D., Lebourgeois, V., Evangelista, B.A., Simões, M., Ferraz, R.P.D., 2018. Improved regional-scale Brazilian cropping system's mapping based on a semi-automatic object-based clustering approach. International Journal of Applied Earth Observation and Geoinformation, 68, 127–138, https://doi.org/10.1016/j.jag.2018.01.019.

Benoit, M., Rizzo, D., Marraccini, E., Moonen, A.C., Galli, M., Lardon, S., Rapey, H., Thenail, C., Bonari, E., 2012. Landscape agronomy: A new field for addressing agricultural landscape dynamics. Landscape Ecology, 27, 1385–1394.

Bisquert, M., Bégué, A., Deshayes, M., 2014. A methodology for delineating landscapes patches at the regional scale using OBIA techniques applied to MODIS time series of vegetation and texture indices. International Journal of Applied Earth Observation and Geoinformation (in press).

Blaschke, T., Hay, G.J., Kelly, M., Lang, S., Hofmann, P., Addink, E., Queiroz Feitosa, R., van der Meer, F., van der Werff, H., van Coillie, F., Tiede, D., 2014. Geographic object-based image analysis: Towards a new paradigm. ISPRS Journal of Photogrammetry and Remote Sensing, 87, 180–191.

Boschetti, L., Flasse, S.P., Brivio, P.A., 2004. Analysis of the conflict between omission and commission in low spatial resolution dichotomic thematic products: The Pareto boundary. Remote Sensing of Environment, 91, 280–292.

Breiman, L., 2001. Random forests. Machine Learning, 45, 5–32.

Bridhikitti, A., Overcamp, T.J., 2012. Estimation of Southeast Asian rice paddy areas with different ecosystems from moderate-resolution satellite imagery. Agriculture Ecosystems & Environment, 146, 113–120.

Brown, J.C., Kastens, J.H., Coutinho, A.C., Victoria, D.D., Bishop, C.R., 2013. Classifying multiyear agricultural land use data from Mato Grosso using time-series MODIS vegetation index data. Remote Sensing of Environment, 130, 39–50.

Burley, T.M., 1961. Land use or land utilization? The Professionnal Geographer, 13, 18–20.

Clinton, N., Holt, A., Scraborough, J., Yan, L., Gong, P., 2010. Accuracy assessment measures for object-based image segmentation goodness. Photogrammetric Engineering and Remote Sensing, 76, 289–299.

Colson, F., Bogaert, J., Ceulemans, R., 2011. Fragmentation in the Legal Amazon, Brazil: Can landscape metrics indicate agricultural policy differences? Ecological Indicators, 11, 1467–1471.

Coltri, P.P., Zullo, J., Goncalves, R.R.D., Romani, L.A.S., Pinto, H.S., 2013. Coffee crop's biomass and carbon stock estimation with usage of high resolution satellites images. Ieee Journal of Selected Topics in Applied Earth Observations and Remote Sensing, 6, 1786–1795.

Comber, A.J., Fisher, P.F., Wadsworth, R.A., 2005. You know what land cover is but does anyone else? . . . an investigation into semantic and ontological confusion. International Journal of Remote Sensing, 26, 223–228.

Congalton, R.G., Green, K., 1999. Assessing the Accuracy of Remotely Sensed Data: Principles and Practices. Boca Raton, FL, 137.

Deffontaines, J.P., Thenail, C., Baudry, J., 1995. Agricultural systems and landscape patterns: How can we build a relationship ? Landscape and Urban Planning, 31, 3–10.

DeFries, R., Rosenzweig, C., 2010. Toward a whole-landscape approach for sustainable land use in the tropics. Proceedings of the National Academy of Sciences of the United States of America, 107, 19627–19632.

Dixon, J., Gulliver, A., Gibbon, D., 2001. Farming Systems and Poverty: Improving farmers' livehoods in a changing world. FAO and World Bank, Rome and Washington, DC. http://documents.worldbank.org/curated/en/126251468331211716/Farming-systems-and-poverty-improving-farmers-livelihoods-in-a-changing-world

FAO, 2011. Looking Ahead in World Food and Agriculture: Perspectives to 2050. FAO, Rome (IT), 539.

Florinsky, I.V., 1998. Combined analysis of digital terrain models and remotely sensed data in landscape investigations. Progress in Physical Geography, 22, 33–60.

Fox, J., Kanter, R., Yarnasarn, S., Ekasingh, M., Jones, R., 1994. Farmer decision-making and spatial variables in northern Thailand. Environmental Management, 18, 391–399.

Fu, R., Yin, L., Li, W.H., Arias, P.A., Dickinson, R.E., Huang, L., Chakraborty, S., Fernandes, K., Liebmann, B., Fisher, R., Myneni, R.B., 2013. Increased dry-season length over southern Amazonia in recent decades and its implication for future climate projection. Proceedings of the National Academy of Sciences of the United States of America, 110, 18110–18115.

Galford, G.L., Mustard, J.F., Melillo, J., Gendrin, A., Cerri, C.C., Cerri, C.E.P., 2008. Wavelet analysis of MODIS time series to detect expansion and intensification of row-crop agriculture in Brazil. Remote Sensing of Environment, 112, 576–587.

Galli, M., Bonari, E., Marraccini, E., Debolini, M., 2010. Characterisation of agri-landscape systems at a regional level: A case study in northern tuscany. Italian Journal of Agronomy, 5, 285–294.

Gobin, A., Campling, P., Deckers, J., Feyen, J., 2001. Integrated land resources analysis with an application to Ikem (south-eastern Nigeria). Landscape and Urban Planning, 52, 95–109.

Guillen-Climent, M.L., Zarco-Tejada, P.J., Villalobos, F.J., 2014. Estimating radiation interception in heterogeneous orchards using high spatial resolution airborne imagery. IEEE Geoscience and Remote Sensing Letters, 11, 579–583.

Hatab, A.A., Cavinato, M.E.G., Lindemer, A., Lagerkvist, C., 2019. Urban sprawl, food security and agricultural systems in developing countries: A systematic review of the literature. Cities, 94, 129–142, ISSN 0264–2751, https://doi.org/10.1016/j.cities.2019.06.001. (www.sciencedirect.com/science/article/pii/S0264275118310485).

Hay, G.J., Castilla, G., 2008. Geographic object-based image analysis (GEOBIA): A new name for a new discipline. In: Blaschke, T., Lang, S., Hay, G.J. (Eds.), Object-Based Image Analysis, Springer, Berlin, Heidelberg, 75–89.

Hay, G.J., Marceau, D., Dube, P., Bouchard, A., 2001. A multiscale framework for landscape analysis: Object-specific analysis and upscaling. Landscape Ecology, 16, 471–490.

Herzog, F., Steiner, B., Bailey, D., Baudry, J., Billeter, R., Bukácek, R., De Blust, G., De Cock, R., Dirksen, J., Dormann, C.F., De Filippi, R., Frossard, E., Liira, J., Schmidt, T., Stöckli, R., Thenail, C., van Wingerden, W., Bugter, R., 2006. Assessing the intensity of temperate European agriculture at the landscape scale. European Journal of Agronomy, 24, 165–181.

Huang, Y., Chen, Z., Yu, T., Huang, X., Gu, X., 2018. Agricultural remote sensing big data: Management and applications. Journal of Integrative Agriculture, 17(9), 1915–1931, ISSN 2095–3119, https://doi.org/10.1016/S2095-3119(17)61859-8. (www.sciencedirect.com/science/article/pii/S2095311917618598).

Hurni, K., Hett, C., Epprecht, M., Messerli, P., Heinimann, A., 2013. A texture-based land cover classification for the delineation of a shifting cultivation landscape in the Lao PDR using landscape metrics. Remote Sensing, 5, 3377–3396.

Hutchinson, C.F., 1991. Uses of satellite data for famine early warning in sub-Saharan Africa. International Journal of Remote Sensing, 12, 1405–1421.

Jasinski, E., Morton, D., DeFries, R., Shimabukuro, Y., Anderson, L., Hansen, M., 2005. Physical landscape correlates of the expansion of mechanized agriculture in Mato Grosso, Brazil. Earth Interactions, 9.

Jones, J.W., Antle, J.M., Basso, B., Boote, K.J., Conant, R.T., Foster, I., Godfray, H.C.J., Herrero, M., Howitt, R.E., Janssen, S., Keating, B.A., Munoz-Carpena, R., Porter, C.H., Rosenzweig, C., Wheeler, T.R., 2017. Brief history of agricultural systems modeling. Agricultural Systems, 155, 240–254, ISSN 0308–521X, https://doi.org/10.1016/j.agsy.2016.05.014. (www.sciencedirect.com/science/article/pii/S0308521X16301585).

Jouve, P., 2006. Cropping systems and farming land space organisation: A comparison between temperate and tropical farming systems. Cahiers Agricultures, 15, 255–260.

Jung, J., Maeda, M., Chang, A., Bhandari, M., Ashapure, A., Landivar-Bowles, J., 2021. The potential of remote sensing and artificial intelligence as tools to improve the resilience of agriculture production systems. Current Opinion in Biotechnology, 70, 15–22, ISSN 0958–1669, https://doi.org/10.1016/j.copbio.2020.09.003. (www.sciencedirect.com/science/article/pii/S0958166920301257).

Karthikeyan, L., Chawla, L., Mishra, A.K., 2020. A review of remote sensing applications in agriculture for food security: Crop growth and yield, irrigation, and crop losses. Journal of Hydrology, 586, 124905, ISSN 0022–1694, https://doi.org/10.1016/j.jhydrol.2020.124905. (www.sciencedirect.com/science/article/pii/S0022169420303656).

Khanal, S., Kushal, K.C., Fulton, J.P., Shearer, S., Ozkan, E., 2020. Remote sensing in agriculture: Accomplishments, limitations, and opportunities. Remote Sensing, 12(22), 3783. https://doi.org/10.3390/rs12223783.

Korting, T.S., Fonseca, L.M.G., Camara, G., 2013. GeoDMA-geographic data mining analyst. Computers & Geosciences, 57, 133–145.

Koschke, L., Fürst, C., Lorenz, M., Witt, A., Frank, S., Makeschin, F., 2013. The integration of crop rotation and tillage practices in the assessment of ecosystem services provision at the regional scale. Ecological Indicators, 32, 157–171.

Kuemmerle, T., Erb, K., Meyfroidt, P., Muller, D., Verburg, P.H., Estel, S., Haberl, H., Hostert, P., Jepsen, M.R., Kastner, T., Levers, C., Lindner, M., Plutzar, C., Verkerk, P.J., van der Zanden, E.H., Reenberg, A., 2013. Challenges and opportunities in mapping land use intensity globally. Current Opinion in Environmental Sustainability, 5, 484–493.

Kumar, S., Meena, R.S., Sheoran, S., Jangir, C.K., Jhariya, M.K., Banerjee, A., Raj, A., 2022. Chapter 5: Remote sensing for agriculture and resource management. In: Jhariya, M.K., Meena, R.S., Banerjee, A., Meena, S.N. (Eds.), Natural Resources Conservation and Advances for Sustainability, Elsevier, 91–135, ISBN 9780128229767, https://doi.org/10.1016/B978-0-12-822976-7.00012-0. (www.sciencedirect.com/science/article/pii/B9780128229767000120).

Kunwar, P.K.T.S., Kumar, A., Agrawal, A.K., Singh, A.N., Mendiratta, N., 2010. Use of high-resolution IKONOS data and GIS technique for transformation of landuse/landcover for sustainable development. Current Science, 98, 204–212.

Lambin, E.F., Geist, H.J., Rindfass, R.R., 2006. Introduction: Local processes with global impacts. In: Lambin, E.F., Geist, H.J. (Eds.), Land-Use and Land-Cover Change: Local Processes and Global Impacts. Springer-Verlag, Berlin, Heidelberg, 1–8.

Leenhardt, D., Angevin, F., Biarnes, A., Colbach, N., Mignolet, C., 2010. Describing and locating cropping systems on a regional scale: A review. Agronomy for Sustainable Development, 30, 131–138.

Lei, Z., Bingfang, W., Liang, Z., Peng, W., 2012. Patterns and driving forces of cropland changes in the Three Gorges Area, China. Regional Environmental Change, 12, 765–776.

Lemettais, L., Alleaume, S., Luque, S., Laques, A.-E., Alim, Y., Demagistri, L., Bégué, A., 2024. Radiometric landscape: A new concept and operational approach for landscape characterisation and mapping. Geo-Spatial Information Science, 1–23, https://doi.org/10.1080/10095020.2024.2314558.

Lobell, D.B., Field, C.B., 2007. Global scale climate-crop yield relationships and the impacts of recent warming. Environmental Research Letters, 2, 014002, 7.

Mahlayeye, M., Darvishzadeh, R., Nelson, A., 2022. Cropping patterns of annual crops: A remote sensing review. Remote Sensing, 14(10), 2404. https://doi.org/10.3390/rs14102404.

Marceau, D.J., Howarth, P.J., Gratton, D.J., 1994. Remote-sening and the measurement of geographical entities in a forested environment: 1. The scale and spatial aggregation problem. Remote Sensing of Environment, 49, 93–104.

McGarigal, K., 2014. FRAGSTATS: Spatial Pattern Analysis Program for Categorical Maps-Documentation.

Monfreda, C., Ramankutty, N., Foley, J.A., 2008. Farming the planet: 2. Geographic distribution of crop areas, yields, physiological types, and net primary production in the year 2000. Global Biogeochemical Cycles, 22.

Montagnon, C., 2006. Coffee: Terroirs and Quality. Cemagref/INRA/CIRAD, Versailles, France.

Morton, D., DeFries, R., Shimabukuro, Y., Anderson, L., Arai, E., del Bon Espirito-Santo, F., Freitas, R., Morisette, J., 2006. Cropland expansion changes deforestation dynamics in the southern Brazilian Amazon. Proceedings of the National Academy of Sciences of the United States of America, 103, 14637–14641.

Mougel, B., Lelong, C., 2008. Classification and information extraction in very high resolution satellite images for tree crops monitoring. 28th EARSel Symposium Remote Sensing for a Changing Europe, Istambul, p. 13.

Newton, A.C., Hill, R.A., Echeverria, C., Golicher, D., Benayas, J.M.R., Cayuela, L., Hinsley, S.A., 2009. Remote sensing and the future of landscape ecology. Progress in Physical Geography, 33, 528–546.

Omia, E., Bae, H., Park, E., Kim, M.S., Baek, I., Kabenge, I., Cho, B.-K., 2023. Remote sensing in field crop monitoring: A comprehensive review of sensor systems, data analyses and recent advances. Remote Sensing, 15(2), 354. https://doi.org/10.3390/rs15020354.

Owusu, A., Kagone, S., Leh, M., Velpuri, N.M., Gumma, M.K., Ghansah, B., Thilina-Prabhath, P., Akpoti, K., Mekonnen, K., Tinonetsana, P., Mohammed, I. 2024. A framework for disaggregating remote-sensing cropland into rainfed and irrigated classes at continental scale. International Journal of Applied Earth Observation and Geoinformation, 126, 103607, ISSN 1569–8432, https://doi.org/10.1016/j.jag.2023.103607. (www.sciencedirect.com/science/article/pii/S1569843223004314).

Ozdogan, M., 2010. The spatial distribution of crop types from MODIS data: Temporal unmixing using Independent Component Analysis. Remote Sensing of Environment, 114, 1190–1204.

Panigrahy, S., Manjunath, K.R., Ray, S.S., 2005. Deriving cropping system performance indices using remote sensing data and GIS. International Journal of Remote Sensing, 26, 2595–2606.

Panigrahy, S., Ray, S.S., Manjunath, K.R., Pandey, P.S., Sharma, S.K., Sood, A., Yadav, M., Gupta, P.C., Kundu, N., Parihar, J.S., 2011. A spatial database of cropping system and its characteristics to aid climate change impact assessment studies. Journal of the Indian Society of Remote Sensing, 39, 355–364.

Peña-Barragán, J.M., Jurado-Expósito, M., López-Granados, F., Atenciano, S., Sánchez-de la Orden, M., Garcia-Ferrer, A., Garcia-Torres, L., 2004. Assessing land-use in olive groves from aerial photographs. Agriculture, Ecosystems and Environment, 103, 117–122.

Plexida, S.G., Sfougaris, A.I., Ispikoudis, I.P., Papanastasis, V.P., 2014. Selecting landscape metrics as indicators of spatial heterogeneity: A comparison among Greek landscapes. International Journal of Applied Earth Observation and Geoinformation, 26, 26–35.

Pocas, I., Cunha, M., Pereira, L.S., 2011. Remote sensing based indicators of changes in a mountain rural landscape of Northeast Portugal. Applied Geography, 31, 871–880.

Pontius, R.G., Millones, M., 2011. Death to Kappa: Birth of quantity disagreement and allocation disagreement for accuracy assessment. International Journal of Remote Sensing, 32, 4407–4429.

Raymond, E., Hunt, J., Daughtry, C.S.T., 2018. What good are unmanned aircraft systems for agricultural remote sensing and precision agriculture? International Journal of Remote Sensing, 39, 15–16, 5345–5376. https://doi.org/10.1080/01431161.2017.1410300.

Rizzo, D., Marraccini, E., Lardon, S., Rapey, H., Debolini, M., Benoit, M., Thenail, C., 2013. Farming systems designing landscapes: Land management units at the interface between agronomy and geography. Geografisk Tidsskrift-Danish Journal of Geography, 113, 71–86.

Rozenstein, O., Cohen, Y., Alchanatis, V., et al., 2024. Data-driven agriculture and sustainable farming: Friends or foes?. Precision Agric, 25, 520–531. https://doi.org/10.1007/s11119-023-10061-5.

Segarra, J., Buchaillot, M.L., Araus, J.L., Kefauver, S.C., 2020. Remote sensing for precision agriculture: Sentinel-2 improved features and applications. Agronom y, 10(5), 641. https://doi.org/10.3390/agronomy10050641.

Sishodia, R.P., Ray, R.L., Singh, S.K., 2020. Applications of remote sensing in precision agriculture: A review. Remote Sensing, 12(19), 3136. https://doi.org/10.3390/rs12193136.

Soumare, M., 2008. Dynamique et Durabilité des Systemes Agraires à Base de Coton au Mali. Université de Paris X Nanterre (FR), Paris (France), 373.

Steele-Dunne, S.C., McNairn, H., Monsivais-Huertero, A., Judge, J., Liu, P.W., Papathanassiou, K., 2017. Radar remote sensing of agricultural canopies: A review. IEEE Journal of Selected Topics in Applied Earth Observations and Remote Sensing, 10(5), 2249–2273, May. https://doi.org/10.1109/JSTARS.2016.2639043.

Strahler, A.H., Woodcock, C.E., Smith, J.A., 1986. On the nature of models in remote sensing. Remote Sensing of Environment, 20, 121–139.

Tantalaki, N., Souravlas, S., Roumeliotis, M., 2019. Data-driven decision making in precision agriculture: The rise of big data in agricultural systems. Journal of Agricultural & Food Information, 20(4), 344–380. https://doi.org/10.1080/10496505.2019.1638264.

Teluguntla, P.R., Thenkabail, P., Oliphant, A., Gumma, M., Aneece, I., Foley, D., McCormick, R., 2023. Landsat-Derived Global Rainfed and Irrigated-Cropland Product 30 m V001. NASA EOSDIS Land Processes DAAC [WWW Document].

Thenkabail, P.S., Hanjra, M.A., Dheeravath, V., Gumma, M., 2010. A holistic view of global croplands and their water use for ensuring global food security in the 21st century through advanced remote sensing and non-remote sensing approaches. Remote Sensing, 2, 211–261.

Thenkabail, P.S., Lyon, G.J., Turral, H., Biradar, C.M., 2009. Remote Sensing of Global Croplands for Food Security. CRC Press, Taylor and Francis Group, Boca Raton, London, New York.

Thenkabail, P.S., Teluguntla, P.G., Xiong, J., Oliphant, A., Congalton, R.G., Ozdogan, M., Gumma, M.K., Tilton, J.C., Giri, C., Milesi, C., Phalke, A., Massey, R., Yadav, K., Sankey, T., Zhong, Y., Aneece, I., Foley, D., 2021. Global cropland-extent product at 30-m resolution (GCEP30) derived from Landsat satellite time-series data for the year 2015 using multiple machine-learning algorithms on Google Earth Engine cloud: U.S. Geological Survey Professional Paper 1868. US Geological Survey, Reston, VI. https://doi.org/10.3133/PP1868.

UN, 2010. World Population Prospects, the 2010 Revision. United Nations, Department of Economic and Social Affairs, Population Division, New York, USA, ST/ESA/SER.A/307; Sales No. E.11.XIII.6.

Ursani, A.A., Kpalma, K., Lelong, C.C.D., Ronsin, J., 2012. Fusion of textural and spectral information for tree crop and other agricultural cover mapping with very-high resolution satellite images. Ieee Journal of Selected Topics in Applied Earth Observations and Remote Sensing, 5, 225–235.

USAID, 2009. Application of the Livelihood Zone Maps and Profiles for Food Security Analysis and Early Warning: Guidance for Famine Early Warning Systems Network (FEWS NET) Representatives and Partners. USAID, 23.

Uuemaa, E., Mander, Ü., Marja, R., 2013. Trends in the use of landscape spatial metrics as landscape indicators: A review. Ecological Indicators, 28, 100–106.

Vaclavik, T., Lautenbach, S., Kuemmerle, T., Seppelt, R., 2013. Mapping global land system archetypes. Global Environmental Change, 23, 1637–1647.

Verburg, P.H., Mertz, O., Erb, K.-H., Haberl, H., Wu, W., 2013. Land system change and food security: Towards multi-scale land system solutions. Current Opinion in Environmental Sustainability, 5, 494–502.

Verburg, P.H., van de Steeg, J., Veldkamp, A., Willemen, L., 2009. From land cover change to land function dynamics: A major challenge to improve land characterization. Journal of Environmental Management, 90, 1327–1335.

Victor, N., et al., 2024. Remote sensing for agriculture in the era of industry 5.0: A survey. IEEE Journal of Selected Topics in Applied Earth Observations and Remote Sensing. https://doi.org/10.1109/JSTARS.2024.3370508.

Vintrou, E., Begue, A., Baron, C., Saad, A., Lo Seen, D., Traore, S.B., 2014. A comparative study on satellite- and model-based crop phenology in West Africa. Remote Sensing, 6, 1367–1389.

Vintrou, E., Desbrosse, A., Begue, A., Traore, S., Baron, C., Lo Seen, D., 2011. Crop area mapping in West Africa using landscape stratification of MODIS time series and comparison with existing global land products. International Journal of Applied Earth Observation and Geoinformation, 14, 83–93.

Vintrou, E., Ienco, D., Begue, A., Teisseire, M., 2013. Data mining, a promising tool for large-area cropland mapping. Ieee Journal of Selected Topics in Applied Earth Observations and Remote Sensing, 6, 2132–2138.

Vintrou, E., Soumare, M., Bernard, S., Begue, A., Baron, C., Lo Seen, D., 2012. Mapping fragmented agricultural systems in the Sudano-Sahelian environments of Africa using random forest and ensemble metrics of coarse resolution MODIS imagery. Photogrammetric Engineering and Remote Sensing, 78, 839–848.

Wästfelt, A., Tegenu, T., Nielsen, M.M., Malmberg, B., 2012. Qualitative satellite image analysis: Mapping spatial distribution of farming types in Ethiopia. Applied Geography, 32, 465–476.

Weiss, M., Jacob, F., Duveiller, G., 2020. Remote sensing for agricultural applications: A meta-review. Remote Sensing of Environment, 236, 111402, ISSN 0034–4257, https://doi.org/10.1016/j.rse.2019.111402. (www.sciencedirect.com/science/article/pii/S0034425719304213).

Wintgens, J.-N., 2004. Factors influencing the quality of green coffee. In: Wintgens, J.-N. (Ed.), Coffee: Growing, Processing, Sustainable Production: A Guidebook for Growers, Processors, Traders, and Researchers. Weihnheim, Germany, 789–809.

You, L.Z., Wood, S., 2006. An entropy approach to spatial disaggregation of agricultural production. Agricultural Systems, 90, 329–347.

You, L.Z., Wood, S., Wood-Sichra, U., 2009. Generating plausible crop distribution maps for Sub-Saharan Africa using a spatially disaggregated data fusion and optimization approach. Agricultural Systems, 99, 126–140.

7 Global Food Security Support Analysis Data (GFSAD) Using Remote Sensing in Support of Food and Water Security in the 21st Century

Current Achievements and Future Possibilities

Pardhasaradhi Teluguntla, Prasad S. Thenkabail, Jun Xiong, Adam Oliphant, Murali Krishna Gumma, Chandra Giri, Cristina Milesi, Mutlu Ozdogan, Russell G. Congalton, James Tilton, Temuulen Tsagaan Sankey, Richard Massey, Aparna Phalke, and Kamini Yadav

ACRONYMS AND DEFINITIONS

ASTER	Advanced Spaceborne Thermal Emission and Reflection Radiometer
CDL	Cropland Data Layer
CEOS	Committee on Earth Observing Satellites
EnMAP	Environmental Mapping and Analysis Program
EO	Earth Observing
GEOSS	Global Earth Observation System of Systems
GIMMS	Global Inventory Modeling and Mapping Studies
GIS	Geographic Information Systems
GLAM	Global Agricultural Monitoring
GPS	Global Positioning System
IKONOS	A commercial Earth observation satellite, typically, collecting sub-meter to 5 m data
JERS	Japanese Earth Resources Satellite
LUC	Land Use Classes
LULC	Land Use, Land Cover
MODIS	Moderate-Resolution Imaging Spectroradiometer
NASA	National Aeronautics and Space Administration
NIR	Near-Infrared
NOAA	National Oceanic and Atmospheric Administration

DOI: 10.1201/9781003541165-9

SAR	Synthetic Aperture Radar
SMTs	Spectral Matching Techniques
SPOT	Satellite Pour l'Observation de la Terre, French Earth Observing Satellites
SRTM	Shuttle Radar Topographic Mission
SSV	Spectral Similarity Value
SWIR	Shortwave Infrared
VHRI	Very High Resolution Imagery

7.1 INTRODUCTION

Agriculture land provides humankind with food, fibers, and raw materials that are vital for human livelihood (Viana et al., 2022, Pereira et al., 2018, Stephens et al., 2018). The accurate estimation of the global agricultural cropland- extents, areas, geographic locations, crop types, cropping intensities (Liu et al., 2023), and their watering methods (irrigated or rainfed; type of irrigation) (Owusu et al., 2024, Teluguntla et al., 2023, Van Tricht et al., 2023) provides a critical scientific basis for the development of water and food security policies (Thenkabail et al., 2021, 2012, 2011, Thenkabail, 2010, Turral et al., 2009). The World Summit on Food Security declared that in 2050, "the world's population is expected to grow to almost 10 billion by 2050, boosting agricultural demand—in a scenario of modest economic growth—by some 50 percent compared to 2013" (FAO, 2017) and by year 2100, the global human population is expected to grow 11 to 12 billion (UN DESA, 2021, Gerland and others, 2014; United Nations [UN], 2019). About 10.4 billion under median fertility variants or higher under constant or higher fertility variants (UNDP, 2012, Table 7.1) with over three quarters living in developing countries and in regions that already lack the capacity to produce enough food. With current agricultural practices, the increased demand for food and nutrition would require about 2 billion hectares of additional cropland, about twice the equivalent to the land area of the United States, and lead to significant increases in greenhouse gas productions associated with agricultural practices and activities (Tillman et al., 2011). For example, during 1960–2010, world population more than doubled from 3 billion to 7 billion. The nutritional demand of the population also grew swiftly during this period from an average of about 2000 calories per day per person in 1960 to nearly 3,000 cal/person/day in 2010 and by the year 2050, global daily average calorie consumption is expected to rise to 3,130 cal/person/day by the year 2050 (World Bank, 2022, Bodirsky et al., 2015). These consumption figures may rise even higher if traditionally low meat consuming nations start increasing meat consumption enabled by economic growth.

The food demand of increased population along with increased nutritional demand during this period was met by the "green revolution" which more than tripled the food production; even though croplands decreased from about 0.43 ha per capita to 0.26 ha per capita (FAO, 2009, Funk and Brown, 2009). The increase in food production during the green revolution was the result of factors such as: (1) cropland expansion from ~300 million hectares to ~1.8 billion hectares (Potapov et al., 2022, Thenkabail et al., 2021); (2) expansion of irrigation croplands from ~130 million hectares in 1960 to ~400 million hectares (Siebert et al., 2006, 2015, Thenkabail et al., 2009a, 2009b, 2009c); (3) increase in yield and per capita production of food (e.g., cereal production from 280 kg/person to 380 kg/person and meat from 22 kg/person to 34 kg/person (McIntyre, 2008) by genetic engineering through high yield, fast growing, short-duration crops (Lenaerts et al., 2019); (4) cropland intensification from single to double or triple cropping, in some irrigated croplands (Hu et al., 2020, Wu et al., 2018); (5) heavy inputs such as fertilizer and nitrogen application (He et al., 2021), and (6) improvements in land management (e.g., leveling, drainage) (Viana et al., 2022).

Although modern agriculture met the challenge to increase food production last century, lessons learned from the 20th century "green revolution" and our current circumstances impact the likelihood of another such revolution. The intensive use of chemicals has adversely impacted the environment in many regions, leading to salinization and decreasing water quality and degrading croplands. From 1960 to 2000, worldwide phosphorous use doubled from 10 million metric tons (MT) to 20 MT, pesticide use tripled from near zero to 3 MT, later by 2020 pesticide use remained stable (FAO, 2022), and nitrogen use as fertilizer increased to a staggering 80 MT from just 10 MT (Foley et al., 2007, Khan and Hanjra, 2008). Diversion of croplands to bio-fuels is taking water away from food production (Khan et al., 2022, Popp et al., 2014, Bindraban et al., 2009), even as the economic, carbon sequestration, environmental, and food security impacts of biofuel production are proving to be a net negative (He et al., 2023, Lal and Pimentel, 2009, Gibbs et al., 2008, Searchinger et al., 2008). Climate models predict that the hottest seasons on record will become the norm by the end of the century in most regions of the world (Nyambariga et al., 2023)—a prediction that bodes ill for feeding the world (Kumar and Singh, 2005). Increasing per capita meat consumption is increasing agricultural demands on land and water (Vinnari and Tapio, 2009). Cropland areas are decreasing in many parts of the world due to urbanization, industrialization, and salinization (Khan and Hanjra, 2008). Ecological and environmental imperatives, such as biodiversity conservation and atmospheric carbon sequestration, have put a cap on the possible expansion of cropland areas to other lands such as forests and rangelands (Gordon et al., 2009). Crop yield increases of the green revolution era have now stagnated (Hossain et al., 2005). Given these factors and limitations, further increases in food production through increases in cropland areas and/or increased allocations of water for croplands are widely considered unsustainable or simply infeasible.

Clearly, our continued ability to sustain adequate global food production and achieve future food security in the 21st century is challenged. So, how does the world continue to meet its food and nutrition needs?

Solutions may come from biotechnology and precision farming. However, developments in these fields are not currently moving at rates that will ensure global food security over the next few decades (Foley et al., 2011). Further, there is a need for careful consideration of possible adverse effects of biotechnology. During the green revolution, the focus was only on getting more yield per unit area (Lynch, 2007). Little thought was given to the serious damage done to our natural environments, water resources, and human health as a result of detrimental factors such as uncontrolled use of herbicides, pesticides, and nutrients, drastic groundwater mining, and salinization of fertile soils due to over-irrigation (Thenkabail, 2010). Currently, there are discussions of a "second green revolution" or even an "ever green revolution," but definitions of what these terms actually mean are still debated and are evolving (e.g., Monfreda et al., 2005, 2008). One of the biggest issues that has not been given adequate focus is the use of large quantities of water for food production. Indeed, an overwhelming proportion (60–90%) of all human water use in India, for example, goes for producing their food (Falkenmark and Rockström, 2006). But such intensive water use for food production is no longer sustainable due to increasing competition for water for alternative uses (EPW, 2008), such as urbanization, industrialization, environmental flows, biofuels, and recreation (Shao et al., 2017, Lark et al., 2015, Smakhtin, 2006). This has brought into sharp focus the need to grow more food per drop of water leading to the need for a "blue revolution" in agriculture (Pennisi, 2008).

A significant part of the solution lies in determining how global croplands are currently used and how they might be better managed to optimize use of resources in increasing food production. This will require development of an advanced Global Cropland Area Database (GCAD) with an ability to map global croplands and their attributes routinely, rapidly, consistently, and with sufficient accuracies. This in turn requires the creation of a framework of best practices for cropland mapping and an advanced global geospatial information system on global croplands. Such a system would need to be consistent across nations and regions by providing information on issues such as the composition and location of cropping, cropping intensities (e.g., single, double crop), crop rotations, crop health/vigor, and irrigation status. Opportunities to establish such a global system can be achieved by fusing advanced remote sensing data from multiple platforms and agencies (e.g., http://eros.usgs.gov/ceos/satellites_midres1.shtml; www.ceos-cove.org/index.php) in combination with national statistics, secondary data (e.g., elevation, slope, soils, temperature, precipitation, evapotranspiration), and the systematic collection of field level observations. An example of such a system on a regional scale is United States Department of Agriculture (USDA), National Agriculture Statistics Service (NASS) Cropland Data Layer (CDL), hosted on CropScape (Han et al., 2012) which is a

TABLE 7.1
World Population (Thousands) under All Variants, 1950–2100

Year	Medium Fertility Variant	High Fertility Variant	Low Fertility Variant	Constant Fertility Variant
1950	2,529,346	2,529,346	2,529,346	2,529,346
1955	2,763,453	2,763,453	2,763,453	2,763,453
1960	3,023,358	3,023,358	3,023,358	3,023,358
1965	3,331,670	3,331,670	3,331,670	3,331,670
1970	3,685,777	3,685,777	3,685,777	3,685,777
1975	4,061,317	4,061,317	4,061,317	4,061,317
1980	4,437,609	4,437,609	4,437,609	4,437,609
1985	4,846,247	4,846,247	4,846,247	4,846,247
1990	5,290,452	5,290,452	5,290,452	5,290,452
1995	5,713,073	5,713,073	5,713,073	5,713,073
2000	6,115,367	6,115,367	6,115,367	6,115,367
2005	6,512,276	6,512,276	6,512,276	6,512,276
2010	6,916,183	6,916,183	6,916,183	6,916,183
2015	7,324,782	7,392,233	7,256,925	7,353,522
2020	7,716,749	7,893,904	7,539,163	7,809,497
2025	8,083,413	8,398,226	7,768,450	8,273,410
2030	8,424,937	8,881,519	7,969,407	8,750,296
2035	8,743,447	9,359,400	8,135,087	9,255,828
2040	9,038,687	9,847,909	8,255,351	9,806,383
2045	9,308,438	10,352,435	8,323,978	10,413,537
2050	9,550,945	10,868,444	8,341,706	11,089,178
2055	9,766,475	11,388,551	8,314,597	11,852,474
2060	9,957,399	11,911,465	8,248,967	12,729,809
2065	10,127,007	12,442,757	8,149,085	13,752,494
2070	10,277,339	12,989,484	8,016,514	14,953,882
2075	10,305,146	13,101,094	7,986,122	15,218,723
2080	10,332,223	13,213,515	7,954,481	15,492,520
2085	10,358,578	13,326,745	7,921,618	15,775,624
2090	10,384,216	13,440,773	7,887,560	16,068,398
2095	10,409,149	13,555,593	7,852,342	16,371,225
2100	10,433,385	13,671,202	7,815,996	16,684,501

Source: UNDP (2012).

raster, geo-referenced, crop-specific land cover data layer with a ground resolution of 30 m (Boryan et al., 2011, Johnson and Mueller, 2010). The GCAD will be a major contribution to Group on Earth Observations (GEO) Global Agricultural Monitoring Initiative (GLAM) (Whitcraft et al., 2015), to the overarching vision of GEO Agriculture and Water Societal Beneficial Areas (GEO Ag. SBAs), G20 Agriculture Ministers initiatives, and ultimately to the Global Earth Observation System of Systems (GEOSS) (Herold et al., 2008). These initiatives are also supported by the Committee on Earth Observing Satellites (CEOS) (Withee et al., 2004) Strategic Implementation Team (SIT).

Within the context of the preceding facts, the overarching goal of this chapter is to provide a comprehensive overview of the state-of-art of global cropland mapping procedures using remote sensing as characterized and envisioned by the "Global Food Security Support Analysis Data @ 30 m (GFSAD30)" project working group team (Thenkabail et al., 2021). First, the chapter will provide an overview of

existing cropland maps and their characteristics along with establishing the gaps in knowledge related to global cropland mapping. Second, **definitions** of cropland mapping along with key parameters involved in cropland mapping based on their importance in food security analysis, and cropland naming conventions for standardized cropland mapping using remote sensing will be presented. Third, **existing methods and approaches** for cropland mapping will be discussed. This will include the type of remote sensing data used in cropland mapping and their characteristics along with discussions on the secondary data, field-plot data, and cropland mapping algorithms. Fourth, currently **existing global cropland products** derived using remote sensing will be presented and discussed. Fifth, a **synthesis** of all existing products leading to a composite global cropland extent version 1.0 (GCE V1.0) is presented and discussed. Sixth, a **way forward** for advanced global cropland mapping is visualized.

7.2 GLOBAL DISTRIBUTION OF CROPLANDS AND OTHER LAND USE AND LAND COVER: BASELINE FOR THE YEAR 2000

General Land Use Land Cover (LULC) datasets provide a holistic picture of all the land uses and covers on Earth, without focusing specifically on any individual land use category. Time series of general LULC datasets at a global scale is useful for understanding global patterns of LUC change and their relation with global processes (García-Álvarez et al., 2022) such as climate change or the loss of biodiversity (Shi et al., 2023). MCD12Q1 (Friedl and Sulla-Menashe, 2022, 2019), also known as MODIS Land Cover, was the first time series of LULC maps to be produced on a global scale. The MCD12Q1 dataset continues to be updated today, providing a series of maps for the period 2001–2022. Since the launch of MCD12Q1, many other historical series of LUC maps have been produced, especially in the last decade. In the past few years, many other historical series of LULC maps have been produced by several research groups with better spatial resolution, for example, 100 m Copernicus Global Land Service Dynamic Land Cover Map (CGLS-LC100) (Tsendbazar et al., 2021, Buchhorn et al., 2019) and 1 m WorldCover10v1 (Zanaga et al., 2022) by the European Space Agency (ESA). However, comparative analyses of many remote-sensing land use and landcover (LULC) products show high disagreement in croplands globally (Owusu et al., 2024, Fritz et al., 2010, Pérez-Hoyos et al., 2017). Fritz et al. (2010) found that the total cropland area worldwide differs by approximately 20% between remote-sensing LULC products. The first comprehensive global map of croplands was created by Ramankutty et al. (1998). For example, a first version of LULC in the beginning of the new millennium for the year 2000 (GLC, 2003, Bartholome and Belward, 2005) shows the spatial distribution of global croplands along with other land use and land cover classes (Figure 7.1). This provides a first view of where global croplands are concentrated and helps us focus on the appropriate geographic locations for detailed cropland studies. Water and snow (Class 8 and 9, respectively) have zero croplands and occupy 44% of the total terrestrial land surface. Further, forests (class 6) occupy 17% of the terrestrial area and deserts (class 7) an additional 12%. In these two classes, <5% of the total croplands exist. Therefore, in order to study croplands systematically and intensively, one must prioritize mapping in the areas of classes 1 to 5 (26% of the terrestrial area) where 95% of all global croplands exist, with the first 3 classes (class 1, 2, 3) having ~75% and the next 2 ~20%. In the future, it is likely some of the non-croplands may be converted to croplands or vice versa, highlighting the need for repeated and systematic global mapping of croplands. Segmenting the world into cropland versus non-cropland areas routinely will help us understand and study change dynamics such as water availability, climate change, and cropland area by region which impacts food security.

7.2.1 Existing Global Cropland Maps: Remote Sensing and Non-Remote Sensing Approaches

Since the year 2000, we have seen several remote-sensing-derived global cropland-extent maps that use satellite-sensor data. However, there are several limitations with existing cropland extent products such as coarse spatial resolution, inadequate temporal and spectral resolutions, varying

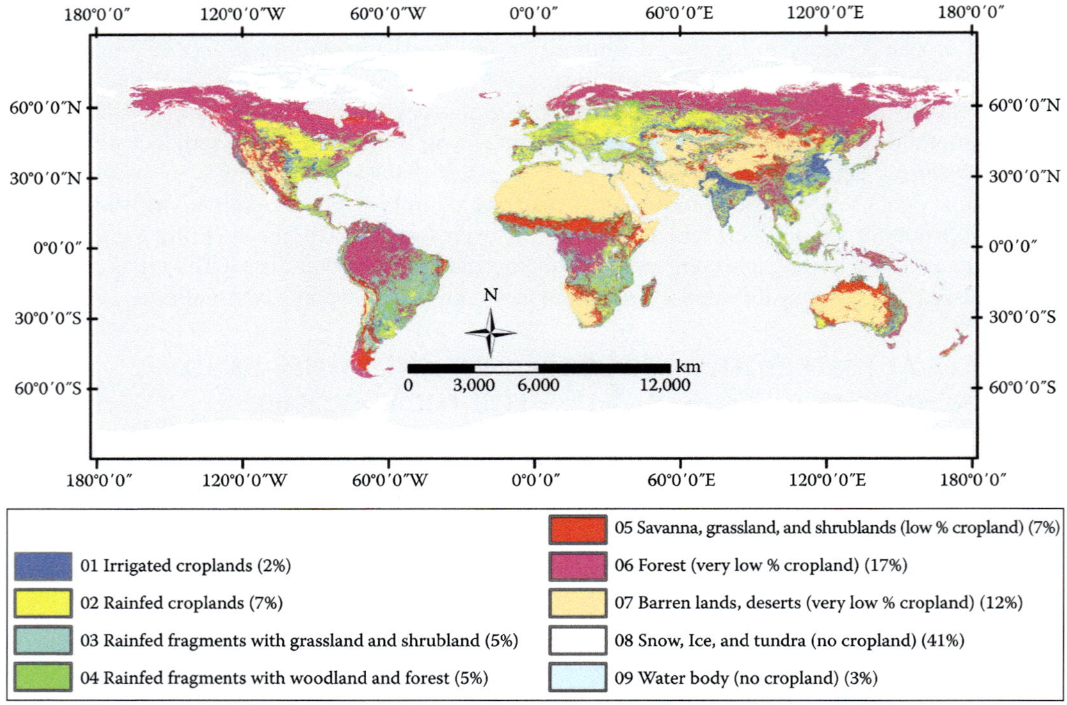

FIGURE 7.1 Global croplands and other land use and land cover classes for Baseline year 2000. (GLC, 2003.)

definitions, varying types of data, varying methods and Inadequate reference-training and validation data. Any one of the factors already mentioned, or combination thereof, could lead to a high degree of uncertainty in cropland-extent products—locally, regionally, and globally.

There are currently ten existing major exclusive global cropland maps: (1) Teluguntla et al. (2023); (2) Thenkabail et al. (2021), (3) Salmon et al. (2015), (4) Thenkabail et al. (2009a, 2009b), (5) Waldner et al. (2016), (6) Goldewijk et al. (2011), (7) Portmann et al. (2008), (8) Pittman et al. (2010), (9) Ramankutty and Foley (1998) and (10) Yu et al. (2013). These studies estimated the total global cropland area anywhere between 1.5 to 1.8 billion hectares between the years 2000–2015. However, there are two significant differences in these products: (1) spatial disagreement on where the actual croplands are, and (2) irrigated to rainfed cropland proportions and their precise spatial locations. Globally, cropland areas have increased from around 265 Mha in year 1700 to around 1,471 Mha in year 1990, whilst the area of pasture has increased approximately sixfold, from 524 to 3,451 Mha (Foley et al., 2011). Ramankutty and Foley (1998) estimated the cropland and pasture to represent about 36% of the world's terrestrial surface (148,940,000 km^2), of which, according to different studies, roughly 12% is croplands and 24% pasture. Multiple studies (Goldewijk et al., 2011, Portmann et al., 2008, Ramankutty et al., 2008) integrated agricultural statistics and census data from the national systems with spatial mapping technologies involving geographic information systems (GIS) to derive global cropland maps.

Thenkabail and others (2011, 2009a, 2009b) produced the first remote sensing-based global irrigated and rainfed cropland maps and statistics through multi-sensor remote sensing data fusion along with secondary data and in situ data. They further used five dominant crop types (wheat, rice, corn, barley, and soybeans) using parcel-based inventory data (Monfreda et al., 2008, Portmann et al., 2008, Ramankutty et al., 2008) to produce a classification of global croplands with crop dominance (Thenkabail et al., 2012). The five crops account for about 60% of the total global cropland areas. he precise spatial location of these crops is only an approximation due to the coarse resolution (approx. 1 km^2) and fractional representation (1 to 100% crop in a pixel) of the crop data in each grid cell of all the maps from which this composite map is produced (Thenkabail et al., 2012). The existing global cropland datasets also differ from each other due to inherent uncertainties in establishing the accurate location of croplands, the watering methods (rainfed *versus* irrigated), cropping intensities, crop types and/or dominance, and crop characteristics (e.g., crop or water productivity measures such as biomass, yield, and water use). Improved knowledge of the uncertainties (Congalton and Green, 2009) in these estimates will lead to a suite of highly accurate spatial data products (Goodchild and Gopal, 1989) to serve as inputs for crop modeling, food security analysis, and decision support.

7.3 KEY REMOTE SENSING DERIVED CROPLAND PRODUCTS: GLOBAL FOOD SECURITY

The production of a repeatable global cropland product requires accurate cropland areas that can be derived consistently across the diverse cropland regions of the world. Four key cropland information systems products that have been identified for global food security analysis and that can be readily derived from remote sensing include (Figure 7.2): (1) cropland extent/areas, (2) watering methods (e.g., irrigated, supplemental irrigated, rainfed), (3) crop types, and (4) cropping intensities (e.g., single crop, double crop, triple/continuous crop) (Thenkabail et al., 2021). Although not the focus of this chapter covering global cropland mapping, many other parameters are also derived in local and regional mapping level, such as (1) precise location of crops, (2) cropping calendar, (3) crop health/vigor, (4) flood and drought information, (5) water use assessments, and (6) yield or productivity (expressed per unit of land and/or unit of water). Remote sensing is specifically suited to derive the four key products over large areas using fusion of multi spectral remote sensing imagery (e.g., Sentinels, Landsat, Resourcesat, MODIS) in combination with national statistics, ancillary data (e.g., slope, elevation, precipitation, evapotranspiration), and field-plot data. Such a system, at the global level, will be complex in data handling and processing and requires coordination between multiple agencies leading to development of a seamless, scalable, transparent, and repeatable methodology. As a result, it is important to have systematic class labeling convention as illustrated in Figure 7.3. A standardized class identifying and labeling process (Figure 7.3) will enable consistent and systematic labeling of classes, irrespective of analysts. First, the area is separated into cropland versus non-cropland. Then, within the cropland class, labeling will involve (Figure 7.3): (1) cropland extent (cropland vs. non-cropland), (2) watering source (e.g., irrigated versus rainfed), (3) irrigation source (e.g., surface water, groundwater), (4) crop type or dominance, (5) scale (e.g., large or contiguous, small or fragmented), and (6) cropping intensity (e.g., single crop, double crop). The detail at which one maps at each stage and each parameter would depend on many factors such as resolution of the imagery, available ground data, and expert knowledge. For example, if there is no sufficient knowledge on whether the irrigation is by surface water or groundwater, but it is clear that the area is irrigated; one could just map it as irrigated without mapping greater details on the type of irrigation. But, for every cropland class, one has the potential to map the details as shown in Figure 7.3.

7.4 DEFINITION OF REMOTE SENSING-BASED CROPLAND MAPPING PRODUCTS

Key to effective mapping is a precise and clear definition of what will be mapped. It is the first step that must be decided because different definitions lead to different products. For example, irrigated areas are defined and understood differently in different applications and contexts. One can define them as areas which receive irrigation at least once during their crop growing period. Alternatively, they can be defined as areas which receive irrigation to meet at least half their crop

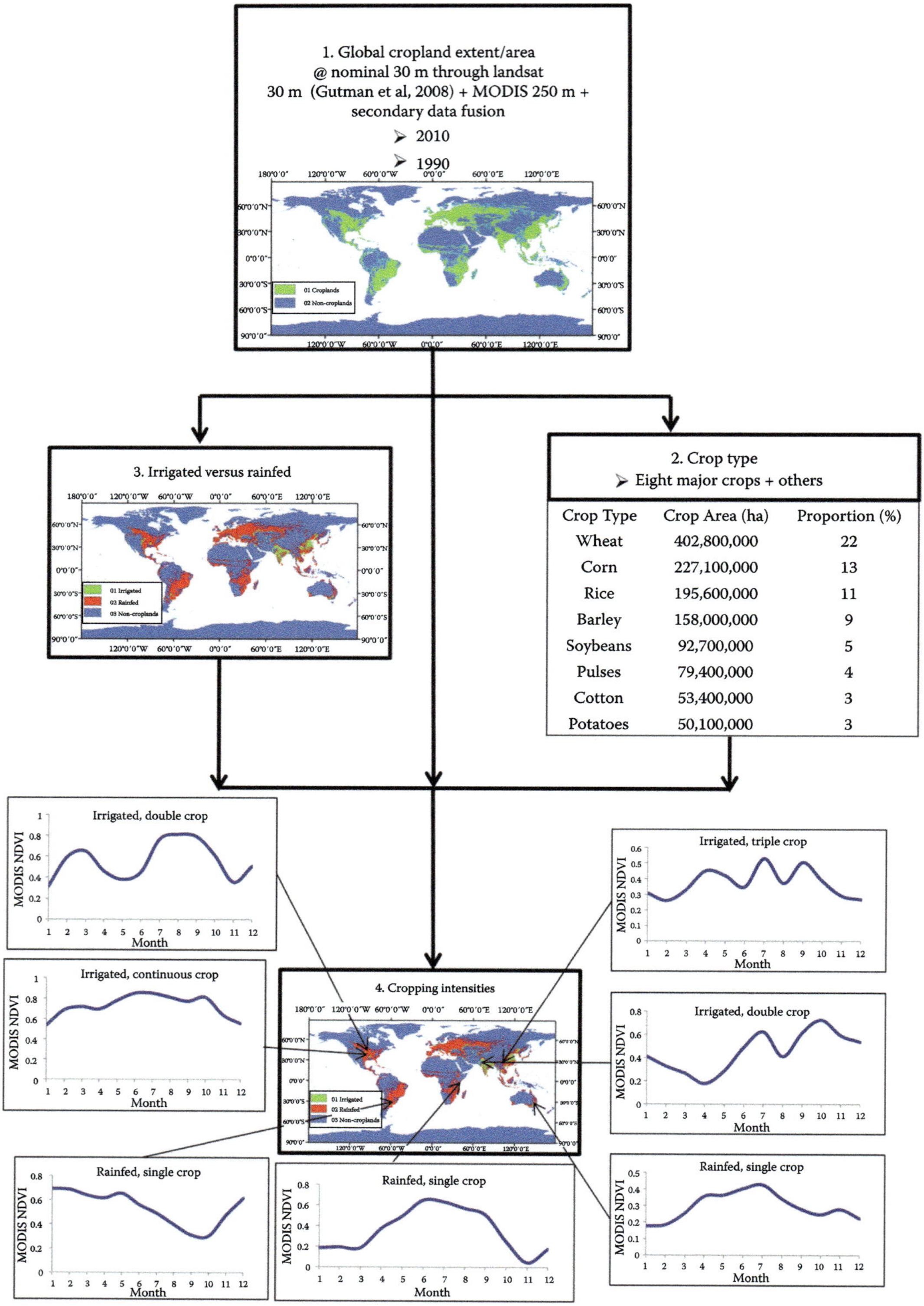

Crop Type	Crop Area (ha)	Proportion (%)
Wheat	402,800,000	22
Corn	227,100,000	13
Rice	195,600,000	11
Barley	158,000,000	9
Soybeans	92,700,000	5
Pulses	79,400,000	4
Cotton	53,400,000	3
Potatoes	50,100,000	3

FIGURE 7.2 Key global cropland area products that will support food security analysis in the 21st century. (Thenkabail et al., 2021.)

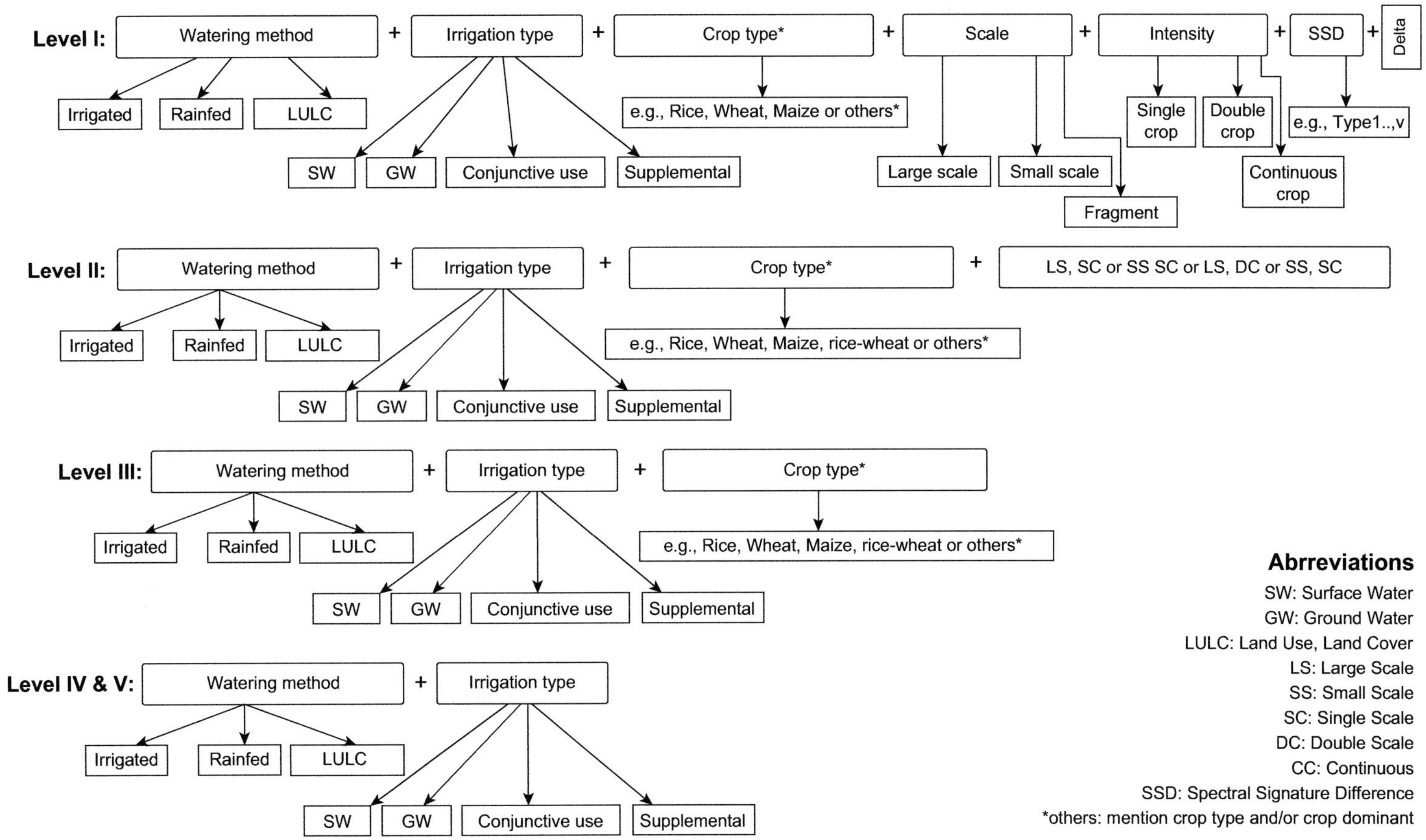

FIGURE 7.3 Cropland class naming convention at different levels. Level I is most detailed, and level IV is least detailed. (Source: Dheeravath et al., 2010.)

water requirements during the growing season. A third definition is that these are areas that are irrigated throughout the growing season. In each of these cases, the irrigated area extent mapped will vary. Similarly, croplands can be defined as all agricultural areas irrespective of type of crops grown or they may be limited to food crops (and not the fodder crops or plantation crops). So, it is obvious that having a clear understanding of the definitions of what we map is extremely important for the integrity of the products developed. We have defined four cropland products as follows:

- **Cropland extent:** All cultivated plants harvested for food, feed, and fiber, including plantations (e.g., orchards, vineyards, coffee, tea, rubber).
- **Irrigated areas:** Irrigation is defined as artificial application of any amount of water to overcome crop water stress. Irrigated areas are those areas which are irrigated one or more times during crop growing season.
- **Rainfed areas:** areas that have no irrigation whatsoever and are precipitation (rainfall, or soil moisture obtained from snow fall) dependent.
- **Cropping intensity:** Number of cropping cycles within a 12-month period, which can be single, double or triple/continuous.
- **Crop type:** Major eight crops of the world Wheat, Corn, Rice, Barley, Soybeans, Pulses, Cotton, Potatoes.

It is also important to define the minimum mapping unit of a particular crop which is here an area of 3 by 3 (0.81 hectares) Landsat pixels identified as having the same crop type. Additionally, a pixel is considered as cropland if greater than 50% of pixel is cropped.

7.5 DATA: REMOTE SENSING AND OTHER DATA FOR GLOBAL CROPLAND MAPPING

Cropland mapping using remote sensing involves multiple types of data: satellite images with a consistent and useful global repeat cycle, secondary data, statistical data, and field plot data. When these data are used in an integrated fashion, the output products achieve highest possible accuracies (Thenkabail et al., 2009b, 2009c).

7.5.1 Primary Satellite Sensor Data

Cropland mapping will require satellite sensor data across spatial, spectral, radiometric, and temporal resolutions from a wide array of satellite/sensor platforms (Table 7.2) throughout the growing season to quantify their growth characteristics. Multispectral, hyperspectral, and hyperspatial data have an important role in characterizing crop growth. The data points per hectare (Table 7.2, last column) will indicate the spatial detail of agricultural information gathered. In addition to satellite-based sensors, it is always valuable to gather ground based hand-held spectroradiometer data from hyperspectral sensors (Thenkabail et al., 2013) and/or imaging spectroscopy, airborne, or space-borne sensors for validation and calibration purposes (Thenkabail et al., 2011).

7.5.2 Secondary Data

There is a wide array of secondary or ancillary data such as the ASTER-derived global digital elevation data (GDEM) (Abrams et al., 2010, 2020), over long time period (50 to 100 year) records of precipitation and temperature from Climatic Research Unit (CRU) (www.uea.ac.uk/groups-and-centres/climatic-research-unit) (Mitchell et al., 2004), digital maps of soil types, and administrative boundaries. Many secondary data are known to improve crop classification accuracies (Thenkabail et al., 2021, 2009a, 2009b). This secondary data will also be critical for the spatial decision support system and final visualization tool in many systems.

TABLE 7.2
Characteristics of Some Satellite Sensor Data Currently Used in Cropland Mapping

Satellite Sensor	Wavelength Range (μm)	Spatial Resolution (m)	Spectral bands (#)	Temporal Resolution (days)	Radiometric Resolution (bits)	Data Points (per hectare)
A. Hyperspectral						
EO-1 Hyperion						
VNIR	0.43–0.93	30	196	16	16	11.1 points for 30 m pixel (0.09 hectares per pixel)
SWIR	0.93–2.40					
DESIS						
VNIR	0.40–1.0	30	235	3–5	13	11.1 points for 30 m pixel
EnMap						
VNIR	0.42–1.0	30	235	4	14	11.1 points for 30 m pixel
SWIR	0.95–2.45					
PRISMA						
VNIR	0.40–1.0	30	238		12	11.1 points for 30 m pixel
SWIR	1.0–2.5					
B. Advanced multispectral						
Sentinel-2	-	-	-	-	-	-
Multi spectral	-	-			-	-
Band 2	0.459–0.525	10	13	10	12	100 points for 10 m pixel
Band 3	0.541–0.577					
Band 4	0.649–0.680					
Band 8	0.779–0.885					
Band 5	0.696–0.711	20				25 points for 20 m pixel
Band 6	0.733–0.748					
Band 7	0.772–0.792					
Band 8a	0.854–0.875					
Band 11	1.568–1.659					
Band 12	2.114–2.289					
Band 1	0.432–0.453	60				2.77 points for 60 m pixel
Band 9	0.935–0.955					
Band 10	1.358–1.389					
Landsat TM						
Multispectral						
Band 1	0.45–0.52	30	7/8	16	8	11.1 points for 30 m pixel
Band 2	0.53–0.61	30				
Band 3	0.63–0.69	30				
Band 4	0.78–0.90	30				
Band 5	1.55–1.75	30				2.77 points for 60 m pixel
Band 6	10.40–12.50	120/60				0.69 points for 120 m pixel
Band 7	2.09–2.35	30				
Panchromatic	0.52–0.90	15				44.4 points for 15 m pixel
Landsat OLI						
Multispectral						

(Continued)

TABLE 7.2 (Continued)
Characteristics of Some Satellite Sensor Data Currently Used in Cropland Mapping

Satellite Sensor	Wavelength Range (μm)	Spatial Resolution (m)	Spectral bands (#)	Temporal Resolution (days)	Radiometric Resolution (bits)	Data Points (per hectare)
Band 1	0.43–0.45	30	11	16	16	11.1 points for 30 m pixel
Band 2	0.45–0.51					
Band 3	0.53–0.59					
Band 4	0.64–0.67					
Band 5	0.85–0.88					
Band 6	1.57–1.65					
Band 7	2.11–2.29					
Band 9	1.36–1.38					
Band 8 Pan	050–0.68	15				44.4 points for 15 m pixel
Band 10	10.6–11.19	100				
Band 11	11.5–12.51					1 point for 100m pixel
EO-1 ALI						
Multispectral						
Band 1	0.43–0.45	30	10	16	16	11.1 points for 30 m pixel
Band 2	0.45–0.52					
Band 3	0.52–0.61					
Band 4	0.63–0.69					
Band 5	0.78–0.81					
Band 6	0.85–0.89					
Band 7	1.20–1.30					
Band 8	1.55–1.75					
Band 9	2.08–2.35					
Panchromatic	0.48–0.69	10				100 points for 10 m pixel
ASTER						
Multispectral						
VNIR			14	16	8	
Band 1	0.52–0.60	15				44.4 points for 15 m pixel
Band 2	0.63–0.69					
Band 3N/3B	0.76–0.86					
SWIR						
Band 4	1.600–1.700	30				11.1 points for 30 m pixel
Band 5	2.145–2.185					
Band 6	2.185–2.225					
Band 7	2.235–2.285					
Band 8	2.295–2.365					
Band 9	2.360–2.430					
TIR						
Band 10	8.125–8.475	90				1.23 points for 90 m pixel
Band 11	8.475–8.825					
Band 12	8.925–9.275					
Band 13	10.25–10.95					
Band 14	10.95–11.65					
MODIS						
Multispectral						
MOD09Q1		250	2	1	12	0.16 points for 250m pixel
Band 1	0.62–0.67					

(Continued)

TABLE 7.2 (*Continued*)
Characteristics of Some Satellite Sensor Data Currently Used in Cropland Mapping

Satellite Sensor	Wavelength Range (μm)	Spatial Resolution (m)	Spectral bands (#)	Temporal Resolution (days)	Radiometric Resolution (bits)	Data Points (per hectare)
Band2	0.84–0.876					
MOD09A1		500	7	1	12	0.04 points for 500m pixel
Band1	0.62–0.67					
Band2	0.84–0.876					
Band3	0.459–0.479					
Band4	0.545–0.565					
Band5	1.23–1.25					
Band6	1.63–1.65					
Band7	2.11–2.16					
C. Hyperspatial						
GeoEye-1						
Multispectral						
Band 1	0.45–0.52	1.65	5	<3	11	59,488 points for 0.41 m pixel
Band 2	0.52–0.60					
Band 3	0.63–0.70					
Band 4	0.76–0.90					
Panchromatic	0.45–0.90	0.41				3673 points for 1.65 m pixel
IKONOS						
Multispectral						
Band 1	0.45–0.52	4	5	3	11	625 points for 4 m pixel
Band 2	0.51–0.60					
Band 3	0.63–0.70					
Band 4	0.76–0.85					
Panchromatic	0.53–0.93	1				10,000 points for 1m pixel
Quickbird						
Multispectral						
Band 1	0.45–0.52	2.44	5	1–6	11	1679 points for 2.44 m pixel
Band 2	0.52–0.60					
Band 3	0.63–0.69					
Band 4	0.76–0.90					
Panchromatic	0.45–0.90	0.61				26,874 points for 0.61 m
Rapideye						
Multispectral						
Band 1	0.44–0.51	5–6.5	5	1–6	16	400 points for 5 m pixel
Band 2	0.52–0.59					
Band 3	0.63–0.68					
Band 4	0.69–0.73					
Band 5	0.76–0.85					
Planet Scope						
Multispectral						
Blue	0.455–515	3	4	1	12	1111 points for 3m pixel
green	0.50–0.59					
Red	0.59–0.67					
NIR	0.78–0.86					

(Modified and adopted from Thenkabail et al., 2004, 2014, 2015, Aneece et al., 2022)

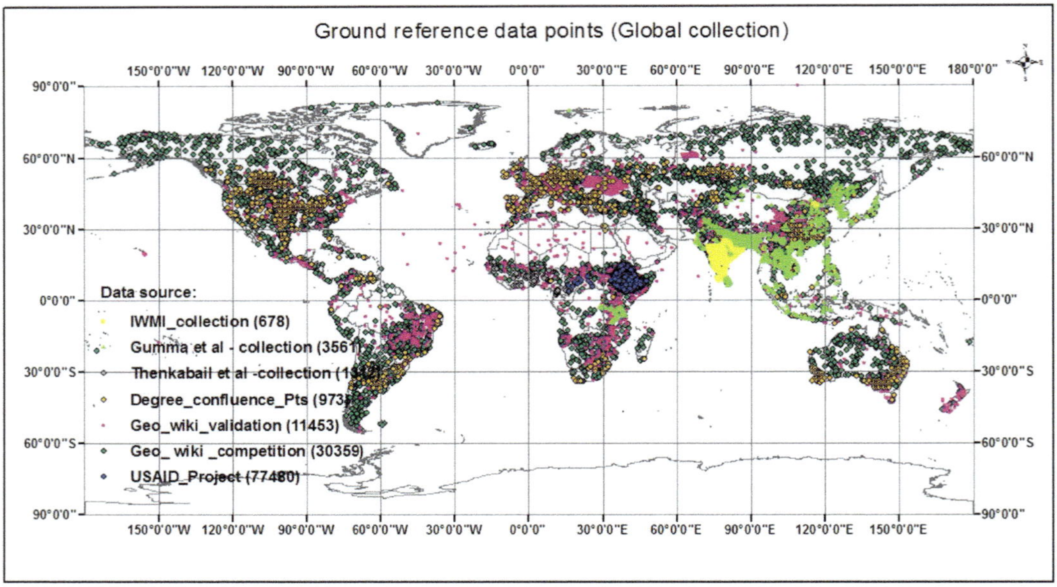

FIGURE 7.4 Field plot data for cropland locations collected over the globe from selected published studies where the number in parenthesis is the number of locations. (Gumma et al., 2011, Thenkabail et al., 2009b, Jarrett et al., 2010, Fritz et al., 2009, Brown and Brickley, 2012.

7.5.3 Field-Plot Data

Field-plot data (e.g., Figure 7.4) can be used for purposes such as (1) class identification and labeling; (2) determining irrigated area fractions; and (3) establishing accuracies, errors, and uncertainties. At each field point (e.g., Figure 7.3), data such as cropland or non-cropland, watering method (irrigated or rainfed), crop type, and cropping intensities are recorded along with GPS locations, digital photographs, and other information (e.g., yield, soil type) as needed. Field plot data also help in gathering an ideal spectral data bank of crop types. One could use the precise locations and the crop characteristics and generate coincident remote sensing data characteristics (e.g., MODIS time-series monthly NDVI).

7.5.4 Very High-Resolution Imagery Data

Very high resolution (sub-meter to 5 m) imagery (VHRI; see hyperspatial data characteristics in Table 7.2) are widely available these days from numerous sources. These data can be used as ground samples in localized areas to classify as well as verify classification results of the coarser resolution imagery. For example, in Figure 7.5, VHRI tiles identify uncertainties existing in cropland classification of coarser resolution imagery. VHRI are specifically useful for identifying croplands versus non-croplands (Figure 7.5). They can also be used for identifying source of irrigation based on associated features such as canals and tanks.

7.5.5 Data Composition: Mega File Data Cube (MFDC) Concept

Data composition requires that all the acquired imagery is harmonized and consistent in known time intervals (e.g., monthly, bi-weekly). All the imagery is acquired is converted to at-sensor reflectance (see Chander et al., 2009, Thenkabail et al., 2004) and then converted to surface reflectance using Landsat Ecosystem Disturbance Adaptive Processing System (LEDAPS) processing system

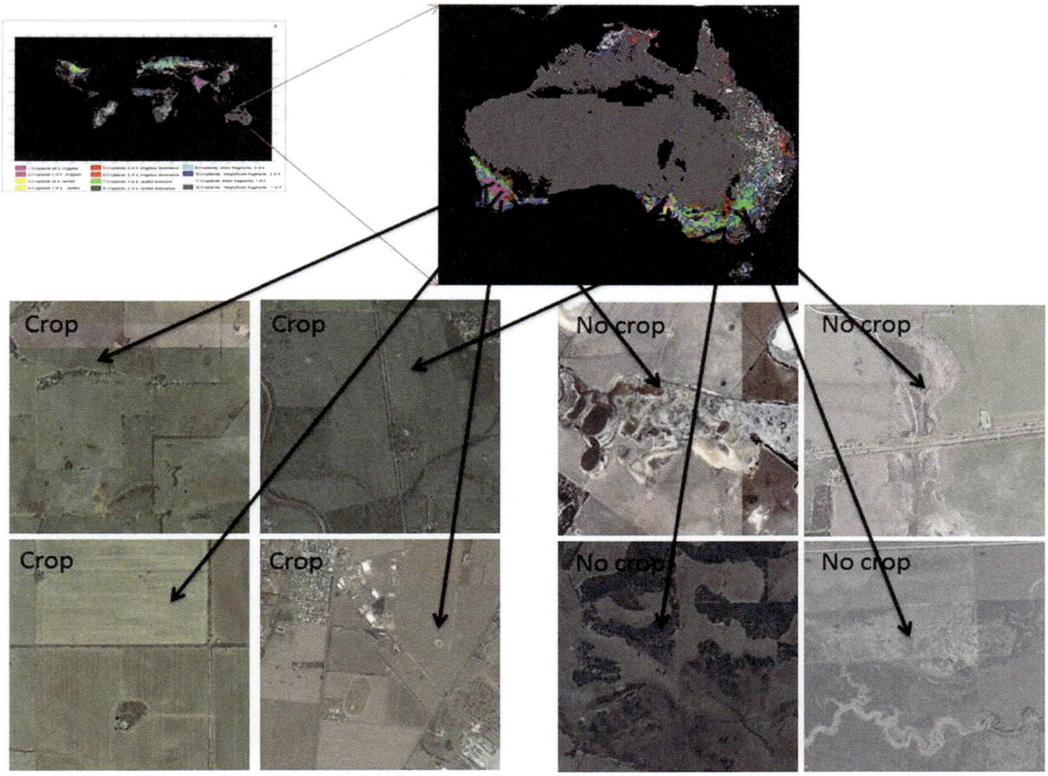

FIGURE 7.5 Very high-resolution imagery used to resolve uncertainties in cropland mapping of Australia. Very high-resolution images were extracted from the Google Earth (Google et al., 2014).

codes for Landsat (Masek et al., 2006) or similar codes for other sensors. All data are processed and mosaicked to required geographic levels (e.g., global, continental).

One method to organize these different but co-located datasets is through the use of a mega-file data cube (MFDC) (Thenkabail et al., 2009b). Numerous secondary datasets are combined in a MFDC, which is then stratified using image segmentation into distinct precipitation-elevation-temperature-vegetation zones. Data within the MFDC can include ASTER-derived refined digital elevation from SRTM (GDEM) (Abrams et al., 2010), monthly long-term precipitation, monthly thermal skin temperature (Mitchell et al., 2004), and forest cover and density. This segmentation allows cropland mapping to be focused; creating distinctive segments of MFDCs and analyzing them separately for croplands which will enhance its accuracy. For example, the likelihood of croplands in a temperature zone of <280 Kelvin is very low. Similarly, croplands in elevation above 1,500 m will be of distinctive characteristics (e.g., patchy, on hilly terrain most likely plantations of coffee or tea). Every layer of data is geo-linked (having precisely same projection and datum and are geo-referenced to one another).

The purpose of mega-file data cube (MFDC; see Thenkabail et al., 2009b for details) is to ensure numerous remote sensing and secondary data layers are all stacked one over the other to form a data cube akin to hyper spectral data cube. The MFDC allows us to have the entire data stack for any geographic location (global to local) as a single file available for analysis. For example, one can classify 10s or 100s or even 1000s of data layers (e.g., monthly MODIS NDVI time series data for a geographic area for an entire decade along with secondary data of the same area) stacked together in a single file and classify the image. The classes coming out of such a mega-file data cube inform us about the phenology along with other characteristics of the crop.

7.6 CROPLAND MAPPING METHODS

7.6.1 Remote Sensing-Based Cropland Classification Mapping Methods for Global, Regional, and Local Scales

There is growing literature on cropland mapping across resolutions for both irrigated and rainfed crops (Teluguntla et al., 2023, 2018, See et al., 2023, Owusu et al., 2024, Zhang et al., 2022, Xie et al., 2021, Gumma et al., 2022, 2011, Friedl et al., 2002, Hansen et al., 2002, Kurz and Seelan, 2007, Loveland et al., 2000, Olofsson et al., 2011, Ozdogan and Woodcock, 2006, Thenkabail et al., 2009a, 2009c, Wardlow and Egbert, 2008, Wardlow et al., 2007, 2006). Based on these studies, an ensemble of methods that is considered most efficient include (1) spectral matching techniques (SMTs) (Teluguntla et al., 2023, Gumma et al., 2022, Thenkabail et al., 2007a, 2009a, 2009c); (2) decision tree algorithms (DeFries et al., 1998); (3) Tassel cap brightness-greenness-wetness (Cohen and Goward, 2004, Crist and Cicone, 1984, Masek et al., 2008); (4) Space-time spiral curves and Change Vector Analysis (Thenkabail et al., 2005); (5) Phenology (Teluguntla et al., 2015, Loveland et al., 2000; Wardlow et al., 2006); and (6) climate data fusion with MODIS time-series spectral indices using decision tree algorithms and sub-pixel classification (Ozdogan and Gutman, 2008).

7.6.2 Spectral Matching Techniques (SMTs) Algorithms

SMTs (Thenkabail et al., 2007a, 2009a, 2011) are innovative methods of identifying and labeling unsupervised land use/cover classes (see illustration in Figures 7.6, 7.7a). For each derived class, this method identifies its NDVI characteristics over time using MODIS time-series data (e.g., Figure 7.6). NDVI time-series or other metrics (Teluguntla et al., 2017, Thenkabail et al., 2005, 2007a, Biggs et al., 2006, Dheeravath et al., 2010) are analogous to spectra, where time is substituted for wavelength. The principle in SMT is to match the shape, or the magnitude or both to an ideal or target spectrum (pure class or "endmember"). The spectra at each pixel to be classified is compared to the endmember spectra and the fit is quantified using the following SMTs (Thenkabail et al., 2007a): (1) Spectral Correlation Similarity (SCS)-a shape measure; (2) Spectral Similarity Value (SSV)-a shape and magnitude measure; (3) Euclidian Distance Similarity (EDS)-a distance measure; and (4) Modified Spectral Angle Similarity (MSAS)-a hyper angle measure.

7.6.2.1 Ideal Spectra Data Bank (ISDB)

The term "ideal or target" spectrum refers to time-series spectral reflectivity or NDVI generated for classes for which we have precise location-specific ground knowledge. From these locations, signatures are extracted using MFDC, synthesized, and aggregated by examining the mean class spectra across time to generate a few hundred signatures that will constitute an ISDB (e.g., Figures 7.6 and 7.7a).

7.6.2.2 Generating Class Spectra

The MFDC (Section 7.5.5) of each of segment (Figures 7.6 and 7.7a) is processed using unsupervised approach with ISOCLASS clustering algorithm (Hexagon, 2021) to produce a large number of class spectra that are then interpreted and labeled. In more localized applications, it is common to undertake a field-plot data collection to identify and label class spectra. However, at the global scale this is not possible due to the enormous resources required to identify and label classes. Therefore, spectral matching techniques (Thenkabail et al., 2007a) to match similar classes or to match class spectra from the unsupervised classification with a library of ideal or target spectra (e.g., Figure 7.6a) will be used to identify and label the classes.

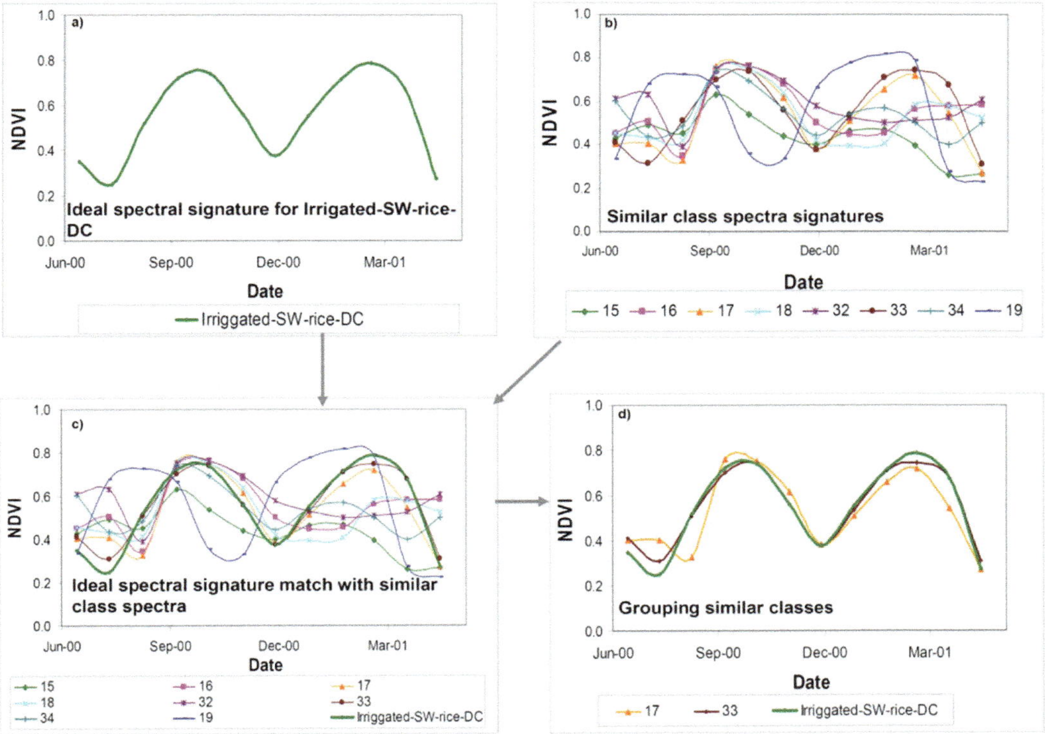

FIGURE 7.6 Spectral matching technique (SMT). In SMTs, the class temporal profile (NDVI curves) are matched with the ideal temporal profile (quantitatively based on temporal profile similarity values) in order to group and identify classes as illustrated for a rice class in this figure. (a) Ideal temporal profile illustrated for "irrigated-surface-water-rice-double crop"; (b) some of the class temporal profile signatures that are similar, (c) ideal temporal profile signature (Figure 7.6a) matched with class temporal profiles (Figure 7.6b), and (d) the ideal temporal profile (Figure 7.6a, in deep green) matches with class temporal profiles of classes 17 and 33 perfectly. Then one can label classes 17 and 33 to be same as the ideal temporal profile ("irrigated-surface-water-rice-double crop"). This is a qualitative illustration of SMTs. For quantitative methods refer to Thenkabail et al. (2007a).

7.6.2.3 Matching Class Spectra with Ideal Spectra Using Spectral Matching Techniques (SMTs)

Once the class spectra are generated, they are compared with ideal spectra to match, identify, and label classes. Often quantitative spectral matching techniques like spectral correlation similarity R-square (SCS R-square) and spectral similarity value (SSV) are used (Thenkabail et al., 2007a).

7.7 AUTOMATED CROPLAND CLASSIFICATION ALGORITHM (ACCA)

Thenkabail et al., 2012, Wu et al., 2014a, 2014b, Xiong et al., 2017, Teluguntla et al., 2017): The first part of the ACCA method involves knowledge-capture to understand and map agricultural cropland dynamics by: (1) identifying croplands versus non-croplands and crop type/dominance based on spectral matching techniques, decision trees tassel cap bi-spectral plots, and very high resolution imagery; (2) determining watering method (e.g., irrigated or rainfed) based on temporal characteristics (e.g., NDVI), crop water requirement (water use by crops), secondary data (elevation, precipitation, temperature), and irrigation source (e.g., canals and wells); (3) identifying

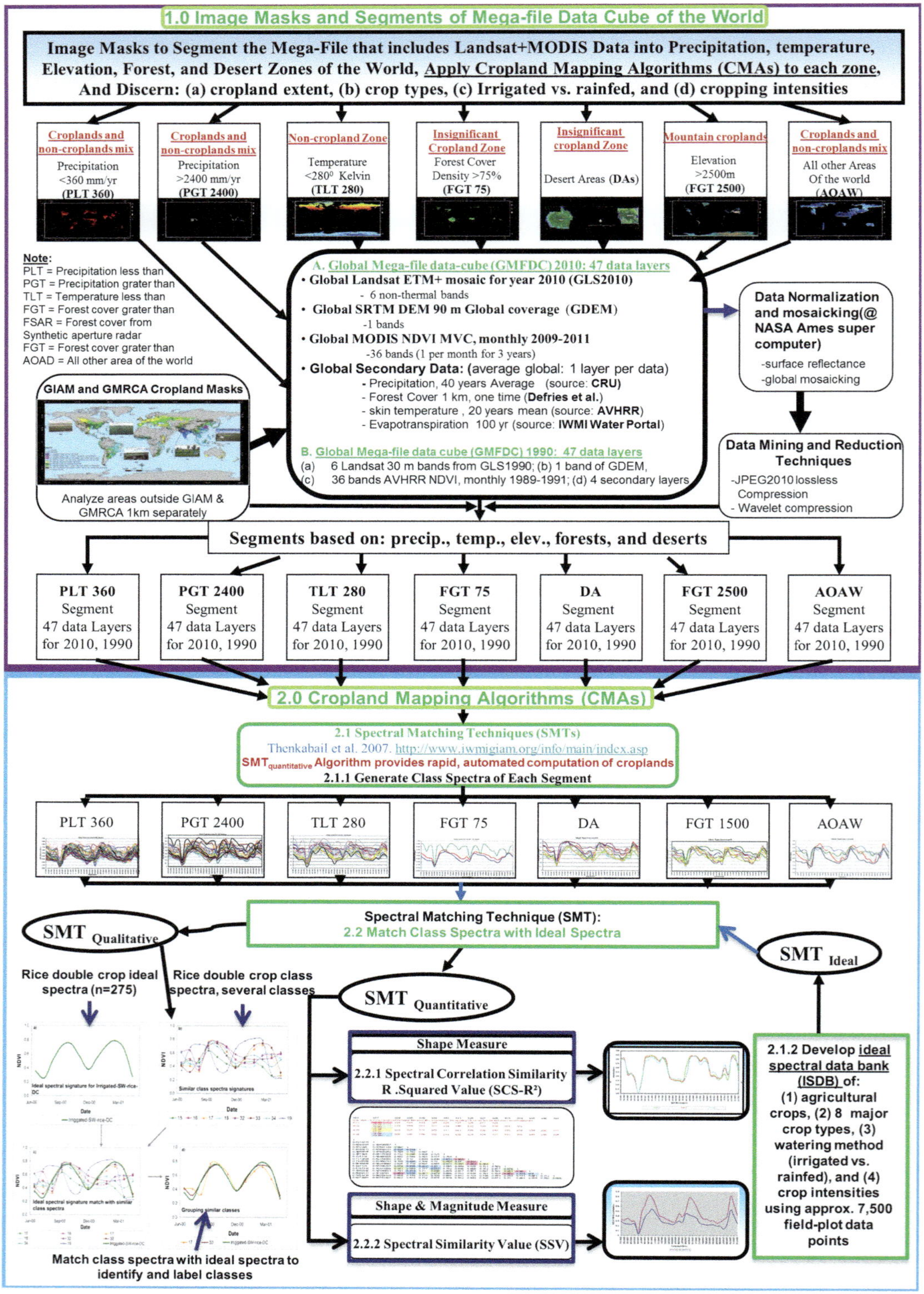

FIGURE 7.7 Cropland mapping method illustrated here for a global scale (see Thenkabail et al., 2009b, 2011). (a) The flowchart demonstrates comprehensive global cropland mapping methods using multi-sensor, multi-date remote sensing, secondary, field plot, and very high-resolution imagery data.

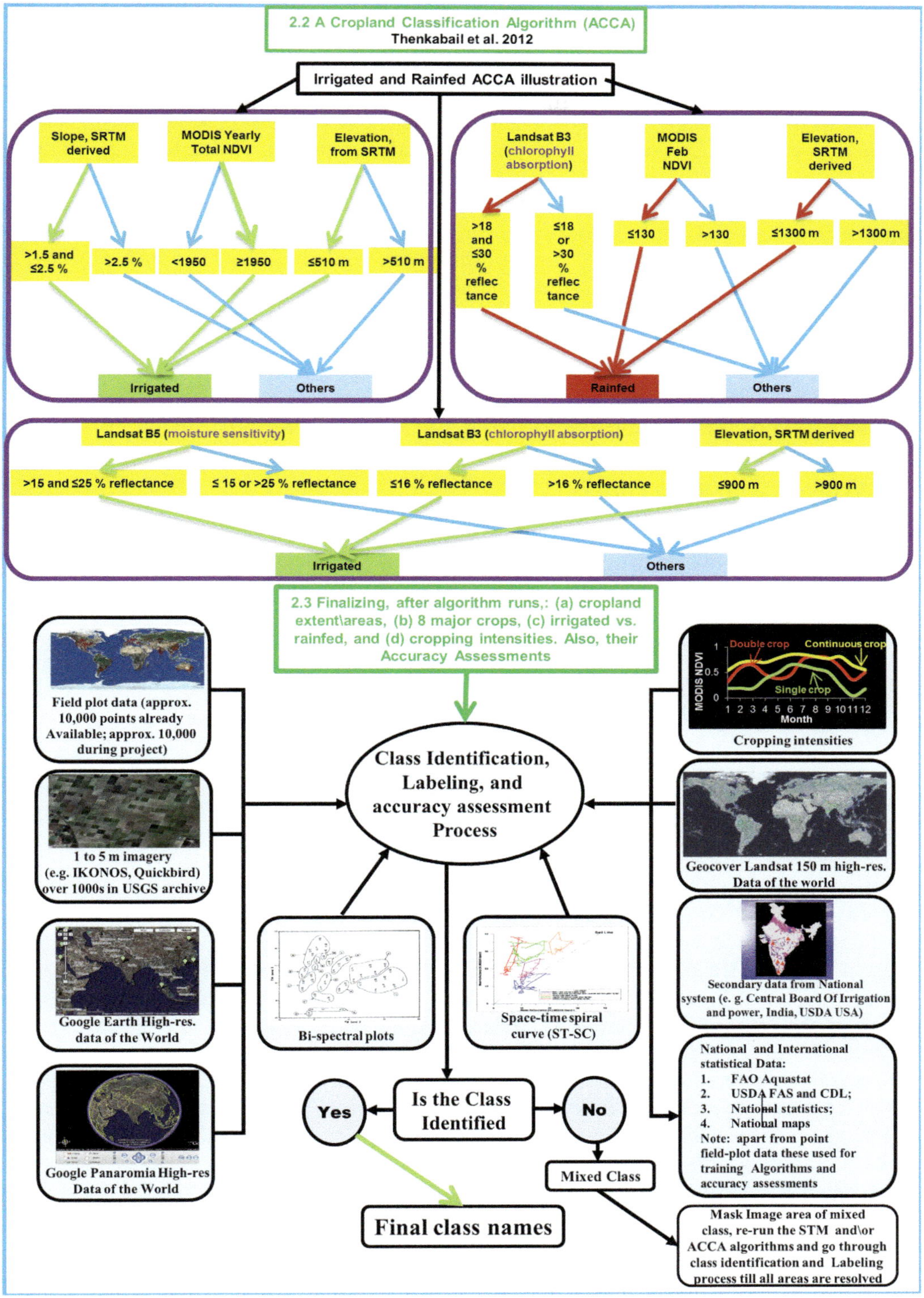

FIGURE 7.7 (Contiuned) (b) Cropland mapping methods illustrated for a global scale. Top half shows automated cropland classification algorithm (see Thenkabail and Wu, 2012; Wu et al., 2014a) and bottom half shows class identification and labeling process (Thenkabail et al., 2005, 2007a, 2009b).

large contiguous and small fragmented; (4) characterizing cropping intensities (single, double, triple, and continuous cropping); (5) interpreting MODIS NDVI Temporal bi-spectral Plots to Identify and Label Classes; and (6) using data from very high resolution imagery, in situ field-plot data, and national statistics (see Figure 7.7b for details). The second part of the method establishes accuracy of the knowledge-captured agricultural map (Congalton, 1991, 2009) and statistics by comparison with national statistics, field-plot data, and very high-resolution imagery. The third part of the method makes use of the captured knowledge to code and map cropland dynamics through an automated algorithm. The fourth part of the method compares the agricultural cropland map derived using an automated algorithm (classified data) with that derived based on knowledge capture (reference map). The fifth part of the method applies the tested algorithm on an independent dataset of the same area to automatically classify and identify agricultural cropland classes. The sixth part of the method assesses accuracy and validates the classes derived from independent dataset using an automated algorithm.

7.8 REMOTE SENSING-BASED GLOBAL CROPLAND PRODUCTS: FOUR SELECTED KEY CROPLAND MAPS, THEIR STRENGTHS, AND LIMITATIONS

Remote sensing offers the best opportunity to map and characterize global croplands most accurately, consistently, and repeatedly. We have selected and used three global remote sensing maps (Thenkabail et al., 2011, Pittman et al., 2010, Yu et al., 2013) that focus exclusively on cropland and a MODIS global land cover and land use map (Friedl et al., 2010) where croplands are included as one class. We examined these maps to identify their strengths and weaknesses, to see how well they compare with each other, and to understand the knowledge gaps that need to be addressed. These maps were produced by:

1. Thenkabail et al. (2011), (Thenkabail et al., 2009b, Biradar et al., 2009)
2. Pittman et al. (2010)
3. Yu et al. (2013)
4. Friedl et al. (2010)

Thenkabail et al. (2009b, 2011); and Biradar et al. (2009) (Figure 7.8, Table 7.3), used a combination of AVHRR, SPOT VGT, and numerous secondary (e.g., precipitation, temperature, and elevation) data to produce a global irrigated area map (GIAM) along with a global map of rainfed cropland areas (GMRCA) (Biradar et al., 2009, Thenkabail et al., 2011; Figure 7.8, Table 7.3). Pittman et al. (2010; Figure 7.9, Table 7.4) used MODIS 250 m data to map global cropland extent. Yu et al. (2013; Figure 7.10, Table 7.5) produced a nominal 30 m resolution cropland extent of the world. Friedl et al. (2010; Figure 7.11, Table 7.6) used 500 m MODIS data in their global land cover and land use product (MCD12Q1) where croplands were one of the 16 land cover classes. The methods, approaches, data, and definitions used in each of these products differ extensively. As a result, the cropland extents mapped by these products also vary significantly. The areas in Tables 7.3–7.6 only show the full pixel areas (FPAs) and not sub-pixel areas (SPAs). Due to the significant difference in the spatial resolution between 1km and 30m products, it can be useful to calculate cropland percentage with in the 1km pixels so the resulting areas can be more accurately compared to 30m pixels. SPAs are actual areas, which can be estimated by re-projecting these maps to appropriate projections and calculating the areas. In this chapter, we did not estimate any SPAs. However, a comparison of the FPAs of the four maps (Figures 7.8 to 7.11) show significant differences in the cropland areas (Tables 7.3 to 7.6) as well as significant differences in the precise locations of the croplands (Figures 7.8 to 7.11), the reasons for which are discussed in the next section.

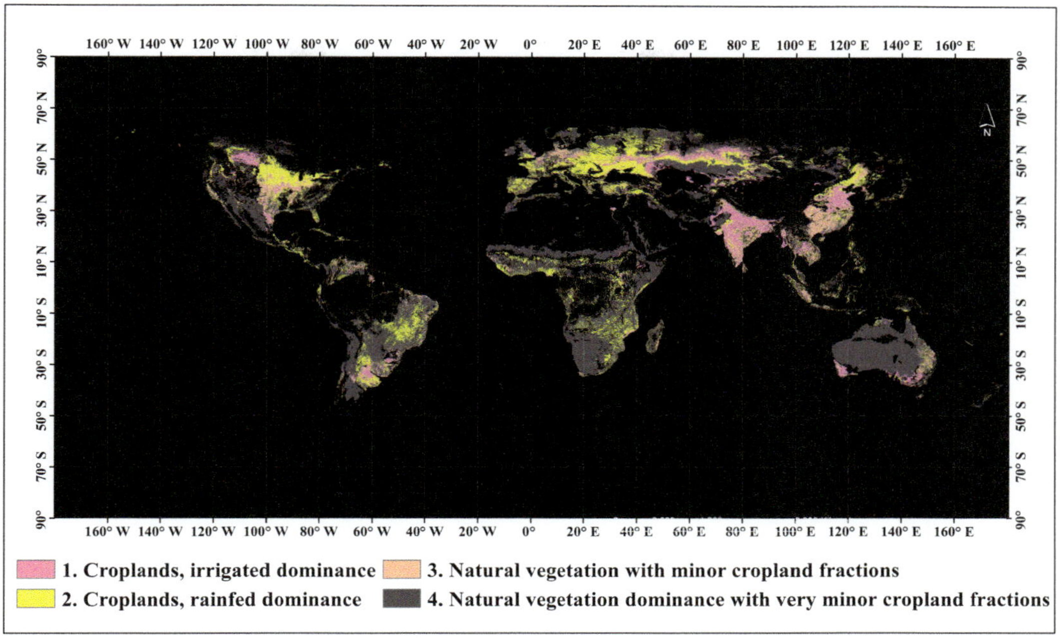

FIGURE 7.8 Global cropland product by Thenkabail et al. (2011, 2009b) using the method illustrated in Figure 7.7 (details in Thenkabail et al., 2011, 2009b). This includes irrigated and rainfed areas of the world. The product is derived using remotely sensed data fusion (e.g., NOAA AVHRR, SPOT VGT, JERS SAR), secondary data (e.g., elevation, temperature, and precipitation), and in situ data. Total area of croplands is 2.3 billion hectares.

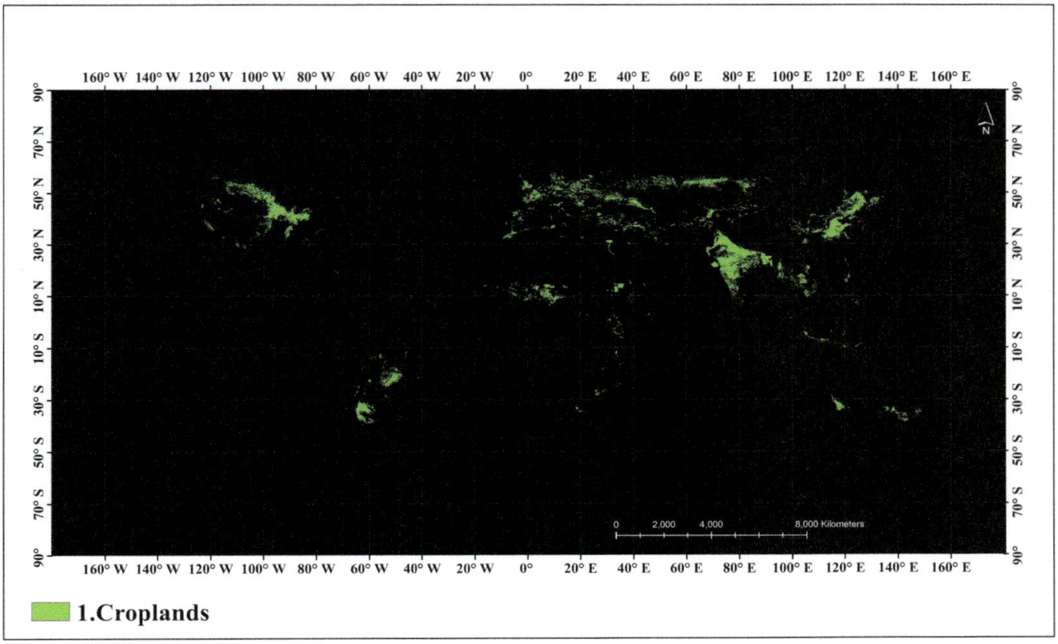

FIGURE 7.9 Global cropland extent map by Pittman et al. (2010) derived using MODIS 250 m data. There is only one cropland class, which includes irrigated and rainfed areas of the world. Total area of croplands is 0.9 billion hectares.

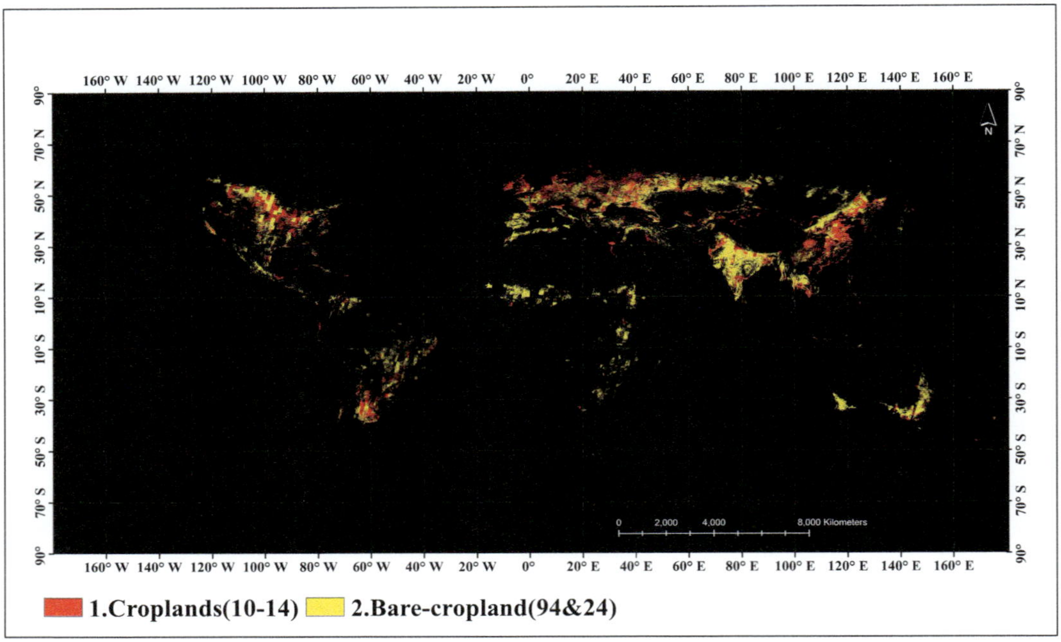

FIGURE 7.10 Global cropland extent map by Yu et al. (2013) derived at nominal 30 m data. Total area of croplands is 2.2 billion hectares. There is no discrimination between rainfed and irrigated areas.

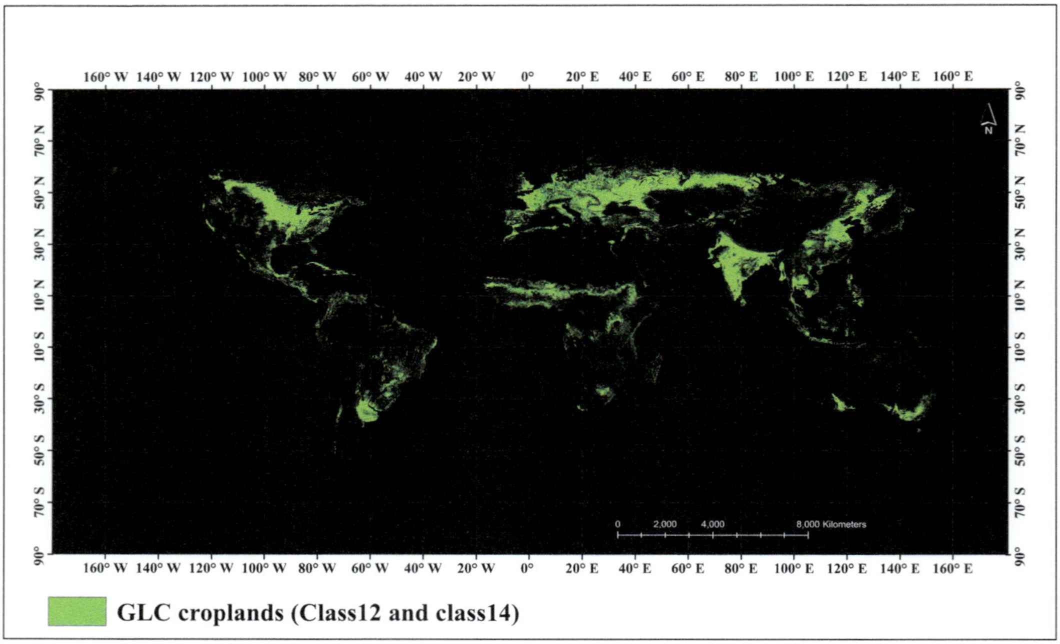

FIGURE 7.11 Global cropland classes (Class 12 and Class 14) extracted from MODIS Global land use and land cover (GLC) 500 m product MCD12Q2 by Friedl et al. (2010). Total area of croplands is 2.7 billion hectares. There is no discrimination between rainfed and irrigated cropland areas.

TABLE 7.3
Global Cropland Extent at Nominal 1-km Based on Thenkabail et al. (2009b, 2011)[1,2] Product shown in Figure 7.8.

Class#	Class Description	Pixels	Percent
#	Names	1 km	%
1	Croplands, irrigated dominance	9359647	40%
2	Croplands, rainfed dominance	14273248	60%
3	Natural vegetation with minor cropland fractions	5504037	
4	Natural vegetation dominance with very minor cropland fractions	44170083	
		23632895	100%

[1] = total of approximately 2.3 billion hectares; Note that these are full pixel areas (FPAs). Actual area is = sub-pixel area (SPA). The SPA is not estimated here. See Thenkabail et al. (2007b) for the methods for calculating SPAs.
[2] = % calculated based on class 1 and 2. Class 3 and 4 are very small cropland fragments

TABLE 7.4
Global Cropland Extent at Nominal 250 m Based on Pittman et al. (2010)[1,2]./ Product shown in Figure 7.9.

Class#	Class Description	Pixels	Percent
#	Names	1km	%
1	Croplands	8948507	100

[1] = total of approximately 0.9 billion hectares. Note that these are full pixel areas (FPAs). Actual area is = sub-pixel area (SPA). The SPA is not estimated here. See Thenkabail et al. (2007b) for the methods for calculating SPAs.
[2] = % calculated based on class 1

TABLE 7.5
Global Cropland Extent at Nominal 30 m Based on Yu et al. (2013)[1,2] Product shown in Figure 7.10.

Class#	Class Description	Pixels	Percent
#	Names	1km	%
1	Croplands (classes 10 to 14)	7750467	35
2	Bare-cropland (classes 94 and 24)	14531323	65
		22281790	100

[1] = total of approximately 2.2 billion hectares. Note that these are full pixel areas (FPAs). Actual area is = sub-pixel area (SPA). The SPA is not estimated here. See Thenkabail et al. (2007b) for the methods for calculating SPAs.
[2] = % calculated based on class 1 and 2.

7.8.1 Global Cropland Extent at Nominal 1-km Resolution

We synthesized the aforementioned four global cropland products and produced a unified Global Cropland Extent map (GCE V1.0) at nominal 1 km (Table 7.7a; Figure 7.12a). The process involved resampling each global cropland product to a common resolution of 1 km and then performing GIS

TABLE 7.6
Global Cropland Extent at Nominal 500 m Based on Friedl et al. (2010)[1] Product shown in Figure 7.11.

Class#	Class Description	Pixels	Percent
#	Names	1km	%
1	Global croplands (Class 12 and 14)	27046084	100

[1] = approximately, total 2.7 billion hectares based on class 12 and class 14. Note that these are full pixel areas (FPAs). Actual area is = sub-pixel area (SPA). The SPA is not estimated here. See Thenkabail et al. (2007b) for the methods for calculating SPAs.

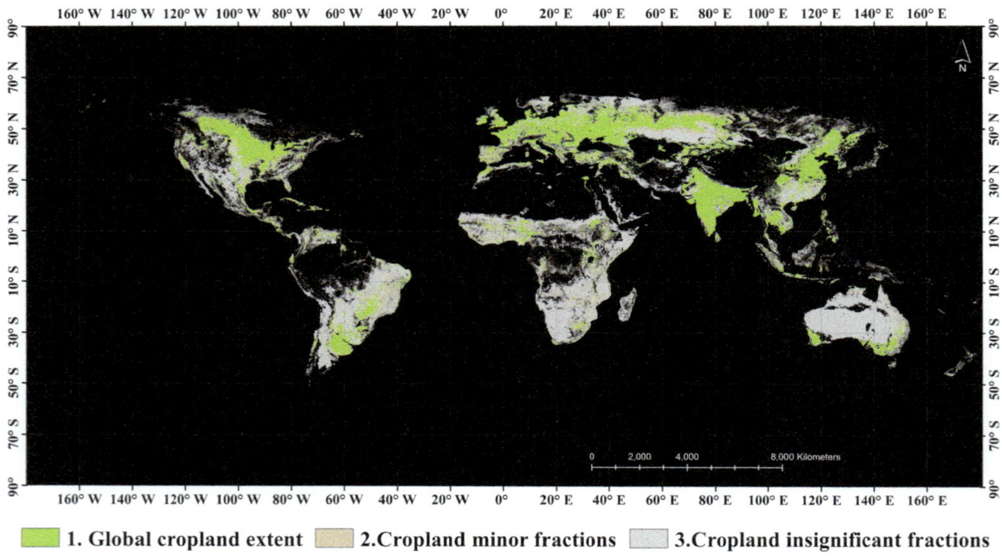

FIGURE 7.12A An aggregated three class global cropland extent map at nominal 1-km based on four major studies: Thenkabail et al. (2009a, 2011), Pittman et al. (2010), Yu et al. (2013), and Friedl et al. (2010). Class 1 is total cropland extent; total cropland extent is 2.3 billion hectares (full pixel areas). Class 2 and Class 3 have ONLY minor fractions of croplands. Refer to Table 7.7a for cropland statistics of this map.

data overlays to determine where the cropland extents matched and where they differed by expert visual interpretation.

Figure 7.12a shows the aggregated global cropland extent map with its statistics in Table 7.7a. Class 1 in Figure 7.12a and Table 7.7a provides the global cropland extent included in all 4 maps. Actual area of this extent is not calculated yet, but it includes approximately 2.3 billion full pixel areas (FPAs) (Table 7.7a). The spatial distribution of these 2.3 billion hectares is demonstrated as class 1 in Figure 7.12a. Class 2 and 3 are areas with minor or very minor cropland fractions. Class 2 and Class 3 are classes with large areas of natural vegetation and/or desert lands and other lands.

Figure 7.12b and Table 7.7b demonstrate where and by how much the 4 products match with one another. For example, 2,802,397 pixels (class 1, Table 7.7b, Figure 7.12b) are croplands that are

TABLE 7.7A

Global Cropland Extent at Nominal 1 km Based on Four Major Studies: Thenkabail et al. (2009b, 2011), Pittman et al. (2010), Yu et al. (2013), and Friedl et al. (2010). Three class map1,2,3 (Figure 7.12a).

Class#	Class Description	Pixels	Percent
#	Names	1 km	%
1	1. Croplands	23493936	100
2	2. Cropland minor fractions	13700176	
3	3. Cropland very minor fractions	44662570	

[1] = approximately 2.3 billion hectares (class 1) of cropland is estimated. But this is full pixel area. Actual area is = sub-pixel area (SPA). The SPA is not estimated here. See Thenkabail et al. (2007b) for the methods for calculating SPAs.

[2] = % calculated based on Class 1.

[3] = Class 2 and 3are minor/very minor cropland fragments

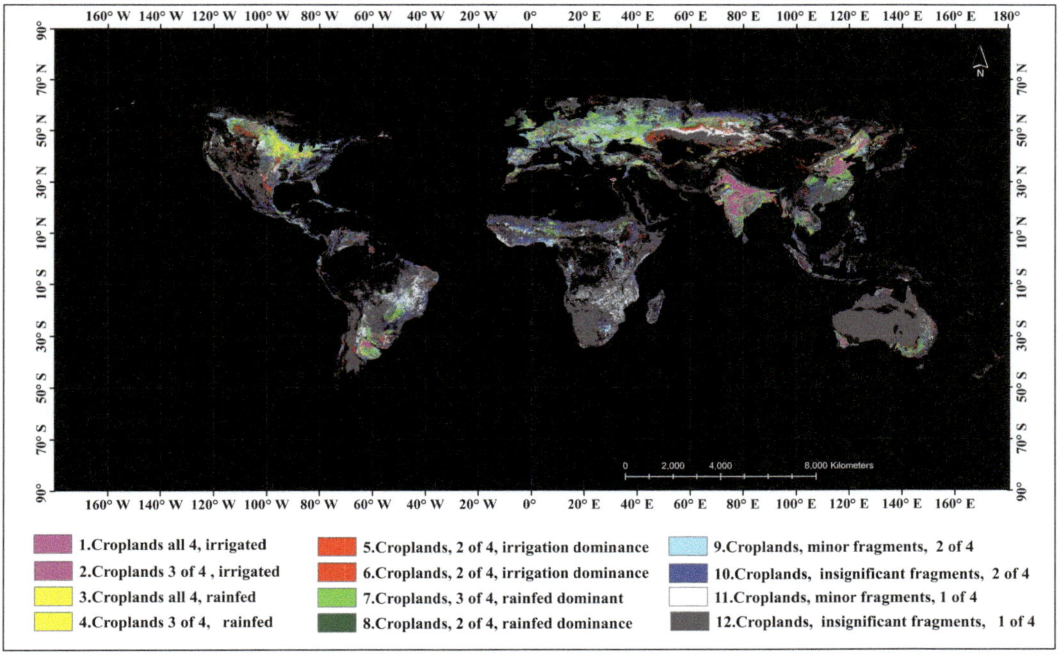

FIGURE 7.12B A disaggregated 12 class global cropland extent map derived at nominal 1-km based on four major studies: Thenkabail et al. (2009a, 2011), Pittman et al. (2010), Yu et al. (2013), and Friedl et al. (2010). Class 1 to Class 9 are cropland classes that are dominated by irrigated and rainfed agriculture. Class 10 to and Class 12 have ONLY minor or very minor fractions of croplands. Refer to Table 7.7b for cropland statistics of this map.

irrigated. Some of the products do not separately classify irrigated vs rainfed croplands, although all 4 products show where croplands are. We first identified where all 4 products match as croplands and then added irrigation status or other indicators (e.g., irrigation dominance, rainfed; Table 7.7b) from the product by Thenkabail et al. (2009b, 2011).

TABLE 7.7B

Global Cropland Extent at Nominal 1 km Based on Four Major Studies: Thenkabail et al. (2009b, 2011), Pittman et al. (2010), Yu et al. (2013), and Friedl et al. (2010). Twelve class map1,2,3,4 (Figure 7.12b).

Class#	Class Description	Pixels	Percent
#	Names	1 km	%
1	Croplands all 4, irrigated	2802397	12
2	Croplands 3 of 4, irrigated	289591	1
3	Croplands all 4, rainfed	1942333	8
4	Croplands 3 of 4, rainfed	427731	2
5	Croplands, 2 of 4, irrigation dominance	3220330	14
6	Croplands, 2 of 4, irrigation dominance	1590539	7
7	Croplands, 3 of 4, rainfed dominance	6206419	26
8	Croplands, 2 of 4, rainfed dominance	3156561	13
9	Croplands, minor fragments, 2 of 4	3858035	17
10	Croplands, very minor fragments, 2 of 4	6825290	
11	Croplands, minor fragments, 1 of 4	6874886	
12	Croplands, very minor fragments, 1 of 4	44662570	
	Class 1 to 9 total	23493936	100

[1]= approximately 2.3 billion hectares (class 1 to 9) of cropland is estimated. But this is full pixel area. Actual area is = sub-pixel area (SPA). The SPA is not estimated here. See Thenkabail et al. (2007b) for the methods for calculating SPAs.

[2]= % calculated based on class 1 to 9

[3]=Class 10,11and 12 are minor cropland fragments

[4]= all four means, all four studies agreed

Table 7.7b and Figure 7.12b show 12 classes of which classes 1 and 2 are croplands with irrigated agriculture, classes 3 and 4 are croplands with rainfed agriculture, classes 5 and 6 are croplands where irrigated agriculture dominates, classes 7 and 8 are croplands where rainfed agriculture dominates, and classes 9 to 12 are areas with minor or very minor cropland fractions. Classes 9 to 12 are those with large areas with natural vegetation and/or desert lands and other lands.

Interestingly, and surprisingly as well, only 20% (class 1 and 3; Table 7.7b, Figure 7.12b) of the total cropland extent matched in all four products. Further, 49% (Class 1, 2, 3, 4, and 7; Table 7.7b, Figure 7.12b) of the total cropland areas match in at least three of the four products. This implies that all the four products have considerable uncertainties in determining the precise location of the croplands. The degree of uncertainty in the cropland products can be attributed to factors including:

1. Coarse resolution of the imagery used in the study
2. Definition of mapping products of interest
3. Methods and approaches adopted
4. Limitations of the secondary data

Table 7.7c and Figure 7.12c show five classes of which classes 1 and 2 are croplands with irrigated agriculture, class 3 is croplands with rainfed agriculture, classes 4 and 5 have ONLY minor or very minor cropland fractions. We recommend the use of this aggregated 5 class global cropland map (Figure 7.12c and Table 7.7c) produced based on the four major cropland mapping efforts [i.e., Thenkabail et al. (2009a, 2011), Pittman et al. (2010), Yu et al. (2013), and Friedl et al. (2010)] using remote sensing. This map (Figure 7.12c, Table 7.7c) is a synthesis of the four prior major studies on global:

TABLE 7.7C

Global Cropland Extent at Nominal 1 km Based on Four Major Studies: Thenkabail et al. (2009b, 2011), Pittman et al. (2010), Yu et al. (2013), and Friedl et al. (2010). Five class map1,2,3 (Figure 7.12c).

Class#	Class Description	Pixels	Percent
#	Names	1 km	%
1	1. Croplands, irrigation major	3091988	13
2	2. Croplands, irrigation minor	4810869	21
3	3. Croplands, rainfed	11733044	50
4	4. Croplands, rainfed minor fragments	3858035	16
5	5. Croplands, rainfed very minor fragments	13700176	
	Class 1 to 4 total	23493936	100.0%

[1]= approximately 2.3 billion hectares (class 1 to 4) of cropland is estimated. But this is full pixel area. Actual area is = sub-pixel area (SPA). The SPA is not estimated here. See Thenkabail et al. (2007b) for the methods for calculating SPAs.

[2] = % calculated based on Class 1 to 4.

[3]= Class 5 is very minor cropland fragments

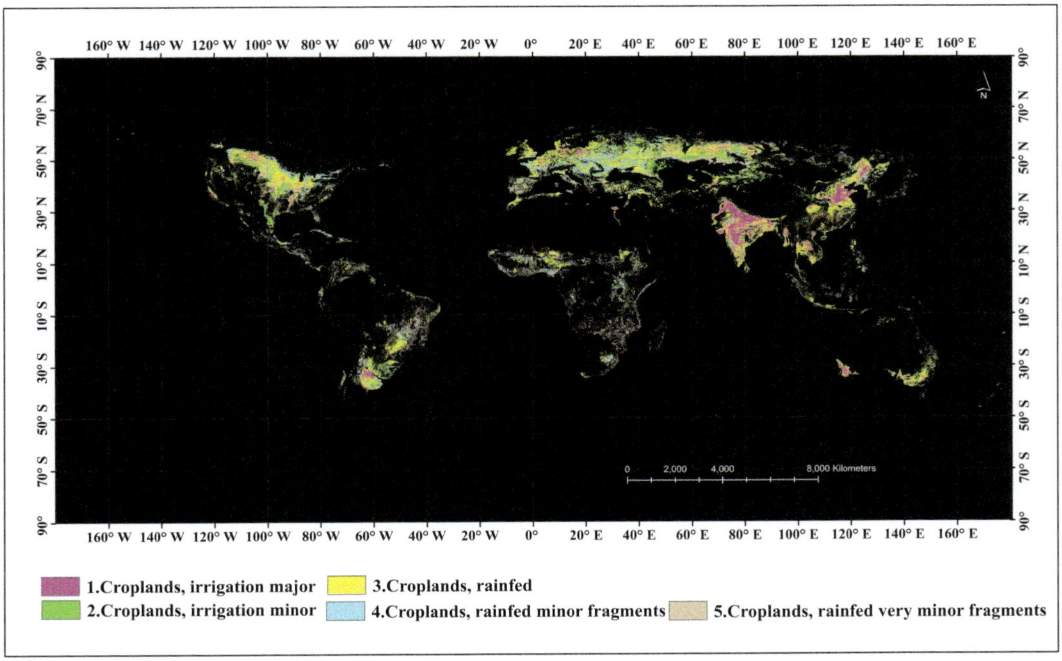

FIGURE 7.12C A disaggregated five class global cropland extent map derived at nominal 1-km based on four major studies: Thenkabail et al. (2009a, 2011), Pittman et al. (2010), Yu et al. (2013), and Friedl et al. (2010). Class 1 to Class 5 are cropland classes, that are dominated by irrigated and rainfed agriculture. However, class 4 and Class 5 have ONLY minor or very minor fractions of croplands. Refer to Table 7.7c for cropland statistics of this map. **Note: Irrigation major**: areas irrigated by large reservoirs created by large and medium dams, barrages and even large groundwater pumping. **Irrigation minor**: areas irrigated by small reservoirs, irrigation tanks, open wells, and other minor irrigation. However, it is very hard to draw a strict boundary between major and minor irrigation and in places there can be significant mixing. So, when major irrigated areas such as the Ganges basin, California's central valley, Nile basin etc. are clearly distinguishable as major irrigation, in other areas major and minor irrigation may inter-mix.

- Cropland extent location
- Cropland watering method (irrigation versus rainfed)

The product (Figure 7.12c, Table 7.7c) does not show where the crop types are or even the crop dominance. However, cropping intensity can be gathered using multi-temporal remote sensing over these cropland areas.

7.9 CHANGE ANALYSIS

Once the croplands are mapped, we can use the time-series historical data such as continuous global coverage of remote sensing data from NOAA Very High Resolution Radiometer (VHRR) and Advanced VHRR (AVHRR), Global Inventory Modeling and Mapping Studies (GIMMS; 1982–2000), MODIS time-series (2001-present) to help build an inventory of historical agricultural development (e.g., Figures 7.13 and 7.14). Such an inventory will provide information including identifying areas that have switched from rainfed to irrigated production (full or supplemental), and non-cropped to cropped (and vice versa). A complete history will require systematic analysis of remotely sensed data as well as a systematic compilation of all routinely populated cropland databases from the agricultural departments of all countries throughout the world. The differences in pixel sizes in AVHRR *versus* MODIS will: (a) influence class identification and labeling, and (b) cause different levels of uncertainties. These issues can be addressed by determining sub-pixel areas and uncertainties involved in class accuracies and area estimation at various spatial resolutions (Thenkabail et al., 2007b, Velpuri et al., 2009, Ozdogan and Woodcock, 2006). Change analyses (Tomlinson, 2003) between products are conducted in order to investigate both the spatial and temporal changes in croplands (e.g., Figure 7.13, 7.14) that will help establish: (1) change in total cropland areas, (2) change in spatial location of cropland areas, (3) expansion on croplands into natural vegetation, (4) expansion of irrigation, (5) change from croplands to bio-fuels, and (6) change from croplands to urban. Massive reductions in cropland areas in certain parts of the world may be detected, including cropland lost as a result of reductions in available groundwater supply due to overdraft (Wada et al., 2012, Jiang, 2009, Rodell et al., 2009).

7.10 UNCERTAINTIES OF EXISTING CROPLAND PRODUCTS

Currently, the main causes of uncertainties in areas reported in various studies (Ramankutty et al., 2008 versus, Thenkabail et al., 2009a, 2009c) can be attributed to, but not limited to: (1) reluctance of national and state agencies to furnish the census data on irrigated area and concerns of their institutional interests in sharing of water and water data; (2) reporting of large volumes of census data with inadequate statistical analysis; (3) subjectivity involved in data collection process of observed data; (4) inadequate accounting of irrigated areas, especially minor irrigation from groundwater, in national statistics; (5) definitional issues involved in mapping using remote sensing as well as national statistics; (6) difficulties in arriving at precise estimates of area fractions (AFs) using remote sensing; (7) difficulties in separating irrigated from rainfed croplands; and (8) imagery resolution in remote sensing. Other limitations include (Thenkabail et al., 2009a, 2011):

1. Absence of precise spatial location of the cropland areas for training and validation
2. Uncertainties in differentiating irrigated areas from rainfed areas
3. Absence of crop types and cropping intensities
4. Inability to generate cropland maps and statistics, routinely
5. Absence of dedicated web/data portal for dissemination cropland products

Further, the need to map accurately specific cropland characteristics such as crop types and watering methods (e.g., irrigated vs. rainfed) is crucial in food security analysis. For example, the

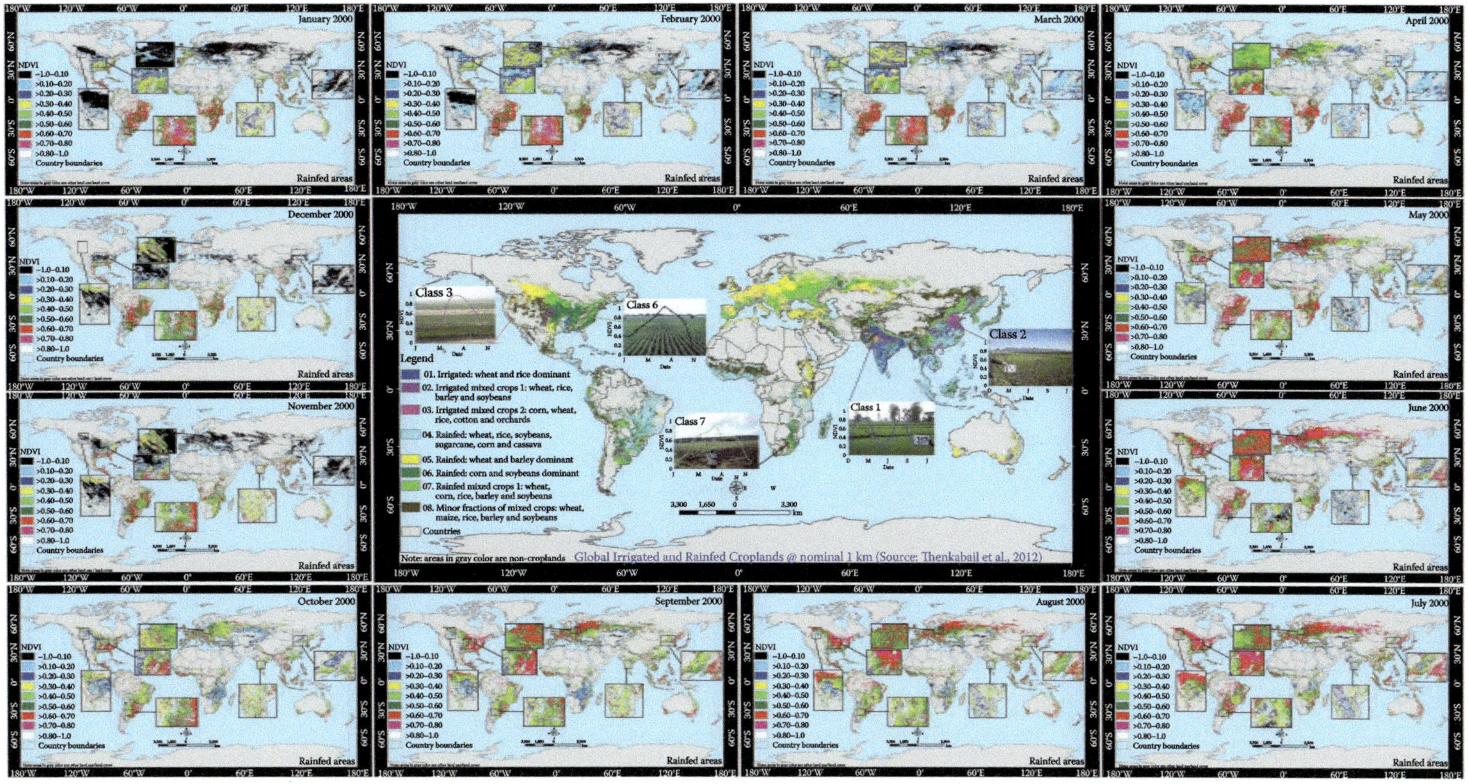

FIGURE 7.13 Center image of global cropland (irrigated and rainfed) areas at 1 km for year 2000 produced by overlying the remote sensing derived product of the International Water Management Institute (IWMI; Thenkabail et al., 2012, 2011, 2009a, 2009b; www.iwmigiam.org) over five dominant crops (wheat, rice, maize, barley, and soybeans) of the world produced by Ramankutty et al. (2008). The five crops constitute about 60% of all global cropland areas. The IWMI remote sensing product is derived using remotely sensed data fusion (e.g., NOAA AVHRR, SPOT VGT, JERS SAR), secondary data (e.g., elevation, temperature, and precipitation), and in situ data. Total area of croplands is 1.53 billion hectares of which 399 million hectares is total area available for irrigation (without considering cropping intensity) and 467 million hectares is annualized irrigated areas (considering cropping intensity). Surrounding NDVI images of irrigated areas: the January to December irrigated area NDVI dynamics is produced using NOAA AVHRR NDVI. The irrigated areas were determined by Thenkabail et al. (2011, 2009a, 2009b).

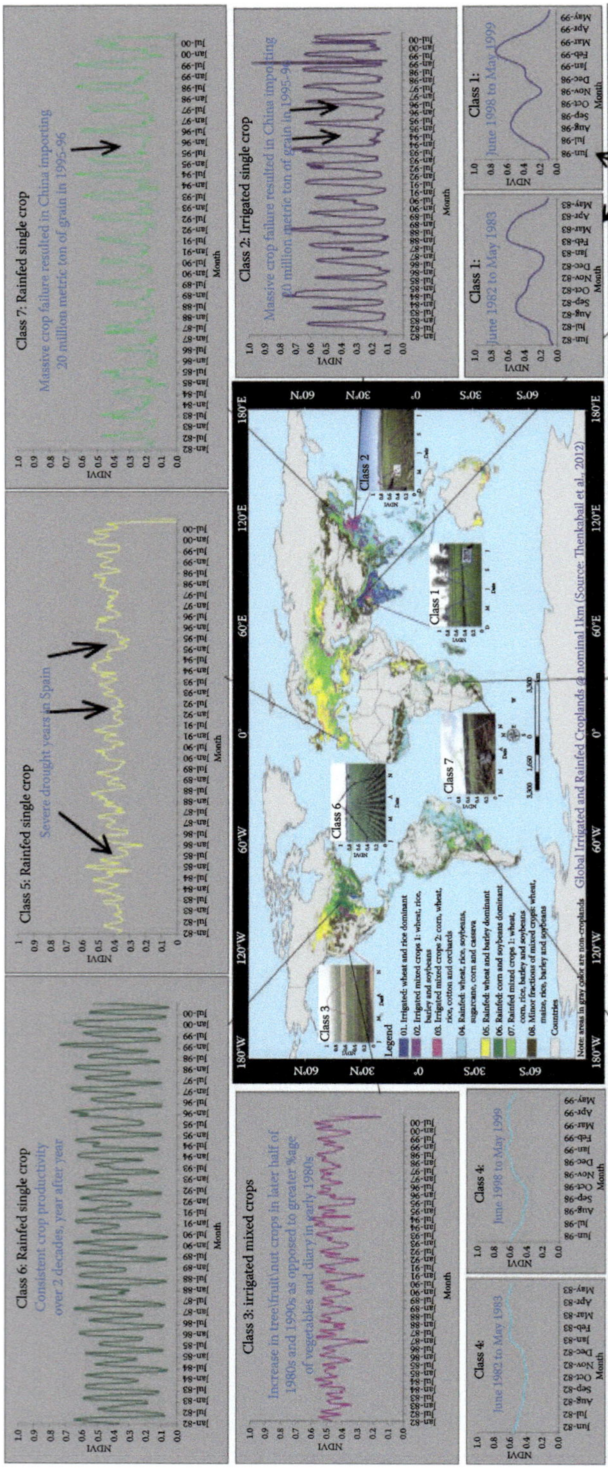

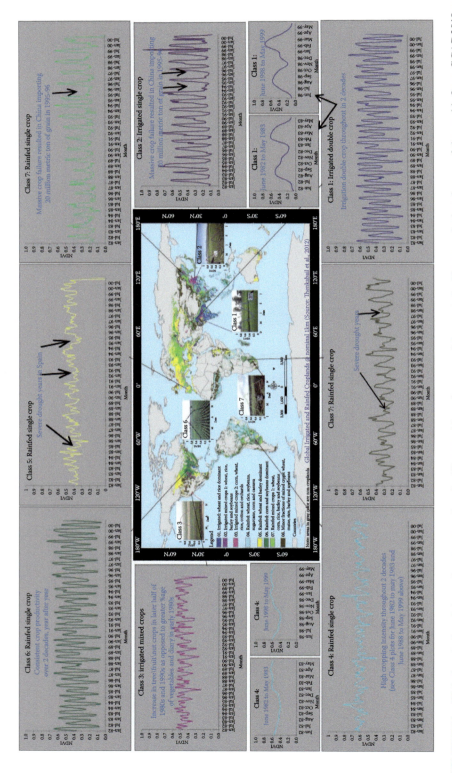

FIGURE 7.14 Global agricultural dynamics over two decades illustrated here for some of the most significant agricultural areas of the world. Once GCAD2010 and GCAD1990 are established at nominal 30 m resolution for the entire world, AVHRR-MODIS monthly MVC NDVI time-series from 1982 to 2017 can be used to provide a continuous time history of global irrigated and rainfed croplands, establish their spatial and temporal changes, and highlight the hot spots of change.

importance of irrigation to global food security is highlighted in a study by Siebert and Döll (2010) who show that without irrigation there would be a decrease in production of various foods including dates (60%), rice (39%), cotton (38%), citrus (32%), and sugarcane (31%) from their current levels. Globally, without irrigation cereal production would decrease by a massive 43%, with overall cereal production, from irrigated and rainfed croplands, decreasing by 20% (Siebert and Döll, 2010).

These limitations are a major hindrance in accurate/reliable global, regional, and country-by-country water use assessments that in turn support crop productivity (productivity per unit of land; kg/m^2) studies, water productivity (productivity per unit of water; kg/m^3) studies, and food security analyses. The higher degrees of uncertainty in coarser resolution data are due to its inherent characteristic of the coarse resolution to capture fragmented, smaller patches of croplands accurately, and the homogenization of both crop and non-crop land within areas of patchy land cover distribution. In either case, there is a strong need for finer spatial resolution to reduce the uncertainty and to produce accurate products.

7.11 WAY FORWARD

Given the aforementioned issues with existing maps of global croplands, the way forward will be to produce global cropland maps at finer spatial resolution and applying a suite of advanced analysis methods. Previous research has shown that at finer spatial resolution the accuracy of irrigated and rainfed area class delineations improve because at finer spatial resolution more fragmented and smaller patches of irrigated and rainfed croplands can be delineated (Ozdogan and Woodcock, 2006, Velpuri et al., 2009). Further, greater details of crop characteristics such as crop types (e.g., Figure 7.15) can be determined at finer spatial resolutions. Crop type mapping will involve use of advanced methods of analysis such as data fusion of higher spatial resolution images from sensors such as Sentinel, Resourcesat, Landsat, AWiFS and MODIS (e.g., Table 7.2) supported by extensive ground surveys and ideal spectral data bank (ISDB) (Thenkabail et al., 2007a). Harmonic analysis is often adopted to identify crop types (Sakamoto et al., 2005) using methods such as the conventional Fourier analysis and adopting a Fourier Filtered Cycle Similarity (FFCS) method (Geerken et al., 2005). Mixed classes are resolved using hierarchical crop mapping protocol based on decision tree algorithm (Wardlow and Egbert, 2008). Irrigated versus rainfed croplands can be distinguished using spectral libraries (Thenkabail et al., 2007a) and ideal spectral data banks (Thenkabail et al., 2009a, 2007a). Similar classes can be grouped by matching class spectra with ideal spectra based on spectral matching techniques (SMTs; Thenkabail et al., 2007a). Details such as crop types are crucial for determining crop water use, crop productivity, and water productivity leading to providing crucial information needed for food security studies. However, the high spatial resolution must be fused with high temporal resolution data in order to obtain time-series spectra that are crucial for monitoring crop growth dynamics and cropping intensity (e.g., single crop, double crop, and continuous year-round crop). Numerous other methods and approaches exist. But usefulness of multi-sensor remote sensing is to produce croplands products such as:

1. Cropland extent/area
2. Crop types (initially focused on eight crops that occupy 70% of global croplands)
3. Irrigated versus rainfed croplands
4. Cropping intensities/phenology (single, double, triple, continuous cropping)
5. Cropped area computation
6. Cropland change over space and time

Recent advances have provided quantum leap in global cropland mapping (Thenkabail et al., 2021, Teluguntla et al., 2023). This involves use of Landsat and Landsat-type (e.g., Sentinel-2A and 2 B) data that have 30 m (1 pixel = 0.09 hectares) or finer spatial resolution. Such data are, typically, acquired in 4-11 spectral bands, over 400-12.500 nm spectral range, in broad band-widths (>30 nm), and in temporal frequency of 5-8 days. The The Harmonized Landsat-8 and 9 and Sentinel-2 (HLS)

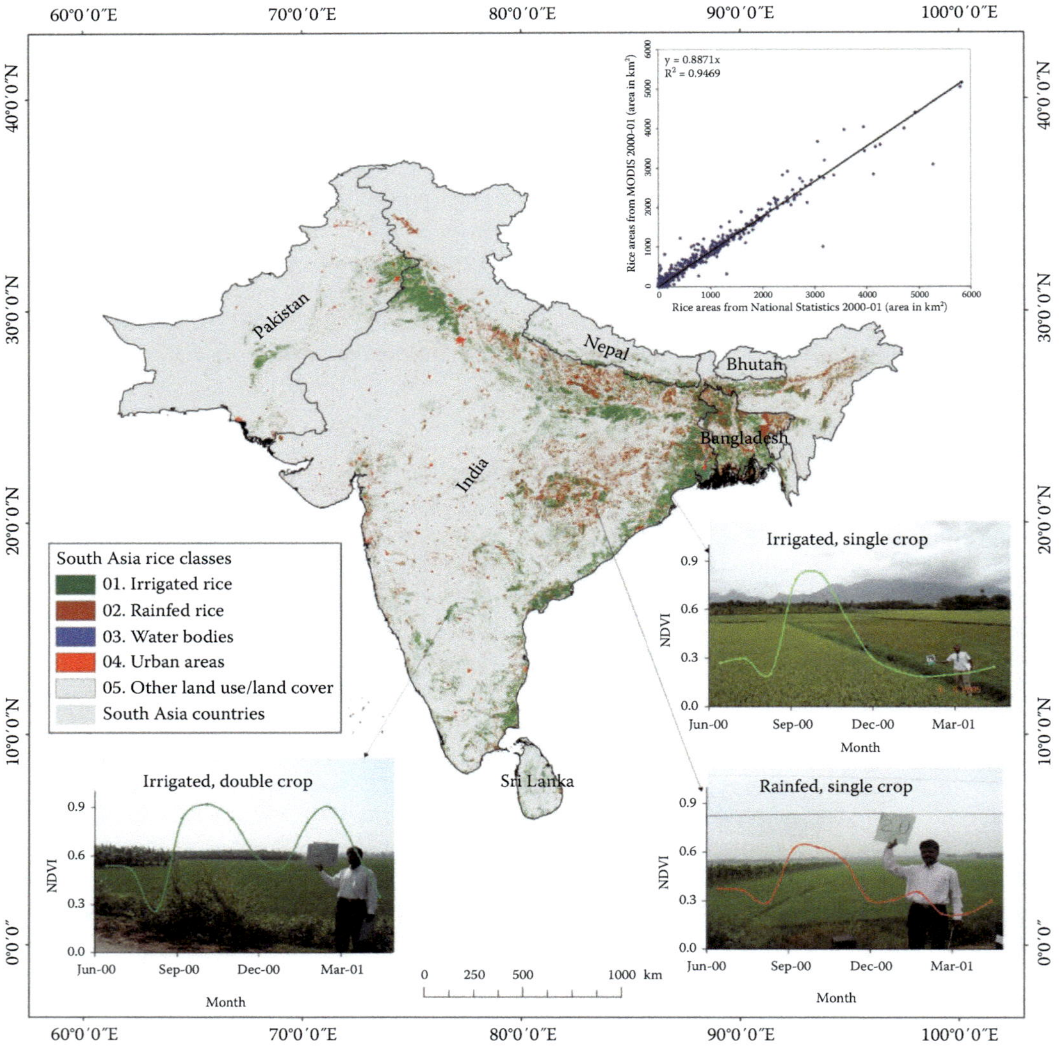

FIGURE 7.15 Rice map of south Asia produced using the method illustrated in Figure 7.6. (Source: From Gumma, M. K. et al., 2011.)

(Masek et al., 2022, 2021) has temporal frequency of 2-3 days covering the entire Planet. This data, covering the world, leads to massively large datasets of several petabytes. This data is seamlessly harmonized, standardized, and synthesized on the cloud platforms such as the Google Earth Engine (GEE) and classified and analyzed using machine learning algorithms by coding in Python and Java Script. The global food-security support analysis data (GFSAD) project generated Landsat-derived global cropland products that are described in detail in our various publications (Thenkabail et al., 2021, Teluguntla et al., 2023, 2018, 2017, Xiong et al., 2018, 2017, Oliphant et al., 2019, Gumma et al., 2020, Phalke et al., 2020) and can be downloaded from NASA's The Land Processes Distributed Active Archive Center (LP DAAC). The GFSAD project produced: 1. Landsat-derived global cropland extent product @ 30m (LGCEP30) (Figure 7.16; Thenkabail et al. 2021; Download data from: https://lpdaac.usgs.gov/news/release-of-gfsad-30-meter-cropland-extent-products/), and 2. Landsat-derived global rainfed and irrigated area product @ 30m (LGRIP30) (Figure 7.17; Teluguntla et al., 2023; download data from: https://lpdaac.usgs.gov/products/lgrip30v001/).

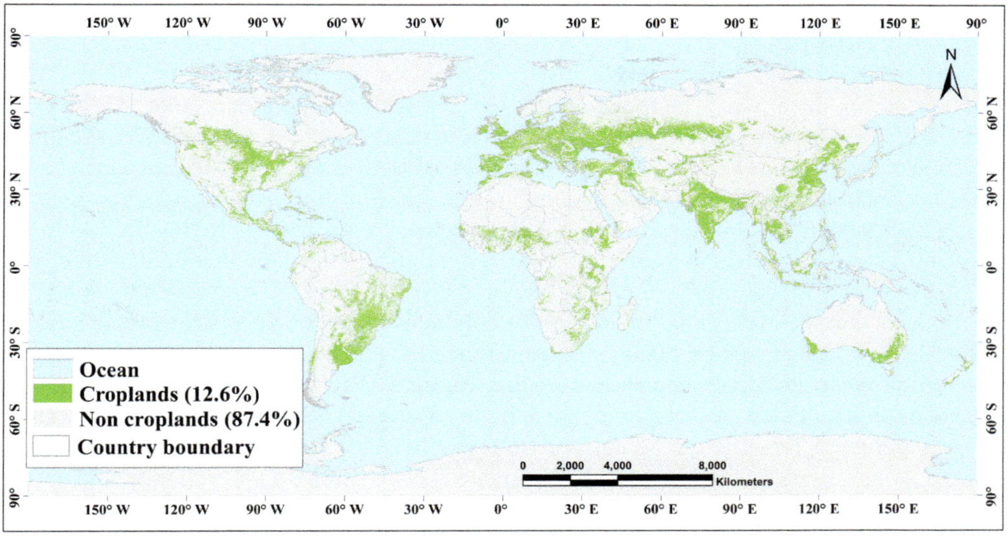

FIGURE 7.16 Global Cropland Extent Product @ 30 m (GCEP30) produced using Landsat time-series 30m data for the nominal year 2015 (Credit: Thenkabail et al., 2021). There was a total of 1.8 billion hectares of croplands in the world (12.6% of the terrestrial area)as of 2015. Download data from: https://lpdaac.usgs.gov/news/release-of-gfsad-30-meter-cropland-extent-products/

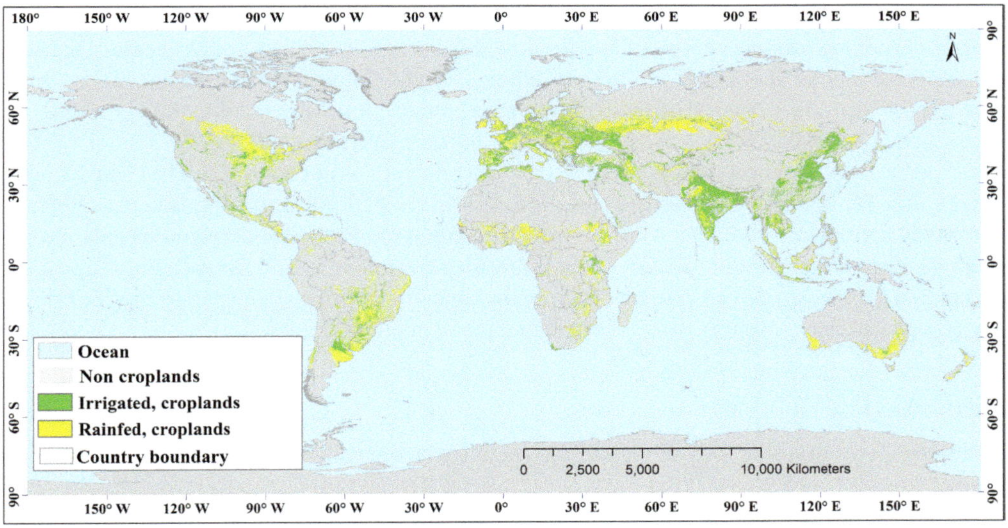

FIGURE 7.17 LGRIP30 V001 2015. Landsat-derived rainfed and irrigated-area product @ 30m Version 1 for the nominal year 2015 (Teluguntla et al., 2023). Of the 1.8 billion hecatres of croplands, there is about 39% irrigated and 61% rainfed croplands. Download data from: https://lpdaac.usgs.gov/products/lgrip30v001/

7.12 CONCLUSIONS

This chapter provides an overview of the importance of global cropland products in food and water security analysis. It is obvious that only remote sensing from Earth Observing (EO) satellites can provide consistent, repeatable, frequent, and objective high-quality data for characterizing and mapping key cropland parameters for global food and water security support analysis. Importance of definitions and class naming conventions in cropland mapping has been re-iterated. Typical EO systems and their spectral, spatial, temporal, and radiometric characteristics useful for cropland mapping have been highlighted including characteristics of old and new generation of satellite sensors over the last 50 years. The chapter provides a review of various cropland mapping methods used at global, regional, and local levels. One of the remote sensing methods for global cropland mapping has been illustrated. The four-selected key global cropland products (e.g., Figures 7.8 to 7.12) derived from remote sensing, based on the work conducted by four major studies (Thenkabail et al., 2009a, 2011, Pittman et al., 2010, Yu et al., 2013, Friedl et al., 2010) are presented and discussed. These studies were conducted using: (1) time-series of multi-sensor data and secondary data, (2) 250 m MODIS time-series data, (3) 30 m Landsat data, and (4) a MODIS 500 m time-series data. These four products were synthesized, at nominal 1 km, to obtain a unified cropland mask of the world (global cropland extent version 1.0 or GCE V1.0). It was demonstrated from these products that the uncertainty in location of croplands in any one given product is quite high and no single product maps croplands particularly well. Therefore, a synthesis identifies where some or all of these products agree and where they disagree. This provides a starting point for the next level of more detailed cropland mapping at higher spatial resolution of 30 m or better. The key cropland parameters identified to be derived from remote sensing are: (1) cropland extent/areas, (2) cropping intensities, (3) watering method (irrigated versus rainfed), (4) crop type, and (5) cropland change over time and space. From these primary products one can derive crop productivity and water productivity. Such products have great importance and relevance in global food security analysis.

The use of a composite global cropland map (see Figure 7.12c, Table 7.7c) that provides clear consensus view on of four major cropland studies on global: Cropland extent location; Cropland watering method (irrigation versus rainfed) may be most useful. The product (Figure 7.12c, Table 7.7c) does not show where the crop types are or even the crop dominance. However, cropping intensity can be gathered using multi-temporal remote sensing over these cropland areas.

Various cropland products are available for download at NASA's The Land Processes Distributed Active Archive Center (LP DAAC). These products include global cropland mask at 1-km (https://lpdaac.usgs.gov/products/gfsad1kcmv001/), global cropland dominance at 1-km (https://lpdaac.usgs.gov/products/gfsad1kcdv001/), Landsat-derived global cropland extent product @ 30 m (LGCEP30) (https://lpdaac.usgs.gov/news/release-of-gfsad-30-meter-cropland-extent-products/), and Landsat-derived global rainfed and irrigated area product @ 30m (LGRIP30) (https://lpdaac.usgs.gov/news/release-of-lgrip30-data-product/). All these products are associated with user guides and algorithm theoretical basis documents (ATBDs), which are also available in LP DAAC. Further, all these products can be viewed at full resolution at: www.usgs.gov/apps/croplands/app/map. For detailed methodology of these products, also refer to Thenkabail et al. (2021).

ACKNOWLEDGMENTS

The authors are grateful for the funding received from National Aeronautics and Space Administration's (NASA's) Making Earth System Data Records for Use in Research Environments (MEaSUREs) through NASA Research Opportunities in Space and Earth Science solicitation (June 1, 2013 to May 31, 2018). This funding was received through NASA MEaSUREs project grant number NNH13AV82I and U.S. Geological Survey (USGS) sales order number 29039. The USGS provided significant direct and indirect supplemental funding through its National Land Imaging (NLI) and Land Change Science (LCS) programs, as well as support from the USGS Land Resources Mission Area, now part of the USGS Core Science Systems. We thank the global food and water security support analysis data @ 30 m (GFSAD30) project team for their valuable inputs. Figures 7.1 and 7.2 were produced by Dr. Zhuoting Wu, USGS. We thank her for it. Any use of trade, firm, or product names is for descriptive purposes only and does not imply endorsement by the U.S. government.

REFERENCES

Abrams, M., Bailey, B., Tsu, H. and Hato, M., 2010. The aster global dem. *Photogrammetric Engineering and Remote Sensing*, *76*(4), pp. 344–348.

Abrams, M., Crippen, R. and Fujisada, H., 2020. ASTER global digital elevation model (GDEM) and ASTER global water body dataset (ASTWBD). *Remote Sensing*, *12*(7), p. 1156.

Aneece, I., Foley, D., Thenkabail, P., Oliphant, A. and Teluguntla, P., 2022. New generation hyperspectral data from DESIS compared to high spatial resolution PlanetScope data for crop type classification. *IEEE Journal of Selected Topics in Applied Earth Observations and Remote Sensing*, *15*, pp. 7846–7858.

Bartholome, E. and Belward, A.S., 2005. GLC2000: A new approach to global land cover mapping from Earth observation data. *International Journal of Remote Sensing*, *26*(9), pp. 1959–1977.

Biggs, T., Thenkabail, P.S., Krishna, M., GangadharaRao, P. and Turral, H., 2006. Vegetation phenology and irrigated area mapping using combined MODIS time-series, ground surveys, and agricultural census data in Krishna River Basin, India. International *Journal of Remote Sensing*, *27*(19), pp. 4245–4266.

Bindraban, P.S., Bulte, E.H. and Conijn, S.G., 2009. Can large-scale biofuel production be sustained by 2020? *Agricultural Systems*, *101*, pp. 197–199.

Biradar, C.M., Thenkabail, P.S., Noojipady, P., Li, Y., Dheeravath, V., Turral, H., Velpuri, M., et al., 2009. A Global Map of Rainfed Cropland Areas (GMRCA) at the end of last millennium using remote sensing. *International Journal of Applied Earth Observation & Geoinformation*, *11*(2), pp. 114–129. https://doi.org/10.1016/j.jag.2008.11.002.

Bodirsky, B.L., Rolinski, S., Biewald, A., Weindl, I., Popp, A. and Lotze-Campen, H., 2015. Global food demand scenarios for the 21st century. *PloS One*, *10*(11), p. e0139201.

Boryan, C., Yang, Z., Mueller, R. and Craig, M., 2011. Monitoring US agriculture: The US department of agriculture, national agricultural statistics service, cropland data layer program. *Geocarto International*, *26*(5), pp. 341–358. https://doi.org/10.1080/10106049.2011.562309.

Brown, M.E. and Brickley, E.B., 2012. Evaluating the use of remote sensing data in the US agency for international development famine early warning systems network. *Journal of Applied Remote Sensing*, *6*(1), pp. 063511–063511.

Buchhorn, M., Smets, B., Bertels, L., Lesiv, M., Tsendbazar, N.E., Herold, M. and Fritz, S., 2019. Copernicus global land service: Land cover 100m: epoch 2015: Globe. *Version V2. 0.2, 10*.

Chander, G., Markham, B.L. and Helder, D.L., 2009. Summary of current radiometric calibration coefficients for Landsat MSS, TM, ETM+, and EO-1 ALI sensors. *Remote Sensing of Environment*, *113*(5), 15 May, pp. 893–903, ISSN 0034–4257, http://dx.doi.org/10.1016/j.rse.2009.01.007. (www.sciencedirect.com/science/article/pii/S0034425709000169).

Cohen, W.B. and Goward, S.N., 2004. Landsat's role in ecological applications of remote sensing. *BioScience*, *54*, 535–545.

Congalton, R., 1991. A review of assessing the accuracy of classifications of remotely sensed data. *Remote Sensing of Environment*, *37*, pp. 35–46.

Congalton, R., 2009. Accuracy and error analysis of global and local maps: Lessons learned and future considerations. In: *Remote Sensing of Global Croplands for Food Security*. P. Thenkabail, J. Lyon, H. Turral, and C. Biradar. (Editors). CRC/Taylor & Francis, Boca Raton, FL, pp. 441–458.

Congalton, R. and Green, K., 2009. *Assessing the Accuracy of Remotely Sensed Data: Principlesand Practices*. 2nd Edition. CRC/Taylor & Francis, Boca Raton, FL, p. 183.

Crist, E.P. and Cicone, R.C., 1984. Application of the tasseled cap concept to simulated Thematic Mapper data. *Photogrammetric Engineering and Remote Sensing*, *50*, 343–352.

DeFries, R., Hansen, M., Townsend, J.G.R. and Sohlberg, R., 1998. Global land cover classifications at 8 km resolution: The use of training data derived from Landsat imagery in decision tree classifiers. *International Journal of Remote Sensing*, *19*, 3141–3168.

Dheeravath, V., Thenkabail, P.S., Chandrakantha, G., Noojipady, P., Biradar, C.B., Turral, H., Gumma, M.L., Reddy, G.P.O. and Velpuri, M., 2010. Irrigated areas of India derived using MODIS 500m data for years 2001–2003. *ISPRS Journal of Photogrammetry and Remote Sensing*, *65*(1), pp. 42–59. http://dx.doi.org/10.1016/j.isprsjprs.2009.08.004.

EPW, 2008. Food security endangered: Structural changes in global grain markets threaten India's food security. *Economic and Political Weekly*, *43*, p. 5.

Falkenmark, M. and Rockström, J., 2006. The new blue and green water paradigm: Breaking new ground for water resources planning and management. *Journal of Water Resource Planning and Management*, 132, pp. 1–15.

FAO, 2009. *Food Outlook: Outlook Global Market Analysis*. FAO, Rome.

FAO, 2017. *The Future of Food and Agriculture: Trends and Challenges*. FAO, Rome.

FAO, 2022. FAOSTAT: Pesticides trade. In: *FAO*. Rome. www.fao.org/faostat/en/#data/RT.

Foley, J.A., DeFries, R., Asner, G.P., Barford, C., Bonan, G., Carpenter, S.R., Chapin, F.S., Coe, M.T., Daily, G.C., Gibbs, H.K., Helkowski, J.H., Holloway, T., Howard, E.A., Kucharik, C.J., et al., 2011. Solutions for a cultivated planet. *Nature*, *478*(7369), pp. 337–342.

Foley, J.A., Monfreda, C., Ramankutty, N. and Zaks, D., 2007. Our share of the planetary pie. *PNAS*, *104*, 12585–12586.

Friedl, M.A., McIver, D.K., Hodges, J.C.F., Zhang, X.Y., Muchoney, D. and Strahler, A.H., 2002. Global land cover mapping from MODIS: Algorithms and early results. *Remote Sensing of Environment*, *83*, pp. 287–302.

Friedl, M., Sulla-Menashe, D., 2019. *MCD12Q1 MODIS/Terra+Aqua Land Cover Type Yearly L3 Global 500m SIN Grid V006*. distributed by NASA EOSDIS Land Processes Distributed Active Archive Center. https://doi.org/10.5067/MODIS/MCD12Q1.006. Accessed 2023–12–28.

Friedl, M. and Sulla-Menashe, D., 2022. MODIS/Terra+ Aqua Land Cover Type Yearly L3 Global 500m SIN Grid V061. *NASAEOSDIS Land Processes DAAC: Sioux Falls, SD, USA*. https://doi.org/10.5067/MODIS/MCD12Q1.061. Accessed 2023–12–28.

Friedl, M.A., Sulla-Menashe, D., Tan, B., Schneider, A., Ramankutty, N.S. and Huang, X.M. 2010. MODIS Collection 5 global land cover: Algorithm refinements and characterization of new datasets. *Remote Sensing of Environment*, *114*(1), pp. 168–182.

Fritz, S., McCallum, I., Schill, C., Perger, C., Grillmayer, R., Achard, F., Kraxner, F. and Obersteiner, M., 2009. Geo-Wiki. Org: The use of crowdsourcing to improve global land cover. *Remote Sensing*, *1*(3), pp. 345–354.

Fritz, S., See, L. and Rembold, F., 2010. Comparison of global and regional land cover maps with statistical information for the agricultural domain in Africa. *International Journal of Remote Sensing*, *31*(9), pp. 2237–2256.

Funk, C. and Brown, M., 2009. Declining global per capita agricultural production and warming oceans threaten food security. *Food Security*. https://doi.org/10.1007/s12571-009-0026-y.

García-Álvarez, D., Lara Hinojosa, J., Jurado Pérez, F.J. and Quintero Villaraso, J., 2022. Global general land use cover datasets with a time series of maps. In: *Land Use Cover Datasets and Validation Tools: Validation Practices with QGIS*. Springer International Publishing, Cham, pp. 287–311.

Geerken, R., Zaitchik, B. and Evans, J.P., 2005. Classifying rangeland vegetation type and coverage from NDVI time series using Fourier Filtered Cycle Similarity. *International Journal of Remote Sensing*, *26*(24), pp. 5535–5554.

Gerland, P., Raftery, A.E., Ševčíková, H., Li, N., Gu, D., Spoorenberg, T., Alkema, L., Fosdick, B.K., Chunn, J., Lalic, N., Bay, G., Buettner, T., Heilig, G.K. and Wilmoth, J., 2014. World population stabilization unlikely this century. *Science*, *346*(6206), pp. 234–237. https://doi.org/10.1126/science.1257469.

Gibbs, H.K., Johnston, M., Foley, J.A., Holloway, T., Monfreda, C., Ramankutty, N. and Zaks, D., 2008. Carbon payback times for crop-based biofuel expansion in the tropics: The effects of changing yield and technology. *Environmental Research Letters*, p. 034001. https://doi.org/10.1088/1748-9326/3/3/034001.

GLC, 2003. Global Land Cover 2000 database. European Commission, Joint Research Centre. Available at: www.gvm.jrc.it/glc2000.

Goldewijk, K., Beusen, A., de Vos, M. and van Drecht, G., 2011. The HYDE 3.1 spatially explicit database of human induced land use change over the past 12,000 years. *Global Ecology and Biogeography*, *20*(1), pp. 73–86. https://doi.org/10.1111/j.1466-8238.2010.00587.x.

Goodchild, M. and Gopal, S. (Editors), 1989. *The Accuracy of Spatial Databases*. Taylor and Francis, New York, p. 290.

Google, CNES, Airbus, and Maxar Technologies. 2014. Imagery Derived from Google Earth. Available at: earth.google.com/web/ Accessed 2014-10-09.

Gordon, L.J., Finlayson, C.M. and Falkenmark, M., 2009. Managing water in agriculture for food production and other ecosystem services. *Agricultural Water Management*, p. 97. https://doi.org/10.1016/j.agwat.2009.03.017.

Gumma, M.K., Nelson, A., Thenkabail, P.S. and Singh, A.N., 2011. Mapping rice areas of South Asia using MODIS multi temporal data. *Journal of Applied Remote Sensing*, *5*, p. 053547, September 1. https://doi.org/10.1117/1.3619838.

Gumma, M.K., Thenkabail, P.S., Teluguntla, P., Oliphant, A., Xiong, J., Giri, C., Pyla, V., Dixit, S., Whitbread, A.M. 2020. *Agricultural cropland extent and areas of South Asia derived using Landsat satellite 30-m time-series big-data using random forest machine learning algorithms on the Google Earth Engine cloud- GIScience & Remote Sensing, 57*(3), 302–322, DOI: 10.1080/15481603.2019.1690780. ISSN: 1548-1603 (Print) 1943-7226. https://doi.org/10.1080/15481603.2019.1690780. IP-111091.

Gumma, M.K., Thenkabail, P.S., Panjala, P., Teluguntla, P., Yamano, T. and Mohammed, I., 2022. Multiple agricultural cropland products of South Asia developed using Landsat-8 30 m and MODIS 250 m data using machine learning on the Google Earth Engine (GEE) cloud and spectral matching techniques (SMTs) in support of food and water security. *GIScience & Remote Sensing*, *59*(1), pp. 1048–1077.

Han, W., Yang, Z., Di, L. and Mueller, R., 2012. CropScape: A web service based application for exploring and disseminating US conterminous geospatial cropland data products for decision support. *Computers and Electronics in Agriculture*, *84*, pp. 111–123. https://doi.org/10.1016/j.compag.2012.03.005.

Hansen, M.C., DeFries, R.S., Townshend, J.R.G., Sohlberg, R., Dimiceli, C. and Carroll, M. 2002. Towards an operational MODIS continuous field of percent tree cover algorithm: Examples using AVHRR and MODIS data. *Remote Sensing of Environment*, *83*, pp. 303–319.

He, G., Liu, X. and Cui, Z. 2021. Achieving global food security by focusing on nitrogen efficiency potentials and local production. *Global Food Security*, *29*, pp. 100536, ISSN 2211–9124, https://doi.org/10.1016/j.gfs.2021.100536.

He, S., Barati, B., Hu, X. and Wang, S., 2023. Carbon migration of microalgae from cultivation towards biofuel production by hydrothermal technology: A review. *Fuel Processing Technology*, *240*, p. 107563.

Herold, M., Woodcock, C.E., Loveland, T.R., Townshend, J., Brady, M., Steenmans, C. and Schmullius, C.C., 2008. Land-cover observations as part of a Global Earth Observation System of Systems (GEOSS): Progress, activities, and prospects. *IEEE Systems Journal*, *2*(3), pp. 414–423.

Hexagon, A.B. 2021. ERDAS field guide. Unsupervised Training • Producer Field Guide • Reader • Documentation Portal. Available at: fluidtopics.net.

Hossain, M., Janaiah, A. and Otsuka, K., 2005. Is the productivity impact of the Green Revolution in rice vanishing? *Economic and Political Weekly*, pp. 5595–9600.

Hu, Q., Xiang, M., Chen, D., Zhou, J., Wu, W. and Song, Q. 2020. Global cropland intensification surpassed expansion between 2000 and 2010: A spatio-temporal analysis based on GlobeLand30. *Science of the Total Environment*, *746*, p. 141035, ISSN 0048–9697, https://doi.org/10.1016/j.scitotenv.2020.141035.

Jarrett, A., et al., 2010. Degree Confluence Project. Available at: www.confluence.org/ Accessed on 2015–06–20.

Jiang, Y., 2009. China's water scarcity. *Journal of Environmental Management*, *90*, pp. 3185–3196.

Johnson, D.M. and Mueller, R., 2010. The 2009 cropland data layer. *Photogrammetric Engineering and Remote Sensing*, *76*(11), pp. 1201–1205.

Khan, S. and Hanjra, M.A. (2008). Sustainable land and water management policies and practices: A pathway to environmental sustainability in large irrigation systems. *Land Degradation and Development*, *19*, pp. 469–487.

Khan, S., Naushad, M., Iqbal, J., Bathula, C. and Ala'a, H., 2022. Challenges and perspectives on innovative technologies for biofuel production and sustainable environmental management. *Fuel*, *325*, p. 124845.

Kumar, M.D. and Singh, O.P., 2005. Virtual water in global food and water policy making: Is there a need for rethinking? *Water Resources Management*, *19*, pp. 759–789.

Kurz, B. and Seelan, S.K. 2007. Use of remote sensing to map irrigated agriculture in areas overlying the Ogallala aquifer, United States. Book Chapter in Global Irrigated Areas. IWMI. Under Review.

Lal, R. and Pimentel, D., 2009. Biofuels from crop residues. *Soil and Tillage Research*, *93*, pp. 237–238.

Lark, T.J., Salmon, J.M. and Gibbs, H.K., 2015. Cropland expansion outpaces agricultural and biofuel policies in the United States. *Environmental Research Letters*, *10*(4), p. 044003.

Lenaerts, B., Collard, B.C.Y. and Demont, M. 2019. Review: Improving global food security through accelerated plant breeding. *Plant Science*, *287*, p. 110207, ISSN 0168–9452, https://doi.org/10.1016/j.plantsci.2019.110207.

Liu, L., Kang, S., Xiong, X., Qin, Y., Wang, J., Liu, Z. and Xiao, X., 2023. Cropping intensity map of China with 10 m spatial resolution from analyses of time-series Landsat-7/8 and Sentinel-2 images. *International Journal of Applied Earth Observation and Geoinformation*, *124*, p. 103504.

Loveland, T.R., Reed, B.C., Brown, J.F., Ohlen, D.O., Zhu, Z. and Yang, L., 2000. Development of global land cover characteristics database and IGBP DISCover from 1 km AVHRR data. *International Journal of Remote Sensing*, *21*, pp. 1303–1330.

Lynch, J.P., 2007. Roots of the second green revolution. *Australian Journal of Botany*, *55*(5), pp. 493–512.

Masek, J.G., Huang, C., Wolfe, R., Cohen, W., Hall, F., Kutler, J. and Nelson, P., 2008. North American forest disturbance mapped from a decadal Landsat record. *Remote Sensing of Environment*, *112*, pp. 2914–2926.

Masek, J.G., Vermote, E.F., Saleous, N., Wolfe, R., Hall, F.G., Huemmrich, K.F., Gao, F., Kutler, J. and Lim, T.K., 2006. A Landsat surface reflectance data set for North America, 1990–2000. *Geoscience and Remote Sensing Letters*, *3*, pp. 68–72.

Masek, J., Ju, J., Roger, J., Skakun, S., Vermote, E., Claverie, M., Dungan, J., Yin, Z., Freitag, B. and Justice, C. 2021. HLS Sentinel-2 MSI surface reflectance daily global 30m v2.0., distributed by NASA EOSDIS Land Processes DAAC, https://doi.org/10.5067/HLS/HLSS30.002.

Masek, J.G., Ju, J., Claverie, M., Skakun, S., Roger, J.C., Vermote, E., Franch, B., Yin, Z., Dungan. J.L. 2022. *Harmonized Landsat Sentinel-2 (HLS) Product User Guide Product Version 2.0*

McIntyre, B.D. (2008). International assessment of agricultural knowledge, science and technology for development (IAASTD): Global report. *Includes Bibliographical References and Index*. ISBN 978-1-59726-538-6, Oxford, UK.

Mitchell, T.D., Carter, T.R., Jones, P.D., Hulme, M. and New, M., 2004. A comprehensive set of high-resolution grids of monthly climate for Europe and the globe: The observed record (1901–2000) and 16 scenarios (2001–2100). *Tyndall Centre for Climate Change Research Working Paper*, *55*(0), p. 25.

Monfreda, C., Patz, J.A., Prentice, I.C., Ramankutty, N. and Snyder, P.K., 2005. Global consequences of land use. *Science*, July 22, *309*(5734), pp. 570–574.

Monfreda, C., Ramankutty, N. and Foley, J.A., 2008. Farming the planet: 2. Geographic distribution of crop areas, yields, physiological types, and net primary production in the year 2000. *Global Biogeochemical Cycles*, *22*(1), p. GB1022.

Nyambariga, F.K., Opere, A.O., Kituyi, E. and Amwata, D.A., 2023. Climate change scenario projections and their implications on food systems in Taita Taveta County, Kenya. *PLOS Climate*, *2*(6), p. e0000114.

Oliphant, A., Thenkabail, P.S., Teluguntla, P., Xiong, J., Gumma, M.K., Congalton, R., and Yadav, K. 2019. Mapping cropland extent of Southeast and Northeast Asia using multi-year time-series Landsat 30-m data using random forest classifier on Google Earth Engine. *International Journal of Applied Earth Observation and Geoinformation*. Vol. 81: 110–124. https://doi.org/10.1016/j.jag.2018.11.014 Accepted. IP-099863.

Olofsson, P., Stehman, S.V., Woodcock, C.E., Sulla-Menasche, D., Sibley, A.M., Newell, J.D., Friedl, M.A. and Herold, M. 2011. A global land cover validation dataset, I: Fundamental design principles. *International Journal of Remote Sensing*. In Review.

Owusu, A., Kagone, S., Leh, M., Velpuri, N.M., Gumma, M.K., Ghansah, B., Thilina-Prabhath, P., Akpoti, K., Mekonnen, K., Tinonetsana, P. and Mohammed, I., 2024. A framework for disaggregating remote-sensing cropland into rainfed and irrigated classes at continental scale. *International Journal of Applied Earth Observation and Geoinformation*, *126*, p. 103607.

Ozdogan, M. and Gutman, G., 2008. A new methodology to map irrigated areas using multi-temporal MODIS and ancillary data: An application example in the continental US. *Remote Sensing of Environment*, *112*(9), pp. 3520–3537.

Ozdogan, M. and Woodcock, C.E., 2006. Resolution dependent errors in remote sensing of cultivated areas. *Remote Sensing of Environment*, *103*, pp. 203–217.

Pennisi, E., 2008. The Blue revolution, drop by drop, gene by gene. *Science*, *320*(5873), pp. 171–173. https://doi.org/10.1126/science.320.5873.171.

Pereira, P., Brevik, E. and Trevisani, S., 2018. Mapping the environment. *Science of the Total Environment*, *610*, pp. 17–23.

Phalke, A.R., Özdoğan, M., Thenkabail, P.S., Erickson, T., Gorelick, N., Yadav, K., Congalton, R.G. 2020. Mapping croplands of Europe, Middle East, Russia, and Central Asia using Landsat, Random Forest, and Google Earth Engine. *ISPRS Journal of Photogrammetry and Remote Sensing, 167*, pp. 104–122, ISSN 0924-2716, https://doi.org/10.1016/j.isprsjprs.2020.06.022. (http://www.sciencedirect.com/science/article/pii/S0924271620301805).

Pérez-Hoyos, A., Rembold, F., Kerdiles, H. and Gallego, J., 2017. Comparison of global land cover datasets for cropland monitoring. *Remote Sensing*, *9*(11), p. 1118.

Pittman, K., Hansen, M.C., Becker-Reshef, I., Potapov, P.V., Justice, C.O., 2010. Estimating global cropland extent with multi-year MODIS data. *Remote Sensing*, *2*(7), pp. 1844–1863.

Popp, J., Lakner, Z., Harangi-Rákos, M. and Fari, M., 2014. The effect of bioenergy expansion: Food, energy, and environment. *Renewable and Sustainable Energy Reviews*, *32*, pp. 559–578.

Portmann, F., Siebert, S. and Döll, P., 2008. MIRCA2000: Global monthly irrigated and rainfed crop areas around the year 2000: A new high-resolution data set for agricultural and hydrological modelling. *Global Biogeochemical Cycles*, p. GB0003435.

Potapov, P., Turubanova, S., Hansen, M.C., . . . and Cortez, J. 2022. Global maps of cropland extent and change show accelerated cropland expansion in the twenty-first century. *Nature Food*, *3*, pp. 19–28. https://doi.org/10.1038/s43016-021-00429-z.

Ramankutty, N., Evan, A.T., Monfreda, C. and Foley, J.A., 2008. Farming the planet: 1. Geographic distribution of global agricultural lands in the year 2000. *Global Biogeochemical Cycles*, p. 22. https://doi.org/10.1029/2007GB002952.

Ramankutty, N. and Foley, J.A., 1998. Characterizing patterns of global land use: An analysis of global croplands data. *Global Biogeochemical Cycles*, *12*(4), pp. 667–685.

Rodell, M., Velicogna, I. and Famiglietti, J.S., 2009. *Nature*, *460*, pp. 999–1002. https://doi.org/10.1038/nature08238.

Sakamoto, T., Yokozawa, M., Toritani, H., Shibayama, M., Ishitsuka, N. and Ohno, H., 2005. A crop phenology detection method using time-series MODIS data. *Remote Sensing of Environment*, *96*(3–4), pp. 366–374.

Salmon, J.M., Friedl, M.A., Frolking, S., Wisser, D. and Douglas, E.M., 2015. Global rain-fed, irrigated, and paddy croplands: A new high resolution map derived from remote sensing, crop inventories and climate data. *International Journal of Applied Earth Observation and Geoinformation*, *38*, pp. 321–334.

Searchinger, T., Heimlich, R., Houghton, R.A., Dong, F., Elobeid, A., Fabiosa, J., Tokgoz, S., Hayes, D. and Yu, T.H., 2008. Use of U.S. croplands for biofuels increases greenhouse gases through emissions from land-use Change. *Science*, *319*, pp. 1238–1240.

See, L., Gilliams, S., Conchedda, G., Degerickx, J., Van Tricht, K., Fritz, S., Lesiv, M., Laso Bayas, J.C., Rosero, J., Tubiello, F.N. and Szantoi, Z., 2023. Dynamic global-scale crop and irrigation monitoring. *Nature Food*, *4*(9), pp. 736–737.

Shao, W., Zhou, Z., Liu, J., Yang, G., Wang, J., Xiang, C., Cao, X. and Liu, H., 2017. Changing mechanisms of agricultural water use in the urbanization and industrialization of China. *Water Policy*, *19*(5), pp. 908–935.

Shi, Q., He, D., Liu, Z., Liu, X. and Xue, J., 2023. Globe230k: A benchmark dense-pixel annotation dataset for global land cover mapping. *Journal of Remote Sensing*, *3*, p. 78.

Siebert, S. and Döll, P., 2010. Quantifying blue and green virtual water contents in global crop production as well as potential production losses without irrigation. *Journal of Hydrology*, *384*(3–4), pp. 198–217. https://doi.org/10.1016/j.jhydrol.2009.07.031.

Siebert, S., Hoogeveen, J. and Frenken, K., 2006. Irrigation in Africa, Europe and Latin America: Update of the digital global map of irrigation areas to version 4. Frankfurt Hydrology Paper 05, p. 134. Institute of Physical Geography, University of Frankfurt, Frankfurt am Main, Germany and Rome, Italy.

Siebert, S., Kummu, M., Porkka, M., Döll, P., Ramankutty, N. and Scanlon, B.R., 2015. A global data set of the extent of irrigated land from 1900 to 2005. *Hydrology and Earth System Sciences*, *19*, pp. 1521–1545. https://doi.org/10.5194/hess-19-1521-2015.

Smakhtin, V.Y., 2006. *An Assessment of Environmental Flow Requirements of Indian River Basins* (Vol. 107). IWMI.

Stephens, E.C., Jones, A.D. and Parsons, D., 2018. Agricultural systems research and global food security in the 21st century: An overview and roadmap for future opportunities. *Agricultural Systems*, *163*, pp. 1–6.

Teluguntla, P., Ryu, D., George, B., Walker, J.P. and Malano, H.M., 2015. Mapping flooded rice paddies using time series of MODIS imagery in the Krishna River Basin, India. *Remote Sensing*, *7*(7), pp. 8858–8882.

Teluguntla, P., Thenkabail, P., Oliphant, A., Gumma, M., Aneece, I., Foley, D. and McCormick, R., 2023. Landsat-derived global rainfed and irrigated-cropland product 30 m V001 (V001), NASA EOSDIS Land Processes DAAC IP148728.

Teluguntla, P., Thenkabail, P.S., Oliphant, A., Xiong, J., Gumma, M.K., Congalton, R.G., Yadav, K. and Huete, A., 2018. A 30-m landsat-derived cropland extent product of Australia and China using random forest machine learning algorithm on Google Earth Engine cloud computing platform. *ISPRS Journal of Photogrammetry and Remote Sensing*, *144*, pp. 325–340.

Teluguntla, P., Thenkabail, P.S., Xiong, J., Gumma, M.K., Congalton, R.G., Oliphant, A., Poehnelt, J., Yadav, K., Rao, M. and Massey, R., 2017. Spectral matching techniques (SMTs) and automated cropland classification algorithms (ACCAs) for mapping croplands of Australia using MODIS 250-m time-series (2000–2015) data. *International Journal of Digital Earth*, *10*(9), pp. 944–977. https://doi.org/10.1080/17538947.2016.1267269.

Thenkabail, P.S., 2010. Global croplands and their importance for water and food security in the twenty-first century: Towards an ever green revolution that combines a second green revolution with a blue revolution. *Remote Sensing*, *2*(9), pp. 2305–2312.

Thenkabail, P.S., Biradar, C.M., Noojipady, P., Cai, X.L., Dheeravath, V., Li, Y.J., Velpuri, M., Gumma, M. and Pandey, S., 2007b. Sub-pixel irrigated area calculation methods. *Sensors Journal*. (special issue: Remote Sensing of Natural Resources and the Environment (Remote Sensing Sensors), Edited by Assefa M. Melesse, *7*, pp. 2519–2538. Available at: www.mdpi.org/sensors/papers/s7112519.pdf.

Thenkabail, P.S., Biradar, C.M., Noojipady, P., Dheeravath, V., Li, Y.J., Velpuri, M., Gumma, M., Reddy, G.P.O., Turral, H., Cai, X.L., Vithanage, J., Schull, M. and Dutta, R., 2009b. Global irrigated area map (GIAM), derived from remote sensing, for the end of the last millennium. *International Journal of Remote Sensing*, *30*(14), July, 20, pp. 3679–3733.

Thenkabail, P.S., Enclona, E.A., Ashton, M.S., Legg, C. and Jean De Dieu, M., 2004. Hyperion, IKONOS, ALI, and ETM+ sensors in the study of African rainforests. *Remote Sensing of Environment*, *90*, pp. 23–43.

Thenkabail, P.S., GangadharaRao, P., Biggs, T., Krishna, M. and Turral, H., 2007a. Spectral matching techniques to determine historical land use/land cover (LULC) and irrigated areas using time-series AVHRR pathfinder datasets in the Krishna River Basin, India. *Photogrammetric Engineering and Remote Sensing*, *73*(9), pp. 1029–1040. (Second Place Recipients of the 2008 John I. Davidson ASPRS President's Award for Practical papers).

Thenkabail, P.S., Gumma, M.K., Teluguntla, P. and Mohammed, I.A., 2014. Hyperspectral remote sensing of vegetation and agricultural crops: Highlight article. *Photogrammetric Engineering and Remote Sensing*, *80*(4), pp. 697–709. IP-052042.

Thenkabail, P.S., Hanjra, M.A., Dheeravath, V. and Gumma, M., 2011. *Global Croplands and Their Water Use from Remote Sensing and Non-Remote Sensing Perspectives*. Taylor and Francis, FL, pp. 383–419.

Thenkabail, P.S., Knox, J.W., Ozdogan, M., Gumma, M.K., Congalton, R.G., Wu, Z., Milesi, C., Finkral, A., Marshall, M., Mariotto, I. and You, S., 2012. Assessing future risks to agricultural productivity, water resources and food security: How can remote sensing help? *Photogrammetric Engineering and Remote Sensing*, *78*(8).

Thenkabail, P.S., Lyon, G.J., Turral, H. and Biradar, C.M., 2009a. *Remote Sensing of Global Croplands for Food Security*. CRC Press, Taylor and Francis Group, Boca Raton, London, New York, p. 556 (48 pages in color). Published in June, 2009.

Thenkabail, P.S., Lyon, G.J., Turral, H. and Biradar, C.M., 2009c. *Remote Sensing of Global Croplands for Food Security*. CRC Press, Taylor and Francis Group, Boca Raton, London, New York, Published in June.

Thenkabail, P.S., Mariotto, I., Gumma, M.K., Middleton, E.M., Landis, D.R. and Huemmrich, F.K., 2013. Selection of hyperspectral narrowbands (HNBs) and composition of hyperspectral twoband vegetation indices (HVIs) for biophysical characterization and discrimination of crop types using field reflectance and Hyperion/EO-1 data. *IEEE Journal of Selected Topics in Applied Earth Observations and Remote Sensing*, *6*(2), APRIL, pp. 427–439. https://doi.org/10.1109/JSTARS.2013.2252601.

Thenkabail, P.S., Schull, M. and Turral, H., 2005. Ganges and Indus River Basin Land Use/Land Cover (LULC) and irrigated area mapping using continuous streams of MODIS data. *Remote Sensing of Environment*, *95*(3), pp. 317–341.

Thenkabail, P.S., Teluguntla, P., Gumma, M.K. and Dheeravath, V., 2015. Hyperspectral remote sensing for terrestrial applications. In: *Land Resources Monitoring, Modeling, and Mapping with Remote Sensing*. CRC Press, Boca Raton, pp. 201–233. ISBN 9781482217957.

Thenkabail, P.S., Teluguntla, P.G., Xiong, J., Oliphant, A., Congalton, R.G., Ozdogan, M., Gumma, M.K., Tilton, J.C., Giri, C., Milesi, C., Phalke, A., Massey, R., Yadav, K., Sankey, T., Zhong, Y., Aneece, I. and Foley, D., 2021. Global cropland-extent product at 30m resolution (GCEP30) derived from Landsat satellite time-series data for the year 2015 using multiple machine-learning algorithms on Google Earth Engine cloud: U.S. Geological Survey Professional Paper 1868, p. 63. https://doi.org/10.3133/pp1868. Available at: https://lpdaac.usgs.gov/news/release-of-gfsad-30meter-cropland-extent-products/. IP-119164.

Thenkabail, P.S. and Wu, Z., 2012. An Automated Cropland Classification Algorithm (ACCA) for Tajikistan by combining landsat, MODIS, and secondary data. *Remote Sensing*, *4*(10), pp. 2890–2918.

Tillman, D., Balzer, C., Hill, J. and Befort, B.L., 2011. Global food demand and the sustainable intensification of agriculture. Proceedings of the National Academy of Sciences of the United States of America, November 21. https://doi.org/10.1073/pnas.1116437108.

Tomlinson, R., 2003. *Thinking about Geographic Information Systems Planning for Managers*. ESRI Press, p. 283.

Tsendbazar, N., Herold, M., Li, L., Tarko, A., De Bruin, S., Masiliunas, D., Lesiv, M., Fritz, S., Buchhorn, M., Smets, B. and Van De Kerchove, R., 2021. Towards operational validation of annual global land cover maps. *Remote Sensing of Environment*, *266*, p. 112686.

Turral, H., Svendsen, M. and Faures, J., 2009. Investing in irrigation: Reviewing the past and looking to the future. *Agricultural Water Management*. https://doi.org/101.1016/j.agwt.2009.07.012.

UN DESA, 2021. United Nations department of economic and social affairs, population division (2021). Global Population Growth and Sustainable Development. UN DESA/POP/2021/TR/NO. 2. Download.

UNDP, 2012. *Human Development Report 2012: Overcoming Barriers: Human Mobility and Development*. United Nations, New York.

United Nations [UN], 2019, Population division: World population prospects 2019: United Nations website. Available at: https://population.un.org/wpp/ Accessed on 2019-08.

Van Tricht, K., Degerickx, J., Gilliams, S., Zanaga, D., Battude, M., Grosu, A., Brombacher, J., Lesiv, M., Bayas, J.C.L., Karanam, S. and Fritz, S., 2023. WorldCereal: a dynamic open-source system for global-scale, seasonal, and reproducible crop and irrigation mapping. *Earth System Science Data*, *15*(12), pp. 5491-5515.

Velpuri, M., Thenkabail, P.S., Gumma, M.K., Biradar, C.B., Dheeravath, V., Noojipady, P. and Yuanjie, L., 2009. Influence of resolution or scale in irrigated area mapping and area estimations. *Photogrammetric Engineering and Remote Sensing (PE&RS)*, *75*(12), December issue.

Viana, C.M., Freire, D., Abrantes, P., Rocha, J. and Pereira, P., 2022. Agricultural land systems importance for supporting food security and sustainable development goals: A systematic review. *Science of the Total Environment*, *806*, p. 150718.

Vinnari, M. and Tapio, P., 2009. Future images of meat consumption in 2030. *Futures*. https://doi.org/10.1016/j.futures.2008.11.014.

Wada, Y., van Beek, L.P.H. and Bierkens, M.F.P., 2012, Nonsustainable groundwater sustaining irrigation: A global assessment. *Water Resources Research*, *48*, p. W00L06. https://doi.org/10.1029/2011WR010562.

Waldner, F., Fritz, S. and Di Gregorio, A., et al., 2016. A unified cropland layer at 250 m for global agriculture monitoring. *Data*, p. 1. https://doi.org/10.3390/data1010003.

Wardlow, B.D. and Egbert, S.L., 2008. Large-area crop mapping using time-series MODIS 250 m NDVI data: An assessment for the U.S. *Central Great Plains. Remote Sensing of Environment*, *112*, pp. 1096–1116.

Wardlow, B.D., Egbert, S.L. and Kastens, J.H., 2007. Analysis of time-series MODIS 250 m vegetation index data for crop classification in the U.S. *Central Great Plains. Remote Sensing of Environment*, *108*, pp. 290–310.

Wardlow, B.D., Kastens, J.H. and Egbert, S.L., 2006. Using USDA crop progress data for the evaluation of greenup onset date calculated from MODIS 250-meter data. *Photogrammetric Engineering and Remote Sensing*, *72*, pp. 1225–1234.

Whitcraft, A.K., Becker-Reshef, I. and Justice, C.O., 2015. A framework for defining spatially explicit earth observation requirements for a global agricultural monitoring initiative (GEOGLAM). *Remote Sensing*, *7*(2), pp. 1461–1481.

Withee, G.W., Smith, D.B. and Hales, M.B., 2004. Progress in multilateral Earth observation cooperation: CEOS, IGOS and the ad hoc Group on Earth Observations. *Space Policy*, *20*(1), pp. 37–43.

World Bank, 2022. GovData360. Available at: https://govdata360.worldbank.org/indicators/h76e4f437.

Wu, W., Yu, Q., You, L., Chen, K., Tang, H. and Liu, J., 2018. Global cropping intensity gaps: Increasing food production without cropland expansion. *Land Use Policy*, *76*, pp. 515–525, ISSN 0264–8377, https://doi.org/10.1016/j.landusepol.2018.02.032.

Wu, Z., Thenkabail, P.S. and Verdin, J., 2014b. An automated cropland classification algorithm (ACCA) using Landsat and MODIS data combination for California. *Photogrammetric Engineering and Remote Sensing*, *80*(1), pp. 81–90.

Wu, Z., Thenkabail, P.S., Zakzeski, A., Mueller, R., Melton, F., Rosevelt, C., Dwyer, J., Johnson, J. and Verdin, J.P., 2014a. Seasonal cultivated and fallow cropland mapping using modis-based automated cropland classification algorithm. *Journal of Applied Remote Sensing*, *8*(1), p. 083685. https://doi.org/10.1117/1.JRS.8.083685.

Xie, Y., Gibbs, H.K. and Lark, T.J., 2021. Landsat-based Irrigation Dataset (LANID): 30 m resolution maps of irrigation distribution, frequency, and change for the US, 1997–2017. *Earth System Science Data*, *13*(12), pp. 5689–5710.

Xiong, J., Thenkabail, P.S., Gumma, M.K., Teluguntla, P., Poehnelt, J., Congalton, R.G., Yadav, K. and Thau, D., 2017. Automated cropland mapping of continental Africa using Google Earth Engine cloud computing. *ISPRS Journal of Photogrammetry and Remote Sensing*, *126*, pp. 225–244.

Xiong, J., Thenkabail, P.S., Gumma, M., Teluguntla, P., Poehnelt, J., Congalton, R., Yadav. K. 2018. Automated Cropland Mapping of Continental Africa using Google Earth Engine Cloud Computing. *The International Society of Photogrammetry and Remote Sensing (ISPRS) Journal of Photogrammetry and Remote Sensing (P&RS).* 126:225-244. http://dx.doi.org/10.1016/j.isprsjprs.2017.01.019. https://www.journals.elsevier.com/isprs-journal-of-photogrammetry-and-remote-sensing/most-downloaded-articles.

Yu, L., Wang, J., Clinton, N., Xin, Q., Zhong, L., Chen, Y. and Gong, P., 2013. FROM-GC: 30 m global cropland extent derived through multisource data integration. *International Journal of Digital Earth.* https://doi.org/10.1080/17538947.2013.822574.

Zanaga, D., Van De Kerchove, R., Daems, D., De Keersmaecker, W., Brockmann, C., Kirches, G., Wevers, J., Cartus, O., Santoro, M., Fritz, S., Lesiv, M., Herold, M., Tsendbazar, N.E., Xu, P., Ramoino, F. and Arino, O., 2022. ESA WorldCover 10 m 2021 v200. https://doi.org/10.5281/zenodo.7254221. Available at: https://developers.google.com/earth-engine/datasets/catalog/ESA_WorldCover_v200.

Zhang, C., Dong, J. and Ge, Q., 2022. Mapping 20 years of irrigated croplands in China using MODIS and statistics and existing irrigation products. ScientificData, 9(1), p. 407.

8 Remote Sensing for Precision Agriculture

Yuxin Miao, David J. Mulla, and Yanbo Huang

ACRONYMS AND DEFINITIONS

AI	Aphid Index
DSSI	Damage Sensitive Spectral Index
GIS	Geographic Information Systems
GPS	Global Positioning System
GNDVI	Green Normalized Difference Vegetation Index
IKONOS	A commercial Earth observation satellite, typically, collecting sub-meter to 5 m data
LAI	Leaf Area Index
LHI	Leaf Hopper Index
LiDAR	Light Detection and Ranging
MERIS	Medium Resolution Imaging Spectrometer
NDVI	Normalized Difference Vegetation Index
NIR	Near-Infrared
PRI	Photochemical Reflectance Index
SAVI	Soil Adjusted Vegetation Index
SWIR	Shortwave Infrared
TIR	Thermal Infrared
UAV	Unmanned Aerial Vehicles
VIS	Visible

8.1 INTRODUCTION

The world population is increasing rapidly, and by 2050 it is estimated that there will be over 9 billion people to feed (Cohen, 2003; UN, 2023). Agricultural production to feed this large population will be severely constrained by a lack of additional arable land combined with a diminishing supply of water and increasing pressure to protect the quality of water resources beyond the edge of agricultural fields. These constraints mean that it will be increasingly imperative to prevent losses in crop productivity due to water stress, nutrient deficiencies, weeds, insects, and crop diseases. These losses in productivity often occur at specific locations within fields, and at critical growth stages. They are not typically uniform in severity across locations within a field. Thus, farmers must take measures to identify where crop stress occurs in a timely fashion, they must identify what is causing crop stress, and they must try to use management practices that overcome crop stress at specific locations and times.

8.2 PRECISION AGRICULTURE

Precision agriculture is one of the top ten revolutions in agriculture (Crookston, 2006). It can be generally defined as doing the right management practices in the right place, with the right rate and at the right time. The International Society of Precision Agriculture recently defined precision agriculture as "a management strategy that gathers, processes and analyzes temporal, spatial and

DOI: 10.1201/9781003541165-10

individual data and combines it with other information to support management decisions according to estimated variability for improved resource use efficiency, productivity, quality, profitability and sustainability of agricultural production." Management practices commonly used in precision agriculture include variable rate application of fertilizers (Diacono et al., 2013; Miao, 2023) or pesticides (Lost Filho et al., 2020), variable rate seeding (da Silva et al., 2022) or tillage (Shamal et al., 2016), and variable rate irrigation (Neupane and Guo, 2019). Precision agriculture offers several benefits, including improved efficiency of farm management inputs, increases in crop productivity, quality or economic returns, and reduced transport of fertilizers and pesticides beyond the edge of field (Mulla et al., 1996; Miao et al., 2007; Erickson and Fausti, 2021).

Precision agriculture is also known as precision farming or site-specific crop management, but precision agriculture can also cover orchard management, pasture management, planation management, forestry, aquaculture, and livestock management. Precision agriculture as it is practiced today had its beginnings in the mid-1980s with two contrasting philosophies, namely, farming by soil (Larson and Robert, 1991) versus grid soil sampling for delineation of management zones (Mulla, 1991; Bhatti et al., 1991; Mulla, 1993).

Precision agriculture aims to improve site-specific agricultural decision making through collection and analysis of data, formulation of site-specific management recommendations and implementation of management practices to correct the factors that limit crop growth, productivity, and quality (Mulla and Schepers, 1997).

Precision agriculture has always relied on technology for data collection and analysis at specific locations and times across agricultural fields. The earliest technology was GIS, followed by variable rate spreaders, yield monitors, GPS, and remote sensing. As technology has improved, the scale at which management actions are implemented has become finer spatial and temporal scales. Ultimately, technology will lead to the ability to manage individual plants within an agricultural field in real-time (Shanahan et al., 2008; Freeman et al., 2007).

Adoption rates of technology in precision agriculture vary widely (McFadden et al., 2023). GPS (including autosteer) and yield monitors are widely used. Variable rate spreaders are moderately popular. Remote sensing has not yet been widely adopted for use in precision agriculture (Moran et al., 1997; Mulla, 2013; Sishodia et al., 2020). The main reasons include the difficulty in interpreting spectral signatures, the slow processing time for data, and the need to collect confirmatory data from ground surveys in order to diagnose causative factors for anomalous spectral reflectance data. Clearly, there is a significant scope for improving the interpretation and utility of remote sensing data for precision agriculture.

Remote sensing in precision agriculture started with Landsat TM imagery for improved mapping of soil fertility patterns across complex agricultural landscapes (Bhatti et al., 1991). Proximal sensing of soil organic matter content or weeds was also developed for early application in precision agriculture, and this approach now includes detection of crop nutrient deficiencies. Commercial satellite imagery was first provided to agricultural users at the beginning of the 21st century with IKONOS and QuickBird. Spatial and spectral resolution and return frequencies of satellite remote sensing platforms have improved rapidly since then with the advent of RapidEye, GeoEye, and WorldView imagery. Satellite imagery is typically unavailable on days with significant cloud cover. During the past few years, Planet has launched over 450 CubeSats (Doves and SuperDoves) that can collect high spatial resolution images (3–5 m) for the entire earth every day, making them ideal for precision agriculture applications.

Interest in remote sensing from airplanes and unmanned aerial vehicles (UAVs) has recently been very intense. One of the most active emerging areas of research in precision agriculture uses cameras mounted on UAVs. The UAVs are relatively inexpensive, can be deployed rapidly at low altitudes when crop stress is starting to appear, and have the flexibility to be flown during windy or partially cloudy conditions. Their limitations include some regulatory restrictions in different countries, inability to carry heavy cameras, mounts and GPS units, and short battery life, etc., although they have been relaxed and improved.

Several companies offer precision farming services that rely on remote sensing. These include companies that are based primarily on satellite imagery, including DigitalGlobe, Satellite Imaging Corp., Geosys SST/GeoVantage and Winfield Solutions. Companies that offer equipment for proximal sensing of crop nutrient deficiencies include Trimble's GreenSeeker (Solie et al., 1996), AgLeader's OptRx (Holland et al., 2012), Topcon's CropSpec (Reusche et al., 2010) and Yara's N-sensor (Link and Reusch, 2006). Trimble also offers equipment for proximal sensing of weeds (WeedSeeker et al., 1998). Numerous companies offer aerial remote sensing services with panchromatic imagery, broadband multispectral imagery or hyperspectral imagery. One example is CERES Imaging, which operates a fleet of airplanes that collect remote sensing imagery for vegetable, cereal, and tree crops. This imagery is used for detecting stresses, guiding crop scouting, and making prescription maps for variable rate growth regulator, irrigation, herbicide, insecticide, or fertilizer applications.

Commercial applications of remote sensing for precision farming have not always been successful. John Deere's Agri-Services division partnered with GeoVantage in 2006 to provide the OptiGro precision remote sensing service to farmers. This service proved to be unprofitable for John Deere, and they sold it to GeoVantage in 2008.

8.3 WAVELENGTHS AND BAND RATIOS OF INTEREST IN PRECISION AGRICULTURE

Remote sensing in precision agriculture has focused on reflectance in the visible (VIS) and near infrared (NIR), emission of radiation in the thermal infrared (TIR), and fluorescence in the visible spectrum. Remote sensing of soil is responsive to spatial patterns in soil moisture and organic matter content, as well as soil carbonate and iron oxide content. Remote sensing of crop canopies in the VIS spectrum responds to plant pigments such as chlorophyll a and b, anthocyanins and carotenoids (Pinter et al., 2003; Blackburn, 2007; Hatfield et al., 2008; Zhou et al., 2018). Plant pigments absorb radiation in narrow wavelength bands centered around 430 nm (blue or B) and 650 nm (red or R) for chlorophyll a and 450 nm (B) and 650 nm for chlorophyll b. Wavelengths with low absorption characteristics, conversely have high reflectance, particularly in the green (550 nm) wavelength. Remote sensing of crops in the NIR spectrum (particularly at 780 nm, 800 nm, and 880 nm) responds to crop canopy biomass and leaf area index (LAI), leaf orientation and leaf size and geometry. Plant pigments and crop canopy architecture in turn respond to many crop stresses, including water stress (Bastiaanssen et al., 2000), nutrient deficiencies (Samborski et al., 2009), crop diseases (West et al., 2003), and infestations of insects (Seelan et al., 2003) or weeds (Lamb and Brown, 2001; Thorp and Tian, 2004). As a result, remote sensing has often proved useful at indirectly detecting crop stresses for applications in precision agriculture.

In contrast to broadband multispectral reflectance imagery collected with older satellite platforms such as Landsat, QuickBird and IKONOS, recent attention in remote sensing has turned to analysis of narrow bands (10 nm wide) collected using hyperspectral imagery (Miao et al., 2009; Thenkabail et al., 2010; Yao et al., 2018). The hyperspectral data cube can be used to represent crop reflectance over large areas at each of these narrow bands (Figure 8.1, Nigon et al., 2014), illustrating the large amount of spatial and spectral information collected with hyperspectral imaging. In theory, hyperspectral imaging offers the capability of sensing a wide variety of soil and crop characteristics simultaneously, including moisture status, organic matter, nutrients, chlorophyll, carotenoids, cellulose, leaf area index, and crop biomass (Haboudane et al., 2002, 2004; Goel et al., 2003). Thenkabail et al. (2000) showed that hyperspectral data can be used to construct three general categories of predictive spectral indices, including (1) optimal multiple narrow band reflectance indices (OMNBR), (2) narrow band normalized difference vegetative indices (NDVI), and (3) soil adjusted vegetative indices (SAVI). Only two to four narrow bands were needed to describe plant characteristics with OMNBR (Table 8.1). The greatest information about plant characteristics in OMNBR includes the longer red wavelengths (650–700 nm), shorter green wavelengths (500–550 nm), red-edge (720 nm), and two

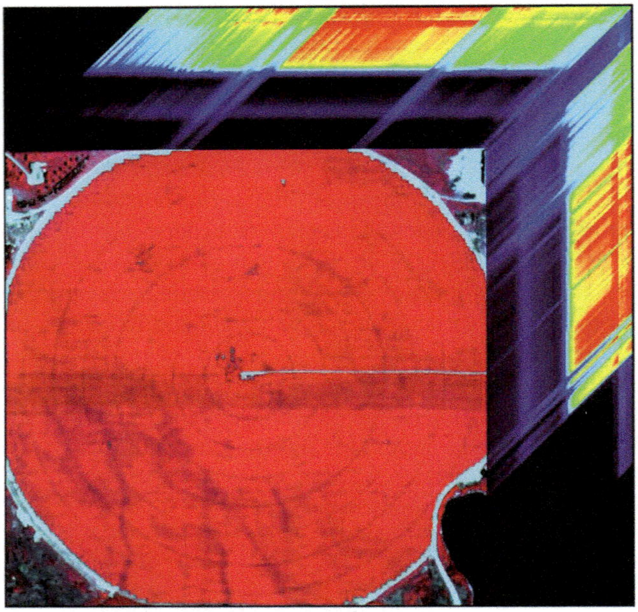

FIGURE 8.1 Hyperspectral data cube for an irrigated Minnesota potato field showing the spatial and spectral resolution available with hyperspectral imaging. The circular slice in front represents a combination of reflectance values at red, green, and blue wavelengths, whereas the cubical slices in back represent narrow band reflectance across a broad range of visible and NIR wavelengths.

TABLE 8.1
Multispectral Broadband Vegetation Indices or Commercial Sensor Midpoint Wavelengths Available for Use in Precision Agriculture

Index	Definition	Reference
GNDVI	(NIR − G)/(NIR + G)	Gitelson et al. (1996)
MSAVI2	0.5 * [2 * (NIR + 1) − SQRT((2 * NIR + 1)^2−8 * (NIR − R))]	Qi et al. (1994)
NDVI	(NIR − R)/(NIR + R)	Rouse et al. (1973)
OSAVI	(NIR − R)/(NIR + R + 0.16)	Rondeaux et al. (1996)
REIP	R/(NIR + R + G)	Sripada et al. (2005)
RVI	NIR/R	Jordan (1969)
SAVI	1.5 * [(NIR − R)/(NIR + R + 0.5)]	Huete (1988)
Crop Circle ACS 430	R_{670}, R_{730}, R_{780}	Holland et al. (2012)
CropSpec	R_{730}, R_{805}	Reusche et al. (2010)
GreenSeeker	R_{650}, R_{770}	Solie et al. (1996)
Yara N Sensor ALS	R_{730}, R_{760}, R_{900}, R_{970}	Link and Reusch (2006)

G refers to green reflectance, NIR to near infrared, and R to red reflectance. For commercial sensors, R_x refers to the center wavelength x of the reflectance band used by the sensor.

NIR (900–940 nm and 982 nm) spectral bands. The information in these bands is only available in narrow increments of 10–20 nm and is easily obscured in broad multispectral bands that are available with older satellite imaging systems. The best combination of two narrow bands in NDVI-like indices was centered in the red (682 nm) and NIR (920 nm) wavelengths but varied depending on the type of crop (corn, soybean, cotton, or potato) as well as the plant characteristic of interest (LAI, biomass, etc.). Analysis of hyperspectral imagery can potentially involve advanced chemometric methods that are not possible with broadband multispectral imagery, including (1) lambda-lambda plots, (2) spectral derivatives, (3) discriminant analysis, and (4) partial least squares analysis (Jain et al., 2007; Alchanatis and Cohen, 2010; Li et al., 2014a; Yuan et al., 2014).

The sharp contrast in reflectance behavior between the red and NIR portions of the spectrum is the motivation for development of spectral indices that are based on ratios of reflectance values in the visible and NIR regions (Sripada et al., 2008). Commonly used spectral reflectance indices (Table 8.1) include NDVI((NIR – Red)/(NIR + Red)), green NDVI, and ratio vegetation index (RVI = NIR/R). These indices, along with indices that are based on reflectance in the red edge spectrum region (700–740 nm) have been found to be very sensitive to crop canopy chlorophyll and nitrogen status due to the rapid change in leaf reflectance caused by the strong absorption by pigments in the red spectrum and leaf scattering in the NIR spectrum (Hatfield et al., 2008; Nguy-Robertson et al., 2012).

Several red edge-based vegetation indices have been identified from hyperspectral imagery (Haboudane et al., 2002) for estimating crop nitrogen status (Table 8.2). For example, red edge inflation point (REIP, Guyot et al., 1988) uses a red band (670 nm), two red edge bands (700 nm and 740 nm) and a NIR band (780 nm). It accurately estimated nitrogen supply to the plant, plant nitrogen concentration and uptake, and the nitrogen nutrition index (NNI), and was not affected significantly by interfering factors (e.g., zenith angle of the Sun, cloud cover, and soil color, etc.) (Heege et al., 2008; Mistele and Schmidhalter, 2008). The canopy chlorophyll content index (CCCI) is an integrated index based on the theory of two-dimensional planar domain illustrated by Clarke et al. (2001) using three bands (red, red edge and NIR). It uses NDVI as a surrogate for ground cover to separate soil signal from plant signal and the normalized difference red edge (NDRE) index as a measure of canopy nitrogen status (Fitzgerald et al., 2010). It is not significantly affected by ground cover (Fitzgerald et al., 2010), and worked well for estimating plant nitrogen status in the early growing season of maize (Li et al., 2014b). Other red edge indices include red edge chlorophyll index (CIred edge) (Gitelson et al., 2005), red edge ratio index (Erdle et al., 2011), Datt (1999) index ((NIR-Red edge)/(NIR-R)), MERIS terrestrial chlorophyll index (MTCI) (Shiratsuchi et al., 2011), red edge Soil Adjusted Vegetation Index (RESAVI), modified RESAVI (MRESAVI), red

TABLE 8.2

Hyperspectral Narrow-Band Vegetation Indices Available for Use in Precision Agriculture

Index	Definition	Reference
Aphid Index (AI)	$(R_{576} - R_{908})/(R_{756} - R_{716})$	Mirik et al. (2007)
CI Red Edge	$(R_{753} / R_{709}) - 1$	Gitelson et al. (2005)
DATT index	$(R_{850} - R_{710})/(R_{850} - R_{680})$	Datt (1999)
Damage Sensitive Spectral Index (DSSI)	$(R_{576} - R_{868} - R_{508} - R_{540})/[(R_{716} - R_{868}) + (R_{508} - R_{540})]$	Mirik et al. (2007)
Leaf Hopper Index (LHI)	$(R_{761} - R_{691})/(R_{550} - R_{715})$	Prabhakar et al. (2011)
MERIS TCI	$(R_{754} - R_{709})/(R_{709} - R_{681})$	Dash and Curran (2004)
NDRE	$(R_{790} - R_{720})/(R_{790} + R_{720})$	Barnes et al. (2000)
REIP	$700 + 40*\{[(R_{667} - R_{782})/2 - R_{702}]/(R_{738} + R_{702})\}$	Guyot et al. (1988)
Red Edge Ratio Index	$(R_{760}) / R_{730})$	Erdle et al. (2011)
PK Index	$(R_{1645} - R_{1715})/(R_{1645} - R_{1715})$	Pimstein et al. (2011)
PRI	$(R_{531} - R_{570})/(R_{531} + R_{570})$	Gamon et al. (1992)
S Index	$(R_{1260} - R_{660})/(R_{1260} + R_{660})$	Mahajan et al. (2014)
TCARI	$3*[(R_{700} - R_{670}) - 0.2*(R_{700} - R_{550})(R_{700}/R_{670})]$	Haboudane et al. (2002)

R refers to reflectance at the wavelength (nm) in subscript. NIR refers to near infrared reflectance.

edge difference vegetation index (REDVI) and red edge re-normalized difference vegetation index (RERDVI) (Cao et al., 2013).

The ultraviolet, violet and blue spectral regions have also been found to be important for estimating plant nitrogen concentration (Li et al., 2010). Wang et al. (2012) developed a new three band vegetation index using NIR, red edge and blue bands $[(R_{924} - R_{703} + 2 \times R_{423})/(R_{924} + R_{703} - 2 \times R_{423})]$, which was found to be closely related to wheat and rice leaf nitrogen concentration. Far NIR (FNIR) and short-wave infrared (SWIR) bands were found to be important for estimating plant aboveground biomass (Gnyp et al., 2014; Thenkabail et al., 2004). These bands are currently missing from the commercial active canopy sensors commonly used in precision agriculture.

The commonly used NDVI can easily become saturated at moderate to high canopy coverage conditions (Figure 8.2, Nigon et al., 2014). One reason is due to the normalization effect embedded in the calculation formula of this index (Nguy-Robertson et al., 2012; Gnyp et al., 2014), and another reason is due to the different transmittance of red and NIR radiation through the crop canopy leaves. The saturation effect of NDVI can be partially addressed by using ratio vegetation indices or wavelengths having similar penetration into the canopy (Gnyp et al., 2014; Li et al., 2014b; Van Niel and McVicar, 2004).

It should be noted that the sensitive spectral reflectance bands for precision agriculture change at different crop growth stages in response to crop growth and development (Li et al., 2010; Gnyp

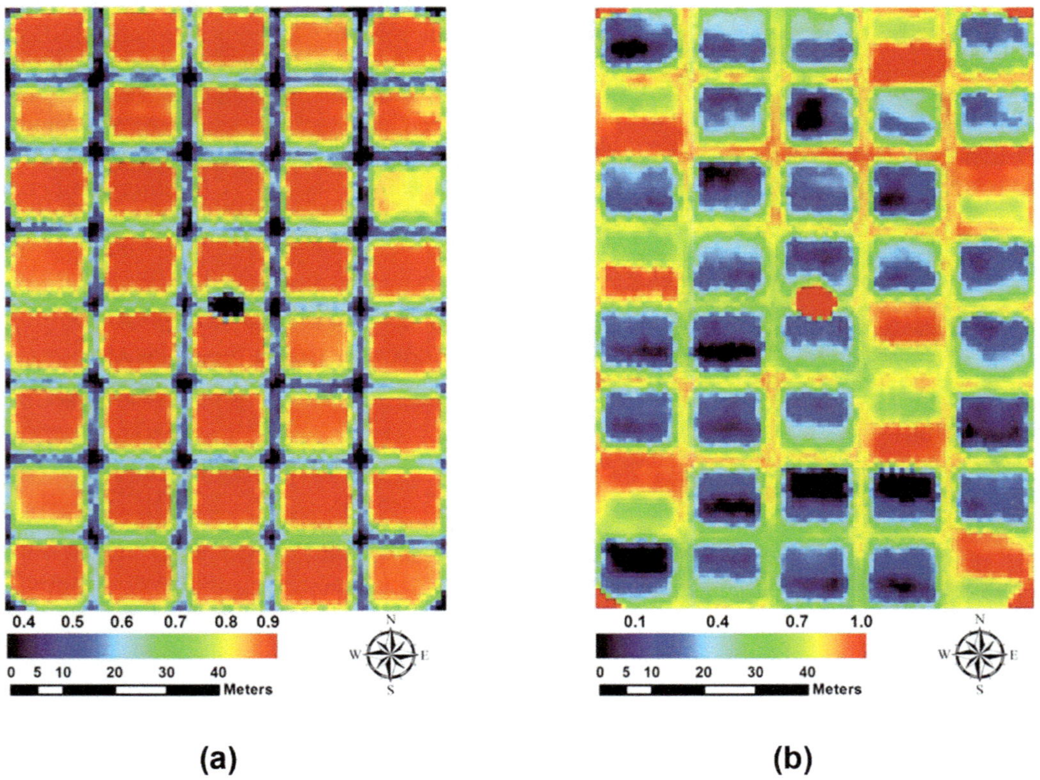

FIGURE 8.2 Hyperspectral estimates of (a) NDVI values and (b) TCARI-OSAVI values for small plots in a Minnesota potato field with two crop varieties receiving a wide range of nitrogen fertilizer application rates and timings. NDVI values exhibit a small range in values due to saturation. In contrast, TCARI-OSAVI values exhibit a large range of values, and are better suited for identifying differences in nitrogen stress for each variety.

et al., 2014). Different vegetation indices are needed for different crops, with different crop growth parameters at different growth stages (Hatfield and Prueger, 2010).

The fluorescence of leaf chlorophyll is an emerging research area in precision agriculture (Tremblay et al., 2012). When leaves that have been in the dark are exposed to UV or blue light, chlorophyll a in photosystem II (PSII) is excited to the first singlet state (Sayed, 2003), and upon decay to the ground energy state, these molecules are capable of fluorescence. Leaf fluorescence is affected by many factors including, the wavelength and intensity of incident light, temperature, canopy structure and leaf chlorophyll content, which may be affected by crop stresses from water, nitrogen, and salinity (Sayed, 2003; Tremblay et al., 2012). On first exposure to light, quinine acceptors in PSII are maximally oxidized (Baker and Rosenqvist, 2004), leading to a minimal fluorescence level (F_o). After further exposure to light, maximal fluorescence (F_m) may be attained, indicating that all electron acceptors are reduced (Baker and Rosenqvist, 2004). Interpretation of plant stress levels is often based on combinations or ratios of these two parameters (Tremblay et al., 2012; Baker and Rosenqvist, 2004; Sayed, 2003). Variable fluorescence (F_v) is defined as $F_m - F_o$, and F_v/F_m represents the photochemical efficiency of PSII (Tremblay et al., 2012). High values of F_o indicate plant stress (Tremblay et al., 2012), whereas low values of F_v/F_m indicate nitrogen stress (Baker and Rosenqvist, 2004). Diagnosis of specific types of crop stress may be facilitated by combining fluorescence spectroscopy with hyperspectral or multispectral imaging (Moshou et al., 2012; Mahlein et al., 2019).

8.4 REMOTE SENSING PLATFORMS

Remote sensing imagery for precision agriculture can be obtained using satellites, airplanes, UAVs, ground robots or agricultural machinery (Moran et al., 1997; Zhang and Kovacs, 2012; Mulla, 2013; Huang et al., 2018). Remote sensing imagery from satellites has improved in spatial resolution, spectral resolution, and the frequency of return visits since the launch of Landsat in the 1970s. Spatial resolution has improved from 30 m to 50 cm. Spectral resolution has improved from broadbands in the blue, green red, and NIR regions to narrow band hyperspectral imagery over blue, green, red, red edge and NIR wavelengths. Return frequencies have improved from several weeks to one day. Despite these improvements, satellite imagery in the visible and NIR regions still suffers from an inability to penetrate cloud cover.

Aerial remote sensing imagery offers excellent capabilities for precision agriculture applications (Yang, 2022). Spatial resolution is typically sub-meter, and spectral resolution ranges from broadband blue, green, red, and NIR to hyperspectral imaging. Aerial imaging can typically be obtained when and where it is needed with high reliability. Cloud cover is a continuing challenge for remote sensing from airplanes. Even though airplanes can fly below cloud cover, shadows from clouds cause difficulties in interpreting imagery. Remote sensing imagery obtained by proximal sensing from agricultural equipment is very popular in precision farming. Examples include on-the-go sensing from fertilizer spreaders for variable rate application of nitrogen fertilizer, and on-the-go sensing from herbicide sprayers for variable rate application of herbicides. Sensors used for proximal sensing are typically limited to two or three narrow bands of reflectance, thereby limiting the number of spectral indices that can be used to diagnose causes of stress. This is particularly limiting in mature crops with leaf area index values greater than three for sensors that calculate NDVI values. The NDVI values are less sensitive to spatial variations in chlorophyll content of leaves in mature crop canopies than at earlier growth stages.

Researchers have increasingly used UAVs for acquisition of remote sensing imagery (Figure 8.3). UAVs typically include fixed wing aircraft or helicopters that fly at altitudes of roughly 100 m (Zhang and Kovacs, 2012). Because of the low altitude, many images are typically acquired, and these must be tiled or mosaiced together to produce a continuous image of the field or farm of interest. Fixed wing UAVs generally have longer flight time (greater power supply) and payload capacity than helicopters. Aircraft have faster flight speeds than helicopters, and this may result in blurring of images due to the

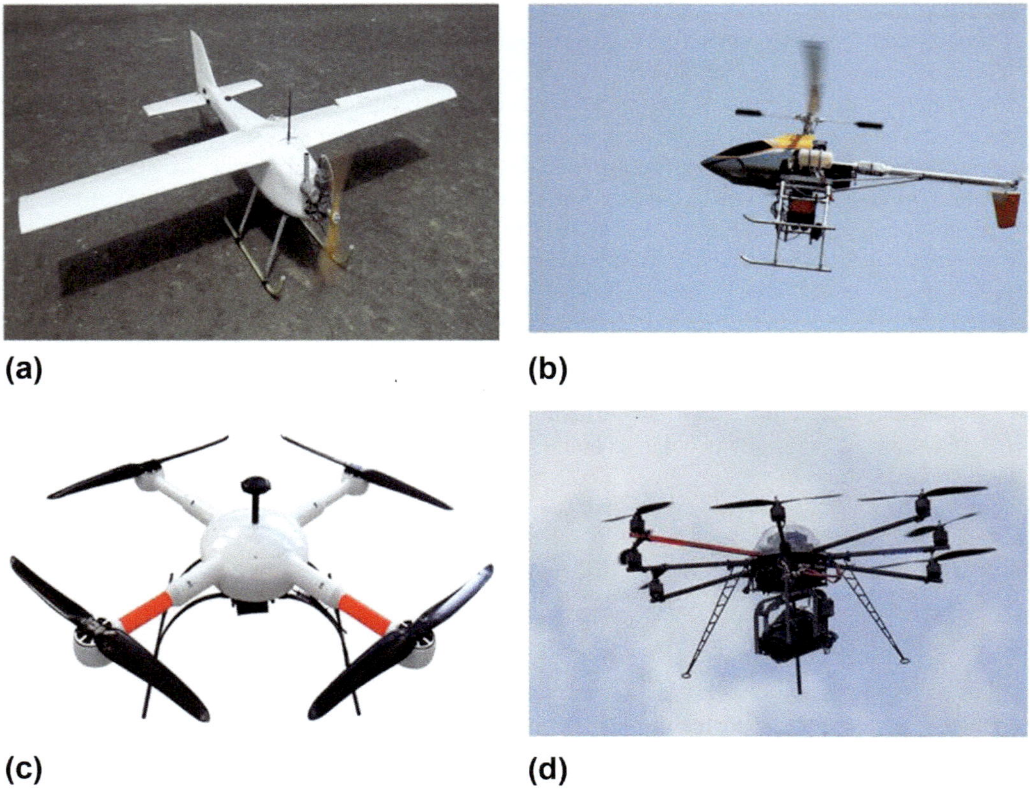

FIGURE 8.3 Different types of unmanned aerial vehicles used in precision farming: (a) fixed wing aircraft, (b) helicopter, (c) quadrocopter, and (d) octocopter.

low altitude. Helicopter UAVs have the advantages of flexibility and less space restriction by allowing vertical take-off and the ability to land vertically, hover, and fly forward, backward, and laterally as compared with fixed wing UAVs, allowing them to inspect isolated small fields closer to obstructions, which may be difficult for fixed-wing UAVs (Huang et al., 2013). Helicopters are generally more stable than aircraft, resulting in fewer problems with variations in viewing angle from one image to another. Remote sensing imagery from UAVs has very high spatial resolution, typically on the order of 10–50 cm. This allows individual plants to be studied. However, it also requires special care in correcting geometric distortion. Cameras used on UAVs range include inexpensive digital cameras that provide panchromatic images, multispectral cameras that provide narrow band reflectance in the blue, yellow, green, red, red-edge and NIR regions of the spectrum, expensive hyperspectral cameras with hundreds of narrow bands, thermal infrared cameras for plant surface temperature and thermal emission measurements and LiDAR sensors (Yao et al., 2019). Promising results have been obtained using UAV-based remote sensing for estimating crop leaf area index, biomass, plant height, nitrogen status, water stress, weed infestation, yield and grain protein content (Bendig et al., 2013; Berni et al., 2009; Samseemoung et al., 2012; Swain et al., 2010; Huang et al., 2018; Yao et al., 2019). It is expected to become a major remote sensing platform for precision agriculture in the future.

8.5 MANAGEMENT ZONES

Conventional agriculture involves uniform management of fields. In contrast, precision agriculture involves customized management in areas that are much smaller than fields. Precision farming has been used to divide fields into management zones (MZ) (Mulla, 1991, 1993) that differ in their requirements for fertilizer, pesticide, irrigation, seed or tillage. MZs are relatively homogeneous units within the field that differ from one another in their response to fertilizer, irrigation or pesticides, etc. They can be delineated based on differences in crop yield, soil type, topography or soil properties (fertility, moisture content, pH, organic matter, etc.) (Miao et al., 2018). Remote sensing has been used to delineate MZs based on variations in soil organic matter content (Mulla, 1997; Fleming et al., 2004; Christy, 2008). Boydell and McBratney (2002) used 11 years of Landsat TM imagery for a cotton field to identify MZs based on yield stability. Cammarano et al. (2020) used Landsat imagery to delineate management zones at village level in a small-scale farming system, with Green normalized difference vegetation index (GNDVI) maps from 2008 to 2018 being used to represent spatial and temporal variability in crop growth conditions and soil brightness index data being used to representation soil variability. Rasmussen et al. (2021) demonstrated and discussed the use of UAV-based remote sensing for MZ delineation in comparison with Sentiniel-2 satellite imagery for mapping heterogeneous crop fields. They found that using UAV and satellite remote sensing images might result in differences in the MZs delineated due to sensor differences.

8.6 IRRIGATION MANAGEMENT

Water stress is one of the major causes for loss of crop productivity (Moran et al., 2004). Irrigation is widely used to overcome crop water stress, but when applied uniformly, can lead to drawdown of water supply and environmental pollution. In precision irrigation, also known as variable rate irrigation (Sadler et al., 2005), sprinkler heads deliver water at rates that are varied using either microprocessors (Stark et al., 1993) or solenoids connected to manifolds (Omary et al., 1997). Nozzle spray rates are varied depending on spatial patterns in soil moisture (Hedley and Yule, 2009), crop stress (Bastiaanssen and Bos, 1999), or soil-landscape patterns, including rock outcroppings (Sadler et al., 2005). Variable rate irrigation uses water more efficiently than uniform irrigation, leading to better water conservation and improved environmental quality, without affecting crop yield.

Remote sensing can be used in variable rate irrigation applications by detecting crop water stress or estimating evapotranspiration or soil moisture through thermal infrared, optical or microwave sensing (Moran et al., 2004; Vereecken et al., 2012; Sishodia et al., 2020). In thermal infrared sensing, crop water stress is inferred by measuring a Crop Water Stress Index (CWSI) that is proportional to the difference between canopy and air temperatures (Moran et al., 2004), but also depends on the atmospheric vapor pressure deficit. CWSI values are estimated relative to the canopy and air temperatures for a non-stressed (well-watered) crop. This method works well for full crop canopies near a well-watered section of the crop field. Meron et al. (2010) developed a simplified approach for estimating CWSI that involves thermal infrared measurements of canopy temperature relative to the temperature of a nearby artificial reference surface consisting of a wet, white fabric covering polystyrene floating in a container of water. Care must be taken to segment thermal images in fields with partial canopy cover in order to eliminate errors due to high soil temperatures. Meron et al. (2010) showed that thermal infrared measurements of CWSI based on the artificial reference surface approach could be used to develop maps showing the spatial patterns in crop water stress to apply variable rates of irrigation.

ET is needed to determine crop water requirement, which is important in precision irrigation management. Four approaches have been taken to use remote sensing data to estimate ET: (1) surface energy balance; (2) crop coefficient, (3) the Penman-Monteith method, and (4) hybrid approach combining crop coefficient and energy balance methods, with each approach having advantages and limitations (Sishodia et al., 2020).

Optical, thermal and microwave remote sensing approaches have been used to estimate soil moisture status, with microwave remote sensing having a greater potential to provide more accurate results (Sishodia et al., 2020). However, microwave remote sensing generally does not provide the spatial resolution or accuracy in soil moisture estimation needed for precision irrigation. Multi-polarization microwave satellite data (e.g., Radarsat-2) can be combined with optical satellite remote sensing data to produce high spatial resolution soil moisture data for precision irrigation (Baghdadi et al., 2015). On the other hand, fast development of UAV-based optical, thermal and microwave remote sensing data provides more new and promising solutions to remote sensing-based precision irrigation management (Khanal et al., 2017; Wigmore et al., 2019; Ye et al., 2024).

8.7 CROP SCOUTING

Crop scouting is used for timely detection of crop stressors that pose an economic risk to production (Mueller and Pope, 2009; Fishel et al., 2001; Linker et al., 1999). If detected at an early stage, management actions can be taken to control crop water stress, nutrient deficiencies, kill weeds or insects and eradicate crop diseases. Crop scouting traditionally involves having a trained professional walk in a pre-determined pattern through an agricultural field in order to conduct a limited and somewhat random sampling to detect and identify crop stress. This approach is time consuming and labor intensive, and it does not guarantee that the sampling strategy covered the right spatial locations or occurred at the right time. Remote sensing offers the potential for improved crop scouting, with better spatial and temporal coverage than would be possible with a trained professional walking through fields. While remote sensing can accurately identify locations where crop stress is occurring, remote sensing alone is often unable to distinguish between crop stress caused by nutrient deficiencies, weed or insect pressure, or crop diseases. However, we can use remote sensing to identify problem areas in a field and guide a crop scout to visit those areas to determine the exact stress(s). A study indicated that UAV multispectral remote sensing-assisted scouting could identify disease foci and concerned areas more rapidly than conventional scouting, with 20% more often early detection of disease in watermelon fields (Kalischuk et al., 2019). A more promising strategy is to use satellite remote sensing to monitor crop fields and identify problem areas, and then a UAV or unmanned ground vehicle (UGV) equipped with multiple sensors can be used to identify the exact stress factor(s).

8.8 NUTRIENT DEFICIENCIES

Crop nutrient deficiencies are a major cause of crop stress and reductions in crop yield or quality. Nutrient deficiencies may be caused by macronutrients such as nitrogen, phosphorus, potassium, sulfur, calcium, and magnesium or micronutrients such as iron, boron, zinc, or manganese. Nutrient deficiencies often cause changes in leaf pigment concentrations, particularly for chlorophyll a and b. Changes in chlorophyll a or b content can be detected using remote sensing in the green (550 nm) and red-edge (710 nm) wavelengths. Nutrient deficiencies from either macro or micronutrients cause spectral reflectance of crop leaves to increase in the green portion of the spectrum. Reflectance spectra of deficient leaves alone are insufficient in many cases to determine which nutrient is responsible for the deficiency, and what rate or formulation of fertilizer is needed to correct the deficiency. Crop deficiencies also cause changes in crop biomass that can be detected using NIR reflectance.

Crop scout professionals have learned to distinguish and identify nutrient deficiencies based on coloration, pattern, location, and timing of the deficiency. Several examples for corn illustrate the approach used by crop scouts (Mueller and Pope, 2009). Nitrogen deficiency in corn appears as a yellowing of leaf color, starting with lower leaves. Deficiencies first appear at leaf tips and progress towards the base of the leaf in a v-shaped pattern. Phosphorus deficiency appears as red to purple leaf tips in the older leaves of young corn plants that appear to have stunted growth. Newly emerged leaves do not show phosphorus deficiencies, and the distinctive coloration associated with phosphorus deficiencies disappears when the crop grows to a meter or more in height. Potassium deficiency appears in corn as a yellowing along the edges of leaves at growth stage V6. It is often associated with conditions that lead to poor rooting depth.

Nutrient deficiencies that are detected and diagnosed in a timely fashion can be corrected using variable rate technology (VRT), which involves applying the right rate of fertilizer, at the right blend, in the right location at the right time. There is a long history of VRT in precision farming, with a primary focus on correcting nutrient deficiencies caused by phosphorus or nitrogen. In the earliest application of remote sensing for precision farming, Landsat TM images were used along with auxiliary data from soil sampling to develop maps showing spatial variability in phosphorus fertilizer recommendations for a wheat farm in Washington state (Bhatti et al., 1991). Landsat imagery was used to estimate spatial patterns in soil organic matter content, which were indirectly correlated with spatial patterns in soil phosphorus.

Proximal sensing of crops is currently the primary tool used to detect nutrient deficiencies for variable rate application of fertilizer. This is based on research that showed nitrogen deficiencies could be detected using spectral reflectance in the green, red, red-edge and near infrared portions of the spectrum. Commercial sensors used in precision farming to detect crop nitrogen deficiencies (Figure 8.4, Table 8.2) are mainly active crop canopy sensors with their own light sources to avoid the influence of different environmental light conditions, including the GreenSeeker, Crop Circle, CropSpec and Yara N-sensor (Shaver et al., 2011; Barker and Sawyer, 2010; Kitchen et al., 2010). GreenSeeker operates with the red (656 nm) and near infrared (774 nm) spectral bands. Crop Circle ACS 210 operates with the green (590 nm) and near infrared (880 nm) bands, while Crop Circle ACS 430 has red (670 nm), red edge (730 nm) and near infrared (780nm) bands. Crop Circle ACS 470 sensor also has three bands but is user configurable with a choice of 6 spectral bands covering blue (450 nm), green (550 nm), red (650 nm, 670 nm), red-edge (730 nm), and near infrared (>760 nm) regions (Cao et al., 2013). CropSpec operates with the red-edge (730 nm) and near infrared (805 nm) bands. Yara's N-sensor operates at 730 (red) and 760 (NIR) nm. A newer version of the Yara N sensor allows the operator to select five reflectance bands between 450–900 nm.

One limitation of the GreenSeeker, Yara N, CropSpec and Crop Circle sensors is that they cannot directly estimate the amount of nitrogen fertilizer needed to overcome crop nitrogen stress

FIGURE 8.4 Active crop canopy sensors commonly used in precision farming in the United States. (GreenSeeker, left; Crop Circle ACS 430, middle; Crop Circle ACS 470, right.)

(Samborski et al., 2009). Instead, sensor readings have to be compared to readings in reference strips receiving sufficient nitrogen fertilizer (Blackmer and Schepers, 1995; Raun et al., 2002; Sripada et al., 2008; Kitchen et al., 2010). These comparisons are the basis for nitrogen fertilizer response functions that relate sensor readings to the amount of nitrogen fertilizer needed to overcome crop nitrogen stress (Scharf et al., 2011). Clay et al. (2012) have shown that for wheat, when both water and nitrogen stress occur simultaneously, nitrogen fertilizer recommendations based on NDVI values are more accurate when reference strips have both sufficient nitrogen and insufficient moisture (water stress) in comparison with reference strips with both sufficient nitrogen and sufficient moisture (no water stress). To detect both nitrogen and water stress simultaneously, integrated sensing systems will be needed. Crop Circle Phenom is an integrated sensing system including a Crop Circle ACS-430 sensor and a Crop Circle DAS43X sensor, enabling the collection of reflectance data of red, red-edge and near-infrared wavelengths, automatically calculated NDVI, NDRE, estimated leaf area index(eLAI), and estimated canopy chlorophyll content (eCCC), atmospheric pressure, relative humidity, incoming and reflected photosynthetically active radiation, canopy temperature, and air temperature (Cummings et al., 2021). It has the potential for simultaneous identification of crop nitrogen and water stress. Integrated multispectral and thermal cameras have also been developed to enable UAV remote sensing-based monitoring of crop nitrogen and water status, like Altum-PT camera by MecaSense (now AgEagle) and 6X Multispectral & Thermal camera by Sentera.

Phosphorus deficiencies typically appear as changes in reflectance in the near infrared and blue portions of the spectrum. There has been little research on remote sensing methods to distinguish nitrogen, phosphorus, and potassium deficiencies in crops (Pimstein et al., 2011). Spectral signatures for nitrogen, phosphorus and potassium deficiency show responses at different wavelengths (Pimstein et al., 2011). NDVI values (such as those estimated using GreenSeeker technology) are often not able to distinguish between nitrogen and phosphorus deficiencies (Grove and Navarro, 2013). To distinguish nitrogen, phosphorus, and potassium deficiencies in wheat, Pimstein et al. (2011) proposed new spectral indices that require collecting reflectance data not only in the NIR region (870 nm), but also in the short-wave infrared (SWIR) region (1450, 1645 and 1715 nm). These new indices were able to predict phosphorus or potassium deficiency with an accuracy ranging from 78–80%, but accuracy levels decreased as variability in crop biomass increased.

In order to distinguish between different types of nutrient deficiencies, remote sensing must rely on more than changes in reflectance at key wavelengths. A diagnosis with remote sensing must also be able to detect where on the plant (upper vs lower leaves, leaf tips or edges, etc.) symptoms of deficiency occur and in what pattern. These patterns change over time, and early detection is important. High RGB resolution imagery at the scale of millimeter size pixels is needed for early detection (Zermas et al., 2020), otherwise it will be difficult to identify whether symptoms of deficiency are in upper or lower leaves, at leaf tips or basal regions, or along the edges or in interveinal regions of the leaf. For deficiencies that tend to occur in young plants, remote sensing must be able to compensate for reflectance from bare soil, hence spectral indices such as soil adjusted vegetation index (SAVI, Huete, 1988), modified SAVI (MSAVI, Qi et al., 1994) or optimized SAVI (OSAVI, Rondeaux et al., 1996) may be useful.

8.9 INSECT DETECTION

Insects cause crop damage by sucking plant sap, eating plant tissue, or damaging crop roots. Examples include European corn borer and Russian wheat aphid. These damages usually result in decreased crop biomass, deformed or stripped leaves. Because decreased biomass also occurs in response to other crop stressors, identifying insect damage via remote sensing has proved challenging.

Insect growth and development is more strongly linked with temperature and growing degree days than crop phenology (Hicks and Naeve, 1998; MacRae, 1998). Insects can first appear in a variety of locations, including along edges of fields, on undersides of leaves, or in the soil. It is difficult to detect insects in soil or on the undersides of leaves with remote sensing. Remote sensing often detects crop damage caused by insects, rather than the insects themselves. Harmful insects should be detected and identified before they can cause significant damage to crops. Proper identification is important because control methods vary by insect species.

Remote sensing is not widely used in precision agriculture for detecting insect infestations. Franke and Menz (2007) used hyperspectral imaging from an airplane in Iowa corn plots inoculated with European corn borer. Spectral indices were largely ineffective at differentiating inoculated plots from control plots during the first generation of insect growth. NDVI was consistently able to identify inoculated plots during the second generation of corn borer growth. These results show that it is difficult to use remote sensing for early detection of European corn borer. Mirik et al. (2007) used a handheld hyperspectral radiometer to measure reflectance in the visible and NIR wavelengths for Texas, Colorado, and Oklahoma winter wheat plots with and without significant Russian wheat aphid infestations. Their results showed that aphid damage resulted in changes in biomass that reduced NIR reflectance in infested plants relative to undamaged plants. They also showed increased reflectance in the green portion of the spectrum due to changes in chlorophyll content of leaves for infested plants relative to un-infested plants. They proposed using an Aphid Index (AI) and a Damage Sensitive Spectral Index (DSSI) to detect Russian wheat aphid damage (Table 8.2). AI is estimated based on $(R_{576} - R_{908})/(R_{756} - R_{716})$, where R is reflectance and the subscript denotes the wavelength (nm) of interest. DSSI is more complicated, and is estimated using $(R_{716} - R_{868} - R_{508} - R_{540})/[(R_{716} - R_{868}) + (R_{508} - R_{540})]$. Because the field of view for the handheld spectrometer was narrow, there was little mixing of pixels from infested and un-infested leaves, something that would be a significant impediment if reflectance measurements were obtained using satellites. More recently, Ribeiro et al. (2023) overcame this impediment by using Sentinel-2 satellite bands 6, 7 and 8a along with Support Vector Machine models to identify commercial soybean fields in the Midwestern USA exceeding an economic damage threshold of 250 aphids per plant with an accuracy of 91%.

Prabhakar et al. (2011) used hyperspectral imaging to detect leaf hopper damage in cotton. They found that leaf hopper damage was associated with decreases in the content of chlorophyll a and b pigments in leaves. The best spectral indices for identifying leaf hopper damage were based on changes in leaf reflectance in the visible (376, 496, 691 nm) and NIR (761, 1124 and 1457 nm) portions of the spectrum. A leaf hopper index (LHI) defined as $(R_{761} - R_{691})/(R_{550} - R_{715})$ could explain 46–82% of the variability in leaf hopper damage across three fields. A number of other spectral indices also performed relatively well, including NDVI, OSAVI, AI, and DSSI (Table 8.2).

In recent years, multispectral and hyperspectral UAV remote sensing has been increasingly used to detect crop pests. They can be more efficient than ground-based handheld sensing systems and fly at lower altitude to collect high spatial resolution images with less mixed pixels than satellite or aircraft-based remote sensing systems for better mapping pest distributions (Lost Filho et al., 2020). It should be noted that the different remote sensing systems mainly detect insect-induced plant stress, not the insects themselves. Field scouting is still needed to determine the exact insects and their damages, which will be more efficient guided by the remote sensing-based stress maps (Lost Filho et al., 2020).

8.10 DISEASE DETECTION

Diseases are caused by infestations of virus, fungi, or bacteria. They can affect any part of the plant, including leaves, stalks, roots, or grain. Damage to leaves often occurs as lesions or pustules that may lead to white, tan, brown or orange leaf colors (Mueller and Pope, 2009). Lesions can occur in shapes as varied as spots, rectangles or strips that vary in size and area. Each disease has a specific location where infection tends to occur, and each is associated with different shapes and colors of infected areas. Infected plants may eventually become stunted, and have chlorotic or necrotic leaves (Mirik et al., 2011). Early detection of disease is essential to limit economic damage (Sankaran et al., 2010).

Spectral characteristics of crops are often affected by disease, as described by West et al. (2003). Disease propagules often influence reflectance in the visible spectrum. Necrotic or chlorotic damage affects chlorophyll content and reflectance in the green and red-edge regions. Senescence affects reflectance in the red to NIR region. Stunting and reduced leaf area influences NIR reflectance. Impacts of disease on photosynthesis affect fluorescence in the spectral region between 450–550 nm and 690–740 nm. Crop disease also affects transpiration rates and water contents of leaves, these effects can be detected in the short-wave and thermal infrared regions.

Remote sensing is not widely used to detect crop disease in precision agriculture; however, research has shown that remote sensing has the potential to be used for more efficient monitor crop disease development, even at early stage of disease development when visible signs are not obvious (Table 8.3) (Sishodia et al., 2020). Different remote sensing technologies using RGB, multispectral, hyperspectral, thermal and fluorescence imaging have been used to detect fungal and viral infections in soybean (Das et al., 2013), wheat (Huang et al., 2007; Muhammed, 2005; Mewes et al., 2011; Mirik et al., 2011) and other crops (Mahlein, 2016; Barbedo, 2019). Yellow rust infections of wheat in China were accurately detected using aerial hyperspectral remote sensing and a photochemical reflectance index (PRI) (Huang et al., 2007). The values of PRI were estimated using reflectance values at 531 and 570 nm. Fluorescence at 550 and 690 nm was also useful for distinguishing wheat leaves infected with yellow rust from uninfected leaves (Bravo et al., 2004). Wheat infected with Septoria *tritici* leaf blotch in France was accurately distinguished from uninfected wheat using a combination of NDVI and thermal infrared measurements (Nicolas, 2004). Infestations of powdery

TABLE 8.3
Spectral Indices or Commercial Sensors Available for Diagnosis of Nutrient Deficiencies, Crop Disease, and Insect or Weed Infestations in Precision Agriculture

Index	N, P, or K	Disease	Insects	Weeds
Aphid Index (AI)			X	
CI Red Edge	X			
DATT index	X			
Damage Sensitive Spectral Index (DSSI)			X	
Fluorescence	X	X		
Leaf Hopper Index (LHI)			X	
MERIS TCI	X			
NDRE	X			
NDVI	X	X	X	X
REIP	X			
Red Edge Ratio Index	X			
RVI	X			X
PK Index	X			
PRI		X		
SAVI (or related)	X		X	X
S Index	X			
TCARI	X			
Crop Circle ACS 430	X			
CropSpec	X			
GreenSeeker	X			
Yara N Sensor	X			
WeedSeeker				X

mildew and leaf rust on wheat in Germany were difficult to detect at early stages of infection with Quickbird-like NDVI values (Franke and Menz, 2007). This is because at early stages of infection, reflectance in the red portion of the spectrum is affected, but NIR reflectance is not (Lorenzen and Jensen, 1989). At more advanced stages of infection, plant canopy structure and biomass are affected, causing changes in NIR reflectance that result in large decreases in NDVI values.

Yuan et al. (2014) used hyperspectral imaging to simultaneously detect and distinguish damage to wheat leaves caused by yellow rust and powdery mildew diseases and Russian wheat aphids. Reflectance in leaves damaged by disease and insects generally increased relative to undamaged leaves at wavelengths between 500–690nm. Distinguishing between disease and insect damage required analysis of reflectance in the NIR portion of the spectrum between 750–1300 nm. Powdery mildew and aphid damage caused reflectance in this region to decrease, whereas reflectance in this region increased for yellow rust damage. Partial least squares regression of reflectance in these regions, along with spectral derivative parameters and conventional spectral indices such as AI (Table 8.3) could explain 73% of the variability in intensity of wheat damage by the three stressors studied. Distinguishing damage from yellow rust versus powdery mildew versus aphids with hyperspectral imaging and Fisher Linear Discriminate Analysis (FLDA) was more challenging, however, especially at low intensities of infestation.

Thermal remote sensing can be used to detect plant diseases that affect plant's water status, such as root infections reducing plant water uptake, diseases reducing water transport within plants, modifying stomatal aperture, or causing changes in cuticular conductance (Oerke, 2020). Chlorophyll fluorescence sensing can be used to assess photosystem II activity, which can be affected by abiotic and biotic factors (Murchie and Lawson, 2013). The attack of pathogens will affect the plant's photosynthetic apparatus directly by causing leaf chlorosis (chlorophyll degradation) or reducing leaf area due to necrosis (Oerke, 2020).

It is challenging to use remote sensing to detect crop diseases in commercial fields due to spatial and temporal variability in environmental conditions, the crop characteristics (e.g., 3D architecture, different growth stages, etc.), and the possibility of multi-disease or multi-stress occurrence (Oerke, 2020). More research is still needed to develop accurate and efficient methods to detect and diagnose different crop diseases under diverse on-farm conditions.

8.11 WEED DETECTION

Weeds compete with crops for light, water, and nutrients. Above critical weed density thresholds, crop yields and quality will decline substantially. In most fields, weed infestations are not uniform, rather, weeds tend to occur in patches or clusters, leaving up to 80% of the field free of weeds (Wiles et al., 1992; Lamb and Brown, 2001). Because of this, there has been quite a bit of interest in precision agriculture (variable rate herbicide application) to control weeds that occur in patches, while avoiding herbicide application in areas without weeds (Stafford and Miller, 1993; Mulla et al., 1996; Hanks and Beck, 1998; Khakural et al., 1999). Variable rate herbicide application is especially of interest in Europe, where genetically modified crops (such as Roundup Ready soybean) are not allowed.

Weeds can be identified using remote sensing based on their spectral signatures, leaf shape, and organization of the weedy plant. Detecting and identifying weeds in a bare soil that is crop free is easier than detecting and identifying weeds in an actively growing crop (Thorp and Tian, 2004; Lópes-Granados, 2011). Detecting weeds that occur in large, dense clusters is easier with aerial remote sensing than identifying small, isolated weeds.

Remote sensing with satellites or airplanes is adequate for detecting weeds that occur in large, dense clusters within a crop or crop free soil (Lamb and Brown, 2001). Ground based proximal sensing is more suited than aerial remote sensing to detect and identify small, isolated weeds in a growing crop (Thorp and Tian, 2004). Proximal sensing has been used for real-time monitoring and spraying of weeds from a field herbicide applicator (López-Granados, 2011). A commercial example of this technology is WeedSeeker (Hanks and Beck, 1998), which uses gallium arsenide photoelectric emitters to detect weeds growing in bare soil or in a crop canopy (Sui et al., 2008). This technology is best suited to detecting weeds at intermediate growth stages that are growing between crop rows. It is not well suited to detecting recently emerged weeds (Thorp and Tian, 2004).

Zwiggelaar (1998) reviewed remote sensing methods for distinguishing weeds from soils or crops. Remote sensing is only useful if weeds have a spectral signature that is uniquely different from surrounding bare soil or crops, and if the spatial resolution of images is fine enough to detect individual weeds or patches of weeds (Lamb and Brown, 2001). Distinguishing weeds from soil is often based on graphing reflectance in the red portion of the spectrum versus reflectance in the NIR portion of the spectrum. A graph of these two reflectance bands for bare soil gives the soil line (Wiegand et al., 1991). For fields with mixtures of bare soil and weeds, the presence of weeds increases with vertical distance above the soil line along the NIR axis. Graphs of red versus NIR reflectance are commonly referred to as Tasseled Cap Transformations. Band ratios have also been used to distinguish weeds from bare soil. The most common approach for detection of weeds in bare soil is to use the NDVI ratio (Table 8.3). This ratio has the advantage of canceling out effects of shadows produced by weeds. Reflectance from bare soil can also be diminished through use of Soil Adjusted Vegetation Indices (SAVI, OSAVI, MSAVI, etc.).

Spectral reflectance patterns of weeds and crops are in general very similar when bare soil is absent (Zwiggelaar, 1998; Lamb and Brown, 2001). When bare soil is present, reflectance values at two wavelengths (e.g., 758 nm and 658 nm) can be used along with discriminant analysis to distinguish crops from weeds from soil (Borregaard et al., 2000). RVI (=NIR/R) and NDVI have also often been used to discriminate between weeds and crops (Table 8.3), especially when crops occur in systematic rows and weeds occur as patches between rows. Detection of weeds at early growth stages is very challenging (López-Granados, 2011), especially if they occur in recently germinated crops with similar physiology (e.g., grassy weeds in cereal crops or broad leaf weeds in dicotyledonous crops). Detection is easier at later growth stages, when spectral differences between weeds and crops are greatest (López-Granados, 2011).

UAV remote sensing can collect high spatial resolution images at centimeter scale, which makes it more suitable for weed detection and mapping, especially for low weed densities (Huang et al., 2018; Wu et al., 2021). Three types of cameras are commonly used in UAV remote sensing of weeds (RGB, multispectral and hyperspectral cameras), which can map weed patches accurately (Esposito et al., 2021).

8.12 MACHINE VISION FOR WEED DISCRIMINATION

Discrimination between weeds and crops requires high spatial resolution of imagery (Zwiggelaar, 1998). Remote sensing images with a spatial resolution of tens of meters will not be sufficient for discrimination of weeds and crops. Images at a spatial resolution of tens of centimeters to a meter are needed to distinguish plants from weeds (Lamb and Brown, 2001; Rasmussen et al., 2013). However, even spectral indices at this fine scale of resolution are often by themselves not sufficient, because crops and weeds often have similar reflectance signatures. Crops and weeds are more easily distinguished based on differences in their canopy or leaf shapes, heights, and structures. These features can be described and distinguished from one another using machine vision analysis of color images or video imagery (Gée et al., 2008; Burgos-Artizzu et al., 2011).

Discrimination of one weed species from another is more challenging than discriminating weeds from crops. Gibson et al. (2004) used supervised classification of weeds in soybean based on aerial remote sensing in the yellow, green, red and NIR bands. While weedy areas could be distinguished from soybeans or bare soil with accuracies of greater than 90%, distinguishing giant foxtail from velvetleaf had accuracy levels ranging from 41% to 83%.

Machine vision is commonly used for precision agriculture applications of discriminating weeds from bare soil or crops (Thorp and Tian, 2004; Wu et al., 2021). There are two basic steps in discriminating weeds (Gée et al., 2008; Burgos-Artizzu et al., 2011). The first is distinguishing regions with vegetation from regions with bare soil (segmentation). The second is distinguishing weeds from crops (discrimination). As an example of this two-step process, Gée et al. (2008) used a RGB color image in various row crops to estimate an excess green index (Gée et al., 2008), which was then reclassified into black (soil) and white (vegetation) components. The reclassified image was then subjected to a double Hough transformation (DHT) to identify the position of the linear crop rows. Blobs of white (vegetation) that were offset from rows were assumed to be weeds. Burgos-Artizzu et al. (2011) use real-time analysis of video imagery to perform these same two steps and were able to accurately identify 85% of the weeds in a field of maize.

Examples of machine vision for precision weed management are numerous (Wang et al., 2019; Wu et al., 2021). The University of Tokyo developed an autonomous vehicle for mechanical weeding and variable rate application of chemicals (Torii, 2000). This vehicle is guided along crop rows based on a Hue, Saturation, and Intensity (HSI) transformation. Tillet et al. (2008) used real-time machine vision in conjunction with a mechanical weeder to reduce weed populations in cabbage by 62–87%. Blasco et al. (2002) used machine vision with a robotic weeder that produced an electrical discharge of 15,000 volts. These studies both show that it is possible to use precision agriculture techniques to avoid using herbicides to control weeds.

8.13 MULTI-SOURCE DATA FUSION AND MACHINE LEARNING

Previous applications of remote sensing in precision agriculture mainly used spectral data alone. With the development of UAV remote sensing allowing collection of high spatial resolution images, structural (e.g., plant height, canopy cover or vegetation fractions) and textural (e.g., mean, variance, homogeneity, contrast, dissimilarity, entropy, second moment, and correlation) data are used together with spectral data to improve prediction of crop traits for applications in precision agriculture (Zhang et al., 2022; Liang et al., 2023). As more remote sensing features are used, traditional multiple linear regression approach is not sufficient, and machine learning algorithms are increasingly adopted, which can consider both linear and nonlinear relationships and handle large datasets (Chlingaryan et al., 2018; Zha et al., 2020). Considering the limitations of different sensing technologies, multi-sensor data fusion using machine learning has been adopted to improve crop growth monitoring, crop stress diagnosis and mapping (Munnaf et al., 2020; Ahmad et al., 2022; Fei et al., 2022). In addition to multi-sensor data fusion, it is important to incorporate multi-source data to improve prediction of crop growth conditions together with proximal or remote sensing data for precision crop management (Wang et al., 2021; Li et al., 2022; Lu et al., 2022). Li et al. (2022) integrated genetics, environmental (soil and weather), and management data together with active canopy sensor data to significantly improve corn N status prediction and diagnosis across the US Midwest. The 3-band Crop Circle ACS 430 sensor performed better than the 2-band GreenSeeker sensor for corn N status diagnosis and recommendation when they were used alone, but their performance differences were minimized when multisource data fusion was used (Wang et al., 2023). In addition to commonly used machine learning algorithms, deep learning algorithms are also increasingly used to analyze remote sensing and other related data to improve crop stress diagnosis and precision crop management decisions (Liu et al., 2021; Zhou et al., 2021).

8.14 KNOWLEDGE GAPS

Remote sensing applications in precision farming have increased dramatically over the last decades (Mulla, 2013; Huang and Thomson, 2015; Yao et al., 2018). This increased adoption is associated with investments in precision farming research, coupled with improvements in the spatial and spectral resolution and return frequency of satellite, aerial and UAV remote sensing imagery, and the development of proximal sensors. Aerial and proximal remote sensing are primarily used for variable rate application of irrigation water, nitrogen fertilizer or for detection of weeds. Remote sensing is not widely used for detection of crop stresses by insects or plant diseases and is rarely used for detection of nutrient deficiencies other than nitrogen.

There is a pressing need for broader use of proximal and remote sensing in precision agriculture. Current applications of remote sensing are rarely able to simultaneously identify locations of a field afflicted with crop stress, and distinguish between stresses caused by water, nutrients, weeds, insects, and disease. Furthermore, remote sensing is rarely able to distinguish between stresses caused by different types of nutrients, different types of diseases or different types of insects. The main reason for this failure is that remote sensing applications typically rely only on spectral signatures at a few important wavelengths (green, red, red-edge, and NIR) or combinations of these wavelengths where different types of crop stress have similar influences on chlorophyll content of leaves and adverse effects on crop biomass or canopy structure (Table 8.3). Distinguishing between stresses caused by water, nutrients, weeds, insects and disease will require fusion of remote sensing information (e.g., hyperspectral and fluorescence spectroscopy) that are sensitive to these influences and effects, combined with machine vision to identify the locations on a plant (stems or leaves, leaf tips or leaf edges, upper leaves, or lower leaves), colors of stress (yellow, purple, red, brown, white, etc.), and the shapes associated with stresses (e.g., monocotyledonous vs dicotyledonous weeds, spots versus stripes).

Further development of remote sensing applications in precision agriculture will require multidisciplinary efforts by experts in crop water, nutrient, weed, insect, and disease stresses working collaboratively with experts in remote sensing and engineering. At present, these types of multidisciplinary team efforts are rare. Further development of remote sensing applications in precision agriculture will require use of high resolution (cm scale) aerial imagery at key wavelengths to identify locations affected by crop stress, coupled with proximal sensing and machine vision to differentiate between different types of crop stress to diagnose the problem. Platforms to collect remote sensing imagery must be capable of deployment at intervals of at least every week during the growth of the crop, and these platforms must be capable of distinguishing between stresses caused by water, nutrients, weeds, diseases, and insects. Unmanned aerial vehicles and proximal sensors offer significant potential to address these capabilities, and further research with these platforms and sensors is encouraged.

Many remote sensing-based models generally work well in the region(s) of their development. However, when they are used in other regions, they may not work well. It is important for researchers in different regions to work together and build large databases to develop proximal or remote sensing-based machine or deep learning models with multisource data fusion that can be used across different regions.

8.15 CONCLUSIONS

Precision agriculture can be generally defined as doing the right management practices at the right location, in the right rate and at the right time. Precision agriculture offers several benefits, including improved efficiency of farm management inputs, increases in crop productivity or quality, and reduced transport of fertilizers and pesticides beyond the edge of field.

Losses in crop productivity often occur non-uniformly at specific locations within fields, and at critical growth stages. Crop stress must be detected in a timely fashion, the type of stress causing it must be identified, and management practices must be implemented at the right locations and times to overcome crop stress.

Research applications of remote sensing in precision agriculture are numerous, and include techniques for detecting water stress, nitrogen stress, weed infestations, fungal disease, and insect damage. Significant advances have been made in identifying key wavelengths and spectral indices at which these stresses influence the reflectance or fluorescence properties of plant pigments and crop canopy architecture. However, little research has been conducted on detecting locations affected by crop stress and simultaneously distinguishing between different types of crop stress. A basic problem is that remote sensing does not typically respond directly to water, nutrient, weed, insect, or disease stresses, rather it responds indirectly to the changes in chlorophyll or crop canopy architecture caused by these crop stresses. For this reason, remote sensing has not yet been widely adopted by farmers for routine use in precision agriculture. The main reasons include the difficulty in interpreting spectral signatures, the slow processing time for data, the high expense, and the need to collect confirmatory data from ground surveys to diagnose causative factors for anomalous spectral reflectance data. Clearly, there is a significant scope for improving the interpretation and utility of remote sensing data for precision agriculture.

Researchers have focused significant effort on identifying key wavelengths at which areas with crop stress can be distinguished from areas without crop stress. These wavelengths, and spectral indices based on them typically occur in the green, red, red-edge and NIR bands. Significant progress has been made in identifying spectral indices that respond to changes in leaf pigmentation or canopy biomass and architecture, or indices that can eliminate interference from shadows and soil background effects. As the spatial resolution of remote sensing imagery used in precision agriculture has improved (from 30 m to sub-meter resolution), techniques for discriminating crops, soils, and weeds have also improved. As spectral bandwidth has decreased (from broadband blue, green, red, and NIR to narrow band hyperspectral and fluorescence spectroscopy), researchers have discovered that crop stress is more easily detected with narrow bands (10–20 nm wide) rather than broadbands (50–100 nm wide) at these key wavelengths. Narrow band hyperspectral imagery is amenable to image analysis with advanced chemometric techniques that allow for better diagnosis of crop stress, including lambda-lambda plots, derivative analysis, and partial least squares analysis.

Less progress has been made in the use of remote sensing coupled with computer vision for differentiating between specific types of crop stress based on the location within the plant where stress occurs, and the shape or color of the stressor. Advances in computer vision are needed that require collaborative research by multidisciplinary teams of agronomists, engineers and remote sensing experts working with high resolution hyperspectral and video imagery capable of viewing individual plants. High resolution imagery is increasingly possible because of improvements in camera technology and proximal sensors deployed on unmanned aerial or ground vehicles that collect imagery at short distances from the growing crop.

Multi-sensor and multi-source data fusion using machine learning and deep learning has greatly improved crop stress detection for precision agriculture. Future research is needed for multi-disciplinary and multi-region scientists to work together to develop machine learning and deep learning models that can diagnose multi-stresses simultaneously across different regions to better support the development of precision agriculture.

REFERENCES

Ahmad, U., A. Nasirahmadi, O. Hensel, and S. Marino. 2022. Technology and data fusion methods to enhance site-specific crop monitoring. Agronomy 12: 555.

Alchanatis, V., and Y. Cohen. 2010. Spectral and spatial methods of hyperspectral image analysis for estimation of biophysical and biochemical properties of agricultural crops. Ch. 13. In: (P.S. Thenkabail, J.G. Lyon and A. Huete, eds.), Hyperspectral Remote Sensing of Vegetation. CRC Press, Boca Raton, FL, p. 705.

Baghdadi, N., M.E. Hajj, M. Zribi, and I. Fayad. 2015. Coupling SAR C-band and optical data for soil moisture and leaf area index retrieval over irrigated grasslands. IEEE J. Sel. Top. Appl. Earth Obs. Remote Sens. 9: 1–15.

Baker, N.R., and E. Rosenqvist. 2004. Applications of chlorophyll fluorescence can improve crop production strategies: An examination of future possibilities. J. Exp. Bot. 55: 1607–1621.

Barbedo, J. 2019. A review on the use of unmanned aerial vehicles and imaging sensors for monitoring and assessing plant stresses. Drones 3: 40.

Barker, D.W., and J.E. Sawyer. 2010. Using active canopy sensors to quantify corn nitrogen stress and nitrogen application rate. Agron. J. 102: 964–971.

Barnes, E.M., T.R. Clarke, S.E. Richards, P.D. Colaizzi, J. Haberland, M. Kostrzewski, and P. Waller. 2000. Coincident detection of crop water stress, nitrogen status and canopy density using ground based multispectral data. In: (P.C. Robert, R.H. Rust and W.E. Larson, eds.), Proc. 5th Int. Conf. Precis Agric. Bloomington, MN.

Bastiaanssen, W.G.M., D.J. Molden, and I.W. Makin. 2000. Remote sensing for irrigated agriculture: Examples from research and possible applications. Agric. Water Manag. 46: 137–155.
Bastiaanssen, W.G.M., and M.G. Bos. 1999. Irrigation performance indicators based on remotely sensed data: A review of literature. Irrig. Drain. Syst. 13: 291–311.
Bendig, J., A. Bolten, and G. Bareth. 2013. UAV-based imaging for multi-temporal, very high resolution crop surface models to monitor crop growth variability. PFG 2013(6): 551–562.
Berni, J.A.J., P.J. Zarco-Tejada, L. Suárez, and E. Fereres. 2009. Thermal and narrowband multispectral remote sensing for vegetation monitoring from an unmanned aerial vehicle. IEEE Trans. Geosci. Remote Sens. 47(3): 722–738.
Bhatti, A.U., D.J. Mulla, and B.E. Frazier. 1991. Estimation of soil properties and wheat yields on complex eroded hills using geostatistics and thematic mapper images. Remote Sens. Environ. 37: 181–191.
Blackburn, G.A. 2007. Hyperspectral remote sensing of plant pigments. J. Exp. Bot. 58: 855–867.
Blackmer, T.M., and J.S. Schepers. 1995. Use of a chlorophyll meter to monitor nitrogen status and schedule fertigation for corn. J. Prod. Agric. 8: 56–60.
Blasco, J., N. Aleixos, J.M. Roger, G. Rabatel, and E. Molto. 2002. Robotic weed control using machine vision. Biosys. Eng. 83(2): 149–157.
Borregaard, T., H. Nielsen, L. Norgaard, and H. Have. 2000. Crop-weed discrimination by line imaging spectroscopy. J. Agric. Eng. Res. 75: 389–400.
Boydell, B., and A. McBratney. 2002. Identifying potential within-field management zones from cotton-yield estimates. Precis. Agric. 3(1): 9–23.
Bravo, C., D. Moshou, R. Oberti, J. West, A. McCartney, L. Bodria, and H. Ramon. 2004. Foliar disease detection in the field using optical sensor fusion. Agric. Eng. Int.: CIGR J. Sci. Res. Dev. Manuscript FP 04 008. 6.
Burgos-Artizzu, X.P., A. Ribeiro, M. Guijarro, and G. Pajares. 2011. Real-time image processing for crop/weed discrimination in maize fields. Comp. Electron. Agric. 75(2): 337–346.
Cammarano, D., H. Zha, L. Wilson, Y. Li, W.D. Batchelor, and Y. Miao. 2020. A remote sensing-based approach to management zone delineation in small scale farming systems. Agronomy 10(11): 1767.
Cao, Q., Y. Miao, H. Wang, S. Huang, S. Cheng, R. Khosla, and R. Jiang. 2013. Non-destructive estimation of rice plant nitrogen status with Crop Circle multispectral active canopy sensor. Field Crops Res. 154: 133–144.
Chlingaryan, A., S. Sukkarieh, and B. Whelan. 2018. Machine learning approaches for crop yield prediction and nitrogen status estimation in precision agriculture: A review. Comput. Electron. Agric. 151: 61–69.
Christy, C.D. 2008. Real-time measurement of soil attributes using on-the-go near infrared reflectance spectroscopy. Comp. Electron. Agric. 61: 10–19.
Clarke, T.R., M.S. Moran, E.M. Barnes, P.J. Pinter, and J. Qi. 2001. Planar domain indices: A method for measuring a quality of a single component in two-component pixels. In: Proc. IEEE International Geosci. Remote Sens. Sympos. [CD ROM], Sydney, Australia, 9–13 July.
Clay, D.E., T.P. Kharel, C. Reese, D. Beck, C.G. Carlson, S.A. Clay, and G. Reicks. 2012. Winter wheat crop reflectance and nitrogen sufficiency index values are influenced by nitrogen and water stress. Agron. J. 104: 1612–1617.
Cohen, J.E. 2003. Human population: The next half century. Science 302: 1172–1175.
Crookston, K. 2006. A top 10 list of developments and issues impacting crop management and ecology during the past 50 years. Crop Sci. 46: 2253–2262.
Cummings, C., Y. Miao, G.D. Paiao, S. Kang, and F.G. Fernández. 2021. Corn nitrogen status diagnosis with an innovative multi-parameter Crop Circle Phenom sensing system. Remote Sens. 13: 401.
Das, D.K., S. Pradhan, V.K. Sehgal, R.N. Sahoo, V.K. Gupta, and R. Singh. 2013. Spectral reflectance characteristics of healthy and yellow mosaic virus infected soybean (Glycine max L.) leaves in a semiarid environment. J. Agrometeor. 15: 37–39.
Dash, J., and P.J. Curran. 2004. The MERIS terrestrial chlorophyll index. Intl. J. Rem. Sensing 25: 5403–5413.
da Silva, E.E., F.H.R. Baio, L.P.R. Teodoro, et al. 2022. Variable-rate seeding in soybean according to soil attributes related to grain yield. Precis. Agric 23: 35–51.
Datt, B. 1999. A new reflectance index for remote sensing of chlorophyll content in higher plants: Tests using eucalyptus leaves. J. Plant Physio. 154: 30–36.
Diacono, M., P. Rubino, and F. Montemurro. 2013. Precision nitrogen management of wheat: A review. Agron. Sustain. Dev. 33: 219–241.
Erdle, K., B. Mistele, and U. Schmidhalter. 2011. Comparison of active and passive spectral sensors in discriminating biomass parameters and nitrogen status in wheat cultivars. Field Crops Res. 124: 74–84.

Erickson, B., and S.W. Fausti. 2021. The role of precision agriculture in food security. Agron. J. 113(6): 4455–4462.

Esposito, M., M. Crimaldi, V. Cirillo, F. Sarghini, and A. Maggio. 2021. Drone and sensor technology for sustainable weed management: A review. Chem. Biol. Technol. 8: 1–11.

Fei, S., M.A. Hassan, Y. Xiao, X. Su, Z. Chen, Q. Cheng, et al. 2022. UAV-based multi-sensor data fusion and machine learning algorithm for yield prediction in wheat. Precis. Agric. 24: 187–212.

Fishel, F.M., W.C. Bailey, M. Boyd, W.G. Johnson, M. O'Day, L.E. Sweets, and W.J. Wiebold. 2001. Introduction to Crop Scouting. IPM Manual. University of Missouri, Columbia, MO.

Fitzgerald, G., D. Rodriguez, and G. O'Leary. 2010. Measuring and predicting canopy nitrogen nutrition in wheat using a spectral index: The canopy chlorophyll content index (CCCI). Field Crops Res. 116: 318–324.

Fleming, K.L., D.F. Heermann, and D.G. Westfall. 2004. Evaluating soil color with farmer input and apparent soil electrical conductivity for management zone delineation. Agron. J. 96(6): 1581–1587.

Franke, J., and G. Menz. 2007. Multi-temporal wheat disease detection by multi-spectral remote sensing. Precis. Agric. 8: 161–172.

Freeman, K.W., K. Girma, D.B. Arnall, R.W. Mullen, K.L. Martin, R.K. Teal, and W.R. Raun. 2007. By-plant prediction of corn forage biomass and nitrogen uptake at various growth stages using remote sensing and plant height. Agron. J. 99: 530–536.

Gamon, J.A., J. Penuelas, and C.B. Field. 1992. A narrow-waveband spectral index that tracks diurnal changes in photosynthetic efficiency. Remote Sens. Environ. 41: 35–44.

Gée, C., J. Bossu, G. Jones, and F. Truchetet. 2008. Crop/weed discrimination in perspective agronomic images. Comp. Electron. Agric. 60: 49–59.

Gibson, K.D., R. Dirks, C.R. Medlin, and L. Johnston. 2004. Detection of weed species in soybean using multispectral digital images. Weed Tech. 18(3): 742–749.

Gitelson, A.A., Y.J. Kaufman, and M.N. Merzlyak. 1996. Use of a green channel in remote sensing of global vegetation from EOS-MODIS. Remote Sens. Environ. 58: 289–298.

Gitelson, A.A., A. Viña, V. Ciganda, D.C. Rundquist, and T.J. Arkebauer. 2005. Remote estimation of canopy chlorophyll content in crops. Geophys. Res. Lett. 32, L08403.1–L08403.4.

Gnyp, M.L., Y. Miao, F. Yuan, S.L. Ustin, K. Yu, Y. Yao, S. Huang, and G. Bareth. 2014. Hyperspectral canopy sensing of paddy rice aboveground biomass at different growth stages. Field Crops Res. 155: 42–55.

Goel, P.K., S.O. Prasher, J.A. Landry, R.M. Patel, R.B. Bonnell, A.A. Viau, and J.R. Miller. 2003. Potential of airborne hyperspectral remote sensing to detect nitrogen deficiency and weed infestation in corn. Comp. Electro. Agric. 38: 99–124.

Grove, J.H., and M.M. Navarro. 2013. The problem is not N deficiency: Active canopy sensors and chlorophyll meters detect P stress in corn and soybean. In: (J.V. Stafford, ed.), Precision Agriculture'13. Wageningen Academic Publishers, Wageningen, The Netherlands, pp. 137–144.

Guyot, G., F. Baret, and D.J. Major. 1988. High spectral resolution: Determination of spectral shifts between the red and infrared. Intl. Arch. Photogram. Remote Sens. 11: 750–760.

Haboudane, D., J.R. Miller, E. Pattey, P.J. Zarco-Tejada, and I.B. Strachan. 2004. Hyperspectral vegetation indices and novel algorithms for predicting green LAI of crop canopies: Modeling and validation in the context of precision agriculture. Remote Sens. Environ. 90: 337–352.

Haboudane, D., J.R. Miller, N. Tremblay, P.J. Zarco-Tejada, and L. Dextraze. 2002. Integrated narrow-band vegetation indices for prediction of crop chlorophyll content for application to precision agriculture. Remote Sens. Environ. 81: 416–426.

Hanks, J.E., and J.L. Beck. 1998. Sensor-controlled hooded sprayer for row crops. Weed Tech. 12: 308–314.

Hatfield, J.L., A.A. Gitelson, S. Schepers, and C.L. Walthall. 2008. Application of spectral remote sensing for agronomic decisions. Agron. J. 100: 117–131.

Hatfield, J.L., and J.H. Prueger. 2010. Value of using different vegetative indices to quantify agricultural crop characteristics at different growth stages under varying management practices. Remote Sens. 2: 562–578.

Hedley, C.B., and I.J. Yule. 2009. Soil water status mapping and two variable-rate irrigation scenarios. Precis. Agric. 10: 342–355.

Heege, H.J., S. Reusch, and E. Thiessen. 2008. Prospects and results for optical systems for site-specific on-the-go control of nitrogen-top-dressing in Germany. Precis. Agric. 9: 115–131.

Hicks, D.R., and S.L. Naeve. 1998. The Minnesota Soybean Field Book. University of Minnesota Extension Service, St. Paul, MN.

Holland, Lamb and Schepers. 2012. Radiometry of proximal active optical sensors (AOS) for agricultural sensing. IEEE J. Sel. Topics Appl. Earth Observ. Remote Sens. 5: 1793–1802.

Huang, W., D.W. Lamb, Z. Niu, L. Liu, and J. Wang. 2007. Identification of yellow rust in wheat by in situ and airborne spectrum data. Precis. Agric. 8 (4–5): 187–197.

Huang, Y., K.N. Reddy, R.S. Fletcher, and D. Pennington. 2018. UAV low-altitude remote sensing for precision weed management. Weed Tech. 32: 2–6.

Huang, Y., and S.J. Thomson, 2015. Remote sensing for cotton farming. In: (D.D. Fang and R.G. Percy, eds.), Cotton, 2nd edition, American Society of Agronomy, Inc., Crop Science Society of America, and Soil Society of America, Inc. Madison, WI, USA, Agronomy Monograph, Vol. 57, pp. 1–26.

Huang, Y., S.J. Thomson, W.C. Hoffmann, Y. Lan, and B.K. Fritz. 2013. Development and prospect of unmanned aerial vehicle technologies for agricultural production management. Int. J. Agric. Bio. Eng. 6(3): 1–10.

Huete, A. 1988. A soil adjusted vegetation index (SAVI). Remote Sens. Environ. 25: 295–309.

Jain, N., S.S. Ray, J.P. Singh, and S. Panigrahy. 2007. Use of hyperspectral data to assess the effects of different nitrogen applications on a potato crop. Precis. Agric. 8: 225–239.

Jordan, C.F. 1969. Derivation of leaf area index from quality of light on the forest floor. Ecology 50: 663–666.

Kalischuk, M., M.L. Paret, J.H. Freeman, D. Raj, S. Da Silva, S. Eubanks, et al. 2019. An improved crop scouting technique incorporating unmanned aerial vehicle-assisted multispectral crop imaging into conventional scouting practice for gummy stem blight in watermelon. Plant Dis. 103: 1642–1650.

Khakural, B.R., P.C. Robert, and D.R. Huggins. 1999. Variability of corn/soybean yield and soil/landscape properties across a southwestern Minnesota landscape. In: (Robert, et al., ed.), Proceedings of the 4th International Conference on Precision Agricultural. St. Paul, MN. 19–22 July, 1998. ASA, CSSA, SSSA, Madison, WI, pp. 573–579.

Khanal, S., J. Fulton, and S. Shearer. 2017. An overview of current and potential applications of thermal remote sensing in precision agriculture. Comput. Electron. Agric. 139: 22–32.

Kitchen, N.R., K.A. Sudduth, S.T. Drummond, P.C. Scharf, H.L. Palm, D.F. Roberts, and E.D. Vories. 2010. Ground-based canopy reflectance sensing for variable-rate nitrogen corn fertilization. Agron. J. 102: 71–84.

Lamb, D.W., and R.B. Brown. 2001. Remote-sensing and mapping of weeds in crops. J. Agric. Eng. Res. 78: 117–125.

Larson, W.E., and P.C. Robert. 1991. Farming by soil. In: (R. Lal and F.J. Pierce, eds.), Soil Management for Sustainability. Soil and Water Conserv. Soc, Ankeny, IA, USA, pp. 103–112.

Li, D., Y. Miao, C.J. Ransom, G.M. Bean, N.R. Kitchen, F.G. Fernández, et al. 2022. Corn nitrogen nutrition index prediction improved by integrating genetic, environmental, and management factors with active canopy sensing using machine learning. Remote Sens. 14: 394.

Li, F., B. Mistele, Y. Hu, X. Chen, and U. Schmidhalter. 2014a. Reflectance estimation of canopy nitrogen content in winter wheat using optimized hyperspectral spectral indices and partial least squares regression. European J. Agron. 52: 198–209.

Li, F., Y. Miao, G. Feng, F. Yuan, S. Yue, X. Gao, Y. Liu, B. Liu, S.L. Ustin, and X. Chen. 2014b. Improving estimation of summer maize nitrogen status with red edge-based spectral vegetation indices. Field Crops Res. 157: 111–123.

Li, F., Y. Miao, S.D. Hennig, M.L. Gnyp, X. Chen, L. Jia, and G. Bareth. 2010. Evaluating hyperspectral vegetation indices for estimating nitrogen concentration of winter wheat at different growth stages. Precis. Agric. 11: 335–357.

Liang, J., W. Ren, X. Liu, H., Zha, X. Wu, C. He, et al. 2023. Improving nitrogen status diagnosis and recommendation of maize using UAV remote sensing data. Agronomy 13: 1994.

Link, A., and S. Reusch. 2006. Implementation of site-specific nitrogen application-status and development of the YARA N-Sensor. In: NJF Seminar 390, Precision Technology in Crop Production Implementation and Benefits. Norsk Jernbaneforbund, Stockholm, Sweden, pp. 37–41.

Linker, H.M., J.S. Bacheler, H.D. Coble, E.J. Dunphy, S.R. Koenning, and J.W. Van Duyn. 1999. Integrated pest management soybean scouting manual. North Carolina Cooperative Extension Service pub. no. AG-385.

Liu, J., J. Xiang, Y. Jin, R. Liu, J. Yan, and L. Wang. 2021. Boost precision agriculture with unmanned aerial vehicle remote sensing and edge intelligence: A survey. Remote Sens. 13: 4387.

López-Granados, F. 2011. Weed detection for site-specific weed management: Mapping and real time approaches. Weed Res. 51: 1–11.

Lorenzen, B., and A. Jensen. 1989. Changes in leaf spectral properties induced in barley by cereal powdery mildew. Remote Sens. Environ. 27: 201–209.

Lost Filho, F.H., W.B. Heldens, Z. Kong, E.S. de Lange. 2020. Drones: Innovative technology for use in precision pest management. J. Econ. Entomol. 113: 1–25.

Lu, J., E. Dai, Y. Miao, and K. Kusnierek. 2022. Improving active canopy sensor-based in-season rice nitrogen status diagnosis and recommendation using multi-source data fusion with machine learning. J. Clean. Prod. 380: 134926.

MacRae, I. 1988. Scouting for Insects in Wheat, Alfalfa and Soybeans. University of Minnesota Extension Service, St. Paul, MN.

Mahajan, G.R., R.N. Sahoo, R.N. Pandey, V.K. Gupta, and D. Kumar. 2014. Using hyperspectral remote sensing techniques to monitor nitrogen, phosphorus, sulphur and potassium in wheat (*Triticum aestivum* L.). Precis. Agric. 15(5): 499–522. https://doi.org/10.1007/s11119-014-9348-7.

Mahlein, A.K. 2016. Plant disease detection by imaging sensors: Parallels and specific demands for precision agriculture and plant phenotyping. Plant Dis. 100: 241–251.

Mahlein, A.-K., E. Alisaac, A. Al Masri, J. Behmann, H.-W. Dehne, and E.-C. Oerke. 2019. Comparison and combination of thermal, fluorescence, and hyperspectral imaging for monitoring fusarium head blight of wheat on spikelet scale. Sensors 19: 2281.

McFadden, J., E. Njuki, and T. Griffin. 2023. Precision agriculture in the digital era: Recent adoption on U.S. farms. US department of agriculture. Economic Research Service 248. Available online: www.ers.usda.gov/publications/pub-details/?pubid=105893.

Meron, M., J. Tsipris, V. Orlov, V. Alchanatis, and Y. Cohen. 2010. Crop water stress mapping for site-specific irrigation by thermal imagery and artificial reference surfaces. Precis. Agric. 11: 148–162.

Mewes, T., J. Franke, and F. Menz. 2011. Spectral requirements on airborne hyperspectral remote sensing data for wheat disease detection. Precis. Agric. 12: 795–812.

Miao, Y. 2023. Precision nutrient management. In: (Q. Zhang, ed.), Encyclopedia of Smart Agriculture Technologies. Springer Nature, Switzerland AG, pp. 1054–1061.

Miao, Y., D.J. Mulla, G. Randall, J. Vetsch, and R. Vintila. 2009. Combining chlorophyll meter readings and high spatial resolution remote sensing images for in-season site-specific nitrogen management of corn. Precision Agric. 10: 45–62.

Miao, Y., D.J. Mulla, J.A. Hernandez, M. Wiebers, and P.C. Robert. 2007. Potential impact of precision nitrogen management on corn yield, protein content, and test weight. Soil Sci. Soc. Am. J. 71: 1490–1499.

Miao, Y., D.J. Mulla, and P.C. Robert. 2018. An integrated approach to site-specific management zone delineation. Front. Agric. Sci. Eng. 5: 432–441.

Mirik, M., G.J. Michels, S. Kassymzhanova-Mirik, and N.C. Elliott. 2007. Reflectance characteristics of Russian wheat aphid (Hemiptera: Aphididae) stress and abundance in winter wheat. Comp. Electron. Agric. 57: 123–134.

Mirik, M., Y. Aysan, and F. Sahin. 2011. Characterization of *Pseudomonas cichorii* isolated from different hosts in Turkey. Int. J. Agric. Bio. 13: 203–209.

Mistele, B., and U. Schmidhalter. 2008. Estimating the nitrogen nutrition index using spectral canopy reflectance measurements. Eur. J. Agron. 29: 184–190.

Moran, M.S., C.D. Peters-Lidard, J.M. Watts, and S. McElroy. 2004. Estimating soil moisture at the watershed scale with satellite-based radar and land surface models. Can. J. Remote Sens. 30: 805–826.

Moran, M.S., Y. Inoue, and E.M. Barnes. 1997. Opportunities and limitations for image-based remote sensing in precision crop management. Remote Sens. Environ. 61: 319–346.

Moshou, D., I. Gravalos, D.K.C. Bravo, R. Oberti, J.S. West, and H. Ramon. 2012. Multisensor fusion of remote sensing data for crop disease detection. In: Geospatial Techniques for Managing Environmental Resources. Springer Netherlands, pp. 201–219.

Mueller, D., and R. Pope. 2009. Corn Field Guide: A Reference for Identifying Diseases, Insect Pests and Disorders of Corn. Iowa State University, University Extension, Ames, IA.

Muhammed, H.H. 2005. Hyperspectral crop reflectance data for characterizing and estimating fungal disease severity in wheat. Biosys. Eng. 91(1): 9–20.

Mulla, D.J. 1991. Using geostatistics and GIS to manage spatial patterns in soil fertility. In: (G. Kranzler, ed.), Automated Agriculture for the 21st Century. ASAE, St. Joseph, MI, USA, p. 336e345.

Mulla, D.J. 1993. Mapping and managing spatial patterns in soil fertility and crop yield. In: (P. Robert, W. Larson and R. Rust, eds.), Soil Specific Crop Management. ASA, Madison, WI, USA, pp. 15–26.

Mulla, D.J. 1997. Geostatistics, remote sensing and precision farming. In: (A. Stein and J. Bouma, eds.), Precision Agriculture: Spatial and Temporal Variability of Environmental Quality: Ciba Foundation Symposium. Wiley, Chichester, UK, Vol. 210, p. 100e119.

Mulla, D.J. 2013. Twenty five years of remote sensing in precision agriculture: Key advances and remaining knowledge gaps. Biosys. Eng. 114: 358–371.

Mulla, D.J., C.A. Perillo, and C.G. Cogger. 1996. A site-specific farm-scale GIS approach for reducing groundwater contamination by pesticides. J. Environ. Qua. 25: 419–425.

Mulla, D.J., and J.S. Schepers. 1997. Key processes and properties for site-specific soil and crop management. In: (F.J. Pierce and E.J. Sadler, eds.), The State of Site Specific Management for Agriculture. ASA/CSSA/SSSA, Madison, WI, pp. 1–18.

Munnaf, M.A., G. Haesaert, M. Van Meirvenne, and A.M. Mouazen. 2020. Site-specific seeding using multisensor and data fusion techniques: A review. Adv. Agron. 161: 241–323.

Murchie, E.H., and T. Lawson. 2013. Chlorophyll fluorescence analysis: A guide to good practice and understanding some new applications. J. Exp. Bot. 64: 3983–3998.

Neupane, J., and W.X. Guo. 2019. Agronomic basis and strategies for precision water management: A review. Agronomy 9: 87.

Nguy-Robertson, A., A. Gitelson, Y. Peng, A. Viña, T. Arkebauer, and D. Rundquist. 2012. Green leaf area index estimation in maize and soybean: Combining vegetation indices to achieve maximal sensitivity. Agron. J. 104: 1336–1347.

Nicolas, H. 2004. Using remote sensing to determine the date of a fungicide application on winter wheat. Crop Prot. 23: 853–863.

Nigon, T.J., D.J. Mulla, C.J. Rosen, Y. Cohen, V. Alchanatis, and R. Rud. 2014. Evaluation of the nitrogen sufficiency index for use with high resolution, broadband aerial imagery in a commercial potato field. Precis. Agric. 15: 202–226.

Oerke, E. 2020. Remote sensing of diseases. Annu. Rev. Phytopathol. 58: 225–252.

Omary, M., C.R. Camp, and E.J. Sadler. 1997. Center pivot irrigation system modification to provide variable water application depths. Applied Eng. Agric. 13(2): 235–239.

Pimstein, A., A. Karnieli, S.K. Bansal, and D.J. Bonfil. 2011. Exploring remotely sensed technologies for monitoring wheat potassium and phosphorus using field spectroscopy. Field Crops Res. 121: 125–135.

Pinter, Jr., P.J., J.L. Hatfield, J.S. Schepers, E.M. Barnes, M.S. Moran, C.S.T. Daughtry, and D.R. Upchurch. 2003. Remote sensing for crop management. Photogr. Engin. Remote Sens. 69: 647–664.

Prabhakar, M., Y.G. Prasad, M. Thirupathi, G. Sreedevi, B. Dharajothi, and B. Venkateswarlu. 2011. Use of ground based hyperspectral remote sensing for detection of stress in cotton caused by leafhopper (Hemiptera: Cicadellidae). Computers Elec. Agric. 79: 189–198.

Qi, J., A. Chehbouni, A.R. Huete, Y.H. Keer, and S. Sorooshian. 1994. A modified soil vegetation adjusted index. Remote Sens. Environ. 48: 119–126.

Rasmussen, J., J. Nielsen, F. Garcia-Ruiz, S. Christensen, and J.C. Streibig. 2013. Potential uses of small unmanned aircraft systems (UAS) in weed research. Weed Res. 53: 242–248.

Rasmussen, J., S. Azim, S.K. Boldsen, T. Nitschke, S.M. Jensen, J. Nielsen, and S. Christensen. 2021. The challenge of reproducing remote sensing data from satellites and unmanned aerial vehicles (UAVs) in the context of management zones and precision agriculture. Precis. Agric. 22: 834–851.

Raun, W.R., J.B. Solie, G.V. Johnson, M.L. Stone, R.W. Mullen, K.W. Freeman, W.E. Thomason, and E.V. Lukina. 2002. Improving nitrogen use efficiency in cereal grain production with optical sensing and variable rate application. Agron. J. 94: 815–820.

Reusche, S., J. Jasper, and A. Link. 2010. Estimating crop biomass and nitrogen uptake using CropSpec ™, a newly developed active crop-canopy reflectance sensor. In: (R. Khosla, ed.), Proc. 10th Intl. Conf. Prec. Ag. Denver, CO.

Ribeiro, A.V., L.N. Lacerda, M.A. Windmuller-Campione, T.M. Cira, Z.P.D. Marston, T.M. Alves, et al. 2023. Economic-threshold-based classification of soybean aphid, Aphis glycines, infestations in commercial soybean fields using Sentinel-2 satellite data. Crop Protection (Online-first). https://doi.org/10.1016/j.cropro.2023.106557.

Rondeaux, G., M. Steven, and F. Baret. 1996. Optimization of soil-adjusted vegetation indices. Remote Sens. Environ. 55: 95–107.

Sadler, E.J., R.G. Evans, K.C. Stone, and C.R. Camp. 2005. Opportunities for conservation with precision irrigation. J. Soil Water Cons. 60(6): 371–379.

Samborski, S.M., N. Tremblay, and E. Fallon. 2009. Strategies to make use of plant sensors-based diagnostic information for nitrogen recommendations. Agron. J. 101: 800–816.

Samseemoung, G., P. Soni, H.P.W. Jayasuriya, and V.M. Salokhe. 2012. Application of low altitude remote sensing (LARS) platform for monitoring crop growth and weed infestation in a soybean plantation. Precis. Agric. 13: 611–627.

Sankaran, S., A. Mishra, R. Ehsani, and C. Davis. 2010. A review of advanced techniques for detecting plant diseases. Comp. Electron. Agric. 72(1): 1–13.

Sayed, O.H. 2003. Chlorophyll fluorescence as a tool in cereal crop research. Photosynthetica 41: 321–330.

Scharf, P.C., D.K. Shannon, H.L. Palm, K.A. Sudduth, S.T. Drummond, N.R. Kitchen, L.J. Mueller, V.C. Hubbard, and L.F. Oliveira. 2011. Sensor-based nitrogen applications out-performed producer-chosen rates for corn in on-farm demonstrations. Agron. J. 103: 1683–1691.

Seelan, S.K., S. Laguette, G.M. Casady, and G.A. Seielstad. 2003. Remote sensing applications for precision agriculture: A learning community approach. Remote Sens. Environ. 88: 157–169.

Shamal, S.A.M., S.A. Alhwaimel, and A.M. Mouazen. 2016. Application of an on-line sensor to map soil packing density for site specific cultivation. Soil Tillage Res. 162: 78–86.

Shanahan, J.F., N.R. Kitchen, W.R. Raun, and J.S. Schepers. 2008. Responsive in-season nitrogen management for cereals. Comp. Electron. Agric. 61: 51–62.

Shaver, T.M., R. Khosla, and D.G. Westfall. 2011. Evaluation of two crop canopy sensors for nitrogen variability determination in irrigated maize. Precis. Agric. 12: 892–904.

Shiratsuchi, L., R. Ferguson, J. Shanahan, V. Adamchuk, D. Rundquist, D. Marx, and G. Slater. 2011. Water and nitrogen effects on active canopy sensor vegetation indices. Agron. J. 103: 1815–1826.

Sishodia, R.P., R.L. Ray, and S.K. Singh. 2020. Applications of remote sensing in precision agriculture: A review. Remote Sens. 12(19): 3136.

Solie, J.B., W.R. Raun, R.W. Whitney, M.L. Stone, and J.D. Ringer. 1996. Optical sensor based field element size and sensing strategy for nitrogen. Trans. ASAE 39: 1983–1992.

Sripada, R.P., J.P. Schmidt, A.E. Dellinger, and D.B. Beegle. 2008. Evaluating multiple indices from a canopy reflectance sensor to estimate corn N requirements. Agron. J. 100: 1553–1561.

Sripada, R.P., R.W. Heiniger, J.G. White, and R. Weisz. 2005. Aerial color infrared photography for determining late-season nitrogen requirements in corn. Agron. J. 97: 1443–1451.

Stafford, J.V., and P.C.H. Miller. 1993. Spatially selective application of herbicide to cereal crops. Comp. Electron. Agric. 9: 217–229.

Stark, J.C., I.R. McCann, B.A. King, and D.T. Westermann. 1993. A two-dimensional irrigation control system for site specific application of water and chemicals. Agron. Abs. 85: 329.

Sui, R., J.A. Thomasson, and J. Hanks. 2008. Ground-based sensing system for weed mapping in cotton. Comp. Electron. Agric. 60(1): 31–38.

Swain, K.C., S.J. Thomson, and H.P.W. Jayasuriya. 2010. Adoption of an unmanned helicopter for low-altitude remote sensing to estimate yield and total biomass of a rice crop. Trans. ASABE 53(1): 21–27.

Thenkabail, P.S., E.A. Enclona, M.S. Ashton, and B. Van Der Meer. 2004. Accuracy assessments of hyperspectral waveband performance for vegetation analysis applications. Remote Sens. Environ. 91: 354–376.

Thenkabail, P.S., J.G. Lyon, and A. Huete. 2010. Hyperspectral remote sensing of vegetation and agricultural crops: Knowledge gain and knowledge gap after 40 years of research. Ch. 28. In: (P.S. Thenkabail, J.G. Lyon and A. Huete, eds.), Hyperspectral Remote Sensing of Vegetation. CRC Press, Boca Raton, FL, p. 705.

Thenkabail, P.S., R.B. Smith, and E. De Pauw. 2000. Hyperspectral vegetation indices and their relationships with agricultural crop characteristics. Rem. Sens. Environ. 71: 158–182.

Thorp, K.R., and L. Tian. 2004. A review on remote sensing of weeds in agriculture. Precis. Agric. 5(5): 477–508.

Tillet, N.D., T. Hague, A.C. Grundy, and A.P. Dedousis. 2008. Mechanical within-row weed control for transplanting crops using computer vision. Biosys. Eng. 99(2): 171–178.

Torii, T. 2000. Research in autonomous agriculture vehicles in Japan. Comp. Electron. Agric. 25(1–2): 133–153.

Tremblay, N., Z. Wang, and Z.G. Cerovic. 2012. Sensing crop nitrogen status with fluorescence indicators: A review. Agron. Sustain. Dev. 32: 451–464.

UN (United Nations). 2023. Globe issues: Population. Available online: www.un.org/en/global-issues/population#:~:text=Our%20growing%20population&text=The%20world's%20population%20is%20expected,billion%20in%20the%20mid%2D2080s. (accessed on Nov. 30, 2023).

Van Niel, T.G., and T.R. McVicar. 2004. Current and potential uses of optical remote sensing in rice-based irrigation systems: A review. Aust. J. Agric. Res. 55(2): 155–185.

Vereecken, H., L. Weihermüller, F. Jonard, and C. Montzka. 2012. Characterization of crop canopies and water stress related phenomena using microwave remote sensing methods: A review. Vadose Zone J. 11. https://doi.org/10.2136/vzj2011.0138ra.

Wang, A., W. Zhang, and X. Wei. 2019. A review on weed detection using ground-based machine vision and image processing techniques. Comput. Electron. Agric. 158: 226–240.

Wang, W., X. Yao, X. Yao, Y. Tian, X. Liu, J. Ni, W. Cao, and Y. Zhu. 2012. Estimating leaf nitrogen concentration with three-band vegetation indices in rice and wheat. Field Crops Res. 129: 90–98.

Wang, X., Y. Miao, R. Dong, H. Zha, T. Xia, Z. Chen, et al. 2021. Machine learning-based in-season nitrogen status diagnosis and side-dress nitrogen recommendation for corn. Eur. J. Agron. 123: 126193.

Wang, X., Y. Miao, R. Dong, and K. Kusnierek. 2023. Minimizing active canopy sensor differences in nitrogen status diagnosis and in-season nitrogen recommendation for maize with multi-source data fusion and machine learning. Precis. Agric 24: 2549–2565.

West, J.S., C. Bravo, R. Oberti, D. Lemaire, D. Moshou, and H.A. McCartney. 2003. The potential of optical canopy measurement for targeted control of field crop disease. Ann. Rev. Phytopath. 41: 593–614.

Wiegand, C.L., A.J. Richardson, and D.E. Escobar. 1991. Vegetation indices in crop assessment. Remote Sens. Environ. 35: 105–119.

Wigmore, O., B. Mark, J. McKenzie, M. Baraerd, and L. Lautz. 2019. Sub-metre mapping of surface soil moisture in proglacial valleys of the tropical Andes using a multispectral unmanned aerial vehicle. Remote Sens. Environ. 222: 104–118.

Wiles, L.J., G.G. Wilkerson, H.J. Gold, and H.D. Coble. 1992. Modeling weed distribution for improved postemergence control decisions. Weed Sci. 40: 546–553.

Wu, Z., Y. Chen, B. Zhao, X. Kang, and Y. Ding. 2021. Review of weed detection methods based on computer vision. Sensors 21: 3647.

Yang, C. 2022. Remote sensing technologies for crop disease and pest detection. In: (M. Li, C. Yang, and Q. Zhang, eds.), Soil and Crop Sensing for Precision Crop Production. Agriculture Automation and Control. Springer, Cham, pp. 159–184.

Yao, H., R. Qin, and X. Chen. 2019. Unmanned aerial vehicle for remote sensing applications: A review. Remote Sens. 11: 1443.

Yao, H., Y. Huang, L. Tang, L. Tian, D. Bhatnagar, and T.E. Cleveland. 2018. Using hyperspectral data in precision farming applications. In: (P.S. Thenkabail and J.G. Lyon, eds.), Hyperspectral Remote Sensing of Vegetation, 2nd Edition. CRC Press, Boca Raton, FL, pp. 3–36.

Ye, N., J.P. Walker, Y. Gao, I. PopStefanija, and J. Hills. 2024. Comparison between thermal-optical and L-band passive microwave soil moisture remote sensing at farm scales: Towards IAV-based near-surface soil moisture mapping. IEEE J. Sel. Top. Appl. Earth Obs. Remote Sens. 17: 633–642.

Yuan, L., Y. Huang, R.W. Loraamm, C. Nie, J. Wang, and J. Zhang. 2014. Spectral analysis of winter wheat leaves for detection and differentiation of diseases and insects. Field Crops Res. 156: 199–207.

Zermas, D., H. Nelson, P. Stanitsas, V. Morellas, D. Mulla, and N. Papanikolopoulos. 2020. A methodology for the detection of nitrogen deficiency in corn fields using high resolution RGB imagery. IEEE Trans. Automation Sci. Eng. 18(4): 1879–1891.

Zha, H., Y. Miao, T. Wang, Y. Li, J. Zhang, W. Sun, Z. Feng, and K. Kusnierek. 2020. Improving unmanned aerial vehicle remote sensing-based rice nitrogen nutrition index prediction with machine learning. Remote Sens. 12: 215.

Zhang, C., and J.M. Kovacs. 2012. The application of small unmanned aerial systems for precision agriculture: A review. Precis. Agric. 13: 693–712.

Zhang, J., T. Cheng, L. Shi, W. Wang, Z. Niu, W. Guo, et al. 2022. Combining spectral and texture features of UAV hyperspectral images for leaf nitrogen content monitoring in winter wheat. Int. J. Remote Sens. 43: 2335–2356.

Zhou, X., W. Huang, J. Zhang, W. Kong, and Y. Huang. 2018. A novel combined spectral index for estimating the ratio of carotenoid to chlorophyll content to monitor crop physiological and phenological status. Eur. J. Agron. 76: 128–142.

Zhou, Z., Y. Majeed, G. Diverres Naranjo, and E.M.T. Gambacorta. 2021. Assessment for crop water stress with infrared thermal imagery in precision agriculture: A review and future prospects for deep learning applications. Comput. Electron. Agric. 182: 106019.

Zwiggelaar, R. 1998. A review of spectral properties of plants and their potential use for crop/weed discrimination in row-crops. Crop Prot. 17: 189–206.

9 Remote Sensing of Tillage Status

Baojuan Zheng, James B. Campbell, Guy Serbin, Craig S.T. Daughtry, Heather McNairn, and Anna Pacheco

ACRONYMS AND DEFINITIONS

ALI	Advanced Land Imager
ALOS	Advanced Land Observing Satellite
ASAR	Advanced Synthetic Aperture Radar
ASTER	Advanced Spaceborne Thermal Emission and Reflection Radiometer
CAI	Cellulose Absorption Index
COSMO-SkyMed	Constellation of Small Satellites for Mediterranean Basin Observation
EnMAP	Environmental Mapping and Analysis Program
JERS	Japanese Earth Resources Satellite
MODIS	Moderate-Resolution Imaging Spectroradiometer
NASA	National Aeronautics and Space Administration
NDI	Normalized Difference Index
NDTI	Normalized Difference Tillage Index
NDVI	Normalized Difference Vegetation Index
NIR	Near-Infrared
PALSAR	Phased Array type L-band Synthetic Aperture Radar
RADARSAT	Radar Satellite
RISAT	Radar Imaging Satellite
SAR	Synthetic Aperture Radar
SPOT	Satellite Pour l'Observation de la Terre, French Earth Observing Satellites
STARFM	Spatial and Temporal Adaptive Reflectance Fusion Model
SWIR	Shortwave Infrared
TerraSAR-X	A radar Earth observation satellite, with its phased array synthetic aperture radar
USDA	United States Department of Agriculture

9.1 INTRODUCTION

Tillage prepares the seedbed by mechanical disturbance to loosen and smooth the soil surface, often mixing topsoil with surface organic debris to aerate soil, assist in weed suppression, control insects and pests, and, in mid-latitudes, promote springtime warming and drying. Tillage has been practiced, in varied forms, throughout the world since antiquity. During the 1700s and 1800s innovations in designs of plowshares greatly increased tillage effectiveness by increasing depth of the disturbed soil, and by turning the surface soil to more completely mix surface crop residue (also referred to as plant litter, senescent vegetation, or non-photosynthetic vegetation) with disturbed soil.

For millennia, mechanical disturbance of the soil was accomplished using hand tools and animal power. By the mid-19th century, steam-powered tractors (later replaced by internal-combustion engines), greatly increased tillage efficiency and speed, and expanded tillage into a wider range of

DOI: 10.1201/9781003541165-11

slopes, topography, and into a wider range of ecosystems. Notable impacts of mechanization are the expansion of tillage into formerly uncultivated environments, especially prairies and steppes of several continents that have since become some of the most productive agricultural systems, but also some of the most susceptible to drought and erosion. Mechanization also led to further innovations in designs of specialized tillage implements, and to increases in tillage operations, which often created the context for soil and water erosion.

Detrimental impacts of tillage include increased wind and water erosion, increased soil compaction, especially in the context of mechanization, decreased soil organic matter, reduced water infiltration, and increased amounts of nutrients reaching streams and rivers. By the 1940s, increased awareness of detrimental aspects of tillage (Faulkner, 1943), combined with availability of herbicides, led to alternative practices to minimize adverse aspects of tillage. Such practices include increased use of tillage instruments that minimize soil disturbance and leave crop residue on the soil surface.

Recognized environmental benefits of conservation tillage systems include reduced soil erosion from wind and water, carbon emission reductions, and improvements of air, soil, and water quality (Papadopoulos et al., 2023, Calcagno et al., 2022). Long-term adoption of conservation tillage practices can increase soil organic matter content, hence, can potentially sequester atmospheric carbon into soils (Lal, 2004). Conservation tillage practices increase soil water infiltration, improve nutrient cycling, and, in general, improve water quality because of improved retention of soil nutrients (Karlen et al., 1994). Soil quality is improved because of accumulation of surface organic matter increases aggregate stability, and higher levels of crop residues provide shelter and food for wildlife. As for economic perspective, conservation tillage practices decrease labor and fuel costs because of reduced tillage operations, and reduced fertilizer requirements as a result of improved soil quality (West et al., 2002). Conservation tillage, especially no-till, requires fewer field operations and reduces the number of field days needed to plant a crop. As a result, it reduces the risk of delayed planting due to unfavorable weather conditions, and also provides possibilities to practice double-cropping.

As alternative tillage practices gained acceptance and were implemented, conservationists needed objective data to gauge the extent and benefits of their use. The Soil Tillage Intensity Rating (STIR), developed by USDA-NRCS, provides a physically based evaluation of tillage systems across the spectrum from true no-till to conventional plow systems (USDA-NRCS, 2014). STIR requires information on 1) each tillage implement used; 2) the operating speed of the implement; 3) the depth of tillage; and 4) the fraction of the total soil surface disturbed by the tillage implement. STIR provides robust evaluations of complex tillage systems and crop rotations for conservation planning. However, STIR is impractical for surveys over many fields and large regions.

Tillage intensity can also be characterized by the fraction of the soil surface covered by crop residue. The Conservation Technology Information Center (CTIC) defined the following categories of tillage based on the crop residue cover on the soil surface shortly after planting: conventional tillage has < 15% residue cover; reduced tillage has 15–30% residue cover; conservation tillage has >30% residue cover (CTIC, 2014). This less robust definition of tillage intensity has a few caveats that must be considered, for example, fields where crop residues were harvested for feed or biofuel may have low crop residue cover without soil disturbing tillage.

Over time, varied efforts to collect information on tillage intensity have included visual assessment, field measurements, agricultural censuses, and remote sensing techniques. Such information is required by a number of agro-ecosystem models and is important for assessing the impacts of tillage practices on soil erosion, soil carbon sequestration, and water quality. Field measurements and agricultural survey to acquire tillage information are time-consuming and difficult. Moreover, it is unrealistic to survey every single field using these methods over large regions and over time. Therefore, it is of great interest to develop techniques that can routinely and systematically map tillage practices. Synoptic remote sensing imagery offers opportunities to provide spatial-temporal information on tillage practices efficiently at low costs (Zhang et al., 2024, Jiang et al.,

2023, Papadopoulos et al., 2023, Calcagno et al., 2022, Liu et al., 2022, Luotamo et al., 2022, Mandal et al., 2022, Xiang et al., 2022, Martins et al., 2021, Rossi et al., 2020, Azzari et al., 2019, Yeom et al., 2019). The first investigation on the potential of using remote sensing imagery to map crop residues can be traced back to 1975 by Gausman et al. (1975). Thereafter, both aerial and satellite imagery were tested to differentiate different tillage practices and estimate crop residue cover. For instance, Airborne Visible/Infrared Imaging Spectrometer (AVIRIS) data were found to be useful for crop residue cover estimation (Daughtry et al., 2005).

Although aerial imagery, properly timed, and collected at suitable resolutions, offers the capability to assess soil tillage status, the broad-scale surveys require the areal coverage, revisit capabilities, and spectral channels that are, as a practical matter, available only through satellite observation systems. Here we discuss the two main classes of satellite systems with the potential for routine broad-scale tillage assessment (Zhang et al., 2024, Jiang et al., 2023, Papadopoulos et al., 2023, Calcagno et al., 2022, Liu et al., 2022, Luotamo et al., 2022, Mandal et al., 2022, Xiang et al., 2022, Martins et al., 2021, Rossi et al., 2020, Azzari et al., 2019, Yeom et al., 2019): (1) optical remote sensing (visible, near infrared, and mid-infrared imaging sensors), and (2) microwave remote sensing (SAR, synthetic aperture imaging sensors).

9.2 FIELD ASSESSMENT OF CROP RESIDUE COVER

Methods appropriate for assessing crop residue cover in fields can be grouped into intercept and photographic techniques (Morrison et al., 1993). Intercept methods use a system of grid points, cross-hairs, or points along a line where the presence or absence of residue is determined. The standard technique used by USDA-NRCS is the line-point transect method where a 15–30 m line with 100 evenly spaced markers along the line is stretched diagonally across the crop rows in the field and markers intersecting crop residue are counted. Accuracy of the line-point transect method depends on the length of the line, the number of points per line, and the skill of the observer. At least 500 points must be observed to determine corn residue cover to within 15% of the mean (Laflen et al., 1981). Significant modifications to the line-point transect method include the use of measuring tapes, meter sticks, and wheels with pointers (Morrison et al., 1993, Corak et al., 1993). However, the line-point transect is impractical for monitoring crop residue cover in many fields over broad areas in a timely manner.

For the photographic method, a color or color infrared digital camera is used to take multiple vertical photographs within a sampling area where residue conditions appear visually homogeneous. A grid or cross-hairs is superimposed on the digital images and the points intersecting residue are counted. Software programs, such as Sample Point, can randomly select sample points within each image for the user to identify and can tabulate the proportion of each class (Booth et al., 2006). Alternatively, the image may be classified into soil and residue classes using objective image analysis procedures (Papadopoulos et al., 2023, Xiang et al., 2022, Yeom et al., 2019). Classification errors occur when the spectral differences between soil and residues classes are not sufficiently large for discrimination. Shortly after harvest, crop residues are often much brighter than soils, but as the residues decompose, the residues may be brighter or darker than the soil. The best time to acquire information of tillage practices in the field is shortly after sowing and before crop emergence, which is also the optimal time window to acquire images to map tillage practices.

The CTIC, established at Purdue University in 1983 as clearinghouse for tillage and conservation information, has conducted field surveys to assess tillage status in the United States (www.ctic.purdue.edu). For the CTIC surveys, trained observers visually assessed tillage status in fields at regular intervals along selected routes through participating counties. The survey provided county-level estimates of overall tillage practices. The roadside assessment task is subject to various degrees of error and uncertainty because it mainly relies on visual interpretation. The quality of the data has also varied from time to time and from county to county due to a variety of reasons, such as unfavorable weather conditions at the time of survey and inconsistent levels of experience among

TABLE 9.1
Tillage Types and Their Corresponding Crop Residue Cover

Tillage Category	Tillage Types	Description	Crop Residue Cover (%)
Conservation	No-till/strip-till	Minimal soil disturbance (< 25%)	>30% (likely > 70%)
	Ridge-till	Residue left on the surface between ridges	>30%
	Mulch-till	100% soil surface disturbance	>30%
Non-conservation	Reduced-till	15%–30%	15–30%
	Conventional-till or intensive-till	<15%	<15%

the observers. Finally, some counties have stopped acquiring tillage data after the national survey program was discontinued in 2004 (CTIC, 2014).

Limited soil tillage information is available for other countries. Canada conducts tillage inventory as part of its 5-year Census of Agriculture. Tillage practices are reported by province in three categories: (1) Tillage incorporating most of the crop residue into the soil; (2) Tillage retaining most of the crop residue on the surface, and (3) No-till seeding or zero-till seeding. Thus, it is difficult and impractical to evaluate tillage practices over time, and by nation, because of wide variations in field-data collection, survey responses, and agricultural censuses (Zheng et al., 2014). The tillage categories defined by Canada are less precise than the CTIC definitions. Definitions of tillage categories may slightly differ from one country to another, and even differ from organization to organization. To evaluate tillage practices for a particular field using visual assessment or remote sensing methodologies, we have to link the ground surface status observed from the ground, air, or space to types of tillage practices. Although soil texture and smoothness can be one of the indicators for different tillage status, the amounts of crop residues left on the ground after planting are often considered as the most reliable indicator. Here, we listed types of tillage practices and their expected crop residue cover according to CTIC and Natural Resource Conservation Services' (NRCS) definitions in Table 9.1. Globally, a systematic monitoring of soil tillage is needed to manage the finite soil resources as demand for food, feed, fiber, and fuel intensities.

9.3 MONITORING WITH OPTICAL REMOTE SENSING

Optical remote sensing imagery is valuable for monitoring biophysical properties of various objects on the Earth (Calcagno et al., 2022, Martins et al., 2021, Rossi et al., 2020). Crop residue, although spectrally similar to soils, has a unique absorption feature near 2,100 nm. The absorption depth becomes deeper as the amount of crop residue increases. Thus, optical remote sensing imagery provides better capability in estimating crop residue cover than radar data. This section first describes spectral properties of soils, green vegetation, and non-photosynthetic vegetation, following with the second subsection (9.3.2) on tillage spectral indices based on spectral differences among soils, green vegetations, and non-photosynthetic vegetation. Section 9.3.3 reviewed tillage assessment using different remote sensing platforms, followed by Section 9.3.4 which discussed current challenges and future possibilities.

9.3.1 Spectral Properties of Soils, Green Vegetation, and Non-Photosynthetic Vegetation

Soil tillage intensity is defined by the proportion of the soil surface covered by crop residue shortly after planting. Green vegetation may also be present in the field as the planted crop or as weeds. This section focuses on the spectral properties of soils, green vegetation, and crop residues.

9.3.1.1 Spectral Properties of Soils

Soil reflectance typically increases monotonically with increasing wavelength (Figure 9.1). Major contributors to the reflectance spectra of soils include moisture content, iron-oxide content, organic matter content, particle-size distribution, mineralogy, and soil structure (Liu et al., 2022, Rossi et al., 2020, Azzari et al., 2019, Yeom et al., 2019, Baumgardner et al., 1986, Ben-Dor, 2002). Stoner and Baumgardner (1981) measured the spectral reflectance of 485 soil samples representing 10 soil taxonomic orders and identified five distinct soil reflectance curve forms. Soil organic matter content and iron oxide content were the primary factors determining shape of the reflectance spectra.

In general, soil reflectance decreased as soil moisture content, organic matter content, and iron oxide content increased. Spectral reflectance is strongly correlated with soil organic matter among soils from the same parent materials (Henderson et al., 1992). Reflectance spectra of soils may also have absorption features near 2210 nm that are associated with Al-OH in phyllosilicate clays (Figures 9.1 and 9.2) (Serbin et al., 2009a). However, mineral absorption features evident in the reflectance spectra of dry soils are often obscured by the strong absorption of water in the reflectance spectra of wet soils (Stoner et al., 1980, Daughtry et al., 2004).

Soil tillage roughens the soil surface and often decreases soil reflectance, but the effect is short-lived and soil reflectance increases as the soil surface is smoothed by precipitation or additional tillage operations (Azzari et al., 2019, Yeom et al., 2019). As water wets the soil surface and fills pore spaces, soil reflectance decreases.

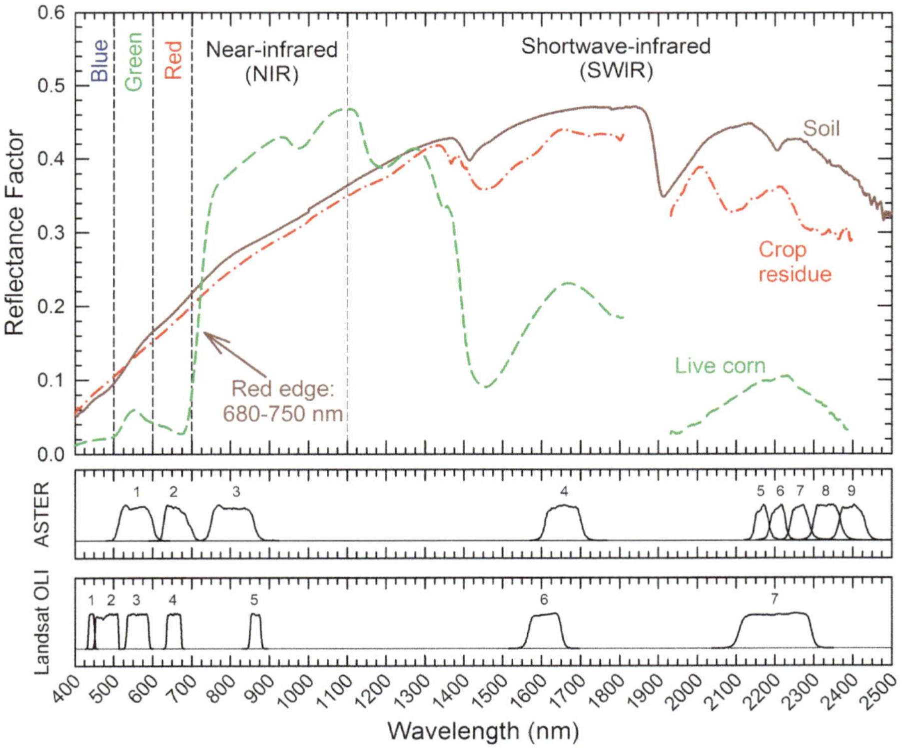

FIGURE 9.1 Spectra of a soil, corn residue, and live corn canopy for the visible through SWIR, and relative spectral response (RSR) for ASTER and Landsat OLI bands. Note that reflectance values vary from sample to sample. (Adapted from Daughtry et al., 2005.)

9.3.1.2 Spectral Properties of Green Vegetation

Reflectance of solar radiation from a dense canopy of actively growing green plants is characterized by three distinct regions: visible, near infrared, and shortwave infrared (Figure 9.1). In the visible wavelength region (400–700 nm), chlorophyll and other leaf pigments strongly absorb blue and red wavelengths which largely determines the reflectance and transmittance spectra (Thomas and Gausman, 1977). In the near-infrared (NIR) wavelength region (700–1200 nm), there is very little absorption and spectral reflectance and transmittance are largely determined by leaf mesophyll structure and cell wall-air interfaces (Slaton et al., 2001). Reflectance and transmittance in the shortwave-infrared (SWIR) wavelength region (1200–2500 nm) are affected primarily by the amount of water in the leaves (Zhang et al., 2024, Luotamo et al., 2022, Azzari et al., 2019, Yilmaz et al., 2008). Thus a distinguishing spectral characteristic of green vegetation is the step-like transition from low reflectance and low transmittance in the visible region to high reflectance and transmittance in the near infrared (Figure 9.1). Soils and non-photosynthetic vegetation lack this spectral feature. Spectral vegetation indices that exploit this fundamental spectral feature are particularly sensitive to green vegetation, for example, the Normalized Difference Vegetation Index (NDVI) (Rouse et al., 1973).

9.3.1.3 Spectral Properties of Non-Photosynthetic Vegetation

Non-photosynthetic vegetation (NPV) broadly refers to any senesced vegetation and includes crop residues, which are the portions of a cultivated crop remaining in the field after harvest. Initially, crop residues may completely cover the soil surface, but when the soil is tilled or the crop residues are harvested for feed or biofuel, crop residue cover decreases. Crop residues on the soil surface decrease soil erosion, increase soil organic matter, and improve soil quality (Ben-Dor, 2002). Quantification of crop residue cover is required to assess the effectiveness and extent of conservation tillage practices.

The reflectance spectra of both soils and crop residues lack the unique spectral signature of green vegetation (Figure 9.1). Crop residues and soils are spectrally similar and differ only in amplitude in the 400 to 1100 nm wavelength region, which makes quantification of crop residue cover by spectral reflectance challenging (Streck et al., 2002). Crop residues may be brighter than the soil shortly after harvest, but as residues weather and decompose, they may become either brighter or darker than the soil (Jiang et al., 2023, Luotamo et al., 2022, Mandal et al., 2022, Azzari et al., 2019, Yeom et al., 2019, Nagler et al., 2000, Daughtry et al., 2010). Residue water content also has impacts on its spectral properties. The presence of water in crop residues decreases reflectance across all wavelengths (Daughtry, 2001). Thus, assessing crop residue cover with broadband multispectral data can be challenging and may require extensive local calibration data.

An alternative approach for discriminating crop residues from soils is based on detecting absorption features in the 2100–2350 nm wavelength region that are associated with cellulose and lignin in crop residues (Workman and Weyer, 2008). High residue water content can obscure the absorption feature at 2100 nm (Daughtry, 2001). Increases in soil moisture content also decrease our ability to separate crop residue from soils (Daughtry, 2001). Thus, it becomes more difficult to discriminate crop residue from soils as residue and soil water content increases. As illustrated in Figure 9.2, these absorption features are not shared by common soil minerals, but are obscured by the strong absorptions of water often present in soils, crop residues, and green vegetation, which can significantly attenuate the cellulose and lignin absorption features (Daughtry and Hunt, 2008, Serbin et al., 2009b).

9.3.2 Spectral Indices for Assessing Crop Residue Cover

Spectral vegetation indices designed for assessing green vegetation, such as NDVI, cannot distinguish soil and crop residues (Zhang et al., 2024, Calcagno et al., 2022, Xiang et al., 2022, Rossi et al., 2020, Azzari et al., 2019, Yeom et al., 2019). Numerous tillage or residue indices use various

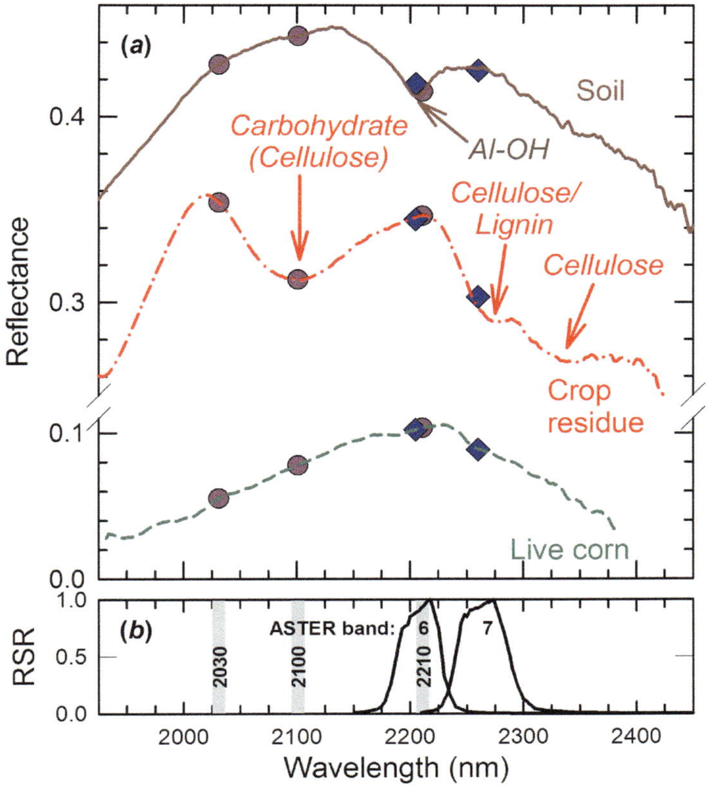

FIGURE 9.2 Spectra of a soil, corn residue, and live corn canopy in the SWIR, and relative spectral response (RSR) for ASTER bands 6 and 7 and 11 nm-wide CAI bands centered at 2030, 2100, and 2210 nm. (Adapted from Serbin et al., 2013.)

combinations of visible, near infrared and shortwave multispectral bands to discriminate crop residues from soils. The index best suited for crop residue cover estimation from single scenes is Cellulose Absorption Index (CAI), which specifically targets this feature. It has the distinct advantage that crop residues always have CAI > 0, live vegetation ≈ 0, and soils ≤ 0 (Figure 9.3). The CAI is defined as the relative intensity of the absorption feature at 2100 nm, which is attributed to an O—H stretching and C—O bending combination in cellulose and other carbohydrates in crop residues. CAI is measured using three relatively narrow (10–30 nm spectral resolution depending on the sensors) spectral bands—two on the shoulders and one near the center of the absorption feature at 2,100 nm (Nagler et al., 2000) (Table 9.2). CAI is effective in discriminating crop residues from soils for dry to moderately moist mixtures of crop residues and soils, but less effective for mixtures of wet crop residues and soils (Rossi et al., 2020).

Additional spectral indices that also target the cellulose and lignin absorption features of crop residues have used the relatively narrow (30–90 nm) shortwave infrared bands of the Advanced Spaceborne Thermal Emission and Reflectance Radiometer (ASTER) on the NASA Terra satellite, that is, the Lignin Cellulose Absorption (LCA) and the Shortwave Infrared Normalized Difference Residue Index (SINDRI) (Daughtry et al., 2005, Serbin et al., 2009c). For two-band normalized difference indices, the ASTER-based SINDRI performs well, and targets a decrease in reflectance associated with cellulose and lignin feature between ASTER SWIR bands 6 and 7 (Serbin et al., 2009c; Table 9.2). However, SINDRI is sensitive to green vegetation (Figures 9.2–9.3) and certain soil minerals (Figure 9.4), which also experience reflectance decreases between these bands, such

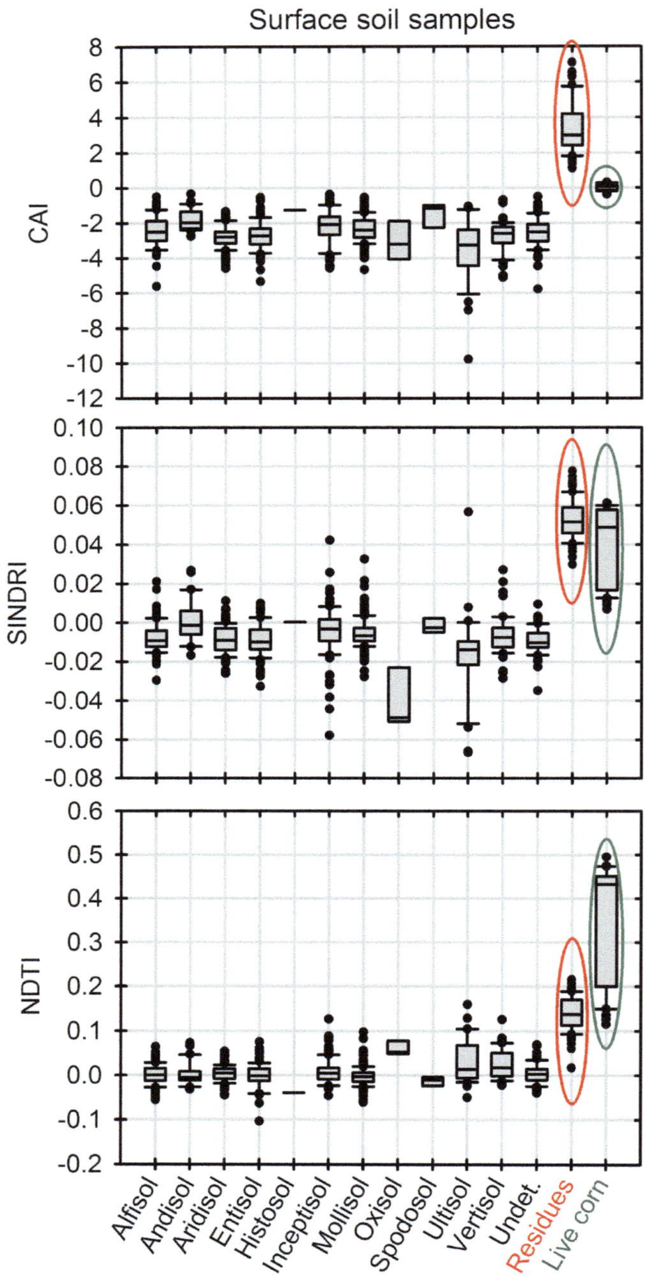

FIGURE 9.3 Spectral index values for surface soils, crop residues, and live corn canopy. (Adapted from Serbin et al. 2009b; Serbin et al., 2013.)

that it may not work well for a limited number of soils or where emerged crops may be present (Serbin et al., 2013).

While Landsat TM/ETM bands 5 and 7 and Landsat 8 OLI bands 6 and 7 are too wide and not properly placed to capture the cellulose absorption feature at 2100 nm, they can be used for tillage estimation via Normalized Difference Tillage Index (NDTI) (van Deventer et al., 1997; Table 9.2).

TABLE 9.2
Selected Tillage Indices and Their Calculation

Sensor	Tillage Indices	Formula	Description	Reference
Landsat TM & ETM+	NDTI	$(B5-B7)/(B5+B7)$	B5, B7: Landsat bands 5 & 7	Van Deventer et al. (1997)
AVIRIS	CAI	$100\times\left[0.5\left(R_{2030}+R_{2210}\right)-R_{2100}\right]$	R_{2030} and R_{2210} are the reflectances of the shoulders at 2030 nm & 2210 nm, R_{2100} is at the center of the absorption	Daughtry et al. (2005)
Hyperion				Daughtry et al. (2006)
ASTER	LCA	$100(2\times B6-B5-B8)$	B5, B6, B7, B8: ASTER shortwave infrared bands 5, 6, 7, & 8	Daughtry et al. (2005)
	SINDRI	$(B6-B7)/(B6+B7)$		Serbin et al. (2009b)

In addition to NDTI, Normalized Difference Index (NDI) (McNairn and Protz, 1993) and Simple Tillage Index (STI) (van Deventer et al., 1997) are Landsat-based tillage indices. Serbin et al. (2009b) showed that NDTI performed the best of several Landsat-based normalized difference tillage indices, but underperformed in comparison to CAI and the ASTER-based LCA. Furthermore, NDTI was found to lack adequate contrast for a number of soils with high contents of kaolinite or smectite, and had a much stronger signal for live vegetation than either crop residues or soil minerals (Figure 9.3). In Figure 9.3, the median values of NDTI for crop residues are consistently higher than the median values of surface soils. However, discrimination of some combinations of soils and crop residues may be difficult without adequate quantities of local data for calibration and validation. For example, the NDTI values of most crop residues may not differ significantly from NDTI values of soils with high contents of kaolinite or smectite (Serbin et al., 2009b). As the fraction of green vegetation in a scene increases, NDTI also increases which alters the estimation of crop residue cover. One approach is to exclude pixels with green vegetation using a NDVI threshold (Papadopoulos et al., 2023, Xiang et al., 2022, Martins et al., 2021, Thoma et al., 2004; Daughtry et al., 2005). Another robust approach to reduce effects of soil and green vegetation on estimates of crop residue cover is to identify the minimum NDTI (minNDTI) values from multi-temporal NDTI data, because the minNDTI values were found to be well-correlated with crop residue cover (Liu et al., 2022, Yeom et al., 2019, Zheng et al., 2012, 2013a). This method was found to be similar in accuracy to single collects using SINDRI or CAI (Figure 9.5) (Zheng et al., 2013a). However, as we can see in Figure 9.5 that minNDTI results in higher root mean squared errors (RMSE), NDTI is more subject to the negative influences of soil moisture and soil organic carbon than SINDRI and CAI (Zheng et al., 2013a).

9.3.3 Tillage Assessment Using Airborne and Satellite Imagery

Until recently, most assessments of crop residue cover and tillage intensity were snapshots of conditions using single dates of multispectral imagery (Papadopoulos et al., 2023, Calcagno et al., 2022, Liu et al., 2022, Mandal et al., 2022, Xiang et al., 2022, Martins et al., 2021). For example, various spectral indices using Landsat TM bands 5 and 7 successfully differentiated conventional tillage from conservation tillage using logistic regression (van Deventer et al., 1997, Gowda et al., 2001). Other classification methods (e.g., minimum distance, Mahalanobis distance, Maximum Likelihood, spectral angle mapping, and cosine of the angle concept), and data mining approaches

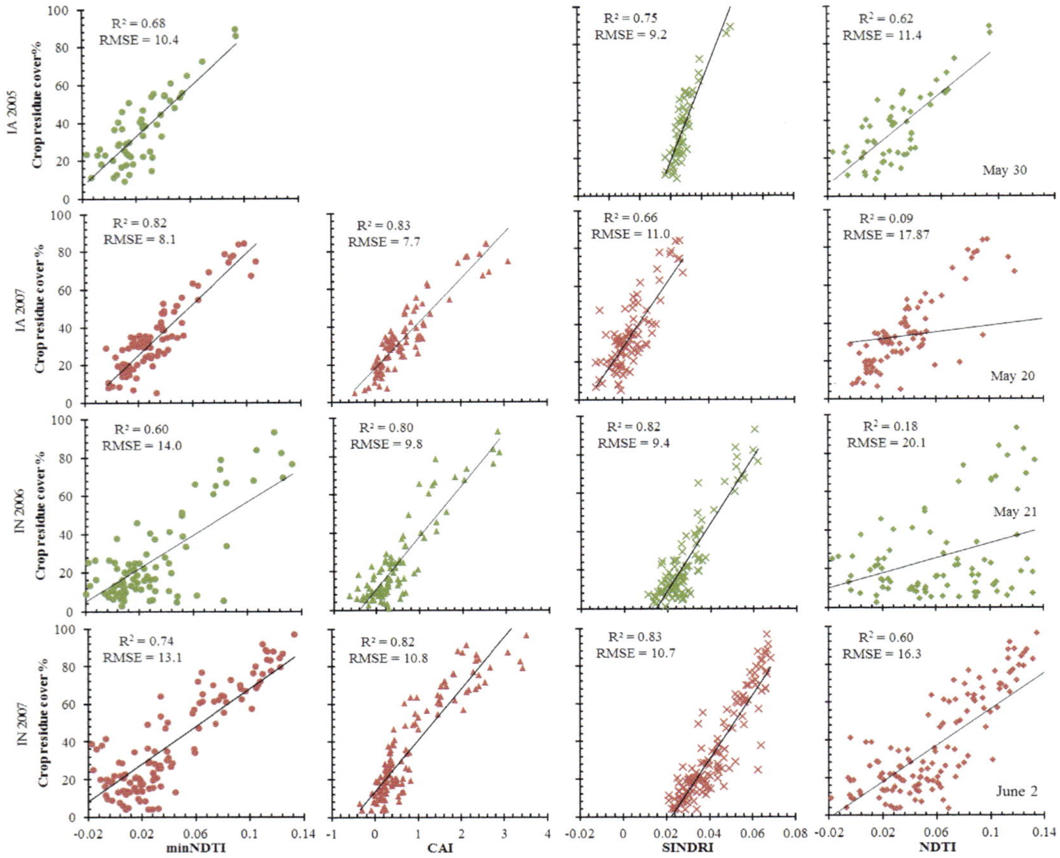

FIGURE 9.4 Spectra of Gibbsite HS423.3B (Kokaly et al., 2017), a gibbsitic Ultisol (Brown et al., 2006), and corn residue with convolved spectral band values. (Adapted from Serbin et al., 2013.)

(e.g., random forest classifier and support vector machine) have been examined for identifying two broad tillage categories (Zhang et al., 2024, Luotamo et al., 2022, Mandal et al., 2022, Xiang et al., 2022, South et al., 2004, Bricklemyer et al., 2006, Sudheer et al., 2010, Samui et al., 2012). These studies demonstrated the capability of Landsat TM imagery to discriminate between two broad tillage categories (i.e., conventional and conservation tillage) (Gowda et al., 2001, van Deventer et al., 1997), but fell short of achieving the reliability and consistency required for operational applications. Based on previous studies, it remains unclear which classification approach performs the best in classifying tillage categories. Research also has been conducted to test the feasibility of estimating crop residue cover using Landsat data (McNairn and Protz, 1993, Thoma et al., 2004, Daughtry et al., 2006). These studies used single-date multispectral images and yielded mixed results. The inconsistent results of these studies may be related the spectral resolution of Landsat TM data, different image preprocessing strategies to correct for atmospheric transmittance, spatial and temporal variations in soils and green vegetation.

Tillage indices developed using hyperspectral and advanced multispectral (e.g., ASTER) data have provided consistent assessments of crop residue cover across years and study sites (Table 9.3; Figure 9.5). These tillage indices (e.g., CAI, SINDRI) detect absorption features associated with cellulose and lignin and are robust for discriminating crop residues from soils and green vegetation. However, the sensor systems with the appropriate spectral bands have very limited spatial

TABLE 9.3
Summary of Studies in Crop Residue Estimation Using Remote Sensing Imagery

Sensor	n*	Image Dates	Indices or Methods	R^2	References
Landsat TM	266	4/18/1990	NDI	0.74	McNairn and Protz (1993)
Landsat ETM+	468	03/28/2000;	NDI	0.38	Thoma et al. (2004)
		06/03/2001;	STI	0.47	
		11/10/2001;	NDTI	0.48	
Landsat TM	54	06/12/2004	NDI	0.14	Daughtry et al. (2006)
			NDTI	0.11	
SPOT	varied	varied	Spectral unmixing	0.58–0.78	Pacheco and McNairn
Landsat TM	39	05/28/2008		0.69	(2010)
Hyperion	54	05/03/2004	CAI	0.85	Daughtry et al. (2006)
Landsat TM§	varied	varied	NDTI	0.004–0.64	
ASTER†	varied	varied	LCA	0.39–0.86	Serbin et al. (2009c)
			SINDRI	0.61–0.87	
Airborne hyperspectral data	varied	varied	CAI	0.72–0.89	
Landsat TM & ETM+	31	multitemporal	minNDTI	0.89	Zheng et al. (2012)
Landsat TM & ETM+	varied	multitemporal	minNDTI	0.66–0.89	Zheng et al. (2013a)

* n denotes number of samples;
§ Data were simulated using ASTER data when Landsat TM imagery was unavailable;
† ASTER data were simulated using airborne hyperspectral data when they are unavailable.

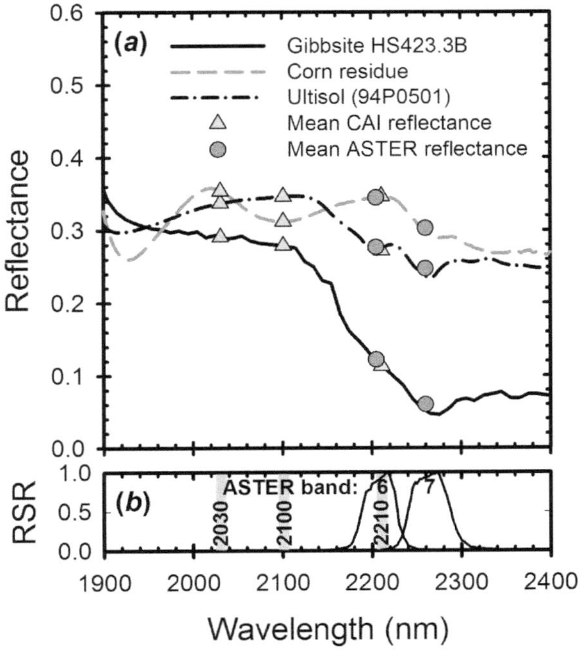

FIGURE 9.5 Comparison of minNDTI, and single scene CAI, SINDRI, and NDTI in two study areas: Ames, Iowa (year 2005 & 2007) and Fulton, Indiana (year 2006 & 2007). (Adapted from Zheng et al., 2013a.)

and temporal coverage, which limits their usefulness for monitoring crop residue cover and tillage intensity over large areas. Finally, the SWIR bands of ASTER needed to characterize residue cover are no longer available due to detector failure in April 2008 (Azzari et al., 2019). Spaceborne multispectral imagery, however, is favorable due to its ability to provide extended repetitive coverage of the Earth. Landsat TM/ETM+ imagery, thus, is extremely attractive for monitoring tillage practices and crop residue cover over large areas because it is freely available and provides a long-term synoptic view of the Earth with a 16-day revisit frequency.

Timing of image acquisition is very important for monitoring agricultural resources because agricultural land surfaces change rapidly as growers prepare soils for planting, and as crops emerge from soils, mature, and are harvested. It is well recognized that soil and residue status changes rapidly during the planting season and varies in space and time (McNairn et al., 2001), but tillage and crop residue mapping have been long treated as a one-time mapping effort using only one image at a time, until Watts et al. (2011) incorporated temporal dimensions into tillage mapping. Zheng et al. (2012) emphasized the need to consider varied timings of tillage and planting in tillage mapping, and significantly improved mapping accuracy using multitemporal Landsat imagery (Table 9.3). Minimum NDTI values extracted from a time-series Landsat image that included images from 1–2 months before expected planting date to 1–2 months after planting date (Zheng et al., 2012). The method was designated as minNDTI and forms an effective way to minimize confounding effects of green vegetation (Zheng et al., 2012). Figure 9.6 shows a tillage map and its corresponding NDTI values of Champaign County, Illinois. The left image in Figure 9.6 is the minimum NDTI values extracted from a time-series NDTI image. Agricultural fields managed with conservation tillage are relatively brighter because higher levels of crop residue cover result in higher NDTI values. The multitemporal approach requires the use of surface reflectance Landsat data products, which are available from EarthExplorer (http://earthexplorer.usgs.gov/) and USGS ESPA (EROS Science Processing Architecture) ordering interface (https://espa.cr.usgs.gov). The minNDTI approach was also applied to six additional datasets collected in different regions of the United States and the

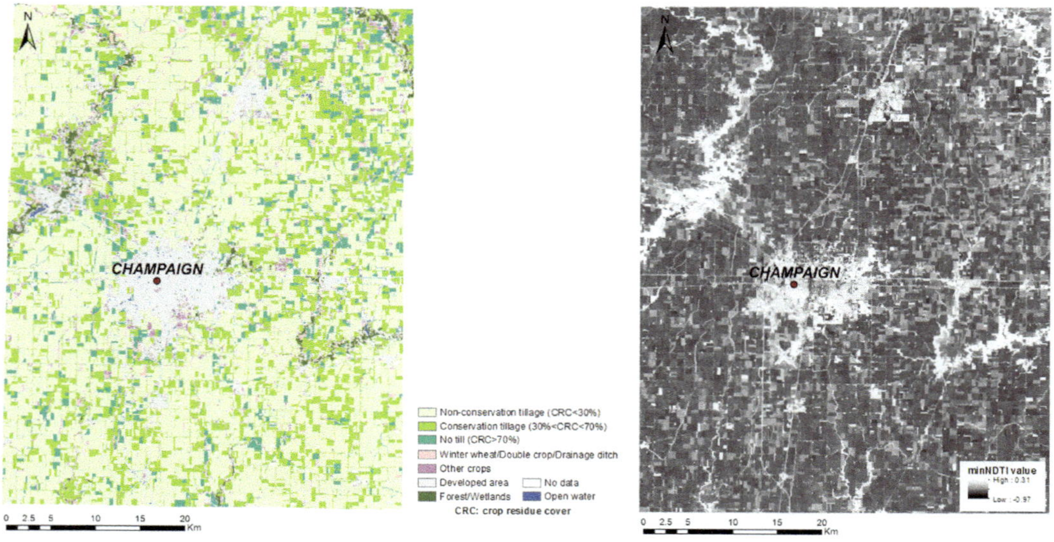

FIGURE 9.6 2006 Tillage map of Champaign County, Illinois (left), and its corresponding minimum NDTI values (right) extracted from a time-series NDTI image. Agricultural fields with brighter tones indicate higher levels of crop residue cover, which corresponds to the conservation tillage category.

technique was comparable to CAI and SINDRI in achieving similar classification accuracy of three tillage categories (Zheng et al., 2013a). Zheng et al. (2013a) reported 68% to 86% overall accuracies for three tillage categories—a significant improvement compared to 42% to 56% accuracies reported by Thoma et al. (2004). However, the minNDTI approach cannot address the effects of surface soil variability as its performance was degraded when applied to a larger geographical area. Nevertheless, multitemporal approach has shown a substantial potential to track changes of tillage practices over time and space using freely available Landsat and Landsat-like data (Watts et al., 2011, Zheng et al., 2012, Zheng et al., 2013a).

9.3.4 Summary

9.3.4.1 Challenges

The primary challenges for operational tillage mapping using optical remote sensing imagery (Papadopoulos et al., 2023, Xiang et al., 2022, Martins et al., 2021, Rossi et al., 2020, Azzari et al., 2019) include (1) revisit rates of moderate-spatial resolution imagery are not frequent enough to capture the rapid changes in agricultural land surfaces during planting season; (2) limited spatial coverage of satellite hyperspectral imagery; (3) confounding effects of soil background and green vegetation; (4) lack of transferability of locally developed models.

Landsat is currently the best satellite system to provide the capabilities for long-term and broad-scale tillage assessment. Although the minNDTI technique showed promises in tillage mapping at large scales, the 8-day revisit rate of combined Landsat 8 OLI and 7 ETM+ cannot guarantee adequate numbers of cloud-free observations to capture the "recently tilled surface." In tropical regions or other areas that have persistent cloud cover, one may be lucky to obtain two or three cloud-free images per year. The data gap issues of Landsat 7 ETM+ imagery also prevent rapid application of the minNDTI technique because additional image preprocessing skills are required to fill the missing data. Zheng et al. (2013b) have presented an easy way to fill the missing data for broad-scale tillage mapping using the multi-scale segmentation method. Landsat images with partial cloud cover can be incorporated into the time series, however, estimation of tillage status for the cloud-contaminated pixels could be less accurate and a quality assessment map should be provided to inform users about locations of cloud and cloud shadow pixels (Zheng et al., 2014).

The Spatial and Temporal Adaptive Reflectance Fusion Model (STARFM) (Zhang et al., 2024, Jiang et al., 2023, Papadopoulos et al., 2023), which produces cloud-free synthetic Landsat images with 30-m spatial resolution at MODIS temporal frequency, could be an alternative option to enhance temporal resolution for tillage mapping. The enhanced STARFM (Zhu et al., 2010), future improvement of data fusion techniques, and the higher quality of Landsat 8 and the European Space Agency (ESA) Sentinel-2 data could open possibilities to provide data optimized in both temporal and spatial resolutions for tillage assessment. However, the potential to incorporate data fusion techniques into minNDTI technique to improve our ability to map tillage practices currently remains unknown and required for future investigation (Zheng et al., 2014).

Locally developed empirical models often show degraded performance when applied to the same location over time or to a broader region. Variations in weather, soil, and terrain conditions across landscape are the main reasons for the degraded performance when a model is extrapolated to new situations. Zheng et al. (2013a) reported superior performance of local models than a "universal" model, and highlighted negative impacts of local variation in terrain, moisture, and soil color upon crop residue estimation. Thus, estimation of crop residue cover with broadband multispectral may require extensive local calibration data. Alternatively, the effects of soil variation can be reduced or minimized using local soil-adjusted tillage indices (Biard and Baret, 1997) or the spectral unmixing approach (Pacheco and McNairn, 2010). The spectral unmixing approach has the potential to map crop residue cover over large geographic regions as the approach is insensitive to variations in soil and residue when endmembers are retrieved directly from the image (Pacheco and McNairn, 2010). However, future work is required to examine how well the unmixing approach performs in the presence of green vegetation.

Much of the research to apply remote sensing to tillage assessment has been developed in the context of mid-latitude agriculture, characterized by distinct seasonal cycles, large field sizes, common use of monoculture, or reduced crop diversity, over large regions. In other regions of the world, or in irrigated regions, there may be a much larger range of crops, with a variety of planting and harvesting dates, not synchronized with each other, and smaller field sizes—in such situations, the tillage assessment task requires different strategies than may be effective in mid-latitude regions.

9.3.4.2 Future Capabilities

At the time of this writing, due to the limited availability of hyperspectral data, the minNDTI approach is probably the most effective method to map tillage practices at broad-scale using optical remote sensing imagery (Jiang et al., 2023, Papadopoulos et al., 2023, Calcagno et al., 2022, Liu et al., 2022, Luotamo et al., 2022). The minNDTI can be applied to Landsat 7 ETM+ and Landsat 8 OLI, which together provides 8-day observation cycle. The OLI imagery has potential to enhance our ability to accurately estimate crop residue cover with its narrower spectral bands and 12 bit dynamic range, as indicated by Galloza et al. (2013), who found that the Advanced Land Imager (ALI) has better capability to discriminate crop residues from soils than Landsat TM data.

The upcoming launch of the European Space Agency (ESA) Sentinel-2 satellite will provide enhanced Landsat-type data with < 5-day revisit time. Sentinel-2 is particularly useful for monitoring the rapid changes of agricultural lands. Operational tillage assessment is likely to involve multi-sensor multi-date image fusion and could be implemented using Landsat and Sentinel-2 data together. The planned hyperspectral satellite missions, including ESA Environmental Mapping and Analysis Program (EnMAP) and NASA Hyperspectral Infrared Imager (HyspIRI), will also make contribution to large-scale tillage assessment. These hyperspectral data can be used to calculate CAI. Fusion of hyperspectral and multispectral images could estimate crop residue cover at the multispectral spatial extent with improved accuracy (Galloza et al., 2013). The WorldView-3 satellite scheduled for launch in late 2014 includes SWIR bands equivalent to ASTER SWIR sensor (www.digitalglobe.com/content/worldview3/), which can be used to derive SINDRI for crop residue estimation. The very high spatial resolution (3.7 m) of WorldView-3 SWIR data will permit fine-scale assessment of crop residue cover, soil texture, and soil roughness.

9.4 MONITORING WITH SYNTHETIC APERTURE RADAR (SAR)

9.4.1 Introduction

Synthetic Aperture Radars (SARs) are considered active remote sensing sensors as they generate pulses of energy which are propagated towards a target. SARs then record the energy scattered by the target, back towards the radar antenna. The strength (intensity) of the received or "backscattered" signal is measured as sigma naught (σ^o), expressed in decibels (dB). Since these sensors provide their own source of energy, SARs are able to collect data day or night. SARs generate energy at microwave frequencies (0.2–300 GHz), with Earth observing SAR satellites typically operating at X-Band (2.40–3.75 cm; 8.0–12.5 GHz), C-Band (3.75–7.5 cm; 4.0–8.0 GHz), and L-Band (15–30 cm; 1.0–2.0 GHz) (Lewis and Henderson, 1998) (Table 9.4). These lower frequencies are unaffected by the presence of cloud and haze. Given this context, and the sensitivity of microwaves to soil conditions, SARs are an important data source for mapping and monitoring tillage and residue.

9.4.2 Critical Variables for Tillage Assessment

The interaction of microwaves with a target and the characteristics of the scatter which results from this interaction are a function of the condition of the target as well as the SAR sensor specifications.

TABLE 9.4
Selected Civilian Spaceborne Radar Sensors

Frequency (in GHz)		Sensor	Polarization*	Incidence Angle	Resolution (m)	Swath (km)	Dates of Operation
X	8.600	Cosmo-SkyMed 1	SP, DP	25–50°	1–100	10–200	2007—
		Cosmo-SkyMed 2	SP, DP				2007—
		Cosmo-SkyMed 3	SP, DP				2008—
		Cosmo-SkyMed 4	SP, DP				2010—
	8.650	TerraSAR-X	SP, DP, QP	15–60°	0.25–40	4–270	2007—
	8.650	TanDEM-X	SP, DP, QP	15–60°	0.25–40	4–270	2010—
C	5.300	RADARSAT-1	SP (HH)	10–60°	8–100	45–500	1995–2013
	5.300	ERS-2	SP (VV)	20–26°	30	100	1995–2011
	5.331	Envisat ASAR	SP, DP	15–45°	10–1000	5–405	2002–2012
	5.350	RISAT-1	SP, DP, QP, CP	12–55°	1–50	25–223	2012—
	5.405	RADARSAT-2	SP, DP, QP	10–60°	3–100	18–500	2007—
	5.405	RADARSAT Constellation	SP, DP, QP, CP	10–60°	1–500	5–500	2018
	5.405	Sentinel 1A	SP, DP	20–45°	5–40	80–400	2014
		Sentinel 1B	SP, DP	20–45°	5–40	80–400	2015
L	1.200	ALOS /PALSAR-1	SP, DP, QP	8–60°	10–100	20–350	2006–2011
	1.200	ALOS/PALSAR-2	SP, DP, QP, CP	8–60°	1–100	25–490	2014
	1.260	SMAP	SP, VV/HH/HV[1]	40°	1–3 (km)	1000	2014
	1.275	SAOCOM 1A	SP, DP, QP, CP	17–51°	10–100	20–350	2014
		SAOCOM 1B					2015

* In the polarization column, SP = Single polarization, DP = Dual polarization, QP = Quadrature polarization, and CP = Compact polarization.

[1] SMAP will acquire radar imagery simultaneously in VV, HH, and HV.

SAR response is driven by the dielectric permittivity, roughness, and structural properties of the target. In the context of tillage monitoring, SARs are sensitive to small scale roughness and large macrostructures produced by farming implements, as well as volumetric soil moisture. In addition to their spatial resolution, SARs are characterized by their frequency, incidence angle, and polarization—configurations that also affect the target interaction.

9.4.2.1 Sensitivity of SAR to Soil Characteristics

Surface Roughness: Random and periodic roughness determines the angular scattering pattern with diffuse scattering increasing as roughness increases. For agricultural fields, roughness is created by land management activities (principally tillage and seedbed preparation) modified over time by water and wind erosion. Roughness is defined by two parameters—the root mean square variance (RMS) and surface correlation length (l). RMS describes the surface's random vertical statistical variability relative to a reference surface while correlation length is an autocorrelation function that measures the statistical independence of surface heights at two points (Ulaby et al., 1986). For very smooth surfaces, as expected from no-till fields, the random roughness (RMS) is small and the height of every point is correlated with the height of every other point (hence l is large). In this case, most microwave energy is forward scattered and backscatter is low. Inversely randomly rough surfaces, created by tillage, result in more diffuse scattering with a greater proportion of the incident energy scattered back to the sensor. These surfaces have higher RMS, short correlation lengths and higher backscatter.

Dielectric Permittivity: The intensity of backscatter from soils is largely determined by the soil permittivity (dielectric constant) while the angular pattern of microwave scattering is governed by the surface roughness. The permittivity ε is a frequency dependent complex quantity [$\varepsilon(f)=\varepsilon'(f)-j\varepsilon''(f)$], where the real component ε' describes the polarizability of a material when an electric field is applied, and the imaginary component ε'' energy losses (Hasted, 1973). Dielectric losses are due to relaxation ε''_{ref} and direct current electrical conductivity σ in Siemens/m: $\varepsilon''(f)=\varepsilon''_{ref}(f)+\frac{\sigma}{2\pi f\varepsilon_0}$, where ε_0 is the permittivity of free space ($8.854 \cdot 10^{-12}$ Farad/m). On agricultural fields (without vegetation cover), scattering occurs at the air/soil boundary as a dielectric discontinuity exists at this interface. The majority of dry soils have ε' of 3–8, and bulk soil permittivity increases with water content. This is due to the much greater, albeit frequency dependent, permittivity of water, which at 1.4 GHz ranges from $84.1-j10.7$ at 5°C to $74.5-j4.1$ at 35°C, $68.0-j32.1 \sim 71.4-j14.6$ at 5.3 GHz, and $48.2-j38.7 \sim 65.1-j23.7$ at 8.6 GHz for pure water where $\sigma=0$ S/m. Increases in either part of the permittivity will increase soil reflectivity. Electromagnetic wavelength is an inverse function of ε', thus wavelength becomes shorter within the soil as it becomes wetter. As backscatter intensity is a function of permittivity, a strong linear relationship exists between soil moisture and backscatter. The depth of sensitivity within the soil volume is dependent upon three parameters: the SAR configuration, soil moisture, and bulk soil ε''. Penetration depth is an inverse function of bulk soil permittivity, and thus, soil moisture and conductivity. Consequently, SARs respond to moisture over deeper volumes as soils dry. Regardless, sensitivity is still near surface with this depth approximately equivalent to the microwave wavelength (Boisvert et al., 1995).

Residue: If vegetation (green or senesced vegetation, or post-harvest residue) is present, SAR response will be affected if water is present in the vegetation. Residue is considered "dead" vegetation and thus its effect on backscatter is often assumed insignificant, effectively transparent to the incident microwaves. This assumption has proven invalid in circumstances where residue retains water. The impact of residue on backscatter varies depending upon the volume of water held, a function of the amount and type of residue (McNairn et al., 2001). Jackson and O'Neill (1991) reported that residue can retain significant moisture with McNairn et al. (2001) measuring up to 60% and 40–50% moisture in corn and barley residue, respectively, following rain events.

Row Direction: Land management practices (planting, harvesting, and tillage) can create row effects and row direction relative to the radar look direction impacts SAR response. When row direction is perpendicular to the look direction, SAR response is stronger when compared to a look direction parallel to rows (Beaudoin et al., 1990, McNairn et al., 1996). Producers follow a rectangular pattern operating parallel to the long and short axes of fields. This practice creates a "bow-tie" effect visible on SAR imagery where, within a single field, backscatter is significantly higher for the axis of the field oriented perpendicular to the sensor.

9.4.2.2 Impact of SAR Configuration

SAR sensors are defined by three configurations—frequency (GHz, or cm, if characterized as free-space wavelength), incidence angle (degrees), and polarization. These configurations affect how microwaves interact with the target in terms of backscatter intensity and scattering characteristics. SAR configurations can be selected to maximize sensitivity to the target property of interest (soil moisture, surface roughness or residue). Alternatively, as these properties are confounded in the microwave signal, multiple configurations can be used together to resolve individual contributions.

Frequency: As well as affecting penetration depth, SAR frequency determines sensitivity to surface roughness. Thus, surface roughness must be considered relative to frequency. Surfaces are defined as "rough" or "smooth" according to the Rayleigh criterion. Surfaces are smooth if $h<\frac{\lambda}{25\sin\tau}$ and rough if $h>\frac{\lambda}{4.4\sin\tau}$ where h is the RMS, λ is the wavelength and τ is the depression angle (Sabins, 1986). Assuming flat terrain, τ is the complement of the incidence angle ($\theta=90-\tau$). In practice this means that a field will appear rougher (higher backscatter) at shorter wavelengths (i.e., X-Band) than at longer wavelengths (i.e., L-Band). With this strong dependency, choice of wavelength is especially important when monitoring tillage. Short wavelength (high frequency) SARs will see many fields as "rough" and thus may not differentiate among tillage classes at the upper ranges of roughness. Several studies (Liu et al., 2022, Azzari et al., 2019, Yeom et al., 2019, Panciera et al., 2013, Aubert et al., 2011, Pacheco et al., 2010) reported that X-Band data from TerraSAR-X was not well suited for roughness mapping when RMS was high. Panciera et al. (2013) found that TerraSAR-X backscatter was sensitive to roughness (RMS) which fell between 0.5 and 1.5 cm, but that the signal saturated beyond 2 cm. Conversely, large wavelength (low frequency) SARs may view even tilled fields as "smooth." Nevertheless, numerous studies have reported sensitivity of C- and L-Band responses to roughness and residue (McNairn et al., 2001, 2002, Baghdadi et al., 2008). Baghdadi et al. (2008) compared three frequencies (X-, C-, and L-band) demonstrating that sensitivity to roughness increased with wavelength.

Incidence Angle: Regardless of the target, backscatter decreases with increasing incidence angle, which is defined as the angle between the radar beam and a line perpendicular to the surface. The rate of decrease is target dependent, with backscatter decreasing with angle at a higher rate when soils are smooth. This differential rate of decrease can be used to separate smooth from rough fields, if fields are imaged at contrasting incidence angles (McNairn et al., 1996). As simultaneous multi-angle data are typically unavailable from spaceborne SARs, a simpler approach is to select an incidence angle which maximizes sensitivity to surface roughness. Steeper (smaller) angles minimize roughness contributions to backscatter and are thus more suited to estimate soil moisture, while shallower (larger) angles maximize roughness effects on backscatter (McNairn et al., 1996). Similarly, larger angles are more sensitive to residue as soil moisture contributions are minimized and more microwave interaction occurs with residue at these angles (McNairn et al., 2001)). Although these larger angles are more suited to roughness and residue applications, contributions from soil moisture are not completely eliminated. Aubert et al. (2011) noted that the range in X-HH backscatter due to surface roughness increased as incidence angle increased, with backscatter varying 3.5 dB and 1.9 dB at angles of 50° and 25°, respectively. Baghdadi et al. (2008) reported a slightly larger range in X-Band backscatter (5.5 dB at 50° – 52° and 4 dB at 26° – 28°).

Polarization: Polarization is defined by the orientation of the electric field vectors of the transmitted and received electromagnetic wave. Polarization should be considered relative to target structure and response interpreted according to characteristics of scattering from the target, including the sources of scattering and the randomness of the scatter. Scattering is categorized as single bounce (surface), multiple (volume) or double-bounce. Targets usually produce more than one type of scattering although typically one source dominates. For smooth soils devoid of residue, surface single bounce scattering dominates. Rough soils result in multiple scattering of microwaves. Residue (depending on amount and water content) also cause multiple scattering and, if residue is vertically oriented, double-bounce events may also contribute.

Most SAR sensors transmit and receive microwaves in the horizontal (H) and/or vertical (V) linear polarizations (Table 9.4). Early satellites transmitted and received microwaves in a single linear polarization (ERS-1 and 2 (VV), JERS-1 (HH) and RADARSAT-1 (HH)). Next generation sensors (i.e., ASAR) transmitted and/or received in both linear polarizations which permitted acquisition of like (HH and/or VV) and cross (HV or VH) polarizations. When targets are physically oriented parallel to the polarization of the incident wave, greater microwave interaction occurs. This is most obvious for targets like crops where their vertical structure aligns well with vertical transmitted waves. Consequently, a VV configuration provides more information on crops than HH. For soils without residue, horizontal or vertical orientation is absent and thus HH and VV backscatter is correlated. A linear cross polarization response (HV or VH) results when the transmitted wave (i.e., H) is repolarized to its orthogonal polarization (i.e., V). Repolarization of H to V (or V to H) occurs as a result of multiple scattering (at least two bounces) and thus a target must be able to cause more than a single scatter event to elicit a HV or VH response. Smooth soils, devoid of structure, are dominated by single bounce forward scattering and produce very low cross polarized backscatter. For soils with random roughness or residue (assuming moisture in the residue), incident waves experience multiple scattering and higher cross polarization response is observed. McNairn et al. (2001) reported that, of all the linear polarizations, the cross-polarization was most sensitive to the amount of residue. The cross polarization has the advantage of being insensitive to planting, harvesting or tillage row direction (McNairn and Brisco, 2004). This is important considering that Brisco et al. (1991) established that row direction from tillage significantly impacted like-polarized backscatter.

Polarimetry: Some satellites (i.e., ALOS PALSAR, RADARSAT-2, and TerraSAR-X) are polarimetric capable. Polarimetric sensors capture the complete characterization of the scattering field meaning that they record all 4 mutually coherent channels (HH, VV, HV, VH), with phase information between orthogonal polarizations retained and processed. Any linear, elliptical or circular polarization can be synthesized from polarimetric data. Circular polarizations are described by their handedness (direction of rotation) relative to the observer. Right-handed circular waves (R) rotate clockwise (relative to observer) while left-handed waves (L) rotate counter-clockwise. The application of circular polarizations for agriculture has received limited attention although for soils, circular and linear backscatter is highly correlated (Sokol et al., 2004). As with linear polarizations, multiple scattering must occur to change the handedness of the transmitted circular polarization. Roughness or residue can cause two or more bounces, changing the handedness and resulting in a higher circular co-polarization (RR or LL) response (recall rotation is defined relative to the observer). Indeed de Matthaeis et al. (1992) observed high circular cross-polarized backscatter (LR) returns for surfaces with dominant surface scattering. Circular co-polarized (RR) backscatter increases when the mechanisms producing volume scattering dominates (McNairn et al., 2002).

Polarimetric data can be processed to extract additional parameters, which characterize scattering and thus tillage and residue conditions. SARs transmit completely polarized waves but with multiple scattering, microwaves become completely or partially depolarized. The degree of depolarization (or proportion of unpolarized energy) is indicative of the randomness of scattering within the target. Smooth soils create little depolarization (Evans and Smith, 1991). Degree of depolarization increases with roughness and residue cover as phase becomes unpredictable from point to point within the target. Degree of depolarization can be measured by pedestal height with height increasing as roughness increases or in the presence of residue (Calcagno et al., 2022, Xiang et al., 2022, Martins et al., 2021, Rossi et al., 2020, de Matthaeis et al., 1991, McNairn et al., 2002, Adams et al., 2013a). Adams et al. (2013a) also reported the dynamic range of the degree of polarization (Δ_{POL}) was sensitive to roughness and residue. Δ_{POL} is the difference between the maximum and minimum degree of polarization and reflects the heterogeneity of scattering mechanisms within the target (Touzi et al., 1992).

Absolute phase (ϕ) of a scattered wave is a function of distance from the target and carries no target scattering information (Langman and Inggs, 1994). However, the difference in the phase between two orthogonal polarizations (i.e., H and V) is of interest for tillage monitoring. Shifts in the phase (characterized by the co-polarized phase difference (PPD) (ϕ_{VV-} ϕ_{HH})) occur due to double bounce or multiple scattering. For smooth soils with minimal contributions from multiple scattering, HH and VV are in phase and mean PPD is close to zero (Evans et al., 1988). Vertical structure can cause a double bounce and here PPD values approach 180° (de Matthaeis et al., 1991). Large phase differences are typically associated with cropped fields although high PPD values have been observed for standing senesced crops (McNairn et al., 2002). Ulaby et al. (1987) reported that ploughed and disked fields, as well as those with corn and soybean residue, had a mean PPD close to zero. However, the standard deviation of the phase difference among the disked, ploughed, residue and standing crops was very different. These results were confirmed by McNairn et al. (2002) where multiple scattering in residue caused a highly varying PPD with a noise-like distribution for these fields. The co-polarized complex correlation coefficient ($\rho_{HH\text{-}VV}$) measures the de-correlation of the phase and some sensitivity to residue has also been reported (Adams et al., 2013a).

Methods which decompose the SAR signal have drawn considerable interest with the Cloude-Pottier (Cloude and Pottier, 1997) and Freeman-Durden (Freeman and Durden, 1998) decompositions showing sensitivity to tillage and residue. Cloude-Pottier decomposes the signal into a set of eigenvectors (which characterize the scattering mechanism) and eigenvalues (which estimate the intensity of each mechanism) (Alberga et al., 2008). From the eigenvalues, entropy (H) and anisotropy (A) are calculated. H measures the degree of randomness of the scattering (from 0 to 1); values near zero are characteristic of single scatter targets (i.e., smooth soils). Rough soils and those with residue have larger contributions from multiple scattering. This increase in randomness of scattering is measured as an increase in H. Anisotropy estimates the relative importance of the dominant scattering mechanism and the contribution from secondary and tertiary scattering mechanisms. Zero A identifies two mechanisms of approximately equal proportions while values approaching 1 indicate that the second mechanism dominates the third (Papadopoulos et al., 2023). The Cloude-Pottier decomposition also calculates the average alpha ($\bar{\alpha}$) angle (0 to 90°) which identifies the dominant scattering source (Alberga et al., 2008). Smooth soils with single bounce scattering have angles close to 0°, volume scatterers close to 45° and double bounce nearing 90°. Adams et al. (2013a) reported that H and $\bar{\alpha}$ were significantly correlated with roughness and percent crop residue. The Freeman-Durden decomposition separates the total power of every SAR resolution cell into contributions from three scattering mechanisms—volume (multiple), double bounce and single bounce (surface) scattering. Adams et al. (2013b) demonstrated that H, $\bar{\alpha}$ and the Freeman-Durden multiple scattering could statistically separate fields with different harvesting, tillage and residue conditions, particularly at higher incidence angles (49°). In addition, the best separability was found between unharvested or fields not tilled, and conventionally tilled fields; fields under conservation tillage were confused with other tillage classes (Adams et al., 2013b).

9.4.3 Methods

9.4.3.1 Change Detection and Classification

Change detection identifies and measures differences between two (or more) images, indicated by a change in SAR response or in derived surface properties (roughness, residue). Several SAR metrics can be used to capture change and include: (1) incoherent SAR backscatter (HH, VV, HV, VH), (2) degree of polarization, (3) co-polarized phase parameters, (4) decomposition parameters, and (5) coherent change. When change is measured directly from SAR response, consideration must be given to the confounding effects of target parameters, the SAR configuration and sensor calibration. To isolate change in SAR response due to roughness (or residue), soil moisture must not vary and thus the period between acquisitions should be minimized. Since frequency, incidence angle and polarization affect target interaction; images must have the exact same SAR configuration. For spaceborne SARs, this means using exact repeat orbits. Constellations of satellites (such as the planned Canadian RADARSAT-Constellation) will be of interest for change detection since repeat acquisitions in the same SAR configuration will be possible within a short period of time. Finally, SARs must be well calibrated; scene to scene calibration of spaceborne sensors is typically well below 1 dB. If changes in derived properties (roughness, residue) are used, errors in methods or model performance will be carried forward in the change detection process. Whatever metric is adopted, interpretation of the change is required. This means that a threshold must be determined, above which change is considered significant. In addition, change must be linked to information meaningful for tillage monitoring (type of implement used, tillage or residue class, change in residue amount).

McNairn et al. (1998) applied a simple change detection approach to a pair of RADARSAT-1 (HH) images acquired one week apart. The incidence angle difference between the Standard Mode 2 and 3 images was limited to 6° and was considered of secondary importance. In the one week separating the first from second acquisition, C-HH backscatter remained stable (average difference

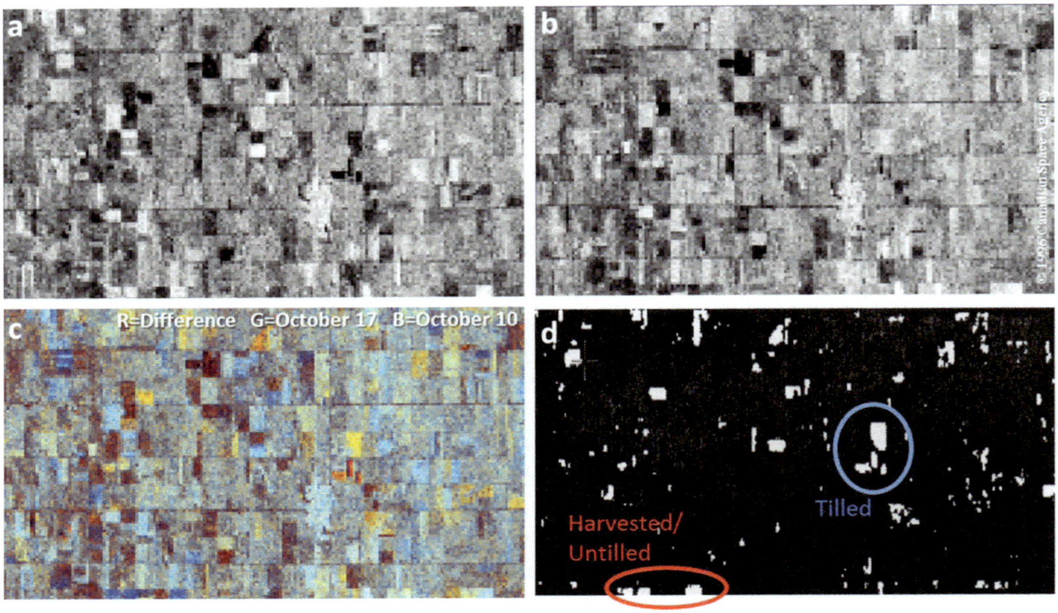

FIGURE 9.7 Detection of tillage and harvesting activities using RADARSAT-1 over an agricultural site in Canada (Altona, Manitoba). Standard beam mode images were collected on October 10 (a) and October 17 (b) in 1996. A difference image (c) and a change detection product (d) were produced from the backscatter. (Taken from McNairn et al., 1998.)

of 0.7 dB) for fields not tilled. No rain fell during the week, and the small difference was attributed to the 6° difference in angles. For fields which were tilled, the average change (increase) in backscatter was 5.6 dB. This technique (Figure 9.7) enabled the identification of broad conservation tillage classes (no-till, intermediate, and tilled) and flagged fields where harvesting and tillage had occurred. Zhang et al. (2024) combined SAR (Envisat ASAR) and optical data to classify broad categories of tillage. The authors used a combination of image thresholding and decision tree classification. Envisat ASAR was especially helpful at differentiating smooth surfaces (no-till) from other rougher (tilled) surfaces.

Coherent change detection (CCD) exploits the coherence between two polarimetric complex images acquired at different times but in the same imaging geometry (Milisavljević et al., 2010). A pixel by pixel correlation of the coherence between the images reveals changes in the target; if no change has taken place, the pixels remain correlated. This technique requires that the target is coherent, allowing changes in coherence from image to image to be measured. Random phase characterizes most distributed natural targets like forests and crops. These targets typically have low coherence and are not ideal candidates for this method. As well, external effects like wind can cause these targets to temporally de-correlate. Polarimetric interferometric (PolInSAR) may be useful in optimizing coherence for detecting change in distributed targets like crops (Li et al., 2014). Although CCD for tillage change detection has not been explored, this approach may be capable of observing changes from tillage activities.

9.4.3.2 Semi-Empirical and Physical Models

Physical scattering models estimate backscatter using the soil's physical properties and sensor configurations. Soil properties include the dielectric constant, RMS and correlation length. The small perturbation and Kirchhoff models (geometrical optics and physical optics models) are two common physical models. However, these models are not suited to targets with multiple sources of scattering and large ranges of roughness, as expected from agricultural fields. The Integral Equation Model (IEM) (Fung et al., 1992) integrates these two models and is better adapted for targets with surface and multiple scattering and with roughness ranging from smooth to rough.

The goodness of fit between backscatter predicted by the IEM and that observed by SARs has varied depending on the roughness, frequency and incidence angle. Speculation has been that in many cases the error in IEM simulated backscatter is due to inaccurate representation of the correlation length (l), a parameter difficult to adequately measure in the field (Merzouki et al., 2010). As a solution Baghdadi et al. (2004) proposed a calibrated version of the IEM, introducing an optimum correlation length (l_{opt}). The optimum correlation length is derived from a set of equations which relates correlation length (l) to RMS, as a function of polarization and incidence angle (Baghdadi et al., 2006). Simulated backscatter from the calibrated IEM has more closely matched backscatter from C-Band SAR backscatter (Merzouki et al., 2010). Figure 9.8a is an example of a surface roughness (RMS height) map derived from this study (Merzouki et al., 2010). Rahman et al. (2008) also derived surface roughness over sparsely vegetated fields using Envisat ASAR and the IEM in a multi-angle approach. The image-derived RMS (2.19 cm) overestimated the field-derived RMS (0.79 cm) (Figure 9.8b). The subsurface rock fragments may have caused multiple bounce interactions, thus increasing response and generating a larger "radar-perceived" roughness (Rahman et al., 2008).

Inversion of the IEM or calibrated IEM is difficult due to the complexity of the model. As well, multiple unknowns in the IEM (dielectric constant, RMS and l) and the calibrated IEM (dielectric constant and RMS) require multiple sources of SAR information. In this case, a look-up table (LUT) approach can be used to estimate roughness or dielectric from SAR response (Merzouki et al., 2011). Forward runs of the model are used to create the LUTs with incremental steps in dielectric, RMS, l and incidence angle, and their modeled backscatter (in HH and VV). Direct search functions are used to find the LUT entry which minimizes the difference between the measured (from SAR sensor) and modeled (from IEM) backscatter. This LUT entry provides the model estimate of soil dielectric and surface roughness. With multiple unknowns, multiple SAR configurations are needed to solve the IEM (3 unknowns) or calibrated IEM (2 unknowns). Typically SAR data acquired at two polarizations (i.e., HH and VV) are used with the calibrated IEM. With a third unknown (l), an additional source of backscatter is needed to implement the original IEM. One approach is to use SAR data acquired at two polarizations (HH and VV) and two contrasting incidence angles.

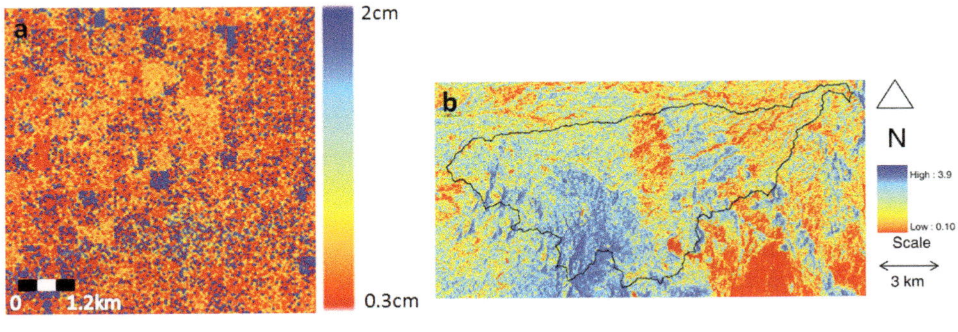

FIGURE 9.8 Surface roughness maps derived from radar images and the IEM model over two agricultural sites: an area within the Red River Watershed in Southern Manitoba in Canada (a) and Walnut Gulch Experimental Watershed in Arizona in the United States (b). The surface roughness map for the Red River watershed is expressed in root mean square (RMS) height in centimeters (Merzouki et al., 2010). The Walnut Gulch Experimental Watershed surface roughness map is defined by the root mean squared height (hRMS) variation of the surface at centimeter scale. The solid line represents the boundary of the Watershed (taken from Rahman et al., 2008).

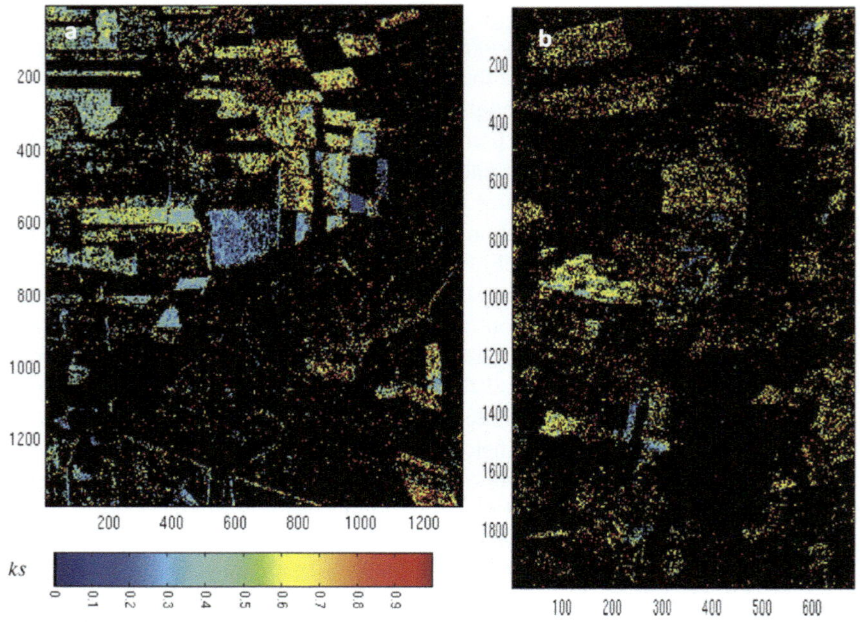

FIGURE 9.9 Estimated surface roughness, ranging from 0 to 1, over two study sites in Germany: Elbe-Auen (a) and Weiherbach (b). Areas in black represent data gaps. (Taken from Hajnsek et al., 2003.)

The Oh (Oh et al., 1992, Oh, 2004) and Dubois (Dubois et al., 1995) models are semi-empirical models created from the collection of large experimental datasets, and subsequently empirically relating soil dielectric (directly or via the Fresnel reflectivity), RMS, the wavelength (through the wavenumber) and the incidence angle to SAR backscatter. Oh modeled backscatter from all three linear polarizations (HH, VV, HV) and for three frequencies (X, C, and L). In contrast, the Dubois model uses only the co-polarized backscatter (HH and VV) and was developed using data collected only at L-Band. As with the IEM, the Oh model can be inverted using a LUT. The Dubois model is easily inverted by solving the model's two backscatter equations. Because these models were created with experimental data, application of these models to target conditions or SAR configurations beyond those of the experimental data used to create them may yield uncertain results. Indeed, Merzouki et al. (2010) found that these models tended to overestimate backscatter when modeled backscatter was compared to that measured by RADARSAT-2, which would lead to an overestimation of RMS. The Oh model resulted in larger errors between modeled and measured backscatter on smoother fields (<2 cm). Conversely, errors were greater on rougher fields (>1.5 cm) for the Dubois model.

Hajnsek et al. (2003) developed a model to invert surface roughness by coupling a Bragg scattering term and a roughness variable derived from the scattering entropy and anisotropy, and the alpha angle. This model was validated against airborne polarimetric L-Band (E-SAR) data and yielded low root mean square errors (19%). Figure 9.9 shows a roughness map created using this approach.

9.4.4 Linking Radar Products to Tillage Information

SAR sensors can provide information on roughness (RMS) and residue, and as well as changes in these conditions. However, to be meaningful, roughness and residue must be linked to information of interest such as tillage implement or tillage class. This linkage is required for applications such as watershed management, soil erosion risk assessment, or estimating carbon sequestration.

Establishing this linkage is not a simple task given the complexity and dynamics of tillage activities. Producers use a combination of tillage implements and tillage occurs periodically and at a range of soil depths and directions. Tillage-induced roughness also varies depending upon soil texture and moisture, and is modified over time by erosion events. Winter crops and weeds present on fields also complicate tillage mapping. How to link SAR-derived products and tillage information will vary depending on the approach used to create these products. For example, if models are used to estimate roughness (RMS), an association between RMS and tillage operation could be established. Such an approach was proposed by Jackson et al. (1997). However, the roughness (RMS) created by each tillage implement, and sequences of tillage applications, is likely to vary field to field due to soil conditions, erosion and characteristics of the implement itself. Consequently a much larger data base of roughness responses to tillage is required, and these data must be acquired over regions with varying tillage systems. For example, Pacheco et al. (2010) found that in eastern Canada some conservation tilled fields (chiseled ploughed) had greater roughness (RMS) than conventional tilled fields (moldboard ploughed). As well, RMS varied greatly within the chisel class, creating confusion when attempting to use backscatter to identify classes. Classifications or change detection approaches typically identify broad tillage classes (untilled, conservation, conventional). While these classes may be useful for some mapping applications (identifying adoption of no-till for carbon sequestration), they may not be adequate for others (erosion modeling).

9.4.5 Summary

Given the dynamics of tillage activities during the pre-seeding and post-harvest seasons, SAR sensors can be a valuable data source for time critical applications (McNairn et al., 1998). With longer wavelengths, SAR data acquisition is unaffected by atmospheric conditions such as cloud cover and haze. The number of SAR satellites in orbit continues to increase and the engineering behind these satellites has led to a greater diversity in SAR configurations. This means that users now have choices in incidence angle and polarization and in some cases, access to polarimetric data (Table 9.4). Research has demonstrated that success in this application will be best achieved when data can be accessed at more than one frequency and polarization. Choice of incidence angle and polarization is clear with researchers agreeing that shallower angles and cross polarizations are best for roughness and residue mapping. The availability of polarimetric-capable sensors is relatively recent, and thus more research is needed to develop methods to exploit these complex data. The primary challenge is the coupling of roughness, soil moisture and residue in the SAR response. This coupling complicates the extraction of tillage information from the signal but can be accomplished by exploiting SAR data acquired at multiple configurations (frequency, angle or polarization). Planned, and soon-to-be-launched satellites, include the C-Band RADARSAT-Constellation (Canada), C-Band Sentinel-1A and B (European Space Agency (ESA)) and L-Band SAOCOM-1A and B (Comisión Nacional de Actividades Espaciales (CONAE)) (Table 9.4). These satellites will provide frequent data at a range of angles and polarizations and promise to provide an important source of data for monitoring tillage.

9.5 REVIEW AND OUTLOOK

This chapter has summarized recent progress to advance applications of remote sensing technologies to broad-scale assessment of tillage status (Zhang et al., 2024, Jiang et al., 2023, Papadopoulos et al., 2023, Calcagno et al., 2022, Liu et al., 2022, Luotamo et al., 2022, Mandal et al., 2022, Xiang et al., 2022, Martins et al., 2021, Rossi et al., 2020, Azzari et al., 2019, Yeom et al., 2019). Nonetheless, important challenges remain. Here we recap some of the key elements of current research to apply remote sensing technologies to broad-scale, site-specific tillage assessment, then highlight some of the principal challenges this effort faces as further research progresses (see also Zheng et al., 2014).

Optical remote sensing and SAR data provide different capabilities for tillage assessment. Whereas optical remote sensing imagery provides the spectral basis for detection of crop residues on soil surface, SAR data provide information on soil physical properties, such as roughness and texture, which can reveal the nature of tillage practices. With the presence of green vegetation, both SAR and optical remote sensing data have difficulties to discriminate different tillage categories. Remotely sensed imagery sensitive to radiation near 2100 nm cellulose absorption bands provide the best opportunity to estimate crop residue cover and to map tillage practices. In this context, the best three tillage indices are CAI, SINDRI, and NDTI. Because current satellite hyperspectral systems cannot provide systematic spatial coverage, at present, multispectral imagery now forms the preferred candidate for a broad-scale tillage assessment. Multi-temporal imagery is required to provide accurate assessment on tillage practices for regions with diverse crop calendars—a range of dates for soil preparation and planting schedules. The upcoming launch of several new satellite systems with optical sensors will offer solid opportunities to enhance our ability to monitor rapid changes of agricultural lands, providing timely, and low-cost, information for monitoring site-specific tillage assessment.

Challenges—Optical systems: As noted previously, optical systems provide capabilities for monitoring tillage in a systematic manner. Yet they are subject to disruptive influences of soil moisture variations and uneven terrain. Possible solutions include: (1) development of terrain and soil data layers that can guide interpretations of image data in such areas, and (2) development of specialized indices or other strategies to detect or adjust for spectral variations caused by these effects. Further, although current systems can provide revisit intervals adequate in key agricultural regions, these capabilities may not be adequate in other regions, where higher cloud cover may require more frequent revisit capabilities to acquire cloud-free coverage necessary for the temporal sequences required for the minNDTI strategy.

From evaluation of the SINDRI and CAI tillage indices, we know that carefully, and narrowly, defined spectral channels are effective in tillage assessment. However, it seems unlikely that future satellite systems are likely to incur the costs of designing and operating new bands to support a single application mission. As a result, future opportunities for optical tillage assessment seem likely to be based on the NDTI model (which relies upon broadly defined spectral channels, but ones that support a range of application missions), and relying upon the sequential imagery to apply strategies, such as the minNDTI.

Challenges—SAR systems: Although specific strategies for application of SAR for monitoring tillage status are still under development, it has great potential for systematic tillage assessment, in part because of its ability to acquire data in the presence of cloud cover, and the potential to extract a suit of terrain measurements as part of a tillage assessment mission. As reported here, current research has been successful in applying radar fundamental to the tillage assessment task, although the multiplicity of system variables that interact with each other and with the landscape offer challenges in isolating tillage information. The SAR tillage effort has yet to scale current findings to examine larger regions, allowing identification of unexpected effects of local terrain, and interactions between agricultural practices and the geometries of varied SAR satellite systems.

Challenges—Sequential Observations: Monitoring tillage status by remote sensing by its nature requires broad-scale observation of very large regions. Within such broad regions, weather, terrain, and local practices vary, necessarily dispersing tillage and planting data operations over intervals of several weeks. Because the tillage event is ephemeral, soon concealed by the foliage of the emerging crop, it must be assessed as it occurs, not at a later date. As a result, a single "snapshot" satellite image can capture only a partial record of a region's tillage pattern. This effect is significant regardless of the sensor system, or tillage assessment strategy—sequential imagery of the entire planting season is necessary to observe the correct tillage status of a landscape. Otherwise, the inventory will record only a portion of the tillage operations within the area. In this context, both SAR and optical satellite systems are challenged to provide reliable coverage in the sense that current revisit intervals of optical systems are subject to disruption by cloud cover, and current

SAR systems are challenged to simultaneously provide the spatial detail, broad-scale coverage, and revisit intervals necessary to observe the full planting season.

Challenges—Global Tillage Monitoring: Current research to apply remote sensing to tillage assessment has been developed largely in mid-latitudes, in regions characterized by large fields, simple crop calendars that apply for very large areas, limited numbers of crops, known crop rotation sequences, and availability of supporting data. These conditions may apply in many of the other major grain-producing regions (e.g., Brazil, China, Argentina, Ukraine, Mexico), where current tillage assessment strategies may transfer. Many of the world's other agricultural regions present much different conditions that do not favor their direct transfer. For irrigated crops, there may be several planting cycles. Many tropical regions are characterized by smaller fields and complicated crop calendars, so investigators may require mastery of detailed knowledge of a diversity of cropping systems and irrigation practices, which may all vary within short distances. Such agricultural systems may exhibit levels of spatial and temporal variability that will greatly complicate applications of remote sensing strategies that have been successful in the context of mid-latitude agricultural systems.

Challenges—Field and Validation Data: Further advances in tillage assessment will require development of additional strategies for collection of field data for preparation of assessment model and for valuation of survey findings. Field data collection campaigns following established and co-coordinated protocols have a role in broad-scale survey, especially when it is feasible to mobilize a network or experienced volunteers to support campaigns. However, such efforts inevitably encounter logistical problems, especially when unfavorable weather creates uncertainties, or prevents acquisition of viable imagery. Work to investigate alternative strategies, including the feasibility of using commercial satellite imagery to collect tillage observations to support model development, and validation of project findings, deserves attention.

REFERENCES

Adams, J. R., A. A. Berg, H. McNairn, and A. Merzouki. 2013a. Sensitivity of C-band SAR polarimetric variables to unvegetated agricultural fields. *Can. J. Remote Sens.* 39 (1):1–16. doi: 10.5589/m13-003.

Adams, J. R., T. L. Rowlandson, S. J. McKeown, A. A. Berg, H. McNairn, and S. J. Sweeney. 2013b. Evaluating the Cloude—Pottier and Freeman—Durden scattering decompositions for distinguishing between unharvested and post-harvest agricultural fields. *Can. J. Remote Sens.* 39 (04):318–327. doi: 10.5589/m13-040.

Alberga, V., G. Satalino, and D. K. Staykova. 2008. Comparison of polarimetric SAR observables in terms of classification performance. *Int. J. Remote Sens.* 29 (14):4129–4150. doi: 10.1080/01431160701840182.

Aubert, M., N. Baghdadi, M. Zribi, A. Douaoui, C. Loumagne, F. Baup, M. El Hajj, and S. Garrigues. 2011. Analysis of TerraSAR-X data sensitivity to bare soil moisture, roughness, composition and soil crust. *Remote Sens. Environ.* 115 (8):1801–1810. doi: 10.1016/j.rse.2011.02.021.

Azzari, G., P. Grassini, J. Ignacio Rattalino Edreira, S. Conley, S. Mourtzinis, D. B. Lobell. 2019. Satellite mapping of tillage practices in the North Central US region from 2005 to 2016. *Remote Sens. Environ.* 221:417–429, ISSN 0034–4257, doi: 10.1016/j.rse.2018.11.010. (www.sciencedirect.com/science/article/pii/S0034425718305157).

Baghdadi, N., I. Gherboudj, M. Zribi, M. Sahebi, C. King, and F. Bonn. 2004. Semi-empirical calibration of the IEM backscattering model using radar images and moisture and roughness field measurements. *Int. J. Remote Sens.* 25 (18):3593–623. doi: 10.1080/01431160310001654392.

Baghdadi, N., M. Zribi, C. Loumagne, P. Ansart, and T. P. Anguela. 2008. Analysis of TerraSAR-X data and their sensitivity to soil surface parameters over bare agricultural fields. *Remote Sens. Environ.* 112 (12):4370–4378. doi: 10.1016/j.rse.2008.08.004.

Baghdadi, N., N. Holah, and M. Zribi. 2006. Calibration of the integral equation model for SAR data in C-band and HH and VV polarizations. *Int. J. Remote Sens.* 27 (4):805–816. doi: 10.1080/01431160500212278.

Baumgardner, M. F., L. F. Silva, L. L. Biehl, and E. R. Stoner. 1986. Reflectance properties of soils. In *Advances in Agronomy*, edited by N. C. Brady, 1–44. San Diego, CA: Academic Press.

Beaudoin, A., T. Le Toan, and Q. H. J. Gwyn. 1990. SAR observations and modeling of the C-band backscatter variability due to multiscale geometry and soil moisture. *IEEE Trans. Geosci. Remote Sens.* 28 (5):886–895. doi: 10.1109/36.58978.

Ben-Dor, E. 2002. Quantitative remote sensing of soil properties. In *Advances in Agronomy*,, 173–243. San Diego, CA: Academic Press.

Biard, F., and F. Baret. 1997. Crop residue estimation using multiband reflectance. *Remote Sens. Environ.* 59 (3):530–536.

Boisvert, J. B., Q. H. J. Gwyn, B. Brisco, D. Major, and R. J. Brown. 1995. Evaluation of soil moisture techniques and microwave penetration depth for radar applications. *Can. J. Remote Sens.* 21:110–123.

Booth, D. T., S. E. Cox, and R. D. Berryman. 2006. Point sampling digital imagery with 'SamplePoint'. *Environ. Monit. Assess.* 123:97–108.

Bricklemyer, R. S., R. L. Lawrence, P. R. Miller, and N. Battogtokh. 2006. Predicting tillage practices and agricultural soil disturbance in north central Montana with Landsat imagery. *Agric., Ecosyst. Environ.* 114 (2–4):210–216. doi: 10.1016/j.agee.2005.10.005.

Brisco, B., R. J. Brown, B. Snider, G. J. Sofko, J. A. Koehler, and A. G. Wacker. 1991. Tillage effects on the radar backscattering coefficient of grain stubble fields. *Int. J. Remote Sens.* 12 (11):2283–2298. doi: 10.1080/01431169108955258.

Brown, D. J., K. D. Shepherd, M. G. Walsh, M. Dewayne Mays, and T. G. Reinsch. 2006. Global soil characterization with VNIR diffuse reflectance spectroscopy. *Geoderma* 132 (3–4):273–290. doi: 10.1016/j.geoderma.2005.04.025.

Calcagno, F., E. Romano, N. Furnitto, A. Jamali, and S. Failla. 2022. Remote sensing monitoring of durum wheat under no tillage practices by means of spectral indices interpretation: A preliminary study. *Sustainability* 14 (22):15012. doi: 10.3390/su142215012.

Cloude, S. R., and E. Pottier. 1997. An entropy based classification scheme for land applications of polarimetric SAR. *IEEE Trans. Geosci. Remote Sens.* 35 (1):68–78. doi: 10.1109/36.551935.

Corak, S. J., T. C. Kaspar, and D. W. Meek. 1993. Evaluating methods for measuring residue cover. *J. Soil Water Conserv.* 48:700–704.

CTIC. 2014. National Crop Residue Management Survey. www.ctic.purdue.edu/CRM/ (accessed January 21, 2014).

Daughtry, C. S. T. 2001. Discriminating crop residues from soil by shortwave infrared reflectance. *Agron. J.* 93 (1):125–131.

Daughtry, C. S. T., and E. R. Hunt, Jr. 2008. Mitigating the effects of soil and residue water contents on remotely sensed estimates of crop residue cover. *Remote Sens. Environ.* 112 (4):1647–1657. doi: 10.1016/j.rse.2007.08.006.

Daughtry, C. S. T., E. R. Hunt, Jr., and J. E. McMurtrey, III. 2004. Assessing crop residue cover using shortwave infrared reflectance. *Remote Sens. Environ.* 90 (1):126–134. doi: 10.1016/j.rse.2003.10.023.

Daughtry, C. S. T., E. R. Hunt, Jr., P. C. Doraiswamy, and J. E. McMurtrey, III. 2005. Remote sensing the spatial distribution of crop residues. *Agron. J.* 97 (3):864–871. doi: 10.2134/agronj2003.0291.

Daughtry, C. S. T., G. Serbin, J. B. Reeves, III, P. C. Doraiswamy, and E. R. Hunt, Jr. 2010. Spectral reflectance of wheat residue during decomposition and remotely sensed estimates of residue cover. *Remote Sens.* 2 (2):416–431.

Daughtry, C. S. T., P. C. Doraiswamy, E. R. Hunt, Jr., A. J. Stern, J. E. McMurtrey, III, and J. H. Prueger. 2006. Remote sensing of crop residue cover and soil tillage intensity. *Soil Tillage Res.* 91 (1–2):101–108.

de Matthaeis, P., P. Ferrazzoli, G. Schiavon, and D. Solimini. 1991. Radar response to vegetation parameters: Comparison between theory and MAESTRO-1 results. Paper presented at the Proceedings of the International Geoscience and Remote Sensing Symposium (IGARSS 1991), Espoo, Finland.

de Matthaeis, P., P. Ferrazzoli, G. Schiavon, and D. Solimini. 1992. Agriscatt and MAESTRO: Multifrequency radar experiments for vegetation remote sensing. Paper presented at the MAESTRO-1/AGRISCATT: Radar Techniques for Forestry and Agricultural Applications, Final Workshop, The Netherlands.

Dubois, P. C., J. Van Zyl, and T. Engman. 1995. Measuring soil moisture with imaging radars. *IEEE Trans. Geosci. Remote Sens.* 33 (4):915–926. doi: 10.1109/36.406677.

Evans, D. L., and M. O. Smith. 1991. Separation of vegetation and rock signatures in thematic mapper and polarimetric SAR images. *Remote Sens. Environ.* 37 (1):63–75. doi: 10.1016/0034-4257(91)90051-7.

Evans, D. L., T. G. Farr, J. J. van Zyl, and H. A. Zebker. 1988. Radar polarimetry: Analysis tools and applications. *IEEE Trans. Geosci. Remote Sens.* 26 (6):774–788. doi: 10.1109/36.7708.

Faulkner, E. H. 1943. *Plowman's Folly*. New York: Grosset & Dunlap.

Freeman, A., and S. L. Durden. 1998. A three-component scattering model for polarimetric SAR data. *IEEE Trans. Geosci. Remote Sens.* 36 (3):963–973. doi: 10.1109/36.673687.

Fung, A. K., Z. Li, and K. S. Chen. 1992. Backscattering from a randomly rough dielectric surface. *IEEE Trans. Geosci. Remote Sens.* 30 (2):356–368. doi: 10.1109/36.134085.

Galloza, M. S., M. M. Crawford, and G. C. Heathman. 2013. Crop residue modeling and mapping using landsat, ALI, hyperion and airborne remote sensing data. *IEEE J. Sel. Topics Appl. Earth Observ.* 6 (2):446–456. doi: 10.1109/jstars.2012.2222355.

Gausman, H. W., A. H. Gerbermann, C. L. Wiegand, R. W. Leamer, R. R. Rodriguez, and J. R. Noriega. 1975. Reflectance differences between crop residues and bare soils. *Soil Sci. Soc. Am. J.* 39 (4):752–75.

Gowda, P. H., B. J. Dalzell, D. J. Mulla, and F. Kollman. 2001. Mapping tillage practices with landstat thematic mapper based logistic regression models. *J. Soil Water Conserv.* 56 (2):91–96.

Hajnsek, I., E. Pottier, and S. R. Cloude. 2003. Inversion of surface parameters from polarimetric SAR. *IEEE Trans. Geosci. Remote Sens.* 41 (4):727–744. doi: 10.1109/tgrs.2003.810702.

Hasted, J. B. 1973. *Aqueous Dielectrics*. London: Chapman and Hall.

Henderson, T. L., M. F. Baumgardner, D. P. Franzmeier, D. E. Stott, and D. C. Coster. 1992. High dimensional reflectance analysis of soil organic matter. *Soil Sci. Soc. Am. J.* 56 (3):865–872. doi: 10.2136/sssaj1992.03615995005600030031x.

Jackson, T. J., H. McNairn, M. A. Weltz, B. Brisco, and R. Brown. 1997. First order surface roughness correction of active microwave observations for estimating soil moisture. *IEEE Trans. Geosci. Remote Sens.* 35 (4):1065–1068. doi: 10.1109/36.602548.

Jackson, T. J., and P. E. O'Neill. 1991. Microwave emission and crop residues. *Remote Sens. Environ.* 36 (2):129–136. doi: 10.1016/0034-4257(91)90035-5.

Jiang, D., J. Du, K. Song, B. Zhao, Y. Zhang, and W. Zhang. 2023. Classification of conservation tillage using enhanced spatial and temporal adaptive reflectance fusion model. *Remote Sens.* 15 (2):508. doi: 10.3390/rs15020508.

Karlen, D. L., N. C. Wollenhaupt, D. C. Erbach, E. C. Berry, J. B. Swan, N. S. Eash, and J. L. Jordahl. 1994. Long-term tillage effects on soil quality. *Soil Tillage Res.* 32 (4):313–327. doi: 10.1016/0167-1987(94)00427-G.

Kokaly, R. F., R. N. Clark, G. A. Swayze, K. E. Livo, T. M. Hoefen, N. C. Pearson, R. A. Wise, W. M. Benzel, H. A. Lowers, R. L. Driscoll, and A. J. Klein. 2017. USGS Spectral Library Version 7: U.S. Geological Survey Data Series 1035, 61 p, ISSN 2327-638X, doi: 10.3133/ds1035.

Laflen, J. M., M. Amemiya, and E. A. Hintz. 1981. Measuring crop residue cover. *J. Soil Water Conserv.* 36 (6):341–343.

Lal, R. 2004. Soil carbon sequestration to mitigate climate change. *Geoderma* 123 (1–2):1–22. doi: 10.1016/j.geoderma.2004.01.032.

Langman, A., and M. R. Inggs. 1994. The use of polarimetry in subsurface radar. Paper presented at the Geoscience and Remote Sensing Symposium, 1994. IGARSS '94. Surface and Atmospheric Remote Sensing: Technologies, Data Analysis and Interpretation, California Institute of Technology in Pasadena, California, 8–12 Aug.

Lewis, A. J., and F. M. Henderson. 1998. Radar fundamentals: The Geoscience perspective. In *Principles and Applications of Imaging Radar. Manual of Remote Sensing*, edited by F. M. Henderson and A. J. Lewis. New York: John Wiley & Sons, Inc.

Li, Y., T. Liu, G. Lampropoulos, H. McNairn, J. Shang, and R. Touzi. 2014. RADARSAT-2 POLInSAR coherence optimization for agriculture crop change detection. Paper presented at the Proceedings in International Geoscience and Remote Sensing Symposium (IGARSS 2014), Quebec City, Quebec, Canada.

Liu, Y, P. Rao, W. Zhou, B. Singh, A. K. Srivastava, S. P. Poonia, et al. 2022. Using Sentinel-1, Sentinel-2, and Planet satellite data to map field-level tillage practices in smallholder systems. *PLoS ONE* 17 (11):e0277425. doi: 10.1371/journal.pone.0277425.

Luotamo, M., M. Yli-Heikkilä, and A. Klami. 2022. Density estimates as representations of agricultural fields for remote sensing-based monitoring of tillage and vegetation cover. *Appl. Sci.* 12 (2):679. doi: 10.3390/app12020679.

Mandal, A., A. Majumder, S. S. Dhaliwal, A. S. Toor, P. K. Mani, R. K. Naresh, R. K. Gupta, and T. Mitran. 2022. Impact of agricultural management practices on soil carbon sequestration and its monitoring through simulation models and remote sensing techniques: A review. *Crit. Rev. Environ. Sci. Technol.* 52 (1):1–49. doi: 10.1080/10643389.2020.1811590.

Martins, R. N., M. F. Portes, H. M. Fialho e Moraes, M. R. Furtado Junior, J. T. Fim Rosas, W. D. A. Orlando, Jr. 2021. Influence of tillage systems on soil physical properties, spectral response and yield of the bean crop. *Remote Sens. Appl.: Soc. Envir.* 22:100517, ISSN 2352–9385, doi: 10.1016/j.rsase.2021.100517. (www.sciencedirect.com/science/article/pii/S2352938521000537).

McNairn, H., and B. Brisco. 2004. The application of C-band polarimetric SAR for agriculture: A review. *Can. J. Remote Sens.* 30 (3):525–542. doi: 10.5589/m03-068.

McNairn, H., C. Duguay, B. Brisco, and T. J. Pultz. 2002. The effect of soil and crop residue characteristics on polarimetric radar response. *Remote Sens. Environ.* 80 (2):308–320. doi: 10.1016/s0034-4257(01)00312-1.

McNairn, H., C. Duguay, J. Boisvert, E. Huffman, and B. Brisco. 2001. Defining the sensitivity of multi-frequency and multi-polarized radar backscatter to post-harvest crop residue. *Can. J. Remote Sens.* 27 (3):247–263.

McNairn, H., D. Wood, Q. H. J. Gwyn, and R. J. Brown. 1998. Mapping tillage and crop residue management practices with RADARSAT. *Can. J. Remote Sens.* 24:28–35.

McNairn, H., J. B. Boisvert, D. J. Major, Q. H. J. Gwyn, R. J. Brown, and A. M. Smith. 1996. Identification of agricultural tillage practices from C-band radar backscatter. *Can. J. Remote Sens.* 22:154–162.

McNairn, H., and R. Protz. 1993. Mapping corn residue cover on agricultural fields in Oxford County, Ontario, using Thematic Mapper. *Can. J. Remote Sens.* 19:152–158.

Merzouki, A., H. McNairn, and A. Pacheco. 2010. Evaluation of the Dubois, Oh, and IEM radar backscatter models over agricultural fields using C-band RADARSAT-2 SAR image data. *Can. J. Remote Sens.* 36 (S2):S274–S86. doi: 10.5589/m10-055.

Merzouki, A., H. McNairn, and A. Pacheco. 2011. Mapping soil moisture using RADARSAT-2 data and local autocorrelation statistics. *IEEE J. Sel. Topics Appl. Earth Observ.* 4 (1):1–10.

Milisavljević, N., D. Closson, and I. Bloch. 2010. Detecting human-induced scene changes using coherent change detection in SAR images. Paper presented at the ISPRS TC VII Symposium-100 Years ISPRS, Vienna, Austria, 5–7 July.

Morrison, J. E., C.-H. Huang, D. T. Lightle, and C. S. T. Daughtry. 1993. Residue measurement techniques. *J. Soil Water Conserv.* 48 (6):478–483.

Nagler, P. L., C. S. T. Daughtry, and S. N. Goward. 2000. Plant litter and soil reflectance. *Remote Sens. Environ.* 71 (2):207–215. doi: 10.1016/S0034-4257(99)00082-6.

Oh, Y. 2004. Quantitative retrieval of soil moisture content and surface roughness from multipolarized radar observations of bare soil surfaces. *IEEE Trans. Geosci. Remote Sens.* 42 (3):596–601. doi: 10.1109/tgrs.2003.821065.

Oh, Y., K. Sarabandi, and F. T. Ulaby. 1992. An empirical model and an inversion technique for radar scattering from bare soil surfaces. *IEEE Trans. Geosci. Remote Sens.* 30 (2):370–381. doi: 10.1109/36.134086.

Pacheco, A., and H. McNairn. 2010. Evaluating multispectral remote sensing and spectral unmixing analysis for crop residue mapping. *Remote Sens. Environ.* 114 (10):2219–2228. doi: 10.1016/j.rse.2010.04.024.

Pacheco, A. M., H. McNairn, and A. Merzouki. 2010. Evaluating TerraSAR-X for the identification of tillage occurrence over an agricultural area in Canada. Proceedings Volume 7824, Remote Sensing for Agriculture, Ecosystems, and Hydrology XII, SPIE Remote Sensing, Toulouse, France. doi: 10.1117/12.868218.

Panciera, R., F. MacGill, M. Tanase, K. Lowell, and J. Walker. 2013. Sensitivity of TerraSAR-X-band data to surface parameters in bare agricultural areas. Paper presented at the IEEE 2013 International Geoscience and Remote Sensing Symposium (IGARSS), 21–26 July.

Papadopoulos, G., A. Mavroeidis, I. Roussis, I. Kakabouki, P. Stavropoulos, and D. Bilalis. 2023. Evaluation of tillage & fertilization in *Carthamus tinctorius* L. using remote sensing. *Smart Agri. Techn.* 4:100158, ISSN 2772–3755, doi: 10.1016/j.atech.2022.100158. (www.sciencedirect.com/science/article/pii/S2772375522001228).

Rahman, M. M., M. S. Moran, D. P. Thoma, R. Bryant, C. D. Holifield Collins, T. Jackson, B. J. Orr, and M. Tischler. 2008. Mapping surface roughness and soil moisture using multi-angle radar imagery without ancillary data. *Remote Sens. Environ.* 112 (2):391–402. doi: 10.1016/j.rse.2006.10.026.

Rossi, F. S., C. A. da Silva, Jr, J. F. de Oliveira, Jr., P. E. Teodoro, L. S. Shiratsuchi, M. Lima, L. P. R. Teodoro, A. V. Tiago, and G. F. Capristo-Silva. 2020. Identification of tillage for soybean crop by spectro-temporal variables, GEOBIA, and decision tree. *Remote Sens. Appl.: Soc. Environ.* 19:100356, ISSN 2352–9385, doi: 10.1016/j.rsase.2020.100356. (www.sciencedirect.com/science/article/pii/S2352938520300641).

Rouse, J. W., Jr., R. H. Haas, D. W. Deering, and J. A. Schell. 1973. Monitoring the vernal advancement and retrogradation (green wave effect) of natural vegetation. In *93*. College Station, TX, USA: Texas A&M University.

Sabins, F. F. 1986. *Remote Sensing: Principles and Interpretation*. San Francisco: Freeman.

Samui, P., P. H. Gowda, T. Oommen, T. A. Howell, T. H. Marek, and D. O. Porter. 2012. Statistical learning algorithms for identifying contrasting tillage practices with Landsat Thematic Mapper data. *Int. J. Remote Sens.* 33 (18):5732–5745. doi: 10.1080/01431161.2012.671555.

Serbin, G., C. S. T. Daughtry, E. R. Hunt, Jr., D. J. Brown, and G. W. McCarty. 2009b. Effect of soil spectral properties on remote sensing of crop residue cover. *Soil Sci. Soc. Am. J.* 73 (5):1545–1558. doi: 10.2136/sssaj2008.0311.

Serbin, G., C. S. T. Daughtry, E. R. Hunt, Jr., J. B. Reeves, III, and D. J. Brown. 2009a. Effects of soil composition and mineralogy on remote sensing of crop residue cover. *Remote Sens. Environ.* 113 (1):224–238.

Serbin, G., E. R. Hunt, Jr., C. S. T. Daughtry, and G. W. McCarty. 2013. Assessment of spectral indices for cover estimation of senescent vegetation. *Remote Sens. Lett.* 4 (6):552–560. doi: 10.1080/2150704x.2013.767478.

Serbin, G., E. R. Hunt, Jr., C. S. T. Daughtry, G. W. McCarty, and P. C. Doraiswamy. 2009c. An improved ASTER index for remote sensing of crop residue. *Remote Sens.* 1 (4):971–991.

Slaton, M. R., E. R. Hunt, Jr., and W. K. Smith. 2001. Estimating near-infrared leaf reflectance from leaf structural characteristics. *Am. J. Bot.* 88 (2):278–284. doi: 10.2307/2657018.

Sokol, J., H. NcNairn, and T. J. Pultz. 2004. Case studies demonstrating the hydrological applications of C-band multipolarized and polarimetric SAR. *Can. J. Remote Sens.* 30 (3):470–483. doi: 10.5589/m03-073.

South, S., J. Qi, and D. P. Lusch. 2004. Optimal classification methods for mapping agricultural tillage practices. *Remote Sens. Environ.* 91 (1):90–97. doi: 10.1016/j.rse.2004.03.001.

Stoner, E. R., and M. F. Baumgardner. 1981. Characteristic variations in reflectance of surface soils1. *Soil Sci. Soc. Am. J.* 45 (6):1161–1165. doi: 10.2136/sssaj1981.03615995004500060031x.

Stoner, E. R., M. F. Baumgardner, R. A. Weismiller, L. L. Biehl, and B. F. Robinson. 1980. Extension of laboratory-measured soil spectra to field conditions1. *Soil Sci. Soc. Am. J.* 44 (3):572–574. doi: 10.2136/sssaj1980.03615995004400030028x.

Streck, N. A., D. Rundquist, and J. Connot. 2002. Estimating residueal wheat dry matter from remote sensing measurements. *Photogramm. Eng. Remote Sens.* 68:1193–1201.

Sudheer, K. P., P. Gowda, I. Chaubey, and T. Howell. 2010. Artificial neural network approach for mapping contrasting tillage practices. *Remote Sens.* 2 (2):579–590.

Thoma, D. P., S. C. Gupta, and M. E. Bauer. 2004. Evaluation of optical remote sensing models for crop residue cover assessment. *J. Soil Water Conserv.* 59 (5):224–233.

Thomas, J. R., and H. W. Gausman. 1977. Leaf reflectance vs. leaf chlorophyll and carotenoid concentrations for Eight crops1. *Agron. J.* 69 (5):799–802. doi: 10.2134/agronj1977.00021962006900050017x.

Touzi, R., S. Goze, T. Le Toan, A. Lopes, and E. Mougin. 1992. Polarimetric discriminators for SAR images. *IEEE Trans. Geosci. Remote Sens.* 30 (5):973–980. doi: 10.1109/36.175332.

Ulaby, F. T., D. Held, M. C. Donson, K. C. McDonald, and T. B. A. Senior. 1987. Relating polaization phase difference of SAR signals to scene properties. *IEEE Trans. Geosci. Remote Sens.* GE-25 (1):83–92. doi: 10.1109/tgrs.1987.289784.

Ulaby, F. T., R. K. Moore, and A. K. Fung. 1986. *MicrowaveRemote Sensing: Active and Passive*. Vol. 3. Reading, MA: Addison-Wesley.

USDA-NRCS. 2014 The Soil Tillage Intensity Rating (STIR). www.nrcs.usda.gov/Internet/FSE_DOCUMENTS/stelprdb1119754.pdf (accessed February 28, 2014).

van Deventer, A. P., A. D. Ward, P. H. Gowda, and J. G. Lyon. 1997. Using thematic mapper data to identify contrasting soil plains and tillage practices. *Photogramm. Eng. Remote Sens.* 63 (1):87–93.

Watts, J. D., S. L. Powell, R. L. Lawrence, and T. Hilker. 2011. Improved classification of conservation tillage adoption using high temporal and synthetic satellite imagery. *Remote Sens. Environ.* 115 (1):66–75. doi: 10.1016/j.rse.2010.08.005.

West, T. O., and G. Marland. 2002. A synthesis of carbon sequestration, carbon emissions, and net carbon flux in agriculture: Comparing tillage practices in the United States. *Agric., Ecosyst. Environ.* 91 (1–3):217–232. doi: 10.1016/S0167-8809(01)00233-X.

Workman, J. J., and L. Weyer. 2008. *Practical Guide to Interpretive Near-Infrared Spectroscopy*. Boca Raton, FL: Taylor & Francis Group.

Xiang, X., J. Du, D. Jacinthe, B. Zhao, H. Zhou, H. Liu, and K. Song. 2022. Integration of tillage indices and textural features of Sentinel-2A multispectral images for maize residue cover estimation. *Soil Till. Res.* 221:105405, ISSN 0167–1987, doi: 10.1016/j.still.2022.105405. (www.sciencedirect.com/science/article/pii/S0167198722000915).

Yeom, J., J. Jung, A. Chang, A. Ashapure, M. Maeda, A. Maeda, and J. Landivar. 2019. Comparison of vegetation indices derived from UAV data for differentiation of tillage effects in agriculture. *Remote Sens.* 11 (13):1548. doi: 10.3390/rs11131548.

Yilmaz, M. T., E. R. Hunt, Jr., and T. J. Jackson. 2008. Remote sensing of vegetation water content from equivalent water thickness using satellite imagery. *Remote Sens. Environ.* 112 (5):2514–2522. doi: 10.1016/j.rse.2007.11.014.

Zhang, W., Q. Yu, H. Tang, J. Liu, and W. Wu. 2024. Conservation tillage mapping and monitoring using remote sensing. *Comput Electron Agr* 218:108705, ISSN 0168–1699, doi: 10.1016/j.compag.2024.108705. (www.sciencedirect.com/science/article/pii/S0168169924000966).

Zheng, B., J. B. Campbell, G. Serbin, and C. S. T. Daughtry. 2013a. Multi-temporal remote sensing of crop residue cover and tillage practices: A validation of the minNDTI strategy in the United States. *J. Soil Water Conserv.* 68 (2):120–131.

Zheng, B., J. B. Campbell, G. Serbin, and J. M. Galbraith. 2014. Remote sensing of crop residue and tillage practices: Present capabilities and future prospects. *Soil Tillage Res.* 138 (0):26–34. doi: 10.1016/j.still.2013.12.008.

Zheng, B., J. B. Campbell, and K. M. de Beurs. 2012. Remote sensing of crop residue cover using multitemporal Landsat imagery. *Remote Sens. Environ.* 117 (0):177–183. doi: 10.1016/j.rse.2011.08.016.

Zheng, B., J. B. Campbell, Y. Shao, and R. H. Wynne. 2013b. Broad-scale monitoring of tillage practices using sequential landsat imagery. *Soil Sci. Soc. Am. J.* 77 (5):1755–1764. doi: 10.2136/sssaj2013.03.0108.

Zhu, X., J. Chen, F. Gao, X. Chen, and J. G. Masek. 2010. An enhanced spatial and temporal adaptive reflectance fusion model for complex heterogeneous regions. *Remote Sens. Environ.* 114 (11):2610–2623. doi: 10.1016/j.rse.2010.05.032.

10 Hyperspectral Remote Sensing for Terrestrial Applications

Prasad S. Thenkabail, Itiya Aneece, Pardhasaradhi Teluguntla, Richa Upadhyay, Asfa Siddiqui, Justin George Kalambukattu, Suresh Kumar, Murali Krishna Gumma, and Venkateswarlu Dheeravath

ACRONYMS AND DEFINITIONS

ASTER	Advanced Spaceborne Thermal Emission and Reflection Radiometer
CDL	Cropland Data Layer
CHRIS	Compact High Resolution Imaging Spectrometer
DLR	German Aerospace Center
EnMAP	Environmental Mapping and Analysis Program
ENVISAT	Environmental Satellite
FAO	Food and Agriculture Organization of the United Nations
GHISA	Global Hyperspectral Imaging Spectral-library of Agricultural crops
GPS	Global Positioning System
HNB	Hyperspectral Narrowbands
HVI	Hyperspectral Vegetation Indices
LAI	Leaf Area Index
LiDAR	Light Detection and Ranging
LST	Land Surface Temperature
LUE	Light Use Efficiency
NASA	National Aeronautics and Space Administration
NDVI	Normalized Difference Vegetation Index
NIR	Near-Infrared
VNIR	Visible and Near-Infrared
OMI	Ozone Monitoring Instrument
PRI	Photochemical Reflectance Index
PROBA	Project for On Board Autonomy
SAR	Synthetic Aperture Radar
SMTs	Spectral Matching Techniques
SVM	Support Vector Machines
SWIR	Shortwave Infrared
TOA	Top of Atmosphere
UASs	Unmanned Aircraft Systems
USDA	United States Department of Agriculture

10.1 INTRODUCTION

Remote sensing data are considered hyperspectral when the data are gathered from numerous wavebands, contiguously over an entire range of the spectrum (e.g., 400–2,500 nm). Goetz (1992) defined

DOI: 10.1201/9781003541165-12

hyperspectral remote sensing as: "The acquisition of images in hundreds of registered, contiguous spectral bands such that for each picture element of an image it is possible to derive a complete reflectance spectrum." More recently, Jensen (2004) defines hyperspectral remote sensing as: "The simultaneous acquisition of images in many relatively narrow, contiguous and/or noncontiguous spectral bands throughout the ultraviolet, visible, and infrared portions of the electromagnetic spectrum."

Overall, the three key factors in considering data to be hyperspectral are:

1. *Contiguity in data collection*
   Data are collected contiguously over a spectral range (e.g., wavebands spread across 400 to 2500 nm).
2. *Number of wavebands*
   The number of wavebands by itself does not make the data hyperspectral. A strict definition of hyperspectral data is having contiguous spectral bands throughout the electromagnetic spectrum (e.g., 400–2,500 nm) in narrow bandwidths.
3. *Bandwidths*
   Hyperspectral data are collected in sufficiently narrow bandwidths (≤ 10 nm). At times, at specific locations, a very narrow band of even 0.5 nm bandwidth may be required for measuring features such as light-use efficiency (531 nm) from field spectroradiometers. However, too narrow a bandwidth (e.g., less than 1 nm) may lead to low signal to noise ratio and provide noisy bands. In summary:

Remote sensing data are called hyperspectral (or imaging spectroscopy from satellites, uncrewed aircraft systems, or airborne platforms) when the data are collected contiguously over a spectral range such as visible, near-infrared, short-wave infrared, and thermal (e.g., 400–12,000 nm), preferably in narrow bandwidths of 10 nm or less and in reasonably high number of bands (e.g., 30, 200, or 1,000). The number of bands is less important than contiguity of narrow bands dispersed across the spectral range. The terms multispectral, hyperspectral, and superspectral can be differentiated as follows: (1) multispectral data consist of a set of spectral bands with broad bandwidth (e.g., 30 nm or wider) along the electromagnetic spectrum (e.g., 400–12,500 nm). For example, Landsat OLI has 11 bands spread across 400 to 12,500 nm. (2) Hyperspectral data are defined earlier. (3) Superspectral data fall between multispectral and hyperspectral. For example, Landsat Next will have data acquired in 26 spectral bands from 400–12,500 nm with varying bandwidths, and is considered superspectral.

Such a definition will meet many requirements and expectations of hyperspectral data.

Hyperspectral remote sensing is also referred to as imaging spectroscopy since data for each pixel are acquired in numerous contiguous wavebands resulting in: (a) 3D image cube, and (b) hyperspectral signatures. The various forms and characteristics of hyperspectral data (imaging spectroscopy) are illustrated in Figures 10.1 to 10.9. The distinction between hyperspectral and multispectral is based on the narrowness and contiguous nature of the measurements, not the "number of bands" (Qi et al., 2012).

The overarching goal of this chapter is to introduce hyperspectral remote sensing, its characteristics, data mining approaches, and methods of analysis for terrestrial application. *First*, hyperspectral sensors from various platforms are noted. *Second*, data mining to overcome data redundancy are enumerated. *Third*, the concept of the Hughes phenomenon and the need to overcome it are highlighted. *Fourth*, hyperspectral data analysis methods are presented and discussed. *Fifth*, the methods section includes approaches to optimal band selection, deriving hyperspectral vegetation indices, and various classification methods.

10.2 HYPERSPECTRAL SENSORS

Hyperspectral data (or imaging spectroscopy) are gathered from various sensors. These are briefly discussed in the following sections.

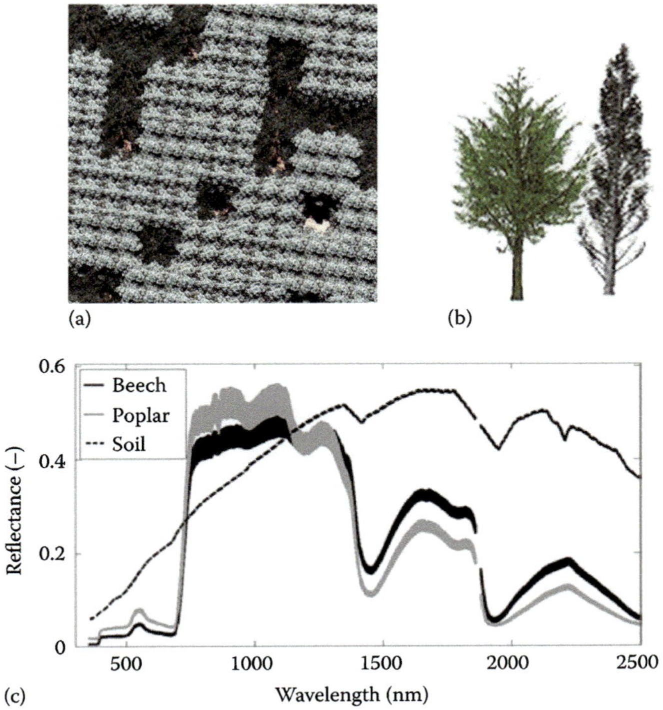

FIGURE 10.1 Tree-spectra collected using Analytic Spectral Devices (ASD) Fieldspec JR spectroradiometer. Hyperspectral shape-based unmixing to improve intra- and interclass variability for forest and agro-ecosystem monitoring. A detail of a 30 × 30 m image pixel of the virtual forest consisting of two species with a different structure, with 10% of the trees removed to include gaps in the canopy (a). An example of a virtual tree for the two species, used to build up the forest (b), while the spectral variability of the two species and the soil is given (c). (Source: Tits et al., 2012.)

10.2.1 Spectroradiometers

The most common and widely used over the last 50 years is the handheld or platform-mounted spectroradiometer (Thenkabail et al., 2021). Typically, spectroradiometers gather hyperspectral data in ~1-nm-wide bands over the entire spectral range (e.g., 400–13,500 nm or 400–2,500 nm). For example, Figure 10.1 illustrates the hyperspectral data gathered for Beech versus Poplar forests (Tits et al., 2012) based on FieldSpec Pro FR spectroradiometer manufactured by Analytical Spectral Devices (ASD) (also see Stagakis et al. (2016); Tanner (2013); Thomas (2012); Zhang (2012)). Data are acquired over 400–2500 nm at every 1 nm bandwidth. Gathering spectra at any given location involves optimizing the integration time (typically set at 17 ms), providing foreoptic information, recording dark current, conducting optimization, collecting white reference reflectance, and then obtaining target reflectance at a set field of view (FOV) such as 18° (Julitta et al., 2016; Thenkabail et al., 2004a). Data are either in radiance (watts per square meter steradian micrometer, W m^{-2} sr^{-1} μm^{-1}) or reflectance factor, as shown in Figure 10.1, or in percentage. Note these spectroradiometer data are not imaging data.

Other hand-held sensors include the Ocean Optic USB4000 (Middleton et al., 2010), UniSpec DC Spectrometer Analysis System (Davidson et al., 2016), and those made by Spectral Evolution (Kandel et al., 2019).

10.2.2 Airborne Hyperspectral Remote Sensing

An airborne hyperspectral remote sensing platform is the next most common hyperspectral data collection format, which has a history of over 30 years. The most common of these is the airborne visible/infrared imaging spectrometer (AVIRIS) by the National Aeronautics and Space Administration's (NASA's) Jet Propulsion Laboratory (JPL). As an imaging spectrometer, AVIRIS gathers data in 614-pixel swath, in 224 bands over 400–2,500 nm. The data can be constituted as an image cube (e.g., Figure 10.2; Guo et al. (2013)). Figure 10.2 shows hyperspectral imaging data gathered by AVIRIS over an agricultural area. The hyperspectral signatures of tilled versus untilled corn and soybeans, and a few other crops, are illustrated by Guo et al. (2013) (Figure 10.2). Spectral reflectivity of no-till corn fields is highest in the red region (around 680 nm, Band 30). In contrast grass/pasture and woods are lowest around 680 nm and reflectivity is highest for these land cover types in the near-infrared region (760–900 nm, Bands 42–56). The healthy grass/pasture and woods also absorb heavily around the 960–970 nm (Bands 62–63) range due to the presence of moisture/water. Many other unique features can be observed qualitatively by individuals trained in imaging spectroscopy. For example, the corn-no-till class has high reflectance in the visible bands due to the soil background whereas Grass/Pasture has high absorption in that spectral region due to the photosynthetic vegetation.

Another frequently used airborne hyperspectral imager is the Australian HyMap (Buzzi et al., 2014; Hajaj et al., 2023; Riaza et al., 2015). It has 126 wavebands over 400–2500 nm. The data captured by HyMap are illustrated in Figure 10.3 (Andrew and Ustin, 2008). Typical characteristics of healthy vegetation for certain plant species are obvious as described earlier for wavelengths centered in red (680 nm) and NIR (760–900 nm). In contrast the soil and litter have comparable spectra, with litter having higher reflectivity than soil in NIR (760–900 nm) and SWIR (900–1,700 nm) bands. Water absorbs heavily in NIR (760–900 nm) and SWIR (900–1700 nm) and hence the reflectances are very low or zero (Figure 10.3).

Other airborne sensors include NASA's Airborne Visible InfraRed Imaging Spectrometer—Next Generation (AVIRIS-NG) (Agrawal et al., 2022; Bhattacharya et al., 2019; Chaube et al., 2019; Jha et al., 2019; Ratheesh et al., 2019), the United States (US) Hyperspectral Digital Imagery Collection Experiment (HYDICE) (Zhang et al., 2006), Hyperspectral Sensor Surveying (AISA-EAGLE) (Abdel-Rahman et al., 2015; Beamish et al., 2020; Lausch et al., 2015a; Mansour et al., 2012), Compact Airborne Spectrographic Imager (CASI) (Legleiter et al., 2016; Vahtmäe et al., 2021; Xu et al., 2018), AisaEAGLET (Doneus et al., 2014), airborne Portable Remote Imaging SpectroMeter (PRISM) (Erickson et al., 2019; Mouroulis et al., 2014; Thompson et al., 2015), and SpecTIR (Hively et al., 2011).

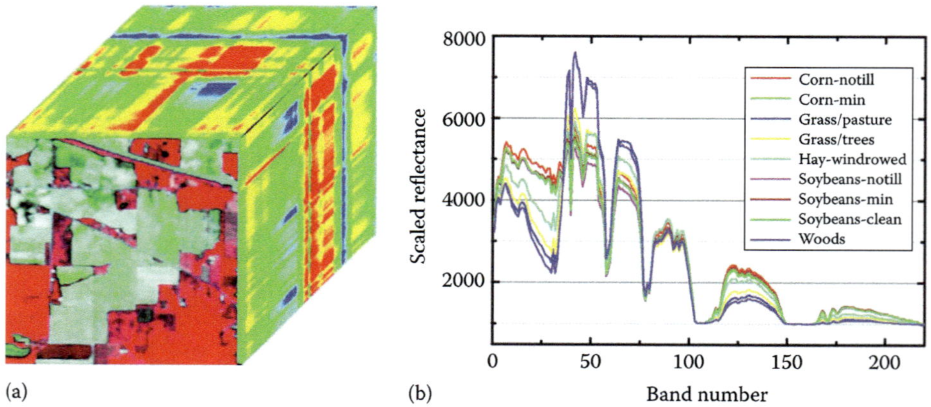

FIGURE 10.2 Land cover spectra. AVIRIS Indian Pines dataset: (a) 3D hyperspectral cube and (b) the scaled reflectance plot. Spectra are for tilled, not tilled (no-till), minimally tilled (min), and weed-managed (clean) fields. (Source: Guo et al., 2013.)

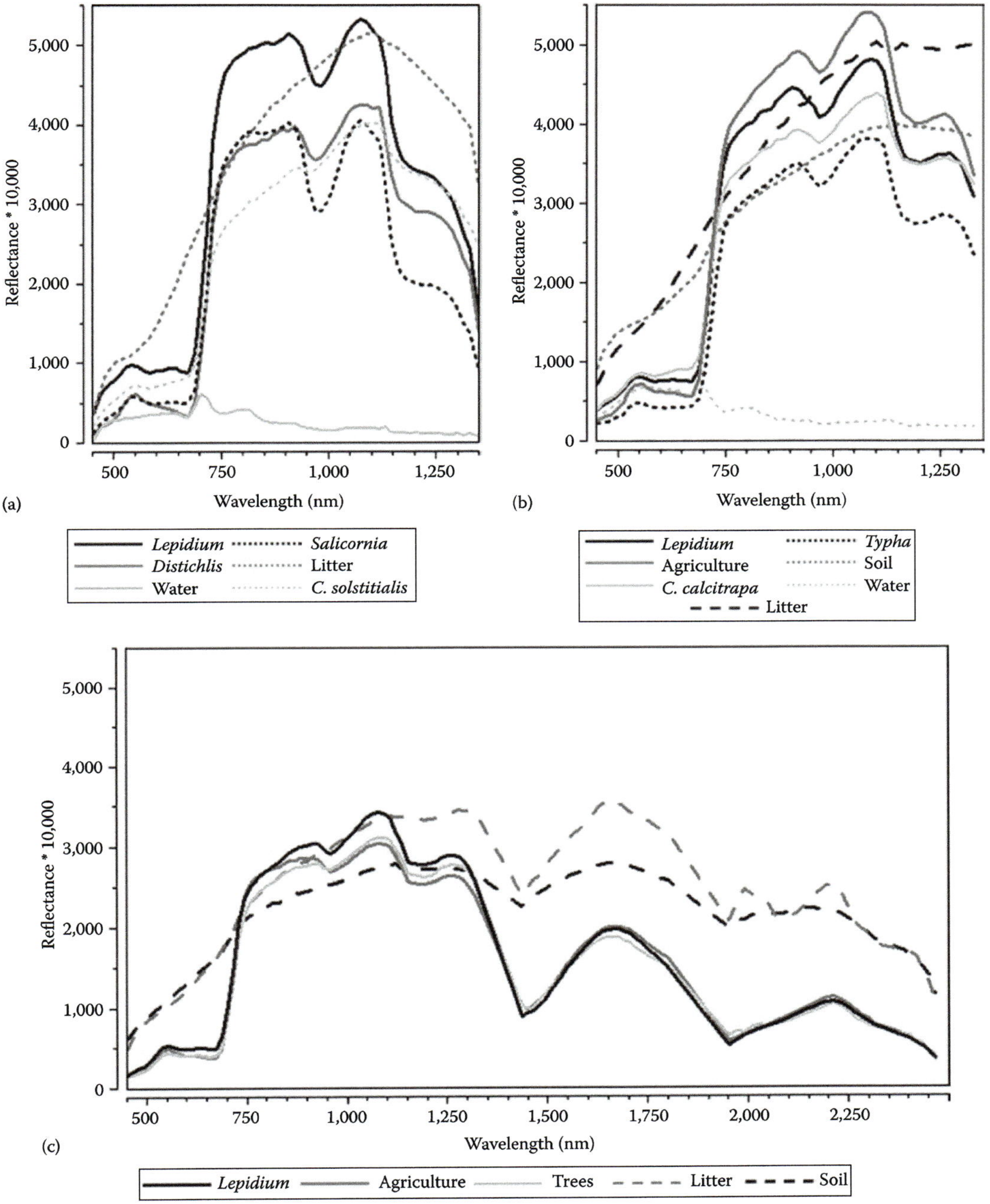

FIGURE 10.3 Reflectance spectra derived from HyMap imagery. Spectra of the dominant species at (a) Rush Ranch, (b) Jepson Prairie, and (c) Consumes River Preserve. These spectra were used as training endmembers for mixture tuned matched filtering (MTMF). (Source: Andrew and Ustin, 2008.)

10.2.3 Spaceborne Hyperspectral Data

Hyperspectral sensors are of increasing interest to the remote sensing community given their inherent advantages over multispectral sensors (Aneece et al., 2022; Marshall and Thenkabail, 2015; Qi et al., 2012; Thenkabail et al., 2012a). In the year 2000, NASA launched the first civilian spaceborne hyperspectral imager called Hyperion onboard the Earth Observing-1 (EO-1) satellite. Hyperion gathered data in 242 bands spread across 400–2,500 nm. Each band is 10 nm wide. Of the original 242 Hyperion bands, 196 are unique and calibrated: bands 8 (427.55 nm) to 57 (925.85 nm) from the visible and near-infrared (VNIR) sensors; and bands 79 (932.72 nm) to 224 (2395.53 nm) from the SWIR sensors (Thenkabail et al., 2004b). The redundant and uncalibrated bands are in the spectral range: 357–417 nm, 936–1,068 nm, and 852–923 nm. The 196 bands are further reduced to 157 bands after removing bands in atmospheric windows affected by such atmospheric components as carbon dioxide and water vapor: 1,306–1,437 nm, 1,790–1,992 nm, and 2,365–2,396 nm ranges, which show low signal and high noise level (Thenkabail et al., 2004b).

Since 2000, Hyperion has acquired over 83,000 images throughout the world (Figure 10.4) that are now freely available from U.S. Geological Survey's (USGS) EarthExplorer (https://earthexplorer.usgs.gov/) and Glovis (https://glovis.usgs.gov/) portals. Each image is 7.5 km by 185 km with a pixel resolution of 30 m. The data cubes composed from these images allow us to derive hyperspectral signature banks of various land cover or cropland themes (e.g., Figure 10.4). Figure 10.5 illustrates two Hyperion images acquired over California and several hyperspectral signatures

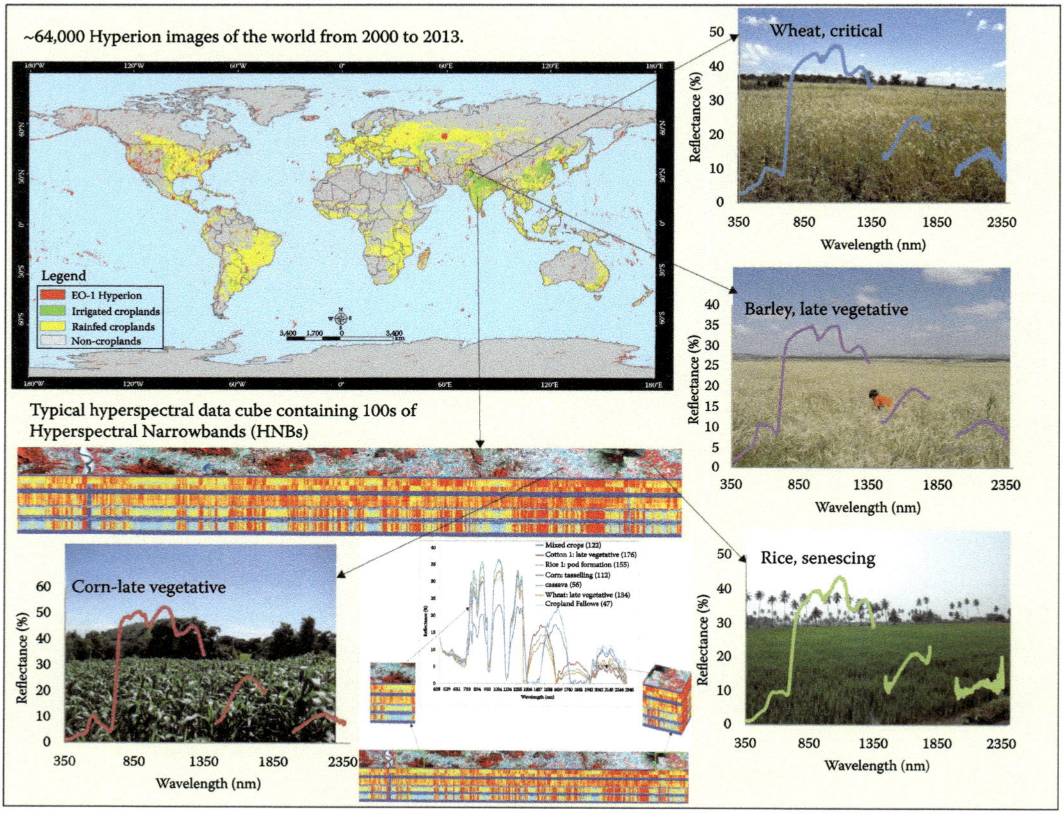

FIGURE 10.4 EO-1 Hyperion. EO-1 Hyperion was the first spaceborne civilian hyperspectral sensor. It was launched in 2000 and acquired ~64,000 land surface images from 2000 to 2014 (see the area covered by Hyperion images marked in red on global image). Each image is 7.5 km by 185 km and has 242 bands over 400–2,500 nm with 30 m pixel resolution. A single such image data cube is shown in the center with spectral signatures derived from the Hyperion sensor shown for a few land cover themes. Typical ASD spectroradiometer gathered hyperspectral data of crops are shown in preceding photos. The gaps in ASD hyperspectral data are in areas of atmospheric windows where data are too noisy and hence deleted. (Author's original contribution.)

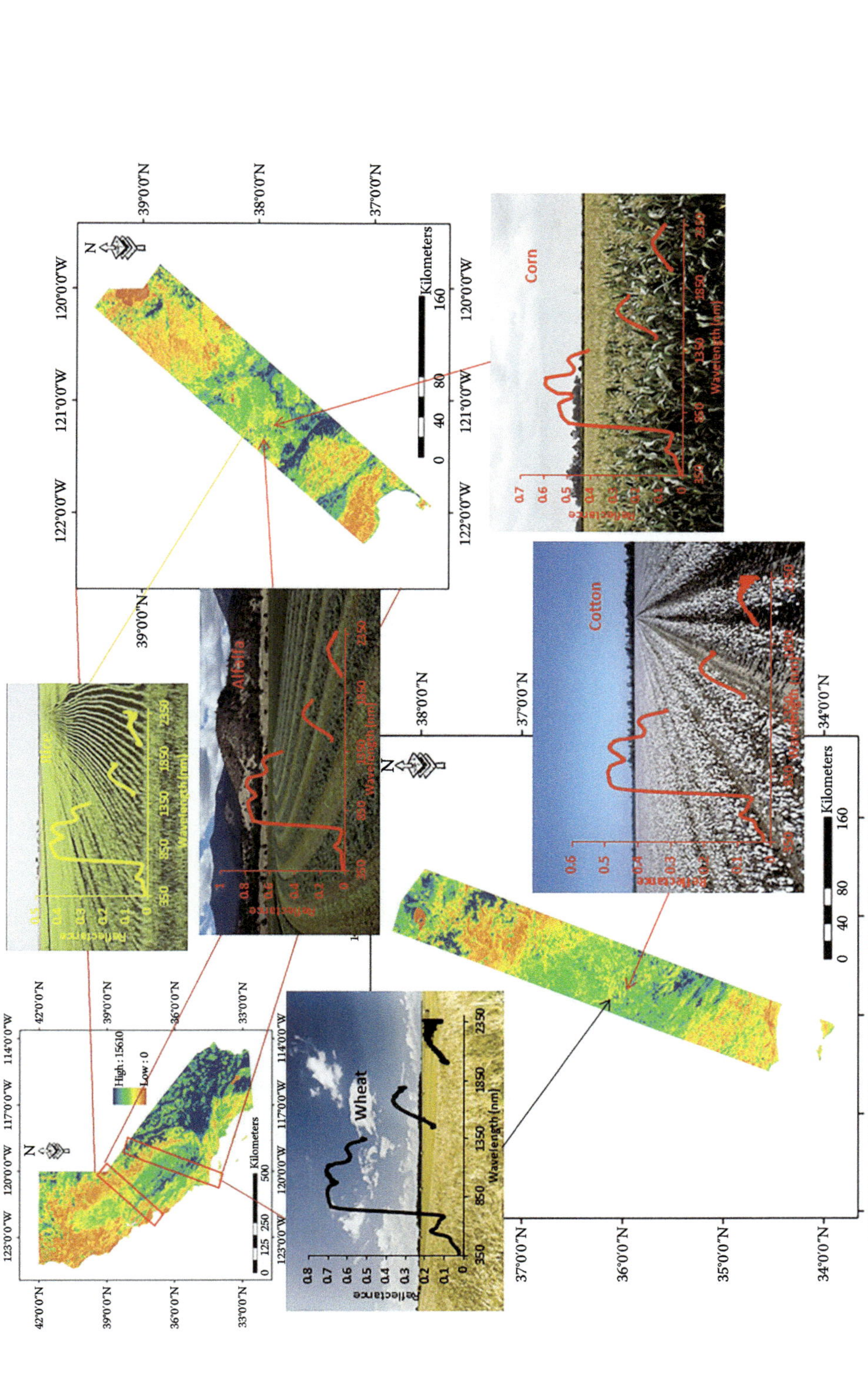

FIGURE 10.5 Hyperspectral signatures of some of the major crops of California. The depicted spectral signatures are representative of the particular crops measured using an ASD spectroradiometer. Two Hyperion images (each of 7.5 km by 185 km) are also illustrated. (Author's original contribution.)

of major crops gathered using an ASD field spectroradiometer. Although decommissioned in 2017, Hyperion has been used for various applications such as agricultural crop classification, chlorophyll and nitrogen estimation, and mineral identification (Aneece and Thenkabail, 2018a; Moharana and Dutta, 2016; Oskouei and Babakan, 2016). Regardless of its low signal to noise ratio, Hyperion was the only publicly available spaceborne hyperspectral sensor available at the time and advanced our understanding of hyperspectral data characteristics and analysis workflows. Similarly, NASA's Hyperspectral Imager for the Coastal Ocean (HICO) and Ozone Monitoring Instrument (OMI), and the European Space Agency's Scanning Imaging Absorption Spectrometer for Atmospheric CHartographY (SCIAMACHY) hyperspectral sensors advanced oceanic and atmospheric research (Table 10.1, Miura and Yoshioka, 2012; Ortenberg, 2012; Qi et al., 2012). Recently launched spaceborne hyperspectral sensors include Germany's Environmental Mapping and Analysis Program (EnMAP) (Bracken et al., 2019; Kokhanovsky et al., 2023; Okujeni et al., 2015), Italy's Compact High Resolution Imaging Spectrometer (CHRIS) onboard the Project for On Board Autonomy (PROBA) satellite (CHRIS PROBA) (Glowienka and Zembol, 2022; Lin et al., 2019; Verrelst et al., 2012), German Aerospace Center (German: Deutsches Zentrum fu¨r Luft- und Raumfahrt-DLR) or DLR's Earth Sensing Imaging Spectrometer (DESIS) (Farmonov et al., 2023; Krutz et al., 2019) onboard the International Space Station (ISS), Italy's PRecursore IperSpettrale della Missione Applicativa (PRISMA) (Pignatti et al., 2013, 2015), NASA's Earth Surface Mineral Dust Source Investigation (EMIT) mission onboard the ISS (Sousa and Small, 2023), and Japanese Hyperspectral Imager Suite also onboard the ISS (HISUI) (Yamamoto et al., 2022). Upcoming hyperspectral sensors include NASA's Surface Biology and Geology mission (formerly known as HyspIRI; Clark (2017); Iqbal et al. (2018); Lee et al. (2015, 2022)), the European Space Agency's (ESA's) Copernicus Hyperspectral Imaging Mission for the Environment (CHIME) (Rast et al., 2021), and the United States' Planet Labs PBC Tanager satellites (www.planet.com/products/hyperspectral/). There are also current initiatives from the private industry and the commercial sector to launch hyperspectral sensors. The spatial, spectral, radiometric, and temporal characteristics of some of the key ocean, atmospheric, and land observation spaceborne hyperspectral data are provided in Table 10.1. For studies discussing the strengths and weaknesses of various sensors for different applications, the reader is referred to Roy et al. (2016), Tripathi and Garg (2023), Marshall et al. (2022), Davies et al. (2023), Aneece and Thenkabail (2021, 2022b), and Muthusamy et al. (2023).

10.2.4 Uncrewed Aircraft Systems (UASs)

Hyperspectral sensors are increasingly carried onboard Uncrewed Aircraft Systems (UASs; Colomina and Molina (2014)), also called proximal remote sensing (Zhao et al., 2021). The UASs are fast evolving as widely used remote sensing platforms. A wide array of UASs (e.g., Figure 10.6) are currently used, including miniature to large UASs with fixed, rotary, flapping, or hybrid wing designs (Liu et al., 2021b). UASs can carry several types of sensors, including RGB, multispectral, hyperspectral, thermal, Light Detection and Ranging (LiDAR), and Synthetic Aperture Radar (SAR) sensors (Liu et al., 2021b).

Some examples of hyperspectral UAS sensors include Micro-Hyperspec X sensors (Dao et al., 2019; Guo et al., 2019), Rikola Hyperspectral camera (Ivushkin et al., 2019; Mozgeris et al., 2018), SOC710-GX (Ad˜ao et al., 2017; Rhee et al., 2018), Specim ImSpector V10 2/3 (Domingues Franceschini et al., 2017; van der Meij et al., 2017), OCI-UAV-1000 (Cahalane et al., 2017; Manfreda et al., 2018), MicroHSI 410-SHARK (Manfreda et al., 2018), and Headwall (Adler-Golden et al., 2017).

10.2.5 Multispectral versus Hyperspectral

While multispectral broadband satellite data from sensors such as the Landsat ETM+ provide a continuous record of the Earth's surface dating back to 1972, hyperspectral sensors such as Hyperion offer many possibilities for visualizations and quantification of terrestrial earth features

TABLE 10.1

Characteristics of Spaceborne Hyperspectral Sensors (Decommissioned, in Orbit, or Planned for Launch) for Ocean, Atmosphere, Land, and Water Applications Compared with ASD Spectroradiometer[a]

Sensor, Satellite[c]	Spatial (m)	Spectral (#)	Swath (km)	Band Range (μm)	Band Widths	Data Points (# per hectares)	Launch (Date)
I. Coastal Hyperspectral Spaceborne Imagers							
1. HICO, ISS USA	90	128	42	353–1080	5.7 μm	0.81	2009–2018
II. Atmosphere/Ozone Hyperspectral Spaceborne Imagers							
1. OMI, Aura USA	13000 × 12000	740	145	270–500	0.45–1 μm	1/16900	2004-present
2. SCIAMACHY, ENVISAT ESA	30000 × 60000	2000	960	212–2384	0.2–1.5 μm	1/180000	2002–2012
III. Land and Water Hyperspectral Spaceborne Imagers							
1. Hyperion, EO-1 NASA USA	30	220 (196[b])	7.5	427.55–925.85 nm	10 nm	11.1	2000–2017
2. CHRIS, PROBA ESA	25	19	17.5	200–1050	1.25–11 μm	16	2001–2022
3. DESIS, ISS Germany	30	235	30	402–1000	2.55 nm (no binning)	11.1	2018-present
4. HISUI, ISS Japan	20m × 30m	185	20	400–2500	10 nm VNIR, 12.5 nm SWIR	11.1	2019-present
5. EnMAP Germany	30	224	30	91 in 420–1000 nm, 133 in 900-2450 nm	6.5, 10 nm	11.1	2022-present
6. PRISMA Italy	30	238	30	400–2505	≤12 nm	11.1	2019-present
7. EMIT, NASA USA	60	285	75	380–2500 nm	7.4 nm	2.77	2022-present
8. SBG NASA USA	30–45, 30–60	210+	30	400–2500 nm, 10.5–11.5 μm and 11.5–12.5 μm	≤10 nm	11.1	2026+

(Continued)

TABLE 10.1 *(Continued)*

Characteristics of Spaceborne Hyperspectral Sensors (Decommissioned, in Orbit, or Planned for Launch) for Ocean, Atmosphere, Land, and Water Applications Compared with ASD Spectroradiometer[a]

Sensor, Satellite[c]	Spatial (m)	Spectral (#)	Swath (km)	Band Range (μm)	Band Widths	Data Points (# per hectares)	Launch (Date)
9. Tanager Planet Labs PBC USA	30	400+		400–2500	5 nm		2023+
10. CHIME, ESA	30	200+	130	400–2500	<10		2028+
IV. Land and Water Handheld Spectroradiometer							
1. ASD spectroradiometer	1134 cm^2 @ 1.2 m Nadir view 18° Field of view	2100	N/A	400–2500 nm	1 nm	88183	last 30+ years

Source: Modified and adopted from Qi et al. (2012); Thenkabail et al. (2014, 2011).

Note:

[a] = information for the table modified and adopted from Qi et al. (2012); Thenkabail et al. (2014, 2011)

[b] = Of the 242 bands, 196 are unique and calibrated. These are: (A) Band 8 (427.55 nm) to band 57 (925.85 nm) that are acquired by visible and near-infrared (VNIR) sensor; and (B) Band 79 (932.72 nm) to band 224 (2395.53 nm) that are acquired by short wave infrared (SWIR) sensor

[c] = HICO = Hyperspectral Imager for the Coastal Ocean onboard International Space Station. OMI = Ozone Monitoring Instrument onboard AURA of NASA; SCIAMACHY (Scanning Imaging Absorption Spectrometer for Atmospheric CHartographY) of ESA; Hyperion EO-1= hyperspectral sensor onboard EO-1= Earth observing 1; CHRIS PROBA = Compact High Resolution Imaging Spectrometer Project for On Board Autonomy satellite of ESA; SBG = Surface Biology and Geology of NASA; DESIS = DLR Earth Sensing Imaging Spectrometer of Germany; EnMAP = Environmental Mapping and Analysis Program of Germany; PRISMA = PRecursore IperSpettrale della Missione Applicativa of Italy; EMIT = Earth Surface Mineral Dust Source Investigation (EMIT) mission of NASA; HISUI = Hyperspectral Imager Suite of Japan, on ISS; CHIME = Copernicus Hyperspectral Imaging Mission for the Environment of ESA. 2= For irradiance ($W\ m^{-2}\ sr^{-1}\ \mu m^{-1}$), see Neckel and Labs (1984).

FIGURE 10.6 Microdrone MD4–1000. Uncrewed Aircraft System (UAS) flying over the experimental crop. (Source: Torres-Sánchez et al., 2014.)

(e.g., Figures 10.7 and 10.8). In Figure 10.7, the depiction of different false color composites (FCC) of Hyperion (e.g., RGB: 843 nm, 680 nm, 547 nm; or RGB: 680 nm, 547 nm, 486 nm and so on) and compared with FCC of Landsat ETM+ bands 4, 3, 2 clearly demonstrates, even by visual observation, the many possibilities that exist with Hyperion. For example, a 7-band Landsat image will provide 21 unique two-band indices (7 × 7= 49 indices—seven indices on diagonal of the matrix divided by two since the values above and below the matrix are transpose of each other). In contrast, a 157 band clean Hyperion image (after being reduced from the original 242 bands by eliminating bands in atmospheric windows and uncalibrated bands) allows for 12,246 unique two-band indices (157 × 157= 24,640 indices—157 indices on diagonal of the matrix divided by two since the values above and below the matrix are the transpose of each other). In Figure 10.8, Hyperion and Landsat spectral data for corn, soybean, and winter wheat are compared from images over Ponca City, Oklahoma acquired on comparable dates during the 2014 growing season. This figure demonstrates the added details provided by the continuous, numerous, narrow bands available in Hyperion data as compared with Landsat broadbands.

10.2.6 Hyperspectral Data: 3D Data Cube Visualization and Spectral Data Characterization

One quick way to visualize hyperspectral data is to create 3D cubes as illustrated by an EO-1 Hyperion image in Figure 10.9. The 3D cube basically is a data layer stack of 242 bands over 400–2,500 nm. Looking through this stack, the same color along the bands 1 to 242 indicates less diversity in data.

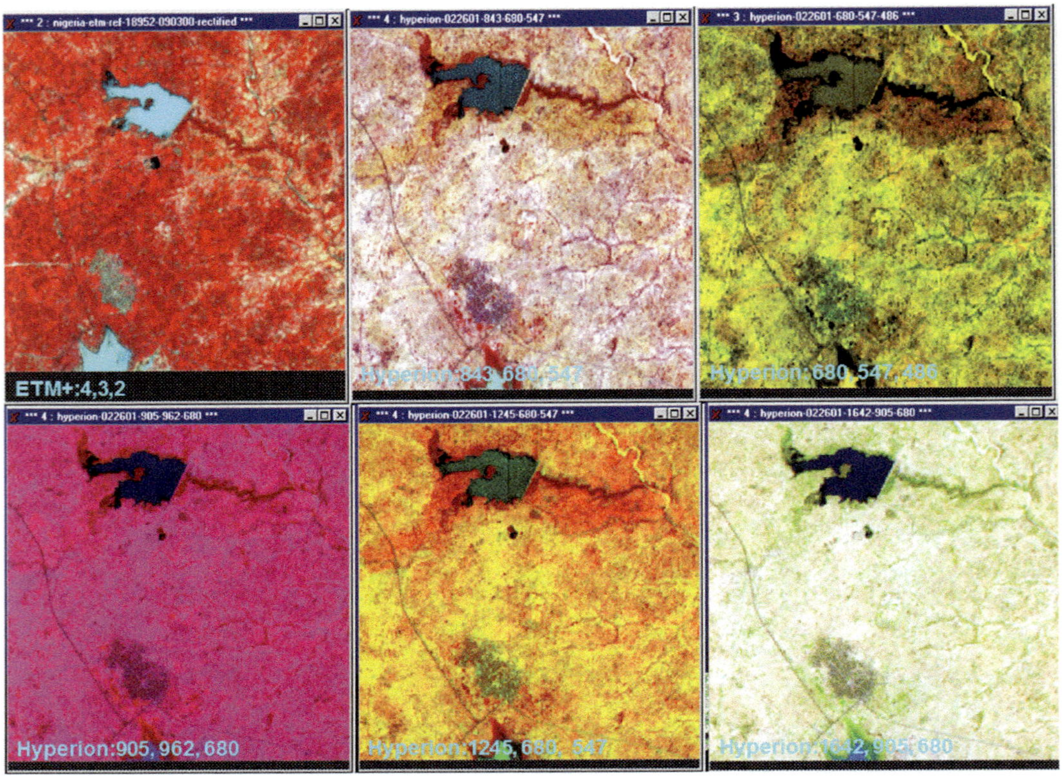

FIGURE 10.7 Hyperion FCCs. Hyperion images displayed in a number of different combinations of false color composites (FCCs) (e.g., wavebands centered at 843 nm, 680 nm, 547 nm, which is NIR, Red, Green as RGB FCC) and compared with classic RGB 4, 3, 2 (NIR, red, green) FCC combination of Landsat ETM+ data on top left. Unlike multispectral data, hyperspectral data offer numerous different opportunities to depict, quantify, and study the Earth. (Author's original contribution.)

The spectral regions with significant diversity are in different colors (e.g., red versus cyan in Figure 10.9). Once preprocessed (see Aneece and Thenkabail, 2018b; Thenkabail et al., 2004b), a click on any pixel in any image analysis software will give surface reflectances in 242 bands, which are then plotted as hyperspectral signatures (e.g., Figure 10.7) and analyzed quantitatively.

10.2.7 Hyperspectral Data Normalization

Here we illustrate hyperspectral data normalization by taking the case of Hyperion data (Aneece and Thenkabail, 2018b; Thenkabail et al., 2004b). The digital numbers (DNs) of the Hyperion level 1 products are 16-bit radiances and are stored as 16-bit signed integers. The DNs were converted to radiances (W m^{-2} sr^{-1} μm^{-1}) using an appropriate scaling factor (e.g., for a Hyperion image dated March 21, 2002, factor: 40 for visible and VNIR, and 80 for SWIR) (Equations 10.1 and 10.2). Users should check the header file of the image they work with to determine the exact scaling factor for their image.

$$\text{Radiance}\left(\text{Wm}^{-2}\text{sr}^{-1}\mu\text{m}^{-1}\right)\text{for VNIR bands} = \frac{DN}{40} \quad (10.1)$$

$$\text{Radiance}\left(\text{Wm}^{-2}\text{sr}^{-1}\mu\text{m}^{-1}\right)\text{for SWIR bands} = \frac{DN}{80} \quad (10.2)$$

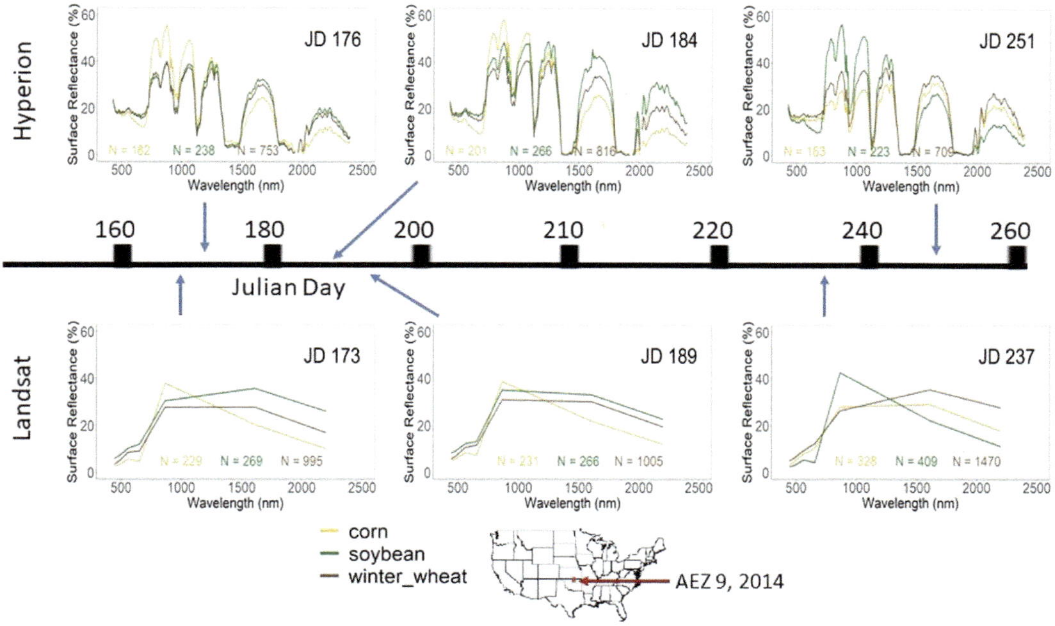

FIGURE 10.8 Spectral-libraries of world crops. Illustrated for crop type and their growth stages for corn crop in agroecological zone (AEZ) 9 of the U.S. during 2014 growing season using hyperspectral narrowband Hyperion and Multispectral broadband Landsat images. (Author's original contribution.)

Radiance to at-sensor top of atmosphere reflectance is then calculated using Equation 10.3:

$$\text{TOA Reflectancce (\%)} = \frac{\pi * L_\lambda * d^2}{ESUN_\lambda * cos\theta_s} \tag{10.3}$$

Where top of atmosphere (TOA) reflectance (at-satellite exo-atmospheric reflectance), $L\lambda$ is the radiance (W m^{-2} sr^{-1} μm^{-1}), d is the Earth to Sun distance in astronomic units at the acquisition date (see Markham and Barker (1987)), $ESUN\lambda$ is irradiance (W m^{-2} sr^{-1} μm^{-1}) or solar flux (Neckel and Labs, 1984), and θs = solar zenith angle

Note: θs is solar zenith angle in degrees (i.e., 90° minus the Sun elevation or Sun angle when the scene was recorded as given in the image header file).

Atmospheric correction methods include: (a) Dark object subtraction technique (Chavez, 1988), (b) Improved dark object subtraction technique (Chavez, 1989), (c) Radiometric normalization technique: Bright and dark object regression (Elvidge et al., 1995), and (d) 6S model (Vermote et al., 2002). Readers with further interest in this topic are referred to Chapters 4 to 8 in Volume I and to Chander et al. (2009).

New generation spaceborne hyperspectral sensors have similar atmospheric correction protocols, with a few crucial differences for the German Aerospace Center (DLR) and Teledyne Brown Engineering DLR Earth Sensing Imaging Spectrometer (DESIS) (Heiden et al., 2019; Krutz et al., 2019; Peschel et al., 2018). DESIS is mounted on the Multi-User System for Earth Sensing (MUSES) platform on the International Space Station (ISS) (Aneece and Thenkabail, 2022b) and is unable to point at the Sun, Moon, or deep space for calibration (Krutz et al., 2019). For this reason, it has in-orbit spectral and radiometric calibration (Krutz et al., 2019). This can make atmospheric correction challenging. Fortunately, these data are available to users in surface reflectance products. The DESIS Level 2A (L2A) data are atmospherically corrected to ground surface reflectance (Heiden et al., 2019) using the DLR's Python Atmospheric COrrection (PACO) algorithm, based on the

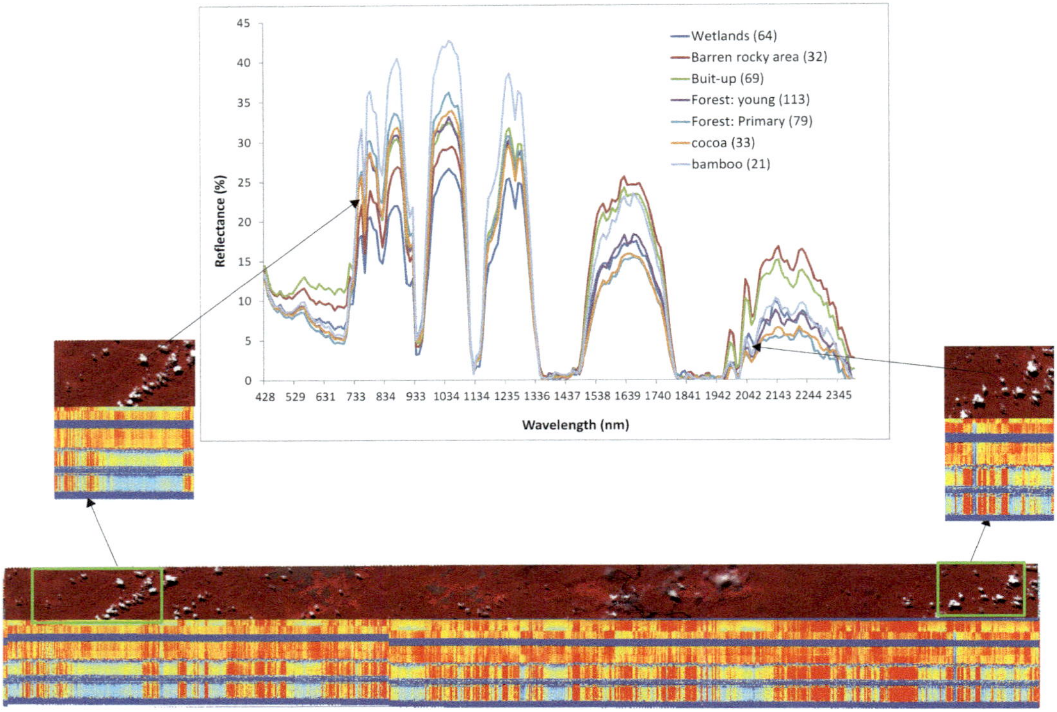

FIGURE 10.9 Hyperspectral signatures. Spectra derived from Hyperion data cube for certain land cover themes. The numbers within parentheses show sample sizes. (Author's original contribution.)

Atmospheric and Topographic CORrection (ATCOR) algorithm (Alonso et al., 2019). The DLR's Environmental Mapping and Analysis Program (EnMAP) polar-orbiting sensor also uses the PACO algorithm for its L2A surface reflectance land product. For a detailed description of the atmospheric correction methods, please see de los Reyes et al. (2022).

The Italian Space Agency's (ASI) PRecursore IperSpettrale della Missione Applicativa (PRISMA) polar-orbiting sensor has an on-board radiometric calibration system including absolute calibration using the Sun, relative calibration using two tungsten lamps, dark calibration using a shutter slit, and on-demand calibration using flat-field calibration and lunar observation (Aneece and Thenkabail, 2022b; Pignatti et al., 2013). Atmospheric correction is done using a method similar to MODerate resolution atmospheric TRANsmission (MODTRAN) (Berk et al., 2008), a simplified radiative transfer equation, and a digital elevation model (Pignatti et al., 2015). The PRISMA Level 2d product consists of geocoded surface reflectance (Loizzo et al., 2016).

Differences in sensor characteristics and preprocessing methods necessitate intersensor comparisons for various applications (Aneece and Thenkabail, 2022b).

10.3 SPECTRAL LIBRARIES

Several spectral libraries exist, like the USGS Spectral Library (Kokaly et al., 2017) with signatures of minerals, rocks, and vegetation. These libraries are useful for various applications including training classification algorithms. Building a spectral library for agricultural crops is challenging due to all the factors that can affect the crop spectra, including crop type, growth stage, growing condition, management, soil type, climate, inputs like nitrogen, potassium, and phosphorous, pests, and diseases (Aneece and Thenkabail, 2019). Information on these factors is needed to make an agricultural

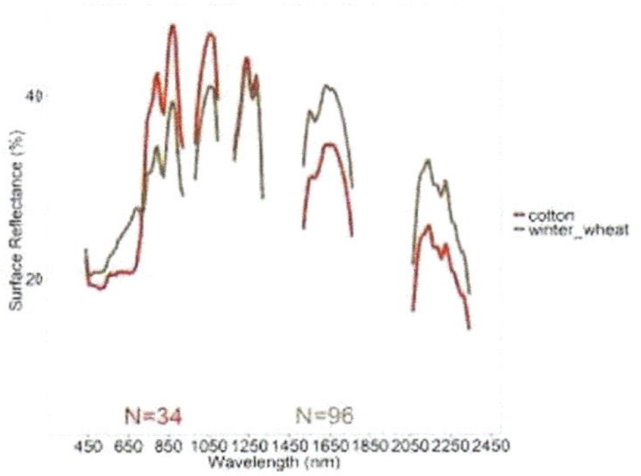

(a) GHISA for cotton and winter wheat in one Hyperion image from AEZ 5.

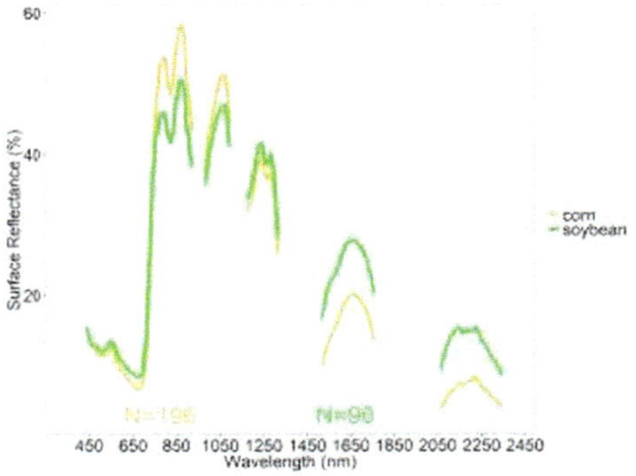

(b) GHISA for corn and soybean in one Hyperion image from AEZ 6.

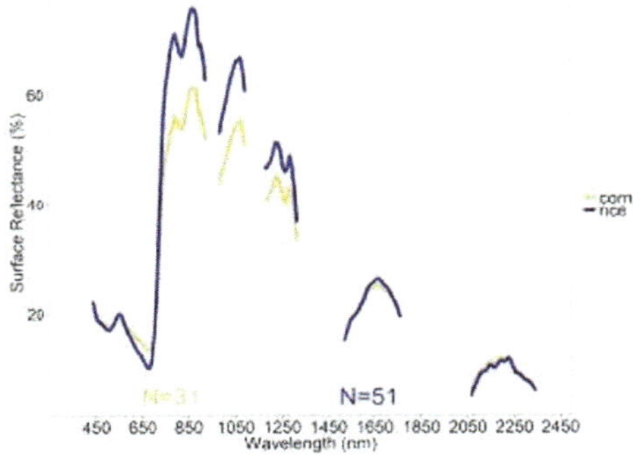

(c) GHISA for corn and rice in one Hyperion image from AEZ 7.

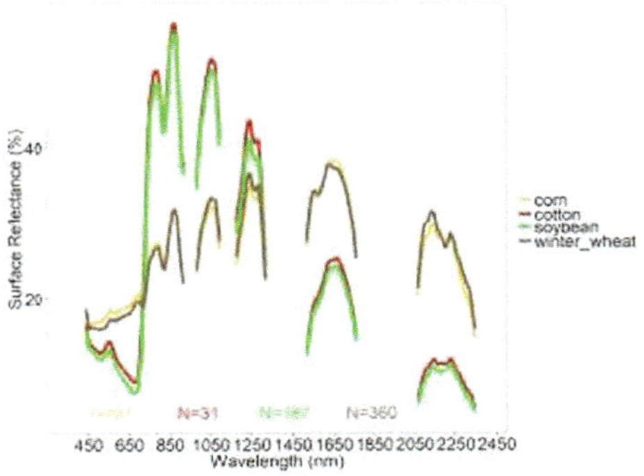

(d) GHISA for corn, cotton, soybean, and winter wheat in one Hyperion image from AEZ 9.

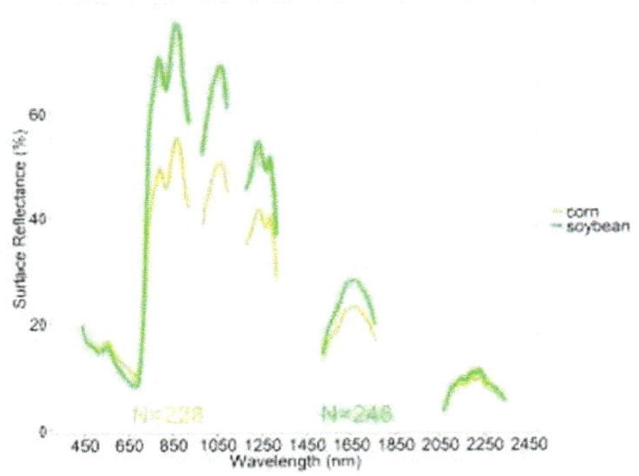

(e) GHISA for corn and soybean in one Hyperion image from AEZ 10.

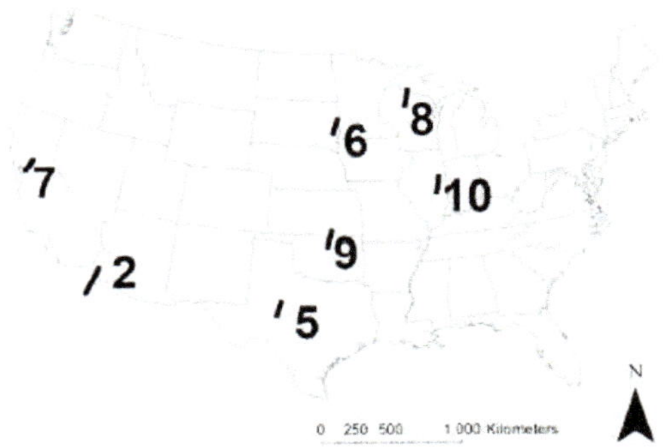

(f) US map showing locations of study areas.

FIGURE 10.10 GHISA CONUS crop types. An illustration of GHISA CONUS, demonstrating EO-1 Hyperion spectra of crop types in different Food and Agriculture (FAO) agroecological zones (AEZs) within the continental USA (CONUS). Hyperion image tracks are shown in (f). (Source: Aneece and Thenkabail, 2019.)

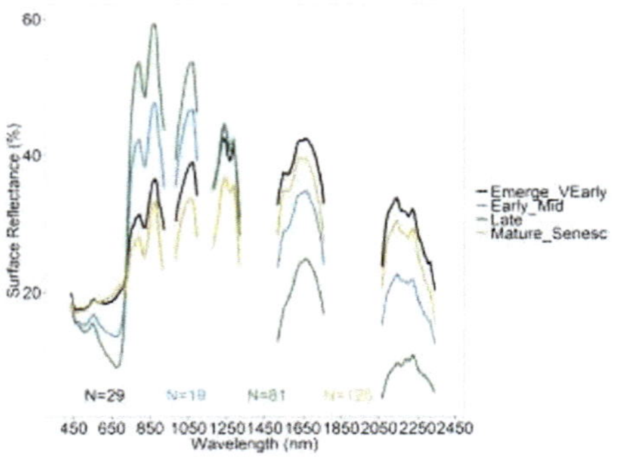

(a) GHISA of corn growth stages in AEZ 9.

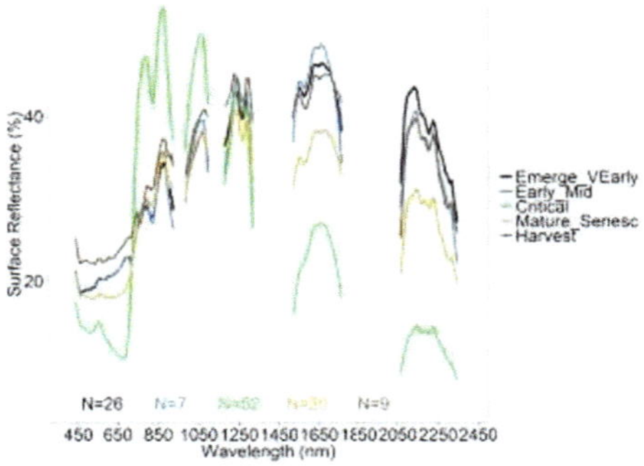

(b) GHISA of cotton growth stages in AEZ 9.

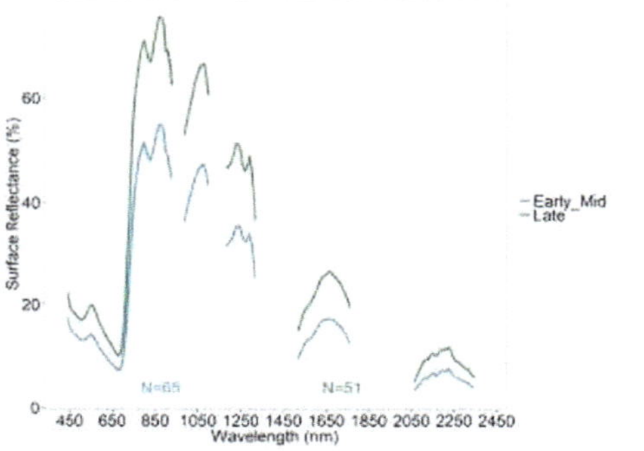

(c) GHISA of rice growth stages in AEZ 7.

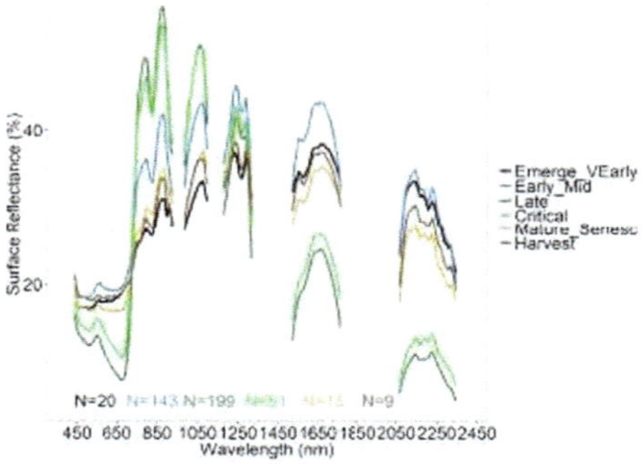

(d) GHISA of soybean growth stages in AEZ 9.

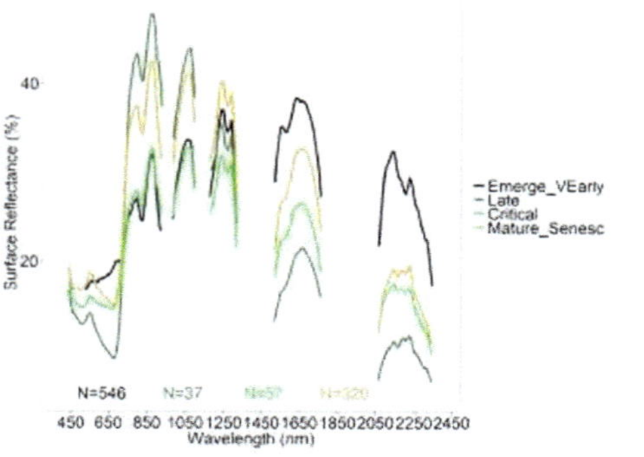

(e) GHISA of winter wheat growth stages in AEZ 9.

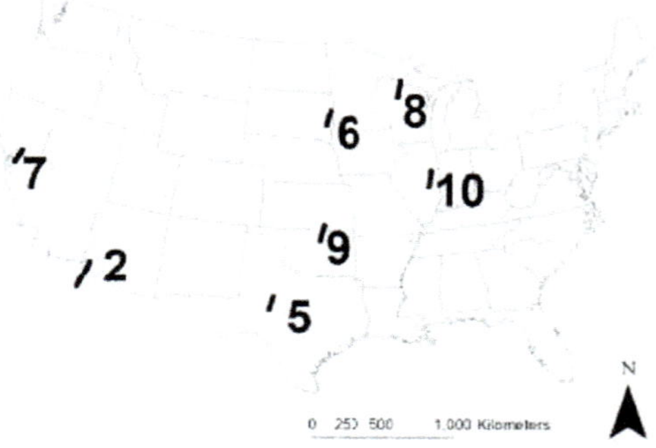

(f) US map showing locations of study areas.

FIGURE 10.11 GHISA CONUS crop growth stages. An illustration of GHISA CONUS, demonstrating EO-1 Hyperion spectra of crop growth stages for different crop types in different Food and Agriculture (FAO) agroecological zones (AEZs) within the continental USA (CONUS). Hyperion image tracks are shown in (f). (Source: Aneece and Thenkabail, 2019.)

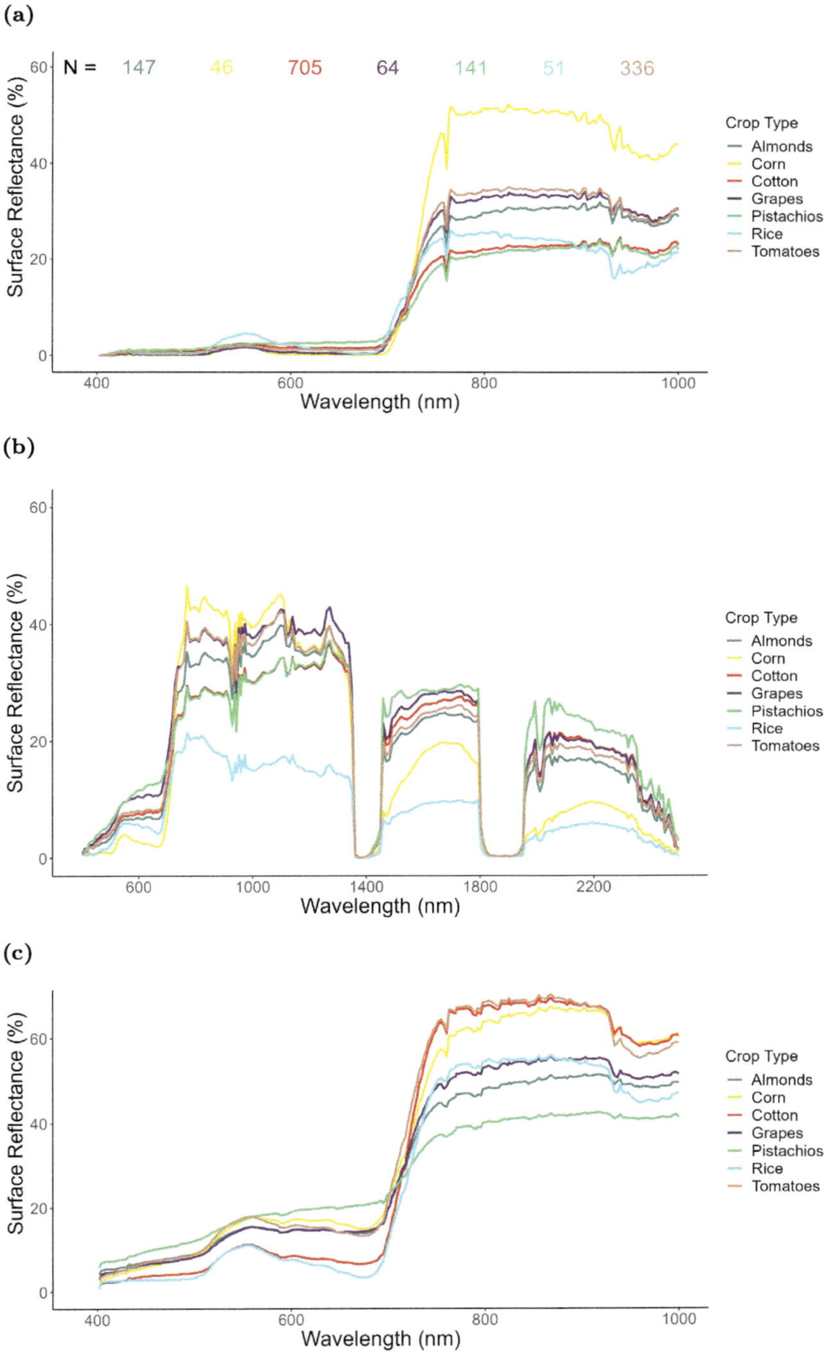

FIGURE 10.12 DESIS and PRISMA Derived Spectral Library of Agricultural Crops in California. Spectral libraries of different crops in California's Central Valley. (a) Deutsches Zentrum fu¨r Luft- und Raumfahrt (DLR) Earth Sensing Imaging Spectrometer (DESIS) June, (b) PRecursore IperSpettrale della Missione Applicativa (PRISMA) June, and (c) DESIS August spectra, averaged by crop type. The number of spectra (N) used to calculate the average is the same for all three plots. (Source: Aneece and Thenkabail, 2022b.)

library robust and useful. Including spectra from various spaceborne, airborne, and ground-based sensors, like those listed in Aneece and Thenkabail (2019), would also make the library more robust.

10.3.1 Spectral Libraries of Agricultural Crops

The Global Hyperspectral Imaging Spectra-library of Agricultural crops (GHISA) platform seeks to collect agricultural spectral data and associated metadata from different sensors, countries, agro-ecological zones, and management regimes (Aneece and Thenkabail, 2019). Such a library would help understand, model, map, and monitor crops and their biophysical and biochemical quantities. Following are examples of GHISA spectra for the continental USA (CONUS, Figures 10.10 and 10.11) and California (Figure 10.12), and examples of EnMAP spectra of different land cover types (Figure 10.13).

The GHISA-CONUS dataset, associated documentation, and code are available for download through LP DAAC (Thenkabail and Aneece, 2019) and described in Aneece and Thenkabail (2018a, 2018b, 2019). To create this dataset, 99 Hyperion images from 2008 to 2015 were used to compile spectral libraries of corn, soybean, winter wheat, cotton, and rice in different growth stages across seven agro-ecological zones throughout CONUS (Aneece and Thenkabail, 2019). The United States Department of Agriculture (USDA) National Agricultural Statistics Service (NASS) Cropland Data Layer (CDL) (USDA National Agricultural Statistics Service Cropland Data Layer, 2022) was used as reference for labeling crop types and crop calendars from the Center for Sustainability and Global Environment (SAGE) at the University of Wisconsin-Madison (Sacks et al., 2010) were used for determining crop growth stages (Aneece and Thenkabail, 2019).

Similarly, DESIS and PRISMA Derived Spectral Library of Agricultural Crops in California and associated documentation are published and available (Aneece and Thenkabail, 2022a) and described in Aneece and Thenkabail (2022b). This dataset contains spectral profiles of almonds, corn, cotton, grapes, pistachios, rice, and tomatoes from DESIS and PRISMA images in California's Central Valley during the growing season of 2020.

Other GHISA datasets are also available through LP DAAC. For example, the GHISA for Central Asia (GHISACASIA) dataset, documentation, and code can be downloaded at Mariotto et al. (2020c), described in Mariotto et al. (2020a, 2020b). GHISACASIA contains 2006–2007 field spectroradiometer and Hyperion data for wheat, rice, corn, alfalfa, and cotton in different growth stages in the Syr Darya river basin in Central Asia. Other GHISA data from different countries, crops, and sensors (e.g., EnMAP, Figure 10.13) are currently being compiled. These data can be used for training classification models that can then be tested and validated using new generation hyperspectral data.

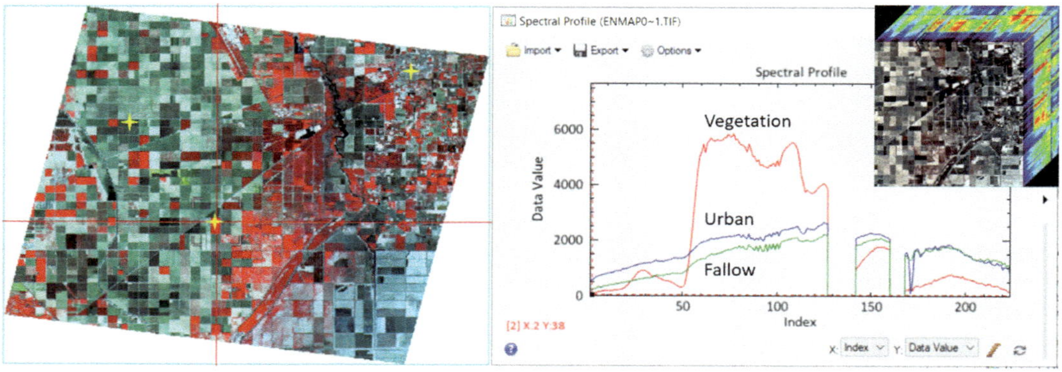

FIGURE 10.13 EnMAP spectra. Examples of EnMAP spectra for different land cover types. (Author's original contribution.)

10.4 DATA MINING AND DATA REDUNDANCY OF HYPERSPECTRAL DATA

Data mining is one of the critical first steps in hyperspectral data analysis. The primary goal of data mining is to eliminate redundant data and retain only the useful data. Data volumes are reduced through data mining methods such as feature selection (e.g., principal component analysis, derivative analysis, wavelets), lambda by lambda correlation plots (Fan et al., 2023; Thenkabail et al., 2000), minimum noise fraction (MNF) (Boardman and Kruse, 1994; Chen et al., 2023; Green et al., 1988) and hyperspectral vegetation indices (Danilov et al., 2023; Thenkabail et al., 2014). Data mining methods lead to (Thenkabail et al., 2012b): (a) reduction in data dimensionality, (b) reduction in data redundancy, and (c) extraction of unique information.

Wavebands adjacent to one another (e.g., 680 nm versus 690 nm or 550 nm versus 560nm) are often highly correlated for a given application. Various research papers (Numata, 2012; Thenkabail et al., 2014, 2012b, 2004a, 2004b, 2010, 2000; Thenkabail and Wu, 2012) showed that in a large stack of 242 bands in a Hyperion image, typically ~10% of the wavebands (~20 bands) are very useful in agricultural cropland or vegetation studies. It means for any given application (e.g., agriculture), a large number of bands are likely to be redundant. So, the goal of data mining is to identify and eliminate redundant bands. This will help eliminate unnecessary processing of redundant data, at the same time retaining the optimal power of hyperspectral data. This process is of great importance in the age of "big data."

However, eliminating redundant bands needs to be done with considerable care and expertise. What is redundant for one application (e.g., agriculture) may be critical for another application (e.g., geology) (Yao et al., 2012).

Data mining requires the merging of different disciplines such as digital imagery, pattern recognition, database management, artificial intelligence, machine learning algorithms, and statistics. There are various models of data mining. The generic concept of data mining is illustrated in Figure 10.14 (Lausch et al., 2015b). Figure 10.15 (Lausch et al., 2015b) shows data mining model applications for studies in soil clay content and soil organic content.

10.5 HUGHES PHENOMENON AND THE NEED FOR DATA MINING

If the number of bands remains high, the number of observations required to train a classifier increases exponentially to maintain classification accuracies; this is called the Hughes phenomenon (Kuang et al., 2023; Thenkabail and Wu, 2012). For example, Thenkabail et al. (2004a, 2004b) used 20 Hyperion bands to classify five crop types and achieve an accuracy of 90%. Relative to this, the seven-band Landsat data provided only an accuracy of 60% in classifying the same five crops. However, the number of observation points (e.g., ground data) to train and test the algorithms will be exponentially higher (depending on number and variability of classes, number of bands used, etc.) for the Hyperion data relative to Landsat data because larger numbers of bands are involved with Hyperion. So, one needs to weigh the higher classification accuracies achieved using a greater number of bands versus the resources required to gather exponentially higher number of observations (e.g., ground data) required to train and test the algorithms. In general, the ratio of variables to samples should be between 0.1 and 0.2 (Marshall et al., 2022). Higher accuracy by as much as 30% using 20 hyperspectral narrowbands (HNBs) when compared with seven bands of Landsat will justify the greater amount of ground data required. Beyond 20 bands, the increase in accuracy per increase in wavebands becomes asymptotic (e.g., Thenkabail et al. (2012b, 2004a, 2004b)). These studies, for example, show that when 40 Hyperion bands were used, the classification accuracies increased only by another 5% (from 90% with 20 bands to 95% with 40 bands). Here, using 20 additional Hyperion bands (from 20 to 40) cannot be justified since the ground observations needed to train and test the algorithm will also increase exponentially for 40 bands relative to 20. The key aim is to balance the higher classification accuracies with an optimal number of bands such as 20 instead of seven or 40. By doing so, we achieve a number of goals:

1. Increased classification accuracies with an optimal number of bands.
2. Significantly reduced data redundancies with optimal number of bands.

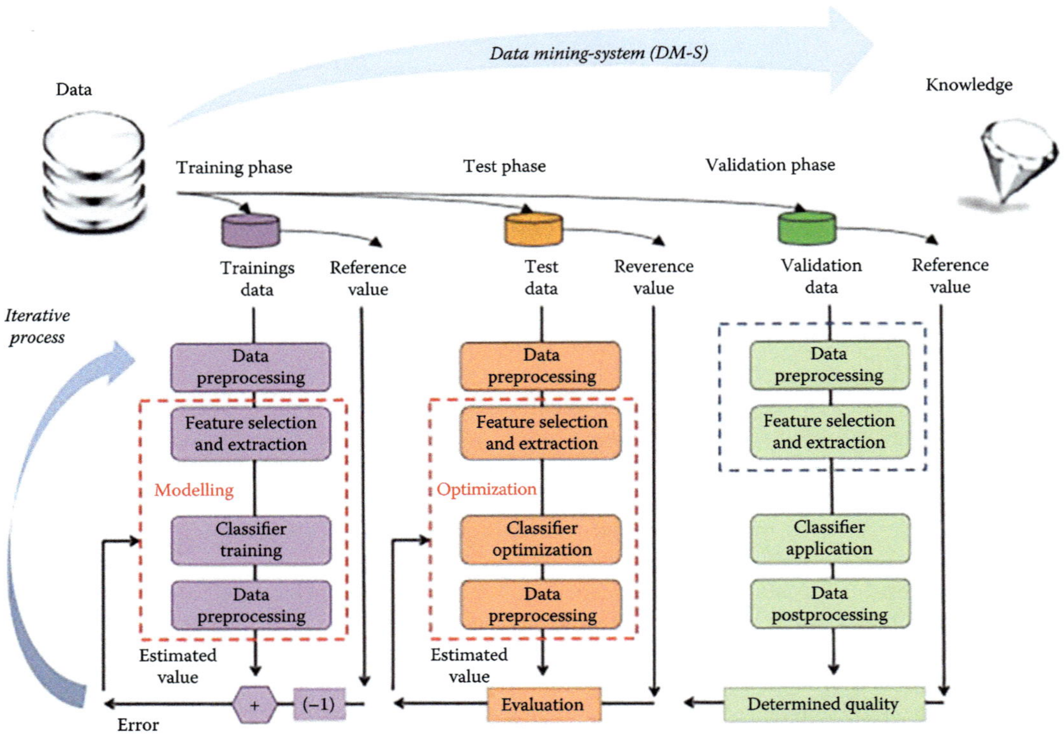

FIGURE 10.14 Data mining process. Data mining and linked open data—New perspectives for data analysis in environmental research. Data mining process with the data mining system (DM-S) in the Phases: (1) Training Phase, (2) Test Phase, and (3) Validation Phase. The data mining process works in a comparable way in all data mining types like text mining or web mining (Fayyad et al. (1996); Tanner (2013)). (Source: Lausch et al., 2015b.)

3. Overcoming the Hughes phenomenon by using optimal number of bands (e.g., 20) in which observation data (ground data) to train and test the algorithms will be kept to reasonable levels.

10.6 METHODS OF HYPERSPECTRAL DATA ANALYSIS

Hyperspectral data analysis methods are broadly grouped under two categories (Bajwa and Kulkarni, 2012):

1. Feature extraction methods
2. Information extraction methods

Under each of these two categories, specific unsupervised and supervised classification approaches exist (Figure 10.16, Bajwa and Kulkarni (2012); Plaza et al. (2012)). Methods of classifying vegetation classes, crop types, or vegetation species using hyperspectral narrowbands are discussed extensively in this chapter and include unsupervised classification, supervised approaches, spectral angle mapper (SAM) (Chakravarty et al., 2021), artificial neural networks (ANN) (Srivastava et al., 2021), support vector machines (SVM) (Ji et al., 2019; Xiang et al., 2020), multivariate or partial least square regressions (Padghan and Deshmukh, 2017; Song et al., 2023), and discriminant analysis (Furlanetto et al., 2020; Thenkabail et al., 2012a; Xia et al., 2019).

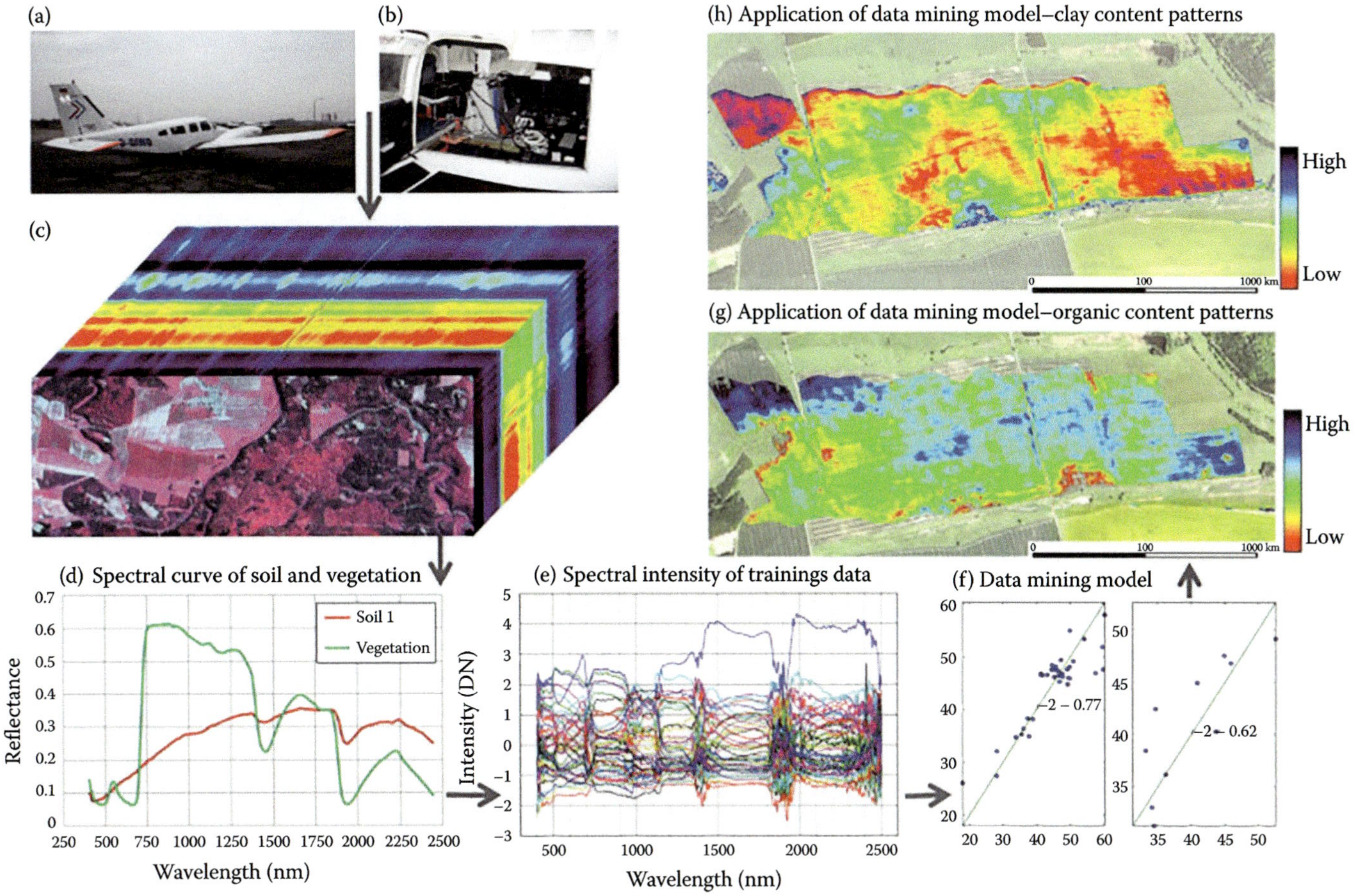

FIGURE 10.15 Data mining examples. (a–b) Data mining and linked open data—New perspectives for data analysis in environmental research. Airborne hyperspectral AISA-Eagle/HAWK remote sensor mounted on Piper, (c) CIR-image from hyperspectral sensors of the AISA-EAGLE/HAWK (AISA-DUAL) 400–2,500 nm with data cube, 367 spectral bands with 2 m recorded ground resolution, date of recording Mai 2012 with a Piper, Region Sch¨afertal—Bode Catchment, (d) Spectral curve of ground truth sampling points for soil and vegetation in the test site, (e) Spectral intensity curves of imaging hyperspectral data, (f) Data Mining Model, (g) Application of the best data mining model on airborne hyperspectral image data for quantification and recognition of organic content patterns, and (h) Pattern of clay content. (Source: Lausch et al., 2015b.)

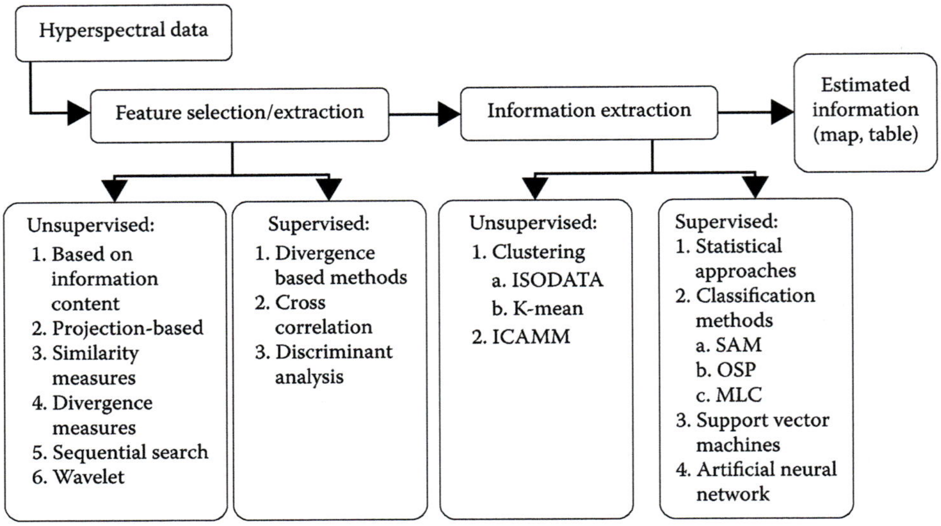

FIGURE 10.16 Hyperspectral data analysis methods. Examples of feature selection/extraction and information extraction methods. (Source: Bajwa and Kulkarni, 2012.)

Within information extraction methods, fundamental philosophies of hyperspectral data analysis involve two approaches:

1. Optimal hyperspectral narrowbands (OHNBs) where only a select number of nonredundant bands are used (e.g., ~20 of Hyperion OHNBs are used); and
2. Full spectral analysis (FSA) where all the bands in the continuum (e.g., all 242 Hyperion bands in 400–2,500 nm) are used.

10.7 OPTIMAL HYPERSPECTRAL NARROWBANDS (OHNBS)

Determining wavebands that are optimal for different studies requires a thorough study in itself. The importance of the wavebands for different studies such as vegetation, geology, and water are all different. So, determining OHNBs requires subject knowledge and considerable experience working with hyper spectral data. Based on the synthesis of the extensive studies conducted by Aneece and Thenkabail (2022b); Mariotto et al. (2013); Marshall and Thenkabail (2014); Thenkabail et al. (2014, 2012b, 2013); Thenkabail (2002); Thenkabail et al. (2004a, 2004b, 2000), the OHNBs for agriculture and vegetation studies are established and presented in Tables 10.2 and 10.3 (also see Ahmad et al. (2021)). Each of these hyperspectral narrowbands (HNBs) are identified for their importance in studying one or more vegetation and crop biophysical and biochemical characteristics. Most of these bands are also very distinct from one another; so none of them are redundant. Using some combination of these bands will help better quantify the biophysical and biochemical characteristics of vegetation and agricultural crops (Alchanatis and Cohen, 2012; Pu, 2012). In the following sections and subsections, we will demonstrate how these HNBs are used in classifying, modeling, and mapping agricultural croplands and other vegetation.

Table 10.2 shows that over the 400–2,500 nm range of the spectrum, there are 28 bands (e.g., ~12% of the 242 Hyperion bands in the 400–2,500 nm range) that are optimal in the study of agriculture and vegetation. However, the redundant bands here (i.e., agriculture and vegetation applications) may be very useful bands in other applications such as geology (Ben-Dor, 2012; Tan et al., 2020). For example, the critical absorption bands for studying minerals like biotite, kaolinite, hematite, and others are shown in Table 10.4. In cases where minerals are mixed with other

TABLE 10.2

Optimal (Nonredundant) Hyperspectral Narrowbands to Study Vegetation and Agricultural Crops[1,2,3]

Waveband Number (#)	Waveband Range (λ)	Waveband Center (λ)	Waveband Width (δλ)	Importance and physical significance of waveband in vegetation and cropland studies
A. Ultraviolet bands				
1	373–377	375	5	Fraction of photosynthetically active radiation (fPAR), leaf water content
B. Blue bands				
2	403–407	405	5	Nitrogen, Senescing: sensitivity to changes in leaf nitrogen. Reflectance changes due to pigments is moderate to low. Sensitive to senescing (yellow and yellow green leaves).
3	491–500	495	10	Carotenoid, light use efficiency (LUE), stress in vegetation, sensitive to senescing and loss of chlorophyll/browning, ripening, crop yield, and soil background effects
C. Green bands				
4	513–517	515	5	Pigments (carotenoid, chlorophyll, anthocyanins), nitrogen, vigor: positive change in reflectance per unit change in wavelength of this visible spectrum is maximum around this green waveband
5	530.5-531.5	531	1	Light use efficiency (LUE), xanthophyll cycle, stress in vegetation, pest and disease: Senescing and loss of chlorophyll/ browning, ripening, crop yield, and soil background effects
6	546–555	550	10	Chlorophyll: Total chlorophyll; Chlorophyll/carotenoid ratio, vegetation nutritional and fertility level; vegetation discrimination; vegetation classification
7	566–575	570	10	Pigments (Anthrocyanins, Chlorophyll), Nitrogen: negative change in reflectance per unit change in wavelength is maximum as a result of sensitivity to vegetation vigor, pigment, and N.
D. Red bands				
8	676–685	680	10	Biophysical quantities and yield: leaf area index, wet and dry biomass, plant height, grain yield, crop type, crop discrimination
E. Red-edge bands				
9	703–707	705	5	Stress and chlorophyll: Nitrogen stress, crop stress, crop growth stage studies
10	718–722	720	5	Stress and chlorophyll: Nitrogen stress, crop stress, crop growth stage studies
11*	700–740	700–740	700–740	Chlorophyll, senescing, stress, drought: first-order derivative index over 700–740 nm has applications in vegetation studies (e.g., blue-shift during stress and red-shift during healthy growth)
F. Near infrared (NIR) bands				
12	841–860	850	20	Biophysical quantities and yield: LAI, wet and dry biomass, plant height, grain yield, crop type, crop discrimination, total chlorophyll
13	886–915	900	20	Biophysical quantities, Yield, Moisture index: peak NIR reflectance. Useful for computing crop moisture sensitivity index, NDVI; biomass, LAI, Yield.
14	961–980	970	20	Plant moisture content, center of moisture sensitive "trough"; water band index, leaf water, biomass;

(Continued)

TABLE 10.2 (*Continued*)

Optimal (Nonredundant) Hyperspectral Narrowbands to Study Vegetation and Agricultural Crops[1,2,3]

Waveband Number (#)	Waveband Range (λ)	Waveband Center (λ)	Waveband Width (δλ)	Importance and physical significance of waveband in vegetation and cropland studies
G. Far near infrared (FNIR) bands				
15	1073–1077	1075	5	Biophysical and biochemical quantities: leaf area index, wet and dry biomass, plant height, grain yield, crop type, crop discrimination, total chlorophyll, anthocyanin, carotenoids
16	1178–1182	1080	5	Water absorption band
17	1243–1247	1245	5	Water sensitivity: water band index, leaf water, biomass. Reflectance peak in 1050-1300 nm.
H. Early short-wave infrared (ESWIR) bands				
18	1448–1532	1450	5	Vegetation classification and discrimination: ecotype classification; plant moisture sensitivity. Moisture absorption trough in early short wave infrared (ESWIR)
19	1516–1520	1518	5	Moisture and biomass: A point of most rapid rise in spectra with unit change in wavelength in SWIR. Sensitive to plant moisture.
20	1648–1652	1650	5	Heavy metal stress, Moisture sensitivity: Heavy metal stress due to reduction in Chlorophyll. Sensitivity to plant moisture fluctuations in ESWIR. Use as an index with 1548 or 1620 or 1690 nm.
21	1723–1727	1725	5	Lignin, biomass, starch, moisture. Discriminating crops and vegetation.
I. Far short-wave infrared (FSWIR) bands				
22	1948–1952	1950	5	Water absorption band: Highest moisture absorption trough in FSWIR. Use as an index with any one of 2025 nm, 2133 nm, and 2213 am. Affected by noise at times.
23	2019–2027	2023	8	Litter (plant litter), lignin, cellulose: litter-soil differentiation: moderate to low moisture absorption trough in FSWIR. Use as an index with any one of 2025 nm, 2133 nm, and 2213 nm.
24	2131–2135	2133	5	Litter (plant litter), lignin, cellulose: typically highest reflectivity in FSWIR for vegetation. Litter-soil differentiation
25	2203–2207	2205	5	Litter, lignin, cellulose, sugar, starch, protein; Heavy metal stress: typically, second highest reflectivity in FSWIR for vegetation. Heavy metal stress due to reduction in Chlorophyll
26	2258–2266	2262	8	Moisture and biomass: moisture absorption trough in far short-wave infrared (FSWIR). A point of most rapid change in slope of spectra based on land cover, vegetation type, and vigor.
27	2293–2297	2295	5	Stress: sensitive to soil background and plant stress
28	2357–2361	2359	5	Cellulose, protein, nitrogen: sensitive to crop stress, lignin, and starch

Source: Modified and adopted from Thenkabail et al. (2014, 2011, 2013); Thenkabail (2002); Thenkabail et al. (2004a, 2004b, 2000). These bands were obtained through extensive studies of agricultural crop biophysical and biochemical quantities using EO-1 Hyperion and field spectroradiometer data.

Note:

[1] = most hyperspectral narrowbands (HNBs) that adjoin one another are highly correlated for a given application. Hence from a large number of HNBs, these non-redundant (optimal) bands are selected.

[2] = these optimal HNBs are for studying vegetation and agricultural crops. When we use some or all of these wavebands, we can attain highest possible classification accuracies in classifying vegetation categories or crop types.

[3] = wavebands selected here are based on careful evaluation of a large number of studies.

[*] = One can obtain an integral spectrum from 700–740 for a derivative index

TABLE 10.3

Across-Sensor Optimal Hyperspectral Narrowbands to Study Vegetation and Agricultural crops.

Band Center (nm)	Significance
498	Carotenoids, LUE, Stress
530	LUE, Stress, Disease
565	Nitrogen, Pigments
714	Stress, Pigments
773	LAI, Biomass/yield
806	Biomass/yield
834	Biomass/yield
916	Biomass/yield
989	Biomass/yield, Moisture, Protein
1177	Biomass/yield, Moisture
1240	Biomass/yield, Moisture
1298	Biomass/yield, Moisture
1712	Biomass/yield, Nitrogen, Lignin, Cellulose
2078	Moisture, Nitrogen, Protein
2188	Lignin, Cellulose, Starch, Protein
2346	Lignin, Cellulose, Protein, Stress

Note: LUE = Light Use Efficiency; LAI = Leaf Area Index. (Aneece and Thenkabail, 2018a, 2021, 2022b; Thenkabail et al., 2014, 2011, 2013; Thenkabail, 2002; Thenkabail et al., 2004a, 2004b, 2000). These bands were obtained from agricultural classification studies using EO-1 Hyperion and new-generation Germany's Deutsches Zentrum fu¨r Luft- und Raumfahrt Earth Sensing Imaging Spectrometer (DESIS) and the Italian Space Agency's (ASI) PRecursore IperSpettrale della Missione Applicativa (PRISMA) data.

TABLE 10.4

Sub-Pixel Mineral Mapping of a Porphyry Copper Belt using EO-1 Hyperion Data

Hyperion Band Number	Wavelength nanometer	Feature	Minerals	Mineral Characteristic
210, 217	2254, 2324	absorption	biotite	potassic-biotitic alteration zone
205	2203	absorption	muscovite and illite	Al-OH vibration in minerals with muscovite deeper absorption than illite
201, 205	2163, 2203	absorption	kaolinite	Al-OH vibration
14, 79, 205	487, 932, 2203	absorption	goethite	
14, 53, 205	487, 884, 2203	absorption	hematite	
79, 211, 205	932, 2264, 2203	absorption	jarosite	
201	2163	absorption	pyrophyllite	Al-OH and Mg-OH
218	2335	absorption	chlorite	Al-OH and Mg-OH

Source: adopted and modified from information in manuscript by Hosseinjani Zadeh et al. (2014).

cover types (e.g., vegetation, rocks, soil) there will be mixed mineral signatures that require specific understanding and analysis (e.g., sub-pixel decomposition methods) for better obtaining pure samples. The HNBs required for mineralogy are quite different from those required for vegetation and cropland studies (Chung et al., 2020; Slonecker, 2012; Vaughan et al., 2012). The preceding fact clearly establishes the need to determine OHNBs that are application-specific. For example, the bands in Table 10.2 were determined from studies focused on photosynthetic vegetation, and differ from those best for non-photosynthetic vegetation (Dennison et al., 2023).

10.8 HYPERSPECTRAL VEGETATION INDICES (HVIS)

One of the most common, powerful, and useful forms of feature selection methods for hyperspectral data is based on the calculation of hyperspectral vegetation indices (HVIs) (Clark, 2012; Colombo et al., 2012; Danilov et al., 2023; Galv˜ao, 2012; Gitelson, 2012a, 2012b; Lian et al., 2023; Roberts, 2012; Sellami et al., 2022). The HVIs achieve two important goals of hyperspectral data analysis:

1. Compute many specific targeted HVIs to help model biophysical and biochemical quantities;
2. Reduce the data volume (mine the data) to eliminate all redundant bands for a given application. There are several approaches to deriving HVIs. These are briefly presented and discussed in the sections that follow.

10.8.1 Two Band Hyperspectral Vegetation Indices (TBHVIs)

The TBHVIs are defined using Equation 10.4 (Thenkabail et al., 2000):

$$TBHVI_{I,j} = \frac{(R_j - R_i)}{(R_j + R_i)} \tag{10.4}$$

where i, j = 1 . . . N, with N = number of narrowbands. As defined in Section 10.1.3, only 157 of the 242 Hyperion bands are useful after removing the wavebands in the atmospheric windows and those that are uncalibrated. This will still leave C^2 = 12,246 unique TBHVIs.

Any one crop biophysical or biochemical quantity (e.g., biomass, leaf area index, nitrogen) can be correlated with each one of the 12,246 TBHVIs (Stroppiana et al., 2012; Zhu et al., 2012). This will result for each crop variable (e.g., biomass) a total of 12,246 unique models, each providing an R-square. Figure 10.17 shows the contour plot of 12,246 R-square values plotted for (1) rice crop wet biomass with TBHVIs (Figure 10.17; above the diagonal), and (2) barley crop wet biomass with TBHVIs (Figure 10.17, below the diagonal). The areas with "bulls-eyes" are regions of rich information having high R-square values whereas the areas in gray are redundant bands with low R-square values. Based on these lambda (λ) versus lambda (λ) plots (Figure 10.17) the optimal wavebands centers (λ) and widths ($\Delta\lambda$) are determined (Tables 10.2 and 10.3). Table 10.2 shows the optimal wavebands (λ), wavebands centers (λ), and widths ($\Delta\lambda$) based on meta-analyses of several studies (Thenkabail et al., 2014, 2013; Thenkabail, 2002; Thenkabail et al., 2004a, 2004b, 2000; Thenkabail and Wu, 2012).

10.8.1.1 Refinement of Two-Band HVIs

Further refinement of each of the two-band HVIs (TBHVIs) is possible by computing: (1) soil adjusted versions of TBHVIs, and (2) atmospheric corrected versions of TBHVIs. Interested readers can read more on this topic in Thenkabail et al. (2000).

10.8.1.2 Multi-Band Hyperspectral Vegetation Indices (MBHVIs)

The MBHVIs are computed using Equation 10.5 (Li et al., 2012; Thenkabail et al., 2000):

$$MBHVI_i = L_{J=1}^{N} a_{i,j} R_j \tag{10.5}$$

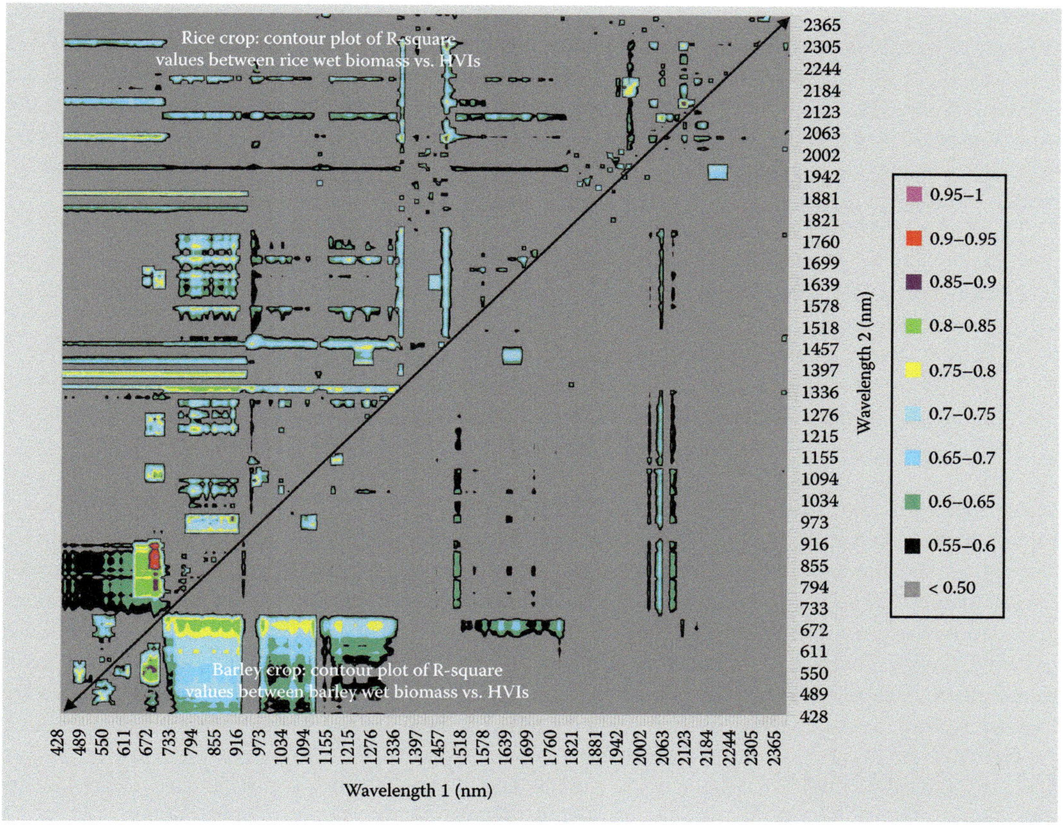

FIGURE 10.17 Lambda (λ) versus Lambda (λ) plot. Plot of R-square values between wet biomass and hyperspectral vegetation indices (HVIs) for the rice crop (above the diagonal) and barley crop (below the diagonal). (Author's original contribution.)

Where $MBHVI_i$ = crop variable i, R = reflectance in bands j (j = 1 to N with N = 242 for Hyperion); a = the coefficient for reflectance in band j for *i*th variable. The process of modeling involves running stepwise linear regression models (e.g., using the MAXR algorithm in Statistical Analysis System or SAS (SAS, 2009)) with any one biophysical or biochemical variable (e.g., biomass) as the dependent variable and the numerous hyperspectral narrowbands as independent variables (e.g., 157 of the 242 useful bands of Hyperion). In this modeling approach we will get the best 1-band, 2-band, 3-band, and so on to the best n-band model. The best 1-band model is the one in which the biomass (taken as an example) has the highest R-square value with a single band out of the total 157 Hyperion hyperspectral narrowbands (HNBs). Then we obtain the best 2-band model, in which 2 HNBs provide the best R-square value with biomass. Similarly, the best 3-band, best 4-band, and best n-band (e.g., all 157 Hyperion bands) models are obtained. Even though theoretically, all 157 bands can be involved in providing a 157-band biomass model that is usually meaningless due to over-fitting. A plot of R-square values (y-axis) versus the number of bands (x-axis) will show us when the increase in R-square values with the addition of wavebands becomes asymptotic. Alternatively, we can also consider additional bands when there is an increase of at least 0.03 or higher in R-square value when additional bands are added. So, the approach we can use is to look at 1-band model and examine its R-square. Then when a 2-band model increases R-square value by at least 0.03 (a threshold we can set), consider the 2-band model; otherwise retain the 1-band model as final. Eventually, we will notice the addition of a band does not increase the R-square value by more than 0.03. Typically, we have noticed that anywhere between 3 and 10 HNBs explain the greatest variability in modeling various biophysical and biochemical quantities in most agricultural crop and vegetation variables. Beyond these 3 to 10 bands, the increase in R-square per additional band is insignificant or asymptotic (Mariotto et al., 2013; Marshall and Thenkabail, 2014; Thenkabail et al., 2004a, 2004b). Yet, which 3 to 10 bands within 400 to 2,500 nm are retained will vary based on the type of crop variable. When we study these variables across crop types and locations with different sensors, we can synthesize outcomes across studies and arrive at frequently repeating wavebands to model and map crop quantities.

Through MBHVIs, we can establish:

1. How many HNBs are required to achieve an optimal R-square for any biophysical or biochemical quantity;
2. Which HNBs are involved in providing optimal R-square;
3. Which are important HNBs and which are redundant. The best approach to achieve this is by a study conducted for many crops, involving several crop variables, and based on data from multiple sites and years. Tables 10.2 and 10.3 provide such a summary.

These MBHVIs take advantage of the key absorptive and reflective portions of the spectrum (e.g., Figure 10.18; Gnyp et al. (2014)). Using four HNBs, two reflective (900 nm and 1050 nm) and two absorptive (955 nm and 1220 nm), Gnyp et al. (2014) constructed an MBHVI for winter wheat biomass (Equation 10.5), and in their paper, clearly demonstrated the significantly higher R-square values provided by such a multi-band HVI when compared with two-band HVIs (e.g., in Figure 10.19 the MBHVI, GnyLi, has a much higher R-square than other indices). Interestingly, while the typical saturation effect (lack of sensitivity) at higher biomass amounts is still present, it is evidently somewhat less severe with GnyLi than the others (except the red-edge position (REP) but it has a lower R-square).

The specific band centers and bandwidths are not definitive. This is because, with crop type and crop growing conditions, the specific reflective maxima (900 nm and 1050 nm) and reflective

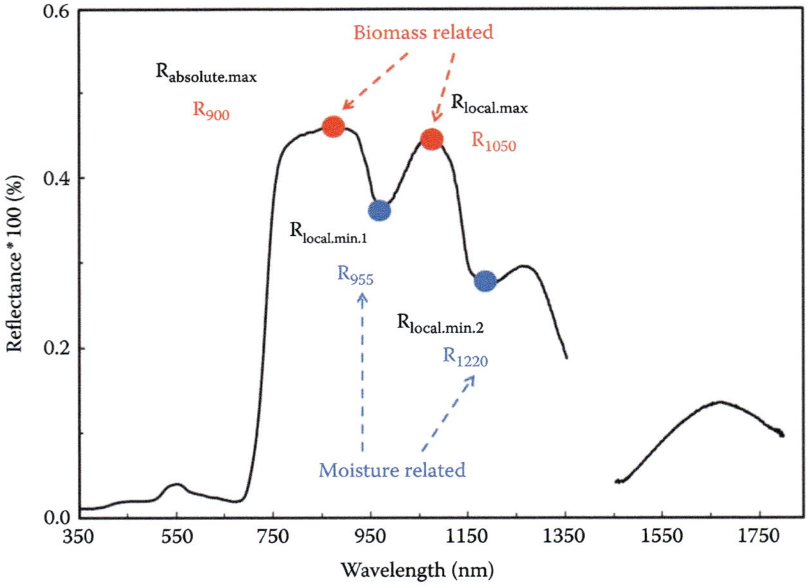

FIGURE 10.18 New index. Development and implementation of a multiscale biomass model using hyperspectral vegetation indices for winter wheat in the North China Plain. Reflectance of winter wheat and its characteristic peaks and troughs with the reflectance maxima and minima in the NIR and SWIR domains. These peaks were used to compute the VI GnyLi. R is the reflectance value (%) at a specific wavelength (nm). (Source: Gnyp et al., 2014.)

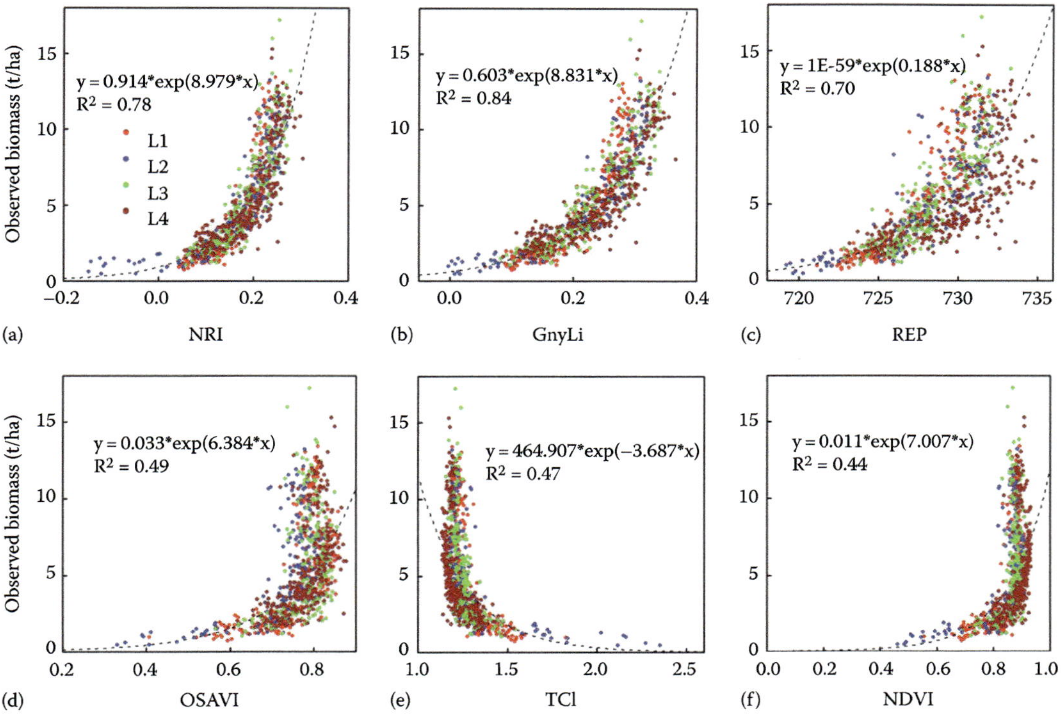

FIGURE 10.19 Biomass estimation. Observed biomass versus various HVIs and also GnyLi (see Figure 10.26). R is the reflectance value (%) at a specific wavelength (nm). (Source: Gnyp et al., 2014.)

minima (955 nm and 1220 nm) shown in Figure 10.18 and Equation 10.6 can vary. For example, the moisture absorption maxima can be at 750 nm, 760 nm, 770 nm, or 780 nm (Thenkabail et al., 2013; Thenkabail and Wu, 2012) or can be at 755 nm as shown in Figure 10.18 and Equation 10.6. As a result, we performed a meta-analysis of several papers for the recommendations of HNB centers and widths (Table 10.3) that are optimal for use in HVI computations across crops and vegetation.

$$GnyLi = \frac{R_{900} * R_{1050} - R_{955} * R_{1220}}{R_{900} * R_{1050} + R_{955} * R_{1220}} \tag{10.6}$$

10.9 HYPERSPECTRAL VEGETATION INDICES (HVIs) AND THEIR CATEGORIES

Based on extensive research over the last few decades (Thenkabail et al., 2014, 2013; Thenkabail, 2002; Thenkabail et al., 2004a, 2004b, 2000; Thenkabail and Wu, 2012), six distinct categories of two-band hyperspectral vegetation indices (TBHVIs; Table 10.5) are considered most significant and important to study specific biophysical and biochemical quantities of agriculture and vegetation. Researchers use these HVIs, derived using HNBs, for their studies to quantify and model biophysical and biochemical quantities of various agricultural crops and vegetation of different types. The values of two such indices are illustrated. These are: (a) Hyperspectral Biomass and Structural Index 1 (HBSI1; Thenkabail et al., 2014; narrowband NDVI), derived using the Hyperion bands centered around 855 nm and 682 nm (each with 10 nm width), which is applied to an agricultural area to determine biomass (Figure 10.20); and (b) Photochemical Reflectance Index (PRI) for stress detection (e.g., Figure 10.21; Hernández-Clemente et al. (2011), Middleton et al. (2012)). Readers are encouraged to compare Figure 10.21 with Tables 10.5, 10.2, and 10.3 for a better understanding of HNBs

TABLE 10.5

Hyperspectral Vegetation Indices or HVIs

Band Number (#)	Plant	Bandwidth ($\Delta\lambda 1$)	Hyperspectral Narrowband ($\lambda 2$)	Bandwidth ($\Delta\lambda 2$)	Hyperspectral Vegetation Index (HVI)	Best Index under Each Category
I. Hyperspectral Biomass and Structural Indices (HBSIs) [to best study biomass, LAI, plant height, and grain yield]						
HBSI1	855	20	682	5	(855−682)/(855+682)	HBSI: Hyperspectral Biomass and Structural Index
HBSI2	910	20	682	5	(910−682)/(910+682)	
HBSI3	550	5	682	5	(550−682)/(550+682)	
II. Hyperspectral BioChemical Indices (HBCIs) [pigments like carotenenoids, anthocyanins nitrogen, and chlorophyll]						
HBCI8	550	5	515	5	(550−515)/(550+515)	HBCI: Hyperspectral BioChemical Index
HBCI9	550	5	490	5	(550−490)/(550+490)	
III. Hyperspectral Red-Edge Indices (HREIs) [to best study plant stress, drought]						
HREI14	700–740	40	first-order derivative integrated over red-edge			HREI: Hyperspectral Red-Edge Index
HREI15	855	5	720	5	(855−720)/(855+720)	
IV. Hyperspectral Water and Moisture Indices (HWMIs) [to best study plant water and moisture]						
HWMI17	855	20	970	10	(855−970)/(855+970)	HWMI: Hyperspectral Water and Moisture Index
HWMI18	1075	5	970	10	(1075−970)/(1075+970)	
HWMI19	1075	5	1180	5	(1075−1180)/(1075+1180)	
HWMI20	1245	5	1180	5	(1245−1180)/(1245+1180)	
V. Hyperspectral Light-use Efficiency Index (HLEI) [to best study light use efficiency or LUE]						
HLUE24	570	5	531	1	(570−531)/(570+531)	HLEI: Hyperspectral Light-use Efficiency Index
VI. Hyperspectral Lignin Cellulose Index (HLCI) [to best study plant legnin, cellulose, and plant residue]						
HLCI25	2205	5	2025	1	(2205−2025)/(2205+2025)	HLCI: Hyperspectral Lignin Cellulose Index

Source: Modified and adopted from Thenkabail et al. (2014), also see wavebands in Table 10.3 used to derive these indices.

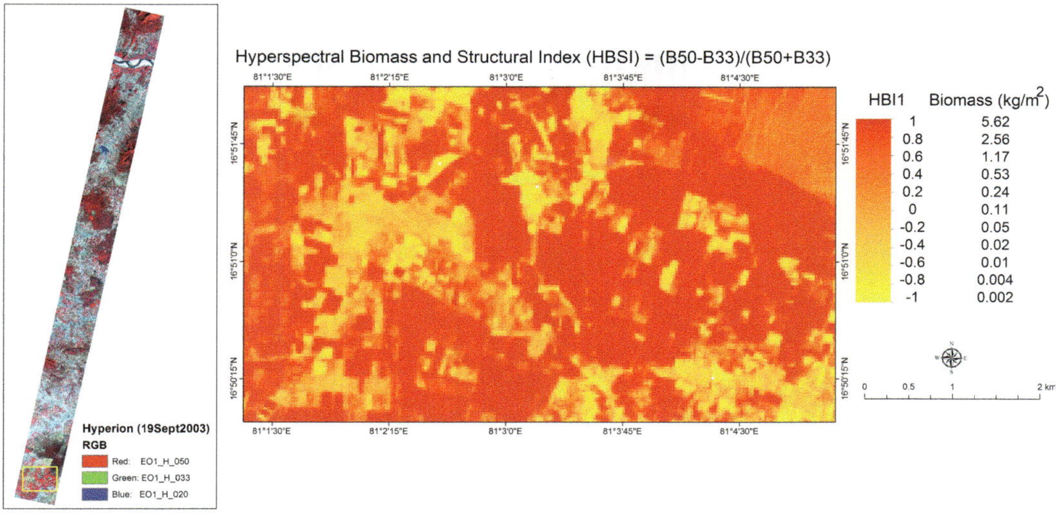

FIGURE 10.20 Spatial depiction of a Hyperspectral Biomass and Structural Index 1 (HBSI1). HBSI1 as applied to an agricultural area. One of the HVIs (HBSI1, unitless) in mapping wet biomass for a study area using Hyperion hyperspectral data. The red area in the z-scale can be stretched further to show better biomass variability with changes in HBSI1. For example, HBSI1 0.4 = 0.53 and HBSI 0.6 = 1.16, HBSI1 0.8 = 2.56, and HBSI1= 5.62. The current stretch does not adequately show these differences (much of the higher end is in red). If we stretch between HBSI1 from 0.4 to 1.0 then the biomass differences in this HBSI1 range, which is 0.53 to 5.62, will show up in better contrast. The relationship between HBSI1 and biomass is nonlinear due to the saturation of indices at the higher end of the biomass. The dynamic range of the narrowband index is higher than that of broadband NDVI, perhaps due to sensor design and radiometric resolution. (Author's original contribution.)

(Tables 10.2 and 10.3), HVIs (Table 10.5), and their importance (Figure 10.22) in studies pertaining to crops and vegetation. The importance of wavebands in computing the indices for various biophysical and biochemical quantities is illustrated in Figure 10.22. For studying crop residues with HVIs, the reader is referred to Dennison et al. (2023), Lamb et al. (2022), and Hively et al. (2021).

10.10 FULL SPECTRAL ANALYSIS (FSA)

Several chapters of this book discuss the usefulness and utility of applying full spectra (e.g., continuous and entire spectra over 400–2,500 nm) for analysis using methods such as partial least squares regression (PLSR) (dos Santos et al., 2023; Song et al., 2023), wavelet analysis (Zhao et al., 2022), continuum removal (Yang and Du, 2021), spectral angle mapper (SAM) (Chakravarty et al., 2021), and spectral matching techniques (SMTs) (Gumma et al., 2022b; Thenkabail and Wu, 2012).

10.10.1 Spectral Matching Techniques (SMTs)

Spectral Matching Techniques (SMTs) (Thenkabail et al., 2007) involve:

1. **Ideal or target spectral library creation:** Collecting ideal or target spectra (e.g., specific crops, specific species, specific minerals) and creating a spectral library.
2. **Class spectra generation:** Class spectra are generated by classifying an image into distinct classes (e.g., crop types or land cover types). Once the image is classified, the spectra of each class can be generated.

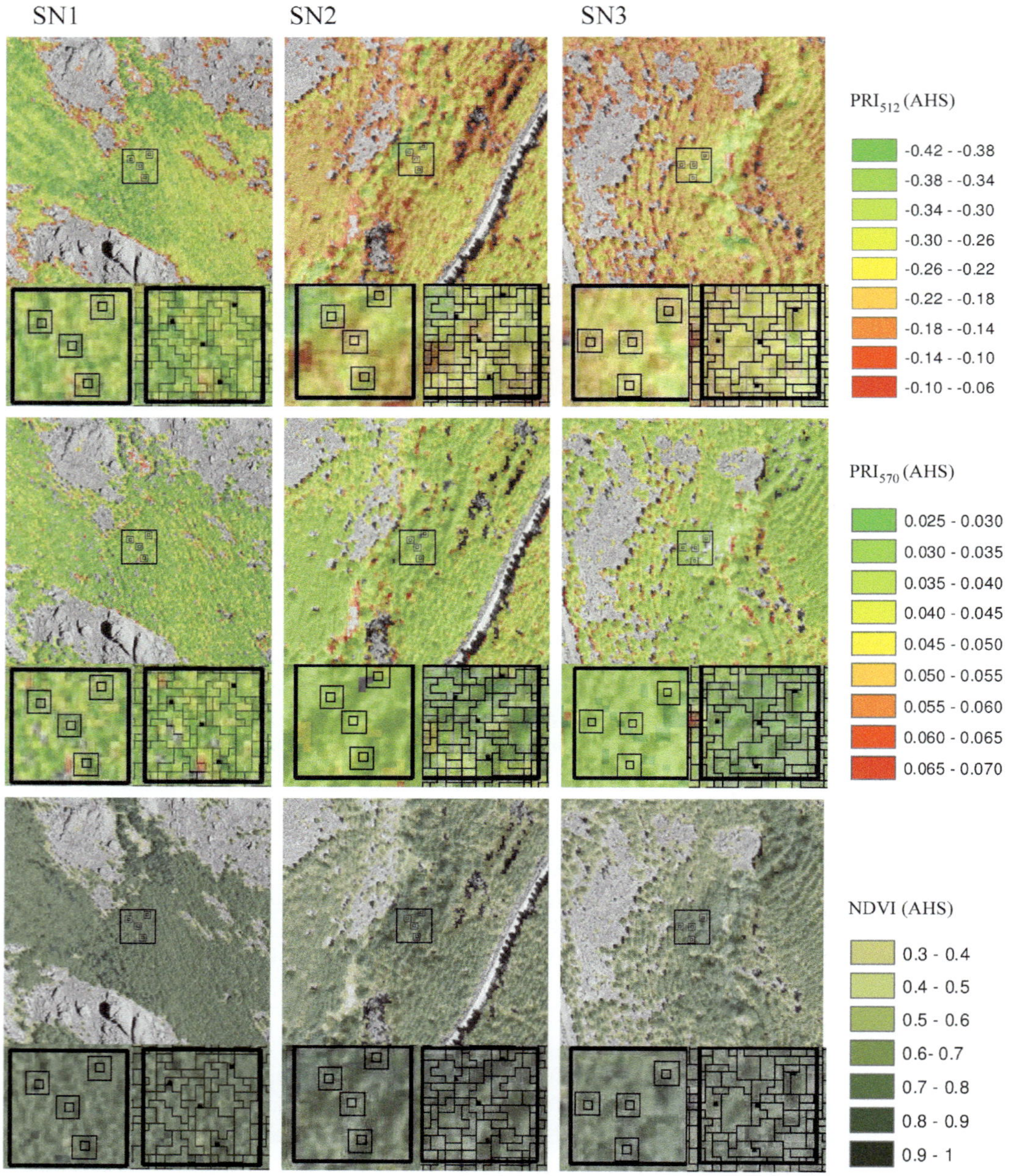

FIGURE 10.21 Assessing structural effects on Photochemical Reflectance Index (PRI). PRI for stress detection in conifer forests. PRI512, PRI570, and NDVI obtained from the AHS airborne sensor from three study areas of *Pinus nigra* with different levels of stress: SN1, SN2, and SN3. At the bottom of each image, two zoom images of a central plot, one pixel-based displaying 1 × 1 and 3 × 3 resolutions and the other at object level. Note: PRI512 is a normalized index involving a waveband centered at 512 nm and 531 nm whereas PRI570 is a normalized index involving a waveband centered at 570 nm and 531 nm. Airborne Hyperspectral Scanner, AHS (Sensytech Inc., currently Argon St. Inc., Ann Arbor, MI, USA), acquiring 2 m spatial resolution imagery in 38 bands in the 0.43–12.5 μm spectral range. (Source: Hernández-Clemente et al., 2011.)

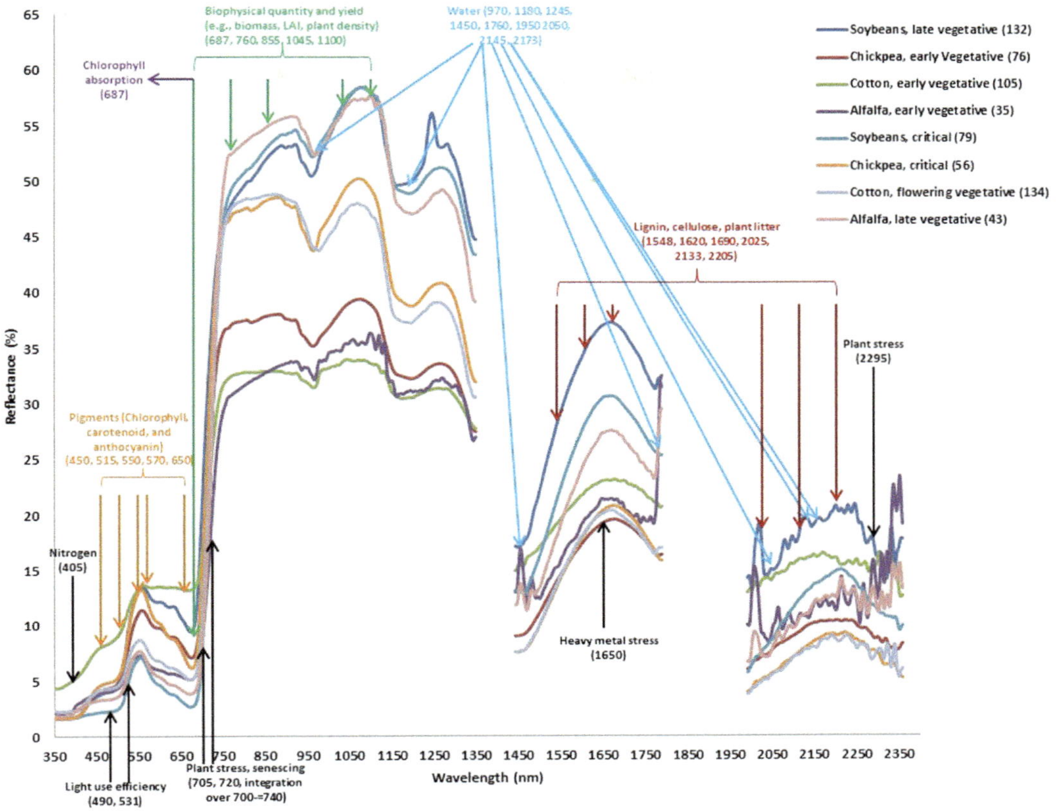

FIGURE 10.22 Hyperspectral features, demonstrated using field spectroradiometer data. Importance of various portions of hyperspectral data in characterizing biophysical and biochemical quantities of crops and vegetation. (Author's original contribution.)

3. **Matching class spectra with ideal spectra** to identify and label classes.
   The principal approach in SMT is to match the shape, magnitude, or (preferably) both to an ideal or target spectrum (pure class or "endmember"). Thenkabail et al. (2007) proposed and implemented SMT for multi-temporal data illustrated in Figure 10.23. The qualitative pheno-SMT approach concept remains the same for hyperspectral data (replace the number of bands of temporal data with the number of hyperspectral bands).

The quantitative SMTs consist of Thenkabail et al. (2007): (a) Spectral Correlation Similarity (SCS)—a shape measure; (b) Spectral Similarity Value (SSV)—a shape and magnitude measure; (c) Euclidian Distance Similarity (EDS)—a distance measure; and (d) Modified Spectral Angle Similarity (MSAS)—a hyper angle measure. Figure 10.23 illustrates the SMT method on time-series spectra; a similar approach can be adopted for hyperspectral data analysis.

10.10.1.1 Continuum Removal through Derivative Hyperspectral Vegetation Indices (DHVIs)

Continuum removal is used to quantify absorption features while removing the effects of the overall reflectance and slopes in a spectral profile (Kokaly et al., 2003). Derivative hyperspectral vegetation indices (DHVIs) are one form of continuum removal (Thenkabail et al., 1999). They are computed

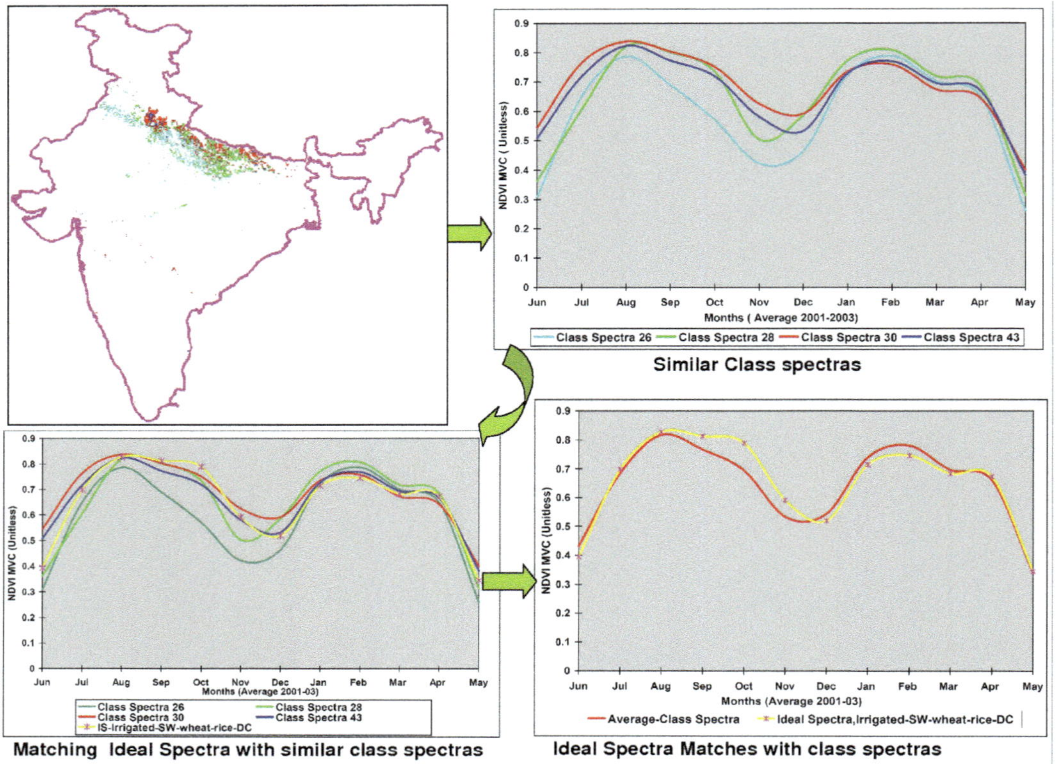

FIGURE 10.23 Pheno-Spectral Matching Techniques (SMTs). In SMTs, the class temporal profiles (NDVI curves) are matched with the ideal temporal profile (quantitatively based on temporal profile similarity values) to group and identify classes as illustrated for a rice class in this figure. Illustration of double-crop (DC) irrigation. The NDVI spectra of the four classes (C-26, C-28, C-30, and C-43) of DC irrigation are "matched" with ideal spectra (shown in yellow) for the same. This is a qualitative illustration of SMTs. For quantitative methods refer to Thenkabail et al. (2007). (Author's original contribution.)

by integrating an index over a certain wavelength range (e.g., 600 to 700 nm or 700 to 760 nm) (Equation 10.7):

$$DHVI = \lambda_1 L \frac{\rho'(\lambda_i) - \rho'(\lambda_j)}{\Delta\lambda_1} \tag{10.7}$$

where i and j are band numbers, λ = center of wavelength. The process of obtaining the DHVI value for 600 nm to 700 nm is as follows: (a) DHVI1 = lambda 1 (e.g., $\lambda_1 = 600$ nm) versus lambda 2 (e.g., $\lambda_2 = 610$ nm). The difference in reflectivity at these two bands is then divided by their bandwidth ($\Delta\lambda_I$ = 10 nm); (b) DHVI2 = the process is repeated for lambda 1 (e.g., $\lambda_1 = 610$ nm) versus lambda 2 (e.g., $\lambda_2 = 620$ nm). The difference in reflectivity of these two bands is then divided by their bandwidth ($\Delta\lambda_I$ = 10 nm); and (c) DHVIn = so on to lambda 1 (e.g., $\lambda_1 = 690$ nm) versus lambda 2 (e.g., λ_2 = 700 nm). The difference in reflectivity of these two bands is then divided by their bandwidth ($\Delta\lambda_I$ = 10 nm). Finally, DHVI1, DHVI2, and so on to DHVIn are added to get a single integrated DHVI value over the entire 600 to 700 nm range.

The DHVIs can be derived over various wavelengths such as 400 to 2500 nm, 500 to 600 nm, 600 to 800 nm, and any other wavelength useful for the particular application. There are opportunities to further investigate the significance of DHVIs over different wavelengths for a wide array of applications.

10.11 PRINCIPAL COMPONENT ANALYSIS (PCA)

Another common, powerful, and useful feature selection method for hyperspectral data analysis is principal component analysis (PCA) (Tian et al., 2020). The PCA performs the following functions:

1. **Reduces data volumes.** This happens since the PCA generates numerous principal components (PCs) (as many as the number of wavebands), but the first few PCs explain almost all the variability of the data. The first PC (PC1) explains the highest, followed by the others. Since each PC is constituted based on information from all the bands (e.g., PC1 = factor loading for band 1 * band 1 reflectivity +. . . . + factor loading for band n * band n reflectivity), the PCs have the explanatory power of the hyperspectral bands without the redundancy;
2. **Provides new single bands of information** (e.g., PC1, PC2), each of which (e.g., PC1) actually has the information derived from all the hyperspectral narrowbands (HNBs). These new bands of information (e.g., PC1) can then be used to classify an area (e.g., to establish crop types) or used to model crop biophysical or biochemical quantities.
3. **The power of principal components** can be used to discriminate crop types, land cover themes, or species (e.g., Figure 10.24).

10.12 SPECTRAL MIXTURE ANALYSIS (SMA) OF HYPERSPECTRAL DATA

Hyperspectral data have a great ability to distinguish specific objects based on their unique signatures. For example, wheat versus barley crops are distinguished based on spectral reflectivity in two HNBs, each 10 nm wide, and centered at 687 nm and 855 nm (e.g., Figure 10.25). Often, we find multiple objects or classes within a single pixel. In such situations, we need to perform Spectral Mixture Analysis (SMA) (Damarjati et al., 2022; Fitzgerald et al., 2019; Masudul Islam et al., 2022) or Independent Component Analysis (ICA) (Jayaprakash et al., 2020; Li et al., 2022) to unmix the spectral signatures within each pixel.

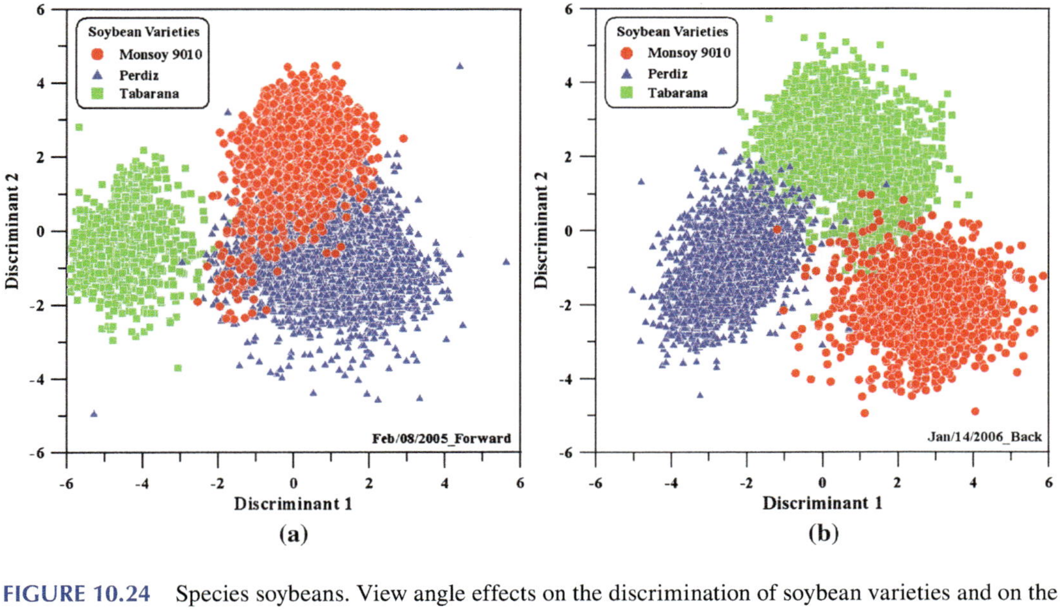

FIGURE 10.24 Species soybeans. View angle effects on the discrimination of soybean varieties and on the relationships between vegetation indices and yield using off-nadir Hyperion data. Projection of the Hyperion discriminant scores of the three soybean varieties in the (a) forward-scattering and (b) back-scattering directions for different years. (Source: Galv˜ao et al., 2009.)

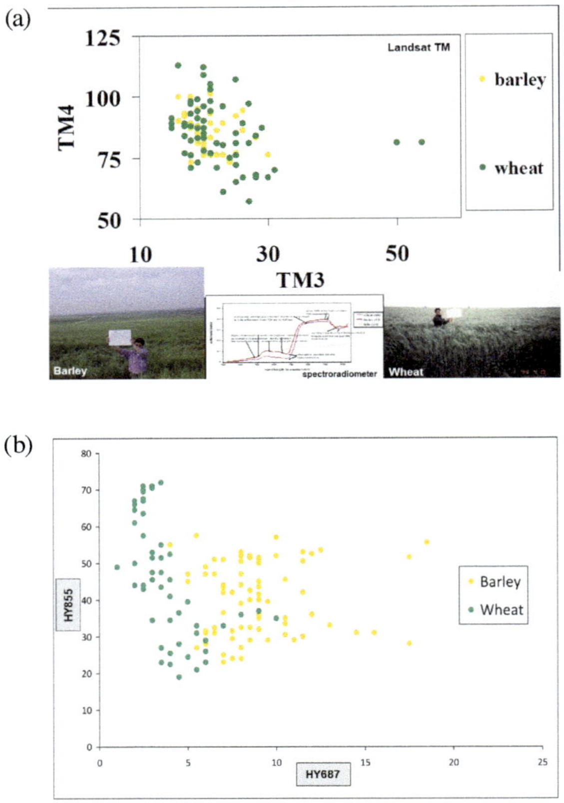

FIGURE 10.25 Crop type differentiation. Differentiating barley from wheat using two Landsat TM broadbands (a) and two hyperspectral narrowbands (b). (Author's original contribution.)

The reference spectra for SMA are derived from "endmembers" (i.e., characteristic spectra of each component). Once all the classes in the image are identified, it is possible to use linear or nonlinear spectral unmixing to determine how much of each material is in each pixel (Figure 10.26).

The concept of unmixing hyperspectral data is illustrated by showing Hyperion spectra of vegetation fractional cover in Figure 10.27 and minerals in Figure 10.28. Sub-pixel mineral mapping of a porphyry copper belt using EO-1 Hyperion data in Figure 10.29 involved mineral spectra extracted from Hyperion compared to convolved spectra from field samples and reference library spectra (Figures 10.26 and 10.27). Extensive discussions on linear and nonlinear SMA can be found in Plaza et al. (2012).

10.13 MACHINE LEARNING ALGORITHMS (MLAS) FOR HYPERSPECTRAL DATA ANALYSIS

Machine Learning Algorithms (MLAs) suitable for hyperspectral analysis can be broadly categorized into two categories: supervised and unsupervised (Aneece and Thenkabail, 2021). Supervised MLAs are those where the number and types of classes are known. A subset of data is used to build a classification model, another subset to test the model and tune model parameters, and a third subset is used to validate the model on unseen data (Aneece and Thenkabail, 2021).

Unsupervised MLAs do not need that a priori information (Aneece and Thenkabail, 2021). This is beneficial when the user does not know which, or how many, classes are present in a dataset. It is

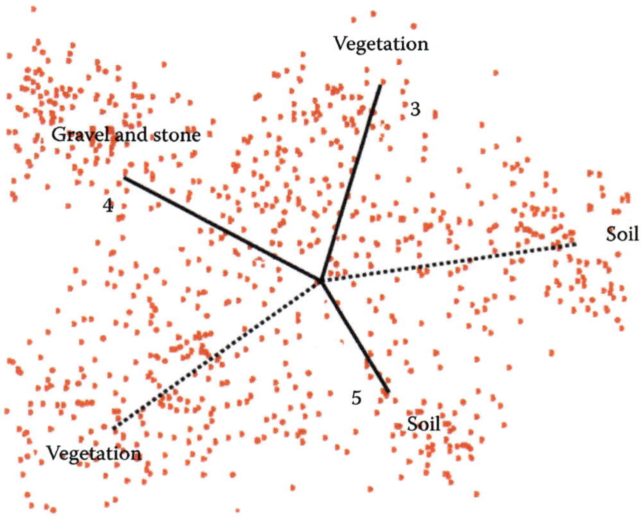

FIGURE 10.26 Endmembers. Arid land characterization with EO-1 Hyperion hyperspectral data. Endmember extraction in n-dimension visualizer using bands 3, 4, and 5 of the Maximum Noise Fraction transform (MNF) Hyperion image. The visualizer allows users to see a three-dimensional point cloud representing individual samples distributed based on spectral similarity across the three bands. The samples that are furthest apart are usually endmembers associated with particular classes. (Source: Jafari and Lewis, 2012.)

also useful in cases with limited ground reference data for labeling classes. Some common examples of supervised MLAs are support vector machines (SVM), random forests (RF), and AdaBoost tree-based ensembles; examples of unsupervised MLAs include isoclass clustering and k-means clustering (Aneece and Thenkabail, 2021).

10.13.1 Supervised Machine Learning Algorithms (MLAs)

10.13.1.1 Support Vector Machines (SVM)

Support vector machines (SVM) is a machine learning supervised classification approach that determines support vectors (samples closest to the hyperplane) to define a hyperplane that best separates classes (Abdi, 2020; Aneece and Thenkabail, 2021; Lu et al., 2020; Lv and Wang, 2020; Sheykhmousa et al., 2020; Vali et al., 2020; Wu and Zhang, 2019). Hyperplanes can be linear or nonlinear (e.g., radial basis function, polynomial); besides kernel type, parameters that need to be selected and optimized include cost, gamma, and other kernel-specific parameters (Aneece and Thenkabail, 2021).

Unlike the feature selection approach, data dimensionality is not an issue here; any number of bands can be used, but more bands can lead to more support vectors and overfitting. The process involves supervised training of classes, based on sufficient and accurate knowledge of the class (e.g., ground data), where one can use all or some of the hyperspectral bands to train the algorithm. Once the algorithm is sufficiently trained, it can be run on the rest of the data to gather the same class occurring in other areas. Figure 10.30a shows the classification performed using all 272 AISA hyperspectral bands based on the SVM algorithm. In Figure 10.30b the same classification is performed using only 51 of the most important AISA hyperspectral bands. The results of the 51-band classification output (Figure 10.30b) are comparable to the 272-band classification output (Figure 10.30a) in most areas; there is significant uncertainty in the Northern portion of the image. Studies have shown that by using only 1% of training pixels per class, almost 90% overall

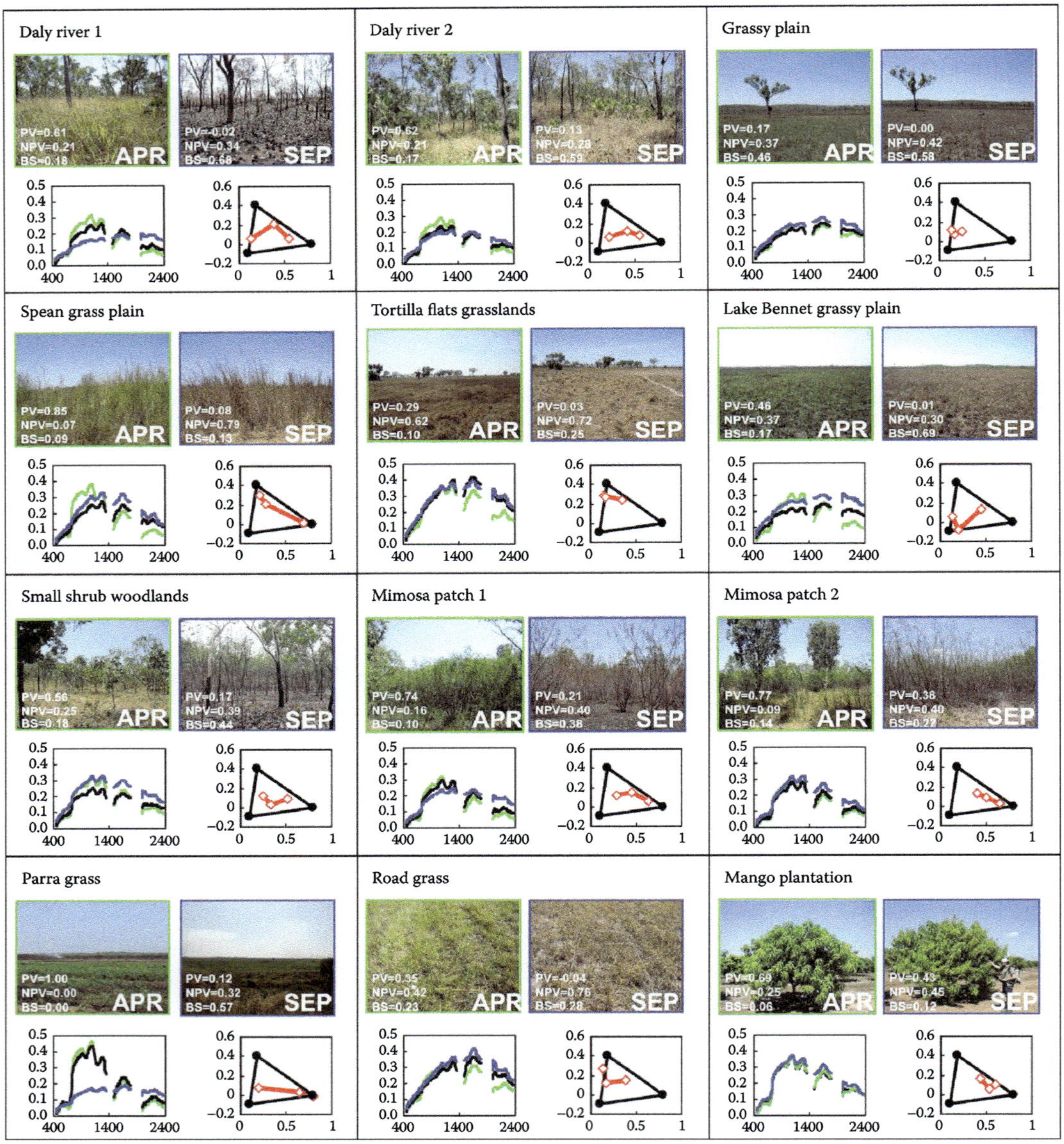

FIGURE 10.27 Unmixing. Qualitative assessment of Hyperion unmixing of vegetation fractional cover. Each set of pictures and graphs corresponds to one of 12 sites visited from 16 to 19 of May and 29 to 31 of August 2005. The left graphs show the reflectance spectra derived from Hyperion images for the April (green curve), July (black curve), and September (blue curve) images from 400 to 2,400 nm. The right graphs show the position of each spectrum in the Normalized Difference Vegetation Index or NDVI (x-axis) (detecting live, green vegetation) and Cellulose Absorption Index or CAI (y-axis) (detecting non-photosynthetic vegetation) space from April to September (red dots and line) and the position of the endmembers (black lines). The derived photosynthetic vegetation (fPV), non-photosynthetic vegetation (fNPV), and bare soil (fBS) fractions are shown in the plots to the right of the respective spectral profiles and are critical for natural resource management and modeling carbon dynamics. (Source: Guerschman et al., 2009.)

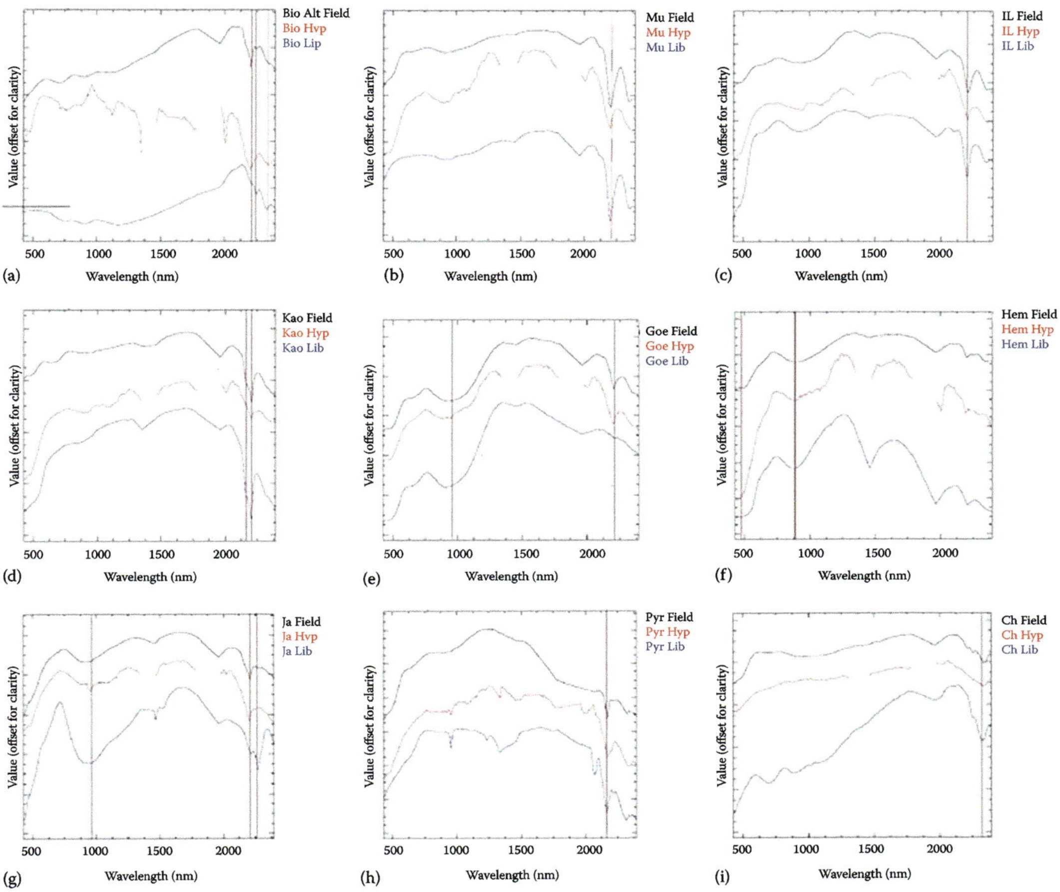

FIGURE 10.28 Mineral mapping spectra. Sub-pixel mineral mapping of a porphyry copper belt using EO-1 Hyperion data. Mineral spectra extracted from Hyperion compared to convolved spectra from field samples and reference library spectra. (a) Biotite (Bio), (b) Muscovite (Mu), (c) Illite (Il), (d) Kaolinite (Kao), (e) Goethite (Goe), (f) Hem (Hem), (g) Jarosite (Ja), (h) Pyrophylite (Pyr), (i) Chlorite (Ch). Hyp and Lib are abbreviations of Hyperion and Library, respectively. The red vertical lines indicate locations of diagnostic absorption features. (Source: Hosseinjani Zadeh et al., 2014.)

classification accuracies are obtained using SVM methods (Bajwa and Kulkarni, 2012; Ramsey III and Rangoonwala, 2012).

10.13.1.2 Random Forest and AdaBoost

Random Forest and AdaBoost (Adaptive Boosting) are two tree-based ensemble classifiers, in which numerous decision trees are generated and a majority vote is used to determine a class label (Aneece and Thenkabail, 2021; Brovelli et al., 2020; Chen et al., 2020; Lu et al., 2020; Sheykhmousa et al., 2020; Vali et al., 2020). Parameters that need tuning include the number of trees, number of samples in a node, out-of-bag fraction, and variables per split (Aneece and Thenkabail, 2021). These classifiers serve two purposes:

1. Help select hyperspectral bands that are important and those that are redundant
2. Classify hyperspectral data through decision tree-based classifiers

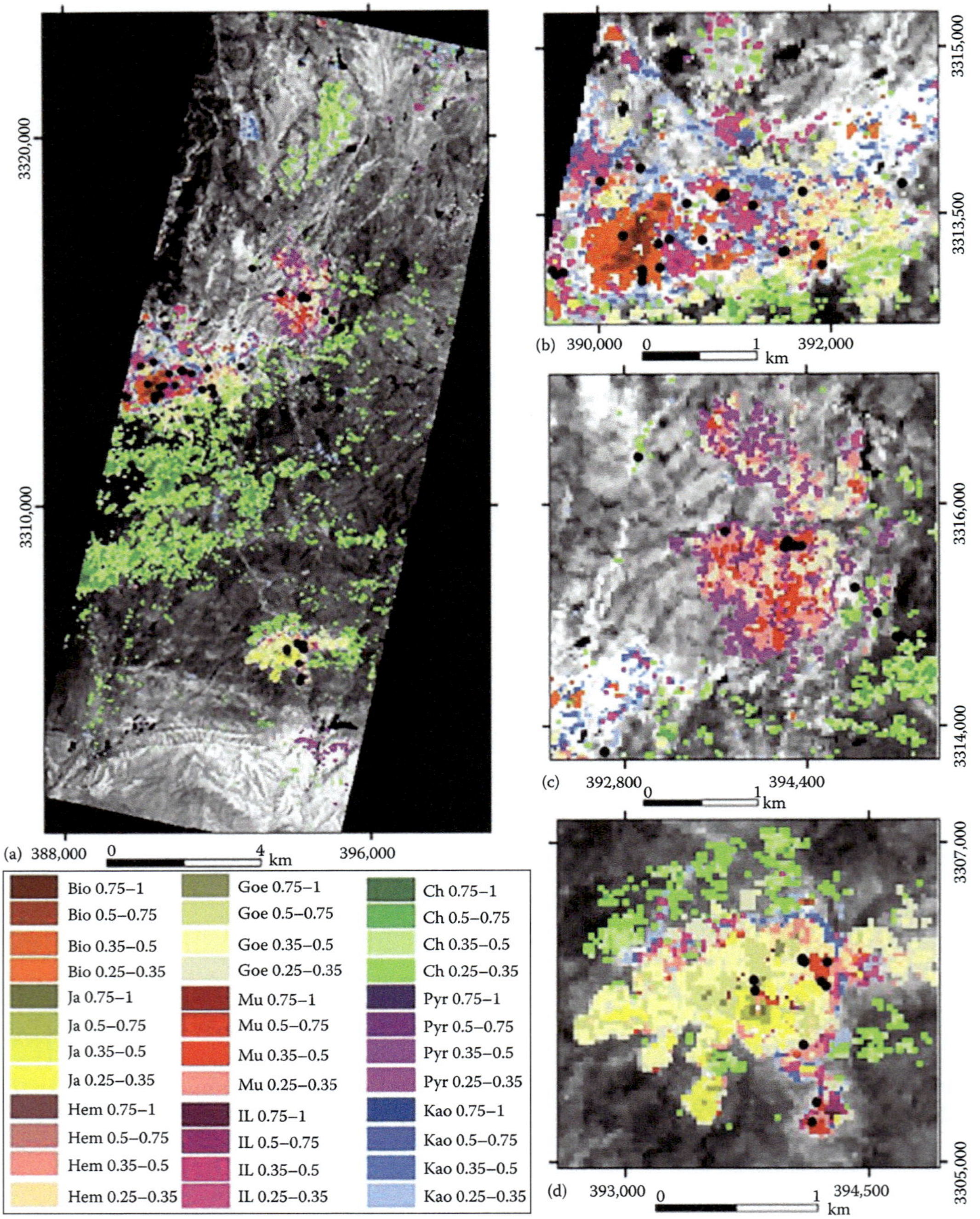

FIGURE 10.29 Mineral Mapping. Sub-pixel mineral mapping of a porphyry copper belt using EO-1 Hyperion data. Thematic mineral maps using sub-pixel Mixture Tuned Matched Filtering (MTMF) method (a) Final classification image map of alteration minerals derived from MTMF algorithm (top left). (b) Sarcheshmeh mine (top right); (c) Sereidun (middle right), and (d) Darrehzar (bottom right).) Bio, Mu, Il, Kao, Goe, Hem, Ja, Pyr, and Ch indicate Biotite, Muscovite, Illite, Kaolinite, Goethite, Hematite, Jarosite, pyrophyllite, and Chlorite, respectively. These values indicate percentages of each mineral at the pixel. For instance, a value of 0.25 shows that 25% of the pixel contains the selected mineral. (Source: Hosseinjani Zadeh et al., 2014.)

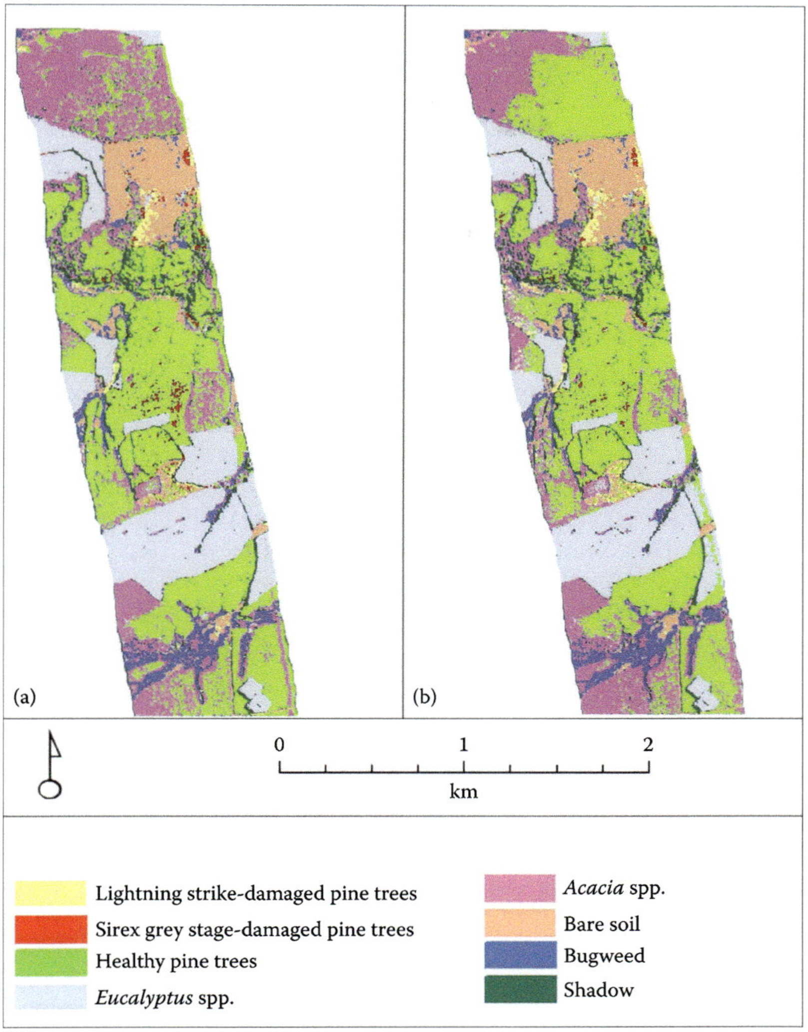

FIGURE 10.30 SVM. Detecting *Sirex noctilio* gray-stage-attacked and lightning-struck pine trees using airborne hyperspectral data, random forest, and support vector machines classifiers. Classification maps obtained using support vector machine (SVM) classification algorithm, including all (a) or only the 51 most important (b) Airborne Imaging System for different Applications (AISA) Eagle spectral bands. The AISA image spatial resolution was about 2 m and there were 272 spectral bands ranging from 393.23 to 994.09 nm (VNIR: Visible Near-Infrared) with bandwidths between 2 and 4 nm. (Source: Abdel-Rahman et al., 2014.)

This approach has been discussed in great detail by Chan and Paelinckx (2008) for the thorough classification of detailed ecotopes using hyperspectral data (Figures 10.31 and 10.32). They gathered extensive hyperspectral data for: (1) six grassland classes and (2) ten tree classes (Figure 10.31). In terms of accuracy, Random Forest and AdaBoost perform similarly and both have outperformed older neural network classifiers (Chan and Paelinckx, 2008). Both feature selection routines, the best-first search and the out-of-bag ranking index under Random Forest, are successful in identifying substantially smaller band subsets that attain almost the same accuracy as all the bands (e.g., Figure 10.32, Chan and Paelinckx (2008)). There are many approaches to selecting spectral wavebands for obtaining the best classification results. For agriculture and vegetation studies, one could use various combinations of band selection (e.g., Table 10.6) depending on the number of bands one decides to use, the classification accuracies desired, and the need to overcome the Hughes phenomenon.

10.13.2 Unsupervised Machine Learning Algorithms (MLAs)

Unsupervised MLAs are advantageous in situations where the classes in an area of interest are unknown (Abbas et al., 2016). They include various clustering algorithms, such as k-means clustering and its variant ISOCLASS clustering, in which data are mapped in n-dimensional space and split based on a selected distance measure between the cluster centers (e.g., Euclidean, Chebyshev, Manhattan) (Aneece and Thenkabail, 2021; Biradar et al., 2009; Duveiller et al., 2015; Teluguntla et al., 2017; Thenkabail et al., 2009; Tsyganskaya et al., 2018).

K-means clustering is easy to implement, computationally efficient, results in high-quality clusters, and is still widely used today (Abbas et al., 2016; Ahmed and Akter, 2017; Tsyganskaya et al., 2018). For example, Tsyganskaya et al. (2018) used k-means clustering for image segmentation, with which object-based classification of flood extent outperformed pixel-based classification. Similarly, Santos et al. (2021) used k-means for segmenting images into distinct crops and soil classes. Wan et al. (2018) used k-means clustering to classify oilseed rape flowers so the flower coverage could be used to predict yield. Ahmed and Akter (2017) used k-means clustering to classify land cover.

Several variations of k-means clustering have also been proposed (Liu et al., 2021a; Sinaga and Yang, 2020).

For example, the number of clusters remains constant in k-means clustering. In contrast, ISOCLASS clustering has the ability to split and merge clusters to obtain the optimal number of clusters based on thresholds such as the minimum number of points in a cluster and the maximum standard deviation of a cluster (Abbas et al., 2016). This method has been used to classify land use/land cover for studying flood-based farming (Gumma et al., 2022a), climate-smart agriculture (Traor´e et al., 2021), soil and water conservation (Birhanu et al., 2019), water availability (Garg et al., 2020), and effects of irrigation projects on crop yields (Gumma et al., 2023).

10.13.3 Advances in Artificial Intelligence and Cloud Computing

Artificial intelligence (AI) informs agriculture in several ways including precision farming and irrigation, crop breeding, crop yield prediction, disease and pest detection, and task automation (Katoch et al., 2023). Jung et al. (2021) list commercially available AI-based tools for agricultural applications. While traditional MLAs have been successfully used for hyperspectral remote sensing, the complex and nonlinear nature of hyperspectral data (e.g., atmospheric effects, intraclass variation, and interclass similarities (Zhu et al., 2017)) could be more efficiently analyzed using Deep Learning (DL) models (Li et al., 2019). For example, Zhao et al. (2019) found a DL method (convolutional neural network) outperformed a traditional ML method (Random Forest) in classifying vineyards with an increase in overall accuracy by 9.87% (Figure 10.33).

Unlike more traditional ML models which still rely on user-directed feature engineering, DL models automatically select the most useful feature space through auto-encoders and numerous hierarchical layers (Audebert et al., 2019; Li et al., 2019; Ma et al., 2019; Zhu et al., 2017). This allows users to put in numerous variables and allow the model to select the best variables and avoid the Hughes Phenomenon. To do this variable selection, these models need enormous sample sizes for training and validation. DL models have outperformed ML models for vegetation studies and agricultural applications (Lee et al., 2020; Liu et al., 2021b). With advances made in DL through open-source frameworks, such as TensorFlow and PyTorch, Caffe, Theano, and Microsoft-CNTK (Zhu et al., 2017), sophisticated neural networks can be built and modified for various applications

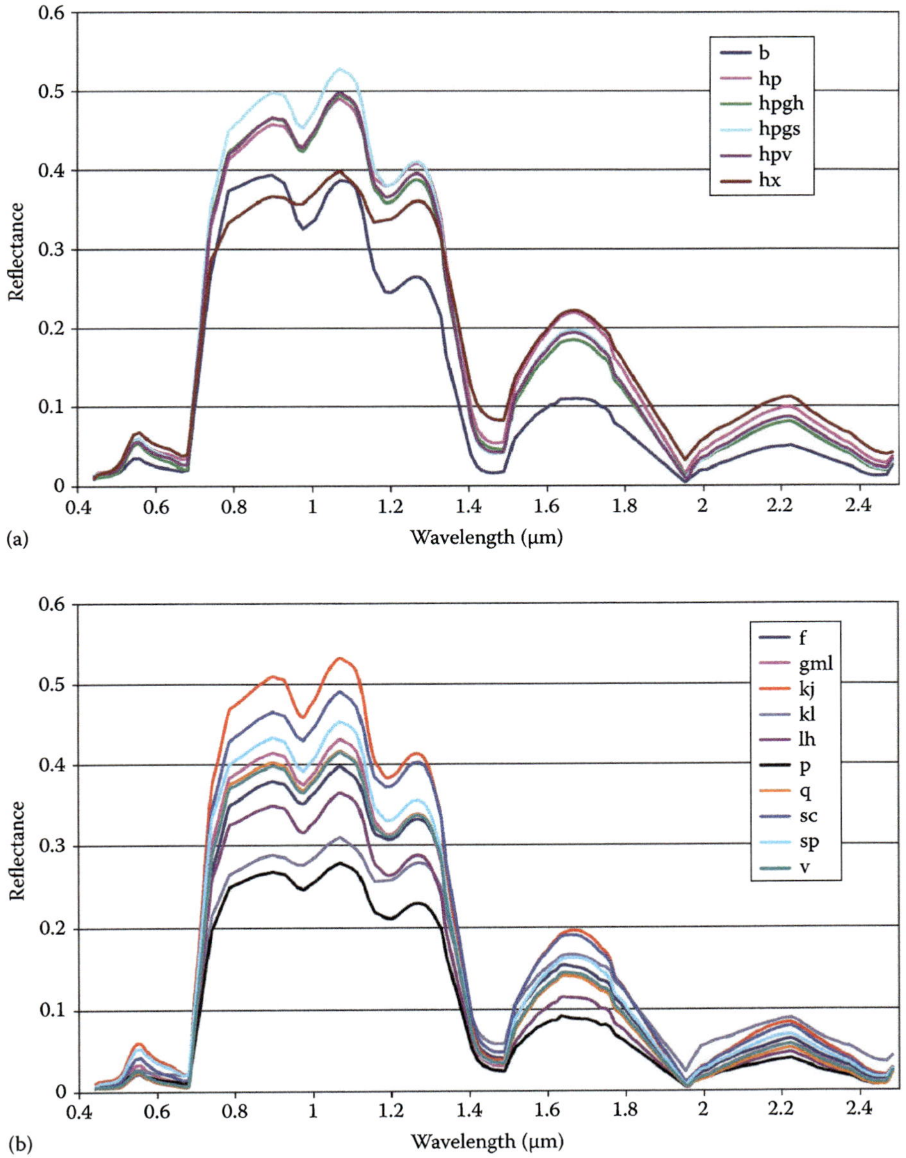

FIGURE 10.31 Tree-based ensembles. Evaluation of Random Forest and AdaBoost tree-based ensemble classification and spectral band selection for ecotope mapping using airborne hyperspectral imagery. Mean spectrum of the (a) six grassland classes and (b) ten tree classes. Note: b = grassland, arable land; hp = grassland, species poor improved grassland (normally more homogenous for the whole parcel); hpgh = grassland, semi-natural grassland; hpgs = grassland, species rich improved grassland (between hpgh and hp); hpv = grassland, grassland with patches hp and either patches hpgs or hpgh; hx = grassland, grass monocultures (equal to arable land sown with grasses of one or more years); f = tree/tall veg, deciduous forest, dominated by beech (Fagus sp.); gml = tree/tall veg, plantation of deciduous tree species other than beech, oak, alder and poplar; kj = tree/tall veg, tall tree orchard; kl = tree/tall veg, low tree orchard; lh = tree/tall veg, poplar plantation; p = tree/tall veg, conifer plantation; q = tree/tall veg, deciduous forest, dominated by oak trees (Quercus sp.); sp = tree/tall veg sc, scrubs of clearings and scrubs on abandoned land; sp = tree/tall veg, thorn thicket; v = tree/tall veg, Woodland of alluvial soil, fens and bogs (mostly dominated by alder = Alnus sp.). (Source: Chan and Paelinckx, 2008.)

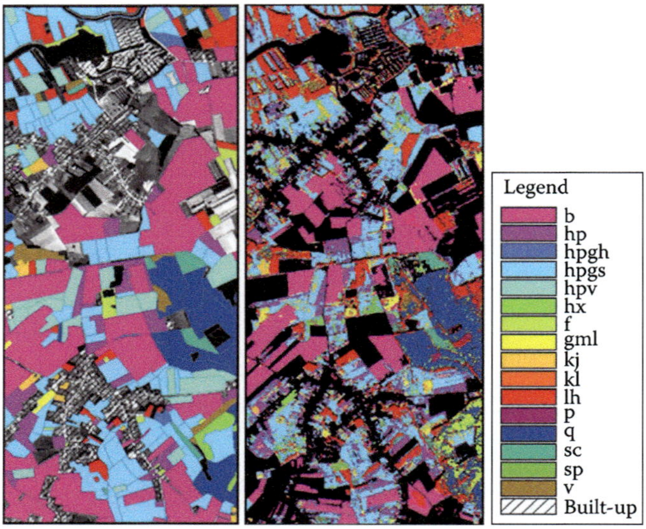

FIGURE 10.32 Mapping with ensembles. Evaluation of Random Forest and AdaBoost tree-based ensemble classification and spectral band selection for ecotope mapping using airborne hyperspectral imagery. On the left is the ground truth image of the Biological Valuation Map. On the right is the Biological Valuation Map classification based on airborne hyperspectral data using 99 trials of AdaBoost with 21 bands selected by the best-first search method. The black areas represent unclassified land covers that have been masked out. (Source: Chan and Paelinckx, 2008.)

TABLE 10.6

Best Hyperspectral Narrowband (HNB) Combinations Based on the Number of Bands Available to Classify Crops or Vegetation.

Best 4 bands	550, 680, 850, 970
Best 6 bands	550, 680, 850, 970, 1075, 1450
Best 8 bands	550, 680, 850, 970, 1075, 1180, 1450, 2205
Best 10 bands	550, 680, 720, 850, 970, 1075, 1180, 1245, 1450, 2205
Best 12 bands	550, 680, 720, 850, 910, 970, 1075, 1180, 1245, 1450, 1650, 2205
Best 16 bands	490, 515, 550, 570, 680, 720, 850, 900, 970, 1075, 1180, 1245, 1450, 1650, 1950, 2205
Best 20 bands	490, 515, 531, 550, 570, 680, 720, 850, 900, 970, 1075, 1180, 1245, 1450, 1650, 1725, 1950, 2205, 2262, 2359

Source: adopted and modified from Thenkabail et al. (2014, 2011, 2013).

including classification (Audebert et al., 2019; Li et al., 2019; Ma et al., 2019), regression (Audebert et al., 2019; Kamilaris and Prenafeta-Boldu´, 2018; Li et al., 2019), pansharpening (Zhu et al., 2017), data fusion (Ma et al., 2019; Zhu et al., 2017), image coregistration (Zhu et al., 2017), and image segmentation (Ma et al., 2019). Within agriculture, DL has been used for leaf, crop, and weed classification, disease detection, object detection, land cover classification, plant and plant phenology recognition, root and soil segmentation, crop yield estimation, soil moisture estimation, fruit counting, monitoring farm animal growth, and weather prediction (Kamilaris and Prenafeta-Boldu´, 2018). For example, Wang et al. (2020) used CNN to map crop types in Southeast India. Yubo et al. (2020) used DL for land use/land cover change reconstruction in Northeast China.

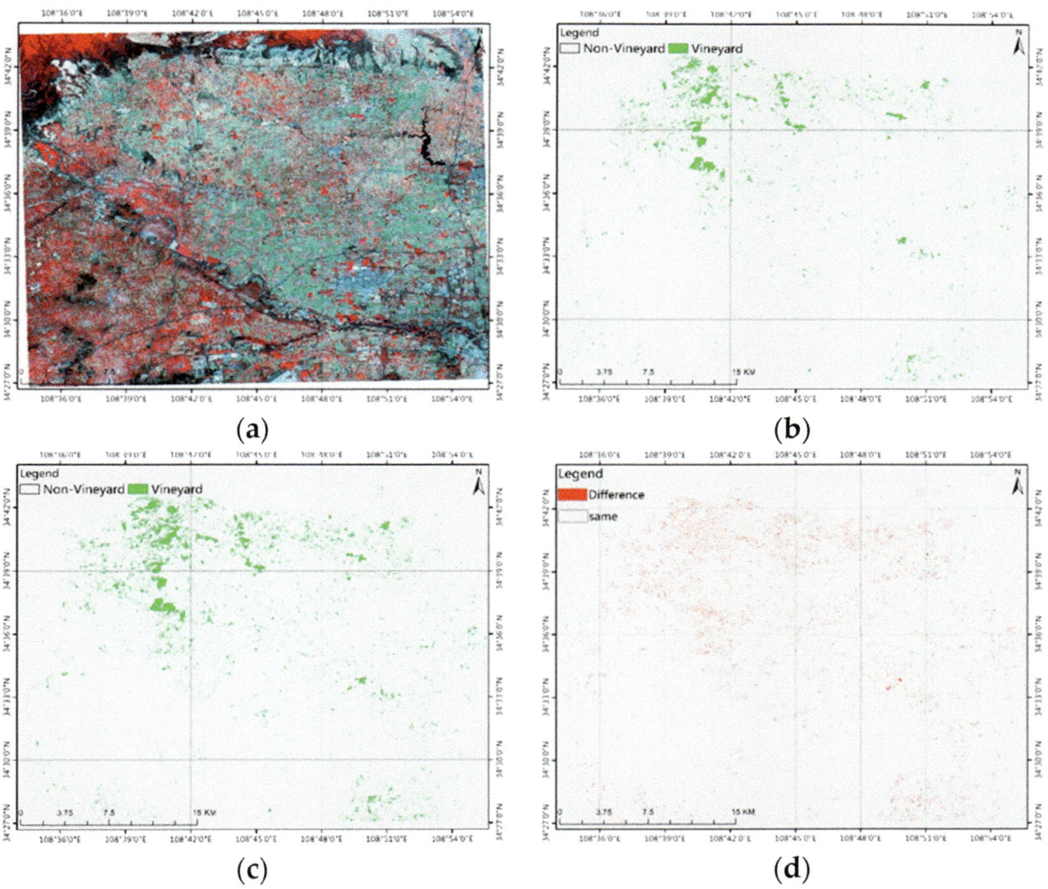

FIGURE 10.33 Mapping vineyards. A comparison of vineyard classification results using the machine learning Random Forest (RF) approach versus the deep learning Convolutional Neural Network (CNN) approach where (a) is the original false-color image, (b) result of the CNN, (c) result of RF, and (d) difference between CNN and RF. (Source: Zhao et al., 2019.)

Convolutional Neural Networks (CNNs) are some of the most commonly used DL models, consisting of convolutional, pooling, and fully connected layers for synthesizing the most relevant information and a classification or regression algorithm. The model would likely use batch normalization, stochastic gradient descent (SGD) as the back-propagation algorithm for training the model, the ReLU activation function to speed up model convergence, dropout to avoid overfitting, and a Softmax optimizer (Li et al., 2019). An example of the CNN architecture is shown in Figure 10.34.

Hyperspectral DL models taking both spectral and spatial information into account are usually more robust, efficient, and accurate than those only using spectral information (Audebert et al., 2019; Li et al., 2019; Ma et al., 2019). Unsupervised learning, data augmentation, and transfer learning can ameliorate issues of limited labeled samples for training and validation (Audebert et al., 2019; Li et al., 2019).

In addition, current cloud computing platforms and High-Performance Computing (HPC) systems (of which there are thousands) enable researchers to run these complex models without relying on local data repositories or compute capabilities (Tamiminia et al., 2020). For example, Figure 10.35 shows the performances of several classifiers in Google Earth Engine across studies (Tamiminia et al. (2020)).

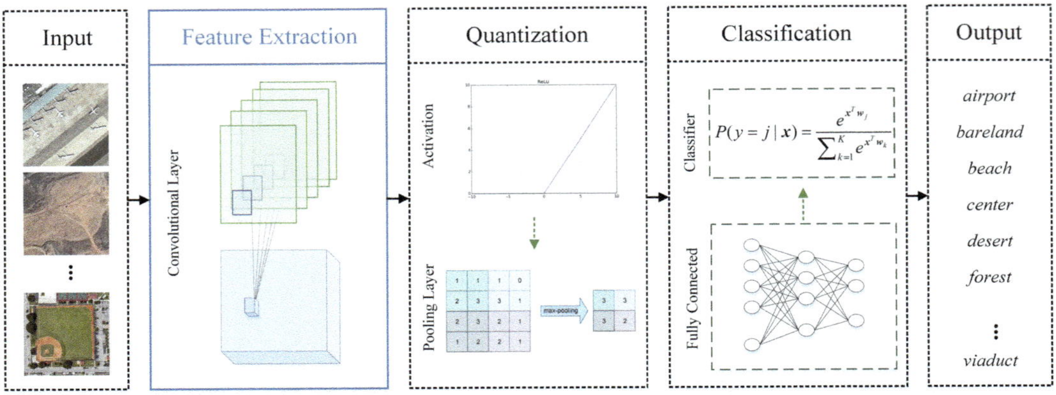

FIGURE 10.34 Convolutional Neural Network. An example of a convolutional neural network architecture. (Source: He et al., 2019.)

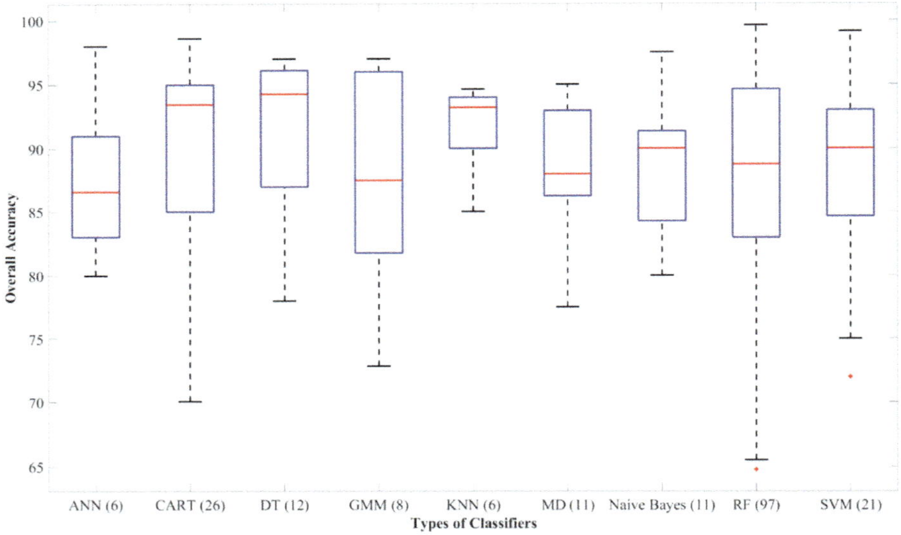

FIGURE 10.35 Classifier comparison. A comparison of performances of different classifiers in Google Earth Engine. (Source: Tamiminia et al., 2020.)

10.14 CASE STUDIES

Hyperspectral remote sensing can be used for a slew of applications ranging from agriculture studies, soil studies, mineral identification and mapping, water quality assessment, urban material characterization, etc. Some of the case studies highlighting the use of the aforementioned techniques are now described.

10.14.1 Hyperspectral Remote Sensing for Mineral Mapping

Geologists have used remote sensing data for a very long time. This technology is used for regional mapping, structural interpretation, and to aid in prospecting for ores and hydrocarbons. Earlier in remote sensing, multispectral images (Landsat Multispectral scanner and Thematic Mapper) were used for mineral mapping (Bedini, 2011; Carrino et al., 2015; Clark et al., 2006; Honarmand, 2016; Parashar et al., 2016; Pour and Hashim, 2012, 2011; Richards and Jia, 2006; Vincheh and Arfania, 2017). Later, geologists developed band ratio techniques and selective principal component analysis (Crosta and McMoore, 1989) to produce iron oxide and hydroxyl images that could be related to hydrothermal alteration. The beginning of the Advanced Spaceborne Thermal Emission and Reflectance Radiometer (ASTER) allowed geologists to map alteration facies as they were now able to produce qualitative surface mineral maps of clay minerals (kaolinite, illite), sulfate minerals (alunite), carbonate minerals (calcite, dolomite) and iron oxides (hematite, goethite) (van der Meer et al., 2012).

Multispectral images have discrete (as opposed to contiguous) spectral bands for pixels (Ting and Fei, 2012), and the majority of the pixels are used for identification of the minerals. However, in some regions, minerals are not exactly identified due to those discrete spectral bands as a result of non-homogeneous composition that involves minerals mixed with other land cover (e.g., vegetation or water) or land features (e.g., soil composition, color, texture, etc.). Minerals of the same group cannot be differentiated in those regions due to the previously described reason (Pippi, 1989).

A step towards quantitative and validated (subpixel) surface mineralogic mapping was taken with the arrival of high spectral resolution hyperspectral remote sensing. Hyperspectral remote sensing works on the basis of specific chemical bonds to identify and map the materials. When a sensor is flown over different geographic regions having the same minerals or rocks, the sensor can identify and map the minerals more accurately (Clark et al., 2006). Hyperspectral sensors provide spatial and spectral contiguous bands of the Earth's surface for mineral mapping (Goetz et al., 1985). This leads to a wealth of techniques to match image pixel spectra to library and field spectra, and to unravel mixed pixel spectra to pure endmember spectra to derive subpixel surface compositional information.

In the year 2000, EO-1 Hyperion was the first spaceborne hyperspectral sensor and was highly used for mineral mapping. Minerals mapped using Hyperion include calcite, dolomite, kaolinite, alunite, buddingtonite, muscovite (several varieties), hydrothermal silica, and zeolites (Clark et al., 2006; Parashar et al., 2016). The availability of spaceborne hyperspectral data is very limited; therefore airborne sensors like AVIRIS are being used by researchers for mineral mapping (Goetz and Srivastava, 1985; Kruse et al., 1993; Kruse, 1988; Kruse, 2002; Kruse et al., 2003). Airborne sensors provide high spatial resolution (2–20 m), and high spectral resolution (10–20 nm) data for various scientific domains (Kruse, 2002). Spectral reflectance curves have absorption features at different positions and those absorption peaks are used for analysis of imagery. Absorption-band parameters, such as position, depth, width, and asymmetry of the feature have been used to quantitatively estimate the composition of samples from hyperspectral field and laboratory reflectance data (van der Meer, 2004). In a nutshell, multispectral remote sensing allows geologists a qualitative assessment of surface composition while hyperspectral remote sensing enables quantitative mapping of surface mineralogy (van der Meer et al., 2012).

Utilizing airborne hyperspectral imagery, the examined case study by Jain and Sharma (2019) mainly focuses on mineral identification and mapping with the help of various algorithms such as Spectral Angle Mapper (SAM), Spectral Feature Fitting (SFF), and Mixture Tuned Matched Filtering (MTMF) using airborne hyperspectral data. The study area, a part of the Bhilwara district of Rajasthan, India, is located at the south-eastern region of the city of Jahazpur (Figure 10.36, Jain and Sharma, 2019). The location of the area is between the latitude 25°26′34.80″ to 25°30′7.20″ in the North and longitude 75°10′22.80″ to 75°14′16.80″ in the East. The area is of great importance for geologists due to the presence of iron ore along with dolomites. The majority of iron ore is mineralized in the form of banded iron formations (BIF). The oldest rocks recorded from the south-central part of the Rajasthan region belong to the Bhilwara Supergroup (BSG) and show the directional trend of NNE-SSW. Gupta et al. (1997) gave the stratigraphic sequence of the Bhilwara Supergroup (Table 10.7).

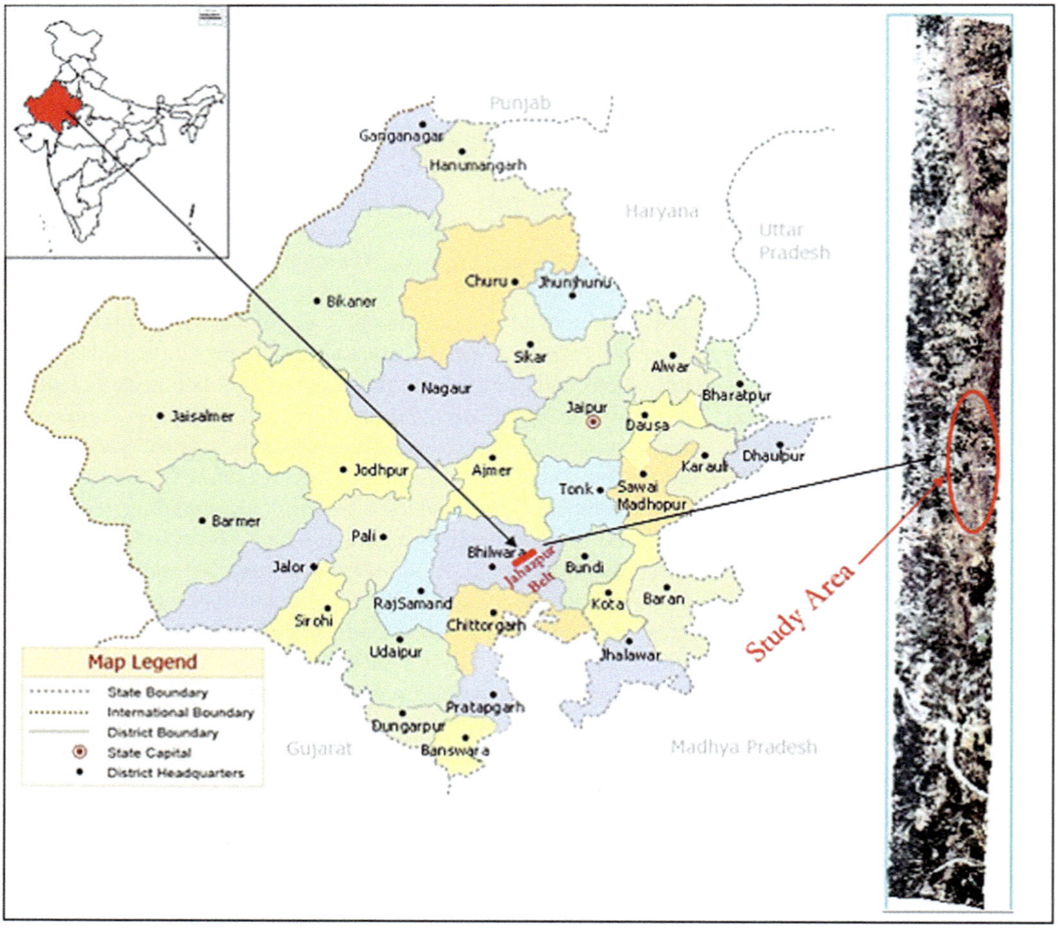

FIGURE 10.36 Study area: Bhilwara district of Rajasthan, India located in south-eastern region of the city of Jahazpur. (Source: Jain and Sharma, 2019.)

AVIRIS-NG L2 (level2) data, acquired in February 2016, with 425 contiguous bands and spectral resolution of 10 nm were used in the study. To extract endmembers, Minimum Noise Fraction (MNF) (Green et al., 1988) was used for the reduction of the dimensionality along with the Pixel Purity Index (PPI) algorithm proposed by Boardman et al. (1995) which uses PPI (van der Meer et al., 2012) for the extraction of spectrally pure pixels. Preprocessing of the data involved the removal of bad bands and bands with no information after which 363 bands remained for further processing. Fourteen MNF components were selected out of 363 components for endmember selection. Table 10.8 provides information about the pure pixel spectra obtained from the image and their score with the predefined spectral library and were further used for the pixel-based and sub pixel-based classification.

Spectra acquired from the imagery matched with the predefined mineral spectral library and characteristic absorption depth were used for the identification of minerals. These minerals belonged to clay (i.e., kaolinite and montmorillonite), metallic (i.e., pyrrhotite and limonite), dolomite categories and others like quartz, talc, augite, vermiculite, and barite. The authors identified 13 minerals (Figure 10.37) and out of 13 different endmembers, four endmembers were identified as clay (kaolinite: doublet absorption, 2159.53 nm—2209.61 nm; montmorillonite: 2194.59 nm). Out of those four clay endmembers, three endmembers fell under the category of kaolinite and

TABLE 10.7

Stratigraphic Sequence of the Study Area (modified after Gupta et al., 1997; GSI, 2004)

Age	Supergroup	Group	Formation	Lithology/Litho Units
Lower Proterozoic	Bhilwara Supergroup	Jahazpur Group	Jawal Formation	Dolomite with BIF Conglomerate & gritty quartzite
Archean		Hindoli Group	Sujanpura Formation	Metagreywacke Phyllite ± garnet

TABLE 10.8

Analysis of the Spectral Characteristics of the Pure Pixels Provided the Information about the Minerals Present in the Region [Author's original contribution.]

S. No.	Class Name (Mineral Name)	Total Score	Image Coordinates	
			X	Y
	Clay Minerals			
1.	Montmorillonite	767	561	3485
2.	Kaolin/Smectite 1	714	679	3673
3.	Kaolin/Smectite 2	723	674	3111
4.	Kaolin/Smectite 4	819	699	3746
	Metallic Minerals			
5.	Pyrrhotite	713	562	3048
6.	Limonite (+Kaolin/Smectite 2)	720 (628)	541	3680
	Carbonate Mineral			
7.	Dolomite	805	404	3105
	Other Minerals			
8.	Vermiculite	704	613	3806
9.	Augite	616	441	3204
10.	Quartz	777	688	3595
11.	Talc	695	609	3618
12.	Orthoclase	787	627	3412
13.	Barite (+Montmorillonite)	745 (703)	617	3074

smectite mixture but all three had variation in the mixture composition. The variation was due to the change in the percentage of potassium (K) in the chemical composition. The other two endmembers belonged to the iron-carrying (metallic) mineral class (pyrrhotite: 511.67 nm; limonite: 681.97 nm) and one endmember was identified for the carbonate (dolomite: 2314.80 nm) class. The remaining counts of endmembers were recognized as silicates, sulfates, oxides, etc., which were less abundant than the aforementioned minerals.

In the region, iron ore mineralization was present in the form of the banded iron formation along with dolomite. Minerals pyrrhotite (Fe(1-x)S; x = 0–0.2) and limonite (FeO(OH)) were outlined for the iron ore from the gossan deposit which was also present in the field. These minerals occurred with the intermixing of clay minerals, so, mapping of these minerals was highly affected by the clay minerals such as kaolinite (Al2Si2O5(OH)4) from kaolin group and montmorillonite ((Na,Ca)0,3 (Al,Mg)2Si4O10 (OH)2·n(H2O)) from smectite group.

Comparison of all three mineral maps has been done to understand and compare the capabilities of the techniques Spectral Angle Mapper (SAM), Mixture Tuned Matched Filtering (MTMF), and Spectral Feature Fitting (SFF) for mineral mapping.

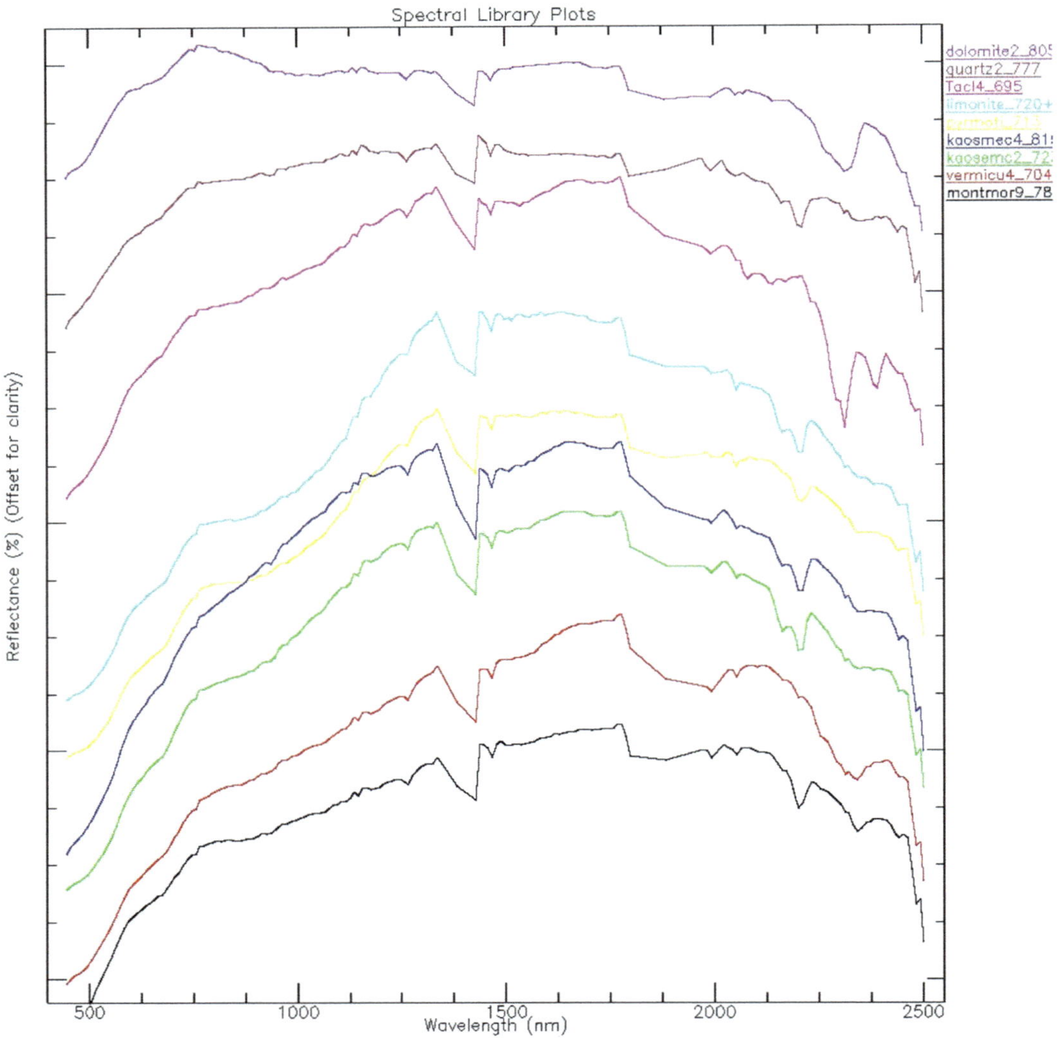

FIGURE 10.37 Spectral profiles: Spectra of the endmembers obtained from AVIRIS—NG data and their diagnostic absorption features. (Source: Jain and Sharma, 2019.)

Three different mineral mapping algorithms SAM, SFF, and MTMF were used for the identification and mapping of the minerals in the study region. The SAM algorithm mapped the minerals based on similarities of the reference and pixel spectra. Most portions of the study area fell under the clay mineral group and the presence of other minerals was lower. The SAM mineral mapping technique mapped the various minerals in the study area and out of those, pyrrhotite mineral was mapped with intermixing of the clay minerals. The pyrrhotite mineral was lower in abundance and most of the region was mapped by the clay minerals (as shown in Figure 10.38 SAM), so, the accuracy of the pyrrhotite mineral was very low.

The SFF algorithm also mapped the clay minerals in large amounts but less than the SAM algorithm, and mixing of clay with other minerals was also reduced. The SFF algorithm also mapped the pyrrhotite mineral with the melding of clay minerals, so, this algorithm was also not very suitable for mapping the iron ore (as shown in Figure 10.38 SFF). The accuracy of pyrrhotite mineral mapping with the SFF algorithm was 53.85%. Here also, clay minerals produced the obstructions for actual mapping of pyrrhotite minerals. The MTMF algorithm mapped the target minerals, lessening the number of false positives. As a result, clay minerals were mapped at genuine places, and intermixing was highly reduced. Mapping of pyrrhotite minerals with the help of the MTMF algorithm generated high abundance of these minerals. Using the MTMF algorithm, mixing of the pyrrhotite mineral with the clay minerals was less (as shown in Figure 10.38 MTMF) and produced better accuracy compared to the aforementioned techniques. The accuracy of the pyrrhotite mineral was 88.23%, which was higher than other techniques. So, the MTMF algorithm was highly useful for the identification and mapping of the minerals with better accuracy. A comparison between the maps was done based on mineral abundance (presence and absence) in the field.

The classified mineral map from the MTMF algorithm produced better results and greater accuracy compared to the other two mineral mapping methods. The overall accuracy for the MTMF classified mineral map was 80.49%

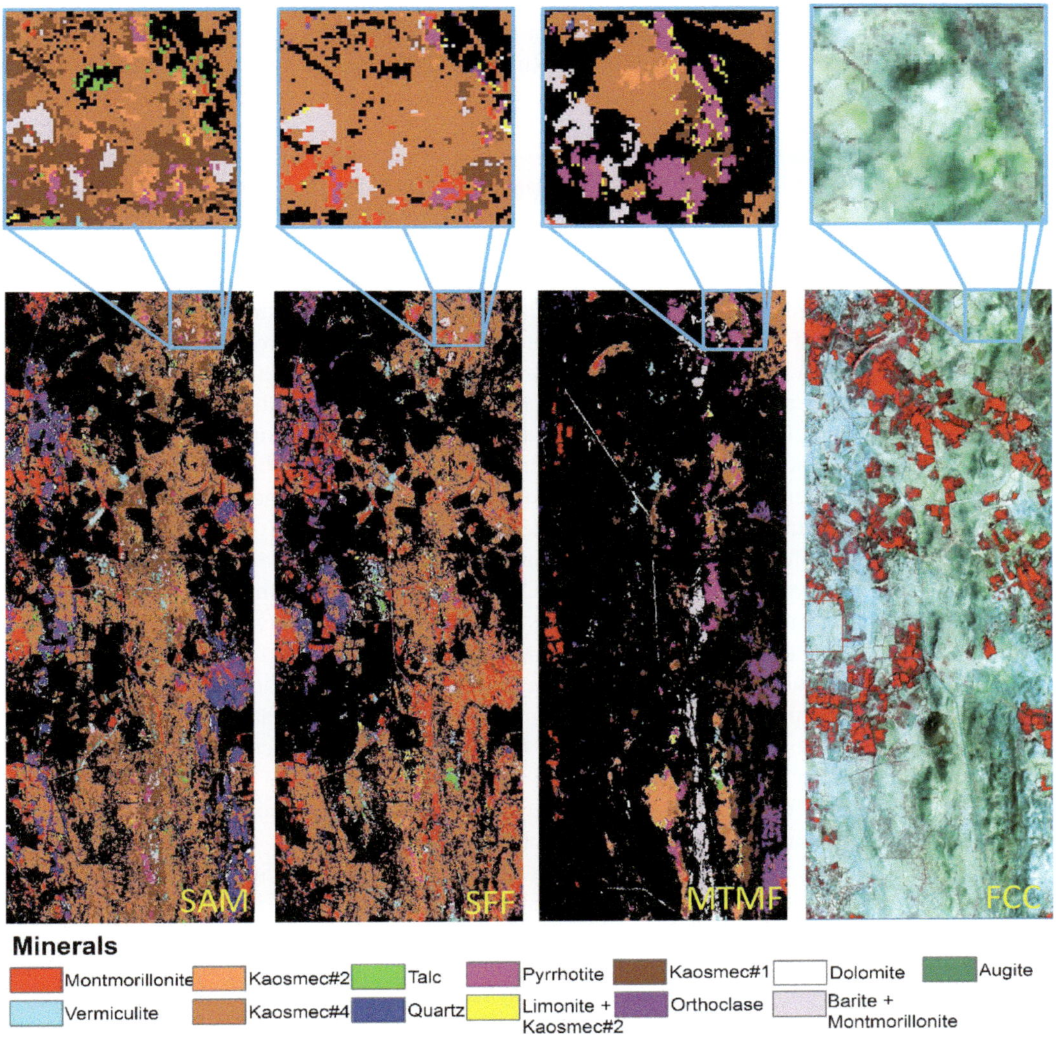

FIGURE 10.38 Mineral maps: Mineral maps showing the abundance of different minerals by using different algorithms (SAM, SFF, and MTMF) and representation of False Color Composite (FCC) imagery of the study area. Zoomed in, the area illustrates the intermixing of pyrrhotite and clay minerals. (Source: Jain and Sharma, 2019.)

10.14.2 Hyperspectral Remote Sensing for Urban Material Characterization

Identification of materials in an urban scenario using Hyperspectral Remote Sensing (HRS) techniques is a promising area of research today. Urban areas are an ensemble of natural and human-made landscapes, suggesting high variability of materials like grass, trees, roof types, road types, water bodies, pavements, etc. Additionally, urban materials have a diverse range of surface properties driven by the chemical composition of materials used during their preparation. The problem of land cover heterogeneity within a pixel, resulting in a mixed pixel, makes it imperative to use higher spatial resolution and appropriate wavelengths for multiple endmember extraction and identification of specific urban materials (Boardman and Kruse 2011). In urban landscapes, different types of land uses, land covers, and surface materials play a vital role in regulating the thermal environment within the city.

Urban material mapping has not been a widely explored area of research and few studies have explored the potential use of hyperspectral data. Hyperspectral data have been used for a few studies for the urban purlieu ranging from extraction of built-up, impervious surface extraction, mapping road types, roof types, etc. (Herold and Roberts, 2005; Herold and Roberts, 2010; Heiden et al., 2001). Hyperspectral data have also been used for mapping several natural targets (Griffin et al., 2005; Kisevic et al., 2016; Garg et al., 2017). Several un-mixing techniques like linear spectral unmixing and mixture tuned matched filtering algorithms are used for identifying materials (Siddiqui et al., 2020, Mishra et al., 2023). Ben-Dor et al. (2001) employed MTMF classification to identify urban materials using Compact Airborne Spectrographic Imager (CASI) data, leveraging the high spectral resolution of the sensor to distinguish subtle variations within each material class.

The scenario of urbanization is having major setbacks towards the thermal comfort scenario of the cities. More and more heat absorbing materials are being used for construction purpose, thereby elevating the problem of urban heating. To scientifically understand the role of hyperspectral data in urban micro-climates within cities, a study was conducted by Kushwaha et al. (2020). This study focused on the identification of major urban materials in the city of Ahmedabad through HRS imagery and analyzing the role of different surface materials on the urban heating scenario in the study area. The hyperspectral imagery was procured from Airborne Visible and Infrared Imaging Spectrometer—Next Generation (AVIRIS-NG) during February 2016 and used to classify urban materials. Landsat-8 data were used to estimate land surface temperature (LST) through Radiative Transfer Model (RTM) for quantification of urban heating.

Field observations were carried out using SVC-HR 1024 spectroradiometer with GPS locations covering all major urban material types such as bitumen, China mosaic, concrete, granite, iron, marble, soil, TAR, terracotta, tin, and water. The field observations were taken at a total of 71 samples, with sample numbers varying for each material. The spectroradiometer has a range of 1024 channels with a range of 348 to 2502.6 nm. Spectra at a given location were gathered after optimizing the integration time which provides fore-optic information, recording dark current, and collecting white reference reflectance. The target reflectance is the ratio of reflected energy to incident energy on the target. The measurements were acquired from 1 m above the material with the sensor facing the target and oriented normal to the surface.

Two classification techniques, Spectral Angle Mapper (SAM) and Support Vector Machine (SVM) were performed to classify the selected study area into five natural materials (water, bare soil, grass, crop, and trees) and six anthropogenic materials (asphalt, PVC, marble, tin, concrete, China mosaic). The overall classification accuracy obtained through SVM (95.41%) was better than that obtained through SAM (70.68%) (see Figure 10.39).

The highest mean LST of human-made urban material was observed for tin (29.68°C) followed by asphalt (29.49°C) and the lowest was observed for China mosaic (28.84°C). The range of LST of human-made surface materials was between 5–11°C; the lowest LST values were observed in areas surrounded by vegetation and water, and the highest in areas surrounded by high heat absorption materials like asphalt and tin. This indicates that a material's heating behavior is not only

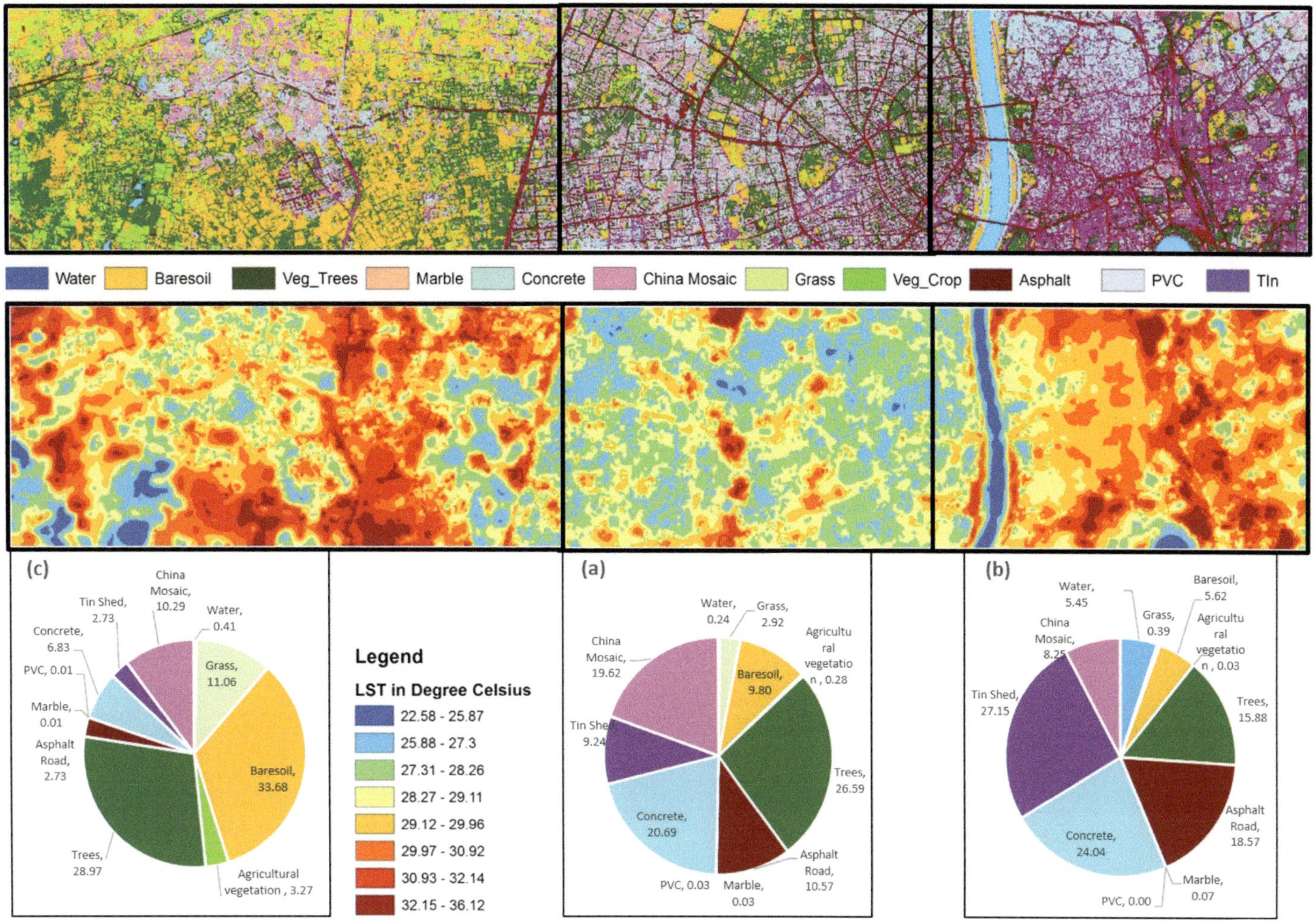

FIGURE 10.39 Material classification from hyperspectral data: SVM classified material map with LST imagery calculated from Landsat. The pie chart shows the proportion of each material in the study area. Section (a) is the high density built up with high surface temperature, (b) shows the low temperature observed due to vegetation, and (c) shows the abundance of bare soil. (Source: Kushwaha et al., 2020.)

dependent on its thermal properties but also depends on the surrounding materials. Concrete, Tin, Asphalt and China Mosaic were major built-up materials in the study area. Specific to the study area, LST ranged between 22°C and 36°C. Asphalt (36.12°C), Tin (35.73°C) and Bare-soil (35.38°C) had the highest temperatures, whereas Water (24°C) and Trees (23.3°C) had the lowest temperatures. Having classified land use and land cover maps, diurnal and seasonal variability in LST of different urban materials can be studied. Hyperspectral thermal images have the potential to give better scenarios of urban material heating.

10.14.3 Hyperspectral Remote Sensing for Soil Property Mapping

Quantitative assessment of various soil properties, including nutrients like soil organic carbon (SOC), is important for our efforts towards planning and implementation of various management measures aimed towards sustainable use of soil resources. Various environmental and agricultural applications, such as precision agriculture, require fine scale detailed mapping of soil properties to better understand environmental processes, study nutrient recycling mechanisms, and make management decisions.

Traditional analytical procedures (dry combustion, wet digestion, distillation, sedimentation analysis, etc.) involve longer time periods, use dangerous chemicals, generate harmful residues, and have higher costs. These procedures are less suitable for the generation of rapid and accurate soil information, especially for large geographical regions (Carmo and Silva, 2012; Apesteguia et al., 2018). Reflectance spectroscopy based techniques are nowadays becoming a promising alternative to laboratory-based analyses and are increasingly being adopted globally for soil property prediction (Sithole et al., 2018; Benedet et al., 2020).

Hyperspectral remote sensing (HRS), owing to its spatially continuous nature and rich spectral information (due to interaction of various soil constituents with electromagnetic radiation) across a large range of wavelengths covering the VIS-NIR spectrum, provides a powerful diagnostic tool for mapping and monitoring of various soil properties in a quantitative manner (Demattê et al., 2003). Airborne HRS data sources, such as AVIRIS-NG (Airborne Visible and Infrared Imaging Spectrometer—Next Generation) sensors, provide a large volume of spectral data at high spatial resolution (less than 5 m) owing to its very fine spectral resolution (https://avirisng.jpl.nasa.gov/). Deciphering valuable information present in the numerous spectral bands necessitates the adoption of different multivariate statistical techniques capable of input dimensionality reduction and information extraction (Liu et al., 2020). Such techniques help in identifying and limiting our analysis only to those wavelengths which are highly related to various soil properties and thus could explain a majority of variance present in the dataset (Curcio et al., 2013).

Among the various multivariate statistical modeling techniques widely adopted for prediction and mapping of soil properties using hyperspectral data, Partial Least Squares Regression (PLSR) needs a special mention (Hutengs et al., 2019; Mahajan et al., 2021) (also see PLS in Hively et al. (2011)). PLSR technique can make use of the correlation between spectra and soil properties during data decomposition. This helps in maintaining the direct relationship between soil properties and the spectral vectors. PLSR is able to deal with a large number of predictor variables, can handle multicollinearity, is robust in handling noisy data and missing values, and explains variances associated with predictor and response variables (Rossel et al., 2006).

Owing to its advantages, this technique has widely been employed for spectral modeling based soil property prediction (Adak et al., 2018; Conforti et al., 2018; Bangelesa et al., 2020). Most studies involve use of laboratory generated spectra only for model development and prediction, thereby lacking spatial mapping. Spatial prediction and mapping of soil properties using multivariate statistical techniques and remote sensing data have also been attempted and reported by various researchers (Nawar et al., 2015; Steinberg et al., 2016). However, such studies using HRS data for quantitative estimation and mapping of different soil properties have rarely been reported from India. In this forthcoming scenario, the study was carried out in the semi-arid tropical region of India with the aim of using high resolution airborne hyperspectral remote sensing data for spatial mapping of different soil attributes including nutrients using PLSR modeling technique (George et al., 2017).

The study region is comprised of two sites located in the state of Telangana, India (Figure 10.40). One site is located near Patancheru area (ICRISAT site) of Sangareddy district (between 17°.438 N and 78°.134 E) and the other is situated in the Shadnagar area of RangaReddy district (between 17°.0715N and 78°.2049 E). Major crops cultivated in the region include rice, maize, cotton, Bengal gram, red gram, and various vegetables like cabbage, tomato, brinjal, etc. The Shadnagar site had more fallow area and was found to be much drier compared to Patancheru site. The major soil types

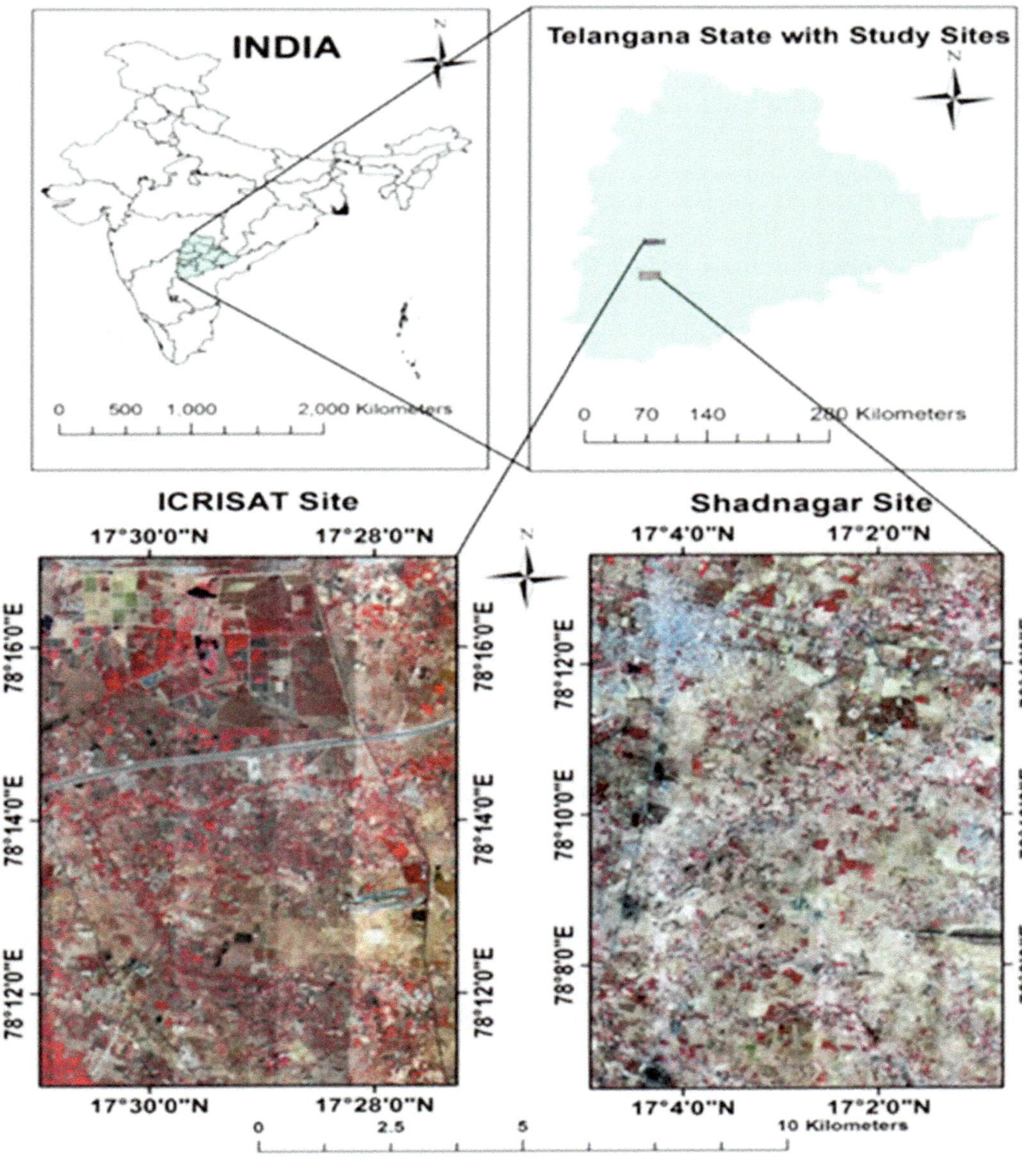

FIGURE 10.40 Study sites: (a) ICRISAT site and (b) Shadnagar site shown in FCC within India and Telangana state. (Author's original contribution.)

present in both the sites are red soil, black soil, and mixed soil, corresponding to the micro variations in topography and drainage conditions.

Airborne HRS data collection was carried out as part of the Indian Space Research Organization (ISRO)-NASA AVIRIS-NG airborne hyperspectral campaign during 2015–2016. AVIRIS—NG images of both the study sites were captured during December 2015 in the forenoon under bright illumination conditions from a flight height of 567 m above ground level. A detailed description regarding the sensor parameters and specifications can be read at NASA's JPL website (https://avirisng.jpl.nasa.gov/). Ground data collection and soil sampling were also conducted under similar conditions during the same period. Field data including crop and soil information and ground spectra using spectroradiometer and soil samples from the different locations, were collected under conditions similar to flight condition. Samples were collected from multiple locations depending on the variability as indicated by differences in soil color, crop type and growth form, and land use/land cover type. Special care was taken to ensure that the soil samples were drawn from uniform bare fields of considerable size with negligible vegetation cover, to ensure the generation of pure soil spectra from the airborne image. Collected soil samples were analyzed to determine the soil properties as per standard analytical procedures—organic carbon and total nitrogen were analyzed using a carbon, hydrogen, nitrogen and sulfur (CHNS) analyzer. Soil texture was identified by using the Bouyoucos hydrometer method and later the textural classes of samples were determined using the USDA soil texture triangle.

AVIRIS-NG radiance data were obtained for both study sites in the form of five strips, which were stitched together for further analysis. The data were examined strip-wise visually, for identification of bad bands with a high amount of noise. The data were then georeferenced and atmospherically corrected to remove atmospheric effects on spectral measurements and converted to surface reflectance using the Fast Line-of-sight Atmospheric Analysis of Spectral Hypercubes (FLAASH) atmospheric correction module in ENVI (ENVI, 2009). The coordinates of sampling locations were overlaid on the airborne HRS data and spectral profiles were generated for spectral modeling using PLSR for spatial mapping of soil properties. The AVIRIS-NG based image spectra and lab analysis generated soil property database were used for further model development and validation.

The spectral data were pre-processed initially to ensure removal of noise signals (de-noising). The data were then smoothened using the Savitzky-Golay algorithm, using 2nd order polynomial function with window size of three and further transformed into 1st derivative spectra. These pre-processing steps helped in diminishing the influence of light scattering, eliminating the background effects, and reducing the variability arising due to differences in optical environment, baseline differences, etc. The processed spectra were used for developing predictive models for estimation of soil properties like carbon, nitrogen, sand, silt, and clay contents in the area using Partial Least Squares Regression (PLSR), which is one of the most common and widely used multivariate statistical technique for VNIR spectra-based soil property estimation, globally.

The laboratory generated soil property database comprising of soil organic carbon (SOC) content (%), total nitrogen (%), and textural composition (percentage values of sand, silt, and clay components) was used for model development and mapping. The descriptive statistics of the soil attributes are given in Table 10.9. Measured and model predicted values of soil properties were compared for

TABLE 10.9
Descriptive Statistics of Different Soil Attributes (N = 73) [Author's original contribution.]

Variable	Min	Max	Mean	SD
Carbon (%)	0.532	2.952	1.193	0.48
Nitrogen (%)	0.041	0.479	0.119	0.062
Sand (%)	27.16	88.4	58.25	15.28
Silt (%)	4.28	55.28	20.43	12.43
Clay (%)	4.40	37.56	21.29	7.51

validation of various PLSR models to find out the most efficient models for soil property prediction. Calibrated models were validated using a subset of soil data before the spatial prediction. R-square, root mean square error (RMSE), and the ratio of the standard deviation of the population to RMSE, also known as RPD, were also calculated to evaluate the accuracy of the models.

In the case of SOC, the PLSR model with eight components gave the best results in comparison to other models during calibration. Based on the loading weights for various wavelengths, produced by the eight-component PLSR model, sensitive wavelengths were selected for further analysis and prediction model development (Figure 10.41). Regression analysis was done using the selected wavelengths to further identify the significant wavelengths (up to 90% level of significance, $p<0.1$) and to generate prediction equations (Table 10.10). The multiple linear regression (MLR) models were also validated for their prediction efficiencies using R^2 and RMSE values. The best prediction model yielded an R^2 value of 0.68, RMSE of 0.33, and an RPD value of 1.45 during validation. The validated model with significant wavelengths was further applied to the entire AVIRIS-NG image (after applying a soil mask to use only bare soil pixels and avoid vegetated pixels) to produce the spatial distribution map of soil organic carbon in both the study areas.

A similar approach was followed for the prediction and mapping of other soil properties such as nitrogen, sand, and clay contents in the region. For total nitrogen, the best PLSR model contained seven components, yielding an R^2 value of 0.58, RMSE of 0.09, and RPD value of 0.68 (Figure 10.41). The validated model was used with AVIRIS-NG bare soil pixels for both sites (Figure 10.42).

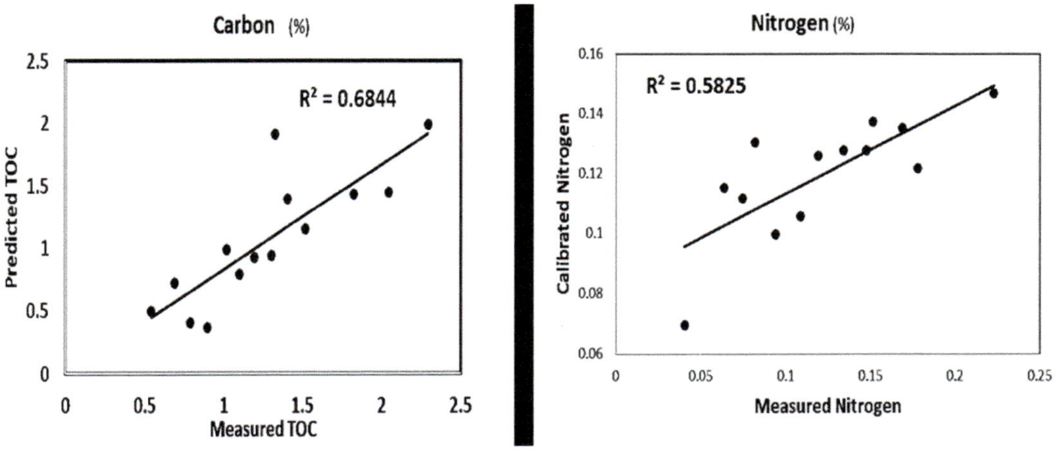

FIGURE 10.41 Validation results: PLSR based models for SOC and Nitrogen calibration. (Author's original contribution.)

TABLE 10.10
Equations for Prediction of Carbon and Nitrogen with the Best Models Constructed Using PLSR with the Respective R^2 [Author's original contribution.]

Variable	Prediction Model	R2
Carbon	Y = 1.9397 + 14.4360 (ρ′ 547 nm) + 25 (ρ′ 762 nm) – 71.7687 (ρ′ 882 nm) + 85.7768 (ρ′ 967 nm) – 11.2455 (ρ′ 1133 nm) – 21.4878 (ρ′ 2004 nm) + 39.4765 (ρ′ 2019 nm) + 23.4351 (ρ′ 2094 nm) – 46.0306 (ρ′ 2305 nm) + 25.3863 (ρ′ 2460 nm)	0.68
Nitrogen	Y = 0.04735–4.89246 (ρ′ 757 nm) + 6.31222 (ρ′ 762 nm) + 4.00632 (ρ′947nm) – 3.75546 (ρ′1133 nm) – 6.55997 (ρ′ 1513 nm) – 2.40539 (ρ′ 2004 nm) + 3.02379 (ρ′ 2039 nm)	0.58

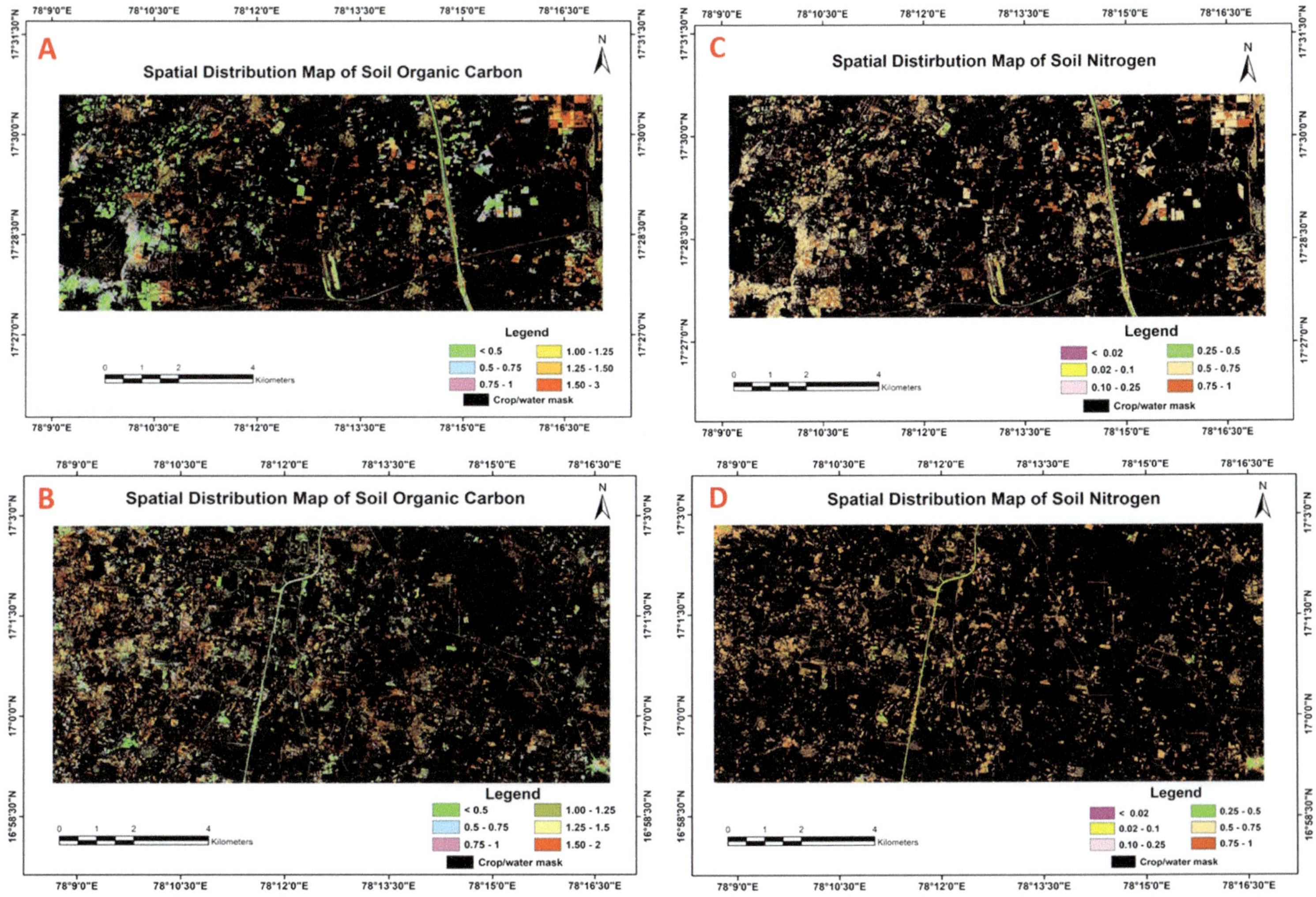

FIGURE 10.42 Spatial distribution maps: Soil Organic Carbon (SOC) at (a) ICRISAT site and (b) Shadnagar site, Nitrogen at (c) ICRISAT site and (d) Shadnagar site. (Author's original contribution.)

Even though the identification of certain sensitive wavelengths was possible, with respect to sand, silt, and clay contents during PLSR analysis, the developed prediction models were not stable and failed during testing. Hence those parameters were excluded from spatial mapping in the region during the study. Although the results obtained were promising, their prediction accuracies were found to be low in comparison to various lab spectra-based studies (Adak et al., 2018; Conforti et al., 2018; Ribeiro et al., 2021) in addition to image spectra-based prediction studies (Nawar et al., 2015). The soil texture results were found to be comparable with the results obtained by Steinberg et al. (2016) for prediction and mapping of different soil properties using airborne hyperspectral (simulated EnMAP) remote sensing data. Identification becomes even more challenging when bare soil is mixed with crop residue and photosynthetic vegetation cover. The adoption of various other spectral pre-processing and spectral decomposition techniques, along with machine learning methods, need to be explored for developing improved prediction models.

10.15 CONCLUSIONS

This chapter provided an overview of hyperspectral remote sensing for terrestrial applications. *First*, hyperspectral remote sensing, imaging spectroscopy, and superspectral remote sensing were defined. *Second*, characteristics of hyperspectral data acquired from three distinct platforms were discussed: (a) ground-based, hand-held, or truck-mounted spectroradiometers, (b) airborne, and (c) spaceborne. *Third*, hyperspectral data acquired from old generation (Hyperion) and new generation (e.g., DESIS, PRISMA, EnMAP) spaceborne hyperspectral sensors were presented and discussed. *Fourth*, the need for data mining to eliminate redundant bands was discussed and various data mining methods were presented. *Fifth*, the importance of understanding the Hughes phenomenon and approaches to overcome this shortcoming were highlighted. *Sixth*, methods of hyperspectral analyses were presented and discussed. These methods included feature extraction methods and information extraction methods. *Seventh*, hyperspectral data analysis philosophies (optimal hyperspectral narrowbands or OHNBs, and full spectral analyses or FSAs) were presented and discussed. *Eighth*, OHNBs best suited for agricultural and vegetation studies were determined from a meta-analysis. *Ninth*, FSAs were presented and discussed. FSAs were performed through spectral matching techniques and continuum removal derivative hyperspectral vegetation indices (HVIs). *Tenth*, HVIs, two-band and multi-band versions, best suited for agricultural and vegetation studies were also determined from a meta-analysis. *Eleventh*, hyperspectral image classification for land cover and species types was performed using methods such as spectral mixture analysis, support vector machines, and tree-based ensemble classifiers such as random forest and AdaBoost. *Twelfth*, new advanced methods of hyperspectral data classifications using AI, deep learning, and cloud computing were discussed in detail.

Thirteenth, case studies were described to demonstrate the previously mentioned concepts. These case studies include: mineral mapping, urban material characterization, and soil property mapping. Finally, it must be noted we are currently in a new era of hyperspectral remote sensing with many spaceborne hyperspectral sensors launched by various countries and private enterprises. These include: Germany's DESIS and EnMAP, Italy's PRISMA, NASA's upcoming SBG, and Planet Lab PBC's Tanager. The advancement in hyperspectral data is backed by equally impressive advances in machine learning, cloud computing, and artificial intelligence. These advances in hyperspectral sensors, tools, and techniques will usher in a new era of advanced scientific applications in the study of the Earth.

ACKNOWLEDGMENTS

Any use of trade, firm, or product names is for descriptive purposes only and does not imply endorsement by the U.S. government. Prasad S. Thenkabail (PST) is grateful to the India-U.S. Professional Engineer and Scientist Exchange Program (PESEP) to enable a PESEP Visit to the Indian Space Research Organization's (ISRO) Indian Institute of Remote Sensing (IIRS) and National Remote

Sensing Center (NRSC) and meet new colleagues, some of whom are coauthors on this paper. PST is very appreciative of longstanding support from the USGS and is especially grateful to Dr. Thomas Cecere (USGS) and Dr. Timothy Stryker (USGS) for securing travel funds for the PESEP Visit. PST thanks lively discussions at IIRS about this chapter and thanks all authors. Thanks also to Mr. Shantanu Bhatawdekar (ISRO Headquarters), Dr. Prakash Chauhan (NRSC), Dr. V. V. Rao (NRSC), Dr. P. V. Raju (NRSC), Dr. Raghavendra Pratap Singh (IIRS), Dr. Pramod Kumar (IIRS), and Dr. Minakshi Kumar (IIRS) for their discussions, insights, and discussions. Further to these meetings and discussions appropriate co-authors from ISRO/IIRS were identified (see authors). They made significant contributions to this manuscript.

REFERENCES

Abbas, A., Minallh, N., Ahmad, N., Abid, S. A. R., and Khan, M. A. A. (2016). K-means and ISODATA clustering algorithms for landcover classification using remote sensing. *Sindh University Research Journal*, 48.

Abdel-Rahman, E. M., Makori, D. M., Landmann, T., Piiroinen, R., Gasim, S., Pellikka, P., and Raina, S. K. (2015). The utility of AISA Eagle hyperspectral data and random forest classifier for flower mapping. *Remote Sensing*, 7:13298–13318.

Abdel-Rahman, E. M., Mutanga, O., Adam, E., and Ismail, R. (2014). Detecting *Sirex Noctilio* grey-attacked and lightning-struck pine trees using airborne hyperspectral data, random forest and support vector machines classifiers. *ISPRS Journal of Photogrammetry and Remote Sensing*, 88:48–59.

Abdi, A. (2020). Land cover and land use classification performance of machine learning algorithms in a boreal landscape using Sentinel-2 data. *GIScience and Remote Sensing*, 57:1–20.

Adak, S., Bandyopadhyay, K. K., Sahoo, R. N., Purakayastha, T. J., Shrivastava, M., and Mridha, N. (2018). Assessment of soil health parameters using proximal hyperspectral remote sensing. *Journal of Agricultural Physics*, 18(1):88–98.

Adão, T., Hruška, J., Pádua, L., Bessa, J., Peres, E., Morais, R., and Sousa, J. J. (2017). Hyperspectral imaging: A review on UAV-based sensors, data processing and applications for agriculture and forestry. *Remote Sensing*, 9(11):1110.

Adler-Golden, S., Sundberg, R., and St Peter, B. (2017). Object classification in hyperspectral imagery based on normalized, whitened reflectance. In *2017 IEEE International Geoscience and Remote Sensing Symposium (IGARSS)*, pages 1324–1327. Fort Worth, Texas, USA.

Agrawal, N., Govil, H., Chatterjee, S., Mishra, G., and Mukherjee, S. (2022). Evaluation of machine learning techniques with AVIRIS-NG dataset in the identification and mapping of minerals. *Advances in Space Research*, 73(2):1517–1534.

Ahmad, S., Pandey, A., Kumar, A., and Lele, N. (2021). Potential of hyperspectral AVIRIS-NG data for vegetation characterization, species spectral separability, and mapping. *Applied Geomatics*, 13:361–372.

Ahmed, K. R., and Akter, S. (2017). Analysis of landcover change in southwest Bengal Delta due to floods by NDVI, NDWI and k-means cluster with Landsat multi-spectral surface reflectance satellite data. *Remote Sensing Applications:Society and Environment*, 8:168–181.

Alchanatis, V., and Cohen, Y. (2012). Spectral and spatial methods for hyperspectral image analysis for estimation of biophysical and biochemical properties of agricultural crops. In Thenkabail, P., Lyon, G., and Huete, A., editors, *Hyperspectral Remote Sensing of Vegetation*, Chapter 13, pages 239–308. CRC Press & Taylor and Francis Group, Boca Raton, London and NY.

Alonso, K., Bachmann, M., Burch, K., Carmona, E., Cerra, D., de los Reyes, R., Dietrich, D., Heiden, U., Holderlin, A., Ickes, J., Knodt, U., Krutz, D., Lester, H., Muller, R., Pagnutti, M., Reinartz, P., Richter, R., Ryan, R., Sebastian, I., and Tegler, M. (2019). Data products, quality and validation of the DLR Earth Sensing Imaging Spectrometer (DESIS). *Sensors*, 19(4471):1–44.

Andrew, M. E., and Ustin, S. L. (2008). The role of environmental context in mapping invasive plants with hyperspectral image data. *Remote Sensing of Environment*, 112(12):4301–4317.

Aneece, I., Foley, D., Thenkabail, P., Oliphant, A., and Teluguntla, P. (2022). New generation hyperspectral data from DESIS compared to high spatial resolution PlanetScope data for crop type classification. *IEEE Journal of Selected Topics in Applied Earth Observations and Remote Sensing*, 15:7846–7858.

Aneece, I., and Thenkabail, P. S. (2018a). Accuracies achieved in classifying five leading world crop types and their growth stages using optimal Earth Observing-1 Hyperion hyperspectral narrowbands on Google Earth Engine. *Remote Sensing*, 10(12):2027.

Aneece, I., and Thenkabail, P. S. (2018b). Spaceborne hyperspectral EO-1 Hyperion data pre-processing: Methods, approaches, and algorithms. In *Hyperspectral Remote Sensing of Vegetation*. CRC Press & Taylor and Francis Inc., Boca Raton, FL.

Aneece, I., and Thenkabail, P. S. (2019). *Global Hyperspectral Imaging Spectral-Library of Agricultural crops (GHISA) for the conterminous United States (CONUS). Algorithm Theoretical Basis Document (ATBD).* NASA Land Processes Distributed Active Archive Center (LP DAAC), page 25.

Aneece, I., and Thenkabail, P. S. (2021). Classifying crop types using two generations of hyperspectral sensors (Hyperion and DESIS) with machine learning on the cloud. *Remote Sensing*, 13:1–24.

Aneece, I., and Thenkabail, P. S. (2022a). *DESIS and PRISMA Spectral Library of Agricultural Crops in California's Central Valley in the 2020 Growing Season*. U.S. Geological Survey Data Release.

Aneece, I., and Thenkabail, P. S. (2022b). New generation hyperspectral sensors DESIS and PRISMA provide improved agricultural crop classifications. *Photogrammetric Engineering and Remote Sensing*, 88(11):715–729.

Apesteguia, M., Plante, A. F., and Virto, I. (2018). Methods assessment for organic and inorganic carbon quantification in calcareous soils of the Mediterranean region. *Geoderma Regional*, 12:39–48.

Audebert, N., Le Saux, B., and Lefevre, S. (2019). Deep learning for classification of hyperspectral data: A comparative review. *IEEE Geoscience and Remote Sensing Magazine*, 7(2):159–173.

Bajwa, S., and Kulkarni, S. (2012). Hyperspectral data mining. In Thenkabail, P., Lyon, G., and Huete, A., editors, *Hyperspectral Remote Sensing of Vegetation*, Chapter 4, pages 93–120. CRC Press & Taylor and Francis Group, Boca Raton, London and NY.

Bangelesa, F., Adam, E., Knight, J., Dhau, I., Ramudzuli, M., and Mokotjomela, T. M. (2020). Predicting soil organic carbon content using hyperspectral remote sensing in a degraded mountain landscape in Lesotho. *Applied and Environmental Soil Science*, 2020:1–11.

Beamish, A., Chabrillat, S., Brell, M., Heim, B., and Sachs, T. (2020). Toolik lake research natural area AISA-eagle hyperspectral mosaic—an EnMAP preparatory flight campaign. *EnMAP Flight Campaigns Technical Report*. Technical Report, GFZ Data Services.

Bedini, E. (2011). Mineral mapping in the Kap Simpson complex, central East Greenland, using HyMap and ASTER remote sensing data. *Advances in Space Research*, 47(1):60–73. https://doi.org/10.1016/j.asr.2010.08.021.

Ben-Dor, E. (2012). Characterization soil properties using reflectance spectroscopy. In Thenkabail, P., Lyon, G., and Huete, A., editors, *Hyperspectral Remote Sensing of Vegetation*, Chapter 22, pages 513–560. CRC Press & Taylor and Francis Group, Boca Raton, London and NY.

Ben-Dor, E., Levin, N., and Saaroni, H. (2001). A spectral based recognition of the urban environment using the visible and near-infrared spectral region (0.4–1.1 lm). A case study over Tel-Aviv, Israel. *International Journal of Remote Sensing*, 22(11):2193–2218.

Benedet, L., Faria, W. M., Silva, S. H. G., Mancini, M., Demattê, J. A. M., Guilherme, L. R. G., and Curi, N. (2020). Soil texture prediction using portable X-ray fluorescence spectrometry and visible near-infrared diffuse reflectance spectroscopy. *Geoderma*, 376:114553.

Berk, A., Anderson, G., Acharya, P., and Shettle, E. (2008). *MODTRAN 5.2.0.0 User's Manual*. Technical Report.

Bhattacharya, B. K., Green, R. O., Rao, S., Saxena, M., Sharma, S., Kumar, K. R. M. A., Srinivasulu, P., Sharma, S., Dhar, D., Bandyopadhyay, S., Bhatwadekar, S., and Kumar, R. (2019). An overview of AVIRIS-NG airborne hyperspectral science campaign over India. *Current Science*, 116(7):1082–1088.

Biradar, C. M., Thenkabail, P., Noojipady, P., Li, Y., Dheeravath, V., Turral, H., Velpuri, M., Gumma, M., Gangalakunta, O., and Cai, X. (2009). A Global Map of Rainfed Cropland Areas (GMRCA) at the end of last millennium using remote sensing. *International Journal of Applied Earth Observation and Geoinformation*, 11(2):114–129.

Birhanu, B. Z., Traore, K., Gumma, M. K., Badolo, F., Tabo, R., and Whitbread, A. M. (2019). A watershed approach to managing rainfed agriculture in the semiarid region of southern Mali: Integrated research on water and land use. *Environment,Development and Sustainability*, 21:2459–2485.

Boardman, J. W., and Kruse, F. A. (1994). Automated spectral analysis: A geologic example using AVIRIS data, north grapevine mountains, Nevada. In *Proc. Tenth Thematic Conference on Geologic Remote Sensing,Environmental Research Institute of Michigan*, pages 1407–1418. San Antonio, Texas, USA.

Boardman, J. W., and Kruse, F. A. (2011). Analysis of imaging spectrometer data using N-dimensional geometry and a mixture-tuned matched filtering approach. *IEEE Transactions on Geoscience and Remote Sensing*, 49(11):4138–4152.

Boardman, J. W., Kruse, F. A., and Green, R. O. (1995). Mapping target signatures via partial unmixing of AVIRIS data. In *Summaries of the Fifth Annual JPL Airborne Earth Science Workshop. Volume 1:AVIRIS Workshop*, pages 23–26. JPL Publication, US. https://ntrs.nasa.gov/search.jsp?R=19950027316.

Bracken, A., Coburn, C., Staenz, K., Rochdi, N., Segl, K., Chabrillat, S., and Schmid, T. (2019). Detecting soil erosion in semi-arid Mediterranean environments using simulated EnMAP data. *Geoderma*, 340:164–174.

Brovelli, M., Sun, Y., and Yordanov, V. (2020). Monitoring forest change in the Amazon using multi-temporal remote sensing data and machine learning classification on Google Earth Engine. *ISPRS International Journal of Geo-information*, 9:1–21.

Buzzi, J., Riaza, A., García-Meléndez, E., Weide, S., and Bachmann, M. (2014). Mapping changes in a recovering mine site with hyperspectral airborne HyMap imagery (Sotiel, SW Spain). *Minerals*, 4(2):313–329.

Cahalane, C., Walsh, D., Magee, A., Mannion, S., Lewis, P., and McCarthy, T. (2017). Sensor pods: Multi-resolution surveys from a light aircraft. *Inventions*, 2(1):2.

Carmo, D. L., and Silva, C. A. (2012). Métodos de quantificação de carbono e matéria orgânica em resíduos orgânicos. *Revista Brasileira de Ciência do Solo*, 36:1211–1220.

Carrino, T. A., Crósta, A. P., Toledo, C. L. B., Silva, A. M., and Silva, J. L. (2015). Geology and hydrothermal alteration of the Chapi Chiara prospect and nearby targets, Southern Peru, using ASTER data and reflectance spectroscopy. *Economic Geology*, 110(1):73–90. https://doi.org/10.2113/econgeo.110.1.73.

Chakravarty, S., Paikaray, B. K., Mishra, R., and Dash, S. (2021). Hyperspectral image classification using spectral angle mapper. In *2021 IEEE International Women in Engineering (WIE) Conference on Electrical and Computer Engineering (WIECON-ECE)*, pages 87–90. San Diego, CA, USA.

Chan, J. C.-W., and Paelinckx, D. (2008). Evaluation of random forest and adaboost tree-based ensemble classification and spectral band selection for ecotope mapping using airborne hyperspectral imagery. *Remote Sensing of Environment*, 112(6):2999–3011.

Chander, G., Markham, B. L., and Helder, D. L. (2009). Summary of current radiometric calibration coefficients for Landsat MSS, TM, ETM+, and EO-1 ALI sensors. *Remote Sensing of Environment*, 113(5):893–903.

Chaube, N., Lele, N., Misra, A., Murthy, T., Manna, S., Hazra, S., Panda, M., and Samal, R. (2019). Mangrove species discrimination and health assessment using AVIRIS-NG hyperspectral data. *Current Science*, 116(7):1136–1142.

Chavez, P. S. (1988). An improved dark-object subtraction technique for atmospheric scattering correction of multispectral data. *Remote Sensing of Environment*, 24(3):459–479.

Chavez, P. S. (1989). Radiometric calibration of Landsat Thematic Mapper multispectral images. *Photogrammetric Engineering and Remote Sensing*, 55:1285–1294.

Chen, G., Krzyzak, A., and en Qian, S. (2023). Noise robust hyperspectral image classification with MNF-based edge preserving features. *Image Analysis andStereology*, 42(2).

Chen, W., Li, Y., Tsangaratos, P., Shahabi, H., Ilia, I., Xue, W., and Bian, H. (2020). Groundwater spring potential mapping using artificial intelligence approach based on kernel logistic regression, random forest, and alternating decision tree models. *Applied Sciences*, 10:1–23.

Chung, B., Yu, J., Wang, L., Kim, N. H., Lee, B. H., Koh, S., and Lee, S. (2020). Detection of magnesite and associated gangue minerals using hyperspectral remote sensing—a laboratory approach. *Remote Sensing*, 12(8).

Clark, M. (2012). Identification of canopy species in tropical forests using hyperspectral data. In Thenkabail, P., Lyon, G., and Huete, A., editors, *Hyperspectral Remote Sensing of Vegetation*, Chapter 18, pages 423–446. CRC Press & Taylor and Francis Group, Boca Raton, London and NY.

Clark, M. L. (2017). Comparison of simulated hyperspectral HyspIRI and multispectral Landsat 8 and Sentinel-2 imagery for multi-seasonal, regional land-cover mapping. *Remote Sensing of Environment*, 200:311–325.

Clark, R. N., Boardman, J., Mustard, J., Kruse, F., Ong, C., Pieters, C., and Swazye, G. A. (2006). Mineral mapping and applications of imaging spectroscopy. In *International Symposium on Geoscience and Remote Sensing, 1986–1989*. Denver, CO, USA. https://doi.org/10.1109/IGARSS.2006.514.

Colombo, R., Lorenzo, B., Michele, M., Micol, R., and Cinzia, P. (2012). Optical remote sensing of vegetation water content. In Thenkabail, P., Lyon, G., and Huete, A., editors, *Hyperspectral Remote Sensing of Vegetation*, Chapter 10, page 227. CRC Press & Taylor and Francis Group, Boca Raton, London and NY.

Colomina, I., and Molina, P. (2014). Unmanned aerial systems for photogrammetry and remote sensing: A review. *ISPRS Journal of Photogrammetry and Remote Sensing*, 92:79–97.

Conforti, M., Matteucci, G., and Buttafuoco, G. (2018). Using laboratory Vis-NIR spectroscopy for monitoring some forest soil properties. *Journal of Soils and Sediments*, 18:1009–1019.

Crosta, A. P., and McMoore, J. (1989). Enhancement of Landsat Thematic Mapper imagery for residual soil mapping in SW Minas Gerais State, Brazil: A prospecting case history in greenstone belt terrain. In *Seventh Thematic Conference on Remote Sensing for Exploration Geology*, Erim, Calgary, AB, Canada.

Curcio, D., Ciraolo, G., D'asaro, F., and Minacapilli, M. (2013). Prediction of soil texture distributions using VNIR-SWIR reflectance spectroscopy. *Procedia Environmental Sciences*, 19:494–503.

Damarjati, S., Nugraha, W. A., and Arjasakusuma, S. (2022). Mapping the invasive palm species *Arenga obtusifolia* using Multiple Endmember Spectral Mixture Analysis (MESMA) and PRISMA hyperspectral data in Ujung Kulon National Park, Indonesia. *Geocarto International*, 37(27):18254–18274.

Danilov, R., Kremneva, O., and Pachkin, A. (2023). Identification of the spectral patterns of cultivated plants and weeds: Hyperspectral vegetation indices. *Agronomy*, 13(3):859.

Dao, P. D., He, Y., and Lu, B. (2019). Maximizing the quantitative utility of airborne hyperspectral imagery for studying plant physiology: An optimal sensor exposure setting procedure and empirical line method for atmospheric correction. *International Journal of Applied Earth Observation and Geoinformation*, 77:140–150.

Davidson, S. J., Santos, M. J., Sloan, V. L., Watts, J. D., Phoenix, G. K., Oechel, W. C., and Zona, D. (2016). Mapping Arctic tundra vegetation communities using field spectroscopy and multispectral satellite data in North Alaska, USA. *Remote Sensing*, 8(12):978.

Davies, B. F. R., Gernez, P., Geraud, A., Oiry, S., Rosa, P., Zoffoli, M. L., and Barillé, L. (2023). Multi- and hyperspectral classification of soft-bottom intertidal vegetation using a spectral library for coastal biodiversity remote sensing. *Remote Sensing of Environment*, 290:113554.

de los Reyes, R., Langheinrich, M., and Bachmann, M. (2022). *EnMAP Ground Segment Level 2A Processor (Atmospheric Correction over Land) ATBD*. Technical Report, DLR.

Demattê, J. A. M., Epiphanio, J. C. N., and Formaggio, A. R. (2003). Influência da matéria orgânica e de formas de ferro na reflectância de solos tropicais. *Bragantia*, 62:451–464.

Dennison, P. E., Lamb, B. T., Campbell, M. J., Kokaly, R. F., Hively, W. D., Vermote, E., Dabney, D., Serbin, G., Quemada, M., Daughtry, C. S. T., Masek, J., and Wu, Z. (2023). Modeling global indices for estimating non-photosynthetic vegetation cover. *Remote Sensing of Environment*, 295:113715.

Domingues Franceschini, M. H., Bartholomeus, H., Van Apeldoorn, D., Suomalainen, J., and Kooistra, L. (2017). Intercomparison of unmanned aerial vehicle and ground-based narrow band spectrometers applied to crop trait monitoring in organic potato production. *Sensors*, 17(6):1428.

Doneus, M., Verhoeven, G., Atzberger, C., Wess, M., and Rus, M. (2014). New ways to extract archaeological information from hyperspectral pixels. *Journal of Archaeological Science*, 52:84–96.

dos Santos, G. L. A. A., Reis, A. S., Besen, M. R., Furlanetto, R. H., Rodrigues, M., Crusiol, L. G. T., de Oliveira, K. M., Falcioni, R., de Oliveira, R. B., Batista, M. A., and Nanni, M. R. (2023). Spectral method for macro and micronutrient prediction in soybean leaves using interval partial least squares regression. *European Journal of Agronomy*, 143:126717.

Duveiller, G., Lopez-Lozano, R., and Cescatti, A. (2015). Exploiting the multi-angularity of the MODIS temporal signal to identify spatially homogeneous vegetation cover: A demonstration for agricultural monitoring applications. *Remote Sensing of Environment*, 166:61–77.

Elvidge, C., Yuan, D., Weerackoon, R., and Lunetta, R. (1995). Relative radiometric normalization of Landsat Multispectral Scanner (MSS) data using an automatic scattergram controlled regression. *Photogrammetric Engineering and Remote Sensing*, 61:1255–1260.

ENVI. (2009). *Atmospheric Correction Module: QUAC and FLAASH User's Guide*. ITT Visual Information Solutions.

Erickson, Z. K., Frankenberg, C., Thompson, D. R., Thompson, A. F., and Gierach, M. (2019). Remote sensing of chlorophyll fluorescence in the ocean using imaging spectrometry: Toward a vertical profile of fluorescence. *Geophysical Research Letters*, 46(3):1571–1579.

Fan, Y., Feng, H., Yue, J., Liu, Y., Jin, X., Xu, X., Song, X., Ma, Y., and Yang, G. (2023). Comparison of different dimensional spectral indices for estimating nitrogen content of potato plants over multiple growth periods. *Remote Sensing*, 15(3):602.

Farmonov, N., Amankulova, K., Szatmári, J., Sharifi, A., Abbasi-Moghadam, D., Mirhoseini Nejad, S. M., and Mucsi, L. (2023). Crop type classification by DESIS hyperspectral imagery and machine learning algorithms. *IEEE Journal of Selected Topics in Applied Earth Observations and Remote Sensing*, 16:1576–1588.

Fayyad, U., Piatetsky-Shapiro, G., and Smyth, P. (1996). The KDD process for extracting useful knowledge from volumes of data. *Communications of the ACM*, 39:27–34.

Fitzgerald, G. J., Perry, E. M., Flower, K. C., Callow, J. N., Boruff, B., Delahunty, A., Wallace, A., and Nuttall, J. (2019). Frost damage assessment in wheat using spectral mixture analysis. *Remote Sensing*, 11(21).

Furlanetto, R. H., Moriwaki, T., Falcioni, R., Pattaro, M., Vollmann, A., Sturion, A. C. Jr., Antunes, W. C., and Nanni, M. R. (2020). Hyperspectral reflectance imaging to classify lettuce varieties by optimum selected wavelengths and linear discriminant analysis. *Remote Sensing Applications:Society and Environment*, 20:100400.

Galvão, L. (2012). Crop type discrimination using hyperspectral data. In Thenkabail, P., Lyon, G., and Huete, A., editors, *Hyperspectral Remote Sensing of Vegetation*, Chapter 17, pages 397–422. CRC Press & Taylor and Francis Group, Boca Raton, London and NY.

Galvão, L. S., Roberts, D. A., Formaggio, A. R., Numata, I., and Breunig, F. M. (2009). View angle effects on the discrimination of soybean varieties and on the relationships between vegetation indices and yield using off-nadir Hyperion data. *Remote Sensing of Environment*, 113(4):846–856.

Garg, K. K., Anantha, K., Nune, R., Akuraju, V. R., Singh, P., Gumma, M. K., Dixit, S., and Ragab, R. (2020). Impact of land use changes and management practices on groundwater resources in Kolar District, southern India. *Journal of Hydrology:Regional Studies*, 31:100732.

Garg, V., Kumar, A. S., Aggarwal, S. P., Kumar, V., Dhote, P., Thakur, P. K., et al. (2017). Spectral similarity approach for mapping turbidity of an inland waterbody. *Journal of Hydrology*. https://doi.org/10.1016/j.jhydrol.2017.05.039.

George, J. K., Kumar, V., Danodia, A., Kumar, S., and Kumar, A. S. (2017). Hyperspectral modelling for prediction of soil texture using ASD spectroradiometer derived soil spectra. In *38th Asian Conference on Remote Sensing—ACRS 2017*. Delhi, India.

Gitelson, A. (2012a). Non-destructive estimation of foliar pigment (chlorophylls, carotenoids, and anthocyanins) contents: Evaluating a semi-analytical three-band model. In Thenkabail, P., Lyon, G., and Huete, A., editors, *Hyperspectral Remote Sensing of Vegetation*, Chapter 6, pages 141–166. CRC Press & Taylor and Francis Group, Boca Raton, London and NY.

Gitelson, A. (2012b). Remote estimation of crop biophysical characteristics at various scales. In Thenkabail, P., Lyon, G., and Huete, A., editors, *Hyperspectral Remote Sensing of Vegetation*, Chapter 15, pages 329–360. CRC Press & Taylor and Francis Group, Boca Raton, London and NY.

Glowienka, E., and Zembol, N. (2022). Forest community mapping using hyperspectral (CHRIS/PROBA) and Sentinel-2 multispectral images. *Geomatics and Environmental Engineering*, 16(4):103–117.

Gnyp, M. L., Bareth, G., Li, F., Lenz-Wiedemann, V. I., Koppe, W., Miao, Y., Hennig, S. D., Jia, L., Laudien, R., Chen, X., and Zhang, F. (2014). Development and implementation of a multiscale biomass model using hyperspectral vegetation indices for winter wheat in the North China Plain. *International Journal of Applied Earth Observation and Geoinformation*, 33:232–242.

Goetz, A. F. H. (1992). Imaging spectrometry for Earth observations. *Episodes*, 15:7–14.

Goetz, A. F. H., and Srivastava, V. (1985). Mineralogical mapping in the cuprite mining district, Nevada. In *Proceedings of the Airborne Imaging Spectrometer (AIS) Data Analysis Workshop*, pages 22–29. JPL Publication, Pasadena, CA, USA.

Goetz, A. F. H., Vane, G., Solomon, J. E., and Rock, B. N. (1985). Imaging spectrometry for Earth remote sensing. *Science*, 228(4704):1147–1153. https://doi.org/10.1126/science.228.4704.1147.

Green, A., Berman, M., Switzer, P., and Craig, M. (1988). A transformation for ordering multispectral data in terms of image quality with implications for noise removal. *IEEE Transactions on Geoscience and Remote Sensing*, 26(1):65–74.

Griffin, M. K., Hsu, S. M., Burke, H. K., Orloff, S. M., and Upham, C. A. (2005). Examples of EO-1 Hyperion data analysis. *Linc Lab Journal*, 15(2):271–298.

GSI. (2004). *Kota Quadrangle*. Retrieved May 19, 2017, from www.portal.gsi.gov.in/pls/gsihub/PKG_PTL_SEARCH_PAGES.pGetImage_PaperMap?inpRecId=1344&inPaperMapImageId=PUB_PAPER_MAP.

Guerschman, J. P., Hill, M. J., Renzullo, L. J., Barrett, D. J., Marks, A. S., and Botha, E. J. (2009). Estimating fractional cover of photosynthetic vegetation, non-photosynthetic vegetation and bare soil in the Australian tropical savanna region upscaling the EO-1 Hyperion and MODIS sensors. *Remote Sensing of Environment*, 113(5):928–945.

Gumma, M. K., Amede, T., Getnet, M., Pinjarla, B., Panjala, P., Legesse, G., Tilahun, G., Van den Akker, E., Berdel, W., Keller, C., Siambi, M., and Whitbread, A. M. (2022a). Assessing potential locations for flood-based farming using satellite imagery: A case study of Afar Region, Ethiopia. *Renewable Agriculture and Food Systems*, 37(S1):S28–S42.

Gumma, M. K., Takashi, Y., Panjala, P., Deevi, K. C., Inthavong, V., Bellam, P. K., and Mohammed, I. (2023). Assessment of cropland changes due to new canals in Vientiane Prefecture of Laos using Earth observation data. *Smart Agricultural Technology*, 4:100149.

Gumma, M. K., Thenkabail, P. S., Panjala, P., Teluguntla, P., Yamano, T., and Mohammed, I. (2022b). Multiple agricultural cropland products of South Asia developed using Landsat-8 30 m and MODIS 250 m data using machine learning on the Google Earth Engine (GEE) cloud and Spectral Matching Techniques (SMTs) in support of food and water security. *GIScience and Remote Sensing*, 59(1):1048–1077.

Guo, L., Zhang, H., Shi, T., Chen, Y., Jiang, Q., and Linderman, M. (2019). Prediction of soil organic carbon stock by laboratory spectral data and airborne hyperspectral images. *Geoderma*, 337:32–41.

Guo, X., Huang, X., Zhang, L., and Zhang, L. (2013). Hyperspectral image noise reduction based on rank-1 tensor decomposition. *ISPRS Journal of Photogrammetry and Remote Sensing*, 83:50–63.

Gupta, S. N., Arora, Y. K., Mathur, R. K., Iqballuddin, P. B., Sahai, T. N., and Sharma, S. B. (1997). *The Precambrian Geology of the AravalliRegion, Southern Rajasthan and North-Eastern Gujarat. Memoir of the Geological Survey of India*, Volume 123, pages 1–262. Government of India, Calcutta.

Hajaj, S., Harti, A. E., Jellouli, A., Pour, A. B., Himyari, S. M., Hamzaoui, A., Bensalah, M. K., Benaouiss, N., and Hashim, M. (2023). HyMap imagery for copper and manganese prospecting in the east of Ameln Valley shear zone (Kerdous Inlier, Western Anti-Atlas, Morocco). *Journal of Spatial Science*, 69(1):81–102.

He, C., Shi, Z., Qu, T., Wang, D., and Liao, M. (2019). Lifting scheme-based deep neural network for remote sensing scene classification. *Remote Sensing*, 11(22):2648.

Heiden, U., Bachmann, M., Alonso, K., Carmona, E., Daniele, C., Dietrich, D., Langheinrich, M., de los Reyes, R., Mueller, R., Pinnel, N., and Ziel, V. (2019). DESIS imaging spectrometer data access and synergistic use with other ISS Earth observing instruments. In *Workshop on International Cooperation in Spaceborne Imaging Spectroscopy,9.–11. July 2019, ESA-ESRIN*, Frascati, Italy. https://elib.dlr.de/129175/.

Heiden, U., Roessner, S., Segl, K., and Kaufmann, H. (2001). Analysis of spectral signatures of urban surfaces for their identification using hyperspectral HyMap data. In *IEEE/ISPRS Joint Workshop on Remote Sensing and Data Fusion over Urban Areas*, pages 173–177. Rome, Italy.

Hernández-Clemente, R., Navarro-Cerrillo, R. M., Suárez, L., Morales, F., and Zarco-Tejada, P. J. (2011). Assessing structural effects on PRI for stress detection in conifer forests. *Remote Sensing of Environment*, 115(9):2360–2375.

Herold, M., and Roberts, D. A. (2005). Spectral characteristics of asphalt road aging and deterioration: Implications for remote-sensing applications. *Applied Optics*, 44(20):4327–4334.

Herold, M., and Roberts, D. A. (2010). The spectral dimension in urban remote sensing. In Rashed, T., and Jürgens, C., editors, *Remote Sensing of Urban and Suburban Areas. Remote Sensing and Digital Image Processing*, Volume 10, pages 47–65. Springer Nature, Dordrecht, The Netherlands.

Hively, W. D., Lamb, B. T., Daughtry, C. S. T., Serbin, G., Dennison, P., Kokaly, R. F., Wu, Z., and Masek, J. G. (2021). Evaluation of SWIR crop residue bands for the Landsat Next mission. *Remote Sensing*, 13:3718.

Hively, W. D., McCarty, G. W., Reeves, J. B., Lang, M. W., Oesterling, R. A., and Delwiche, S. R. (2011). Use of airborne hyperspectral imagery to map soil properties in tilled agricultural fields. *Applied and Environmental Soil Science*, 358193:13.

Honarmand, M. (2016). Application of airborne geophysical and ASTER data for hydrothermal alteration mapping in the Sar-Kuh Porphyry Copper Area, Kerman Province, Iran. *Open Journal of Geology*, 6:1257–1268. https://doi.org/10.4236/ojg.2016.610092.

Hosseinjani Zadeh, M., Tangestani, M. H., Roldan, F. V., and Yusta, I. (2014). Sub-pixel mineral mapping of a porphyry copper belt using EO-1 Hyperion data. *Advances in Space Research*, 53(3):440–451.

Hutengs, C., Seidel, M., Oertel, F., Ludwig, B., and Vohland, M. (2019). In situ and laboratory soil spectroscopy with portable visible-to-near infrared and mid-infrared instruments for the assessment of organic carbon in soils. *Geoderma*, 355:113900.

Iqbal, A., Ullah, S., Khalid, N., Ahmad, W., Ahmad, I., Shafique, M., Hulley, G. C., Roberts, D. A., and Skidmore, A. K. (2018). Selection of HyspIRI optimal band positions for the Earth compositional mapping using Hytes data. *Remote Sensing of Environment*, 206:350–362.

Ivushkin, K., Bartholomeus, H., Bregt, A. K., Pulatov, A., Franceschini, M. H., Kramer, H., van Loo, E. N., Jaramillo Roman, V., and Finkers, R. (2019). UAV based soil salinity assessment of cropland. *Geoderma*, 338:502–512.

Jafari, R., and Lewis, M. (2012). Arid land characterisation with EO-1 Hyperion hyperspectral data. *International Journal of Applied Earth Observation and Geoinformation*, 19:298–307.

Jain, R., and Sharma, R. U. (2019). Airborne hyperspectral data for mineral mapping in Southeastern Rajasthan, India. *International Journal of Applied Earth Observation and Geoinformation*, 81:137–145.

Jayaprakash, C., Damodaran, B. B., Viswanathan, S., and Soman, K. P. (2020). Randomized independent component analysis and linear discriminant analysis dimensionality reduction methods for hyperspectral image classification. *Journal of Applied Remote Sensing*, 14(3):036507.

Jensen, J. R. (2004). *Introductory Digital Image Processing: A Remote Sensing Perspective*, 3rd ed. Prentice-Hall, NJ.

Jha, C., Fararoda, R., Singhal, J., Reddy, C., Rajashekar, G., Maity, S., Patnaik, C., Das, A., Misra, A., Singh, C., Mohapatra, J., Krishnayya, N., Kiran, S., Townsend, P., and Martinez, M. (2019). Characterization of species diversity and forest health using AVIRIS-NG hyperspectral remote sensing data. *Current Science*, 116(7):1124–1135.

Ji, Y., Sun, L., Li, Y., Li, J., Liu, S., Xie, X., and Xu, Y. (2019). Non-destructive classification of defective potatoes based on hyperspectral imaging and support vector machine. *Infrared Physics andTechnology*, 99:71–79.

Julitta, T., Corp, L. A., Rossini, M., Burkart, A., Cogliati, S., Davies, N., Hom, M., Mac Arthur, A., Middleton, E. M., Rascher, U., Schickling, A., and Colombo, R. (2016). Comparison of sun-induced chlorophyll fluorescence estimates obtained from four portable field spectroradiometers. *Remote Sensing*, 8(2):122.

Jung, J., Maeda, M., Chang, A., Bhandari, M., Ashapure, A., and Landivar-Bowles, J. (2021). The potential of remote sensing and artificial intelligence as tools to improve the resilience of agriculture production systems. *Current Opinion in Biotechnology*, 70:15–22.

Kamilaris, A., and Prenafeta-Boldú, F. X. (2018). Deep learning in agriculture: A survey. *Computers and Electronics in Agriculture*, 147:70–90.

Kandel, T. P., Gowda, P. H., Northup, B. K., and Rocateli, A. C. (2019). Impacts of tillage systems, nitrogen fertilizer rates and a legume green manure on light interception and yield of winter wheat. *Cogent Food andAgriculture*, 5(1):1580176.

Katoch, R., Kumar, P., and Sankhyan, N. (2023). Artificial intelligence equipped remote sensing and GIS techniques for sustainable agriculture. *Vigyan Varta*, 4(3):169–172.

Kisevic, M., Morovic, M., and Andricevic, R. (2016). The use of hyperspectral data for evaluation of water quality parameters in the River Sava. *Fresenius Environmental Bulletin*, 25(11):4814–4822.

Kokaly, R. F., Clark, R., Swayze, G., Livo, K., Hoefen, T., Pearson, N., Wise, R., Benzel, W., Lowers, H., Driscoll, R., and Klein, A. (2017). *USGS Spectral Library Version 7*. USGS Publications Warehouse.

Kokaly, R. F., Despain, D. G., Clark, R. N., and Livo, K. E. (2003). Mapping vegetation in Yellowstone National Park using spectral feature analysis of AVIRIS data. *Remote Sensing of Environment*, 84:437–456.

Kokhanovsky, A. A., Brell, M., Segl, K., Bianchini, G., Lanconelli, C., Lupi, A., Petkov, B., Picard, G., Arnaud, L., Stone, R. S., and Chabrillat, S. (2023). First retrievals of surface and atmospheric properties using EnMAP measurements over Antarctica. *Remote Sensing*, 15(12):3042.

Kruse, F. A. (1988). Use of airborne imaging spectrometer data to map minerals associated with hydrothermally altered rocks in the northern grapevine mountains, Nevada, and California. *Remote Sensing of Environment*, 24(1):31–51. https://doi.org/10.1016/0034-4257(88)90004-1.

Kruse, F. A. (2002). Comparison of AVIRIS and Hyperion for hyperspectral mineral mapping. In *Proceedings of the 11th JPL Airborne Geoscience Workshop*. Pasadena, California, USA.

Kruse, F. A., Boardman, J. W., and Huntington, J. F. (2003). Comparison of airborne hyperspectral data and EO-1 Hyperion for mineral mapping. *IEEE Transactions on Geoscience and Remote Sensing*, 41(6):1388–1400. https://doi.org/10.1109/TGRS.2003.812908.

Kruse, F. A., Lefkoff, A. B., and Dietz, J. B. (1993). Expert system-based mineral mapping in Northern Death Valley, California/Nevada, using the airborne visible/infrared imaging spectrometer (AVIRIS). *Remote Sensing of Environment*, 44(2–3):309–336. https://doi.org/10.1016/0034-4257(93)90024-R.

Krutz, D., Muller, R., Knodt, U., Gunther, B., Walter, I., Sebastian, I., Sauberlich, T., Reulke, R., Carmona, E., Eckardt, A., Venus, H., Fischer, C., Zender, B., Arloth, S., Lieder, M., Neidhardt, M., Grote, U., Schrandt, F., Gelmi, S., and Wojtkowiak, A. (2019). The instrument design of the DLR Earth Sensing Imaging Spectrometer (DESIS). *Sensors*, 19(1622):1–16.

Kuang, X., Guo, J., Bai, J., Geng, H., and Wang, H. (2023). Crop-planting area prediction from multi-source Gaofen satellite images using a novel deep learning model: A case study of Yangling district. *Remote Sensing*, 15(15):3792.

Kushwaha, G., Siddiqui, A., and Kumar, V. (2020). Assessing the thermal behaviour of Urban Materials using Hyperspectral Data in Ahmedabad city, India. In *National Seminar on Recent Advances in Geospatial Technology and Applications*, March 02, 2020. Dehradun, India.

Lamb, B. T., Dennison, P. E., Hively, W. D., Kokaly, R. F., Serbin, G., Wu, Z., Dabney, P. W., Masek, J. G., Campbell, M., and Daughtry, C. S. T. (2022). Optimizing Landsat Next shortwave infrared bands for crop residue characterization. *Remote Sensing*, 14(23):6128.

Lausch, A., Salbach, C., Schmidt, A., Doktor, D., Merbach, I., and Pause, M. (2015a). Deriving phenology of barley with imaging hyperspectral remote sensing. *Ecological Modelling*, 295:123–135.

Lausch, A., Schmidt, A., and Tischendorf, L. (2015b). Data mining and linked open data—new perspectives for data analysis in environmental research. *Ecological Modelling*, 295:5–17.

Lee, B., Kim, N., Kim, E.-S., Jang, K., Kang, M., Lim, J.-H., Cho, J., and Lee, Y. (2020). An artificial intelligence approach to predict gross primary productivity in the forests of South Korea using satellite remote sensing data. *Forests*, 11(9):1000.

Lee, C. M., Cable, M., Hook, S., Green, R., Ustin, S., Mandl, D., and Middleton, E. (2015). An introduction to the NASA Hyperspectral InfraRed Imager (HyspIRI) mission and preparatory activities. *Remote Sensing of Environment*, 167(2015):6–19.

Lee, C. M., Glenn, N. F., Stavros, E. N., Luvall, J., Yuen, K., Hain, C., and Schollaert Uz, S. (2022). Systematic integration of applications into the Surface Biology and Geology (SBG) Earth mission architecture study. *Journal of Geophysical Research: Biogeosciences*, 127(4):e2021JG006720.

Legleiter, C., Overstreet, B., Glennie, C., Pan, Z., Fernandez-Diaz, J., and Singhania, A. (2016). Evaluating the capabilities of the CASI hyperspectral imaging system and Aquarius bathymetric LiDAR for measuring channel morphology in two distinct river environments. *Earth Surface Processes and Landforms*, 41:344–363.

Li, J., He, L., Liu, M., Chen, J., and Xue, L. (2022). Hyperspectral dimension reduction and navel orange surface disease defect classification using independent component analysis-genetic algorithm. *Frontiers in Nutrition*, 9:993737. https://doi.org/10.3389/fnut.2022.993737.

Li, J., Li, C., Xhao, D., and Gang, C. (2012). Hyperspectral narrow-bands and their indices on assessing nitrogen contents of cotton crop applications. In Thenkabail, P., Lyon, G., and Huete, A., editors, *Hyperspectral Remote Sensing of Vegetation*, Chapter 24, pages 579–590. CRC Press & Taylor and Francis Group, Boca Raton, London and NY.

Li, S., Song, W., Fang, L., Chen, Y., Ghamisi, P., and Benediktsson, J. A. (2019). Deep learning for hyperspectral image classification: An overview. *IEEE Transactions on Geoscience and Remote Sensing*, 57(9):6690–6709.

Lian, S., Guan, L., Peng, Z., Zeng, G., Li, M., and Xu, Y. (2023). Retrieval of leaf chlorophyll content in Gannan navel orange based on fusing hyperspectral vegetation indices using machine learning algorithms. *Ciencia Rural*, 53(3).

Lin, J., Pan, Y., Lyu, H., Zhu, X., Li, X., Dong, B., and Li, H. (2019). Developing a two-step algorithm to estimate the leaf area index of forests with complex structures based on CHRIS/PROBA data. *Forest Ecology and Management*, 441:57–70.

Liu, D. Z., Yang, F. F., and Liu, S. P. (2021a). Estimating wheat fractional vegetation cover using a density peak k-means algorithm based on hyperspectral image data. *Journal of Integrative Agriculture*, 20(11):2880–2891.

Liu, J., Han, J., Xie, J., Wang, H., Tong, W., and Ba, Y. (2020). Assessing heavy metal concentrations in Earth-cumulic-orthicanthrosols soils using VIS-NIR spectroscopy transform coupled with chemometrics. *Spectrochimica Acta,Part A: Molecular and Biomolecular Spectroscopy*, 226:117639.

Liu, J., Xiang, J., Jin, Y., Liu, R., Yan, J., and Wang, L. (2021b). Boost precision agriculture with unmanned aerial vehicle remote sensing and edge intelligence: A survey. *Remote Sensing*, 13(21):4387.

Loizzo, R., Ananasso, C., Guarini, R., Lopinto, E., Candela, L., and Pisani, A. (2016). The PRISMA hyperspectral mission. In *Workshop on Living Planet Symposium, 9.–13. May 2019, ESA*, Prague, Czech Republic. https://lps16.esa.int/.

Lu, B., Dao, P., Liu, J., He, Y., and Shang, J. (2020). Recent advances of hyperspectral imaging technology and applications in agriculture. *Remote Sensing*, 12:1–44.

Lv, W., and Wang, X. (2020). Overview of hyperspectral image classification. *Journal of Sensors*, 2020:1–13.

Ma, L., Liu, Y., Zhang, X., Ye, Y., Yin, G., and Johnson, B. A. (2019). Deep learning in remote sensing applications: A meta-analysis and review. *ISPRS Journal of Photogrammetry and Remote Sensing*, 152:166–177.

Mahajan, G. R., Das, B., Gaikwad, B., Murgaonkar, D., Desai, A., Morajkar, S., Patel, K. P., and Kulkarni, R. M. (2021). Monitoring properties of the salt-affected soils by multivariate analysis of the visible and near-infrared hyperspectral data. *Catena*, 198:105041.

Manfreda, S., McCabe, M. F., Miller, P. E., Lucas, R., Pajuelo Madrigal, V., Mallinis, G., Ben Dor, E., Helman, D., Estes, L., Ciraolo, G., Müllerová, J., Tauro, F., De Lima, M. I., De Lima, J. L. M. P., Maltese, A., Frances, F., Caylor, K., Kohv, M., Perks, M., Ruiz-Pérez, G., Su, Z., Vico, G., and Toth, B. (2018). On the use of unmanned aerial systems for environmental monitoring. *Remote Sensing*, 10(4):641.

Mansour, K., Mutanga, O., Everson, T., and Adam, E. (2012). Discriminating indicator grass species for rangeland degradation assessment using hyperspectral data resampled to AISA Eagle resolution. *ISPRS Journal of Photogrammetry and Remote Sensing*, 70:56–65.

Mariotto, I., Thenkabail, P. S., Huete, A., Slonecker, E. T., and Platonov, A. (2013). Hyperspectral versus multispectral crop-productivity modeling and type discrimination for the HyspIRI mission. *Remote Sensing of Environment*, 139:291–305.

Mariotto, I., Thenkabail, P., and Aneece, I. (2020a). *Global Hyperspectral Imaging Spectral-Library of Agricultural Crops (GHISA) Area of Study: Central Asia. Algorithm Theoretical Basis Document (ATBD)*. NASA Land Processes Distributed Active Archive Center (LP DAAC), page 28.

Mariotto, I., Thenkabail, P., and Aneece, I. (2020b). *Global Hyperspectral Imaging Spectral-Library of Agricultural Crops (GHISA) Area of Study: Central Asia. User Guide*. NASA Land Processes Distributed Active Archive Center (LP DAAC), page 7.

Mariotto, I., Thenkabail, P., and Aneece, I. (2020c). *Global Hyperspectral Imaging Spectral-Library of Agricultural Crops (GHISA) for Central Asia*. NASA Land Processes Distributed Active Archive Center (LP DAAC).

Markham, B., and Barker, J. (1987). Thematic Mapper bandpass solar exoatmospheric irradiances. *International Journal of Remote Sensing*, 8(3):517–523.

Marshall, M., and Thenkabail, P. (2014). Biomass modeling of four leading world crops using hyperspectral narrowbands in support of HyspIRI mission. *Photogrammetric Engineering and Remote Sensing*, 80(8):757–772.

Marshall, M., and Thenkabail, P. (2015). Advantage of hyperspectral EO-1 Hyperion over multispectral IKONOS, Geoeye-1, WorldView-2, Landsat ETM+, and MODIS vegetation indices in crop biomass estimation. *ISPRS Journal of Photogrammetry and Remote Sensing*, 108:205–218.

Marshall, M., Belgiu, M., Boschetti, M., Pepe, M., Stein, A., and Nelson, A. (2022). Field-level crop yield estimation with PRISMA and Sentinel-2. *ISPRS Journal of Photogrammetry and Remote Sensing*, 187:191–210.

Masudul Islam, S., Kumar, V., Kumar, S., and Agrawal, S. (2022). Spectral mixture analysis of AVIRIS-NG hyperspectral data for material identification and classification for the part of Kolkata City. *Advances in Space Research*, 73(2):1560–1572.

Middleton, E. M., Cheng, Y. B., Corp, L. A., Campbell, P. K. E., and Kustas, W. P. (2010). Diurnal and directional responses of chlorophyll fluorescence and the PRI in a cornfield. In *4th International Workshop on Remote Sensing of Vegetation Fluorescence*, Valencia, Spain.

Middleton, E. M., Huemmrich, K., Cheng, Y. B., and Margolis, H. (2012). Spectral bio-indicators of photosynthetic efficiency and vegetation stress. In Thenkabail, P., Lyon, G., and Huete, A., editors, *Hyperspectral Remote Sensing of Vegetation*, Chapter 12, pages 265–288. CRC Press & Taylor and Francis Group, Boca Raton, London and NY.

Mishra, K., Siddiqui, A., Kumar, V., Pandey, K., and Dev Garg, R. (2023). Examining effect of super-resolution on AVIRIS-NG data: A precursor to generation of large-scale urban material and natural cover maps. *Advances in Space Research*. https://doi.org/10.1016/j.asr.2023.05.020.

Miura, T., and Yoshioka, H. (2012). Hyperspectral data in long-term cross-sensor vegetation index continuity for global change studies. In Thenkabail, P., Lyon, G., and Huete, A., editors, *Hyperspectral Remote Sensing of Vegetation*, Chapter 26, pages 611–636. CRC Press & Taylor and Francis Group, Boca Raton, London and NY.

Moharana, S., and Dutta, S. (2016). Spatial variability of chlorophyll and nitrogen content of rice from hyperspectral imagery. *ISPRS Journal of Photogrammetry and Remote Sensing*, 122:17–29.

Mouroulis, P., Gorp, B. V., Green, R. O., Dierssen, H., Wilson, D. W., Eastwood, M., Boardman, J., Gao, B.-C., Cohen, D., Franklin, B., Loya, F., Lundeen, S., Mazer, A., McCubbin, I., Randall, D., Richardson, B., Rodriguez, J. I., Sarture, C., Urquiza, E., Vargas, R., White, V., and Yee, K. (2014). Portable remote imaging spectrometer coastal ocean sensor: Design, characteristics, and first flight results. *Applied Optics*, 53(7):1363–1380.

Mozgeris, G., Juodkienė, V., Jonikavičius, D., Straigytė, L., Gadal, S., and Ouerghemmi, W. (2018). Ultra-light aircraft-based hyperspectral and colour-infrared imaging to identify deciduous tree species in an urban environment. *Remote Sensing*, 10(10):1668.

Muthusamy, A., Thiel, F., Pham, V., Hellmann, C., and van der Linden, S. (2023). *Mapping Vegetation Cover on Rewetted Fen Peatlands Using Hyperspectral Spaceborne Images from DESIS and PRISMA*. EGU General Assembly, EGU23-5352, pages 1–2.

Nawar, S., Buddenbaum, H., and Hill, J. (2015). Digital mapping of soil properties using multivariate statistical analysis and ASTER data in an arid region. *Remote Sensing*, 7(2):1181–1205.

Neckel, H., and Labs, D. (1984). The solar radiation between 3300 and 12500 Å. *Solar Physics*, 90:205–258.

Numata, I. (2012). Characterization on pastures using field and imaging spectrometers. In Thenkabail, P., Lyon, G., and Huete, A., editors, *Hyperspectral Remote Sensing of Vegetation*, Chapter 9, pages 207–226. CRC Press & Taylor and Francis Group, Boca Raton, London and NY.

Okujeni, A., van der Linden, S., and Hostert, P. (2015). Extending the vegetation—impervious—soil model using simulated EnMAP data and machine learning. *Remote Sensing of Environment*, 158:69–80.

Ortenberg, F. (2012). Hyperspectral sensor characteristics airborne, spaceborne, hand-held, and truck-mounted and integration of hyperspectral data with LiDAR. In Thenkabail, P., Lyon, G., and Huete, A., editors, *Hyperspectral Remote Sensing of Vegetation*, Chapter 2, pages 39–68. CRC Press & Taylor and Francis Group, Boca Raton, London and NY.

Oskouei, M., and Babakan, S. (2016). Detection of alteration minerals using Hyperion data analysis in Lahroud. *Journal of the Indian Society of Remote Sensing*, 44(5):713–721.

Padghan, M., and Deshmukh, R. (2017). Spectroscopic determination of aboveground biomass in grass using partial least square regression model. *International Journal of Scientific Engineering and Technology*, 6(9):332–335.

Parashar, C., Sharma, R. U., Chattoraj, S. L., Sengar, V. K., and Champati Ray, P. K. (2016). Identification and mapping of minerals by using imaging spectroscopy in Southeastern Region of Rajasthan. In *SPIE9880, Multispectral, Hyperspectral, and Ultraspectral Remote Sensing Technology, Techniques and Applications, VI, 988013-1-988013-11*. https://doi.org/10.1117/12.2223579.

Peschel, T., Beier, M., Damm, C., Hartung, J., Jende, R., Muller, S., Rohde, M., Gebhardt, A., Risse, S., Walter, I., Sebastian, I., and Krutz, D. (2018). Integration and testing of an imaging spectrometer for Earth observation. In *International Conference on Space Optics*, page 7, Chania, Greece.

Pignatti, S., Acito, N., Amato, U., Casa, R., Castaldi, F., Coluzzi, R., De Bonis, R., Diani, M., Imbrenda, V., Laneve, G., Matteoli, S., Palombo, A., Pascucci, S., Santini, F., Simoniello, T., Ananasso, C., Corsini, G., and Cuomo, V. (2015). Environmental products overview of the Italian hyperspectral PRISMA mission: The SAP4PRISMA project. In *IEEE International Geoscience and Remote Sensing Symposium*, page 5, Milan, Italy.

Pignatti, S., Palombo, A., Pascucci, S., Romano, F., Santini, F., Simoniello, T., Amato, U., Cuomo, V., Acito, N., Diani, M., Matteoli, S., Corsini, G., Casa, R., De Bonis, R., Laneve, G., and Ananasso, C. (2013). The PRISMA hyperspectral mission: Science, activities and opportunities for agriculture and land monitoring. In *IEEE International Geoscience and Remote Sensing Symposium*, page 5, Melbourne, Australia.

Pippi, I. (1989). Mineral identification by the AVIRIS data. IGARSS'89. In *12th Canadian Symposium on Geoscience and Remote Sensing944–947*. https://doi.org/10.1109/IGARSS.1989.579045.

Plaza, A., Plaza, J., Martin, G., and Sanchez, S. (2012). Hyperspectral data processing algorithms. In Thenkabail, P., Lyon, G., and Huete, A., editors, *Hyperspectral Remote Sensing of Vegetation*, Chapter 5, pages 121–140. CRC Press & Taylor and Francis Group, Boca Raton, London and NY.

Pour, A. B., and Hashim, M. (2011). Identification of hydrothermal alteration minerals for exploring of porphyry copper deposit using ASTER data, SE Iran. *Journal of Asian Earth Sciences*, 42(6):1309–1323. https://doi.org/10.1016/j.jseaes.2011.07.017.

Pour, A. B., and Hashim, M. (2012). The application of ASTER remote sensing data to porphyry copper and epithermal gold deposits. *Ore Geology Reviews*, 44:1–9. https://doi.org/10.1016/j.oregeorev.2011.09.009.

Pu, R. (2012). Detecting and mapping invasive plant species by using hyperspectral data. In Thenkabail, P., Lyon, G., and Huete, A., editors, *Hyperspectral Remote Sensing of Vegetation*, Chapter 19, pages 447–468. CRC Press & Taylor and Francis Group, Boca Raton, London and NY.

Qi, J., Inoue, Y., and Wiangwang, N. (2012). Hyperspectral sensor systems and data characteristics in global change studies. In Thenkabail, P., Lyon, G., and Huete, A., editors, *Hyperspectral Remote Sensing of Vegetation*, Chapter 3, pages 69–92. CRC Press & Taylor and Francis Group, Boca Raton, London and NY.

Ramsey, E. III, and Rangoonwala, A. (2012). Hyperspectral remote sensing of wetlands. In Thenkabail, P., Lyon, G., and Huete, A., editors, *Hyperspectral Remote Sensing of Vegetation*, Chapter 21, pages 487–512. CRC Press & Taylor and Francis Group, Boca Raton, London and NY.

Rast, M., Nieke, J., Adams, J., Isola, C., and Gascon, F. (2021). Copernicus hyperspectral imaging mission for the environment (CHIME). In *2021 IEEE International Geoscience and Remote Sensing Symposium IGARSS*, pages 108–111. Brussels, Belgium.

Ratheesh, R., Chaudhury, N. R., Rajput, P. K., Arora, M., Gujrati, A., Arunkumar, S., Shetty, A., Baral, R., Patel, R. M., Joshi, D. M., Patel, H., Pathak, B. J., Jayappa, K. S., Samal, R. N., and Rajawat, A. S. (2019). Coastal sediment dynamics, ecology and detection of coral reef macroalgae from AVIRIS-NG. *Current Science*, 116(7):1157–1165.

Rhee, D., Kim, Y., Kang, B., and Kim, D. (2018). Applications of unmanned aerial vehicles in fluvial remote sensing: An overview of recent achievements. *KSCE Journal of Civil Engineering*, 22(2):588–602.

Riaza, A., Buzzi, J., García-Meléndez, E., Carrère, V., Sarmiento, A., and Müller, A. (2015). Monitoring acidic water in a polluted river with hyperspectral remote sensing (HyMap). *Hydrological Sciences Journal*, 60(6):1064–1077.

Ribeiro, S. G., Teixeira, A. D. S., de Oliveira, M. R. R., Costa, M. C. G., Araújo, I. C. D. S., Moreira, L. C. J., and Lopes, F. B. (2021). Soil organic carbon content prediction using soil-reflected spectra: A comparison of two regression methods. *Remote Sensing*, 13(23):4752.

Richards, J. A., and Jia, X. (2006). *Remote Sensing Digital Image Analysis: An Introduction*, 4th ed. Springer Verlag, Berlin Heidelberg, Germany.

Roberts, D. (2012). Hyperspectral vegetation indices. In Thenkabail, P., Lyon, G., and Huete, A., editors, *Hyperspectral Remote Sensing of Vegetation*, Chapter 14, pages 309–328. CRC Press & Taylor and Francis Group, Boca Raton, London and NY.

Rossel, R. V., Walvoort, D. J. J., McBratney, A. B., Janik, L. J., and Skjemstad, J. O. (2006). Visible, near infrared, mid infrared or combined diffuse reflectance spectroscopy for simultaneous assessment of various soil properties. *Geoderma*, 131(1–2):59–75.

Roy, D., Kovalskyy, V., Zhang, H., Vermote, E., Yan, L., Kumar, S., and Egorov, A. (2016). Characterization of Landsat-7 to Landsat-8 reflective wavelength and normalized difference vegetation index continuity. *Remote Sensing of Environment*, 185:57–70.

Sacks, W., Deryng, D., Foley, J., and Ramankutty, N. (2010). Crop planting dates: An analysis of global patterns. *Global Ecology and Biogeography*, 19:607–620.

Santos, J. F. B., Dias, J. D. Jr., Backes, A. R., and Escarpinati, M. C. (2021). Segmentation of agricultural images using vegetation indices. In *Proceedings of the 16thInternational Joint Conference on Computer Vision,Imaging and Computer Graphics Theory and Applications—Volume 4: VISAPP, (VISIGRAPP 2021)*, pages 506–511. INSTICC, SciTePress, Vienna, Austria.

SAS Institute Inc. (2009). *SAS/ACCESS® 9.4 Interface to ADABAS: Reference*. SAS Institute Inc., Cary, NC.

Sellami, M. H., Albrizio, R., Čolovíc, M., Hamze, M., Cantore, V., Todorovic, M., Piscitelli, L., and Stellacci, A. M. (2022). Selection of hyperspectral vegetation indices for monitoring yield and physiological response in sweet maize under different water and nitrogen availability. *Agronomy*, 12(2):489.

Sheykhmousa, M., Mahdianpari, M., Ghanbari, H., Mohammadimanesh, F., Ghamisi, P., and Homayouni, S. (2020). Support vector machine vs. random forest for remote sensing image classification: A meta-analysis and systematic review. *IEEE Journal of Selected Topics in Applied Earth Observations and Remote Sensing*, 13:6308–6325.

Siddiqui, A., Chauhan, P., Kumar, V., Jain, G., Deshmukh, A., and Kumar, P. (2020). Characterization of urban materials in AVIRIS-NG data using a mixture tuned matched filtering (MTMF) approach. *Geocarto International*, 37(1):332–347.

Sinaga, K. P., and Yang, M.-S. (2020). Unsupervised k-means clustering algorithm. *IEEE Access*, 8:80716–80727.

Sithole, N. J., Ncama, K., and Magwaza, L. (2018). Robust VIS-NIRS models for rapid assessment of soil organic carbon and nitrogen in Feralsols Haplic soils from different tillage management practices. *Computers and Electronics in Agriculture*, 153:295–301.

Slonecker, T. (2012). Hyperspectral analysis of the effects of heavy metals on vegetation reflectance. In Thenkabail, P., Lyon, G., and Huete, A., editors, *Hyperspectral Remote Sensing of Vegetation*, Chapter 23, pages 561–578. CRC Press & Taylor and Francis Group, Boca Raton, London and NY.

Song, G., Wang, Q., and Jin, J. (2023). Fractional-order derivative spectral transformations improved partial least squares regression estimation of photosynthetic capacity from hyperspectral reflectance. *IEEE Transactions on Geoscience and Remote Sensing*, 61:1–10.

Sousa, D., and Small, C. (2023). Topological generality and spectral dimensionality in the Earth mineral dust source investigation (EMIT) using joint characterization and the spectral mixture residual. *Remote Sensing*, 15(9):2295.

Srivastava, P., Gupta, M., Singh, U., Prasad, R., Pandey, P., Raghubanshi, A., and Petropoulos, G. (2021). Sensitivity analysis of artificial neural network for chlorophyll prediction using hyperspectral data. *Environment,Development and Sustainability*, 23:5504–5519.

Stagakis, S., Vanikiotis, T., and Sykioti, O. (2016). Estimating forest species abundance through linear unmixing of CHRIS/PROBA imagery. *ISPRS Journal of Photogrammetry and Remote Sensing*, 119:79–89.

Steinberg, A., Chabrillat, S., Stevens, A., Segl, K., and Foerster, S. (2016). Prediction of common surface soil properties based on Vis-NIR airborne and simulated EnMAP imaging spectroscopy data: Prediction accuracy and influence of spatial resolution. *Remote Sensing*, 8(7):613.

Stroppiana, D., Fava, F., Boschetti, M., and Brivio, P. (2012). Estimation of nitrogen content in crops and pastures using hyperspectral vegetation indices. In Thenkabail, P., Lyon, G., and Huete, A., editors, *Hyperspectral Remote Sensing of Vegetation*, Chapter 11, pages 245–264. CRC Press & Taylor and Francis Group, Boca Raton, London and NY.

Tamiminia, H., Salehi, B., Mahdianpari, M., Quackenbush, L., Adeli, S., and Brisco, B. (2020). Google Earth Engine for geo-big data applications: A meta-analysis and systematic review. *ISPRS Journal of Photogrammetry and Remote Sensing*, 164:152–170.

Tan, Y., Lu, L., Bruzzone, L., Guan, R., Chang, Z., and Yang, C. (2020). Hyperspectral band selection for lithologic discrimination and geological mapping. *IEEE Journal of Selected Topics in Applied Earth Observations and Remote Sensing*, 13:471–486.

Tanner, R. (2013). Data mining—das etwas andere eldorado. *Technologie IT-Methoden*, 8:37–42.

Teluguntla, P., Thenkabail, P. S., Xiong, J., Gumma, M. K., Congalton, R. G., Oliphant, A., Poehnelt, J., Yadav, K., Rao, M., and Massey, R. (2017). Spectral Matching Techniques (SMTs) and Automated Cropland Classification Algorithms (ACCAs) for mapping croplands of Australia using MODIS 250-m time-series (2000–2015) data. *International Journal of Digital Earth*, 10(9):944–977.

Thenkabail, P. S. (2002). Evaluation of narrowband and broadband vegetation indices for determining optimal hyperspectral wavebands for agricultural crop characterization. *Photogrammetric Engineering and Remote Sensing*, 68:607–622.

Thenkabail, P. S., and Aneece, I. (2019). *Global Hyperspectral Imaging Spectral-Library of Agricultural Crops for Conterminous United States v001*. NASA EOSDIS Land Processes DAAC.

Thenkabail, P. S., Aneece, I., Teluguntla, P., and Oliphant, A. (2021). Hyperspectral narrowband data propel gigantic leap in the Earth remote sensing. *Photogrammetric Engineering and Remote Sensing*, 87(7):461–467.

Thenkabail, P. S., Biradar, C. M., Noojipady, P., Dheeravath, V., Li, Y., Velpuri, M., Gumma, M., Gangalakunta, O. R. P., Turral, H., Cai, X., Vithanage, J., Schull, M. A., and Dutta, R. (2009). Global Irrigated Area Map (GIAM), derived from remote sensing, for the end of the last millennium. *International Journal of Remote Sensing*, 30(14):3679–3733.

Thenkabail, P. S., Enclona, E. A., Ashton, M. S., Legg, C., and De Dieu, M. J. (2004a). Hyperion, IKONOS, ALI, and ETM+ sensors in the study of African rainforests. *Remote Sensing of Environment*, 90(1):23–43.

Thenkabail, P. S., Enclona, E. A., Ashton, M. S., and van der Meer, B. (2004b). Accuracy assessments of hyperspectral waveband performance for vegetation analysis applications. *Remote Sensing of Environment*, 91(3):354–376.

Thenkabail, P. S., Gumma, M., Teluguntla, P., and Mohammed, I. (2014). Hyperspectral remote sensing of vegetation and agricultural crops. *Photogrammetric Engineering and Remote Sensing*, 80(4):697–709.

Thenkabail, P. S., Hanjra, M. A., Dheeravath, V., and Gumma, M. (2010). A holistic view of global croplands and their water use for ensuring global food security in the 21st century through advanced remote sensing and non-remote sensing approaches. *Remote Sensing*, 2(1):211–261.

Thenkabail, P. S., Hanjra, M. A., Dheeravath, V., and Gumma, M. (2011). Global croplands and their water use from remote sensing and nonremote sensing perspectives. In Weng, Q., editor, *Advances in Environmental Remote Sensing: Sensors,Algorithms, and Applications*, Chapter 16, pages 383–419. CRC Press, Boca Raton, FL.

Thenkabail, P. S., Krishna, M., and Turral, H. (2007). Spectral matching techniques to determine historical Land-Use/Land-Cover (LULC) and irrigated areas using time-series 0.1-degree AVHRR pathfinder datasets. *Photogrammetric Engineering and Remote Sensing*, 73:1029–1040.

Thenkabail, P. S., Lyon, G., and Huete, A. (2012a). Advances in hyperspectral remote sensing of vegetation and agricultural crops. In Thenkabail, P., Lyon, G., and Huete, A., editors, *Hyperspectral Remote Sensing of Vegetation*, Chapter 1, pages 3–29. CRC Press & Taylor and Francis Group, Boca Raton, London and NY.

Thenkabail, P. S., Lyon, G., and Huete, A. (2012b). Hyperspectral remote sensing of vegetation and agricultural crops: Current status and future possibilities. In Thenkabail, P., Lyon, G., and Huete, A., editors, *Hyperspectral Remote Sensing of Vegetation*, Chapter 28, pages 663–668. CRC Press & Taylor and Francis Group, Boca Raton, London and NY.

Thenkabail, P. S., Mariotto, I., Gumma, M., Middleton, E., Landis, D., and Huemmrich, K. (2013). Selection of Hyperspectral Narrowbands (HNBs) and composition of Hyperspectral two band Vegetation Indices (HVIs) for biophysical characterization and discrimination of crop types using field reflectance and Hyperion/EO-1 data. *IEEE Journal of Selected Topics in Applied Earth Observations and Remote Sensing*, 6(2):427–439.

Thenkabail, P. S., Smith, R. B., and De Pauw, E. (1999). Hyperspectral vegetation indices for determining agricultural crop characteristics. In *CEO Research Publication No.1*, page 45, ISBN: 0-9671303-0-1. Yale University, New Haven, CT.

Thenkabail, P. S., Smith, R. B., and De Pauw, E. (2000). Hyperspectral vegetation indices and their relationships with agricultural crop characteristics. *Remote Sensing of Environment*, 71(2):158–182.

Thenkabail, P. S., and Wu, Z. (2012). An Automated Cropland Classification Algorithm (ACCA) for Tajikistan by combining Landsat, MODIS, and secondary data. *Remote Sensing*, 4(10):2890–2918.

Thomas, V. (2012). Hyperspectral remote sensing for forest management. In Thenkabail, P., Lyon, G., and Huete, A., editors, *Hyperspectral Remote Sensing of Vegetation*, Chapter 20, pages 469–486. CRC Press & Taylor and Francis Group, Boca Raton, London and NY.

Thompson, D., Seidel, F., Gao, B., Gierach, M., Green, R., Kudela, R., and Mouroulis, P. (2015). Optimizing irradiance estimates for coastal and inland water imaging spectroscopy. *Geophysical Research Letters*, 42(10):4116–4123.

Tian, X., Fan, S., Huang, W., Wang, Z., and Li, J. (2020). Detection of early decay on citrus using hyperspectral transmittance imaging technology coupled with principal component analysis and improved watershed segmentation algorithms. *Postharvest Biology and Technology*, 161:111071.

Ting-ting, Z., and Fei, L. (2012). Application of hyperspectral remote sensing in mineral identification and mapping. In *2nd International Conference on Computer Science and Network Technology*, pages 103–106. Changchun, China.

Tits, L., De Keersmaecker, W., Somers, B., Asner, G. P., Farifteh, J., and Coppin, P. (2012). Hyperspectral shape-based unmixing to improve intra- and interclass variability for forest and agro-ecosystem monitoring. *ISPRS Journal of Photogrammetry and Remote Sensing*, 74:163–174.

Torres-Sánchez, J., Peña, J., de Castro, A., and López-Granados, F. (2014). Multi-temporal mapping of the vegetation fraction in early-season wheat fields using images from UAV. *Computers and Electronics in Agriculture*, 103:104–113.

Traoré, B., Birhanu, B. Z., Sangaré, S., Gumma, M. K., Tabo, R., and Whitbread, A. M. (2021). Contribution of climate-smart agriculture technologies to food self-sufficiency of smallholder households in Mali. *Sustainability*, 13(14):7757.

Tripathi, P., and Garg, R. (2023). Potential of DESIS and PRISMA hyperspectral remote sensing data in rock classification and mineral identification: A case study for Banswara in Rajasthan, India. *Environmental Monitoring and Assessment*, 195(575).

Tsyganskaya, V., Martinis, S., Marzahn, P., and Ludwig, R. (2018). Detection of temporary flooded vegetation using Sentinel-1 time series data. *Remote Sensing*, 10(8):1286.

USDA National Agricultural Statistics Service Cropland Data Layer. (2022). *Published Crop-Specific Data Layer*. Retrieved September 06, 2022, from https://nassgeodata.gmu.edu/CropScape/.

Vahtmäe, E., Kotta, J., Lõugas, L., and Kutser, T. (2021). Mapping spatial distribution, percent cover and biomass of benthic vegetation in optically complex coastal waters using hyperspectral CASI and multispectral Sentinel-2 sensors. *International Journal of Applied Earth Observation and Geoinformation*, 102:102444.

Vali, A., Comai, S., and Matteucci, M. (2020). Deep learning for land use and land cover classification based on hyperspectral and multispectral Earth observation data: A review. *Remote Sensing*, 12:1–31.

van der Meer, F. D. (2004). Analysis of spectral absorption features in hyperspectral imagery. *International Journal of Applied Earth Observation and Geoinformation*, 5(1):55–68. https://doi.org/10.1016/j.jag.2003.09.001.

van der Meer, F. D., van der Werff, H. M. A., van Ruitenbeek, F. J. A., Hecker, C. A., Bakker, W. H., Noomen, M. F., van der Meij, B., Carranza, E. J. M., de Smeth, J. B., and Woldai, T. (2012). Multi- and hyperspectral geologic remote sensing: A review. *International Journal of Applied Earth Observation and Geoinformation*, 14(1):112–128. https://doi.org/10.1016/j.jag.2011.08.002.

van der Meij, B., Kooistra, L., Suomalainen, J., Barel, J. M., and De Deyn, G. B. (2017). Remote sensing of plant trait responses to field-based plant—soil feedback using UAV-based optical sensors. *Biogeosciences*, 14(3):733–749.

Vaughan, R., Titus, T., Johnson, J., Hagerty, J., Gaddis, L., Soderblom, L., and Geissler, P. (2012). Hyperspectral analysis of rocky surfaces on the Earth and other planetary systems. In Thenkabail, P., Lyon, G., and Huete, A., editors, *Hyperspectral Remote Sensing of Vegetation*, Chapter 27, pages 637–662. CRC Press & Taylor and Francis Group, Boca Raton, London and NY.

Vermote, E. F., El Saleous, N. Z., and Justice, C. O. (2002). Atmospheric correction of MODIS data in the visible to middle infrared: First results. *Remote Sensing of Environment*, 83(1):97–111.

Verrelst, J., Romijn, E., and Kooistra, L. (2012). Mapping vegetation density in a heterogeneous river floodplain ecosystem using pointable CHRIS/PROBA data. *Remote Sensing*, 4(9):2866–2889.

Vincheh, Z. H., and Arfania, R. (2017). Lithological mapping from OLI and ASTER multispectral data using matched filtering and spectral analogues techniques in the Pasab-e-Bala Area, Central Iran. *Open Journal of Geology*, 7:1494–1508. https://doi.org/10.4236/ojg.2017.

Wan, L., Li, Y., Cen, H., Zhu, J., Yin, W., Wu, W., Zhu, H., Sun, D., Zhou, W., and He, Y. (2018). Combining UAV-based vegetation indices and image classification to estimate flower number in oilseed rape. *Remote Sensing*, 10(9):1484.

Wang, S., Di Tommaso, S., Faulkner, J., Friedel, T., Kennepohl, A., Strey, R., and Lobell, D. B. (2020). Mapping crop types in Southeast India with smartphone crowdsourcing and deep learning. *Remote Sensing*, 12(18):2957.

Wu, Y., and Zhang, X. (2019). Object-based tree species classification using airborne hyperspectral images and LiDAR data. *Forests*, 11:1–25.

Xia, C., Yang, S., Huang, M., Zhu, Q., Guo, Y., and Qin, J. (2019). Maize seed classification using hyperspectral image coupled with multi-linear discriminant analysis. *Infrared Physics andTechnology*, 103:103077.

Xiang, P., Zhou, H., Li, H., Song, S., Tan, W., Song, J., and Gu, L. (2020). Hyperspectral anomaly detection by local joint subspace process and support vector machine. *International Journal of Remote Sensing*, 41(10):3798–3819.

Xu, Q., Liu, S., Ye, F., Zhang, Z., and Zhang, C. (2018). Application of CASI/SASI and Fieldspec4 hyperspectral data in exploration of the Baiyanghe Uranium Deposit, Hebukesaier, Xinjiang, NW China. *International Journal of Remote Sensing*, 39(2):453–469.

Yamamoto, S., Tsuchida, S., Urai, M., Mizuochi, H., Iwao, K., and Iwasaki, A. (2022). Initial analysis of spectral smile calibration of Hyperspectral Imager Suite (HISUI) using atmospheric absorption bands. *IEEE Transactions on Geoscience and Remote Sensing*, 60:1–15.

Yang, H., and Du, J. (2021). Classification of desert steppe species based on unmanned aerial vehicle hyperspectral remote sensing and continuum removal vegetation indices. *Optik*, 247:167877.

Yao, H., Tang, L., Tian, L., Brown, R., Bhatnagar, D., and Cleveland, T. (2012). Using hyperspectral data in precision farming applications. In Thenkabail, P., Lyon, G., and Huete, A., editors, *Hyperspectral Remote Sensing of Vegetation*, Chapter 25, pages 591–610. CRC Press & Taylor and Francis Group, Boca Raton, London and NY.

Yubo, Z., Zhuoran, Y., Jiuchun, Y., Yuanyuan, Y., Dongyan, W., Yucong, Z., Fengqin, Y., Lingxue, Y., Liping, C., and Shuwen, Z. (2020). A novel model integrating deep learning for land use/cover change reconstruction: A case study of Zhenlai County, Northeast China. *Remote Sensing*, 12(20):3314.

Zhang, J., Rivard, B., Sanchez-Azofeifa, A., and Castro-Esau, K. (2006). Intra- and inter-class spectral variability of tropical tree species at La Selva, Costa Rica: Implications for species identification using HYDICE imagery. *Remote Sensing of Environment*, 105:129–141.

Zhang, Y. (2012). Forest leaf chlorophyll content study using hyperspectral remote sensing. In Thenkabail, P., Lyon, G., and Huete, A., editors, *Hyperspectral Remote Sensing of Vegetation*, Chapter 7, pages 167–186. CRC Press & Taylor and Francis Group, Boca Raton, London and NY.

Zhao, L., Li, Q., Zhang, Y., Wang, H., and Du, X. (2019). Integrating the continuous wavelet transform and a convolutional neural network to identify vineyard using time series satellite images. *Remote Sensing*, 11(22):2641.

Zhao, X., Zhang, J., Huang, Y., Tian, Y., and Yuan, L. (2022). Detection and discrimination of disease and insect stress of tea plants using hyperspectral imaging combined with wavelet analysis. *Computers and Electronics in Agriculture*, 193:106717.

Zhao, Y., Sun, Y., Chen, W., Zhao, Y., Liu, X., and Bai, Y. (2021). The potential of mapping grassland plant diversity with the links among spectral diversity, functional trait diversity, and species diversity. *Remote Sensing*, 13:3034.

Zhu, X. X., Tuia, D., Mou, L., Xia, G.-S., Zhang, L., Xu, F., and Fraundorfer, F. (2017). Deep learning in remote sensing: A comprehensive review and list of resources. *IEEE Geoscience and Remote Sensing Magazine*, 5(4):8–36.

Zhu, Y., Wang, W., and Yao, X. (2012). Estimating Leaf Nitrogen Concentration (LNC) of cereal crop with hyperspectral data. In Thenkabail, P., Lyon, G., and Huete, A., editors, *Hyperspectral Remote Sensing of Vegetation*, Chapter 8, pages 187–206. CRC Press & Taylor and Francis Group, Boca Raton, London and NY.

Part III

Rangelands

11 A Global View of Remote Sensing of Rangelands

Evolution, Applications, Future Pathways

Matthew Reeves, Robert Washington-Allen, Jay Angerer, E. Raymond Hunt Jr., Wasantha Kulawardhana, Lalit Kumar, Tatiana Loboda, Thomas Loveland, Graciela Metternicht, R. Douglas Ramsey, Joanne V. Hall, Trenton Benedict, Pedro Millikan, Angus Retallack, Arjan J.H. Meddens, William K. Smith, and Wen Zhang

ACRONYMS AND DEFINITIONS

ATSR	Along-Track Scanning Radiometer
ENVISAT	Environmental Satellite
ERTS	Earth Resources Technology Satellite
ESRI	Environmental Systems Research Institute
ETM+	Enhanced Thematic Mapper Plus
EVI	Enhanced Vegetation Index
FAO	Food and Agriculture Organization of the United Nations
fPAR	Fraction of Photosynthetically Active Radiation
GIMMS	Global Inventory Modeling and Mapping Studies
GPP	Gross Primary Production
ICESat	Ice, Cloud, and Land Elevation Satellite
LAI	Leaf Area Index
MERIS	Medium Resolution Imaging Spectrometer
MSS	Multispectral Scanner
MODIS	Moderate-Resolution Imaging Spectroradiometer
NASA	National Aeronautics and Space Administration
NDBR	Normalized Difference Burn index
NDVI	Normalized Difference Vegetation Index
NIR	Near-Infrared
NOAA	National Oceanic and Atmospheric Administration
NPP	Net Primary Productivity
OLI	Operational Land Imager
SAR	Synthetic Aperture Radar
SAVI	Soil Adjusted Vegetation Index
S-NPP	Suomi National Polar-orbiting Partnership
SST	Sea Surface Temperature
SWIR	Shortwave infrared

DOI: 10.1201/9781003541165-14

TIRS Thermal Infrared Sensor
TNDVI Transformed Normalized Difference Vegetation Index
UAV Unmanned Aerial Vehicles
VIIRS Visible Infrared Imaging Radiometer Suite

11.1 INTRODUCTION

The term "rangeland" is rather nebulous and there is no single definition of rangeland that is universally accepted by land managers, scientists, or international bodies (Reeves and Mitchell, 2011; Lund, 2007). Dozens and possibly hundreds (Lund, 2007) of definitions and ideologies exist because various stakeholders often have unique objectives requiring different information. To describe the role of remote sensing in a global context it is, however, necessary to provide definitions to orient the reader. The Food and Agricultural Organization of the United Nations (FAO) convened a conference in 2002 and again in 2013 to begin addressing the issue of harmonizing definitions of forest-related activities. Based on this concept, here rangelands are considered lands usually dominated by non-forest vegetation. The Society for Range Management defines rangelands as (SRM, 1998):

> Land on which the indigenous vegetation (climax or natural potential) is predominantly grasses, grass-like plants, forbs, or shrubs and is managed as a natural ecosystem. If plants are introduced, they are managed similarly. Rangelands include natural grasslands, savannas, shrublands, many deserts, tundra, alpine communities, marshes, and wet meadows.

Rangelands occupy a wide diversity of habitats and are found on every continent except Antarctica. Excluding Antarctica and barren lands, rangelands occupy 52% of the Earth's surface based on the land cover analysis presented in Figure 11.1. Figure 11.1 is based on the 2018 MODIS

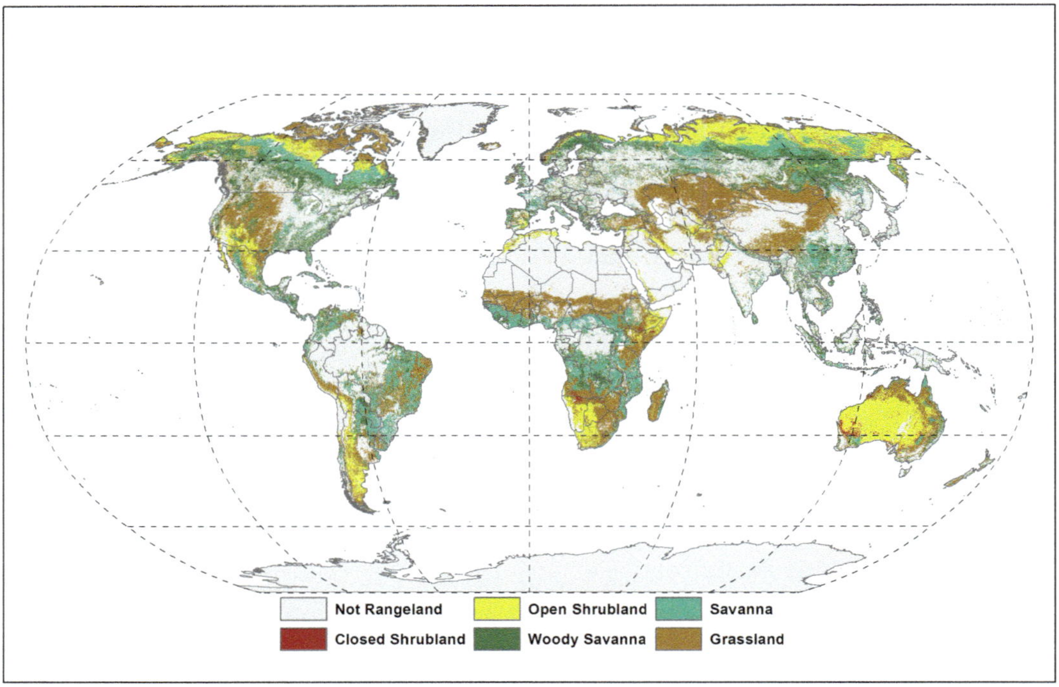

FIGURE 11.1 Global distribution of rangeland land cover types (collection 6 MODIS land cover (MCD12Q1 and MCD12C1) considered rangelands for this chapter. (Sulla-Menashe and Fiedl, 2018.)

Collection 6.0, 1 km land cover (the University of Maryland "UMD" classification) and suggested rangeland classes for this dataset are closed shrubland, open shrubland, woody savanna, savanna, and grassland. Using these classes, Russia, Australia, and Canada are the top three countries with the most rangelands (Table 11.1) representing 18, 10, and 8% of the global extent, respectively. The large areal extent of rangelands, high cost of ground data collection and quest for societal well-being have, for decades, provided rich opportunity for remote sensing to aid in answering pressing questions. Collectively, these lands provide nearly 18 trillion dollars to the global economy each year (Costanza et al., 2014).

11.2 HISTORY AND EVOLUTION OF GLOBAL REMOTE SENSING

The application of digital remote sensing to rangelands is as long as the history of digital remote sensing itself. Before the launch of the Earth Resources Technology Satellite (ERTS)—later renamed Landsat, scientists were evaluating the use of multispectral aerial imagery to map soils and range vegetation (Yost and Wenderoth, 1969). During the late 1960s, the promise of ERTS, designed to drastically improve our ability to update maps and study Earth resources, particularly in developing countries, was eagerly anticipated by a number of government agencies (Carter, 1969). With the ERTS launch on July 23, 1972, a flurry of research activity aimed at the application of this new data source to map Earth resources began. Practitioners who pioneered the use of satellite based digital remote sensing found the new data source a significant value for rangeland assessments (e.g., Rouse et al., 1973, Rouse et al., 1974, Bauer, 1976). This early work established many of the basic techniques still in use today to assess and monitor global rangelands. The following subsections discuss the evolution of remote sensing data, methods, and approaches in various decades.

11.2.1 Beginning of Landsat MSS Era, 1970s

In this first decade of satellite-based digital remote sensing, rangeland scientists quickly assessed the capabilities of this new tool across the globe (Graetz et al., 1976, Rouse et al., 1973). Work

TABLE 11.1

Global Area of Rangeland Vegetation Types Estimated Using MODIS Land Cover Data (Mod12Q1, Collection 6.1) for the Top 10 Countries with the Most Rangeland. CSL is closed shrubland, OSL is open shrubland and rangeland proportion is the rangeland area column divided by the Area column divided by 100.

	Area	CSL	Grassland	OSL	Savanna	Woody Savanna	Rangeland area	Rangeland Proportion
Country				Km²				%
Russia	16,911,018	19,331	2,376,004	3,341,197	2,879,614	2,819,843	11,435,990	68
Australia	7,701,679	277,052	1,859,416	4,190,891	475,195	101,371	6,903,926	90
Canada	9,901,655	398	2,028,688	893,386	1,733,698	1,863,027	6,519,196	66
United States	9,449,630	19,604	2,999,071	890,464	898,098	1,265,172	6,072,409	64
China	9,403,880	1,601	2,786,941	8,019	960,887	833,217	4,590,665	49
Brazil	8,504,556	19,493	1,937,084	1,425	2,096,610	370,273	4,424,885	52
Kazakhstan	2,834,000	179	2,318,399	5,103	9,374	12,748	2,345,802	83
Argentina	2,781,182	1,297	873,035	667,471	413,937	86,635	2,042,375	73
Mexico	1,961,256	7,658	513,387	398,736	323,783	218,373	1,461,938	75
Angola	1,252,148	16,052	453,782	19,959	421,079	217,768	1,128,640	90

by Rouse et al. (1973), in what would later become the Normalized Difference Vegetation Index (NDVI) (Rouse et al., 1974), applied multi-temporal ERTS (Landsat 1) at 79 m spatial resolution data to the grasslands of the central Great Plains of the United States and documented that the normalized ratio of the Multispectral Scanner (MSS) near infrared (NIR) (band 7) and Red band (band 5) was sensitive to vegetation dry biomass, percent green, and moisture content (Figure 11.2). They also determined that within uniform grasslands, ground-based estimates of moisture content and percent green cover accounted for 99% of the variation in their "Transformed Vegetation Index" (TVI). The TVI was later renamed to the Transformed Normalized Difference Vegetation Index (TNDVI) (Deering et al., 1975) and is calculated as the square-root of the NDVI plus an arbitrary constant (0.5 in their case). This transformation of the NDVI was done to avoid negative values.

The NDVI is, to date, one of the most widely used vegetation index on a global basis. Figure 11.2 shows the graphic published by Rouse et al. (1973) identifying the tight relationship between ground derived green biomass and the TVI. The significance of Figure 11.2 is demonstration of potential to track vegetation growth across time, thus documenting the ability for remote sensing instruments to monitor vegetation dynamics and the importance of systematic and uninterrupted collection of remotely sensed imagery.

Another significant development during this first decade of satellite based remote sensing was the "Tasseled Cap Transformation" (Kauth and Thomas, 1976). The Tasseled Cap (or "Kauth-Thomas Transformation" to some) employed principal component analysis to understand the covariate nature of the four MSS spectral bands and extract from those data the primary ground features, or components, influencing the spectral signature. The Tasseled Cap and its eventual successor—the Brightness-Greenness-Wetness Transform (Crist and Cicone, 1984) applied to the Landsat Thematic Mapper sensor, has been a widely used tool for many land resource applications (Todd et al., 1998, Hacker, 1980, Graetz et al., 1986). The NDVI and the Tasseled-Cap provided the ability to convert reflectance values collected across multiple spectral bands into biophysically focused data layers, thus giving range managers and ecologists a tool by which to directly assess and monitor vegetation growth.

11.2.2 Multiple Sensor Era, 1980s

With the development of the NDVI and the launch of the Television Infrared Observation Satellite Next Generation (TIROSN) satellite carrying the Advanced Very High Resolution Radiometer

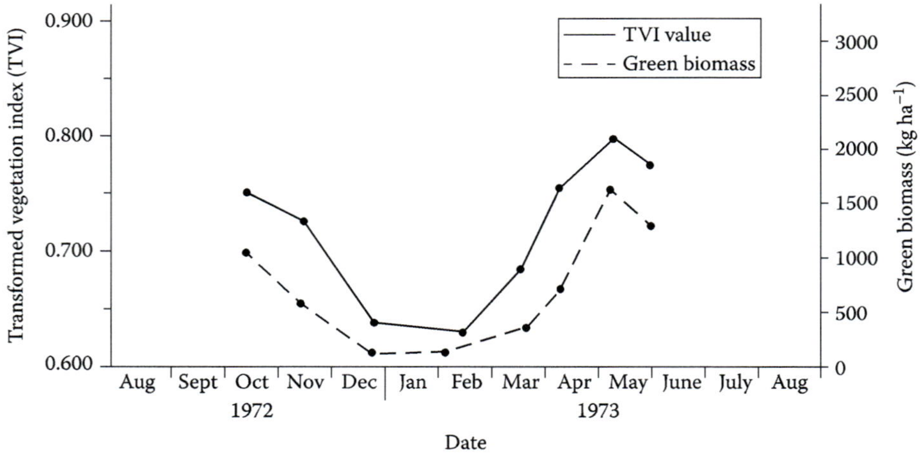

FIGURE 11.2 Earth Resources Technology Satellite (ERTS)-1 Transformed Vegetation Index values versus green biomass. (Original from Rouse et al., 1973.)

(AVHRR) in October of 1978, remote sensing practitioners now had the means to monitor temporal vegetation dynamics across very large areas (Tucker, 1979). The 1 km resolution of the AVHRR was ideal for continental scale monitoring, which was not possible with Landsat images given the computing power and data storage capacities of that era. Further, a 1day global repeat cycle provided the ability to track phenological changes in vegetation growth within and between years—a feature also not possible with the 18 and 16 day repeat cycles of the Landsat platforms. Gray and McCrary (1981) showed the utility of the AVHRR for vegetation mapping and noted that vegetation indices derived from this sensor could be related to plant growth stress due to water deficits. This relationship, coupled with the high temporal repeat interval of the TIROSN, led to the use of the NDVI to monitor the impact of drought on grasslands across the Sahel region of Africa (Tucker et al., 1983) and by direct inference predict the impact of drought to local human populations (Prince and Tucker, 1986).

The application of the NDVI to semiarid landscapes was somewhat problematic due to generally low vegetation canopy cover in these environments and the fact that background soil brightness tended to influence the resulting NDVI values (Elvidge and Lyon, 1985). Moreover, Reeves et al. (2021) estimate that the rather linear relationship between NDVI and aboveground annual productivity became asymptotic above approximately 2,300 kg ha^{-1} suggesting the need for other types of indices in higher productivity environments. The soil adjusted vegetation index (SAVI) (Huete, 1988) was developed as a simple modification to the NDVI to account for the influence of soil on the reflectance properties of green vegetation. The SAVI has been used widely within semiarid environments where vegetation cover is low. The 1980s also saw great strides in satellite based terrestrial remote sensing with the launch of Landsat 4 in July of 1982 and Landsat 5 in March of 1985, as well as the launch of the French Satellite Pour l'Observation de la Terre (SPOT) in 1986. Each platform carried sensors with slightly different capabilities, but each focused their spectral resolution on the red and NIR portions of the electromagnetic spectrum, save one. The Landsat TM was a significant improvement over its predecessor, the MSS. Not only were the spatial and radiometric resolutions improved, but also the TM supported two additional spectral bands calibrated to the shortwave infrared portion of the electromagnetic (EM) spectrum. This significant addition provided the ability to monitor leaf moisture (Tucker, 1980, Hunt and Rock, 1989) as well as identify and map recent wildfires (Chuvieco and Congalton, 1988, Key and Benson, 1999). While the work with AVHRR in Africa expanded and new sensors were becoming readily available, researchers in Australia were evaluating the applicability of Landsat images to monitoring and assessment of rangelands. Work by Dean Graetz, now retired from the Commonwealth Scientific and Industrial Organisation (CSIRO) of Australia, was instrumental in fostering use of satellite remote sensing to monitor rangelands (Graetz et al., 1983, 1986, 1988; Pech et al., 1986; Graetz, 1987). This work, coupled with other CSIRO scientists such as Geoff Pickup (Pickup and Nelson, 1984; Pickup and Foran, 1987; Pickup and Chewings, 1988), firmly established Australia as a leader in the use of remote sensing for rangeland monitoring and assessment. Researchers in Australia had similar problems applying digital imagery to semiarid rangelands as did the United States and Africa teams. The difficulty in applying imagery collected by the Landsat sensors to rangeland assessment is documented by Tueller et al. (1978) and McGraw and Tueller (1983), who found that the spectral differences among semiarid range plant communities were so small that they approached the noise level of the imagery. Even with these limitations, Robinove et al. (1981) and Frank (1984) developed methodologies for using albedo to measure soil erosion on rangelands. Pickup and Nelson (1984) developed the soil stability index (SSI) by using the ratio of the MSS green band divided by the NIR, plotted against the ratio of the red divided by the NIR. This comparison between the two ratios provided a quantitative measure of soil stability. Further, a temporal sequence of SSI images could be used as a monitoring tool to identify changes in landscape state (Pickup and Chewings, 1988). As research progressed in the use of imagery on rangelands through the 1980s, the US civilian remote sensing program began a transition to private sector management of the Landsat program. Issues of data cost and data licensing arose placing financial and legal limitations on research and data sharing. Still, research and application continued into the 1990s with an increased demand by federal land managers for landscape level information.

11.2.3 Advanced Multisensor Era, 1990s

In 1989 and throughout the 1990s, the US Fish and Wildlife Service (USFWS) and the US Geological Survey (USGS) embarked on several largescale land cover mapping efforts across the United States. The Gap Analysis Program initiated by the USFWS and later absorbed into the USGS was designed as a spatial database to identify landscapes of high biological diversity and evaluate their management status (Scott et al., 1993). The Gap Analysis was built around the linkage between wildlife habitat relationship (WHR) models and a detailed land cover map. This linkage allowed the WHR database to be spatially visualized by relating habitat parameters to land cover. The significance of this effort to remote sensing is that at the time, no one had attempted to map vegetation across landscapes requiring multiple frames of radiometrically normalized satellite imagery. The first digitally produced land cover map derived from a statistical classification of a 14image mosaic of radio metrically normalized Landsat TM imagery was completed for the state of Utah in 1995 by Utah State University (Homer et al., 1997). Programs like the Gap Analysis, coupled with the advent of the publicly available internet in 1991, provided the impetus for a new brand of remote sensing centering on large data and improved data access and product delivery. During the late 1980s, the National Aeronautics and Space Administration (NASA) was envisioning the need to provide rapid data access to users. At the time, image acquisition and delivery to the end user required a minimum of a few weeks. There was a need for time critical imagery by users and to meet that demand; NASA set a goal of data delivery to within 24 h of acquisition. Even with the advent of data transfer through the internet, a 24-hour lag between acquisition and delivery is a relatively new phenomenon of the mid-2000s.

11.2.4 New Millennium Era, 2000s

In this era, noteworthy changes to the remote sensing community, including dramatic improvements in data availability, spatial and spectral resolution and temporal frequency (Figure 11.3) were made. Commonly used high spatial resolution sensors launched during this time include Ikonos, QuickBird, GeoEye1, and WorldView2 exhibit spatial resolutions in the multispectral domain of 4, 2.4, 1.65, and 2 m, respectively.

These sensors have enabled improvements in species discrimination (e.g., Everitt et al., 2008) and stand level attributes such as canopy cover (e.g., Sant et al., 2014). Use of Quickbird for identifying giant reed (*Arundo donax*) improved both user's and producer's accuracy by an average of 12% over the use of Spot 5 alone (Everitt et al., 2008). Similarly, Sant et al. (2014) used Ikonos imagery to quantify percent vegetation cover and explained 5% more variation than using Landsat (r^2 of 0.79 vs 0.84) alone. Hyperspectral data emanating from this era also enable greater discrimination of many biophysical features than multispectral sensors alone especially in the realm of invasive species mapping. Parker and Hunt (2004) distinguished leafy spurge with AVIRIS data with overall accuracy of 95% while Oldeland et al. (2010) detected bush encroachment by Acacia spp. ($r^2 = 0.53$). Gaffney et al. (2021) were able to differentiate closely related plant community classes concluding that compositional changes were owed to growing season weather and not the management regime. These improved capabilities emanate not only from improved sensor characteristics in the 2000s but also greatly improved data availability.

In 1999, the launch of Landsat 7, coupled with new sensors from a host of other countries as well as commercial, high spatial resolution sensors, ushered a new era of global assessment and monitoring of natural and human landscapes. With the end of private sector management of the Landsat program in 1999, imagery was again placed in the public domain and costs for Landsat imagery was

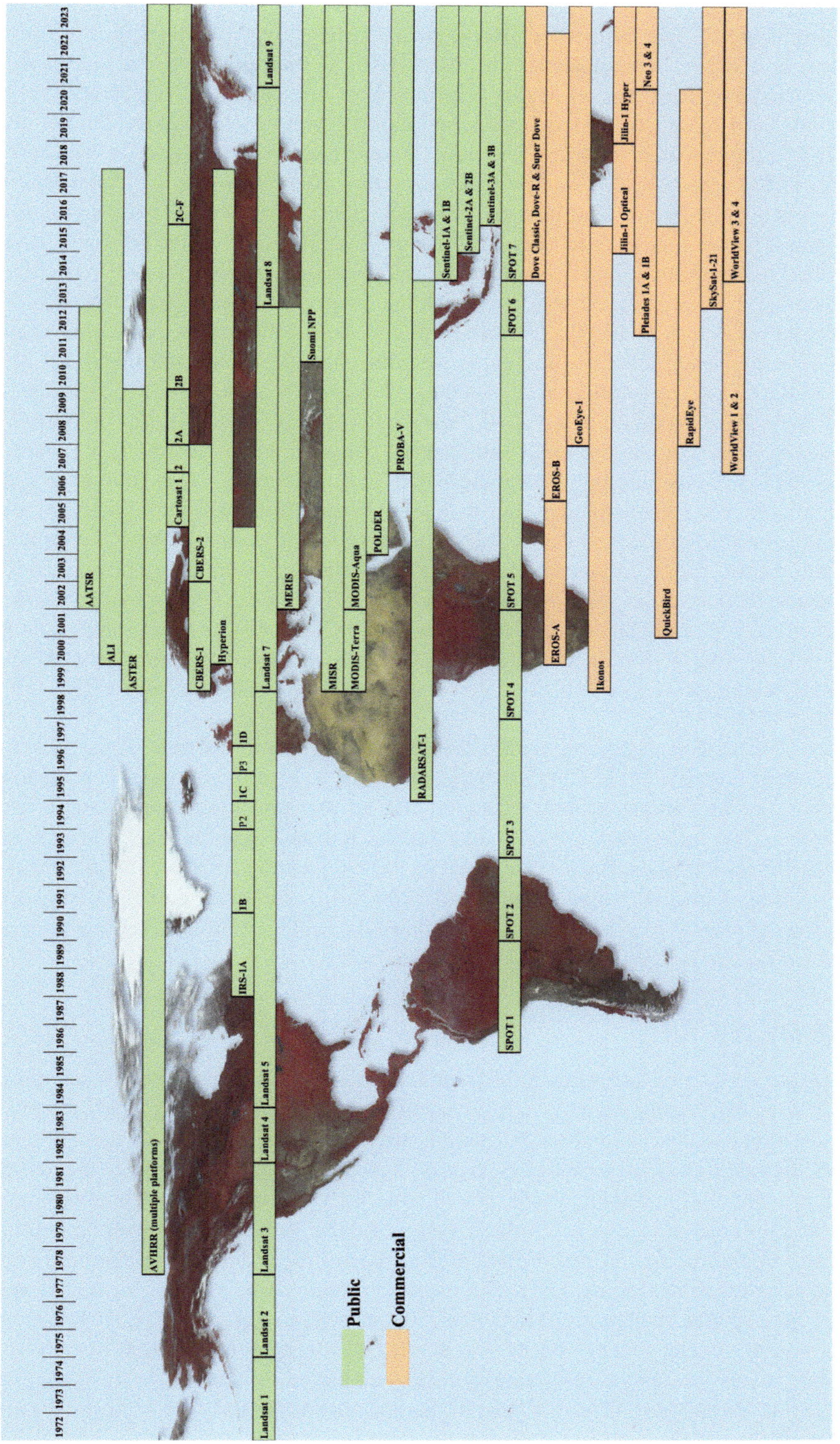

FIGURE 11.3 Temporal sequence of common remote sensing platforms relevant to mapping vegetation attributes such as cover.

reduced to $600 per scene (previously set at $4,400 per scene) for Landsat 7 Enhanced Thematic Mapper Plus (ETM+) imagery and $450 per scene for Landsat 5 TM. This reduction in cost coupled with the free exchange of data between collaborators, boosted research and application of satellite remote sensing. Further, the replacement of the AVHRR as the primary global sensor with the much-advanced Moderate Resolution Imaging Spectroradiometer (MODIS) with 36 spectral bands provided the ability for scientists to model, map, and monitor not only land cover, but net primary productivity (NPP) among other metrics. The now 23-year history of the MODIS sensor aboard two platforms (TERRA and AQUA) has provided an unprecedented source of global land cover dynamics data freely available to land managers and scientists. However, the MODIS sensors are being retired and a newer sensor called Visible Infrared Imaging Radiometer Suite (VIIRS), carried onboard the NASA-NOAA Suomi National Polar-orbiting Partnership (S-NPP) satellite, was launched in October 2011 was developed having similar sensor characteristics to MODIS. VIIRS provides moderate resolution imagery at 375m–1 km and the S-NPP orbital properties align well with MODIS sensors carried on Terra and Aqua satellites. The data collected by VIIRS are best matched to the MODIS imagery collected from the Aqua sensor (Benedict et al., 2021; Roger et al., 2017; Skakun et al., 2018; Vermote et al., 2016). For data continuity missions, key properties (e.g., equatorial crossing time, spatial, temporal, and spectral resolutions) are essential to be similar to maintain historical continuity among MODIS and VIIRS data collections. Even having the property similarities, application of a mathematical transformation between two sensors might be needed for compatible data values. Multiple studies have shown data values are higher in VIIRS products than MODIS (Benedict et al., 2021; Skakun et al., 2018). A translation formula may be applied to sensor spectral bands (Skakun et al., 2018) or calculated vegetation index (i.e., normalized difference vegetation index (NDVI)) (Benedict et al., 2021; Ji and Gallo, 2006), to transform sensor data values for VIIRS continuity in MODIS trained models. The United States Geological Survey (USGS) produces weekly NDVI composites of expedited versions of MODIS (eMODIS) and VIIRS (eVIIRS) (Benedict et al., 2021).

In 2008, The USGS made all Landsat data accessed through the internet free of charge. With this policy change, scene requests at the USGS EROS Center jumped from 53 images per day to about 5,800 images per day. This increase in data demand and delivery has arguably resulted in research in the 2000s centered on the copious use of imagery across multiple temporal and landscape scales. Commercial satellites such as the Ikonos, launched in 1999, Quickbird in 2001, and the WorldView and GeoEye satellites launched between 2007 and 2009 has provided on demand access to high spatial resolution (sub-meter to a few meters) that allows data integration between a wide array of platforms and spatial scales (Sant et al., 2014).

11.3 STATE OF THE ART

Millions of people depend on rangelands for their livelihood. This dependence raises numerous concerns about the health, maintenance, and management of rangelands from local to global perspectives. Discerning and describing how rangelands are changing at multiple spatial and temporal scales requires the integration of sensors that possess specific characteristics. The current suite of government-sponsored and commercial sensors suitable for regional to global analysis span the spatial range of sub-meter to 1 km, a temporal range of daily to bimonthly (temporal resolution is inversely proportional to spatial resolution), and all have the capacity to image landscapes in the visible and near infrared (Figure 11.4). The most commonly used sensors for global applications, however, have spatial resolutions of between 250 and 1000 m (e.g., MODIS, AVHRR, Visible Infrared Imaging Radiometer Suite (VIIRS)) and exhibit high temporal frequency, numerous spectral bands, but relatively low spatial resolution. Sensors best suited for regional to local applications (e.g., Landsat, Spot, WorldView, GeoEye) have higher spatial resolutions (sub-meter to 30 m), and lower temporal repeat cycles.

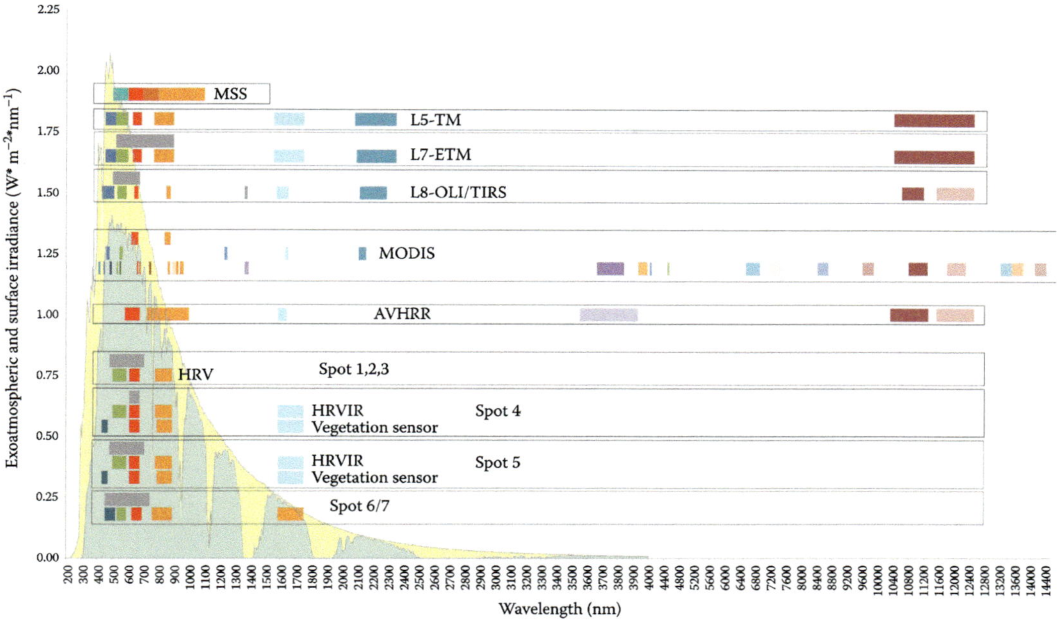

FIGURE 11.4 Comparison of the spectral bands associated with common Earth remote sensing instruments superimposed over exoatmospheric (yellow) and surface (green) solar irradiance. The list is not exhaustive and does not include coverage of radar or lidar instruments.

The present role of remote sensing for characterizing five globally significant phenomena are discussed hereafter, including land degradation, fire, food security, land cover, and vegetation response to global change (Table 11.2). These factors are not mutually exclusive and often exhibit significant interaction. Using remote sensing at global scales provides insight to what may be anticipated in the future, and indicates regions where ecological thresholds have been crossed, beyond which, decreased goods and services from rangelands can be expected.

11.3.1 Rangeland Degradation

Land and soil degradation are accelerating, and drought is escalating worldwide. At the UN Conference on Sustainable Development (Rio+20), world leaders acknowledged that desertification, land degradation, and drought (DLDD) are challenges of a global dimension affecting the sustainable development of all countries, especially developing countries. Drylands are often identified and classified according to the aridity index (AI), which is defined as P/PET where P is the annual precipitation and PET is the potential evapotranspiration. Drylands yield AI values ≤0.65. Despite decades of research, standards to measure progression of land degradation (e.g., global mapping and monitoring systems) remain elusive, but remote sensing plays a significant role.

11.3.1.1 Soil and Land Degradation and Desertification: What Is the Difference?

Land degradation and desertification have been sometimes used synonymously. Land degradation refers to any reduction or loss in the biological or economic productive capacity of the land (UNCCD, 1994) caused by human activities, exacerbated by natural processes, and often magnified by the impacts of climate change and biodiversity loss. In contrast, desertification only occurs in drylands and is considered as the last stage of land degradation (Safriel, 2009).

TABLE 11.2

The Four Most Common Sensors for Regional and Global Applications, Their Characteristics, and Example Applications. Many sensors which may have use for evaluating rangeland are not included. Svoray et al. (2013) provide a larger number of example applications in rangeland environments, but this table focuses largely on globally applicable sensors and global applications.

Satellite [sensors]	Characteristics (a is spatial resolution, b is launch date, c is swath width and d is revisit time.)	Rangeland Application Examples	References
Landsat (5, 7, 8) [Thematic Mapper, Enhanced Thematic Mapper Plus, Optical Land Imager]	a) 15 (panchromatic), 30 (multispectral), 100 (thermal), b) 1999 (Enhanced Thematic Mapper plus) (ETM+) and 2013 (Optical Land Imager) (OLI), c) 185 km × 170 km, d) 16 days	Fire (often DNBR, NBR, LWCI) 1) Burn severity (DNBR, RDNBR, Tasseled Cap Brightness) 2) Burned area mapping (Eidenshenk et al., 2007) 3) Fuel moisture (variety of indices such as NDVI, NDII, LWCI etc.)	Key and Benson (2006); Miller and Thode (2007); Loboda et al. (2013) Eidenshink et al. (2007) Chuvieco et al. (2002)
		Vegetation attributes 1) Land Cover (varied methods) 2) Leaf area index (LAI)/fraction of photo-synthetically active radiation absorbed by vegetation (fpar) (radiative transfer and vegetation indices) 3) Net primary production (multi-sensor fusion and process modeling[100]) 4) Degradation (change detection and residual trend analysis)	Gong et al. (2013); Fry et al. (2011); Rollins (2009) Shen et al. (2014) Li et al. (2012) Jabbar and Zhou (2013)
SPOT [VEGETATION]	a) 1000 b) 1998 c) 2250 km, d) 1–2 days	Fire 1) Burned area mapping Dnbr (NDVI, NDWI) 2) Fuel moisture (primarily NDVI, NDWI)	Silva et al. (2005); Tansey et al. (2004) Verbesselt et al. (2007)
		Vegetation attributes 1) Land Cover (GLC2000) 2) Net primary production/abundance (NDVI, process modeling) 3) Degradation (Trend analysis)	Bartholomé and Belward (2005) Telesca and Lasaponara (2006); Geerken et al. (2005); Jarlan et al. (2008) Fang and Ping (2010); Tchuenté et al. (2011)
Aqua and Terra [Moderate Resolution Imaging Spectroradiometer]	a) 250 (red, near-IR), 500 (multispectral), 1000 (multispectral) b) 2000 (Terra), 2002 (Aqua) c) 2230 km, d) 1–2 days	Fire (often DNBR, NBR, LWCI) 1) Active fire detection (Thermal anomalies and Fire Radiative Potential) 2) Burned area evaluation (SWIR VI and change detection) 3) Burn severity (Time integrated dNBR) 4) Fuel moisture (Empirical relations and radiative transfer modeling; many vegetation indices [GVMI, NDWI, MSI etc.])	Giglio et al. (2003); Giglio et al. (2009) Roy et al. (2008) Veraverbeke et al. (2011) Yebra et al. (2008); Sow et al. (2013)

TABLE 11.2 *(Continued)*

The Four Most Common Sensors for Regional and Global Applications, Their Characteristics, and Example Applications. Many sensors which may have use for evaluating rangeland are not included. Svoray et al. (2013) provide a larger number of example applications in rangeland environments, but this table focuses largely on globally applicable sensors and global applications.

Satellite [sensors]	Characteristics (a is spatial resolution, b is launch date, c is swath width and d is revisit time.)	Rangeland Application Examples	References
		Vegetation attributes 1) Land Cover (varied methods) 2) Leaf area index (LAI)/fraction of photo-synthetically active radiation absorbed by vegetation (fpar) (radiative transfer modeling) 3) Net primary production (process modeling) 4) Degradation (Rain use efficiency, local NPP scaling, trend and condition analysis) 5) Livestock Early Warning System (time series analysis of NDVI, and biomass)	Friedl et al. (2010) Myneni et al. (2002); Wenze et al. (2006) Running et al. (2004); Reeves et al. (2006); Zhao et al. (2011) Bai et al. (2008); Prince et al. (2009); Reeves and Bagget (2014) Angerer (2012); Yu et al. (2011)
National Oceanic and Atmospheric Administration [Advanced Very High Resolution Radiometer]	a) 1000 m, b) NOAA-15 (1998), NOAA-16 (2000), NOAA-18 (2005), NOAA-19 (2009) satellite series (1980 to present). The approximate scene size is 2400 km × 6400 km	Fire 1) Active fire detection (Thermal anomalies and NDVI) 2) Burned area evaluation (multi-temporal multi-threshold approach) 3) Fuel moisture (NDVI) Vegetation attributes 1) Land Cover (unsupervised and supervised time series analysis) 2) Leaf area index (LAI)/fraction of photo-synthetically active radiation absorbed by vegetation (fpar) (radiative transfer modeling, Feed-Forward Neural Network) 3) Net primary production (time-integrated NDVI) 4) Degradation (NDVI and rainfall use efficiency)	Pu et al. (2004); Flasse and Ceccato (1996); Dwyer et al. (2000) Barbosa et al. (1999) Paltridge and Barber (1988); Eidenshink et al. (2007) Loveland et al. (2000); Hansen et al. (2000) Myneni et al. (2002); Ganguly et al. (2008); Zhu and Southworth (2013) An et al. (2013) Wessels et al. (2004); Bai et al. (2008)

11.3.1.2 Role of Remote Sensing for Monitoring Rangeland Degradation

Much research conducted over the last decade has been on remotely sensed biophysical indicators of land degradation processes (e.g., soil salinization, soil erosion, waterlogging, and flooding), without integration of socioeconomic indicators (Metternicht and Zinck, 2003, 2009; Allbed and Kumar, 2013). Studies from the 1970s onward have related soil erosion severity to variations in spectral response. Good reviews of spectrally based mapping of land degradation are found in

Metternicht and Zinck (2003), Bai et al. (2008), Marini and Talbi (2009), and Shoshanya et al. (2013). Moreover, research work from the 1990s and 2000s (Metternicht, 1996; Vlek et al., 2010; Le et al., 2012; Shoshanya et al., 2013) reports the benefits of a synergistic use of satellite and/or airborne remote sensing with ground-based observations to provide consistent, repeatable, cost-effective information for land degradation studies at regional and global scales. Hereafter follows a brief description of some of the most frequent applications of remote sensing applied in "global or sub global assessments" of land degradation. These remotely sensed products include biomass and vegetation health modeling via NDVI and NPP, rain use efficiency (RUE), and local NPP scaling.

11.3.1.3 Biomass and Vegetation Health Modeling As an Indicator of Degradation

The biomass produced by soil and other natural resources can be a proxy for land health (Nkonya et al., 2013). In this vein, Bai et al. (2008) framed land degradation in the context of the Land Degradation Assessment in Drylands (LADA) program as long-term loss of ecosystem function and productivity and used trends in 8 km NDVI from the Global Inventory Modeling and Mapping Studies (GIMMS) as a "proxy indicator" of changes in NPP. Figure 11.5 represents changes in NPP from 1981 to 2003 resulting from fusion of GIMMS NDVI and MODIS 1 km NPP (Bai et al., 2008). The NDVI is related to variables such as leaf area index (LAI) (Myneni et al., 1997), the fraction of photosynthetically active radiation (fPAR) absorbed by vegetation, and NPP. This explains why many NPP estimates derived from remote sensing approaches are based on LAI, and fPAR commonly from the AVHRR onboard the National Oceanic and Atmospheric Administration (NOAA) satellite, and the MODIS on the Terra and Aqua satellites (Ito, 2011). One caveat to remotely sensed estimates of NPP for degradation analyses is the need for comparison with ground measured biophysical parameters such as NPP, LAI, or soil erosion (or salinization) for accuracy assessment (Bai et al., 2008; Le et al., 2012).

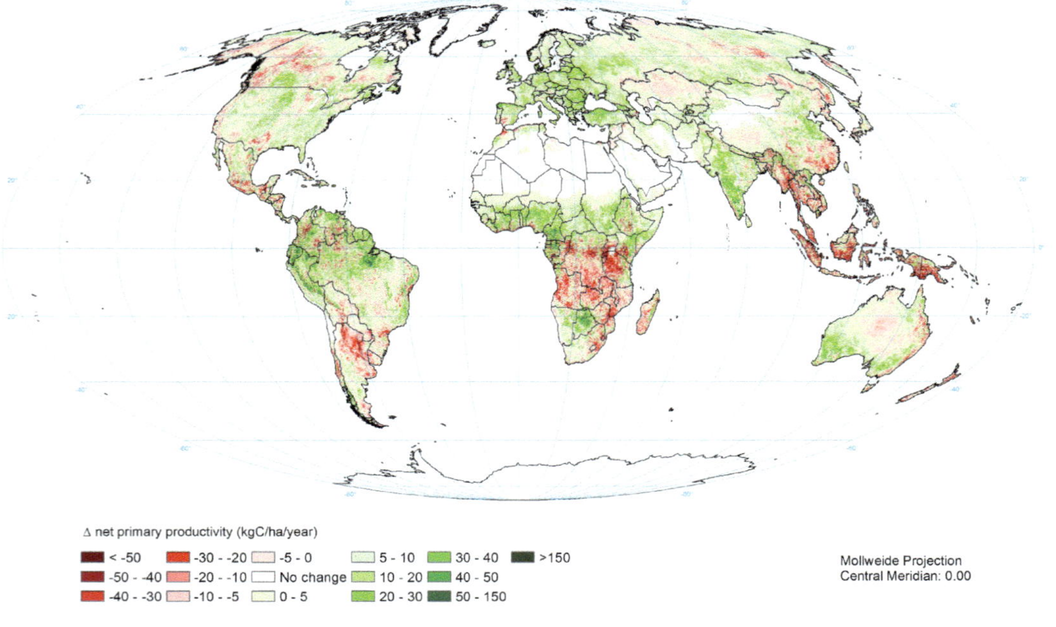

FIGURE 11.5 Global change in NDVI scaled in terms of NPP using MODIS 1 km^2 8-day composite net photosynthesis data (1981–2003). NDVI is a proxy indicator of changes in NPP. (Source: Bai et al. 2008.)

11.3.1.4 Rain Use Efficiency

RUE (ratio of NPP to rainfall) can be used to distinguish between the relatively low NPP of drylands associated with inherent moisture deficit and the additional decline in primary production due to land degradation (Le Houérou, 1984; Le Houérou et al., 1988; Pickup, 1996). In the context of the LADA project, Bai et al. (2008) estimated RUE from the ratio of the annual sum of NDVI (derived from MODIS and NOAA AVHRR) to annual rainfall and used it to identify and isolate areas where declining productivity was a function of drought (Figure 11.6). Figure 11.6 was produced using the same GIMMS NDVI data as Figure 11.5 in concert with Variability Analyses of Surface Climate Observations (VASClimO) gridded precipitation data at 0.5° resolution. This recalibration process was thought to yield a proxy index for land degradation, if a decline in vegetation for any other reason than rainfall (and temperature) differences would be an expression of some form of degradation. Statistical analysis showed 2% of the land area exhibited a negative trend at the 99% confidence level, 5% at the 95% confidence level, and 7.5% at the 90% confidence level (Bai et al., 2008).

A drawback of this mapping approach is that an area of land degradation much smaller than 8 km (pixel size of the GIMMS AVHRR) must be severe to significantly change the signal from a much larger surrounding area. In addition, the application of RUE to identify degraded landscapes has been somewhat controversial and misinterpreted as an indicator of degradation (Prince et al., 2007) since the RUE is highly variable (Fensholt and Rasmussen, 2011). In addition, errors in gridded precipitation data can add significant uncertainty, and noise to a degradation analysis suggesting analyses based solely on remotely sensed data may be beneficial (Reeves and Bagget, 2014).

11.3.1.5 Local NPP Scaling

Prince (2002) developed the local net primary productivity scaling (LNS) approach. Though the LNS approach can be applied to data of any resolution, derived from a host of sensors yielding visible and infrared bandpasses, AVHRR and Terra MODIS are commonly used. The LNS approach compares seasonally summed NDVI (ΣNDVI) of a single pixel to that of highest pixel value (or, commonly, the 90th percentile) observed in homogeneous biophysical land units (e.g., similar soils,

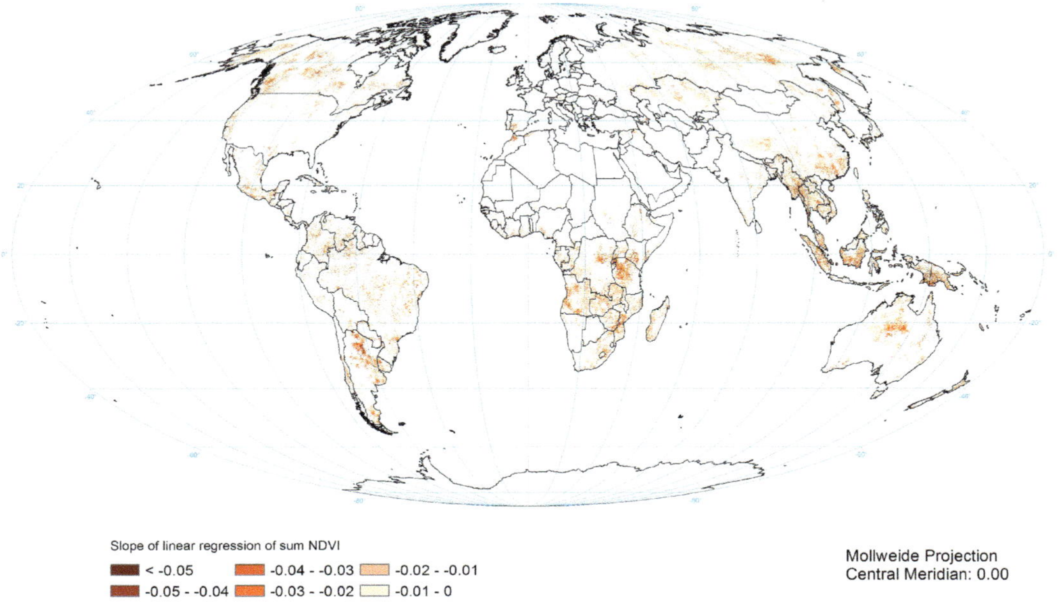

FIGURE 11.6 Average rain use efficiency (RUE) adjusted NDVI, from GIMMS-AVHRR 8 km² and VASClimO at 0.5° spatial resolution. (Source: Bai et al. 2008.)

climate, and landforms). The highest ΣNDVI value is assumed as a proxy for the potential aboveground NPP (ANPP) for each unit, and the other ΣNDVI values are rescaled accordingly. Prince et al. (2009) applied the LNS approach at national scales in Zimbabwe using MODIS 250 m NDVI and concluded that 17.6 Tg C year–1 were lost due to degradation. Similarly, Wessels et al. (2007) used 1 km time-integrated NDVI in northeastern South Africa. More recently, Fava et al. (2012) used annual summations of MODIS 250 m NDVI resolution in an LNS study for assessing pasture conditions in the Mediterranean resulting in a mean agreement of 65% with ground-based classes of degradation. In a variant of the LNS approach, Reeves and Bagget (2014) used the mean 250 m MODIS NDVI response of like-kind sites compared with reference conditions using a time series analysis to identify degradation on the northern and southern Great Plains, United States. With this approach, 11.5% of the region was estimated to be degraded.

11.3.1.6 Global Assessment of Land Degradation: The Evolution of Remote Sensing

The use of remote sensing data in global programs of land degradation assessment is related to the history of the global assessment of human induced soil degradation (GLASOD), the global LADA (Global Assessment of Land Degradation and Improvement [GLADA]), and the Global Land Degradation Information System (GLADIS) programs, funded by the global organizations such as United Nations Environment Program (UNEP), the UN FAO, and the Global Environmental Facility (GEF). Table 11.3 summarizes the objectives, methods, and main outputs derived from these programs, including the use of remote sensing technologies in their implementation. The GLASOD, an expert opinion based study (Table 11.3), and Oldeman (1998) had two follow up assessments, namely, the regional assessments of soil degradation status in South and Southeast Asia (Assessment of the Status of Human induced Soil Degradation in South and Southeast Asia [ASSOD]) and Central and Eastern Europe (Soil and Terrain Vulnerability in Central and Eastern Europe [SOVEUR]) and the global LADA project, under UNEP/FAO. The LADA had the objectives of developing and testing effective methodological frameworks land degradation assessment, at global, national, and subnational scales. The global component of LADA (i.e., GLADA) provided a baseline assessment of global trends in land degradation using a range of indicators collected by processing satellite data and existing global databases (NPP, RUE, AI, rainfall variability, and erosion risk) as described in Bai et al. (2008). The GLADA was implemented between 2006 and 2009, based on 22 years (1981–2003) of fortnightly NDVI data collection and processing (Table 11.3). The project developed and validated a harmonized set of methodologies for the assessment of land use, land degradation, and land management practices at global, national, subnational, and local levels.

The GLADIS was developed by FAO, UNEP, and the GEF using preexisting data and newly developed global databases to inform decision makers on all aspects of land degradation. The GLADIS developed a global land use system (LUS) classification and mapping using a set of pressures and threat indicators at the global level, allowing access to information at country, LUS, and pixel (5 arcminute resolution) levels. It accounts for socioeconomic factors of land degradation, using a variety of ancillary data to this end. Lastly, Zika and Erb (2009) produced a global estimate of NPP losses caused by human induced dryland degradation using existing datasets from GLASOD and other sources. Table 11.3 shows an evolution in the use of remote sensing technology from the first global assessments (GLASOD), expert based, with no use of remote sensing imagery, to the latest GLADIS, heavily reliant on remote sensing derived data coupled with an ecosystem approach. The GLASOD estimated that 20% of drylands ("excluding" hyperarid areas) was affected by soil degradation. A study commissioned by the Millennium Assessment based on regional datasets (including hyperarid drylands) derived from literature reviews, erosion models, field assessments, and remote sensing found lower levels of land degradation in drylands, to be around 11% (although coverage was not complete) (Lepers et al., 2005). The LADA project reported that over the period of 1981–2005, 23.5% of the global land area was being degraded. On the other hand, Zika and Erb (2009) report that approximately 2% of the global terrestrial NPP is lost each year due to dryland

TABLE 11.3
Cursory Comparison between Major Global Rangeland Degradation Efforts

Program	Objective	Methodology—Remote Sensing Usage
GLASOD-Global Assessment of Human Induced Soil Degradation (UNEP) (1987–1990)	Produce a world map of human-induced *soil degradation*, on the basis of incomplete knowledge, in the shortest possible time.	No remote sensing; expert-based approach; distinguishes "types" of soil degradation, based on perceptions; it is **not a measure** of land degradation.
LADA-GLADA Land Degradation Assessment *in Drylands* (LADA)—Global project, under UNEP/FAO. (2006–2009)	Assess (quantitative, qualitative and georeferenced) land degradation at global, national and sub-national levels to identify status, driving forces and impacts, and trends of land degradation in drylands; identify *hot* (degradation) and *bright* (improvement) spots:	The Global LADA was based on 22 years (1981–2003) of fortnightly NDVI data, derived from GIMMS and MODIS-related NPP (MOD 17). Method: 1. Identify degrading areas (negative trend in sum of NDVI) 2. Eliminate false alarms of productivity decline by: masking out urban areas; areas with a positive correlation between rainfall and NDVI, and a positive NDVI-RUE 3. Produce RUE-adjusted NDVI map 4. Calculate NDVI trends for remaining areas.
LADA-GLADIS Global Land Degradation Information System FAO-UNEP-GEF (2006–2010)	Focus on land degradation as a process resulting from pressures on a given status of the ecosystem resources.	Remote sensing is used for biomass status and trends, based on a correction factor to the GLADA-RUE-adjusted NDVI, to present trends in NDVI (1981–2006) translated in greenness losses and gains distinguished by climatic and human-induced (e.g., deforestation from FAO-FRA dataset) causes. Outputs are a series of global maps on the **status and trends** of the main ecosystem services considered, and radar graphs.

Prepared by Metternicht, G. *Sources:* Oldeman (1998); Bai et al. (2008); Nachtergaele et al. (2010).

degradation, or between 4% and 10% of the potential NPP in drylands. Figure 11.7 is a compilation of the global extent of drylands and human induced dryland degradation, produced for the fifth Global Environment Outlook (GEO5) based on research of Zika and Erb (2009) who express dryland degradation in croplands and grasslands as a function of NPP losses.

The three dryland area zones (top of the figure) are derived on basis of the AI. Only dryland areas (arid, semiarid, and dry subhumid), characterized by an AI between 0.05 and 0.65, are considered. Degradation is assessed by calculating the difference of the potential NPP (NPP_0) and current NPP (NPP_{act}). NPP losses due to human induced degradation amount to 965 Tg C year^{-1}, giving evidence that about 4%–10% of the potential pro duction in drylands is lost every year due to human induced soil degradation. The largest losses are occurring in the Sahelian and Chinese arid and semiarid regions, followed by the Iranian and Middle Eastern drylands and to a lesser extent the Australian and Southern African regions (UNEP, 2012) (Table 11.4) (Figure 11.5). A loss of NPP in the range of 20%–30% means reductions of potential productivity in that range; in most pixels of Figure 11.7, productivity losses range between 0% and 5% of their $NPP_{0.}$ The results presented in Figures 11.5 and 11.7 illustrate the scope and patterns of degradation but must only be considered as rough estimates (Zika and Erb, 2009). Major uncertainties related to the results arise from three assumptions:

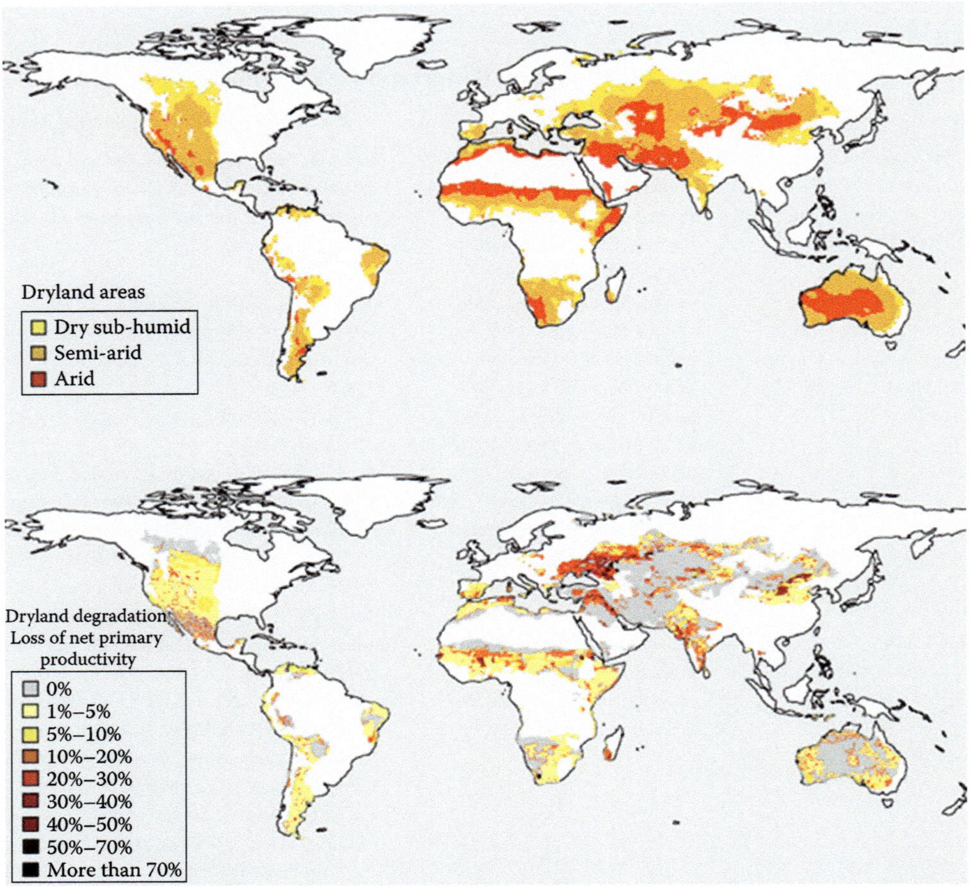

FIGURE 11.7 Global extent of drylands and human-induced dryland degradation. (Source: UNEP, 2012.) Redrawn from Zika and Erb (2009). We thank UNEP and the GEO-5 process for use of the figure.

(1) estimates of degradation extent, (2) assumptions on NPP losses due to degradation processes, and (3) potential NPP as a proxy for production potential. In recognition of the scope of degradation globally, the UN Conference on Sustainable Development (Rio+20) prompted the international community to develop universal sustainable development goals providing a timely opportunity to respond to the threat of soil and land degradation (Koch et al., 2013). Despite over 30 years of applied research in this area, however, the need to provide a baseline and method from which to measure degradation still remains (Gilbert, 2011). For regional refinements to degradation analyses, radar satellite based aboveground biomass estimations by Carreiras et al. (2012), or regional vegetation cover (Dong et al., 2014), could aid degradation analyses since cloud issues faced by LADA GLADA and GLADIS could be mitigated. Additionally, Blanco et al. (2014) propose ecological site classification of semiarid rangelands enabling more refined spatial units across which remote sensing can be conducted. Finally, engaging citizens in knowledge production (including ground verification of remotely sensed derived information), as fostered by current global (UNEPLive, Future Earth, Group on Earth Observations Biodiversity Observation Network) and sub-global initiatives (Eionet of the European Environmental Agency), could address the significant lack of ground truthing of previous global land degradation studies.

TABLE 11.4
Estimates of NPP Losses Due to Dryland dDgradation, Regional Breakdown. Source: Zika and Erb (2009).

	Degraded Dryland[a]		NPP Loss[b]	
Region	**1000 km²**	%	**Tg C yr⁻¹**	%
Central Asia and Russian Federation	1,432	19.5	250	26
Eastern and Southeastern Europe	391	55.5	73	8
Eastern Asia	1,887	45.3	50	5
Latin America and the Caribbean	1,206	18.8	98	10
Northern Africa and Western Asia	1,207	33.8	70	7
Northern America	607	11.3	51	5
Oceania and Australia	866	13.2	24	2
South-Eastern Asia	45	40.4	10	1
Southern Asia	1,437	30.9	106	11
Sub-Saharan Africa	2,597	22.8	215	22
Western Europe	128	24.7	18	2
Total	11,802	23.2	965	100

[a] % of dryland area
[b] Estimated NPP losses associated with dryland degradation. (see Zika and Erb, 2009 for more detail)

11.3.2 Fire in Global Rangeland Ecosystems

The extremely wide range of rangeland environments makes it virtually impossible to develop generalized statements about global fire regimes. However, the general composition of fuel and fuel characteristics defines some specifics of fire occurrence common for these ecosystems. Vegetation of rangelands is characterized by fast growth and slow decomposition rates (Vogl, 1979) leading to considerable buildup of surface litter. Most fuels in these ecosystems, except for chaparral systems, are flash and fine fuels (<0.25 in diameter), which dry out rapidly (i.e., 1 hour time lag fuels) and burn readily (National Wildfire Coordinating Group, 2012). Therefore, it is not unusual for these ecosystems to transition from low fire danger state to extreme fire danger state over a comparatively short period. Contiguity and loading of fuel in these ecosystems are highly variable both spatially and temporally: interannual variation in fuel loading often exceeds 110% (Ludwig, 1987). While fire is currently a common and widespread disturbance agent globally in rangelands, its prominence is expected to rise under projected climate change. Past and ongoing satellite monitoring and mapping of rangeland fire extent provide a much-needed baseline for assessment of potential future change in fire occurrence and its impact on ecosystem functioning.

11.3.2.1 Satellite Monitoring of Ongoing Burning

The hotspot detections from the nighttime top of atmosphere radiance data from the Along Track Scanning Radiometer (ATSR2) and Advanced ATSR (AATSR) were used to build the first World Fire Atlas (Jenkins et al., 1997). Neither of the source instruments was designed to support fire detection specifically, and therefore, the algorithms were based on suboptimal ranges of electromagnetic radiation (at brightness temperature [BT] centered on 3.7 and 11.8 μm) using a suite of simple thresholds (Arino et al., 2012). The MODIS was, however, designed with a specific goal to enhance fire mapping capabilities (Kaufman et al., 1998). MODIS collects daily global observations from Terra ~ 11:30 a.m. and 11:30 p.m. and Aqua at ~1:30 a.m. and 1:30 p.m. equatorial crossing time. In addition, several "fire" channels were included in the instrument to support fire monitoring:

two 4 μm channels (channel 21 with 500 K saturation level and channel 22 with 331 K saturation level) and 11 μm channel (channel 31 with 400 K saturation level) at 1 km nominal resolution (Giglio et al., 2003). The flexibility of switching the high and low saturation 4 μm channels in the contextual active fire detection algorithm is particularly important for tropical savanna environments.

The MODIS active fire product is the first product to include fire characterization metrics in addition to the binary "fire/no fire" masks. Fire radiative power (FRP), expressed in watts (W) is an instantaneous measurement of power released by ongoing burning during the satellite overpass (Kaufman et al., 1996a, 1996b) and are estimated using an empirical relationship established in Kaufman et al. (1998). FRP is directly related to the intensity of biomass burning and, when integrated overtime to fire radiative energy (FRE) expressed in joules (J), is linearly related to biomass consumption (Wooster et al., 2005).

11.3.2.2 Satellite Estimates of Burned Area

Unlike active fire detection, which is primarily based on BT in mid and long infrared spectrum, burned area estimates are most frequently based on changes in surface reflectance due to burning observable within the visible (0.4–0.6 μm), NIR (0.7–1.0 μm), and shortwave infrared (SWIR 1.1–2.4 μm) spectrum. The relatively short wavelength of radiation in this range determines that burned area mapping relies on clear surface observations and is strongly limited by considerable aerosol contamination from smoke during the burning process and high cloud cover in high northern latitudes. The first multiyear global burned area products were developed from data acquired by VEGETATION (VGT) (onboard SPOT), ATSR2 (onboard ERS2), Medium Resolution Imaging Spectrometer (MERIS), and AATSR (onboard Environmental Satellite [ENVISAT]) instruments (Plummer et al., 2006) within the GLOBCARBON initiative.

The suite of fire products developed from the MODIS 500 m data included two global burned area algorithms. The MCD45 algorithm (Roy et al., 2008), which is no longer in production, was based on the detection of rapid changes in surface reflectance within a MODIS 500 m pixel. The MCD64 algorithm (Giglio et al., 2009, 2018), the official NASA operational MODIS burned area product, relies on the detection of persistent changes in vegetation state and subsequent attribution of the change to burning by comparison to active fire occurrence within a specified spatiotemporal window (Figure 11.8).

A detailed study in Central Asia (Loboda et al., 2012) has shown that MODIS based products deliver spatially accurate estimates of burned area in Central Asia, with MCD64A1 estimates in close agreement to the Landsat based assessments (~18% underestimation). The independent accuracy assessment results within drylands of Central Asia are similar to those in North America (Giglio et al., 2009). This makes MODIS based products appear to deliver a reasonable estimate of fire impact on grasslands and shrublands of the world despite the overall higher mapping uncertainty relative to other land covers. Studies have also explored the capabilities of synthetic aperture radar (SAR) from the Sentinel-1 mission (e.g., Hosseini and Lim, 2023). However, the accuracy within rangelands is lower than optical burned area products such as MCD64A1 since the SAR backscatter variations in pre- and post-fire imagery tend to be greater in forests land covers than in grassland and shrubland landscapes (Belenguer-Plomer et al., 2019). In recent years, machine-learning and artificial intelligence have grown in popularity. At present, the majority of these studies are focused on localized analyses (e.g., Knopp et al., 2020) as these methods are still premature and require a large volume of training and validation data.

11.3.2.3 Remote Sensing Methods for Fire Impact Characterization

The characteristics of grass and shrubland fires (i.e., the large footprint of savanna fires, the remote locations of tundra fires, and the short-lived nature of these burn scars), make remote sensing the only viable source of data for consistent, global post-fire characterization of burned area. While a healthy debate about what constitutes burn severity and how much the ecological definition ranges across ecosystems is still ongoing in the fire science community (French et al., 2008; Chen et al.,

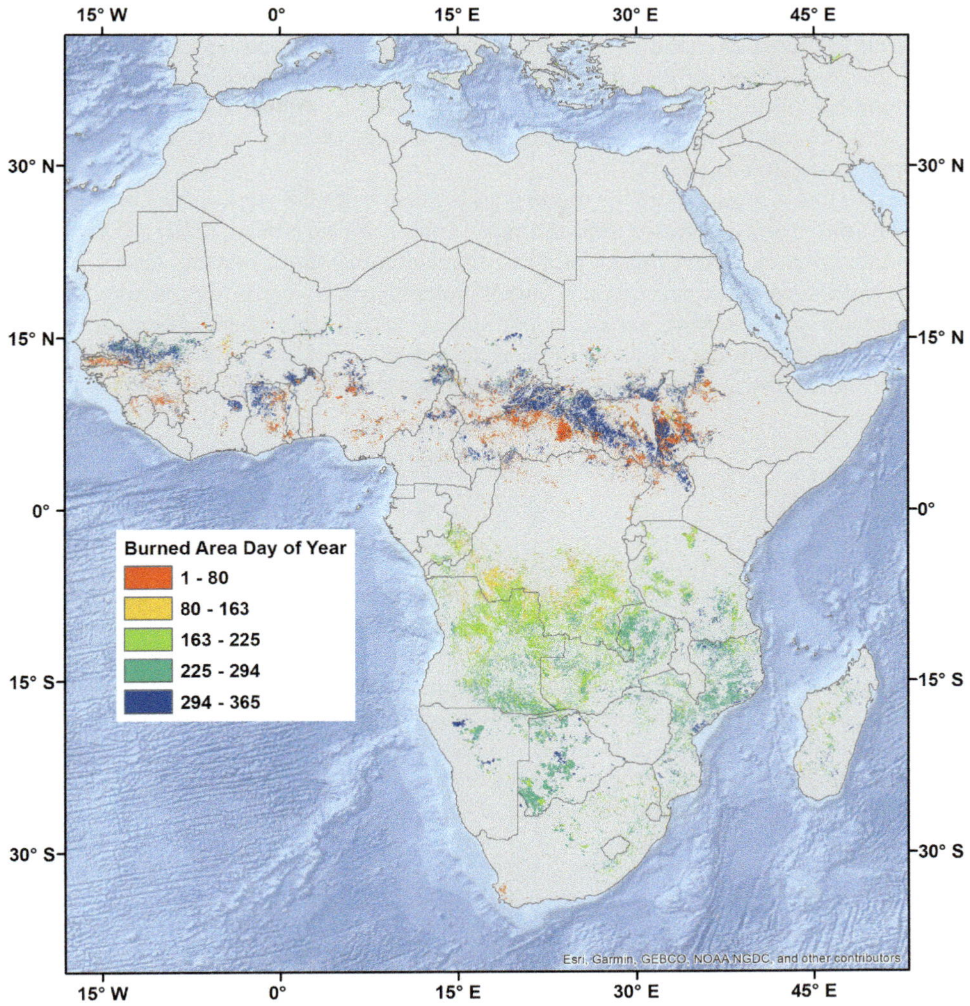

FIGURE 11.8 Example of the 2021 MCD64A1 C6.1 MODIS burned area product depicting approximate date of fire in rangelands globally. (Prepared by Joanne V. Hall.)

2021), the Monitoring Trends in Burn Severity (MTBS) program established the baseline definition. This includes the assumption that this parameter can be mapped from remotely sensed data and is ultimately based on a combination of "visible changes in living and nonliving biomass, fire byproducts (scorch, char, and ash), and soil exposure" among other components. The same ranges of electromagnetic spectrum (visible—NIR—SWIR), therefore, constitute the basis for the strongest differentiation between soil, vegetation, char, and ash components characterizing burn severity as those used most for burned area mapping. It is not surprising that the first widely applied index for mapping and quantifying burn severity is based on the normalized difference of NIR and SWIR in 2.2 μm range ($SWIR_{2.2}$) originally developed by LopezGarcia and Caselles (1991) for burned area mapping. The Normalized Difference Burn index (NDBR), as it was subsequently named by Key and Benson (1999), is calculated as follows:

$$NBR = \frac{NIR - SWIR_{2.2}}{NIR + SWIR_{2.2}}.$$

where NIR refers to the TM band 4 (0.76–0.90 μm) $SWIR_{2.2}$ refers to band 7 (2.08–2.35 μm) Key and Benson (1999) aimed to capture the fire induced changes to the proportions of soil, char, ash, and vegetation through differencing the pre-burn and post-burn NDBR measurement within a fire perimeter. This approach (differenced normalized burn ratio [dNBR], calculated as dNBR = $NBR_{pre\text{-}burn}$ − $NBR_{Post\text{-}burn}$ has become the most widely applied index of burn severity across all ecosystems in the United States (Eidenshink et al., 2007).

Compared to forest cover, where the original assessment of dNBR were closely related to ground measurements of burn severity expressed through a composite burn index (CBI) (Key and Benson, 2006, Allen and Sorbel, 2008), these grass and shrub dominated ecosystems have a low amount of aboveground biomass and are spatially highly heterogeneous. Thus, the magnitude of change between pre-burn and post-burn surface conditions is considerably more muted and uneven. To account for the initial lower fuel loading in these ecosystems, an adjustment to dNBR, named relativized dNBR (RdNBR), was developed by Miller and Thode (2007). This index is calculated as follows:

$$RdNBR = \frac{dNBR}{\sqrt{\left|NBR_{pre-burn} / 1100\right|}}$$

Although RdNBR versus CBI assessments show that RdNBR is more robust in assessing burn severity compared to dNBR in grass and shrub dominated ecosystems (Miller and Thode, 2007; Loboda et al., 2013), it does not overcome a major limitation of spectral signature change due to fire in NIR/SWIR spectral space within these ecosystems.

It is likely that the success rate of any one spectral index in mapping and quantifying burn severity depends strongly on the specific proportions of grass, woody biomass, exposed soil, geo graphic location (as related to frequency of observation allowing for a wider range of mapping days and different Sun sensor geometries), moisture status during image acquisition, and the timing of mapping. A recent study highlighted the importance of standardizing the pre- and post-fire image methodology for the tundra region as changes in the image acquisition seasonality (i.e., the impact of phenological differences in the image pairs) has an impact on the dNBR values (Chen et al., 2020).

11.3.3 Food Security: Role of Remote Sensing in Forage Assessment

Climate change, land degradation, landuse changes, and population growth are causing grand challenges for land managers that utilize rangelands for food production (Roche, 2016). Global demand for food is estimated to increase by 60% by 2050 with the demand for meat expected to increase by 70% (Alexandratos and Bruinsma, 2012). Meeting the food demands of a growing global population will become more difficult when coupled with the environmental and socio-economic factors (i.e., climate change, increasing land values, and limitations on land use). Changes in vegetation structure and forage quantity on rangelands due to a changing climate will have operational and economic consequences for industries reliant on these resources, such as cattle production (Klemm et al., 2020; Wold et al., 2023).

On rangelands, quantifying the amount of forage available to livestock on a near real-time basis using traditional methods (e.g., clipping vegetation along transects) can be costly, time consuming, and logistically challenging. A lack of information for making livestock management decisions at critical times could lead to loss of livestock due to lack of forage, or lead to vegetation over use, which, in turn, could result in rangeland degradation (Weber et al., 2000). Therefore, having an objective means of setting stocking rates on rangelands based on productivity will allow range land managers to better adapt to changing weather conditions. Because of the large areal cover that remote sensing products provide, in addition to the greater temporal of collection compared to traditional ground sampling over large areas, the use of remote sensing imagery is attractive for assessing vegetation production on rangelands. Multiple satellite platforms exist that are useful for rangeland forage assessments and early warning systems. Two approaches have generally been used for assessing rangeland forage conditions using remote sensing imagery. These include (1) empirical approaches that estimate the forage biomass or quality based on a statistical relationship between the spectral bands (or some combination of bands) in the imagery and ground collected vegetation data and (2) process models that use remote sensing data as inputs for predicting vegetation biomass or quality.

11.3.3.1 Empirical Approaches

Empirical approaches for assessing rangeland forage conditions using remote sensing products generally involve the use of a statistical relationship between the remote sensing spectral response or product variable and data collected from field measurements (Dungan, 1998). Using the empirical approach example in Figure 11.9, a MODIS 250 m maximum value composite and NDVI value of 7500 correspond to approximately 3414 kg ha^{-1} of annual production, after accounting for unavailability (ϕ = 0.15) and suggested utilization (υ = 0.5) results in stocking rate of 5.3 animal unit month's (AUM) ha^{-1}.

In a similar manner, Tucker et al. (1983) used both a linear and logarithmic regression between the ground collected biomass data in the Sahel region and AVHRR NDVI to predict biomass on a regional scale. AlBakri and Taylor (2003) used a linear regression approach to predict shrub biomass production for rangelands in Jordan using 7.6 km AVHRR NDVI. Both these studies reported accounting for >60% of the variation in herbaceous biomass with AVHRR NDVI alone using linear regression against biomass. In the Xilingol steppe of Inner Mongolia, Kawamura et al. (2005) used 500 m MODIS enhanced vegetation index (EVI) to predict live biomass and total biomass of livestock forage with linear regression models, which accounted for 80% of the variation in live

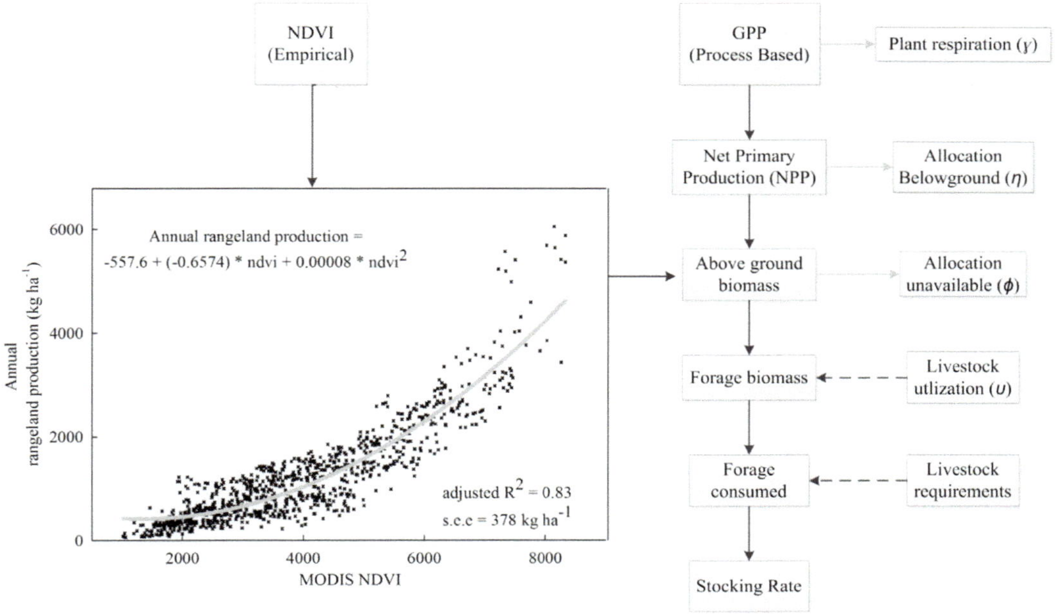

FIGURE 11.9 Process for estimating stocking rate from remote sensing data, either empirically or using a process model. Gross primary production (GPP) is determined from land cover type, spectral vegetation indices, incident photosynthetically active radiation, and climate-dependent radiation use efficiencies or empirically. Solid arrows represent reductions based on physiology while dashed arrows represent critical management decisions are determined.

biomass and 77% of the variation in total biomass. In the Tibetan Autonomous Prefecture of Golog, Qinghai, China, Yu et al. (2011) used the 250 m resolution MODIS NDVI to estimate aboveground green biomass using regression relationships between the NDVI and ground collected biomass data (r^2 of 0.51) from sites across the region.

As with forage biomass, empirical approaches can be used for forage quality assessments generally involving examining statistical relationships between forage quality variables such as crude protein or energy and spectral information from remote sensing imagery. For example, Thoma et al. (2002) used simple linear regression with AVHRR NDVI as the independent variable to predict forage quality and quantity on rangelands in Montana, United States. Their analysis indicated reasonable relationships between NDVI and live biomass ($r^2 = 0.68$) and nitrogen in standing biomass ($r^2 = 0.66$). Similarly, Kawamura et al. (2005) used regression relationships between ground collected data and MODIS EVI to predict live and dead biomass and crude protein in standing biomass. They found good predictability between standing live biomass and total biomass (live + dead) ($r^2 =$ 0.77–0.80), but correlations with crude protein were poor ($r^2 = 0.11$).

Remote sensing imagery provides a dense and exhaustive dataset that can serve as a secondary variable for geostatistical interpolation given that a correlation exists (both direct and spatial) between the primary and secondary variable (Dungan, 1998). Use of MODIS NDVI in the cokriging analysis of forage crude protein provides reasonable during the dry season ($r^2 = 0.69$) but less so during the wet season ($r^2 = 0.51$) (Awuma et al., 2007) likely because the amount of unpalatable shrub cover increased the greenness signal in the NDVI in some of the sampling areas that did not contribute to the available forage.

11.3.3.2 Process Models Using Remote Sensing Inputs

One problem that has been noted for regression models that use remote sensing variables is that they violate the regression assumption of no autocorrelation in the predictor variable(s) (Dungan, 1998; Foody, 2003). Since most remote sensing data are inherently autocorrelated, violation of this assumption may reduce the effectiveness of the regression model (Dungan, 1998). One way of overcoming the autocorrelation problems is to use process models that are driven by remotely sensed input variables on a pixel-by-pixel basis. Reeves et al. (2001) describe such an approach for predicting rangeland biomass using remote sensing products from the MODIS system and a light use efficiency model for plant growth. Hunt and Miyake (2006) used a similar light use efficiency model approach for estimating stocking rates for livestock at 1 km resolution in Wyoming, United States (Figure 11.9). Using the approach of Hunt and Miyake (2006), the stocking rate is estimated as gross primary production (GPP) $(1 - \chi)(1 - \eta)(1 - \phi)\ \upsilon$ (AUM/273 kg month^{-1}). From Hunt and Miyake (2006), the parameters for grasslands are approximately $\chi = 0.48$, $\eta = 0.79$, $\phi = 0.15$, and $\upsilon = 0.5$ where χ is autotrophic respiration, η is belowground carbon allocation, ϕ is carbon allocation to nonpalatable stems and other vegetation, and υ is an estimated accepted level of utilization. Therefore, a monthly GPP of 11,000 kg ha^{-1} month^{-1} is about 1.7 AUM's ha^{-1}, but this is just one method of using process models parameterized with remote sensing inputs. An example of a process-based modeling approach for forage quantity assessment at the regional level is the Livestock Early Warning Systems (LEWS) in East Africa (Stuth et al., 2003a, 2005) and Mongolia (Angerer, 2012) (Figure 11.10). In East Africa, the LEWS has evolved into the Predictive Livestock Early Warning System (PLEWS; Matere et al., 2019) and now encompasses Sudan, South Sudan, Ethiopia, Djibouti, Somalia, Kenya, Tanzania, and Uganda. The PLEWS platform has been incorporated into the Food and Agriculture Organization's East Africa Animal Feed Action Plan with the goal of using the system to provide information for informing drought declarations and responses, as well as providing inputs on rangeland forage availability needed for national feed inventories in each county (FAO and IGAD, 2019; Opio et al., 2020).

Coupling remotely sensed data in process models that also include economic information may provide improvements to global food security efforts in the future, For example, Wold et al. (2023) combined estimates of remotely sensed forage production on a ranch level with an economic model and assessed potential effects of climate change effects to a ranching operation. This small-scale example (coupling remote sensing estimates to an economic model) provides a framework for how we can use economic modeling in combination with remote sensing to provide a deeper understanding of land use changes, vegetation productivity, and environmental dynamics related to rangeland management. Future research could couple remotely sensed rangeland productivity at larger regions (e.g., western United States) to economic models that assess market fluctuations.

Figure 11.10 presents results of the LEWS applied in Mongolia in 2013. Note the significant decline of forage in southwestern Mongolia in 2013. The LEWS was developed to provide near real-time estimates of forage biomass and deviation from average conditions (anomalies) to provide pastoralists, policy makers, and other stakeholders with information on emerging forage conditions to improve risk management decision making. The LEWS combines MODIS 250 m NDVI, ground data collection from a series of monitoring sites, simulation model outputs, and statistical forecasting, to produce regional maps of current and forecast forage conditions and anomalies. The system uses the Phytomass Growth Simulation model (PHYGROW) (Stuth et al., 2003b), parameterized with the MODIS 250 m NDVI, as the primary tool for estimating available forage. Model verification indicates the model performs well in estimating forage biomass (Stuth et al., 2005). For example, model verification across monitoring sites in Mongolia indicated a good correspondence between the PHYGROW predicted biomass and observed ground data ($r^2 = 0.76$) with forage biomass ranging from 3 to 1230 kg ha^{-1}. PHYGROW tended to underestimate forage biomass across sites by 14% with an overall mean bias error of −18 kg ha^{-1} (Angerer, 2008).

11.3.4 Rangeland Vegetation Response to Global Change: The Role of Remote Sensing

Monitoring global change is an increasingly important endeavor (Running et al., 1999) since ecosystem goods and services, essential to human survival, are directly linked to the health of the biosphere (Fox et al., 2009). The Earth is a dynamic system with many interacting components that are complex and highly variable in space and time. Though change has always been present, human activities have influenced rates and extent of change beyond historical ranges (Vitousek, 1992; Levitus et al., 2000; Foley et al., 2005). Global change involves terrestrial, aquatic, oceanic, and atmospheric systems and cycles and is not limited to climate change alone (Beatriz and Valladares, 2008). Other factors such as invasive species, habitat change, overexploitation, and pollution are equally or even more important to the Earth's future (Millennium Ecosystem Assessment, 2005). Thus, the goal of global monitoring is aimed at characterizing "human habitability" through evaluation of vegetation that provides food, fiber, and fuel (Running et al., 1999) to a rapidly growing population. In the burgeoning field of global change monitoring, satellite remote sensing is increasingly more important. Only remote sensing offers a truly synoptic perspective of our surroundings and is therefore a critical tool for describing the type, rate, and extent of change unfolding across the globe. This is especially true for rangeland ecosystems that experienced losses of about 700 million ha by 1983 due to agriculture. In the United States alone, an estimated 75 million ha of former rangelands have been converted to agricultural land use since Euro-American settlement (Reeves and Mitchell, 2011) (Figure 11.11). The impacts of global change, such as climate impacts and land conversion, are often quantified through evaluation of vegetation cover and NPP in the context of the global carbon budget (Running et al., 1999).

11.3.4.1 Vegetation Productivity

Given the lack of ground referenced data available for determining productivity for rangelands globally, ecosystem modeling, remote sensing (Hunt and Miyake, 2006; Fensholt et al., 2006; Reeves et al., 2006), or a combination of both (Jinguo et al., 2006; Wylie et al., 2007; Xiao et al., 2008) can be used to estimate spatial and temporal trends across large areas. Many studies have evaluated the growth, total production, and health of rangeland vegetation, but two general approaches are normally applied that are very similar to the procedures outlined in the food security section. The

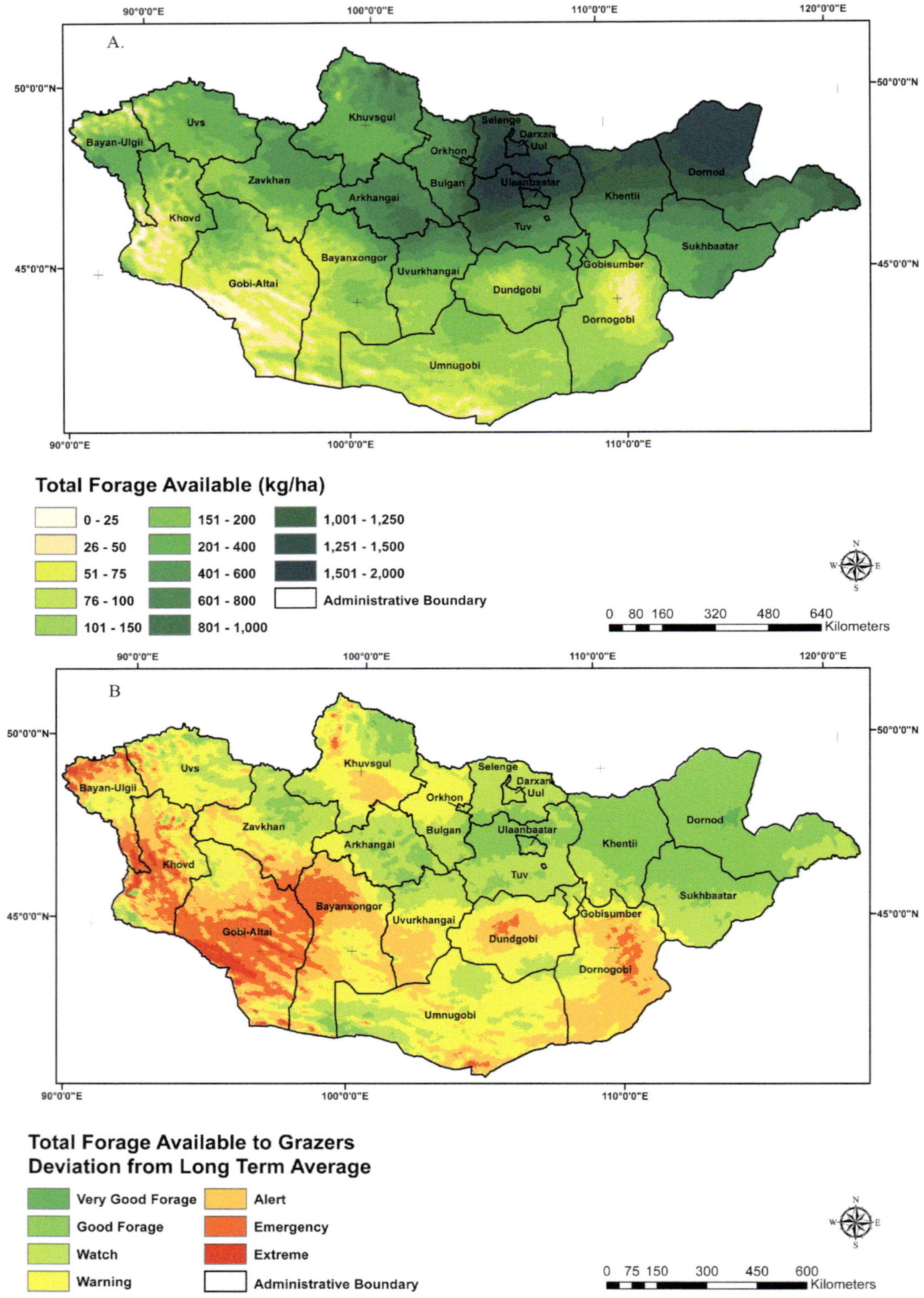

FIGURE 11.10 Panel A represents total forage available (kg ha^{-1}) during August 2013 for the Mongolia Livestock Early Warning System (LEWS). Panel B represents a map of forage deviation from long-term average (i.e., forage anomaly) for August 2013. Note areas in southwestern Mongolia experiencing Emergency to Extreme drought conditions.

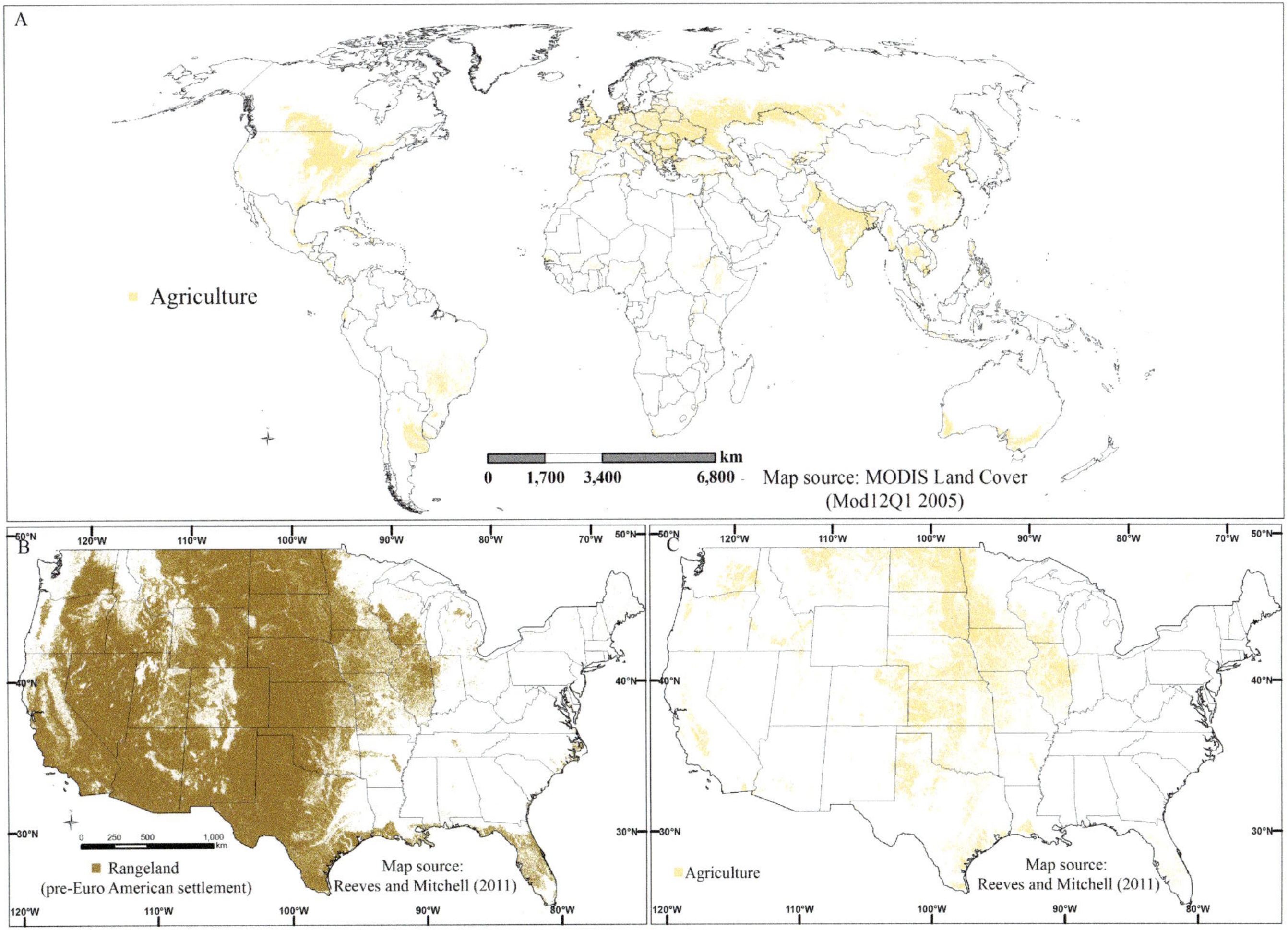

FIGURE 11.11 Panel A represents the estimated distribution of agricultural land use globally derived from MODIS MOD12Q1, 2006; University of Maryland Classification. Also shown is the hypothesized pre Euro-American extent of rangeland (Reeves and Mitchell, 2011) is shown in Panel B, while Panel C demonstrates areas of former rangeland now in agricultural production (estimated using the Biophysical Settings data product from the LANDFIRE Project (Rollins, 2009).

first approach involves directly sensing, via radiometric measurement, the amount of growth that has occurred over a given time. Direct quantification of biomass across rangeland vegetation types requires a set of spatially explicit ground samples describing the amount of peak biomass or annual production. Once ground data are collected and properly scaled, statistical models can be developed to describe the relationship between NDVI and biomass (Figure 11.12) that can, in turn, be used to monitor the response of vegetation through time. If peak biomass is estimates are sought, the annual maximum NDVI value should work reasonably well, but if annual production estimates are desired, a time integration of NDVI is usually employed (e.g., Paruelo et al., 1997).

Though NDVI has been widely used for monitoring global vegetation conditions, it exhibits well known saturation characteristics at relatively higher levels of biomass. The EVI can be used, with some success to overcome the saturation limitations inherent in NDVI. The saturation component of the NDVI signal, however, does not render it less useful for most applications. The reason for this is that across the range of productivity levels expected in most rangeland environments, the response is linear (Skidmore and Ferwerda, 2008).

The second approach for monitoring growth, total production, and health of vegetation involves use of remote sensing for quantifying canopy parameters, such as LAI, and f PAR, which, in turn, become part of a vegetation modeling system (Figure 11.12). Such a system is exemplified by the MODIS NPP algorithm (MOD17), which provides gross and net primary productivity (GPP and NPP) products at 1 km resolution for the entire globe. This approach is more sophisticated than direct sensing of biomass but enables carbon accounting for the global extent of rangelands. The modeling approach also requires a good deal more information including biome specific physiological parameters (Running et al., 2004). In addition, since this type of modeling approach requires meteorological and land cover information, it is directly informed by land cover/land use changes associated with global change. The NPP of rangeland vegetation from 2000 to 2012 is depicted in Figure 11.13, which demonstrates the type of ecosystem analysis possible with the MODIS NPP product. Figure 11.13

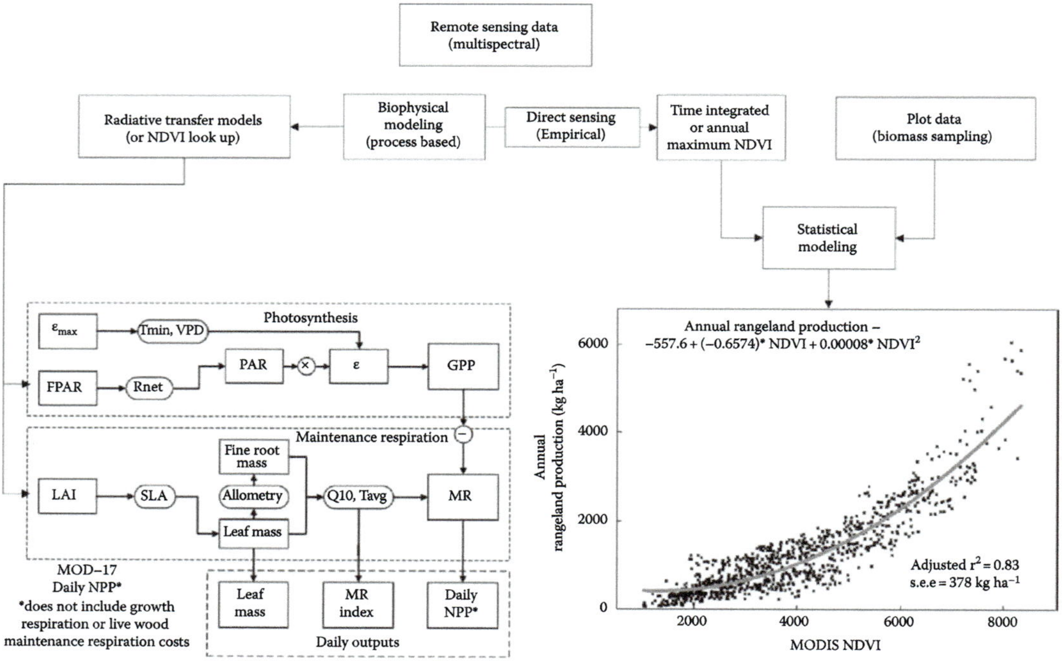

FIGURE 11.12 Direct sensing and biophysical modeling (process modeling) are two methods for estimating productivity of rangeland landscapes.

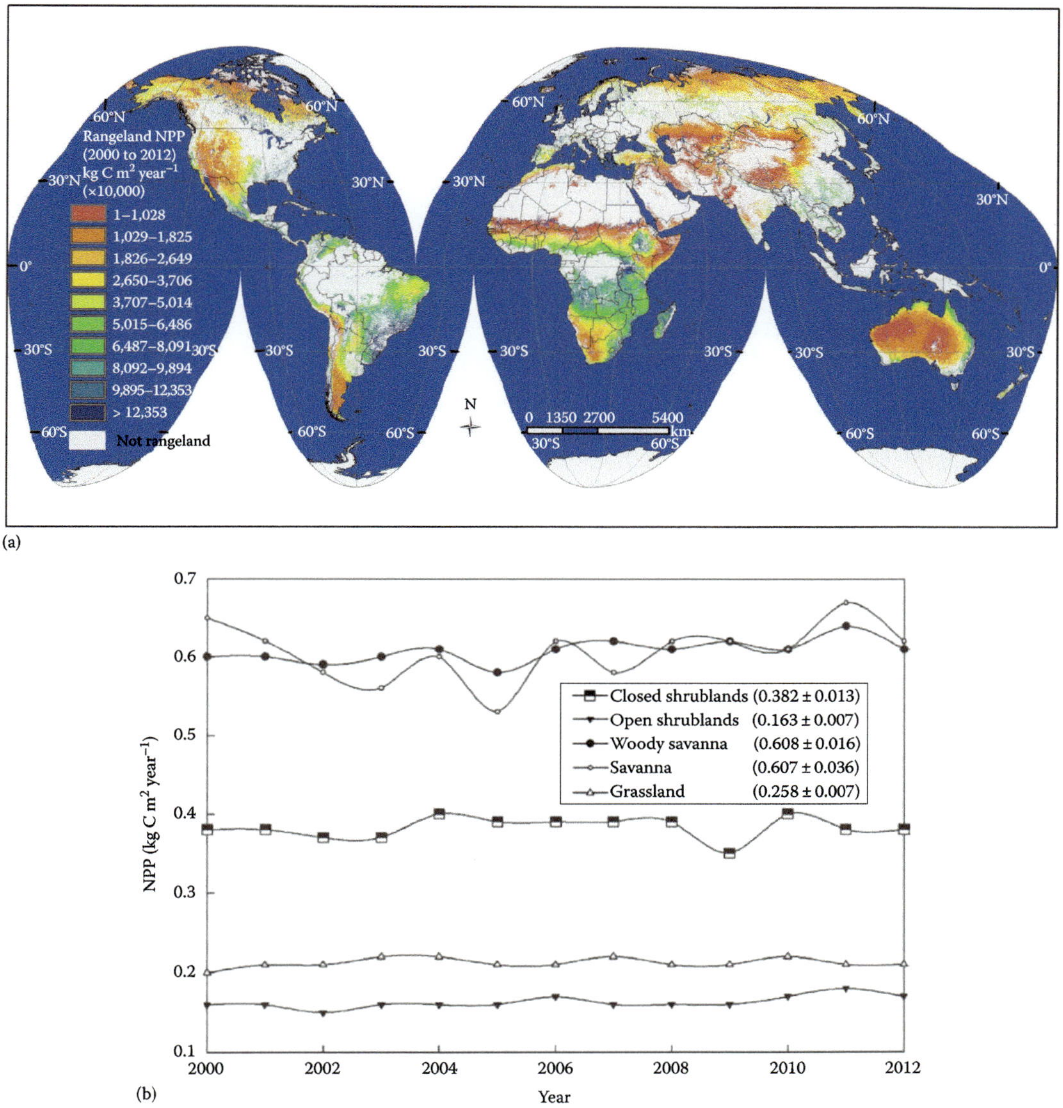

FIGURE 11.13 Panel A represents the mean (2000–2012) global distribution of rangeland NPP from the MODIS NPP (MOD17) product. Panel B represents the time series (2000–2012) of global rangeland NPP from the MOD17 product.

was created using a time series analysis from 2000 to 2012 of the MODIS derived annual NPP and Collection 4.5 land cover products. From this analysis, significant overlap and similarities between the savanna and woody savanna land cover classes are evident. These similarities suggest similar biophysical and bioclimatic conditions are present in these two classes or confusion exists between the classes. The close relationship between woody savanna and savanna could also be related to spatial commingling of the two types, which could be alleviated using higher resolution imagery. More recent analyses of (GPP) and leaf area index (LAI) suggest all major rangeland types have been greening at a significant rate. The latest long-term satellite-based leaf area index (LAI4g; Cao et al., 2023) and gross primary productivity (CEDAR-GPP; Kang et al., 2023) observations reveal widespread increases across global drylands from 1982–2020 (Figure 11.14.).

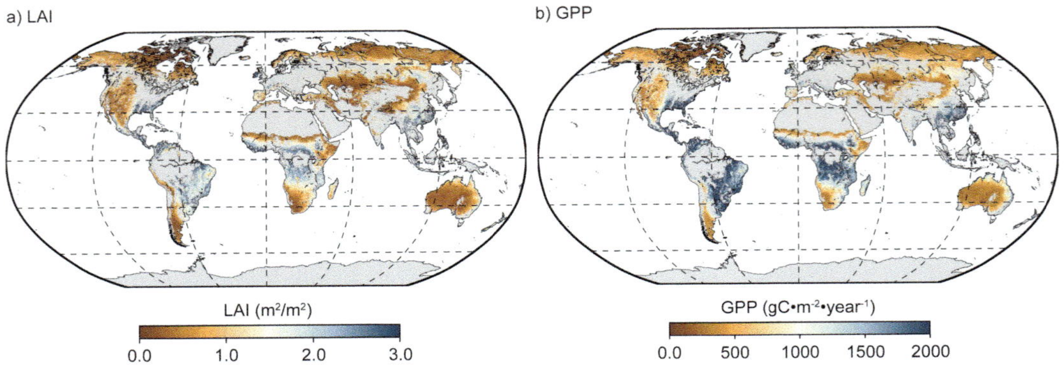

FIGURE 11.14 Long-term annual average LAI and GPP over 1982–2020. Global Inventory Modeling and Mapping Studies (GIMMS) leaf area index product (GIMMS LAI4g) is used as the proxy for vegetation greenness. GIMMS LAI4g product (version 1.1), is developed by Peking University ranging from 1982 to 2020 with a 15-day temporal resolution and ~ 8 km spatial resolution (Cao et al., 2023). GIMMS LAI4g addresses major uncertainties presented in current global long-term LAI products (e.g., GIMMS LAI3g) from NOAA satellite orbital drift and AVHRR sensor degradation and (2) insufficient LAI validation data to build robust LAI particularly before the late 1990s. For GPP dataset, upsCaling Ecosystem Dynamics with ARtificial intelligence GPP (CEDAR-GPP) is used, which incorporates the direct CO2 fertilization effect on photosynthesis. CEDAR-GPP ranging from 1982 to 2020 with monthly temporal resolution and at 0.05° resolution from 1982 to 2020 (Kang et al., 2023).

LAI has increased by an average of 0.0017 ± 0.025 m^2 m^{-2} y–1 across global rangelands. GPP has increased by an average of 1.7 ± 2.4 g C m^{-2} y–1 across global rangelands. LAI data indicate statistically significant greening ($p < 0.05$) across more than 50.1 % of rangelands, with a significant browning trend observed over 5.3 % of rangelands (Figure 11.15C). Similarly, GPP data indicates statistically significant greening ($p < 0.05$) across more than 56.5% of rangelands, with a significant browning trend observed over 3.5% of rangelands (Figure 11.15d). Notably, regions with the most pronounced increasing trend, as observed in both LAI and GPP datasets, include Central South America, Eastern China, the high-latitude Russia and Siberia landmass, and Central and South Africa. In contrast, regions exhibiting decreasing trends are found in Western US, Australia, Central Asia, and Southern South America. Generally, the spatial patterns of CEDAR-GPP consistent with those observed in GIMMS LAI4g. When considering land cover types, woody savanna experienced the largest increases in in LAI and GPP, whereas closed and open shrublands experiences relatively small increases in LAI and GPP. Often, using remotely sensed data alone, it can be difficult to assign causality but directly incorporating effects from changing climate, land cover, and associated vegetation responses simultaneously enables improved analysis of global change effects on rangeland environments. Using long time series, however, requires cross-calibration between sensors. to ensure continuity across new sensors with varying bandpasses and associated target atmospheric effects, drifts in calibration, and filter degradation (Huete et al., 2002).

11.3.4.2 Extending Remote Sensing Time Series Using Cross-Sensor Calibration

Recent ecological research has shown that declines in dryland productivity (often estimated measured using trends in NDVI and/or NPP), and increases in soil loss are due to the synergistic effects of extreme climatic events and land management practices. Livestock grazing and El Niño and La Niña events have 3 to 7 year return intervals (Holmgren and Scheffer, 2001; Holmgren et al., 2006; Washington-Allen et al., 2006) indicating that 10–20 years of continuous data is required to replicate, monitor, and assess the influence of land use practices and these extreme events (Washington-Allen et al., 2006). Sensors have finite life spans, and developing long-term observations often

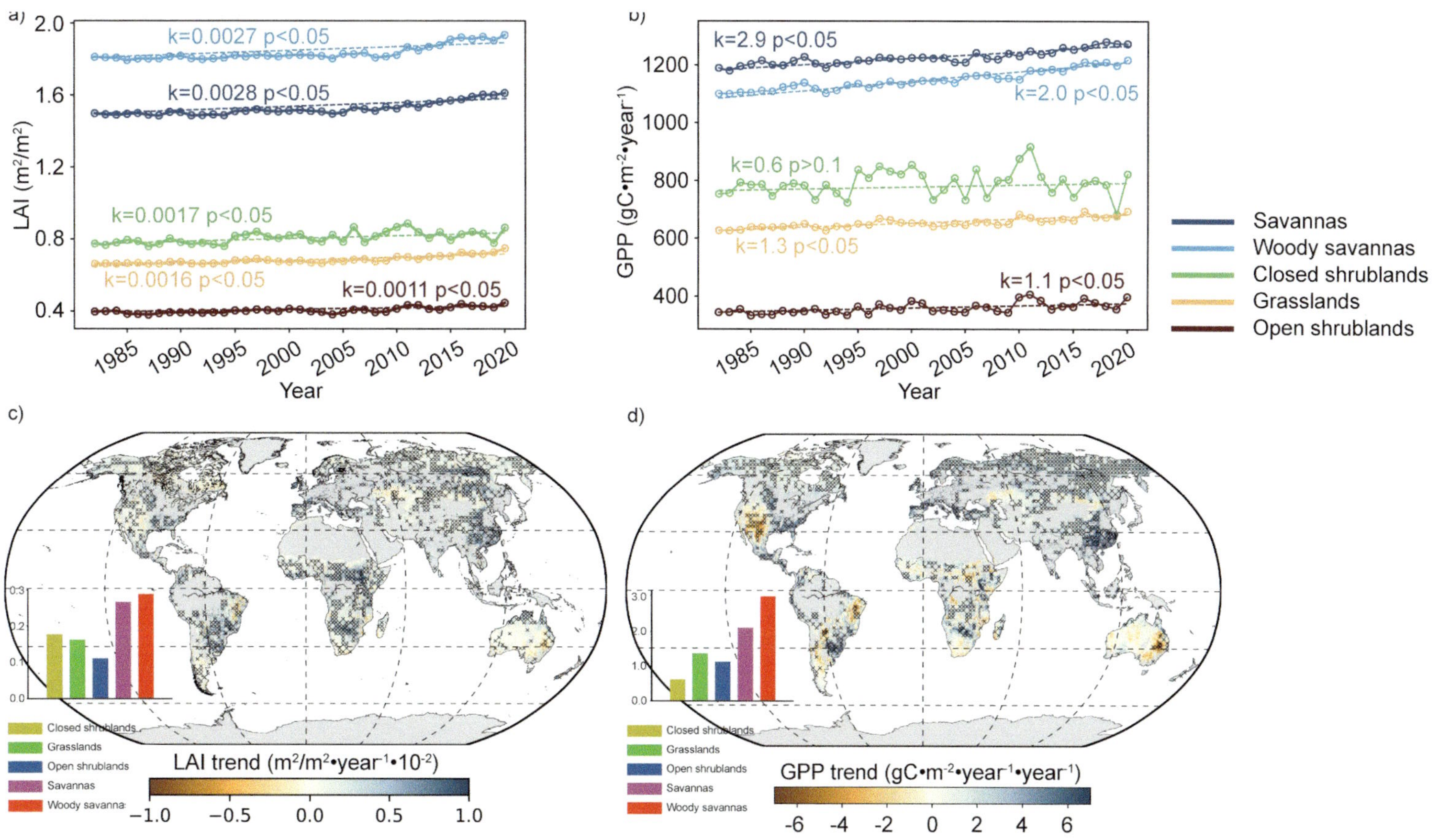

FIGURE 11.15 Interannual changes in LAI estimated by GIMMS(a) and GPP estimated by CEDAR for the period 1982–2020 over five landcovers. Spatial distribution pattern of the annual trend in LAI (c) and GPP (d) for the period 1982–2020. Regions labeled by dots have trends that are statistically significant ($p < 0.05$). The trend is calculated and evaluated using the Mann–Kendall test at the 5% significance level.

requires using multiple sources of data to develop a continuous, compatible dataset. The extension of time series is challenging due to drifts in calibration, filter degradation, and band locations (Miura et al., 2006). These characteristics create errors and uncertainties that vary with the landscape and sensors being evaluated. As examples, red and NIR spectral channels from AVHRR are relatively broad occupying the spectral space between 580–680 and 730–1,000 nm, respectively. In contrast, MODIS provides more narrow bands in the red and NIR space at 620–670 and 841–876 nm, respectively. The broader AVHRR red channel incorporates a portion of the green reflectance region (500–600 nm) (Figure 11.4) inevitably yielding a different spectral response of vegetation than MODIS.

The approaches for extending a satellite data time series via sensor (or product) cross-calibration involve remote sensing data fusion that accounts for multi-sensory, multi-temporal, multi-resolution, and multi-frequency image data from operational satellites (Pohl and Van Genderen, 1998; Zhang et al., 2010). Extension of satellite data records to produce time series of NDVI or NPP data typically involve:

1. Development of equations to simulate the spectral responses of individual channels (e.g., Suits et al., 1988)
2. Development of calibration equations to simulate the vegetation indices derived from other sensors (e.g., Steven et al., 2003; Tucker et al., 2005)
3. Crosscalibration of NDVI (e.g., from AVHRR) and NPP data products (e.g., from the MODIS sensor) to back cast the NPP record.

These techniques have been explored in a good number of studies and indicate suitable relations between sensors, but results are often inconclusive (Fensholt et al., 2009). Suits et al. (1988) determined that multiple regression analysis compared to principal component analysis was the best approach for spectral response substitution between Landsat and AVHRR sensors. Steven et al. (2003) found that vegetation indices from Landsat, SPOT, AVHRR, and MODIS were strongly linearly related, which allowed them to develop a table of conversion coefficients that allowed simulation of NDVI and SAVI across these sensors within a 1%–2% margin of error. Except for AVHRR, which was designed for other purposes, most high temporal resolution sensors have similar sensitivity to green vegetation. In addition, vegetation indices from many global platforms can be calibrated to within approximately ± 0.02 units if surface reflectance (as opposed to top of atmosphere) is used (Steven et al., 2003). Fensholt and Proud (2012) compared the GIMMS 3 g 8 km NDVI archive with MODIS 1 km NDVI and showed that global trends exhibit similar tendencies but significant local and regional differences were present, especially in more xeric environments. A comprehensive analysis of four long-term AVHRR based NDVI datasets with MODIS and SPOT NDVI datasets for the common period (from 2001 to 2008) clearly demonstrated lower correlations in more xeric regions such as the southwest and Great Basin of the United States (Scheftic et al., 2014). Similarly, Gallo et al. (2005) reported that 90% of the variation between 1 km MODIS and AVHRR NDVI can be explained by a simple linear relationship, while Miura et al. (2006) developed translation equations to emulate MODIS NDVI from AVHRR resulting in an r^2 of 0.97. Despite these successes, trend analyses from AVHRR can differ strongly from those estimated with MODIS and SPOTVGT (Steven et al., 2003) and lead to spurious conclusions. Unlike MODIS, AVHRR does not provide additional necessary channels permitting analysis of atmospheric composition for suitable atmospheric correction (Yin et al., 2012). Therefore, cross sensor calibration must be carefully planned and should leverage the strengths of previous efforts. Most efforts aimed for extending time series to improve trend analyses involve spectral calibration, either of individual band passes or indices. For monitoring global change and ecosystem performance, however, it is useful to quantify NPP trends given its link with the global carbon cycle and paramount importance to maintaining goods and services. Bai et al. (2008, 2009) developed a 23-year time series of global NPP data from 1982 to 2003 using the overlap period (2000–2003) between 1 km MODIS NPP and the mean annual sum of 8 km AVHRR GIMMS, for LADA program of FAO. Next, linear regression was applied to four-year mean, global, annual sum of NDVI from the GIMMS dataset and MODIS NPP to generate a single empirical equation between these two datasets. The resulting equation was then used to produce an 8 km NPP time series from 1982 to 2003. Wessels (2009) critiqued the approach of Bai et al. (2008) arguing that spatial variability was reduced and unaccounted for by using a single mean equation rather than a pixel-by-pixel approach. As a result, the following case study used a pixelwise regression approach for establishing relationships between 8 km GIMMS NDVI and 1 km MODIS NPP. The goal of this case study was to produce a continuous, compatible dataset describing annual NPP from 1982 to 2009 using both 8 km AVHRR GIMMS from Tucker et al. (2005) and 1 km MODIS net photosynthesis. A more recent version of GIMMS AVHRR NDVI (GIMMS 3g) data is available from 1981 to 2011 at 1/12th° spatial resolution.

11.3.4.2.1 Case Study

The strategy suggested by Steven et al. (2003) and Wessels (2009) was followed for calibrating 8 km pixel resolution GIMMS annual ΣNDVI from 1982 to 2006 to MODIS NPP data aggregated from 1 to 8 km using the 2000 to 2006 overlap period between these two separate time series. Collection 5 annual estimates of MODIS NPP from 2000 to 2006 and GIMMS ΣNDVI time series were subset to the rangeland portion of the contiguous United States and classified according to varying levels of aridity using AI (Figure 11.16). The AI of drylands (AI ≤ 0.65) is partitioned into four classes including the hyper-arid, arid, semi-arid, and dry sub-humid classes.

11.3.4.2.1.1 Application and Validation of Linear Regression Approach The Taiga Earth Trend Modeler from IDRISI was used to conduct a simple linear regression on a pixel-by-pixel basis between the two time series using the years 2000, 2002, 2004, and 2006. This was done so that a holdout dataset could be retained for comparing predicted and observed NPP. Across all pixels in

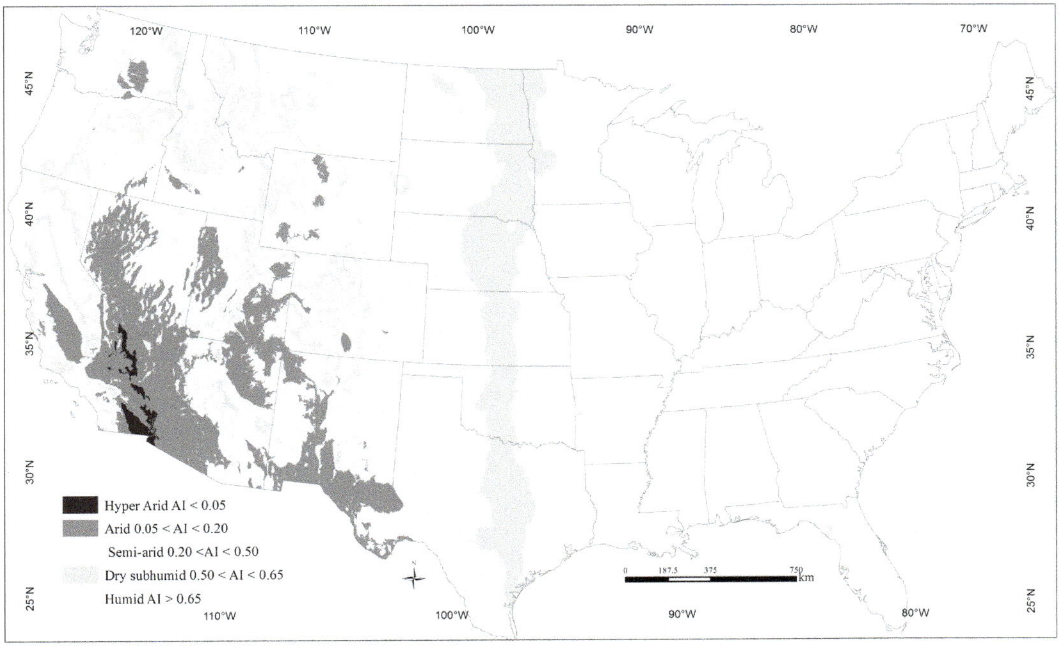

FIGURE 11.16 Distribution of aridity index (annual precipitation/potential evapotranspiration) classes throughout the United States used for aggregating NPP estimates for coterminous U.S. rangelands.

the rangeland domain, the mean NDVI was 0.03 and mean NPP was 281.6 g C m^{-2} $year^{-1}$. The mean equation across all pixels was

$$Y = 0.03 * X + (-31.7)$$

where
X is the annual GIMMS ΣNDVI
Y is the predicted 8 km MODIS NPP and $r^2 = 0.41$ (Figure 11.17)

Panels A, B, and C in Figure 11.17 represent the estimated slope, intercept, and r^2 of a linear regression for each pixel in the study area between GIMMS NDVI and MODIS NPP for the years 2000, 2002, 2004, and 2006. Predicted MODIS NPP was subsequently compared to the observed MODIS NPP (Table 11.5).

Figure 11.18 indicates a strong relationship between monthly integrated 8 km GIMMS NDVI and monthly integrated 1 km MODIS net photosynthesis (PSNnet) over the domain of coterminous US rangelands.

The net photosynthesis is a major component of the annual NPP product. To derive the final model to extend the NPP time series, the pixel level regressions developed were applied to the annual GIMMS ΣNDVI from 1982 to 1999. To these data, the MODIS NPP time series from 2000 to 2009 were added, thus extending the final time series from 1982 to 2009. Using the final time series, temporal and spatial variations in NPP response can be quantified. The mean NPP for each class from 1982 to 2009 was 95 ± 28 for hyper-arid, 115 ± 47 for arid, 218 ± 114 for semi-arid, and 370 ± 117 (g C m^{-2} $year^{-1}$) for the dry sub-humid class. In addition, the temporal trend (not accounting for temporal autocorrelation) of NPP within each AI class was as follows: hyper-arid ($r^2 = 0.08$, $p = 0.08$), arid ($r^2 = 0.01$, $p = 0.37$), semi-arid ($r^2 = 0.25$, $p = 0.004$), and dry sub-humid ($r^2 = 0.22$, $p = 0.006$) (Figure 11.19). Using this approach, significant carbon gains were detected for both semi-arid and arid systems. In addition, the positive response in arid and semiarid systems agrees with conclusions by Reeves and Bagget (2014) that significant increasing trends have been observed from 2000 to 2012 across much of the US rangeland domain, owed mostly to increased precipitation. The results portrayed in Figure 11.19 demonstrate improved chances for successfully interpreting vegetation response to global change through increasing the time series of satellite observation.

11.3.5 Remote Sensing of Global Land Cover

Global land cover data are essential to most global change research objectives, including the assessment of current global environmental conditions and the simulation of future environmental scenarios that ultimately lead to public policy development. In addition, land cover data are applied in national and sub-continental scale operational environmental and land management applications (e.g., weather forecasting, fire danger assessments, resource development planning, and the establishment of air quality standards). Land cover characteristics are integral to many Earth system processes (Hansen et al., 2000), in addition to providing information for carbon exchange and general circulation models. A common and important application of global land cover information is inference of biophysical parameters, such as LAI and fPAR, which influence global scale climate and ecosystem process models. Use of these models and monitoring the state of the Earth's rangelands is needed for global change research, especially given the influence of growing anthropogenic disturbances (Lambin et al., 2001; Jung et al., 2006; Xie et al., 2008).

One of the remote sensing community's grand challenges is to provide globally consistent but locally relevant land cover information (Estes et al., 1999). Global mapping presents special challenges since the geographic variability of both land cover and remote sensing inputs add complexity that can lead to inconsistent results. The evolution of global land cover datasets over the past 30 years has attempted to meet the grand challenge while adhering to general remote sensing land

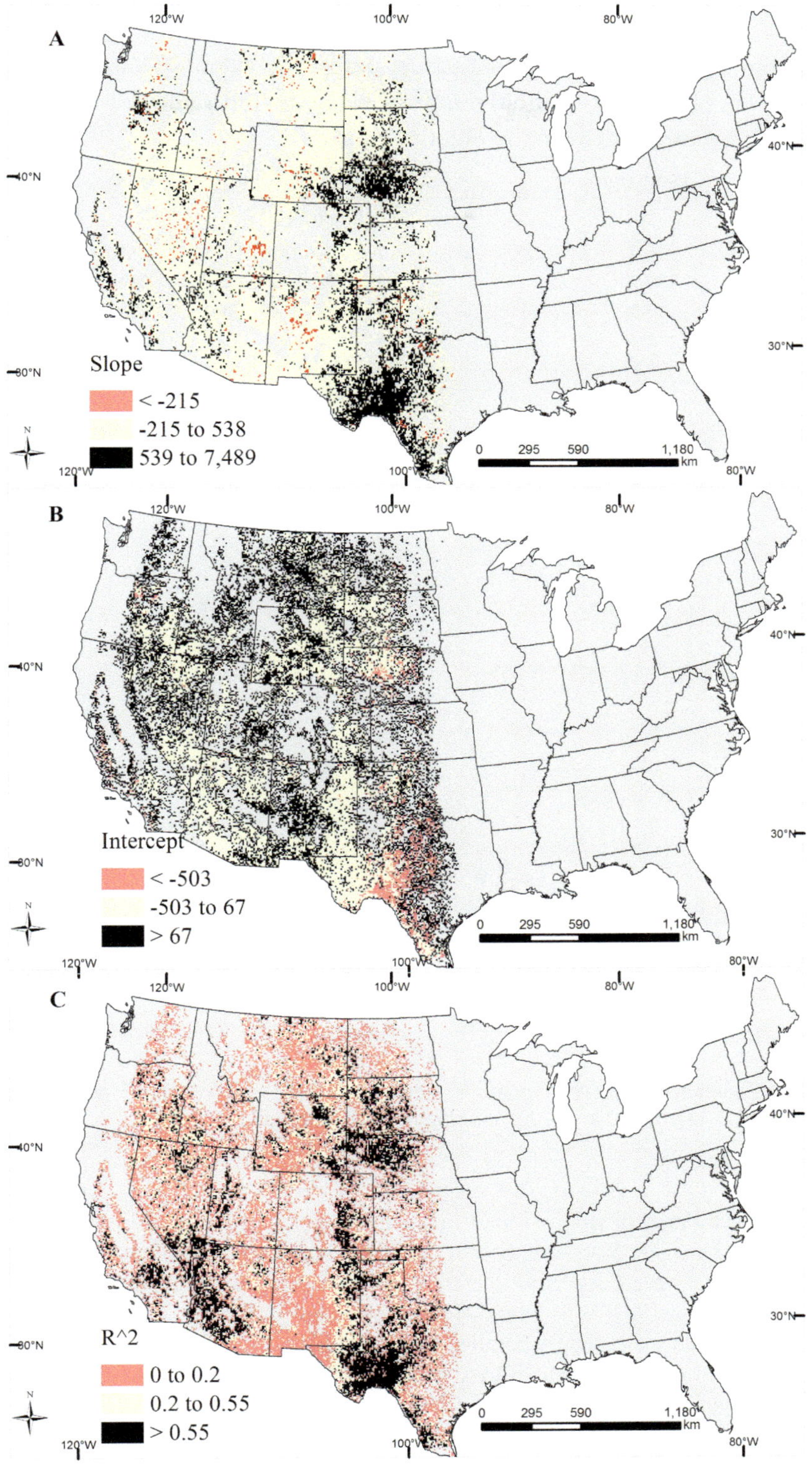

FIGURE 11.17 The resulting pixel-to-pixel linear regression models that were developed to calibrate GIMMS annual ΣNDVI to the MODIS NPP time series for the years 2000, 2002, 2004, and 2006. Panels A, B, and C represent the slope, intercept, and R^2 values for each pixel.

TABLE 11.5

Comparison of Predicted and Observed Values across the Extent of Rangelands in the Coterminous U.S. (g C m2 yr−1). Bold numbers are predicted values based on the pixel level regression equations depicted in Figure 11.15.

	Minimum	Median	Mean	SD
Year		g C m² yr⁻¹		
2001	16.5	179	211	123
	0.1	**186**	**220**	**137**
2003	15.3	189	218	131
	2.4	**184**	**217**	**131**
2005	19.8	236	126	126
	0.1	**210**	**157**	**157**

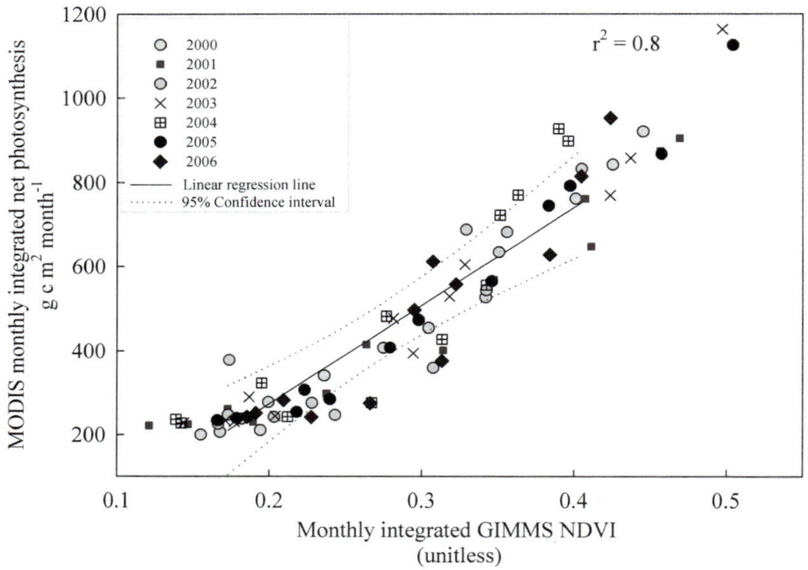

FIGURE 11.18 Relationship between monthly integrated MODIS derived 1 km² net photosynthesis and GIMMS 4 km² NDVI aggregated across all coterminous U.S. rangelands.

cover—mapping standards dealing with accuracy, consistency, and repeatability. The earliest contemporary efforts to provide global land cover data did not rely on remote sensing inputs but instead was based on the developer's expertise and the quality of information from best available sources (Matthews, 1983; Olson, 1983; Wilson and Henderson-Sellers, 1985). These maps were coarse (i.e., 1° × 1°) in resolution but thematically detailed. Global land cover mapping based on remote sensing advanced rapidly in the 1990s when NOAA polar-orbiting data from the AHVRR were compiled into global coverage. Initially, 4 km AVHRR Global Area Coverage Pathfinder data aggregated to 1° × 1° (DeFries and Townshend, 1994) and later to 8 km resolution (DeFries et al., 1998) were inputs to the first remote sensing—based global land cover products. The International Geosphere Biosphere Programme (IGBP) served as the catalyst for a worldwide effort led by the USGS to generate a 1992–1993 set of 1 km resolution AVHRR global 11-day maximum NDVI composites

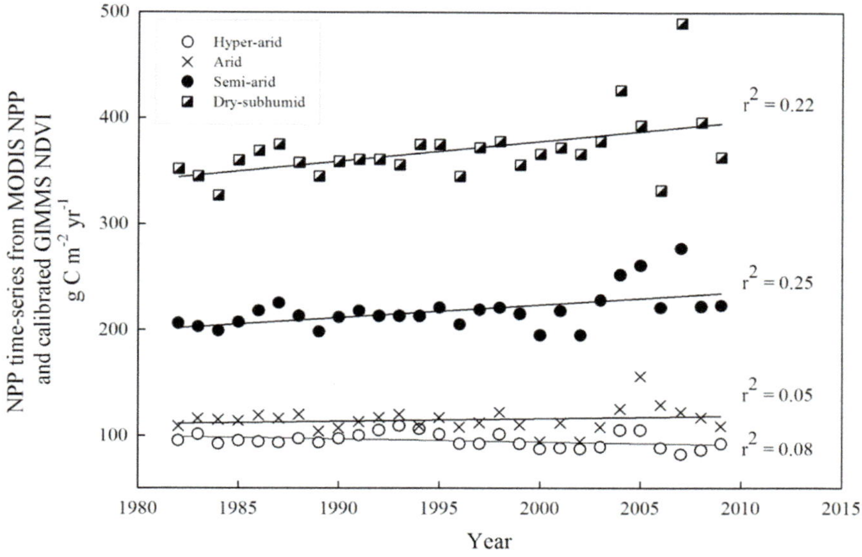

FIGURE 11.19 Rangeland mean net primary productivity (NPP) from 1982 to 2009 across four zones of aridity index (AI) including hyper arid, arid, semi-arid, and dry subhumid.

(Eidenshink and Faundeen, 1994). Also under IGBP auspices, these data were used to produce the first 1 km resolution global land cover dataset using the 17 class International Geosphere-Biosphere Programme Global Land Cover Classification (IGBP DISCover) legend (Loveland et al., 2000) (Figure 11.20). Hansen et al. (2000) followed with the completion of a 1 km, 12 class land cover dataset (UMD land cover map).

These two maps served as the foundation for future global mapping initiatives since their development experiences and map strength and weaknesses provided valuable lessons for the next generation of maps. The NASA Earth Observing System's ambitious global land product program based on multiresolution MODIS data established a new state of the art in global land cover mapping. MODIS global land cover based on 500 m resolution imagery and the 17 class IGBP DISCover legend started in the 2001 and since then has been updated annually (Friedl et al., 2002). This ongoing activity represents the only sustained global land cover initiative. In the 2000s, European global land cover projects contributed significantly to advancing global land cover understanding. The Global Land Cover 2000 (GLC2000) project used SPOT vegetation instrument data to produce a 22 class 1 km resolution land cover dataset (Bartholomé and Belward, 2005). In a follow-on effort, the European Space Agency sponsored a follow-on project, GlobCover, that used ENVISAT MERIS imagery to generate the highest resolution (300 m) global land dataset ever. The MERIS-based map contained 22 land cover classes based on the United Nations sponsored international standard land cover classification system (LCCS).

The China led Fine Resolution Observation and Monitoring Global Land Cover (FROMGLC) dataset was a groundbreaking effort to map global land cover based on Landsat 5 and 7 Thematic Mapper/ETM+ and other high resolution Earth observation data spanning the first decade of the 21st century (Gong et al., 2013). The FROMGLC dataset established new standards for high resolution land cover mapping and monitoring, with 29 land cover classes.

With new accessibility to free remote sensing data in the cloud and cloud-based computing that facilitates rapid generation of land cover across broad regional extents, there has been a proliferation of new global land cover datasets at high resolution. With these new tools, the academic community

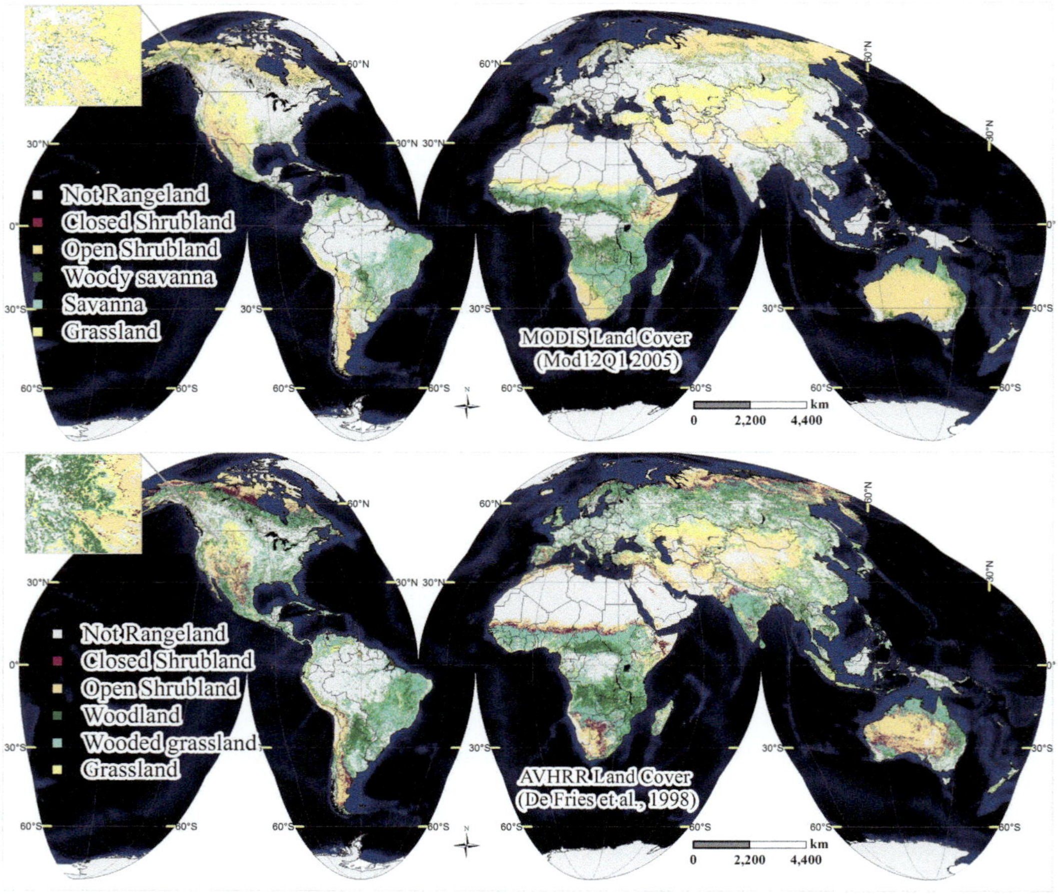

FIGURE 11.20 Comparison between 1 km^2 MODIS land cover data (Mod12Q1; UMD classification) and AVHRR derived land cover (DeFries et al., 1998) using the Simple Biosphere Model legend (Table 11.6).

produced a global land cover, land use, and ecozone map using Landsat and topographic data and defined a potential approach for multitemporal monitoring of change (Hansen et al., 2022), The European Space Agency initiated the WorldCover project in 2017 and released the first global land cover product at 10m resolution for 2020 and 2021, based on Sentinel1 and Sentinel2 data (Zanaga et al., 2020, 2021). Not only was the product the first to map land cover at 10 m resolution, but it marked a paradigm shift in the use of multimodal platforms (optical with Sentinel2, Sentinel1 based radar) for global mapping of land cover. By the early 2020s, for the first time the commercial sector began development of independent global, high resolution land cover. The Environmental Systems Research Institute (ESRI) in partnership with Impact Observatory (IO) created a Sentinel2 based 10 m global land cover product, including mapping change from 2017 to 2022 (ESRI, 2023). Google partnered with the World Resources Institute to produce Dynamic World, a global, 10 m, nine land cover class land cover dataset based on Sentinel2 (Brown et al., 2022).

In addition to the thematic land cover mapping efforts described earlier (Table 11.6), global "continuous fields" products provide quantitative estimates of the percent tree cover within each grid cell. DeFries et al. (1999) developed global percent tree cover data using 1 km AVHRR imagery, and Hansen et al. (2003) created similar products using MODIS.

TABLE 11.6
Summary of Characteristics of the Major Remote Sensing Global Land Cover Datasets

Database	Source	Vintage	Resolution	Land Cover Content (Suggested Rangeland Classes)	Strengths	Weaknesses
Global AVHRR NDVI Land Cover (DeFries and Townshend, 1994)	AVHRR	1987	1.0°^{2} Latitude	11 (3) land cover classes—based on Simple Biosphere Model	First remote sensing-based depiction of global land cover	Coarse resolution, applications limited to Global Circulation Model applications
Global AVHRR Land Cover (DeFries et al., 1998)	AVHRR Global Area Coverage Pathfinder	1987	8 km^2	14 (5) land cover classes—based on Simple Biosphere Model	Improved spatial resolution provided more realistic view of global land cover	Land cover classes were general and specific to one application requirement
IGBP DISCover (Loveland et al., 2000)	AVHRR Local Area Coverage	1992–93	1 km^2	17 (5) IGBP DISCover land cover classes, and other land cover legends	Highest resolution global land cover to date, validated based on statistical design	Variable image quality contributed to unevenness of land cover accuracy
UMD Global Land Cover (Hansen et al., 2000)	AVHRR Local Area Coverage	1992–93	1 km^2	12 (5) land cover classes	Based on an automated analysis strategy	Not validated, affected by variable image quality
MODIS Global Land Cover (Friedl et al., 2002)	MODIS	2001-present, produced annually	500 m^2	17 (5) IGBP DISCover land cover	Uses highest quality remotely sensed inputs available, based on rigorous automated methods	Unknown accuracy due to the lack of a design-based map validation
GLC2000 (Bartholomé and Belward, 2005)	SPOT 4 VEGETATION	2000	1 km^2	22 (5) land cover classes	Based on standardized land cover legend, validated results	Affected by variable image quality
GlobCover (Arino et al., 2007)	ENVISAT MERIS	2005–2006	300 m^2	22 (4) land cover classes, UN Land Cover Classification System	Based on standardized land cover legend, validated results, and highest resolution imagery to-date	Regional variability in image quality increased uncertainty of results in some part of the world
Fine Resolution Global Land Cover (Gong et al., 2013)	Landsat 5 and 7	Nominally 2005–2006	30 m^2	29 (6) land cover classes	Highest resolution dataset ever produced	Limited temporal inputs resulted in regional inconsistencies
GLC2000 (Bartholomé and Belward, 2005)	SPOT 4 Vegetation	2000	1 km^2	22 (5) land cover classes	Based on standardized land cover legend, validated results	Affected by variable image quality
GlobCover (Arino et al., 2007)	ENVISAT MERIS	2005–2006	300 m^2	22 (4) land cover classes, UN Land Cover Classification System	Based on standardized land cover legend, validated results, and highest-resolution imagery to date	Regional variability in image quality increased uncertainty of results in some parts of the world

(Continued)

TABLE 11.6 (*Continued*)
Summary of Characteristics of the Major Remote Sensing Global Land Cover Datasets

Database	Source	Vintage	Resolution	Land Cover Content (Suggested Rangeland Classes)	Strengths	Weaknesses
Fine resolution global land cover (Gong et al., 2013)	Landsat 5 and 7	Nominally 2005–2006	30 m^2	29 (6) land cover classes	Highest-resolution data produced to date	Limited temporal inputs resulted in regional inconsistencies
Global Land Cover and Land use 2019 (Hansen et al., 2022)	Landsat 7 and 8	2019	30 m^2	19 (5) land cover and use classes	Hybrid of land cover and land use; Validated with probability sample	Land use does not include pastures despite acknowledgment of their status as most widespread human landuse
WorldCover (Zanaga et al., 2020, 2021)	Sentinel-2, Sentinel 1	2020, 2021	10 m^2	11 (3) land cover classes	First 10 m^2 global land cover product; Multi-modal with optical and active remote sensing systems	Thematically coarse
ESRI Land Cover (ESRI, 2023)	Sentinel 2	2017–2022	10 m^2	9 (1) land cover classes	Multi-temporal mapping of global land cover at high resolution	Thematically coarse with 1 rangeland class, Questionable representation of change
Dynamic World (Brown et al., 2022)	Sentinel 2	On Demand (limited to Sentinel 2 data availability)	10 m^2	9 (2) land cover classes	First near real-time capability for global land cover	Thematically coarse, questionable representation of change

11.3.5.1 Comparative Investigations of Global Land Cover Datasets

With a relatively large number of global land cover datasets available, users face a challenge in understanding which one is best suited for their application. The differences in spatial resolution, temporal properties, land cover legend, and quality complicate the selection. Land cover legend and quality are particularly significant factors. Accuracy assessments that provide insights into data quality are available for some of the global products. For example, both the IGBP DISCover and GLC2000 datasets were evaluated using an independent accuracy assessment. DISCover accuracy was measured at 66.9% (Scepan, 1999). Mayaux et al. (2006) determined that the overall GLC2000 product accuracy was 68.6%. The MODIS land cover dataset accuracy was assessed based on a comparison with training data, with the results showing 78.3% agreement (Friedl et al., 2002). The more recent GlobCover land cover dataset's (Table 11.6) independent accuracy was measured to be 73.0%. Finally, the China led fine resolution global land cover product was determined to have an overall accuracy of 71.5% (Gong et al., 2013). Accuracy assessments were not produced for UMD global land cover datasets.

The overall accuracies mask the significant variations in per class accuracies (e.g., Scepan, 1999 estimates that the DISCover individual class accuracies varied from 40% to 100%). The class accuracy variations, as well as variations in land cover legends and class definitions, make cover specific applications problematic. As a response to this problem, several global dataset comparison studies have been undertaken, which focus on determining dataset strengths and weaknesses. Some have used independent datasets to look at regions or continents, such as Achuete et al.'s (2011) evaluation of GLC2000, GlobCover, and MODIS land cover for Africa and Frey and Smith's (2007) evaluation of IGBP DISCover and MODIS land cover over western Siberia. Other comparisons have looked at agreement between datasets across the globe. For example, Hansen and Reed (2000) compared UMD and IGBP DISCover products; Giri et al. (2005) compared MODIS and GLC2000; McCullum et al. (2006) compared IGBP, UMD, GLC2000, and MODIS products; and Fritz and See (2007) compared MODIS, GLC2000, and GlobCover. McCullum et al. (2006) concluded that while there is general agreement at the global level in total area and general land cover patterns; there is limited agreement when looking at specific spatial distributions.

A more comprehensive and definitive effort to understand the difference in global datasets comes from Wang and Mountrakis (2023). They assessed the accuracy of 11 global land cover products and several national scale products in the conterminous US. The different datasets had resolutions from 10m to 1km, and mapping dates from 1992 through 2023 and were validated using a consistent reference dataset of 25,000 manually interpreted reference points over the conterminous United States. Differences in thematic class definitions make precise comparisons difficult, but in general the study found wide discrepancies in landcover accuracies. Thematic class accuracies were highest on average for a "treed" cover class, at 78%, but average accuracies for other classes ranged from 55% to 70%. Spatial discrepancies were apparent as well, with the best performing datasets often differing among different geographies. The regional products generally performed better than global products at the scale of the conterminous United States, demonstrating the challenge of achieving regional scale accuracy with a global scale product. The study also found that accuracy across all products decreased when assessing "edge" samples that may represent boundaries among multiple land cover classes.

One of the remote sensing community's grand challenges is to provide globally consistent but locally relevant land cover information (Estes et al., 1999). Evaluations of remote sensing based global land cover datasets have shown general agreement of patterns and total area of different land covers at the global level, but have more limited agreement in spatial patterns at local to regional levels (McCullum et al., 2006) As the quality and resolution of remotely sensed data used for global land cover mapping improves, the logical expectation is that overall and individual class accuracies will also improve. However, Fritz et al. (2011a,b) emphasized the continued uncertainty in global land cover products, especially in land cover classes associated with agriculture and some forest groups. They suggest that increased use of in situ data is the key to improving global land cover datasets. Wang and Mountrakis (2023) similarly noted general room for improvement in mapping accuracy for global scale land cover products. As global high resolution products such as the ESRI and WorldCover products become increasingly available for multiple dates, there is also a need for rigorous validation of not only land cover, but land cover change.

11.4 FUTURE PATHWAYS OF GLOBAL SENSING IN RANGELAND ENVIRONMENTS

Remote sensing has created unprecedented capacity to study the Earth by providing repeated measurements of biological phenomena at global scales. Since the first regional applications of NDVI (one of the earliest regional applications found is Rouse et al., 1973) (Section 10.2), the study of the global rangeland situation has benefitted greatly from advancements made in a relatively short period of time. Degradation of global rangelands because of climate change and anthropogenic land use requires accurate, objective and repeatable measures of land condition across broad spatial scales. As presented in this chapter, the use of remote sensing over previous decades has largely focused on broad-scale satellite-based products that have been used in many international and national monitoring programs, based on trust built over many years of use. While satellite remote sensing has provided the backbone of rangeland assessment and monitoring at global and broad regional scales, condition assessments based on ground-referenced data have widely served as the dominant approach for many rangeland managers at regional and localized scales (Kachergis et al., 2022; Lawley et al., 2016; Sparrow et al., 2020). Tueller (1989) suggested that the majority of monitoring data in rangelands would be collected using remote sensing by 2010, but this still not the case, with a notable lack of uptake for fine-scale remote sensing methods for condition assessment (Retallack et al., 2023). However, recent increases in the accessibility of ultra-high-resolution data have led to the development of corresponding methods for extracting fine-scale indicators of ecosystem condition. Though future uses of remotely sensed data will be used in unexpected ways, obvious areas of enhancement and progress are anticipated. These future pathways can be expressed in distinct areas including improvements in data availability and processing, biophysical product improvement with emphasis on operational conversion of data to information.

The design and intended application of spaceborne sensors will continue to evolve, and a wider variety of satellite systems including radar and lidar could be quite beneficial in the future. If the past provides a glimpse into the future, new sensors with improved capabilities will be developed, but it is unclear, however, whether improved spatial, spectral, and temporal resolution of satellite remote sensing will provide the greatest advancements in the evaluations of rangelands on a global scale. The ability to extract surface features and quantify biophysical properties will still be limited by the same factors presently hindering remote sensing of rangelands. Characteristics such as bright soil background, leaf anatomy and physiology, and relatively low biomass conspire to hinder remote sensing of rangelands. Very little can be done to change these situations, and as a result, future pathways should include a focus on data continuity, increased data availability, better computer processing systems, and global campaigns for collecting ground referenced data and greater use of unmanned aerial vehicles for scaling ground referenced data.

Remote-sensing data continuity is important to monitoring global rangelands, and loss of this critical aspect will significantly weaken our ability to understand what the biosphere is indicating. The need for continuity is recognized in the Land Remote Sensing Policy Act of 1992, which states:

> The continuous collection and utilization of land remote sensing data from space are of major benefit in studying and understanding human impacts on the global environment, in managing the Earth's resources, in carrying out national security functions, and in planning and conducting many other activities of scientific, economic, and social importance.

Since the first civilian spaceborne missions (e.g., Landsat 1), the global monitoring community and government agencies have been reasonably successful in providing the needed continuity. The

Landsat program is a good example of the flow and continuity with incremental improvements with each successive launch generally maintaining a 30 m resolution benchmark. If archive data from Landsat 4 (deployed in 1982) are included, 41 years of 30 m spatial resolution from the TM sensor in visible and NIR (at the minimum) are available. Landsat 8, launched on February 11, 2013, is the most recent addition to the suite of Landsat satellite launches and provides an example of maintaining continuity with previous missions while improving capability. Landsat 8 contains the Operational Land Imager (OLI) and the Thermal Infrared Sensor (TIRS), which provide global coverage at varying resolutions. The OLI provides two new spectral bands for detecting cirrus clouds and the other for coastal zone observations. Now that the entire archive of Landsat data has been made freely and publicly available, usage has increased exponentially. The unprecedented data availability has and will continue to lead to new algorithmic and ecological discoveries. Increased data usage may signal greater interest in remote sensing but certainly tracks the increased microprocessor speed over the last decade (Figure 11.21). As processing speed and memory have increased so has the level of algorithmic sophistication and spatial domain for analysis. Indeed, the global remote-sensing community is poised for improved characterization capabilities, due to new data policies and concurrent advances in computing (Hansen and Loveland, 2011).

Even a decade ago, it would have been unthinkable to regularly process and store a global time series of satellite imagery with a pixel resolution of less than about 250 m. Although it is certainly possible to monitor rangelands globally at 30 m, it will be a monumental task. Each TM path/row contains 0.534 GB in the seven multispectral and thermal channels and approximately 0.234 GB for the panchromatic band. Since roughly 16,396 scenes are required for global coverage (including oceans), that is an estimated 12.3 TB of data for a single 16-day period. The repeat frequency or revisit cycle is 16 days (~22 periods per year), so the total amount of data since 1999 is near 6224 TB. Based on an online storage price of $0.05 per month per GB (https://cloud.google.com/products/cloud-storage/), the storage cost is tantamount to roughly 311,000 dollars per year. While this represents a significant amount of data and resources, a growing number of global applications at 30 m spatial resolution can be expected. Sexton et al. (2013) produced a Landsat-based global database of tree cover at 30 m resolution (Sexton et al., 2013), while Potapov et al.

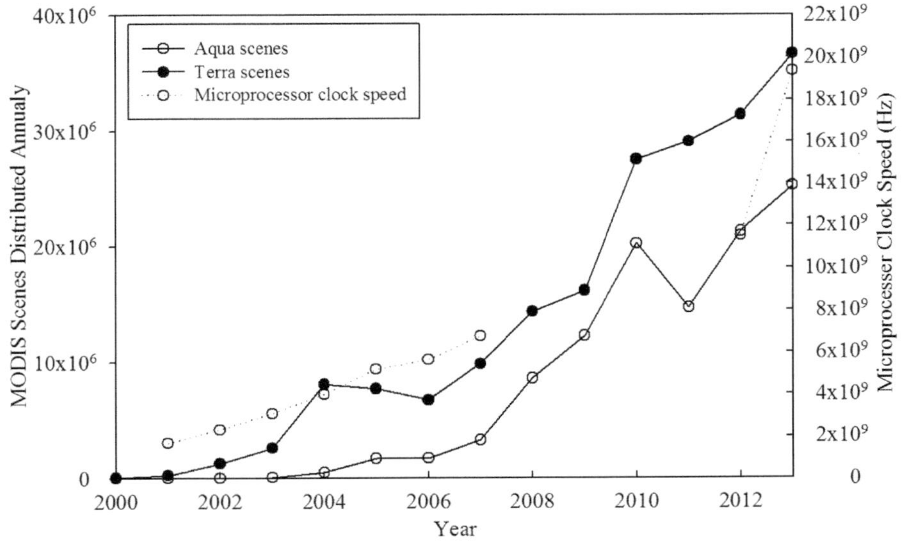

FIGURE 11.21 Microprocessor speed and MODIS data usage (Microprocessor speed data courtesy of www.raptureready.com/ (accessed 1, April 2014), MODIS usage data courtesy of B. Ramachandran NASA Earth Observing System I LP DAAC).

(2022) created the Global Land Cover and Land Use Change (GLAD) dataset quantifying changes in forest extent and height, cropland, built-up lands, surface water, and perennial snow and ice extent from the year 2000 to 2020 at 30 m spatial resolution. Moreover the fractional cover of lifeforms including those in the rangeland domain, have been quantified and made available as Google Earth Engine assets (projects/glad/GLCLU2020/). Global efforts like these will be further enhanced with Landsat 9 (27 September, 2021) as fusion between L8/OLI and S2/MSI enables revisit cycles of about 2 days (Wulder et al., 2019) compared with the nominal revisit of 3 days prior to 2021 (Bolton et al., 2020). This is important for global applications in rangeland domains when one considers the prevalence of cloud cover derived from monsoon periods across much the Earth. Presently, most efforts aimed at global remote sensing of rangelands are based on MODIS sensors aboard the Terra and Aqua satellites. Since 2006, the number of scenes annually distributed from MODIS data from both Aqua and Terra has increased by 7.6 million per year (about 181 TB year^{-1}) (Figure 11.21). This use is a testament to the breadth of vetted science data products offered globally.

Satellite sensors degrade with age, therefore new sensors are essential to continue collecting data within the same quality and usability ranges. Recently, data collected using MODIS sensors have shown degradation caused by time and orbital drift. This degradation has led to the decision to retire these satellites with continuity being supplied by the VIIRS through 2031. A visual comparison between expedited MODIS and VIIRS can be seen in Figure 11.22.

A transformation may lengthen the historical data compatible with VIIRS as it continues to monitor Earth's surface. NOAA leadership plans to continue VIIRS through 2031 with four missions called NASA-NOAA Joint Polar Satellite System (JPSS). The JPSS is a joint program between NOAA, NASA, and the Defense Weather Satellite System, tasked with developing the next generation requirements for environmental research, weather forecasting, and climate monitoring (npp.gsfc.nasa.gov/viirs.html). The JPSS provides operational continuity of satellite-based observations and products through a series of advanced spacecraft of which Suomi NPoP is a member. Suomi NPoP was launched in 2011 with five key instruments, but the instrument with greatest application, to rangelands globally, and similarity with the AVHRR and MODIS predecessors is the VIIRS. The VIIRS instrument observes the Earth and atmosphere at 22 visible and infrared wavelengths (Table 11.7). The continuity of land remote-sensing instruments is well established and provides a critical component to researchers involved with global change research in rangeland environments. Most future global issues will emulate present concerns. In other words, the problems, or area of focus, today (e.g., vegetation trends, land degradation, and fire processes) will continue and perhaps intensify in the future.

Regardless of the increasingly important roles remote sensing will play, georeferenced ground data will play an equally critical aspect of biospheric monitoring (Baccini et al., 2007). Fritz and See (2007) suggest that increased use of in situ data is the key to improving global datasets. The collection, maintenance, analysis, and distribution of georeferenced ground data, however, are a time-consuming and resource-intensive exercise, especially over regional or global domains. In this vein, the citizen scientist is an underutilized concept that can be cheaply and effectively employed to globally collect biospheric observations.

Citizen science can be defined as

> the systematic collection and analysis of data; development of technology; testing of natural phenomena; and the dissemination of these activities by researchers on a primarily avocational basis.
>
> *OpenScientist (2011)*

These open networks promote interactions between scientists, society, and policymakers leading to decision making by scientific research conducted by amateur or nonprofessional scientists (Socientize, 2013). Advancements in communication and technology are credited with aiding the growth of citizen scientists (Silverton, 2009). Collectively, citizen science efforts from around the

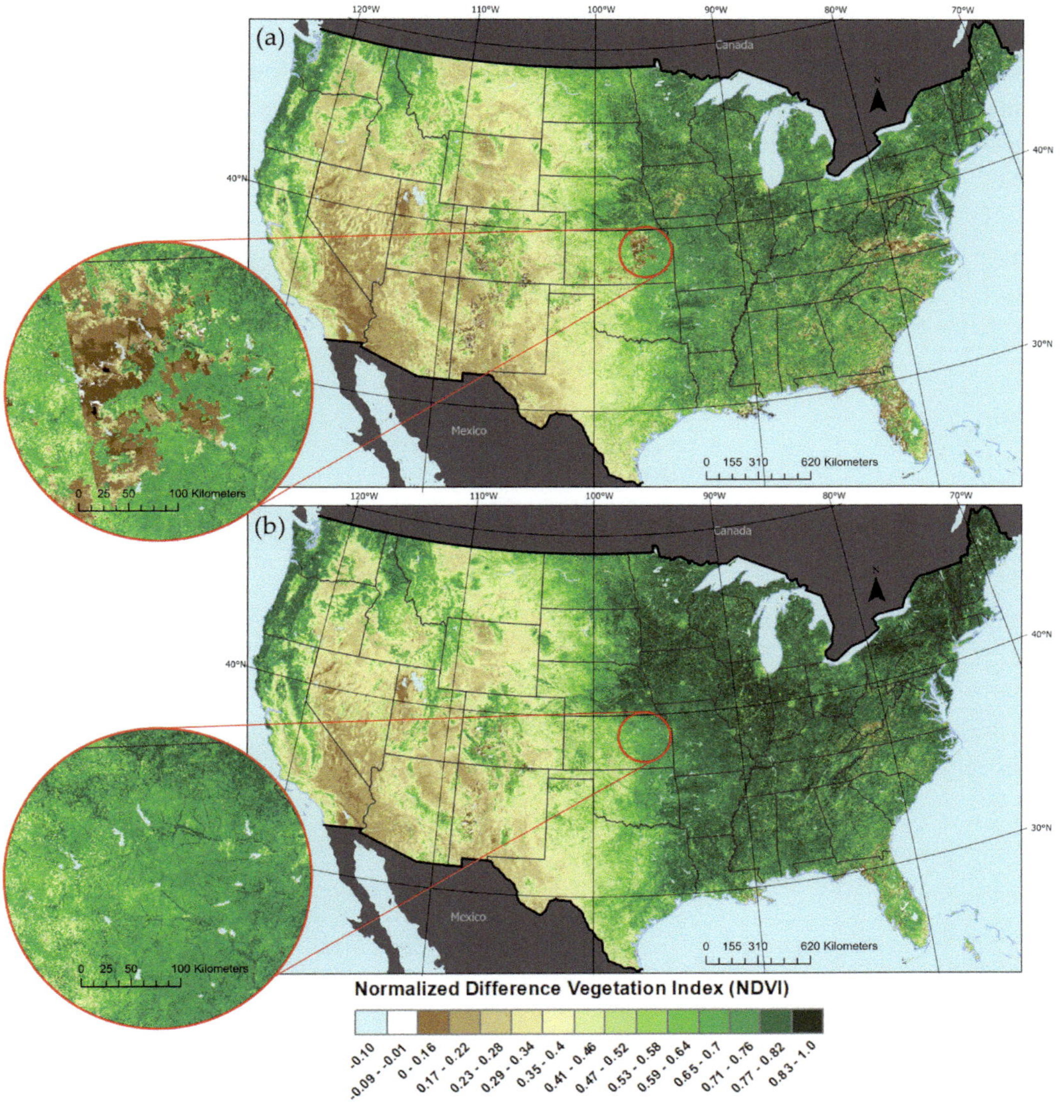

FIGURE 11.22 Expedited Moderate Resolution Imaging Spectroradiometer (eMODIS) 250-m Normalized Difference Vegetation Index (NDVI) (a) and expedited Visible Infrared Imaging Radiometer Suite (eVIIRS) 375-m NDVI (b) for composite (Julian day of year 215–221), 2016. Highlighted location in eastern Kansas shows an example of how eVIIRS composites have less noise artifacts than eMODIS composites. (Benedict et al., 2021.)

globe could possibly provide powerful venues for validating and calibrating future remote-sensing efforts. Both citizen science and crowdsourcing are valuable sources of data for calibrating or validating rangeland remote sensing but both have yet to be fully exploited Fritz et al. (2017). Fritz et al. (2017) provided a special issue dedicated to citizen science and Earth observation. A recent example of using citizen science to aid remote sensing of nonforest landscapes include mapping presence and phenological status of Buffelgrass in southern Arizona where the species is considered a major threat to desert ecosystems (Wallace et al., 2016). The Geo-Wiki project, established in 2010 by the International Institute for Applied Systems Analysis (IIASA), enables anyone with the means to

TABLE 11.7
VIIRS Spectral Channels and Suggested Usefulness. The LWIR are long wave infrared bands while the S/MWIR are short to mid wave infrared bands (adapted from http://cimss.ssec.wisc.edu/itwg/itsc/itsc13/proceedings/session10/10_2_schueler.pdf) (Schueler et al., 2003).

		Band No.	Driving EDR(s)	Spectral Range (um)	Horiz. Sample Interval (km) (Track × Scan) Nadir	End of Scan
Reflective Bands	VisNIR	M1	Ocean Color Aerosol	0.402–0.422	0.742 × 0.259	1.60 × 1.58
		M2	Ocean Color Aerosol	0.436–0.454	0.742 × 0.259	1.60 × 1.58
		M3	Ocean Color Aerosol	0.478–0.498	0.742 × 0.259	1.60 × 1.58
		M4	Ocean Color Aerosol	0.545–0.565	0.742 × 0.259	1.60 × 1.58
		I1	Imagery EDR	0.600–0.680	0.371 × 0.387	0.80 × 0.789
		M5	Ocean Color Aerosol	0.662–0.682	0.742 × 0.259	1.60 × 1.58
		M6	Atmospheric Correction	0.739–0.754	0.742 × 0.776	1.60 × 1.58
		I2	NDVI	0.846–0.885	0.371 × 0.387	0.80 × 0.789
		M7	Ocean Color Aerosol	0.846–0.885	0.742 × 0.259	1.60 × 1.58
	S/WMIR	M8	Cloud Particle Size	1.230–1.250	0.742 × 0.776	1.60 × 1.58
		M9	Cirrus/Cloud Cover	1.371–1.386	0.742 × 0.776	1.60 × 1.58
		I3	Binary Snow Map	1.580–1.640	0.371 × 0.387	0.80 × 0.789
		M10	Snow Fraction	1.580–1.640	0.742 × 0.776	1.60 × 1.58
		M11	Clouds	2.225–2.275	0.742 × 0.776	1.60 × 1.58
Emissive Bands		I4	Imagery Clouds	3.550–3.930	0.371 × 0.387	0.80 × 0.789
		M12	SST	3.660–3.840	0.742 × 0.776	1.60 × 1.58
		M13	SST Fires	3.973–4.128	0.742 × 0.259	1.60 × 1.58
	LWIR	M14	Cloud Top Properties	8.400–8.700	0.742 × 0.776	1.60 × 1.58
		M15	SST	10.263–11.263	0.742 × 0.776	1.60 × 1.58
		I5	Cloud Imagery	10.500–12.400	0.371 × 0.387	0.80 × 0.789
		M16	SST	11.538–12.488	0.742 × 0.776	1.60 × 1.58

classify satellite, drone or ground-level imagery. This novel system has grown rapidly and currently has 22,000 contributors and over 18 million crowdsourced image classifications have been uploaded. In addition to Geo-Wiki, the Global Geo-Referenced Field Photo Library (Xiao et al., 2011) enables visualization and archival of geo-referenced photos. This system, managed by the Center for Earth Observation and Monitoring, also includes a smartphone app. Similarly, the NASA-funded Global Learning and Observations to Benefit the Environment (GLOBE) Observer (GO) program is a smartphone app that enables citizen-collected land cover photos to be tagged and archived with location, date and time, and, in some cases, land cover type (Kohl et al., 2021). Despite a promising future, Fritz et al. (2022) identified grand challenges facing environmental citizen science spanning methodological, technological, political and ethical domains. Lastly, while remote sensing of rangelands has undergone major advances since inception, the need for quickly accessible information targeted for land managers is increasing. Tueller (1989) stated:

> Remote sensing has been recommended for at least 30 years for assisting with rangeland resources development and management on a worldwide basis.

That was 35 years ago and yet while numerous examples of remote sensing to improve our understanding and management rangelands are discussed in this chapter, there remains a great need for operational products for meeting specific end user's needs. The suite of global biophysical products derived from MODIS since 2000 is closely aligned with this idea. The relatively coarse resolution and academic, not managerial, focus limits their usefulness for land management endeavors. However, in recent years, deep learning-based classifications have proven valuable for extracting conservation-relevant information from this high-resolution imagery, including accurate classifications of certain plants to the species-level (Retallack et al., 2023). A new example of this is the inception of the Rangeland Analysis Platform (RAP; https://rangelands.app/) has begun to change this situation as it offers fast information for user defined areas of interest, for most rangelands in the United States regarding plant functional types from 1986 to 2022 and beyond (Allred et al., 2021). While not global the RAP, leveraging the Google Earth Engine, has demonstrated the capability of converting data to information at 30 m since it emanates from the Landsat time series. Similar efforts by Rigge et al. (2020) have used machine learning techniques to classify sagebrush to the genus for sagebrush (*Artemisia* spp.) cover and these products are available for the United States through LandCART which led to development of the Rangeland Condition Monitoring Assessment and Projection data product (Zhou et al., 2020). Potapov et al. (2023) provide an excellent overview of techniques for evaluating rangeland condition using remote sensing.

These recent developments have also led to the development of novel analytical methods, where machine deep-learning has been employed to great effect for extracting information from data collected at all spatial scales, which may prove to be exceptionally useful for scaling ground-referenced data. The analysis of high-resolution UAV or aerial imagery has often been approached using object based image analysis (Lu et al., 2016; Oldeland et al., 2021; Wilson et al., 2022). The collection of this high-resolution data has been enabled by a rapid rise in the accessibility of UAV (uncrewed aerial vehicle) technology over the past decade. UAV platforms have also enabled the collection of high-resolution three-dimensional data using methods such as lidar and structure from motion (SfM). This three-dimensional data has proven invaluable in measuring and mapping changes in erosion at fine scales compared with previous aerial or satellite-based approaches (Chakrabortty and Pal, 2023). UAV-derived SfM and lidar have also been used for the prediction of aboveground biomass in rangeland systems (Anderson et al., 2018; Barnetson et al., 2020; Grüner et al., 2019; Ku and Popescu, 2019), as well as canopy structural attributes (Hillman et al., 2021). While UAV-derived lidar data have proven useful in rangelands, in terms of global applications with spaceborne lidar, the picture is less clear. The launch of the new spaceborne laser altimeter missions ICESat-2 (Ice, Cloud, and Land Elevation Satellite-2) and GEDI (Global Ecosystem Dynamics Investigation) in September and December 2018, has significantly improved observational coverage, providing more accurate estimates with footprint sizes better suited for forest observations, particularly when combined with optical or radar data. The application of spaceborne lidar systems to measure ecosystems characterized by short-stature vegetation, including rangelands (e.g., grasslands, shrublands, and savannas) and wetlands (e.g., mangroves, bogs, and swamps) is still in its infancy, as ICESat-2 and GEDI sensors are not optimized to these types of ecosystems. Ilangakoon (2020), using GEDI measurements from rangeland sites in Idaho, highlighted its capability to capture trends and patterns of functional diversity in semi-arid ecosystems and potential to identify critical biophysical and ecological shifts in similar ecosystems, which can be useful for monitoring changes in carbon-cycle dynamics, habitats and biodiversity. In African savanna ecosystems, Li et al. (2023) concluded that GEDI data products could not estimate canopy heights of shrubs below 2.34 m but accurately estimated the canopy height of trees between 3 and 15 m. In comparison studies ICESat-2 exhibited greater precision in terrain height retrieval than GEDI, contradicting previous observations favoring GEDI in forested regions over ICESat-2. This disparity can be attributed to the challenges faced by GEDI in distinguishing ground and canopy returns, especially in areas with short-stature vegetation (Zhu et al., 2023; Latifi, 2023; Singh et al., 2023). According to Liu et al. (2021) GEDI still outperforms

ICESat-2 for canopy height measurements, because ICESat-2 tends to overestimate the canopy height of dwarf shrublands and underestimate the canopy height of forest ecosystems.

11.5 CONCLUSIONS

Rangelands are found extensively throughout the world covering about 50% of the global land mass. The remoteness, harsh conditions, and high interannual variation in productivity make remote sensing the most cost effective and efficacious tool for evaluating the status and health of rangelands globally. Global remote sensing has unique constraints from a remote sensing perspective and spatial resolution is often sacrificed in place of temporal resolution. A broad suite of sensors possessing various spectral channels, revisit times, and spatial resolutions are available for regional to global rangeland applications. However, most global applications, especially those sponsored for national or international applications (e.g., LADA, IGBP), use AVHRR, SPOTVGT, MODIS, and to a lesser degree TM. Additionally, many biophysical phenomena can be investigated with the myriad of sensors, but as discussed in this chapter, we focused on the globally relevant issues of degradation, fire, land cover, food security, and global change. In this chapter, we demonstrate sensors, data, algorithms, strengths, and limitations of various methods to address these globally significant issues. Though estimates vary, the proportion of degraded rangelands is around 23% globally (Table 11.4). The use and interpretation of RUE for evaluating degradation patterns is controversial (Prince et al., 2007), but alternative techniques are subject to similar issues and assumptions. Thus, when considering degradation, especially in a global context, a model ensemble approach (e.g., combine local NPP scaling, rainfall use efficiency, and NPP trend analysis) may be most useful to indicate trends and identify where action is needed to lessen detrimental effects on goods and services.

Most global land cover efforts have limited thematic resolution of rangeland classes (average number of rangeland classes is 4.75; Table 11.6). However, computational resources and algorithmic complexity is sufficient to produce higher spatial and thematic resolution land cover maps as inaugurated by studies such as Gong et al. (2013) and Hansen et al. (2013). Land cover and land use will continue to evolve in response to broad-scale disturbance and global change. As a result, monitoring global change and extent and severity of fire has been the focus of many algorithms, national programs, and sensors. As an example, the MODIS sensor aboard both the Terra and Aqua platforms was designed with fire monitoring in mind with channels 21, 22, 31, and 33, but given the sensor drift and lifespan, MODIS is being replaced by VIIRS and other sensors. Burn severity evaluation is a relatively new capability since the AVHRR and SPOTVGT sensors lack the spectral channels necessary for contemporary algorithms. Likewise, the advent of the MODIS-derived NPP product (Running et al., 2004)—has spawned numerous studies aimed at evaluating NPP patterns globally. More recently, regional and national tools that enable visualization of aboveground annual production and fractional abundance of various lifeforms has been made possible with Google Earth Engine through the Rangeland Analysis Platform. In this chapter, we demonstrate that from 1982 to 2020, rangelands have become more productive, on average increasing in GPP by 1.7 ± 2.4 g C m^{-2} $y^{-1.}$ Despite increasing temporal trajectories globally, drought and degradation are detrimental on a regional basis and regularly threaten the security of food derived from rangelands, especially since much of the greening around the Earth is due to increasing abundance of woody species. In this vein, tools like the LEWS, driven by MODIS derived 250 m NDVI, is a useful program to provide guidance local governments and international aid organizations. As world population continues to grow, it is likely that the programs like LEWS will become increasingly important. These issues emphasize the critical importance of mission and spectral continuity.

REFERENCES

AlBakri, J. T., J. C. Taylor. 2003. Application of NOAA AVHRR for monitoring vegetation conditions and biomass in Jordan. *Journal of Arid Environments*, 54, 579–593.

Alexandratos, N., J. Bruinsma. 2012. *World Agriculture Towards 2030/2050: The 2012 Revision Ch.4(ESA/12-03, FAO,2012)*. www.fao.org/3/ap106e/ap106e.pdf.

Allbed, A., L. Kumar. 2013. Soil salinity mapping and monitoring in arid and semiarid regions using remote sensing technology: A review. *Advances in Remote Sensing*, 2, 373–385.

Allen, J. L., B. Sorbel. 2008. Assessing the differenced Normalized Burn Ratio's ability to map burn severity in the boreal forest and tundra ecosystems of Alaska's national parks. *International Journal of Wildland Fire*, 17, 463–475.

Allred, B. W., B. T. Bestelmeyer, C. S. Boyd, C. Brown, K. W. Davies, M. C. Duniway, L. M. Ellsworth, T. A. Erickson, S. D. Fuhlendorf, T. V. Griffiths, V. Jansen, M. O. Jones, J. Karl, A. Knight, J. D. Maestas, J. J. Maynard, S. E. McCord, D. E. Naugle, H. D. Starns, D. Twidwell, D. R. Uden. 2021. Improving Landsat predictions of rangeland fractional cover with multitask learning and uncertainty. *Methods in Ecology and Evolution*. https://doi.org/10.1111/2041-210X.13564.

An, N., K. P. Price, J. M. Blair. 2013. Estimating aboveground net primary productivity of the tallgrass prairie ecosystem of the Central Great Plains using AVHRR NDVI. *International Journal of Remote Sensing*, 34, 3717–3735.

Anderson, K. E., N. F. Glenn, L. P. Spaete, D. J. Shinneman, D. S. Pilliod, R. S. Arkle, S. K. McIlroy, D. R. Derryberry. 2018. Estimating vegetation biomass and cover across large plots in shrub and grass dominated drylands using terrestrial lidar and machine learning. *Ecological Indicators*, 84, 793–802. https://doi.org/10.1016/j.ecolind.2017.09.034.

Angerer, J. P. 2008. Examination of high-resolution rainfall products and satellite greenness indices for estimating patch and landscape forage biomass. PhD Dissertation. College Station, TX: Texas A&M University.

Angerer, J. P. 2012. Gobi forage livestock early warning system. In *Conducting National Feed Assessments*, eds. M. B. Coughenour, H. P. S. Makkar, pp. 115–130. Rome, Italy: Food and Agriculture Organization.

Arino, O., S. Casadio, D. Serpe. 2012. Global nighttime fire season timing and fire count trends using the ATSR instrument series. *Remote Sensing of Environment*, 116, 226–238.

Arino, O., D. Gross, F. Ranera, L. Bourg, M. Leroy, P. Bicheron, R. Whit. 2007. Globcover: ESA service for global land cover from MERIS. In *International Geoscience and Remote Sensing Symposium*, pp. 2412–2415. Barcelona, Spain.

Awuma, K. S., J. W. Stuth, R. Kaitho, J. Angerer. 2007. Application of Normalized Differential Vegetation Index and geostatistical techniques in cattle diet quality mapping in Ghana. *Outlook on Agriculture*, 36, 205–213.

Baccini, A., M. A. Friedl, C. E. Woodcock, Z. Zhu. 2007. Scaling field data to calibrate and validate moderate spatial resolution remote sensing models. *Photogrammetric Engineering and Remote Sensing*, 73, 945–954.

Bai, E., T. W. Boutton, X. B. Wu, F. Liu, S. R. Archer. 2009. Landscape scale vegetation dynamics inferred from spatial patterns of soil δ13C in a subtropical savanna parkland. *Journal of Geophysical Research*, 114, G01019.

Bai, Z. G., D. L. Dent, L. Olsson, M. E. Schaepman. 2008. Proxy global assessment of land degradation. *Soil Use and Management*, 24, 223–234.

Barbosa, P. M., J. M. Grégoire, J. M. C. Pereira. 1999. An algorithm for extracting burned areas from time series of AVHRR GAC data applied at a continental scale. *Remote Sensing of Environment*, 69, 253–263.

Barnetson, J., S. Phinn, P. Scarth. 2020. Estimating plant pasture biomass and quality from UAV imaging across queensland's rangelands. *AgriEngineering*, 2, 523–543.

Bartholomé, E., A. S. Belward. 2005. GLC2000: A new approach to global land cover mapping from Earth observation data. *International Journal of Remote Sensing*, 26, 1959–1977. https://doi.org/10.1080/01431160412331291297.

Bauer, M. E. 1976. Technological basis and applications of remote sensing of the Earth's resources. *IEEE Transactions on Geoscience Electronics*, 14, 3–9.

Beatriz, A., F. Valladares. 2008. International efforts on global change research. In *Earth Observation of Global Change: The Role of Satellite Remote Sensing in Monitoring the Global Environment*, ed. E. Chuvieco, pp. 1–21. Dordrecht, the Netherlands: Springer.

Belenguer-Plomer, M. A., A. Tanase Fernandez-Carrillo, E. Chuvieco. 2019. Burned area detection and mapping using Sentinel-1 backscatter coefficient and thermal anomalies. *Remote Sensing of Environment*, 233, 111345.

Benedict, T. D., J. F. Brown, S. P. Boyte et al. 2021. Exploring VIIRS continuity with MODIS in an expedited capability for monitoring drought-related vegetation conditions. *Remote Sensing*, 13. https://doi.org/10.3390/rs13061210.

Blanco, P., H. D. Valle, P. Bouza, G. Metternicht, L. Hardtke. 2014. Ecological site classification of semiarid range lands: Synergistic use of Landsat and Hyperion imagery. *International Journal of Applied Earth Observation and Geoinformation*, 29, 11–21.

Bolton, D. K., J. M. Gray, E. K. Melaas, M. Moon, L. Eklundh, M. A. Friedl. 2020. Continental scale land surface phenology from harmonized Landsat 8 and Sentinel-2 imagery. *Remote Sensing ofEnvironment*, 240, 111685.

Brown, C. F., S. P. Brumby, B. Guerrillas, T. Birch, S. B. Hyde, J. Mazzariello, W. Czerwinski, V. J. Pasquarella, R. Haertel, S. Ilyushchenko, K. Schwehr, M. Weisse, F. Stolle, C. Hanson, O. Guinan, R. Moore, A. M. Tait. 2022. Dynamic world, near realtime global 10m land use land cover mapping. *ScientificData*, 9, 251.

Cao, S., M. Li, Z. Zhu, J. Zha, W. Zhao, Z. Duanmu, Y. Chen. 2023. Spatiotemporally consistent global dataset of the GIMMS Leaf Area Index (GIMMS LAI4g) from 1982 to 2020. *Earth System Science Data Discussions*, 1–31. https://doi.org/10.5194/essd-2023-68.

Carreiras, A., M. Vasconcelos, R. Lucas. 2012. Understanding the relationship between aboveground biomass and ALOS PALSAR data in the forests of Guinea-Bissau (West Africa). *Remote Sensing of Environment*, 121, 426–442.

Carter, L. J. 1969. Earth resources satellite: Finally off the ground? *Science*, 163, 796–798.

Chakrabortty, R., S. C. Pal. 2023. Systematic review on gully erosion measurement, modelling and management: Mitigation alternatives and policy recommendations. *Geological Journal*, 58, 3544–3576. https://doi.org/10.1002/gj.4709.

Chen, D., C. Fu, J. Hall, E. E. Hoy, T. V. Loboda. 2021. Spatio-temporal patterns of optimal Landsat data for burn severity index calculations: Implications for high northern latitudes wildfire research. *Remote Sensing of Environment*, 258, 112393.

Chen, D., T. V. Loboda, J. V. Hall. 2020. A systematic evaluation of influence of image selection process on remote sensing-based burn severity indices in North American boreal forest and tundra ecosystems. *ISPRS Journal of Photogrammetry and Remote Sensing*, 159, 63–77. https://doi.org/10.1016/j.isprsjprs.2019.11.011.

Chuvieco, E., R. G. Congalton. 1988. Mapping and inventory of forest fires from digital processing of TM data. *Geocarto International*, 3, 41–53.

Chuvieco, E., D. Riano, I. Aguado, D. Cocero. 2002. Estimation of fuel moisture content from multitemporal analysis of Landsat Thematic Mapper reflectance data: Applications in fire danger assessment. *International Journal of Remote Sensing*, 23, 2145–2162.

Costanza, R., R. Groot, P. Sutton, S. Van der Ploeg, S. Anderson, S. Farber, R. Turner. 2014. Changes in the global value of ecosystem services. *Global Environmental Change*, 26, 152–158.

Crist, E. P., R. C. Cicone. 1984. Application of the tasseled cap concept to simulated Thematic Mapper data. *Photogrammetric Engineering and Remote Sensing*, 50, 343–352.

Deering, D. W., J. W. Rouse, R. H. Haas, J. A. Schell. 1975. Measuring forage production of grazing units from Landsat MSS data. In *Tenth International Symposium on Remote Sensing of Environment*, pp. 1169–1178. Ann Arbor, MI: University of Michigan.

DeFries, R. S., M. Hansen, J. R. G. Townshend, R. Solberg. 1998. Global land cover classifications at 8 km spatial resolution: The use of training data derived from Landsat imagery in decision tree classifiers. *International Journal of Remote Sensing*, 19, 3141–3168.

DeFries, R. S., J. R. G. Townshend. 1994. NDVI derived land cover classifications at a global scale. *International Journal of Remote Sensing*, 15, 3567–3586.

DeFries, R. S., J. R. G. Townshend, M. C. Hansen. 1999. Continuous fields of vegetation characteristics at the global scale at 1km resolution. *Journal of Geophysical Research:Atmospheres*, 104, 16911–16923.

Dong, J., X. Xiao, S. Sheldon, C. Biradar, G. Zhang, N. D. Duong, M. Hazarika, K. Wikantika et al. 2014. A 50m forest cover map in southeast Asia from ALOS/PALSAR and its application on forest fragmentation assessment. *PLoS One*, 9, e8580. https://doi.org/10.1371/journal.pone.0085801.

Dungan, J. 1998. Spatial prediction of vegetation quantities using ground and image data. *Remote Sensing*, 19, 267–285.

Dwyer, E., S. Pinnock, J. M. Gregorie, J. M. C. Pereira. 2000. Global spatial and temporal distribution of vegetation fire as determined from satellite observations. *International Journal of Remote Sensing*, 21, 1289–1302.

Eidenshink, J. C., J. L. Faundeen. 1994. The 1 km AVHRR global land data set: First stages in implementation. *International Journal of Remote Sensing*, 104, 3443–3462.

Eidenshink, J. C., B. Schwind, K. Brewer, Z. Zhu, B. Quayle, S. Howard. 2007. A project for monitoring trends in burn severity. *Fire Ecology*, 3, 3–19.

Elvidge, C. D., R. J. P. Lyon. 1985. Influence of rock soil spectral variation on the assessment of green biomass. *Remote Sensing of Environment*, 17, 265–279.

Environmental Systems Research Institute (ESRI). 2023. *Sentinel2 10meter Land Use/Land Cover*. https://livingatlas.arcgis.com/landcover/ [accessed 13 Oct., 2023].

Estes, J., A. Belward, T. Loveland, J. Scepan, A. Strahler, J. Townshend, C. Justice. 1999. The way forward. *Photogrammetric Engineering and Remote Sensing*, 65, 1089–1093.

Everitt, J. H., C. Yang, R. Fletcher, C. J. Deloach. 2008. Comparison of QuickBird and SPOT 5 satellite imagery for mapping giant reed. *Journal of Aquatic Plant Management*, 46, 77–82.

Fang, H., W. Ping. 2010. Vegetation change of ecotone in west of northeast China plain using timeseries remote sensing data. *Chinese Geographical Society*, 20, 167–175.

Fava, F., R. Colombo, S. Bocchi, C. Zucca. 2012. Assessment of Mediterranean pasture condition using MODIS Normalized Difference Vegetation Index time series. *Journal of Applied Remote Sensing*, 6, 063530106353012.

Fensholt, R. I., S. R. Proud. 2012. Evaluation of Earth observation based global long term vegetation trends—comparing GIMMS and MODIS global NDVI time series. *Remote Sensing of Environment*, 119, 131–147.

Fensholt, R. I., K. Rasmussen. 2011. Analysis of trends in the Sahelian 'rain use efficiency' using GIMMS NDVI, RFE and GPCP rain fall data. *Remote Sensing of Environment*, 115, 438–451.

Fensholt, R. I., K. Rasmussen, T. T. Nielsen, C. Mbow. 2009. Evaluation of earth observation based long term vegetation trends inter-comparing NDVI time series trend analysis consistency of Sahel from AVHRR GIMMS, Terra MODIS and SPOT VGT data. *Remote Sensing of Environment*, 113, 1886–1898.

Fensholt, R. I., M. S. Sandholt, S. Rasmussen, S. Stisen, A. Diou. 2006. Evaluation of satellite based primary production modeling in the semiarid Sahel. *Remote Sensing of Environment*, 105, 173–188.

Flasse, S., P. Ceccato. 1996. A contextual algorithm for AVHRR fire detection. *International Journal of Remote Sensing*, 17, 419–424.

Foley, J., R. DeFries, G. P. Asner. 2005. Global consequences of land use. *Science*, 309, 570–574.

Food and Agriculture Organization (FAO), Intergovernmental Authority on Development (IGAD). 2019. *East Africa Animal Feed Action Plan*, p. 50. Rome: Food and Agriculture Organization of the United Nations (FAO) and Intergovernmental Authority on Development (IGAD). www.fao.org/3/ca5965en/ca5965en.pdf [accessed 31 Oct., 2023].

Foody, G. M. 2003. Geographical weighting as a further refinement to regression modelling: An example focused on the NDVI—rainfall relationship. *Remote Sensing of Environment*, 88, 283–293.

Fox, W. E., D. W. McCollum, J. E. Mitchell, L. E. Swanson, U. P. Kreuter, J. A. Tanaka, G. R. Evans, H. T. Heintz. 2009. An Integrated Social, Economic, and Ecologic Conceptual (ISEEC) framework for considering rangeland sustainability. *Society and Natural Resources*, 22, 593–606.

Frank, T. D. 1984. Assessing change in the surficial character of a semiarid environment with Landsat residual images. *Photogrammetric Engineering and Remote Sensing*, 50, 471–480.

French, N. H. F., J. L. Allen, R. J. Hall, E. E. Hoy, E. S. Kasischke, K. A. Murphy, D. L. Verbyla. 2008. Using Landsat data to assess fire and burn severity in the North American boreal forest region: An overview. *International Journal of Wildland Fire*, 17, 443–462.

Frey, K. E., L. C. Smith. 2007. How well do we know northern land cover? Comparison of four global vegetation and wet land products with a new ground truth database for West Siberia. *Global Biogeochemical Cycles*, 21, 1–15.

Friedl, M. A., D. K. McIver, J. C. F. Hodges, X. Y. Zhang, D. Muchoney, A. H. Strahler, C. E. Woodcock, S. Gopal et al. 2002. Global land cover mapping from MODIS: Algorithms and early results. *Remote Sensing of Environment*, 83, 287–302.

Friedl, M. A., D. Sulla-Menashe, B. Tan, A. Schneider, N. Ramankutty, A. Sibley, X. M. Huang. 2010. MODIS Collection 5 global land cover: Algorithm refinements and characterization of new datasets. *Remote Sensing of Environment*, 114, 168–182.

Fritz, S., C. C. Fonte, L. See. 2017. The role of citizen science in earth observation. *Remote Sensing*, 9, 357. https://doi.org/10.3390/rs9040357.

Fritz, S., I. McCallum, C. Schill, C. Perger, L. See, D. Schepaschenko, M. van der Velde, F. Kraxner, M. Obersteiner. 2011a. GeoWiki: An online platform for improving global land cover. *Environmental Modelling andSoftware*, 31, 110123.

Fritz, S., L. See. 2007. Identifying and quantifying uncertainty and spatial disagreement in the comparison of Global Land Cover for different applications. *Global Change Biology*, 14, 1057–1075.

Fritz, S., L. See, F. Grey. 2022. The grand challenges facing environmental citizen science. *Frontiers of Environmental and Science*, 10. https://doi.org/10.3389/fenvs.2022.1019628.

Fritz, S., L. See, I. McCallum, C. Schill, M. Obersteiner, M. van der Velde, H. Boettcher, P. Havlík et al. 2011b. Highlighting continued uncertainty in global land cover maps for the user community. *Environmental Research Letters*, 6, 1–7.

Fry, J., G. Xian, S. Jin, J. Dewitz, C. Homer, L. Yang, C. Barnes, N. Herold et al. 2011. Completion of the 2006 national land cover database for the conterminous United States. *Photogrammetric Engineering and Remote Sensing*, 77, 858–864.

Gaffney, R., D. J. Augustine, S. P. Kearney, L. M. Porensky. 2021. Using hyperspectral imagery to characterize rangeland vegetation composition at process-relevant scales. *Remote Sensing*, 13, 4603. https://doi.org/10.3390/rs13224603.

Gallo, K., L. Li, B. Reed, J. Eidenshink, J. Dwyer. 2005. Multiplatform comparisons of MODIS and AVHRR Normalized Difference Vegetation Index data. *Remote Sensing of Environment*, 99, 221–231.

Ganguly, S., A. Samanta, M. A. Schull, N. V. Shabanov, C. Milesi, R. R. Nemani, Y. Knyazikhin, R. B. Myneni. 2008. Generating vegetation leaf area index Earth system data record from multiple sensors. Part 2: Implementation, analysis and validation. *Remote Sensing of Environment*, 112, 4318–4332.

Geerken, R., N. Batikha, D. Celisj, E. Depauw. 2005. Differentiation of rangeland vegetation and assessment of its status: Field investigations and MODIS and SPOT VEGETATION data analyses. *International Journal of Remote Sensing*, 26, 4499–4526.

Giglio, L., L. Boschetti, D. P. Roy, M. L. Humber, C. O. Humber. 2018. The Collection 6 MODIS burned area mapping algorithm and product. *Remote Sensing of Environment*, 217, 72–85.

Giglio, L., J. Descloitres, C. O. Justice, Y. J. Kaufman. 2003. An enhanced contextual fire detection algorithm for MODIS. *Remote Sensing of Environment*, 87, 272–282.

Giglio, L., T. Loboda, D. P. Roy, B. Quayle, C. O. Justice. 2009. An active fire based burned area mapping algorithm for the MODIS sensor. *Remote Sensing of Environment*, 113, 408–420.

Gilbert, N. 2011. Science enters desert debate. *Nature*, 447, 262.

Giri, G., Z. L. Zhu, B. Reed. 2005. A comparative analysis of the global land cover 2000 and MODIS land cover data sets. *Remote Sensing of Environment*, 94, 123–132.

Gong, P., J. Wang, L. Yu, Y. Zhao, Y. Zhao, L. Liang, Z. Niu, X. Huang et al. 2013. Finer resolution observation and monitoring of global land cover: First mapping results with Landsat TM and ETM+ data. *International Journal of Remote Sensing*, 34, 2607–2654.

Graetz, R. D. 1987. Satellite remote sensing of Australian range lands. *Remote Sensing of Environment*, 23, 313–331.

Graetz, R. D., D. M. Carneggie, R. Hacker, C. Lendon, D. G. Wilcox. 1976. A qualitative evaluation of Landsat imagery of Australia rangelands. *Australian Rangeland Journal*, 1, 53–59.

Graetz, R. D., M. R. Gentle, R. P. Pech, J. F. O'Callaghan, G. Drewien. 1983. The application of Landsat image data to rangeland assessment and monitoring: An example from South Australia. *Australian Rangeland Journal*, 5, 63–73.

Graetz, R. D., R. P. Pech, A. W. Davis. 1988. The assessment and monitoring of sparsely vegetated rangelands using calibrated Landsat data. *International Journal of Remote Sensing*, 9, 1201–1222.

Graetz, R. D., R. P. Pech, M. R. Gentle, J. F. Ocallaghan. 1986. The application of Landsat image data to rangeland assessment and monitoring—the development and demonstration of a Land Image-Based Resource Information System (LIBRIS). *Journal of Arid Environments*, 10, 53–80.

Gray, T. I., D. G. McCrary. 1981. The environmental vegetation index, a tool potentially useful for arid land management. *AgRISTAR Report No. EWN104076*, p. 7. Houston, TX: Johnson Space Center, 17132.

Hacker, R. 1980. Prospects for satellite applications in Australian rangelands. *Tropical Grasslands*, 14, 289.

Hansen, M. C., R. S. DeFries, J. R. G. Townshend, M. Carroll, C. Dimiceli, R. A. Dimiceli. 2003. Global percent tree cover at a spatial resolution of 500 meters: First results of the MODIS vegetation continuous fields algorithm. *Earth Interactions*, 7, 1–15.

Hansen, M. C., R. S. DeFries, J. R. G. Townshend, R. Sohlberg. 2000. Global land cover classification at the 1 km spatial resolution using a classification tree approach. *International Journal of Remote Sensing*, 21, 1331–1364.

Hansen, M. C., T. R. Loveland. 2011. A review of large area monitoring of land cover change using Landsat data. *Remote Sensing of Environment*, 122, 66–74.

Hansen, M. C., P. V. Potapov, R. Moore, M. Hancher, S. A. Turubanova, A. Tyukavina, D. Thau, S. V. Stehman et al. 2013. High resolution global maps of 21stcentury forest cover change. *Science*, 342, 850–853.

Hansen, M. C., P. V. Potapov, A. H. Pickens, A. Tyukavina, A. Hernandez-Serna, V. Zalles, S. Turubanova, I. Kommareddy, S. Stehman, X. P. Song. 2022. Global land use extent and dispersion within natural land cover using Landsat data. *Environmental Research Letters*, 17, 034050. https://doi.org/10.1088/1748-9326/ac46ec.

Hansen, M. C., B. Reed. 2000. A comparison of the IGBP DISCover and University of Maryland 1 km global land cover products. *International Journal of Remote Sensing*, 21, 1365–1373.

Hillman, S., L. Wallace, A. Lucieer, K. Reinke, D. Turner, S. Jones. 2021. A comparison of terrestrial and UAS sensors for measuring fuel hazard in a dry sclerophyll forest. *International Journal of Applied Earth Observation and Geoinformation*, 95, 102261. https://doi.org/10.1016/j.jag.2020.102261.

Holmgren, M., M. Scheffer. 2001. El Niño as a Window of Opportunity for the Restoration of Degraded arid ecosystems. *Ecosystems*, 4, 151–159.

Holmgren, M., P. Stapp, C. R. Dickman, C. Gracia, S. Graham, J. R. Gutiérrez, C. Hice, F. Jaksic, D. A. Kelt, M. Letnic, M. Lima, B. C. López, P. L. Meserve, W. B. Milstead, G. A. Polis, M. A. Previtali, M. Richter, S. Sabaté, F. A. Squeo. 2006. Extreme climatic events shape arid and semiarid ecosystems. *Frontiers in Ecology and the Environment*, 4, 87–95.

Homer, C. G., R. D. Ramsey, T. C. Edwards, A. Falconer. 1997. Landscape covertype modeling using a multi scene thematic mapper mosaic. *Photogrammetric Engineering and Remote Sensing*, 63, 59–67.

Hosseini, M., S. Lim. 2023. Burned area detection using Sentinel-1 SAR data: A case study of Kangaroo Island, South Australia. *Applied Geography*, 151, 102854.

Huete, A. R. 1988. A Soil Adjusted Vegetation Index (SAVI). *Remote Sensing of Environment*, 25, 295–309.

Huete, A. R., K. Didan, T. Miura, E. P. Rodriguez, X. Gao, L. G. Ferreira. 2002. Overview of the radiometric and biophysical performance of the MODIS vegetation indices. *Remote Sensing of Environment*, 83, 195–213.

Hunt, E. R., B. A. Miyake. 2006. Comparison of stocking rates from remote sensing and geospatial data. *Rangeland Ecology andManagement*, 59, 11–18.

Hunt, E. R., B. N. Rock. 1989. Detection of changes in leaf water content using near and middle infrared reflectances. *Remote Sensing of Environment*, 30, 43–54.

Ilangakoon, G. Y. M. N. T. 2020. Complexity and dynamics of semi-arid vegetation structure, function and diversity across spatial scales from full waveform lidar. PhD Dissertation. Boise, ID: Boise State University.

Ito, A. 2011. A historical meta-analysis of global terrestrial net primary productivity: Are estimates converging? *Global Change Biology*, 17, 3161–3175.

Jabbar, M. T., J. Zhou. 2013. Environmental degradation assessment in arid areas: A case study from Basra Province, southern Iraq. *Environmental Earth Sciences*, 70, 2203–2214.

Jarlan, L., S. Mangiarotti, E. Mougin, P. Mazzega, P. Hiernaux, V. L. Dantec. 2008. Assimilation of SPOT/VEGETATION NDVI data into a Sahelian vegetation dynamics model. *Remote Sensing of Environment*, 112, 1381–1394.

Jenkins, G., K. Mohr, V. Morris, O. Arino. 1997. The role of convective processes over the Zaire Congo basin to the southern hemispheric ozone maximum. *Journal of Geophysical Research*, 102, 18963–18980.

Ji, L., K. Gallo. 2006. An agreement coefficient for image comparison. *Photogramm Engineering and Remote Sensing*, 72, 823–833.

Jinguo, Y., N. Zheng, C. Wang. 2006. Vegetation NPP distribution based on MODIS data and CASA model—a case study of northern Hebei Province. *Chinese Geographical Science*, 16, 334–341.

Jung, M., K. Henkel, M. Herold, G. Churkina. 2006. Exploiting synergies of global land cover products for carbon cycle modeling. *Remote Sensing of Environment*, 101, 534–553.

Kachergis, E., S. W. Miller, S. E. McCord, M. Dickard, S. Savage, L. V. Reynolds, N. Lepak, C. Dietrich, A. Green, A. Nafus, K. Prentice, Z. Davidson. 2022. Adaptive monitoring for multiscale land management: Lessons learned from the Assessment, Inventory, and Monitoring (AIM) principles. *Rangelands*, 44, 50–63. https://doi.org/10.1016/j.rala.2021.08.006000609802.

Kang, Y., M. Gaber, M. Bassiouni, X. Lu, T. Keenan. 2023. CEDAR-GPP: Spatiotemporally upscaled estimates of gross primary productivity incorporating CO 2 fertilization. *Earth System Science Data Discussions*, 2023, 1–51.

Kaufman, Y. J., C. O. Justice, L. P. Flynn, J. D. Kendall, E. M. Prins, L. Giglio, D. E. Ward, W. P. Menzel et al. 1996a. Potential global fire monitoring from EOSMODIS. *Journal of Geophysical Research: Atmospheres*, 103, 32215–32238.

Kaufman, Y. J., C. O. Justice, L. P. Flynn, J. D. Kendall, E. M. Prins, L. Giglio, D. E. Ward, W. P. Menzel, A. W. Setzer. 1998. Potential global fire monitoring from EOS-MODIS. *Journal of Geophysical Research*, 103(D24), 32215–32238. https://doi.org/10.1029/98JD01644.

Kaufman, Y. J., L. A. Remer, R. D. Ottmar, D. E. Ward, R. R. Li, R. Kleidman, R. S. Fraser, L. Flynn et al. 1996b. Relationship between remotely sensed fire intensity and rate of emission of smoke: SCARC experiment. In *Global Biomass Burning*, ed. J. Levin, pp. 685–696. Cambridge, MA: The MIT Press.

Kauth, R. J., G. S. Thomas. 1976. The tasseled cap—a graphic description of the spectral temporal development of agricultural crops as seen by Landsat. In *LARS Symposia*, Paper 159, pp. 4B41–4B51. West Lafayette, IN: Purdue ePubs, Purdue University.

Kawamura, K., T. Akiyama, H. Yokota, M. Tsutsumi. 2005. Monitoring of forage conditions with MODIS imagery in the Xilingol steppe, Inner Mongolia. *International Journal of Remote Sensing*, 26, 1423–1436.

Key, C. H., N. C. Benson. 1999. Measuring and remote sensing of burn severity. In *Proceedings of the Joint Fire Science Conference*, eds. L. F. Neuenschwander, K. C. Ryan, G. E. Goldberg, Vol. II, p. 284. Boise, ID: University of Idaho and the International Association of Wildland Fire.

Key, C. H., N. C. Benson. 2006. Landscape assessment: Ground measure of severity, the Composite Burn Index, and remote sensing of severity, the Normalized Burn Index. In *FIREMON: Fire Effects Monitoring and Inventory System,For. Serv. Gen. Tech. Rep. RMRSGTR164*, eds. D. C. Lutes et al., pp. CD: LA1–LA51. Ogden, UT: USDA For. Serv. Rocky Mt. Res. Stn.

Klemm, T., D. D. Briske, M. C. Reeves. 2020. Vulnerability of rangeland beef cattle production to climate-induced NPP fluctuations in the US Great Plains. *Global Change Biology*, 26, 4841–4853.

Knopp, L., M. Wieland, M. Rättich, S. Martinis. 2020. A deep learning approach for burned area segmentation with Sentinel-2 data. *Remote Sensing*, 12(15), 2422.

Koch, A., A. McBratney, M. Adams, D. Field, R. Hill, J. Crawford, B. Minasny, R. Lal et al. 2013. Soil security: Solving the global soil crisis. *Global Policy*, 4, 434–444.

Kohl, H. A., P. V. Nelson, J. Pring, K. L. Weaver, D. M. Wiley, A. B. Danielson, R. M. Cooper, H. Mortimer, D. Overoye, A. Burdick, S. Taylor, M. Haley, S. Haley, J. Lange, M. E. Lindblad. 2021. GLOBE observer and the GO on a trail data challenge: A citizen science approach to generating a global land cover land use reference dataset. *Frontiers in Climate*, 3. https://doi.org/10.3389/fclim.2021.620497.

Ku, N.-W., S. C. Popescu. 2019. A comparison of multiple methods for mapping local-scale mesquite tree aboveground biomass with remotely sensed data. *Biomass and Bioenergy*, 122, 270–279. https://doi.org/10.1016/j.biombioe.2019.01.045.

Lambin, E. F., B. L. Turner, J. Helmut. 2001. The causes of land use and landcover change: Moving beyond the myths. *Global EnvironmentalChange*, 11, 261–269.

Latifi, H., R. Valbuena, C. A. Silva. 2023. Towards complex applications of active remote sensing for ecology and conservation. *Methods in Ecology and Evolution*, 14(7), 1578–1586.

Lawley, V., M. Lewis, K. Clarke, B. Ostendorf. 2016. Site-based and remote sensing methods for monitoring indicators of vegetation condition: An Australian review. *Ecological Indicators*, 60, 1273–1283. https://doi.org/10.1016/j.ecolind.2015.03.021.

Le, Q. B., L. Tamene, P. L. G. Vlek. 2012. Multipronged assessment of land degradation in West Africa to assess the importance of atmospheric fertilization in masking the processes involved. *Global PlanetaryChange*, 92, 71–81.

Le Houérou, H. N. 1984. Rain use efficiency: A unifying concept in arid land ecology. *Journal of Arid Environments*, 7, 213–247.

Le Houérou, H. N., R. L. Bingham, W. Skerbek. 1988. Relationship between the variability of primary production and the variability of annual precipitation in world arid lands. *Journal of Arid Environments*, 15, 1–18.

Lepers, E., E. F. Lambin, A. C. Janetos, R. DeFries, F. Achard, N. Ramankutty, R. J. Scholes. 2005. A synthesis of information on rapid landcover change for the period 1981–2000. *BioScience*, 55, 115–124.

Levitus, S., J. I. Antonov, T. P. Boyer, C. Stephens. 2000. Warming of the world ocean. *Science*, 287, 2225–2229.

Li, S., C. Potter, C. Hiatt. 2012. Monitoring of net primary production in California rangelands using Landsat and MODIS satellite remote sensing. *Natural Resources*, 3(2), 10.

Li, X., K. J. Wessels, S. Armston, P. Hancock, R. Mathieu, R. Main, R. Scholes. 2023. First validation of GEDI canopy heights in African savannas. *Remote Sensing of Environment*, 285, 113402.

Liu, A., X. Cheng, Z. Chen. 2021. Performance evaluation of GEDI and ICESat-2 laser altimeter data for terrain and canopy height retrievals. *Remote Sensing of Environment*, 264, 112571.

Loboda, T. V., N. H. F. French, C. Hight-Harf, L. Jenkins, M. E. Miller. 2013. Mapping fire extent and burn severity in Alaskan tussock tundra: An analysis of the spectral response of tundra vegetation to wildland fire. *Remote Sensing of Environment*, 134, 194–209.

Loboda, T. V., L. Giglio, L. Boschetti, C. O. Justice. 2012. Regional fire monitoring and characterization using global NASA MODIS fire products in dry lands of Central Asia. *Frontiers in Earth Science*, 6, 196–205.
Lopez-Garcia, M. J., V. Caselles. 1991. Mapping burns and natural reforestation using Thematic Mapper data. *Geocarto International*, 6, 31–37.
Loveland, T. R., B. C. Reed, J. F. Brown, D. O. Ohlen, L. Yang, J. W. Merchant. 2000. Development of a global land cover characteristics database and IGBP DISCover from 1 km AVHRR data. *International Journal of Remote Sensing*, 29, 3–10.
Lu, B., Y. He, H. Liu. 2016. Investigating species composition in a temperate grassland using unmanned aerial vehicle-acquired imagery. In *20164th International Workshop on Earth Observation and Remote Sensing Applications (EORSA). Presented at the 2016 4th International Workshop on Earth Observation and Remote Sensing Applications (EORSA)*, pp. 107–111. Guangzhou, China: IEEE. https://doi.org/10.1109/EORSA.2016.7552776.
Ludwig, J. 1987. Primary productivity in arid lands: Myths and realities. *Journal of Arid Environments*, 13, 1–7.
Lund, G. H. 2007. Accounting for the worlds rangelands. *Rangelands*, 29, 3–10.
Marini, A., M. Talbi. 2009. *Desertification and Risk Analysis Using High and Medium Resolution Satellite Data*, p. 271. Dordrecht, the Netherlands: Springer.
Matere, J., P. Simpkin, J. Angerer, E. Olesambu, S. Ramasamy, F. Fasina. 2019. Predictive Livestock Early Warning System (PLEWS): Monitoring forage condition and implications for animal production in Kenya. *Weather and Climate Extremes*, 27, 1–8.
Matthews, E. 1983. Global vegetation and land use: New high resolution data bases for climate studies. *Journal of Climatology and Applied Meteorology*, 22, 474–487.
Mayaux, P., H. Eva, J. Gallego, A. H. Strahler, M. Herold, S. Agrawal, S. Naumov, E. E. D. Miranda et al. 2006. Validation of the global land cover 2000 map. *IEEE Transactions on Geoscience and Remote Sensing*, 44, 1728–1738.
McCullum, I., M. Obersteiner, S. Nilsson, A. Shvidenko. 2006. A spatial comparison of four satellite derived 1 km global land cover datasets. *International Journal of Applied Earth Observation and Geoinformation*, 8, 246–255.
McGraw, J. F., P. T. Tueller. 1983. Landsat computer aided analysis techniques for range vegetation mapping. *Journal of Range Management*, 36, 627–631.
Metternicht, G. 1996. *Detecting and Monitoring Land Degradation Features and Processes in the Cochabamba Valleys, Bolivia*. Enschede, the Netherlands: International Institute for GeoInformation Science and Earth Observation, ITC Publication No.36.
Metternicht, G., J. Zinck. 2003. Remote sensing of soil salinity: Potentials and constraints. *Remote Sensing of Environment*, 85, 1–20.
Metternicht, G., J. Zinck. 2009. *Remote Sensing of Soil Salinization—Impact on Land Management*. London, U.K.: CRC Press.
Millennium Ecosystem Assessment. 2005. *Ecosystems and Human Well-Being: Desertification Synthesis*. Washington, DC: Island Press.
Miller, J. D., A. E. Thode. 2007. Quantifying burn severity in a heterogeneous landscape with a relative version of the delta Normalized Burn Ratio (dNBR). *Remote Sensing of Environment*, 109, 66–80.
Miura, T., A. Huete, H. Yoshioka. 2006. An empirical investigation of cross sensor relationships of NDVI and red/near infrared reflectance using EO1 Hyperion data. *Remote Sensing of Environment*, 100, 223–236.
Myneni, R. B., S. Hoffman, Y. Knyazikhin, J. K. Privette, J. Glassy, Y. Tian, Y. Wang, X. Song, Y. Zhang, G. R. Smith, A. Lotsch, M. Friedl, J. T. Morisette, P. Votava, R. R. Nemani, S. W. Running. 2002. Global products of vegetation leaf area and fraction absorbed PAR from year one of MODIS data. *Remote Sensing of Environment*, 83, 214–231.
Myneni, R. B., R. R. Nemani, S. Running. 1997. Estimation of global leaf area index and absorbed PAR using radiative transfer models. *IEEE Transactions on Geoscience and Remote Sensing*, 35, 1380–1393.
Nachtergaele, F., M. Petri, R. Biancalani, G. Van Lynden, H. Van Velthuizen. 2010. Global Land Degradation Information System (GLADIS). Beta Version. An information database for land degradation assessment at global level. *Land Degradation Assessment in Drylands Technical Report*, 17, 1563.
National Wildfire Coordinating Group (NWCG). 2012. *Glossary of Wildland Fire Terminology*. www.nwcg.gov/pms/pubs/glossary/ [accessed 13 Oct., 2023].
Nkonya, E., J. von Braun, A. Mirzabaev, Q. B. Le, H. Y. Kwon, O. Kurui. 2013. Economics of land degradation initiative: Methods and approach for global and national assessments. *ZEF Discussion Papers on Development Policy No. 183*, pp. 40. Bonn, Germany.

Oldeland, J., W. Dorigo, D. Wesuls, N. Jürgens. 2010. Mapping bush encroaching species by seasonal differences in hyper spectral imagery. *Remote Sensing*, 2, 1416–1438.

Oldeland, J., R. Revermann, J. Luther-Mosebach, T. Buttschardt, J. R. K. Lehmann. 2021. New tools for old problems—comparing drone- and field-based assessments of a problematic plant species. *Environmental Monitoring and Assessment*, 193. https://doi.org/10.1007/s10661-021-08852-2.

Oldeman, L. R. 1998. Soil degradation: A threat to food security? *Report 98/01*. Wageningen: International Soil Reference and Information Centre.

Olson, J. S. 1983. *Carbon in Live Vegetation of Major World Ecosystems*, ORNL5862. Environmental Sciences Division Publication No. 1997. Oak Ridge, TN: Oak Ridge National Laboratory.

OpenScientist. 2011. *Finalizing a Definition of "Citizen Science"and "Citizen Scientists."* www.openscientist.org/2011/09/finalizingdefinitionofcitizen.html [accessed 13 Oct., 2023].

Opio, P., H. P. Makkar, M. Tibbo, S. Ahmed, A. Sebsibe, A. M. Osman, E. Olesambu, C. Ferrand, S. Munyua. 2020. Regional animal feed action plan for East Africa: Why, what, for whom, how used and benefits. *CABI Reviews*, 15(44), 1–16.

Paltridge, G. W., I. Barber. 1988. Monitoring grassland dry ness and fire potential in Australia with NOAA/AVHRR data. *Remote Sensing of Environment*, 25, 381–394.

Parker, A. E., E. R. Hunt. 2004. Accuracy assessment for detection of leafy spurge with hyperspectral imagery. *Journal of Range Management*, 57, 106–112.

Paruelo, J. M., H. E. Epstein, W. K. Lauenroth, I. C. Burke. 1997. ANPP estimates from NDVI for the central grassland region of the United States. *Ecology*, 78, 953–958.

Pech, R. P., R. D. Graetz, A. W. Davis. 1986. Reflectance modeling and the derivation of vegetation indexes for an Australian semiarid shrubland. *International Journal of Remote Sensing*, 7, 389–403.

Pickup, G. 1996. Estimating the effects of land degradation and rainfall variation on productivity in rangelands, an approach using remote sensing and models of grazing and herbage dynamics. *Journal of Applied Ecology*, 33, 819–832.

Pickup, G., V. H. Chewings. 1988. Forecasting patterns of soil erosion in arid lands from Landsat MSS data. *International Journal of Remote Sensing*, 9, 69–84.

Pickup, G., B. D. Foran. 1987. The use of spectral and spatial variability to monitor cover change on inert landscapes. *Remote Sensing of Environment*, 23, 351–363.

Pickup, G., D. J. Nelson. 1984. Use of Landsat radiance parameters to distinguish soil erosion, stability, and deposition in arid central Australia. *Remote Sensing of Environment*, 16, 195–209.

Plummer, S., O. Arino, M. Simon, W. Steffen. 2006. Establishing an earth observation product service for the terrestrial carbon community: The GLOBCARBON initiative. *Mitigation and Adaptation Strategies for Globalchange*, 11, 97–111.

Pohl, C., J. L. Van Genderen. 1998. Multi-sensor image fusion in remote sensing: Concepts, methods and applications. *International Journal of Remote Sensing*, 19, 823–854. https://doi.org/10.1080/014311698215748.

Potapov, P., M. C. Hansen, A. Pickens, A. Hernandez-Serna, A. Tyukavina, S. Turubanova, V. Zalles, X. Li, A. Khan, F. Stolle, N. Harris, X.-P. Song, A. Baggett, I. Kommareddy, A. Kommareddy. 2022. The global 2000–2020 land cover and land use change dataset derived from the Landsat archive: First results. *Frontiers in Remote Sensing*, 3. https://doi.org/10.3389/frsen.2022.856903.

Potapov, P., M. C. Hansen, A. Pickens, A. Hernandez-Serna, A. Tyukavina, S. Turubanova, V. Zalles, X. Li, A. Khan, F. Stolle, N. Harris, X.-P. Song, A. Baggett, I. Kommareddy, A. Kommareddy, A. Retallack, G. Finlayson, B. Ostendorf, K. Clarke, M. Lewis. 2023. Remote sensing for monitoring rangeland condition: Current status and development of methods. *Environmental and Sustainability Indicators*, 19, 100285. https://doi.org/10.1016/j.indic.2023.100285.

Prince, S. D. 2002. Spatial and temporal scale for detection of desertification. In *Do Humans Create Deserts*, eds. J. F. Reynolds, M. Stafford-Smith, pp. 23–40. Berlin, Germany: Dahlem University Press.

Prince, S. D., I. Becker-Reshef, K. Rishmawi. 2009. Detection and mapping of longterm land degradation using local net production scaling: Application to Zimbabwe. *Remote Sensing of Environment*, 113, 1046–1057.

Prince, S. D., C. J. Tucker. 1986. Satellite remote sensing of rangelands in Botswana 2 NOAA AVHRR and herbaceous vegetation. *International Journal of Remote Sensing*, 7, 15551570.

Prince, S. D., K. J. Wessels, C. J. Tucker, S. E. Nicholson. 2007. Desertification in the Sahel: A reinterpretation of a reinterpretation. *Global Change Biology*, 13, 1308–1313.

Pu, R., P. Gong, Z. Li, J. Scarborough. 2004. A dynamic algorithm for wildfire mapping with NOAA/AVHRR data. *International Journal of Wildland Fire*, 13, 275–285.

Reeves, M. C., L. S. Bagget. 2014. A remote sensing protocol for identifying rangelands with degraded productive capacity. *Ecological Indicators*, 43, 172–182.

Reeves, M. C., B. B. Hanberry, H. Wilmer, N. E. Kaplan, W. K. Lauenroth. 2021. An assessment of production trends on the Great Plains from 1984 to 2017. *Rangeland Ecology andManagement*, 78, 165–179.

Reeves, M. C., J. E. Mitchell. 2011. Extent of coterminous US rangelands: Quantifying implications of differing agency perspectives. *Rangeland Ecology and Management*, 64, 1–12.

Reeves, M. C., J. C. Winslow, S. W. Running. 2001. Mapping weekly rangeland vegetation productivity using MODIS algorithms. *Journal of Range Management*, 54, 90–105.

Reeves, M. C., M. Zhao, S. W. Running. 2006. Applying improved estimates of MODIS productivity to characterize grassland vegetation dynamics. *Journal of Rangeland Ecology and Management*, 59, 1–10.

Retallack, A., G. Finlayson, B. Ostendorf, K. Clarke, M. Lewis. 2023. Remote sensing for monitoring rangeland condition: Current status and development of methods. *Environmental and Sustainability Indicators*, 19, 100285. https://doi.org/10.1016/j.indic.2023.100285.

Rigge, M. B., C. Homer, L. Cleeves, D. K. Meyer, B. Bunde, H. Shi, G. Xian, S. Schell, M. Bobo. 2020. Quantifying western U.S. rangelands as fractional components with multi-resolution remote sensing and in situ data. *Remote Sensing*, 12, 412.

Robinove, C., J. P. S. Chavez, D. Gehring, R. Holmgren. 1981. Arid land monitoring using Landsat albedo difference images. *Remote Sensing of Environment*, 11, 133–156.

Roche, L. M. 2016. Adaptive rangeland decision-making and coping with drought. *Sustainability*, 8, 1334.

Roger, J. C., E. F. Vermote, S. Devadiga et al. 2017. *Suomi-NPP VIIRS Surface Reflectance User's Guide*. V1 Re-processing (NASA Land SIPS).

Rollins, M. G. 2009. LANDFIRE: A nationally consistent vegetation, wildland fire, and fuel assessment. *International Journal of Wildland Fire*, 18(3), 235–249.

Rouse, J. W., R. H. Haas, J. A. Schell, D. W. Deering. 1973. Monitoring vegetation systems in the Great Plains with ERTS. In *Proceedings of the Third ERTS Symposium*, pp. 309–317. Washington, DC.

Rouse, J. W. Jr., R. H. Hass, J. A. Schell, D. W. Deering, J. C. Harlan. 1974. *Monitoring the Vernal Advancement and Retrogradation(Greenwave Effect) I of Natural Vegetation*, p. 390. College Station, TX: Remote Sensing Center, Texas A&M University.

Roy, D. P., L. Boschetti, C. O. Justice, J. Ju. 2008. The collection 5 MODIS burned area product—global evaluation by comparison with the MODIS active fire product. *Remote Sensing of Environment*, 112, 3960–3707.

Running, S. W., D. D. Baldocchi, D. P. Turner, S. T. Gower, P. S. Bakwin, K. A. Hibbard. 1999. A global terrestrial monitoring network integrating tower fluxes, flask sampling, ecosystem modeling and EOS satellite data. *Remote Sensing of Environment*, 70, 108–127.

Running, S. W., R. R. Nemani, F. A. Heinsch, M. Zhao, M. Reeves, H. Hashimoto. 2004. A continuous satellite derived measure of global terrestrial primary production. *Bioscience*, 54, 547–560.

Safriel, U. 2009. Deserts and desertification: Challenges but also opportunities. *Land Degradation and Development*, 20, 353–366.

Sant, E. D., G. E. Simonds, R. D. Ramsey, R. T. Larsen. 2014. Assessment of sagebrush cover using remote sensing at multiple spatial and temporal scales. *Ecological Indicators*, 43, 297–305.

Scepan, J. 1999. Thematic validation of high resolution global landcover data sets. *Photogrammetric Engineering and Remote Sensing*, 65, 1051–1060.

Scheftic, W., X. Zeng, P. Broxton, M. Brunke. 2014. Intercomparison of seven NDVI products over the United States and Mexico. *Remote Sensing*, 6, 1057–1084.

Schueler, C., J. E. Clement, L. Darnton, F. DeLuccia, T. Scalione, H. Swenson. 2003. VIIRS sensor performance. In *International Geoscience and Remote Sensing Symposium Proceedings*, Vol. I, p. 369–372. Toulouse, France.

Scott, J. M., F. Davis, B. Csuti, R. Noss, B. Butterfield, C. Groves, H. Anderson, S. Caicco et al. 1993. Gap analysis—a geo graphic approach to protection of biological diversity. *Wildlife Monographs*, 123, 1–41.

Sexton, J. O., X. P. Song, M. Feng, P. Noojipady, A. Anand, C. Huang, D. Kim, K. M. Collins et al. 2013. Global, 30m resolution continuous fields of tree cover: Landsat-based rescaling of MODIS vegetation continuous fields with lidar based estimates of error. *International Journal of Digital Earth*, 6(5), 427–448.

Shen, L., Z. Li, X. Guo. 2014. Remote sensing of Leaf Area Index (LAI) and a spatiotemporally parameterized model for mixed grasslands. *International Journal of Applied Science and Technology*, 4, 46–61.

Shoshanya, M., N. Goldshleger, A. Chudnovsky. 2013. Monitoring of agricultural soil degradation by remote sensing methods: A review. *International Journal of Remote Sensing*, 34, 6152–6181.

Silva, J. M. N., A. C. L. Sá, J. M. C. Pereira. 2005. Comparison of burned area estimates derived from SPOT-VEGETATION and Landsat ETM+ data in Africa: Influence of spatial pattern and vegetation type. *Remote Sensing of Environment*, 96, 188–201.

Silverton, J. 2009. A new dawn for citizen science. *Trends in Ecological Evolution*, 24, 467–201.

Singh, J., P. B. Boucher, E. G. Hockridge, A. B. Davies. 2023. Effects of long-term fixed fire regimes on African savanna vegetation biomass, vertical structure and tree stem density. *Journal of Applied Ecology*, 60, 1223–1238. https://doi.org/10.1111/1365-2664.14435.

Skakun, S., C. O. Justice, E. Vermote et al. 2018. Transitioning from MODIS to VIIRS: An analysis of inter-consistency of NDVI data sets for agricultural monitoring. *International Journal of Remote Sensing*, 39, 971–992.

Skidmore, A. K., J. G. Ferwerda. 2008. Resource distribution and dynamics. In *Resource Ecology: Spatial and Temporal Dynamics of Foraging*, eds. H. H. T. Prins, F. van Langevelde. Wageningen, the Netherlands: Springer.

Socientize Project (2013–12–01). 2013. *Green Paper on Citizen Science: Citizen Science for Europe: Towards a Better Society of Empowered Citizens and Enhanced Research*. Socientize Consortium.

Society for Range Management [SRM], Glossary Update Task Group. 1998. *Glossary of Terms Used in Range Management* (4th edn.), p. 32. Denver, CO: Society for Range Management.

Sow, M., C. Mbow, C. Hély, R. Fensholt, B. Sambou. 2013. Estimation of herbaceous fuel moisture content using vegetation indices and land surface temperature from MODIS data. *Remote Sensing*, 5, 2617–2638.

Sparrow, B., J. Foulkes, G. Wardle, E. Leitch, S. Caddy-Retalic, S. van Leeuwen, A. Tokmakoff, N. Thurgate, G. R. Guerin, A. J. Lowe. 2020. A vegetation and soil survey method for surveillance monitoring of rangeland environments. *Frontiers in Ecology and Evolution*, 8, 157.

Steven, M. D., T. J. Malthus, F. Baret, H. Xu, M. J. Chopping. 2003. Intercalibration of vegetation indices from different sensor systems. *Remote Sensing of Environment*, 88, 412–422.

Stuth, J. W., J. Angerer, R. Kaitho, A. Jama, R. Marambii. 2005. Livestock early warning system for Africa rangelands. In *Monitoring and Predicting Agricultural Drought: A Global Study*, eds. V. K. Boken, A. P. Cracknell, R. L. Heathcote, pp. 283–294. New York: Oxford University Press.

Stuth, J. W., J. Angerer, R. Kaitho, K. Zander, A. Jama, C. Heath, J. Bucher, W. Hamilton, R. Conner, D. Inbody. 2003a. The Livestock Early Warning System (LEWS): Blending technology and the human dimension to support grazing decisions. *Arid Lands Newsletter*, University of Arizona. http://cals.arizona.edu/OALS/ALN/aln53/stuth.html [accessed 13 Oct., 2023].

Stuth, J. W., D. Schmitt, R. C. Rowan, J. P. Angerer, K. Zander. 2003b. *PHYGROW Users Guide and Technical Documentation*, Texas A&M University. http://cnrit.tamu.edu/physite/PHYGROW_userguide.pdf [accessed 13 Oct., 2023].

Suits, G., W. Malila, T. Weller. 1988. Procedures for using signals from one sensor as substitutes for signals of another. *Remote Sensing of Environment*, 25, 395–408.

Sulla-Menashe, D., M. A. Friedl. 2018. *User Guide to Collection 6 MODIS Land Cover (MCD12Q1 and MCD12C1) Product*, Vol. I, p. 18. Reston, VA, USA: US Geological Survey (USGS). https://lpdaac.usgs.gov/documents/101/MCD12_User_Guide_V6.pdf [accessed 30 Oct., 2023].

Svoray, T., A. Perevolotsky, P. M. Atkinson. 2013. Ecological sustainability in rangelands: The contribution of remote sensing. *International Journal of Remote Sensing*, 34, 6216–6242.

Tansey, K., J. M. Gregoire, E. Binaghi, L. Boschetti, P. A. Brivio, D. Ershov, S. Flasse, R. Fraser et al. 2004. A global inventory of burned area at 1 km resolution for the year 2000 derived from SPOT Vegetation data. *Climatic Change*, 67, 345–377.

Tchuenté, K., A. Thibaut, J. L. Roujean, S. M. D. Jong. 2011. Comparison and relative quality assessment of the GLC2000, GLOBCOVER, MODIS and ECOCLIMAP land cover data sets at the African continental scale. *International Journal of Applied Earth Observation and Geoinformation*, 13, 207–219.

Telesca, L., R. Lasaponara. 2006. Quantifying intra annual persistent behaviour in SPOT VEGETATION NDVI data for Mediterranean ecosystems of southern Italy. *Remote Sensing of Environment*, 101, 95–103.

Thoma, D. P., D. W. Bailey, D. S. Long, G. A. Nielsen, M. P. Henry, M. C. Breneman, C. Montagne. 2002. Short term monitoring of rangeland forage conditions with AVHRR imagery. *Journal of Range Management*, 55, 383–389.

Todd, S. W., R. M. Hoffer, D. G. Milchunas. 1998. Biomass estimation on grazed and ungrazed rangelands using spectral indices. *International Journal of Remote Sensing*, 19, 427–438.

Tucker, C. J. 1979. Red and photographic infrared linear combinations for monitoring vegetation. *Remote Sensing of Environment*, 8, 127–150.

Tucker, C. J. 1980. Remote sensing of leaf water content in the near infrared. *Remote Sensing of Environment*, 10, 23–32.

Tucker, C. J., J. E. Pinzon, M. E. Born, D. A. Slayback, E. W. Pak, R. Mahoney, E. F. Vermote, N. E. Saleous. 2005. An extended AVHRR 8km NDVI dataset compatible with MODIS and SPOT vegetation NDVI data. *International Journal of Remote Sensing*, 26, 4485–4498.

Tucker, C. J., C. Vanpraet, E. Boerwinkel, A. Gaston. 1983. Satellite remote sensing of total dry matter production in the Senegalese Sahel. *Remote Sensing of the Environment*, 13, 461–474.

Tueller, P. T. 1989. Remote sensing technology for rangeland management applications. *Journal of Range Management*, 42, 442–453.

Tueller, P. T., F. R. Honey, I. J. Tapley. 1978. Landsat and photo graphic remote sensing for arid land applications in Australia. In *International Symposium on Remote Sensing of Environment, Proceedings*, pp. 2177–2191. Ann Arbor, MI.

UNEP. 2012. *Global Environment Outlook: GEO5*, p. 551. Nairobi, Kenya: United Nations Environment Programme.

United Nations Convention to Combat Desertification (UNCCD). 1994. *Article 2 of the Text of the United Nations Convention to Combat Desertification.* https://catalogue.unccd.int/936_UNCCD_Convention_ENG.pdf [accessed 13 Oct., 2023].

Veraverbeke, S., S. Lhermitte, W. W. Verstraeten, R. Goossens. 2011. Time-integrated MODIS burn severity assessment using the multitemporal differenced Normalized Burn Ratio (dNBR(MT)). *International Journal of Applied Earth Observation and Geoinformation*, 13, 52–58.

Verbesselt, J., B. Somers, S. Lhermitte, I. Jonckheere, J. van Aardt, P. Coppin. 2007. Monitoring herbaceous fuel moisture content with SPOT VEGETATION timeseries for fire risk prediction in savanna ecosystems. *Remote Sensing of Environment*, 108, 357–368.

Vermote, E., B. Franch, M. Claverie. 2016. *VIIRS/NPP Surface Reflectance 8-Day L3 Global 500m SIN Grid V001 [Digital Data Set].* https://lpdaac.usgs.gov/products/vnp09h1v001/#using [accessed 30 Oct., 2023].

Vitousek, P. M. 1992. Global environmental change: An introduction. *Annual Review of Ecology,Evolution,and Systematics*, 23, 1–14.

Vlek, P., Q. B. Le, L. Tamene. 2010. Assessment of land degradation, its possible causes and threat to food security in Sub-Saharan Africa. In *FoodSecurityand Soil Quality*, eds. R. Lal, B. A. Stewart, pp. 57–86. Boca Raton, FL: CRC Press.

Vogl, R. 1979. Some basic principles of grassland fire management. *Environmental Management*, 3, 51–57.

Wallace, C. S. A., J. J. Walker, S. M. Skirvin, C. Patrick-Birdwell, J. F. Weltzin, H. Raichle. 2016. Mapping presence and predicting phenological status of invasive buffelgrass in southern Arizona using MODIS, climate and citizen science observation data. *Remote Sensing*, 8, 524. https://doi.org/10.3390/rs8070524.

Wang, Z., G. Mountrakis. 2023. Accuracy assessment of eleven medium resolution global and regional land cover land use products: A case study over the conterminous United States. *Remote Sensing*, 15, 3186.

Washington-Allen, R. A., N. E. West, R. D. Ramsey, R. A. Efroymson. 2006. A protocol for retrospective remote sensing based ecological monitoring of rangelands. *Rangeland Ecology Management*, 59, 19–29.

Weber, G. E., K. Moloney, F. Jeltsch. 2000. Simulated long term vegetation response to alternative stocking strategies in savanna rangelands. *Plant Ecology*, 150, 77–96.

Wenze, Y., B. Tan, D. Huang, M. Rautiainen, N. V. Shabanov, Y. Wang, J. L. Privette, K. F. Huemmrich et al. 2006. MODIS leaf area index products: From validation to algorithm improvement. *IEEE Transactions on Geoscience and Remote Sensing*, 44, 1885–1898.

Wessels, K. J. 2009. Letter to the Editor: Comments on 'Proxy global assessment of land degradation' by Bai et al. 2008. *Soil Use and Management*, 25, 91–92.

Wessels, K. J., S. D. Prince, P. E. Frost, D. van Zyl. 2004. Assessing the effects of human-induced land degradation in the former homeland of northern South Africa with a 1 km AVHRR NDVI timeseries. *Remote Sensing of Environment*, 91, 47–67.

Wessels, K. J., S. D. Prince, J. Malherbe, J. Small, P. E. Frost, D. van Zyl. 2007. Can human induced land degradation be distinguished from the effects of rainfall variability? A case study in South Africa. *Journal of Arid Environments*, 68, 271–297.

Wilson, M. F., A. Henderson-Sellers. 1985. A global archive of land cover and soils data for use in general circulation climate models. *International Journal of Climatology*, 5, 119–143.

Wilson, L., R. Van Dongen, S. Cowen, T. P. Robinson. 2022. Mapping restoration activities on Dirk Hartog Island using remotely piloted aircraft imagery. *Remote Sensing*, 14, 1402.

Wold, A. N., A. J. Meddens, D. Lee, V. S. Jansen. 2023. Quantifying the effects of vegetation productivity and drought scenarios on livestock production decisions and income. *Rangelands*, 45(2), 21–32.

Wooster, M. J., G. Roberts, G. Perry, Y. J. Kaufman. 2005. Retrieval of biomass combustion rates and totals from fire radiative power observations: Calibration relationships between biomass consumption and fire radiative energy release. *Journal of Geophysical Research:Atmospheres*, 110, 1–24.

Wulder, M. A., T. R. Loveland, D. P. Roy, C. J. Crawford, J. G. Masek, C. E. Woodcock, R. G. Allen, M. C. Anderson, A. S. Belward, W. B. Cohen et al. 2019. Current status of Landsat program, science, and applications. *Remote Sensing of Environment*, 225, 127–147.

Wylie, B. K., E. A. Fosnight, T. G. Gilmanov, A. B. Frank, J. A. Morgan, M. R. Haferkamp, T. P. Meyers. 2007. Adaptive data driven models for estimating carbon fluxes in the northern Great Plains. *Remote Sensing of Environment*, 106, 399–413.

Xiao, X., P. Dorovskoy, C. Biradar, E. Bridge. 2011. A library of georeferenced photos from the field. *Eos, Transactions American Geophysical Union*, 92:453. https://doi.org/10.1029/2011EO490002.

Xiao, J., Q. Zhuang, D. D. Baldocchi, B. E. Law, A. D. Richardson, J. Chen, R. Oren, G. Starr et al. 2008. Estimation of net eco system carbon exchange for the conterminous United States by combining MODIS and AmeriFlux data. *Agricultural and Forest Meteorology*, 148, 1827–1847.

Xie, Y., Z. Sha, M. Yu. 2008. Remote sensing imagery in vegetation mapping: A review. *Journal of Plant Ecology*, 1, 9–23.

Yebra, M., E. Chuvieco, D. Riaño. 2008. Estimation of live fuel moisture content from MODIS images for fire risk assessment. *Agricultural and Forest Meteorology*, 148, 523–536.

Yin, H., T. Udelhoven, R. Fensholt, D. Pflugmacher, P. Hostert. 2012. How Normalized Difference Vegetation Index (NDVI) trends from Advanced Very High Resolution Radiometer (AVHRR) and Systeme Probatoire d'Observation de la Terre VEGETATION (SPOT VGT) time series differ in agricultural areas: An inner mongolian case study. *Remote Sensing*, 4, 3364–3389.

Yost, E., S. Wenderoth. 1969. Ecological applications of multispectral colour aerial photography. In *Remote Sensing in Ecology*, ed. P. L. Johnson, pp. 46–62. Athens, GA: University of Georgia Press.

Yu, L., L. Zhou, W. Liu, H. K. Zhou. 2011. Using remote sensing and GIS technologies to estimate grass yield and livestock carrying capacity of alpine grasslands in Golog Prefecture China. *Pedosphere*, 17, 419–424.

Zanaga, D., R. Van De Kerchove, D. Daems, W. De Keersmaecker, C. Brockmann, G. Kirches, J. Wevers, O. Cartus, M. Santoro, S. Fritz, M. Lesiv, M. Herold, N. E. Tsendbazar, P. Xu, F. Ramoino, O. Arino. 2020. *ESA WorldCover 10 m 2021 v200*. https://doi.org/10.5281/zenodo.7254221.

Zanaga, D., R. Van De Kerchove, W. De Keersmaecker, N. Souverijns, C. Brockmann, R. Quast, J. Wevers, A. Grosu, A. Paccini, S. Vergnaud, O. Cartus, M. Fritz, S. Santoro, I. Georgieva, M. Lesiv, S. Carter, M. Herold, L. Linlin, N. E. Tsendbazar, F. Ramoino, O. Arino. 2021. *ESA WorldCover 10 m2020v100.*

Zhang, L., D. J. Jacob, X. Liu, J. A. Logan, K. Chance, A. Eldering, B. R. Bojkov. 2010. Intercomparison methods for satellite measurements of atmospheric composition: Application to tropospheric ozone from TES and OMI. *Atmospheric Chemistry and Physics*, 10, 4725–4739.

Zhao, M., S. Running, F. Heinsch, R. R. Nemani. 2011. MODIS-derived terrestrial primary production. In *Land Remote Sensing and Global Environmental Change*, eds. B. Ramachandran, C. O. Justice, M. J. Abrams, pp. 635–660. New York: Springer.

Zhou, B., G. S. Okin, J. Zhang. 2020. Leveraging Google Earth Engine (GEE) and machine learning algorithms to incorporate in situ measurement from different times for rangelands monitoring. *Remote Sensing of Environment*, 236, 111521. https://doi.org/10.1016/j.rse.2019.111521.

Zhu, X., S. Nie, Y. Zhu, Y. Chen, B. Yang, W. Li. 2023. Evaluation and comparison of ICESat-2 and GEDI data for terrain and canopy height retrievals in short-stature vegetation. *Remote Sensing*, 15(20), 4969.

Zhu, L., J. Southworth. 2013. Disentangling the relationships between net primary production and precipitation in Southern Africa savannas using satellite observations from 1982 to 2010. *Remote Sensing*, 5, 3803–3825.

Zika, M. E., K. H. Erb. 2009. The global loss of net primary production resulting from human induced soil degradation in drylands. *Ecological Economics*, 69, 310–388.

12 Remote Sensing of Rangeland Biodiversity

E. Raymond Hunt Jr., Cuizhen Wang, D. Terrance Booth, Samuel E. Cox, Lalit Kumar, and Matthew Reeves

ACRONYMS AND DEFINITIONS

ASTER	Advanced Spaceborne Thermal Emission and Reflection Radiometer
ETM+	Enhanced Thematic Mapper Plus
EVI	Enhanced Vegetation Index
GIS	Geographic Information System
IKONOS	A commercial Earth observation satellite, typically, collecting sub-meter to 5 m data
ISODATA	Iterative Self-Organizing Data Analysis Technique
LiDAR	Light Detection and Ranging
MERIS	Medium Resolution Imaging Spectrometer
MODIS	Moderate-Resolution Imaging Spectroradiometer
NASA	National Aeronautics and Space Administration
NDVI	Normalized Difference Vegetation Index
NIR	Near-Infrared
NOAA	National Oceanic and Atmospheric Administration
OLI	Operational Land Imager
PROSAIL	Combination of PROSPECT and SAIL, the two nondestructive physically based models to measure biophysical and biochemical properties
PROSPECT	Radiative transfer model to measure leaf optical properties spectra
SAIL	Scattering by Arbitrary Inclined Leaves
SAVI	Soil Adjusted Vegetation Index
SVM	Support Vector Machines
SWIR	Shortwave Infrared
TM	Thematic Mapper
UAS	Unmanned Aircraft Systems
USDA	United States Department of Agriculture
VIIRS	Visible Infrared Imaging Radiometer Suite

12.1 INTRODUCTION

Rangelands are a type of land cover dominated by grasses, grass-like plants, broadleaf herbaceous plants (forbs), shrubs, and isolated trees, usually in which large herbivores evolved as part of the ecosystem. In many rangelands, the large herbivores were replaced by livestock, and thus livestock grazing represents a major land use for production of food and fiber. Sustainability is maintained by species diversity (Kouba et al., 2024; Retallack et al., 2023; Hooper et al., 2005; Tilman et al., 2006, 2012; Zaveleta et al., 2010; Reich et al., 2012) and reduction of biodiversity is expected to be one of the major consequences of global climatic change (Soussanna and Lüscher, 2007; Janetos et al., 2008; McKeon et al., 2009; Pereira et al., 2010, 2012; Belgacem and Louhaichi, 2013; Joyce et al., 2013; Polley et al., 2013). Along with climatic change, invasions of

DOI: 10.1201/9781003541165-15

non-native species are threatening rangeland sustainability by decreasing native species diversity (Retallack et al., 2023; Thornley et al., 2023; Matongera et al., 2021; Sawalhah et al., 2018; Jafari et al., 2017; McCord et al., 2017; Bradley et al., 2009; Lavergne et al., 2010; Ziska et al., 2011).

Rangelands cover large, sparsely populated areas; thus, remote sensing is becoming much more important for rangeland monitoring, whether the objectives are sustainable livestock foraging or maintenance of other ecosystem goods and services with climatic change. Different stakeholders at the national, state/province, and landscape scales have different needs, which may require remotely sensed data at different spatial, temporal or spectral resolutions. Reeves et al. (Chapter 11 of this volume) examined the relationship between rangeland productivity and climate, which highlights some of the direct mechanisms on how rangelands may respond to increased atmospheric CO_2, global warming, and changes in precipitation regimes. Methods and techniques of rangeland characterization using remote sensing are covered by Kumar et al. (Chapter 12 of this volume).

Monitoring the biodiversity of rangelands is critical for maintaining sustainability; therefore, one of the major challenges for remote sensing is how to use imagery to estimate biodiversity (Kouba et al., 2024; Ocholla et al., 2024; Fenetahun et al., 2022; Matongera et al., 2021; Sawalhah et al., 2018; Jafari et al., 2017; McCord et al., 2017; Ludwig et al., 2004; John et al., 2008; Gillespie et al., 2008; Huang and Asner, 2009; Ward and Kutt, 2009). Diversity of plant functional types (PFTs) may be a good indicator of biodiversity, and determining the diversity of PFTs by remote sensing may be easier than determining plant species richness (Ustin and Gamon, 2010). However, land-cover and land-use maps based on global-scale PFTs (Running et al., 1995; Friedl et al., 2010) do not provide sufficient information for conserving and managing rangelands.

This chapter examines data types and methods for remote sensing of biodiversity at different resolutions: spectral, temporal and spatial. Medium resolution sensors, such as the Landsat 8 Operational Line Imager (OLI), generally have resolutions on the order of 10–60 m, 10–20 days, and 4–10 bands for spatial, temporal and spectral resolutions, respectively. High spectral resolution (hyperspectral) sensors generally have 100 or more contiguous bands, which are used to determine a reflectance spectrum of the land surface. High temporal resolution is usually provided by satellites, such as the MODerate resolution Imaging Spectroradiometer (MODIS), that have a broad swath (>400 km) in order to cover the Earth frequently. Commercial vendors provide satellite data with panchromatic bands of about 0.5 m and multispectral bands of about 2–3 m spatial resolution, but for this chapter, high spatial resolution (HSR) data have pixel sizes of 10 cm or less, and are acquired by aircraft or ground-based imaging.

To be useful for rangeland managers, remote sensing data must be able to estimate biodiversity at a landscape scale, and the information must be compatible with rangeland management systems based on ecological sites, rangeland state-and-transition models, and rangeland health. With over 40 years of data acquired from the Landsat satellites, and extensive programs of research, Landsat imagery have not been used routinely to provide information necessary to affect managers' decisions on which land areas are used for a given purpose at a given time. Remote sensing data providers and analysts must adapt to the needs of rangeland managers, and the needs are increasingly being driven by issues of sustainability, based on maintaining biodiversity. Therefore, we briefly describe rangeland management concepts and then describe remote sensing data and methods related to rangeland biodiversity. From this perspective, we conclude that HSR data acquired from aircraft provide the necessary data and will have the highest impact on management decisions. However, HSR data have their own unique challenges.

12.2 BIODIVERSITY AND RANGELAND MANAGEMENT

A plant community is a set of interacting species co-occurring at a given site, with dominant species identified by mass and typical longevity of an individual. Management of U.S.

rangelands was built on Clements' (1916) theory of plant succession where the dominant plant species changed over time to a set of species in equilibrium with climate, which is defined as the climax plant community (Brown, 2010). These communities were assumed to have the highest sustainable productivity and greatest resistance to invasive species. Overgrazing was assumed to reverse the plant community to an earlier stage of succession (Sampson, 1919). For monitoring and management, rangeland sites were defined by assuming the climax plant community was the dominant plant community at the onset of European immigration and settlement into the Western United States.

However, it was recognized that with natural disturbances such as fire, drought and grazing, plant succession was much more dynamic and variable. Furthermore, after severe overgrazing and soil erosion, it was unlikely that the climax community could re-establish itself without costly interventions. In response to the deficiencies of Clementsian succession, two ideas emerged in parallel (Briske et al., 2005): state-and-transition models (Brown, 1994, 2010), and rangeland health (NRC, 1994). To better guide range management based on these two ideas, the concept of rangeland sites was replaced with the concept of ecological sites in which the various plant communities are in a dynamic equilibrium determined by the natural disturbance regime.

12.2.1 Ecological Sites and State-and-Transition Models

> *An ecological site is a conceptual division of the landscape, defined as a distinctive kind of land based on recurring soil, landform, geological, and climate characteristics that differs from other kinds of land in its ability to produce distinctive kinds and amounts of vegetation and in its ability to respond similarly to management actions and natural disturbances.*
>
> *(Caudle et al., 2013)*

In the U.S., the top-levels of a hierarchical classification system based on soil, landform, geology, and climate are the Land Resource Region (LRR) and Major Land Resource Area (MLRA), defined by the USDA NRCS (2006). Subdividing MRLAs based on finer scale differences in climate, geomorphology and soils, the next level down in the hierarchy is the Land Resource Unit (LRU), LLRs, MLRAs, LRUs, and ecological sites are provisional and may be revised with more information (Moseley et al., 2010; Caudle et al., 2013). In a MRLA or LRU, ecological sites are identified by a reference state (State 1) and reference plant community (Community Phase 1.1), based on the dominant vegetation thought to be present at the time of European immigration and settlement (Figure 12.1). Ecological sites are divided into a series of alternative states (States 2, 3 . . . to some number *N*), which recognize new sets of stable communities that become self-perpetuating (Figure 12.1). Land areas in most alternative states are not barren wastes; with appropriate management, these areas may be used sustainably by preventing further degradation.

The definition of an ecological site raises an important drawback for remote sensing – the reference plant community is usually inferred from a wide variety of sources, and mapped using geographic information data. An ecological site is not defined by the plant community that is currently occupying it; therefore, remotely sensed land cover, at any resolution, cannot be used to define an ecological site. Whereas ecological sites are not defined by the plant communities, ecological states may be mapped using characteristic plant communities determined from high-resolution aerial photographs (Steele et al., 2012).

Within an ecological state (for example, Reference State 1, Figure 12.1), various plant community phases result from natural disturbances (including grazing by livestock) and subsequent plant succession leads back to the reference community phase (Figure 12.1). Together, all of the community phases in Reference State 1 represent the natural range of variation found over time (Caudle et al., 2013). Land areas in Community Phases 1.2, 1.3, and 1.4 are not considered degraded simply because the plant community is not in Phase 1.1, even though there may be fewer species and less biomass. Change detection must distinguish between changes within a state compared to a change of state. For example, there is less foliar mass in grasslands either after a drought (a frequent occurrence within the range of natural variability) or after severe overgrazing by livestock (which frequently entails a state change; Stafford Smith et al., 2007). Demonstrating a change with remote sensing is not a demonstration that a state-change has occurred (Thornley et al., 2023; Matongera et al., 2021; Sawalhah et al., 2018; Jafari et al., 2017; McCord et al., 2017 (Bastin et al., 2012; Bradley, 2013).

State-and-transition models are provisional hypotheses that describe the ecological processes leading to transitions (T1.A, T1.B, Figure 12.1) from one state to another (Caudle et al., 2013).

For example, invasion of downy brome (*Bromus tectorum* L.) into Great Basin sagebrush ecosystems increased the frequency of fire, which enhanced the dominance of downy brome in a feedback loop (Balch et al., 2013).

Thresholds are the conceptual boundaries dividing alternative states which are crossed during transitions (Briske et al., 2005, 2006; Bestelmeyer, 2006). Ecological resilience describes how much disturbance an ecological site can withstand without crossing a threshold into an alternative state; operationally, resilience may be defined as multiple species having the same ecological function (Questad et al., 2022; McCord et al., 2017; Elmqvist et al., 2003; Folke et al., 2004; Allen et al., 2005; Briske et al., 2008). For most stakeholders, the goal for monitoring rangelands is to determine if an ecological site is in the process of transitioning to alternative state. These transitions are not distributed evenly over an area or over time, so monitoring plant communities requires a larger and longer perspective (Bestelmeyer et al., 2011; Williamson et al., 2012). Determining the amount of species diversity, or at least an array of PFTs, is a primary objective for developing science-based, cost-effective tools for rangeland monitoring based on remote sensing.

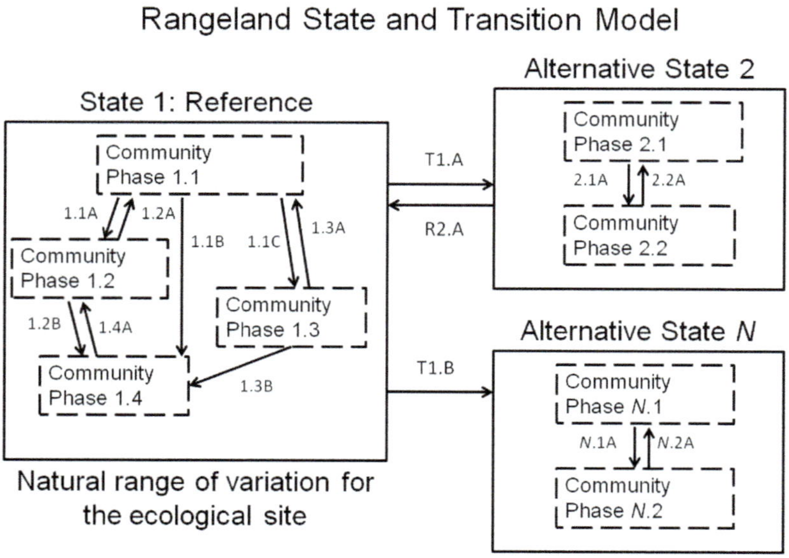

FIGURE 12.1 State-and-transition model for an ecological site in the USA. Outer boxes (solid lines) represent stable ecological states (A, B, . . . , *N*) within which changes in community phases (dashed lines) result from natural disturbances and succession (arrows). State 1 is the reference state and community phase 1 represents the historic climax plant community. Other community phases represent the range of natural variation and may be identified by dominant plant species. Transitions from one state to another stable state occur when a threshold is crossed (bold arrows), which is usually irreversible without intensive inputs.

12.2.2 Biodiversity Metrics for Managing Australian Rangelands

Rangelands, including tropical savannas, woodlands, shrublands and grasslands, make up 75% of Australia, with 55% of these rangelands being grazed by livestock. The Australian Collaborative Rangelands Information System (ACRIS) was set up in 2002 to support the Commonwealth and state governments in better managing the rangelands (Bastin et al., 2009; Eyre et al., 2011a; Oliver et al., 2014). In Australia, each state has its own regulations and methods of assessing and managing rangelands, and ACRIS is the overarching body that supports this by collating and synthesizing the monitoring data and making these data available to interested parties. This information assists natural resource management organizations, state governments, and the commonwealth in planning and reporting obligations, and evaluating the effectiveness of investments.

For determining the health, condition and biodiversity of rangelands, extensive surveys on a repeated basis are essential. Surveys of species presence and abundance are time consuming and expensive if undertaken at broad scales. For rapid monitoring, three states developed multi-metrics for biodiversity to help in the assessment of site conditions:

(1) BioCondition (in Queensland),
(2) Habitat Hectares (in Victoria), and
(3) BioMetric (in New South Wales).

BioCondition is a vegetation and biodiversity assessment framework developed by the Queensland Department of Resource Management to provide on-site guidance to beginners and document the assessment process for future revision and comparison (Eyre et al., 2011b). Various surrogates are used to represent the health and condition of the environment being assessed. The BioCondition Assessment Tool uses cameras on mobile devices to take visual evidence of flora and uses an interactive means to identify the flora. Keeping a visual record and associating this with a description can be used by experts at a later stage for validation purposes. Repeated measurements and comparisons of the BioCondition Index provides a measure of how well a terrestrial ecosystem is functioning and this can then be linked to biodiversity values. It should be noted that BioCondition is more geared to be used at the local or property scale.

In Victoria, the Department of Sustainability and Environment has developed the Habitat Hectares method for estimating the quality of an area of vegetation (Parkes et al., 2003). It is mainly geared towards native vegetation and includes site-based measures of quality and quantity of vegetation and condition within the landscape context. Habitat Hectares provides a step-by-step approach to habitat and landscape assessment in the field and includes useful tips for ensuring consistency of application. Vegetation condition is determined by utilizing variables such as presence and amounts of weeds, amounts of log and leaf litter, cover and diversity of understory, and canopy cover and presence of older trees. Repeat measurements and comparisons against pre-determined benchmarks allows for the calculation of native vegetation losses and gains.

BioMetric is a terrestrial biodiversity assessment tool used in New South Wales (Gibbons et al., 2008, 2009). It is mainly applicable for assessment at the paddock or property scale and assesses losses of biodiversity from proposed activities, gains in biodiversity from proposed offsets, or gains in biodiversity as a result of management actions. In BioMetric, the vegetation is assessed against benchmarks, which are quantitative measures of the range of variability in the condition compared to pre-European settlement. Vegetation condition benchmarks are available by vegetation class and BioMetric compares the current or predicted future condition against this benchmark to denote scores that are then converted to a metric.

Currently, no satellite or aircraft data products are used to determine BioCondition, Habitat Hectares, or BioMetric, although research is being conducted in this area. To conclude, monitoring rangeland biodiversity is the basis of management in Australia, so the question is how biodiversity could be measured more efficiently and accurately by remote sensing.

12.2.3 Assessing Rangeland Health by Remote Sensing

Rangeland health is the degree to which soils and vegetation are maintained which would sustain the kinds and amounts of vegetation that would typically occur for that site (NRC, 1994; Pyke et al., 2002). A series of 17 qualitative indicators (with two optional indicators, Table 12.1) are intended to help people with some training to determine rangeland health on the ground in a consistent manner. An overall rating of rangeland health is determined by a preponderance of evidence (Pellant et al., 2005).

Multiple indicators of rangeland health (Table 12.1) are related to the ecological processes leading to transitions between alternative states at an ecological site (Caudle et al., 2013). These indicators may show that a site is either at risk for a transition or has crossed the threshold to an alternative state (Figure 12.1) when monitored on the ground (Herrick et al., 2005a, 2005b). Probably, not all indicators need to be determined (MacKinnon et al., 2011); the three most important in Table 12.1 are:

1. Bare ground cover (Indicator 4),
2. Vegetation composition (Indicator 12), and
3. Presence of invasive species (Indicator 16).

Gaps of bare ground may be the single most-important indicator for rangeland health (Booth and Tueller, 2003).

TABLE 12.1

Indicators of Rangeland Health (Pellant et al., 2005). By examining several qualitative indicators to establish a "preponderance of evidence," an overall assessment of soil and site stability, hydrologic function, and biotic integrity may be made. In the third column, we suggest the potential remote sensing methods suitable for monitoring the indicator.

Rangeland Health Indicator	Assessment	Potential Data Type*
1. Rills in soil	Active soil erosion by water	HSR
2. Water flow patterns	Water infiltration/runoff	HSR
3. Pedestals/terracettes	Active soil erosion by wind or water	
4. Bare ground cover/gap sizes	Potential soil erosion by wind or water	HSR, Hysp, Medium resolution
5. Gullies	Active soil erosion by water	HSR, LiDAR
6. Wind-scoured/deposition areas	Active soil erosion by wind	HSR
7. Litter movement	Soil erosion by wind or water	HSR
8. Soil surface resistance to erosion	Soil quality	
9. Soil surface loss/degradation	Soil quality	
10. Community composition and distribution	Water infiltration/runoff	HSR
11. Compaction layer	Water infiltration/runoff	
12. Vegetation composition/functional groups	Biogeochemical cycles	HSR, Hysp
13. Plant mortality/decadence	Population dynamics	HSR
14. Litter amount	Soil quality	Hysp
15. Net primary production/green leaves	Biogeochemical cycles	fPAR
16. Invasive plants	Population dynamics and biogeochemical cycles	HSR, Hysp, fPAR
17. Perennial plant reproduction	Population dynamics	HSR
18. Biological crusts on soil (optional)	Soil quality	Hysp
19. Vertical vegetation structure (optional)	Animal communities	HSR, LiDAR

* Abbreviations are: HSR—high spatial resolution, Hysp—high spectral resolution (hyperspectral), LiDAR—light detection and ranging, fPAR—high temporal resolution estimating the fraction of absorbed photosynthetically active radiation.

Table 12.1 (third column) lists the potential data sources for 15 of the 19 indicators. The data resolution, which provides the information about an ecological process, is given instead of the sensor name, because the spectral, temporal, and spatial resolutions overlap among different sensors. For detecting gaps of bare ground (Indicator 4), HSR provides direct measurements of cover, whereas hyperspectral (Hysp) and medium resolution data could be used either for direct estimates using spectral unmixing or indirect estimates using spectral indices.

Image classification for vegetation composition (Indicator 12) is one of the primary applications and research areas in remote sensing (Lu and Weng, 2007; Franklin, 2010). In general, the two methods for classification are supervised and unsupervised (Maphanga et al., 2022, Questad et al., 2022, Dube et al., 2021). Supervised classification uses known areas on the ground to create rules (training) for assigning a pixel to a specific category. Unsupervised classification groups similar pixels, which are assigned to different categories afterwards. Independent areas on the ground are then used for assessing the accuracy of the classification (Congalton and Green, 2008; Olofsson et al., 2013). Based on the PROSAIL model (Jacquemoud et al., 2009), the major interacting variables affecting spectral reflectance from a plant canopy are: (1) leaf area index; (2) plant structure and leaf angle distribution; (3) soil background reflectance; (4) positions of the Sun, target, and sensor; and (5) leaf spectral reflectance and transmittance. A plant community or functional type will have similar structural and spectral properties, and to the extent they have constant leaf area index and plant density on similar landscapes, the community or functional type will comprise a single class on an image.

Remote sensing of invasive species (Indicator 16) is based on some detectable difference between the invasives and native species (Ocholla et al., 2024, Thornley et al., 2023, Questad et al., 2022, Hunt et al., 2003; Underwood et al., 2003; Asner, 2004; Madden, 2004; Franklin, 2010; He et al., 2011; Pu, 2012; Bradley, 2013), either spectrally (Section 11.4.2), temporally (Section 11.5.3), or spatially (Section 11.6.3). Most invasive plant species need to be detected at the initial stages of infestation for control, perhaps limiting the usefulness of medium resolution and high temporal resolution sensors. In the Western United States alone, there are more than 300 species of invasive weeds. According to DiTomaso (2000), the five most problematic are:

(1) downy brome (*B. tectorum* also called cheatgrass);
(2) yellow star thistle (*Centaurea solstitialis* L.);
(3) spotted knapweed [*C. stoebe* L., the synonym *C. maculosa* is more common in the literature];
(4) diffuse knapweed (*C. diffusa* Lam.); and
(5) leafy spurge (*Euphorbia esula* L.).

Added to this list is tamerisk (*Tamerix* spp., also called salt cedar), a shrub spreading along rivers and streams in the western United States (Nagler et al., 2011).

12.2.4 Remote Sensing for Animal Biodiversity

The use of remote sensing to estimate animal diversity is increasing. Leyequien et al. (2007) listed five variables that relate to animal needs for food and shelter (Table 12.2). Habitat suitability uses land-cover class determined with medium resolution satellite data, particularly bird species (Gottschalk et al., 2005). Productivity and phenology (Variables 2 and 3, Table 12.2) are important variables that are determined from high temporal resolution data which are related to animal biodiversity (Pettorelli et al., 2011). Correlations between animal species diversity and plant production were established before satellite data were available (Gaston, 2000), although the underlying causes for the correlations are being debated. Habitat structure (Variable 4, Table 12.2) combines several attributes in a given area: variation the amount of bare soil and vegetation, variation in the occurrence of PFTs, and variation of shadows related to the vertical structure of vegetation. Estimates

TABLE 12.2
Variables for the Remote Sensing of Animal Biodiversity According to Leyequien et al. (2007)

Variable	Potential Data Type*
1. Habitat suitability	Medium resolution
2. Photosynthetic productivity	fPAR
3. Multi-temporal patterns	fPAR
4. Habitat structure	Medium resolution, LiDAR, HSR
5. Forage quality	Hysp

* Abbreviations defined in Table 12.1.

of image heterogeneity are strongly related to habitat structure and biodiversity (Section 11.3.2). Vertical habitat structure is measured directly at high spatial resolution using either Light Detection and Ranging (LiDAR) or stereo HSR. Forage quality (Variable 5, Table 12.2) depends on the protein and fiber consumed by herbivores, and may be detectable with hyperspectral remote sensing (Section 11.4.2) or with commercial satellite sensors that have bands at the red-edge of the chlorophyll absorption spectrum.

12.3 MEDIUM RESOLUTION REMOTE SENSING

The Landsat series of satellites are the archetype of medium resolution remote sensing and have provided data globally for over 40 years. Landsat's 4 and 5 carried the Thematic Mapper (TM) sensor (Figure 12.2), Landsat 7 carries the Enhanced Thematic Mapper Plus (ETM+), and the recently launched Landsat 8 carries the Operational Land Imager (OLI) with two new bands (Figure 12.2).

Red and near-infrared spectral indices have been a standard method in analysis of multispectral data since the Landsat 1 was launched to enhance differences between soil and vegetation (Figure 12.2) and to reduce effects of atmospheric transmittance and solar irradiance from either time of year or topography. The spectral indices used most frequently in remote sensing are the Normalized Difference Vegetation Index (NDVI, Rouse et al., 1974; Tucker, 1979), the Soil Adjusted Vegetation Index (SAVI, Huete, 1988), and the Enhanced Vegetation Index (EVI, Huete et al., 2002).

Seasonal and annual precipitation totals in rangelands have high variability, so drought is relatively frequent. Vegetation index differences among images acquired on the same day of the year for different years may be from: (1) drought, (2) recent grazing, (3) fire, or (4) a state change of the ecological site. Long-term monitoring is the only way to distinguish among these possibilities to account for the natural range of variation at a single ecological site (Parracciani et al., 2024; Angerer et al., 2023; Maphanga et al., 2022; Questad et al., 2022; Washington-Allen et al., 2006; Bastin et al., 2014). Comparisons of vegetation indices for a specific area in relation to the average for all areas of that ecological site will highlight areas that have changed or in the process of changing to another ecological state (Maynard et al., 2007; Williamson et al., 2012).

Wildlife and livestock grazing patterns in rangelands are not random; they selectively graze areas with high forage quality (Angerer et al., 2023; Kleinhesselink et al., 2023; Sawalhah et al., 2018; Jafari et al., 2017; McCord et al., 2017; Ramoelo et al., 2012; Zengeya et al., 2013). There are several management changes that affect livestock grazing patterns at the landscape scale, such as fences and locations for water (Kleinhesselink et al., 2023; Questad et al., 2022; Jafari et al., 2017; McCord et al., 2017; Washington-Allen et al., 2004; Bastin et al., 2012). In contrast, wildlife feed in areas with sufficient cover from predators. In the future, determining the non-random grazing patterns at a landscape scale may be important information to manage biodiversity.

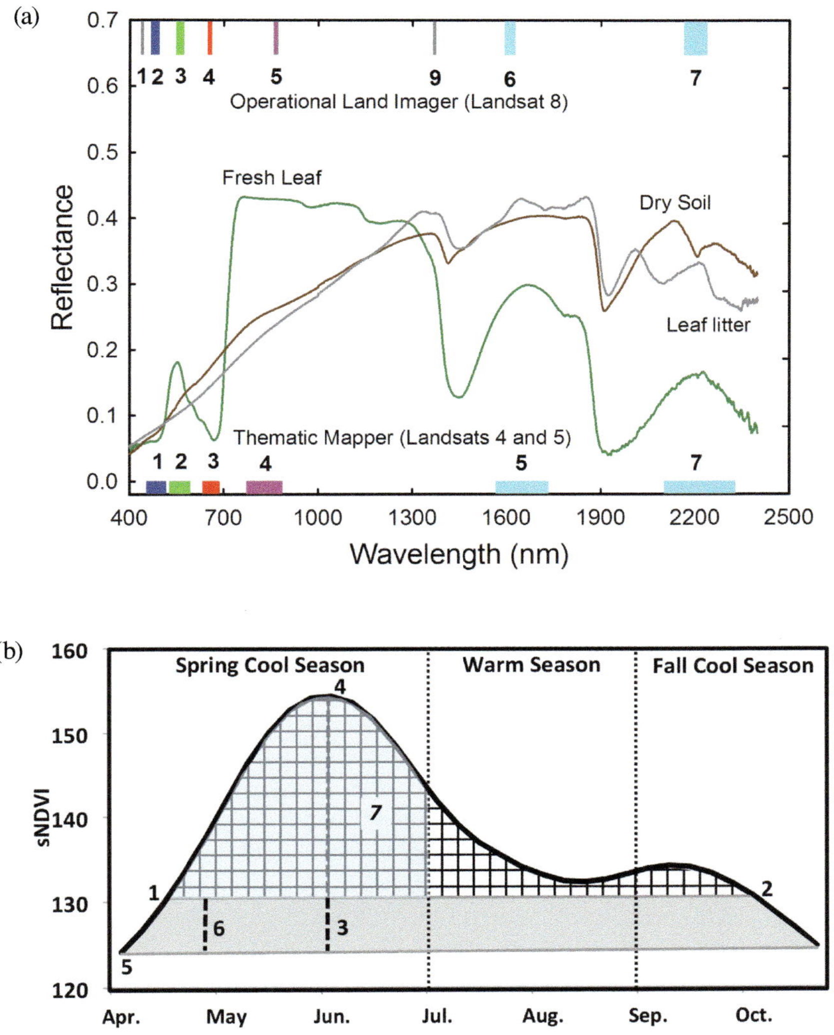

FIGURE 12.2 Spectral reflectances of a healthy green leaf (green line), old leaf litter (gray), and dry soil (brown line). Along the bottom are the wavebands for the Landsats 4 & 5 Thematic Mapper and along the top are the wavebands for the Landsat 8 Operational Line Imager. The width of each bar shows the spectral width of each waveband.

12.3.1 Spectral Unmixing

There is a large spectral difference between vegetation and soils, primarily at red and near-infrared wavelengths (Figure 12.2). Differences in the short-wave infrared region are small if the vegetation and soil are either dry or moist, and are large if the leaves are moist and the soils are dry (Figure 12.2). Linear spectral mixture models are generally thought of as a method for analyzing hyperspectral data (Roberts et al., 1993); however, Adams et al. (1986) developed this method using multispectral data. There are several important assumptions for linear spectral unmixing:

1. The fractional covers are non-negative and sum to one,
2. No multiple scattering among the spectral components (endmembers), and
3. All of the spectral components are known.

Distinct patches of vegetation, litter, and bare soil meet these assumptions; whereas within a patch of vegetation, multiple scattering between individual plants and soil creates nonlinear mixing.

The simplest case of a linear spectral mixture model has two spectral components (S_1 and S_2):

$$S_\lambda = f\,S_1 + (1-f)\,S_2 \tag{12.1}$$

where S_λ is the sensor measurement, f is the percentage of the first component, and $(1-f)$ is the percentage of the second component. If the two components are vegetation and soils (Figure 12.2), Equation 12.1 may be rearranged and solved for f:

$$f = (S_\lambda - S_2)/(S_1 - S_2) \tag{12.2}$$

thus, the percentage of bare soil may be calculated directly from the sensor data, and an indicator of rangeland health becomes unambiguously measured.

From Equation 12.2, vegetation indices based on near-infrared (NIR) and red wavebands are nonlinearly related to the fractional cover of vegetation and soil (Jiang et al., 2006; Montandon and Small, 2008). Empirical relationships between fractional cover and vegetation indices have problems because there is a large amount of variability in soil spectra, particularly in rangelands where plant cover is low. Equation 12.2 is also used to normalize NDVI for a site (S_λ) based on maximum NDVI for the year (S_1) and minimum NDVI for the year (S_2), where f is called the vegetation condition index (Kogan et al., 2003).

Equation 12.1 may be extended to include other spectral components such as plant litter, other non-photosynthetic vegetation, and crop residue (Retallack et al., 2023; Maphanga et al., 2022; Sawalhah et al., 2018; Davidson et al., 2008; Guerschman et al., 2009). In some areas, broadband spectral indices may be used to calculate plant litter cover based on statistical regressions (Thornley et al., 2023; Fenetahun et al., 2022; Sawalhah et al., 2018; McNairn and Protz, 1993; van Deventer et al., 1997; Marsett et al., 2006). However, it is often difficult to distinguish bare soil from plant litter with medium-resolution data, particularly when the soil and litter are moist (Daughtry and Hunt, 2008).

12.3.2 Habitat Heterogeneity and Structure

In general, landscapes with a large diversity of cover types and vertical structure also have a high amount of species diversity (Kouba et al., 2024; Kleinhesselink et al., 2023; Maphanga et al., 2022; Questad et al., 2022; Jafari et al., 2017; McCord et al., 2017; Fuhlendorf and Engle, 2001; Tews et al., 2004; Gillespie et al., 2008). Image texture is a general term for the amount of heterogeneity in gray-scale values surrounding a pixel, and there are different statistical formulae (randomness, variance, skewness, entropy, correlation, and more) used for calculating texture (Ocholla et al., 2024; Retallack et al., 2023; Questad et al., 2022; Dube et al., 2021; Haralick et al., 1973; Franklin and Wulder, 2002; Wood et al., 2012). It is recognized that heterogeneity is a function of scale (Haralick et al., 1973; Franklin and Wulder, 2002), so image texture needs to be evaluated with both medium resolution data and higher-resolution commercial satellite data (Johansen and Phinn, 2006).

Plant species richness is strongly related to remotely sensed texture for different regions (Gould, 2000; Dobrowski et al., 2008). Commercial satellite sensors such as DigitalGlobe's WorldView-2 have a panchromatic band with about 0.5-m pixel resolution and multispectral bands with about 2-m pixel resolution. Plant species richness can be determined with these data using various methods of analysis (Parracciani et al., 2024; Thornley et al., 2023; Dube et al., 2021; Hall et al., 2012; Mansour and Mutanga, 2012; Adelabu et al., 2013; Dalmayne et al., 2013; Müllerová et al., 2013). When pixels are smaller than image features, object-based classification creates clusters from adjacent pixels based on spectral information and texture (Yu et al., 2006; Dobrowski et al., 2008; Blaschke, 2010).

Remotely sensed image texture may have more information when flow direction of water and other resource flows are included. Ludgwig et al. (2000, 2002) developed a spatial index (Leakiness) to examine the connections among patches of bare soil, which is related to water flow patterns and the potential to lose soil and nutrients. Later, Ludwig et al. (2006, 2007), using ideas based on spectral unmixing, extended the Leakiness index to landscapes using Landsat Thematic Mapper data.

Habitat suitability for a given bird species is usually known, so land cover data from medium resolution satellites have been used to predict species distribution (Gottshalk et al., 2005). Small scale disturbances create mosaics of different vegetation types and the resulting heterogeneity was associated with greater species richness (Fuhlendorf et al., 2006). Image texture is sensitive to small variations of both vertical structure and vegetation mosaics for determination of bird species richness in a variety of habitats (Retallack et al., 2023; Questad et al., 2022; Sawalhah et al., 2018; Jafari et al., 2017; McCord et al., 2017; St. Louis et al., 2009; Bellis et al., 2008; Culbert et al., 2012; Wood et al., 2013).

12.3.3 Assessment of Medium Resolution

During most of the operational life of Landsats 4 and 5, the data were licensed commercially and thus had limited availability for rangeland management. The economic value per area of rangeland is low, but the total value is high because of the large areas of rangelands on Earth. The U.S. Geological Survey has made the entire archive of Landsat data available free of charge (http://landsat.usgs.gov), and analysis of this long-term archive will provide important insights into the patterns and processes of ecological states (Retallack et al., 2023; Thornley et al., 2023; Fenetahun et al., 2022; Maphanga et al., 2022; Jafari et al., 2017; McCord et al., 2017; Washington-Allen et al., 2006; Hernandez and Ramsey, 2013; Bastin et al., 2014).

We included the new commercial satellites (IKONOS, GeoEye, and WorldView) with pixel sizes between 2 and 5 m for the multispectral bands in this section, because the commercial sensors will be analyzed in large part with the same techniques that are currently used to analyze medium resolution data. Specifically, commercial satellite data will be used to estimate vegetation biomass using spectral vegetation indices, classify habitat suitability from land cover, and estimate vegetation structure with image texture. Some commercial satellite sensors include a band at the red edge of the chlorophyll absorption spectrum, which provides better information on forage quality compared to current medium resolution sensors. But the main disadvantage of commercial satellites is that these are pointed to specific areas, thereby missing adjacent areas on that satellite orbit.

12.4 HIGH SPECTRAL RESOLUTION

Acquisition and analysis of high spectral resolution data, properly called imaging spectrometer data, but ubiquitously called hyperspectral remote sensing, obtains radiance data in numerous, narrow, contiguous bands over a target (Green et al., 1998). With atmospheric correction, the reflectance spectrum of the target is calculated (Gao et al., 1993). From the reflectance spectrum, the identity of the target is determined based on chemical composition.

Linear spectral unmixing may be able to distinguish one spectral component (endmembers) more than the number of bands (Equation 12.1), if the components are known. However, the reflectances at one waveband are highly correlated with those nearby, so the effective number of spectral components is much less than the number of bands (Thorp et al., 2013). Usually, the number of spectral components is not known, or there is multiple, nonlinear scattering among them, which increases the complexity of the spectral unmixing.

Spectral matching compares a pixel's spectrum with some reference spectrum acquired from: a spectral library, field-acquired spectra, or the image itself. The advantage of spectral matching is that the total number of component spectra in the mixture does not need to be known. It must be decided *a priori* the amount of similarity for a match; a typical value for a match is 5.7 °

or 0.1 radians. The common spectral matching algorithms are: the Spectral Angle Mapper (SAM, Kruse et al., 1993), Mixture Tuned Matched Filter (MTMF, Boardman and Kruse, 2011), Spectral Information Divergence (SID, Chang, 2000), Spectral Correlation Measure (SCM, van der Meer, 2006), and Tetracorder (Clark et al., 2003).

SAM calculates a vector angle (Θ) between a reference spectrum $\mathbf{R} = (R_{\lambda 1}, R_{\lambda 2}, \ldots, R_{\lambda n})$ and a target spectrum $\mathbf{T} = (T_{\lambda 1}, T_{\lambda 2}, \ldots, T_{\lambda n})$, where λ1 to λn are the spectral wavelengths:

$$\Theta = \arccos\left[(\mathbf{R} \cdot \mathbf{T})/(\|\mathbf{R}\|\ \|\mathbf{T}\|)\right] \tag{12.3}$$

where $\|\mathbf{R}\|$ and $\|\mathbf{T}\|$ are vector normalizations and $\mathbf{R} \cdot \mathbf{T}$ is the vector dot product. The spectral angle between the leaf and soil spectra in Figure 12.2 is 28 /° or 0.49 radians. A major advantage (or disadvantage in some instances) of SAM is that differences in brightness are removed by the vector normalization, so large values of Θ are from spectral differences only.

12.4.1 Spectral Separability of Plant Species

The challenge using SAM and other spectral matching algorithms is determining the threshold value for determining a match, so that there are not large numbers of false positives and false negatives. A large threshold value will increase the number of false negatives and a small threshold value will increase the number of false positives. The value also depends on the variability of the target and reference spectra; therefore, how variable are the spectra from different species? To minimize the variation within a species, 10 leaves were acquired from four species growing in a common garden.

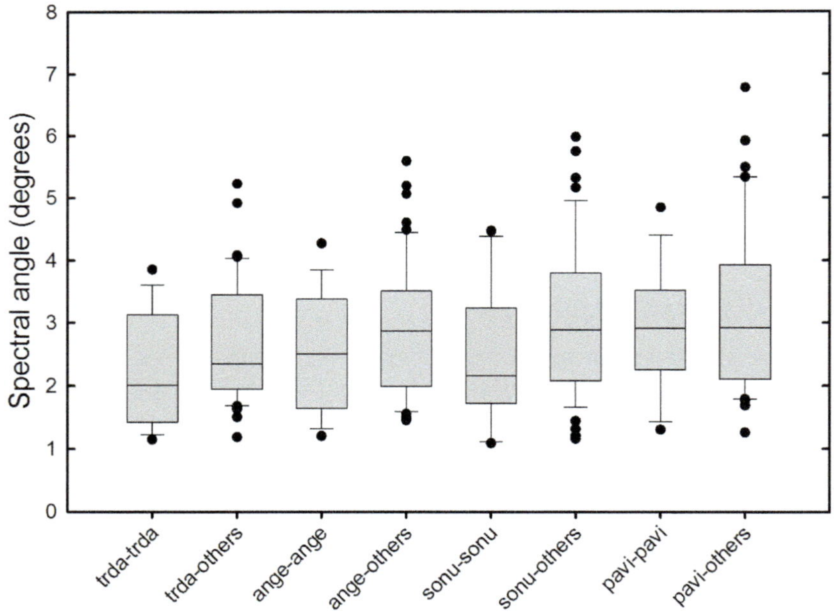

FIGURE 12.3 Spectral angles from leaf spectral reflectances (400–2,400 nm) for four grass species: *Tripsacum dactyloides* (L.) L. (trda, gamagrass); *Andropogon gerardii* Vitman (ange, big bluestem); *Sorghastrum nutans* (L.) Nash (sonu); and *Panicum virgatum* L. (pavi, switch grass). Mean spectra were calculated for each species and used for the reference spectrum, which was compared to the individual leaf spectra of that species (e.g., trda-trda) and spectra from the other species (e.g., trda-others). The center line is the median, the boxes show the range from the 25th to 75th percentiles, the error bars show the range from the 10th and 90th percentiles, and outliers are shown as single points.

Spectral reflectances were measured using a field portable spectrometer collecting light from an integrating sphere (Hunt et al., 2004).

The median spectral angles within a species were less than the median angles compared to the other species (Figure 12.3). The largest separations of spectral angles between within-species and among-species were for *Tripsacum dactyloides* (L.) L. and *Sorghastrum nutans* (L.) Nash. Using the default angle for classification (5.7° or 0.1 radians), only 5 leaves of three species were not spectrally similar to the others.

These results were expected based on previous results (Kouba et al., 2024; Kleinhesselink et al., 2023; Fenetahun et al., 2022; Jafari et al., 2017; McCord et al., 2017; Carter et al., 2005; Clark et al., 2005; Irisarri et al., 2009; Cho et al., 2010; Martin et al., 2011). Furthermore, the PROSPECT leaf optics model (Jacquemoud et al., 1996, 2009; Feret et al., 2008) accurately predicts leaf reflectance and transmittance with only four parameters: the chlorophyll content, the liquid water content, the dry matter content, and a leaf structure parameter equivalent to the number of parallel plates reflecting and transmitting radiation. Therefore, there may not be sufficient degrees of freedom to differentiate species using SAM or other methods (Mansour et al., 2012). Spectral absorption of radiation is based on the types and amounts of chemical bonds (Shenk et al., 2008), so while there may be unique organic compounds in different plant species, there will be few differences in the types and amounts of the chemical bonds. Differences in leaf structure are also highly variable and depend on the conditions during leaf development. Leaf water and chlorophyll contents are affected by current environmental conditions; so much of the variation in leaf reflectance spectra is not related to species.

12.4.2 Plant Chemical Composition

High spectral resolution data have potential for determining forage quality (Variable 5, Table 12.2), which is important for estimating animal biodiversity and livestock distribution (Parracciani et al., 2024; Fenetahun et al., 2022; Jafari et al., 2017; McCord et al., 2017; Leyequien et al., 2007; Skidmore et al., 2010; Knox et al., 2011). Near-infrared (NIR) spectroscopy is a standard laboratory method for analysis of dried plant materials for protein and fiber contents (Shenk et al., 2008), and hyperspectral sensors provide coverage of the same wavelength regions. One problem is determination of forage quality when the vegetation has high water content, because the spectral absorption of water dominates the SWIR reflectance spectrum (Ramoelo et al., 2011; Ustin et al., 2012). As the spectral absorption coefficients for liquid water are reasonably well known (Ustin et al., 2012), it is possible to remove the spectral features caused by water to estimate proteins and dry matter (Ramoelo et al., 2011; Wang et al., 2011b). Extensive studies with field spectrometers show strong potential for determination of forage quality (Starks et al., 2004; Knox et al., 2011), but the algorithms require extensive calibration and may not be appropriate for other species at other locations.

The areal cover of bare soil and litter are important indicators of rangeland health (Indicators 4 and 14, respectively, Table 12.1) and are readily detectable using high spectral resolution data (Figure 12.1). Cellulose and lignin are synthesized by plants and these substances are not found in soil, so narrow-band indices emphasizing spectral features in the shortwave infrared (SWIR) are linearly related to the cover of litter over bare soil (Ocholla et al., 2024; Kleinhesselink et al., 2023; Maphanga et al., 2022; Jafari et al., 2017; McCord et al., 2017; Daughtry, 2001; Nagler et al., 2003; Serbin et al., 2009, 2013). SWIR absorption features of bare soil (Figure 12.1) exposed from erosion are also detectable the Advanced Spaceborne Thermal Emission and Reflection Radiometer (ASTER) on NASA's Terra satellite (Maphanga et al., 2022; Questad et al., 2022; Dube et al., 2021; Matongera et al., 2021; Vrieling et al., 2007; Gill and Phinn, 2008; Serbin et al., 2009).

Biological soil crusts (Indicator 18, Table 12.1) are mixtures of cyanobacteria, fungi, and bacteria that form on the soil surface in arid and semiarid ecosystems. Chlorophyll a in the cyanobacteria creates a small absorption feature at 680 nm wavelength distinct from soil (Karnieli et al., 2003; Weber et al., 2008; Ustin et al., 2009). The problem is that the chlorophyll-a feature is also found in algae, lichens, moss, and plants; reflectances in the SWIR allow separation of biological soil crusts from small amounts of plants and plant-like classes (Ustin et al., 2009).

12.4.3 Detection of Invasive Plant Species

Whereas leaves of most rangeland species may not have much spectral diversity, distinctive leaves and flowers of invasive plants may result in characteristic spectral signatures that may be detected with high spectral resolution data (Table 12.3). Leafy spurge is a noxious invasive weed infesting large areas of the U.S. Great Plains. It is clonal, forming dense stands of genetically identical aboveground shoots spreading from a single root system. Flowers are clustered at or near the top of the shoot (umbels) and the flower cover is frequently over 20% in a single clone (Hunt et al., 2007). The flower bracts' yellow-green color is from a small amount of chlorophyll and a large ratio of carotenoids to chlorophylls (Hunt et al., 2004). When leafy spurge is found, one management option is the release of insects that specifically feed on the roots (larvae) and shoots (adult) for biological control (Thornley et al., 2023; Questad et al., 2022; Dube et al., 2021; Jafari et al., 2017; Anderson et al., 2003; Lym, 2005; Samuel et al., 2008; Lesica and Hanna, 2009).

Hyperspectral sensors are successful in detecting leafy spurge based on the yellow-green bracts (Retallack et al., 2023; Dube et al., 2021; Jafari et al., 2017; McCord et al., 2017; O'Neill et al., 2000; Parker Williams and Hunt, 2002, 2004; Dudek et al., 2004; Glenn et al., 2005; Lawrence et al., 2006). Furthermore, medium-resolution satellite sensors generally were not very successful (Retallack et al., 2023; Maphanga et al., 2022; Matongera et al., 2021; Sawalhah et al., 2018; McCord et al., 2017; Hunt and Parker Williams, 2006; Mladinich et al., 2006; Stitt et al., 2006; Hunt et al., 2007). The flower bracts of leafy spurge are readily detectable with aerial photography and videography (Anderson et al., 1996).

Parker Williams and Hunt (2002, 2004) analyzed two Airborne Visible Infrared Imaging Spectrometer (AVIRIS) scenes acquired on July 6, 1999 over Devils Tower National Monument and surrounding areas in Crook County, Wyoming. Two sets of plots were acquired, one with the cover of leafy spurge estimated ($N = 66$) and one with a simple classification of presence or absence ($N = 146$). Two scenes were atmospherically corrected to land-surface reflectance using the ATREM version 3.1 program (Gao et al., 1993) and processed using the "Spectral Hourglass" approach (Boardman and Kruse, 2011). Dimensionality was reduced using the Minimum Noise Fraction (MNF) in the Environment for Visualizing Images (ENVI version 3.2, Research Systems Inc.), and spectral signatures for leafy spurge were picked from the AVIRIS images using the Pixel Purity Index and n-Dimensional Visualizer (Boardman and Kruse, 2011).

The Mixture Tuned Matched Filter (MTMF, Boardman and Kruse, 2011) was used to identify pixels with possible leafy spurge. The MTMF score was related to leafy spurge cover with an R^2 of 0.66 (Parker Williams and Hunt, 2002), and the classification accuracy was 87% (Parker Williams and Hunt, 2004). However, accuracies of leafy spurge detection with MTMF appear to be dependent on the number of other spectral signatures in a scene (Fenetahun et al., 2022; Maphanga et al., 2022; Questad et al., 2022; Dudek et al., 2004; Glenn et al., 2005; Mitchell and Glenn, 2009). Using the data from Parker Williams and Hunt (2002, 2004), 11 AVIRIS scenes were combined into a single image for analysis; the overall classification accuracy using MTMF was no better than chance (Hunt et al., 2007). The type of spectral mixing needs to be considered, differences in plant density may change the amount of nonlinear mixing (Mitchell and Glenn, 2009).

There were many studies using high spectral resolution data to detect invasive species; two species commonly studied were tamarisk and yellow star thistle (Table 12.3). Detection of these invasive species may seem at odds with Figure 12.3 and related studies. In Table 12.3, the reflectance spectra of invasive species were in context with native species, soils, and subtle variations in spatial arrangement, leaf area index or leaf angle distribution. The leaf spectra in Figure 12.3 were isolated from any context. Successful use of hyperspectral data for detection of invasive species were based on knowing the invasive was present and having areas for training of supervised classification algorithms (Underwood et al., 2007).

TABLE 12.3
Invasive Plant Species in Rangelands Identified with High Spectral Resolution Data

Study	Species	Methods*
O'Neill et al., 2000; Parker Williams and Hunt, 2002, 2004; Dudek et al., 2004; Glenn et al., 2005; Hunt et al., 2007; Lawrence et al., 2006; Mitchell and Glenn, 2009	Leafy spurge (*Euphorbia esula* L.)	MNF, MTMF, SAM, supervised classification
Lass et al., 2002	Spotted knapweed (*Centaurea stoebe* L., *C. maculosa* is a commonly used synonym)	SAM
Underwood et al., 2003, 2007	Ice plant [*Carpobrotus edulis* (L. N. E. Br.)], Jubata grass/Pampas grass [*Cortaderia jubata* (Lemoine ex Carrière) Stapf], Blue gum (*Eucalyptus globules* Labill.)	MNF, continuum removal, spectral indices, supervised classification
Anderson et al., 2005	Tamarisk (*Tamarix chinensis Lour.*; T. gallica L.; and *T. ramosissima* Ledeb.)	PC, spectral indices
Lass et al., 2005	Yellow starthistle (*C. solstitialis* L.) and Babysbreath (*Gypsophila paniculata* L.)	SAM
Mundt et al., 2005	Hoary cress (*Cardaria draba* L.)	MNF, MTMF, SAM
Andrew and Ustin, 2006, 2008	Perennial pepperweed (*Lepidium latifolium* L.)	MNF, MTMF
Ge et al., 2006, 2007	Yellow star thistle	Spectral indices, SAM
Miao et al., 2006, 2007	Yellow star thistle	Band selection, feature extraction
Mirik et al., 2006, 2013	Musk thistle (*Carduus nutans* L.)	Regression, SVM
Narumalani et al., 2006, 2009	Tamerisk, Musk thistle, Canada thistle (*Cirsium arvense* L.), Reed canary grass (*Phalaris arundinacea* L.), Russian olive (*Elaeagnus augustifolia* L.)	ISODATA, MNF, SAM
Cheng et al., 2007	Kudzu [*Pueraria montana* (Lour.) Merr.]	MNF, MTMF, SAM
Hamada et al., 2007	Tamarisk	MNF, spectral indices, MTMF, stepwise discriminant analysis, hierarchical clustering, root sum squared differential area
Hestir et al., 2008	Perennial pepperweed, Water hyacinth [*Eichhornia crassipes* (Mart.) Solms], Brazilian waterweed (*Egeria densa* Planch.)	MNF, band indices, spectral mixture analysis, SAM, continuum removal
Noujdina and Ustin, 2008	Downy brome (*Bromus tectorum* L.)	MNF, MTMF
Pu, 2008; Pu et al., 2008	Tamarisk	PC, spectral indices
Wang et al., 2008	Sericea lespedeza [*Lespedeza cuneata* (Dum. Cours.) G. Don]	ISODATA, spectral derivatives
Carter et al., 2009	Tamarisk	MNF, supervised classification
Yang et al., 2009; Yang and Everitt, 2010a, 2010b	Ashe juniper (Juniperus ashei J. Buchholz), Broom snakeweed [*Gutierrezia sarothrae* (Pursh.) Britt. and Rusby], water hyacinth	MNF, SAM, supervised classification
Wang et al., 2010a; Bentivegna et al., 2012	Cut-leaved teasel (*Dipsacus laciniatus* L.)	SAM, supervised classification, spectral information divergence

(Continued)

TABLE 12.3 (*Continued*)
Invasive Plant Species in Rangelands Identified with High Spectral Resolution Data

Study	Species	Methods*
Fletcher et al., 2011; Yang et al., 2013	Tamarisk	SVM
Miao et al., 2011	Tamarisk	MNF, PC, supervised classification, SAM
Olsson and Morisette, 2014	Buffelgrass [*Pennisetum ciliare* (L.) Link]	MNF, MTMF, random forest

* Abbreviations are: minimum noise fraction (MNF), mixture tuned matched filtering (MTMF), spectral angle mapper (SAM), Iterative Self Organizing Data Analysis Technique Algorithm (ISODATA), principal components (PC), and Support Vector Machine (SVM)

12.4.4 Assessment of High Spectral Resolution

There are only two indicators of rangeland health (Table 12.1) and one variable for animal biodiversity (Table 12.2) for which the preferred method of remote sensing is high spectral resolution. The large spectral difference between vegetation and soils allows accurate assessments of relative cover, so one of the three most important indicators is reliably determined with hyperspectral sensors. Compared to other methods, the suitability of hyperspectral remote sensing for rangeland monitoring depends on detection of the other two critical indicators, invasive species and composition of plant functional types. The extensive discussion in Section 11.4.3 shows that more research is required, perhaps using simulated Hyperspectral Infrared Imager (HyspIRI) mission data produced with high-altitude AVIRIS imagery (http://hyspiri.jpl.nasa.gov; last checked June 12, 2014).

Compared to other sensors, it is more expensive to acquire an airborne hyperspectral sensor, and it takes more time and expertise to analyze hyperspectral imagery, so mapping the distribution of an invasive species over a landscape may not be cost effective. Geospatial species-distribution models classify suitable habitat over larger areas based on combinations of climate, vegetation, soils, topography and other variables within a geographic information system. There are too many combinations of factors to establish field plots for complete model testing. Classified hyperspectral imagery for small areas should be used to test species geospatial niche models, and then the geospatial models should be used to classify suitable habitat over larger areas (Kleinhesselink et al., 2023; Questad et al., 2022; Dube et al., 2021; Matongera et al., 2021; Sawalhah et al., 2018; Jafari et al., 2017; McCord et al., 2017; Underwood et al., 2004; Andrew and Ustin, 2009; Hunt et al., 2010).

Global biodiversity is one of the important science questions for the NASA HyspIRI mission (Roberts et al., 2012). The total number of species in a small area (alpha diversity) and the total number of species in a large area (gamma diversity) are more encompassing assessments of biodiversity. Alpha and gamma diversities are inter-related by beta diversity, which is a measure of the proportion of shared species between two points along an environmental gradient. Rocchini et al. (2010a, 2010b) hypothesized spectral diversity is an important method for estimating beta diversity. If this hypothesis is validated, then high spectral resolution data will become invaluable for estimating biodiversity, even though invasive species or plant functional types may not be detected.

12.5 HIGH TEMPORAL RESOLUTION

Repeated satellite acquisitions reveal seasonal patterns and growth of vegetation; however, cloud cover may obscure vegetation during critical time periods of active plant growth. High temporal resolution data acquire data every few days, which are composited to form weekly to monthly cloud-free time series. Even in rangelands, "where the skies are not cloudy all day" (Higley and Kelly, 1873), medium resolution imagery are frequently unusable because of cloud occurrence during the onset of the plant growing season. Meteorological satellites, such as NOAA's Advanced Very High Resolution Radiometer (AVHRR) and NASA's Moderate Resolution Imaging Spectroradiometer (MODIS), use the change in NDVI over the year to determine phenology and the fraction of absorbed photosynthetically active radiation (fPAR). Gross and net primary production are calculated from the total absorbed photosynthetically active radiation and light use efficiency (Hunt et al., 2003). Productivity is an indicator of rangeland health (Indicator 15, Table 12.1) and a variable for predicting animal biodiversity (Variable 2, Table 12.2).

Over a small region, interannual variation in phenology is related to climatic variability (Reed et al., 1994; Schwartz et al., 2002), whereas consistent spatial variation in phenology is related to plant functional type (Kouba et al., 2024; Retallack et al., 2023; Dube et al., 2021; Loveland et al., 1991; Kremer and Running, 1993; Peters et al., 1997; Gu et al., 2010). Grasses (Poaceae) have two functional types based on photosynthetic pathway: cool-season grasses with C_3 photosynthesis, and warm-season grasses with C_4 photosynthesis. Physiologically, C_3 photosynthesis is greater at cool temperatures and C_4 photosynthesis is higher at warm temperatures, but temperature affects many different processes, not just photosynthesis (Sage and Kubien, 2007). With global climatic change, warmer temperature may favor C_4 grasses whereas elevated atmospheric CO_2 may promote C_3 plants (Morgan et al., 2007). Therefore, C_3/C_4 relative abundance is a complex response of climate change, and will affect ecosystem biogeochemical cycles at regional and global scales.

Temperate grasslands in North America, South America, Australia, Africa, and Asia have mixtures of C_3 and C_4 grasses with the ratio depending on the amount of rainfall during the warm and cool seasons (Winslow et al., 2003). Determining the ratios of C_3 and C_4 grasses is a challenge because the reflectance spectra may not be separable (Adjorlolo et al., 2012a, 2012b). High temporal resolution AVHRR or Medium Resolution Imaging Spectrometer (MERIS) data was used to separate C_3 and C_4 grasses (Ocholla et al., 2024; Thornley et al., 2023; Questad et al., 2022; Jafari et al., 2017; McCord et al., 2017; Goodin and Henebry, 1997; Tieszen et al., 1997; Ricotta et al., 2003; Foody and Dash, 2007, 2010). Medium resolution imagery acquired two or three times per year also showed differences based on temperature (Davidson and Csillag, 2001; Peterson et al., 2002; Guo et al., 2003). High temporal resolution data may add more information by subdividing both C_3 and C_4 functional types into tall and short functional types (Wang et al., 2013).

12.5.1 Phenology Metrics

The first step for determining phenological metrics from high temporal resolution data is applying a low-pass filter to the time series of NDVI to remove the effects of sub-pixel clouds, atmospheric conditions, and view angles. Several smoothing algorithms have been used: a nonlinear median filter (Reed et al., 1994), an upper-envelope three-point filter (Gu et al., 2006), and a 2nd-order polynomial filter (Savitzky and Golay, 1964).

Most studies define a series of metrics from the smoothed NDVI time series (Table 12.4). These metrics may be determined by a sequence of logical statements; tools have been developed to automate calculation of the various metrics: fitting the NDVI time series into piecewise logistic functions (Zhang et al., 2003), fitting the time series using a quadratic function (de Beurs and Henebry, 2004), and in the open-source TIMESAT program (Jönsson and Eklundh, 2004). In rangelands, these mathematical approaches may produce errors because grasses often have prolonged growing season and asymmetric trajectories (de Beurs and Henebry, 2010). For this reason, some studies directly use the polynomial-smoothed time series to extract phenological metrics directly (Parracciani et al., 2024; Thornley et al., 2023; Questad et al., 2022; Dube et al., 2021; Matongera et al., 2021; McCord et al., 2017; Wang et al., 2010b, 2011a, 2013).

TABLE 12.4

Common Phenology Metrics for Determining Plant Functional Types. These metrics have to be determined with a complete time series of NDVI.

Metric	Definition
Start of season (*SOS*)	Before Peak NDVI, DOY* when NDVI > threshold
End of season (*EOS*)	After peak NDVI, DOY when NDVI < threshold
Length of season (*LOS*)	*EOS* – *SOS*
Peak NDVI	Maximum NDVI in time series
Peak date	DOY of Peak NDVI
Cumulative NDVI (Σ-*NDVI*)	Sum NDVI when *SOS* < NDVI < *EOS*

* Day of the Year (DOY)

12.5.2 Grass Functional Types

Dividing the U.S. Great Plains at 100° West Longitude, east of this meridian is primarily sub-humid tallgrass prairie with 500–750 mm of annual precipitation (Figure 12.4a). West of this meridian are the southern shortgrass steppe and the northern mixed-grass prairie (Figure 12.4a), which get about 250–500 mm of precipitation annually. Precipitation strongly controls grassland productivity resulting in the designations, tallgrass and shortgrass. The latitudinal C_3/C_4 variations and longitudinal tall/short differences are combined into tallgrass C_3, tallgrass C_4, shortgrass C_3 and shortgrass C_4. The latter three PFTs correspond to naturally occurring ecological regions; however, tallgrass C_3 may be the result of seeding cool-season species (e.g., tall fescue) into pastures for higher forage production (Fenetahun et al., 2022; Sawalhah et al., 2018; Jafari et al., 2017; McCord et al., 2017; Wang et al., 2013).

Grassland areas were determined from the USGS National Land-Cover Database (Homer et al., 2004); the grassland areas were not divided into functional types. Time series of the 500-m, eight-day Terra MODIS Surface Reflectance products (MOD09A1) for the years 2000 to 2009 show phenological variations of C_3 and C_4 grasses (Figure 12.5). Tallgrass C_3 and C_4 have higher NDVI values than shortgrass C_3 and C_4. Summer dormancy and fall growth for tallgrass C_3 was seen in some years as a second NDVI peak in the autumn (Figure 12.5), but this phenological feature was much less important than an earlier start of growing season (*SOS*), later end of growing season (*EOS*), and therefore longer length of growing season (*LOS*). Shortgrass C_3 have much earlier *SOS* than shortgrass C_4 and both have shorter *LOS* than the tallgrass PFTs (Figure 12.5).

The four grass PFTs in the U.S. Great Plains were mapped over the ten-year period (2000–2009) with a decision tree approach (Figure 12.4b). Because of interannual variability in temperature and precipitation, a given pixel would be classified as one PFT for some years and would be classified as another PFT for other years. For example, during drought, tallgrass C_3 and shortgrass C_3 *SOS* may be delayed and *EOS* may be earlier, so that the phenological features become similar to those of tallgrass C_4 and shortgrass C_4, respectively.

Therefore, delineation of grass C_3/C_4 PFTs is problematic using data from only a single year. A long time series of data were required for reliable separation (Winslow and Hunt, 2003; Wang et al., 2013). The final classification in Figure 12.4b displays the distributions where one PFT was selected for at least 6 of the 10 years (Wang et al., 2013). The four PFTs followed the longitudinal transition of ecological regions based on precipitation. However, latitudinal trends caused by temperature were not present, because east of 100 ° West, tallgrass C_3 areas were south of tallgrass C_4 areas. Splitting grasses into tallgrass and shortgrass highlighted the occurrence of a new ecological

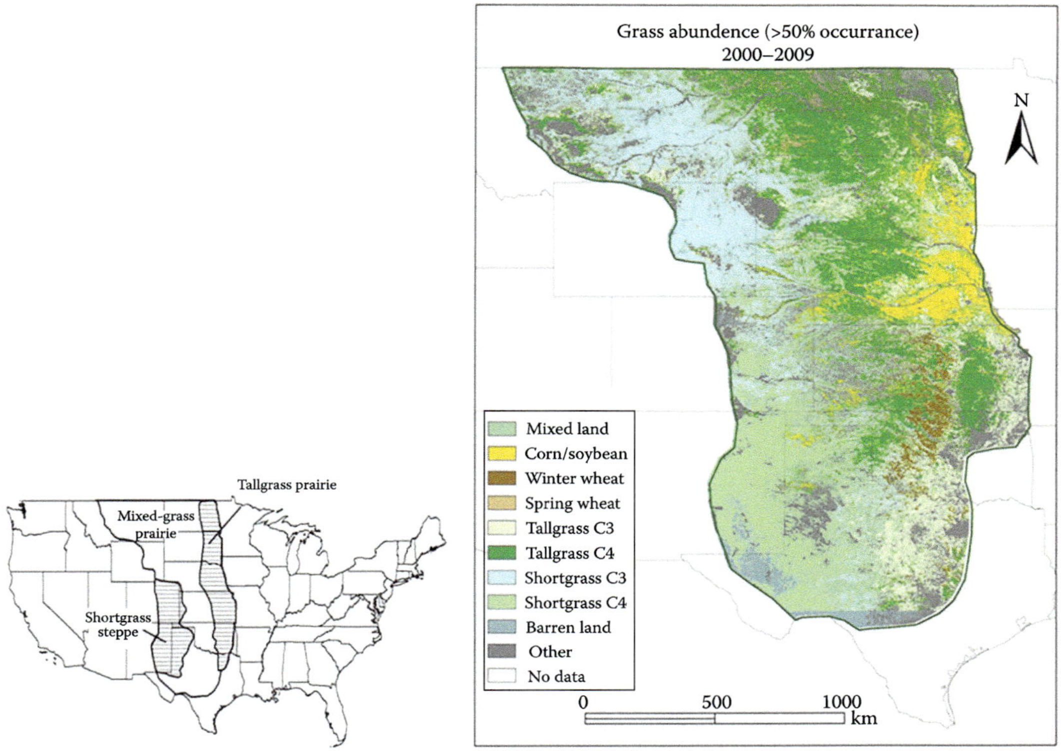

(a) (b)

FIGURE 12.4 (a) Ecological regions of the tallgrass prairie, shortgrass steppe, and northern mixed-grass prairie. (b) Distributions of four grass functional types in the U.S. Great Plains: tallgrass C_3, tallgrass C_4, shortgrass C_3, and shortgrass C_3. Tallgrass C_3 is not a naturally occurring ecological region in the U.S. Great Plains and may be the result of seeding C_3 species into pastures. Interannual variability in precipitation and temperature affected phenology, so a functional type had to be selected at least six years from 2000–2009 to be classified that functional type.

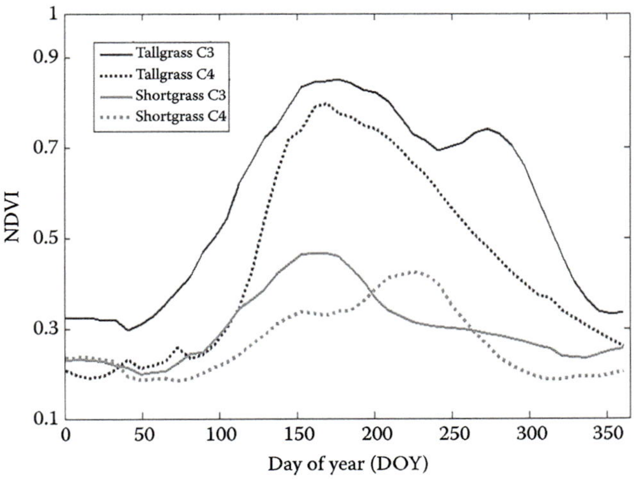

FIGURE 12.5 Examples of NDVI time series of the four grass functional types: tallgrass C_3, tallgrass C_4, shortgrass C_3, and shortgrass C_3. The NDVI values of pure pixels are extracted from the 46 MODIS scenes acquired in 2007.

region; the areas classified as tallgrass C_3 are located in areas generally considered to be pastures (Maphanga et al., 2022; Dube et al., 2021; McCord et al., 2017; Wang et al., 2013). Because more of the variation in temporal profiles was now explained, the resulting classification had somewhat increased accuracy.

12.5.3 Invasive Plant Species

Few invasive species are detected from the large pixels of high temporal resolution data (e.g., AVHRR and MODIS). One species is downy brome in the western United States, Bradley and Mustard (2005, 2006) showed that the temporal variation between downy brome and co-occurring species was large enough to classify downy brome invasion using AVHRR. Singh and Glenn (2009) and Clinton et al. (2010) replicated this study using a time-series of Landsat and MODIS data, respectively.

Two other species that may be detected using high temporal resolution data are broom snakeweed [*Gutierrezia sarothrae* (Pursh) Britton & Rusby; Peters et al., 1992] and Lehmann lovegrass (*Eragrostis lehmanniana* Nees; Huang and Geiger, 2008; Huang et al., 2009). Broom snakeweed is a C_3 shrub surrounded by C_4 grasses, so it was detected based on its earlier *SOS*. Lehmann lovegrass is a C_4 grass that crowds out native C_4 grasses; during senescence it produces a litter layer that is dense, bright and yellow which is detectable even with large pixel sizes (Huang and Geiger, 2008; Huang et al., 2009).

Two important questions about invasive species are addressed using ecological niche or habitat suitability models: (1) what areas on Earth may be susceptible to an invasive species, and (2) how will the areas expand or contract with anthropogenic global warming (Angerer et al., 2023; Fenetahun et al., 2022; Sawalhah et al., 2018; Jafari et al., 2017; Peterson, 2003; Thuiller et al., 2005; Andrew and Ustin, 2009; Stohlgren et al., 2010). MODIS vegetation index data are important for predicting the potential distribution of tamarisk (Morisette et al., 2006) and purple loosestrife (*Lythrum salicaria* L.; Anderson et al., 2006). In these studies, time series of NDVI are used as a surrogate for meteorological data, because even with 1-km pixel size, the data are much more densely distributed compared to available weather station data.

12.5.4 Assessment of High Temporal Resolution

The specifications for NASA MODIS and the recently launched Suomi National Polar-orbiting Partnership Visible Infrared Imaging Radiometer Suite (VIIRS) were developed in part based on the long history of AVHRR for global remote sensing of interannual climatic variability. Most land-cover classifications using AVHRR data were designed to be used by global ecosystem models, so fewer PFTs were defined. Because of limited computing power, there were limitations placed on the spatial resolution in order to provide high-quality data at high temporal resolution (Townshend and Justice, 1988). Therefore, variations in phenology related to taxonomy, plant functional types, and other considerations at smaller scales were rarely explored because of the large pixel sizes.

Two satellite systems have broad swaths and high temporal resolution with much smaller pixel sizes: (1) the Indian Space Research Organization's Advanced Wide Field Sensor (AWiFS) and (2) the international Disaster Monitoring Constellation (Wang et al., 2010c). However, data from these two sensors are only available from commercial venders and are not freely available from public archives (Goward et al., 2012).

With more powerful computers available, tradeoffs between high temporal resolution and high spatial resolution are no longer required. For example, the United States Department of Agriculture (USDA) National Agricultural Statistics Service (NASS) previously used AWiFS data (56–70 m pixels) and currently uses Deimos-1 data (22 m pixels) to produce annual Cropland Data Layers for the USA (Boryan et al., 2011). The potential of high temporal resolution data for monitoring rangeland biodiversity is simply not known. Making AWiFS, Deimos, and similar multispectral sensor data available worldwide is an important investment for monitoring global biodiversity (Retallack et al., 2023; Questad et al., 2022; Dube et al., 2021; Wang et al., 2010c).

12.6 HIGH SPATIAL RESOLUTION

While ground and aerial photographs have been used for natural resource management since the last century (Cooper, 1924), advances in digital sensors and computer processing have made high spatial resolution (HSR) the newest dimension in remote sensing. The term "high spatial resolution" is subjective, changes with time, and can mean pixel sizes from 0.25 mm to 2 m with the current state of the art. Smaller pixel sizes must be acquired from aircraft flying at lower altitudes, usually resulting in smaller areas covered per image (< 1 ha compared to 27,000 ha for WorldView-2 data). HSR images are also referred to as "very-large scale" in reference to the map scale (e.g., 1:500 compared to 1:24,000 for a 7.5 ft × 7.5 ft topographic map).

HSR imagery are useful in most aspects of rangeland resource management including assessments of ground cover, riparian condition, wildlife habitat, woodlands and other resource concerns. One of the characteristics of HSR images is very high pixel to pixel variability, but each pixel is spectrally pure (Figure 12.6). With larger pixel sizes, each pixel is a heterogeneous mixture of different spectral components (Figure 12.6). At the scale of 1 m, spectral unmixing or spectral indices are required to measure the amount of bare ground, so the advantages of HSR satellite data compared to medium resolution satellite data are not clear. On the other hand, at the scale of 0.4-mm HSR aircraft data, the established techniques of remote sensing are inadequate and new techniques

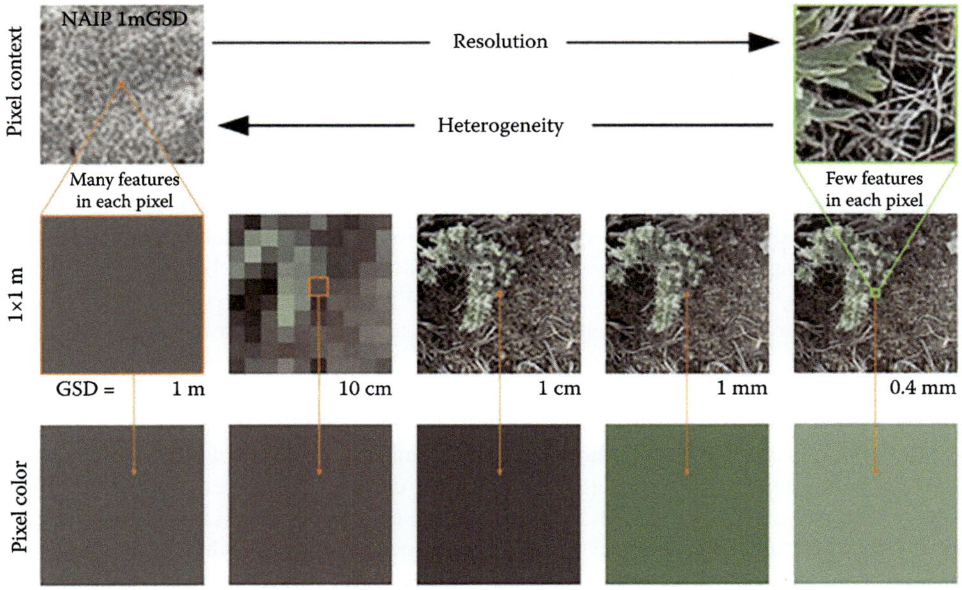

FIGURE 12.6 Effect of spatial resolution (also called ground sample distance) on pixel color and visual interpretation. The top left panel shows a 100 × 100-m area of rangeland in southeast Idaho at 1-m resolution to illustrate the heterogeneity of the imaged area. The middle row shows a 1 × 1-m patch of rangeland in all panels with the lowest spatial resolution on the left (1-m pixel size) and highest resolution on the right (0.4-mm pixel size). The top right panel shows an enlargement of the 0.4-mm resolution image; the red dot in the center (shown by the blue arrow) is one 0.4-mm pixel situated on a sagebrush leaf. The lower row shows pixel color change from increasing resolution from 1 m to 0.4 mm. (Figure adapted from Weber et al. (2013), and used with permission.)

are being developed and tested. Alternatively, established techniques of visual photo-interpretation can often be updated for use with digital imagery. As shown by Figure 12.6, the resolution of HSR data needs to be from 0.25 mm to 10 mm in order to visually interpret the data effectively.

12.6.1 Ground Imaging

Most ground assessments continue to be interpretations and judgments based on monitoring a limited number of non-randomly selected sites. It simply has not been practical to do otherwise. However, ground and aerial digital photography are replacing or augmenting time-consuming field methods, and are increasing the objectivity of rangeland monitoring (Booth et al., 2008; Moffet, 2009; Sadler et al., 2010; Booth and Cox, 2011). Digital photography and subsequent photo interpretation can increase sampling rates over traditional field sampling methods (Kouba et al., 2024; Retallack et al., 2023; Questad et al., 2022; Jafari et al., 2017; McCord et al., 2017 Luscier, 2006; Cagney et al., 2011). Reduced monitoring time and associated labor costs can increase monitoring precision by allowing for increased sample numbers.

Methods for obtaining non-aerial, nadir-looking digital images include a free-hand method (Cagney et al., 2011) and the use of staffs, tripods, stands (Booth et al., 2004; Booth and Cox, 2011), booms, and gantries (Louhaichi et al., 2010), vehicle mounts and other aids to reduce camera motion. Furthermore, these tools allow for a range of camera positions above ground level and a range of image resolutions. For a permanent record, the digital images are archived at much lower cost compared to film negatives or prints. Even where sampling will be done primarily from aircraft, acquisition of ground images is an important part of monitoring since it provides the opportunity to capture nadir images with less motion blur than can be obtained from the air.

12.6.2 Aerial Imaging

Some image users require boundary-to-boundary coverage, and HSR mosaics may be made from airborne data at considerable expense. However, many users do not require mosaics, and can manage effectively at lower cost by capturing single-image samples that are systematically distributed across watersheds, allotments, and other landscape-scale management units (Figure 12.7). Geographic Information Systems (GIS) should be used to draft sampling plans that define intended locations for ground or aerial image acquisition to acquire a statistically adequate sample that represents the natural range of variation in the areas monitored (Blumenthal et al., 2007; Booth and Cox, 2008).

Platforms used for HSR aerial-image acquisition include manned conventional and light sport airplanes (LSAs), manned and unmanned helicopters, and unmanned fixed-wing aircraft (Booth and Cox, 2011). Surprisingly, contracting for a piloted LSA was the least expensive, because fewer personnel were required to be on site for safety. Furthermore, only LSAs and helicopters were able to fly slow enough to avoid excessive motion blur with image resolutions of 1-mm resolution or less (Figure 12.8). Contract costs for manned helicopters are usually more expensive than for LSAs, but helicopters (manned or unmanned) can fly during windy conditions when LSAs cannot safely operate. To mitigate risk, LSAs need to be equipped with rocket-deployed, whole-plane parachutes in the event of mechanical failure. Frequently in windy areas, flights are made early in the day before strong winds and atmospheric turbulence develops. A potential problem with early morning flights is presence of shadows that may affect estimates of bare ground cover.

Manned helicopters and LSAs can carry multiple cameras allowing for nested, simultaneously acquired, multi-resolution images; for example: 1-, 10-, and 20-mm pixel sizes. This capability allowed users to acquire wide fields of view *and* high resolution, a capability that proved important for the study of fire intervals in shrub ecosystems (Moffet, 2009), and monitoring invasive species (Kouba et al., 2024; Retallack et al., 2023; Maphanga et al., 2022; Questad et al., 2022; Dube et al., 2021; Blumenthal et al., 2007, 2012; Booth et al., 2010).

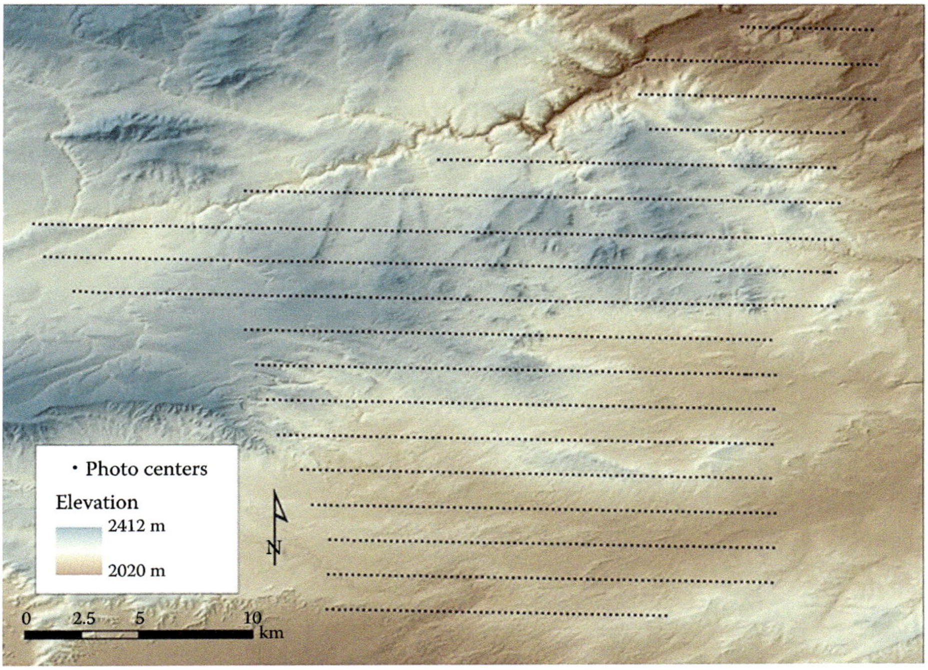

FIGURE 12.7 Flight plan in Wyoming's Red Desert (center: 43.37°N, 108.40°W) showing 1,457 image acquisition points which were 200 m apart along east-west flight lines spaced 1600 m apart. Three images were collected at each acquisition point: (1) an area of 3 × 4 m with 0.9-mm pixels; (2) an area of 24 × 36 m, with7-mm pixels; and (3) an area of 48 × 72 m with 14-mm pixels.

Aerial surveys acquired imagery with 1-mm pixels that overcame the need to depend on subjective selection of "representative" study areas. Summarizing, aerial surveys acquiring multi-resolution imagery that includes 1-mm pixel data have repeatedly demonstrated:

1. Lower cost than extensive conventional ground sampling for areas greater than about 200 ha;
2. Practical acquisition of large sample numbers;
3. Reduced sample-collection time;
4. Collection of many samples within short phenological windows;
5. Creation of permanent records that may be examined any time in the future; and
6. Capability of capturing sub-meter details for detecting ecologically important changes.

Measurement of bare ground from 1-mm pixel images is just one example of the different indicators of rangeland health that may be monitored.

12.6.3 Visual Analysis

For measuring ground cover, image resolution should be at least 1-mm resolution (Booth and Cox, 2009). Attempts to accurately measure ground-cover from 10-, 20-, and 50-mm resolution imagery were unsuccessful (Booth and Cox, 2009; Duniway et al., 2012; Weber et al., 2013). Figure 12.8 shows that 1-mm resolution images are essential to capturing the detail needed for identifying species. Current methods for analysis of this type of imagery include visual-analysis-facilitating

FIGURE 12.8 Portion of a 3 × 4-m aerial image (0.9-mm pixel size) acquired from a light sport airplane traveling 84 km/h at an altitude of 103 m above ground level in the Red Desert of central Wyoming. The main image shows bare soil, litter, shadow, grasses, sagebrush (*Artemisia tridentata* Nutt.) and yellow rabbitbrush (*Chrysothamnus viscidiflorus* (Hook.) Nutt.), while the circular inset (circled area on main image enlarged twofold) shows goldenweed (*Stenotus acaulis* (Nutt.) Nutt.).

software programs (Booth et al., 2006a, 2006b). These programs are accurate when used with appropriate image resolutions (you can't measure what you can't see), and by people with adequate field experience.

SamplePoint is free software (available online at www.SamplePoint.org; last accessed 30 May 2014) that facilitates point sampling of digital images. Because the sample point is always a single pixel of the image (Figure 12.9a), where the spatial resolution is equal or less than 1 mm, the analysis has a potential accuracy of 92%. To use the program, images are loaded from a computer directory, and a database is created to store the data entered. The number and pattern of sample points (grid or random) are determined by the user (Figure 12.9b). The predetermined classes are associated with buttons along the bottom of the image. *SamplePoint* automatically moves from point to point when the class information is entered by clicking one of the buttons. The software allows image magnification (zoom), and it will support up to three monitors, each of which displays the image with different levels of magnification.

Images need not be orthorectified or georeferenced for use in SamplePoint. Thus, users with no GIS experience can use the software effectively. This is an important point because other point-sampling tools like Image Interpreter Tool (Duniway et al., 2012) and Digital Mylar (Clark et al., 2004; RSAC, 2014) do require georeferenced imagery. That adds a workflow step that makes the software more difficult and time-consuming to use for those who have GIS experience, and virtually impossible for those who don't. The cited tools function very well with orthoimagery, as intended. Nevertheless, for ground- or aerial-based photo plot monitoring, a georeferenced image prerequisite restricts usage of these tools because (1) it is impractical to georeference hundreds of disconnected plot-based images and (2) commercial sub-meter orthoimagery is too expensive to acquire frequently. To illustrate, over the last six years the Bureau of Land Management contracted

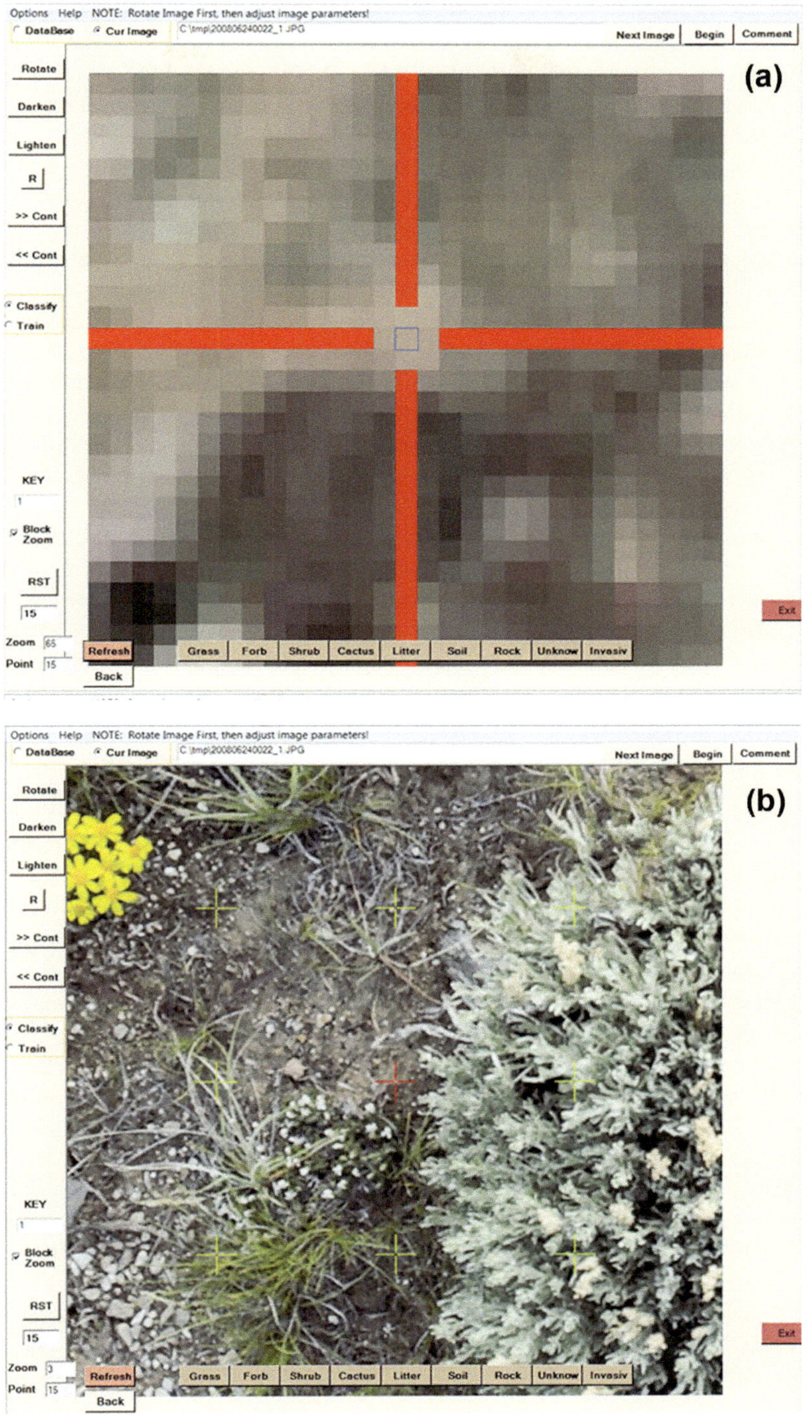

FIGURE 12.9 SamplePoint analysis of a ground photograph in Wyoming's Red Desert associated with aerial images (Figure 12.8) collected using the flight plan in Figure 12.7. Classification is performed on a single pixel within a 9-pixel array framed by the red crosshair by clicking one of the user-defined buttons along the bottom of the screen (a). The same sampling point is shown zoomed out (b) to show the context of the point that is hitting bare ground in between spiny phlox (*Phlox hoodii* Richardson) and sagebrush (*Artemesia tridentata* Nutt.). Note the systematic grid sampling pattern of the classification crosshairs.

only one sub-meter orthoimage collection in the state of Wyoming: 30-cm resolution at \$18/km². Collection of the 2–3-cm orthoimagery that Duniway et al. (2012) reported as yielding reliable estimates for many cover types will cost substantially more than \$18/km², limiting usage among agencies where even contracting for 30-cm orthoimagery is usually cost-prohibitive. As internal collection capabilities improve, cost may become less of a factor.

First-time users of 1-mm resolution imagery are often surprised at the number of species they can identify in an image. This leads to an initial presumption that they can use *SamplePoint* to measure cover by species. While this may be true for appropriately spaced woody species, measuring cover by plant functional type (grasses, grass-like plants, and forbs) is usually more practical except where and when plant species are distinctive. *SamplePoint* does *not* correct for user biases that may occur due to personal interpretations of protocol (e.g., what is litter), or correct for conditions such as age that can influence color perception. In fact, the variation among *SamplePoint* users was found to be about equal with that of users of the line-point intercept (Moffet, 2009). Additionally, multiple people used *SamplePoint* to analyze the same set of images showing there are user biases in class selection. When biases are found, the image dataset can be re-analyzed. Thus, data verifiability and the capability to significantly increase sampling are key advantages of using *SamplePoint* with image-based monitoring.

Two other programs are also freely available at www.SamplePoint.org, *SampleFreq* and *ImageMeasurement*. *SampleFreq* functions similarly to *SamplePoint*, but it facilitates nested frequency sampling. *ImageMeasurement* is used to measure areas and lengths from nadir imagery of known scale. Originally, this program was used to measure stream and channel widths with other ecological indicators for riparian areas. *ImageMeasurement* is unique among similar programs by incorporating exact image resolution for every image in a multi-image dataset.

12.6.4 Unmanned Aircraft Systems (UAS)

Unmanned aircraft systems (UAS) may also be effective platforms for remote sensing (Hardin and Jackson, 2005; Rango et al., 2006, 2009; Hardin and Hardin, 2010). Some of the first UAS that were used in agricultural and natural resource management were made from radio-control model aircraft (Parracciani et al., 2024; Fenetahun et al., 2022; Matongera et al., 2021; Tomlins and Lee, 1983; Quilter and Anderson, 2000, 2001). Now, there are many types of advanced civilian UAS, from high-altitude long-endurance to low-altitude short-endurance aircraft (Watts et al., 2012), so few generalizations may be made about current UAS flight capabilities. LSAs may fly in steady wind conditions of 32 km/h when there are no wind gusts. Two particular UAS (UX5 and Gatewing X100, Trimble Navigation Limited, Westminster, CO, USA) are claimed to be able to fly in 65 km/h winds (Trimble Navigation, 2013), because the automatic pilot updates aircraft controls at 100 Hz (Southard, 2013). The ability of UAS to fly in moderate to high winds may facilitate timely collection of aerial data for biodiversity.

Like manned aircraft, UAS may be configured with medium-resolution, high spectral resolution, or high spatial resolution sensors. Furthermore, LiDAR, thermal infrared, radar, and other sensors are available for UAS platforms. Most of the research uses small low-flying UAS with digital cameras to measure ground cover (Ocholla et al., 2024; Dube et al., 2021; McCord et al., 2017; Hardin et al., 2007; Laliberte et al., 2007, 2010; Breckenridge and Dakins, 2011; Breckenridge et al., 2011). Hardin et al. (2007) using small UAS to detect squarrose knapweed found that overall detection accuracy was low but the rate of false positives was miniscule. Breckenridge et al. (2011) also attempted to determine presence and sex of sage grouse (*Centrocercus urophasianus*) using decoys to represent grouse during spring mating season. At a height of 73 m, 100% of the male and 80% of the female decoys could be detected by a skilled observer, whereas at 305 m height, 90% of the male and only 10% of the female decoys could be detected by the skilled observer.

With small UAS and digital cameras, it is easy to acquire many more images than would be possible to analyze using aids such as *SamplePoint*, so automated processing or preprocessing is required.

Many people using these systems have backgrounds in image processing, in which the typical workflow would be production of orthorectified image mosaics. Two general methods used in orthorectification are the scale invariant feature transform (Lowe, 2004) and structure from motion (Turner et al., 2012; Westoby et al., 2012). Classification of land cover types from these image mosaics is not very accurate when using methods developed for multispectral satellite data, because the spectral properties of a small pixel have little connection with its cover type (Figure 12.6). Instead, the spatial relationships among pixels are used in an object-based classification (Parracciani et al., 2024; Retallack et al., 2023; Questad et al., 2022; Dube et al., 2021; McCord et al., 2017; Luscier et al., 2006; Laliberte et al., 2007, 2010; Laliberte and Rango, 2011).

12.6.5 Assessment of High Spatial Resolution

Remote sensing with high spatial resolution sensors can be used for monitoring numerous indicators of rangeland health (Table 12.1) in order to get a preponderance of evidence for evaluation. Aircraft are necessary for HSR data acquisition; and image analysis currently requires human inputs. The typical image-processing workflow for aerial photography is to geo- and ortho-rectify the large number of photographs to produce a single image with boundary-to-boundary coverage. We suggest that this time-consuming effort is not necessary for rangeland monitoring. By assuming each photograph is one plot in a statistical sample, accurate conclusions may be determined regarding the land area as a whole. Furthermore, from the geographic coordinates of each photograph, the plot-scale results may be combined for landscape-scale information by kriging or co-kriging.

Unsupervised or supervised classification of remote sensed imagery requires areas of a known category for training, which is a problem for early detection of invasive species. Computer-facilitated interpretation of HSR data may be used to detect plant species new to an area, because photographs from other areas would be used as examples. Furthermore, with low-cost data storage, HSR acquisitions may be archived for re-examination.

Most of the airborne HSR data used for method development were acquired using a manned light sport aircraft (Booth and Cox, 2009, 2011), and the costs for LSA remote sensing were less than other methods. The decade-long LSA effort included a strong emphasis on safety and there were no incidents during the thousands of LSA flight-hours. Even though the availability of high-quality components have made UAS much more reliable than a few years ago, there will be flight and control failures. The expanse of rangelands is large, so the number of flight hours required for monitoring will be large; using either LSA or UAS as low-altitude platforms should facilitate the acquisition of HSR imagery for rangeland monitoring.

12.7 CONCLUSIONS

Rangeland state-and-transition models account for differences in plant communities occurring at an ecological site, and these models are being adopted in different countries for rangeland management (Kouba et al., 2024; Ocholla et al., 2024; Parracciani et al., 2024; Angerer et al., 2023; Kleinhesselink et al., 2023; Retallack et al., 2023; Thornley et al., 2023; Fenetahun et al., 2022; Maphanga et al., 2022; Questad et al., 2022; Dube et al., 2021; Matongera et al., 2021; Sawalhah et al., 2018; Jafari et al., 2017; McCord et al., 2017). Based on the natural range of variation, recently disturbed ecological sites in the reference state may have lower diversity and biomass compared to an alternative state. Therefore, after accounting for interannual variation in precipitation, detecting lower biomass with spectral vegetation indices is not necessarily equal to detecting rangeland degradation. Instead, degraded rangelands crossed a transition into an alternative ecological state such that the area will not recover over time to its former diversity, biomass, and structure. Operational monitoring of rangelands by remote sensing needs to be based on the information required for management in order to prevent degradation.

As discussed in this chapter, there are many examples showing how sensors with different combinations of high spectral resolution, high temporal resolution, and high spatial resolution may be used to measure biodiversity, usually by detecting various indicators of rangeland health. We attempted to make the best possible case for rangeland monitoring with each data type. *At the current time, we conclude that facilitated image interpretation of high spatial resolution data is the best method for rangeland monitoring for management based on state-and-transition models.* This conclusion is based on one narrowly focused objective, providing information required for modifying management practices in order to protect rangeland diversity and prevent rangeland degradation. Other objectives, such as estimating biomass forage production, require other sources of remotely sensed data.

Medium resolution satellite data have been available for decades, and the only rangeland health indicator that may be reliably estimated is the amount of bare ground (Table 12.1). Transitions to alternative states may be recognized afterwards, but perhaps too late to prevent further degradation. High temporal resolution data from operational meteorological satellites are necessary for monitoring vegetation phenology, functional types, and effects of drought. Phenology using higher spatial resolution data from satellites currently in orbit is an extremely important new source of information, and there should be an international effort to archive these data for world-wide availability. Multiple years' worth of data need to have been acquired and available in order to account for interannual variability in temperature and precipitation.

High spectral resolution data may be used to detect and measure bare soil, many different invasive species, and most plant functional types, the three most important indicators of rangeland health (MacKinnon et al., 2011). Combined hyperspectral sensors and LiDAR in aircraft (Kouba et al., 2024; Ocholla et al., 2024; Parracciani et al., 2024; Angerer et al., 2023; Kleinhesselink et al., 2023; Retallack et al., 2023; Thornley et al., 2023; Fenetahun et al., 2022; Maphanga et al., 2022; Questad et al., 2022; Dube et al., 2021; Matongera et al., 2021; Sawalhah et al., 2018; Jafari et al., 2017; McCord et al., 2017; Asner et al., 2007; Asner and Martin, 2009; Kampe et al., 2010) are promising because it is much easier to identify pure spectral components. Airborne sensors have pixel sizes from 5 to 30 m, so image texture would help identify areas of spatial heterogeneity for animal habitat. Finally, litter cover, biological crusts, and forage quality may be determined by chemical composition.

Hyperspectral remote sensing does well when either the analyst has some prior information on where the target is located (such as invasive species) or the spectral signature is invariant (cellulose and lignin for litter cover). Soil minerals have spectral signatures, whereas soil types are classified based on pedogenesis. Plant species do not have signatures in the way that soil minerals and biological compounds do, but this conclusion is being re-evaluated using more-advanced airborne sensors (Asner and Martin, 2009, 2011; Asner et al., 2011). Furthermore, Rocchini et al. (2010a, 2010b) hypothesized spectral diversity *per se* is important for estimating biodiversity, which is parallel to the claim that spatial diversity from image texture is important for estimating biodiversity. Object-based classification combines both spectral and spatial diversity and may become an effective method.

Rangeland monitoring with high spatial resolution data provides information directly relevant for management based on rangeland health, to avoid transitions to more degraded states. There may be a large difference between the scales at which information is useful compared to the scales at which the data is acquired to produce the information. The scale for information required by rangeland managers is the landscape, but acquisition of high spatial resolution data may be the most cost effective method of obtaining the required information.

12.8 DISCLAIMER

Mention of product names is for information only, does not imply government endorsement and other products may be equally suitable.

REFERENCES

Adams, J. B., Smith, M. O., & Johnson, P. E. 1986. Spectral mixture modeling: A new analysis of rock and soil types at the Viking Lander 1 site. Journal of Geophysical Research, 91, 8098–8112.

Adelabu, S., Mutanga, O., Adam, E., & Cho, M. A. 2013. Exploiting machine learning algorithms for tree species classification in a semiarid woodland using RapidEye image. Journal of Applied Remote Sensing, 7, 073480–1.

Adjorlolo, C., Cho, M. A., Mutanga, O., & Ismail, R. 2012a. Optimizing spectral resolutions for the classification of C_3 and C_4 grass species, using wavelengths of known absorption features. Journal of Applied Remote Sensing, 6, 063560.

Adjorlolo, C., Mutanga, O., Cho, M. A., & Ismail, R. 2012b. Challenges and opportunities in the use of remote sensing for C_3 and C_4 grass species discrimination and mapping. African Journal of Range and Forage Science, 29, 47–61.

Allen, C. R., Gunderson, L., & Johnson, A. R. 2005. The use of discontinuities and functional groups to assess relative resilience in complex systems. Ecosystems, 8, 958–966.

Anderson, G. L., Carruthers, R. I., Ge, S., & Gong, P. 2005. Monitoring of invasive *Tamarix* distribution and effects of biological control with airborne hyperspectral remote sensing. International Journal of Remote Sensing, 26, 2487–2489.

Anderson, G. L., Everitt, J. H., Escobar, D. E., Spencer, N. R., & Andrascik, R. J. 1996. Mapping leafy spurge (*Euphorbia esula*) infestations using aerial photography and geographic information systems. Geocarto International, 11, 81–89.

Anderson, G. L., Prosser, C. W., Wendel, L. E., Delfosse, E. S., & Faust, R. M. 2003. The ecological area-wide management (TEAM) of leafy spurge program of the United States Department of Agriculture – Agricultural Research Service. Pest Management Science, 59, 609–613.

Anderson, R. P., Peterson, A. T., & Egbert, S. L. 2006. Vegetation-index models predict areas vulnerable to purple loosestrife (*Lythrum salicaria*) invasion in Kansas. Southwestern Naturalist, 51, 471–480.

Andrew, M. E., & Ustin, S. L. 2006. Spectral and physiological uniqueness of perennial pepperweed (*Lepidium latifolium*). Weed Science, 54, 1051–1062.

Andrew, M. E., & Ustin, S. L. 2008. The role of environmental context in mapping invasive plants with hyperspectral image data. Remote Sensing of Environment, 112, 4301–4317.

Andrew, M. E., & Ustin, S. L. 2009. Habitat suitability modeling of an invasive plant with advanced remote sensing data. Diversity and Distributions, 15, 627–640.

Angerer, J. P., Fox, W. E., Wolfe, J. E., Tolleson, D. R., & Owen, T. 2023. Chapter 20—Land degradation in rangeland ecosystems. Pages 394–434 in *Hazards and Disasters Series, Biological and Environmental Hazards, Risks, and Disasters* (Second Edition) (Ramesh Sivanpillai and John F. Shroder, eds.). Elsevier, ISBN 9780128205099, https://doi.org/10.1016/B978-0-12-820509-9.00007-1. (www.sciencedirect.com/science/article/pii/B9780128205099000071)

Asner, G. P. 2004. Biophysical remote sensing signatures of arid and semiarid ecosystems. Pages 53–109 in *Remote Sensing for Natural Resource Management and Environmental Monitoring. Manual of Remote Sensing, 3rd Edition, Volume 4* (S. L. Ustin, ed.). Hoboken, NJ: John Wiley and Sons.

Asner, G. P., Knapp, D. E., Kennedy-Bowdoin, T., Jones, M. O., Martin, R. E., Boardman, J., & Field, C. B. 2007. Carnegie airborne observatory: In-flight fusion of hyperspectral imaging and waveform light detection and ranging (wLiDAR) for three-dimensional studies of ecosystems. Journal of Applied Remote Sensing, 1, 013536.

Asner, G. P., & Martin, R. E. 2009. Airborne spectranomics: Mapping canopy chemical and taxonomic diversity in tropical forests. Frontiers in Ecology and Environment, 7, 269–276.

Asner, G. P., & Martin, R. E. 2011. Canopy phylogenetic, chemical and spectral assembly in a lowland Amazonian Forest. New Phytologist, 189, 999–1012.

Asner, G. P., Martin, R. E., Knapp, D. E., Tupayachi, R., Anderson, C., Carranza, L., Martinez, P., Houcheime, M., Sinca, F., & Weiss, P. 2011. Spectroscopy of canopy chemicals in humid tropical forests. Remote Sensing of Environment, 115, 3587–3598.

Balch, J. K., Bradley, B. A., D'Antonio, C. M., & Gómez-Dans, J. 2013. Introduced annual grass increases regional fire activity across the arid western USA. Global Change Biology, 19, 173–183.

Bastin, G. N., Denham, R., Scarth, P., Sparrow, A., & Chewings, V. 2014. Remotely-sensed analysis of ground-cover change in Queensland's rangelands, 1988–2005. Rangeland Journal, 36, 191–204.

Bastin, G. N., Scarth, P., Chewings, V., Sparrow, A., Denham, R., Schmidt, M., O'Reagain, P., Shepard, R., & Abbot, B. 2012. Separating grazing and rainfall effects at regional scale using remote sensing imagery: A dynamic reference-cover method. Remote Sensing of Environment, 121, 443–457.

Bastin, G. N., Stafford Smith, D. M., Watson, I. W., & Fisher, A. 2009. The Australian collaborative rangelands information system: Preparing for a climate of change. Rangeland Journal, 31, 111–125.

Belgacem, A. O., & Louhaichi, M. 2013. The vulnerability of native rangeland plant species to global climate change in the West Asia and North African regions. Climatic Change, 119, 451–463.

Bellis, L. M., Pidgeon, A. M., Radeloff, V. C., St-Louis, V., Navarro, J. L., & Martella, M. B. 2008. Modeling habitat suitability for greater rheas based on satellite image texture. Ecological Applications, 18, 1956–1966.

Bentivegna, D. J., Smeda, R. J., & Wang, C. 2012. Detecting cutleaf teasel (*Dipsacus laciniatus*) along a Missouri highway with hyperspectral imagery. Invasive Plant Science and Management, 5, 155–163.

Bestelmeyer, B. T. 2006. Threshold concepts and their use in rangeland management and restoration: The good, the bad, and the insidious. Restoration Ecology, 14, 325–329.

Bestelmeyer, B. T., Goolsby, D. P., & Archer, S. R. 2011. Spatial perspectives in state-and-transition models: A missing link to land management. Journal of Applied Ecology, 48, 746–757.

Blaschke, T. 2010. Object-based image analysis for remote sensing. ISPRS Journal of Photogrammetry and Remote Sensing, 65, 2–16.

Blumenthal, D. M., Booth, D. T., Cox, S. E., & Ferrier, C. E. 2007. Large-scale aerial images capture details of invasive plant populations. Rangeland Ecology & Management, 60, 523–528.

Blumenthal, D. M., Norton, A. P., Cox, S. E., Hardy, E. M., Liston, G. E., Kennaway, L., Booth, D. T., & Derner, J. D. 2012. *Linaria dalmatica* invades south-facing slopes and less grazed areas in grazing-tolerant mixed-grass prairie. Biological Invasions, 14, 395–404.

Boardman, J. W., & Kruse, F. A. 2011. Analysis of imaging spectrometer data using N-dimensional geometry and a mixture-tuned matched filter approach. IEEE Transactions on Geoscience and Remote Sensing, 49, 4138–4152.

Booth, D. T., & Cox, S. E. 2008. Image-based monitoring to measure ecological change in rangeland. Frontiers of Ecology and Management, 6, 185–190.

Booth, D. T., & Cox, S. E. 2009. Dual-camera, high-resolution aerial assessment of pipeline revegetation. Environmental Monitoring and Assessment, 158, 23–33.

Booth, D. T., & Cox, S. E. 2011. Art to science: Tools for greater objectivity in resource monitoring. Rangelands, 33(4), 27–34.

Booth, D. T., Cox, S. E., & Berryman, R. D. 2006a. Point sampling digital imagery with 'Samplepoint'. Environmental Monitoring and Assessment, 123, 97–108.

Booth, D. T., Cox, S. E., & Berryman, R. D. 2006b. Precision measurements from very-large scale aerial digital imagery. Environmental Monitoring and Assessment, 112, 293–307.

Booth, D. T., Cox, S. E., Louhaichi, M., & Johnson, D. E. 2004. Lightweight camera stand for close-to-earth remote sensing. *Journal of Range Management*, 57, 675–678.

Booth, D. T., Cox, S. E., Meilke, T., & Zuuring, H. R. 2008. Ground-cover measurements: Assessing correlation among aerial and ground-based measurements. Environmental Management, 42, 1091–1100.

Booth, D. T., Cox, S. E., & Teel, D. 2010. Aerial assessment of leafy spurge (*Euphorbia esula* L.) on Idaho's Deep Fire Burn. Native Plants Journal, 11, 327–339.

Booth, D. T., & Tueller, P. T. 2003. Rangeland management using remote sensing. Arid Land Research and Management, 17, 455–467.

Boryan, C., Yang, Z., Mueller, R., & Craig, M. 2011. Monitoring US agriculture: The US department of agriculture, national agricultural statistics service, cropland data layer program. Geocarto International, 26, 341–358.

Bradley, B. A. 2013. Remote detection of invasive plants: A review of spectral, textural and phenological approaches. Biological Invasions, DOI: 10.1007/s10530-013-0578-9.

Bradley, B. A., Blumenthal, D. M., Wilcove, D. S., & Ziska, L. H. 2009. Predicting plant invasions in an era of global change. Trends in Ecology and Evolution, 25, 310–318.

Bradley, B. A., & Mustard, J. F. 2005. Identifying land cover variability distinct from land cover change: Cheatgrass in the Great Basin. Remote Sensing of Environment, 94, 204–213.

Bradley, B. A., & Mustard, J. F. 2006. Characterizing the landscape dynamics of an invasive plant and risk of invasion using remote sensing. Ecological Applications, 16, 1132–1147.

Breckenridge, R. P., & Dakins, M. 2011. Evaluation of bare ground on rangelands using unmanned aerial vehicles: A case study. GIScience & Remote Sensing, 48, 74–85.

Breckenridge, R. P., Dakins, M., Bunting, S., Harbour, J. L., & White, S. 2011. Comparison of unmanned aerial vehicle platforms for assessing vegetation cover in sagebrush steppe ecosystems. Rangeland Ecology & Management, 64, 521–532.

Briske, D. D., Bestelmeyer, B. T., Stringham, T. K., & Shaver, P. L. 2008. Recommendations for development of resilience-based state-and-transition models. Rangeland Ecology & Management, 61, 359–367.

Briske, D. D., Fuhlendorf, S. D., & Smeins, F. E. 2005. State-and-transition models, thresholds, and rangeland health: A synthesis of ecological concepts and perspectives. Rangeland Ecology & Management, 58, 1–10.

Briske, D. D., Fuhlendorf, S. D., & Smeins, F. E. 2006. A unified framework for assessment and application of ecological thresholds. Rangeland Ecology & Management, 59, 225–236.

Brown, J. R.1994. State and transition models for rangelands. 2. Ecology as a basis for rangeland management: Performance criteria for testing models. Tropical Grasslands, 28, 206–213.

Brown, J. R. 2010. Ecological sites: Their history, status and future. Rangelands, 32(6), 5–8.

Cagney, J., Cox, S. E., & Booth, D. T. 2011. Comparison of point intercept and image analysis for monitoring rangeland transects. Rangeland Ecology & Management, 64, 309–315.

Carter, G. A., Knapp, A. K., Anderson, J. E., Hoch, G. A., & Smith, M. D. 2005. Indicators of plant species richness in AVIRIS spectra of a mesic grassland. Remote Sensing of Environment, 98, 304–316.

Carter, G. A., Lucas, K. L., Blossom, G. A., Lassitter, C. L., Holiday, D. M., Mooneyhan, D. S., Fastring, D. R., Holcombe, T. R., & Griffith, J. A. 2009. Remote sensing and mapping tamarisk along the Colorado River, USA: A comparative use of summer-acquired hyperion, thematic mapper, and quickbird data. Remote Sensing, 1, 318–329.

Caudle, D., Sanchez, H., DiBenedetto, J., Talbot, C., & Karl, M. 2013. *Interagency Ecological Site Handbook for Rangelands*. Washington, DC: US DOI Bureau of Land Management, USDA Forest Service, and USDA Natural Resource Conservation Service. http://www.ars.usda.gov/Research/docs.htm?docid=18502, last accessed 28 Feb. 2014.

Chang, C.-I. 2000. An information-theoretic approach to spectral variability, similarity, and discrimination for hyperspectral image analysis. IEEE Transactions on Information Theory, 46, 1927–1932.

Cheng, Y. B., Tom, E., & Ustin, S. L. 2007. Mapping an invasive species, kudzu (*Pueraria montana*), using hyperspectral imagery in western Georgia. Journal of Applied Remote Sensing, 1, 013514.

Cho, M. A., Debba, P., Mathieu, R., Naidoo, L., van Aardt, J., & Asner, G. P. 2010. Improving discrimination of savanna tree species through a multiple-endmember spectral angle mapper approach: Canopy level analysis. IEEE Transactions on Geoscience and Remote Sensing, 48, 4133–4142.

Clark, J., Finco, M., Warbington, R., & Schwind, B. 2004. Digital Mylar: A tool to attribute vegetation polygon features over high-resolution imagery. In: *Proceedings of the Tenth Forest Service Remote Sensing Applications Conference* Salt Lake City, Utah. www.fs.fed.us/r5/rsl/publications/, last accessed 9 March 2014.

Clark, M. L., Roberts, D. A., & Clark, D. B. 2005. Hyperspectral discrimination of tropical rainforest tree species at leaf to crown scales. Remote Sensing of Environment, 96, 375–398.

Clark, R. N., Swayze, G. A., Livo, K. E., Kokaly, R. F., Sutley, S. J., Dalton, J. B., McDougal, R. R., & Gent, C. A. 2003. Imaging spectroscopy: Earth and planetary remote sensing with the USGS Tetracorder and expert systems. Journal of Geophysical Research, 108, E12, 5131.

Clements, F. E. 1916. *Plant Succession:An Analysis of the Development of Vegetation*. Carnegie Institute Publication Volume 242. Washington, DC: Carnegie Institute of Washington.

Clinton, N. E., Potter, C., Crabtree, B., Genovese, V., Gross, P., & Gong, P. 2010. Remote sensing-based time-series analysis of cheatgrass (*Bromus tectorum* L.) phenology. Journal of Environmental Quality, 39, 955–963.

Congalton, R. G., & Green, K. 2008. *Assessing the Accuracy of Remotely Sensed Data: Principles and Practices*. Boca Raton, FL: CRC press, Taylor and Francis Group, 183 pp.

Cooper, W. S. 1924. An apparatus for photographic recording of quadrats. Journal of Ecology, 12, 317–321.

Culbert, P. D., Radeloff, V. C., St-Louis, V., Flather, C. H., Rittenhouse, C. D., Albright, T. P., & Pidgeon, A. M. 2012. Modeling broad-scale patterns of avian species richness across the Midwestern United States with measures of satellite image texture. Remote Sensing of Environment, 118, 140–150.

Dalmayne, J., Möckel, T., Prentice, H. C., Schmid, B. C., & Hall, K. 2013. Assessment of fine-scale plant species beta diversity using WorldView-2 satellite spectral dissimilarity. Ecological Informatics, 18, 1–9.

Daughtry, C. S. T. 2001. Discriminating crop residues from soil by shortwave infrared reflectance. Agronomy Journal, 93, 125–131.

Daughtry, C. S. T., & Hunt Jr., E. R. 2008. Mitigating the effects of soil and residue water contents on remotely sensed estimates of crop residue cover. Remote Sensing of Environment, 112, 1647–1657.

Davidson, A., & Csillag, F. 2001. The influence of vegetation index and spatial resolution on a two-date remote sensing-derived relation to C4 species coverage. Remote Sensing of Environment, 75, 138–151.

Davidson, E. A., Asner, G. P., Stone, T. A., Neill, C., & Figueiredo, R. O. 2008. Objective indicators of pasture degradation from spectral mixture analysis of Landsat imagery. Journal of Geophysical Research, 113, G00B03.

de Beurs, K. M., & Henebry, G. M. 2004. Land surface phenology, climatic variation, and institutional change: Analyzing agricultural land cover change in Kazakhstan. Remote Sensing of Environment, 89, 497–509.

de Beurs, K. M., & Henebry, G. M. 2010. Spatio-temporal statistical methods for modeling land surface phenology. Pages 177–208 in *Phenological Research* (I. L. Hudson and M. R. Keatley, eds.). New York, NY: Springer Science and Business Media.

DiTomaso, J. M. 2000. Invasive weeds in rangelands: Species, impacts, and management. Weed Science, 48, 255–265.

Dobowski, S. Z., Safford, H. D., Cheng, Y. B., & Ustin, S. L. 2008. Mapping mountain vegetation using species distribution modeling, image-based texture analysis, and object-based classification. Applied Vegetation Science, 11, 499–508.

Dube, T., Shoko, C. and Gara, T. W. 2021. Remote sensing of aboveground grass biomass between protected and non-protected areas in savannah rangelands. African Journal of Ecology, Wiley online library. 59:3, 687–695.

Dudek, K. B., Root, R. R., Kokaly, R. F., & Anderson, G. L. 2004. Increased spatial and temporal consistency of leafy spurge maps from multidate AVIRIS imagery: A modified, hybrid linear spectral mixture analysis/mixture-tuned matched filtering approach. In *Proceedings of the 13th JPL Airborne Earth Science Workshop* (R. O. Green, ed.). Pasadena, CA, USA: Jet Propulsion Laboratory.

Duniway, M. C., Karl, J. W., Schrader, S., Baquera, N., & Herrick, J. E. 2012. Range and pasture monitoring using high resolution aerial imagery: A repeatable image interpretation approach. Environmental Monitoring and Assessment, 184, 789–804.

Elmqvist, T., Folke, C., Nyström, M., Peterson, G., Bengtsson, J., Walker, B., & Norberg, J. 2003. Response diversity, ecosystem change, and resilience. Frontiers of Ecology and Management, 1, 488–494.

Eyre, T. J., Fisher, A., Hunt, L. P., & Kutt, A. S. 2011a. Measure it to better manage it: A biodiversity monitoring framework for the Australian rangelands. Rangeland Journal, 33, 239–253.

Eyre, T. J., Kelly, A. L., Neldner, V. J., Wilson, B. A., Ferguson, D. J., Laidlaw, M. J., & Franks, A. J. 2011b. *BioCondition: A Condition Assessment Framework for Terrestrial Biodiversity in Queensland.* Assessment Manual. Version 2.1. Brisbane, Queensland, Australia: Department of Environment and Resource Management (DERM), Biodiversity and Ecosystem Sciences.

Fenetahun, Y., Yuan, Y., Xu, X., Wang, Y. 2022. Borana rangeland of southern Ethiopia: Estimating biomass production and carrying capacity using field and remote sensing data. Plant Diversity, 44:6, 598–606, ISSN 2468–2659, https://doi.org/10.1016/j.pld.2022.03.003. (www.sciencedirect.com/science/article/pii/S2468265922000361)

Feret, J. B., François, C., Asner, G. P., Gitelson, A. A., Martin, R. E., Bidel, L. P. R., Ustin, S. L., le Maire, G., & Jaquemoud, S. 2008. PROSPECT-4 and 5: Advances in the leaf optical properties model separating photosynthetic pigments. Remote Sensing of Environment, 112, 3030–3043.

Fletcher, R. S., Everitt, J. H., & Yang, C. 2011. Identifying saltcedar with hyperspectral data and support vector machines. Geocarto International, 26, 195–209.

Folke, C., Carpenter, S., Walker, B., Scheffer, M., Elmqvist, T., Gunderson, L., & Holling, C. S. 2004. Regime shifts, resilience, and biodiversity in ecosystem management. Annual Review of Ecology, Evolution, and Systematics, 35, 557–581.

Foody, G. M., & Dash, J. 2007. Discriminating and mapping the C3 and C4 composition of grasslands in the northern Great Plains, USA. Ecological Informatics, 2, 89–93.

Foody, G. M., & Dash, J. 2010. Estimating the relative abundance of C_3 and C_4 grasses in the Great Plains from multi-temporal MTCI data: Issues of compositing period and spatial generalizability. International Journal of Remote Sensing, 31, 351–362.

Franklin, S. E. 2010. *Remote Sensing for Biodiversity and Wildlife Management.* New York: McGraw-Hill, 346 pp.

Franklin, S. E., & Wulder, M. A. 2002. Remote sensing methods in medium spatial resolution satellite data land cover classification of large areas. Progress in Physical Geography, 26, 173–205.

Friedl, M. A., Sulla-Menashe, D., Tan, B., Schnieder, A., Ramankutty, N., Sibley, A., & Huang, X. 2010. MODIS collection 5 global land cover: Algorithm refinements and characterization of new datasets. Remote Sensing of Environment, 114, 168–182.

Fuhlendorf, S. D., & Engle, D. M. 2001. Restoring heterogeneity on rangelands: Ecosystem management based on evolutionary grazing patterns. Bioscience, 51, 625–632.

Fuhlendorf, S. D., Harrell, W. C., Engle, D. M., Hamilton, R. G., Davis, C. A., & Leslie Jr., D. M. 2006. Should heterogeneity be the basis for conservation? Grassland bird response to fire and grazing. Ecological Applications, 16, 1706–1716.

Gao, B. C., Heidebrecht, K. B., & Goetz, A. F. H. 1993. Derivation of scaled surface reflectances from AVIRIS data. Remote Sensing of Environment, 44, 145–163.

Gaston, K. J. 2000. Global patterns in biodiversity. Nature, 405, 220–227.

Ge, S., Everitt, J., Carruthers, R., Gong, P., & Anderson, G. 2006. Hyperspectral characteristics of canopy components and structure for phonological assessment of an invasive weed. Environmental Monitoring and Assessment, 120, 109–126.

Ge, S., Xu, M., Anderson, G. L., & Carruthers, R. I. 2007. Estimating yellow starthistle (*Centaurea solstitialis*) leaf area index and aboveground biomass with the use of hyperspectral data. Weed Science, 55, 671–678.

Gibbons, P., Briggs, S. V., Ayers, D. A., Doyle, S., Seddon, J., McElhinny, C., Jones, N., Sims, R., & Doody, J. S. 2008. Rapidly quantifying reference conditions in modified landscapes. Biological Conservation, 141, 2483–2493.

Gibbons, P., Briggs, S. V., Ayers, D. A., Seddon, J., Doyle, S., Cosier, P., McElhinny, C., Pelly, V., & Roberts, K. 2009. An operational method to assess impacts of land clearing on terrestrial biodiversity. Ecological Indicators, 9, 26–40.

Gill, T. K., & Phinn, S. R. 2008. Estimates of bare ground and vegetation cover from Advanced Spaceborne Thermal Emission and Reflection Radiometer (ASTER) short-wave-infrared reflectance imagery. Journal of Applied Remote Sensing, 2, 023511.

Gillespie, T. W., Foody, G. M., Rocchini, D., Giorgi, A. P., & Saatchi, S. 2008. Measuring and modeling biodiversity from space. Progress in Physical Geography, 32, 203–221.

Glenn, N. F., Mundt, J. T., Weber, K. T., Prather, T. S., Lass, L. W., & Pettingill, J. 2005. Hyperspectral data processing for repeat detection of small infestations of leafy spurge. Remote Sensing of Environment, 95, 399–412.

Goodin, D. G., & Henebry, G. M. 1997. A technique for monitoring ecological disturbance in tallgrass prairie using seasonal NDVI trajectories and a discriminant function mixture model. Remote Sensing of Environment, 61, 270–278.

Gottschalk, T. K., Huettmann, F., & Ehlers, M. 2005. Thirty years of analyzing and modeling avian habitat relationships using satellite imagery data: A review. International Journal of Remote Sensing, 26, 2631–2656.

Gould, W. 2000. Remote sensing of vegetation, plant species richness, and regional biodiversity hotspots.
Ecological Applications, 10, 1861– 1870.

Goward, S. N., Chander, G., Pagnutti, M., Marx, A., Ryan, R., Thomas, N., & Tetrault, R. 2012. Complementarity of ResourceSat-1 AWiFS and Landsat TM/ETM+ sensors. Remote Sensing of Environment, 123, 41–56.

Green, R. O., Eastwood, M. L., Sarture, C. M., Chrien, T. G., Aronsson, M., Chippendale, B. J., Faust, J. A., Pavri, B. E., Chovit, C. J., Solis, M., Olah, M. R., & Williams, O. 1998. Imaging Spectroscopy Environment, 65, 227–248.

Gu, Y., B´elair, S., Mahfouf, J., & Deblonde, G. 2006. Optional interpolation analysis of leaf area index using MODIS data. Remote Sensing of Environment, 104, 283–96.

Gu, Y., Brown, J. F., Miura, T., van Leeuwen, W. J. D., & Reed, B. C. 2010. Phenological classification of the United States: A geographic framework for extending multi-sensor time-series data. Remote Sensing, 2, 526–544.

Guerschman, J. P., Hill, M. J., Renzullo, L. J., Barrett, D. J., Marks, A. S., & Botha, E. J. 2009. Estimating fractional cover of photosynthetic vegetation, non-photosynthetic vegetation and bare soil in the Australian tropical savanna region upscaling the EO-1 Hyperion and MODIS sensors. Remote Sensing of Environment, 113, 928–945.

Guo, X., Price, K. P., & Stiles, J. 2003. Grasslands discriminant analysis using Landsat TM single and multitemporal data. Photogrammetric Engineering & Remote Sensing, 69, 1255–1262.

Hall, K., Reitalu, T., Sykes, M. T., & Prentice, H. C. 2012. Spectral heterogeneity of QuickBird satellite data is related to fine-scale plant species spatial turnover in semi-natural grasslands. Applied Vegetation Science, 15, 145–157.

Hamada, Y., Stow, D. A., Coulter, L. L., Jafolla, J. C., & Hendricks, L. W. 2007. Detecting Tamarisk species (*Tamarix* spp.) in riparian habitats of Southern California using high spatial resolution hyperspectral imagery. Remote Sensing of Environment, 109, 237–248.

Haralick, R. M., Shanmugam, K., & Dinstein, I. 1973. Textural features for image classification. IEEE Transactions on Systems, Man and Cybernetics, 3, 610–621.

Hardin, P. J., & Hardin, T. J. 2010. Small-scale remotely piloted vehicles in environmental research. Geography Compass, 4:9, 1297–1311.

Hardin, P. J., & Jackson, M. W. 2005. An unmanned aerial vehicle for rangeland photography. Rangeland Ecology & Management, 58, 439–442.

Hardin, P. J., Jackson, M. W., Anderson, V. J., & Johnson, R. 2007. Detecting squarrose knapweed (*Centaurea virgata* Lam. ssp. *squarrosa* Gugl.) using a remotely piloted vehicle: A Utah case study. GIScience & Remote Sensing, 44, 203–219.

He, K. S., Rocchini, D., Neteler, M., & Nagendra, H. 2011. Benefits of hyperspectral remote sensing for tracking plant invasions. Diversity and Distributions, 17, 381–392.

Hernandez, A. J., & Ramsey, R. D. 2013. A landscape similarity index: Multitemporal remote sensing to track changes in big sagebrush ecological sites. Rangeland Ecology & Management, 66, 71–81.

Herrick, J. E., Van Zee, J. W., Havstad, K. M., Burkett, L. M., & Whitford, W. G. 2005a. *Monitoring Manual for Grassland, Shrubland, and Savanna Ecosystems, Volume I: Quick Start*. Las Cruces, NM: USDA-ARS Jornada Experimental Range. 36 pp.

Herrick, J. E., Van Zee, J. W., Havstad, K. M., Burkett, L. M., & Whitford, W. G. 2005b. *Monitoring Manual for Grassland, Shrubland, and Savanna Ecosystems, Volume II: Design, supplementary methods and interpretation*. Las Cruces, NM: USDA-ARS Jornada Experimental Range, 200 pp.

Hestir, E. L., Khanna, S., Andrew, M. E., Santos, M. J., Viers, J. H., Greenberg, J. A., Rajapakse, S. S., & Ustin, S. L. 2008. Identification of invasive vegetation using hyperspectral remote sensing in the California Delta ecosystem. Remote Sensing of Environment, 112, 4034–4047.

Higley, B. M., & Kelly, D. E. 1873. Home on the range. Kansas: Smith County Pioneer.

Homer, C., Huang, C., Yang, L., Wylie, B., & Coan, M. 2004. Development of a 2001 National Landcover Database for the United States. Photogrammetric Engineering & Remote Sensing, 70, 829–840.

Hooper, D. U., Chapin III, F. S., Ewel, J. J., Hector, A., Inchausti, P., Lavorel, S., Lawton, J. H., Lodge, D. M., Loreau, M., Naeem, S., Schmid, B., Setälä, H., Symstad, A. J., Vandermeer, J., & Wardle, D. A. 2005. Effects of biodiversity on ecosystem functioning: A consensus of current knowledge. Ecological Monographs, 75, 3–35.

Huang, C., & Asner, G. P. 2009. Applications of remote sensing to alien invasive plant studies. Sensors, 9, 4869–4889.

Huang, C., & Geiger, E. L. 2008. Climate anomalies provide opportunities for large-scale mapping of non-native plant abundance in desert grasslands. Diversity and Distributions, 14, 875–884.

Huang, C., Geiger, E. L., van Leeuven, W. J. D., & Marsh, S. E. 2009. Discrimination of invaded and native species sites in a semi-desert grassland using MODIS multi-temporal data. International Journal of Remote Sensing, 30, 897–917.

Huete, A. R. 1988. A soil-adjusted vegetation index (SAVI). Remote Sensing of Environment, 25, 295–309.

Huete, A. R., Didan, K., Miura, T., Rodriguez, E. P., Gao, X., & Ferreira, L. G. 2002. Overview of the radiometric and biophysical performance of the MODIS vegetation indices. Remote Sensing of Environment, 83, 195–213.

Hunt Jr., E. R., Daughtry, C. S. T., Kim, M. S., & Parker Williams, A. E. 2007. Using canopy reflectance models and spectral angles to assess potential of remote sensing to detect invasive weeds. *Journal of Applied Remote Sensing*, 1: 013506.

Hunt Jr., E. R., Everitt, J. H., Ritchie, J. C., Moran, M. S., Booth, D. T., Anderson, G. L., Clark, P. E., & Seyfried, M. S. 2003. Applications and research using remote sensing for rangeland management. Photogrammetric Engineering & Remote Sensing, 69, 675–693.

Hunt Jr., E. R., Gillham, J. H., & Daughtry, C. S. T. 2010. Improving potential geographic distribution models for invasive plants by remote sensing. Rangeland Ecology & Management, 63, 505–513.

Hunt Jr., E. R., McMurtrey III, J. E., Parker Williams, A. E., & Corp, L. A. 2004. Spectral characteristics of leafy spurge (*Euphorbia esula*) leaves and flower bracts. Weed Science, 52, 492–497.

Hunt Jr., E. R., & Parker Williams, A. E. 2006. Detection of flowering leafy spurge with satellite multispectral imagery. Rangeland Ecology & Management, 59, 494–499.

Irisarri, J. G. N., Oesterheld, M., Vernón, S. R., & Paruelo, J. M. 2009. Grass species differentiation through canopy hyperspectral reflectance. International Journal of Remote Sensing, 30, 5959–5975.

Jacquemoud, S., Ustin, S. L., Verdebout, J., Schmuck, G., Andreoli, G., & Hosgood, B. 1996. Estimating leaf biochemistry using the PROSPECT leaf optical properties model. Remote Sensing of Environment, 56, 194–202.

Jacquemoud, S., Verhoef, W., Baret, F., Bacour, C., Zarco-Tejada, P. J., Asner, G. P., François, C., & Ustin, S. L. 2009. PROSPECT+SAIL models: A review of use for vegetation characterization. Remote Sensing of Environment, 113(suppl), S56–S66.

Jafari, F., Jafari, R., & Bashari, H. 2017. Assessing the performance of remotely sensed landscape function indices in semi-arid rangelands of Iran. The Rangeland Journal, 39, 253–262.

Janetos, A., Hansen, L., Inouye, D., Kelly, B. P., Meyerson, L., Peterson, B., & Shaw, R. 2008. Chapter 5. Biodiversity. Pages 151–182 in: *The Effects of Climate Change on Agriculture, Land Resources, Water Resources, and Biodiversity in the United States*. Synthesis and Assessment Product 4.3. Washington, DC: U.S. Climate Change Science Program and the Subcommittee on Global Change Research.

Jiang, Z., Huete, A. R., Chen, J., Chen, Y., Li, J., Yan, G., & Zhang, X. 2006. Analysis of NDVI and scaled difference vegetation index retrievals of vegetation fraction. Remote Sensing of Environment, 101, 366–378.

Johansen, K., & Phinn, S. 2006. Mapping structure parameters and species composition of riparian vegetation using IKONOS and Landsat ETM+ data in Australian tropical savannahs. Photogrammetric Engineering & Remote Sensing, 72, 71–80.

John, R., Chen, J., Lu, N., Guo, K., Liang, C., Wei, Y., Noormets, A., Ma, K., & Han, X. 2008. Predicting plant diversity based on remote sensing products in the semi-arid region of Inner Mongolia. Remote Sensing of Environment, 112, 2018–2032.

Jönsson, P., & Eklundh, L. 2004. TIMESAT—A program for analyzing time-series of satellite sensor data. Computers & Geoscience, 30, 833–845.

Joyce, L. A., Briske, D. D., Brown, J. R., Polley, H. W., McCarl, B. A., & Bailey, D. W. 2013. Climate change and North American Rangelands: Assessment of mitigation and adaption strategies. Rangeland Ecology & Management, 66, 512–528.

Kampe, T. U., Johnson, B. R., Kuester, M., & Keller, M. 2010. NEON: The first continental-scale observatory with airborne remote sensing of vegetation biochemistry and structure. Journal of Applied Remote Sensing, 4, 043510.

Karnieli, A., Kokaly, R. F., West, N. E., & Clark, R. N. 2003. Remote sensing of biological soil crusts. Pages 431–455 in *Biological Soil Crusts: Structure,Function, and Management* (J. Belnap and O. L. Lange, eds.). Ecological Studies Volume 150. Berlin, Germany: Springer-Verlag.

Kleinhesselink, A. R., Kachergis, E. J., McCord, S. E., Shirley, J., Hupp, N. R., Walker, J., Carlson, J. C., Morford, S. L., Jones, M. O., Smith, J. T., Allred, B. W., Naugle, D. E., 2023. Long-term trends in vegetation on bureau of land management rangelands in the Western United States. Rangeland Ecology & Management, 87, 1–12, ISSN 1550–7424, https://doi.org/10.1016/j.rama.2022.11.004. (www.sciencedirect.com/science/article/pii/S1550742422001075)

Knox, N. M., Skidmore, A. K., Prins, H. H. T., Asner, G. P., van der Werff, H., de Boer, W. F., van der Waal, C., de Knegt, H. J., Kohi, E. M., Slotow, R., & Grant, R. C. 2011. Dry season mapping of savanna forage quality, using the hyperspectral Carnegie Airborne Observatory sensor. Remote Sensing of Environment, 115, 1478–1488.

Kogan, F., Gitelson, A., Zakarin, E., Spivak, L., & Lebed, L. 2003. AVHRR-based spectral vegetation index for quantitative assessment of vegetation state and productivity: Calibration and validation. Photogrammetric Engineering & Remote Sensing, 69, 899–906.

Kouba, Y., Doghbage, A., Merdas, S., & Bagousse-Pinguet, Y. E. 2024. Grazing exclusion modulates the effects of different components of plant diversity on biomass production in semiarid rangeland. Agriculture, Ecosystems & Environment, 365, 108914, ISSN 0167–8809, https://doi.org/10.1016/j.agee.2024.108914. (www.sciencedirect.com/science/article/pii/S016788092400032X)

Kremer, R. G., & Running, S. W. 1993. Community type differentiation using NOAA/AVHRR data within a sagebrush-steppe ecosystem. Remote Sensing of Environment, 46, 311–318.

Kruse, F. A., Lefkoff, A. B., Boardman, J. W., Heidebrecht, K. B., Shapiro, A. T., Barloon, P. J., & Goetz, A. F. H. 1993. The Spectral Image Processing System (SIPS)—interactive visualization and analysis of imaging spectrometer data. Remote Sensing of Environment, 44, 145–163.

Kumar, L., Sinha, P., Brown, J. F., Ramsey, R. D., Rigge, M., Stam, C. A., Hernandez, A. J., Hunt, E. R., & Reeves, M. C. In press. Rangeland, grassland, shrublands monitoring: Methods and approaches. Chapter 12 in *Remote Sensing Handbook. Volume II: Land Resources: Monitoring, Modeling, and Mapping* (P. S. Thenkabail, ed.). Boca Raton, FL: CRC Press Taylor & Francis Group.

Laliberte, A. S., Herrick, J. E., Rango, A., & Winters, C. 2010. Acquisition, orthorectification, and object-based classification of unmanned aerial vehicle (UAV) imagery for rangeland monitoring. Photogrammetric Engineering & Remote Sensing, 76, 661–672.

Laliberte, A. S., & Rango, A. 2011. Image processing and classification procedures for analysis of sub-decimeter imagery acquired with an unmanned aircraft over arid rangelands. GIScience & Remote Sensing, 48, 4–23.

Laliberte, A. S., Rango, A., Herrick, J. E., Fredrickson, E. L., & Burkett, L. 2007. An object-based image analysis approach for determining fractional cover of senescent and green vegetation with digital plot photography. Journal of Arid Environments, 69, 1–14.

Lass, L. W., Prather, T. S., Glenn, N. F., Weber, K. T., Mundt, J. T., & Pettingill, J. 2005. A review of remote sensing of invasive weeds and example of the early detection of spotted knapweed (*Centaurea maculosa*) and babysbreath (*Gypsophila paniculata*) with a hyperspectral sensor. Weed Science, 53, 242–251.

Lass, L. W., Thill, D. C., Shafii, B., & Prather, T. S. 2002. Detecting spotted knapweed (*Centaurea maculosa*) with hyperspectral remote sensing technology. Weed Technology, 16, 426–432.

Lavergne, S., Mouquet, N., Thuiller, W., & Ronce, O. 2010. Biodiversity and climate change: Integrating evolutionary and ecological responses of species and communities. Annual Review of Ecology, Evolution, and Systematics, 41, 321–350.

Lawrence, R. L., Wood, S. D., & Sheley, R. L. 2006. Mapping invasive plants using hyperspectral imagery and Breiman Cutler classifications (Random Forest). Remote Sensing of Environment, 100, 356–362.

Lesica, P., & Hanna, D. 2009. Effect of biological control on leafy spurge (*Euphorbia esula*) and diversity of associated grasslands over 14 years. Invasive Plant Science and Management, 2, 151–157.

Leyequien, E., Verrelst, J., Slot, M., Schaepman-Strub, G., Heitkönig, I. M. A., & Skidmore, A. 2007. Capturing the fugitive: Applying remote sensing to terrestrial animal distribution and diversity. International Journal of Applied Earth Observation and Geoinformation, 9, 1–20.

Louhaichi, M., Johnson, M. D., Woerz, A. L., Jasra, A. W., & Johnson, D. E. 2010. Digital charting technique for monitoring rangeland vegetation cover at local scale. International Journal of Agriculture and Biology, 12, 406–410.

Loveland, T. R., Merchant, J. W., Ohlen, D. O., & Brown, J. F. 1991. Development of a land-cover characteristics database for the conterminous U.S. Photogrammetric Engineering & Remote Sensing, 57, 1453–1463.

Lowe, D. G. 2004. Distinctive image features from scale-invariant keypoints. International Journal of Computer Vision, 60, 91–110.

Lu, D., & Weng, Q. 2007. A survey of image classification methods and techniques for improving classification performance. International Journal of Remote Sensing, 28, 823–870.

Ludwig, J. A., Bastin, G. N., Chewings, V. H., Eager, R. W., & Liedloff, A. C. 2007. Leakiness: A new index for monitoring the health of arid and semiarid landscapes using remotely sensed vegetation cover and elevation data. Ecological Indicators, 7, 442–454.

Ludwig, J. A., Eager, R. W., Bastin, G. N., Chewings, V. H., & Liedloff, A. C. 2002. A leakiness index for assessing landscape function using remote sensing. Landscape Ecology, 17, 157–171.

Ludwig, J. A., Eager, R. W., Liedloff, A. C., Bastin, G. N., & Chewings, V. H. 2006. A new landscape leakiness index based on remotely sensed ground-cover data. Ecological Indicators, 6, 327–336.

Ludwig, J. A., Tongway, D. J., Bastin, G. N., & James, C. D. 2004. Monitoring ecological indicators of rangeland functional integrity and their relation to biodiversity at local and regional scales. Austral Ecology, 29, 108–120.

Ludwig, J. A., Wiens, J. A., & Tongway, D. J. 2000. A scaling rule for landscape patches and how it applies to conserving soil resources in savannas. Ecosystems, 3, 84–97.

Luscier, J. D., Thompson, W. L., Wilson, J. M., Gorham, B. E., & Dragut, L. D. 2006. Using digital photographs and object-based image analysis to estimate percent ground cover in vegetation plots. Frontiers of Ecology and Management, 4, 408–413.

Lym, R. G. 2005. Integration of biological control agents with other weed management technologies: Success from the leafy spurge (*Euphorbia esula*) IPM program. Biological Control, 35, 366–375.

MacKinnon, W. C., Karl, J. W., Toevs, G. R., Taylor, J. J., Karl, M., Spurrier, C. S., & Herrick, J. E. 2011. *BLM Core Terrestrial Indicators and Methods*. Tech Note 440. Denver, CO: U.S. Department of the Interior, Bureau of Land Management, National Operations Center.

Madden, M. 2004. Remote sensing and geographic information system operations for vegetation mapping of invasive exotics. Weed Technology, 18, 1457–1463.

Mansour, K., & Mutanga, O. 2012. Classifying increaser species as an indicator of different levels of rangeland degradation using WorldView-2 imagery. Journal of Applied Remote Sensing, 6, 063558–1.

Mansour, K., Mutanga, O., & Everson, T. 2012. Remote sensing based indicators of vegetation species for assessing rangeland degradation: Opportunities and challenges. African Journal of Agricultural Research, 7, 3261–3270.

Maphanga, T., Dube, T., Shoko, C., & Sibanda, M. 2022. Advancements in the satellite sensing of the impacts of climate and variability on bush encroachment in savannah rangelands. Remote Sensing Applications: Society and Environment, 25, 100689, ISSN 2352–9385, https://doi.org/10.1016/j.rsase.2021.100689. (www.sciencedirect.com/science/article/pii/S2352938521002251)

Marsett, R. C., Qi, J., Heilman, P., Biedenbender, S. H., Watson, M. C., Amer, S., Weltz, M., Goodrich, D., & Marset, R. 2006. Remote sensing for grassland management in the arid southwest. Rangeland Ecology & Management, 59, 530–540.

Martin, M. P., Barreto, L., Riaño, D., Fernandez-Quintanilla, C., & Vaughan, P. 2011. Assessing the potential of hyperspectral remote sensing for the discrimination of grass weeds in winter cereal crops. International Journal of Remote Sensing, 31, 49–67.

Matongera, T. N., Mutanga, O., Sibanda, M., & Odindi, J. 2021. Estimating and monitoring land surface phenology in Rangelands: A review of progress and challenges. Remote Sensing, 13:11, 2060. https://doi.org/10.3390/rs13112060

Maynard, C. L., Lawrence, R. L., Nielson, G. A., & Decker, G. 2007. Ecological site descriptions and remotely sensed imagery as a tool for rangeland evaluation. Canadian Journal of Remote Sensing, 33, 109–115.

McCord, S. E., Buenemann, M., Karl, J. W., Browning, D. M., & Hadley, B. C. 2017. Integrating remotely sensed imagery and existing multiscale field data to derive Rangeland indicators: Application of Bayesian additive regression trees. Rangeland Ecology & Management, 70:5, 644–655, ISSN 1550–7424, https://doi.org/10.1016/j.rama.2017.02.004. (www.sciencedirect.com/science/article/pii/S1550742417300222)

McKeon, G. M., Stone, G. S., Syktus, J. I., Carter, J. O., Flood, N. R., Ahrens, D. G., Bruget, D. N., Chilcoty, C. R., Cobon, D. H., Cowley, R. A., Crimp, S. J., Fraser, G. W., Howden, S. M., Johnson, P. W., Ryan, J. G., Stokes, C. J., & Day, K. A. 2009. Climate change impacts on northern Australian rangeland livestock carrying capacity: A review of issues. Rangeland Journal, 31, 1–29.

McNairn, H., & Protz, R. 1993. Mapping corn residue cover on agricultural fields in Oxford County, Ontario, using Thematic Mapper. Canadian Journal of Remote Sensing, 19:152, 159.

Miao, X., Gong, P., Swope, S., Pu, R., Carruthers, R., & Anderson, G. L. 2007. Detection of yellow starthistle through band selection and feature extraction from hyperspectral imagery. Photogrammetric Engineering & Remote Sensing, 73, 1005–1015.

Miao, X., Gong, P., Swope, S., Pu, R., Carruthers, R., Anderson, G. L., Heaton, J. S., & Tracy, C. R. 2006. Estimation of yellow starthistle abundance through CASI-2 hyperspectral imagery using linear spectral mixture models. Remote Sensing of Environment, 101, 329–341.

Miao, X., Patil, R., Heaton, J. S., & Tracy, R. C. 2011. Detection and classification of invasive saltcedar through high spatial resolution airborne hyperspectral imagery. International Journal of Remote Sensing, 32, 2131–2150.

Mirik, M., Ansley, R. J., Steddom, K., Jones, D. C., Rush, C. M., Michels Jr., G. J., & Elliot, N. C. 2013. Remote distinction of a noxious weed (musk thistle: *Carduss nutans*) using airborne hyperspectral imagery and the support vector machine classifier. Remote Sensing, 5, 612–630.

Mirik, M., Steddom, K., & Michels Jr., G. J. 2006. Estimating biophysical characteristics of musk thistle (*Carduus nutans*) with three remote sensing instruments. Rangeland Ecology & Management, 59, 44–54.

Mitchell, J. J., & Glenn, N. F. 2009. Subpixel abundance estimates in mixture-tuned matched filtering classifications of leafy spurge (*Euphorbia esula* L.). International Journal of Remote Sensing, 30, 6099–6119.

Mladinich, C. S., Ruiz Bustos, M., Stitt, S., Root, R., Brown, K., Anderson, G. L., & Hager, S. 2006. The use of Landsat 7 Enhanced Thematic Mapper Plus for mapping leafy spurge. Rangeland Ecology & Management, 59, 500–506.

Moffet, C. A. 2009. Agreement between measurements of shrub cover using ground-based methods and very large scale aerial imagery. Rangeland Ecology & Management, 62, 268–277.

Montandon, L. M., & Small, E. E. 2008. The impact of soil reflectance on the quantification of the green vegetation fraction from NDVI. Remote Sensing of Environment, 112, 1835–1845.

Morgan, J. A., Milchunas, D. G., LeCain, D. R., West, M., & Mosier, A. R. 2007. Carbon dioxide enrichment alters plant community structure and accelerates shrub growth in the shortgrass steppe. Proceedings of the National Academy of Sciences, 104, 14724–14729.

Morisette, J. T., Jarnevich, C. S., Ullah, A., Cai, W., Pedelty, J. A., Gentle, J. E., Stohlgren, T. J., & Schnase, J. L. 2006. A tamarisk habitat suitability map for the continental United States. Frontiers of Ecology and Management, 4, 11–17.

Moseley, K., Shaver, P. L., Sanchez, H., & Bestelmeyer, B. T. 2010. Ecological site development: A gentle introduction. Rangelands, 32:6, 16–22.

Müllerová, J., Pergl, J., & Pyšek, P. 2013. Remote sensing as a tool for monitoring plant invasions: Testing the effects of data resolution and image classification approach on the detection of a model plant species *Heracleum mantegazzianum* (giant hogweed). International Journal of Applied Earth Observation and Geoinformation, 25, 55–65.

Mundt, J. T., Glenn, N. F., Weber, K. T., Prather, T. S., Lass, L. W., & Pettingill, J. 2005. Discrimination of hoary cress and determination of its detection limits via hyperspectral image processing and accuracy assessment techniques. Remote Sensing of Environment, 96, 509–517.

Nagler, P. L., Glenn, E. P., Jarnevich, C. S., & Shafroth, P. B. 2011. Distribution and abundance of saltcedar and Russian olive in the Western United States. Critical Reviews of Plant Science, 30, 508–523.

Nagler, P. L., Inoue, Y., Glenn, E. P., Russ, A. L., & Daughtry, C. S. T. 2003. Cellulose absorption index (CAI) to quantify mixed soil-plant litter scenes. Remote Sensing of Environment, 87, 310–325.

Narumalani, S., Mishra, D. R., Burkholder, J., Merani, P. B. T., & Wilson, G. 2006. A comparative evaluation of ISODATA and Spectral Angle Mapping for the detection of saltcedar using airborne hyperspectral imagery. Geocarto International, 21, 59–66.

Narumalani, S., Mishra, D. R., Wilson, R., Reece, P., & Kohler, A. 2009. Detecting and mapping four invasive species along the floodplain of North Platte River, Nebraska. Weed Technology, 23, 99–107.

National Research Council (NRC). 1994. *Rangeland Health, New Methods to Classify, Inventory, and Monitor Rangelands*. Washington, DC: National Academy Press, 180 pp.

Noujdina, N. V., & Ustin, S. L. 2008. Mapping downy brome (*Bromus tectorum*) using multidate AVIRIS data. Weed Science, 56, 173–179.

Ocholla, I. A., Pellikka, P., Karanja, F. N., Vuorinne, I., Odipo, V., & Heiskanen, J. 2024. Livestock detection in African rangelands: Potential of high-resolution remote sensing data. Remote Sensing Applications: Society and Environment, 33, 101139, ISSN 2352–9385, https://doi.org/10.1016/j.rsase.2024.101139. (www.sciencedirect.com/science/article/pii/S235293852400003X)

Oliver, I., Eldridge, D. J., Nadolny, C., & Martin, W. K. 2014. What do site condition multi-metrics tell us about species biodiversity? Ecological Indicators, 38, 262–271.

Olofsson, P., Foody, G. M., Stehman, S. V., & Woodcock, C. E. 2013. Making better use of accuracy data in land change studies: Estimating accuracy and area and quantifying uncertainty using stratified estimation. Remote Sensing of Environment, 129, 122–131.

Olsson, A. D., & Morisette, J. T. 2014. Comparison of simulated HyspIRI with two multispectral sensors for invasive species mapping. Photogrammetric Engineering & Remote Sensing, 80, 217–227.

O'Neill, M., Ustin, S. L., Hager, S., & Root, R. 2000. Mapping the distribution of leafy spurge at Theodore Roosevelt National Park using AVIRIS. In *Proceedings of the 10th AVIRIS Airborne Geoscience Workshop* (R. O. Green, ed.). Pasadena, CA: Jet Propulsion Laboratory.

Parker Williams, A. E., & Hunt Jr., E. R. 2002. Estimation of leafy spurge cover from hyperspectral imagery using mixture tuned matched filtering. Remote Sensing of Environment, 82, 446–456.

Parker Williams, A. E., & Hunt Jr., E. R. 2004. Accuracy assessment for detection of leafy spurge with hyperspectral imagery. Journal of Range Management, 57, 106–112.

Parkes, D., Newell, G., & Cheal, D. 2003. Assessing the quality of native vegetation: The 'habitat hectares' approach. Ecological Management & Restoration, 4(suppl), S29–S38.

Parracciani, C., Gigante, D., Mutanga, O., Bonafoni, S. & Vizzari, M. 2024. Land cover changes in grassland landscapes: Combining enhanced Landsat data composition, LandTrendr, and machine learning classification in google earth engine with MLP-ANN scenario forecasting. GIScience & Remote Sensing, 61:1, DOI: 10.1080/15481603.2024.2302221

Pellant, M., Shaver, P., Pyke, D. A., & Herrick, J. E. 2005. *Interpreting Indicators of Rangeland Health, Version 4.* Technical Reference 1734–6. Denver, CO: United States Department of Interior, Bureau of Land Management, 122 pp.

Pereira, H. M., Leadley, P. W., Proença, V., Alkemade, R., Scharlemann, J. P. W., Fernandez-Manjarrés, J. F., Araújo, M. G., Balvanera, P., Biggs, R., Cheung, W. W. L., Chini, L., Cooper, H. D., Gilman, E. L., Guénette, S., Hurtt, G. C., Huntington, H. P., Mace, G. M., Oberdorff, T., Revenga, C., Rodrigues, P., Scholes, R. J., Sumaila, U. R., & Walpole, M. 2010. Scenarios for global biodiversity in the 21st Century. Science, 330, 1496–1501.

Pereira, H. M., Navarro, L. M., & Martins, I. S. 2012. Global biodiversity change: The bad, the good, and the unknown. Annual Review of Environment and Resources, 37, 25–50.

Peters, A. J., Eve, M. D., Holt, E. H., & Whitford, W. G. 1997. Analysis of desert plant community growth patterns with high temporal resolution satellite spectra. Journal of Applied Ecology, 34, 418–432.

Peters, A. J., Reed, B. C., Eve, M. D., & McDaniel, K. C. 1992. Remote sensing of broom snakeweed (*Gutierrezia sarothrae*) with NOAA-10 spectral image processing. Weed Technology, 6, 1015–1020.

Peterson, A. T. 2003. Predicting the geography of species' invasions via ecological niche modeling. Quarterly Review of Biology, 78, 419–433.

Peterson, D. L., Price, K. P., & Martinko, E. A. 2002. Discriminating between cool season and warm season grassland cover types in northeastern Kansas. International Journal of Remote Sensing, 23, 5015–5030.

Pettorelli, N., Ryan, S., Mueller, T., Bunnefeld, N., Jędrejewska, B., Lima, M., & Kausrud, K. 2011. The normalized difference vegetation index (NDVI): Unforeseen successes in animal ecology. Climate Research, 46, 15–27.

Polley, H. W., Briske, D. D., Morgan, J. A., Wolter, K., Bailey, D. W., & Brown, J. R. 2013. Climate change and North American Rangelands: Trends, projections, and implications. Rangeland Ecology & Management, 66, 493–511.

Pu, R. 2008. Invasive species change detection using artificial neural networks and CASI hyperspectral imagery. *Environmental Monitoring and Assessment*, 140, 15–32.

Pu, R. 2012. Detecting and mapping invasive plant species by using hyperspectral data. Pages 447–465 in *Hyperspectral Remote Sensing of Vegetation* (P. S. Thenkabail, J. G. Lyon and A. Huete, eds.). Boca Raton, FL: CRC Press-Taylor Francis Group.

Pu, R., Gong, P., Tian, Y., Miao, X., Carruthers, R. I., & Anderson, G. L. 2008. Using classification and NDVI differences for monitoring sparse vegetation coverage: A case study of saltcedar in Nevada, USA. International Journal of Remote Sensing, 29, 3987–4011.

Pyke, D. A., Herrick, J. E., Shaver, P., & Pellant, M. 2002. Rangeland health attributes and indicators for qualitative assessment. Journal of Range Management, 55, 584–597.

Questad, E. J., Antill, M., Liu, N., Natasha Stavros, E., Townsend, P. A., Bonfield, S., & Schimel, D. 2022. A camera-based method for collecting rapid vegetation data to support remote-sensing studies of Shrubland biodiversity. Remote Sensing, 14:8, 1933. https://doi.org/10.3390/rs14081933

Quilter, M. C., & Anderson, V. J. 2000. Low altitude/large scale aerial photographs: A tool for range and resource managers. Rangelands, 22:2, 13–17.

Quilter, M. C., & Anderson, V. J. 2001. A proposed method for determining shrub utilization using (LA/LS) imagery. Journal of Range Management, 54, 378–381.

Ramoelo, A., Skidmore, A. K., Cho, M. A., Schlerf, M., Mathieu, R., & Heitkönig, I. M. A. 2012. Regional estimation of savanna grass nitrogen using the red-edge band of the spaceborne RapidEye sensor. International Journal of Applied Earth Observation and Geoinformation, 19, 151–162.

Ramoelo, A., Skidmore, A. K., Schlerf, M., Mathieu, R., & Heitkönig, I. 2011. Water-removed spectra increase the retrieval accuracy when estimating savanna grass nitrogen and phosphorus concentrations. ISPRS Journal of Photogrammetry and Remote Sensing, 66, 408–417.

Rango, A., Laliberte, A., Herrick, J. E., Winters, C., Havstad, K., Steele, C., & Browning, D. 2009. Unmanned aerial vehicle-based remote sensing for rangeland assessment, monitoring and management. Journal of Applied Remote Sensing, 3, 033542.

Rango, A., Laliberte, A., Steele, C., Herrick, J. E., Bestelmeyer, B., Schmugge, T., Roanhorse, A., & Jenkins, V. 2006. Using unmanned aerial vehicles for rangelands: Current applications and future potentials. Environmental Practice, 8, 159–168.

Reed, B. C., Brown, J. F., VanderZee, D., Loveland, T. R., Merchant, J. W., & Ohlen, D. O. 1994. Measuring phenological variability from satellite imagery. Journal of Vegetation Science, 5, 703–714.

Reeves, M. C., Angerer, J., Hunt, E. R., Kulawardhana, W., Kumar, L., Loboda, T., Loveland, T., Metternicht, G., Ramsey, R. D., & Washington-Allen, R. A. In press. A global view of remote sensing of rangelands: Evolution, applications, and future pathways. Chapter 10 in *Remote Sensing Handbook. Volume II: Land Resources: Monitoring, Modeling, and Mapping* (P. S. Thenkabail, ed.). Boca Raton, FL: CRC Press Taylor & Francis Group.

Reich, P. B., Tilman, D., Isbell, F., Mueller, K., Hobbie, S. E., Flynn, D. F. B., & Eisenhauer, N. 2012. Impacts of biodiversity loss escalate through time as redundancy fades. Science, 336, 587–592.

Retallack, A., Finlayson, G., Ostendorf, B., Clarke, K., & Lewis, M. 2023. Remote sensing for monitoring rangeland condition: Current status and development of methods. Environmental and Sustainability Indicators, 19, 100285, ISSN 2665–9727, https://doi.org/10.1016/j.indic.2023.100285. (www.sciencedirect.com/science/article/pii/S2665972723000624)

Ricotta, C., Reed, B. C., & Tieszen, L. T. 2003. The role of C_3 and C_4 grasses to interannual variability in remotely sensed ecosystem performance over the US Great Plains. International Journal of Remote Sensing, 24, 4421–4431.

Roberts, D. A., Quattrochi, D. A., Hulley, G. C., Hook, S. J., & Green, R. O. 2012. Synergies between VSWIR and TIR data for the urban environment: An evaluation of the potential for the Hyperspectral Infrared Imager (HyspIRI) Decadal Survey mission. Remote Sensing of Environment, 117, 83–101.

Roberts, D. A., Smith, M. O., & Adams, J. B. 1993. Green vegetation, nonphotosynthetic vegetation, and soils in AVIRIS data. Remote Sensing of Environment, 44, 255–269.

Rocchini, D., Balkenhol, N., Carter, G. A., Foody, G. M., Gillespie, T. W., He, K. S., Kark, S., Levin, N., Lucas, K., Luoto, M., Nagendra, H., Oldeland, J., Ricotta, C., Southworth, J., & Neteler, M. 2010a. Remotely sensed spectral heterogeneity as a proxy of species diversity: Recent advances and open challenges. Ecological Informatics, 5, 318–329.

Rocchini, D., He, K. S., Oldeland, J., Wesuls, D., & Neteler, M. 2010b. Spectral variation versus species β-diversity at different spatial scales: A test in African highland savannas. Journal of Environmental Monitoring, 12, 825–831.

Rouse, J. W., Haas, R. H., Schell, J. A., & Deering, D. W. 1974. Monitoring vegetation systems in the Great Plains with ERTS. Pages 309–317 in *Third Earth Resources Technology Satellite-1Symposium, Volume 1: Technical Presentations*, NASA SP-351 (S. C. Freden, E. P. Mercanti and M. Becker, eds.). Washington, DC: National Aeronautics and Space Administration.

RSAC (Remote Sensing Applications Center). 2014. *Digital Mylar*. www.fs.fed.us/eng/rsac/digitalmylar/, last accessed 9 June 2014.

Running, S. W., Loveland, T. R., Pierce, L. L., Nemani, R., & Hunt Jr., E. R. 1995. A remote sensing based vegetation classification logic for global land cover analysis. Remote Sensing of Environment, 51, 39–48.

Sadler, R. J., Hazelton, M., Boer, M., & Grierson, P. F. 2010. Deriving state-and-transition models from an image series of grassland pattern dynamics. Ecological Modelling, 221, 433–444.

Sage, R. F., & Kubien, D. S. 2007. The temperature response of C_3 and C_4 photosynthesis. Plant, Cell & Environment, 30, 1086–1106.

Sampson, A. W. 1919. *Plant Succession Relation to Range Management*. USDA Technical Bulletin No. 791. Washington, DC: United States Department of Agriculture, 76 p.

Samuel, L. W., Kirby, D. R., Norland, J. E., & Anderson, G. L. 2008. Leafy spurge suppression by flea beetles in the Little Missouri Drainage Basin, USA. Rangeland Ecology & Management, 61, 437–443.

Savitzky, A., & Golay, M. J. E. 1964. Smoothing and differentiation of data by simplified least squares procedures. Analytical Chemistry, 36, 1627–1639.

Sawalhah, M. N., Al-Kofahi, S. D., Othman, Y. A., & Cibils, A. F. 2018. Assessing rangeland cover conversion in Jordan after the Arab spring using a remote sensing approach. Journal of Arid Environments, 157, 97–102, ISSN 0140–1963, https://doi.org/10.1016/j.jaridenv.2018.07.003. (www.sciencedirect.com/science/article/pii/S0140196318304415)

Schwartz, M. D., Reed, B. C., & White, M. A. 2002. Assessing satellite-derived start-of-season measures in the Conterminous USA. International Journal of Climate, 22, 1793–1805.

Serbin, G., Hunt Jr., E. R., Daughtry, C. S. T., & McCarty, G. W. 2013. Assessment of spectral indices for cover estimation of senescent vegetation. Remote Sensing Letters, 4, 552–560.

Serbin, G., Hunt Jr., E. R., Daughtry, C. S. T., McCarty, G. W., & Doraiswamy, P. C. 2009. An improved ASTER index for remote sensing of crop residue. Remote Sensing, 1, 971–991.

Shenk, J. S., Workman Jr., J. J., & Westerhaus, M. O. 2008. Application of NIR spectroscopy to agricultural products. Pages 347–386 in *Handbook of Near-Infrared Analysis, 3rd Edition* (D. A. Burns and E. W. Ciurczak, eds.). Boca Raton, FL: CRC Press.

Singh, N., & Glenn, N. F. 2009. Multitemporal spectral analysis for cheatgrass (*Bromus tectorum*) classification. International Journal of Remote Sensing, 30, 3441–3462.

Skidmore, A. K., Ferwerda, J. G., Mutanga, O., Van Wieren, S. E., Peel, M., Grant, R. C., Prins, H. H. T., Balcik, F. B., and Venus, V. 2010. Forage quality of savannas – simultaneously mapping foliar protein and polyphenols for trees and grass using hyperspectral imagery. Remote Sensing of Environment, 114, 64–72.

Soussanna, J.-F., & Lüscher, A. 2007. Temperate grasslands and global atmospheric change: A review. Grass and Forage Science, 62, 127–134.

Southard, G. 2013. A little UAV grows up. Presented on 27 March 2013 at the American Society of Photogrammetry and Remote Sensing 2013 Annual Conference, Baltimore, MD, USA.

Stafford Smith, D. M., McKeon, G. M., Watson, I. W., Henry, B. K., Stone, G. S., Hall, W. B., & Howden, S. M. 2007. Learning from episodes of degradation and recovery in variable Australian rangelands. Proceedings of the National Academy of Sciences, 104, 20690–20695.

Starks, P. J., Coleman, S. W., & Phillips, W. A. 2004. Determination of forage chemical composition using remote sensing. Rangeland Ecology & Management, 57, 635–640.

Steele, C. M., Bestelmeyer, B. T., Burkett, L. M., Smith, P. L., & Yanoff, S. 2012. Spatially explicit representation of state-and-transition models. Rangeland Ecology & Management, 65, 213–222.

Stitt, S., Root, R., Brown, K., Hager, S., Mladinich, C., Anderson, G. L., Dudek, K., Ruiz Bustos, M., & Kokaly, R. 2006. Classification of leafy spurge with Earth Observing-1 Advanced Land Imager. Rangeland Ecology & Management, 59, 507–511.

St-Louis, V., Pidgeon, A. M., Clayton, M. K., Locke, B. A., Bash, D., & Radeloff, V. C. 2009. Satellite image texture and a vegetation index predict avian biodiversity in the Chihuahuan desert of New Mexico. Ecography, 32, 468–480.

Stohlgren, T. J., Ma, P., Kumar, S., Rocca, M., Morisette, J. T., Jarnevich, C. S., & Benson, N. 2010. Ensemble habitat mapping of invasive plant species. Risk Analysis, 30, 224–235.

Tews, J., Brose, U., Grimm, V., Tielbörger, K., Wichmann, M. C., Schwager, M., & Jeltsch, F. 2004. Animal species diversity driven by habitat heterogeneity/diversity: The importance of keystone structures. Journal of Biogeography, 31, 79–92.

Thornley, R. H., Gerard, F. F., White, K., and Verhoef, A. 2023. Prediction of Grassland biodiversity using measures of spectral variance: A meta-analytical review. Remote Sensing, 15(3), 668. https://doi.org/10.3390/rs15030668

Thorp, K. R., French, A. N., & Rango, A. 2013. Effect of image spatial and spectral characteristics on mapping semi-arid rangeland vegetation using multiple endmember spectral mixture analysis (MESMA). Remote Sensing of Environment, 132, 120–130.

Thuiller, W., Richardson, D. M., Pyšek, P., Midgley, G. F., Hughes, G. O., & Rouget, M. 2005. Niche-based modeling as a tool for predicting the risk of alien plant invasions at a global scale. Global Change Biology, 11, 2234–2250.

Tieszen, L. L., Reed, B. C., Bliss, N. B., Wylie, B. K., & DeJong, D. D. 1997. NDVI, C_3 and C_4 production and distributions in Great Plains grassland land cover classes. Ecological Applications, 7, 59–78.

Tilman, D., Reich, P. B., & Isbell, F. 2012. Biodiversity impacts ecosystem productivity as much as resources, disturbance, or herbivory. Proceedings of the National Academy of Sciences, 109, 10394–10397.

Tilman, D., Reich, P. B., & Knops, J. M. H. 2006. Biodiversity and ecosystem stability in a decade-long grassland experiment. Nature, 441, 629–632.

Tomlins, G. F., & Lee, Y. J. 1983. Remotely piloted aircraft—an inexpensive option for large-scale aerial photography in forestry applications. Canadian Journal of Remote Sensing, 9, 76–85.

Townshend, J. R. G., & Justice, C. O. 1988. Selecting the spatial resolution of satellite sensors required for global monitoring of land transformations. International Journal of Remote Sensing, 9, 187–236.

Trimble Navigation Limited. 2013. *Trimble Unmanned Aircraft Systems for Surveying and Mapping*. Westminster, CO: Trimble Navigation Limited. www.trimble.com/Survey/unmanned-aircraft-systems.aspx, last accessed 9 June 2014.

Tucker, C. J. 1979. Red and photographic infrared linear combinations for monitoring vegetation. Remote Sensing of Environment, 8, 127–150.

Turner, D., Lucieer, A., & Watson, C. 2012. An automated technique for generating georectified mosaics from ultra-high resolution unmanned aerial vehicle (UAV) imagery, based on structure from motion (StM) point clouds. Remote Sensing, 4, 1392–1410.

Underwood, E. C., Klinger, R., & Moore, P. E. 2004. Predicting patterns of non-native plant invasions in Yosemite National Park, California, USA. Diversity and Distributions, 10, 447–459.

Underwood, E. C., Ustin, S., & DiPietro, D. 2003. Mapping nonnative plants using hyperspectral imagery. Remote Sensing of Environment, 86, 150–161.

Underwood, E. C., Ustin, S. L., & Ramirez, C. M. 2007. A comparison of spatial and spectral image resolution for mapping invasive plants in coastal California. Environmental Management, 39, 63–83.

USDA NRCS. 2006. United States Department of Agriculture, Natural Resource Conservation Service. *Land Resource Regions and Major Land Resource Areas of the United States, the Caribbean, and the Pacific Basin*. Handbook 206. Washington, DC: United States Department of Agriculture. www.nrcs.usda.gov/wps/portal/nrcs/detail/soils/survey/?cid=nrcs142p2_053624, last accessed 28 Feb. 2014.

Ustin, S. L., & Gamon, J. A. 2010. Remote sensing of plant functional types. New Phytologist, 186, 795–816.

Ustin, S. L., Riaño, D., & Hunt Jr., E. R. 2012. Estimating canopy water content from spectroscopy. Israel Journal of Plant Science, 60, 9–23.

Ustin, S. L., Valko, P. G., Kefauver, S. C., Santos, M. J., Zimpfer, J. F., & Smith, S. D. 2009. Remote sensing of biological soil crust under simulated climate change manipulations in the Mojave Desert. Remote Sensing of Environment, 113, 317–328.

van der Meer, F. 2006. The effectiveness of spectral similarity measures for the analysis of hyperspectral imagery. International Journal of Applied Earth Observation and Geoinformation, 8, 3–17.

van Deventer, A. P., Ward, A. D., Gowda, P. H., & Lyon, J. G. 1997. Using thematic mapper data to identify contrasting soil plains to tillage practices. Photogrammetric Engineering & Remote Sensing, 63, 87–93.

Vrieling, A., Rodrigues, S. C., Bartholomeus, H., & Sterk, G. 2007. Automatic identification of erosion gullies with ASTER imagery in the Brazilian Cerrados. International Journal of Remote Sensing, 28, 2723–2738.

Wang, C., Bentivegna. D. J., Smeda, R. J., & Swanigan, R. E. 2010a. Comparing classification approaches for mapping cut-leaved teasel in highway environments. *Photogrammetric Engineering & Remote Sensing*, 76, 567–575.

Wang, C., Fritschi, F. B., Stacey, G., & Yang, Z. 2011a. Phenology-based assessment of perennial energy crops in North American Tallgrass Prairie. Annals of the Association of American Geographers, 101, 741–751.

Wang, C., Hunt Jr., E. R., Zhang, L., & Guo, H. 2013. Spatial distributions of C_3 and C_4 grass functional types in the U.S. Great Plains and their dependency on inter-annual climate variability. Remote Sensing of Environment, 128, 90–101.

Wang, C., Jamison, B. E., & Spicci, A. A. 2010b. Trajectory-based warm season grassland mapping in Missouri prairies with multi-temporal ASTER imagery. Remote Sensing of Environment, 114, 531–539.

Wang, C., Zhou, B., & Palm, H. L. 2008. Detecting invasive sericea lespedeza (*Lespedeza cuneata*) in Mid-Missouri pastureland using hyperspectral imagery. Environmental Management, 41, 853–862.

Wang, K., Franklin, S. E., Guo, X., & Cattet, M. 2010c. Remote sensing of ecology, biodiversity and conservation: A review from the perspective of remote sensing specialists. Sensors, 10, 9647–9667.

Wang, L., Hunt Jr., E. R., Qu, J., Hao, X., & Daughtry, C. S. T. 2011b. Estimating dry matter content of fresh leaves from the residuals between leaf and water reflectance. Remote Sensing Letters, 2, 137–145.

Ward, D. P., & Kutt, A. S. 2009. Rangeland biodiversity assessment using fine scale on-ground survey, time series of remotely sensed ground cover and climate data: An Australian savanna case study. Landscape Ecology, 24, 495–507.

Washington-Allen, R. A., Van Niel, T. G., Ramsey, R. D., & West, N. E. 2004. Remote sensing-based piosphere analysis. GIScience & Remote Sensing, 41, 136–154.

Washington-Allen, R. A., West, N. E., Ramsey, R. D., & Efromson, R. A. 2006. A protocol for retrospective remote sensing-based ecological monitoring of rangelands. Rangeland Ecology & Management, 59, 19–29.

Watts, A. C., Ambrosia, V. G., & Hinckley, E. A. 2012. Unmanned aircraft systems in remote sensing and scientific research: Classification and considerations of use. Remote Sensing, 4, 1671–1692.

Weber, B., Olehowski, C., Knerr, T., Hill, J., Deutschewitz, K., Wessels, D. C. J., Eitel, B., & Büdel, B. 2008. A new approach for mapping of biological soil crusts in semidesert areas with hyperspectral imagery. Remote Sensing of Environment, 112, 2187–2201.

Weber, K. T., Chen, F., Booth, D. T., Raza, M., Serr, K., & Gokhale, B. 2013. Comparing two ground-cover measurement methodologies for semiarid rangelands. Rangeland Ecology & Management, 66, 82–87.

Westoby, M. J., Brasington, J., Glasser, N. F., Hambrey, M. J., & Reynolds, J. M. 2012. 'Structure-from-Motion' photogrammetry: A low-cost, effective tool for geosciences applications. Geomorphology, 179, 300–314.

Williamson, J. C., Bestelmeyer, B. T., & Peters, D. P. C. 2012. Spatiotemporal patterns of production can be used to detect state change across an arid landscape. Ecosystems, 15, 34–47.

Winslow, J. C., Hunt Jr., E. R., & Piper, S. C. 2003. The influence of seasonal water availability on global C_3 vs C_4 grassland biomass and its implications for climate change research. Ecological Modelling, 163, 153–173.

Wood, E. M., Pidgeon, A. M., Radeloff, V. C., & Keuler, N. S. 2012. Image texture as a remotely sensed measure of vegetation structure. Remote Sensing of Environment, 121, 516–526.

Wood, E. M., Pidgeon, A. M., Radeloff, V. C., & Keuler, N. S. 2013. Image texture predicts avian density and species richness. PLoS-One, 8, e63211.

Yang, C., & Everitt, J. H. 2010a. Comparison of hyperspectral imagery with aerial photography and multispectral imagery for mapping broom snakeweed. International Journal of Remote Sensing, 31, 5423–5438.

Yang, C., & Everitt, J. H. 2010b. Mapping three invasive weeds using airborne hyperspectral imagery. Ecological Informatics, 5, 429–439.

Yang, C., Everitt, J. H., & Fletcher, R. S. 2013. Evaluating airborne hyperspectral imagery for mapping saltcedar infestations in west Texas. Journal of Applied Remote Sensing, 7, 073556–1.

Yang, C., Everitt, J. H., and Johnson, H. B. 2009. Applying image transformation and classification techniques to airborne hyperspectral imagery for mapping Ashe juniper infestations. International Journal of Remote Sensing, 30, 2741–2758.

Yu, Q., Gong, P., Clinton, N., Biging, G., Kelly, M., & Schirokauer, D. 2006. Object-based detailed vegetation classification with airborne high spatial resolution remote sensing imagery. Photogrammetric Engineering & Remote Sensing, 72, 799–811.

Zavaleta, E. S., Pasari, J. R., Hulvey, K. B., & Tilman, G. D. 2010. Sustaining multiple ecosystem functions in grassland communities requires higher biodiversity. Proceedings of the National Academy of Sciences, 107, 1443–1446.

Zengeya, F. M., Mutanga, O., and Murwira, A. 2013. Linking remotely sensed forage quality estimates from WorldView-2 multispectral data with cattle distribution in a savanna landscape. International Journal of Applied Earth Observation and Geoinformation, 21, 513–524.

Zhang, X., Friedl, M. A., Schaaf, C. B., Strahler, A. H., Hodges, J. C. F., Gao, F., Reed, B. C., & Huete, A. 2003. Monitoring vegetation phenology using MODIS. Remote Sensing of Environment, 84, 471–75.

Ziska, L. H., Blumenthal, D. M., Runion, G. B., Hunt Jr., E. R., & Diaz-Soltero, H. 2011. Invasive species and climate change: An agronomic perspective. Climatic Change, 105, 13–42.

Part IV

Phenology and Food Security

13 Characterization, Mapping, and Monitoring of Rangelands

Methods and Approaches

Lalit Kumar, Priyakant Sinha, Jesslyn F. Brown, R. Douglas Ramsey, Matthew Rigge, Carson A. Stam, Alexander J. Hernandez, E. Raymond Hunt Jr., and Matthew Reeves

ACRONYMS AND DEFINITIONS

ASTER	Advanced Spaceborne Thermal Emission and Reflection Radiometer
DEM	Digital Elevation Model
ETM+	Enhanced Thematic Mapper Plus
EVI	Enhanced Vegetation Index
GIS	Geographic Information System
ISODATA	Iterative Self-Organizing Data Analysis Technique
LAI	Leaf Area Index
LiDAR	Light Detection and Ranging
MSS	Multispectral Scanner
MODIS	Moderate-Resolution Imaging Spectroradiometer
NDVI	Normalized Difference Vegetation Index
NDWI	Normalized Difference Water Index
NIR	Near-Infrared
NPP	Net Primary Productivity
OLI	Operational Land Imager
PAR	Photosynthetically Active Radiation
SAVI	Soil Adjusted Vegetation Index
SPOT	Satellite Pour l'Observation de la Terre, French Earth Observing Satellites
TM	Thematic Mapper
USDA	United States Department of Agriculture

13.1 INTRODUCTION

While there are many definitions of rangeland, the central theme of all these is that it is land on which the dominating vegetation is mainly grasses, grass-like plants, forbs, shrubs and isolated trees (De Cauwer et al., 2024; Arogoundade et al., 2023; Feizizadeh et al., 2023; Sharifi and Felegari, 2023; McCord and Pilliod, 2022; Matongera et al., 2021; Zhou et al., 2020; Hadian et al., 2019; Al-bukhari et al., 2018; Eddy et al., 2017; Gökalp et al., 2017). Rangelands include shrublands, natural grasslands, woodlands, savannahs, tundra and many desert regions. A distinguishing factor of rangelands from pasture lands is that they grow primarily native vegetation, rather than plants established by humans. Rangelands are also managed mainly through extensive practices such as managed livestock grazing and prescribed fire rather than more intensive agricultural practices and the use of fertilizers. Rangelands worldwide are known to provide a wide range of desirable goods

DOI: 10.1201/9781003541165-17

and services, including but not limited to livestock forage, wildlife habitat, wood products, mineral resources, water and recreation space. Large populations depend on rangelands for their livelihoods, hence effective monitoring and management is crucial for sustainable production, health and biodiversity of these systems (McCord and Pilliod, 2022; Matongera et al., 2021; Zhou et al., 2020).

Effective monitoring of rangelands has proven logistically and statistically difficult using field based monitoring methods alone due to the sheer size, range and complexity of rangelands (Hadian et al., 2019; Al-bukhari et al., 2018; Eddy et al., 2017; Gökalp et al., 2017). Mapping and monitoring of rangelands, especially those in a disturbed state or under rapid change, requires data that is extensive, accurate, timely and with regular repeat coverage. All this makes remote sensing an ideal platform for rangeland monitoring as recent developments in sensor capabilities means we have repeat coverage from multiple satellites, spectral resolutions sufficient to distinguish many rangeland vegetation species and communities, spatial resolutions allowing monitoring and management at micro-scales and costs that are a small fraction of a few decades ago. Remote Sensing data is more easily available and the systems cover almost the entire world, and certainly all the regions where rangelands occur.

There is a plethora of literature available that describe various uses of remote sensing data for rangeland monitoring, ranging from mapping species distribution, biomass, degradation, woody cover, net primary production, biodiversity, change detection, fuel loads, fire extents and frequency, invasive species encroachment, livestock foraging, etc. New methods of image classification and interpretation are regularly published, as well as novel techniques of incorporating satellite data with ancillary data to better understand rangeland dynamics (Arogoundade et al., 2023; Feizizadeh et al., 2023). Some of these applications have been discussed in the previous two chapters of this volume of the *Remote Sensing Handbook* (Second Edition, six volumes), with Reeves et al. (Chapter 11) looking at the relationship between rangeland productivity and climate, food security, fire and rangeland degradation, and Hunt et al. (Chapter 12) exploring rangeland biodiversity using different spectral, spatial and temporal satellite data.

The focus of this chapter is to present and discuss methods and approaches used in the mapping and monitoring of rangelands. We do this by presenting characterization, mapping, and monitoring of rangelands in specific applications such as: 1. Phenology and Productivity studies; 2. Fuel Analysis; 3. Biodiversity and Gap Analysis; 4. Vegetation Continuous Fields, and 5. Change Detection Analysis. We summarize some oft-used traditional means of mapping and monitoring rangelands but concentrate on newer developments in this field.

13.2 RANGELAND MONITORING METHODS USING VEGETATION INDICES

13.2.1 Rangeland Phenology

Rangeland phenology for vegetation types (shrublands, grasslands, steppes, deserts, and woodlands) is affected by environmental drivers (temperature, precipitation, sunshine) and factors such as topography (elevation, slope, aspect), edaphic conditions (variations in soil type, texture, nutrients) and latitude (Feizizadeh et al., 2023; Gökalp et al., 2017). Rangeland vegetation growth is the result of overall influence of all environmental drivers and factors and their interaction, and responds more rapidly to the environmental variations as compared to other kinds of vegetation (Reed et al., 1994a, 1994b). Since rangelands are mainly located in dry areas characterized by low and variable annual rainfall (Grice and Hodgkinson, 2002; Zhou et al., 2020), precipitation regime has a much more significant influence on rangeland vegetation among all the environmental drivers and factors (Reed et al., 1994a, 1994b). They usually respond to precipitation in a pulsed way where their phenology is dependent on discrete rainfall events in terms of productivity, density, and abundance (Rauzi and Dobrenz, 1970). Temperature has also been observed to have direct influence on phenological phases and a large number of studies have been conducted to determine the effects of temperature on the phonological timings of plants (Badeck et al., 2004; Sparks et al., 2000). Livestock grazing is another notable factor that influences rangeland phenology and has impact on rangeland vegetation (Desalew et al., 2010). The grazing-induced vegetation change is dependent on the type of livestock and the composition of vegetation types and hence an important consideration in monitoring or predicting rangeland plant phenology (Sharifi and Felegari, 2023; McCord and Pilliod, 2022).

The advent of remote sensing technology induced great changes in vegetation phenology studies by providing temporal data at regular intervals. Time-series data have been used to predict phenophase in terms of onsets and offsets of the vegetation growing season as well as budburst, flowering or leaf color changing dates. A simple way to predict the onset and/or offset of the growing season is to analyze the time series of vegetation indices (VIs). There have been many methods to determine the onset/offset dates, including thresholds, maximum rate of change, or a certain percentage of the greatest VI increase (McCord and Pilliod, 2022; Al-bukhari et al., 2018). For example, Rigge et al. (2013) evaluated the productivity and phenology of western South Dakota mixed-grass prairie in the period from 2000 to 2008 using the Normalized Difference Vegetation Index (NDVI) derived from MODIS data. They used growing-season NDVI images on a weekly basis to produce time-integrated NDVI, a proxy of total annual biomass production, and also integrated seasonally to represent annual production by cool- and warm-season species (C3 and C4, respectively). Heumann et al. (2007) studied phenological change in the Sahel and Soudan, Africa from 1982–2005. They used TIMESAT software to estimate phenological parameters from the AVHRR NDVI dataset and have found significant positive trends for the length of the growing and end of the growing season for the Soudan and Guinean regions. Kumar et al. (2002) showed that soil type has a significant impact on the early season growth variation of annual vegetation on sandy and clay soils, a fact that is utilized in the movement patterns of graziers in the Sahel.

13.2.2 Vegetation Indices in Rangeland Monitoring

13.2.2.1 What to Measure?

To be effective estimators of biomass, Leaf Area Index (LAI), or percentage cover, spectral indices must be able to differentiate vegetation features from soil features (Todd et al., 1998). For such differentiation, it is a requirement that the soils and vegetation have different reflectance patterns and the spectral index should be sensitive enough to detect the differences. Green vegetation has characteristically low reflectance in the visible portion of the spectrum (lowest in red portion of the spectrum) with a sharp increase in reflectance in the near-infrared portion (Figure 13.1). Most of the commonly used vegetation indices exploit this expected difference in near-infrared and red reflectance for vegetation discrimination and in separating them from non-vegetated areas (Sharifi and Felegari, 2023; McCord and Pilliod, 2022). For example, the Normalized Difference Vegetation Index (NDVI), computed as: NDVI = NIR–Red/NIR + Red, has been used in a wide range of practical remote sensing applications (e.g., Tucker et al., 1985). The NDVI values range between –1 to +1, with dense vegetation having a high NDVI while soil values are low but positive, and water is negative due to its strong absorption of NIR. Tucker (1979) tested various combinations of the red, NIR, and green bands to predict biomass, water and chlorophyll content of grass plots. A strong correlation was observed between NDVI values and chlorophyll content and crop characteristics such as green biomass and leaf water content. Sellers (1985) used a canopy radiative transfer model to show that NDVI is near-linearly related to area-averaged net carbon assimilation and plant transpiration, even at different values of fractional vegetation cover (fc) and LAI over an area of interest.

The TM Tasseled Cap green vegetation index (GVI) is a linear combination of the six reflecting wavebands of Landsat TM (Crist, 1983). The GVI coefficients with the highest values are for the red (negatively loaded) and the near-infrared (positively loaded) wavebands.

13.2.2.2 Soil Reflectance Variation

As compared to vegetation, soil reflectance patterns are usually quite different and generally increase linearly with increasing wavelength: from visible to near-infrared to mid-infrared (Figure 13.1).

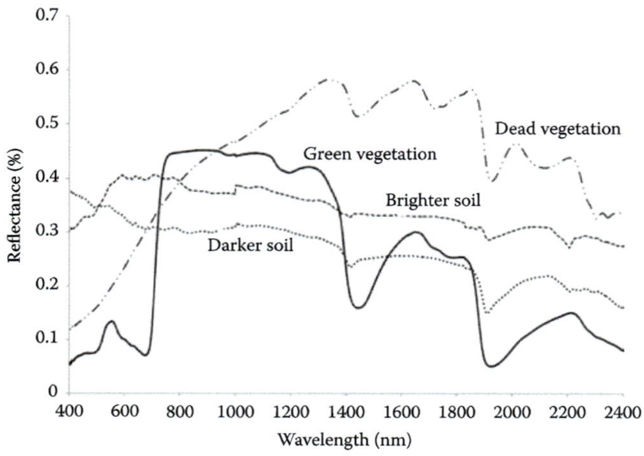

FIGURE 13.1 Idealized representation of spectral reflectance curves for green and dead vegetation, and light and dark soils.

Unlike vegetation, soil usually has high reflectance in the visible wavelengths and low reflectance in the near-infrared wavelengths. Figure 13.1 illustrates the contrast in reflectance patterns between dark (low reflecting) and light (high reflecting) soils. The difference between reflectance of soil and vegetation, in case of high reflecting soils, can be small in the near-infrared wavelengths (Eddy et al., 2017; Gökalp et al., 2017).

Soil reflectance properties vary considerably with soil type, texture, moisture content, organic matter content, color, and the presence of iron oxide (McCord and Pilliod, 2022). Low reflecting soils are usually dark, high in organic matter or moisture, containing iron oxides, and/or coarse textured. Dry soils show high reflectance values in visible, near-infrared, and mid-infrared regions of the spectrum. In contrast wet soils show low reflectance values for these regions (Bowers and Hanks, 1965).

13.2.2.3 Soil Background Impacts on Spectral Indices

In rangelands and semi-arid regions, background soil effects lead to soil-vegetation spectral mixing, a major concern in vegetation identification from spectral indices using remote sensing data. Soil, plant and shadow reflectance components mix interactively to produce composite reflectance (Richardson and Wiegand, 1990). The knowledge on soil reflectance variations for different soil types and conditions and their interaction with vegetation reflectance is essential to differentiate the soil and vegetation reflectance patterns. This helps in assessing the potential effectiveness of remote sensing techniques to map spatial distribution of plant species or communities and estimate biomass (De Cauwer et al., 2024; Sharifi and Felegari, 2023; McCord and Pilliod, 2022; Al-bukhari et al., 2018).

Generally, the values of ratio based indices, such as the normalized vegetation index, tend to increase with dark or low reflecting soil backgrounds (De Cauwer et al., 2024; Hadian et al., 2019; Al-bukhari et al., 2018; Elvidge and Lyon, 1985; Huete et al., 1985; Todd and Hoffer, 1998, 1997). However, a few studies have shown that the Tasseled Cap green vegetation index (GVI) decreases with low reflecting soil backgrounds (Huete et al., 1985; Huete and Jackson, 1987). The influence of soil type and moisture content on vegetation index was found less for Landsat TM derived GVI as compared to NDVI when estimating percentage green vegetation cover, based on a two component soil and vegetation model (Todd and Hoffer, 1997). Where soil effects on NDVI are a problem, alternative VIs such as the Soil Adjusted Vegetation Index (e.g., Huete, 1988) or the scaled difference vegetation index (Jiang et al., 2006) can be used.

13.2.2.4 Vegetation Reflectance Variation

The relationship between biomass and spectral indices can also be affected by vegetation condition, distribution and structure (Sharifi and Felegari, 2023; McCord and Pilliod, 2022; Todd et al., 1998). The loss in chlorophyll content due to drying of vegetation alters spectral reflectance characteristics in the visible and infrared regions. This phenomena is a very common occurrence in semi-arid rangelands such as the shortgrass steppe (Hadian et al., 2019). In case of drying vegetation, the reflectances in both the visible and in the mid-infrared regions of the spectrum increase significantly. The reflectance patterns of dead or dry plant material are more similar to soil than to healthy green vegetation (Matongera et al., 2021). Both dry vegetation and most soils have high reflectance in the visible and mid-infrared regions (Hoffer and Johannsen, 1969) and hence the spectral similarities introduce difficulties in remote sensing applications in rangelands.

For a given region, if dry or senescent biomass forms a significant portion of total vegetation present, the spectral distinction between vegetation and soil background is altered (Todd et al., 1998). For bright soil, spectral bands such as Red or indices such as Tasseled Cap brightness index, which are responsive to scene brightness, should help in vegetation discrimination in the presence of dry and/or senescent vegetation which appear less bright than the soil background.

Todd et al. (1998) found that the RED index, which responds to surface brightness, estimated vegetation cover on high reflecting soils more accurately than vegetation indices because dry vegetation was less bright than the soil background.

The grazing pattern and intensity also change the relative proportions of the standing dead and green biomass (Matongera et al., 2021; Sims and Singh, 1978a, 1978b) and cover characteristics (Milchunas et al., 1989) of plant communities within the rangelands. With increasing grazing intensity, the amount of standing dead plant material as well as litter decreases; however, in the absence of grazing, there will be considerable amount of both green and dry vegetation present. The presence of dead and/or drying plant material causes spectral confusion between this dead/drying plant material and the soil background and hence creates difficulties in estimating biomass using remote sensing techniques. Therefore, comparisons between grazed and ungrazed sites can provide insights into the impacts of senescence on biomass detection. In addition, grazing intensity varies considerably with topography (Milchunas et al., 1989), therefore information on the effectiveness of various remote sensing indices in detecting biomass under different grazing intensities could be useful in developing models across landscapes with heterogeneous grazing patterns. The field methods used to sample vegetation may also affect possible relationships between remote sensing indices and biomass.

13.2.2.5 Overview of Vegetation Indices

Vegetation indices combine reflectance measurements from different portions of the electromagnetic spectrum to provide information about vegetation cover on the ground (De Cauwer et al., 2024; Matongera et al., 2021; Zhou et al., 2020). Healthy green vegetation has distinctive reflectance in the visible and near-infrared regions of the spectrum. In the visible region, and in particular red wavelengths, plant pigments strongly absorb the energy for photosynthesis, whereas in the near-infrared region the energy is strongly reflected by the internal leaf structures. This strong contrast between red and near-infrared reflectance has formed the basis of many different vegetation indices. When applied to multispectral remote sensing images, these indices involve numeric combinations of the sensor bands that record land surface reflectance at various wavelengths.

One of the most promising applications of satellite data is the estimation of net primary productivity over time and space. The use of satellite derived vegetation indices has been useful for estimating net productivity (Al-bukhari et al., 2018). Vegetation indices are not a direct measure of biomass or primary productivity, but establish empirical relationships with a range of vegetation/climatological parameters (Weiser et al., 1986) such as: (a) Fraction of Absorbed Photosynthetically Active Radiation (FAPAR); (b) Leaf Area Index (LAI); and (c) Net Primary

Productivity (NPP). They are simple to understand and their implementation is fast as most of them use spectral band values in a mathematical formulation (e.g., ratio, difference, etc.). While maintaining sensitivity to vegetation temporal characteristics and seasonality, most of these indices reduce sensitivity to topographic effects, soil background, view/Sun angle and atmosphere to some extent. While many vegetation indices have been developed over the years, Table 13.1 lists a few of these and provides references for further investigation (McCord and Pilliod, 2022; Matongera et al., 2021).

TABLE 13.1
Vegetation Indices Commonly Used in Rangeland Studies

	Vegetation Index	Formula
1	SR—Simple Ratio	$\left[SR = \frac{\rho_{NIR}}{\rho_R}\right]$
2	EVI—Environmental Vegetation Index (Birth and McVey, 1968)	$\left[EVI = \frac{\rho_{IR}}{\rho_R}\right]$
3	NDVI—Normalized Difference Vegetation Index (Rouse et al., 1974)	$\left[NDVI = \frac{\rho_{IR} - \rho_R}{\rho_{IR} + \rho_R}\right]$
4	PVI—Perpendicular Vegetation Index (Richardson and Wiegand, 1977)	$\left[PVI = \sqrt{\left(\rho R_{soil} - \rho R_{veg}\right)^2 + \left(\rho IR_{soil} - \rho IR_{veg}\right)^2}\right]$
5	SAVI—Soil-Adjusted Vegetation Index (Huete, 1988)	$\left[SAVI = (1+L)\frac{(\rho_{IR} - \rho_R)}{(\rho_{IR} + \rho_R + L)}\right]$ L is a correction factor whose values range from 0 (high vegetation cover) to 1 (low vegetation).
6	TSAVI—Transformed SAVI (Baret et al., 1989)	$\left[TSAVI = \frac{a^*(\rho_R - a^*\rho_{IR} - b)}{(\rho_R + a^*\rho_{IR} - a^*b))}\right]$
7	Kauth-Thomas transformation [Tasseled cap, K-T] (Kauth and Thomas, 1976)	GVI= – 0′2728(TM1) – 0′2174(TM2) – 0′5508(TM3) + 0′722(TM4) + 0′0733(TM5) – 0′1648(TM7)
8	GNDVI—Green NDVI (Gitelson et al., 1996)	$\left[NDVI_g = \frac{(\rho_{NIR} - \rho_{green})}{(\rho_{NIR} + \rho_{green})}\right]$
9	GEMI—Normalized Difference Vegetation Index (Pinty and Verstraete, 1992)	$\left[GEMI = \eta(1 - 0.25\eta)\frac{(\rho_R - 0.125)}{(1 - \rho_R)}\right]$
10	EVI—Enhanced Vegetation Index (MODIS) Huete et al., 2002)	$\left[EVI = G\frac{\rho_{NIR} - \rho_{red}}{\rho_{NIR} + C_1 * \rho_{red} - C_2 * \rho_{blue} + L}\right]$ L = canopy background adjustment that addresses nonlinear, differential NIR and red radiant transfer through a canopy, and C1, C2 are the coefficients of the aerosol resistance term, which uses the blue band to correct for aerosol influences in the red band.
11	II—Infrared Index Hardisky et al., 1983)	$\left[II = \frac{(TM4 - TM5)}{(TM4 + TM5)}\right]$
12	MSI—Moisture Stress Index (Rock et al., 1985)	$\left[MSI = \frac{TM5}{TM4}\right]$

13.2.2.6 Vegetation Indices (VIs) as Proxies for Other Canopy Attributes

13.2.2.6.1 VIs and LAI

The leaf area index (LAI) is an estimate of the total leaf area per unit area (Glenn et al., 2008). LAI links vegetation indices to photosynthesis through the absorbed photosynthetically active radiation (Tucker and Sellers, 1986). When light travels through a series of leaves the transmission and therefore reflectance in near infrared of the electromagnetic radiation decreases (McCord and Pilliod, 2022). On the other hand, the uppermost leaf layers of a green vegetation canopy strongly absorb the red portion of the spectrum and therefore reduce their transmittance to successive leaf layers. The relation between LAI and vegetation indices varies among vegetation types (Peterson et al., 1987). In the case of saturated LAI, the ability to estimate biomass using vegetation indices is constrained, however, in rangelands, such as the shortgrass steppe, LAI saturation is improbable (Sharifi and Felegari, 2023; McCord and Pilliod, 2022).

LAI is derived mathematically and has no direct relationship to fPAR or processes that depend on fPAR (Glenn et al., 2008). LAI is related to light interception by a canopy (Ri) by:

$$Ri = Rs(1 - \exp - kLAI)$$

where k is a factor that accounts for leaf angles and other factors that affect absorption of Rs within a canopy (Monteith and Unsworth, 1990). Vertical leaved (erectophiles) plants typically absorb less light per unit leaf area than plants with relatively horizontal leaves (planophiles). Within a stand, the coefficient k also varies with respect to plant arrangements. Single stand plants receive light from all sides of their canopy as compared to dense stands of plants which receive light from the top of the canopy. The fraction of light absorbed by the canopy (fPAR) depends not only on Ri but also on the spectral properties of the leaves. The reflective surfaces present in some leaves reduce the heat gain while other surfaces absorb nearly all of the incident radiation between the visible bands (400 and 700 nm). Therefore, a correlation can easily be established between LAI and NDVI for single plant species grown under uniform conditions (Asrar et al., 1985). However, the same is not the case with mixed canopies, often the case in remote sensing data, in which a single pixel contains several landscape units.

13.2.2.6.2 VIs and Fractional Cover (fc)

Models use fractional vegetation cover to divide the landscape into areas of vegetation and bare soil (Zhou et al., 2020; Hadian et al., 2019; Glenn et al., 2007; Kustas and Norman, 1996; Anderson, 1997; Timmerman et al., 2007). Different methods are used in these models to estimate carbon and moisture fluxes from vegetation and bare soil and the fraction of the landscape that is vegetated (Glenn et al., 2008). Typically, a landscape unit is partitioned into bare soil and vegetation through the use of vegetation indices (VIs). For example, NDVI values derived from satellite images and ranging from −1 to +1 are rescaled between 0 and 1 to represent bare soil at values near 0 and 100% vegetation cover at values near to 1 to get fc for a given pixel or area of interest in the scene. Depending on the models used, the scaling is done linearly or in a nonlinear way to represent the vegetation type of interest. Some models require both LAI and fc often estimated by VIs (Anderson, 1997). In a few cases, ground-based information relating to vegetation species and canopy characteristics are also included to improve the estimates. Sometimes average leaf angles for a particular type of landscape are used to predict both fc and LAI from VIs (Anderson, 1997).

For partially vegetated scenes with LAI in the range of 1–3, Hadian et al. (2019) found VIs to be much more closely related to fc than to LAI in case of clumped vegetation, and the relationships between NDVI and fc to be nonlinear. They showed that, for partially vegetated scenes of uniform vegetation type, VIs were a good measure of fc. Other studies on a variety of landscape types have also reported strong linear (McCord and Pilliod, 2022) or nonlinear (Al-bukhari et al., 2018) relationships between VIs and fc. However, at 100% cover, different plant species may have different VIs due to differences in chlorophyll content and canopy structure, and thus create a potential practical problem in using VIs to estimate fc over mixed scenes. Amiri and Shariff (2010), in their study of vegetation cover assessment in semi-arid rangelands of Iran, used 26 different VIs to determine suitable indices for vegetation cover and production assessment, and found a significant relationship between NDVI derived from ASTER data with the vegetation cover. In another study, Ajorlo and Abdullah (2007) examined four vegetation indices (NDVI, SAVI, PVI and RVI) to assess rangeland degradation in semi-arid parts of the Qazvin province, Iran. The results showed that NDVI was a more powerful index for assessing the rangeland degradation as compared to other indices. McCord and Pilliod (2022) used NDVI and ratio vegetation index (RVI) in grassland study and found good correlation between fresh herbage yields and RVI and NDVI ($P < 0.01$) in four grassland types with correlation coefficient (r) >0.679. Fresh herbage yields correlated better with RVI than with NDVI for lowland meadow, hill desert steppe, and mountain meadow, but not for plains desert steppe. Matongera et al. (2021) monitored grassland health with remote sensing approaches and assessed the effectiveness of remote sensing in grassland monitoring. They found that it was challenging to use remotely sensed data in mixed grasslands because the large proportion of dead material complicated analysis for indices that were not developed for heterogeneous landscapes, especially in conservation areas. They investigated the relationship between remote sensing data and grassland biophysical measurement, including aboveground biomass and plant moisture content, in the native mixed prairie ecosystem with its high litter component. Their results indicated that the NDVI was not suitable for biomass estimation although a moderate relationship was found between NDVI and plant moisture content. Compared to NDVI, leaf area index (LAI) provided promising results on both biomass and plant moisture content estimation. John et al. (2008) evaluated the utility of MODIS-based productivity (GPP and EVI) and surface water content (NDSVI and LSWI) in predicting species richness in the semi-arid region of Inner Mongolia, China. They found that these metrics correlated well with plant species richness and could be used in biome- and life form-specific models (De Cauwer et al., 2024; Matongera et al., 2021; Zhou et al., 2020).

Although, different vegetation indices are used in assessing the rangeland degradation, there are still challenges facing the classification of vegetation species in degraded areas where the reflectance is strongly affected by the soil background as a result of relatively sparse vegetation and atmospheric conditions.

13.2.3 Case Study: Rangeland Phenology and Productivity in the Northern Mixed-Grass Prairie, North America

13.2.3.1 Introduction

Rangelands across the northern plains of North America are predominantly mixed-grass prairie communities and can be dominated by either C_3 (cool-season) or C_4 (warm-season) grass species. These mixed-grass communities exhibit different growth dynamics or phenological patterns that can be detected from satellite remote sensing (Sharifi and Felegari, 2023; McCord and Pilliod, 2022; Matongera et al., 2021; Zhou et al., 2020; Gökalp et al., 2017; Rigge et al., 2013; Tieszen et al., 1997). Although remote sensing cannot detect traditional phenological events such as budding and flowering in individual plants (Tieszen et al., 1997; White et al., 2009), it can detect important landscape-level phenological measures such as the onset of the growing season, the end of the growing season, rate of green-up, and peak vegetation vigor based on satellite image time series over a growing season (De Cauwer et al., 2024; Matongera et al., 2021; Gökalp et al., 2017; Kovalskyy and Henebry, 2012; Reed et al., 1994b; Tao et al., 2008; van Leeuwen et al., 2010). In this case study, conducted in the Bad River watershed in western South Dakota, USA, the spatial and temporal dynamics of rangelands as measured by remote sensing indicators or "phenological metrics" varied related to climate, management, and plant photosynthetic pathway (Rigge et al., 2013).

Land management, livestock grazing, invasive species, and prolonged droughts have the potential to change rangeland plant community structure and therefore contribute to altered phenological patterns (De Cauwer et al., 2024; Matongera et al., 2021; Gökalp et al., 2017; Foody and Dash, 2007; Tieszen et al., 1997)). These factors also have the potential to modify ecosystem goods and services (Matongera et al., 2021). One benefit in monitoring rangelands is to detect and mitigate any damaging trends caused by land management (Paruelo and Lauenroth, 1998).

Biomass production in rangelands by C_3 and C_4 plants displays significantly different phenological timing that is detectable utilizing satellite remote sensing. Information on rangeland phenological dynamics has been applied to assessing rangeland health monitoring, due to inferences that can be made about community composition and presence of invasive species (Arogoundade et al., 2023; McCord and Pilliod, 2022; Eddy et al., 2017; Boyte et al., 2014; Rigge et al., 2013; Tateishi and Ebata, 2004). For example, rangeland plant communities dominated by either C_3 or C_4 species can be identified through their unique and asynchronous phenological time series signals (Foody and Dash, 2007). Grasses with C_3 pathways are most active during the cooler spring and fall seasons, while many C_4 grasses are adapted to the hot and dry summer months (Arogoundade et al., 2023; Zhou et al., 2020; Gökalp et al., 2017; Foody and Dash, 2007; Tieszen et al., 1997; Wang et al., 2010). State-and transition models indicate that cool-season (C_3) grasses tend to dominate historic plant climax communities in many ecological sites of the northern mixed prairie, while shortgrass (C_4) species generally increase under disturbance such as heavy, continuous, season-long grazing (U.S. Department of Agriculture, 2008). Furthermore, phenological differences are useful for identifying vegetation types. For example, in western South Dakota, riparian vegetation tends to be dominated by C_4 species while upland vegetation communities consist primarily of C_3 plants. In late summer, this difference allows for the detection of a clear riparian vegetation signal (Kamp et al., 2013).

In the northern mixed-grass prairie, the majority of C_3 production occurs in spring and fall, and most C_4 production occurs in summer, although there is both spatial and temporal overlap in production (Ode et al., 1980). For example, both C_3 and C_4 plants actively produce biomass in mid- to late June with no clear separation between the timing of their production (Ode et al. 1980). In midsummer (1 July to 31 August) production is typically dominated by C_4 grasses in the study area and was therefore used to define the warm-season period (Ode et al., 1980; Wang et al., 2010).

13.2.3.2 Methods

13.2.3.2.1 Study Site

The Bad River watershed of western South Dakota (~lat 45°N, long 101°W) is dominated by the Clayey ecological site description in the Major Land Resource Area classification (U.S. Department of Agriculture, 2008). The topography of this region is generally typified by long, smooth slopes, with steeper slopes along well-defined waterways. Bedrock throughout the watershed is Pierre Shale, resulting in soils with a high clay content and low permeability. The climate is semiarid, receiving an average of 398 mm precipitation annually over the 2000–2008 period of which 80% occurred during the growing season of April to September. Annual precipitation is highly variable, with drought and insufficient moisture common (Matongera et al., 2021). The daily mean temperature ranges from 32°C in July to 14°C in January, with a yearly mean of 8°C (Smart et al., 2007). Analysis was constrained to pixels classified in the National Land Cover Database 2006 as herbaceous cover (Fry et al., 2011).

The study area vegetation is mixed-grass rangeland dominated by C_3 grasses including western wheatgrass (*Pascopyrum smithii Rybd.*) and green needlegrass (*Stipa viridula Trin. & Rupr.*; Smart et al. 2007). Shortgrasses (C_4) include buffalograss (*Bouteloua dactyloides Nutt.*) and blue grama (*Bouteloua gracilis H.B.K.*). Midgrasses are typically C_4 species such as little bluestem (*Schizachyrium scoparium [Michx.] Nash*) and sideoats grama (*Bouteloua curtipendula [Michx.] Torr.*). Little forb and succulent cover exists (Sims and Singh et al., 1978a, 1978b).

13.2.3.2.2 Input Data

Moderate Resolution Imaging Spectroradiometer (MODIS) satellite time series data provided an ideal balance between spatial resolution and temporal repeat, and are therefore well-suited for

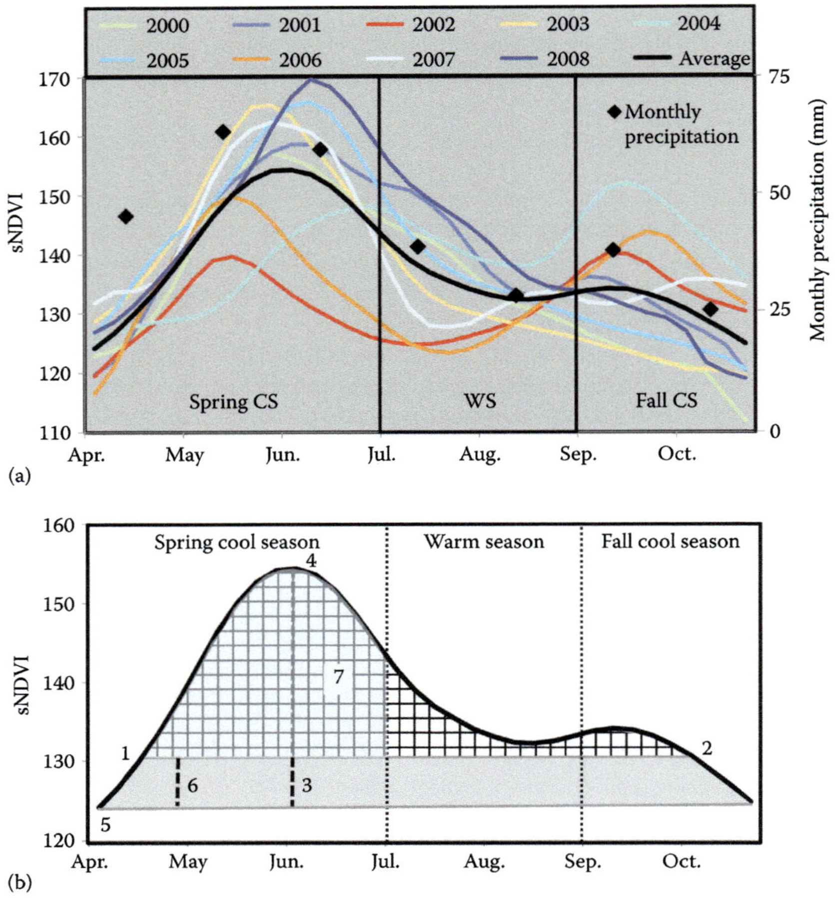

FIGURE 13.2 (a) Phenological profiles for Bad River watershed, South Dakota, USA, from 2000–2008 where years are colored according to growing-season precipitation (shown on the secondary vertical axis). Colors range from red (for dryer growing seasons to blue (for relatively wetter growing seasons). The left third of the graph shows the spring cool season (CS), the middle shows the warm season (WS), and the right thirds shows the fall CS. (b) Illustration of idealized phenological profile and main phenological indicators used in the study. The entire gridded area corresponds to the growing season time integrated NDVI (TIN) and the blue gridded area corresponds to the spring CS integrated NDVI. (Based on Figure 13.1 and 13.3 in Rigge et al. 2013.)

phenology studies across relatively large regions. This study is based on 9 years (2000–2008) of eMODIS weekly composite Terra MODIS imagery at 250-m resolution (van Leeuwen et al., 2010). NDVI values were calculated and rescaled to a range of 0 to 200 to simplify calculations by eliminating negative values. Hereafter, rescaled NDVI will be referred to as sNDVI (scaled NDVI).

Annual variability in rangeland phenology and productivity as shown in MODIS sNtDVI time series plots (Figure 13.2a) is fairly common (Tieszen et al., 1997), where complex herbaceous communities respond to highly variable inter- and intra-annual precipitation (Lauenroth and Sala, 1992; Smart et al., 2007).

13.2.3.2.3 Phenological Analysis

Several measures calculated on a pixel by pixel basis were utilized in this study to describe the phenology of rangeland communities (Table 13.2). The start of the season was calculated for each pixel as the first point in time each year when sNDVI reached 20% of the total growing-season sNDVI

TABLE 13.2
Phenological Measures Calculated from Input MODIS Time Series Profiles

	Remote Sensing Phenological Metric	Description/Notes
1	SOS: Start of season	Time of first occurrence of sNDVI >= 20% of growing season amplitude (within a calendar year)
2	EOS: End of season	Time of last occurrence of sNDVI >= 20% growing season amplitude (within a calendar year)
3	AMP: Amplitude	Difference between the growing season minimum sNDVI and the growing season peak NDVI
4	$sNDVI_{peak}$	Value of peak growing season NDVI
5	$sNDVI_{min}$	Value of minimum growing season NDVI
6	Baseline	Values of <=20% of growing season amplitude (within a calendar year) that are eliminated
7	TIN_{GS}	Summation of sNDVI from Apr 1– Oct 1 where sNDVI is >= 20% of growing season amplitude

amplitude of that year (van Leeuwen et al., 2010). The total growing season amplitude was calculated as the difference between the peak sNDVI value (during the April 1 to October 31 growing season) and the minimum sNDVI value during this period. Similarly, the end of the season was the time at which the sNDVI value dropped below 20% of the total growing-season sNDVI amplitude. This approach better approximated the total seasonal biomass production than simply averaging NDVI across the entire growing season. TIN served as a proxy for growing-season total biomass production. Because TIN is influenced by the magnitude and duration of mNDVI values, the saturation effects that occur at larger leaf area index values (Huete et al., 2002) are minimized. The TIN of the spring cool season (start of season to 30 June) was also used in this study (Figure 13.2b).

Since phenological profiles can reveal community species composition and vegetation communities (McCord and Pilliod, 2022; Matongera et al., 2021; Zhou et al., 2020; Gökalp et al., 2017; Reed et al., 1996; Tateishi and Ebata, 2004) and phenology can strongly influence biomass production (Smart et al., 2007), the combined geospatial data on cool-season grasses and TIN was combined to create a watershed-level vegetation map (Figure 13.3). The per-pixel 2000–2008 average TIN was grouped into three classes: (1) within one standard deviation of the study area mean, (2) below one standard deviation of the mean, and (3) above one standard deviation of the mean. A similar approach was used to group the 2000–2008 average cool-season percentages into three classes. The TIN and cool season percentage classes were integrated to form nine vegetation classes, representing the average of the 2000–2008 conditions.

Table 13.3 provides vegetation class numbers, generally indicative of the plant community state, with higher numbers closely approaching the western wheatgrass/green needlegrass community (historic climax plant community of the Clayey ecological site) and lower numbers generally indicative of the blue grama/buffalograss plant community (Smart et al., 2007; U.S. Department of Agriculture, 2008). Classes marked with "+" indicated TIN or CS% over one standard deviation higher than the study area average, "N" indicated TIN or CS% within one standard deviation of the average, and "–" indicated TIN or CS% below one standard deviation of the average.

Spatial patterns of vegetation classes are useful for describing topographic, edaphic, and land management influences on plant communities (Figure 13.3). For example, areas with both low cool-season percentage and low TIN (class 1) were likely dominated by buffalograss and blue grama, both low-producing C_4 species. This community was reported to result from strong grazing pressure (Smart et al., 2007; U.S. Department of Agriculture, 2008) in the Clayey ecological site, but might stem from lower than normal warm-season moisture in Pierre Shale/badland outcrops.

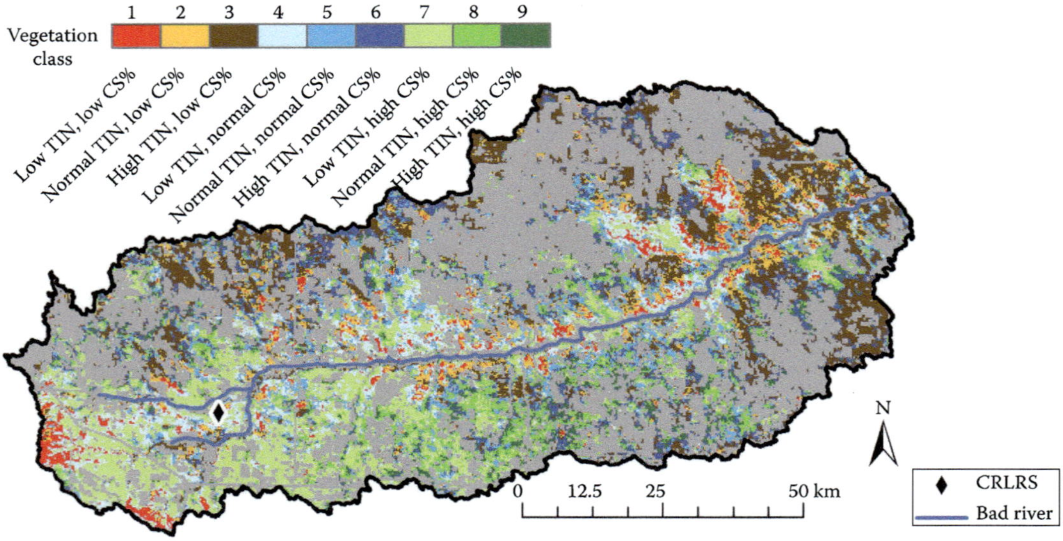

FIGURE 13.3 Nine vegetation classes derived from growing season time-integrated NDVI (TIN) and average cool season percentage (CS%) in the Bad River watershed, South Dakota, USA, shaded gray in the inset map. Light gray areas denote non-rangeland vegetation types that were excluded from this analysis. The Cottonwood Range and Livestock Research Station (CRLRS) is located in the southwest. (Based on Figure 13.6 in Rigge et al. 2013.)

TABLE 13.3
Bad River Watershed Rangeland Vegetation Classes

Vegetation Class	TIN	CS%	Area (%)	Mean TIN	Mean CS%
1	–	–	4.7	268.2	74.5
2	N	–	8.8	329.0	74.7
3	+	–	19.1	389.4	73.7
4	–	N	10.2	279.2	80.3
5	N	N	12.8	327.3	80.0
6	+	N	10.3	376.7	79.8
7	–	+	18.1	270.1	85.7
8	N	+	11.5	325.0	85.0
9	+	+	4.5	370.7	84.7

13.2.3.2.4 Validation Methods

Validation is a critical, yet challenging, component of phenological studies based on remote sensing. Methods were performed based on field plot data and carbon flux tower data. The field plot data were collected for two distinct plant communities (mixed-grass and midgrass-dominated) in fields that were retained under long-term grazing management practices (Dunn et al., 2010). Field-measured vegetation data estimated species composition by weight in late June (peak cool-season biomass) and early August (peak warm-season biomass). A nonrandom sampling scheme consisted of 14 plots per field represented local variation in soil type, slope, and aspect. The remotely sensed yearly average TIN and cool-season percentage for each field were regressed against the corresponding field-measured average biomass production and cool-season percentage data. The comparison of remotely sensed growing season TIN and phenology to field vegetation data collected at the Cottonwood Range and Livestock Research Station (CRLRS) in the Bad River watershed suggested that the remote observations were successful in describing actual conditions (Rigge et al., 2013). Field-measured cool-season biomass production and cool-season TIN were greater on the midgrass-dominated pastures than the mixed-grass pastures, following the expected pattern (Sims and Singh, 1978a, 1978b; Smart et al., 2007). Overall, the relationship between field-measured annual biomass production and TIN by pasture was strong ($R^2 = 0.69$, $P < 0.01$, $n = 24$).

A carbon flux tower located on a mixed-grass plant community pasture at the CRLRS was used for validation at that site. Thirty-minute flux tower quality-controlled ecosystem data for 2007 and 2008 were modeled using nonlinear light response curves driven by photosynthetically active radiation (PAR), soil temperature, and vapor pressure deficit to partition carbon fluxes associated with PAR (i.e., gross primary productivity) from those associated with total ecosystem respiration (Gilmanov et al., 2010). Gross photosynthesis (PG) data were calculated as the sum of day CO_2 flux and respiration minus the rate of change in atmospheric CO_2 storage below the tower. Phenological metrics from the flux tower data were generated using the same methods employed for MODIS data, making the results generated from both datasets more directly comparable. Flux tower PG data were strongly related to sNDVI values ($R^2 = 0.67$, $P < 0.01$) at the overlapping pixel. Similarly, the accumulation of TIN and PG throughout both growing seasons were related ($R^2 > 0.90$ in both years).

13.2.3.2.5 Conclusions

Information on the timing and magnitude of biomass production can be a useful tool for assessment of rangeland health (De Cauwer et al., 2024; McCord and Pilliod, 2022; Eddy et al., 2017; Gökalp et al., 2017). Species diversity and variable weather across the northern mixed-grass prairie make modeling native rangelands problematic. However, the MODIS-based phenological indicators were successful in capturing important characteristics of plant community phenology and productivity. The results of this study clarify the spatial and temporal dynamics (inter and intra-annual) of phenology and biomass production in response to precipitation in this region.

This approach could be useful to land managers in adjusting the stocking rate and season of grazing to 1) maximize rangeland productivity and profitability (Dunn et al., 2010) and 2) achieve conservation objectives, through improving understanding of management impacts to rangeland phenology and production. Maps such as these might also be useful to identify areas where land degradation may be occurring or is likely to occur if current management practices are continued. Further, if the phenological metrics of a patch (e.g., pasture) of land is dramatically different than the surrounding landscape or contrasts with the general landscape gradient, it can be presumed that land management practices on that parcel, not climate, soils, or topography, are primarily responsible. These specific parcels can then be the focus of best management practices, field examination, and data collection.

13.2.4 Rangeland Fuel Load Assessment

Knowledge of the spatial distribution of fuels is critical when characterizing susceptibility of landscapes to wildfire and for estimating expected fire behavior (Arroyo et al., 2008). In the simplest terms the susceptibility of a rangeland landscape to wildfire is a function of expected fire weather, topography, probability of ignition, juxtaposition of the landscape, and fuel bed characteristics. Many rangelands are dominated by small diameter fuel components, which respond very quickly to environmental conditions such as heating, relative humidity and precipitation (Cheney and Sullivan, 1997). In addition, the high surface area to volume ratio, high packing ratios and small sizes make rangeland fuels characteristically prone to ignition and very high rates of spread. For example, under extreme fire weather conditions, grassland fires can exceed 28 km hr^{-1} making rangeland fires unpredictable. The temporal and spatial variability of fuels add to this uncertainty.

Productivity of rangeland varies much more on an inter-annual, proportional basis than that of forests (Zhou et al., 2020; Hadian et al., 2019; Briggs and Knapp, 1995; Le Houerou and Hoste, 1977; Teague et al., 2008; Zhang et al., 2010). As a result, the fuel bed properties associated fire behavior characteristics and potential risk can change commensurately. In addition, rangelands are the most extensive kind of land cover, occupying nearly 33% of ice-free land globally (Ellis and Ramankutty, 2008). In the coterminous U.S.A. alone there is an estimated 268 million ha that often occur across large, open, remote expanses prone to high and gusty winds. The high variability and large areas inherent in rangeland systems suggest that regular monitoring is needed for properly addressing the fire danger situation.

Regular monitoring of fuels is needed for both strategic (Kaloudis et al., 2005) and tactical purposes (Reeves et al., 2009). Strategic assessment involves planning and resource allocation modeling. In contrast, tactical uses of wildland fuel data involve specific fire behavior projections and assessment for determining things such as arrival time, risks to resources, and suppression tactics.

A complete evaluation of wildland fuels requires accounting for all fuel bed characteristics including such components as fuel bed depth, vegetation structure and species composition, litter, fuel moisture, fuel quantity of various size classes (Sharifi and Felegari, 2023; Gökalp et al., 2017; Anderson, 1982; Scott and Burgan, 2005; Ottmar et al., 2007; Sandberg et al., 2001). The diameters of the fuel particles can be classified based on their time lag. Larger diameter fuels have longer time-lags because they respond more slowly to changes in environmental conditions. The time lag categories most often used for fire behavior and danger assessment include 1hr, 10hr, 100hr, and 1000hr corresponding to diameter ranges of 0–0.635 cm, 0.635–2.54 cm, 2.54–7.62 cm, and 7.62–20.32 cm, respectively. These fuel bed components are most strongly influenced by the kinds and amounts of vegetation. The pressing need for timely information, dynamic nature of rangeland vegetation and fuels, and paucity of plot data suggest that remote sensing can play a significant role in the evaluation of rangeland fuels.

Remote sensing has long been used to evaluate vegetation conditions at spatial scales and spatial resolutions varying from plot level evaluation (Blumenthal et al., 2007) to global assessment at 8 km^2. Mere vegetation classification, however, is generally not sufficient for quantifying fuel conditions. As a result, many approaches have been derived for using remote sensing to evaluate wildland fuels (Riaño et al., 2002; Riaño et al., 2003; Rollins, 2009; Reeves et al., 2009; Arroyo et al., 2008). Most methods for quantifying fuels across the landscape involve a combination of remote sensing techniques (e.g., spectral analysis for species composition, structure and production; LiDAR for vegetation structure) and modeling or expert systems (Keane and Reeves, 2011). Use of satellite remote sensing for capturing temporal dynamics of fuel bed properties is usually most successful for characterizing the herbaceous biomass response. These biomass estimates can subsequently be converted to fuel loadings (1-hour time lag category). With woody vegetation the situation is more complicated and quantifying fuels can be accomplished by first determining vegetation structure (height and cover) (e.g., McCord and Pilliod, 2022; Eddy et al., 2017; Gökalp et al., 2017; Chopping et al., 2006; Vierling et al., 2012). Yet another method is to first determine stand structure (stand cover and height) and species composition and then using allometric relationships to estimate individual fuel components (Means et al., 1996). Once the appropriate fuel bed attributes have been estimated it is often necessary to invoke expert systems to crosswalk stand level fuel attributes into a fuel model depending on the intended use of the data. This is a critical step because describing all fuel characteristics in an area is exceedingly difficult given the extreme spatial and temporal variation of fuel bed components (Arroyo et al., 2008; Keane, 2013). As a result, description of fuel properties relevant to fire behavior or effects is normally based on classification schemes, which summarize large groups of fuel characteristics. These classes are expressed as "an identifiable association of fuel elements of distinctive species, form, size arrangement, and continuity that will exhibit characteristic fire behavior under defined burning conditions" (Merrill and Alexander, 1987).

Commonly used fuel classification systems include Surface Fire Behavior Fuel Models (Sharifi and Felegari, 2023; Zhou et al., 2020; Eddy et al., 2017; Gökalp et al., 2017; Anderson, 1982; Scott and Burgan, 2005) and Fuel Loading Models (Lutes et al., 2009). The reason for this is that most fire behavior or fire affects processors such as Farsite (Lutes et al., 2009), FlamMap (Lutes et al., 2009), Behave (Andrews and Bevins, 2003) and Promethius (Lutes et al., 2009) require stylized fuel models or standardized classifications of fuel components. In the United States, many decision support systems and tactical evaluations utilize FarSite and Flammap which generally require stylized fuel models from either Anderson (1982) or Scott and Burgan (2005). Prometheus, on the other hand, is widely used in Canada and components of it have been used in the U.S.A., Spain, Portugal, Sweden, Argentina, Mexico, Fiji, Indonesia and Malaysia (www.nrcan.gc.ca/forests/fires/14470) to estimate fire spread and uses fuel models from the Canadian Forest Fire Danger Rating System (CFFDRS) (Wotton, 2009).

13.2.5 Case Study: Using Remote Sensing to Aid Rangeland Fuel Analysis

This case study focuses on the rangelands of the coterminous U.S.A. where, in places, annual aboveground rangeland production can vary by more than 150% in extreme years. Prominent fire decision support systems in the U.S.A. presently require the Surface Fire Behavior Fuel Models (FBFM) (Anderson, 1982; Scott and Burgan, 2005). As a result, here we present a case study using remote sensing to inter-annually quantify 1-hour time lag fuels for informing mapping processes designed to predict surface FBFMs for all coterminous U.S.A (Figure 13.4).

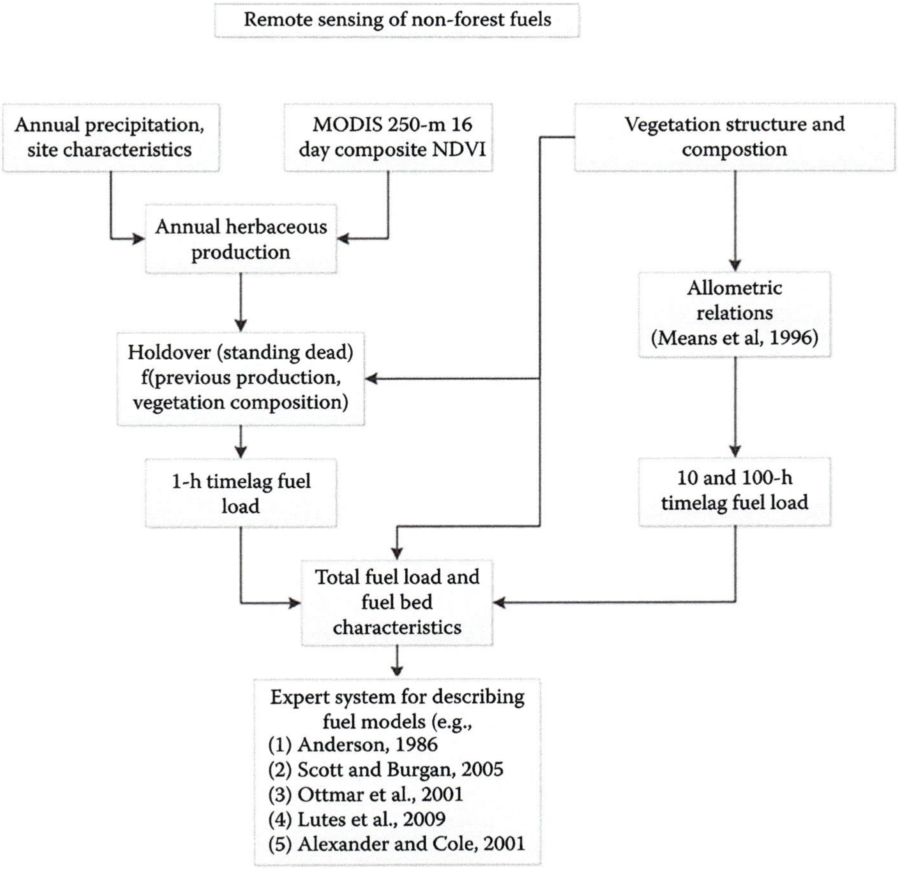

FIGURE 13.4 Fuel flowchart.

The fine fuels of rangelands are driven primarily by grasses and forbs. Capturing these inter-annual fuel dynamics requires a relatively high repeat cycle and reasonably good spatial resolution, both of which are inherent in the 250 m^2 16-day composite MODIS Normalized Difference Vegetation Index (NDVI) from the Moderate Resolution Imaging Spectroradiometer (MODIS). From the 23 periods in each year from 2000 to 2012 the annual maximum NDVI was chosen to correlate with ground observations of aboveground productivity. Since spatially explicit, consistent, comprehensive ground data do not exist, production estimates from the Soil Survey Geographic database (SSURGO; www.nrcs.usda.gov/wps/portal/nrcs/detail/soils/survey/?cid=nrcs142p2_053627) were used as a surrogate for plot based ground data. For each soil type in the SSURGO database, the expected annual production based on normal (mean), drought (low) and above average (high) growing conditions is available. Thus, for each soil site, represented as a soil map unit, there are three data points representing expected annual production. Annual maximum NDVI, gridded annual precipitation from the PRSIM project, and the Biophysical Settings (BPS) data layer were used to develop a model for predicting annual production and, ultimately, 1-hr time lag fuels through time. For precipitation, the maximum, minimum, and mean annual total precipitation from 2000 to 2012 was selected to correspond with the high, mean and low productivity for each BPS type. Likewise, the maximum, minimum, and mean annual maximum from 2000 to 2012 was also selected to correspond with the rangeland productivity from the SSURGO soil types. For each BPS type, the range of annual production, range of annual precipitation, and annual maximum NDVI was spatially averaged. For example, average mean annual maximum NDVI, annual summation of precipitation and rangeland productivity were averaged for all shortgrass prairie BPS. The annual productivity model resulted in a simple combination of these predictors and their interactions as

$$\text{AnnualProduction} = (\text{Y int}) + (\text{Precip}_{\text{annsum}} * 0.141) + (\text{NDVI}_{\text{annmax}} * 3.0056) + \text{BPS} + (\text{Precip}_{\text{annsum}} * \text{BPS} * -0.1138) + (\text{NDVI}_{\text{annmax}} * \text{BPS} * -1.2961)$$

where AnnualProduction is the estimated annual production of rangeland biomass, $\text{Precip}_{\text{annsum}}$ is the annual sum of precipitation for a BPS unit and the $\text{NDVI}_{\text{annmax}}$ is the average annual maximum NDVI. This model resulted in an R^2 value of 0.94, a bias estimate of 1.43, and a mean absolute error (MAE) of 164 kg $ha^{-1.}$

This relationship between NDVI, precipitation and site specific (BPS) coefficients enabled estimates of rangeland annual production from 2000 to 2012.

After quantifying annual production, estimates of standing dead residue ("holdover"), which add considerably to the total fuel load on a rangeland site, were estimated. This was accomplished using a simple decay function represented in Figure 13.4. As a result, for a given year, the total 1-hr fuel load can be described as function of the following form f(annual production + holdover $_{t-1}$ + holdover $_{t-2}$, holdover $_{t-3}$). These fine fuel loads were combined with estimates of 10- and 100-hour time lag fuels of the woody components (e.g., shrubs) of each stand. This was accomplished by combining the cover, height, and species composition from the LANDFIRE project, and quantifying woody biomass and 10- and 100-hour time lag fuels using allometric equations from Means et al. (1996) (Figure 13.4). Once the full suite of fuel characteristics were known, surface FBFMs from the Scott and Burgan (2005) suite were estimated for average fuel conditions and then modulated through time based on the inter-annual change in 1-hour time lag fuel loads (Figure 13.5). Figure 13.5 demonstrates how remote sensing is used with climate and site characteristics to estimate 1-hr time lag fuels every year. In addition in Figure 13.5, the amount of deviation (as a percent of the 12-year average) in 2000, 2005 and 2011 are shown for the southwestern Ecological Province (Bailey and Hogg, 1986). Finally, the resulting change in the distribution of surface FBFMs can be seen in panel C for all three years presented. Note the large increase in 1-hr time lag fuels in 2005 in the Southwestern Ecoregion. This corresponds to a large increase in rainfall and a strong NDVI signal resulting in very high biomass and subsequent changes to surface FBFMs suggesting much

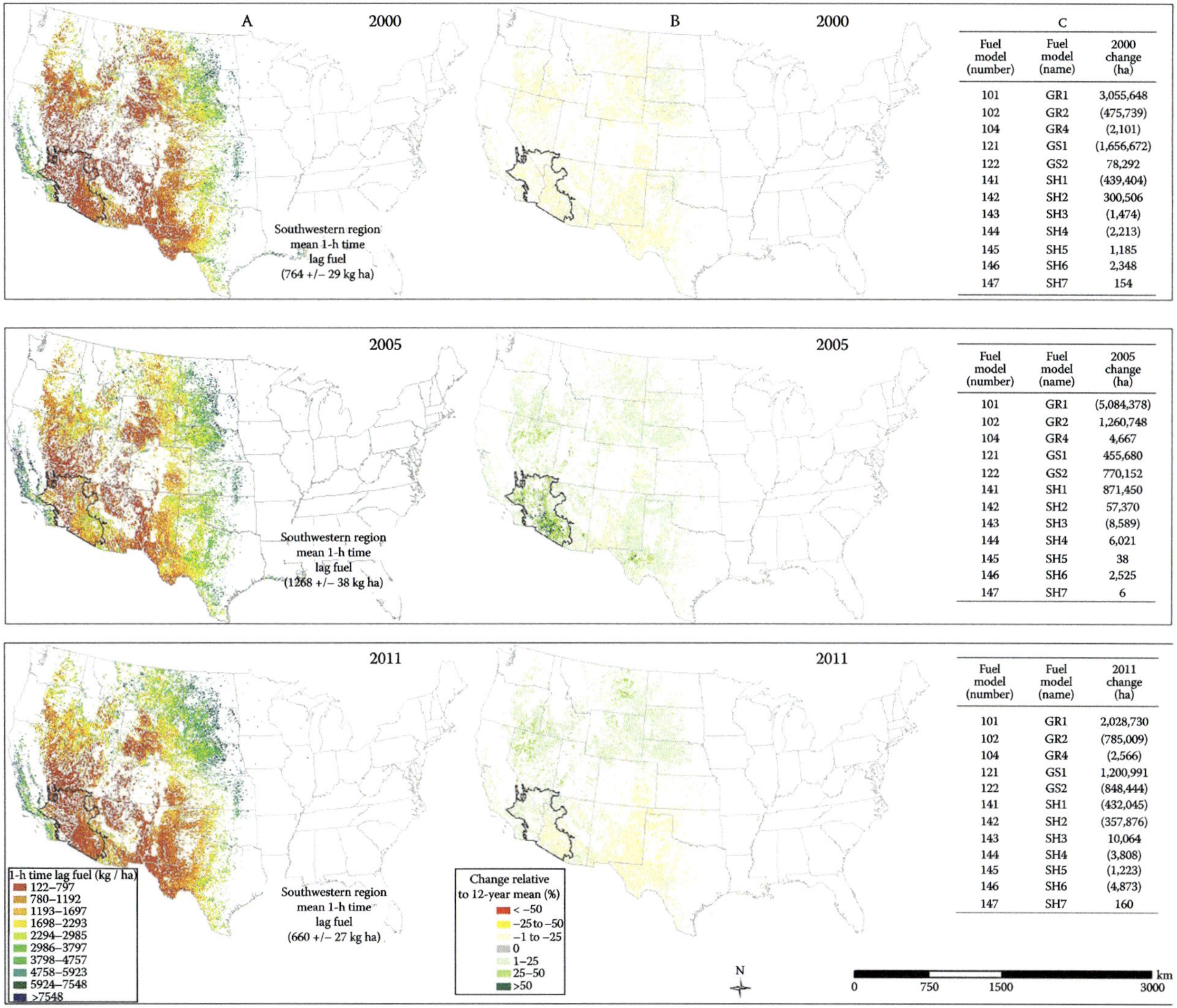

Fuel model (number)	Fuel model (name)	2000 change (ha)
101	GR1	3,055,648
102	GR2	(475,739)
104	GR4	(2,101)
121	GS1	(1,656,672)
122	GS2	78,292
141	SH1	(439,404)
142	SH2	300,506
143	SH3	(1,474)
144	SH4	(2,213)
145	SH5	1,185
146	SH6	2,348
147	SH7	154

Fuel model (number)	Fuel model (name)	2005 change (ha)
101	GR1	(5,084,378)
102	GR2	1,260,748
104	GR4	4,667
121	GS1	455,680
122	GS2	770,152
141	SH1	871,450
142	SH2	57,370
143	SH3	(8,589)
144	SH4	6,021
145	SH5	38
146	SH6	2,525
147	SH7	6

Fuel model (number)	Fuel model (name)	2011 change (ha)
101	GR1	2,028,730
102	GR2	(785,009)
104	GR4	(2,566)
121	GS1	1,200,991
122	GS2	(848,444)
141	SH1	(432,045)
142	SH2	(357,876)
143	SH3	10,064
144	SH4	(3,808)
145	SH5	(1,223)
146	SH6	(4,873)
147	SH7	160

FIGURE 13.5 Fuel modulation.

greater flame lengths and spread rates (the GR1 is reduced and GR2 is substantially increased). As an example, under the identical environmental parameters with an open wind speed of 24 km hr^{-1}, a GR2 surface FBFM results in a spread rate of 9 times greater than a GR1.

This brief case study demonstrates a method for using high temporal resolution satellite remote sensing for quantifying fuels in rangeland landscapes. The important characteristics of a sensor for characterizing fine fuels across large landscapes in a timely manner suggest satellites with relatively high repeat frequencies and appropriate spectral channels for modeling changes in aboveground biomass. Application of the methods outlined here are limited to regions where sufficient numbers of spatially explicit measures (or estimates as in the present work) of biomass or fuels are available.

13.3 RANGELAND VEGETATION CHARACTERIZATION

13.3.1 Rangeland Biodiversity and Gap Analysis

Rangeland biological diversity (biodiversity) refers to the variety and variability among living organisms and the environments in which they occur and is recognized at species, ecosystem and landscape level (Al-bukhari et al., 2018; Eddy et al., 2017, Gökalp et al., 2017). The goal of biodiversity conservation is to reverse the processes of biotic impoverishment at each of these levels of organization. Gap analysis provides a quick overview of the distribution and conservation status of several components of biodiversity (Scott et al., 1993). Gap Analysis is the process by which the distribution of species and vegetation types are compared with the distribution of different land management and land ownership classifications. It seeks to identify gaps (vegetation types and species that are not represented in the network of biodiversity management) that may be filled through establishment of changes in land management practices (Scott et al., 1993). The goal here is to ensure that all ecosystems and areas rich in species diversity are represented adequately in biodiversity management areas. Gap Analysis uses vegetation types and vertebrates as indicators of biodiversity. Maps of existing vegetation are prepared from satellite imagery and other sources and entered into a geographic information system (GIS). Because the mapping is carried on a regional scale, the smallest area identified on vegetation maps is generally 100 ha. Vegetation maps are verified through field checks and aerial photographs. Predicted species distributions are based on existing range maps and other distributional data, combined with information on the habitat of each species. Distribution maps for individual species are overlaid in the GIS to produce maps of species richness. An additional GIS layer of land ownership and management status allows identification of gaps in the representation of vegetation types and centers of species richness in biodiversity management areas through a comparison of the vegetation and species richness maps with ownership and management status maps. Gap Analysis is a powerful and efficient first step towards setting land management priorities. It provides focus, direction, and accountability for conservation efforts. Areas identified as important through Gap Analysis can then be examined more closely for their biological qualities and management needs (Scott et al., 1993).

13.3.2 Case Study: Vegetation Continuous Fields with Regression Trees

13.3.2.1 Introduction

The sagebrush ecosystem is an important and distinguishing natural system of the U.S. Intermountain West. However, an estimated 40% of its pre-European settlement distribution has been reduced due to conversion to agriculture, urban/suburban growth, energy development, invasion by exotic plants, and encroachment by woodlands among others (Wisdom et al., 2005). The sagebrush ecosystem is an example of many other globally distributed natural systems that have been impacted by similar disturbances. Monitoring land cover change, across vast landscapes, in an efficient manner is touted by many to be a significant strength of digital remote sensing. The holistic view of remotely sensed data allows us to evaluate landscapes in their entirety. Traditional, field-based, monitoring systems are limited to specific locations whose characteristics are then extrapolated across the entire landscape. The number of field-based samples required to characterize landscapes with statistical certainty is often cost prohibitive and in some cases impossible to acquire due to access restrictions. The temporally systematic, landscape level monitoring that remote sensing offers coupled with modern statistical modeling approaches can provide land managers with much of the information necessary for effective planning and management.

Traditional image interpretation techniques that convert digital remote sensing data into discrete land cover maps have been used to monitor landscapes (Jin et al., 2013; Vogelmann et al., 2012). These techniques rely on the ability to accurately classify and map vegetation types at multiple time intervals to determine how change has occurred at the vegetation community level. A limitation of this technique is the assumption that adjacent vegetation community types have discrete boundaries (sharp ecotones) and that variation within the community is typically ignored. The reality is that sharp ecotones between community types, while they exist, are typically not the norm. Further, the spatial variation of vegetation cover within a mapped community type is often an important diagnostic of community condition. In traditional classification of imagery, this information is lost.

This case study describes the use of digital remote sensing data coupled with field-based measurements and advanced statistical modeling techniques to map vegetation continuous fields (VCF) in the sagebrush ecosystem. Vegetation Continuous Fields (VCF) is a relatively new concept that attempts to model percent canopy cover of specific vegetation types using remotely sensed imagery as its primary input. This technique has been used to estimate canopy cover of woody and herbaceous vegetation, as well as bare ground, on a global basis (De Cauwer et al., 2024; McCord and Pilliod, 2022; Defries et al., 2000; Sexton et al., 2013). A series of VCFs consist of several continuous response surfaces (one for each cover type) in which every pixel value corresponds to a percent canopy cover estimate predicted through a regression model. The VCF offers an advantage over traditional discrete classifications because areas of heterogeneity within vegetation community types are better represented (Hansen et al., 2002).

The objectives of the case study were to: (a) Test the effectiveness of a regression tree algorithm (Random Forest) to model estimates of percent cover, and (b) Develop a series of Vegetation Continuous Fields models for shrubs, woodland, herbaceous vegetation and bare ground for a semiarid shrub-steppe landscape.

13.3.2.2 Methods

13.3.2.2.1 Study Area

Our research was conducted in the northwest corner of Box Elder County, Utah (114°2′31. 2″ – 112°43′40. 8″ West longitude and 41°6′ 27. 36″–41°59′ 59. 64″ North latitude) depicted in Figures 13.6 and 13.7. The area covers 1,742,860 ha, with approximately 60% of the county occupied by the Great Salt Lake and barren playa bottoms. Of the remaining area, salt desert scrub occupies about one fifth of the area while big sagebrush shrubland and steppe covers nearly the same amount (19%). Pinyon-juniper ecosystems are an important part of the landscape making up 12% of the area. The remainder consists of greasewood flats, montane sagebrush steppe, xeric mixed sagebrush shrubland and invasive annual grasslands (Program, 2004). The elevation ranges from 1,278 m in the lowlands close to the Great Salt Lake to 3,027 m at the peak of the Raft River range. The mean elevation is 1,520 m.

13.3.2.2.2 Field Data

Field-based estimates of percent canopy cover for shrubs, trees, herbaceous (grasses and forbs) vegetation, and bare-ground were used to develop the VCF models. These data were prepared as

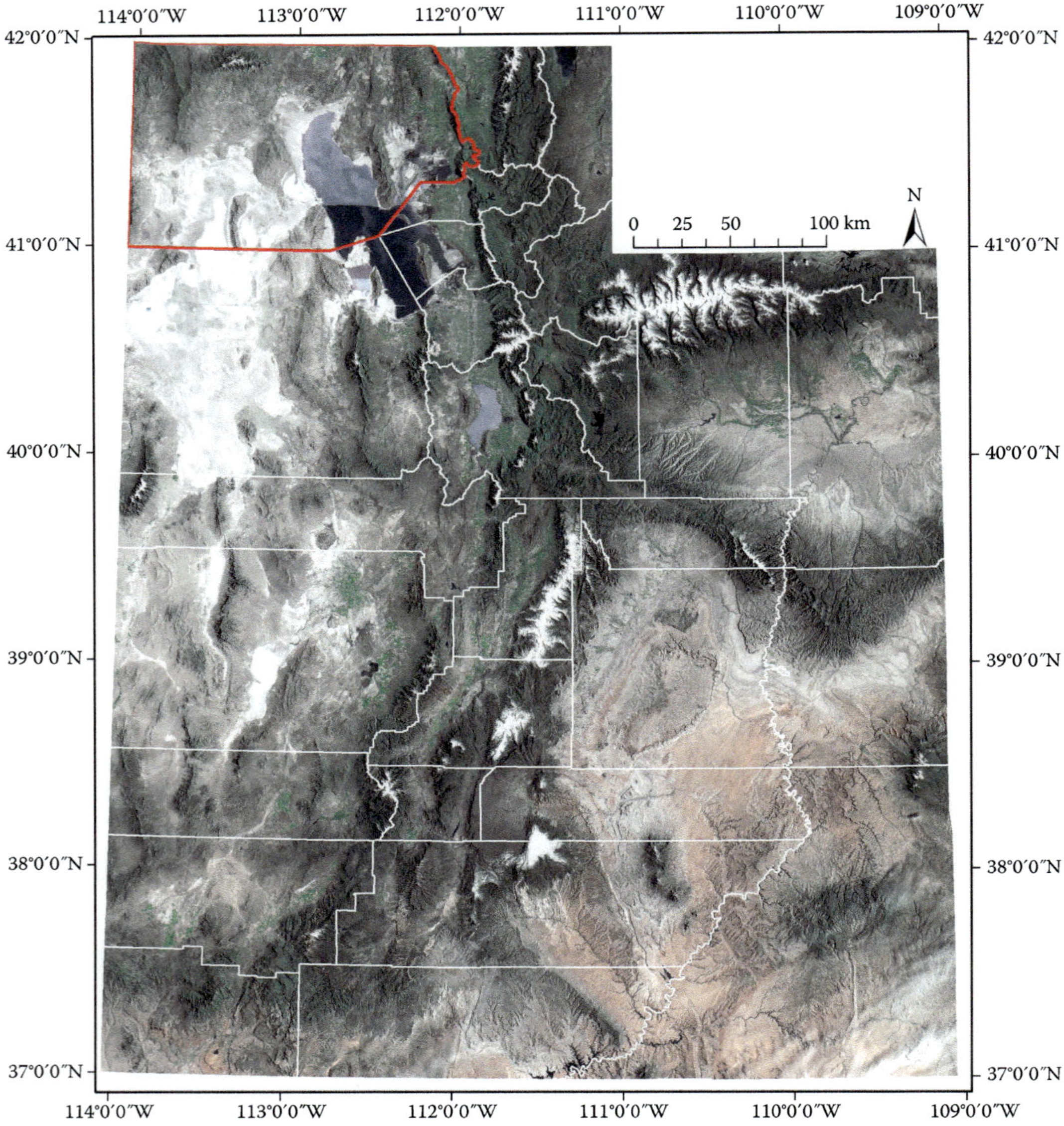

FIGURE 13.6 Utah image mosaic of 2013 Landsat 8 OLI imagery with Box Elder County highlighted in red.

geo-referenced field points and obtained from different sources: (a) 482 points collected by the South West Regional GAP (SWREGAP) project during 2001 (Lowry et al., 2007), (b) Points collected by The Nature Conservancy (TNC) for the Northwest Utah Landscape modeling project in 2007 (Conservancy, 2009), and (c) Field points that we collected during a field season in 2007. In total, 135 field observations were available for the year 2007. A fourth dataset was available from the Utah Division of Wildlife Resources (UDWR). Figure 13.8 contains the spatial distribution of the different datasets across the study area.

With the exception of the UDWR dataset, which are permanent sample plots and follow a standardized and quantitative data gathering method, the rest of the field points were visually assessed (qualitative assessment) in terms of percent canopy cover for shrubs, trees, herbaceous vegetation, and bare-ground on an area that resembled a 3 × 3 Landsat TM pixel (approximately 90 × 90 m). Cover estimates were taken independently at four cardinal directions from the center of the plot in the NE, SE, SW, and NW directions. These estimates were averaged to represent the entire plot.

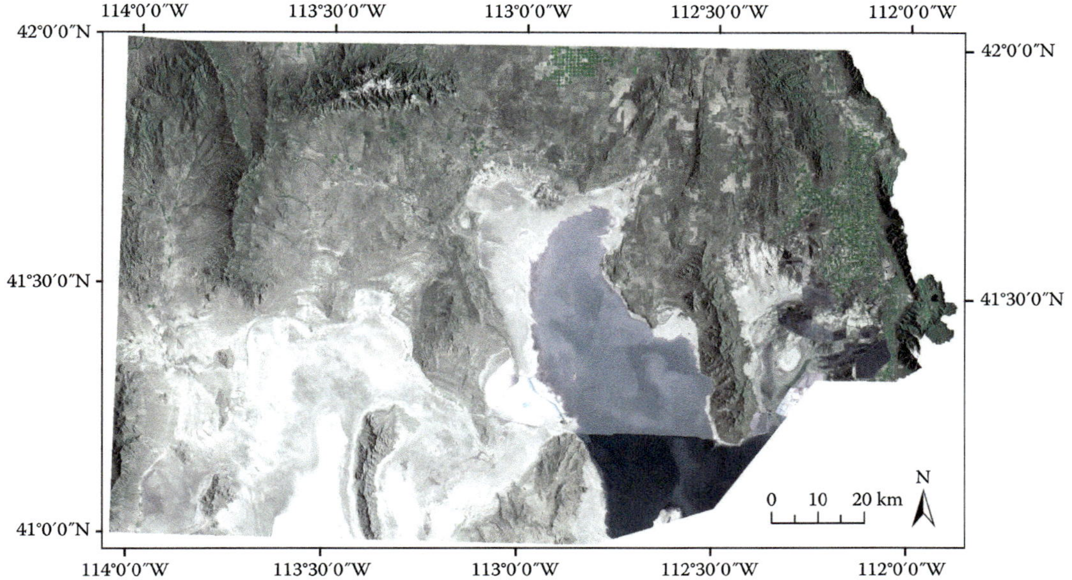

FIGURE 13.7 Landsat 8 OLI image mosaic (2013) of Box Elder County, Utah.

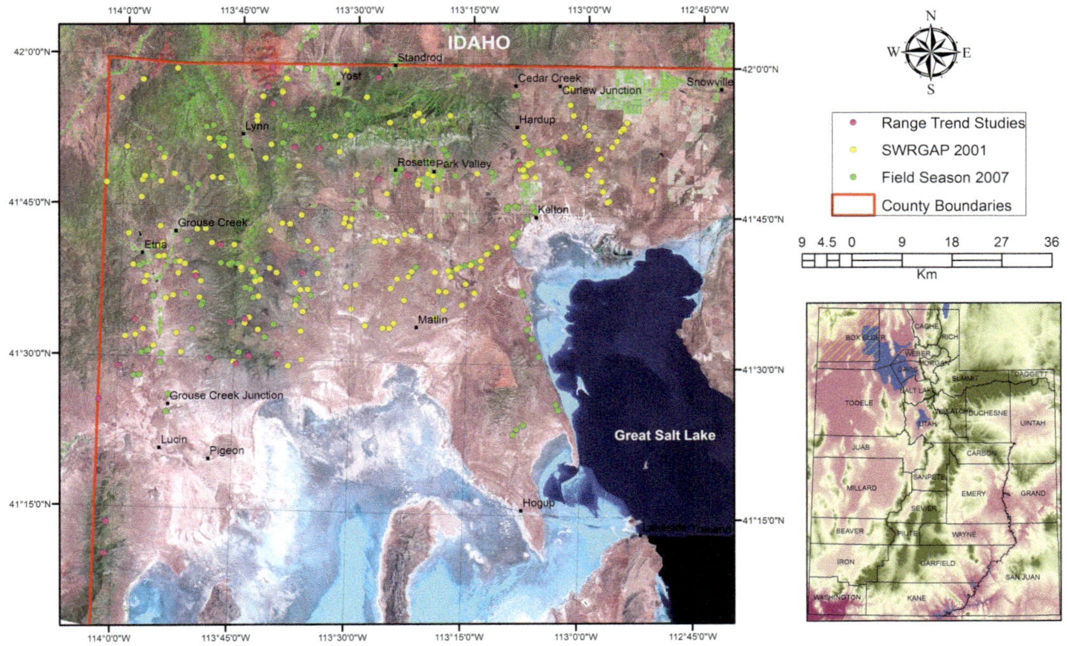

FIGURE 13.8 Distribution of field observations to model multi-temporal CVF.

The percent cover estimates were recorded using 5% increments for each life form and bare ground (i.e. 0%, 5%, 10%, etc.). The sum of the percent cover for shrubs, herbaceous vegetation, trees, and bare ground totaled 100% at each point. The sampling scheme consisted of locating sites along an elevation range that included Wyoming big sagebrush (Artemisia tridentata ssp. wyomingensis), basin big sagebrush (Artemisia tridentata ssp. tridentata), and mountain big sagebrush (Artemisia tridentata ssp. vaseyana) communities.

13.3.2.2.3 Explanatory Variables: Remote Sensing and Topography

Remotely sensed images and topographic datasets were used as explanatory variables for this modeling. With regards to the remotely sensed data, scenes from the Thematic Mapper (TM) sensor of the Landsat 5 satellite, Path 39 Row 31 were obtained. Due to the underlying differences in phenology that most vegetation types exhibit in semiarid landscapes (Bradley and Mustard, 2008), imagery from multiple dates during the summer months of 2001 were acquired. Within year seasonal imagery allowed us to capture major phenological variations that occur during the growing season. Landsat TM imagery collection was concentrated during late spring, mid-summer and early fall. An effort to obtain only imagery with the best quality (i.e., minimum cloud cover) was made. Three images were selected for this study. These were collected on April 28, July 01, and October 5, 2001.

Where necessary, imagery was rectified and resampled to a common map projection UTM Zone 12 WGS 1984. Standardization of the imagery was performed by converting the raw digital numbers to exo-atmospheric reflectance values using an image-based atmospheric correction procedure (Chavez, 1996) with the most up-to-date calibration coefficients for the Landsat TM sensor (Chander et al., 2009).

The Soil Adjusted Vegetation Index (SAVI) was computed for each date. SAVI may be calculated as follows:

$$SAVI = \frac{NIR - RED}{(NIR + RED + L)} * (1 + L) \tag{13.1}$$

In Equation 13.1, NIR = near-infrared reflected radiant flux, RED = red reflected radiant flux, and L = adjustment factor (typically a value of 0.5 is used).

SAVI has been reported to work well in semiarid ecosystems because it minimizes soil background effects that are known to affect other indices such as the Normalized Difference Vegetation Index (NDVI) (Huete, 1988; Jensen, 2007). It has been widely reported that a vegetation index such as SAVI may be used to follow the phenological trajectory or seasonal and inter-annual change in vegetation growth and activity (Jensen, 2007).

A new variable was created from the multi-temporal SAVI and named as NDSAVI or the normalized difference SAVI. The NDSAVI takes advantage of the contrast between the spring and the summer SAVI and may be used to improve our understanding of the phenological dynamics of grasses on the landscape. Higher values in the NDSAVI would correspond with higher greenness during the early spring relative to summer whereas low values of NDSAVI would relate to areas that become green later in the growing season. This new variable conveys a multi-temporal signature of greenness variation that may be used to discriminate among different land cover types and particularly focus on non-native grasses such as cheatgrass that follows this phenological pattern. Within this environment, this index allows us to identify areas where cheatgrass, a common and significant invasive annual grass, is a major component of the plant community.

NDSAVI was estimated as follows:

$$\frac{SAVIspr - SAVIsum}{SAVIspr + SAVIsum} \tag{13.2}$$

The Normalized Difference Water Index (NDWI; Gao, 1996) was also calculated for each image. NDWI takes advantage of the contrast found between the near and middle infrared bands to provide information about water content. Forest disturbances have been successfully detected using the NDWI (Jin and Sader, 2005), and thus it was appropriate to test its performance in our regression models. The brightness, greenness, and wetness (BGW) transformation (Crist and Kauth, 1986) was also derived for each image. This transformation has been used extensively to monitor condition and changes in soil brightness, vegetation, and moisture content, respectively (Jensen, 2007; Lowry et al., 2007).

In addition to the remotely sensed information (Landsat TM spectral bands, SAVI, NDWI, BGW), derivatives from a 30 m digital elevation model (DEM) including slope, aspect, and landform were obtained. Two transformations of aspect, namely southness and westness indices (Chang et al., 2004) and a modification to the original topographic relative moisture index TRMI (Parker, 1982) were generated. An existing land cover map from the SWRGAP project (Lowry et al., 2007) was included as an explanatory variable. The inclusion of this type of ancillary information has been shown to greatly improve classification and regression modeling in rangelands (Peterson, 2005).

13.3.2.2.4 Regression with Random Forests

13.3.2.2.4.1 Background Random Forests (RF) is a relatively new statistical method that emerged from the machine learning literature, and is based on the same philosophy as CARTs (De Cauwer et al., 2024; Arogoundade et al., 2023; Feizizadeh et al., 2023; Sharifi and Felegari, 2023; McCord and Pilliod, 2022; Matongera et al., 2021; Zhou et al., 2020; Hadian et al., 2019; Al-bukhari et al., 2018; Eddy et al., 2017; Gökalp et al., 2017). In RF, multiple bootstrapped regression trees without pruning are created. In a typical bootstrap sample, approximately 63% of the original observations occur at least once (Cutler et al., 2007). The data that are not used in the training set are termed "out-of-bag" observations and are customarily used to provide estimates of errors (Prasad et al., 2006). Out-of-bag samples are also used to calculate variable importance (Cutler et al., 2007). In RF, each tree is grown with a randomized subset of predictors, which equal the square root of the number of variables. In general, 500 to 2000 trees are grown, and averaging aggregates the results. The method is very effective in reducing variance and error in multi-dimensional datasets. One of the strengths of RF is that because it grows a large number of trees, the method tends not to overfit the data, and because the selection of predictors is random, the bias can be kept low (Prasad et al., 2006). More comprehensive descriptions of the method may be found in Sutton (2005), Lawler et al. (2006), Prasad et al. (2006), and Cutler et al. (2007). The application of RF was done in two phases. First, the best subset of variables to model each response variable: shrubs, trees, herbaceous vegetation, and bare-ground were identified. Second, the best subset of variables identified for each response variable was used to model the VCF for that variable.

13.3.2.2.5 Variable Importance and Parsimony

The underlying principle that the phenological pattern of a given vegetation type should dictate which remotely sensed datasets to use (Bradley and Mustard, 2008) was followed. For example, it is sensible to use only one scene (mid-summer for instance) to model bare ground percent cover due to its relatively constant spectral response throughout the year. On the other hand, it makes sense to utilize two to three images (i.e., mid-summer and early fall) to model herbaceous vegetation due to its conspicuous phenological signature which peaks during the summer and then significantly decreases during the fall.

In order to develop a simple yet effective model the concept of variable importance (Cutler et al., 2007) was used. This is based on the mean decrease in accuracy concept, and is assessed based on how much poorer the predictions would be if the data for that predictor were permuted arbitrarily. This provides a measure of the impact that a specific variable has in decreasing the precision of prediction. This is a somewhat subjective approach to choosing the most important variables since thresholds of decreasing precision tend to be arbitrary. Once the most important variables are chosen, the correlation coefficients between each variable are evaluated and further used to eliminate highly correlated variables (choosing the variable that makes most ecological/spectral sense) to arrive at a parsimonious set of predictor variables.

13.3.2.2.6 Regression

The R package RandomForest (Liaw and Wiener, 2002) was used to develop regression trees to calculate the VCF. Regressions were run separately for each of the four response variables (i.e., shrubs, trees, etc.) using the selected subset of variables determined to be most important. The R package YaImpute (Crookston and Finley, 2008) was used to extract the model for each VCF run, and then applied a predict function to generate a continuous geospatial response surface for the entire study area.

13.3.2.2.7 Validation and Comparison Metrics

For independent validation purposes, 20% of the field observations were withheld during model development. Pearson's correlation coefficients were calculated for each of the VCF predictions using this withheld set of data and two metrics were further calculated: mean absolute error (MAE) and root mean square error (RMSE). MAE is the average absolute difference of the predicted value from the field-observed estimate, while RMSE is the square root of the mean squared error (Prasad et al., 2006; Walton, 2008).

13.3.2.3 Results and Discussion

Table 13.4 contains the Pearson's Correlation Coefficients (PCC), mean absolute error (MAE), and root mean square error (RMSE). A global average of the PCC was 0.65. Our highest PCC was

TABLE 13.4
Validation Metrics between MRTS and RF

VCF	Pearson's Correlation	MAE*	RMSE**
Shrubs	0.72	7.81	10.00
Trees	0.52	12.56	16.69
Herbaceous	0.77	9.39	12.02
Bare ground	0.62	8.15	11.05
Average	0.65		

* Mean Absolute Error
** Root Mean Square Error

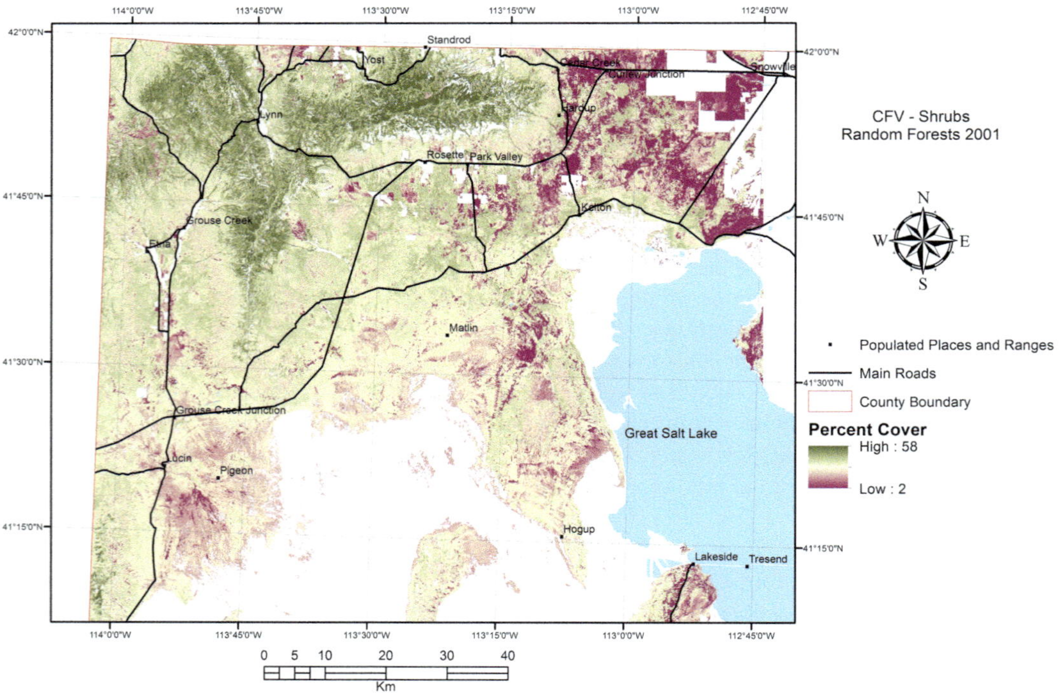

FIGURE 13.9 Percent canopy cover of shrubs as modeled using Random Forests.

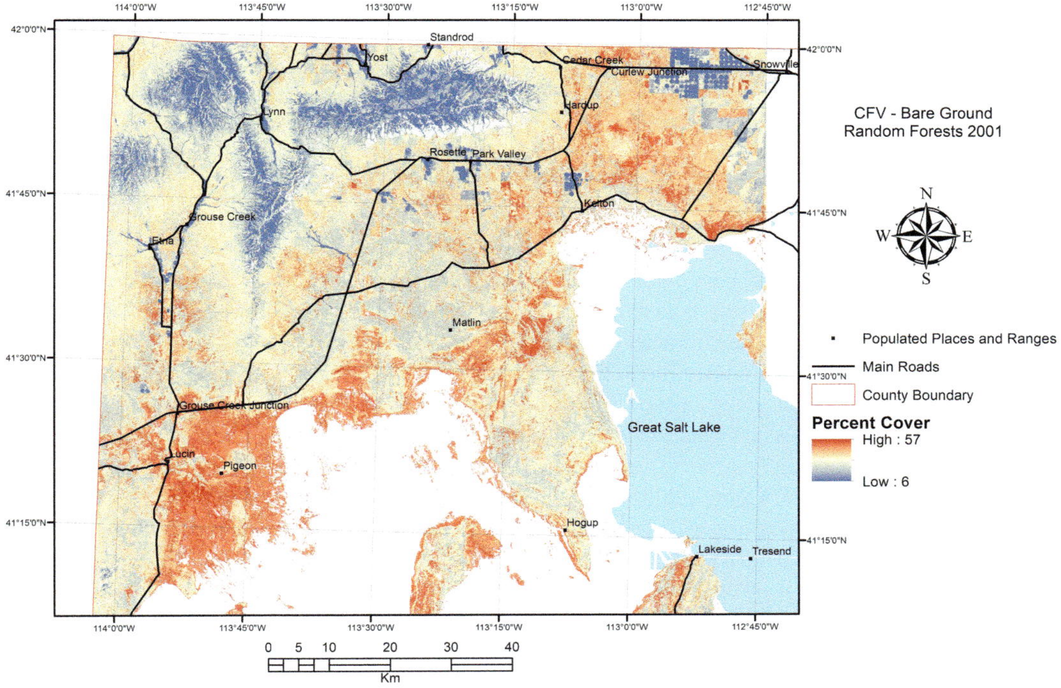

FIGURE 13.10 Percent canopy cover of bare ground as modeled using Random Forests.

for herbaceous cover at 0.77 and the lowest was for trees at 0.52. Figures 13.9 and 13.10 illustrate samples of the VCF spatial layers generated using RF.

The use of regression trees to depict sub-pixel heterogeneity has been widely reported in the literature. In rangeland environments, work has been conducted to model woody vegetation cover (Danaher et al., 2004), bare ground cover (Weber et al., 2009), and shrub cover and encroachment (Laliberte et al., 2004). With the exception of the MODIS global continuous vegetation maps (Danaher et al., 2004), the preceding examples dealt with only one response variable. In this work there were four response variables. Since a pixel is in essence an integrated multi-dimensional spectral response of vegetation, bare ground and other features, it makes sense to attempt to decompose that response to understand the land cover dynamics of a given pixel in relation to the surrounding landscape.

With regards to the ease of understanding, RF has been frequently described as a "black box" (Prasad et al., 2006) because the individual trees cannot be examined separately due to the sheer number of trees that may be generated. This can limit the ability of an analyst to understand the underlying dynamics of the resulting model. Random Forests does provide metrics to aid in interpretation. One metric is variable importance, which can be used to compare relative importance among predictor variables. Such a feature is not available in other regression tree tools and therefore the importance of variables must be determined with a careful data mining process.

13.3.2.4 Implications

The development of a multi-temporal collection of VCF may be used to update information about the status or condition of a particular ecological site as well as characterizing the states and transitions for that site. For instance, a specific spatial unit of an ecological site may be characterized in terms of its occupancy by shrubs, grasses, trees, and bare ground using modeled VCF. Knowledge about the relative dominance of these life forms in a particular unit may shed light about its current condition relative to a reference condition. The VCF process may provide knowledge about usage of the ground by major life forms and bare ground and in this way pinpoint areas that are diverging from a desired condition.

13.4 RANGELAND CHANGE DETECTION ANALYSIS

13.4.1 Rangeland Indicators

Numerous landscape metrics have been developed to characterize the patterns and configurations of different land cover types in a landscape. Categorical maps generated from remote sensing data can then be used for landscape characterization to better understand spatial arrangements between different cover classes, particularly forest fragments (Read and Lam, 2002). These spatial arrangements are expressed numerically in the form of landscape indices or pattern metrics and have been used in many studies to assess land cover change and its impact, ecosystem health, or as variables for models that support environmental assessment and planning efforts (e.g., Botequilha Leitão and Ahern, 2002; Fuller, 2001; Gergel, 2005; Griffith et al., 2000; Liu and Cameron, 2001).

More than 100 pattern metrics can be computed using freeware such as FRAGSTAT (McGarigal and Marks, 1995), Patch Metrics (Rempel et al., 1999) and others (Crews-Meyer, 2002; Cumming and Vervier, 2002; Stanfield et al., 2002). However, many pattern metrics are highly correlated (Cain et al., 1997; Riitters et al., 1995). Efforts have been made to identify a minimum set of pattern metrics that describe landscape patterns adequately. Multivariate data analysis using principal component analysis (PCA) and factor analysis (FA) are the most commonly used methods to reduce pattern metrics data (Griffith et al., 2000; Honnay et al., 2003; McAlpine and Eyre, 2002; Stanfield et al., 2002). These methods identify a small number of components, which are then interpreted in terms of their dominant characteristics and underlying causes (Griffith et al., 2000). Multivariate data analysis requires large datasets and several landscape units to be statistically consistent (e.g., Cumming and Vervier, 2002; Schmitz et al., 2003). A few empirical landscape studies apply pattern analysis to only one landscape (e.g., Griffith et al., 2000) and are useful because they tackle problems at relevant scales, but the validity of making statistical inferences with such an approach is seriously compromised (Li and Wu, 2004). Gustafson (1998) overcame this by generating artificial landscapes (called neutral models), but the technique was difficult to relate to pattern metrics in real landscapes (Li and Wu, 2004). Therefore, the behavior of pattern metrics in real landscapes over time needs further investigation (Griffith et al., 2003). Irrespective of the landscape unit used, pattern metrics require rigorous validation in order to be interpreted and applied with confidence (McAlpine and Eyre, 2002).

13.4.2 Pattern Metrics to Measure Landscape Attributes

Area: Area distribution pattern metrics include basic attributes of the landscape such as the number of patches (*NP*) and the total area (*CA*) of all class patches, expressed in hectares. Mean patch size (*AREA_MN* in ha) is an intuitive index for measuring aggregation and is particularly suitable for categorical maps (Matongera et al., 2021; Zhou et al., 2020). Other measures are the standard deviation of the mean patch size (*AREA_SD* in ha) and the coefficient of variation (*AREA_CV* in %), determined by Equation 13.3.

$$AREA\_CV = \frac{AREA\_SD}{AREA\_MN} * 100 \tag{13.3}$$

$AREA\_CV$ indicates patch area distribution by determining the difference among patches within one landscape class. Lower values indicate a more uniform class distribution (Batistella et al., 2003; McAlpine and Eyre, 2002), that is if a landscape class is dominated by big patches both $AREA\_SD$ and $AREA\_CV$ values are large. All of these pattern metrics assign equal weights to each patch. For measuring landscape resistance to fragmentation, large forest patches are most important (Batistella et al., 2003), therefore metrics such as largest patch index (LPI) are useful. LPI is equal to the percent of the total landscape made up by the largest patch (McAlpine and Eyre, 2002).

Edge: Total edge (TE) refers to the length of edge that exists at the interface between land classes (McGarigal and Marks, 1995) while mean patch edge (MPE) indicates the amount of edge per patch (Equation 13.4).

$$MPE = TE / NP \tag{13.4}$$

Patch Shape Complexity: Mean perimeter—area ratio (*PAR_MN*) is a simple index of patch shape complexity, computed as the mean ratio between patch perimeter and area; it describes the amount of edge per unit area of landscape unit (such as forest) (McGarigal and Marks, 1995). Since the simple ratio is usually affected by patch size, a modified perimeter—area ratio called Landscape index (*LSI*), implied as shape of landscape, is computed using Equation 13.5.

$$LSI = Perimeter / 2\left(area * \pi\right) \tag{13.5}$$

Higher LSI values indicate higher complexity (McAlpine and Eyre, 2002). Because *LSI* is influenced both by shape complexity and number of patches (*NP*), preference is sometimes given to the mean shape index (*SHAPE_MN*) or the area-weighted mean shape index (*SHAPE_AM*), which weights larger patches more heavily than smaller patches (Batistella et al., 2003).

Fractal dimension (*FRAC*) has equally been used to describe patch shape complexity across a range of spatial scales (patch sizes). This overcomes one of the major limitations of the perimeter—area ratio as a measure of shape complexity. The value of *FRAC* ranges between 1 and 2 and is computed as:

$$FRAC = 2 \log P / \log A \tag{13.6}$$

where P is the perimeter and A is the area of the patch. Fractal dimension approaches 1 for shapes with very simple perimeters such as squares, and approaches 2 for shapes with highly complex perimeters (Read and Lam, 2002). Mean patch fractal dimension (*FRAC_MN*) and the area-weighted mean patch fractal dimension (*FRAC_AW*) are derived measures of shape complexity, interpreted similarly to *FRAC* with the addition of an individual patch area weighting applied to each patch in the case of *FRAC_AW* (McGarigal and Marks, 1995).

Interior-to-edge (core area): Core area (CA) represents the area of patch greater than a specified depth-of-edge distance from the perimeter. It is determined by defining a vanishing distance as the distance from a patch boundary inward, to a point where edge effects are eliminated (Baskent and Jordan, 1995). The higher the ratio between the core area and the total area, the lesser the degree of fragmentation (Batistella et al., 2003; Tang et al., 2005). The vanishing distance is arbitrarily set. The number of core areas (*NCA*) is the number of distinct core areas contained within a patch boundary. Total core area (*TCA*) is the same as *CA* except that core area is aggregated (summed) over all patches, approaching *CA* as patch shapes are simplified. Core area distribution in the landscape is measured using the parameters mean core area per patch (ha) (*CORE_MN*), patch core area standard deviation (*CORE_SD*) and patch core area coefficient of variation (*CORE_CV*).

Isolation: The distance of patch to its nearest neighbor measures the degree of isolation of that patch. Averaging this distance for all individual land cover classes or patches in a landscape gives the mean nearest neighbor distance (*ENN_MN*). Mean proximity index (*PROX_MN*) is also computed (Equation 13.7) and is a measure of isolation and fragmentation:

$$PROX\_MN = \sum \frac{\sum_{i}^{N} \sum_{j}^{m'} \frac{a_{ij}}{h_{ij}}}{N} \tag{13.7}$$

where a_{ij} is the area of patch ij in the neighborhood of patch i, h_{ij} the distance between patch i and patch ij, N is the number of patches and m' is the number of patches in the neighborhood of patch i. This index allows comparison among landscapes of any size, as long as the search buffer around each patch is the same (Gustafson and Parker, 1992).

Contagion/Interspersion: Contagion (*CONTAG*) provides an effective summary of overall clumpiness of categorical data where lower values indicate fragmentation of larger patches into smaller patches, that is the patch types are maximally disaggregated and interspersed (equal proportions of all pairwise adjacencies) (Equation 13.8):

$$CONTAG = 1 + \frac{\sum_{i=1}^{m}\sum_{k=1}^{m}(P_i)\left(\frac{g_{ik}}{\sum_{k=1}^{m} g_{ik}}\right)*ln(P_i)\left(\frac{g_{ik}}{\sum_{k=1}^{m} g_{ik}}\right)}{2ln(m)}(100) \qquad (13.8)$$

where P_i is the proportion of the landscape occupied by path type (class) i, g_{ik} is the number of adjacencies (joins) between pixels of patch types (classes) i and k based on the double-count method, and m is the number of patch types (classes) present in the landscape, including the landscape border if present.

Aggregation index (AI) is calculated from an adjacency matrix at the class level, AI equals 0 when the patch types are maximally disaggregated. *SPLIT* and *MESH* are two subdivision metrics computed as isolation measures (McGarigal and Marks, 1995). The interspersion and juxtaposition index (*IJI*), a configuration metric, is a measure of relative interspersion of each class at class level and each patch at landscape level (Crews-Meyer, 2002; McAlpine and Eyre, 2002). *IJI* approaches 100 when all patch types are equally adjacent to all other patch types, and as the distribution of class types depart from evenness, the *IJI* approaches 0 (Equation 13.9).

$$IJI = \frac{-\sum_{i=1}^{m}\sum_{k=i+1}^{m}\left(\frac{e_{ik}}{E}\right)*ln\left(\frac{e_{ik}}{E}\right)}{ln(0.5m(m-1))}(100) \qquad (13.9)$$

where e_{ik} is total length (m) of edge in the landscape between patch types (classes) i and k, E is total length (m) of edge in the landscape, excluding background, and m is the number of patch types (classes) present in the landscape, including the landscape border, if present.

Spatial heterogeneity: Spatial diversity representing the extent to which all patches are equally adjacent to each other, number of different land cover classes, and interspersion metrics determine the spatial arrangements of land cover classes (Trani and Giles, 1999). Patch richness (*PR*) is a simple metric equal to the number of different patch types present within the landscape boundary. Shannon's diversity index (*SHDI*) assesses the diversity of patches, and the extent to which one or a few patch types dominate the landscape (McAlpine and Eyre, 2002). *SHDI* starts at zero and increases without limit and is more sensitive to the number of patch types than evenness. *SHDI* is given by Equation 13.10 (Read and Lam, 2002):

$$SHDI = -\sum_{i=1}^{m}(P_i*lnP_i) \qquad (13.10)$$

where P_i is the proportion of landscape occupied by patch type (class) i, and m is the number of patch types (classes) present in the landscape, excluding the landscape border, if present.

The Shannon's evenness index (*SHEI*) is expressed such that an even distribution of area among patch types results in maximum evenness (Equation 13.11):

$$SHEI = -\frac{\sum_{i=1}^{m}(P_i*lnP_i)}{\ln m} \qquad (13.11)$$

SHEI varies between 0 and 1, and the highest value is reached when the distribution among patch types is perfectly even (Wickham and Rhtters, 1995). The modified Simpson's diversity (*MSIDI*) is similar to the *SHEI* (Equation 13.12):

$$MSIDI = -\ln \sum_{i=1}^{m} \left(P_i^2\right) \tag{13.12}$$

where P_i is the proportion of the landscape occupied by patch type (class) *i*. Diversity metrics represent landscape composition in terms of the relative proportions of each patch types (evenness), together with the number of patch types (richness). Detailed information on these indices and their computation can be found in Wickham and Rhtters (1995).

Ludwig et al. (2004) Monitored ecological indicators of rangeland functional integrity and their relation to biodiversity at local to regional scales. Functional integrity is the intactness of soil and native vegetation patterns and the processes that maintain these patterns. The integrity of these patterns and processes has been modified by clearing, grazing and fire. Intuitively, biodiversity should be strongly related to functional integrity. Ludwig et al. (2004), based on published work on Australian rangelands, identified several indicators of landscape functional integrity at finer patch and hillslope scales. These indicators, based on the quantity and quality of vegetation patches and inter patch zones, are related to biodiversity. These vegetation-cover and bare-soil patches can be measured using remote-sensing data at various resolutions depending on specific rangeland areas. De Cauwer et al. (2024) used Landsat MSS for prevailing dry periods before the major rainfall events and about six weeks after them when vegetation growth had peaked for the assessment of biodiversity condition in a rangelands in central Australia. They computed Contagion and Interspersion scores for land systems and vegetation types. The Contagion and Interspersion values, as indices of the spatial arrangement of patches in the landscape, indicated the extent to which both water development and land degradation associated with pastoralism had fragmented habitat types. Non-degraded and water-remote patches of pastorally more productive land systems or vegetation types were more dispersed and more fragmented relative to their original state, than for patches of pastorally less productive land systems or vegetation types.

13.4.2.1 Resilience

Ecological resilience is defined as the degree, manner, and pace of the restoration of vegetation attributes after a disturbance (Westman, 1985). Ecological resilience and its characteristics have been quantified statistically in ecosystem simulations (O'Neill, 1976; Westman and O'Leary, 1986), plant community field studies (Westman and O'Leary, 1986), and regional analyses of land degradation (Wessels et al., 2004, 2007). O'Neill (1976) simulated energy flow through a three-compartment (autotrophs, heterotrophs, and detrivores) model for six different biomes to test their ability to recover from a 10% reduction in plant biomass over a 25-year period. A recovery index, that is, a measure of the amount of deviation from a reference equilibrium value or initial state after a disturbance (a measure of malleability), was computed for each year over the 25-year period. The mean malleability of each biome for this period was interpreted to indicate which biomes were least and most resistant to disturbance (O'Neill, 1976). Westman and O'Leary (1986) developed four measures of ecological resilience: elasticity, that is, the rate of recovery from a disturbance; amplitude, that is, the threshold beyond which recovery to a previous reference state no longer occurs; damping, that is, the extent and duration of an ecosystem parameter following disturbance, and malleability. These were used to estimate the responses of various plant functional types within a coastal sage scrub plant community at 5–6 years after fire using field data and for 200 years after fire using a simulation model. Wessels et al. (2004, 2007) used an inter-annual time series of AVHRR- and Moderate Resolution Imaging Spectroradiometer (MODIS)-derived NDVI (or ANPP) from 1985 to 2005 to examine the resilience of a landscape in northeastern South Africa that had been subject to both overpopulation and apparent overgrazing by the forced settlement of pastoralists into "homelands" during the apartheid era from 1910 to 1994. A spatially aggregated approach was used in which the mean annual ΣNDVI for the period of 1985–2005 was compared between non-degraded benchmark areas and degraded areas within the homelands as

$$RDI\ or\ PD = \frac{non-degraded\sum NDVI - degraded\sum NDVI}{non-degraded\sum NDVI} \times 100 \tag{13.13}$$

where RDI is the relative degradation index or the percent difference (PD) between the mean NDVIs of non-degraded and degraded sites (Wessels et al., 2004, 2007). Wessels et al. (2004, 2007) found that these paired sites were significantly different from each other and that degraded sites had significantly lower mean annual NDVI than did non-degraded sites. Also, there was no indication of recovery (malleability) of degraded sites toward the mean conditions of non-degraded sites from the end of the apartheid era in 1994 to 2005 (Wessels et al., 2004, 2007). This provided evidence in support of the hypothesis that the observed migrations of former rural homeland populations to jobs in major urban centers, such as Johannesburg and Durban, were due in part to environmental degradation of the homelands. However, at the coarse resolution of MODIS (0.25 km^2) and AVHRR (1 km^2), it was observed that finer-scale phenomena at the community and species levels were masked or averaged out (Wessels et al., 2007). A number of scaling studies in drylands suggest characteristic length scales for vegetation and bare soil patches of approximately 1–100 m; these scales are probably not amenable to analysis using spatially coarse-resolution sensors such as MODIS or AVHRR (Hudak and Wessman, 1998; Rietkerk et al., 2004). To account for this disparity, Wessels et al. (2007) suggested a staged remote sensing approach to monitoring at regional and national scales in which both coarse-resolution sensors such as MODIS are used in conjunction with relatively finer-resolution datasets such as Landsat.

13.4.3 Change Detection Methods

It is important to monitor and understand change in rangelands so that effective action can be taken to maintain ecological, economic and social values. Both seasonal conditions and grazing management play a role in vegetation dynamics on pastoral areas. For example, De Cauwer et al. (2024) determined the effect of seasonal spectral variability on vegetation and land cover classification of Landsat TM by comparing accuracies in different seasons in a mid-latitudinal (29°30′–31°0′S) region with summer and winter rainfall, a broad altitudinal range, a temperate to subtropical climate and diverse land uses (e.g., summer and winter crops, and nature conservation). By comparing the observed changes with those expected under the prevailing seasonal conditions and by investigating the response of species known to be adversely or positively affected by livestock grazing, it was possible to conclude that at least some of the positive changes observed could be attributed to the type of grazing management, rather than seasonal conditions alone. In a similar study, De Cauwer et al. (2024) used three-date composite land cover maps through a process called referential refinement and aggregation for understanding the level of land exploitation (extensive vs intensive) activities carried out in the region. Landscape-scale monitoring is important for providing regional to national-scale intelligence on habitat quality and trends in threats to or drivers of biodiversity, with data obtained using systematic ground-based and remote methods. Landscape-scale monitoring is typically based on remotely obtained or extrapolated data that can be mapped, and use temporal scales appropriate to the indicator (Ludwig et al., 2004). The advantage of landscape-scale monitoring is that it is often relatively cheap to undertake over space and time and provides a broader context for national reporting.

There are a variety of opinions and suggestions about the selection of the most effective and appropriate techniques for change detection studies (Lu et al., 2004). Therefore, it is difficult to specify which change detection algorithm will suit a specific problem. A review of different change detection techniques used in previous research would be of great help in selecting methods to aid in producing good quality change detection results. For any change detection project, the following conditions must be satisfied (Lu et al., 2004): (1) accurate and precise registration between multi-temporal images; (2) precise radiometric and atmospheric normalization between multi-temporal images; (3) similar phenological or seasonal conditions between multi-temporal images; and (4) selection of the same spatial and spectral resolution images if possible.

The aim of change detection studies is to compare the spatial representation of two points in time by controlling all variance caused by differences in variables that are not of interest and to measure changes caused by differences in the variables of interest (Green et al., 1994). Since there are many change detection techniques that can be used in one study, the selection of the most suitable method for a given project is not easy since different approaches, with the same environmental conditions, produce different results (Coppin et al., 2004).

13.4.3.1 Classification Based Change Detection

Broadly, image classification is a process of drawing meaningful information by differentiation and extraction of different classes or types (e.g., land use types, vegetation species) from remote sensing data through a number of image processing procedures including image pre-processing. Image is classified either using traditional methods or improved or modified techniques.

The classification group of change detection techniques includes supervised and unsupervised classification followed by post-classification comparison of results of change/no-change identification. The aim of these methods is to produce high quality, accurate classification results from remote sensing data through the use of adequate numbers of accurate training sample data, to produce accurate change results after comparison. However, the selection of sufficient numbers of high quality training samples to truly represent each land use/land cover class is laborious and time consuming and requires a thorough knowledge of the study area in order to obtain high quality classification results. The situation is more difficult in the classification of historical data when there is no or insufficient ground or area information, making accurate classification a challenge and often leading to unsatisfactory change detection results (Lu et al., 2004). The change detection results are often represented in the form of matrix showing land use/land cover dynamics of pixel changes from one class to another, providing a detailed description of changes over a specified period of time and minimizing the effect of atmospheric and environment difference between the multi-date images. A detailed review of quality assessment of different image classification algorithms for land cover mapping and accuracy assessment has been summarized by Arogoundade et al., (2023) and Feizizadeh et al., 2023. These techniques have been used by many researches in different types of land cover change detection analysis with good results (e.g., Xiuwan, 2002; Petit and Lambin, 2002; Li and Zhou, 2009).

Among traditional methods of classifying remote sensing data, unsupervised and supervised techniques are the most commonly used algorithms. ISODATA clustering and K-means methods are mainly used in unsupervised techniques while the Maximum Likelihood Classification (MLC) and Minimum Distance to Mean (MDM) are commonly applied techniques in supervised methods since the beginning of digital image feature extraction using statistical software. Unsupervised approaches, such as ISODATA clustering and K-means algorithms, are easy to apply for thematic mapping of landcover or vegetation classification using statistical software packages (Hadian et al., 2019 and Al-bukhari et al., 2018) based on iterative learning of remote sensing data defined by the user. In this classification approach, an arbitrary initial cluster vector is assigned first, based on which each pixel is classified as closest to that cluster value. Finally, based on all the pixels present in one cluster, the new cluster mean vector is calculated and the process is repeated until the gap between the iterations becomes smaller than the threshold value (Xie et al., 2008). This method does not require any prior knowledge of the area or theme being studied (Tso and Olsen, 2005) and classifies the image based on natural grouping or spectrally homogenous thematic classes using spectral values of each pixel. The analyst then assigns the spectral classes into thematic information classes of interest (Jensen, 2005). It has the benefit of automatically converting the raw image data into useful information so long as higher accuracy is achieved (Tso and Olsen, 2005). However, since each pixel is treated as spatially independent, a traditional unsupervised classification based on spectral data alone results in poor accuracy. To overcome this, Tso and Olsen (2005) introduced

Hidden Markov Model (HMM) as a fundamental framework to incorporate both the spectral and contextual information for unsupervised classification and achieved higher classification accuracies.

In supervised classification, a prior knowledge about some cover types are obtained in advance through a combination of field work, aerial photo interpretation, existing maps and other sources. Based on this, the analyst attempts to locate specific sites (class representation) for these cover classes in remote sensing data. These sites are called training sites and used to train the classification algorithm based on spectral characteristics of each site to classify the rest of the image. Statistical parameters such as mean, standard deviation, covariance matrices, correlation matrices etc. are computed for each training site, based on these statistics, every pixel is then evaluated and assigned to a class of which it has the highest likelihood of being a member of (Jensen, 2005). Various supervised classification algorithms such as Parallelpiped (PAR), Minimum Distance (MID), Nearest Neighborhood (NN), and Maximum Likelihood (MLC) may be used to assign an unknown pixel to one of the possible classes; therefore the choice of a particular classifier or decision rule depends on the nature of input data and desired output (Jensen, 2005). MLC is considered to be classic and most widely used supervised classification technique, applied by many researchers in different studies for satellite image classification (Laba et al., 1997; Langford and Bell, 1997; Rogan et al., 2002; Xiuwan, 2002; Abdulaziz et al., 2009). In rangeland studies, Abdulaziz et al. (2009) evaluated the potential of multispectral Landsat images (MSS and TM) and applied MLC to classify vegetation cover in the degraded land of MuUs sandy land in China. Torahi (2012), in monitoring rangeland dynamics between 1990–2006, used MLC for classifying TM and ASTER data and generated detailed rangeland maps and also separated grazing intensity levels in rangelands.

However, in complex areas, the assumption of MLC that data follow Gaussian distribution is not always applicable, resulting in less satisfactory results. Strahler (1980) demonstrated the effective use of modified prior probabilities into MLC to improve image classification. In addition to MLC alone, there are studies conducted by researchers using a combination of other supervised classification algorithms, for example Miller et al. (1998) applied parallelepiped technique to classify water, open land, clouds and cloud shadows, assuming them as discrete classes in spectral space, while other forest classes were classified using MLC in landcover change study in Northern Forest of New England. Xiuwan (2002) compared several classification methods such as PAR, MID, Mahalanobis Distance (MAD), MLC, ISODATA and a method combining an unsupervised algorithm and training data (CUT) to develop a new classification method and suitable change detection technique for analysis of landcover change and its regional impacts on sustainable development in Korea. J. Im et al. (2008) used nearest neighborhood classification techniques for landuse and landcover classification and compared the accuracy with other techniques. Munyati et al. (2011) used the hybrid image classification approach to monitor savanna rangeland deterioration in Mokopane, South Africa. For classification, initial clustering was undertaken by unsupervised classification using the ISODATA algorithm. On the resulting cluster image, the field data sites were then located and the spectral signature clusters were assigned names. The named spectral signatures were then used for a final supervised classification.

13.4.3.1.1 Improved/Modified Classifiers

Depending upon different geographic conditions and other factors, there is a possibility that same ground covers or vegetation types show different spectral response or different covers show similar spectral response on remote sensing imagery due to spectral mixing, therefore an accurate classification result is very difficult to obtain from traditional supervised or unsupervised classification techniques. Hence, there is always a need for improvement in the classification methods for better classification results. Some of the new methods have been based on traditional supervised or unsupervised approaches through a process of combination, extension or as an expansion to provide better classification results from remote sensing data. Sohn and Rebello (2002) developed the new spectral angle classifier (SAC), a combination of supervised spectral angle classifier (SSAC) and unsupervised spectral angle classifiers (USAC), based on the fact that the spectra of same types of surface objects are approximately linearly scaled variations of one another due to atmospheric and topographic effects. SAC allows the spectral angle to be used as a metric for measuring angular distances in feature space for classification and clustering of multispectral satellite data and was successfully applied in biotic and landcover classification (Sohn and Qi, 2005). Collado et al. (2002) found use of spectral mixture analysis (SMA) to be valuable in monitoring desertification processes in the crop-rangeland boundary of Argentina. Savanna rangeland degradation in Namibia was classified by Vogel and Strohbach (2009), who used Landsat TM and ETM+ data. The decision tree classifier was also used. Their results showed that savanna degradation could be classified into the following six classes: vegetation densification, vegetation decrease, complete vegetation loss, long-term vegetation patterns, the recovery of vegetation on formerly bare soils, and no change with an overall accuracy of 73.4%, with class pair accuracies ranging from 80 to 100% for producer and user accuracies. Okin et al. (2001) assessed the utility of AVIRIS satellite imagery for accurately discriminating among vegetation types in the Mojave Desert, USA. Multiple Endmember Spectral Mixture Analysis (MESMA) and Spectral Mixture Analysis (SMA) were performed to estimate the proportion of each ground pixels area that fitted with different cover types. They concluded that AVIRIS showed low potential for classifying vegetation types, with an overall accuracy of only 30% due to low vegetation cover. This is a common problem in rangelands, where the overall classification accuracies are generally lower than for forest or well vegetated areas.

James et al. (2003) suggested that ecological condition of rangelands is a major factor in their environmental quality, their overall performance as watersheds and in wildlife and livestock production. They pointed out that to maintain the quality of rangelands, they must be monitored over time and space and also must take into account topography, climate, soils, plant communities, and animal population. Several studies have shown that extensive field knowledge and support data improves classification accuracy, for example, Wang et al. (2009) used stratified classification approaches based on segmentation of an image into focused area and categories based on existing GIS landcover data in order to improve classification accuracy. Franklin and Wilson (1992) developed a three-stage classifier that incorporated a quadtree-based segmentation operator, a Gaussian minimum-distance to mean test and final test involving ancillary geomorphometric data and a spectral curve measure and attained significant increase in classification accuracy in less time and with minimum field training data as compared to MLC. Jianlong et al. (1998) used green herbage yield data, environment, and remote sensing data recorded in different grassland types in Fukang County, Xinjiang from 1991 to 1996. They explored the methods of processing images, analyzing information, and linking of remote sensing data with ground grassland data. Tagestad and Downs (2007) studied landscape measures of rangeland condition using texture methods, highlighting the apparent roughness in the visible surface due to drastic changes in brightness between adjacent pixels. The texture model reduces the color signal in the image and maximizes the texture signal. The field-measured shrub canopy cover in each plot was compared to the corresponding texture ratio values for pixels representing that plot to develop a simple linear regression relationship between shrub canopy cover and image texture. Franklin et al. (2001) used spatial co-occurrence texture measures and MLC to generate higher forest species composition classification accuracies in New Brunswick forest stand than the use of spectral patterns alone. Gong and Howarth (1992) developed and evaluated a contextual method of landuse classification using SPOT data involving two steps: gray-level vector reduction and frequency based classification and found that the frequency based classification method was comparatively fast, efficient and could improve landuse classification accuracies over MLC and was effective in identification of spatially heterogeneous landuse classes. Pin˜eiro et al. (2006) estimated seasonal variation in aboveground production and radiation-use efficiency of temperate rangelands using remote sensing. They evaluated, at a seasonal scale, the relationship between ANPP and the NDVI and estimated the seasonal variations in the coefficient of conversion of absorbed radiation into aboveground biomass and also identified the environmental controls on such temporal changes. Their results indicated that NDVI produced good, direct estimates of ANPP only if NDVI, PAR, and aboveground biomass were correlated throughout the seasons and hence suggested seasonal variations of aboveground biomass associated with temperature and precipitation to be taken into account to generate seasonal ANPP estimates with acceptable accuracy.

Fuzzy methods in the classification of remote sensing images have become popular because of their ability to deal with situations where the geographical boundary is inherently fuzzy or heterogeneous due to the presence of mixed pixels and traditional methods of classification are often incapable of performing satisfactorily (Okeke and Karnieli, 2006). For example, Tagestad and Downs (2007) investigated the fuzzy approach for the classification of sub-urban landcover from remote sensing imagery and reported to have advantages over both conventional hard method and partially fuzzy approach. Other similar applications have been by Okeke and Karnieli (2006) for vegetation change study, and Sha et al. (2008) for grassland classification. Discriminating and mapping vegetation degradation at Fowlers Gap Arid Zone Research Station in Western New South Wales, Australia, Okeke and Karnieli (2006) used Random Forest method to classify perennial vegetation, chenopod shrubs and trees using hyperspectral imaging (CASI). An area of less than 25% was discriminated and mapped. Okeke and Karnieli (2006) concluded that high-spectral resolution imagery had potential for the discrimination of vegetation cover in arid regions.

Recently, decision tree (DT) classifiers have been used for landcover and vegetation classification from remote sensing data based on the concept of splitting a complex decision into several simpler decisions which may be easier to interpret. DT is based on multistage or hierarchal decision scheme or a tree like structure composed of a root node (containing all data), a set of internal nodes (splits) and a set of terminal nodes (leaves) and processes from top to bottom by moving down the tree where each node of the decision tree structure makes a binary decision that separates either one class or some of the classes from the remaining classes (Xu et al., 2005; Chen and Rao, 2008). DT is relatively simple, explicit, computationally fast, makes no statistical assumptions and can handle data on different measurement scales (Friedl and Brodley, 1997; Pal and Mather, 2003). Chen and Rao (2008) determined the rate and status of grassland degradation and soil salinization based on DT and field investigation with overall classification accuracy of more than 85%. Xu et al. (2005) employed a decision tree regression approach to determine class proportions within a pixel so as to produce a soft classification and the accuracy achieved by DT regression was found to be significantly higher as compared to MLC applied in soft mode and supervised version of fuzzy-c-soft classification, especially when data contained a large proportion of mixed pixels.

Pal and Mather (2003) assessed the utility of DT classifier for landcover classification using multispectral and hyperspectral data and compared the performance of univariate and multivariate DT with that of ANN and MLC. They concluded that DT performance was always affected by training data size, univariate DT was more systematic than multivariate DT for common training and test datasets and, in the case of univariate DT, a minimum of 300 training pixels per class were needed to the achieve most suitable classification accuracy. They further concluded that DT classifiers were not recommended for high-dimensional datasets. Friedl and Brodley (1997) tested univariate DT, multivariate DT and a hybrid DT on three different remote sensing datasets and compared the classification results from each DT algorithm with those of MLC and linear discriminant function classifier, and reported that DT, hybrid DT in particular, consistently outperformed MLC and linear function discriminant classifier in regard to classification accuracy. Rogan et al. (2002) compared MKT, MSMA MLC and DT to accurately identify changes in vegetation cover in south California and showed that DT classification approach outperformed MLC by nearly 10% regardless of enhancement technique used and using DT classification, MSMA change fractions outperformed MKT change features by nearly 5%. Borak and Strahler (1999) developed a tree-based model for landcover identification from satellite data for a semiarid region in Cochise County, Arizona and compared the results with other classifiers such as fuzzy ARTMAP, MLC and reported that DT could reduce a high-dimension dataset to a manageable set of inputs that retained most of the information of the original database. However fuzzy ARTMAP achieved the highest accuracy in comparison to MLC or DT classifier. Muchoney et al. (2000) compared the classification results obtained from MODIS data for vegetation and landcover mapping using DT, Gaussian ART and Fuzzy ART ANN algorithms in Central America and attained high accuracies from DT (88%), Gaussian ART (83%), and Fuzzy ART NN (79%).

Jarman et al. (2011) studied rangeland conditions in the regions of Kvemo Kartli and Samtskhe-Javakheti in southern Georgia, USA using remote sensing data. They used (NDVI) as the indicator of condition and used object based image analysis. The Landsat scenes and ancillary datasets were collated within eCognition following a two tiered (multi-level) hierarchical approach. The first level brought together the non-rangeland datasets and the second level classified the rangeland area into relative states of rangeland condition (good, moderate, poor). A rule based classification was undertaken within eCognition to map the classes. Laliberte et al. (2011) applied IHS transformation on remote sensing data followed by object based segmentation and classification for structure and species level rangeland mapping. They developed specific rules to define threshold for broader classes and nearest neighborhood classification for finer species level classification. They reported that classification accuracies were highly dependent on the level of detail, number of classes, size of area and specific mapping objectives.

13.4.3.2 Image Differencing and Image Ratioing

The algebraic technique of change detection analysis is based on the selection of a threshold to determine the changed areas (Lu et al., 2004). In this group of change detection techniques, many researchers have applied image differencing and image ratioing with different combinations of spectral bands as their first choice for identifying change, and derived satisfactory results. Conclusions and recommendations vary about whether image differencing and regression, vegetation index differencing or image ratioing is the best change detection technique. Since each method has been applied to different areas, with different datasets and under different environmental conditions, the decision on the selection of a suitable method is not an easy task (Coppin et al., 2004). For example, Sinha and Kumar (2013a) tested 11 different binary change detection methods and compared their capability in detecting land-cover change/no-change information in different seasons using multi-date Thematic Mapper (TM) data. They proposed a relatively new approach for optimal threshold value determination for separation of change/no-change areas and found improved results with this as compared to traditional thresholding (Sinha and Kumar, 2013b).

13.4.3.2.1 Transformation

Various linear data transformation techniques such as Principal Component Analysis (PCA), Tasselled Cap (TC), Gramm-Schmidt (GS), and chi-square transformations have been applied to multi-temporal remote sensing data to identify changes in various land use/land cover classes. In PCA, only two bands of the multi-date image are used as input instead of all bands (Richards, 1986), and hence reducing the data volume and redundancy between bands. After transformation using two bands, the derived components contain information about change and no change areas as the first and second components, respectively, based on information common or unique in the two input bands (Chavez and Kwarteng, 1989). PCA is based on three steps: calculation of a variance—covariance matrix, computation of eigenvectors and linear transformation of datasets (Richards, 1986). Two types of PCA, standardized PCA (uses correlation matrix) and non-standardized PCA (uses covariance matrix), have been used for change detection (De Cauwer et al., 2024; Arogoundade et al., 2023). PCA has certain disadvantages in not providing detailed change matrices and difficulty in interpretation, identification and labelling of changed areas.

TC transformation is carried out by assigning tasseled cap coefficients to spectral bands of two dates, a positive coefficient to the first date and a negative coefficient to the second date, as explained by Crist and Cicone (1984). This is followed by Gramm-Schmidt transformation to make the derived vectors orthogonal to each other. The three transformed images thus obtained contain information about differences in greenness, brightness and wetness, with highest classification accuracy in the greenness change image. Maynard et al. (2007) classified Landsat 7 ETM+ to identify spectrally anomalous locations on satellite data and their correlation with corresponding ground locations for rangeland evaluation. Their classification was carried out using TC brightness, greenness and wetness components stratified by ecological site descriptions. PCA and TC have been the most used approaches for detecting change/no-change information. An additional advantage of TC transformation is that the TC coefficients are scene independent compared to PCA coefficients, which are scene-dependent.

13.4.4 Case Study: Spectral-Spatial Characteristics of Selected Ecological Sites

13.4.4.1 Introduction

Ecological sites (ES) characterize land of specific biophysical and plant community properties. Spatially distinct land areas of the same ecological site respond similarly to management actions and natural disturbances (U.S. Department of Agriculture, NRCS 2012). Therefore, the identification and mapping of ecological sites across large landscapes can be an important tool for rangeland managers. Ecological sites, as they are applied by the United States Department of Agriculture Natural Resources Conservation Service (NRCS) are defined on the basis of soils, geomorphology, hydrology, and the plant species composition that occur on those soils. The NRCS spatially ties ESs to soil components mapped within soil map units (SMU) contained in their spatial digital soil surveys.

Bestelmeyer et al. (2009) formulated an approach to identify ESs from Landsat (or similar platforms) interpreted into land cover maps to identify vegetation distribution. These mapped vegetation areas are used to infer possible ecological sites. In addition to this effort, Maynard et al. (2007) found that there was a high correlation between field measures of productivity and exposed soil when compared to the tasseled cap brightness component extracted from Landsat Thematic Mapper (TM) imagery. Differences in brightness have been shown to discriminate between deciduous shrubs (or harvested forest stands) and closed canopy forests (Dymond et al. 2002).

Gamon et al. (1995) discussed the usefulness of the NDVI as an indicator of photosynthetic activity as well as canopy structure and plant nitrogen content. Jensen (2000) showed that NDVI was sensitive to canopy variations including soil visible through canopy openings. While the sensitivity to soil background has typically been seen as a disadvantage of NDVI for vegetation assessment, it could prove useful for studying ESs because areas of the same ES may have a similar amount and type of bare soil. The NDVI values within a polygon, such as an SMU, and the variation in the NDVI within that polygon has also been used to distinguish between cover types (Pickup and Foran, 1987).

Accurately classifying and identifying the spatial extent of ESs on a landscape level is a time consuming process involving extensive field-work to characterize soils. While remotely sensed data cannot yet be used to obtain detailed data about soils, it can be used to identify the unique vegetation components of ESs. Being able to accurately identify the vegetation component of ESs should provide a means by which soil field sample locations can be identified more efficiently. Based on past research, the use of satellite derived NDVI and brightness, coupled with biophysical geospatial data (elevation, slope, and aspect) should allow areas of the same ES vegetation components to be mapped.

13.4.4.2 Methods

13.4.4.2.1 Study Area

This research was conducted in Rich County, Utah, U.S. (Figure 13.11), located in the northeastern corner of the state (long 111°30′38.5″—long 111°2′42.2″ West and lat 42°0′0″- lat 42°08′24.3″ North). The western portion of the study area is characterized by high elevations with vegetation consisting of aspen forests, subalpine conifer forests, and scattered mountain sagebrush steppe. Moving east, elevation decreases, and the mountain sagebrush steppe becomes dominant. Central and eastern Rich County is made up of relatively lower elevations with vegetation consisting of basin big sagebrush steppe and shrubland, subalpine grasslands, and agriculture (Figure 13.12). The average elevation is 2093 m. The highest point is Bridger Peak at 2821 m and the lowest point is about 1800 m. The climate is variable and is affected by the changing topography of the county.

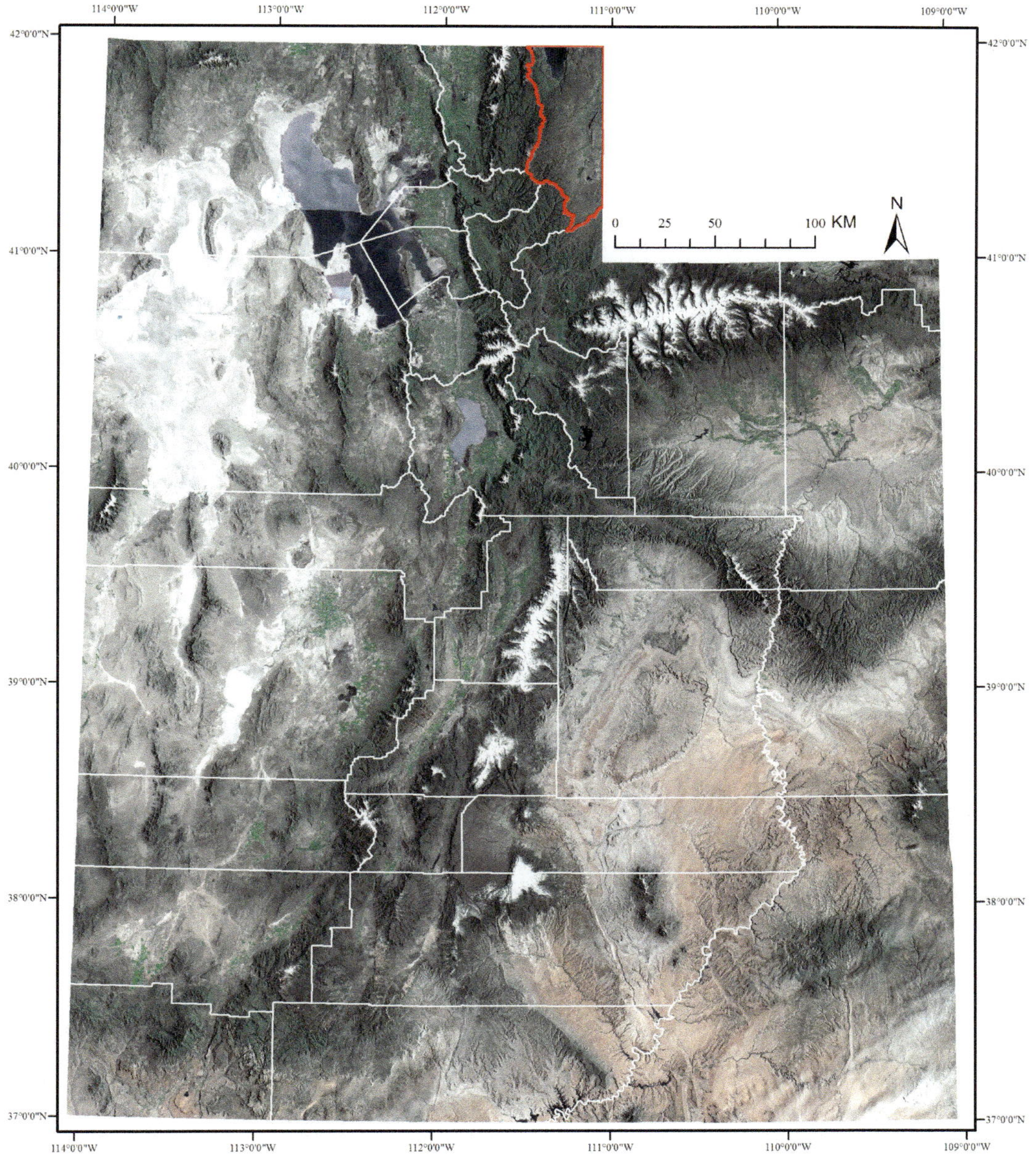

FIGURE 13.11 Utah natural color mosaic from the Landsat 8 Operational Land Imager (OLI) showing Rich County outlined in red.

13.4.4.2.2 Biophysical Geospatial Datasets

A series of Landsat 5 TM images (Path 38/Row 31) for each year between 1984–2011 with Julian date as close to 207 (July 26th) as possible (given acceptable cloud cover) were collected from the U.S. Geological Survey Global Visualization Viewer (GLOVIS). The Julian date of 207 was chosen by averaging the date for each year that displayed the greatest variance in NDVI between different land cover types. The dates were obtained by examining line graphs of mean NDVI values collected by the Moderate Resolution Imaging Spectroradiometer (MODIS) of evergreen forests, shrubs, and deciduous forests. Figure 13.13 is an example of one of these graphs from 2009.

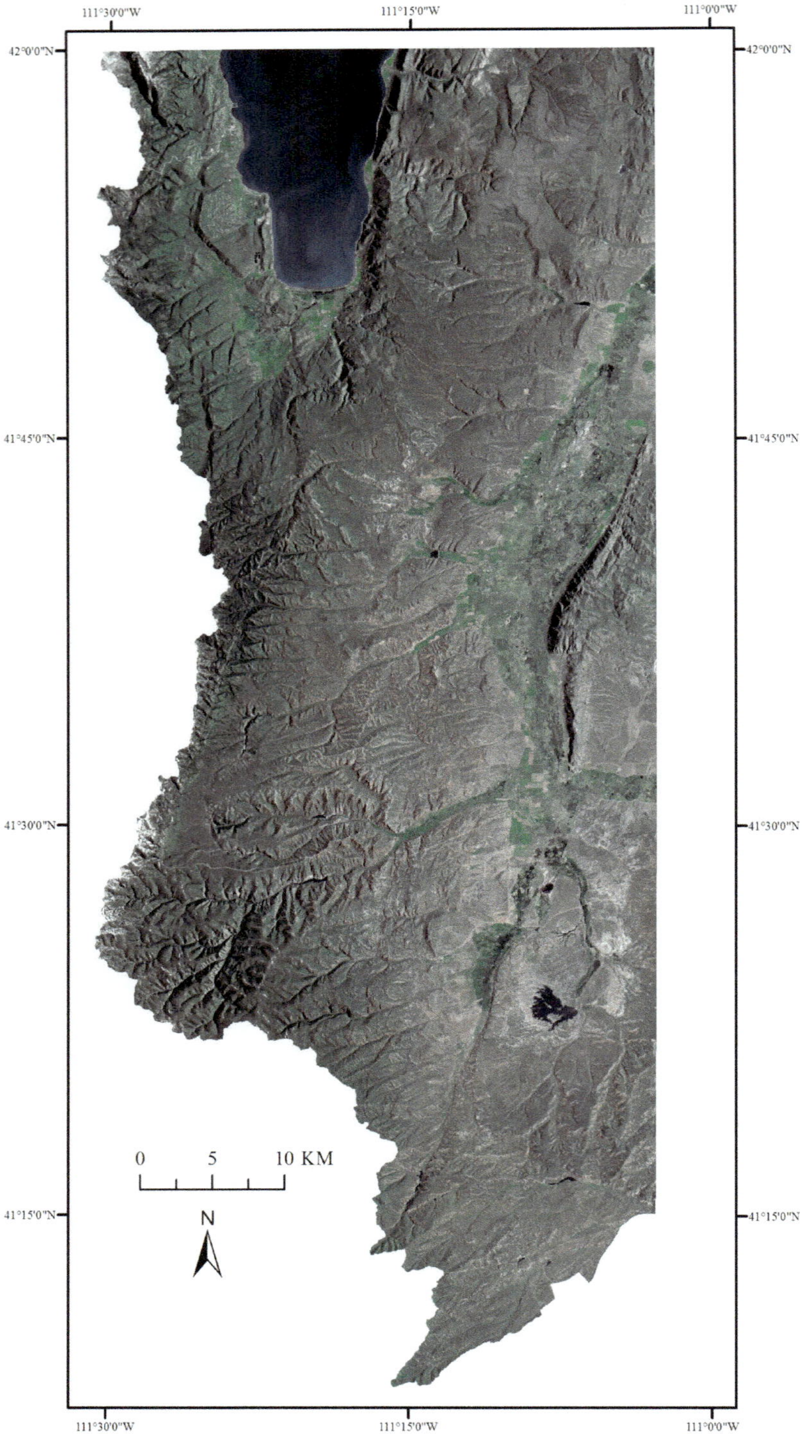

FIGURE 13.12 Rich County, Utah, natural color image from Landsat 8 OLI.

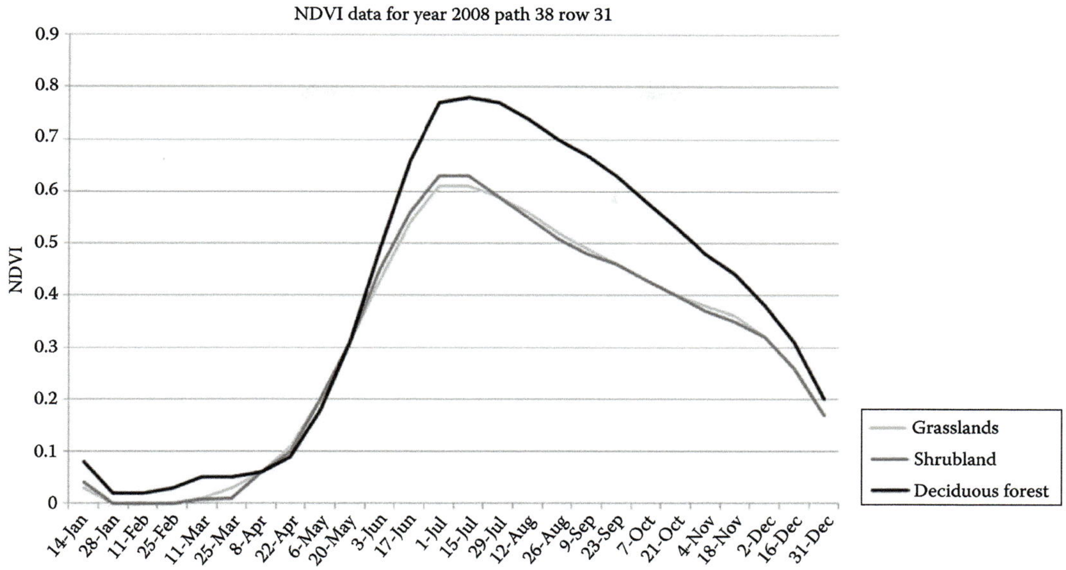

FIGURE 13.13 Line graph of annual fluctuations in NDVI for evergreen forests, shrublands, and deciduous forests. The largest differences in NDVI can be seen in mid-summer.

Of the 28 years' images, 18 were within 20 days of 207, five more were within 30 days of 207, and three more were within 40 days of 207. The cloud free scene closest to Julian date 207 from 1987 had a Julian date of 153 and was 54 days off. The year 2001 was the only year that an image was not available due to cloud cover.

Raw pixel values were converted to reflectance using an image-based atmospheric correction (Chavez, 1996) using appropriate calibration coefficients for Landsat 5 TM (Chander et al., 2009). The NDVI was calculated for each image (Rouse et al., 1974) and for each NDVI product we calculated the spatial standard deviation of the NDVI using a 5 × 5 pixel focal window. The brightness component for each year was calculated using the published transformation coefficients for Landsat 5 TM (Crist and Cicone, 1984). Images from every year were utilized, based on literature indicating that longer time series of remotely sensed data were necessary to adequately characterize different ecological states due to inherent year-to-year variance (Hernandez, 2011).

A 30 m digital elevation model (DEM) produced by the USGS National Elevation Dataset program was used to extract slope and aspect. Elevation, slope, and aspect have been shown to drive microclimatic variation and therefore the spatial distribution and patterns of vegetation (Jin et al., 2008).

13.4.4.2.3 Ecological Sites

To test the process, land cover types representing five ESs were selected. These were Wyoming big sagebrush (*Artemisia tridentata* ssp. *wyomingensis*) steppe, mountain big sagebrush (*Artemisia tridentata* ssp. *vaseyana*), Utah juniper (*Juniperus osteosperma*), Douglas-fir (*Pseudotsuga menziesii*), and aspen (*Populus tremuloides*). With the exception of Utah juniper, these vegetation components were selected because of their prevalence in the county. Wyoming big sagebrush accounts for much of the foothill vegetation in the study area, and aspen is prevalent in the mountainous section, with Douglas-fir, and mountain big sagebrush as secondary and tertiary types. Utah juniper is not prevalent within the study area; however, it is an important vegetation component due to its potential to encroach into sagebrush steppe communities (Miller and Rose, 1999). Together, these vegetation components represent approximately 71% of the county by area.

Twenty polygons were digitized for each of the five ESs using 2009 National Agricultural Imagery Project (NAIP) 1m resolution aerial orthoimagery. In total, one-hundred polygons were created (20 for each ES vegetation component).

Polygons were intersected with the topographic data layers, yearly NDVI imagery, and yearly brightness component images. For each polygon, the mean values of topographic and brightness variables were extracted along with the mean and standard deviation of each NDVI image for each year. From these data, a 28-year mean of the average brightness as well as the mean of the average NDVI and the mean NDVI standard deviation (sdNDVI) were produced. This was done to minimize the effects of inter-annual climate variability and clouds and provide long-term average values for each polygon of each land cover type. Inter-annual climate variability has been shown to affect plant species productivity (Goulden et al., 1996; Arain et al., 2002) and ecological processes (Westerling and Swetnam, 2003). The resulting data matrix was therefore composed of the ES vegetation component name followed by three columns for the DEM derivatives, and three columns for the NDVI, sdNDVI, and brightness.

13.4.4.2.4 Results

Figure 13.14 shows the 28-year mean of the average NDVI value for each polygon plotted against the 28-year mean of the sdNDVI for each polygon showing that the five ESs occupied unique NDVI mean and spatial variance "niches." Some overlap occurred between Wyoming big sagebrush and Utah juniper and between Douglas-fir and aspen ESs. Figure 13.15 shows the same data points with 28-year mean of the average NDVI value plotted against the 28-year mean of the average brightness. The brightness component was able to cleanly separate Aspen polygons from the Douglas-fir polygons. However, brightness provided little separation between Utah juniper and Wyoming big sagebrush.

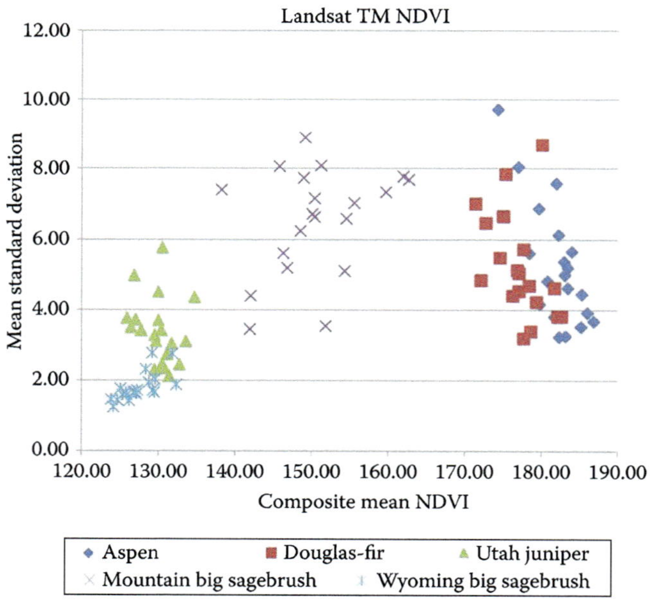

FIGURE 13.14 Scatter-plot showing the distribution of each ecological site vegetation component in our study with average NDVI value on the x-axis and the average standard deviation in NDVI on the y-axis. Both of these variables provide separation between vegetation classes.

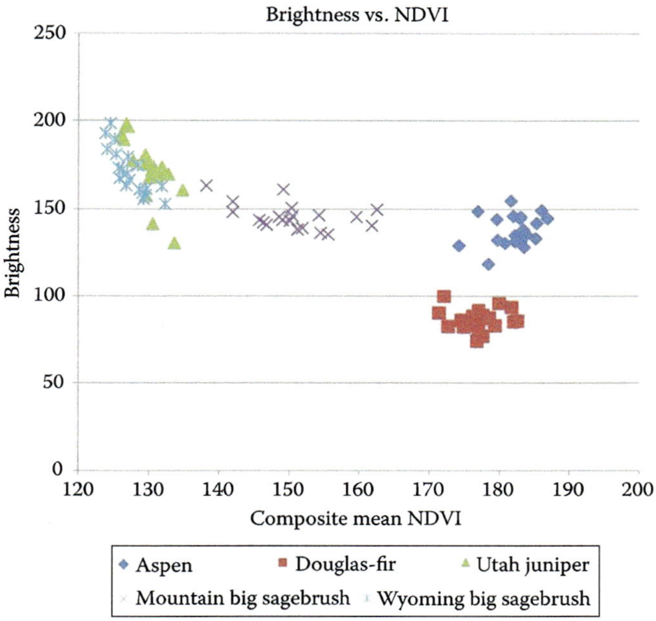

FIGURE 13.15 Scatter-plot showing the distribution of each ecological site vegetation component in our study with average NDVI value on the x-axis and the average brightness component (obtained from the tasseled cap transformation) value on the y-axis.

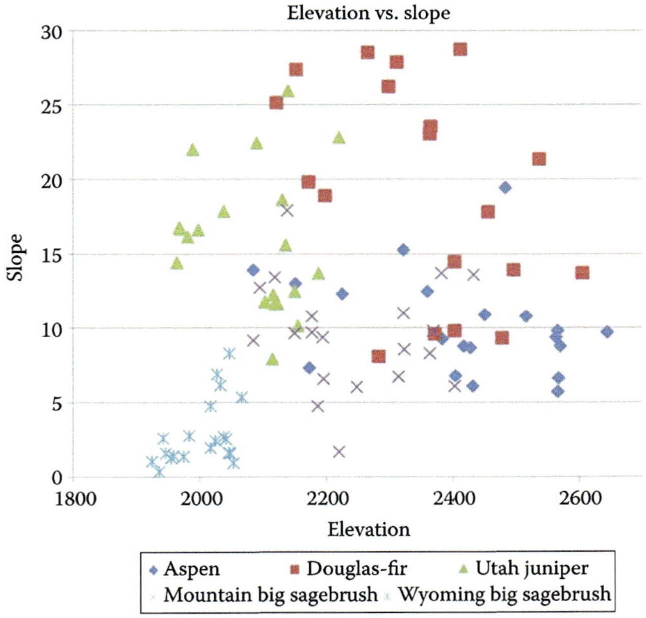

FIGURE 13.16 Scatter-plot showing the distribution of each ecological site vegetation component in our study with elevation on the x-axis and slope on the y-axis. Most vegetation classes overlap one another. However, slope does help with separating Wyoming big sagebrush from Utah juniper.

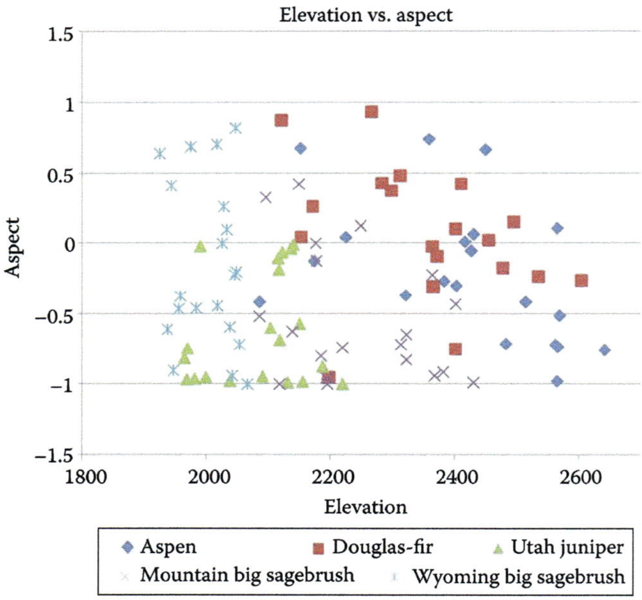

FIGURE 13.17 Scatter-plot showing the distribution of each ecological site vegetation component in our study with elevation on the x-axis and aspect on the y-axis. Aspect does not appear to separate any vegetation classes.

Each polygon was plotted against elevation and slope (Figure 13.16) and also against elevation and the cosine of aspect (Figure 13.17). Topographic variables alone were able to somewhat separate vegetation components along an elevation gradient (as expected). Slope seemed to be a good variable to separate Utah Juniper from Wyoming big sagebrush and Douglas-fir from Aspen. Aspect was not useful for distinguishing between any vegetation types.

The Wyoming big sagebrush polygons collectively had low NDVI and low spatial variation in NDVI. Utah juniper sites had similarly low average NDVI, but due to high contrast between green juniper trees and a relatively larger amount of bare ground, these sites had higher spatial variation in NDVI. Mountain big sagebrush had higher average NDVI values. This was expected since mountain big sagebrush occurs at higher elevations that receive more precipitation than either Wyoming big sagebrush or Utah juniper and therefore is associated with higher plant production. Aspen polygons tended to have higher NDVI values compared to Douglas-fir polygons with both ESs having a similar, relatively large distribution of spatial variance. Where the aspen sites are concerned, there was a slight but distinct trend of decreasing composite mean NDVI with increasing mean standard deviation (sdNDVI). Indeed considering the four aspen sites with the highest sdNDVI, three sites consisted of lower aspen canopy cover and the site with the highest sdNDVI contained a mix of immature aspen trees and shrubs. The remaining 16 aspen sites were all similar in canopy cover.

13.4.4.2.5 Discussion

This case study has shown that using variables derived from remotely sensed images as well as biophysical geospatial data, selected ESs can be discriminated on a per-pixel basis. Prediction of the spatial distribution of ESs on a pixel basis has been suggested as the next step in remote sensing applications to rangeland conservation (Hernandez, 2011). This research has shown that selected ESs can be somewhat cleanly discriminated by utilizing spectral (NDVI and Brightness) and spectral-spatial (sdNDVI) variables and therefore could be mapped by utilizing remotely sensed imagery. This is, of course, not surprising since the remote sensing community has clearly shown the ability to map land cover. What is important here is that by identifying the spatiotemporal spectral nature of selected land cover types, land cover change can be tracked. Ecological sites are used by the USDA-NRCS as a benchmark condition to develop state and transition models (Briske et al., 2005; Bestelmeyer et al., 2009). Ecological sites linked to soil map units provide a means to compare current land cover condition to historic conditions (as defined by the ES). By establishing the spectral nature of ESs, the spectral/spatial signature of soil map polygons can be compared with an expected ES for that polygon. Deviations from the expected response could indicate type and directionality of change, thus providing managers with a powerful land management tool.

13.5 CONCLUSION

Rangelands are dynamic environments that probably exhibit change more frequently and profoundly than many other systems (De Cauwer et al., 2024; Arogoundade et al., 2023; Sharifi and Felegari, 2023; McCord and Pilliod, 2022; Matongera et al., 2021; Hadian et al., 2019; Al-bukhari et al., 2018; Eddy et al., 2017). Land management, livestock grazing, invasive species, prolonged droughts and fire have the potential to alter rangeland community structure at both the micro and macro levels. Vegetation communities can be dominated by either C_3 or C_4 grass species or have permanent shrublands; thus productivity is very phenology dependent. Biomass production in rangelands by C_3 and C_4 plants displays significantly different phenological timing that is detectable using satellite remote sensing. The phenological differences are also useful for identifying vegetation types. Such information about rangeland productivity, vegetation species and timing are useful for rangeland managers in adjusting stocking rate, broad-scale movement of stock, land degradation monitoring and long-term trends in the health and condition of the systems. This chapter has highlighted numerous methods of achieving the preceding by using remotely sensed imagery.

Accurately classifying and identifying the spatial extent of Ecological Sites on a landscape scale is a laborious task involving extensive fieldwork to characterize soils. Case studies presented in this chapter show that variables extracted from remote sensing images, together with biophysical geospatial data, can be used to discriminate Ecological Sites on a per pixel basis with a relatively high degree of confidence. Ecological Sites are widely used as a benchmark condition to develop state and transition models and provide a means to compare current land cover condition with historic conditions, thus indicating type and directionality of change and helping in rangeland management.

Monitoring of fuel in rangelands is also important and this chapter has provided examples of how high temporal resolution satellite remote sensing can be used for quantifying fuel loads in rangeland landscapes. We have also summarized a range of change detection techniques relevant to rangeland monitoring and discussed, through a case study, the effectiveness of a Regression Tree algorithm (Random Forest) in modeling estimates of percent cover and developing Vegetation Continuous Fields for a range of vegetation communities in a rangeland environment.

Remote Sensing has undergone major transformations since its mainstream application began about 50 years ago. Spatial resolutions of images have shrunk from a few kilometers range to the sub-meter range, spectral resolutions have moved to the hyperspectral domain and temporal resolutions enable images to be available almost on a daily basis. In conjunction with this development, image classification techniques have made major progress and continue to be developed. New techniques include neural networks, random forest and object based classifications, and research cited earlier show that they have led to a marked increase in classification accuracies. All these developments have greatly enhanced our ability to map and monitor rangelands to a degree not possible in the past and no doubt will continue to be improved into the future as new techniques and sensor capabilities become mainstream. This chapter has covered some of these developments and highlighted the importance and usefulness of remote sensing for rangeland mapping and monitoring (De Cauwer et al., 2024; Arogoundade et al., 2023; Sharifi and Felegari, 2023; McCord and Pilliod, 2022; Matongera et al., 2021; Hadian et al., 2019; Al-bukhari et al., 2018; Eddy et al., 2017).

REFERENCES

Abdulaziz, A. M., Hurtado, J. M., & Al-Douri, R. 2009. Application of multitemporal Landsat data to monitor land cover changes in Eastern Nile Delta region, Egypt. International Journal of Remote Sensing, 30, 2977–2996.

Ajorlo, M., & Abdullah, B. R. 2007. Develop an Appropriate Vegetation Index for Assessing Rangeland Degradation in Semi-Arid Areas. In: Proceedings of 28th Asian Conference on Remote Sensing, 12–16 November, Kuala Lumpur, Malaysia.

Al-bukhari, A., Hallett, S., & Brewer, T. 2018. A review of potential methods for monitoring rangeland degradation in Libya. Pastoralism, 8(13). https://doi.org/10.1186/s13570-018-0118-4

Amiri, F., & Shariff, A. R. B. M. 2010. Using remote sensing data for vegetation cover assessment in semi-arid rangeland of center province of Iran. World Applied Sciences Journal, 11(12), 1537–1546.

Anderson, H. 1982. Aids to Determining Fuel Models for Estimating Fire Behavior. General Technical Report GTR INT-122. USDA Forest Service, Intermountain Forest and Range Experiment Station: Ogden, Utah. p. 22.

Anderson, M. A. 1997. Two-source time-integrated model for estimating surface fluxes using thermal infrared remote sensing.Remote Sensing of Environment, 60, 195– 216.

Andrews, P. L., & C. D. Bevins. 2003. BehavePlus Fire Modeling System, Version 2.0: Overview. In: Proceedings of the Second International Wildland Fire Ecology and Fire Management Congress and Fifth Symposium on Fire and Forest Meteorology, 16–20 November, American Meteorological Society, Orlando, FL. P5.11.

Arain, M. A., Black, T. A., Barr, A. G., Jarvis, P. G., Massheder, J. M., Verseghy, D. L., & Nesic, Z. 2002. Effects of seasonal and interannual climate variability on net ecosystem productivity of boreal deciduous and conifer forests. Canadian Journal of Forest Research, 32, 878–891.

Arogoundade, A. M., Mutanga, O., Odindi, J., & Odebiri, O. 2023. Leveraging Google Earth Engine to estimate foliar C: N ratio in an African savannah rangeland using Sentinel 2 data. Remote Sensing Applications: Society and Environment, 30, 2023, 100981, ISSN 2352–9385, https://doi.org/10.1016/j.rsase.2023.100981. (www.sciencedirect.com/science/article/pii/S2352938523000630)

Arroyo, L. A., Pascual, C., & Manzanera, J. A. 2008. Fire models and methods to map fuel types: The role of remote sensing. Forest Ecology and Management, 256, 1239–1252.

Asrar, G., Kanemasu, E., & Yoshida, M. 1985. Estimates of leaf area index from spectral reflectance of wheat under different cultural practices and solar angles. Remote Sensing of Environment, 17, 1–11.

Badeck, F.-W., Bondeau, A., Bottcher, K., Doktor, D., Lucht, W., Schaber, J., & Sitch, S. 2004. Responses of spring phenology to climate change. New Phytologist, 162(2), 295–309.

Bailey, R. G., & Hogg, H. C. 1986. A world ecoregions map for resource reporting. Environmental Conservation, 13, 195–202. https://doi.org/10.1017/S0376892900036237.

Baret, F., Guyot, G., & Major, D. J. 1989. TSAVI: A Vegetation Index Which Minimizes Soil Brightness Effects on LAI and APAR Estimation. 12th Canadian Symp. on Remote Sensing and IGARSS' 90, Vancouver, Canada, 10–14 July, 1989.

Baskent, E. Z., & Jordan, G. A. 1995. Characterizing spatial structure of forest landscapes. Canadian Journal of Forest Research, 25, 1830–1849.

Batistella, M., Robeson, S., & Moran, E. F. 2003. Settlement design, forest fragmentation, and landscape change in Rondônia, Amazônia. Photogrammetric Engineering and Remote Sensing, 69, 805–811.

Bestelmeyer, B. T., Tugel, A. J., Peacock, G. L., Robinett, D. G., Shaver, P. L., Brown, J. R., Herrick, J. E., Sanchez, H., & Havstad, K. M. 2009. State-and-transition models for heterogeneous landscapes: A strategy for development and application. Rangeland Ecology & Management, 52, 1–15.

Birth, G. S., & McVey, G. 1968. Measuring the color of growing turf with a reflectance spectrophotometer. Agronomy Journal, 60, 640–643.

Blumenthal, D., Booth, D. T., Cox, S. E., & Ferrier, C. E. 2007. Large-scale aerial images capture details of invasive plant populations. Rangeland Ecological Management, 60, 523–528.

Borak, J. S., & Strahler, A. H. 1999. Feature selection and land cover classification of MODIS-like data set for a semiarid environment. International Journal of Remote Sensing, 20, 919–938.

Bowers, S. A., & Hanks, R. J. 1965. Reaction of radiant energy from soil. Soil Science, 100, 130–138.

Boyte, S. P., Wylie, B. K., Major, D. J., & Brown, J. F. 2014. The integration of geophysical and eMODIS NDVI data into a rule-based, piecewise regression-tree model to estimate cheatgrass beginning of spring growth. International Journal of Digital Earth, 8, 118–132.

Bradley, B. A., & Mustard, J. F. 2008. Comparison of phenology trends by land cover class: A case study in the Great Basin, USA. Global Change Biology, 14, 334–346.

Briggs, J. M., & Knapp, Z. A. K. 1995. Interannual variability in primary production in tallgrass prairie: Climate, soil moisture, topographic position, and fire as determinants of aboveground biomass. American Journal of Botany, 82, 1024–1030.

Briske, D. D., Fuhlendorf, S. D., & Smeins, F. E. 2005. State-and-transition models, thresholds, and rangleland health: A synthesis of ecological concepts and perspectives. Rangeland Ecology & Management, 58, 1–10.

Cain, D. H., Riitters, K., & Orvis, K. 1997. A multi-scale analysis of landscape statistics. Landscape Ecology, 12, 199–212.

Chander, G., Markham, B. L., & Helder, D. L. 2009. Summary of current radiometric calibration coefficients for Landsat MSS, TM, ETM+, and EO-1 ALI sensors. Remote Sensing of Environment, 113, 893–903.

Chang, C. R., Lee, P. F., Bai, M. L., & Lin, T. T. 2004. Predicting the geographical distribution of plant communities in complex terrain—a case study in Fushian Experimental Forest, northeastern Taiwan. Ecography, 27, 577–588.

Chavez, P. S. 1996. Image-based atmospheric corrections revisited and improved. Photogrammetric Engineering and Remote Sensing, 62, 1025–1036.

Chavez, P. S. Jr., & Kwarteng, A. Y. 1989. Extracting spectral contrast in Landsat Thematic Mapper image data using selective principal component analysis. Photogrammetric Engineering and Remote Sensing, 55, 339–348.

Chen, S., & Rao, P. 2008. Land degradation monitoring using-multi-temporal Landsat TM/ETM data in a transition zone between grassland and cropland of northeast China. International Journal of Remote Sensing, 29, 2055–2073.

Cheney, P., & Sullivan, A. 1997. Grassfires: Fuel, Weather and Fire Behaviour. CSIRO Publishing, Collingwood, Australia.

Chopping, M., Lihong, S., Rango, A., Martonchik, J. V., Peters, D. P. C., & Laliberte, A. 2006. Remote sensing of woody shrub cover in desert grasslands using MISR with a geometric-optical canopy reflectance model. Remote Sensing of Environment, 112, 19–34.

Collado, A. D., Chuvieco, E., & Camarasa, A. 2002. Satellite remote sensing analysis to monitor desertification processes in the crop-rangeland boundary of Argentina. Journal of Arid Environment, 52, 121–133.

Conservancy, T. N. 2009. Spatial Modeling of the Cumulative Effects of Land Management Actions on Ecological Systems of the Grouse Creek Mountains—Raft River Mountains Region, Utah. Salt Lake City, Utah. p. 243.

Coppin, P., Jonckheere, I., Nackaerts, K., & Muys, B. 2004 Digital change detection methods in ecosystem monitoring: A review. International Journal of Remote Sensing, 25, 1565–1596.

Crews-Meyer, K. A. 2002. Characterizing landscape dynamism using pane-pattern metrics. Photogrammetric Engineering & Remote Sensing, 68, 1031–1040.

Crist, E. P. 1983. The Thematic Mapper Tasseled Capped a Preliminary Formulation. In: Proceedings of 1983 Machine Processing of Remotely Sensed Data Symposium, Environmental Research Institute of Michigan, Ann Arbor, Michigan, pp. 357–363.

Crist, E. P., & Cicone, R. C. 1984. Application of the tasseled cap concept to simulated Thematic Mapper data. Photogrammetric Engineering and Remote Sensing, 50, 343–352.

Crist, E. P., & Cicone, R. C. 1984. A physically-based transformation of thematic mapper data—the TM tasseled cap. IEEE Transactions on Geosciences and Remote Sensing, 22, 256–263.

Crist, E. P., & Kauth, R. J. 1986. The Tasseled Cap De-Mystified. Photogrammetric Engineering and Remote Sensing, 52, 81–86.

Crookston, N. L., & Finley, A. O. 2008. yaImpute: An R package for kNN imputation. Journal of Statistical Software, 23, 1–16.

Cumming, S., & Vervier, P. 2002. Statistical models of landscape pattern metrics, with applications to regional scale dynamic forest simulations. Landscape Ecology, 17, 433–444.

Cutler, D. R., Edwards, T. C., Beard, K. H., Cutler, A., & Hess, K. T. 2007. Random forests for classification in ecology. Ecology, 88, 2783–2792.

Danaher, T., Armston, J., & Collett, L. 2004. A Regression Model Approach for Mapping Woody Foliage Projective Cover Using Landsat Imagery in Queensland, Australia. In: Geoscience and Remote Sensing Symposium, 2004. IGARSS '04. Proceedings. 2004 IEEE International, Anchorage, AK, pp. 523–527.

De Cauwer, V., Colace, M. P., Mendelsohn, J., Antonio, T., & Van Der Waal, C. 2024. A Rangeland Management-Oriented Approach to Map Dry Savanna—Woodland Mosaics. Available at SSRN: https://ssrn.com/abstract=4703486 or https://doi.org/10.2139/ssrn.4703486

Defries, R. S., Hansen, M. C., & Towsend, J. R. G. 2000. Global continuous fields of vegetation characteristics: A linear mixture model applied to multi-year 8 km AVHRR data. International Journal of Remote Sensing, 21, 1389–1414.

Desalew, T., Tegegne, A., Nigatu, L., & Teka, W. 2010. Rangeland Condition and Feed Resources in Metema District, North Gondar Zone, Amhara Region, Ethiopia, IPMS (Improving Productivity and Market Success) of Ethiopian Farmers Project Working Paper 25: Nairobi, Kenya.

Dunn, B. H., Smart, A. J., Gates, R. N., Johnson, P. S., Beutler, M. K., Diersen, M. A., & Janssen, L. L. 2010. Long-term production and profitability from grazing cattle in the northern mixed grass prairie. Rangeland Ecology and Management, 63, 233–242.

Dymond, C. C., Mladenoff, D. J., & Radeloff, V. C. 2002. Phenological differences in tasseled cap indices improve deciduous forest classification. Remotes Sensing of Environment, 80, 460–472.

Eddy, I. M. S., Gergel, S. E., Coops, N. C., Henebry, G. M., Levine, J., Zerriffi, H., & Shibkov, E. 2017. Integrating remote sensing and local ecological knowledge to monitor rangeland dynamics. Ecological Indicators, 82, 2017, 106–116, ISSN 1470-160X, https://doi.org/10.1016/j.ecolind.2017.06.033. (www.sciencedirect.com/science/article/pii/S1470160X17303746)

Ellis, E. C., & Ramankutty, N. 2008. Putting people in the map: Anthropogenic biomes of the world. Frontier in Ecology and the Environment, 6, 439–447.

Elvidge, C. D., & Lyon, R. J. P. 1985. Influence of rock-soil spectral variation on the assessment of green biomass. Remote Sensing of Environment, 17, 37–53.

Feizizadeh, B., Omarzadeh, D., Garajeh, K. M., Lakes, T., & Blaschke, T. (2023) Machine learning data-driven approaches for land use/cover mapping and trend analysis using Google Earth Engine. Journal of Environmental Planning and Management, 66(3), 665–697, https://doi.org/10.1080/09640568.2021.2001317.

Foody, G. M., & Dash, J. 2007. Discriminating and mapping the C3 and C4 composition of grasslands in the northern Great Plains, USA. Ecological Informatics, 2, 89–93.

Franklin, S. E., & Wilson, B. A. 1992. A three-stage classifier for remote sensing of mountain environments. Photogrammetric Engineering and Remote Sensing, 58, 449–454.

Franklin, S. E., Maudie, A. J., & Lavigne, M. B. 2001. Using spatial co-occurrence texture to increase forest structure and species composition classification accuracy. Photogrammetric Engineering and Remote Sensing, 67, 849–855.

Friedl, M. A., & Brodley, C. E. 1997. Decision tree classification of land cover from remotely sensed data. Remote Sensing of Environment, 61, 399–409.

Fry, J. A., Xian, G., Jin, S., Dewitz, J. A., Homer, C. G., Yang, L., Barnes, C. A., Herold, N. D., & Wickham, J. D. 2011. Completion of the 2006 national land cover database for the conterminous united states. Photogrammetric Engineering and Remote Sensing, 77, 858–864.

Fuller, D. O. 2001. Forest fragmentation in Loudoun County, Virginia, USA evaluated with multitemporal Landsat imagery. Landscape Ecology, 16, 627–642.

Gamon, J. A., Field, C. B., Goulden, M. L., Griffin, K. L., & Hartley, A. E. 1995. Relationships between NDVI, canopy structure, and photosynthesis in three Californian vegetation types. Ecological Applications, 5, 28–41.

Gao, B. C. 1996. NDWI—a normalized difference water index for remote sensing of vegetation liquid water from space. Remote Sensing of Environment, 58, 257–266.

Gergel, S. E. 2005. Spatial and non-spatial factors: When do they affect landscape indicators of watershed loading? Landscape Ecology, 20, 177–189.

Gilmanov, T. G., Aires, L., Barcza, Z., Baron, V. S., Belelli, L., Beringer, J., Billesbach, D., Bonal, D., Bradford, J., Ceschia, E., Cook, D., Corradi, C., Frank, A., Gianelle, D., Gimeno, C., Gruenwald, T., Guo, H., Hanan, N., Haszpra, L., Heilman, J., Jacobs, A., Jones, M. B., Johnson, D. A., Kiely, G., Li, S., Magliulo, V., Moors, E., Nagy, Z., Nasyrov, M., Owensby, C., Pinter, K., Pio, C., Reichstein, M., Sanz, M. J., Scott, R., Soussana, J. F., Stoy, P. C., Svejcar, T., Tuba, Z., & Zhou, G. 2010. Productivity, respiration, and light-response parameters of world grassland and agroecosystems derived from flux-tower measurements. Rangeland Ecology and Management, 63, 16–39.

Gitelson, A. A., Kaufman, Y., & Merzlyak, M. N. 1996. Use of green channel in remote sensing of global vegetation from EOS-MODIS. Remote Sensing of Environment, 58, 289–298.

Glenn, E. P., Huete, A. R., Nagler, P. L., & Nelson, S. G. 2008. Relationship between remotely-sensed vegetation indices, canopy attributes and plant physiological processes: What vegetation indices can and cannot tell us about the landscape. Sensors, 8, 2136–2160.

Glenn, E., Huete, A., Nagler, P., Hirschboeck, K., & Brown, P. 2007. Integrating remote sensing and ground methods to estimate evapotranspiration. Critical Reviews in Plant Sciences, 26, 139–168.

Gökalp, A. E., Akar, Ö., & Yılmaz, V. 2017. Improving classification accuracy of spectrally similar land covers in the rangeland and plateau areas with a combination of WorldView-2 and UAV images. Geocarto International, 32(9), 990–1003, https://doi.org/10.1080/10106049.2016.1178816.

Gong, P., & Howarth, P. J. 1992. Frequency based contextual classification and gray-level vector reduction for land use identification. Photogrammetric Engineering & Remote Sensing, 58, 423–437.

Goulden, M. L., Munger, J. W., Fan, S., Daube, B. C., & Wofsy, S. C. 1996. Exchange of carbon dioxide by a deciduous forest: Response to interannual climate variability. Science, 271, 1576–1578.

Graetz, R. D., Pech, R. P., & Davis, A. W. 1988. The assessment and monitoring of sparsely vegetated rangelands using calibrated Landsat data. International Journal of Remote Sensing, 9, 1201–1222.

Green, K., Kempka, D., & Lackey, L. 1994. Using remote sensing to detect and monitor land-cover and land-use change. Photogrammetric Engineering & Remote Sensing, 60, 331–337.

Grice, A. C., & Hodgkinson, K. C. 2002. Global Rangelands: Progress and Prospects. CABI Publishing, Wallingford, Oxfordshire, UK.

Griffith, J., Stehman, S., Sohl, T., & Loveland, T. 2003. Detecting trends in landscape pattern metrics over a 20-year period using a sampling-based monitoring programme. International Journal of Remote Sensing, 24, 175–181.

Griffith, J. A., Martinko, E. A., & Price, K. P. 2000. Landscape structure analysis of Kansas at three scales. Landscape and Urban Planning, 52, 45–61.

Gustafson, E. J. 1998. Quantifying landscape spatial pattern: What is the state of the art? Ecosystems, 1, 143–156.

Gustafson, E. J., & Parker, G. R. 1992. Relationships between landcover proportion and indices of landscape spatial pattern. Landscape Ecology, 7, 101–110.

Hadian, F., Jafari, R., Bashari, H. et al. 2019. Estimation of spatial and temporal changes in net primary production based on Carnegie Ames Stanford Approach (CASA) model in semi-arid rangelands of Semirom County, Iran. Journal of Arid Land 11, 477–494. https://doi.org/10.1007/s40333-019-0060-3.

Hansen, M. C., DeFries, R. S., Townshend, J. R. G., Carroll, M., Dimiceli, C., & Sohlberg, R. A. 2003a. Development of 500 Meter Vegetation Continuous Field Maps Using MODIS Data. In: Geoscience and Remote Sensing Symposium, 2003. IGARSS '03. Proceedings. 2003 IEEE International, pp. 264–266.

Hansen, M. C., DeFries, R. S., Townshend, J. R. G., Carroll, M., Dimiceli, C., & Sohlberg, R. A. 2003b. Global percent tree cover at a spatial resolution of 500 meters: First results of the MODIS vegetation continuous fields algorithm. American Meteorological Society, 7, 1–15.

Hansen, M. C., DeFries, R. S., Townshend, J. R. G., Sohlberg, R., Dimiceli, C., & Carroll, M. 2002. Towards an operational MODIS continuous field of percent tree cover algorithm: Examples using AVHRR and MODIS data. Remote Sensing of Environment, 83, 303–319.

Hardisky, M. A., Klemas, V., & Smart, R. M. 1983. The influence of soil salinity, growth form, and leaf moisture on the spectral radiance of Spartina alterniflora canopies. Photogrammetric Engineering & Remote Sensing, 49, 77–83.

Hernandez, A. 2011. Spatiotemporal modeling of threats to big sagebrush ecological sites in northern Utah [dissertation]. Logan, UT, USA: Utah State University. 163 p.

Heumann, B. W., Seaquis, J. W., Eklundh, L., & Jönsson, P. 2007. AVHRR derived phenological change in the Sahel and Soudan, Africa, 1982–2005. Remote Sensing of Environment, 108, 385–392.

Hoffer, R. M., & Johannsen, C. J. 1969. Ecological potentials in spectral signature analysis. Chapter 1 in Remote Sensing in Ecology, edited by P. C. Johnson (Athens: University of Georgia Press), 1–6.

Honnay, O., Piessens, K., Van Landuyt, W., Hermy, M., & Gulinck, H. 2003. Satellite based land use and landscape complexity indices as predictors for regional plant species diversity. Landscape and Urban Planning, 63, 241–250.

Hudak, A. T., & Wessman, C. A. 1998. Textural analysis of historical aerial photography to characterize woody plant encroachment in South African savanna. Remote Sensing of Environment, 66(3), 317–330.

Huete, A. R. 1988. A soil-adjusted vegetation index (Savi). Remote Sensing of Environment, 25, 295–309.

Huete, A. R., & Jackson, R. D. 1987. Suitability of spectral indices for evaluating vegetation characteristics on arid rangelands. Remote Sensing of Environment, 25, 89–105.

Huete, A. R., Jackson, R. D., & Post, D. F. 1985. Spectral response of a plant canopy with different soil backgrounds. Remote Sensing of Environment, 17, 37–53.

Huete, A., Didan, K., Miura, T., Rodriguez, E. P., Gao, X., & Ferreira, L. G. 2002. Overview of the radiometric and biophysical performance of the MODIS vegetation indices. Remote Sensing of Environment, 83, 195–213.

Im, J., Jensen, J. R., & Tylliss, J. A. 2008. Object based change detection using correlation image analysis and image segmentation. International Journal of Remote Sensing, 29, 399–423.

James, L. F., Young, J. A., & Sanders, K. 2003. A new approach to monitoring rangelands. Arid Land Research and Management, 17, 319–328.

Jensen, J. R. 2000. Remote Sensing of the Environment: An Earth Resources Perspective. Prentice-Hall, Upper Saddle River, NJ, USA. 592 p.

Jensen, J. R. 2007. Remote Sensing of the Environment: An Earth Resource Perspective. Pearson Prentice Hall, Upper Saddle River, NJ.

Jensen, J., 2005. Introductory Digital Image Processing: A Remote Sensing Perspective. Pearson Prentice Hall, Upper Saddle River, NJ.

Jiang, Z., Huete, A. R., Chen, J., Chen, Y., Li, J., & Yan, G. 2006. Analysis of NDVI and scaled difference vegetation index retrievals of vegetation fraction. Remote Sensing of Environment, 101, 366–378.

Jianlong, L., Tiangang, L., & Quangong, C. 1998. Estimating grassland yields using remote sensing and GIS technologies in China. New Zealand Journal of Agricultural Research, 41, 31–380028–8233/98/4101–0031 $7.00/0 © The Royal Society of New Zealand 1998.

Jin, S. M., & Sader, S. A. 2005. Comparison of time series tasseled cap wetness and the normalized difference moisture index in detecting forest disturbances. Remote Sensing of Environment, 94, 364–372.

Jin, S., Yang, L., Danielson, P., Homer, C., Fry, J., & Xian, G. 2013. A comprehensive change detection method for updating the National Land Cover Database to circa 2011. Remote Sensing of Environment, 132, 159–175.

Jin, X. M., Zhang, Y. K., Schaepman, M. E., Clevers, J. G., & Su, Z. 2008. Impact of elevation and aspect on the spatial distribution of vegetation in the Qilian mountain area with remote sensing data. The International Archives of the Photogrammetry, Remote Sensing and Spatial Information Sciences, 37, 1385–1390.

John, R., Chen, J., Lu, N., Guo, K., Liang, C., Wei, Y., Noormets, A., Ma, K., & Han, X. 2008. Predicting plant diversity based on remote sensing products in the semi-arid region of Inner Mongolia. Remote Sensing of Environment, 112, 2018–2032.

Kaloudis, S. T., Lorentzos, N. A., Sideridis, A. B., & Yialouris, C. P. 2005. A decision support system for forest fire management. Operational Research, 5(1), 141–152.

Kamp, K. V., Rigge, M., Troelstrup Jr, N. H., Smart, A. J., & Wylie, B. 2013. Detecting channel riparian vegetation response to best-management-practices implementation in ephemeral streams with the use of spot high-resolution visible imagery. Rangeland Ecology and Management, 66, 63–70.

Kauth, R. J., & Thomas, G. S. 1976. The Tasseled Cap: A Graphic Description of the Spectratemporal Development of Agricultural Crops as Seen by Landsat. In: Proceedings of the Symposium on Machine Processing of Remotely Sensed Data, Purdue University, West Lafayette, Indiana, pp. 4B41–4B51.

Keane, R. E. 2013. Describing wildland surface fuel loading for fire management: A review of approaches, methods and systems. International Journal of Wildland Fire, 22, 51–62.

Kovalskyy, V., & Henebry, G. M. 2012. A new concept for simulation of vegetated land surface dynamics-Part 1: The event driven phenology model. Biogeosciences, 9, 141–159.

Kumar, L., Rietkerk, M., van Langevelde, F., van de Koppel, J., van Andel, J., Hearne, J., de Ridder, N., Stroosnijder, L., Prins, H. H. T., & Skidmore, A. K. 2002. Relationship between vegetation recovery rates and soil type in the Sahel of Burkina Faso: Implications for resource utilization at large scales. Ecological Modelling, 149, 143–152.

Kustas, W., & Norman, J. 1996. Use of remote sensing for evapotranspiration monitoring over land surfaces. Hydrological Sciences Journal—Journal des Sciences Hydrologiques, 41, 495–516.

Laba, M., Smith, S. D., & Degloria, S. D. 1997. Landsat based land cover mapping in the lower Yuna river watershed in the Dominican Republic. International Journal of Remote Sensing, 18, 3011–3025.

Laliberte, A. S., Rango, A., Havstad, K. M., Paris, J. F., Beck, R. F., McNeely, R., & Gonzalez, A. L. 2004. Object-oriented image analysis for mapping shrub encroachment from 1937 to 2003 in southern New Mexico. Remote Sensing of Environment, 93, 198–210.

Laliberte, A. S., Winters, C., & Rango, A. 2011. UAS remote sensing missions for rangeland applications. Geocarto International, 26(2), 141–156.

Langford, M., & Bell, W. 1997. Land cover mapping in a tropical hillsides environment: A case study in the Cauca region of Colombia. International Journal of Remote Sensing, 18, 1289–1306.

Lauenroth, W. K., & Sala, O. E. 1992. Long-term forage production of North American shortgrass steppe. Ecological Applications, 2, 397–403.

Lawler, J. J., White, D., Neilson, R. P., & Blaustein, A. R. 2006. Predicting climate-induced range shifts: Model differences and model reliability. Global Change Biology, 12, 1568–1584.

Le Houerou, H. N., & Hoste, C. H. 1977. Rangeland production and annual rainfall relations in the mediterranean basin and in the African Sahelo-Sudanian Zone. Journal of Range Management, 30, 181–189.

Leitão, B., & Ahern, A., J. 2002. Applying landscape ecological concepts and metrics in sustainable landscape planning. Landscape and Urban Planning, 59, 65–93.

Li, B., & Zhou, Q. 2009. Accuracy assessment on multi-temporal land-cover change detection using a trajectory error matrix. International Journal of Remote Sensing, 30, 1283–1296.

Li, H., & Wu, J. 2004. Use and misuse of landscape indices. Landscape Ecology, 19, 389–399.

Liaw, A., & Wiener, M. 2002. Classification and Regression by randomForest. R News, 2, 5.

Liu, A. J., & Cameron, G. N. 2001. Analysis of landscape patterns in coastal wetlands of Galveston Bay, Texas (USA). Landscape Ecology, 16, 581–595.

Lowry, J., Ramsey, R. D., Thomas, K., Schrupp, D., Sajwaj, T., Kirby, J., Waller, E., Schrader, S., Falzarano, S., Langs, L., Manis, G., Wallace, C., Schulz, K., Comer, P., Pohs, K., Rieth, W., Velasquez, C., Wolk, B., Kepner, W., Boykin, K., O'Brien, L., Bradford, D., Thompson, B., & Prior-Magee, J. 2007. Mapping moderate-scale land-cover over very large geographic areas within a collaborative framework: A case study of the Southwest Regional Gap Analysis Project (SWReGAP). Remote Sensing of Environment, 108, 59–73.

Lu, D., Mausel, P., Brondizio, E., & Moran, E. 2004. Change detection techniques. International Journal of Remote Sensing, 25, 2365–2407.

Ludwig, J. A., Tongway, D. J., Bastin, G. N., & James, C. D. 2004. Monitoring ecological indicators of rangeland functional integrity and their relation to biodiversity at local to regional scales. Austral Ecology, 29, 108–120.

Lutes, D. C., Keane, R. E., & Caratti, J. F. 2009. A surface fuels classification for estimating fire effects. International Journal of Wildland Fire, 18, 802–814.

Matongera, T. N., Mutanga, O., Sibanda, M., & Odindi, J. 2021. Estimating and monitoring land surface phenology in rangelands: A review of progress and challenges. Remote Sensing, 13, 11, 2060. https://doi.org/10.3390/rs13112060.

Maynard, C. L., Lawrence, R. L., Nielsen, G. A., & Decker, G. 2007. Ecological site descriptions and remotely sensed imagery as a tool for rangeland evaluation. Canadian Journal of Remote Sensing, 33(2), 109–115.

McAlpine, C. A., & Eyre, T. J. 2002. Testing landscape metrics as indicators of habitat loss and fragmentation in continuous eucalypt forests (Queensland, Australia). Landscape Ecology, 17, 711–728.

McCord, S. E., & Pilliod, D. S. 2022. Adaptive monitoring in support of adaptive management in rangelands. Rangelands, 44(1), 2022, 1–7, ISSN 0190–0528, https://doi.org/10.1016/j.rala.2021.07.003. (www.sciencedirect.com/science/article/pii/S0190052821000663)

McGarigal, K., & Marks, B. J. 1995. Spatial Pattern Analysis Program for Quantifying Landscape Structure. Gen. Tech. Rep. PNW-GTR-351. US Department of Agriculture, Forest Service, Pacific Northwest Research Station: Portland, Oregon.

Means, J. E., Krankina, O. N., Jiang, H., & Li, H. 1996. Estimating Live Fuels for Shrubs and Herbs With BIOPAK. Gen. Tech. Rep. PNW-GTR-372. U.S. Department of Agriculture, Forest Service, Pacific Northwest Research Station: Portland, OR. p. 28.

Merrill, D. F., & Alexander, M. E. 1987. Glossary of Forest Fire Management Terms, 4th Edition. National Research Council of Canada, Canadian Committee on Forest Fire Management, Ottawa, Ontario.

Milchunas, D. G., Laurenroth, W. K., Chapman, P. L., & Kazempour, M. K. 1989 Effects of grazing, topography, and precipitation on the structure of a semiarid grassland. Vegetation, 80, 11–23.

Miller, A. B., Bryant, E. S., & Birnie, R. W. 1998. An analysis of land cover changes in the northern forest of New England using multi temporal Landsat MSS data. International Journal of Remote Sensing, 19, 245–265.

Miller, R. F., & Rose, J. A. 1999. Fire history and western juniper encroachment in sagebrush steppe. Journal of Range Management, 52, 550–559.

Monteith, J., & Unsworth, M. 1990. Principles of Environmental Physics. 2nd. Edition, Edward Arnold, London.

Muchoney, D. M., Borak, J., Chi, H., Friedl, M., Gopal, S., Hodges, J., Morrow, N., & Strahler, A. 2000. Application of the MODIS global supervised classification model to vegetation and land cover mapping of Central America. International Journal of Remote Sensing, 21, 1115–1138.

Munyati, C., Shaker, P., & Phasha, M. G. 2011. Using remotely sensed imagery to monitor savanna rangeland deterioration through woody plant proliferation: A case study from communal and biodiversity conservation rangeland sites in Mokopane, South Africa. Environmental Monitoring Assessment, 176, 293–311.
O'Neill, R. V. 1976. Ecosystem persistence and heterotrophic regulation. Ecology, 57(6), 1244–1253.
Ode, D. J., Tieszen, L. L., Lerman, J. C. 1980. The seasonal contribution of C3 and C4 plant species to primary production in a mixed prairie. Ecology, 61, 1304–1311.
Okeke, F., & Karnieli, A. 2006. Methods for fuzzy classification and accuracy assessment of historical aerial photographs for vegetation change analyses. Part I: Algorithm development. International Journal of Remote Sensing, 27, 153–176.
Okin, G. S., Roberts, D. A., Murray, B., & Okin, W. J. 2001. Practical limits on hyperspectral vegetation discrimination in arid and semiarid environments. Remote Sensing of Environment, 77(2), 212–225.
Ottmar, R. D., Sandberg, D. V., Riccardi, C. L., & Prichard, S. 2007. An overview of the fuel characteristic classification system—quantifying, classifying, and creating fuelbeds for resource planners. Canadian Journal of Forest Research, 37, 2381–2382.
Pal, M., & Mather, P. M. 2003. An assessment of the effectiveness of decision tree methods for landcover classification. Remote Sensing of Environment, 86, 554–565.
Parker, A. J. 1982.The topographic relative moisture index: An approach to soil-moisture assessment in mountain terrain. Physical Geography, 3, 160–168.
Paruelo, J. M., & Lauenroth, W. K. 1998. Interannual variability of NDVI and its relationship to climate for North American shrublands and grasslands. Journal of Biogeography, 25, 721–733.
Peterson, D. L., Spanner, M. A., Running, S. W., & Teuber, K. B. 1987. Relationships of Thematic Mapper simulator data to leaf area index of temperate coniferous forests. Remote Sensing of Environment, 22, 323–341.
Peterson, E. B. 2005. Estimating cover of an invasive grass (Bromus tectorum) using tobit regression and phenology derived from two dates of Landsat ETM plus data. International Journal of Remote Sensing, 26, 2491–2507.
Petit, C. C., & Lambin, E. F. 2002. Impact of data integration technique on historical land-use/landcover change: Comparing historical maps with remote sensing data in the Belgian Ardennes. Landscape Ecology, 17, 117–132.
Pickup, G., & Foran, B. D. 1987. The use of spectral and spatial variability to monitor cover change on inert landscapes. Remote Sensing of Environment, 23, 351–363.
Pinty, B., & Verstraete, M. M. 1992. GEMI: A non-linear index to monitor global vegetation from satellites. Vegetation, 101, 15–20.
Prasad, A. M., Iverson, L. R., & Liaw, A. 2006. Newer classification and regression tree techniques: Bagging and random forests for ecological prediction. Ecosystems, 9, 181–199.
Program, U. N. G. A. 2004.Provisional Digital Land Cover Map for the Southwestern United States. Version 1.0. United States Geological Survey.
Rauzi, F., & Dobrenz, A. K. 1970. Seasonal variation of chlorophyll in Western wheatgrass and blue grama. Journal of Range Management, 23, 372–373.
Read, J., & Lam, N. S. 2002. Spatial methods for characterising land cover and detecting land-cover changes for the tropics. International Journal of Remote Sensing, 23, 2457–2474.
Reed, B. C., Brown, J., VanderZee, D., Loveland, T. R., Merchant, J. W., & Ohlen, D. O. 1994a. Measuring phenological variability from satellite imagery. Journal of Vegetation Science, 5, 703–714.
Reed, B. C., Brown, J. F., Vanderzee, D., Loveland, T. R., Merchant, J. W., & Ohlen, D. O. 1994b. Measuring phenological variability from satellite imagery. Journal of Vegetation Science, 5, 703–714.
Reed, B. C., Loveland, T. R., & Tieszen, L. L. 1996. An approach for using AVHRR data to monitor U.S. Great Plains grasslands. Geocarto International, 11, 13–22.
Reeves, M., Ryan, K. C., Rollins, M. G., & Thompson, T. 2009. Spatial fuel data products of the LANDFIRE Project. International Journal of Wildland Fire, 18, 250–267.
Rempel, R. S., Carr, A., & Elkie, P. C. 1999. Patch Analyst User's Manual: A Tool for Quantifying Landscape Structure. Ontario Ministry of Natural Resources, Boreal Science, Northwest Science & Technology, Ontario, Canada.
Riaño, D., Chuvieco, E., Salas, J., Palacios-Orueta, A., & Bastarrika, A. 2002. Generation of fuel type maps from Landsat TM images and ancillary data in Mediterranean ecosystems. Canadian Journal of Forest Research, 32(8), 1301–1315.

Riaño, D., Chuvieco, E., Ustin, Salas, S., Rodríguez-Pérez, J., Ribeiro, L., Viegas, D., Moreno, J., & Fernández, H. 2007. Estimation of shrub height for fuel-type mapping combining airborne LiDAR and simultaneous color infrared ortho imaging. International Journal of Wildland Fire, 16(3), 341–348.
Riaño, D., Meier, E., Allgöwer, B., Chuvieco, E., & Ustin, S. L. 2003. Modeling airborne laser scanning data for the spatial generation of critical forest parameters in fire behavior modeling. Remote Sensing of Environment, 86(2), 177–186.
Richards, J. A. 1986, Thematic mapping from multitemporal image data using the principal components transformation. Remote Sensing of Environment, 16, 35–46.
Richardson, A. J., & Wiegand, C. L. 1990. Comparison of two models for simulating the soil-vegetation composite reflectance of a developing cotton canopy. International Journal of Remote Sensing, 11, 447–459.
Richardson, A. J., & Wiegand, C. L. 1977. Distinguishing vegetation from soil background information. Photogrammetric Engineering & Remote Sensing, 43, 1541–1552.
Rietkerk, M., Dekker, S. C., de Ruiter, P. C., & van de Koppel, J. 2004. Self-organized patchiness and catastrophic shifts in ecosystems. Science, 305(5692), 1926–1929.
Rigge, M., Smart, A., Wylie, B., Gilmanov, T., & Johnson, P. 2013. Linking phenology and biomass productivity in South Dakota mixed-grass prairie. Rangeland Ecology and Management, 66, 579–587.
Riitters, K. H., O'neill, R., Hunsaker, C., Wickham, J. D., Yankee, D., Timmins, S., Jones, K., & Jackson, B. 1995. A factor analysis of landscape pattern and structure metrics. Landscape Ecology, 10, 23–39.
Rock, B. N., Williams, D. L., & Vogelmann, J. E. 1985. Field and Airborne Spectral Characterization of Suspected Acid Deposition Damage in Red Spruce (Picea Rubens) from Vermont. Proc., Symp. on Machine Processing of Remotely Sensed Data, Purdue Univ., West Lafayette, IN, pp. 71–81.
Rogan, J., Franklin, J., & Roberts, D. A. 2002. A comparison of methods for monitoring multitemporal vegetation change using Thematic Mapper imagery. Remote Sensing of Environment, 80, 143–156.
Rollins, M. 2009. LANDFIRE: A nationally consistent vegetation, wildland fire and fuel assessment. International Journal of Wildland Fire, 18, 235–249.
Rouse, J. W., Haas, R. H., Schell, J. A., & Deering, D. W. 1974. Monitoring vegetation systems in the Great Plains with ERTS. Proc. Third ERTS-1 Symposium, NASA Goddard, NASA SP-351, pp. 309–317.
Sandberg, D. V., Ottmar, R. D., & Cushon, G. H. 2001. Characterizing fuels in the 21st Century. International Journal of Wildland Fire, 10, 381–387.
Sharifi, A., & Felegari, S. 2023. Remotely sensed normalized difference red-edge index for rangeland biomass estimation. Aircraft Engineering and Aerospace Technology, 95(7), 1128–1136. https://doi.org/10.1108/AEAT-07-2022-0199.
Schmitz, M., De Aranzabal, I., Aguilera, P., Rescia, A., & Pineda, F. 2003. Relationship between landscape typology and socioeconomic structure: Scenarios of change in Spanish cultural landscapes. Ecological Modelling, 168, 343–356.
Scott, J. H., & Burgan, R. E. 2005. Standard fire behavior fuel models: A comprehensive set for use with Rothermel's surface fire spread model. General Technical Report RMRS-GTR-153. U.S. Department of Agriculture, Forest Service, Rocky Mountain Research Station: Fort Collins, CO. p. 72.
Scott, J. M., Davis, F., Csuti, B., Noss, R., Butterfield, B., & Groves, C. 1993. Gap analysis: A geographic approach to protection of biological diversity. Wildlife Monographs, 123, 3–41.
Sellers, P. 1985. Canopy reflectance, photosynthesis and transpiration. International Journal of Remote Sensing, 6, 1335–1372.
Sexton, J. O., Song, X. P., Feng, M., Noojipady, P., Anand, A., Huang, C., Kim, D. H., Collins, K. M., Channan, S., DiMiceli, C., & Townshend, J. R. 2013. Global, 30-m resolution continuous fields of tree cover: Landsat-based rescaling of MODIS vegetation continuous fields with lidar-based estimates of error. International Journal of Digital Earth, 6, 427–448.
Sha, Z., Baiy, Y., & Xie, Y. 2008. Using a hybrid fuzzy classifier (HFC) to map typical grassland vegetation in Xilinhe River Basin, Inner Mongolia, China. International Journal of Remote Sensing, 29(8), 2317–2337.
Sims, P. L., & Singh, J. S. 1978a. The structure and function of ten western North American grasslands. II. Intra-seasonal dynamics in primary producer compartments. Journal of Ecology, 66, 547–572.
Sims, P. L., & Singh, J. S. 1978b. Primary production of the central grassland region of the United States. Ecology 69, 40–45.
Sinha, P., & Kumar, L. 2013a. Binary images in seasonal land-cover change identification: A comparative study in parts of NSW, Australia. International Journal of Remote Sensing, 34(6), 2162–2182.
Sinha, P., & Kumar, L. 2013b. Independent two-step thresholding of binary images in inter-annual land cover change/no-change identification. ISPRS Journal of Photogrammtery and Remote Sensing, 81(7), 31–43.

Sinha, P., Kumar, L., & Reid, N. 2012a. Seasonal variation in landcover classification accuracy in diverse region. Photogrammetric Engineering & Remote Sensing, 78(3), 770–781.
Sinha, P., Kumar, L., & Reid, N. 2012b. Three-date Landsat TM composite in seasonal land-cover change identification in a mid-latitudinal region of diverse climate and land-use. Journal of Applied Remote Sensing, 6(1), 063595, https://doi.org/10.1117/1.JRS.6.063595.
Smart, A. J., Dunn, B. H., Johnson, P. S., Xu, L., & Gates, R. N. 2007. Using weather data to explain herbage yield on three great plains plant communities. Rangeland Ecology Management, 60, 146–153.
Sohn, Y., & Qi, J. 2005. Mapping detailed biotic communities in the upper San Pedro Valley of southeastern Arizona using Landsat 7 ETM+ data and supervised spectral angle classifier. Photogrammetric Engineering & Remote Sensing, 71, 709–718.
Sohn, Y., & Rebello, N. S. 2002. Supervised and unsupervised spectral angle classifiers. Photogrammetric Engineering & Remote Sensing, 68, 1271–1280.
Sparks, T. H., Jeffree, E. P., & Jeffree, C. E. 2000. An examination of the relationship between flowering times and temperature at the national scale using long-term phenological records from the UK. International Journal of Biometeorology, 44(2), 82–87.
Stanfield, B. J., Bliss, J. C., & Spies, T. A. 2002. Land ownership and landscape structure: A spatial analysis of sixty-six Oregon (USA) Coast Range watersheds. Landscape Ecology 17, 685–697.
Strahler, A. H., 1980. The use of prior probabilities in maximum likelihood classification of remotely sensed data. Remote Sensing of Environment, 10, 1135–1163.
Sutton, C. D. 2005. Classification and regression trees, bagging, and boosting. Handbook of Statistics: Data Mining and Data Visualization, edited by C. R. Rao, E. J. Wegman & J. L. Solka (Elsevier), 303–329.
Tagestad, J. D., & Downs, J. L. 2007. Landscape Measures of Rangeland Condition in the BLM Owyhee Pilot Project: Shrub Canopy Mapping, Vegetation Classification, and Detection of Anomalous Land Areas. Report prepared for U.S. Department of Interior Bureau of Land Management under a Related Services Agreement with the U.S. Department of Energy Contract DE-AC05–76RL01830 Pacific Northwest National Laboratory Richland, Washington 99352.
Tang, J., Corresponding, L. W., & Zhang, S. 2005. Investigating landscape pattern and its dynamics in Daqing, China. International Journal of Remote Sensing, 26, 2259–2280.
Tao, F., Yokozawa, M., Zhang, Z., Hayashi, Y., & Ishigooka, Y. 2008. Land surface phenology dynamics and climate variations in the North East China Transect (NECT), 1982–2000. International Journal of Remote Sensing, 29, 5461–5478.
Tateishi, R., & Ebata, M. 2004. Analysis of phenological change patterns using 1982–2000 advanced very high resolution radiometer (AVHRR) data. International Journal of Remote Sensing, 25, 2287–2300.
Teague, W. R., Ansley, J. R., Pinchak, W. E., Dowhower, S. L., Gerrard, S. A., & Waggoner, J. A. 2008. Interannual herbaceous biomass response to increasing honey mesquite cover on two soils. Rangeland Ecological Management, 61, 496–508.
Tieszen, L. L., Reed, B. C., Bliss, N. B., Wylie, B. K., & DeJong, D. D. 1997. NDVI, C3 AND C4 production, and distributions in Great Plains grassland land cover classes. Ecological Applications, 7, 59–58.
Timmerman, W., Kustas, W., Anderson, M., & French, A. 2007. An intercomparison of the surface energy balance algorithm for land (SEBAL) and the two-source energy balance (TSEB) modeling schemes. Remote Sensing of Environment, 108, 369–384.
Todd, S. W., & Hoffer, R. M. 1997. Responses of spectral indices to variations in vegetation cover and soil background. Photogrammetric Engineering & Remote Sensing, 64(9), 915–921.
Todd, S. W., Hoffer, R. M., & Milchunas, D. G. 1998. Biomass estimation on grazed and ungrazed rangelands using spectral indices. International Journal of Remote Sensing, 19(3), 427–438.
Torahi, A. A. 2012. Rangeland Dynamics Monitoring Using Remotely-Sensed Data, in Dehdez Area, Iran. In: International Conference on Applied Life Sciences (ICALS2012), 10–12 September, Turkey.
Trani, M. K., & Giles, R. H. 1999. An analysis of deforestation: Metrics used to describe pattern change. Forest Ecology and Management, 114, 459–470.
Tso, B., & Olsen, R. C. 2005. Combining spectral and spatial information into hidden Markov models for unsupervised image classification. International Journal of Remote Sensing, 26, 2113–2133.
Tucker, C. 1979. Red and photographic infrared linear combinations for monitoring vegetation. Remote Sensing of Environment, 8, 127–150.
Tucker, C., Townshend, J., & Goff, T. 1985. African land cover classification using satellite data. Science, 227, 229–235.

Tucker, C. J., & Sellers, P. J. 1986. Satellite remote sensing of primary production. International Journal of Remote Sensing, 7, 1395–1416.

Tymstra, C., Bryce, R. W., Wotton, B. M., Taylor, S. W., & Armitage, O. B. 2010. Development and Structure of Prometheus: The Canadian Wildland Fire Growth Simulation Model. Natural Resources Canada, Canadian Forest Service, Northern Forestry Centre, Edmonton, AB. Information Report NOR-X-417. p. 88. Available at: http://cfs.nrcan.gc.ca/bookstore_pdfs/31775.pdf. Accessed 16 March 2011.

U.S. Department of Agriculture, National Resources Conservation Service. 2008. Field Office Technical Guide, Major Land Resource Area 063A: Northern Rolling Pierre Shale Plains, Ecological Site Description: Clayey.

U.S. Department of Agriculture, NRCS. 2012. Ecological Sites. Available at: www.nrcs.usda.gov/wps/portal/nrcs/detail/national/landuse/rangepasture/?cid=STELPRDB1043235. Accessed 17 October 2011.

van Leeuwen, W. J. D., Davison, J. E., Casady, G. M., & Marsh, S. E. 2010. Phenological characterization of desert sky island vegetation communities with remotely sensed and climate time series data. Remote Sensing, 2, 388–415.

Vierling, L. A., Xu, Y., Eitel, J. U. H., & Oldow, J. S. 2012. Shrub characterization using terrestrial laser scanning and implications for airborne LiDAR assessment. Canadian Journal of Remote Sensing, 38(6), 709–722.

Vogel, M., & Strohbach, M. 2009. Monitoring of savanna degradation in Namibia using Landsat TM/ETM+ data, Int. Geosci. Remote Sensing (IGARSS) IEEE, Cape Town, S. Afr., pp. III-931–III-934.

Vogelmann, J. E., Xian, G., Homer, C., & Tolk, B. 2012. Monitoring gradual ecosystem change using Landsat time series analyses: Case studies in selected forest and rangeland ecosystems. Remote Sensing of Environment, 122, 92–105.

Walton, J. T. 2008. Subpixel urban land cover estimation: Comparing cubist, random forests, and support vector regression. Photogrammetric Engineering & Remote Sensing, 74, 1213–1222.

Wang, C., Jamison, B. E., & Spicci, A. A. 2010. Trajectory-based warm season grassland mapping in Missouri prairies with multi-temporal ASTER imagery. Remote Sensing of Environment, 114, 531–539.

Wang, Y., Mitchell, B. R., Nugranad-Marzilli, J., Bonynge, G., Zhou, Y., & Shriver, G. 2009. Remote sensing of land-cover change and landscape context of National Parks: A case study of northeast temperate network. Remote Sensing of Environment, 113, 1453–1461.

Weber, K. T., Alados, C. L., Bueno, C. G., Gokhale, B., Komac, B., & Pueyo, Y. 2009. Modeling bare ground with classification trees in Northern Spain. Rangeland Ecology & Management, 62, 452–459.

Weiser, R. L., Asrar, G., Miller, G. P., & Kanemasu, E. T. 1986 Assessing grassland biophysical characteristics from spectral measurements. Remote Sensing of Environment, 20, 141–152.

Wessels, K. J., Prince, S. D., Carroll, M., & Malherbe, J. 2007. Relevance of rangeland degradation in semiarid northeastern South Africa to the nonequilbrium theory. Ecological Applications, 17(3), 815–827.

Wessels, K. J., Prince, S. D., Frost, P. E., & Van Zyl, D. 2004. Assessing the effects of human-induced land degradation in the former homelands of northern South Africawith a 1km AVHRR NDVI time-series. Remote Sensing of Environment, 91, 47–67.

Westerling, A. L., & Swetnam, T. W. 2003. Interannual to decadal drought and wildfire in the western United States. Transactions American Geophysical Union, 84, 545–560.

Westman, W. E. 1985. Ecology, Impact Assessment, and Environmental Planning. Wiley, New York, USA.

Westman, W. E., & O'Leary, J. F. 1986. Measures of resilience: The response of coastal sage scrub to fire. Vegetation, 65(3), 179–189.

White, M. A., de Beurs, K. M., Didan, K., Inouye, D. W., Richardson, A. D., Jensen, O. P., O'Keefe, J., Zhang, G., Nemani, R. R., van Leeuwen, W. J. D., Brown, J. F., de Wit, A., Schaepman, M., Lin, X., Dettinger, M., Bailey, A. S., Kimball, J., Schwartz, M. D., Baldocchi, D. D., Lee, J. T., & Lauenroth, W. K. 2009. Intercomparison, interpretation, and assessment of spring phenology in North America estimated from remote sensing for 1982–2006. Global Change Biology, 15, 2335–2359.

Wickham, J., & Rhtters, K., 1995. Sensitivity of landscape metrics to pixel size. International Journal of Remote Sensing, 16, 3585–3594.

Wisdom, M., Rowland, M., Suring, L., Shueck, L., Meinke, C., & Knick, S. 2005. Evaluating species of conservation concern at regional scales. Habitat Threats in the Sagebrush Ecosystem: Methods of Regional Assessment and Applications in the Great Basin, edited by M. Wisdom, M. Rowland & L. Suring (Lawrence, Kansas: Alliance Communications Group), 5–74.

Wotton, B. M. 2009. Interpreting and using outputs from the Canadian Forest Fire Danger Rating System in research applications. Environmental and Ecological Statistics, 16(2), 107–131.

Xie, Y., Sha, Z., & Yu, M. 2008. Remote sensing imagery in vegetation mapping: A review. Journal of Plant Ecology, 1, 9–23.

Xiuwan, C. 2002. Using remote sensing and GIS to analyse land cover change and its impacts on regional sustainable development. International Journal of Remote Sensing, 23, 107–124.

Xu, M., Watanachaturaporn, P., Varshney, P. K., & Arora, M. K. 2005. Decision tree regression for soft classification of remote sensing data. Remote Sensing of Environment, 97, 322–336.

Zhang, L., Wylie, B. K., Ji, L., Gilmanov, T. G., & Tieszen, L. L. 2010. Climate-driven interannual variability in net ecosystems exchange in the northern Great Plains grasslands. Rangeland Ecology and Management, 63, 40–50.

Zhou, B., Okin, G. S., & Zhang, J. 2020. Leveraging Google Earth Engine (GEE) and machine learning algorithms to incorporate in situ measurement from different times for rangelands monitoring. Remote Sensing of Environment, 236, 2020, 111521, ISSN 0034–4257. https://doi.org/10.1016/j.rse.2019.111521. (www.sciencedirect.com/science/article/pii/S0034425719305401)

Part V

Soils

14 Global Land Surface Phenology and Implications for Food Security

Molly E. Brown, Kirsten de Beurs, and Kathryn Grace

ACRONYMS AND DEFINITIONS

CDL	Cropland Data Layer
LSP	Land Surface Phenology
MODIS	Moderate-Resolution Imaging Spectroradiometer
NPP	Net Primary Productivity
SPOT	Satellite Pour l'Observation de la Terre, French Earth Observing Satellites
USDA	United States Department of Agriculture

14.1 INTRODUCTION

Phenology is the scientific study of periodic biological phenomena in relation to climate conditions and habitat factors. Phenology varies by species and is influenced by many factors, such as soil temperature, solar illumination, day length, and soil moisture. Land surface phenology (LSP) is the study of the spatiotemporal patterns in the vegetated land surface as observed by satellite sensors. Agriculture and food production are linked inextricably to the seasonal effects of rainfall and temperature changes. Land surface phenology can be used to estimate agriculturally important changes in the start, length and strength of the growing season, which controls how much food is produced in rainfed agricultural systems (Bolton and Friedl 2013; Koetse and Rietveld 2009). Since the supply of food in many countries is strongly affected by how much food is grown locally, understanding land surface phenology is a critical part of assessing food availability.

Food security is defined as the ability of all people to acquire enough culturally relevant food for an active and healthy life (FAO 2012). Roughly 850 million people, most of whom live in the developing world suffer from undernourishment, an outcome of food insecurity (FAO 2012). To help explain the reasons underlying undernourishment, food security is commonly examined by focusing on four underlying pillars:

- Availability—the availability of sufficient quantities of food of appropriate quality, supplied through domestic production or imports (including food aid).
- Access—access by individuals to adequate resources for acquiring appropriate foods for a nutritious diet.
- Utilization—utilization of food through adequate diet, clean water, sanitation and health care to reach a state of nutritional wellbeing where all physiological needs are met.
- Stability—to be food secure, a population, household, or individual must have access to adequate food at all times. They should not risk losing access to food as a consequence of sudden shocks or cyclical events (FAO 2008; Godfray et al. 2010; Schmidhuber and Tubiello 2007).

DOI: 10.1201/9781003541165-19

Vegetative variability, as measured with land surface phenology, can impact each of these elements. Extreme events can affect food production directly (Vrieling et al. 2011); affect distribution of food and thus the price of food in regions where supply is low and demand is high (de Beurs and Brown 2013); affect utilization through increased spread of disease (Myers and Patz 2009) and impact stability through extreme events that reduce the ability of farmers to predict weather conditions from one year to the next. Farming is becoming even more risky because of heat stress, lack of water, pests and diseases that interact with ongoing pressures on natural resources. Lack of predictability in the start and length of the growing season affects the ability of farmers to invest in appropriate fertilizer levels or improved, high yielding varieties (Zaal et al. 2004).

We can use land surface phenology models to better predict interannual variability of food production. Land surface phenology models rely on remote sensing observations of vegetation, such as datasets derived from the Advanced Very High Resolution Radiometer (AVHRR) and the newer Moderate resolution Imaging Spectroradiometer (MODIS) sensors on Aqua and Terra satellites. Vegetation and rainfall data can assess variables such as the start of season, growing season length and overall growing season productivity (Brown and De Beurs 2008; de Beurs and Henebry 2004, 2010). These metrics are common inputs to crop models that estimate the impact of weather on agricultural area and yield (Bolton and Friedl 2013; Funk and Budde 2009). LSP metrics have a strong relationship with regional food production, particularly those with sufficiently long records to capture local variability. This chapter will focus on reviewing how LSP analysis is done, how it can be used to monitor agriculture and food production, and the links between these observations and food security (FCPN 2007; Zaal et al. 2004).

14.2 CHARACTERIZING LAND SURFACE PHENOLOGY (LSP)

The International Biological Program (IBP) defined phenology as "the study of the timing of recurrent biological events, the causes of their timing with regard to biotic and abiotic forces, and the interrelation among phases of the same or different species (Lieth 1974). The importance of phenological observations to understanding the impact of global environmental change has been increasingly recognized (Mu et al. 2013). The impact of the weather on plant phenology can serve as a biological indicator of the impacts of climate change on terrestrial ecosystems, including agroecosystems that support the production of food (Bradley et al. 2011; Schwartz 1992).

The vigor and development of vegetation depends on available moisture and nutrients for plant development. The health of crops can be studied by looking at their phenological characteristics including germination, leaf emergence, and start of senescence (Vrieling et al. 2011). Land surface phenology is defined as the spatiotemporal development of the vegetated land source as observed by synoptic satellite sensors (de Beurs and Henebry 2004). Datasets from satellite remote sensing of vegetation can approximate phenological stages and thus characterize the general vegetation behavior (Justice et al. 1985; Reed et al. 1994). A derived metric of particular interest is the seasonally cumulated vegetation index as it is related to net primary productivity (NPP) (Awaya et al. 2004).

Phenology models produce annual metrics that describe the growing season, including the start of season (SOS), length of season (LOS), maximum NDVI value (maxNDVI), and cumulated NDVI over the season (cumNDVI) (Figure 14.1) (Brown et al. 2010; de Beurs and Henebry 2010). Satellite remote sensing can be used to study the spatiotemporal development of the vegetated land surface. When using LSP in assessing food production, the measurement needs to be focused on the agriculturally productive areas of the land surface, including row crops, pasture, and gardens to maximize the representativeness of the satellite-derived assessment of the start of season (White et al. 2009).

In the context of food security, land surface phenology is used to provide a remote estimate of the timing of the start of the agricultural growing season. The start of season metric is a critical parameter for food security assessment, and is monitored remotely by many food security and agriculture organizations (Brown 2008). Staple cereal crops in semi-arid agricultural zones such as millet and sorghum are often photoperiod sensitive and thus a sowing delay can translate into a

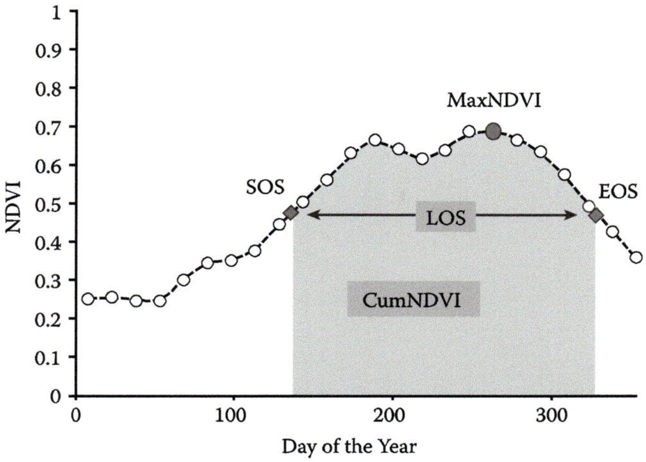

FIGURE 14.1 Start of Season (SOS), Length of Season (LOS), End of Season (EOS), maximum NDVI, and cumulative NDVI for a sample curve. (fFrom Brown, de Beurs and Vrieling 2010 RSE.)

reduction in yield (Brown and de Beurs 2008; Buerkert et al. 2001). Changes in the start of season also may reduce the overall length of the growing season, further reducing the yields obtained in marginal semiarid agro-ecosystems.

Figure 14.2 provide the growing season length for West Africa for the years 2009 (average), 2011 (very dry) and 2012 (relatively wet after a dry winter). The growing season length is calculated with the threshold percentage method (White et al. 1997) based on MODIS bi-directional reflectance function (BRDF) adjusted reflectance derived NDVI data (MCD43C4) at the Global Climate Modeling Grid (0.05°). The maps show a North-South gradient in the length of the growing season with much shorter growing seasons further north. In addition, 2011 reveals much shorter growing seasons, especially compared to 2012. As an example we calculated the average length of the growing season for the agricultural regions (cropland percentage according to the MODIS Land Cover Classification > 0) in Mali. We found that the average growing season length ranged from 101 days in the driest year of 2011, to 125 days in 2012. The year 2009 had an average growing season length of 110 days. Figure 14.3 also reveals that the length of the growing season was very short in 2011, more than 48% of the cropped pixels had a growing season of less than 100 days and the median growing season length was 101 days. The short growing seasons in 2001 resulted in failure of most harvests in Mali and other countries in West Africa. In comparison, the growing season for 2012 was much longer. Only 16.5% of the pixels showed a growing season with fewer than 100 days and the median growing season was 123 days. Unfortunately, there was civil unrest in Mali in 2012, which still resulted in reduced food insecurity.

White et al. (2009) explored the methodology that is used to estimate phenology metrics with remote sensing data. They found that care must be taken with the modeling approach used and matching the method to the region where the agricultural assessment is done. In the context of food security, tropical semi-arid regions dominate the countries where food security is monitored (Figure 14.2) thus the quadratic and multiple-model fit approaches are necessary to correctly assess variations in phenology relevant to agriculture (White et al. 2009).

Henebry and de Beurs (2013) provide a basic overview of land surface phenology and the different satellite sensors that are used in studying land surface phenology going back to Landsat 1 in 1972. Indeed, one of the very first land surface phenology studies was performed based on 80m spatial resolution Landsat data (Dethier 2016). However, most land surface phenology studies have used observations from the Advanced Very High-Resolution Radiometer (AVHRR) which has provided

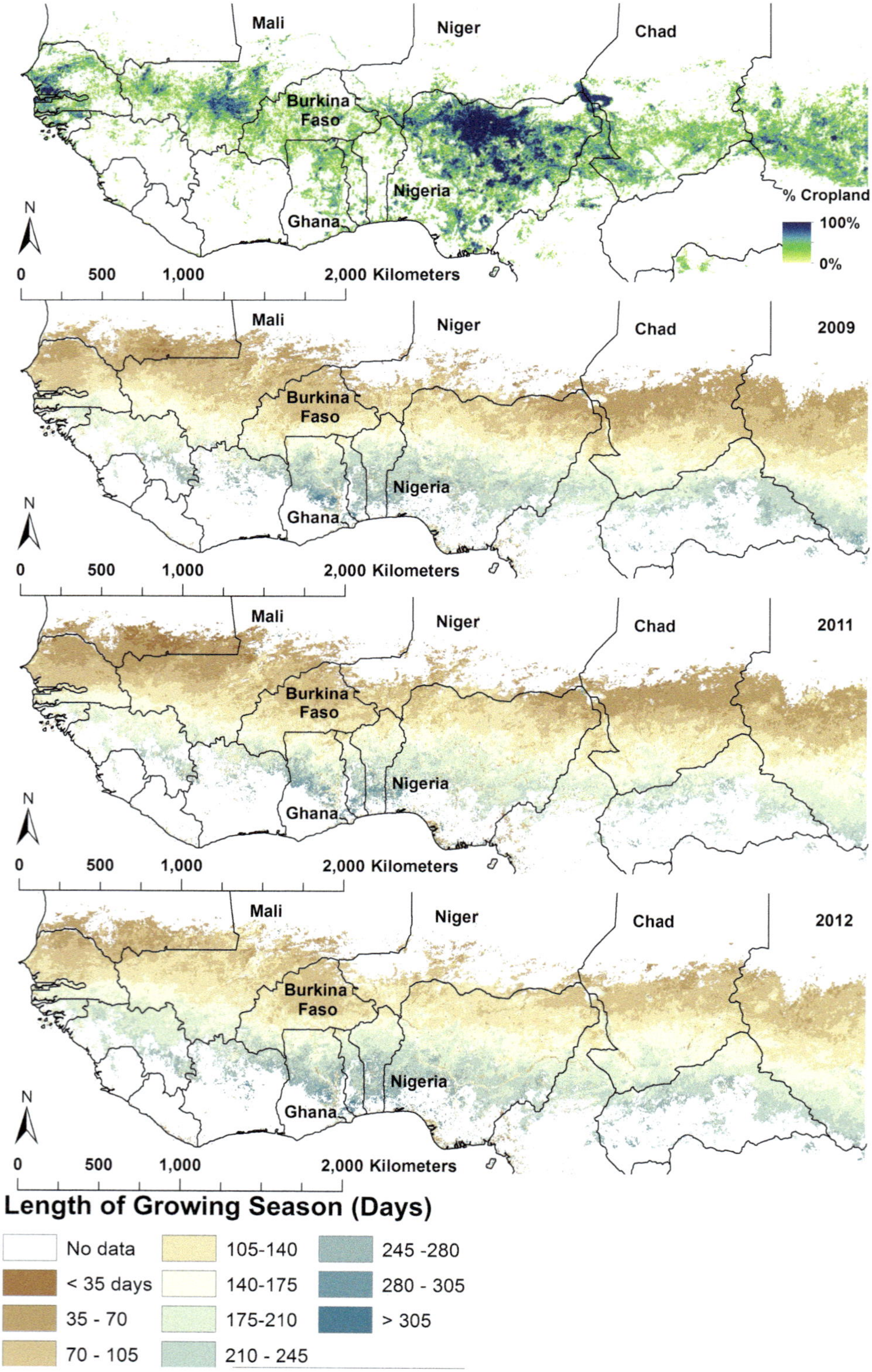

FIGURE 14.2 Top: Percentage of cropland. Bottom three figures: Length of the growing season in West Africa for 2009 (average), 2011 (dry), and 2012 (relatively wet).

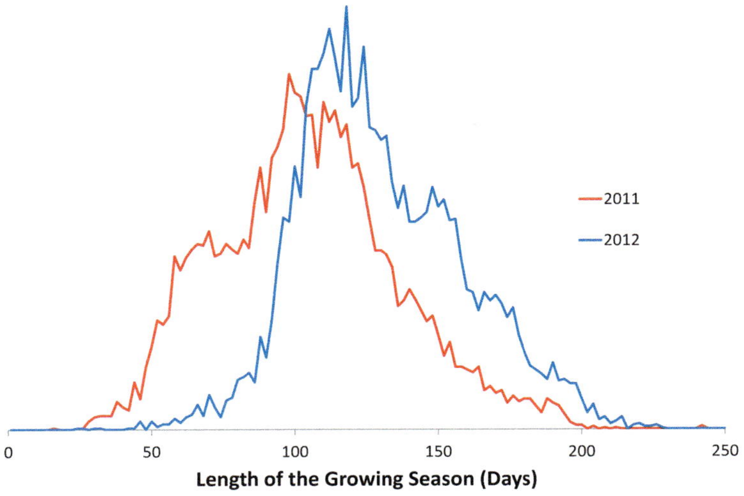

FIGURE 14.3 Histogram of the length of the growing season for 2011 and 2012 for the cropped areas in Mali.

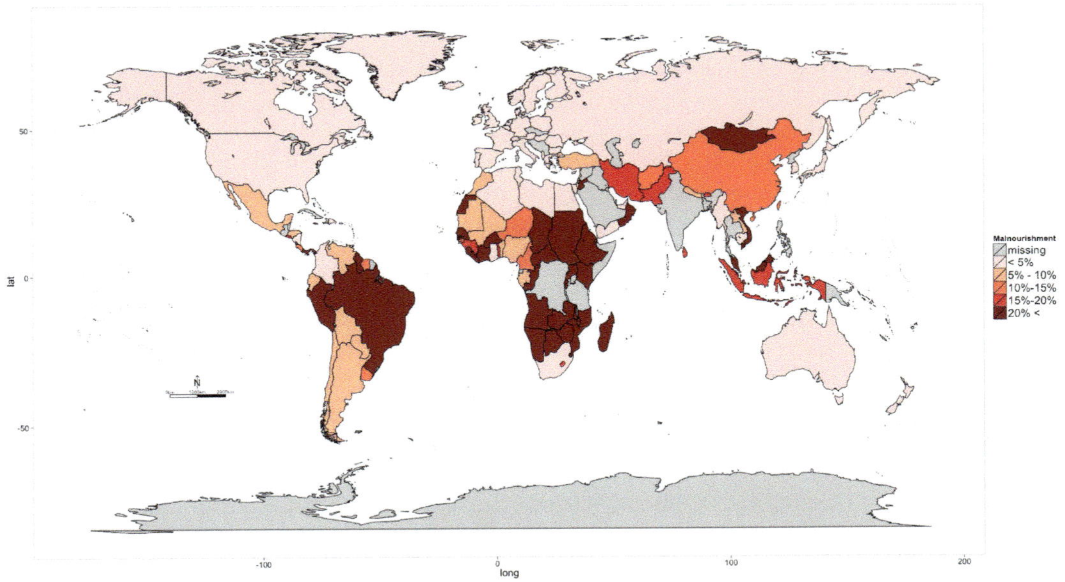

FIGURE 14.4 Map of the percent of the population who are undernourished 2011–2013, World Food Program data.

time series of observations starting in 1982, allowing multi-decade analysis. The AVHRRs have truly been the workhorse platform for the study of land surface phenology (Henebry and de Beurs 2013 and citations within). Since the launch of the MODIS sensors on the Terra and Aqua satellites in 2000 and 2002, MODIS data are also regularly applied in land surface phenology studies as a result of their increased spatial resolution and general accuracy as compared with AVHRR data. While there has been a tremendous increase in the number of published studies that use the term land surface phenology, few of these studies specifically use higher spatial resolution (30 m) Landsat data (Fisher et al. 2006; Fisher and Mustard 2007). However, some recent efforts have focused on the use of fused data products (e.g., fused data based on MODIS and Landsat) for land surface phenology studies (Walker et al. 2014) as well as web-enabled Landsat data (Kovalskyy et al. 2012). More recently, there has been a proliferation of Sentinel 2a/b sensors, which were launched in 2015 (Phiri et al. 2020) and has accelerated the analysis of phenology and land cover in cropped systems.

While White et al. (2009) compared a host of different land surface phenology methods for one dataset (AVHRR) to determine which method resulted in data closest to a variety of field observations, we are aware of just one study that compares land surface phenology metrics based on different datasets (Brown et al. 2008). This study applied one methodology, quadratic regression models, to AVHRR, SPOT-Vegetation and several MODIS products to determine the start of the growing season. They then compare the results with fields observed SOS. The results showed that 8 km MODIS data at eight-day temporal resolution resulted in SOS measurements closest to field observations. It is important to note that the field observations were specifically designed for large-scale satellite validation and thus were better matched with 8km data as opposed to data at finer spatial resolutions.

14.3 AGRICULTURE AND PHENOLOGY METRICS

LSP metrics have been used to better connect observed variability in the weather to impacts on food production. Based on agronomic research on the development and response of different cereal varieties to moisture and temperature stress at various stages of development, LSP assessment can provide a quantitative link between remote sensing observations and yield outcomes (Nafziger 2009).

Funk and Budde (2009) used phenology to adjust the start of season across multiple years to increase the reliability of vegetation-derived metrics to assess yield. They noted that Rasmussen (1992) found that early season vegetation data bore no significant relationship to millet yields in Burkina Faso ($r^2 = 0.1$), while values from 30 days after the mid-season maxima until the end of season explained 93% of the variation in yields (Funk and Budde 2009; Rasmussen 1992). They found that when removing non-agriculture regions and normalizing the start of season across multiple years, they were able to predict 90% of the variability in yields from 2001–2008. This period included the significant change in management of agriculture in the area, which caused a massive decline in productivity outside of weather-induced changes (Funk and Budde 2009).

Bolton and Friedl (2013) used the Funk and Budde (2009) method to derive yield estimates from three different vegetation products and phenology estimates using 500m MODIS. Their results showed that remotely sensed information related to crop phenology is useful for agricultural monitoring. The best times to predict crop yields were 65–75 days after green-up for maize and 80 days after green-up for soybeans (Bolton and Friedl 2013). For relatively homogenous pixels, the timing of maize and soybeans start of season derived from satellite data appeared to be both detectable and separable (Bolton and Friedl 2013). However, because MODIS acquires data at relatively coarse spatial resolution (250m), most pixels will include mixtures of crops. Thus, separation of crops based on satellite-derived phenological information is likely to be challenging in many areas (Ozdogan 2010). Thus most LSP information is used as a measure of moisture availability instead of crop type.

Nieto et al. (2022) describe the role of high resolution, high temporal frequency imagery such as Planet Fusion that can have nearly daily overpass time which can be used to estimate at the field-level crop phenology. The study presented a model that included surface reflectance, vegetation indices, weather and day of the year to create a high quality maize crop phenology estimate in the United States. Robust classifiers such as random forest and machine learning approaches accelerates the modeling of field-level crop maturity, which can be useful for farmers and to forecast phenology across regions. More temporal observations was found to be important in order to capture phenology given the tendency of agricultural regions to also have significant cloudiness during the growing season.

Most satellite-derived agricultural yield estimates using phenology data require a base map of land cover and land use that is used to distinguish land in cultivation from non-agricultural land cover.

Bolton and Friedl (2013) tested two data sources for this purpose: (1) the MODIS Land Cover Type product, which provides a 500m spatial resolution representation of land cover, and (2) the Cropland Data Layer (CDL) created by the USDA, which provides a much higher resolution representation but is only available for the United States. They found that in regions with high intensity, industrial agriculture such as in the United States, the MODIS land cover and the CDL were equally effective. In regions with small field size this will not be the case, however, and the ability of the remote sensing data to identify farm management and moisture-related differences in yield in small fields will be severely limited.

Jain et al. (2013) focused on evaluating the impact of high-resolution satellite data to estimate cropping intensity, or the number of crops planted annually, in regions with small fields. Subsistence agricultural systems are often characterized by small, irregular fields, and these areas often are poorly characterized by methods based on coarse resolution due to the mixing of natural and cropped vegetation in the same pixel (Rasmussen 1992). Using a multi-scalar method where 30m resolution Landsat data is combined with 250m daily MODIS imagery, Jain et al. (2013) show that using a hierarchical training method that uses the high resolution spatial information together with the lower spatial but higher temporal information from MODIS allows for an accurate mapping of crop intensity over these small fragmented agricultural systems (Jain et al. 2013). Other authors have also shown that merging high-resolution imagery with lower resolution imagery in quantitative fusion approaches can improve the results of agricultural assessment (Jin et al. 2010).

Outside of smallholder agricultural regions in the tropics, there is significant food insecurity in the lesser-developed regions of central and eastern Asia. Land abandonment resulting in reduced agricultural production, large changes in economic and governmental institutions and a crumbling physical infrastructure has resulted in increased vulnerability to moisture conditions.

De Beurs and Henebry (2004) used land surface phenology measurements at 8km spatial resolution to demonstrate the effect of the collapse of the Soviet Union on agriculture in Kazakstan. De Beurs and Ioffe (2014) used land surface phenology measurements to determine the number of times agricultural fields were cropped over a ten-year period in European Russia, which is an indicator of potential future cropland abandonment. Prishchepov et al. (2012) found the highest land abandonment was found in Latvia, where 42% of its agricultural land was abandoned from 1990 to 2000, followed by Russia (31%), Lithuania (28%), Poland (14%) and Belarus (13%) (Prishchepov et al. 2012). When abandoned areas are reforested or native grasslands are restored, water quality improves and carbon can be sequestered, but unless economic growth in other sectors occurs, food security of the area could suffer (Gellrich and Zimmermann 2007).

14.4 FOOD SECURITY AND PHENOLOGY

Seasonality in food production, where food is produced in one primary growing season, is a common characteristic across many climates. The impact of seasonality in food availability can translate to seasonality in food prices and food security in poor and food insecure countries with insufficient storage and poorly functioning markets (Alderman et al. 1997; Chen 1991; Crews and Silva 1998; Haddad et al. 1997; Handa and Mlay 2006; Hillbruner and Egan 2008). As Devereux (2012, p. 111) states, “not only does seasonality generate short-term hunger and seasonal food crises, it is also responsible for various ‘poverty ratchets’ that can have irreversible long-term consequences for household well-being and for productive capacity in rural areas.” Coping strategies that have been developed in response to regular reductions in availability and affordability of food involve transfers of assets from poorer households to richer ones at less than full value (Cekan 1992; Devereux 2012).

Understanding the impact of local declines in food production due to the weather relies on understanding the livelihood approach of the people living in the area affected. Livelihood is defined as “the capabilities, assets (including both material and social resources) and activities required for a means of living” (Chambers and Conway 1991). LSP analysis can be used to determine when agricultural production is likely to fall below the needs for farmers in the region. Its relevance to food security depends on the ability of a community to access food if a production decline occurs.

Access to food requires that all individuals have the income sufficient to purchase food that is personally and culturally acceptable (Sen 1981). The concept of access focuses on the ability of households to maintain food consumption in the face of a wide variety of shocks that increase the gap between available income and entitlements and the amount of food that can be purchased with that income at a particular time and place.

Shocks to food security can come from many sources. Droughts can greatly reduce the supply of food, increasing the local food prices (Brown et al. 2012). Personal shocks, such as death of a member of the family, poor health, and loss of employment, can reduce household income and access to food (Gazdar and Mallah 2013). Economic shocks, such as inflation, changes in government policies, changes in public safety nets, and international commodity prices, can reduce the ability of households to attain enough food, despite the fact that the amount of food they produce or their household income has not changed (Brown 2014). Issues of access are described most clearly in social science concepts of individual and household well-being, which capture stress and coping strategies at a variety of scales (Barrett 2010).

14.5 APPROACHES TO MEASURING FOOD INSECURITY

There are two comprehensive approaches to measuring food insecurity. One approach involves measurement of the anthropometrics such as body weight, height, and age of representative samples of the population (e.g., Demographic and Health Surveys) (Brown et al. 2014). The other involves measurements of aggregate household (and/or) individual consumption of food per day of representative samples of the population (e.g., the World Bank's Living Standard Measurement Surveys). However, both these methods require extensive surveys of large numbers of people and households. This data collection strategy is costly in terms of time and money and as a result, these surveys are rarely conducted more than a few times in a decade for a given country. Additionally, a non-trivial limitation of using anthropometry in assessing child nutritional status is the lack of specificity. In other words, changes in body measurements are sensitive to many unobserved factors including intake of essential nutrients, infection, altitude, stress, and genetic background (Barrett 2010).

Thus for measuring and monitoring food security, many organizations use food prices in both urban and rural areas as a proxy for food access, since large, rapid increases in food prices can result in widespread reduction in food consumption, and resulting from widespread declines in food availability and thus food supply in a market (Brown et al. 2009). Many rural dwellers buy food and sell food so they are sensitive to price changes in nearby markets (Brown 2014) Food prices are collected across small towns, regional capitals and in the capital city on a monthly basis for this use. Because food prices can be influenced by a number of international and domestic policies or events that are unrelated to local production (Brown 2014), the links between phenology assessment and food security outcomes can quickly become complex. Including LSP assessments with food prices at multiple scales of analysis can help to identify potential "hot-spots" of food insecurity.

14.6 THE USE OF LSP IN FOOD SECURITY ASSESSMENT IN NIGER

The use of LSP data can be important in food security assessment, particularly in poor agropastoral regions that are remote and frequently food insecure as illustrated here for Niger. Niger consistently ranks in the bottom five poorest countries in the world for gross domestic product per capita (WorldBank 2013). Over half of its 13 million residents are engaged in the agriculture sector, despite the country being semi-arid with a short annual growing season and that only 12% of its land is arable (CIA 2012).

In 2013, Ouallam, a region north of Niamey in the Department of Tillaberi, and other agropastoral areas of Niger had 1.2 million people that had difficulty getting enough to eat with the high price of cereal and limited livelihood strategies (FEWSNET 2013a). Niger's Ouallam Department in the Tillaberi Region has an economy that is based mainly on agriculture (harvests of winter millet, sorghum, cowpeas, groundnuts, and beans) and on raising cattle, sheep, and goats. Annual rainfall in this area ranges from 400 to 600 mm. Local crop production normally covers over 40% of household food needs and accounts for 18% of the income of very poor and poor households. Migration and sales of bush products (wood and straw) are also important sources of cash income for local households. The types of foods normally consumed by households in this area are furnished by on-farm production or purchased with income from livestock sales, farm labor, and sales of wood and straw (Senahoun et al. 2011). Most household spending is on cereal purchases by poor and cereal-short households and on school fees for their children's education. An average to high local demand is helping to generate normal to above-normal levels of income from the farm activities (FEWSNET 2013b).

In 2013, Ouallam Department of Niger experienced a delay in the start of its growing season, restricting the total length of the season and resulting in a production deficit of 68,000 metric tons (FEWSNET 2013b). Although the 2012 cropping season was above average, the poor and very poor households, who make up 61% of the population of the region, depleted their food stocks from the previous year before the start of the 2013 growing season thus are vulnerable to food insecurity due to the low productivity (FEWSNET 2013b).

Given this situation, food security analysts were looking at the remote sensing data products available to determine the likely impact of the reported late start of the growing season. Here we will show the remotely sensed vegetation data that the Famine Early Warning Systems Network or FEWS NET uses operationally to identify and quantify the likely impact of variations in the start of season. FEWS NET uses the eMODIS data product derived from daily MODIS reflectance data and re-composited into 10 day observations for ease of comparison with rainfall and other datasets (Jenkerson et al. 2010; Ji et al. 2010). Figure 14.3 shows the Ouallam Region on an NDVI anomaly image calculated from a ten-year mean (from 2001–2010) used by FEW NET in its Early Warning Explorer tool for the first ten days of July. Figures 14.5a and 14.5b show the time series of the NDVI as it compared to the previous season, and the anomaly of the NDVI. Although the anomaly image showed very little impact of the late start, the time series does show the late start, and then a subsequent robust response to significant rains in August.

The late start in July 2013 resulted in below average yields, despite the later robust response to rainfall in August. A shorter growing season due to a late start has long been recognized as a risk factor for crop development (Brown and de Beurs 2008). Because crops in the Sahel are often photoperiod sensitive, a late start cannot be made up later in the season and typically translates into a yield reduction (Buerkert et al. 2001). Thus the food security analysts using the information from vegetation were able to quantitatively link the late start to identifying regions that were more likely to experience food insecurity as a result in yield declines.

14.7 USING LSP TO CONTEXTUALIZE THE RELATIONSHIP BETWEEN MAIZE PRICE AND HEALTH IN KENYA

Kenya's malnutrition rate has remained high despite a number of improvements in other socioeconomic and health indicators (KNBS 2010). Additionally, Kenya faces important shifts in climate and weather patterns that may be linked to food insecurity (Grace et al. 2013). Recent research has sought to examine the effect of price changes (a measure of food access) on household-level food insecurity outcomes, specifically the birth weight of babies (Grace et al. 2014). A key component of this type of approach to food insecurity and food access is food availability.

Kenya has no annually updated measures of local, household-relevant food production. Instead, to measure local production, the authors calculated the maximum NDVI in a small area (10 km radius) around each community with household-level health data. Given the long time series of available MODIS NDVI data, the authors were able to match births to the relevant growing season's NDVI. Using the MODIS NDVI value as a proxy for local production, the authors identified the importance of local prices on infant birth weights (Grace et al. 2014). The results further suggested

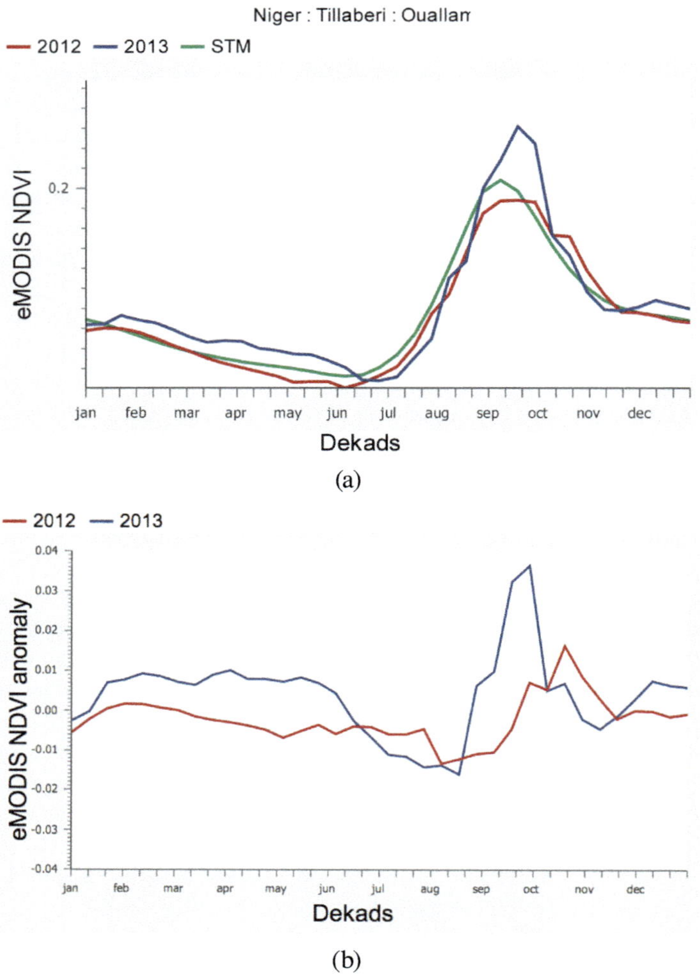

FIGURE 14.5 Comparison of Niger's Ouallam Department in Tillaberi Region's MODIS NDVI. (a) The actual NDVI from 2012 and 2013, along with the short-term mean, (b) the right NDVI anomaly from 2012 and 2013 using the short-term mean.

that price impacts were dependent on NDVI. In general, when NDVI was high and food prices were low, relative to the areas under study—households were less likely to experience food insecurity, but when MODIS NDVI was low and prices were low the likelihood of food insecurity increased. The results suggest that food prices alone cannot predict a household's risk for food insecurity, rather the combination of local production and food prices should be evaluated (Brown 2014).

Because the majority of impoverished, food insecure countries are not able to collect and disseminate fine-resolution estimates of food production, the potential for NDVI and other measures of LSP to support analyses similar to the Kenya analyses just described, is high and very relevant. Ultimately, the use of this type of data can provide an improved understanding of community- and household-level response to food price volatility.

14.8 DISCUSSION

Improving scientific understanding of land surface phenology through the use of remotely sensed data and climate models is necessary for anticipating areas of potential food shortage. Land surface phenology models can be combined with data on food prices, health outcomes and household economics at micro- and macro-levels to more fully examine the links between the physical environment and the people who are most likely impacted by weather changes. This chapter highlighted the work that has been done by physical scientists to advance LSP modeling using remote sensing and climate models. We also highlighted the impacts of food insecurity and undernutrition on human health outcomes. Interdisciplinary research that includes LSP and human health is an important next step in applying the physical science to relevant issues related to development and health.

Having a high quality, remote sensing-based phenology model can accelerate anticipation of production deficits. These can be integrated with socioeconomic and food price drivers of food security, particularly as the resolution and time step improves with additional satellite remote sensing systems such as Sentinel 2a/b. Anticipatory Action is a risk-informed approach that leverages modeling and mapping of geospatial risks and hazards to support communities to take action to reduce the impact of the hazards. Anticipatory actions rely on forecasts of the hazard, are linked to short-term actions that aim to prevent or reduce impacts and require finance to be identified ahead of time to fund the interventions. In this context, determining phenology baselines and then predicting phenology anomalies is a critical first step to identifying and integrating cropping outcomes with these approaches (Choularton and Montier 2023).

There are some drawbacks to phenology as a measure of the progress of the growing season. Most LSP models are most effective when most of the growing season has happened, as the curve-fitting algorithms do best when the beginning, middle and end of the season are present. When only the start is known, estimating the yield impacts from the beginning of the season is very challenging due to uncertainties in the model. Thus although LSP tools derived from remote sensing of vegetation are quite effective as a retrospective analysis, continued reliance on rainfall-based estimation of the start and peak of the season through crop models is likely for famine early warning organizations interested in the current month's growing conditions.

Another challenge to LSP and other new metrics of remote sensing such as soil moisture from new sensors such as Soil Moisture Active Passive (SMAP) and evapotranspiration derived from satellite-derived temperature, is the challenge of communicating the value of such new tools to the widespread and diverse food security and humanitarian communities (Brown and Brickley 2012). FEWS NET, for example, is comprised of a central office, four organizations that are experts on remote sensing and biophysical modeling, and over 20 offices in food insecure countries. This makes the organization extremely susceptible to center-periphery problems where new ways of looking at drought and its impact on crop yield move only extremely slowly from the research centers, located mostly in the United States or Europe, to where food security analysis is actually conducted, in the food insecure countries themselves. Thus although LSP analysis could be extremely helpful to identify and respond to food insecurity, moving the analysis approach from research to the operational context will take a lot of investment and effort (Brown and Brickley 2012).

14.9 CONCLUSIONS

Land surface phenology is the study of the changes in start, peak and end of the growing season on the land surface as observed by satellite sensors. LSP can be used to determine when weather-related events are agriculturally important and may result in impacts on food production, distribution, and cost that may result in changes in food security in vulnerable communities. Food security assessment uses satellite remote sensing to determine how agriculture is changing, the impact of weather on food production, and how these changes affect food availability and access. Research to improve our understanding of environmental and weather drivers of production can provide new and valuable information for the food security community.

REFERENCES

Alderman, H., Bouis, H., & Haddad, L. (1997). Aggregation, flexible forms, and estimation of food consumption parameters: Comment. *American Journal of Agricultural Economics*, *79*, 267

Awaya, Y., Kodani, E., Tanaka, K., Liu, J., Zhuang, D., & Meng, Y. (2004). Estimation of the global net primary productivity using NOAA images and meteorological data: Changes between 1988 and 1993. *International Journal of Remote Sensing*, *25*, 1597–1613

Barrett, C. (2010). Measuring food insecurity. *Science*, *327*, 825–828

Bolton, D.K., & Friedl, M.A. (2013). Forecasting crop yield using remotely sensed vegetation indices and crop phenology metrics. *Agricultural and Forest Meteorology*, *173*, 74–84

Bradley, A.V., Gerard, G.G., Barbier, N., Weedon, G.P., Anderson, L.O., Huntingford, C., Aragao, L.E.O.C., Zelazowski, P., & Arai, E. (2011). Relationships between phenology, radiation and precipitation in the Amazon region. *Global Change Biology*, *17*, 2245–2260

Brown, M.E. (2008). *Famine Early Warning Systems and Remote Sensing Data*. Heidelberg: Springer Verlag

Brown, M.E. (2014). *Food Security, Food Prices and Climate Variability*. London: Routedge Press

Brown, M.E., & Brickley, E.B. (2012). Evaluating the use of remote sensing data in the USAID famine early warning systems network. *Journal of Applied Remote Sensing*, *6*, 063511

Brown, M.E., & de Beurs, K. (2008). Evaluation of multi-sensor semi-arid crop season parameters based on NDVI and rainfall. *Remote Sensing of Environment*, *112*, 2261–2271

Brown, M.E., de Beurs, K., & Vrieling, A. (2010). The response of African land surface phenology to large scale climate oscillations. *Remote Sensing of Environment*, *114*, 2286–2296

Brown, M.E., Grace, K., Shively, G., Johnson, K., & Carroll, M. (2014). Using satellite remote sensing and household survey data to assess human health and nutrition response to environmental change. *Population and Environment*, in press

Brown, M.E., Hintermann, B., & Higgins, N. (2009). Markets, climate change and food security in West Africa. *Environmental Science and Technology*, *43*, 8016–8020

Brown, M.E., Tondel, F., Essam, T., Thorne, J.A., Mann, B.F., Leonard, K., Stabler, B., & Eilerts, G. (2012). Country and regional staple food price indices for improved identification of food insecurity. *Global Environmental Change*, *22*, 784–794

Buerkert, A., Moser, M., Kumar, A.K., Furst, P., & Becker, K. (2001). Variation in grain quality of pearl millet from Sahelian West Africa. *Field Crops Research*, *69*, 1–11

Cekan, J. (1992). Seasonal coping strategies in Central Mali: Five villages during the 'Soudure'. *Disasters*, *16*, 66–73

Chambers, R., & Conway, G. (1991). Sustainable rural livelihoods: Practical concepts for the 21st century. In, *IDS Discussion Paper 296*. London: Institute of Development Studies

Chen, M.A. (1991). *Coping with Seasonality and Drought*. London: Sage Publications

Choularton, R., & Montier, E. (2023) *Windows of Opportunity for Risk-Informed Assistance: An Anticipatory, Early Action, andDisaster Risk Finance Framework*. Report of the USAID Climate Adaptation Support Activity implemented by Tetra Tech and funded by the U.S. Agency for International Development (p. 49).

CIA (2012). *Central Intelligence Agency World Factbook*. In, Washington DC: United States Government

Crews, D.E., & Silva, H.P. (1998). Seasonality and human adaptation: Current reviews and trends. *Reviews in Anthropology*, *27*, 1–15

de Beurs, K.M., & Brown, M.E. (2013). The effect of agricultutral growing season change on market prices in Africa. In A. Tarhule (Ed.), *Climate Variability—Regional and Thematic Patterns*. in press

de Beurs, K.M., & Henebry, G.M. (2004). Land surface phenology, climatic variation, and institutional change: Analyzing agricultural land cover change in Kazakhstan. *Remote Sensing of Environment*, *89*, 497–509

de Beurs, K.M., & Henebry, G.M. (2010). Spatio-temporal statistical methods for modeling land surface phenology. In I.L.H.a.M.R. Keatley (Ed.), *Phenological Research: Methods for Environmental and Climate Change Analysis*. Springer

de Beurs, K.M., & Ioffe, G. (2014). Use of landsat and MODIS data to remotely estimate Russia's sown area. *Journal of Land Use Science*, *9*. https://doi.org/10.1080/1747423X.2013.798038

Dethier, B. (2016). Phenology Satellite Experiment. *On Significant Results . . .*, March. https://books.google.com/books?hl=en&lr=&id=CthRAQAAMAAJ&oi=fnd&pg=PA157&dq=(Dethier+1973+vegetation&ots=z_nqxfmsn1&sig=TpQyBAl3Rc2cNDxxe5bSapcfwYU

Devereux, S. (2012). Seasonal food crises and social protection. In B. Harriss-White, & J. Heyer (Eds.), *The Comparative Political Economy of Development: Africa and South Asia* (p. 358). Routledge Press

FAO (2008). *Assessment of the World Food Security and Nutrition Situation.* In, Rome, Italy: United Nations FAO

FAO (2012). *The State of Food Insecurity in the World.* In, Rome, Italy: United Nations Food and Agriculture Organization

FCPN (2007). *Food Situation in the Sahel and West Africa: Should Satisfactory Agricultural and Food Prospects be Expected*? Food Crises Prevention Network (FCPN)

FEWSNET (2013a). *Niger Food Security Outlook.* In, (p. 8). Washington DC: US Agency for International Development

FEWSNET (2013b). *Niger Food Security Outlook: Auspicious Cropping Season Conditions through August Deteriorate in September.* In, (p. 8). Washington DC: US Agency for International Development Famine Early Warning Systems Network

Fisher, J.I., & Mustard, J.F. (2007). Cross-scalar satellite phenology from ground, landsat, and MODIS data. *Remote Sensing of Environment*, *109*, 261–273. https://doi.org/10.1016/j.rse.2007.01.004

Fisher, J.I., Mustard, J.F., & Vadeboncoeur, M.A. (2006). Green leaf phenology at landsat resolution: Scaling from the field to the satellite. *Remote Sensing of Environment*, *100*, 265–279

Funk, C.C., & Budde, M.E. (2009). Phenologically-tuned MODIS NDVI-based production anomaly estimates for Zimbabwe. *Remote Sensing of Environment*, in press

Gazdar, H., & Mallah, H.B. (2013). Inflation and food security in Pakistan: Impact and coping strategies. *IDS Bulletin*, *44*, 31–37

Gellrich, M., & Zimmermann, N.E. (2007). Investigating the regional-scale pattern of agricultural land abandonment in the Swiss mountains: A spatial statistical modeling approach. *Landscape and Urban Planning*, *79*, 65–76

Godfray, H.C.J., Beddington, J.R., Crute, I.R., Haddad, L., Lawrence, D., Muir, J.F., Pretty, J., Robinson, S., Thomas, S.M., & Toulmin, C. (2010). Food security: The challenge of feeding 9 billion people. *Science*, *327*, 812–818

Grace, K., Brown, M.E., & McNally, A. (2014). Examining the link between food price and food insecurity: A multi-level analysis of maize price and birthweight in Kenya. *Food Policy*, in press

Grace, K., Davenport, F., Funk, C., & Lerner, A. (2013). Child malnutrition and climate conditions in Kenya. *Applied Geography*, *11*, 164–177

Haddad, L., Hoddinott, J., & Alderman, H. (1997). *Intrahousehold Resource Allocation in Developing Countries: Models,Methods and Policies.* Washington DC: IFPRI/JHU Press

Handa, S., & Mlay, G. (2006). Food consumption patterns, seasonality and market access in Mozambique. *Development Southern Africa*, *23*, 541–560

Henebry, G.M., & de Beurs, K.M. (2013). Remote sensing of land surface phenology: A prospectus. In *Phenology: An Integrative Environmental Science* (pp. 385–411). Dordrecht: Springer Netherlands. https://doi.org/10.1007/978-94-007-6925-0_21

Hillbruner, C., & Egan, R. (2008). Seasonality, household food security, and nutritional status in Dinajpur, Bangladesh. *Food and Nutrition Bulletin*, *29*, 221–231

Jain, M., Mondal, P., DeFries, R.S., Small, C., & Galford, G.L. (2013). Mapping cropping intensity of smallholder farms: A comparison of methods using multiple sensors. *Remote Sensing of Environment*, *134*, 210–223

Jenkerson, C., Maiersperger, T., & Schmidt, G. (2010). eMODIS: A user-friendly data source. In, *Open-File Report 2010-1055* (p. 10). Sioux Falls, SD: US Geological Survey

Ji, L., Wylie, B., Ramachandran, B., & Jenkerson, C. (2010). A comparative analysis of three different MODIS NDVI datasets for Alaska and adjacent Canada. *Canadian Journal of Remote Sensing*, *36*, S149–S167

Jin, H., Wang, J., Bo, Y., Chen, G., & Xue, H. (2010). Data assimilation of MODIS and TM observations into CERES-Maize model to estimate regional maize yield. In *Remote Sensing and Modeling of Ecosystems for Sustainability VII.* San Diego, CA: SPIE-Int Soc Optical Engineering

Justice, C.O., Townshend, J.R.G., Holben, B.N., & Tucker, C.J. (1985). Analysis of the phenology of global vegetation using meteorological satellite data. *International Journal of Remote Sensing*, *6*, 1271–1318

KNBS (2010). *Kenya Demographic and Health Survey 2008–09.* In, Calverton, Maryland: Kenya National Bureau of Statistics (KNBS) and ICF Macro

Koetse, M.J., & Rietveld, P. (2009). The impact of climate change and weather on transport: An overview of empirical findings. *Transportation Research Part D:Transport and Environment*, *14*, 205–221

Kovalskyy, V., Roy, D.P., Zhang, X.Y., & Ju, J. (2012). The suitability of multi-temporal web-enabled landsat data NDVI for phenological monitoring—a comparison with flux tower and MODIS NDVI. *Remote Sensing Letters*, *3*, 325–334. https://doi.org/10.1080/01431161.2011.593581

Lieth, H. (1974). *Phenology and Seasonality Modeling* (p. 444). Berlin: Springer Verlag
Mu, J.E., McCarl, B.A., & Wein, A. (2013). Adaptation to climate change: Changes in farmland use and stocking rate in the U.S. *Mitigation andAdaptation Strategies for Global Change*, *18*, 713–730
Myers, S., & Patz, J. (2009). Emerging threats to human health from global environmental change. *Annual Review of Environmental Resources*, *34*, 223–252
Nafziger, E. (2009). Corn. *Illinois Agronomy Handbook* (pp. 13–26). Urbana, IL: University of Illinois
Nieto, L., Houborg, R., Zajdband, A., Jumpasut, A., Vara Prasad, P.V., Olson, B.J.S.C., & Ciampitti, I.A. (2022). Impact of high-cadence earth observation in maize crop phenology classification. *Remote Sensing*, *14*, 469
Ozdogan, M. (2010). The spatial distribution of crop types from MODIS data: Temporal unmixing using independent component analysis. *Remote Sensing of Environment*, *114*, 1190–1204
Phiri, D., Simwanda, M., Salekin, S., Nyirenda, V.R., Murayama, Y., & Ranagalage, M. (2020) Sentinel-2 data for land cover/use mapping: A review. *Remote Sensing*, *12*, 2291
Prishchepov, A.V., Radeloff, V.C., Baumann, M., Kuemmerle, T., & Müller, D. (2012). Effects of institutional changes on land use: Agricultural land abandonment during the transition from state-command to market-driven economies in post-Soviet Eastern Europe. *Environmental Research Letters*, *7*, 024021
Rasmussen, M.S. (1992). Assessment of millet yields and production in northern Burkina Faso using integrated NDVI from the AVHRR. *International Journal of Remote Sensing*, *13*, 3431–3442
Reed, B.C., Brown, J.F., Vanderzee, D., Loveland, T.R., Merchant, J.W., & Ohlen, D.O. (1994). Measuring phenological variability from satellite imagery. *Journal of Vegetation Science*, *5*, 703–714
Schmidhuber, J., & Tubiello, F.N. (2007). Global food security under climate change. *Proceedings of the National Academy of Sciences*, *104*, 19703–19708
Schwartz, M.D. (1992). Phenology and springtime surface-layer change. *Monthly Weather Review*, *120*, 2570–2578
Sen, A.K. (1981). *Poverty and Famines: An Essay on Entitlements and Deprivation*. Oxford: Clarendon Press
Senahoun, J., Ndiaye, C.I., Haido, A.M., Saidou, O., Tahirou, L., Akakpo, K., & Kountche, B.I. (2011). *Special Report: Inter-Agency Crop and Food Security Assessment Mission to Niger*. In, Rome: The World Food Program and the Food and Agriculture Organization
Vrieling, A., de Beurs, K.M., & Brown, M.E. (2011). Variability of African farming systems from phenological analysis of NDVI time series. *ClimaticChange*, *109*, 455–477
Walker, J.J., de Beurs, K.M., & Wynne, R.H. (2014). Dryland vegetation phenology across an elevation gradient in Arizona, USA, investigated with fused MODIS and landsat data. *Remote Sensing of Environment*, *144*, 85–97
White, M.A., de Beurs, K., Didan, K., Inouye, D., Richardson, A., Jensen, O., O'Keefe, J., Zhang, G., Nemani, R., van Leeuwen, W., Brown, J., de Wit, A., Schaepman, M., Lin, X., Dettinger, M., Baily, A., Kimball, J., Schwartz, M., Baldocchi, D., Lee, J., & Lauenroth, W. (2009). Intercomparison, interpretation and assessment of spring phenology in North America estimated from remote sensing for 1982 to 2006. *Global Change Biology*, *15*, 2335–2359
White, M.A., Thornton, P.E., & Running, S.W. (1997). A continental phenology model for monitoring vegetation responses to interannual climatic variability. *Global Biogeochemical Cycles*, *11*, 217–234. https://doi.org/10.1029/97GB00330
WorldBank (2013). *Data: The World Bank*. In, Washington DC: The World Bank
Zaal, F., Dietz, T., Brons, J., Van der Geest, K., & Ofori-Sarpong, E. (2004). Sahelian livelihoods on the rebound: A critical analysis of rainfall, drought index and yields in Sahelian agriculture. In A.J. Dietz, R. Ruben, & A. Verhagen (Eds.), *The Impact of Climate Change on Drylands: With a Focus on West Africa* (pp. 61–77). Dordrecht, The Netherlands: Kluwer Academic Publishers

15 Spectral Sensing from Ground to Space in Soil Science

State of the Art, Applications, Potential, and Perspectives

José A.M. Demattê, Cristine L.S. Morgan, Sabine Chabrillat, Rodnei Rizzo, Marston H.D. Franceschini, Fabrício da S. Terra, Gustavo M. Vasques, Johanna Wetterlind, Henrique Bellinaso, Letícia G. Vogel

ACRONYMS AND DEFINITIONS

ASTER	Advanced Spaceborne Thermal Emission and Reflection Radiometer
CAI	Cellulose Absorption Index
CHRIS	Compact High Resolution Imaging Spectrometer
DEM	Digital Elevation Model
EnMAP	Environmental Mapping and Analysis Program
ENVISAT	Environmental Satellite
ETM+	Enhanced Thematic Mapper Plus
EVI	Enhanced Vegetation Index
FAO	Food and Agriculture Organization of the United Nations
GIS	Geographic Information System
GPS	Global Positioning System
GSS	Ground Spectral Sensing
HyspIRI	Hyperspectral Infrared Imager
IKONOS	A commercial Earth observation satellite, typically, collecting sub-meter to 5 m data
LIDAR	Light Detection and Ranging
LSS	Laboratory Spectral Sensing
MIR	Middle Infrared
MSS	Multispectral Scanner
MODIS	Moderate-Resolution Imaging Spectroradiometer
NASA	National Aeronautics and Space Administration
NDII	Normalized Difference Infrared Index
NDVI	Normalized Difference Vegetation Index
NIR	Near-Infrared
PROBA	Project for On Board Autonomy
PV	Photosynthetic Vegetation
RADAR	Radio Detection and Ranging
RS	Remote Sensing
SAR	Synthetic Aperture Radar
SPOT	Satellite Pour l'Observation de la Terre, French Earth Observing Satellites

DOI: 10.1201/9781003541165-20

SVM	Support Vector Machines
SWIR	Shortwave Infrared
TIR	Thermal Infrared
UAV	Unmanned Aerial Vehicles

15.1 PREFACE

Industrial and agricultural activities are developing faster than the public policy on the use of soil resources. The world needs more information about soil for land use planning and interpretative purposes. Spectral Sensing (SS) has emerged as a major discipline in Remote Sensing science in the past years providing important tools to assist in soil information gathering, mapping, and monitoring. This chapter aims to discuss the role of SS (covering the visible, infrared, thermal, microwave and gamma ranges of the spectrum) in soil science, based on different sensors, scales, and platforms (laboratory, field, aerial, and orbital). We review the state of the art and provide guidance on how to use SS for several purposes, for example, soil classification and mapping, attribute quantification, soil management, conservation and monitoring. Research has shown that SS has the capability to quantify soil attributes, such as clay, sand, soil organic matter (SOM), soil organic carbon (SOC), cation exchangeable capacity (CEC), Fe_2O_3, carbonates and mineralogy, with reliable and repeatable results. Other soil attributes including pH, Ca, Mg, K, N, P and heavy metals have also been evaluated with variable outcomes. Laboratory and field-based measurements are more accurate than aerial or space-based measurements as they are taken under more controlled environments that are less affected by external factors, such as mixing in the field-of-view, vegetation cover, stone cover, water content and atmospheric conditions. Nevertheless, soil is typically evaluated from space using multispectral sensors on board satellites, which offer many options in terms of temporal and spatial coverage and resolution, and are commonly available free-of-charge. On the other hand, hyperspectral images are less commonly applied due to their more limited choices of temporal and spatial resolutions, and difficulty of processing, despite their great potential to correlate with various soil properties. Other SS techniques, such as passive gamma spectroscopy, provide data for surface and below-surface soil inference, primarily relating to the clay content and types of soil minerals, while microwave (i.e., radar) spectroscopy is mainly used in study of soil moisture. In soil science, there are promising results and growing interest for visible-near-infrared (VIS-NIR) and middle infrared (MIR) spectroscopy as they allow quick, non-destructive, and cost-effective estimation of soil properties, reducing the need for sample preparation and the use of reagents, minimizing pollution. It has been observed that MIR spectroscopy can quantify properties, such as clay, clay-sized mineralogy, SOC and inorganic carbon (C), more accurately than VIS-NIR. Both physical (descriptive interpretation of spectral information) and statistical (mathematical approach) methods proved to be useful depending on soil and environmental conditions under study. We observed that the most important limitation of VIS-NIR and MIR spectroscopy for soil classification are their inability to detect soil morphological properties (e.g., soil structure). In the case of VIS-NIR space SS, the limitation is that the radiation only penetrates a few centimeters into the soil surface. On the other hand, satellite-based VIS-NIR data can be used for delineation of soils boundaries supporting soil survey and mapping. Spectral sensing applicability is also increasing in Precision Agriculture (PA), coupled with on-the-go sensors that measure soil properties with high sampling density and in real time. Future advances in SS include: 1) extraction of moisture effects from intact and field moist spectra, allowing a comparison with laboratory measurements, 2) development of local, regional, or global soil Spectral Libraries and their appropriate use, and 3) combining multiple sources of sensed data for better soil inference. Country-based soil spectral libraries started in the early 1980s and today we are moving towards a global spectral library with contribution from as many as 90 countries. Soil spectral libraries, from global to local, will be the future of soil analysis carrying both spatial and hyperspectral data to derive soil information. Spectral sensing has the advantage of providing quantitative data, and thus reducing the subjectivity of soil spatial information for decision making. Spectral sensing techniques are powerful when combined with geoprocessing, landscape modeling, geology, and geomorphology. The past and new studies on soil ground spectral sensing indicate strong information with a great perspective on all spectral sensing platforms, specially for hyperspectral aerial and orbital ones that are currently working and the new ones being develop and launched soon (2017–2020). The goal of all SS techniques is to deliver spatially and spectrally accurate, reliable, and transferable information on soil properties. In order to achieve this, SS applications need to properly account for specific advantages and limitations of each sensor, depending on the overall aim. In summary, it is clear that spectral sensing can be applied in any field of interest of soil science, depending only on the user's creativity.

Index words: Image spectroscopy, soil analysis, soil mapping, proximal sensing, precision agriculture, remote sensing.

15.2 SOILS

15.2.1 Definition and Classification

Soil might be defined as the non-consolidated part of the terrestrial crust, or more specifically as "a continuous and three-dimensional natural body, in constant development, formed by organic and mineral constituents, including solid, liquid and gaseous phases organized in specific structure on a certain pedological medium" (IUSS Working Group WRB, 2014). A "soil body," as defined in Pedology, is governed by the interaction of soil factors and formation processes that "sculpt" the soils, and is usually represented by basic units called pedons. As stated by Buol et al. (2011), soil reveals a vertical arrangement of components that change, often gradually, as one traverse the landscape. Our understanding of soil is limited without the use of chemical, mineralogical, biological and physical quantification techniques, which characterize samples. Besides, soils can be dismembered, sampled and autopsied, but this analysis will only help if we understand what a soil is and how it functions in the ecosystem. Identifying the horizontal and vertical arrangement of soil attributes, that is, across space and at the topsoil, subsoil, and genetic diagnostic horizons, is essential for soil classification. **Perhaps, no single problem has plagued soil scientists more than the identification of the spatial boundaries of an individual soil on the landscape**. Spectral sensing may contribute in identifying horizontal and vertical soil boundaries.

Due to the complex interaction that takes place in the pedogenetic processes, the grouping of similar pedons in soil mapping units, by defined boundaries, is normally based on landscape and soil profile descriptions. To recognize the soil as a distinct individual is very efficient from several points of view. First, it allows us to structure our knowledge in the form of individual groups, also known as classes. Second, it facilitates the drawing of thematic maps where classes are spatially delineated (Legros, 2006). For better communication between members of the soil science community, various soils classification systems have been established, such as Soil Taxonomy (Soil Survey Staff, 2014) and World Reference Base for Soil Resources (IUSS Working Group WRB, 2014).

15.2.2 How Does Soil Form?

Soil formation starts with the weathering of the parent material (e.g., the original rock) through physical and chemical processes over time, where climate and organisms have a major role and all are influenced by relief (Figure 15.1). The soil forming factors described here represent soil formation processes that occur under different time and spatial scales and with variable intensities. This concept can be summarized by the conceptual equation proposed by Jenny (1941): S = f (c, o, r, p, t); *c* (or cl), climate; *o*, organisms; *r*, relief; *p*, parent materials; *t*, time. With the advent of new methods and technologies for soil evaluation, such as global positioning systems (GPS) and geographic information systems (GIS), it has been possible to comprehend soils from a different point of view. McBratney et al. (2003) proposed a renewed model, known as "scorpan," where two factors are

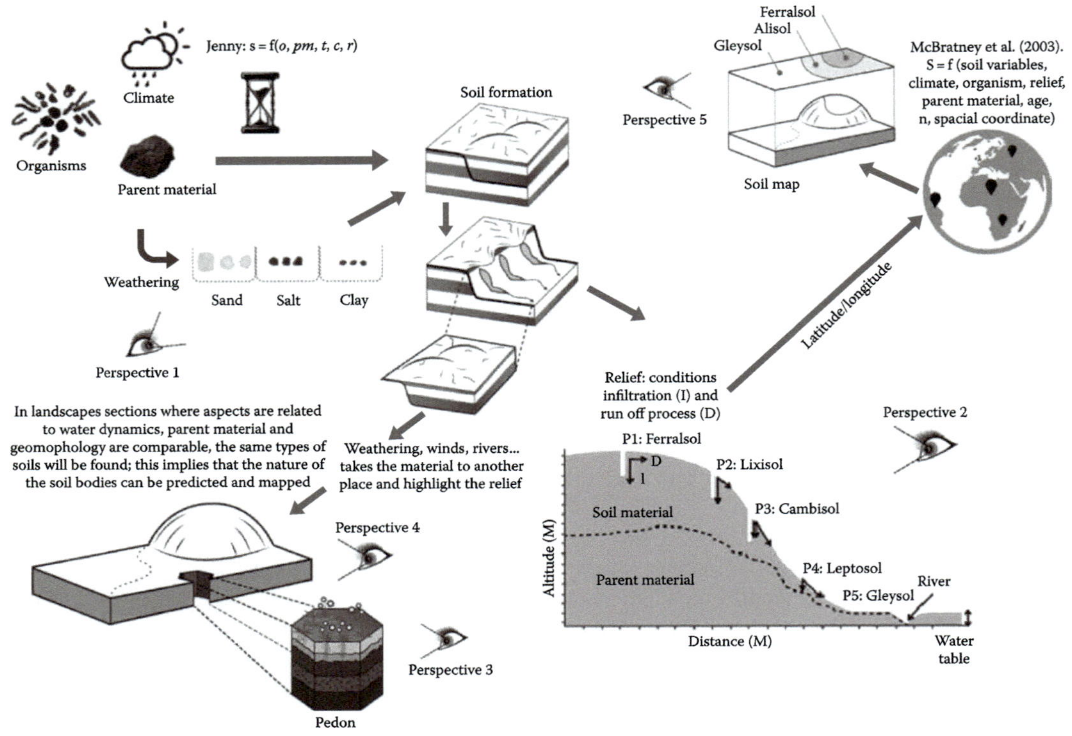

FIGURE 15.1 Conceptual soil and landscape formation. Perspective 1: micro-perspective, intrinsic soil characteristics, such as mineralogy; Perspective 2: longitudinal vision; Perspective 3: soil profile or site perspective; Perspective 4: related with the soil surface composed by the landscape elements, such as landforms, slopes, and drainage patterns; Perspective 5: spatial vision and distribution of the soil.

added to Jenny's equation, namely: the *s* factor corresponding to soil data available at the beginning of the mapping process, including soil maps, data acquired by means of remote or proximal sensing, and expert pedological knowledge; and the *n* factor representing the geographic position of the soil. The collection of soil forming factors (s,c,o,r,p,a,n) represent the underlying landscape characteristics that have allowed the formation of a particular soil class. To completely understand the soil, considering this conceptual soil formation process, and to be able to model this process and map soils, it is important to have different views (Figure 15.1). Thus, we can "look" at soil from five different perspectives, as suggested (Figure 15.1): perspective 1—micro-perspective, intrinsic soil characteristics, such as mineralogy; perspective 2—longitudinal vision; perspective 3—soil profile or site perspective; perspective 4—related with the soil surface composed by the landscape elements, such as landforms, slopes, and drainage patterns; perspective 5—spatial vision and distribution of the soil from space. All of these points of view will assist users in understanding and visualizing the soils as a complete body, in which SS can greatly contribute and which we will discuss in this chapter.

15.3 WHY IS SOIL IMPORTANT?

The world population is expected to exceed 8 billion people by 2050 (FAO, 2013a). Today we have about 867 million chronically undernourished people in the world. At present, more than 1.5 billion ha (about 12%) of the globe's land surface (13.4 billion ha) is or can be used for crop production,

28% (3.7 billion ha) is under forest, and 24–35% (4.6 billion ha) comprises grasslands and woodland ecosystems (FAO, 2011; 2013a). To accomplish the task of feeding almost 9 billion people, one of the most important strategies is closing the yield gap (Godfray et al., 2010), which requires good land use planning for the existent fields and also the upcoming ones. What has been observed is that farmers who seek higher yields (although with higher costs), promote a better use of natural resources (e.g., fertilizers, water, and soil) and consequently are responsible for a lower impact in the environment. In 2006 the World Conference for Structural Geospatial Data Base indicated the necessity to develop soil maps, which are basic information for crop management planning and consequently has the potential to improve food production.

If on one hand we need to use soils, on the other we are degrading them. Mahmood (1987) calculated by 1986 that around 1,100 km^3 of sediment had been lost from soils and accumulated in the world's reservoirs, consuming almost one fifth of the global soil storage capacity. Lal (2003) determined a rate of 201.1 Pg $year^{-1}$ of soil erosion. Indiscriminate use of herbicides and pesticides all over the world leads to soil and water pollution (Center for Food Safety, 2008; Singh and Ghoshal, 2010). Some herbicides, such as Glyphosate and Atrazine might reduce enzyme activity and populations of organisms in soil (Sannino and Gianfreda, 2001). Microbial community structure, often used as an indicator in monitoring soil quality, is affected by various environmental and plant growth factors, such as moisture, temperature, nutrient availability, and management practices (Petersen et al., 2002; Ratcliff et al., 2006).

Soils are relevant not only for food security or environmental quality, but also in climate change issues. In fact, according to Lang (2008) the atmospheric abundance of carbon dioxide (CO_2) has increased by 36%, from 280 ppm in 1750 to 381 ppm in 2006 (Canadell et al., 2007) and in 2014 reached 398.87 ppm (Dlugokencky and Pieter, 2014).

Land use change contributed 158 Pg C, where the deforestation and the attendant biomass burning, and soil tillage along with erosion, contributed with an estimated emission of 78 ±12 Pg C from world soils (Lal, 1999). Soils can sequester C by increasing stable SOC and soil inorganic C (SIC) stocks through judicious land use and recommended management practices. Research has shown a clear link between SOC (stock and change) and the type of soil, which stresses the need for soil maps. Bruinsma (2009) indicate that until 2050, "arable land would expand by some 70 million ha (or less then 5%), the expansion of land in developing countries by about 120 million ha (or 12%) being offset by a decline of some 50 million ha (or 8%) in the develop countries." In fact, recent work (Spera et al., 2014) indicates agriculture expansion in one of the most important states in Brazil, that is, Mato Grosso. Authors observed that 3.3 million hectares of mechanized agriculture in 2001, of which 500,000 hectares had two commercial crops per growing season (double cropping). By 2011, Mato Grosso had 5.8 million hectares of mechanized agriculture, of which 2.9 million hectares were double cropped, an increase of 76%. This is a clear indication of the necessity to have strategies on land use planning to reach a high quality use of soils.

Studies that address comprehension of soils, soil function, and soil in agriculture and society have several applications such as soil classification, pedological and attributes mapping, soil monitoring, soil conservation, experimental design, spatial allocation for agriculture, politics, food improvement, local sustainability, environment productivity systems, agriculture systems, specific plant and soil interactions, precision agriculture, irrigation systems, determination of land use capacity and land aptitude, and many others. Today, to improve agricultural productivity while securing future water quality (and quantity) as well with environment quality, it is necessary to understand soil function in natural and agricultural ecosystems that are tied together over landscapes, watersheds, and larger spatial extents. Nonetheless, soil management occurs at the 1 to 100-meter scale, while soil policy occurs at broader spatial scales (political boundaries). Hence knowledge of soils that includes detailed spatial information is imperative to understand the impact of management and policy decisions on soils, water, and the environment at large. In the following sections, we show how SS can assist in all these aspects.

15.4 THE ROLE OF SPECTRAL SENSING IN SOIL SCIENCE

15.4.1 Concepts

Remote Sensing (RS) has copious definitions in the literature. The most simple and effective one refers to the "acquisition of information about an object by detecting its reflected or emitted energy without being in direct physical contact with it" (Colwell, 1997; Jensen, 2006). Therefore, the term RS can be applied for a radiation detecting sensor installed at any platform or level of acquisition, for instance, laboratory, field, aerial, or orbital (Jensen, 2006). Recently, the expression Proximal Sensing (PS) was proposed for soil studies (Viscarra Rossel et al., 2011a). This concept focuses on detecting information about an object in a short distance, primarily in the field, using not only spectral sensors, but any measurement device. For spectral sensor applications, the concepts of PS and RS, although obvious in some contexts, can be confused. Generally speaking, the term "remote" means that the sensor is "far" from the target, while the term "proximal" means the sensor is "near" the target, but a specific distance that distinguishes one from the other (1 cm, 1 m, or 1 km) is not defined. In fact the word takes in consideration arbitrary "distance." Commonly (by convention or tradition) RS is applied mostly for orbital and aerial acquisition levels, whereas PS is used for laboratory and field sensors. In fact, as stated by Jensen (2006), all of these concepts are correct in their appropriate context. Thus, in this chapter we define Spectral Sensing as: 1) "Spectral": related to the electromagnetic radiation spectrum coming from an object which is dependent on its characteristics and composition; and 2) "Sensing": referring to the acquisition of this spectral information without directly touching the object. This definition removes the relativity from terms proximal and remote. The SS terminology can be divided into Space Spectral Sensing (SSS), which includes aerial and orbital (also known as RS), and Ground Spectral Sensing (GSS), divided into Field Spectral Sensing (FSS) when spectra are taken in natural conditions in the field, and Laboratory Spectral Sensing (LSS) when spectra are read in the lab using benchtop instruments. Ground spectral sensing is also known as proximal sensing (PS). Thus, by definition, SS can be applied using different acquisition distances (scales) and sources, from ground to space. This is our suggestion to better discuss the chapter and does not have the pretension to substitute the traditional terms remote sensing and proximal sensing.

15.4.2 How Does Spectral Sensing Contribute to Soil Science? Why Is It a New Perspective of Science?

The usual way to study soils is by discretizing the knowledge into basic disciplines such as physics, microbiology, pedology, and chemistry. Each discipline has accepted methodologies for analysis, which will allow them to understand and quantify soil properties. The interpretation is related to the results that are commonly obtained from traditional soil laboratory analysis involving chemical, physical and biological assessments. Spectral readings are physical information which, in many cases, requires models to relate to soil properties. The five perspectives (or points of view) to "look" at soils indicated in Figure 15.1 can also be observed by a Spectral Sensing point of view. In fact, these perspectives can be related with Figure 15.2. The micro-scale perspective, related with the soil analysis, can be related with the ground, laboratory, or field sensors. SS can study soils by the longitudinal perspective, where can *see* the relief from only one perspective, along a toposequence. The soil profile is related to the observation of a single point, which can be detected by ground sensing on surface, inside a pit or a borehole, or a pixel from an image. In the case of the surface landscape perspective, the perception of SS corresponds to the combination of all elements of the relief (shape, slope and height). Finally, the spatial perspective, which is the visualization of soil with its boundaries, can be detected from space. The pedologist combines different perspectives to analyze soils depending on the objective. Each one of these will add information about the soil.

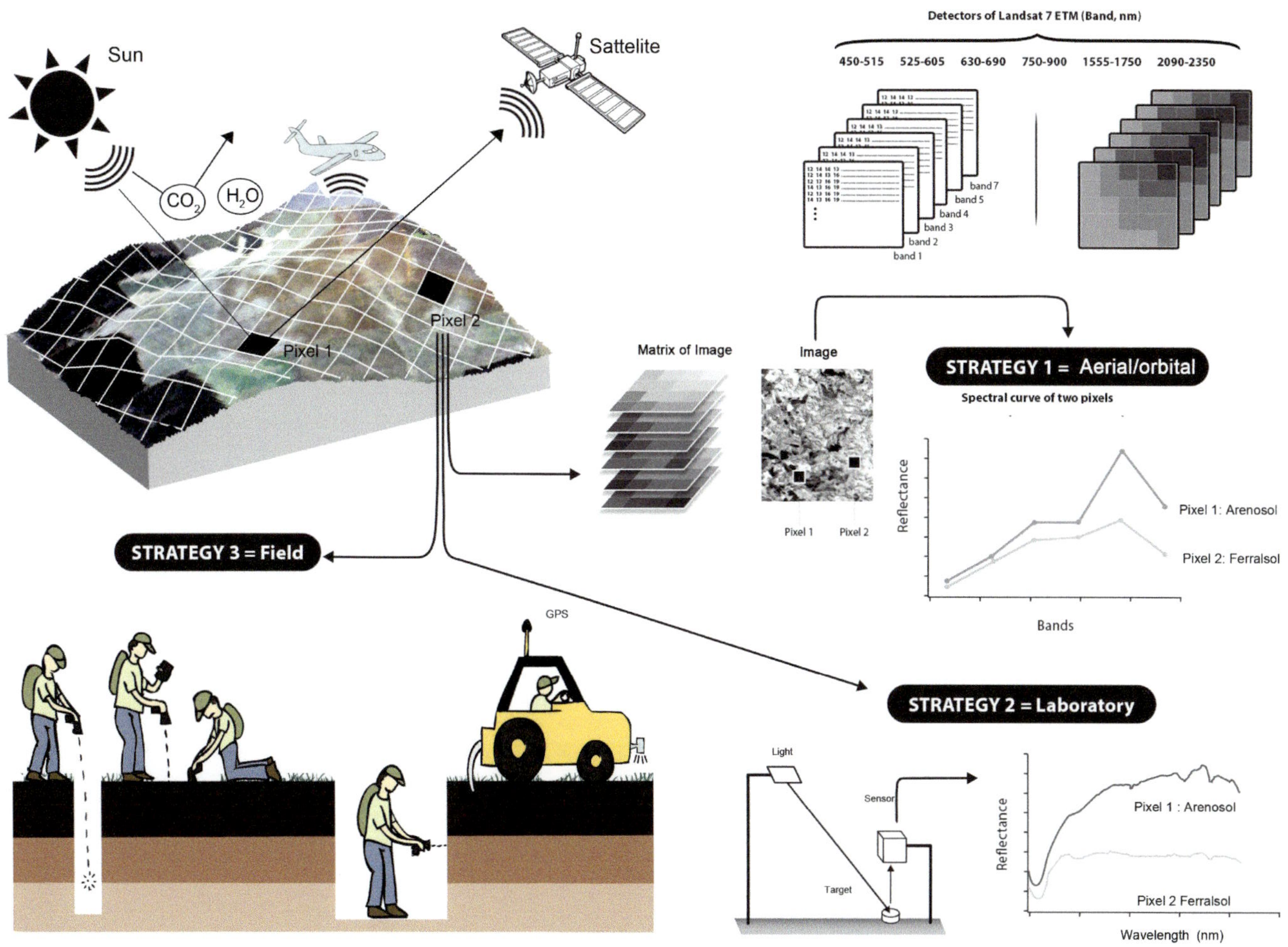

FIGURE 15.2 Alternatives for soil evaluation from ground to space: Strategy 1: Surface spectral curve from orbital source; Strategy 2: Ground (laboratory) spectral curve of a soil sample taken from the same pixel (location) of the image; Strategy 3: Ground spectral data in field (at the topsoil or subsoil measured inside a pit or with a sensor inside the hole to measure underneath).

15.4.3 History and Evolution

We can consider the beginning of SS when people saw and distinguished objects (in our case, soils) by their shape or color (using the eye). The NIR spectral range from 780 to 2,500 nm was discovered by Herschel in the year 1800 (Hershel, 1800). According to Stark et al. (1986), the NIR was discovered before the MIR region due to the instruments available at the time. But both spectral regions were neglected during years and only started their studies in the 1960s (Stark et al., 1986). Before this period, spectral sensing was applied using aerial photographs. In 1826, Joseph Nicéphore Niépce (1765–1833) took the first photography, in France (Gernsheim and Gersheim, 1952). In 1890, Arthur Batut published a book *Aerial Photography by Kites*, which shows the use of aerial photography for agronomy, archaeology, exploration and military uses (Batut, 1890). Aerial photographs for soil survey and applications were determined by Buringh (1960). Later, Vink (1964) published a very important work, which demonstrated how to use the aerial photographs for a soil survey, with examples given by Goosen (1967) and afterwards Hilwig et al. (1974).

In the early 1900s, Bernad Keen and William Haines built the first on-the-go soil sensor (Keen and Haines, 1925). Four decades later, Obukhov and Orlov (1964) selected the 750-nm spectral region to estimate SOC, and Bowers and Hanks (1965) published a paper about the correlation between soil reflectance, soil moisture and particle size. These pioneering studies were followed by Hunt and Salisbury (1970) and Hunt (1982), which proved that water and minerals in the soil have unique spectral fingerprints that can be identified, quantified and further used for the assessment of other soil properties. This took soil spectroscopy to another level, and the first scientists to systematically gather soil spectral information and publish it in the form of a soil spectral atlas were Stoner et al. (1980). These works guided spectroscopy for decades. In fact, findings from Vink (1964) and Stoner and Baumbgardner (1981) took Demattê et al. (2001, 2004c) to integrate and apply these technologies with the first soil mapping system using spectral information combined with aerial images.

The combination of different sensors and platforms for soil assessment started with Krinov (1947), who combined laboratory data, field spectrographs, and aerial spectrographs adapted to operate from aircraft, in the VIS-NIR region (400 to 910 nm) for soil research. In fact, maybe the most important event for SS happened in 1957, when the former USSR managed to keep the first satellite in Earth orbit (Sputnik). In the 1960s, a new era began for SS with new sensors on board satellites capable of taking pictures of Earth (McDonald, 1997). In 1965, the United States (conterminous, Alaska, and Hawaii) was mapped by the Corona satellite (Clark, 1999a). The first Landsat was launched in 1972, called Landsat-MSS (Later Landsat-1). In 1999, a new generation of multispectral satellite sensors was launched, including Landsat-7 ETM+, Terra MODIS and ASTER, CBERS IRS (Brazil and China), and IKONOS-2. In 2001, QuickBird was launched, in 2002 NASA launched the Aqua satellite carrying a MODIS sensor, and in 2005 Google started using data from satellites creating the Google Earth—a great upgrade on SS for humanity. Almost in parallel, important advances in hyperspectral sensing occurred, starting with an airborne platform—the AVIRIS sensor with 224 bands (400–2,500 nm)—operational in 1989 (Vane et al., 1993), followed by the spaceborne Hyperion sensor with 220 bands (400–2,500 nm) in the Earth-Observing 1 satellite, launched in 2000 (USGS, 2014).

15.5 THE THEORY BEHIND SOIL SPECTRAL SENSING

15.5.1 Visible, Near-Infrared, Shortwave-Infrared, and Mid-Infrared

The spectral ranges most used in soil science include the Ultraviolet (UV: 200–380 nm), Visible (VIS: 350–700), Near Infrared (NIR: 700–1,000 nm), Short-wave Infrared (SWIR: 1,000–2,500 nm) and Mid Infrared (MIR: 2,500–25,000 nm). The region 700–2,500 is generally referred to as NIR, whereas the MIR region at 2,500–25,000 can be divided into Thermal Infrared (TIR: 8,000–14,000 nm) and Far Infrared (14,000–25,000 nm). Fundamental studies on soil and spectral sensing in these ranges were conducted in the laboratory. Spectral reflectance is extremely complex, as it is affected by soil constituents that interact and mix to produce the spectrum, with different aspects of interactions between them. Among the factors that influence reflectance are the concentration, size and arrangement of the components that occur in soil samples, soil moisture, soil aggregation, soil texture and the roughness of the soil surface. Studies have reported that the smaller the soil particle size, the larger the surface area and the higher the reflectance (Ben-Dor et al., 2008; Bowers and Hanks, 1965; Kuang and Mouazen, 2013; Wu et al., 2021). The measured soil spectra result from the combination of the intrinsic spectral behavior of different soil constituents including organic materials, mineral materials and water, which interact differently with the incident light (Clark, 1999b). The percent of incident light that is reflected by the soil at different wavelengths constitutes the soil's reflectance behavior and is represented by the soil's spectral curve, or simply soil spectrum. Thus, soil spectra are a result of microscopic interactions between the atoms and molecules of the soil and the incident light, which penetrates the first 10 to 50 μm of soil. These interactions generate specific absorption features at different wavelengths and for different soils, and are mainly affected by soil particle size, porosity for energy dispersion, SOM, and soil water (Jensen, 2006). Therefore, each soil sample has a unique spectral behavior and unique spectral curve.

Microscopic interactions with light occur in mineral (siloxane, oxyhydroxy, silanol, aluminol, and ferrol surfaces) and organic (carboxyl and phenolic hydroxyl) functional groups of soils at the same place where other soil physical-chemical reactions take place as well (Alleoni et al., 2009). Thus, these interactions are dependent on soil composition with respect to types of atomic and molecular structures, strength of bonds among atoms and molecules, and ionic impurities in soils, such as chemical elements adsorbed on colloidal particles (Hollas, 2004). Spectral absorption features related to microscopic effects will only appear if enough energy is available to promote atomic transitions and/or molecular vibrations (Bernath, 2005). There are two levels of electromagnetic energy absorption that must be considered in soil microscopic interaction: atomic and molecular. Atomic refers to radiation required to produce changes in the energetic level of electrons linked to ions in the soil, causing electronic transitions and rearrangement of charges in atoms (Sherman and Waite, 1985). Atomic interactions usually take place from VIS to NIR, except for metals (Cu, Fe, Mn, Cr, and Ti) and other elements (Si, Al, K, Ca, and OH-) whose electronic transitions do not produce spectral features at this range.

Light interactions at the molecular level are responsible for the vibrational processes of molecules with its functional groups (e.g., -COOH, $-CH_x$, NH_x) in the soil and usually take place after 1,100 nm. Molecular vibrations need less energy, but produce more intense and defined absorption features (Stenberg et al., 2010). Molecular vibrations include stretching, bending and torsion, and are divided into fundamental and non-fundamental vibrations. Non-fundamental vibrations are considered secondary vibrations of lower intensity, such as propagations (or reverberations) of fundamental vibrations, which can be overtones or combination tones. They usually occur in the NIR as a result of interactions of radiation with molecular functional groups, for example, $-CO_x$, $-PO_x$, $-CH_x$, and H_2O (Viscarra Rossel et al., 2011b). Fundamental vibrations occur in the MIR and are usually derived from associations of Fe^{3+}, Al^{3+}, Mn^{2+}, and Si^{4+} with O in soil oxides (Viscarra Rossel et al., 2008). The vibrational absorptions in VIS-NIR are overtones and combination frequencies of absorption bands that occur in the MIR (Figure 15.2).

Table 15.1 summarizes the most important wavebands and respective soil attributes. Absorption features have been coming from interaction between electromagnetic energy and soil functional groups in colloidal surfaces. It has been observed a great variety of organic and inorganic functional groups, such as carboxyl, alkyl, methyl, phenolic, aliphatic groups, and others for organic ones, and, mainly, hydroxyl (-OH), silanol (Si]-OH) and aluminol (Al]-OH) for inorganic ones on outer surfaces of clay minerals (phyllosilicates) (Table 15.1). Overlaps of bands and positioning modifications can be observed because molecules of soil constituents are all mixed in a complex system and do not behave harmoniously (Bishop et al., 1994). Vis-NIR-SWIR spectral ranges have showed features related to electronic transition of iron from oxides (hematite) and hydroxides (goethite) in the visible and near infrared, and other features due to non-fundamental vibrations (overtones and combination tones) on molecular structures of 2:1 (vermiculite, smectite, illite, and others) and 1:1 (hematite) clay mineral and Al oxide (gibbsite) have been identified from shortwave infrared. Features due to soil organic compounds have also been identified along VIS-NIR-SWIR spectral ranges (Table 15.1). Due to the stronger microscopic interaction of MIR radiation and soil organic and mineral particles by fundamental vibrations (Janik et al., 1998), a greater number of absorption features can be observed in this range when compared to Vis-NIR-SWIR. This amount of features results in much more information about soil properties in MIR, except for iron contents which have not been identified in this range. Besides the 2:1 and 1:1 clay minerals, features of quartz (from sand particles), phosphates, carbonates, borates, nitrates, and a great variety of organic components has been identified along MIR (Table 15.1). Non-fundamental and fundamental absorptions from Vis-NIR-SWIR and MIR ranges were compiled from: Oinuma and Hayashi (1965), Bowers and Hanks (1965), Hunt and Salisbury (1970), White (1971), van der Marel and Beutelspacher (1975), Stoner and Baumgardner (1981), Baumgardner et al. (1985), Sherman and Waite (1985), White and Roth (1986), Nguyen et al. (1991), Salisbury and D'aria (1992a), Salisbury and D'aria (1994), Srasra et al. (1994), Janik et al. (1998), Fernandes et al. (2004), Madejova and Komadel (2001), Viscarra Rossel et al. (2006a), Stevens et al. (2008), Viscarra Rossel et al. (2008), Richter et al. (2009), and Viscarra Rossel and Behrens (2010).

One of the applications of MIR spectroscopy is that, as well as being a fast and non-invasive physicochemical method in the field, it can be used to detect and identify microorganisms. The region considered typical for identifying the fingerprints of microorganisms is between the wavelengths of 650 cm^{-1} (15385 nm) and 1800 cm^{-1} (5556 nm), coming from cellular compounds of carbohydrates and proteins. Fatty acids and cellular lipids show peaks between 2800 and 3000 cm^{-1} (3571 and 3333 nm). The best classifications are obtained using spectral differences in the 1500 to 1650 cm^{-1} (6667 to 6061 nm) regions (Ammann and Brandl, 2011; Naumann et al., 1991; Subramanian et al., 2007).

15.5.2 Microwaves

Microwaves constitute the portion of the electromagnetic radiation (spectrum) with wavelengths ranging from 0.1 to 100 cm (frequencies from 300 MHz to 300 GHz), and photon energy from 1.24 meV to 1.24 μeV. The Earth's surface emits low levels of microwave radiation which is modeled by the Rayleigh Jeans' law for black bodies (Jensen, 2006). The radiometric magnitude detected by the sensors in microwave spectral range is called brightness temperature and is expressed by $TB = \varepsilon T$, where T is an object's physical temperature and ε the emissivity. Passive microwave radiometers detect energy in the spectral range between 0.15 and 0.30 cm, with frequencies from 1 to 200 GHz. These sensors have poor spectral and spatial resolutions due to low microwave energy emitted from the Earth. Nevertheless, there is growing interest in the measurement of passive microwave radiation to monitor soil moisture conditions (Liu et al., 2012; Mladenova et al., 2014). Active microwave systems, also known as RADARs (RAdio Detection and Ranging), differ from passive systems because they provide an energy source. This energy (incident radiation) interacts with the target, and the response is detected by the sensor. Because they provide their own energy, applications are possible during the day and night. Moreover, differently from VIS-NIR-MIR, microwaves can pass through clouds and plant canopy (depending on the spectral range), and thus can be used under bad weather condition (Woodhouse, 2005) and for ground and understory examination.

15.5.3 Thermal Infrared

Any object with temperature greater than absolute zero (0 K, −273.26°C, −459.69°F) emits some amount of radiant energy, which is generally correlated to the object's temperature (Jensen, 2005). Energy emissions occur in the thermal infrared (TIR) range, but remotely sensed data

TABLE 15.1

Absorption Features (Band Assignments) of Soil Constituents in the VIS-NIR-SWIR (350 to 2,500 nm) and MIR (4,000 to 400 cm^{-1}) Spectral Ranges

Spectral Range	λ (nm)	Functional Groups	Soil Compounds	Microscopic Interactions	Interaction Modes	Types of Propagation
visible	404	Fe^{+3} ion	hematite	atomic level	electronic transition	
	430		goethite			
	444		hematite			
	480		goethite			
	520		hematite			
	650		goethite			
			hematite			
near infrared	751	N-H	organic compounds (amine)			$4\nu_7$ overtone
	825	C-H	organic compounds (aromatic carbon)			$2\nu_6$ overtone
	850	Fe^{+3} ion	hematite	atomic level	electronic transition	
	853	C-H	organic compounds (alkyl)			$4\nu_8$ overtone
	877					$4\nu_9$ overtone
	940	Fe^{+3} ion	goethite	atomic level	electronic transition	
	1000					
		N-H	organic compounds (amine)	molecular level	non-fundamental vibration	$3\nu_7$ overtone
shortwave infrared	1100	C-H	organic compounds (aromatic carbon)			$2\nu_6$ overtone
	1135	(H-O-H)+(O-H)	soil water			$\nu_1 + \nu_2 + \nu_3$ combination tone
	1138	C-H	organic compounds (alkyl)			$3\nu_8$ overtone
	1170					$3\nu_9$ overtone
	1380	(H-O-H)+(O-H)	soil water			$\nu_1 + \nu_3$ combination tone
	1395	2(O-H)	kaolinite			$2\nu_{1a}$ overtone
	1414					$2\nu_{1b}$ overtone
			2:1 vermiculite			$2\nu_4$ overtone
			smectite			$2\nu_4$ overtone
			mica (illite)			$2\nu_4$ overtone

(Continued)

TABLE 15.1 (*Continued*)

Absorption Features (Band Assignments) of Soil Constituents in the VIS-NIR-SWIR (350 to 2,500 nm) and MIR (4,000 to 400 cm^{-1}) Spectral Ranges

Spectral Range	λ (nm)	Functional Groups	Soil Compounds	Microscopic Interactions	Interaction Modes	Types of Propagation
	1449	C=O	organic compound (carboxylic acids)			$4v_{10}$ overtone
	1455	(H-O-H)+(O-H)	soil water			$2v_2 + v_3$ combination tone e overtone
			hygroscopic water			
	1500	N-H	organic compounds (amine)			$2v_7$ overtone
	1524	C=O	organic compounds (amide)			$4v_{11}$ overtone
	1650	C-H	organic compounds (aromatic carbon)			$2v_6$ overtone
	1706	C-H	organic compounds (alkyl)			$2v_8$ overtone
			organic compounds (aliphatics)			$4v_{12}$ overtone
	1730		organic compounds (methyl)			$4v_{13}$ overtone
	1754		organic compounds (alkyl)			$2v_9$ overtone
	1800	Ca, Mg, Fe, Mn, Sr, Ba—BO_3^{-2}	borate			$3v_5$ overtone
		Ca, Mg, Fe, Mn, Sr, Ba—CO_3^{-2}	carbonate			
	1852	C-H	organic compounds (methyl)			$4v_{13}$ overtone
	1915	(H-O-H)+(O-H)	soil water			$v_2 + v_3$ combination tone
			hygroscopic water			
	1930	C=O	organic compound (carboxylic acids)			$3v_{10}$ overtone
	1961	C-OH	organic compound (phenolic)			$4v_{14}$ overtone
	1980	Ca, Mg, Fe, Mn, Sr, Ba—BO_3^{-2}	borate			$3v_5$ overtone
		Ca, Mg, Fe, Mn, Sr, Ba—CO_3^{-2}	carbonate			
	2033	C=O	organic compound (amide)			$3v_{11}$ overtone

	2060	N-H	organic compounds (amine)			$v_7 + \delta_b$ combination tone
	2135	Ca, Mg, Fe, Mn, Sr, Ba—BO_3^{-2}	borate			$3v_5$ overtone
		Ca, Mg, Fe, Mn, Sr, Ba—CO_3^{-2}	carbonate			
	2137	C-O	organic compound (polysaccharides)			$4v_{15}$ overtone
	2160	(O-H)+(Al-OH)	kaolinite			$v_{1a} + \delta$ combination tone
	2205					$v_{1b} + \delta$ combination tone
			2:1 vermiculite			
			smectite			
			mica (illite)			
	2230	AlFe-OH	smectite			$v_{1b} + \delta_a$ combination tone
	2260	(O-H)+(Al-OH)	gibbsite			$v_{1b} + \delta$ combination tone
	2275	C-H	organic compounds (aliphatics)			$4v_{12}$ overtone
	2316	C-H	organic compounds (methyl)			$3v_5$ overtone
		$[3.(CO_3^{2-})]$	carbonate			
	2307	C-H	organic compounds (methyl)			$3v_{13}$ overtone
	2336	Ca, Mg, Fe, Mn, Sr, Ba—BO_3^{-2}	borate			$3v_5$ overtone
		Ca, Mg, Fe, Mn, Sr, Ba—CO_3^{-2}	carbonate			
	2350	(O-H)+(Al-OH)	mica (illite)			$v_{1b} + \delta$ combination tone
	2381	C-O	organic compounds (carbohydrate)			$4v_{16}$ overtone
	2382					$4v_5$ overtone
	2450	(O-H)+(Al-OH)	mica (illite)			$v_{1b} + \delta$ combination tone
	2469	C-H	organic compounds (methyl)			$3v_{13}$ overtone
Spectral range	Wavenumber (cm^{-1})	Functional groups	Soil compounds	Microscopic interactions	Interaction modes	Types of propagation
mid infrared	3695 (v_{1a})	(Al--O-H)OH	kaolinite	molecular level	fundamental vibration	stretching
			halloysite			
			smectite			
			mica (illite)			
	3670		kaolinite			
	3653					
			smectite			

(Continued)

TABLE 15.1 (*Continued*)

Absorption Features (Band Assignments) of Soil Constituents in the VIS-NIR-SWIR (350 to 2,500 nm) and MIR (4,000 to 400 cm^{-1}) Spectral Ranges

Spectral Range	λ (nm)	Functional Groups	Soil Compounds	Microscopic Interactions	Interaction Modes	Types of Propagation
			mica (illite)			
	3620 (v_{1b})		kaolinite			
			halloysite			
			chlorite Al-rich			
			smectite			
			mica (biotite)			
			mica (muscovite)			
			mica (illite)			
	3575 (v_4)	O-H	hydroxyl			
	3560	O-Al-OH	nontronite			
	3550		mica (biotite)			
			chlorite Al-rich			
	3529		2:1 vermiculite			
			kaolinite			
			gibbsite			
	3484 (v_1)	O-H	water			
	3448 (v_3)	PO_4^{3-}	phosphate			
		O-Al-OH	2:1 vermiculite			
			kaolinite			
			gibbsite			
	3400		halloysite			
	3394	(H-O-H)+(O-H)	2:1 vermiculite			
			kaolinite			
	3340		chlorite Al-rich			
	3330 (v^7)	N-H	organic compounds (amine)			
	3278 (v_1)	O-H	water			

3330–3030	N-H	ammonium (nitrate)			
3030 (v_6)	C-H	organic compounds (aromatic carbon)			
2930 (v_8)		organic compounds (alkyl)			
		asymmetric-symmetric doublet			
2924		organic compounds (aliphatic)			
2850 (v_9)		organic compounds (alkyl)			
		asymmetric-symmetric doublet			
2843		organic compounds (aliphatic)			
2341		CO_2 (breath)			
2233	(- C Ξ C—H,—C Ξ C—)	organic compounds (alkyne groups)	molecular level	fundamental vibration	stretching
	Si-O	quartz			
2133	(- C Ξ C—H,—C Ξ C—)	organic compounds (alkyne groups)			
	Si-O	quartz			
1975	Si-O	quartz			
1867					
1790					
1725 (v_{10})	C=O	organic compound (carboxylic acids)			
1678	Si-O	quartz			
1645 (v_2)	H-O-H	water			
1640	O-Al-OH	halloysite			
		smectite			
		nontronite			
		2:1 vermiculite			
1640 (v_{11})	C=O	organic compound (amides—protein)			

(Continued)

TABLE 15.1 (*Continued*)
Absorption Features (Band Assignments) of Soil Constituents in the VIS-NIR-SWIR (350 to 2,500 nm) and MIR (4,000 to 400 cm^{-1}) Spectral Ranges

Spectral Range	λ (nm)	Functional Groups	Soil Compounds	Microscopic Interactions	Interaction Modes	Types of Propagation
	1628	O-H	kaolinite			deformation
			smectite			
			mica (illite)			
			2:1 vermiculite			
		Si-O	quartz			stretching + deformation
	1610 (δ_b)	N-H	organic compound (amine)			
	1527	Si-O	quartz			
		COO-	organic compound (symmetric)			
		N-H + C=N	organic compound (amide)			
		C=C	organic compound (aromatic carbon)			
	1497 (v_4)	Si-O	quartz			
		COO-	organic compound (symmetric)			
		N-H + C=N	organic compound (amide)			
		C=C	organic compound (aromatic carbon)			
	1490–1410	C-O	carbonate			
	1485–1390	N-H	ammonium (nitrate)			
	1465 (v_{12})	C-H	organic compound (aliphatics)			
	1445 (v_{13})		organic compound (methyl)			
	1435	Ca—CO3–2	calcite			
	1415 (v_5)	Ca, Mg, Fe, Mn, Sr, Ba—BO3–2	borate			

		Ca, Mg, Fe, Mn, Sr, Ba—CO3–2	carbonate	
	1362 (v_4)	OH + C-O	organic compound (phenolic)	
		C-O	CH_2 and CH_3 groups (methyl)	
		COO- and -CH	organic compound (aliphatics)	
		Si-O	quartz	
	1350 (v_{13})	C-H	organic compounds (methyl)	
thermal infrared	1275 (v_{14})	C-OH	organic compounds (phenolic)	
	1170 (v_{15})	C-O	organic compounds (polysaccharide)	
	1157			
		C-OH	organic compounds (aliphatic)	
		O-Al-OH	smectite	deformation
			mica (illite)	
			2:1 vermiculite	
	1111	Si-O-Si	kaolinite	stretching
			halloysite	
			smectite	
			mica (illite)	
			mica (muscovite)	
			2:1 vermiculite	
			microcline	
			nontronite	deformation
		O-Al-OH	gibbsite	
	1100–1000	P-O	phosphate	
	1085–1050	C-O	carbonate	
	1050 (v_{16})	C-O	organic compound (carbohydrate)	

(Continued)

TABLE 15.1 (*Continued*)

Absorption Features (Band Assignments) of Soil Constituents in the VIS-NIR-SWIR (350 to 2,500 nm) and MIR (4,000 to 400 cm^{-1}) Spectral Ranges

Spectral Range	λ (nm)	Functional Groups	Soil Compounds	Microscopic Interactions	Interaction Modes	Types of Propagation
	1018 (v_5)	C-C	organic compound (aliphatics)			stretching
		C-O	organic compounds (polysaccharide)			
			organic compounds (carbohydrates)			
		Si-O-Si	kaolinite			
			smectite			
			mica (illite)			
			2:1 vermiculite			
		Si-O	quartz			
		Al-O	gibbsite			
	926 (δ)	Al--O-H	kaolinite			deformation
		((Al, Al)--O-H)	smectite			
			mica (illite)			
			2:1 vermiculite			
			gibbsite			
	885 (δ_a)	AlFe-OH	smectite			stretching
	875–860	C-O	carbonate			
	877	Ca—CO_3^{-2}	calcite			
	814	Si-O	quartz			
			kaolinite			deformation
		((Al, Fe)--O-H)	smectite			
		((Al, Mg)--O-H)	mica (illite)			
			2:1 vermiculite			
	791	Si-O-Si	kaolinite			stretching
			mica (illite)			

			2:1 vermiculite	
	752		kaolinite	
			mica (illite)	
			2:1 vermiculite	
mid infrared	750–680	C-O	carbonate	
	712	Ca—CO_3^{-2}	calcite	
	702	Si-O-Si	kaolinite	
			smectite	
			mica (illite)	
			2:1 vermiculite	
	635–5000	P-O	phosphate	
	517	Si-O-Si	kaolinite	deformation
			smectite	
			mica (illite)	
			2:1 vermiculite	
		Si-O	quartz	
	436		quartz	

λ: wavelength in nm; ν and δ: energy levels of fundamental vibration in microscopic interactions acquired aiming to quantify radiant flux are generally restricted to ranges near 8,000 to 9,000 nm or near 10,000 to 12,000 nm, since these ranges correspond to "atmospheric windows," in which water vapor (H_2O), carbon dioxide (CO_2) and ozone (O_3), among other components, absorb less energy.

Assuming the imaged surface as an ideal black body, thermal infrared radiance measured by the sensor would be proportional to the energy given by Planck's radiation law (Richards, 2013). However, real surfaces are not ideal black bodies, but instead, radiate selectively and emit a certain proportion of the energy that would be emitted by a black body, at the same temperature. The ratio between real radiance and black body radiance, at a given temperature, is the concept of emissivity. This is an intrinsic property of the material, and thus, the radiant flux that leaves a surface in thermal infrared spectroscopy is proportional to the material's emissivity and temperature (Derenne, 2003). In the case of soils, different factors can affect the spectral response, such as soil composition and moisture. Minerals such as quartz, present absorption features in the thermal infrared range, called reststrahlen spectral features, which occur from 8,000 to 10,000 nm (Salisbury and D'Aria, 1992a). However, an increase in SOM can reduce reststrahlen features, that is increase emissivity, especially in soils with more than 2% of organic matter (Breunig et al., 2008). In cases of clay coatings on quartz grains and increases in soil water content, the reststrahlen spectral features are attenuated as well (Salisbury and D'Aria, 1992a).

15.5.4 Gamma-Ray

Gamma ray has initiated studies in rocks, such as performed by Gregory and Horwood (1961) and can be better understood in Grasty (1979). The electromagnetic energy in the γ-ray range is characterized by short wavelengths (on the order of 0.01 nm or less), with high frequency, and high energy photons (more than 0.04 MeV). Ground or airborne γ-ray spectrometry uses the fact that γ-ray photons have discrete energies, which are characteristic of radioactive isotopes from which they originate (passive system). Consequently it is possible to determine the source of the radiation by measuring the energies of γ-ray photons (IAEA, 2003).

Radioactive isotopes of elements that emit gamma radiation are called radionuclides. Many naturally occurring elements have radionuclides, but only potassium (^{40}K), cesium (^{137}Cs), and the decay series of uranium (^{238}U and ^{235}U and their daughters) and thorium (^{232}Th and its daughters) are abundant in the environment, and produce γ-rays of sufficient energy and intensity to be measured by γ-ray spectrometry (Mahmood et al., 2013).

An energy spectrum is typically measured by γ-ray spectrometers over the 0 to 3 MeV range, generally in 256 channels or more (Dierke and Werban, 2013). In the conventional approach, γ-ray measurements are used to monitor four spectral regions of interest, corresponding to the energy levels of K (1.460 MeV), U (1.765 MeV) and Th (2.614 MeV), and to the total radioactivity over the 0.4 to 2.81 MeV range (Viscarra Rossel et al., 2007a). The γ-ray measurements taken by spectrometers are counts of the decay rate (intensity) at the specific energy. The measured intensity (counts per second) can be converted into the activity of the nuclides (Bq/kg), the concentrations (% for K, ppm for U and Th) or into the dose rate (nGy/h) using a calibration method (IAEA, 2003).

The concentration of γ-ray emitting radionuclides in soils depends on different soil properties, which are the result of physical and chemical composition of parent rock, as well as soil genesis under different climatic conditions (Dierke and Werban, 2013). Mineralogy of source rocks, clay content, and type of clay minerals are properties which directly influence the radionuclides concentration in sediments and soils (Serra, 1984). When measured from the top, approximately 50% of the observed γ-rays originate from the top 0.10 m of dry soil, 90% from the top 0.30 m, and 95% from the upper 0.5 m of the profile (Taylor et al., 2002). Soil water content and bulk density can attenuate γ-rays. For example, radiation attenuation increases by approximately 1% for each 1% increase in volumetric water content; while a dry soil with a bulk density of 1.6 Mg m^{-3} causes a decrease in the radiation to half its value at each 10 cm compared to a moist soil (Cook et al., 1996). The radiation decrease caused by air is much smaller, for example, 121 m are need to reduce the radiation to half its value considering a 2 MeV source, hence detection of γ-rays is possible from airborne platforms (Viscarra Rossel et al., 2007a).

15.6 STRATEGIES FOR SOIL EVALUATION BY SPECTRAL SENSING

Strategy is the art of applying knowledge and capabilities to reach a goal. In the case of spectral sensing, an effective strategy requires knowledge of the limitations and capabilities of the instrument making measurements. The goal is detecting and identifying soil properties for the purpose of making an inference regarding the capability of the soil for a specific function. Thus, there are several strategies to use on soil evaluation based on SS, but each strategy is different and should depend on the objectives and tools available. The strategies described here are related to research with the missions of creating, determining, and investigating methods for mapping soil attributes. In fact, quantification of SOM, SOC, pH, clay, sand and other soil constituents for the purpose of converting to other soil properties using pedotransfer functions and ultimately making a soil inference is the goal of SS, and a challenging one. Strategies may focus on Ground SS (GSS) or Space SS (SSS) and the spatial modeling of the soil constituents can be complemented by terrain models (i.e., Digital Elevation Models—DEMs) and other sources of information (Figure 15.2). As we observe in Figure 15.2, there are several strategies to reach soils information. Strategy 1 implies on using aerial or orbital data. Both will only get soil surface information since the energy penetrates a few centimeters into the soil. Orbital information has to be well atmospheric corrected and transformed into reflectance (as will be seen in subsequent sections). On the other hand, when properly detected an area with bare soil, the spectra of the pixel can have a great relationship with a soil type. A case study indicated in Figure 15.2, illustrate the location of a pixel with an arenosol and another with an ferralsol, reaching spectra completely different. Thus, the importance of this strategy is to detect boundaries and spots of a specific soil type (related only with the surface information) and/or a specific soil attribute (i.e., clay or iron content). If this strategy has the limitation to detect soils in depth, we can change to strategy 2. In this case, soil samples can be collected in field and prepared in laboratory in a controlled condition. As we can see in Figure 15.2, the sample collected in the center of the same pixel (surface sample) has similar spectral intensities. This is the basis of the importance of laboratory spectra information as a pattern to understand aerial or orbital data. Despite this, strategy 2 allows the user to collect soil samples in different depths and determine spectra from horizons for future evaluation. Thus, it is possible to make the relationship between the soil surface information obtained in strategy 1 with the soil surface and subsurface of strategy 2, thus making a link between these pieces of information. Although the laboratory strategy has inconveniences, since we have to go to field and collect samples. In many cases, soil scientists need the information in situ. In this case, we have strategy 3, which allows the user to get soil information *inside* a hole to reach underneath information. Another method is to get the surface information. In this case we can use two methods: with the fiber and with the contact probe. The first use the natural light of the Sun and has as inconvenience the alteration of radiation intensity and the loose of water bands. The use of the contact probe does not have these inconveniences, since has a own source. On the other hand, we have to take more care with the contact probe in relation with contamination since the equipment gets into contact with each sample. Still in strategy three, to read spectra undersurface of soils, we can use the probe inside a hole, or direct inside a pit. Another important approach are on-the-go sensors in tractors or vehicles, where can use sensors inside the soil as made by implements or sensors that measures the surface of the soils. Thus, Figure 15.3 demonstrates that there are several situations which sensors can assist soil evaluation, all of them with advantages and limitations and can be used as the necessities of the user. The following sections will show works that use these strategies for several approaches.

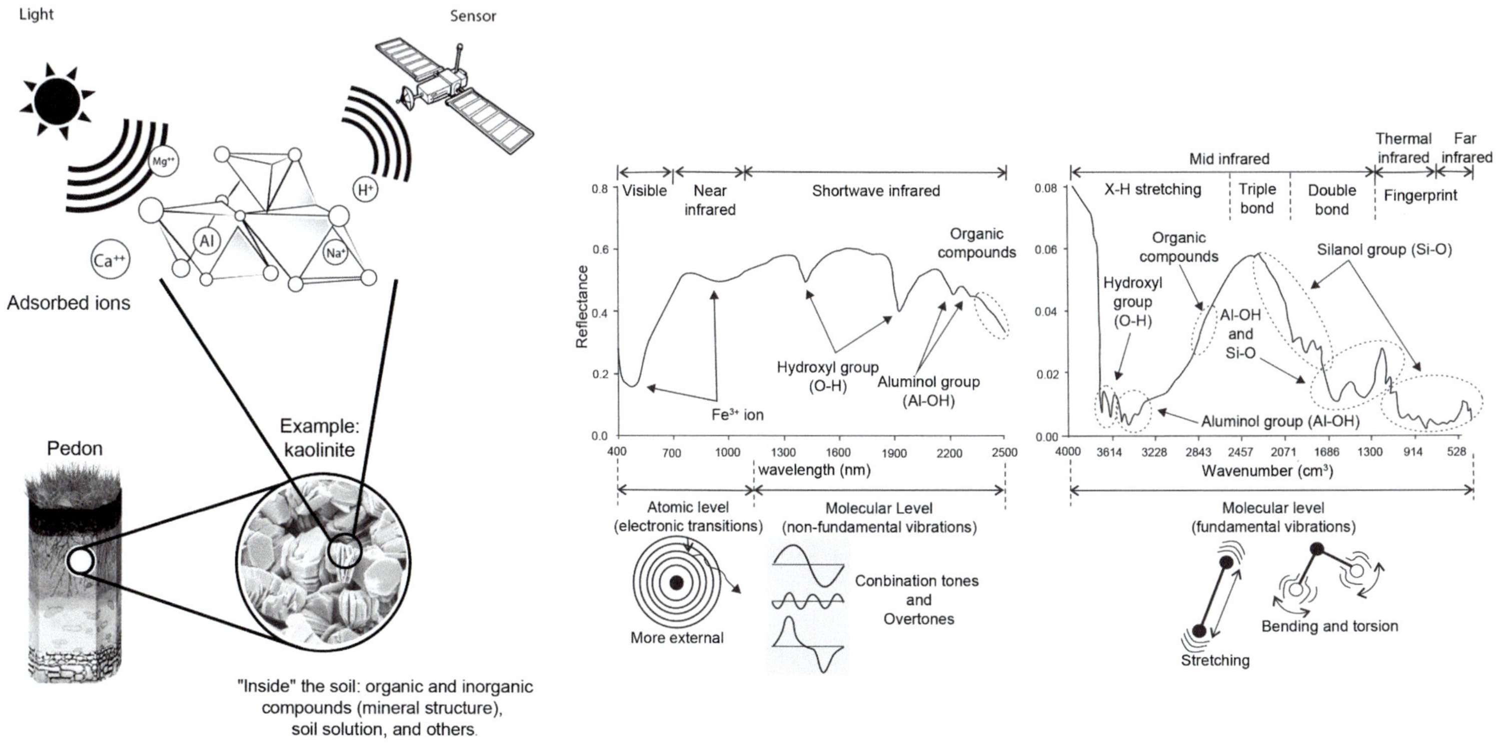

FIGURE 15.3 Illustration of soil spectral curve showing fundamental absorption features of selected functional groups in the MIR (2,500–25,000 nm), and overtones and combination tones in the visible and near infrared (400–2,500 nm).

15.6.1 Strategies for Soil Sampling

A key set of questions when developing a SS strategy involves deciding where and how to collect soil samples for calibration of an instrument or validation of a derived map. Regardless of the answers to the where and how questions, it must be recognized that collecting soil samples in the field is expensive and time consuming. Once collected, the samples still must be prepared and analyzed in the laboratory, which also has costs. Hence, many times a key question is: How can the number of samples reduce? To collect a minimum of representative samples, probably the most used type of sampling is stratified sampling, when landscapes are divided into homogeneous sectors, and each sector is sampled. The usual approach is to collect in each sector a number of samples proportional to the area of the sector. Another common approach is to define the number of samples proportional to the complexity of the sector, with more heterogeneous sectors receiving more samples.

A simple and effective form of stratification in based on the theory of soil change along a toposequence, and thus, a toposequence (meaning positions along a slope) can be used as the spatial stratifier. In general, the concept is summarized as follows: (1) get images from the area; (2) correlate them with an elevation map; (3) mark the places to collect soils from top to bottom, using the topossequence system. With several toposequences (transects along slopes), the most different soils in the area can be likely identified. Other types of data can be used as strata, particularly SS data plotted across a landscape.

A more automated stratification system that can use multiple SS and terrain data is the conditioned Latin Hypercube (cLHS) algorithm (Minasny and Mcbratney, 2006), which, according to the authors, is a stratified random procedure that provides an efficient way of sampling variables from their multivariate distributions. In fact, Mulder et al. (2013) used the cLHS to assess the variation of soil properties at a regional scale by using the first three principal components of an ASTER satellite image and a DEM. According to them, the sampling approach was successful in representing major soil variation. The cLHS algorithm has been used to map multiple soil properties and included GSS data such as electrical conductivity and γ-ray data (Viscarra Rossel et al., 2007a; Adamchuk et al., 2011). In one case, Viscarra Rossel et al. (2008) reduced 1,878 samples with MIR information to 213 samples chosen for laboratorial analysis using cLHS. Other sampling methods have been proposed, and they all stratify the area based on auxiliary information. A method by Simbahan and Dobermann (2006) uses many variables (soil series, relative elevation, slope, electrical conductivity and soil surface reflectance) for sample optimization. A Variance Quadtree Algorithm successively divides the study area into strata and each stratum has a similar variation (Minasny et al., 2007; Rongjiang et al., 2012).

15.6.2 Strategies for Soil Attribute Prediction and Mapping

Soil analysis is the basic source of data for many applications, notably agriculture and environmental monitoring. The conventional laboratory analysis based on wet chemistry is the most important source of this information with more than 100 years of knowledge. On the other hand, the need for faster information with environmental care has taken research to find other technologies. Since many soil attributes have strong relationship with spectra, their quantification using chemometric techniques constitutes one of the main goals of spectral sensing analysis. Since the pioneering work of Schreier (1977), who made quantitative measurements of soil attributes from ground to space sensors, this line of research has evolved to more advanced methods including the NIRA method performed by Ben-Dor and Banin (1995a, 1995b). However, SS still relies on wet chemistry analysis as the basic source of data and greatly benefits from existing databanks. Notably, at the beginning of this decade, soil attribute prediction using SS has been more important and many papers were published on the quantification of several soil elements by SS, including N, Ca, Mg, K, P, Na, pH, clay, sand, silt, mineralogy, organic matter, carbon, carbonates, CEC, micronutrients, Fe_2O_3, Al_2O_3, heavy metals and others, with prediction quality varying for different attributes. History shows how we are going forward on this task: Scherier (1977) used simple regression for each band; which later evolved to the use of band depth (Clark et al., 1990)—although with AVIRIS data; then we reached Coleman et al. (1991) using a multivariable analysis with selected bands; afterwards using the Reflexion Inflexion Difference (RID) (Nanni and Demattê, 2006a); until the use of all spectra with PLS method (Viscarra Rossel et al., 2006). Today, Vohland et al. (2014) raises the discussion about the effects of selected variables on the quantification of soil attributes. Thus, papers indicate that is still needed discussion on this subject to reach the best results, and are discussed in this section as summarized in Tables 15.2 (ground data) and 15.3 (aerial and space data).

15.6.2.1 Strategies for Ground Spectral Sensing

Soil properties assessment using spectroscopic approaches in UV-VIS-NIR-SWIR-MIR ranges (from 100 to 25,000 nm) has gained in importance in the last 20 years (Ben-Dor, 2011). Soil properties with direct relation to the spectral signature have better prediction performances, especially: SOC, calcium carbonate, clay content, clay minerology, Fe and Al oxides, and water content (Brown, 2007; Waiser et al. 2007; Morgan et al., 2009; Reeves III, 2010). Other soil chemical attributes, such as, pH, Ca, CEC, N (and all its forms) and Mg, have been predicted with variable accuracy. These properties are not directly related to the soil spectral response, that is, they do not cause absorption features, and their prediction depend on good correlations with spectrally active properties (Kuang et al., 2012).

The correlation between spectra and attributes is generally found using an array of mathematical and statistical procedures. Early in the use of spectroscopy for quantifying soil constituents, linear methods, such as stepwise multiple linear regression (MLR), principal components regression (PCR) and partial least squares regression (PLSR) were used. Principal components and PLSR can deal with a large number of predictor variables that are highly collinear (this is the case of spectral curves) (Varmuza and Filzmoser, 2009). The PLSR provides better results than PCR, and is by far the most used method for predicting soil properties; however, non-normal distributions of soil properties, for example, SOC, hinder the use of linear models like stepwise MLR and PLSR (Gobrecht et al., 2014). Nonlinear modeling methods offer an alternative in these cases, and include support vector machines (SVM), regression trees (RT) and its derivations, and artificial neural networks (ANN). They are sometimes more complicated to interpret, but can deal with any non-normal distributions, collinearities and missing data better than linear methods (Brown et al., 2006; Viscarra Rossel and Beherns, 2010; Gobrecht et al., 2014).

Usually, spectral preprocessing methods are applied to correct nonlinearities, measurement and sample variations, and for noise reduction in spectra (Stenberg and Viscarra Rossel, 2010). Examples of these approaches are transformation from reflectance (R) to log 1/R (absorbance), and the Kubelka-Munk transformation. Wavelet transformations have been used to reduce the number of variables in a multiple regression model and clarify interpretation of predictors, performing as well as PLSR (Ge et al., 2007). Other pre-treatments are suggested to mitigate scattering and path length variation between samples such as multiplicative scatter correction (MSC) (Geladi et al., 1985), standard normal variate (SNV) transformation (Barnes et al., 1989), smoothing, and first and second derivatives of spectra (Stenberg and Viscarra Rossel, 2010), including Savitzky-Golay transformation (Savitzky and Golay, 1964).

Because of its importance to the soil condition, hydrologic cycle (infiltration and water retention), and global C cycle, interest in estimating SOC with spectral sensing is very high. Bellon-Maurel and McBratney (2011) reviewed the spectroscopic prediction of SOC, and its use in carbon stock evaluation; they concluded that MIR had better results than VIS-NIR. The difficulty is that MIR requires laborious sample preparation (soil ground and sieved at a 200 mesh, and sometimes diluted in KBr) and is not practical for field measurements (Reeves III, 2010). Some studies report predictions of SOC using spectroscopic techniques with rather large R^2 (> 0.90) (Stenberg et al., 2010; Kuang et al., 2012; McDowell et al., 2012). Gmur et al. (2012) characterized soil attributes by spectral

TABLE 15.2

Summary of Revision Related with Quantification of Soil Attributes by Ground Spectral Sensing

Element*	Spectral range**	Element range***	N. cal.	Cal. Scale	R^2 range	R^2 median	R^2 N	RMSEP/RMSECV, *SEP/SECV* (%) range	RMSEP/RMSECV, *SEP/SECV* (%) median	RMSEP/RMSECV, *SEP/SECV* (%) N	Authors
A. Lab dried											
Clay (%)	UV-vis-NIR	70–90	121–207	R/N	0.61–0.72	0.67	2	12.3, *8.9*	12.3, *8.9*	2	(Islam et al., 2003; Pirie et al., 2005)
	Vis-NIR	70->90	457–4184	G	0.73–0.78	0.77	3	7.5–12	9.5	3	(Brown et al., 2006; Ramirez-Lopez et al., 2013; Shepherd and Walsh, 2002)
	Vis-NIR	20->90	30->1000	R/N	0.02–0.91	0.78	22	2–8.5, *4.9–13.7*	5.3, *7.8*	26	(Demattê and Garcia, 1999; Nanni and Demattê, 2006; Bricklemyer and Brown, 2010; Chang et al., 2005; Curcio et al., 2013; Genot et al., 2011; Gogé et al., 2014; Gogé et al., 2012; Islam et al., 2003; Kinoshita et al., 2012; Knadel et al., 2013; Ramirez-Lopez et al., 2013; Ramirez-Lopez et al., 2014; Sankey et al., 2008; Stenberg, 2010; Summers et al., 2011; Thomsen et al., 2009; Waiser et al., 2007; Vendrame et al., 2012; Viscarra Rossel and Behrens, 2010; Viscarra Rossel et al., 2009; Viscarra Rossel and Lark, 2009; Viscarra Rossel and Webster, 2012)
	Vis-NIR	<5–80	16->100	F	0.39–0.82	0.70	11	0.36–6.3	1.8	11	(Debaene et al., 2014; Mahmood et al., 2012; McCarty and Reeves, 2006; Ramirez-Lopez et al., 2014; Wetterlind and Stenberg, 2010)
	NIR	30–90	35->400	R/N	0.50–0.94	0.71	10	3.9–16, *5–10.3*	5.3, *8.6*	11	(Ben-Dor and Banin, 1995; Chang et al., 2001; Islam et al., 2003; Malley et al., 2000; Stenberg et al., 2002; Wang et al., 2013b; Waruru et al., 2014)
	NIR	20–50	20–52	F	0.47–0.86	0.63	6	1.9–5.7	3.1	6	(Igne et al., 2010; Sudduth et al., 2010; Wetterlind et al., 2008; Viscarra Rossel et al., 2006)

(Continued)

TABLE 15.2 (*Continued*)
Summary of Revision Related with Quantification of Soil Attributes by Ground Spectral Sensing

Element*	Spectral range**	Element range***	N. cal.	Cal. Scale	R^2 range	R^2 median	R^2 N	RMSEP/RMSECV, *SEP/SECV* (%) range	RMSEP/RMSECV, *SEP/SECV* (%) median	RMSEP/RMSECV, *SEP/SECV* (%) N	Authors
	MIR	30->90	60–663	R/N	0.55–0.94	0.86	11	3.7–10.5	6.6	8	(D'Acqui et al., 2010; Ge et al., 2014b; Janik et al., 2009; Janik and Skjemstad, 1995; Minasny and McBratney, 2008; Minasny et al., 2008; Minasny et al., 2009; Pirie et al., 2005; Viscarra Rossel and Lark, 2009)
	MIR	20	116–209	F	0.67–0.77	0.70	3	1.6-~2	1.7	3	(Igne et al., 2010; McCarty and Reeves, 2006; Viscarra Rossel et al., 2006)
CEC (cmol kg^{-1})	UV-vis-NIR	13–97	49–205	R/N	0.47–0.64	0.52	3	1.3–5.8, *4.3*	3.5, *4.3*	3	(Islam et al., 2003; Lu et al., 2013; Pirie et al., 2005)
	Vis-NIR	55–165	740–4184	G	0.74–0.88	0.81	2	3.8–6.7	5.3	2	(Brown et al., 2006; Shepherd and Walsh, 2002)
	Vis-NIR	14–92	94->2000	R/N	0.68–0.91	0.81	7	1–7.8, *3.9*	2.1, *3.9*	8	(Chang et al., 2001; Dunn et al., 2002; Genú et al., 2011; Gogé et al., 2014; Islam et al., 2003; Kinoshita et al., 2012; Vendrame et al., 2012; Viscarra Rossel and Webster, 2012)
	Vis-NIR	16	50–299	F	0.73–0.83	0.78	2	1.4	1.4	3	(van Groenigen et al., 2003; Nanni and Demattê, 2006; Sudduth et al., 2010;)
	NIR	29–91	35–1100	R/N	0.64–0.73	0.69	4	9.6, *3.3–8.5*	9.6, *4.1*	4	(Bendor and Banin, 1995; Genot et al., 2011; Islam et al., 2003; Waruru et al., 2014)
	NIR	2–7	49	F	0.13	0.13	1	1.04	1.04	1	(Viscarra Rossel et al., 2006)
	MIR	36–94	60–662	R/N	0.77–0.92	0.86	5	2–4.8	4.6	3	(D'Acqui et al., 2010; Janik et al., 2009; Minasny and McBratney, 2008; Minasny et al., 2008; Minasny et al., 2009; Pirie et al., 2005)
	MIR	5–16	50	F	0.34–0.56	0.45	2	0.9–2.1	1.5	2	(van Groenigen et al., 2003; Viscarra Rossel et al., 2006)

OC (%)	UV-Vis-NIR	1–5	49–207	R/N	0.76–0.83	0.76	3	0.12–0.5, *0.44*	0.31, *0.44*	3	(Islam et al., 2003; Lu et al., 2013; Pirie et al., 2005)
	Vis-NIR	5–50	250->12000	G	0.68–0.82	0.79	10	0.03–5.11	0.90	14	(Brown et al., 2006; Nocita et al., 2014; Ramirez-Lopez et al., 2013; Shepherd and Walsh, 2002; Stevens et al., 2013)
	Vis-NIR	2–33	<50->3000	R/N	0.31–0.96	0.84	29	0.06–0.88, *0.28–2.94*	0.35, *0.72*	27	(Bricklemyer and Brown, 2010; Chang and Laird, 2002; Conforti et al., 2013; Daniel et al., 2003; Deng et al., 2013; Doetterl et al., 2013; Dunn et al., 2002; Fystro, 2002; Gogé et al., 2014; Gogé et al., 2012; Kinoshita et al., 2012; Knadel et al., 2013; Liu and Liu, 2013; Moron and Cozzolino, 2002; Nocita et al., 2011; Ramirez-Lopez et al., 2013; Sankey et al., 2008; Stenberg, 2010; Summers et al., 2011; Tian et al., 2013; Wang et al., 2013a; Viscarra Rossel and Behrens, 2010; Viscarra Rossel and Lark, 2009; Vohland and Emmerling, 2011; Vohland et al., 2014)
	Vis-NIR	<1–14	25–287	F	0.12–0.99	0.75	15	0.07–1.6, *0.33*	0.3, *0.33*	20	(Debaene et al., 2014; Fontan et al., 2011; He et al., 2007; Heinze et al., 2013; Kuang and Mouazen, 2011; Kuang and Mouazen, 2013b; McCarty and Reeves, 2006; Wetterlind and Stenberg, 2010; Yang et al., 2012)
	NIR	2–60	39->2000	R/N	0.45–0.99	0.83	21	0.14–0.83, *0.16–2.95*	0.32, *0.61*	21	(Bendor and Banin, 1995; Cambule et al., 2012; Chen et al., 2011; Dalal and Henry, 1986; Dunn et al., 2002; Fidêncio et al., 2002; Genot et al., 2011; Islam et al., 2003; Malley et al., 2000; McCarty et al., 2010; McCarty et al., 2002; Miltz and Don, 2012; Nocita et al., 2011; Rabenarivo et al., 2013; Reeves et al., 2006; Stenberg et al., 2002; Todorova et al., 2009; Zornoza et al., 2008)
	NIR	1–10	20–299	F	0.34–0.95	0.93	7	0.1–0.77	0.18	8	(Reeves and McCarty, 2001; Viscarra Rossel et al., 2006) (Guerrero et al., 2014; Sudduth et al., 2010; Wetterlind et al., 2008; Xie et al., 2011)
	MIR	3–8	87–1545	G	0.93–0.95	0.94	2	0.02–0.25	0.14	2	(Bornemann et al., 2008; Kamau-Rewe et al., 2011)

(Continued)

TABLE 15.2 *(Continued)*

Summary of Revision Related with Quantification of Soil Attributes by Ground Spectral Sensing

Element*	Spectral range**	Element range***	N. cal.	Cal. Scale	R^2 range	R^2 median	R^2 N	RMSEP/RMSECV, *SEP/SECV* (%) range	RMSEP/RMSECV, *SEP/SECV* (%) median	RMSEP/RMSECV, *SEP/SECV* (%) N	Authors
	MIR	2–30	31–560	R/N	0.77–0.98	0.93	14	0.11–0.61	0.35	13	(D'Acqui et al., 2010; Ge et al., 2014b; Grinand et al., 2012; Janik and Skjemstad, 1995; Masserschmidt et al., 1999; McCarty et al., 2010; McCarty et al., 2002; Minasny et al., 2009; Pirie et al., 2005; Rabenarivo et al., 2013; Reeves et al., 2006; Viscarra Rossel and Lark, 2009; Vohland et al., 2014)
	MIR	1–3	118–217	F	0.73–0.96	0.95	3	0.11–0.15, *0.67*	0.14, *0.67*	3	(McCarty and Reeves, 2006; Viscarra Rossel et al., 2006; Xie et al., 2011)
TC (%)	Vis-NIR	2–55	30->400	R/N	0.66–0.95	0.88	6	0.02–2.8, *0.3*	0.54, *0.3*	6	(Chang and Laird, 2002; Chang et al., 2005; McDowell et al., 2012; Sørensen and Dalsgaard, 2005; Thomsen et al., 2009; Vendrame et al., 2012)
	Vis-NIR	1–17	16–429	F	0.01–0.93	0.76	11	0.16–1.1, *0.1*	0.19, *0.1*	10	(Fontan et al., 2011; Kuang and Mouazen, 2011; Mahmood et al., 2012; van Groenigen et al., 2003; Wetterlind et al., 2010; Yang et al., 2012)
	NIR	4–5	83–91	G	0.85–0.94	0.90	2	*0.35–0.41*	*0.38*	2	(Barthes et al., 2006; Brunet et al., 2007)
	NIR	10–28	150->700	R/N	0.85–0.87	0.86	3	0.54–0.79	0.71	3	(Chang et al., 2001; McCarty et al., 2002; Reeves et al., 2006)
	NIR	1	209	F	~0.65	~0.65	1	~0.16	~0.16	1	(Igne et al., 2010)
	MIR	10–50	56–660	R/N	0.79–0.97	0.95	7	0.24–2.28	0.34	7	(Janik et al., 2009; Ludwig et al., 2008; McDowell et al., 2012; Minasny et al., 2008; Minasny et al., 2009; Reeves et al., 2006) (McCarty et al., 2002; Minasny and McBratney, 2008)
	MIR	1	50–209	F	0.01-~0.85	0.43	2	~0.1	~0.1	1	(Igne et al., 2010; van Groenigen et al., 2003)
IC (%)	Vis-NIR	13	4184	G	0.83	0.83	1	0.62	0.62	1	(Brown et al., 2006)
	Vis-NIR	3–4	76–86	R/N	0.96–0.98	0.97	2	0.15–0.19	0.17	2	(Chang and Laird, 2002; Chang et al., 2005)

	Vis-NIR	1–2	90–492	F	0.53–0.95	0.74	2	0.08, *0.15*	0.08, *0.15*	2	(Fontan et al., 2011; Yang et al., 2012)
	NIR	7	177	R/N	0.87	0.87	1	0.31	0.31	1	(McCarty et al., 2002)
	MIR	7–10	60–418	R/N	0.82–0.97	0.96	6	0.02–0.42, *0.28*	0.12, *0.28*	6	(D'Acqui et al., 2010; Ge et al., 2014b; Grinand et al., 2012; McCarty et al., 2002; Reeves et al., 2006)
B. Lab—field moist											
Clay (%)	Vis-NIR	40–90	187–2000	R/N	0.76–0.77	0.77	2	5.25	9	2	(Chang et al., 2005; Ge et al., 2014a)
	MIR	20	209	F	~0.7	~0.7	1	~2	~2	1	(Igne et al., 2010)
OC (%)	Vis-NIR	3–9	75–2000	R/N	0.53–0.94	0.86	6	0.38–0.73, *0.15–0.67*	0.68, *0.41*	6	(Chang et al., 2005; Fystro, 2002; Ge et al., 2014a; Mouazen et al., 2010; Mouazen et al., 2007; Terhoeven-Urselmans et al., 2008)
	Vis-NIR	1–7	47–104	F	0.58–0.82	0.72	4	0.2–1.7	1.23	9	(Kuang and Mouazen, 2012; Kuang and Mouazen, 2013b)
TC (%)	Vis-NIR	4	162	R/N	0.85	0.85	1	0.42	0.42	1	(Chang et al., 2005)
	MIR	1	209	F	~0.6	~0.6	1	~0.15	~0.15	1	(Igne et al., 2010)
C. In field (including at site and on-the-go measurements)											
Clay (%)	Vis-NIR	40–60	311->1000	R/N	0.17–0.83	0.78	3	6.1–7.9, *9.0*	7.0, *9.0*	3	(Bricklemyer and Brown, 2010; Waiser et al., 2007; Viscarra Rossel et al., 2009)
	Vis-NIR	20	209	F	~0.7	~0.7	1	~2	~2	1	(Igne et al., 2010)
	Gamma	30–40	13–660	R/N	0.86	0.86	1	4–12	6	5	(Petersen et al., 2012; van der Klooster et al., 2011)
	Gamma	<5–40	7–70	F	0.63–0.94	0.72	8	0.81–6.56	3.1	11	(Mahmood et al., 2013; Piikki et al., 2013; Priori et al., 2014; Taylor et al., 2010; van der Klooster et al., 2011; van Egmond et al., 2010; Viscarra Rossel et al., 2007a)
OC (%)	Vis-NIR	2–11	28–765	R/N	0.0–0.91	0.84	7	0.31–1.16, *0.35*	0.42, *0.35*	6	(Gomez et al., 2008b; Gras et al., 2014; Kusumo et al., 2008) (Bricklemyer and Brown, 2010; Daniel et al., 2003; Denis et al., 2014; Nocita et al., 2011)

(Continued)

TABLE 15.2 (*Continued*)

Summary of Revision Related with Quantification of Soil Attributes by Ground Spectral Sensing

					R^2			RMSEP/RMSECV, *SEP/SECV* (%)			
Element*	**Spectral range****	**Element range*****	**N. cal.**	**Cal. Scale**	**range**	**median**	**N**	**range**	**median**	**N**	**Authors**
	Vis-NIR	1–37	15->400	F	0.65–0.90	0.75	3	0.07–7.15, *0.56*	1.74, *0.56*	8	(Knadel et al., 2011; Kuang and Mouazen, 2013a; Kuang and Mouazen, 2013b; Reeves et al., 2010; Wijaya et al., 2001)
	NIR	2–3	24–120	R/N	0.53–0.86	0.74	4	0.18–0.52, *0.42–0.53*	0.36, *0.48*	5	(Christy, 2008; Cozzolino et al., 2013; McCarty et al., 2010; Nocita et al., 2011; Sudduth and Hummel, 1993)
	NIR	1–3	11–216	F	0.61–0.95	0.82	4	0.09–0.27	0.18	4	(Kweon and Maxton, 2013; Reeves et al., 2010; Schirrmann et al., 2012)
TC (%)	Vis-NIR	1–3	38–209	F	~0.6–0.89	0.86	3	0.16–0.19	0.16	3	(Igne et al., 2010; Kodaira and Shibusawa, 2013; Kusumo et al., 2011)
	NIR	1–2	45–78	F	0.46–0.68	0.57	2	0.15–0.48	0.18	2	(Huang et al., 2007; Munoz and Kravchenko, 2011)

* CEC, cation exchangeable capacity, OC, organic carbon, IC, inorganic carbon, TC, total carbon

** The exact spectral range in the individual studies may deviate to some extent but will stay within ranges as described later.

*** This is included to give an idea of the variation in the element range (i.e., clay content) in the studies used to calculate the prediction statistics since this has a large impact on the prediction statistics. The individual ranges are presented as percentage units and "Element range" in the table is the range of these for the studies used.

TABLE 15.3
Summary of Revision Related with Quantification of Soil Attributes by Aerial and Orbital Sensors

Soil Properties	Sensor	Nr. Samples (Total/ Validation)	Modeling Approach	Validation	Validation/Calibration Results R^2 (P)/R^2 (CV)	RMSEP/RMSECV	SEP/SECV	Author
				Multispectral sensors				
SOC	ATLAS[a]	31–40	MLR	-	0.63–0.91	0.11–0.22%	-	Chen et al. (2008)
	SPOT[b]	10–11/17–27	MLR	Pred.	0.55–0.72	4.06–6.57 g kg^{-1}	-	Vaudour et al. (2013)
OM	LANDSAT 5[b]	378/95	MLR	Pred.	0.41	-	-	Fiorio et al. (2010)
	LANDSAT 5[b]	184	MLR	-	0.561	-	-	Nanni and Demattê (2006a)
	LANDSAT 5[b]	164	LR	-	0.79	1,5 g kg^{-1}	-	Dogan and Kılıç (2013)
	LANDSAT 7[b]	110/155	MLR	Pred.	0.27	-	-	Demattê et al. (2007a)
Total Carbon	IKONOS[b]	144–222/14–24	MLR	Pred.	0.11–0.61	0.11–0.24 %	-	Sullivan et al. (2005)
	LANDSAT 7[b]	78	MLR	Jack-knifing (10-fold)	0.33–0.46	0.27–0.25 %	-	Huang et al. (2007)
Total N	LANDSAT 7[b]	164	LR	-	0.612	-	0.06%	Dogan and Kılıç (2013)
Clay	LANDSAT 5[b]	378/95	MLR	Pred.	0.61	-	-	Fiorio et al. (2010)
	LANDSAT 5[b]	184	MLR	-	0.675	-	-	Nanni and Demattê (2006a)
	LANDSAT 7[b]	110/155	MLR	Pred.	0.63	-	-	Demattê et al. (2007a)
Sand	LANDSAT 5[b]	378/95	MLR	Pred.	0.63	-	-	Fiorio et al. (2010)
	LANDSAT 5[b]	184	MLR	-	0.525	-	-	Nanni and Demattê (2006a)
	LANDSAT 7[b]	110/155	MLR	Pred.	0.67	-	-	Demattê et al. (2007a)
	ASTER[b]	22/5	ER	Pred.	0.63	19,33 g kg^{-1}	-	Breunig et al. (2008)
Silt	LANDSAT 5[b]	378/95	MLR	Pred.	0.54	-	-	Fiorio et al. (2010)
	LANDSAT 5[b]	184	MLR	-	0.508	-	-	Nanni and Demattê (2006a)
	LANDSAT 7[b]	110/155	MLR	Pred.	0.29	-	-	Demattê et al. (2007a)
CEC	LANDSAT 5[b]	378/95	MLR	Pred.	0.45	-	-	Fiorio et al. (2010)
	LANDSAT 5[b]	184	MLR	-	0.551	-	-	Nanni and Demattê (2006a)
	LANDSAT 7[b]	44	MLR	-	0.57	-	-	Ghaemi et al. (2013)
	LANDSAT 7[b]	110/155	MLR	Pred.	0.26	-	-	Demattê et al. (2007a)

(Continued)

TABLE 15.3 (*Continued*)
Summary of Revision Related with Quantification of Soil Attributes by Aerial and Orbital Sensors

Soil Properties	Sensor	Nr. Samples (Total/ Validation)	Modeling Approach	Validation	Validation/Calibration Results: R^2 (P)/R^2 (CV)	RMSEP/RMSECV	SEP/SECV	Author
	LANDSAT 5[b]	113	MLR-RK/RT/ GAM-RK/KED	Jack-knifing (100-fold)	-	6.48–5.53/7.27/ 6.94–6.18/5.41 cmolc kg^{-1}	-	Bishop and McBratney (2001)
Fe_2O_3 (total)	LANDSAT 5[b]	184	MLR	-	0.725	-	-	Nanni and Demattê (2006a)
SiO_2				-	0.598	-	0.60%	
TiO_2				-	0.727	-	-	
$CaCO_3$	LANDSAT 7[b]	164	LR	-	0.716	-	-	Dogan and Kılıç (2013)
pH	LANDSAT 7[b]	164	LR	-	0.659	-	0.3	Dogan and Kılıç (2013)
Total P	LANDSAT 7[b]	111	MLR/CK/RK	CV	0.46/—/-	-	356.1/279.2/ 238.8 mg kg^{-1}	Rivero et al. (2007)
	ASTER[b]	111	MLR/CK/RK	CV	0.39/—/-	-	281.8/238.2/ 200.1 mg kg^{-1}	
K	LANDSAT 7[b]	110/155	MLR	Pred.	0.11	-	-	Demattê et al. (2007a)
Ca	LANDSAT 7[b]	110/155	MLR	Pred.	0.13	-	-	Demattê et al. (2007a)
Mg	LANDSAT 7[b]	110/155	MLR	Pred.	0.19	-	-	Demattê et al. (2007a)
H+Al	LANDSAT 7[b]	110/155	MLR	Pred.	0.18	-	-	Demattê et al. (2007a)
Basis (CEC saturation)	LANDSAT 5[b]	378/95	MLR	Pred.	0.01	-	-	Fiorio et al. (2010)
	LANDSAT 7[b]	110/155	MLR	Pred.	0.21	-	-	Demattê et al. (2007a)
Al (CEC saturation)	LANDSAT 5[b]	378/95	MLR	Pred.	0.13	-	-	Fiorio et al. (2010)
EC	LANDSAT 7[b]	164	LR	-	0.498	-	131.3	Dogan and Kılıç (2013)
				Hyperspectral sensors				
SOC	AHS-160[a]	68/16	PLSR	Pred.	0.53/0.75	2.42/1.18 g kg^{-1}	-	Bartholomeus et al. (2011)
	CHRIS-PROBA[b]	72/24	RK/PLSR/PLSR-K	Pred.	-	1.06–1.16 g kg^{-1}	-	Casa et al. (2013a)
	AHS-160[a]	88–91	PLSR	CV (LOO)	0.93–0.96	3.68–4.9 g kg^{-1}	-	Denis et al. (2014)
	AVNIR[a]	321	MLR	-	0.2692	0.081%	-	DeTar et al. (2008)
	HyMap[a]	67/29	PLSR	Pred.	0.71	-	2.07/1.64 g kg^{-1}	Gerighausen et al. (2012)
	EO-1 Hyperion[b]	72	PLSR	CV (LOO)	0.51	0.73%	-	Gomez et al. (2008b)

	HyMap[a]	95	PLSR	CV (LOO)	0.02	2.6 g kg^{-1}	-	Gomez et al. (2012)
	HyMap[a]	204	PLSR	CV (LOO) and Pred.	0.83	1.10/1.05 g kg^{-1}	-	Hbirkou et al. (2012)
	SpecTIR[a]	269	PLSR	CV (LOO)	0.65	0.0019	-	Hively et al. (2011)
	EO-1 Hyperion[b]	49/14	PLSR/MLR	CV (LOO—PLSR)/CV (LOO—MLR) and Pred. (MLR)	0.63 (PLSR)/ 0.50 (P—MLR)/0.63 (CV—MLR)	1.6 g kg^{-1} (PLSR)	-	Lu et al. (2013)
	HyMap[a]	72	PLSR/MLR	CV (LOO)	0.90 (PLSR)/0.86 (MLR)	0.29 (PLSR)/0.22% (MLR)	-	Selige et al. (2006)
	CASI[a]	227/57	PLSR	Pred.	0.85	-	5.1/4.8 g kg^{-1}	Stevens et al. (2006)
	AHS-160[a]	197/102	PLSR	Pred.	0.86/0.89	3.56/3.01 g kg^{-1}	-	Stevens et al. (2010)*
		188/101	PSR	Pred.	0.88/0.91	3.20/2.54 g kg^{-1}	-	
		201/101	SVMR	Pred.	0.84/0.99	4.20/0.43 g kg^{-1}	-	
	AHS-160[a]	400/126	PLSR/PSR/SVMR	Pred.	0.56–0.73/ 0.85–0.98	4.74–6.11/1.37–3.45 g kg^{-1}	-	Stevens et al. (2012)*
	CASI[a]	47	MLR-PCA/ ANN-PCA	CV (10-fold CV)	0.745 (MLR)/0.590 (ANN)	0.49 (MLR)/0.592% (ANN)	-	Uno et al. (2005)
	EO-1 Hyperion[b]	28/08	PLSR	Pred.	0.48/0.56	0.33/0.43%	-	Zhang et al. (2013)
OM	DAIS-7915[a]	62	MLR	Pred.	0.827	-	0.015/0.003%	Ben-Dor et al. (2002)
	AVNIR[a]	321	MLR	-	0.4857	0.08%	-	DeTar et al. (2008)
	SpecTIR[a]	269	PLSR	CV (LOO)	0.75	0.4%	-	Hively et al. (2011)
	EO-1 Hyperion[b]	28/08	PLSR	Pred.	0.74/0.72	0.66/0.72%	-	Zhang et al. (2013)
POM	EO-1 Hyperion[b]	18	PLSR	CV (LOO)	0.67	4.56%	-	Anne et al. (2014)
MAOM					0.74	2.450%	-	
Labile C					0.93	6.72 mg g^{-1}	-	
Stable C					0.71	31.51 mg g^{-1}	-	

(Continued)

TABLE 15.3 (*Continued*)
Summary of Revision Related with Quantification of Soil Attributes by Aerial and Orbital Sensors

Soil Properties	Sensor	Nr. Samples (Total/Validation)	Modeling Approach	Validation	Validation/Calibration Results: R^2 (P)/R^2 (CV)	RMSEP/RMSECV	SEP/SECV	Author
N total	CHRIS-PROBA[b]	73/24	RK/PLSR/PLSR-K	Pred.	-	0.144–0.139 g kg^{-1}	-	Casa et al. (2013a)
	HyMap[a]	72	PLSR/MLR	CV (LOO)	0.92/0.87	0.03/0.02%	-	Selige et al. (2006)
	EO-1 Hyperion[b]	28/08	PLSR	Pred.	0.70/0.63	0.032/0.033%	-	Zhang et al. (2013)
Labile N	EO-1 Hyperion[b]	18	PLSR	CV (LOO)	0.96	0.34 mg g^{-1}	-	Anne et al. (2014)
Stable N					0.69	2.58 mg g^{-1}	-	
Clay	MIVIS[a]	80/29	PLSR	Pred.	0.48	7.20%	-	Casa et al. (2013b)
	CHRIS-PROBA[b]				0.52	6.87%	-	
	CHRIS-PROBA[b]	132/44	RK/PLSR/PLSR-K	Pred.	-	5.33–5.82%	-	Casa et al. (2013a)
	AVNIR[a]	321	MLR	-	0.6708	3.23%	-	DeTar et al. (2008)
	SIM-GA[a]	40/11	Regression with band depth (2210 nm)	Pred.	0.5994	-	-	Garfagnoli et al. (2013)
	HyMap[a]	67/29	PLSR	Pred.	0.85	-	19.41/20.34 g kg^{-1}	Gerighausen et al. (2012)
	HyMap[a]	52	PLSR	CV (LOO)	0.64	49.60 g kg^{-1}	-	Gomez et al. (2008a)
	HyMap[a]	95	PLSR	CV (LOO)	0.67	42.15 g kg^{-1}	-	Gomez et al. (2012)
	SpecTIR[a]	269	PLSR	CV (LOO)	0.66	2.2%	-	Hively et al. (2011)
	HyMap[a]	33	Regression with CR data (2206 nm)	CV (LOO)	0.61	54 g kg^{-1}	-	Lagacherie et al. (2008)
		19			0.60	130 g kg^{-1}	-	
		33 + 19			0.58	82 g kg^{-1}	-	
	AISA-Dual Vis—NIR[a]	152/30	PLSR/Regression trees	CV (LOO) and Pred.	0.81/0.78/0.78/0.77	87/67/93/69 g kg^{-1}	-	Lagacherie et al. (2013)
	HyMap[a]	72	PLSR/MLR	CV (LOO)	0.71/0.65	4.2/3.8%	-	Selige et al. (2006)
	EO-1 Hyperion[b]	28/08	PLSR	Pred.	0.51/0.83	5.46/2.21%	-	Zhang et al. (2013)
Sand	EO-1 Hyperion[b]	18	PLSR	CV (LOO)	0.58	7.02%	-	Anne et al. (2014)
	MIVIS[a]	80/29	PLSR	Pred.	0.64	7.82%	-	Casa et al. (2013b)
	CHRIS-PROBA[b]				0.45	9.32%	-	
	CHRIS-PROBA[b]	132/44	RK/PLSR/PLSR-K	Pred.	-	6.80–7.40%	-	Casa et al. (2013a)
	AVNIR[a]	321	MLR	-	0.8063	4.83%	-	DeTar et al. (2008)
	HyMap[a]	95	PLSR	CV (LOO)	0.20	90.47 g kg^{-1}	-	Gomez et al. (2012)
	SpecTIR[a]	269	PLSR	CV (LOO)	0.79	7.9%	-	Hively et al. (2011)
	AISA-Dual Vis—NIR[a]	152/30	PLSR/Regression trees	CV (LOO) and Pred.	0.75/0.83/0.78/0.77	119/97/111/107 g kg^{-1}	-	Lagacherie et al. (2013)
	HyMap[a]	72	PLSR/MLR	CV (LOO)	0.95/0.87	9.7/12.9%	-	Selige et al. (2006)

Silt	MIVIS[a]	80/29	PLSR	Pred.	0.32	3.28%	-	Casa et al. (2013b)
	CHRIS-PROBA[b]				0.23	3.43%	-	
	CHRIS-PROBA[b]	132/44	RK/PLSR/PLSR-K	Pred.	-	3.12–3.67%	-	Casa et al. (2013a)
	AVNIR[a]	321	MLR	-	0.7518	3.34%	-	DeTar et al. (2008)
	HyMap[a]	95	PLSR	CV (LOO)	0.17	74.84 g kg^{-1}	-	Gomez et al. (2012)
	SpecTIR[a]	269	PLSR	CV (LOO)	0.79	6.9%	-	Hively et al. (2011)
Silt + Clay	EO-1 Hyperion[b]	18	PLSR	CV (LOO)	0.82	1.95%	-	Anne et al. (2014)
Soil Moisture (gravimetric)	EO-1 Hyperion[b]	18	PLSR	CV (LOO)	0.82	3.02%	-	Anne et al. (2014)
	DAIS-7915[a]	62	MLR	Pred.	0.645	-	0.14/0.045%	Ben-Dor et al. (2002)
	HyMap[a]	205	Linear regression with NSMI index	-	0.82	2.30%	-	Haubrock et al. (2008a)
	EO-1 Hyperion[b]	28/08	PLSR	Pred.	0.40/0.79	5.12/3.13%	-	Zhang et al. (2013)
Saturated Soil Moisture (gravimetric)	DAIS-7915[a]	62	MLR	Pred.	0.759	-	0.021/0.019%	Ben-Dor et al. (2002)
	AVNIR[a]	321	MLR	-	0.4859	3.07%	-	DeTar et al. (2008)
Available Soil Water	CHRIS-PROBA[b]	132/44	RK/PLSR/PLSR-K	Pred.	-	2.50–2.79%	-	Casa et al. (2013a)
Iron oxides	ROSIS[a]	35/16	Redness index	Pred.	-	-	4.97–10.59/ 4.38–11.39 %	Bartholomeus et al. (2007)**
			Spectral feature area (550 nm)		-	-	5.73–9.08/ 6.41–7.26 %	
			Standard deviation after CR		-	-	5.79–7.71/ 6.39–6.43 %	
	HyMap[a]	95	PLSR	CV (LOO)	0.78	0.28 g $100g^{-1}$	-	Gomez et al. (2012)
Fe_2O_3 (total)	AVIRIS[a]	22	Regression with band depth (1710 nm)	-	0.83	-	-	Galvão et al. (2001)
TiO_2				-	0.74	-	-	
Al_2O_3			Regression with band depth (2200 nm)	-	0.68	-	-	
$CaCO_3$	HyMap[a]	52	PLSR	CV (LOO)	0.77	76.67 g kg^{-1}	-	Gomez et al. (2008a)
	HyMap[a]	95	PLSR	CV (LOO)	0.76	64.32 g kg^{-1}	-	Gomez et al. (2012)
	HyMap[a]	33	Regression with CR data (2341 nm)	CV (LOO)	0.59	113 g kg^{-1}	-	Lagacherie et al. (2008)
		19			0.61	133 g kg^{-1}	-	
		33 + 19			0.47	132 g kg^{-1}	-	

(Continued)

TABLE 15.3 (*Continued*)

Summary of Revision Related with Quantification of Soil Attributes by Aerial and Orbital Sensors

Soil Properties	Sensor	Nr. Samples (Total/ Validation)	Modeling Approach	Validation	Validation/Calibration Results			Author
					R^2 (P)/R^2 (CV)	RMSEP/RMSECV	SEP/SECV	
RCGb	HyMap[a]	30	Spectral index	-	0.75	-	-	Baptista et al. (2011)
CEC	HyMap[a]	95	PLSR	CV (LOO)	0.62	1.84 cmolc kg^{-1}	-	Gomez et al. (2012)
	AISA-Dual Vis—NIR[a]	147/30	PLSR/RT	CV (LOO) and Pred.	0.71/0.79/ 0.72/0.79	4.6/3.3/4.6/3.4 Meq 100 g^{-1}	-	Lagacherie et al. (2013)
	EO-1 Hyperion[b]	49	PLSR	CV (LOO)	0.40	1.55 cmolc kg^{-1}	-	Lu et al. (2013)
pH	DAIS-7915[a]	62	MLR	Pred.	0.528	-	0.26/0.146	Ben-Dor et al. (2002)
	AVNIR[a]	321	MLR	-	0.6164	0.076	-	DeTar et al. (2008)
	HyMap[a]	95	PLSR	CV (LOO)	0.31	0.37	-	Gomez et al. (2012)
	SpecTIR[a]	269	PLSR	CV (LOO)	0.51	0.40	-	Hively et al. (2011)
	EO-1 Hyperion[b]	49/14	PLSR/MLR	CV (LOO—PLSR)/CV (LOO—MLR) and Pred. (MLR)	0.68 (PLSR)/ 0.65 (P—MLR)/0.83 (CV—MLR)	0.19 (PLSR)	-	Lu et al. (2013)
Extractable P	AVNIR[a]	321	MLR	-	0.6975	4.32 mg kg^{-1}	-	DeTar et al. (2008)
Total P	EO-1 Hyperion[b]	49/14	PLSR/MLR	CV (LOO—PLSR)/CV (LOO—MLR) and Prediction (Indep. data—MLR)	0.62 (PLSR)/0.54 (P—MLR)/0.74 (CV—MLR)	0.2 g kg^{-1} (PLSR)	-	Lu et al. (2013)
	EO-1 Hyperion[b]	28/08	PLSR	Pred.	0.25/0.44	176.67/141.86 mg kg^{-1}	-	Zhang et al. (2013)

K	AVNIR[a]	321	MLR	-	0.6391	45.03 mg kg^{-1}	-	DeTar et al. (2008)
	SpecTIR[a]	269	PLSR	CV (LOO)	0.59	89.5 mg kg^{-1}	-	Hively et al. (2011)
Ca	AVNIR[a]	321	MLR	-	0.6188	9.51 meq L^{-1}	-	DeTar et al. (2008)
	SpecTIR[a]	269	PLSR	CV (LOO)	0.69	166.1 mg kg^{-1}	-	Hively et al. (2011)
Mg	AVNIR[a]	321	MLR	-	0.582	4.32 meq L^{-1}	-	DeTar et al. (2008)
	SpecTIR[a]	269	PLSR	CV (LOO)	0.69	50.3 mg kg^{-1}	-	Hively et al. (2011)
Na	AVNIR[a]	321	MLR	-	0.6224	9.11 meq L^{-1}	-	DeTar et al. (2008)
Al	SpecTIR[a]	269	PLSR	CV (LOO)	0.76	104.7 mg kg^{-1}	-	Hively et al. (2011)
Mn	SpecTIR[a]	269	PLSR	CV (LOO)	0.62	19.6 mg kg^{-1}	-	Hively et al. (2011)
Zn	SpecTIR[a]	269	PLSR	CV (LOO)	0.64	1.4 mg kg^{-1}	-	Hively et al. (2011)
Cl	AVNIR[a]	321	MLR	-	0.7376	12.24 meq L^{-1}	-	DeTar et al. (2008)
Fe	SpecTIR[a]	269	PLSR	CV (LOO)	0.75	49.3 mg kg^{-1}	-	Hively et al. (2011)
EC	DAIS-7915[a]	62	MLR	Pred.	0.665	-	4.58/4.36 ms cm^{-1}	Ben-Dor et al. (2002)
	CHRIS-PROBA[b]	74/24	RK/PLSR/PLSR-K	Pred.	-	123.74–129.6 µS cm^{-1}	-	Casa et al. (2013a)
	AVNIR[a]	321	MLR	-	0.6696	1.96 dS m^{-1}	-	DeTar et al. (2008)
Bulk Density	EO-1 Hyperion[b]	18	PLSR	CV (LOO)	0.82	0.03 g cm^{-3}	-	Anne et al. (2014)

[a]—airborne sensor; [b]—orbital sensor; CK—CoKriging; CR—Continuum Removed; KED—Kriging with external drift; MAOM—Mineral Associated Organic Matter; NSMI—Normalized Soil Moisture Index; PLSR-K—PLSR with kriging interpolation; POM—Particulate Organic Matter; PSR—Penalized-spline Signal Regression; RCGb—Index related to soil weathering; RK—Regression-kriging; RT—Regression trees; SVMR—Support Vector Machine Regression; * These results concern models fitted locally using soil types to stratify the area; ** Quantifications were done using linear regression approach and indices as predictive variables.

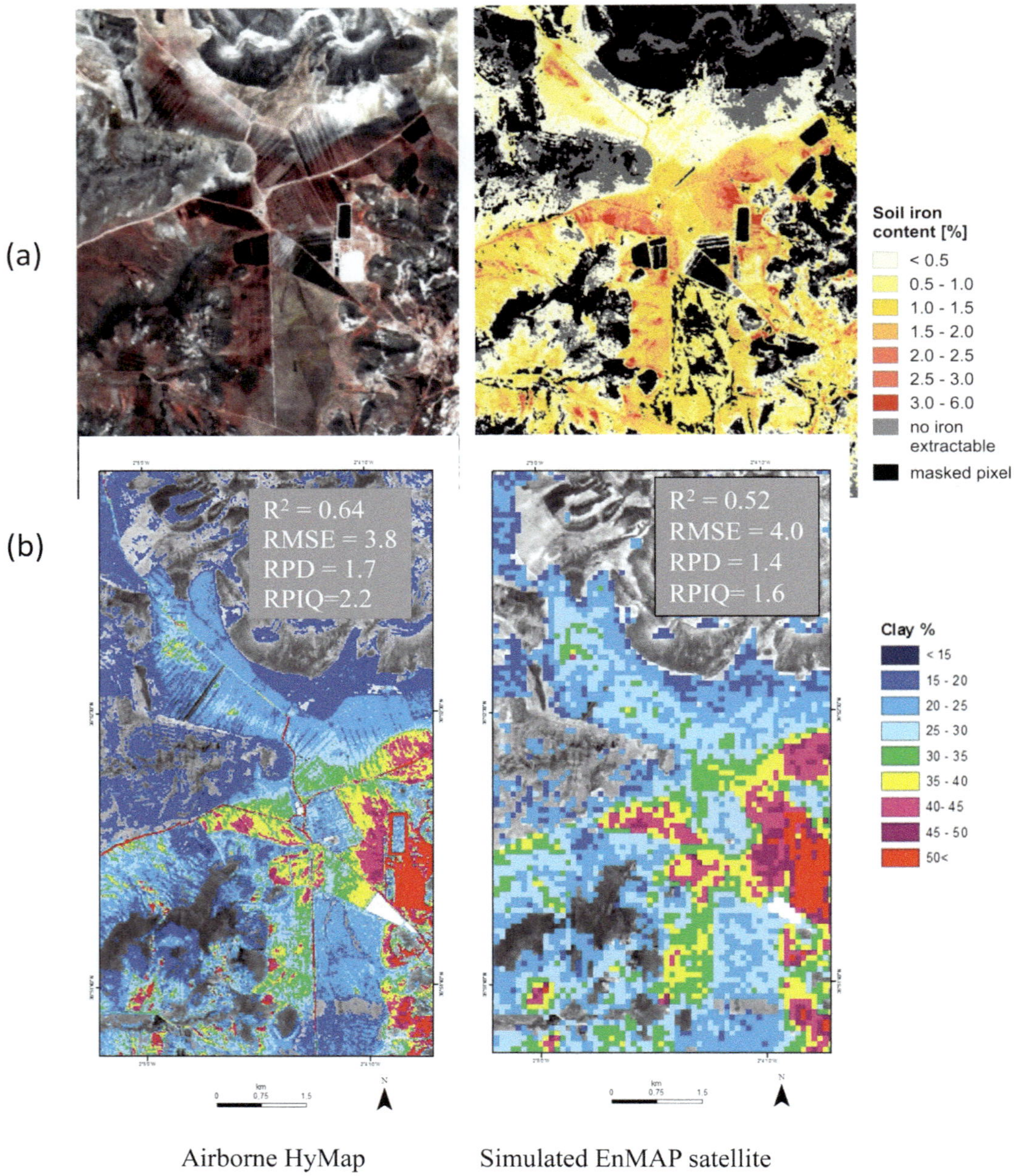

FIGURE 15.4 Cabo de Gata-Nijar Natural Park, southern Spain: (a) HyMap true-color image (left) and associated soil iron map (right) (Richter et al., 2007); (b) Potential of upcoming EnMAP hyperspectral satellite for quantitative surface soil mapping. Soil clay content maps (Cabo de Gata natural Park, Spain) based on airborne HyMap (left) and spaceborne simulated EnMAP hyperspectral images (right) (Chabrillat et al., 2014). Model error metrics calculated using independent validation data.

signatures measurement with ground sensor using classification and regression tree and obtained predictions of N (R^2 0.91), carbonate (R^2 0.95), total carbon (R^2 0.93) and OM (R^2 0.98) in the soil.

Through using classification and regression tree statistical methods was obtained concentrations of N (R^2 0.91), carbonate (R^2 0.95), total carbon (R^2 0.93) and OM (R^2 0.98).

When evaluating the performance of these prediction models it is important to analyze the geographical extent and the availability and independence of validation and calibration data (Bellon-Maurel and McBratney, 2011). Soriano-Disla et al. (2014) reviewed the performance of LSS and FSS for predicting SOC. They concluded that overall MIR (20 studies) performed better (coefficient of determination—R^2—of 0.93) than NIR (23 studies; R^2 of 0.85). Adding UV, VIS or multiple combinations of wavelength ranges, that is, VIS-NIR, NIR-MIR, UV-VIS-NIR, or VIS-NIR-MIR, did not necessarily improve prediction accuracy. The UV range has been used for SOM prediction because it allows detection of any molecule with alternating double and single bonds (Schulthess, 2011).

Nitrogen is part of SOM, relating to SOC, since the absolute majority of this element is in the form of organic N at a C:N rate of about 10:1 in SOM. Organic N has specific absorption features, such as those originated by amide groups, although these features are expected to be very weak due to the very low N content in soil (generally below 1%) (Stenberg et al., 2010). Thus, N may be better predicted based on a correlation with spectrally active soil components, such as SOC. Mean R^2 values for predictions of total N in various soil studies were about 0.9 using MIR, and 0.86 using VIS-NIR (Soriano-Disla et al., 2014).

Prediction of soil particle fractions (clay, silt and sand contents) using spectroscopic techniques is reported as feasible in literature. Soil texture is especially related to the mineralogical composition of soils, mainly of clay minerals and the quartz content, and thus it directly influences the soil spectra. Indeed Araújo et al. (2014a) indicated that when a spectral library was clustered into smaller datasets based on mineralogy and geology prior to model calibration, the R^2 increased. Mean values of R^2 for clay and sand predictions, in validations datasets, using MIR (0.80 and 0.83, considering 14 and 11 studies, respectively) were almost equivalent for those obtained using VIS-NIR wavelengths (Soriano-Disla et al., 2014).

Another physical soil property of interest in spectroscopy is soil moisture. The different forms of water in soil (hydration, hygroscopic, and free water) are all active regarding electromagnetic energy absorption (Ben-Dor, 2011). In a given area, water (soil moisture) is related with quartz, clay minerals and SOM, which are also spectrally active soil properties (Kuang et al., 2012). In general, Soriano-Disla et al. (2014) reported mean R^2 values for volumetric water content predictions of: 0.89 for UV-VIS-NIR; 0.86 for NIR and VIS-NIR; and 0.83 for MIR wavelength ranges. The most important wavelengths for predicting soil water are 1,350 to 1,450, 1,890 to 1,990, and 2,220 to 2,280 nm (Zhu et al., 2010). Table 15.1 indicates the relation of these and other wavebands with respective soil attributes.

Iron and aluminum oxides, hydroxides, and oxi-hydroxides (most commonly hematite, goethite, and gibbsite) together with clay minerals (kaolinite, illite, and smectites) originate the main absorption features in the VIS-NIR range (Ben-Dor, 2011). Vendrame et al. (2012) predicted kaolinite, gibbsite, goethite, and hematite contents, measured by the sulfuric acid method, for Brazilian soils, using radiometric data in the NIR, and obtained R^2 of 0.83, 0.78, 0.56, and 0.60, respectively. Sellitto et al. (2009) reported R^2 values of 0.46 and 0.80 for goethite and hematite prediction in soils using a simple spectral band depth calculation in the VIS range, as corroborated by a similar method performed by Richter et al. (2009) for iron oxides. Summers et al. (2011) quantified carbonates and iron oxides contents through VIS-NIR, and obtained R^2 values of 0.66 and 0.61, respectively.

An important soil property that affects soil fertility and soil management is CEC, which in most cases have good prediction models, with reported mean R^2 values of 0.85, 0.81 and 0.84 for MIR, NIR and VIS-NIR, considering 13, 9 and 8 studies, respectively (Soriano-Disla et al., 2014). Plant nutrients (Ca, Mg, K and P) do not present absorption features in UV-VIS-NIR-MIR, except when these elements are present as constituents in molecular groups (Kuang et al., 2012). For predictions of exchangeable Ca content mean R^2 values of 0.82, 0.75 and 0.80 were reported considering MIR, NIR and VIS-NIR ranges in 8, 11 and 7 studies, respectively, with similar trends for Mg (Soriano-Disla et al., 2014). On the other hand, highly weathered soils from the tropics usually have very low carbonate and Ca contents, which can explain the low values obtained by Nanni and Demattê (2006a). Predictions of exchangeable K are generally less accurate (Kuang et al., 2012; Soriano-Disla et al., 2014). Inaccurate predictions are reported, in most cases, for available P content in soils, with mean R^2 of 0.35, 0.48 and 0.49 for MIR, NIR and VIS-NIR, respectively (Soriano-Disla et al., 2014).

Predictions of soil pH depend mainly on correlation with spectrally active attributes, such as SOM and clay content. An important aspect to be considered is that soil acidity can be developed from different sources (organic or inorganic), generating instabilities in models created over large soil variabilities or geographical areas (Stenberg and Viscarra Rossel, 2010). Despite this, relatively accurate results have been reported with mean R^2 values as great as 0.79 (VIS-NIR) considering 18, 21, and 15 studies for MIR, NIR, and VIS-NIR wavelengths, respectively (Soriano-Disla et al., 2014).

Despite the indications of papers in this section, Table 15.2 aggregates more information about the subject. The quantification of clay, CEC, OC, IC and total carbon (TC) certainly has strong results. In laboratory conditions we observe an average of R^2 between 0.67 to 0.78 depending on the scale of the work, and with low error (RMSEP/RMSECV, SEP/SECV). In this exploratory evaluation, we found 25 papers with regional or national focus, and only 3 for global scale. Despite this, the global reached 0.77 of R^2 very similar to the regional or national, with less or more number of samples. This shows that number of samples, although important is not a prerogative, if global or regional, but the representation of the distribution. Anyway, all results indicate the feasibility and repeatable strong results on the quantification of clay in laboratory conditions.

It is important to observe that there are much more works related with VIS-NIR (about 58 in this revision) and only 11 with MIR. Similar differences are observed for the other elements. This is mostly due to the lack of MIR equipment (Stark et al., 1986) and the difficulty on use them (Reeves III, 2010). On the other hand, the results reached with MIR are higher than VIS-NIR, going from an average of 0.71 to 0.78 of R^2, where MIR reached a maximum of 0.86. For CEC we go from a maximum of 0.81 in VIS-NIR to 0.86 in MIR. For organic C the differences are even more important, where we have a R^2 range of 0.75–0.93 and 0.93–0.95 for VIS-NIR and MIR, respectively. Thus, although not so great, in laboratory conditions MIR presents better results than VIS-NIR. In an important approach Viscarra Rossel et al. (2010) performed thematic maps of iron oxides and color by spectra information of all Australia.

15.6.2.2 Strategies for Space Spectral Sensing

Evaluating soils from aerial or orbital sensor is certainly a difficult but interesting task. Imagine trying to seek a soil property such as SOM with a sensor about 1 to 800 km away from the target! There are numerous factors interfering on the signal. Despite the challenge, new methods of atmospheric correction and sensors provide the possibility of improving current results. The key for detecting reliable soil attribute information from a pixel is directly related to the correct detection of areas of bare soil in the image. Opportunities for further research are plentiful; there are still issues to be studied, such as the alterations of soils due to agricultural management and contamination, for example. Despite these, this section describes several works with this approach and are summarized in Table 15.3.

A considerable amount of studies have evaluated the use of multispectral data to predict soil properties. Sullivan et al. (2005) evaluated high spatial resolution IKONOS multispectral reflectance data for successfully mapping soil properties in regions of Alabama. Chen et al. (2008) used the ATLAS sensor (2 m of spatial resolution, and 8 bands in the VIS-NIR-SWIR) data to predict SOC, reporting an R^2 of 0.63 and root mean square error (RMSE) of 0.22%. Löhrer et al. (2013) identified and mapped mineralogical composition of Pleistocene sediments using ASTER and SPOT-5 images associated with spectral laboratory, obtaining good results.

Several studies in the tropics were successful in quantifying soil properties using a technique to detect bare soil (Demattê et al., 2009a—see Section 15.8). An average R^2 of 0.67 was obtained for clay content estimation using MLR (Demattê and Nanni, 2003; Demattê et al., 2005; Nanni and

Demattê, 2006a, 2006b; Fiorio et al., 2010). Using the same methodology, Demattê et al. (2007a) studied pixels with two different information of soil with a high variance on texture and reached a 0.86 R^2 for clay. Using Landsat, Demattê et al. (2009b) and Fiorio and Demattê (2009) predicted clay, Al_2O_3, Fe_2O_3, and weathering indicators, Ki, SiO_2 and TiO_2, with similar R^2 values of 0.61, 0.68, 0.67, 0.54, 0.65 and 0.72, respectively. To monitor the spatial variation of CEC, Ghaemi et al. (2013) developed a model based on Landsat ETM data in a semi-arid area in the Neyshaboor province, Iran, where the model correctly classified the spatial variation of CEC in 45 to 65% of the cases. Additionally with Landsat, Masoud (2014) predicted salt abundance in soils of an area in Burg Al-Arab, Egypt, using Landsat-based spectral mixture analysis and soil spectroscopy, achieving an R^2 of 0.88. Still with Landsat, Dogan and Kiliç (2013) studied the relationship between soil variables and the Landsat 7 bands using digital number values, and found good correlation for pH, OM, $CaCO_3$, and N with band 5.

Using ASTER with 8 bands, Vicente and Souza Filho (2011) observed R^2 of 0.65 to 0.79 between spectral signatures measured from soil samples and ASTER pixels to map kaolinite and iron oxides using a mixture-tuned matched filtering (MTMF) approach. Similarly, Genú et al. (2013a) combined ground and ASTER pixel reflectance with multiple endmember spectral mixture analysis (MESMA) for mapping SOM, reaching a 60% agreement between them. As to aggregate all information of ASTER, including its 2–14 μm range, Hewson et al. (2012) studied the sensor Spectral library (SWIR and TIR spectral signature) to evaluate the spectral indices for composition and texture of natural soil samples.

Despite the multispectral data, hyperspectral has great importance due to the high number of bands. On the prediction of SOC, Gomez et al. (2008b) reported an R^2 of 0.66, RMSE of 0.61% and residual prediction deviation (RPD) of 1.69 for laboratory measurements, and an R^2 of 0.51, RMSE of 0.73% and RPD of 1.43 for Hyperion data, while Hbirkou et al. (2012) reached 0.83 of R^2 with HyMap. Stevens et al. (2012) using the AHS-160 hyperspectral sensor obtained high accuracy with R^2 of 0.73 and RMSE of 4.7 g kg^{-1} applying a soil type calibration strategy and PLSR to derive the prediction model for SOC contents. For N, Selige et al. (2006) reported similar performance for PLSR and MLR models, using HyMap data, to predict total N content, with R^2 of 0.92 and RMSE of 0.03% for PLSR. Afterwards, Zhang et al. (2013) also performed total N content quantification but used a Hyperion image and reported an R^2 of 0.70.

The influence of soil particle distribution, for example, clay and sand, on spectra is great and studies such as Selige et al. (2006) evaluated the potential of ASS (HyMap sensor) to predict these properties. The authors found R^2 values of 0.71 and 0.95 for clay and sand contents, respectively. De Tar et al. (2008) used MLR with AVNIR hyperspectral imagery (60 spectral bands from 429 to 1,010 nm), and reported R^2 values of 0.81, 0.75 and 0.67 for sand, silt and clay contents, respectively. Gomez et al. (2008) and Gomez et al. (2008a) performed clay content quantification through HyMap sensor data and obtained R^2 of 0.64 and 0.67, with RMSE of 49.6 and 42.2 g kg^{-1}, respectively, using PLSR. Hively et al. (2011) using HyperSpecTIR sensor data obtained R^2 of 0.79, 0.79 and 0.66 for sand, silt and clay contents predictions, respectively. Casa et al. (2013b) compared different hyperspectral sensors—the airborne MIVIS (VIS-NIR-SWIR), and the spaceborne CHRIS-PROBA (VIS-NIR)—with spatial resolutions of 4.8 and 17 m, respectively, and obtained R^2 values of 0.48 and 0.52, and RMSE of 7.20 and 6.87, for clay prediction with both sensors, respectively.

Moisture is highly related to the soil spectra because water content affects the base line height (albedo) and causes several spectral absorption features (Lobel and Asner, 2002). Whiting et al. (2004) proposed a spectral technique to estimate SM content, in which an inverted Gaussian function is fitted centered on the assigned fundamental water absorption region at 2800 nm, and the area of the inverted function accurately estimated the water content within an RMSE of 2.7% and R^2 of 0.94, for laboratory spectral data. Ben-Dor et al. (2002) applied MLR to predict soil gravimetric moisture using DAIS-7915 hyperspectral sensor and obtained an R^2 of 0.64 for the calibrated model. Later, Zhang et al. (2012) used the entire spectral range measured by the Hyperion to model soil gravimetric moisture using PLSR. The authors reported an R^2 of 0.79, RMSE of 3.13% and RPD of

2.22 for the calibration dataset (cross-validation), and R^2 of 0.40, RMSE of 5.12% and RPD of 1.15 for the validation dataset.

The soil mineralogy has also been studied using Space SS. Galvão et al. (2001) evaluated AVIRIS data to quantify TiO_2, Fe_2O_3 and Al_2O_3 contents and reported R^2 values of 0.74, 0.83, and 0.68, respectively. Bartholomeus et al. (2007) observed correlation between spectral bands of ROSIS hyperspectral sensor and iron content determined by the dithionate-citrate method, as high as 0.5, for spectral measurements near 650 nm. Lagacherie et al. (2008) and Gomez et al. (2012) predicted $CaCO_3$ content using band depth measurements (near 2341 nm) from HyMap data and obtained R^2 values between 0.47 and 0.77. Baptista et al. (2011) applied a spectral index (RCGb) to estimate soil weathering degree using HyMap and reported a 0.75 R^2 between the proportions of kaolinite and gibbsite in soil samples. Alterations of minerals are also evaluated, for example, Molan et al. (2014) used HyMap to map the distribution of altered clay minerals in Iran.

In terms of the soil chemical attributes predicted by Space SS, De Tar et al. (2008) estimated Ca, Mg, K, P and pH using a MLR approach and reported R^2 values of 0.62, 0.58, 0.64, 0.70 and 0.62, and RMSE values of 9.51 meq L^{-1}, 4.32 meq L^{-1}, 45.03 mg kg^{-1}, 4.32 mg kg^{-1} and 0.076, respectively, using AVNIR hyperspectral data. Hively et al. (2011) also evaluated predictions of Ca, Mg, K and pH using HyperSpecTIR sensor data and obtained R^2 values from 0.51 to 0.69. Gomez et al. (2012) report a less optimistic result for pH reaching a maximum R^2 of 0.31. Lu et al. (2013) used Hyperion imagery and PLSR to predict soil total P and pH reporting R^2 of 0.62 and 0.68, respectively. Using hyperspectral images (AISA-Dual VIS—NIR), Lagacherie et al. (2013) predicted soil attributes such as clay, sand and CEC, in profiles with R^2 between 0.71 and 0.81 for topsoil samples. Gomez et al. (2012) report a less optimistic result reaching a maximum R^2 of 0.31. Lu et al. (2013) used Hyperion imagery to predict soil total P and pH reporting R^2 of 0.62 and 0.68, respectively. Using hyperspectral images (AISA-Dual VIS—NIR), Lagacherie et al. (2013) predicted soil attributes such as clay, sand and CEC, in profiles with R^2 between 0.55 and 0.81. Gomez et al. (2012) quantified CEC and obtained an RMSE of 1.84 $cmol_c$ kg^{-1} using PLSR with HyMap data. Using Hyperion, Lu et al. (2013) reported cross-validation RMSE of 1.55 $cmol_c$ kg^{-1} for CEC quantification. Figure 15.4a demonstrates the potential of Hymap data to develop a soil map of iron (Richter et al., 2007). The masked and no-iron extractable areas are associated with other components in the image, such as water surfaces or high vegetation cover areas. As another example, Figure 15.4b shows EnMAP sensor-based predictions for clay content in topsoils of exposed surfaces in agricultural areas of southern Spain, where error estimates were about $R^2 \sim 0.5$ (Chabrillat et al., 2014). Another interesting work (van der Meer and de Jong, 2003) takes in account a discussion about hyperspectral sensing and band-depth on the quantification of soil attributes. The authors indicate a simple linear interpolation method introduced to estimate absorption-band parameters from hyperspectral image data. By applying this hyperspectral data it has been demonstrated that absorption feature maps correspond favorably with the main alteration phases characterizing the systems studied. Thus the derived feature maps allow enhancing the analysis of airborne hyperspectral image data for surface compositional mapping. Despite the quantification of soil attributes, images have great importance to provide information of their spatial distribution as observed by Seid et al. (2013).

The soil properties prediction and mapping based on spectral sensing data have been done for a while. By far, the most studied soil properties are clay content and SOC (Table 15.3). This is related to the fact that clay and SOC, as well as mineralogy and water content, are spectrally active attributes in soil (Stenberg et al., 2010). Moreover, these attributes may potentially be predicted with high accuracy. Although attributes like soil cations are not spectrally active, these might be predicted correctly using remotely sensed radiometric data. In these cases, Stenberg et al. (2010) states that these attributes variation on an area is related to spectrally active attributes and consequently satisfactory coefficients of determination (R^2) can be obtained. Considering the R^2 reported for clay and SOC prediction, it is highly variable with a range of 0.01–0.9. This fact proves that although the technique (Multi and hyperspectral) is capable of predicting soil attributes, other conditions need to be accomplished in order to obtain models considered robust. One of these conditions is to correct the radiometric data regarding the influence of atmospheric components, which significantly improves the quantitative potential of soil spectra. Besides, soil moisture contents and non-photosynthetic vegetation mixture with soil in the imaged pixels reduces the predictive potential of algorithms.

Many techniques have been applied to soil properties prediction, but these ones related to multivariate statistics are predominant. When considering multispectral sensors the multiple linear regression is the most applied. On the other hand, the Partial Least Square Regression (PLSR) are recurrent in hyperspectral data modeling (Table 15.3). Furthermore, there is currently a trend in using multispectral images as secondary data in the mapping process, using concomitant other information easily accessible, such as DEM and terrain attributes derived from these. Due to the constrained spectral resolution present in these sensors, using multispectral images as a primary source on soil modeling might be a limiting factor. Moreover, with the availability of new hyperspectral sensors, the diffusion of more robust statistical algorithms, in addition to greater capacity for computational processing, turned the hyperspectral sensing more reliable as a primary source for soil prediction, considering the great quantity of information provided by this kind of data.

In a further and detailed observation of Table 15.3 we observe the following: (1) R^2 results varies from very low (under 0.3) to high (over 0.8) and very strong (over 0.9); (2) in a general vision, multispectral sensors with low number of bands such as IKONOS shows the lowest results; (3) others multispectral data, and the most used is Landsat satellite, has variable, but rather good results with a strong background on processing and understanding over the last 44 years; (4) hyperspectral information should be the best results, as previously discussed, on the other hand, many of them are similar to Landsat data. This can be due to the younger experience with these sensors when related with Landsat, thus these will probably be improve in the future with upcoming experience and these techniques have to be encourage; (5) interesting to see that several soil attributes have been achieved since, clay, sand, silt, carbon, organic matter, Ca, Mg, K, P, N, carbonates, iron, and many others; (6) many airborne hyperspectral sensors have been used and tested such as Hymap and Spectir, which indicates that the community is going toward these equipment; (7) considering repeatable results, the most reliable soil attributes to be quantified are clay, OM, OC, carbonates and Fe_2O_3; (8) there are great results for other elements such as N, Al, Mn, Zn, Cl, and Fe, but these have very low content and low effect directly in spectra, and thus, it is suggested more studies to comprove that the results are reliable.

Finally, it is interesting to observe that, for example, clay soil attribute has great results from aerial or orbital, multi or hyperspectral sensing and, in many cases, have similar R^2 when compared with observations when quantified with ground sensing (Table 15.2). The point is that theoretically, results should present the following *sequence* from the best to the worse results: laboratory sensing—field sensing—aerial hyperspectral sensing—orbital hyperspectral—orbital multispectral sensing. On the other hand, the results observed in literature, as indicated in these sections (and some summarized in Tables 15.2 and 15.3), takes us for the following points: (1) there are overlapping results between the *sequence*, b(2) we can find a multispectral sensor with a better result of a hyperspectral or vice versa; (3) we can have a laboratory result worse than an orbital multispectral data; (4) we can have very low results for chemical elements such as K with laboratory sensor and a good result from a hyperspectral airborne sensor. This takes us to some conclusions: (1) the quantification of soil attributes is a reality and can be done from any platform; (2) some soil attributes have more literature consistency because are repeatable and have theoretical background that explains the results, not lying only on R^2 values; (3) the differences have several factors such as: number and homogeneity or heterogeneity of data, quality of data, atmospheric correction factors, date and quality of images, quality of soil analysis, soil moisture effects when analyzing aerial or orbital data, quality of the sensor data, statistical analysis, spectra data processing methods, and others. Thus, it is important to take these in account in future studies looking towards reaching better results.

15.6.2.3 Strategies for Thermal Infrared

Thermal information has been widely used for geology (e.g., van der Meer et al., 2012), and now its importance to soil evaluation is being explored. On the other hand, it has been very poorly explored for soil studies, despite its exciting approach. The determination of soil temperature can be related with soil organisms, organic matter alterations, soil moisture, culture development, and others. Some works have explored this approach. Zhan et al. (2014) shows that thermal space spectral sensing can be used to estimate soil temperatures. The results was generated using data from the MODIS satellite and demonstrated that soil temperatures with spatial resolution of 1 km under snow-free conditions can be generated at any time of a clear-sky day. Comparison between the MODIS and ground-based soil temperatures shows that the accuracy lies between 0.3 and 2.5 K with an average of approximately 1.5 K.

Using another strategy, Zhao et al. (2014) used time-series remotely sensed data, including thermal imagery extracted from MODIS combined with vegetation indices (Soil Adjusted Vegetation Index—SAVI) calculated from Landsat ETM+ to estimate the spatial variation of SOM by land surface diurnal temperature difference (DTD) in China. They suggested that time-series remotely sensed data can provide tools for mapping SOM. Soliman et al. (2011) described a method by thermal inertia using standardized principal component analysis applied to a time series of thermal infrared images. The images were taken from a camera mounted at a height of 17 m above ground level using a mobile hydraulic boom lift and results were well related with intrinsic soil physical properties, such as soil bulk density and porosity.

15.6.3 Strategies for Soil Classification

Soil classification is a dynamic procedure which requires knowledge on soil science and spatial sciences. The data derived from the soil profile (control section) analysis is the most important information for soil classification of a pedon (smallest, three-dimensional unit at the surface of the Earth that is considered as an individual soil). Many classification systems require soil characterization from the laboratory as well as a complete morphological description of the soil horizons. The morphological evaluation of soil structure in the field, for example, requires determination of the type, shapes and size of soil structural peds, where the soil structure is determined by the activity of soil biota (macro- and microorganisms), clay content, mineralogy, organic matter and soil aggregation. This requires opening soil pits of variable dimensions where the soil is described and classified based on the upper 2 m profile, This is a time consuming procedure. Thus, the advantage of using SS on soil classification is to infer soil properties using sensors.

The radiometric data can be used to aid soil classification following three main approaches: (a) by quantifying soil properties and then using them as input data for the taxonomic classification; (b) by analyzing the spectral curves of single horizons in search for class-specific spectral patterns, since every spectrum constitutes a unique fingerprint that relates to the soil horizon properties; and (c) by analyzing the spectral curves of all soil horizons simultaneously, combining and interpreting this information to achieve a soil taxonomic classification. The last strategy can only be done using ground-based sensors, directly in field or by bringing sampled horizons into the laboratory. Some methods have been proposed to do this type of analysis, for example: going inside a pit with the spectroradiometer to collect radiometric data (Viscarra Rossel et al., 2009); inserting an optic fiber in a hole and evaluating the spectra at different depths or horizons (Ben-Dor et al., 2008); or collecting samples with an auger or hydraulic core and analyzing them using spectral data measured in laboratory (Demattê et al., 2004a; Waiser et al., 2007; Morgan et al., 2009)—see Figure 15.2. When using soil surface sensors (orbital, aerial or ground) the approach is different, since only surface information can be assessed this way (Figure 15.2). Hartemink and Minasny (2014) have recently discussed the importance and future of soil morphometrics, where they describe several soil sensors that can assist in soil classification, including VIS-NIR-SWIR-MIR radiometers, ground penetrating radar, electrical resistivity meters, cone penetrometer, hyperspectral core scanner, X-ray fluorescence meter, and others. Despite these possibilities there is one important limitation: until today no equipment was able to indicate directly the shapes and sizes of peds, and thus, technologies still rely on other (chemical and physical) qualities of the soil profile to infer soil classification. On the other hand, to reach a soil classification we also need to understand the behavior of the attributes in different soils, such as made for class texture (Franceschini et al., 2013), clay activity (Demattê et al., 2007b), weathering indexes (Galvão et al., 2008), mineralogy (Madeira Netto, 2001) and electrical conductivity (Ucha et al., 2002). Other interesting techniques have evoluted in this area. Schuler et al. (2011) indicate that the gamma spectrometry is a promising potential as a tool to distinguish Reference Group WRB soil profile, field, and the landscape scale, but needs to be verified in other regions of the world.

15.6.3.1 Strategies for Ground Spectral Sensing

One of the first attempts to classify soils using spectra was performed by Condit (1970) and complemented by Stoner and Baumgardner (1981). In general, evaluation is done based on the spectrum of each horizon individually (Galvão et al., 1997). Later on, Demattê (2002) proposed the analysis of the spectral morphology considering the following aspects of the curve: complete shape; reflectance intensity (albedo); and absorption features, which were used to characterize several different soils from the tropics (Nanni et al., 2011, 2012; Nanni et al., 2014). This has recently had an upgrade with the development of the M Morphological Interpretation of Reflectance Spectrum (MIRS), Demattê et al. (2014), which indicate a detailed system for users to look at spectra shapes looking towards soil classification.

On the other hand, quantitative methods have been used to directly predict soil classes from soil spectra. For example, Vasques et al. (2014) introduced a system where the spectra of different soil layers (depths) are evaluated at the same time, and reached 90% accuracy for some classes. Bellinaso et al. (2010) reached an 80% agreement when between classified soils by spectra and soils classified in the field following traditional soil survey protocols, as corroborated by Viscarra Rossel and Webster (2011). Thus, it is important to aggregate quantitative and descriptive analysis to reach the best result on soil classification.

15.6.3.2 Strategies for Space Spectral Sensing

Unlike in Ground Spectral Sensing, data acquired by Space Spectral Sensing are usually not enough to classify soils accurately. These aerial or orbital sensors only measure the superficial layer of soil, and consequently do not represent the entire profile. Despite that, Space SS can still have a great contribution in soil discrimination and mapping by allowing to detect spatial patterns in soil surface variation that relate to the spatial distribution of soil classes. Several studies using Landsat data (Demattê et al., 2004c, 2005, 2007a, 2009a; Nanni et al., 2011, 2012, 2014; Genú et al., 2013b) demonstrated a high potential for soil discrimination. Basically these studies used images atmospherically corrected and transformed into reflectance, detected bare soil spots, and correlated these pixels with the field soil classification. They found high correlation among the spectral curves of similar soil classes (see example in Figure 15.2, Strategy 1). The authors indicated that the soil surface patterns are related to underlying soil variation and dynamics within the soil profile that are specific to each soil class, allowing this correlation. On the other hand, authors emphasized that this correlation is not present in all cases. It is possible that two very different soils (e.g., an Arenosol and a Lixisol) have a sandy surface which could present similar spectra and thus not be discriminated. In this case, we would have to aggregate a new strategy to outline this problem, such as the relief information. We have to underline that this strategy can allow the *separation* of different soils without necessarily giving them *names* (classes). This is helpful to find similar soils and delineate soil mapping or management units, for example. Their names (the classification itself) can be given in the field using other strategies (Figure 15.2, Strategies 2 and 3).

15.6.4 Strategies for Soil Class Mapping

Soil class (survey) maps, when used at the appropriate cartographic scale, allow the improvement of the land productive capacity preventing environmental degradation. New methodologies and strategies are needed to map soils in unmapped areas efficiently fostering sustainable food production. Soil SS have the potential to assist users in efficient soil mapping. Like the other applications, the user has to understand that each SS technique, based on field, laboratory, airborne or orbital sensors, needs to be explored and adapted to the different objectives presented, considering their advantages and limitations, the desired characteristics of the final products, including the type of product (model, map, or both), scale and level of accuracy, the final users of the information, etc.

15.6.4.1 Strategies for Ground Spectral Sensing

When spatial data is available a soil map can be created by following this suggested sequence: (1) assign the sampling points where soil samples will be collected in the field, based on methodology described in Section 15.6.1; (2) describe the soil and collect the soil samples at the sampling sites; (3) take samples to the laboratory and prepare them for spectra readings; (4) acquire, process and interpret spectral data; (5) based on spectra information, choose the most representative which will go to traditional wet analysis; (6) wet versus spectra ones, will create models for quantification. The other samples will be quantified by spectra information; (7) relate spectral data to known soil properties and classes; and (8) interpolate the known soil information across the area of interest using spatial interpolation techniques. With portable equipment and appropriate calibration models, the third step can be replaced by spectral measurements made directly in the field. Field SS measurements produce less accurate predictions of soil properties than Laboratory SS. However, more observations can be made because there is no need for transport or sample preparation. Nonetheless, Ground SS measurements can be augmented by traditional soil profile evaluation in the laboratory for a more accurate classification of the soil unit, as suggested in Section 15.6.4. Demattê et al. (2001, 2004a) have demonstrated how soil spectra data measured at sampling points can be used to derive soil maps. These authors used ground-based spectroscopy for pedological purposes, collecting spectra (VIS-NIR-SWIR) in soil profiles along landscapes and relating spectra to soil classes using a spectral pedological databank with descriptive and quantitative information associated to the radiometric measurements. These methods were used to create a pedological map that was well correlated (79%) with the results of the traditional mapping approach.

15.6.4.2 Strategies for Space Spectral Sensing

Soil sensing data are valuable for providing either primary or auxiliary information, which can be used to interpolate soil properties assessed using other techniques. Preferably, the spectral imagery must be atmospherically corrected and transformed into reflectance before use. Generally only pixels with bare soil are considered to study soils using SSS, and techniques to identify these pixels must be applied before the soil characterization. Reflectance values at each pixel of designated soil observations can be related to image data bases of surface spectra. Though the use of orbital/aerial SS data does not provide direct measurements beyond topsoil spectra, this information is a first indication of soil spatial variability and can be combined with undersurface soil properties from another data source to create complete soil maps. For example, Demattê et al. (2012b) combined aerial photographs, radar and laboratory spectral information to develop soil maps. In fact, the use of orbital images in bare soil mapping is restricted in areas under haze or clouds, under shades, and under vegetation (crop fields or natural vegetation). Also, in situations where legacy soil data are scarce or unavailable, remotely sensed soil data can be an important source to fill data gaps, for example, by interpolation (McBratney et al., 2003; Mulder et al., 2011).

Boettinger et al. (2008) used Landsat data as environmental covariates for digital soil mapping in arid and semiarid regions. In this example, data successfully represented environmental covariates for vegetation (e.g., normalized difference vegetation index—NDVI, fractional vegetation cover) and parent material (e.g., band ratios diagnostic for gypsic and calcareous materials). Browning and Duniway (2011) presented a semi-automated method to map soils with Landsat ETM+ imagery and high-resolution (5 m) terrain (IFSAR) data. Later, Grinand et al. (2008) predicted soil distribution using Landsat ETM imagery, terrain factors (e.g., elevation and slope), land cover and lithology maps. Hengl et al. (2007) employed multiple covariates including six terrain parameters, MODIS enhanced vegetation index (EVI) images, and a polygon map of 17 physiographic regions of Iran, to map soil classes. Hansena et al. (2009) used coarse resolution soil maps, combined with NDVI, normalized difference infrared index (NDII), SWIR reflectance, slope, and two relative elevation layers to map soils in Uganda. Although with low spatial resolution (1 km), Hou et al. (2011) used MODIS to identify different soil types and reached a Kappa index of 0.75.

The importance of images in soil mapping is more than simplifying aiding in soil classification. They help to detect geographic boundaries among soils to delineate soil mapping units. In fact, Nanni et al. (2014) tested several SS sources, from ground sensors and spectra from images, to evaluate which source could reach better performance for soil mapping. The traditional soil survey map had 53 polygons, while the ground spectra determined 22 polygons, and Landsat gave 35 polygons. Working with aerial photographs, Demattê et al. (2001) defined 12 polygons versus 15 polygons delineated by traditional field soil survey.

Considering the use of hyperspectral imagery, Galvão et al. (2008) described the use of AVIRIS data to map several soil properties related with soil classification such as Fe_2O_3 and Ki (weathering index)—Figure 15.5. They mapped the selected soil properties using hyperspectral images and found good agreement between weathering indexes and elevation. This Chabrillat et al. (2002) also used AVIRIS data, but mapped soil expansive clays (Figure 15.6a) considering differences between spectral signatures of mineral endmembers (Figure 15.6b). Chabrillat et al. (2002) examined the potential of optical remote sensing and in particular hyperspectral imagery for the detection and mapping of expansive clays in the Front Range Urban Corridor in Colorado, USA. Here spectral endmembers were extracted from the images without field knowledge and were implemented with different algorithms for clay mapping. Results showed that spectral discrimination and identification of variable clay mineralogy (smectite, illite/smectite, kaolinite) related to variable swelling potential was possible using different algorithms, in presence of significant vegetation cover. Field checks have shown that the maps of clay type derived from the imagery and interpretation in terms of swell hazard are accurate. The main limitations for the expansive clay soils detection were in case of a heavy vegetation cover (forest or green grass), and when the reflectance of the soils is approximately 10% or less.

15.6.5 Integrating Strategies

Mapping soils accurately and at detailed enough spatial resolutions require merging multiple types of spatial data and statistical techniques to maximize soil inference from these datasets. Spectral Sensing data from orbital and airborne platforms can provide information supporting Digital Soil Mapping (DSM). Spectral Sensing involving field and laboratory passive optical sensors and active microwave instruments have been used at regional and coarser scales to map soil mineralogy, texture, moisture, SOC, salinity and carbonate content. Ballabio et al. (2012) employed a vegetation-based approach for mapping SOC in alpine grassland. They used a map of the properties of the plant communities created by combining high-resolution multispectral images and LIght Detection And Ranging (LIDAR) data. Additionally, McBratney et al. (2003) discuss methods to find quantitative relationships between soil properties or classes and their environment. One issue of particular difficulty in DSM is the combination of data collected at different spatial resolutions and with different accuracies. For example, a soil pedon description and subsequent laboratory analysis contains the highest quality information but the spatial extent is extremely sparse. Ground SS data, for example, collected across a soil profile, can be more detailed, but still done on a point support. Moreover, if

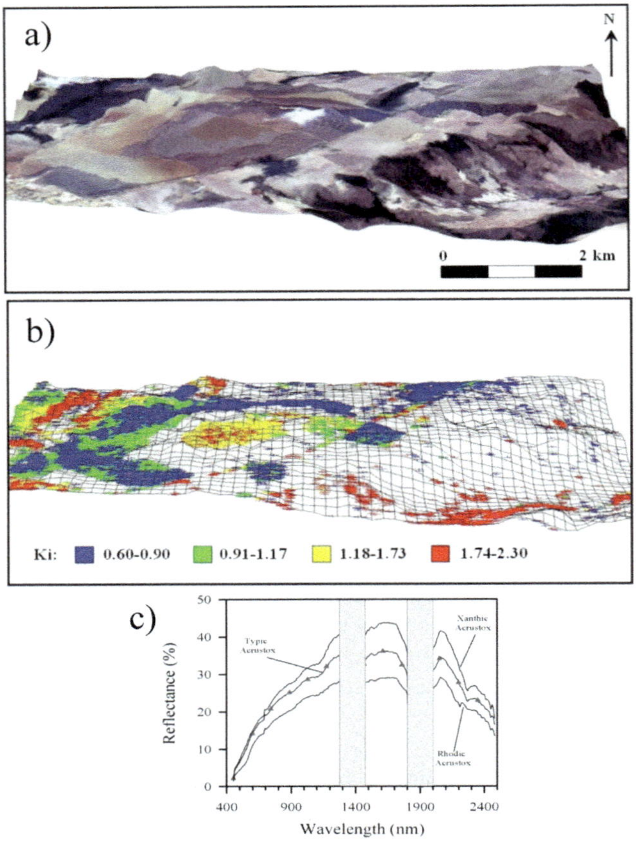

FIGURE 15.5 (a) Digital elevation model, (b) map of weathering index Ki (Al_2O_3/SiO_2) from surface, with higher values indicating more weathered soils, and (c) spectral curves generated by the AVIRIS hyperspectral sensor. Ki: higher values lower weathered soils. (Extracted from Galvão et al. 2008.)

ground SS data is collected without proper soil description, its use is limited for soil mapping. On the other hand, space SS data provides seamless ground coverage to assess soil spatial variation but their spectral resolution and signal-to-noise ratio are usually poorer than laboratory and field sensors. These three types of data acquisition can be combined using a neighborhood-type statistical technique (Zhu et al., 2004). Additionally filtering techniques can be employed to overlay multiple scales of remotely sensed soil data as suggested by Behrens et al. (2010).

15.6.6 Strategies to Infer Soils from Vegetation

Spectral Sensing data obtained directly from bare soils are not always available to allow soil characterization. Often soil, photosynthetic vegetation (PV) and non-photosynthetic vegetation (NPV) occur associated on the landscape and the fractions they occupy in pixels vary with land use and environmental characteristics of the studied area. For this reason, the use of SS to study soils in densely vegetated areas relies on indirect relations between vegetation and soil properties (Mulder et al., 2011). Vegetation has been used to assess soil characteristics and variability in a number of studies (Bartholomeus et al., 2011). For example, Asner et al. (2003) related SOC and N to fractional cover by PV and NPV and were able to show the trends in these soil properties at the ecosystem level. Kooistra et al. (2003, 2004) used vegetation spectral characteristics obtained by a handheld spectrometer in the field, to estimate Zn, Ni, Cd, Cu, and Pb in a floodplain, indicating the potential

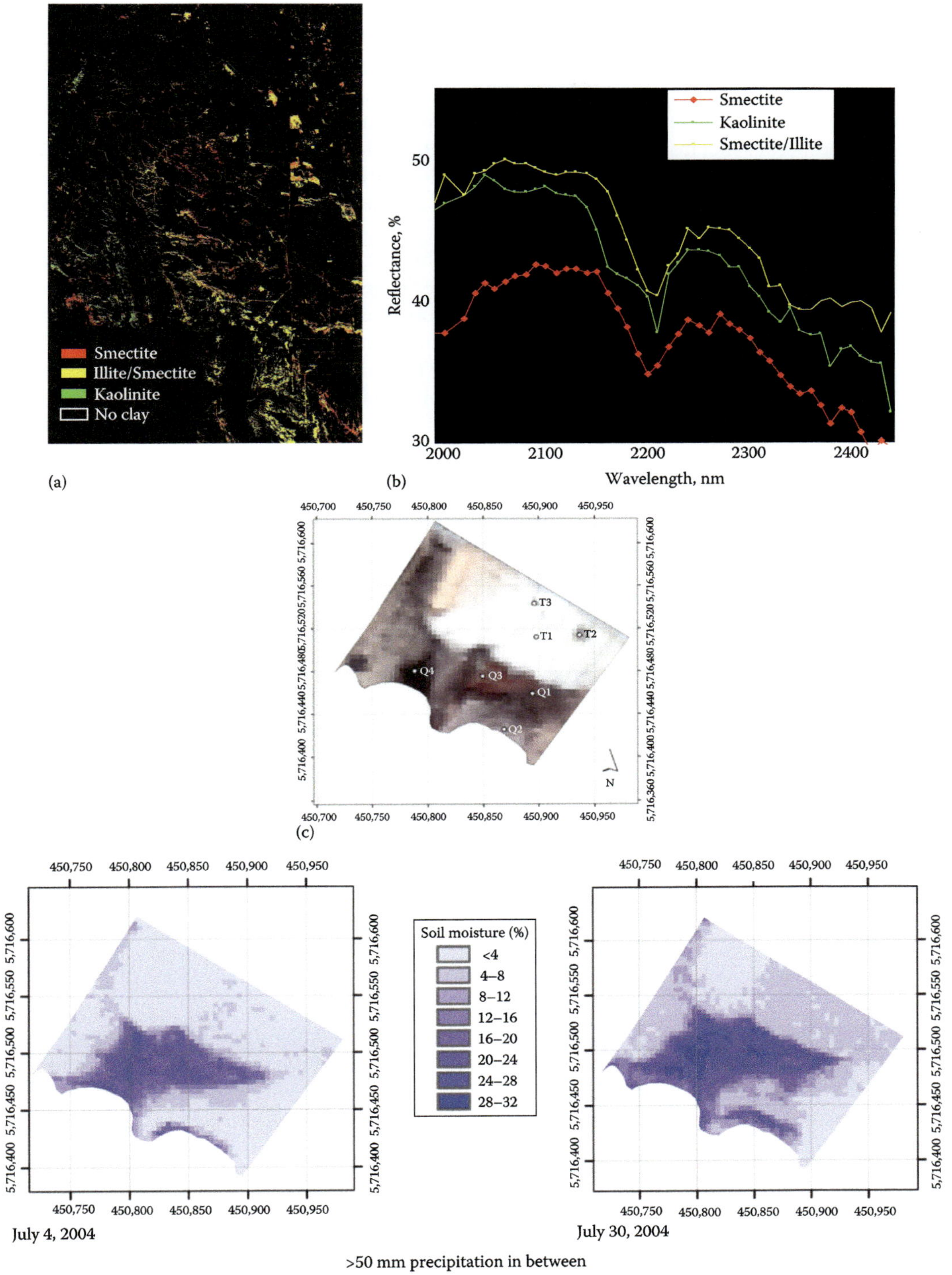

FIGURE 15.6 (a) Map of soil clay type in Colorado, USA, produced from the RGB image of three abundance maps, extracted from AVIRIS high-altitude images covering an area of ~20 × 37 km (Chabrillat et al., 2002); (b) Clay endmembers spectra (Chabrillat et al., 2002); (c) Multitemporal surface soil moisture maps in Welzow Bio-monitoring Recultivation Area, Germany, produced from HyMap airborne images based on NSMI method (Haubrock et al., 2008a, 2008b).

of using SS techniques for the classification of contaminated areas. Vegetation indices, like NDVI, have been applied as indicators of crop growth and site quality (Sommer et al., 2003). Sumfleth and Duttmann (2008) found, in a study carried out in paddy soil landscapes in southeastern China, that NDVI values are related to SOC, N and silt contents, and to some terrain attributes. In specific cases, the combination of several spectral indices can minimize the vegetation influence on estimated soil properties like iron content, as verified by Bartholomeus et al. (2007). Mann et al. (2011) used NDVI, among other covariates, to map the productivity of a citrus grove, and stated that the productivity zones could be used successfully to plan soil sampling and characterize soil variation in new fields.

Mulder et al. (2011) indicated two possible approaches to retrieve soil properties from vegetation SS data: through Plant Functional Types (PFT), and Ellenberg indicator values. The former states that resource limitations, response to disturbance, biotic factors and other environmental aspects are related to morphological and physiological adaptations in plants, and the ability to detect functional types with remote sensing relies on the extent to which such relationships are generalized (Ustin and Gamon, 2010). As an example, Schmidtlein (2004) used the spectral characterization of PFT to discriminate different soil units. In the same way, Ellenberg indicator values can be used to retrieve soil properties, like soil moisture, pH and fertility, and hyperspectral imagery can be used to derive these values (Schmidtlein, 2004; Mulder et al., 2011).

15.6.7 Strategies for Soil Management and Precision Agriculture

The main goal of Precision Agriculture (PA) is site-specific management, considering soil and crop factors to maximize the efficiency of the applied resources. These resources include seed (seeding rates and site-specific genetics), fertilizers, soil amendments such as lime, herbicides, pesticides, and growth regulators. Information needed by PA include high resolution (< 10 m) maps of soil properties, such as plant available water and nutrient status in the rooting zone. Thus, accurate indication of soil properties with dense spatial coverage is important, and this can be done with SS. For example, VIS-NIR models were used by Wetterlind et al. (2010) to assist field soil characterization using small local calibrations and national libraries, reaching good results. Spectral Sensing is particularly useful to PA because it allows, besides posterior soil characterization, real time assessment of soil properties and their variation. This is the most successful data source that might relate the wellbeing of the crop to nutrient needs. Many PA users desire soil sensing strategies that assess nutrient content in the soil linking to plant needs, although this is a difficult task as already demonstrated in Section 15.6.2.1. Looking at this approach, Tekin et al. (2013) used AgroSpec (Tec5) mobile, fiber type, VIS—NIR spectrophotometer to measure on-line soil pH (field measurement). They produce variable rate lime recommendation maps with results $R^2 = 0.81$, RMSEP = 0.20 and RPD = 2.14.

Today, Spectral Sensing can assist in PA implementation providing data on soil properties such as clay content, SOM, SOC and CEC, which are related to nutrient supply and retention as well as productivity potential. Nonetheless, laboratory and controlled experiments are important to understand the crop response to the environment, including soil variability, and perhaps can provide a better basis for relating measurable soil properties to nutrient availability and soil water storage, among others. One attempt to combine lab knowledge with spectroradiometer data is described by Demattê et al. (2003a). They identified alterations in soil spectra due to fertilizer application but concluded that it was not possible to identify P in soils. However, lime application in the soil affected soil spectra (Demattê et al., 2003a). This was corroborated by Araújo et al. (2013), who reached a 0.90 R^2 predicting liming requirement for a sandy and a clayey Oxisol using spectroscopy.

Most SS studies focusing on PA use Ground SS combined with lab-measured soil properties (clay, SOC, SIC, and others). A review on GSS is provided by Kuang et al. (2012). Mouazen et al. (2007) developed a soil sensing system consisting of an optical probe mounted on a soil penetrometer to measure SOC, SM, pH and P. Calibration models were developed using laboratory soil data, and were validated using spectra from field measurements. Estimation of soil moisture content was satisfactory ($R^2 = 0.89$), whereas the estimates of SOC, pH and P were not as well matched to the corresponding reference values (R^2 of 0.73, 0.71, and 0.69, respectively). Later, Kodaira and Shibusawa (2013) found a strong correlation between CEC ($R^2 = 0.89$) and on-the-go spectroscopy data.

To actually measure the soil solution, Viscarra Rossel and Walter (2004) built a soil analysis system comprising a batch type mixing chamber with two inlets for 0.01 M $CaCl_2$ solution and water, respectively. The system was tested in the laboratory using soil solutions of 91 Australian soils, and in a 17-ha agricultural field to estimate lime requirements. The system produced an RMSE of 0.2 for pH ($R^2 = 0.66$). However, the coefficient of determination for pH buffer estimates was not high ($R^2 = 0.49$). Despite this, results indicate a great potential and research is still required. Due to this, Ballari et al. (2012) propose the use of a network of mobile sensors associated with the expected value of information (Evoi) and mobility restrictions to reduce the costs of monitoring phenomena such as soil and natural radioactivity. Christy (2008) used a field spectrophotometer to provide several soil attributes in real time, obtaining RMSE of 0.52% and 0.67 R^2 to soil organic matter.

The success of on-the-go spectral sensing, thus, has several tasks. Gras et al. (2014) focused at optimizing the acquisition procedure of topsoil VNIR spectra in the field with the view to predict soil properties. Obtained good VNIR cross-validation for calcium carbonate, total nitrogen, organic matter and exchangeable potassium, RPD reached up to 9.1, 2.9, 2.8 and 3.0, respectively.

15.6.8 Strategies for Soil Conservation

To perform land use planning and promote soil conservation, information about a specific area is needed such as relief and slope data, erosion susceptibility, soil classes, vegetation cover, among management factors. Additionally, it is also necessary to identify areas that have already exhibited problems, so that corrective measures can be taken. Spectral Sensing can be used to identify possible problems in the soil, and monitor the effects of management decisions by looking at soil and plant spectral responses. Brodský et al. (2013) Mapped the soil organic carbon through VNIR spectroscopy with R^2 over 0.7 and RPD over 1.5 in eroded areas at the farm level. King et al. (2005) and Vrieling (2006) presented an interesting review describing erosion mapping techniques integrating inputs derived from Space SS and additional data sources into runoff and erosion prediction models. Similarly, Shruthi et al. (2014) detected erosion effects and monitored variations in erosion dynamics and degradation levels using Ikonos-2 and GeoEye-1. Both D'Oleire-Oltmanns et al. (2012) and Peter et al. (2014) used an Unmanned Aerial Vehicle for monitoring soil erosion in the Souss Basin (Morocco) where the imagery data were used to quantify gully and badland erosion in 2D and 3D and to analyze the erosion susceptibility of surrounding areas.

Different acquisition levels of SS have been combined for soil conservation studies. Garfagnoli et al. (2013) used a hyperspectral dataset acquired with an airborne Hyper SIM-GA sensor from Selex Galileo simultaneously with ground soil spectral signatures to monitor soil degradation processes. Liberti et al. (2009) assessed how accurately a badland area can be identified from Landsat TM and ETM data. The authors found that the combined use of SS and auxiliary morphological information significantly improved the mapping of badlands over large areas with heterogeneous landscape features. With another approach, Nadal-Romero et al. (2012) assessed badland dynamics using multi-temporal Landsat TM and ETM imagery for the period 1984–2006 in Spain, and the results showed that NDVI helped in revealing degraded areas.

Martínez-Casasnovas (2003) presented a method to compute the rate of retreat of gully walls and the associated rate of sediment production caused by erosion by integrating multitemporal aerial photos and multi-resolution DEMs in Catalonia, Spain. Ries and Marzolff (2003) designed a hot-air blimp as a platform sensor to obtain large-scale aerial photographs from the Barranco de Las Lenas (Spain) with very high spatial and temporal resolution for monitoring the development and dynamics of erosion. Using aerial photo interpretation in southern Italy, Conforti et al. (2013) by, as combined with GSS in the VIS-NIR spectral range with aerial photo interpretation and geostatistics, used to predict SOM content and mapping and relate with water erosion process. Vågen et al. (2013) combined Landsat ETM+ imagery, systematic field methodologies, infrared (IR) spectroscopy and ensemble modeling techniques for landscape-level assessments of land degradation risk and soil condition. The Landsat prediction was robust, with R-squared values of 0.86 for pH and 0.79 for soil organic carbon and were used to create maps for these soil properties. Moreover they developed models for mapping soil erosion and root depth restrictions, with an accuracy of about 80% for both variables.

15.6.9 Strategies for Soil Monitoring

Soil monitoring implies observing the soil over time providing data/information to assure that it stays healthy (chemically, physically and biologically) and secured against environmental or human degradation. Soil and food security are priorities in today's global agenda, and soil monitoring is an essential activity for achieving sustainable food production and sustainable development. Specifically, soil monitoring studies are also interested in observing the temporal, spatial, and concentration changes in organic and inorganic contaminants in the soil. However, perhaps the most common soil monitoring applications relate to the maintenance of soil fertility over time. In this context, SS can be used to monitor soil condition (i.e., quality) spatially and temporally to support management decisions that enable soil stabilization and improvement.

15.6.9.1 Strategies for Ground Spectral Sensing

Kemper and Stefan (2002) used VIS-NIR reflectance spectroscopy to estimate concentrations of contaminants in soils in Spain with R^2 of 0.82 for As, 0.96 for Hg, 0.95 for Pb, and 0.87 for S. Vohland et al. (2009) performed a spectroscopy approach to quantify the same previous elements in floodplain soils using VIS-NIR laboratory data and reported R^2 values between 0.60 and 0.71. Jean-Philippe et al. (2012) detected several heavy metals, in particular Hg, with a performance prediction (R^2) of around 0.91. Song et al. (2012) observed relationships between Cr, Cu, and As, and absorption features caused by iron oxides, clay minerals, and SOM, suggesting they are strongly bounded to these soil constituents. Araújo et al. (2014b) used VIS-NIR-MIR spectroscopy to assess soil contamination with Cr by tannery sludge. They observed strong alterations of absorption features in selected wavelengths (500–600 nm and 2,600 wavelength cm^{-1}) and an overall decrease of reflectance intensity across the spectrum promoted by the sludge.

Brunet et al. (2009) monitored Chlordecone (a toxic insecticide) used in banana plantations in the French West Indies and determined its content by NIR spectroscopy in Andosols, Nitisols, and Ferralsols. Conventional analyses and spectral predictions were poorly correlated for Chlordecone contents higher than 12 mg kg-1. However, 80% of samples were correctly predicted when the dataset was divided into three or four classes of Chlordecone content. Chakraborty et al. (2010) quantified total petroleum hydrocarbons (TPH) in contaminated soils in situ by using VIS-NIR diffuse reflectance spectroscopy, and later mapped them in the field using the same spectral range (Chakraborty et al., 2012a, 2012b). To determine TPH's content in soil, PLSR and boosted RT models were used, and the best performance for validation showed R2 of 0.64 and RPD of 1.70. Schwartz et al. (2012) and Reuben and Mouazen (2013) detected diesel-contaminated soils by spectral sensing and assessed the relationships between petroleum hydrocarbon concentrations, soil moisture and clay content. Forrester et al. (2013) also developed calibration models based on PLSR using NIR and MIR data from a diffuse reflectance infrared Fourier-transform (DRIFT) sensor for predicting TPH concentrations in contaminated soils in southeastern Australia. The authors confirmed DRIFT spectroscopy associated with PLSR as capable to provide accurate models for TPH prediction, where the MIR range outperformed NIR deriving high quality predictions (RPD = 3.7, R2 = 0.93). A common by-product of sugarcane industry is the vinasse (fermentation residue), which can be used as an important K fertilizer. On the other hand, if used in high quantities, can pollute soils and make them very saline. Looking on monitoring this by product, Demattê et al. (2004b) observed differences by VIS-NIR information. Another interesting approach was performed by Shi et al. (2014b) which monitored arsenic in agricultural soils by spectral reflectance of rice plants. Other SS techniques has been used such as by Radu et al. (2013), which clarified that portable XRF can be used to track and quickly identify possible hot spots of pollution and trends in the elementary distributions.

15.6.9.2 Strategies for Space Spectral Sensing

In a review about Image Spectroscopy to study soil properties and applications, Ben-Dor et al. (2009) provided some case studies in which different aerial and orbital sensors, with distinct resolutions, were used. The cases addressed by the authors included a variety of soil science applications, such as soil degradation, salinity, erosion and deposition, mapping and classification, pedogenesis, contamination, water content, and swelling. Reschke and Hüttich (2014) used multitemporal Landsat data combined with high spatial resolution satellite data to extract sub-pixel information of coastal and inland wetland classes for environmental goals. Aerial or orbital detection of reflected radiation from solids has been widely employed in soil monitoring applications (Schwartz et al., 2013). These authors developed steps to identify hydrocarbons in soil.

Adar et al. (2014) studied the automatic identification of soil changes in different surface soil types by HySpex SS in the VIS-NIR and SWIR spectral ranges. This study identified a gradual change over time in SOM, soil crusting and compaction, and demonstrated that the SWIR range allowed better change detection than the VIS-NIR. Ghosh et al. (2012) used EO-1 Hyperion data to identify salt-affected soils, correctly identifying highly affected soils in 84.4% of the cases.

Regarding hyperspectral SS monitoring of heavy metals in soils, Shi et al. (2014a) discussed the applicability of SS for mapping soil contamination over large areas. The authors also presented methodologies to estimate heavy metal concentrations using VIS-NIR imaging spectroscopy and reported good results. In a similar research, Choe et al. (2008) used combined data of geochemistry, field spectroscopy, and hyperspectral sensing (HyMAP) to map heavy metal pollution in stream sediments in Rodalquilar mining area (Spain). In fact, when narrow bands are used they have the opportunity to provide greater focus on targeting specific waveband spots and is likely to provide greater accuracies or R-squares compared to broadbands and their indices, as shown in these papers.

15.6.10 Strategies for Microwave (RADAR) and Gamma-Ray

According to a review made by Mulder et al. (2011), the feasibility of determining soil texture, moisture, and salinity by active and passive microwave SS was scaled from medium to high, whereas the feasibility of determining land cover and degradation ranged from low to medium. Microwave SS has been applied to measure the thermal radiation emitted by bare soil which is mainly affected by its moisture content and temperature (Parrens et al., 2014), where soil temperature is also dependent on soil mineralogy and SOM (van Lier, 2010).

Alternatively, gamma-ray spectroscopy is related with the detection of three basic elements: uranium (^{238}U), thorium (^{232}Th) and potassium (^{40}K) (Minty, 1997). One of the advantages of microwave or gamma-ray sensors, compared with VIS-NIR-MIR reflectance meters, is that they are less influenced by vegetation cover and climate. One exception, for example, is very dense vegetation cover (as in tropical rainforest), where microwave and gamma rays might be attenuated by the vegetation.

15.6.10.1 Ground Penetrating Radar (GPR)

Ground penetrating radar (GPR) has been used since the 1970s. In fact, Johnson et al. (1979) determined the first information botu GPR in soil survey. The GPR emits electromagnetic radiation pulses to the soil, and variations in some soil properties such as moisture, porosity, salinity, SOM, texture, and mineralogical content (iron oxides, high clay activity, and others) affect this radiation

(Pozdnyakova, 1999). GPR uses wavelengths with high frequencies in the microwave spectral range (from 10 MHz to 2.5 GHz) (Daniels, 2004; Jol, 2009). Typically, interfaces between horizons (as in soil layering) produce changes in reflectance that are recognizable in GPR images (Stroh et al., 2001; Doolittle and Butnor, 2009). Since the 1980s, studies have demonstrated the applicability of GPR as a tool for the characterization of organic and mineral undersurface horizons regarding their thickness and lateral variability (Collins et al., 1990), and for the identification of lithic contact (Doolittle et al., 1988). GPR is primarily limited to soils with coarser texture, low electrical conductivity (Ucha et al., 2002), and higher (or high enough) moisture contents (Ardekani, 2013).

The use of GPR to identify argillic and cambic B-horizons has been useful for understanding pedogenetic processes (Inman et al., 2002) that are important for soil mapping. Doolittle et al. (2007) produced the "Ground-Penetrating Radar Soil Suitability Map of the Conterminous United States," which limited areas rated as being "Unsuited" for GPR to saline and sodic soils, reassessed calcareous and gypsiferous soils, and provided a mineralogy override for soils with low activity clay, where the efficiency of the equipment is restrict to a clay content of 35% (Mahmoudzadeh et al., 2012). The great majority of GPR applications in soil science are concentrated in hydropedology (Doolittle et al., 2012; Zhang and Doolittle, 2014), with studies on the variations in water table depths and groundwater flow patterns (Doolittle et al., 2006), vertical moisture dynamics in a soil profile (Steelman and Endres, 2012), and quantification of soil water content in the vadose zone (Minet et al., 2012; Yochim et al., 2013) at the field scale. Transport of contaminants and agrochemicals in the subsurface has also been investigated with GPR (Glaser et al., 2012; McGlashan et al., 2012). In fact, Yoder et al. (2001) identified offsite movement of waterborne agrochemicals using conventional soil survey combined with electromagnetic induction (EMI) and ground-penetrating radar (GPR). They concluded EMI mapping provides rapid identification of areas of high potential for offsite movement of subsurface water, GPR mapping of areas identified by EMI mapping provides a means to identify features that are known to conduct concentrated lateral flow of water, and combining the capabilities of EMI and GPR instrumentation makes possible the surveys of large areas that would otherwise be impossible or unfeasible to characterize.

Other uses of GPR include the assessment of soil porosity (Causse and Sénéchal, 2006), soil compaction (Tosti et al., 2013), and delineation of agriculture management zones (André et al., 2012).

15.6.10.2 Aerial and Orbital Radar

A good review about microwave Space SS and soil salinity was written by Metternicht and Zinck (2003). Bell et al. (2001) used an Airborne Polarimetric Synthetic Aperture Radar for mapping soil salinity in the Alligator River Region of the Northern Territory in Australia. According to Hasan et al. (2014), vegetation cover is still a main factor in the attenuation, scattering and absorption of the microwave emissions from the soil which impacts its brightness. These authors used the airborne Polarimetric L-band Multibeam Radiometer 2 (PLMR2) and the L-band Microwave Emission of the Biosphere (L-MEB) model to simulate microwave emissions from the soil-vegetation layer and to retrieve surface soil moisture in Germany (moisture retrieval with a RMSD of 0.035 m^3 m^{-3} when compared to ground-based measurements). NASA's Soil Moisture Active Passive (SMAP) mission will carry in 2014 the first combined spaceborne L-band radiometer and Synthetic Aperture Radar (SAR) system with the objective of mapping near-surface soil moisture with high (~3 km), low (~36 km), and intermediate resolutions (~9 km). For that, Panciera et al. (2014) conducted three experiments combining field and airborne sources to provide prototype data for the development and validation of soil moisture retrieval algorithms applicable to the SMAP mission.

The sensitivity of spaceborne SAR is well established for soil moisture. However, soil moisture monitoring can be confounded by the effects of vegetation and surface roughness. In this context, Singh and Kathpalia (2007) proposed an approach based on a genetic algorithm with inclusion of empirical modeling to determine the soil moisture, texture and roughness with backscattered data from ERS-2 SAR. Kornelsen and Coulibaly (2013) presented a critical review about technical and methodological advances, limitations and potential of SAR. They concluded that soil moisture estimation can be retrieved with multi-angular SAR without in situ measurements. Fatras et al. (2012) analyzed the potential of the radar altimeter aboard ENVISAT to successfully estimate the surface SM in a semi-arid region in Northern Mali, with correlation coefficients (r) higher than 0.8 between SM and the backscattering coefficient, and SM predictions with RMSE < 2%.

15.6.10.3 Aerial and Ground Gamma-Ray

Passive gamma-ray spectrometry is a fast and cost-efficient tool for developing a spatial map of soil properties related to clay content, mineralogy and soil weathering. It can be used proximally or through an airborne sensor. Initial airborne gamma surveys started with geologists for mineral and lithological exploration (Graham and Bonham-Carter, 1993). Later on, Cook et al. (1996) examined the ability of ground and airborne systems to detect the spatial distribution of soil-forming materials across the landscape, and distinguished highly weathered from fresher soil materials, which was also done by Wilford (2011). This usefulness was recently observed by Gooley et al. (2014). They used gamma-ray spectrometry from aerial and ground sensing to develop a Digital Soil Mapping (DSM) of the available water content (QWC) across an irrigated area through map soil properties.

Approximately 50% of the observed gamma-rays originate from the top 0.10 m of dry soil, 90% from the top 0.30 m reaching 95% from the upper 0.5 m of the profile (Taylor et al., 2002). Different factors can attenuate gamma-rays through the soil, such as moisture and bulk density. Radiation attenuation increases by approximately 1% for each 1% increase in volumetric water content, while a dry soil with a bulk density of 1.6 Mg m^{-3} causes a decrease in the radiation to half its value at each 10 cm (Cook et al., 1996). The radiation decrease caused by air is much smaller, for example 121 m are needed to reduce the radiation to half its value considering a 2 MeV source, thus making possible the detection of gamma-rays from airborne platforms (Viscarra Rossel et al., 2007a). Pracilio et al. (2006) demonstrated the use of gamma-ray radiometric mapping of clay and plant available potassium contents at the farm scale, obtaining R^2 up to 0.68 for clay and 0.60 for potassium.

The gamma region of the electromagnetic spectrum has been applied successfully for studying the properties of cultivated soils (Medhat, 2012), such as field capacity, porosity, moisture content, and bulk density. There is a rising interest in this spectral range for applications in DSM. Gamma-ray spectrometry was found to be an accurate predictor of topsoil clay content in alluvial soils (Piikki et al., 2013). Also van der Klooster et al. (2011) investigated the prediction of soil clay contents in three marine clay districts in the Netherlands with R-squared varied between 0.50 and 0.70.

Vulfson et al. (2013) merged data from microwave and gamma ranges for monitoring soil water content in the root zone and showed strong correlations ($R^2 > 0.9$) between field and laboratory measurements. Evidence suggests that gamma-ray spectrometry can be used for assessing SOC (Dierke and Werban, 2013), which is commonly associated with clay content. Dent et al. (2013) obtained gamma radiometric data from an airborne to imply soil properties. This investigation confirmed that airborne radiometric has the capability to map different parent material and indirectly can infer on soil texture.

Table 15.2 indicates some results for clay content with an on-the-go sensor, reaching 0.86 of R^2. On the other hand, Table 15.4 indicates more soil attributes so we can have an idea of its utility. Attributes such as clay, silt, sand, fine sand, coarse sand, gravel, PAWC (Plant Available Water Capacity), EC, OC, pH, K, obtained variable R^2 results. The best results were for carbon and clay attributes with R^2 0.89 and 0.83, respectively.

Soderstrom and Eriksson (2013) studied the use of aerial and ground-based gamma radiometry to assess Cd contamination risk in food production. In a field with Cd content in the fluvial sediment, gamma-ray measurements allowed to improve mapping of contamination risk in relation to a general soil map. Their results show that geological maps and gamma radiation mapping, calibrated with a few analyses of Cd concentrations in soils and crops, can be used for risk classification of soils at the regional scale. In fact, from aerial gamma there are several that can be evaluated as is shown in Table 15.4. Results of R^2 are variable since 0.02 for Ca until 0.93 for Sr.

TABLE 15.4

Summary of Revision Related with Quantification of Soil Attributes by Ground and Aerial Gamma Platforms

	R^2			RMSE			
	Range	Median	N	Range	Median	N	Authors
A. Aerial**							
Clay (%)	0.53–0.68	0.59	5	2.40–13.65	3.1	5	(Pracilio et al., 2006; Martelet et al., 2013)*
Silt (%)	0.34	0.34	1	10.41	10.41	1	(Martelet et al., 2013)*
Sand (%)	0.56	0.56	1	18.53	18.53	1	(Martelet et al., 2013)*
Gravel (%)	0.13	0.13	1	6.73	6.73	1	(Martelet et al., 2013)*
Bic-K (mg/kg)	0.04–0.55	0.53	4	103–145	127	4	(Pracilio et al., 2006)
Al (g/kg)	0.50	0.50	1	1.18	1.18	1	(Martelet et al., 2013)*
Si (g/kg)	0.56	0.56	1	2.18	2.18	1	(Martelet et al., 2013)*
Ca (g/kg)	0.02	0.02	1	2.98	2.98	1	(Martelet et al., 2013)*
Fe (g/kg)	0.43	0.43	1	0.74	0.74	1	(Martelet et al., 2013)*
Mn (g/kg)	0.24	0.24	1	279.80	279.80	1	(Martelet et al., 2013)*
Mg (g/kg)	0.35	0.35	1	0.16	0.16	1	(Martelet et al., 2013)*
Na (g/kg)	0.35	0.35	1	0.19	0.19	1	(Martelet et al., 2013)*
Pb (g/kg)	0.48	0.48	1	7.33	7.33	1	(Martelet et al., 2013)*
Sr (g/kg)	0.93	0.93	1	4.19	4.19	1	(Martelet et al., 2013)*
V (g/kg)	0.83	0.83	1	11.13	11.13	1	(Martelet et al., 2013)*

* Models include morphological variables

** Bic-K—Plant available Potassium; Al—Aluminum; Si—Silicon; Ca—Calcium; Fe—Iron; Mn—Manganese; Mg—Magnesium; Na—Sodium; Pb—Lead; Sr—Strontium; V—Vanadium

	R^2			RMSE			
	Range	median	N	range	median	N	Authors
B. Proximal*							
Clay (%)	0.17–0.95	0.83	19	1–8.40	2	18	(Viscarra Rossel et al., 2007b; Wong and Harper, 1999; Priori et al., 2014; Van der Klooster et al., 2011)
Silt (%)	0.40–0.89	0.40	3	2.29–2.46	2.38	2	(Viscarra Rossel et al., 2007b; Wong and Harper, 1999)
Sand (%)	0.65–0.85	0.77	4	6.7–7.9	7.4	1	(Priori et al., 2014)
Fine Sand (%)	0.05–0.31	0.18	2	3.23–3.96	3.60	2	(Viscarra Rossel et al., 2007b)
Coarse Sand (%)	0.3–0.73	0.55	2	6.73	9.31	2	(Priori et al., 2014)
Gravel (%)	0.49–0.58	0.51	4	0.11	0.10–0.11	4	(Priori et al., 2014)
PAWC (mm)	0.50	0.50	1	11.4	11.4	1	(Wong et al., 2009)
EC (ms/m)	0.30–0.60	0.45	2	27.96–46.55	37.26	2	(Viscarra Rossel et al., 2007a)
OC (g/kg)	0.89	0.89	1				(Wong et al., 1999)
$\mathbf{pH_{Ca}}$	0.40–0.63	0.52	2	0.43–0.48	0.46	2	(Viscarra Rossel et al., 2007a)
Colwell-P (mg/kg)	0.68	0.68	1			1	(Wong et al., 1999)
K (mg/kg)	0.61	0.61	1	83.57	83.57	1	(Viscarra Rossel et al., 2007a)

* PAWC—Plant Available Water Capacity, EC—Electrical Conductivity in units milliSiemens per meter, OC—organic carbon, pH_{Ca}—, pH using 0.01 M $CaCl_2$, Colwell-P—Phosphorous using the $NaHCO_3$ method, K—Bicarbonate-extractable Potassium

A comparison between ground and airborne information can be extracted from Table 15.4. We observed that clay had R^2 maximum and average of 0.95 and 0.83 respectively, for ground information. On the other hand, aerial presented 0.68 and 0.59 of R^2, lower than field data. This is true since in the field the sensor has a higher spatial resolution and less interference factor. But we rather have to consider that aerial data is good. Another important information given by gamma is the great R^2 for parent material elements such as strontium and vanadium that showed R^2 0.93 and 0.83, which can certainly assist in relation with soils.

15.6.11 Strategies for in Situ Spectral Sensing

Pre-treating the soil samples before scanning, that is, drying and sieving, is common and improves the quality and repeatability of the spectral data acquired in the laboratory, reducing the negative influence of variable soil moisture or soil particle size in data acquisition (Tekin et al., 2012). However, the goal is to scan the soils in the field without pre-treatment. However, field scans are subject to other sources of variation, including variations in viewing angle, illumination, soil roughness, soil moisture, soil temperature, soil structure, in the presence of specific features (e.g., redoximorphic features and clay films), among other, all affecting the quality of the measurements.

Applying SS to field studies reduces costs associated with collection, transport, preparation, and analysis of soil samples. As already stated, laboratory SS measurements are made under controlled conditions with standard protocols, which provide a minimum of interference in the acquired radiometric data (Ben-Dor, 2011), whereas for in situ measurements there are more possibilities for interference due to soil and environmental conditions that cause variations affecting data acquisition. Other in situ factors that affect data quality include noise associated with tractor vibration, sensor-to-soil distance variation (Mouazen et al., 2007), stones, plant roots, and difficulties of matching the position of soil samples collected for validation with corresponding spectra collected from the same position (Kuang et al., 2012).

Water, in the form of soil moisture and soil moisture variability, is a fundamental concern to consider when using SS measurements of unprocessed soil surfaces or soil cores from the field (Waiser et al., 2007; Nduwamungu et al., 2009). Many working in spectroscopy have attempted to address the problem of soil moisture on SS data. However, the solutions usually include linear transformations and or require prior knowledge of the soil moisture, posing difficulties for applications in the field. An effective strategy is to create calibration models using only field-collected spectra (Waiser et al., 2007; Morgan et al., 2009; Bricklemyer and Brown, 2010). However this approach does not contribute to available soil spectral libraries that require accompanying soil samples. Aiming to reduce the effects of water in the soil spectra without prior knowledge, a technique called "External Parameter Orthogonalization (EPO)" has been developed and will be discussed in Section 15.10.

As previously discussed, several factors limit field application of VIS-NIR spectroscopy, such as temperature, luminosity, climatic conditions, sample surface roughness, organic residues, and appropriate equipment to obtain subsurface spectra. Some solutions have been proposed to address these issues, for example, by Ben-Dor et al. (2008), and Ge et al. (2014a). Field spectral sensing has obtained good results for clay content (Viscarra Rossel et al., 2009) and SOC (Gomez et al., 2008b) prediction in Australian soils, as well as in Texas for clay content (Waiser et al., 2007), and both SOC and SIC (Morgan et al., 2009). Chakraborty et al. (2012a, 2012b) demonstrated the feasibility of VIS-NIR analysis in the field to map TPH contamination in soil in an extensive area. Field measurements can also have issues when working with non-contact equipment (see Figure 15.2, Strategy 3) due to atmospheric interference. For example, the 1400- and 1900-nm bands cannot be detected hindering the interpretation of soil mineralogy, as observed by Fiorio et al. (2014). Thus, it is better to use a contact probe in the field, avoiding atmospheric interference and sunlight variations.

Soil MIR analysis is not widely used in the field mainly due to: (1) the need for sample preparation; and (2) strong water absorption in naturally moist soils leading to spectral distortion and total absorption (Ge et al., 2014b). As an alternative, Ge et al., (2014b) used attenuated total reflectance (ATR) as a technique to obtain MIR spectra of neat soil samples. Accordingly, MIR-ATR can be a promising and powerful tool for soil characterization combining the advantages of both VIS-NIR (minimum sample preparation and high analysis throughput) and diffuse reflectance MIR (better model performance) (Ge et al., 2014b).

Table 15.2 indicates quantification of soil attributes in field conditions. The lab-field moist situation indicates high values for all elements going from 0.7 to 0.85 of R^2. Specifically in field conditions (or on-the-go) we still have important results for clay (R^2 0.78), OC with 0.84 and TC with 0.86. Gamma also has interesting results reaching 0.86 of R^2 for clay measurement.

15.6.12 Soil Spectral Libraries

The first soil spectral library (SL) was built by Condit (1970) and complemented by Stoner et al. (1980) as an atlas, published afterwards by Stoner and Baumgardner (1981) with soils mainly from United States and some samples from Paraná state, Brazil. After that, Epiphanio et al. (1992) and Formaggio et al. (1996) constructed a SL comprising 14 soil classes for one Brazilian state, while Bellinaso et al. (2010) reached six Brazilian states with ~8,000 soil samples. Even though these and other SL (e.g., Clark, 1999b) are promising, there are still few examples including a wide diversity of soil classes (Chang et al., 2001; Malley et al., 2004). The first publication using a SL with global samples was presented by Brown et al. (2006).

After 2000 several SL initiatives appeared, including the ICRAF-ISRIC (ICRAF-World Agroforestry; ISRIC-International Soil Reference and Information Centre) (https://data.isric.org/geonetwork/srv/api/records/1081ac75-78f7-4db3-b8cc-23b78a3aa769) world soil spectral library, composed of 785 soil profiles from 58 countries from Africa, Europe, Asia, and the Americas (Shepherd and Walsh, 2002). Recently, through the initiative of numerous researchers, the Global Soil Spectral Library was created (Viscarra Rossel et al., 2016). The Global Soil Laboratory Network—GLOSOLAN, belonging to FAO's Global Soil Partnership, launched the project for a global spectral library with spectral data in the mid-IR range (www.fao.org/global-soil-partnership/glosolan/soil-analysis). Other global spectral library initiatives are the Open Soil Spectral Library (OSSL) (https://soilspectroscopy.github.io/ossl-manual/). Other initiatives, with continental or multiple countries data are: LUCAS SL (Land Use/Cover Area Frame Survey—https://esdac.jrc.ec.europa.eu/projects/lucas) has around 20,000 vis-NIR-SWIR spectral data, obtained from soil samples (0–20 cm) collected in 30 countries on the European continent (Stevens et al., 2013; Orgiazzi et al., 2018). Another important example is ASTER SL, composed of 2,400 spectra of soils, rocks, minerals and other materials (Baldridge et al., 2009). This SL is a compilation of 2,400 spectra of soils, rocks, minerals and other related materials. GEOCRADLE SL has spectral data from samples from 9 countries in the Balkans, Middle East and North Africa regions (Tziolas et al., 2019)

Spectral Libraries have been developed by several countries. Viscarra Rossel and Webster (2011) described a large spectral library with ~4,000 soil profiles covering the Australian continent. A spectral library covering the United States has been collected under the Rapid Carbon Assessment (RaCA) project (Soil Survey Staff, 2013) with 144,833 VIS-NIR spectral data for 32,084 soil profiles. The Brazilian Soil Spectral Library (BSSL) (www.besbbr.com.br) began in 1995, creating a protocol to gather soil samples from different locations in Brazil. The BSSL reached 39,284 soil samples from 65 contributors representing 41 institutions from all 26 states of Brazil (Demattê et al., 2019). Soil SL initiatives in other countries include: Brazil (Mendes et al., 2022; Santana et al., 2019), Czech Republic (Brodský et al., 2011), France (Gogé et al., 2012; Barthès et al., 2020), Denmark (Knadel et al., 2012; Peng et al., 2013), Mozambique (Cambule et al., 2012), Spain (Bas et al., 2013), China (Shi et al., 2014c; Ji et al., 2016; Liu et al., 2018), United States (Wijewardane et al., 2018), Switzerland (Baumann et al., 2021), New Zealand (Baldock et al., 2019; Ma et al., 2023), Central and East African countries (Shepherd and Walsh, 2002; Summerauer et al., 2021), Tajikistan (Hergarten et al., 2013), Hungary (Mohammedzein et al., 2023), Indonesia (Ng et al., 2020), Austria (Sandén et al., 2022), Poland (Debaene, 2019), India (NBSS and LUP, 2005) and

Costa Rica (Perret et al., 2020). In Brazil, spectral libraries have been developed for different states of the federation, such as: Santa Catarina state (Silva et al., 2019), Piauí (Mendes et al., 2021), Roraima (Marques et al., 2019), Rondônia (Tavares et al., 2022), Maranhão (Demattê et al., 2020; Greschuk et al., 2022).

Soil spectral libraries can be applied for many purposes, including: (a) modeling of soil attributes; (b) soil survey, classification, and mapping; (c) soil contamination and monitoring, by extracting the baseline electromagnetic properties of soils, which can be compared with any contaminated samples; (d) communication among researchers (soil classification has several systems, but spectra are the same!); and (e) development of field, aerial, and space sensors, among others. To understand the usefulness of soil SL, consider the following example: the interested parties (farmers or researchers) could send their soil samples to a central spectral library (e.g., a national or global SL) where they would be scanned and the spectral curves stored, or they could send already acquired soil spectral curves that compose their local spectral libraries. Local SL can be explored for personal interests (e.g., soil monitoring), and also feed global SL, growing a global repository. Once having a global SL, spectral curves from a profile of an unknown could be compared with other spectra from the global SL and a preliminary soil classification or the SOC or clay content, could be estimated.

The ideal scale (global, continental, regional, local or farm) for a soil spectral library application has had much inquiry, and the general result is that the spatial scale of coverage and application depends. This topic was initially raised by Coleman et al. (1993), which concluded that there was evidence that regional/local scale is the most reliable. Later on, Demattê and Garcia (1999) observed better results for soil modeling with a local soil spectral library than with a regional one. Brown (2007) used a global VIS-NIR spectral library for local soil characterization and landscape modeling in a 2nd-order Uganda watershed. In brief, a soil attribute may be estimated in a farm using samples collected at the farm to constitute a local SL; or the soil spectra collected at the farm can be compared against other spectra in a global SL to retrieve predictions. One strategy, called "spiking," combines global library with local samples to get better predictions at the local level (Brown et al., 2006; Wetterlind and Stenberg, 2010). Others have tried using pedological knowledge to subsample a geographically more extensive spectral library (Ge et al., 2011). In the subsampling context, parent material filter improved predictions of clay content; however, for SOC subsampling, the first three principal components coupled with Mahalanobis distance was more effective (Ge et al., 2011). Similar results were found by Araújo et al. (2014a) when analyzing spectra of 7,172 tropical soil samples. They found that separating the global dataset into more mineralogically uniform clusters improved predictive performance of clay content regardless of the geographical origin, showing that probably physically based, soil-related stratification criteria in libraries offer better results. An interesting study Ramirez-Lopez et al. (2013) developed the Spectrum based learner, which indicates the best performance of data to reach high quality of quantification when using complex data. Another important discussion is how to use the dataset to reach best results. Debaene et al. (2014) found little significant increase in prediction capacity of soil attributes with use of an entire dataset, watching increase on the R^2 of 0.63 to 0.72 for SOC and R^2 of 0.71 to 0.73 for clay. The point if local, regional or global, is certainly an actual discussion. Genot et al. (2011) built a methodological framework for the use of NIR spectroscopy on a local and global scale by spectral treatment and regression methods. In addition, evaluated the ability of NIR spectroscopy to predict TOC ($R^2 = 0.91$ local and $R^2 = 0.70$ global), TN ($R^2 = 0.73$ local and $R^2 = 0.61$ global), clay ($R^2 = 0.64$ local and $R^2 = 0.61$ global) and CEC ($R^2 = 0.73$ local and $R^2 = 0.43$ global) above several soils conditions.

Despite these discussions, results have suggested that the best scale for a spectral library is very much application dependent. The application will define the precision needed. Generally, developing a global library does not exclude embracing a local or physically based library, and the user will need to decide which scale to use. Figure 15.7 suggests the sequence on how to construct and use soil spectral libraries.

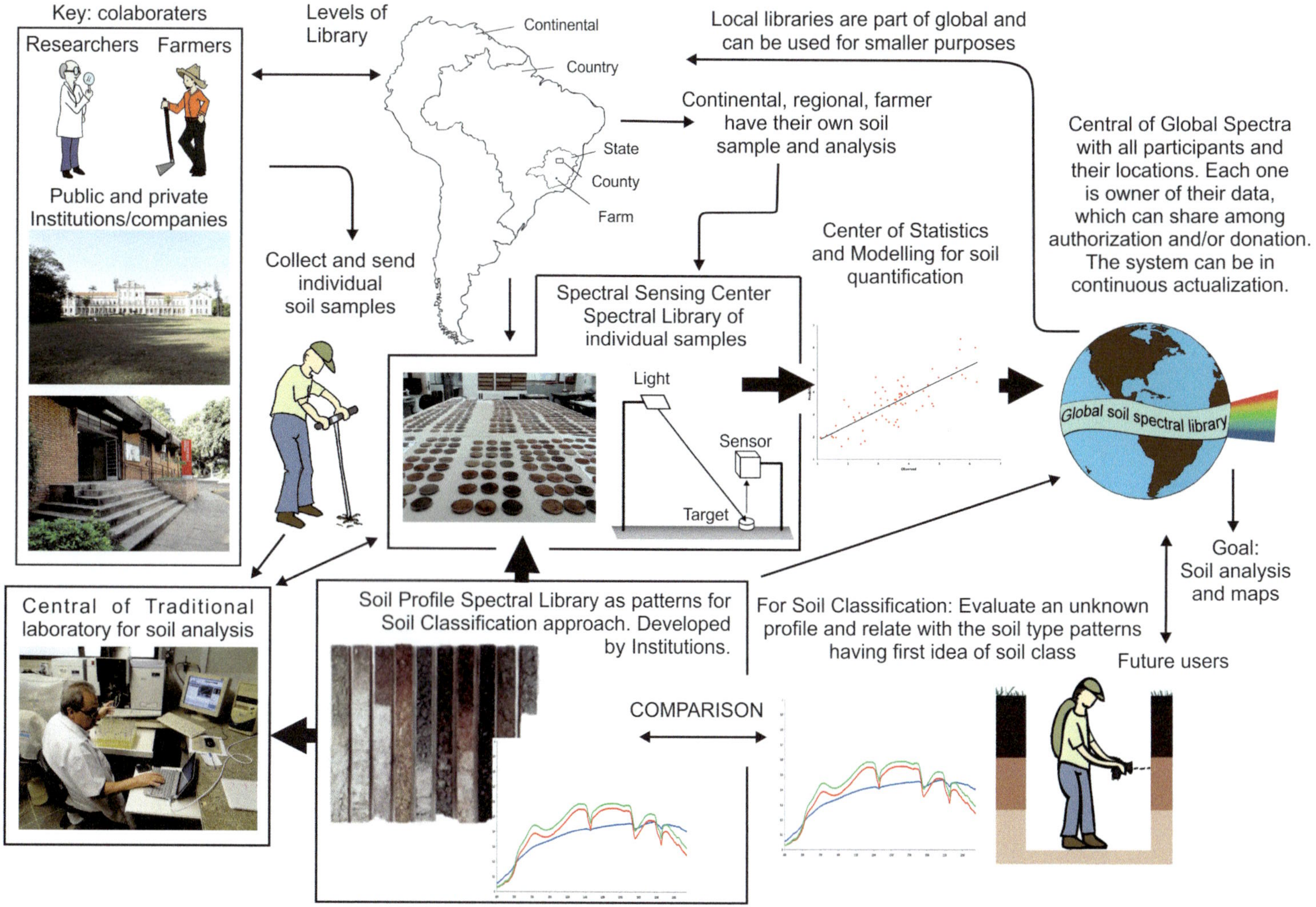

FIGURE 15.7 Illustration on how to construct and use soil spectral libraries.

Recently, several works have been developed with the aim of standardizing data acquisition and correcting spectra collected by different equipment or under different environmental conditions (Kopačková and Ben-Dor, 2016; Romero et al., 2018; Chabrillat et al., 2019; Ogen et al., 2019; Knadel et al., 2023). Francos et al. (2023) proposed a spectral transfer function to harmonize existing SSls generated by different protocols. Francos and Ben-Dor (2022) developed a transfer function to predict the field spectral measurements from the laboratory spectra, this approach has also been tested by Karyotis et al. (2023). Another approach tested in recent works is the application of predictive models generated from soil spectral data from one country to data from other countries (Briedis et al., 2020; Gomez et al., 2020).

To disseminate some applications of SLs, the Brazilian Soil Spectral Service (BraSpecS) (Demattê et al., 2022) was recently launched, carried out by the study group GeoCIS—Geotechnology in Soil Science of the Department of Soil Science at ESALQ/USP (www.besbbr.com.br). On the initiative's portal, the user can upload spectral data and receive the result of estimating the sample's attributes. Additionally, it is possible to view spectral curves of soil samples from different states of Brazil, or average spectral curves of samples with similar characteristics, such as soil class and texture. The platform also allows the user to spontaneously make their spectral data available so that they can be added to the Brazilian Soil Spectral Library database. Shepherd et al. (2022) propose, in a similar way, the creation of an attribute estimation service based on data from a global SL.

15.7 SOIL SPECTRAL BEHAVIOR AT DIFFERENT ACQUISITION LEVELS

Soil spectra result from the interaction of many soil attributes with the electromagnetic energy. According to Demattê and Terra (2014), descriptive spectral analyses of soil horizons are important because their shapes have direct relation with mineralogical and organic constituents that, in turn, result from the pedogenetic processes acting in the soil profile. This took them to define the term "spectral pedology" which can be summarized as

> a detailed and accurate evaluation of soil spectral behavior obtained by proximal and/or remote sensing, and analyzed by its qualitative (shape, absorption features and reflectance intensity/albedo) and/or quantitative information, where the convergence of evidences guides to a probable soil classification or behavior.

This section aims to indicate some important spectral features of soils that relate to their attributes and taxonomic classes.

15.7.1 Spectral Behavior at the Ground Level

Minerals from soils have different and specific spectral shapes (Figure 15.8a), and each one contributes to the total soil spectrum. Features at 1300–1400, 1800–1900 and 2200–2500 nm (hydroxyl groups) are strong indicatives of clay content and type (Zhu et al., 2010). Soil particle size has great influence on spectra. In this case, a coarser texture increases the scatter of energy (reduces reflection) and the apparent absorbance increases as path length increases. In fact, Figure 15.8b presents great differences of energy reflection from clay to sandy soils with different angles (shapes) and intensities of energy with highest peaks occurring after 2000 nm (Demattê, 2002; Franceschini et al., 2013).

The SOM is another important attribute affecting not only specific features in the NIR-MIR region but also the overall spectrum in the VIS region. Udelhoven et al. (2003a) showed the relation between soil brightness and SOC and developed several systems for its analysis (Udelhoven et al., 2003b, 2003c). Demattê (2002) indicated that spectra of superficial soil layers had lower reflectance intensities than undersurface ones, which was related to lower SOC contents (Figure 15.8c). Stenberg et al. (2010) stated that, although soil samples tend to get darker colors with increasing SOM, other

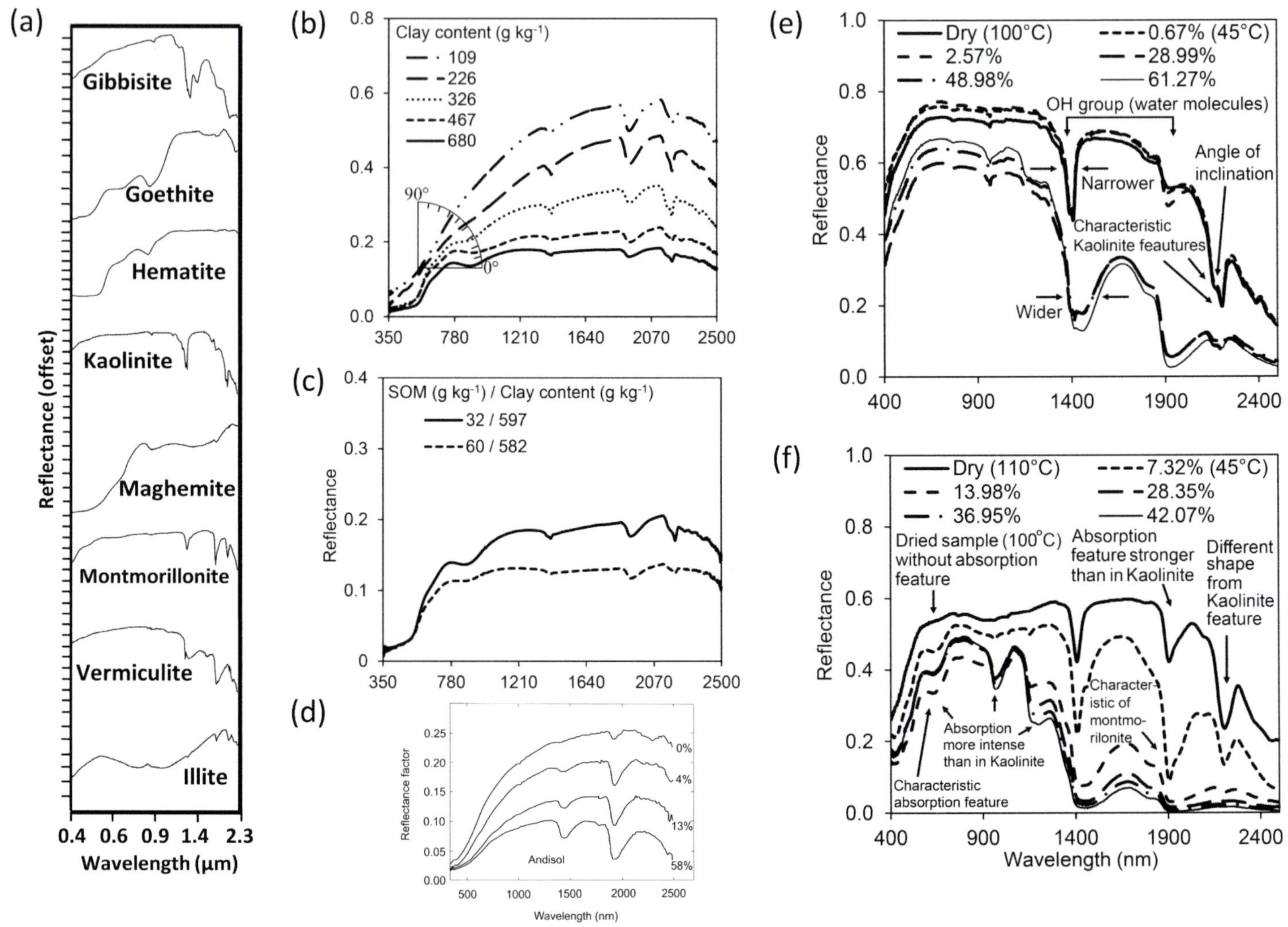

FIGURE 15.8 VIS-NIR spectra of (a) soil minerals (Clark et al., 2007); (b) soils with different clay contents (Franceschini et al., 2013), indicating angular variations (Demattê, 2002); (c) soil with variable organic matter contents (Demattê et al., 2003a); (d) soils with different moisture contents (Lobel and Asner, 2002); and pure kaolinite (e) and pure montmorillonite (f) with varying moisture contents (Demattê et al., 2006).

soil properties, such as texture and moisture, also influence the soil brightness, thus SOM would only be a useful indicator in a specific situation. The same authors have suggested that there are specific features in spectra that would provide a better correlation with SOM, for example, bands around 1,100, 1,600, 1,700 to 1,800, 2,000, 2,200, and 2,400 nm. The spectral features of SOM in soil VIS-NIR spectra are explained by combination and vibration modes of organic functional groups (Chen and Inbar, 1994).

The VIS-NIR-SWIR ranges are strongly influenced by the soil water content. Water in soils can be incorporated to the clay mineral lattice, filling pore spaces as free liquid water or adsorbed on a surface as hygroscopic water. In the first case, the water is related to the mineralogy of the soil samples, and directly affects features near 1400 and 1900 nm bands (Hunt et al., 1971). Bishop et al. (1994) also related the vibrations of bound water in the interlayer mineral lattice to the features in 1400 and 1900 nm. In fact, Stenberg et al. (2010) stated that the bands related to vibrations of bound water occurs at shorter wavelengths close to 1400 and 1900 nm, but the hygroscopic water appear as shoulders at 1468 and 1970 nm. Besides, hygroscopic and free pore water are responsible for reducing the albedo of soil spectrum (Figure 15.8d; Lobel and Asner, 2002). Once the surrounding of soil particles are changed from air to water the refractive index decreases, in other words, changes in the medium surrounding soil particles affect the average degree of forward scattering (Ishida et al., 1991), reducing the reflectance.

Differences in mineral spectra are related to their chemical composition and structure. For example, kaolinite has well delineated features at 1,400 and 2,200 nm, but a weak signal in 1,900 nm, compared to other minerals (Figure 15.8e). For instance, the spectra of montmorillonite have a strong feature at 1,900 nm and a different shape close to 2,200 nm (Figure 15.8f; Demattê et al., 2006). Another important mineral in tropical soils is the gibbsite, which has an aluminum octahedral structure with spectral features occurring mainly close to 1,400 and 2,265 nm (Figure 15.8a, top curve).

Hunt et al. (1971) summarized the physical mechanisms responsible for Fe^{2+} (ferrous) and Fe^{3+} (ferric) spectral activity in the VIS-NIR range, and indicated that iron oxides and hydroxides are spectrally active attributes due to the electronic transition of iron cations. Due to their absorption features and overall spectra shapes, the presence of goethite and hematite alter the shape of spectra of soils (e.g., Figures 15.8a, b, c, d, and 15.9a). Demattê and Garcia (1999) compared the visible range from spectra of oxidic soils, and indicated that soils with predominance of goethite have a narrower feature with higher reflectance between 400–570 nm, whereas the opposite occurs with the predominance of hematite (Figures 15.8 and 15.9). In soils derived from carbonate rocks, that is, rocks composed generally by calcite and dolomite minerals, the carbonate groups ($-CO_3$) are spectrally active causing specific absorption features due to the C-O bonds (Ben-Dor, 2011). A mineral commonly found in temperate soils is montmorillonite, which is a highly enriched Al smectite and consequently presents Al-OH bonds, resulting in a feature at 2160–2170 nm in soil spectra (Figures 15.8f and 15.9c). Other smectites affect the soil spectra in specific regions, so we suggest the referred studies for further information (Stenberg et al., 2010; Ben-Dor, 2011).

As stated by Demattê and Garcia (1999) it is possible to discriminate and identify soils with different weathering conditions. For example, compare the spectra of an Oxisol (highly weathered soil) (Figure 15.9a, Typic Hapludox) with that of a Vertisol (less weathered soil) (Figure 15.9a, Aquic Hapludert). In fact, the degree of weathering can determine the relative amounts of kaolinite and montmorillonite present in the soil, with important differences in spectral features (Figure 15.9d). Moreover, soils with higher weathered mineralogy have higher iron oxide contents, which add other specific spectral features. Demattê et al. (2003b) found that crystalline iron minerals (hematite and goethite) are responsible for the spectral features at 400–850 nm. On the other hand, when these iron forms were extracted from the soil in the laboratory, the concave shapes disappeared from this region (400–850 nm) (Figure 15.9e). For clayey soils, reflectance decreases in intensity around 1200 nm, due to the remaining presence of magnetite. Several wavebands and respective soil attributes can be seen in Table 15.1.

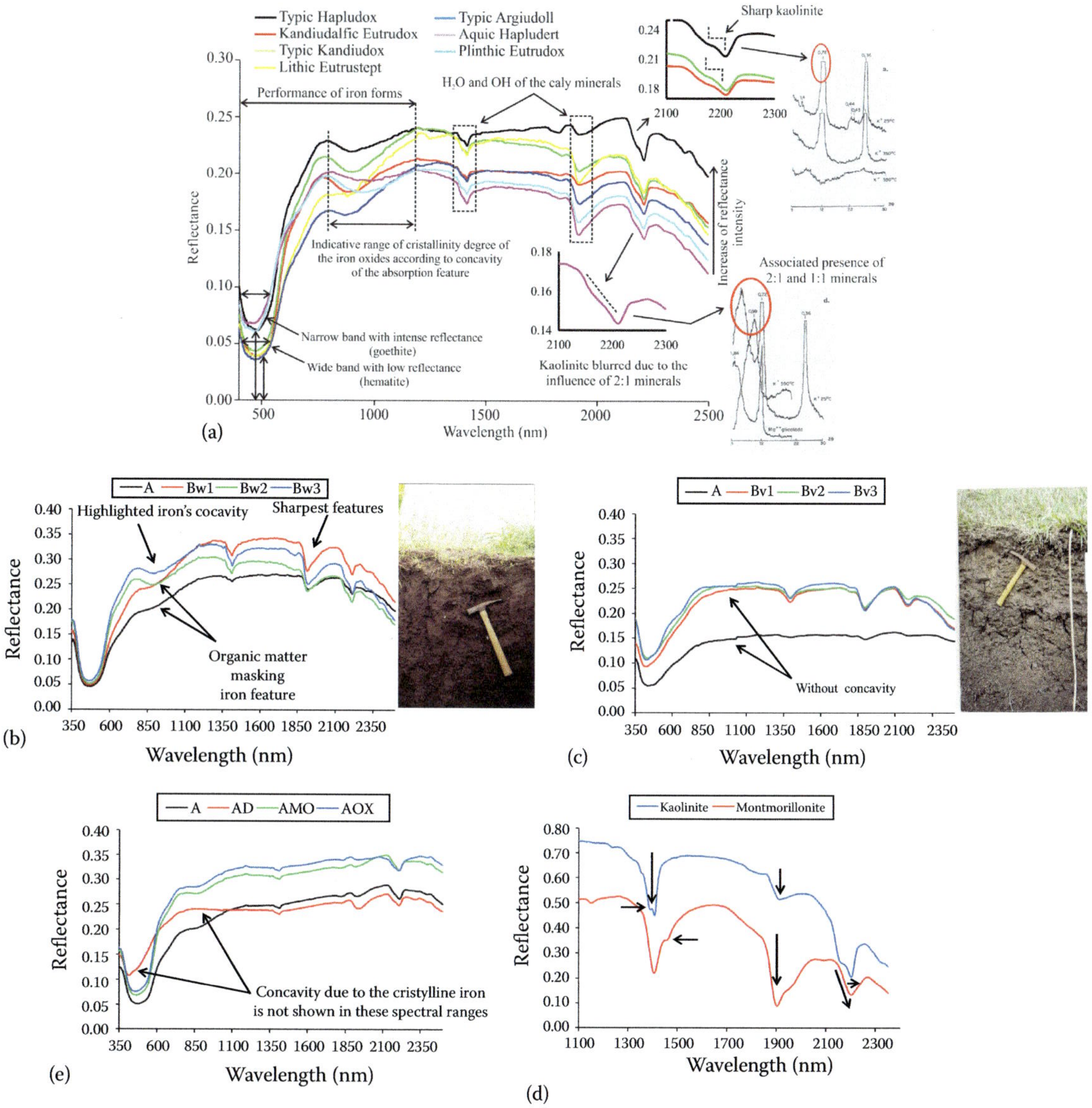

FIGURE 15.9 Examples of soil spectra and their interpretation, from (a) Demattê and Terra (2014); (b, c) Demattê et al. (2012a); (d) Demattê et al. (2006); and (e) Demattê et al. (2003b). A, raw soil (testimony); AMO, soil without organic matter; AOX, soil without amorphous (oxalate-extracted) iron; AD, soil without any type of iron.

Classification of soils by their spectral behavior must be made by interpretation of all horizons. For example, a Ferralsol (Oxisol) has only minor differences regarding reflectance intensity and spectral features among spectra collected at different depths, that is, from surface versus subsurface horizons (Figure 15.10a, b). On the other hand, spectra of a Lixisol (Ultisol) show differences between these horizons, mainly after 2000 nm, due to differences in sand content (quartz mineral). In fact, the absorption features of quartz can be observed in the MIR spectra, mostly between 1,111 and 1,300 cm^{-1} (in wavenumber) (Figure 15.10d). Observe that there is a great difference in the quartz peak (~1,300 cm^{-1}) between the A and B horizons of the Lixisol (Figure 15.10d), and no difference in the Ferralsol (Figure 15.10b), which is in line with the characteristics of these soils. In fact, Ferralsols do not have great differences in clay contents with depth, as opposite to Lixisols. These findings allow to relate the shapes and trends in spectra to specific soil classes, aiding in soil classification.

15.7.2 Spectral Behavior at the Space Level: From Aerial to Orbital Platforms

Compared to ground-based sensors, aerial and orbital ones have low signal-to-noise ratio (SNR) due to the larger atmospheric path length, decreased spatial and spectral resolution, geometric distortions, and spectral ambiguity caused by recording multiple signals from adjacent targets. On the other hand, space-based sensors have larger ground coverage (Obade and Lal, 2013). Data from some satellite sensors, such as Landsat ETM, ASTER (Advanced Spaceborne Thermal Emission and Reflection Radiometer), and MODIS (Moderate Resolution Imaging Spectroradiometer) may be freely downloaded, not requiring a portable radiometer which is an expensive tool.

The first aspect to be analyzed is the feasibility of ground and space information for soil assessment (attribute prediction, soil classification and mapping). Figure 15.11 illustrates the study of Demattê et al. (2009a), and indicates the position on the landscape of Nitisols, Arenosols, and Oxisols in a Landsat TM image (RGB = 5,4,3). Darker and lighter colors are related with clayey and sandy soils, respectively. These and other areas were evaluated and the spectra of pixels were compared with ground spectra by correlating 294 laboratory spectra (Figure 15.11c) simulating Landsat TM spectra (Figure 15.11d) with Landsat TM spectra obtained directly from the pixel in the image (Figure 15.11e). They observed similar trends between spectra simulated from the ground sensor and collected from space. Clayey and oxidic soils (Nitisols) are very different from sandy soils (Arenosols) in many aspects. Of course ground information is more accurate due to the higher spectral resolution. The main differences were related to the reflectance intensities from bands 5 to 7, where drops in the Landsat TM spectra are related to moisture. Other studies proved the capabilities of Landsat data to differentiate moisture content in soils (Shih and Jordan, 1992; Vicente-Serrano et al., 2004). Satellite images can provide a good discrimination of soil classes and consequently support soil mapping either visually or through DSM. In the given example, the image colors range from a dark blue to magenta corresponding to a sequence of Nitisol, Rhodic Oxisol and Typic Oxisol.

As another example, Figure 15.12 presents soil spectra of the same spot taken from ground (2,151 bands), Landsat TM (6 bands), and Hyperion (220 bands). From ground to aerial or orbital sensors, water bands differ among spectra due to atmospheric absorption. Compared to the multispectral sensors, many inferences can be made, although the quality of soil attribute predictions and spectral characterization are limited by the SNR and spectral resolution. Comparing the spectral data acquired from different levels, such as Hyperion, at the field or in the laboratory, the influence of the SNR is clearly observed. Due to the lower spectral resolution, multispectral sensors, such as Landsat TM, do not present important features and consequently many spectrally active attributes cannot be predicted (e.g., minerals). Despite this, Landsat data have a good temporal resolution, that is, information about soils is continuously generated. In this example, older soils (Oxisols and Nitisols) presented lower reflectance intensities, while Arenosols and Inceptisols (younger soils) presented higher reflectance intensity. Some similarities among spectra from different sensors

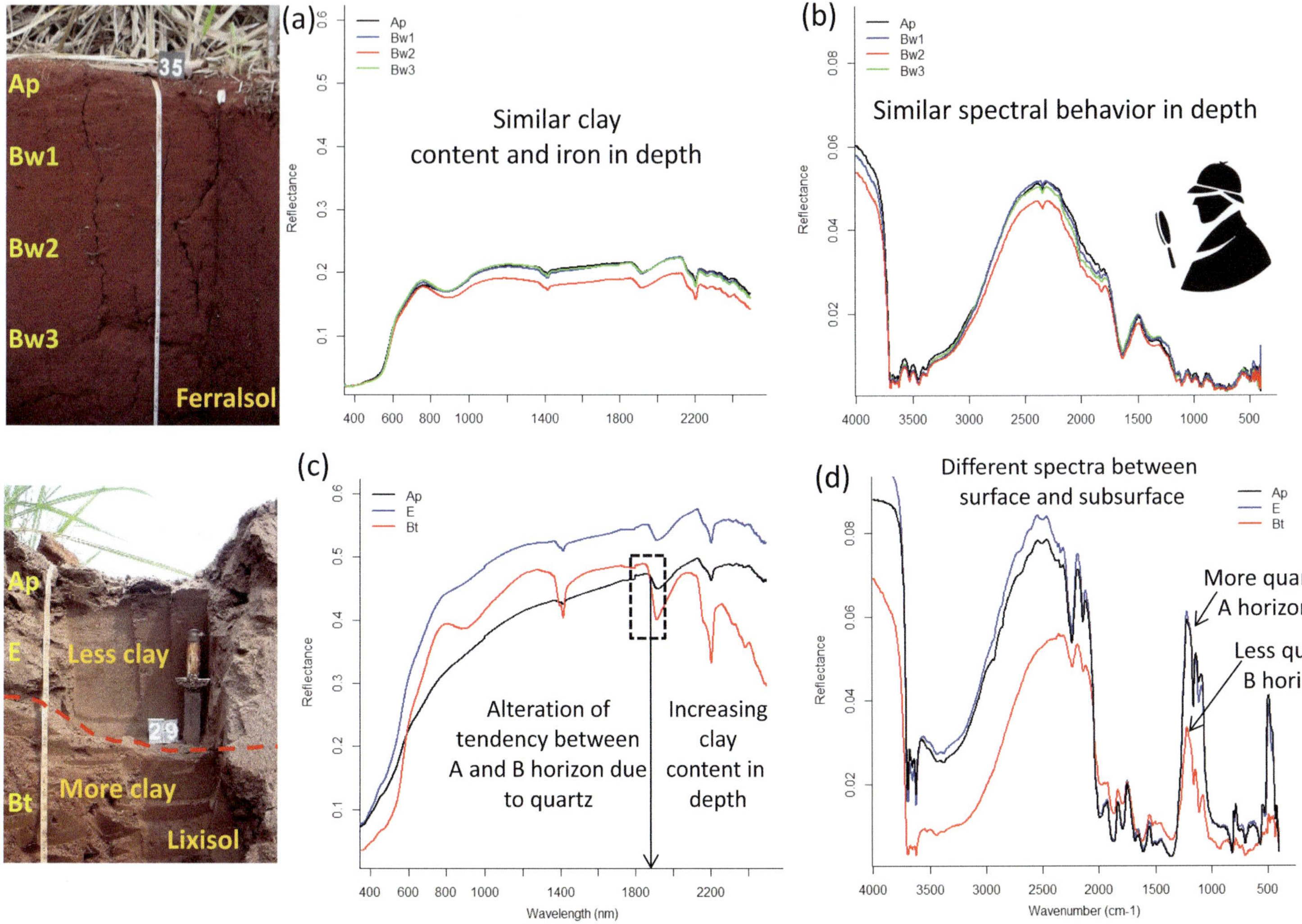

FIGURE 15.10 Methods of soil profile spectral assessment for soil classification: Ferralsol (Oxisol) in VIS-NIR (a), and MIR (b); and Lixisol (Ultisol) in VIS-NIR (c), and MIR (d).

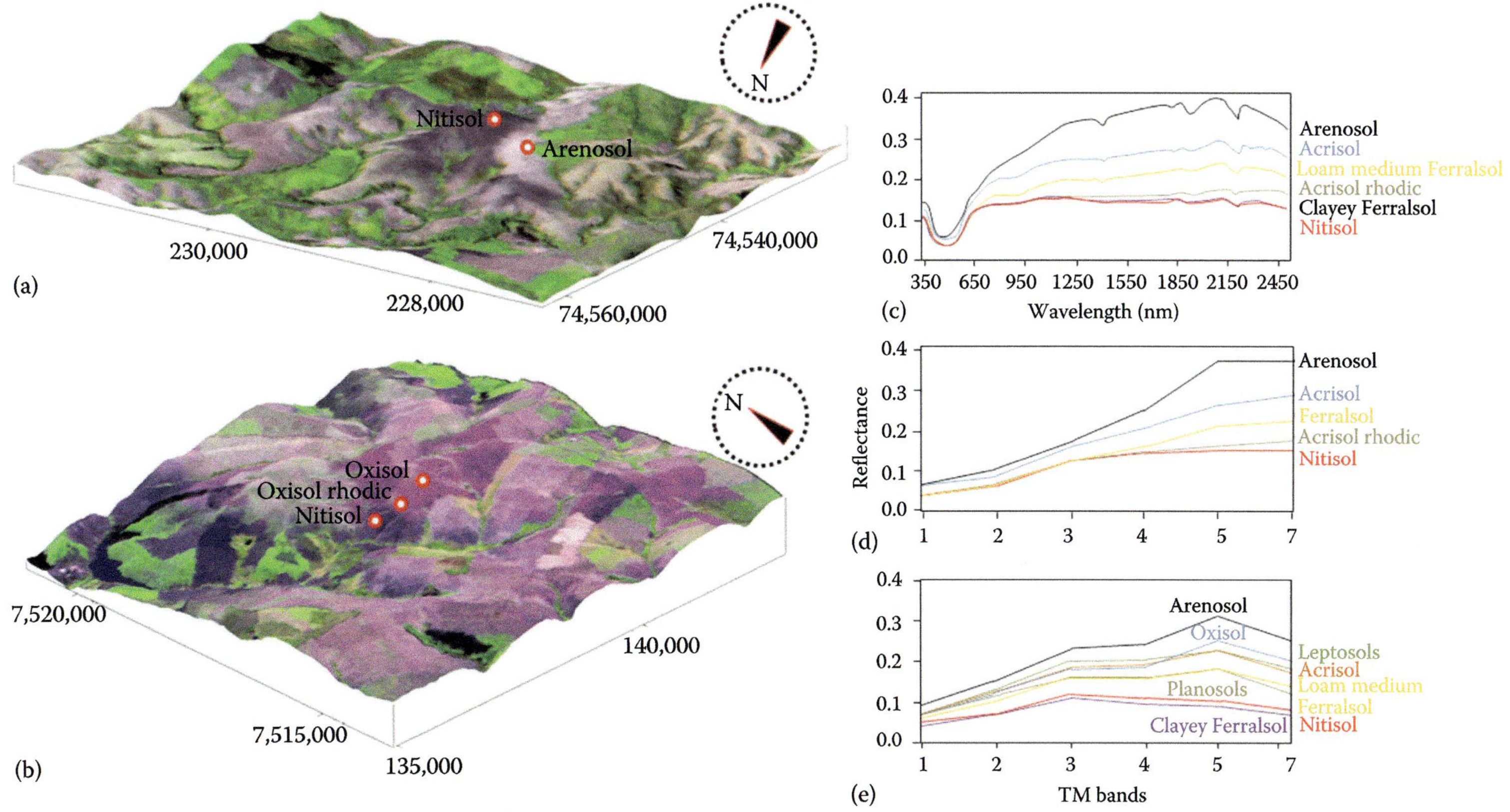

FIGURE 15.11 Landsat TM image compositions (RGB = 5,4,3) with some parts with bare soil indicating areas with clayey and sandy soils (a, b), and soil spectral curves from laboratory (c), laboratory convoluted to Landsat TM (d), and Landsat TM directly (e) (Demattê et al., 2009a).

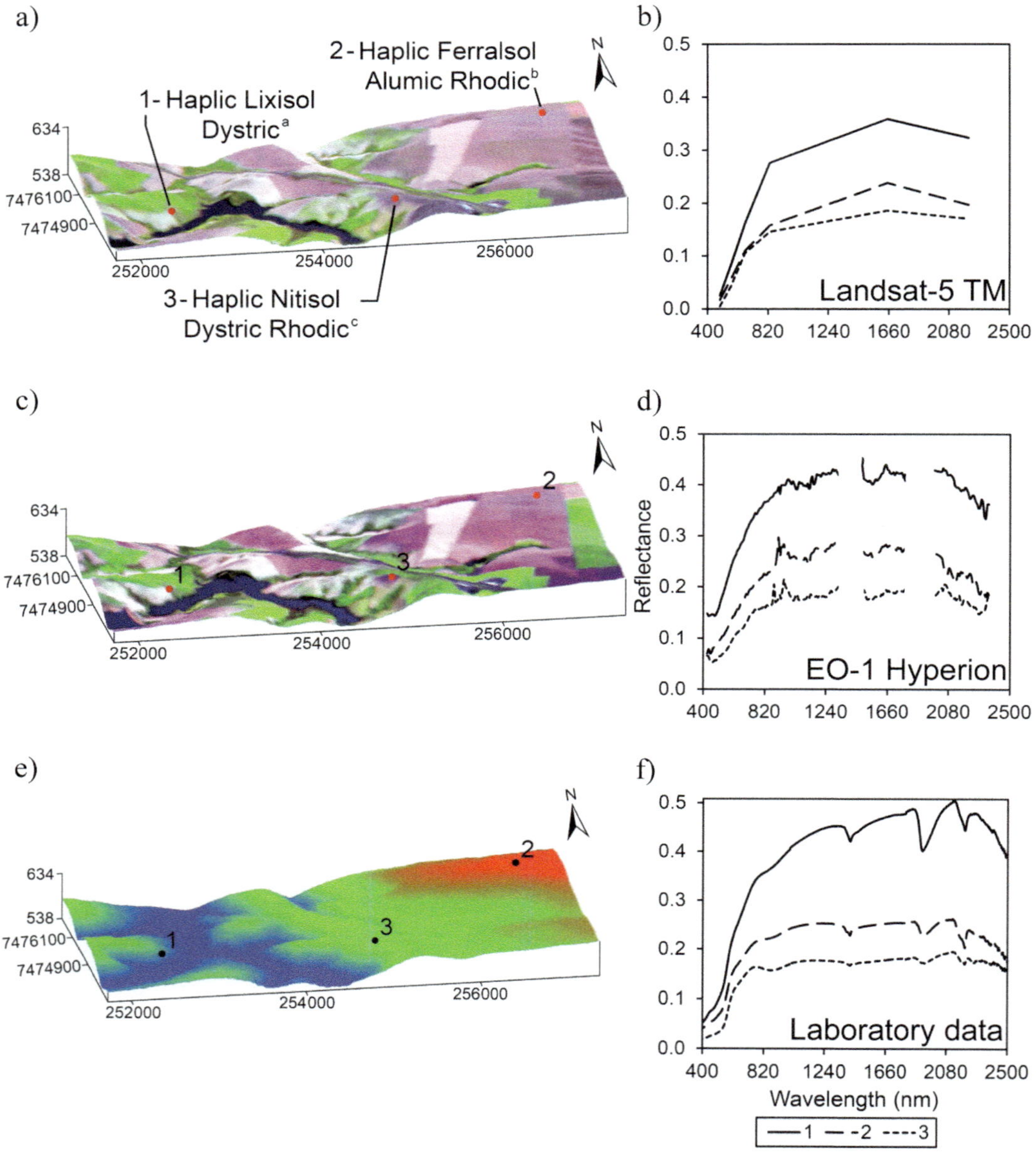

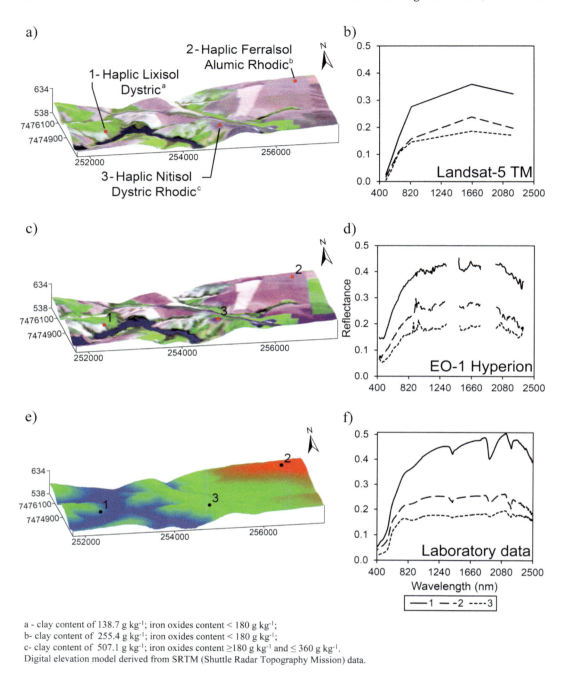

a - clay content of 138.7 g kg^{-1}; iron oxides content < 180 g kg^{-1};
b- clay content of 255.4 g kg^{-1}; iron oxides content < 180 g kg^{-1};
c- clay content of 507.1 g kg^{-1}; iron oxides content ≥180 g kg^{-1} and ≤ 360 g kg^{-1}.
Digital elevation model derived from SRTM (Shuttle Radar Topography Mission) data.

FIGURE 15.12 Landsat-5 TM image composition (R: 1,650 nm, G: 830 nm, B: 660 nm) showing the position of soils (a), and their respective Landsat TM spectra (b); Hyperion image composition (R: 1648 nm, G: 823 nm; B: 661 nm) (c), and Hyperion spectra of the same soils (d); digital elevation model (e), and laboratory VIS-NIR spectra of the same soils (f).

are observed, for example the reflectance intensity and convexity which are greater in the Typic Kandiudult, and lower in the Rodic Hapludox. According to Demattê (2002), Arenosols and some Acrisols present higher reflectance intensities at the infrared band (Landsat TM band 5) and increasing albedo from VIS to NIR due to their lower clay content. It is important to notice that, despite the great spectral resolution of Hyperion (Figure 15.12b) spectra are very noisy which makes its interpretation difficult.

Soil monitoring based on orbital and aerial sensors is another useful method, although interferences due to the soil moisture content in the field, atmospheric conditions, and intensity of illumination in different periods, need to be overcome.

15.8 SPACE SPECTRAL SENSING: FACTORS TO BE CONSIDERED FOR SOIL STUDIES

15.8.1 Data Used for Soil Characterization

An important characteristic of radiometric data acquired by remote sensors is that it is generally recorded and delivered to users as Digital Numbers (DNs). Since the image formed by each spectral band is a monochrome image it is usual to refer to DNs as gray levels (Varshney and Arora, 2004). These DNs or gray levels are scaled integer numbers obtained from quantization of the electromagnetic energy that reaches the sensor, and they are not a physical energy measure (Liang, 2004). Generally, DNs have a linear relation with values of radiance at the top of atmosphere, which are a physical energy measure, and they can be converted to these values using available algorithms that take in parameters related to the sensor and to the environment at the time of acquisition (Lillesand et al., 2007).

Digital number values represent different reflectance intensities (brightness) in distinct images, that is, these values are not comparable between images. However, they can still be used to compare visually different images or to analyze the relative brightness in the same image (Campbell and Wynne, 2011). Although the radiometric data can be statistically analyzed using DNs or radiance at the top of atmosphere, these data can only be compared among sensors and acquisition levels after processing them to surface reflectance, especially when using hyperspectral data (Ustin et al., 2004). Surface reflectance is achieved by correcting the radiometric data regarding the influence of atmospheric conditions, sunlight and viewing angle. Radiative transfer models are usually used for this task, It is a necessary procedure if the objective of the study includes the characterization of biophysical properties of the imaged target, since suitable differences in the signal can have great influence in the results obtained, especially in studies of quantitative nature (Liang, 2004; Jensen, 2005; Lillesand et al., 2007). Only after radiometric data correction to surface reflectance, it is possible to compared space-based spectra with spectra measured in the field or laboratory (Ustin et al., 2004). Figure 15.13 shows examples of these different radiometric products. Besides radiometric correction, an important pre-processing step is to apply geometric correction, assuring that the image pixels are correctly georeferenced, that is, that they match their true position on the surface of the Earth.

To make the image as representative as possible of the scene being recorded is necessary rectify and correct the data measured (Richards, 2013). This is necessary geometric and radiometric distortions hamper an accurate representation of the surface reflectance in the spectral bands measured. Geometric distortions are shape and scale alterations of the obtained pixels, while radiometric distortions concern to the inaccurate transformation of surface reflectance to DNs (Varshney and Arora, 2004). Taken in account that great part of spectral sensing data is provided to the user after complete or partial registration and correction for errors caused by sensor malfunction, the radiometric distortions, concerning the brightness values assigned to the pixels, are an important obstacle between the user and the accurate spectral information. Radiative transfer models are generally used to perform radiometric correction of remotely sensed data. These models consider the scattering and absorption properties of atmospheric components, and in this way make it possible to transform radiance at the top of atmosphere to surface reflectance (Figure 15.13).

Looking at Figure 15.13, is possible to notice that the radiance data (Figure 15.13b) preserve the general irradiance spectra of the energetic source, that is, the Sun, since these data are not corrected

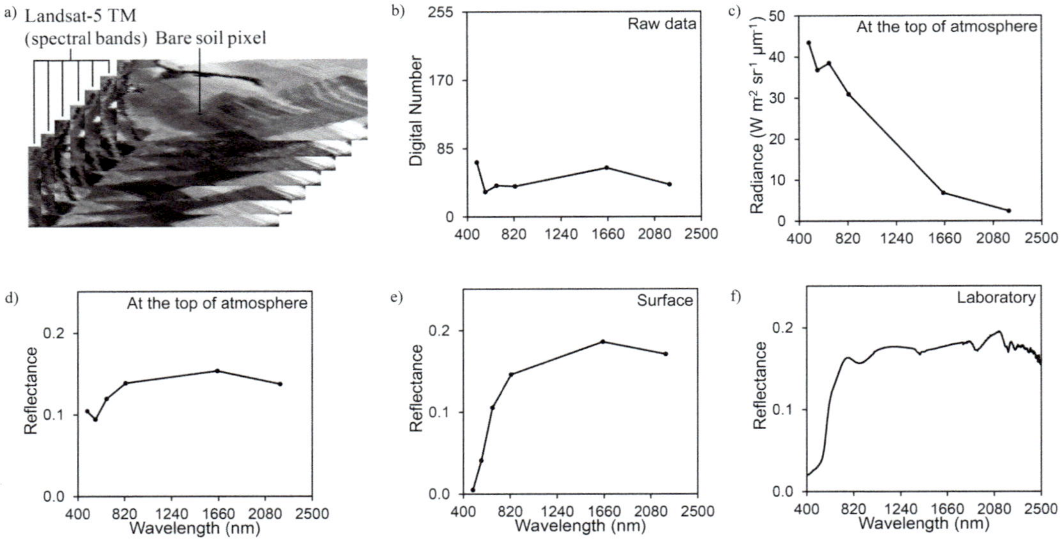

FIGURE 15.13 Landsat-5 TM images from spectral bands 1 to 5 and 7, showing a pixel predominantly with bare soil (a), with reflectance values plotted as: digital number (b), radiance at the top of the atmosphere (c), reflectance at the top of the atmosphere, and (d) reflectance at the target (surface reflectance) (e); and corresponding laboratory VIS-NIR spectrum of a soil sample collected in the same pixel (f).

for wavelength dependence of radiation that reaches the Earth or for atmospheric influences. After transformation to reflectance at the top of the atmosphere (Figure 15.13d), data no more resembles the Sun irradiance spectra, but still are influenced by atmospheric components, especially in the visible wavelengths (from 400 to 700 nm), where absorption and scattering are stronger. Finally, the surface reflectance spectrum (Figure 15.13e) includes correction to the effects caused by atmospheric components, and resembles the spectrum taken in the laboratory (Figure 15.13f) for a soil sample collected in the same spot. In fact, Demattê and Nanni (2003) observed great differences between spectra in DN and surface reflectance, arguing that different soil types were only distinguishable when DNs were transformed into surface reflectance.

15.8.2 Temporal and Spatial Variation: Implications to the Spectral Sensing of Soil in Croplands and Natural Areas

Variations in soils and their surrounding landscapes result from natural and anthropic processes occurred in the past and present, which operate at diverse temporal and spatial scales. Thus, the occurrence and variation of lithological formations, soil bodies, natural or cultivated vegetation, can be gradual or abrupt with great complexity in their development in space and time (De Jong and van der Meer, 2004).

Soil characterization through remotely sensed data can be hampered by the landscape complexity as well as by imagery resolution limitations, since the detail level that can be assessed by images is determined by its resolution (Campbell and Wynne, 2011). The components mixture in the pixels has always to be considered when using remote sensors to study soils. Even images with great spatial resolution will generally contain mixture at some degree in its pixels. To illustrate the influence of mixing different materials on soil spectra, in this case photosynthetic vegetation (PV), and non-photosynthetic vegetation (NPV), and lime were added to bare soil in gradual amounts (Figure 15.14). Considering the sequence of soil+PV spectra, it is important to highlight that chlorophyll and

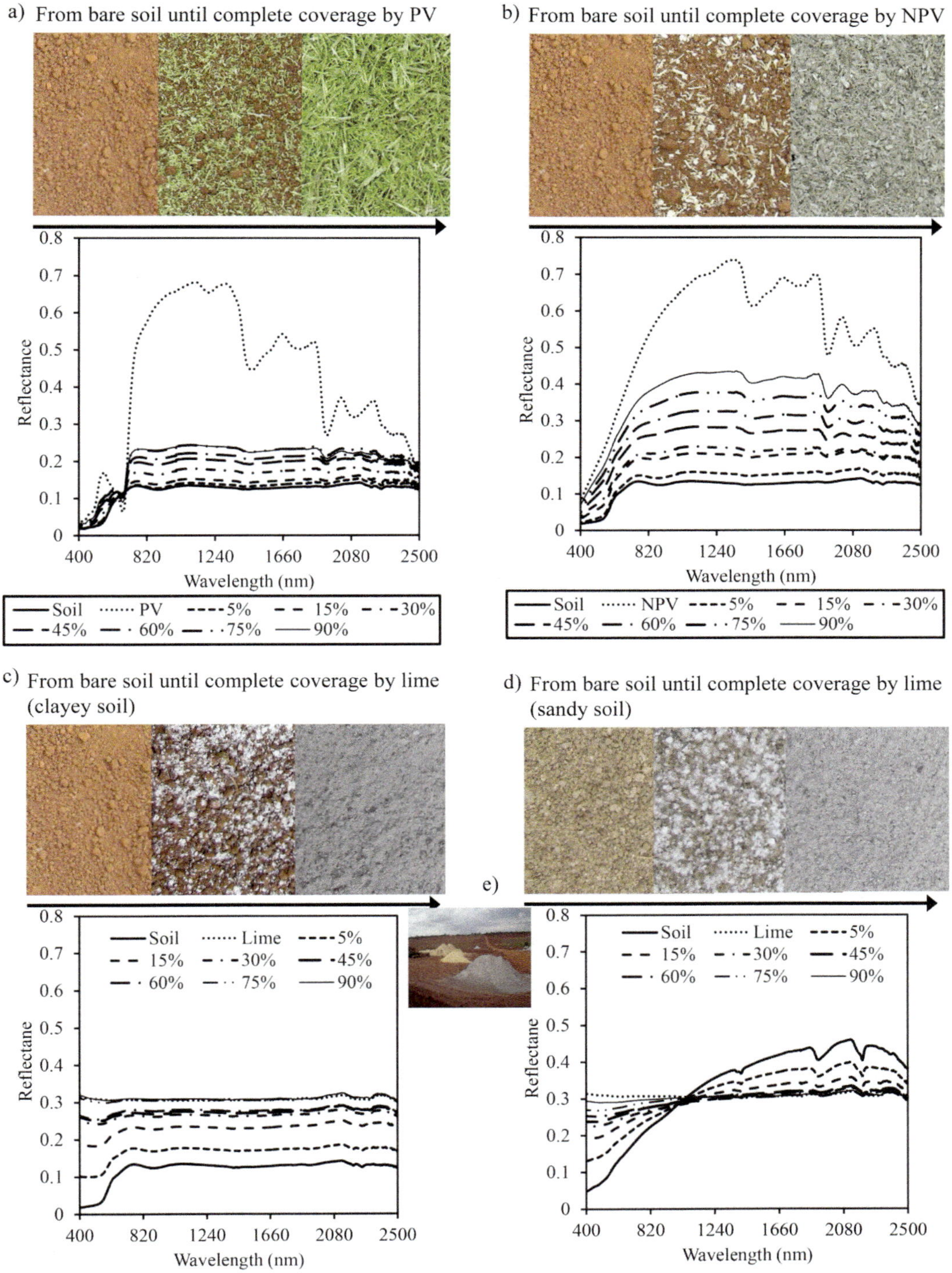

FIGURE 15.14 Some mixing materials that affect soil spectra collected in the field or by airborne or space-borne sensors: photosynthetic vegetation—PV (a); non-photosynthetic vegetation—NPV (b); and lime applied to a clayey soil (560 g kg^{-1} of clay) (c) and a sandy soil (30g kg^{-1} of clay) (d); and (e) fertilizers at the field.

other pigments cause absorption in the visible region of the electromagnetic spectrum, with characteristic features from 350 to 700 nm, and that water in the leaves cause absorption in wavelengths near 970, 1200, 1450, 1950, and 2250 nm (Kokaly et al., 2009; Ustin et al., 2009). In the case of NPV, besides absorption near 1400 and 1900 nm due to water, the main feature occurs at 2100 nm caused by cellulose and other structural components, since sugars and non-structural components are readily degraded by microorganisms (Nagler et al., 2003).

More specifically, adding PV to soil (Figure 15.14a) alters the albedo and attenuates or suppresses soil spectral features. This happens especially in VIS-NIR (from 400 to 950 nm), where characteristic absorption features of iron oxides, hydroxides and oxi-hydroxides are present in soil, and also in wavelengths near 2,200 nm, where absorption features of phyllosilicates and gibbsite occur (Stenberg et al., 2010). As already said, strong absorption features caused by pigments influence the VIS reflectance. However, it is also important to emphasize the effect that PV cover has from 680 to 700 nm, a region called "red edge," to longer wavelengths. In the "red edge" region, reflectance increases in the edge between the chlorophyll absorption feature, in the red wavelengths, and the multiple scattering caused by the cell wall-air interface within leaves, in the NIR (Treitz and Howarth, 1999). For wavelengths longer than the "red edge," a pronounced alteration in albedo occurs as the fraction of PV increases in relation to the soil fraction in the field-of-view.

Soil mixture with NPV (Figure 15.14b) causes soil spectral features attenuation or suppression together with increases in albedo, the same trend observed in the mixture of soil and PV (Figure 15.14a). However, the influence of this mixture in soil spectra is stronger, since 45% of NPV cover will mask all soil absorption features almost completely; however, the spectral feature near 880 nm, for example, caused mainly by hematite (Stenberg et al., 2010), is still visible, although attenuated, even with high PV. The mixture with NPV influences soil spectra markedly near 2100 nm, where absorption features related to cellulose and other structural components in NPV occur, thus masking the phyllosilicate absorption band near 2200 nm. The example presented in Figure 15.14 follows the same principle of the study made by Nagler et al. (2003), corroborating their results, as well as the conclusions of Serbin et al. (2009) and Hbirkou et al. (2012), which state that the VIS-NIR-SWIR spectral signature of PV is distinct from those obtained for soil and NPV. NPV spectra are very similar to soil spectra and can severely affect soil characterization and soil attribute prediction using spectral data. Along the same lines, we observed different changes of soil spectral behavior between clayey and sandy soils upon adding lime, which will affect aerial/orbital image interpretation as well (Figure 15.14c, d). Thus, accurate ground laboratory measurements are always recommended as a basis to correctly interpret aerial/orbital images.

Chabrillat et al. (2002) observed that clay minerals can be identified and mapped even when the PV fraction in a pixel is between 40 and 50% or the NPV is between 20 to 30%. On the other hand, Bartholomeus et al. (2011) found that a PV fraction greater than 5% can already decrease accuracy, and a fraction greater than 20% can hamper soil properties prediction using spectral information. Hbirkou et al. (2012) reported higher performance of SOC prediction by HyMap hyperspectral airborne sensor in recently tilled areas with bare soil ($R^2 = 0.73$), when compared to areas containing 10% of weeds and crop residues ($R^2 = 0.61$) or about 30% of straw residues ($R^2 = 0.34$).

Thus, in studies searching to describe or predict soil properties through remotely sensed data, it is usual to use masks and remove pixels with mixture levels that can hamper an accurate analysis. A methodology developed to isolate pixels containing mostly exposed soil in imagery obtained by remote sensors is described by Demattê et al. (2009a). In brief, they used a semi-automated approach consisting of the following steps: (1) visually evaluate different spectral band compositions, for example, in the case of Landsat-5 TM, they used bands 3 (660 nm), 2 (560 nm) and 1 (485 nm), and bands 5 (1650 nm), 4 (830 nm) and 3 in RGB compositions, respectively; (2) evaluate the pixels to be analyzed using the Soil Adjusted Vegetation Index (SAVI; Huete, 1989); (3) use the soil line concept (Baret et al., 1993) to check if the pixels have characteristic behavior of bare soil areas; and (4) compare pixels spectra with a Spectral Library of soils obtained from laboratory and orbital sensors for surface samples (see Figure 15.11 for examples of spectral curves). The pixels that correspond to bare soil in all the steps are classified as so. Another example of bare soil detection methodology was applied by Chabrillat et al. (2011) and Gomez et al. (2012) using HyMap imagery analyzed in the HYperspectral SOil MApper (HYSOMA) software (available free-of-charge at www.gfz-potsdam.de/hysoma). The approach is similar to that described by Madeira Netto et al. (2007) in which NDVI (Tucker, 1979) is used to mask out pixels containing a high proportion of PV. For this, a threshold was assigned to the NDVI values (0.3) evaluating known bare soil areas in the image and the coherence of the resulting mask. As suggested by Madeira Netto et al. (2007), Chabrillat et al. (2011) and Gomez et al. (2012) evaluated absorption features near 2100 nm, caused by cellulose and other structural components in NPV, in areas covered by significant amounts of NPV. Chabrillat et al. (2011) used the Cellulose Absorption Index (CAI; Nagler et al., 2003) and an threshold-based method to select areas of bare soil for soil attribute estimation (Figure 15.15). In the end of this analysis, depending on the environmental conditions in the study sites, variable results are obtained for pixel classification as bare soil or mixed cover, based on the predominance of soil features in the pixel spectral signature.

One of the most commonly used techniques for pixel unmixing is the Spectral Mixture Analysis (SMA), introduced by Horwitz et al. (1971). The SMA models consider that each pixel is a mix of different components (endmembers), and thus, the pixel spectral signature can be decomposed as a

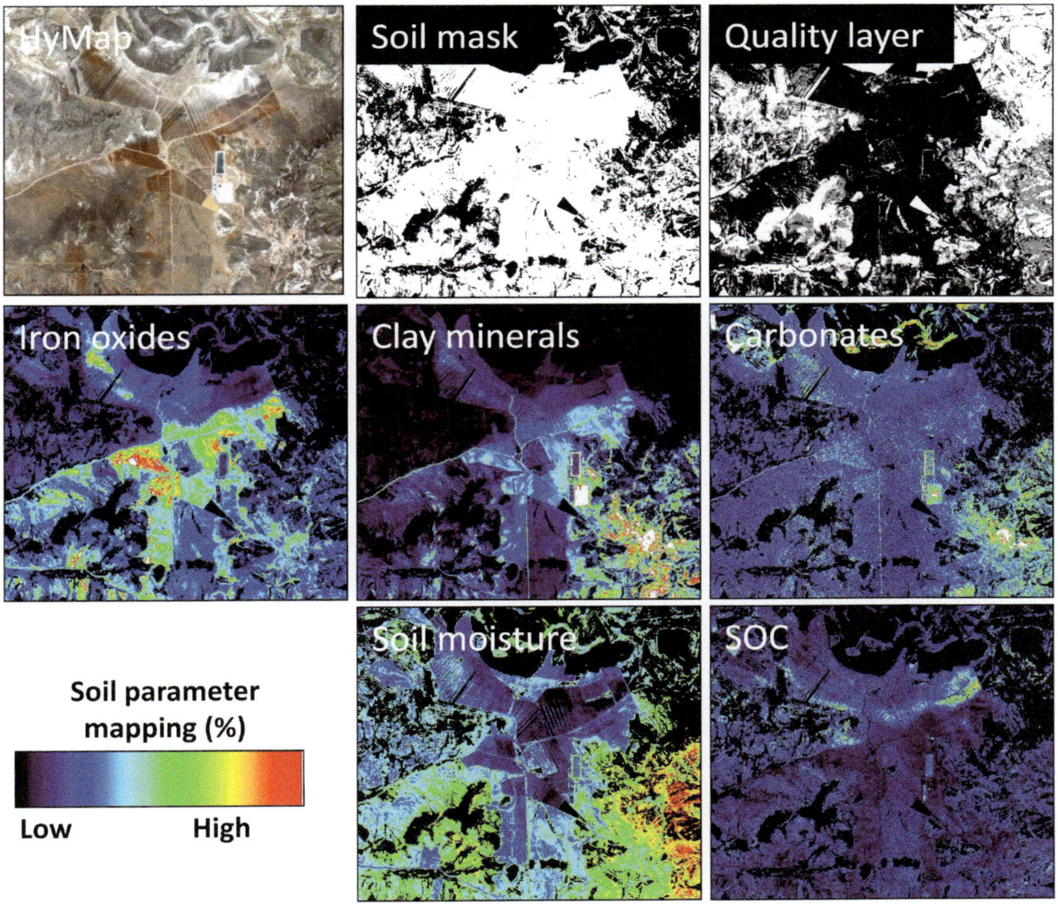

FIGURE 15.15 Soil attribute mapping from airborne HyMap imagery over Cabo de Gata Natural Park, Spain. Map outputs were created using the HYSOMA software (Chabrillat et al., 2011).

linear or nonlinear combination of the spectra of the endmembers (e.g., soil, vegetation, and shape), considering that these represent all the individual components present in the scene. To compose the pixel spectral signature, the endmembers spectra are weighted according to the, respectively, percent fraction of each endmember in the pixel. Guerschman et al. (2009) described an approach to determine fractional cover of bare soil, PV and NPV using a linear SMA of NDVI and CAI images adapted for hyperspectral sensors.

Due to the simplicity of linear SMA models, they are applied more frequently in remotely sensed data analysis, although they are suitable especially to mixtures in which the components are segregated spatially. However, when the components to be analyzed are found intimately associated in a pixel, nonlinear models are more indicated. For example, a bare soil pixel could be roughly decomposed into three intimately associated endmembers: (a) the soil matrix composed of "solid" mineral and organic materials; (b) water in pores and associated with particles of the soil matrix; and (c) air taking the remaining soil volume.

Spectral unmixing is only possible because the electromagnetic energy interacts with each pixel component differently depending on their characteristics as it is multiple scattered (Keshava and Mustard, 2002). The choice between linear and nonlinear SMA models is still controversial and it depends on the desired level of accuracy of the fractions assessment (Somers et al., 2011), since nonlinear models are more complex and difficult to implement. One of the most important limitations of the conventional linear SMA is that only one spectrum (endmember) is assigned to represent a specific scene (pixel) component, when in reality each endmember might have variable spectral signatures in a remote sensing image. For example, in the case of vegetation, biochemical composition and physical structure can vary between different species and phenological stages, among other factors, creating a wide variability in the spectral signatures of this component (Varshney and Arora, 2004). To deal with spatial and temporal endmembers variability, approaches such as Multiple Endmember Spectral Mixture Analysis (MESMA; Roberts et al., 1998) were proposed. In this case, multiple endmembers are considered for each component in an iterative way in the SMA. This methodology was one of the first attempts to manage the endmember variability for each component considered in the SMA. For this, spectral libraries that represent components variability are used. In a review, Somers et al. (2011) indicated other methodologies to deal with spatial and temporal endmembers variability.

To implement the conventional linear SMA it is assumed that all the components (endmembers) in the scene have their spectral signatures known and available to be used in the analysis. However, partial unmixing techniques have been developed in order to avoid the necessity to know all endmembers to estimate the abundance for the components of interest. The Mixture-Tuned Matched Filtering (MTMF; Boardman and Kruse, 2011) is one of the most widespread partial unmixing techniques. An example of the MTMF application is given in Chabrillat et al. (2002), who estimated the clay minerals abundance using hyperspectral imagery. The methodology allowed to quantify the amount of different types of clay with different swelling potentials (2:1 clay minerals—smectite and illite—versus 1:1 clay minerals—kaolinite) exposed at the soil surface in AVIRIS hyperspectral airborne imagery (Figure 15.6a, b).

Sophisticated methodologies to estimate components abundance in pixels have been made available and tested, for example, the approach based on neural networks described by Licciardi and Del Frate (2011) for hyperspectral data. Although bare soil areas identification for subsequent processing is a viable alternative, it limits the extent of the analysis and mapping to the identified bare soil pixels, resulting in information loss in the other areas. This motivated the use of information about vegetation to describe soil characteristics and variability, as detailed in Section 15.6.6. In addition, Bartholomeus et al. (2011) developed a technique entitled Residual Spectral Unmixing (RSU) which removes the vegetation influence in the pixels reflectance spectra. The authors stated that RSU can be applied to obtain continuous spectral information in the imaged area. Table 15.5 presents the main differences between airborne and orbital sensors and will be discussed later.

15.9 COMPARISON BETWEEN CLASSICAL AND SPECTRAL SENSING TECHNIQUES FOR SOIL ANALYSIS

Chemical analyses are largely used for the evaluation of the fertility of crop soils and environment monitoring. Nevertheless, these analyses demand high costs and long periods to obtain results, which are the main obstacles in producing soil information. It is estimated that in Brazil the number of chemical analyses of soils reached one million in 2001, demanding huge amounts of reagents and generating chemical waste (Raij et al., 2001). If such waste is mishandled or inadequately disposed of, it can result in soil and water contamination. These analyses have the support of strong background and reliable results. For example, Cantarella et al. (2006) found consistent results for chemical analysis in support of fertilizer application. Lime recommendation for a soil with 32% of base saturation reached the target value of 70±8% in 74% of the cases. In about 90% of the cases, fertilizer recommendations were on or close to the target rates. Sizeable deviations of the fertilizer recommendations for P and K that could affect profit occurred in less than 5% of the results reported. In terms of mineralogical properties, Lugassi et al. (2014) studied the potential of reflectance spectroscopy across the VIS-NIR-SWIR spectral region in combination with thermal analysis for soil mineralogy assessment. They concluded that the sensitivity of SS is higher than that of x-ray diffractometry (XRD).

Spectral sensing can overcome the issues related to time, cost and environmental pollution of classical soil analysis. However, whether these analyses can be definitely substituted by spectral sensing, still needs to be answered. In general, soil reflectance spectra are directly affected by chemical and physical chromophores, as indicated by Ben-Dor et al. The spectral response is also a product of the interaction between soil constituents, calling for a precise understanding of all chemical and physical reactions in soils. A review of soil attributes already available from reflectance spectroscopy can be found in Malley et al. (2004), Viscarra Rossel et al. (2007a, 2007b). Janik et al. (1998) already questioned if spectra could replace soil extractions. This intriguing question raised several papers along years. In fact, Brown et al. (2006) and Demattê and Nanni (2006) made comparisons between wet laboratory and spectra and stated that the first could not be substituted, but optimized, using spectral information. Despite the great relationship between spectra and soil properties, soil wet analysis cannot be completely substituted. This is because the analytical procedure to derive chemical information from soil spectroscopy is based on models between wet chemistry (as the independent variable) and reflectance data (as the dependent variables). Thus, the accuracy of the spectral data cannot exceed the reference wet chemistry information, or, in other words, these comparison papers assumed wet chemistry as the standard reference method for soil analysis. Nonetheless, the need for rapid, simultaneous and accurate analysis of many soil properties in many soil samples favors the adoption of soil spectroscopy. In fact, Sousa Junior et al. (2011) determined that a measurement with a spectrometer in the laboratory takes 10 minutes including sample preparation (the sensor takes only 1 minute to acquire 100 spectral readings), whereas a granulometric analysis takes 48 hours considering the Bouhoucus method, for example. O'Rourke and Holden (2011) calculated the costs per sample, analytical accuracy and time involved in SOC analysis, in order to identify the best method compared against Walkley-Black (Walkley and Black, 1934). The conclusion indicated that MIR spectroscopy and laboratory hyperspectral imaging were the cheapest techniques, with a cost of €0.45 and €1.26 per sample, respectively. In comparison, samples measured in a total organic carbon (TOC) analyzer costed €15.15 per sample.

An important point to use spectra is related with variances in wet analysis. Schwartz et al. (2012) observed a 20% difference in soil analysis inside the same laboratory and a 103% difference between laboratories, although for very specific analysis (petroleum contamination). Demattê et al. (2010) compared traditional soil analysis and its variations, and the relationship with spectra information. They concluded that there were 84% to 89% agreement obtained for sand content and 74% to 87% for clay between traditional laboratory variations and the estimated spectral data. Despite of this, agriculture needs faster information. Looking towards this goal, Viscarra Rossel et al. (2006a) showed the great advantages of SS analysis. In fact, Nanni and Demattê (2006a) already stated that the analysis of some soil attributes, such as clay and CEC, could already be optimized by spectral analysis, without the complete substitution of traditional laboratory methods. Their data reached high R^2 values (> 0.8), rectified by several other publications (e.g., Shepherd and Walsh, 2002; Araújo et al., 2014a). Recent studies have started using SS as a primary analytical method to measure soil properties. For example, Bradák et al. (2014) used NIR spectroscopy to quantify SOC in a paleoenvironmental investigation, as Gomez et al. (2008b) used MIR spectra to measure SOC in an investigation in Australia.

Another important point would be to define a protocol or method for collecting spectra in the laboratory. For example, the system geometry, number of readings, sample preparation, etc., should be the same to compare spectra among users. In reality, spectroscopy research has been done using different light sources, distances of sensors to soil samples and to the light source, sample preparation, and so on. Thus, data collection in SS has been done accordingly to individual preferences. To overcome this issue, in 2014 Ben-Dor et al. (personal communication; 2014, in press) proposed the first protocol for laboratory SS data collection. As an analogy, SS is going on the same track of traditional soil analysis, with a good perspective to soon become commercially attractive and accurate enough for routine soil analysis in laboratories.

Recently, in view of the growing soil spectral community, Viscarra Rossel (2009) generated an initiative (Soil World Spectral Group, http://groups.google.com/group/soil-spectroscopy) in which all members of the soil spectral community were asked to join together and contribute their local spectral libraries in order to generate a worldwide spectral library that would be accessible to all. This initiative, besides being the first attempt to gather spectral information on the world's soils, is an important step towards establishing a standard protocol and quality indicators that will be accepted by all members of the community. To that end, it is important to mention that special sessions dealing with soil spectroscopy have been organized in several leading conferences on Earth and soil sciences (e.g., EGU 2007, 2008, WSC 2010, WD 2014), and in specific workshops (e.g., EUFAR-2 2014). These meetings expose many scientists from soil and related sciences to these new technologies.

In summary, the main purpose of SS is not to substitute traditional soil analyses, but to optimize them. Samples could be summarized in a primary evaluation by spectra and then follow to wet analysis. The good prediction results for many soil properties observed in the literature (see Section 15.6.2) show the potential of SS, attracting the attention of soil scientists and making SS an interesting and constantly evolving field. Soil spectroscopy is a relatively new science with much ongoing research and the multitude of alternative tools of spectral sensing for soil characterization and attribute quantification make it an intriguing line of research with promising perspectives for soil and environmental analysis. Some important questions to be answered in the future include: (a) Should we need a databank to estimate a soil property by spectra?; (b) If yes, should it be local, regional or global?; (c) How should data be processed and in what statistical packages?; and (d) Should we test universal prediction models or should they be regionalized depending on the types of soils? We hope this chapter provides some guidance on how to achieve these answers.

15.10 MOISTURE EFFECTS IN SPECTRAL SENSING

Soil moisture is a very important property in remote and proximal sensing spectroscopy because it is an important target soil property to detect, but also because it has a nonlinear interference on spectra (Lobel and Asner, 2002; Haubrock et al., 2008a). Water molecules alter spectra in shape and intensity. The wavebands at 1400 and 1900 nm are pronounced under higher moisture contents while at 2,200 nm the opposite is true. Even though variable soil moisture affects our ability to accurately predict soil properties, it is the way in which water and soil minerals bond that provide spectral information that allows prediction of many soil constituents (Demattê et al., 2006).

For field soil spectra to be collected and properly used at its full potential, it is imperative to account for soil moisture. Aiming to reduce the effects of water in the soil spectra, without prior knowledge of the soil water and in a way that spectral libraries of dried and ground samples can be used, Minasny et al. (2011) proposed a technique called External Parameter Orthogonalization (EPO), previously conceived by Roger et al. (2003) to eliminate the effect of temperature in SS data. The EPO technique requires a calibration using a set of soils scanned moist (intact or ground) so that a transformation can be applied to the spectral library in use, and to any subsequent field scans of soils whose properties are to be predicted (Ge et al., 2014a). Most recently, Ge et al. (2014a) tested the EPO concept on intact field moist soil cores and removed the effect of soil moisture and "intactness" so that a VIS-NIR spectral library of dried and ground could be used to predict clay content and SOC (Ge et al., 2014a). This promising technique has the goal of making in situ field prediction of soil constituents using a combination of field VIS-NIR spectroscopy and spectral libraries of dried and ground samples.

From ground to space, Haubrock et al. (2008a, 2008b) (Figure 15.6c) developed a solid technique for topsoil moisture retrieval at the field and space levels, based on the influence of soil available water capacity (AWC), a proxy for soil moisture, on the edges of the spectral absorption band at 1900 nm, which is currently used in many SS applications. Their method, called Normalized Soil Moisture Index (NSMI), and the method from Whiting et al. (2004), based on the analysis of the water absorption feature at 2700 nm (Soil Moisture Gaussian Model—SMGM), were both automatized and implemented in the HYSOMA toolbox (Chabrillat et al., 2011). SMGM seems to deliver slightly better estimates, although in general both methods deliver similar SM retrieval performance, for example, with an R^2 of 0.7 for airborne HyMap images with 4 m pixel size (Chabrillat et al., 2012).

Sobrino et al. (2012) estimated soil moisture at the aerial level using an Airbone Hyperspectral Scanner (AHS) sensor, and at the orbital level using ASTER, by combining remotely sensed images with in situ measurements. Their methodology considered the correlation between surface temperature, NDVI, and emissivity, and allowed SM predictions with RMSE of 0.05 and 0.06 $m^3\ m^{-3}$ from AHS and ASTER, respectively, compared with ground measurements.

15.11 THE BASIC AND INTEGRATED STRATEGY—HOW TO MAKE A SOIL MAP INTEGRATING SPECTRAL SENSING AND GEOTECHNOLOGIES

Why do we need to use geotechnologies for soil mapping? During the field work the pedologist starts to create a mental picture of the soil boundaries envisioning a soil class map. In this task, several tools can be used, including remote sensing. Aerial photographs have been extensively used in the past (and still today). Vink (1964) proved that the use of aerial photographs added efficiency to soil mapping, requiring less field work compared to mapping procedures done without this product. Later on, Campos and Demattê (2004) highlighted the importance of using a colorimeter to quantify soil color in substitution to the visual comparison with Munsell soil color charts. They compared data from five pedologists that performed soil color for the same sample using the Munsell color chart approach. They observed a 17.5 and 8.7% agreement among pedologists for dried and moist samples, respectively. All pedologists superestimated the hue, with consequences for soil classification. Given that field light conditions are highly variable, and the eye sensitivity changes by person and with age, among other factors, we argue that automatic systems should be used for color determination. Bazaglia Filho et al. (2013) compared soil maps of the same area produced by four experienced pedologists, and observed important differences and inconsistencies among maps. These findings prove the necessity to aggregate other technologies in the soil mapping activity, not only to improve the accuracy of the information but to minimize the subjectivity of pedologists. In this aspect, we can integrate Spectral Sensing and digital soil mapping methods for soil map production. Here, we will suggest a sequence that aims to assist users and guide future research on how to integrate SS and DSM. The sequence can be altered depending on user goals and tools available.

Also, some references are cited in the sequence, but there are several other methods that could be applied presented in this paper or elsewhere.

The success of soil class mapping starts by understanding the classical soil survey technique (Legros, 2006). Afterwards, the first step is to define the characteristics and objectives of the map (i.e., soil taxonomic level, map scale/spatial resolution, target users). Second, define the data, information and tools you have (legacy soil data, spectrometers, images, equipment), and understand their advantages and limitations (for example looking at Table 15.5). One of the first used and most important SS tool employed in soil mapping was aerial photography, as described by Vink (1964). Since the stereoscope until the new aerial 3D visualization of the landscape greatly improve the delineation of soil boundaries (Figure 15.16). Third, use orbital and aerial images with color compositions integrated with aerial photographs and elevation maps (e.g., DEMs) to achieve several goals, including: (1) to define spots to collect soil samples (see Section 15.6.1); (2) to define toposequences to study and determine the soils distribution patterns in the area; (3) to define boundaries based on different physiographies (looking at the landscape), and colors (looking at the images), without attributing soil classes at this time; (4) to analyze quantitative information of soil surface by images; and (5) to relate landscape shapes and patterns with soil classes. Fourth, go to field and collect samples, bring them to the laboratory and pass through spectral sensors (mainly VIS-NIR-SWIR-MIR); choose which samples should go to wet laboratory analysis. Another approach would be taking spectra directly in field, although several issues such as moisture still remain. During field collection, measurements should be taken at different depths (Ben-Dor et al., 2008). Fifth, with the laboratory soil data and spectra of samples acquired in the laboratory (Nanni and Demattê, 2006a, 2006b) or field (Waiser et al., 2007), the following activities can be pursued: (1) relate the spectra with landscape (Galvão et al., 2001) and weathering (Demattê and Garcia, 1999) patterns; (2) understand the soil alterations along toposequences (Demattê and Terra, 2014) based on spectra; (3) analyze spectra using quantitative (Brown et al., 2006) and qualitative (Demattê, 2002) methods to group samples and determine soil mapping units (Demattê et al., 2004a); (4) compare soil spectra with available minerals libraries (Clark et al., 2007; Baldridge et al., 2009) to estimate the presence and content of minerals; (5) relate all samples with a spectral library containing pedological data, and analyze spectra for soil classification qualitatively (Bellinaso et al., 2010; Demattê et al., 2014) or quantitatively (Vasques et al., 2014), and/or for soil attribute estimation (Nocita et al., 2013); (6) in the field you can use other equipment such as gamma ray sensors (Piikki et al., 2013) or electrical conductivity meters (Aliah et al., 2013); (7) incorporate laboratory soil data, SS data and other available resources into DSM models (Behrens et al., 2010) and derive soil maps; (8) combine image radiometric data with elevation models; (9) derive models to map related geographical phenomena/properties influencing soil formation (e.g., geomorphic structures; Behrens et al., 2014); (10) finally, organize all available data and derived products in a GIS for publication. In all phases until the final results are achieved, including models and maps, human expertise and interpretation is required. Several skills are necessary, such as photopedology, spectral pedology, chemometrics, geoprocessing, pedometrics, and field experience. The methods described earlier are a suggestion and can be modified depending on the situation (objectives, scale, costs, equipment and products available), or other strategies can always be proposed. The most important message is that today soil mapping is more complex for a pedologist since they have to integrate the classical experience with all these equipment, statistics and modeling. In fact, there has been a great task to professionals to have all these skills, in addition with basic principles of pedology. On the other hand, technology is in constant evolution and pedologists certainly need to learn this new information, and thus, their background have to have these disciplines.

15.12 POTENTIAL OF SPECTRAL SENSING, PERSPECTIVES, AND FINAL CONSIDERATIONS

The potential of SS in soil and environmental sciences is constantly growing and can be perceived based on the impressive developments of hardware and analytical tools going from ground to space

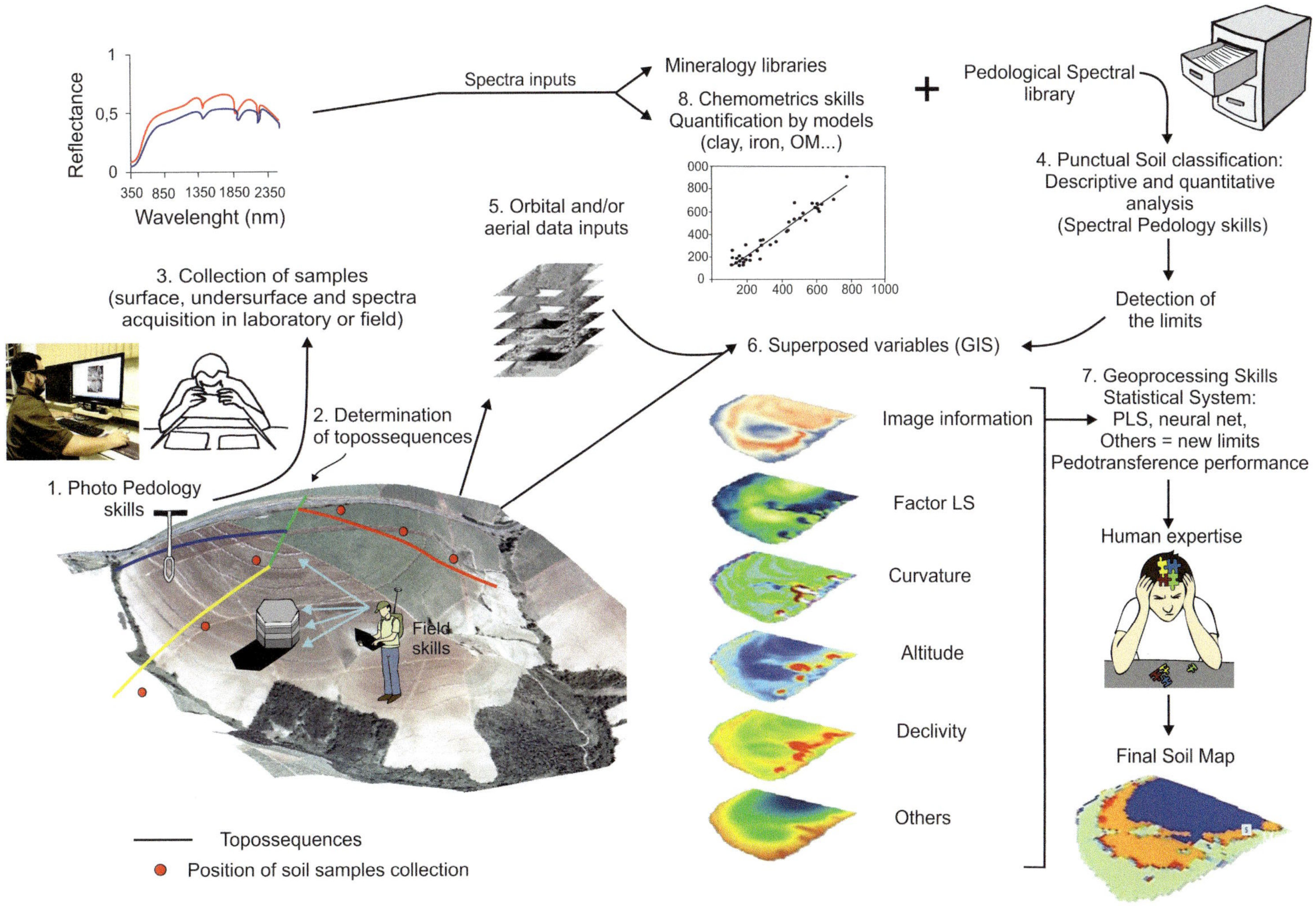

FIGURE 15.16 Suggested framework to produce a soil map by merging several techniques: photopedology; spectral sensing; spectral libraries; digital soil mapping; and statistical soil-landscape modeling.

sensors. In relation with Ground SS that started in the laboratory with, for example, Bowers and Hanks (1965), half a century later we have advanced to extraterrestrial sensors exploring the soils from planet Mars (e.g., Murchie, 2009). Ground spectral sensing has reached early maturity with more than 60 years of papers and a strong theoretical background, but still new opportunities for ground SS may arise with the development of better and less expensive sensors. Basic science in ground sensors has developed useful and fundamental relationships with several soil constituents such as water, granulometry, silicate clay type and content, mineralogy, Fe oxides, SOM, SOC, SIC, CEC and others. The applications toward soil inference are several, covering soil classification, mapping, quantification, conservation and monitoring. The ground sensors are becoming more portable for the field, for example, VIS-NIR-MIR, gamma, radar, and apparent electrical conductivity, while computer technology and software are becoming cheaper and more powerful.

The perspectives to achieve undersurface information with sensor-mounted penetrometers have great importance. Indeed, this will not substitute the usual soil "pit" and the morphological description and sampling of soil horizons, but can certainly reduce the number of soil pits and samples necessary for fine resolution (scale) mapping. Additionally, Ground SS has been adopted by PA management systems to optimize soil quality and crop production. It is still a difficult task to quantify attributes related with soil nutritional information (Ca, Mg, K, pH) because this information is primarily associated with soil water chemistry. However, correlations between soil water chemistry and soil physical (i.e., spectral or electromagnetic) properties, which can be sensed, are possible. Despite this weakness in detecting plant-available nutrients, Ground SS can assist PA with other information such as clay content, SOM, SOC, and CEC for indicating and optimizing soil sampling, assessing soil fertility, and quantifying soil water storage potential, among other properties. Spectral sensing is not a panacea, but certainly can assist on the solution of some issues by providing continuous, quantitative and accurate soil (spectral) data and mapping products, needed to assess and monitor soil status. Since 2000, an exponential increase of publications (e.g., Hartemink and McBratney, 2008; Ben-Dor et al., 2009; Chabrillat et al., 2013; Vasques et al., 2014) has promoted soil spectroscopy as a hot topic in sol science, with new interesting developing ideas. In fact, the question made by Janik et al. (1998) in their title "Can Mid Infrared Diffuse Reflectance Analysis Replace Soil Extractions?" was an indication that a revolution of new ideas on how to quantify soil properties was imminent. In fact, many papers in this research line have merged since then as observed in a review by Soriano-Disla et al. (2014). Today we can conclude that laboratory wet analysis cannot be substituted but certainly can be optimized by spectral sensing.

In 1993, Coleman et al. raised another question in "Spectral Differentiation of Surface Soils and Soil Properties: Is It Possible from Space Platforms?" They answered this question with models that achieved 0.40 and 0.28 of R^2 for clay and iron quantification, respectively, using data from a sensor located 800 km from the target (Landsat). The authors admitted the results were not very accurate, considering the use of a multispectral sensor, but that space-based spectroscopy had *potential*. This stimulated researchers. More recently, Nanni and Demattê (2006a) reached R^2 values of 0.67 and 0.95 also with Landsat and for the same elements, differing on the method to detect bare soil on the image, proving that space SS is no longer a *potential*, but a reality. The perspective is to improve these results with the future generation of optical satellite sensors, such as Copernicus sensors (Sentinel-2, to be launched in 2015), and hyperspectral EnMAP (launch planned for 2017) and other airborne or spaceborne hyperspectral sensors. Also, we expect that new methodologies on how to "isolate" the soil information in image pixels emerge, thus making soil quantification and mapping more accurate.

Considering the actual development in aerial and orbital sensors, which now have a better spectral resolution, for example, hyperspectral sensors, we foresee improvements in soil properties prediction using remote sensing images, and thus, spectral libraries will be important to understand and decomposed the spectra in image pixels. Along the same lines, great opportunities for digital soil mapping applications are expected in the near future with the upcoming availability of satellite hyperspectral sensors that will routinely deliver high spectral resolution images for the entire globe, for example, EnMAP (Germany; Kaufmann et al., 2006), HISUI (Japan), HyspIRI (USA),

HypXIM (France), PRISMA (Italy), SHALOM (Israel-Italy). These hyperspectral satellite sensors are in development, with the earliest (EnMAP, HISUI) presently in phase D with a launch date around 2017, and the others (e.g., HyspIRI, HypXIM, PRISMA, SHALOM) planned for launching around 2020. They will have high SNR and pixel sizes from 8 (HypXIM) to 30 m (EnMAP) up to 60 m (HyspIRI), thus providing a range of options and image products for the soil science community. Table 15.5 summarizes these sensors and is described as follows.

With the development of the imaging spectroscopy concept in the early 1980s, the development of imaging systems was first focused on airborne instruments. Based on the expertise in the development of such systems, many terrestrial spaceborne missions have been under study, such as NASA's High Resolution Imaging Spectrometer (HIRIS) (Goetz and Davis, 1991), the Australian Resource Information and Environment Satellite (ARIES) (Roberts et al., 1997), and ESA's Ecosystem Changes Through Response Analysis (SPECTRA) (Tobehn et al., 2002), to list just a few initiatives. Unfortunately, only a few, such as Hyperion (Middleton et al., 2013), made it into space the last decade. Table 15.5 provides an overview of current and upcoming civilian multispectral and imaging spectroscopy (hyperspectral) sensors currently operating for the imaging of the Earth's soil surface. A survey of spaceborne missions currently operating or ready for launch is provided, completed with a survey of hyperspectral missions under development and a list of new initiatives currently in a planning stage. The latter is probably not a complete list of missions, but provides a good cross-section of sensors, which might be in space around the 2020 time frame. With the launch in 2013 of Landsat 8 (former Landsat Continuity Mission LDCM) and the expected launch in 2015 of the first one of the Sentinel-2 satellites series the new ESA flagship with a repeat rate of 10 days (5 days when the second sensor will be launched two years later), global Earth coverage needs will be covered with additional spectral capabilities than previous multispectral sensors. In complement, with the expected launch of the HSE on the Resurs-P spacecraft in 2013, the next generation of imaging spectroscopy sensors is emerging. It will be followed by PRISMA, HISUI, and EnMAP, the three missions targeting a 30-m spatial resolution resulting in a swath width of 15–30 km and a 10-nm spectral resolution covering the VNIR and SWIR. These missions will replace the existing sensors currently in space in the 2016 to 2018 time frame, with increased data acquisition capacity and superior data quality compared to the technology demonstrators Hyperion and CHRIS. Although it is difficult to predict which of the hyperspectral sensors in the planning stage for a launch around 2020 or beyond will eventually be built and put in space, we can expect that the development of missions will continue and become more operational.

This overview of system shows the technical difficulties linked with the design and development of hyperspectral sensors. Due to the high requirements of hyperspectral systems toward higher spectral resolution and spectral coverage (up to 600 bands for airborne systems, up to 200 bands for spaceborne systems), then a compromise always have to be realized between the 4 pillars of space systems development: Spectral resolution, Spatial resolution, Signal-to-noise ratio and Temporal resolution. In hyperspectral missions, temporal resolution is sacrificed versus spectral resolution, so that all hyperspectral spaceborne systems have low revisit rate in comparison to multispectral missions with ~10 spectral bands. To avoid this difficulty, additional pointing capabilities such as in the EnMAP hyperspectral mission allow a higher revisit rate (4 days instead of 23 days in nadir mode) but then at the expense of adding the effect of different viewing angle. Then the triangle (1) spectral resolution- (2) spatial resolution- (3) signal-to-noise ratio is the determining factor for system performances. The smaller the pixel the lower the signal-to-noise ratio. Signal-to-noise ratio can only be improved by increasing spectral bandwidth or increasing pixel size. As can be seen in the resulting design of the planned hyperspectral missions this compromise leads to higher pixel size (Landsat-equivalent) for many missions (30m EnMAP, HISUI, PRISMA, 60m HypsIRI), and only two missions are considering smaller pixels (8m HypXIM, 10m SHALOM) at the expense of signal quality or lower spectral resolution. Data storage capacity/satellite downlink capabilities are the next limiting factors that prevent global coverage for hyperspectral missions.

Table 15.5 presents additionally a list of currently operating and new airborne hyperspectral sensors. The latter is probably not complete but shows the extensive development of airborne systems

TABLE 15.5
Current and Upcoming Sensors Systems Providing Optical Data for Soil Attributes Mapping

						Spectral Characteristics			
Platform	**Sensor**	**Origin**	**Start of Operation**	**Subsystem**	**Spatial Resolution (m)**	**Number of Bands**	**Wavelength Coverage (μm)**	**Spectral Resolution (nm)***	**Spatial Coverage**
Spaceborne optical sensors**									
Operating/Ready for launch	Landsat 8m	USA	2013	VNIR-TIR	30/100	9	0.45–12.50	-	Global
	MODISm	USA	1999	VNIR-TIR	250/1000	36	0.40–14.40	-	Global
	ASTERm	Japan/USA	1999	VNIR-TIR	15/30/90	14	0.52–11.65	-	Global
	MERISm	ESA	2002 (until 2012)	VNIR	300/1200	15	0.41–1.05	-	Global
	Hyperionh	USA	2000	VNIR-SWIR	30	242	0.36–2.58	10	Regional
	CHRISh	ESA	2001	VNIR	17/34	6/18/37	0.40–1.05	5.6–32.9	Regional
	AVNIR-2m	Japan	2006	VNIR	10	4	0.42–0.89	-	Global
	HJ-1Ah	China	2008	VNIR	100	128	0.45–0.95	5	Regional
	HySIh	India	2008	VNIR	506	64	0.40–0.95	~10	Global
	HICOh	USA	2009	VNIR	90	102	0.35–1.08	5.7	Regional
	HSE Resurs-Ph	Russia	2013	VNIR	30	192	0.40–0.96	5–10	Regional
	Sentinel-2m	ESA	2015	VNIR-SWIR	10/20/60	13	0.44–2.28	-	Global
Under development	EnMAPh	Germany	2017	VNIR-SWIR	30	242	0.42–2.45	6.5/10	Regional
	HISUIh	Japan	2017	VNIR-SWIR	30	185	0.40–2.50	10/12.5	Regional
	PRISMAh	Italy	2017	VNIR-SWIR	30	237	0.40–2.50	~12	Regional
Planned	HIPXIM-Ph	France	~2019	VNIR-SWIR	8	>200	0.40–2.50	10	Regional
	HyspIRIh	USA	≥2020	VNIR-TIR	60	>200/6	0.38–12.30	10/530	Global
	Shalomh	Israel/Italy	TBD	VNIR-SWIR	10	200	0.40–2.50	10	Regional

Airborne hyperspectral sensors									
	AHS-160	Daedalus, Spain	~2004	VNIR-TIR	2.5–10	48	0.45–13	12–550	Local
	aisaEAGLE	Specim, Finland	~2004	VNIR	~0.5–5	~488	0.40–0.97	3.3	Local
	aisaHAWK	Specim, Finland	~2004	SWIR	~0.5–5	254	0.93–2.5	12	Local
	aisaFENIX	Specim, Finland	2013	VNIR-SWIR	~0.5–5	620	0.38–2.5	3.5–12	Local
	aisaOWL	Specim, Finland	2013	TIR	tbd	84	8–12	100	Local
	APEX	VITO, Belgium	2009	VNIR-SWIR	2–10	300	0.38–2.50	5–10	Local
	AVIRIS	JPL, USA	1987	VNIR-SWIR	4–20	224	0.38–2.500	10	Local
	CASI	ITRES, Canada	~1985	VNIR	0.25–1.5	288	0.38–1.05	>3.5	Local
	HyMap	HyVista	1996	VNIR-SWIR	2–10	128	0.45–2.480	13–17	Local
	HySpex-1600	Norsk, Norway	2010	VNIR	0.5–5	160	0.40–1.0	3.7	Local
	HySpex-320	Norsk, Norway	2010	SWIR	2–20	256	1.0–2.5	6	Local
	MIVIS	CNR, Italy	1993	VNIR-TIR	3–10	112	0.43–12.7	20-50-9-450	Local
	ROSIS	DLR, Germany	1993	VNIR	2	115	0.43–0.96	5	Local
	ProspecTIR	specTIR, USA	~2008	VNIR-SWIR	~0.5–5	653	0.40–2.5	3.3–12	Local
	SASI-600	ITRES, Canada	~1985	SWIR	~1–5	160	0.95–2.45	10	Local
	SEBASS	Aerospace, USA	~1985	TIR	~1–10	128	2.5–13.5	~5	Local
	TASI-600	ITRES, Canada	~1985	TIR	~1–5	32	8–11.4	125	Local
	HyperCam	Telops, Lux.	2014	TIR	tbd	256	3–12	tbd	Local
	TRWIS-III	TRW, Canada	1996	VNIR-SWIR	~0.5–11	384	0.4–2.45	5.2/6.2	Local

* full-width half-maximum (provided for hyperspectral sensors only), ** modified from Staenz et al. (2013), m—multispectral sensor; h—hyperspectral sensor, tbd—to be demonstrated.

toward compacter and more portable systems (separated cameras for VIS-NIR and SWIR, e.g., Hyspex and aisa systems), with higher spectral resolution and more variable spectral coverage. One can note the recent development of thermal infrared (TIR) airborne hyperspectral sensors that will open new frontier of soil science as this spectral domain has been few available from afar until now, as it was mainly available only from laboratory instrumentation.

It is difficult to understand the spectra of a soil type from space without knowing basic ground information. The link between field observations (e.g., from portable field sensors or from samples analyzed in the laboratory) and space sensors will be possible through Soil Spectral Libraries. After their start in the 1980s (Stoner and Baumgardner, 1981), soil SL have evolved to a Global Spectral Library soon to be published, with the participation of about 90 countries and coordinated by Dr. Raphael Viscarra Rossel (http://groups.google.com/group/soil-spectroscopy). The available EPO method to remove moisture effects from field samples will allow faster evaluation of soil properties (Ge et al., 2014a) in conjunction with soil SL. Irrespective of the acquisition level, from ground to space, hyperspectral ground information is still the basis for understanding and interpreting soil spectra generated by the new upcoming aerial or orbital sensors. An experienced "spectral" pedologist will be able to directly take measurements in situ and have a mental picture of the soil constituents, and thus, soil class boundaries, in real-time. Removing or minimizing the need of laboratory analyses will create opportunities for the soil professional to cover larger areas with higher observation density and at a higher spatial resolution, reducing the time and cost of this activity.

Probably the most important contribution of SS to soil science (since aerial photographs and now with many available ground and space sensors) relates to the identification, characterization and delineation of soil bodies and mapping units across the landscape. Whereas ground sensors offer the opportunity to observe soils with great detail and fine sampling density, airborne and spaceborne sensors allow to extrapolate soil-spectra and soil-landscape relationships across large areas. Moreover, nowadays other regions of the electromagnetic spectrum are being studied, such as thermal infrared, microwaves and gamma, with promising contributions to soil sensing. These sensors are already available at different acquisition levels, from ground to aerial and orbital.

In precision agriculture, the use of spectral sensing techniques has been part of the concept from the very beginning, and these techniques continue to develop, with diverse applications from soil status evaluation to crop yield monitoring. The demand for detailed information in farms stresses the need for fast and cheap methods which includes both remote (airborne and spaceborne) and proximal (ground) sensing techniques. There is a growing interest in tractor-mounted sensors for real-time soil property quantification by spectral sensing in agriculture. Also, small unmanned aerial vehicles (UAV) which can carry several types of electromagnetic sensors, bringing the airborne techniques closer to the ground, will probably have great impact on soil sensing. Although most of the research so far using UAV has been on vegetation (Zhang and Kovacs, 2012), the possible applications are the same as those in traditional airborne or orbital based SS. Both tractor-mounted and UAV-based sensing will require much research for future commercial applications.

Data integration and interpolation techniques that maximize information contained in multi-scale and multi-accuracy data sources need more development to achieve reliable and repeatable soil property predictions for digital soil mapping applications. For example, radar is still an important method mainly for soil moisture quantification, whereas gamma has been used for clay content estimation, among others. These sensors can be combined to assess both (and other) soil properties simultaneously.

These SS technologies are very attractive to soil scientists, especially younger ones, and care should be taken to focus the attention on the real "patient"—the soil—and not on the technology itself. The technology can help, but the real notion on how to use them comes from the human expertise and creativity, which the SS community has so far demonstrated. Thus, we have to avoid the paradox of the technology driving the science and the questions, because in reality the opposite must be true. We recommend multidisciplinary work that includes classical pedology (i.e., knowledge of soils), statistics, and understanding of the sensors (knowing their benefits and limitations) and their derived SS information. Effective and accurate SS must merge with terrain modeling and soil genesis knowledge to reach the highest levels of accuracy and detail for soil quantification, classification and mapping. The inherent flexibility of SS allows merging micro and macro scale knowledge of soil properties, combining ground to space-based data.

In conclusion, Spectral Sensing provides a general and flexible set of instruments and tools for users who need soil information, but also constitutes a real science domain under rapid development. Relevant SS applications that advance our knowledge of soils can contribute to understand soil forming processes and soil patterns, both horizontally and vertically. Soil spectral sensing equipment is in constant evolution, and, together with powerful computing and statistical tools, will support soil researchers reach the "next level" in soil assessment and mapping. It is a matter of time to coalesce technology and information to reach the main goal of soil science: to understand and preserve soils for the future through sustainable land use and management.

ACKNOWLEDGMENTS

The authors thank Dr. Igo F. Lepsch for revision of Section 15.2; the students: Arnaldo Barros e Souza, Bruna Cristina Gallo, Caio Troula Fongaro, Danilo Jefferson Romero, Luis Gustavo Bedin, Marcus Vinicius Sato, Ana Paula Zanibão, João Paulo Brasiliano Camargo, Julia Antedomenico Cardoso De Morais, Lethícia Magno, Matheus Vinicius Rodrigues and Veridiana Maria Sayão participants of the Geotechnologies in Soil Science Group (GEOSS, http://esalqgeocis.wix.com/english) and Gregory S. Rouze from Texas A&M University, for the assistance on part of revision of this chapter.

REFERENCES

Adamchuk, V. I., Viscarra Rossel, R. A., Marx, D. B., & Samal, A. K. 2011. Using targeted sampling to process multivariate soil sensing data. Geoderma, 163, 63–73.

Adar, S., Shkolnisky, Y., & Ben-Dor, E. 2014. Change detection of soils under small-scale laboratory conditions using imaging spectroscopy sensors. Geoderma, 216, 19–29.

Aliah, B. S. N., Kodaira, M., & Shibusawa, S. 2013. Potential of visible-near infrared spectroscopy for mapping of multiple soil properties using real-time soil sensor. Proceedings of SPIE, 8881, 8881071–88810710.

Alleoni, L. R. F., Mello, J. W. V., & Rocha, W. S. D. 2009. Eletroquímica, adsorção e troca iônica no solo. In: Melo, V. F., & Alleoni, L. R. F. (Ed.), Química e mineralogia do solo: Parte II—Aplicações. Viçosa: Sociedade Brasileira de Ciência do Solo, cap. 7, pp. 69–129.

Ammann, A. B., & Brandl, H. 2001. Detection and differentiation of bacterial spores in a mineral matrix by Fourier transform infrared spectroscopy (FTIR) and chemometrical data treatment. BMC Biophys, 4, pp. 14. https://doi.org/10.1186/2046-1682-4-14.

André, F., van Leeuwen, C., Saussez, S., Van Durmen, R., Bogaert, P., Moghadas, D., Resseguier, L., Delvaux, B., Vereecken, H., & Lambot, S. 2012. High-resolution imaging of a vineyard in south of France using ground-penetrating radar, electromagnetic induction and electrical resistivity tomography. Journal of Applied Geophysics, 78, 113–122.

Anne, N. J. P., Abd-Elrahman, A. H., Lewis, D. B., & Hewitta, N. A. 2014. Modeling soil parameters using hyperspectral image reflectance insubtropical coastal wetlands. International Journal of Applied Earth Observation and Geoinformation, 33, 47–56.

Araújo, S. R., Demattê, J. A. M., & Bellinaso, H. 2013. Analyzing the effects of applying agricultural lime to soils by VNIR spectral sensing: A quantitative and quick method. International Journal of Remote Sensing, 34, 4750–4785.

Araújo, S. R., Demattê, J. A. M., & Vicente, S. 2014b. Soil contaminated with chromium by tannery sludge and identified by Vis-Nir-Mid spectroscopy techniques. International Journal of Remote Sensing and Remote Sensing Letters, 35, 3579–3593.

Araújo, S. R., Wetterlind, J., Demattê, J. A. M., & Stenberg, B. 2014a. Improving the prediction performance of a large national Vis-NIR spectroscopic library by clustering into smaller subsets or the use of data mining calibration techniques. European Journal of Soil Science, in press.

Ardekani, M. R. M. 2013. Off- and on-ground GPR techniques for field-scale soil moisture mapping Original. Geoderma, 200–201, 55–66.

Asner, G. P., Borghi, C. E., & Ojeda, R. A. 2003. Desertification in Central Argentina: Changes in ecosystem carbon and nitrogen from imaging spectroscopy. Ecological Applications, 13, 629–648.

Baldock, J. A., McNally, S. R., Beare, M. H., Curtin, D., & Hawke, B. 2019. Predicting soil carbon saturation deficit and related properties of New Zealand soils using infrared spectroscopy. Soil Research, 57, 835–844.

Baldridge, A. M., Hook, S. J., Grove, C. I., & Rivera, G. 2009. The ASTER spectral library version 2.0. Remote Sensing of Environment, 113, 711–715.

Ballabio, C., Fava, F., & Rosenmund, A. 2012. A plant ecology approach to digital soil mapping, improving the prediction of soil organic carbon content in alpine grasslands. Geoderma, 187, 102–116.

Ballari, D., de Bruin, S., & Bregt, A. K. 2012. Value of information and mobility constraints for sampling with mobile sensors. Computers & Geosciences, 49, 102–111.

Baptista, G. M. M., Corrêa, R. S., Santos, P. F., Madeira Netto, J. S., & Meneses, P. R. 2011. Use of imaging spectroscopy for mapping and quantifying the weathering degree of tropical soils in central Brazil. Applied and Environmental Soil Science, Article ID 641328, 7 p.

Baret, F., Jacquemound, S., & Hanocoq, J. F. 1993. The soil line concept in remote sensing. Remote Sensing of Environment, 7, 1–18.

Barnes, R. J., Dhanoa, M. S., & Lister, S. J. 1989. Standard normal variate transformation and de-trending of near-infrared diffuse reflectance spectra. Applied Spectroscopy, 43, 772–777.

Barthes, B. G., Brunet, D., Ferrer, H., Chotte, J. L., & Feller, C., 2006. Determination of total carbon and nitrogen content in a range of tropical soils using near infrared spectroscopy: Influence of replication and sample grinding and drying. Journal of near Infrared Spectroscopy, 14(5), 341–348.

Barthès, B. G., Kouakoua, E., Coll, P., Clairotte, M., Moulin, P., Saby, N. P. A., Le Cadre, E., Etayo, A., & Chevallier, T. 2020. Improvement in spectral library-based quantification of soil properties using representative spiking and local calibration—The case of soil inorganic carbon prediction by mid-infrared spectroscopy. Geoderma, 369, 114272. https://doi.org/10.1016/j.geoderma.2020.114272

Bartholomeus, H., Epema, G., & Schaepman, M. 2007. Determining iron content in Mediterranean soils in partly vegetated areas, using spectral reflectance and imaging spectroscopy. International Journal of Applied Earth Observation and Geoinformation, 9, 194–203.

Bartholomeus, H., Kooistra, L., Stevens, A., van Leeuwen, M., van Wesemael, B., Ben-Dor, E., & Tychon, B. 2011. Soil organic carbon mapping of partially vegetated agricultural fields with imaging spectroscopy. International Journal of Applied Earth Observation and Geoinformation, 13, 81–88.

Bas, M. V., Meléndez-Pastor, I., Navarro-Pedreño, J., Gómez, I., Mataix-Solera, J., & Hernández, E. 2013. Saline soils spectral library as a tool for digital soil mapping. Geophysical Research Abstracts, 15, EGU 2013-9738.

Batut, A. 1890. La Photographie Aerinne par Cerf-Volant. Gauthier-Villars. Paris: Gauthiers-Villars et Fils.

Baumann, P., Helfenstein, A., Gubler, A., Keller, A., Meuli, R. G., Wächter, D., Lee, J., Viscarra Rossel, R., & Six, J. 2021. Developing the Swiss mid-infrared soil spectral library for local estimation and monitoring. SOIL, 7, 525–546, https://doi.org/10.5194/soil-7-525-2021

Baumgardner, M. F., Silva, L. F., Biehl, L. L., & Stoner, E. R. 1985. Reflectance properties of soils. Advances in Agronomy, Amsterdam, 38, 1–43.

Bazaglia Filho, O., Rizzo, R., Lepsch, I. F., Prado, H. do, Gomes, F. H., Mazza, J. A., & Demattê, J. A. M. 2013. Comparison between detailed digital and conventional soil maps of an area with complex geology. Brazilian Journal of Soil Science, 37, 1136–1148.

Behrens, T., Schmidt, K., Ramirez-Lopez, L., Gallant, J., Zhu, A., & Scholten, T. 2014. Hyper-scale digital soil mapping and soil formation analysis. Geoderma, 213, 578–588.

Behrens, T., Zhu, A. X., Schmidt, K., & Scholten, T. 2010. Multi-scale digital terrain analysis and feature selection for digital soil mapping. Geoderma, 155, 175–185.

Bell, D., Menges, C., Ahmad, W., & van Zyl, J. J. 2001. The application of dielectric retrieval algorithms for mapping soil salinity in a Tropical Coastal environment using airborne polarimetric SAR. Remote Sensing of Environment, 75, 375–384.

Bellinaso, H., Demattê, J. A. M., & Araújo, S. R. 2010. Spectral library and its use in soil classification. Brazilian Journal of Soil Science, 34, 861–870.

Bellon-Maurel, V., & McBratney, A. 2011. Near-infrared (NIR) and mid-infrared (MIR) spectroscopic techniques for assessing the amount of carbon stock in soils—Critical review and research perspectives. Soil Biology & Biochemistry, 43, 1398–1410.

Ben-Dor, E. 2011. Characterization of soil properties using reflectance spectroscopy. In: Thenkabail, P. S., Lyon, J. G., & Huete, A. (Eds.), Hyperspectral Remote Sensing of Vegetation. Boca Raton: CRC Press, pp. 513–557.

Ben-Dor, E., & Banin, A. 1995a. Near infrared analysis (NIRA) as a rapid method to simultaneously evaluate, several soil properties. Soil Science Society of American Journal, 59, 364–372.

Ben-Dor, E., & Banin, A. 1995b. Near infrared analysis (NIRA) as a simultaneously method to evaluate spectral featureless constituents in soils. Soil Science, 159, 259–269.

Ben-Dor, E., Chabrillat, S., Demattê, J. A. M., Taylor, G. R., Hill, J., Whiting, M. L., & Sommer, S. 2009. Using Imaging Spectroscopy to study soil properties. Remote Sensing of Environment, 113, S38–S55.

Ben-Dor, E., Heller, D., & Chudnovsky, A. 2008. A novel method of classifying soil profiles in the field using optical means. Soil Science Society of American Journal, 72, 1113–1123.

Ben-Dor, E., Ong, C., & Lau, I. 2014. Reflectance Measurements of Soils in the Laboratory: Standards and Protocols. Sensors (personal communication and in submission).

Ben-Dor, E., Patkin, K., Banin, A., & Karnieli, A. 2002. Mapping of several soil properties using DAIS-7915 hyperspectral scanner data—a case study over clayey soils in Israel. International Journal of Remote Sensing, 23, 1043–1062.

Bernath, P. F. 2005. Spectra of Atoms and Molecules. New York: Oxford University Press, chap. 1, pp. 1–32.

Bishop, J. L., Pieters, C. M., & Edwards, J. O. 1994. Infrared spectroscopic analyses on the nature of water in montmorillonite. Clays and Clay Minerals, 42, 701–715.

Bishop, T. F. A., & McBratney, A. B. 2001. A comparison of prediction methods for the creation of field-extent soil property maps. Geoderma, 103(1–2), 149–160.

Boardman, J. W., & Kruse, F. A. 2011. Analysis of imaging spectrometer data using N-dimensional geometry and a mixture-tuned matched filtering (MTMF) approach. IEEE Transactions on Geoscience and Remote Sensing, 49, 4138–4152.

Boettinger, J. L., Ramsey, R. D., Bodily, J. M., Cole, N. J., Kienast-Brown, S., Nield, S. J., Saunders, A. M., & Stum, A. K. 2008. Landsat spectral data for digital soil mapping. In: Hartemink, A. E., McBratney, A. B., & Mendonça-Santos, M. L. (Eds.), Digital Soil Mapping with Limited Data, pp. 193–202.

Bornemann, L., Welp, G., Brodowski, S., Rodionov, A., & Amelung, W. 2008. Rapid assessment of black carbon in soil organic matter using mid-infrared spectroscopy. Organic Geochemistry, 39(11), 1537–1544.

Bowers, S. A., & Hanks, J. R. 1965. Reflection of radiant energy from soil. Soil Science, Philadelphia, 100(2), 130–138.

Bradák, B., Kiss, K., Barta, G., Varga, G., Szeberényi, J., Józsa, S., Novothny, Á., Kovács, J., Markó, A., Mészáros, E., & Szalai, Z. 2014. Different paleoenvironments of Late Pleistocene age identified in Verőce outcrop, Hungary: Preliminary results. Quaternary International, 319, 119–136.

Breunig, F. M., Galvão, L. S., & Formaggio, A. R. 2008. Detection of sandy soil surfaces using ASTER-derived reflectance, emissivity and elevation data: Potential for the identification of land degradation. International Journal of Remote Sensing, 29, 1833–1840.

Bricklemyer, R. S., & Brown, D. J. 2010. On-the-go VisNIR: Potential and limitations for mapping soil clay and organic carbon. Computers and Electronics in Agriculture, 70, 209–216.

Briedis, C., Baldock, J., de Moraes Sá, J. C., dos Santos, J. B., & Milori, D. M. B. P. 2020. Strategies to improve the prediction of bulk soil and fraction organic carbon in Brazilian samples by using an Australian national mid-infrared spectral library. Geoderma, 373, 114401. https://doi.org/10.1016/j.geoderma.2020.114401

Brodský, L., Klement, A., Penizek, V., Kodesova, R., & Boruvka, L. 2011. Building soil spectral library of the Czech soils for quantitative digital soil mapping. Soil & Water Research, 6, 165–172.

Brodský, L., Vašát, R., Klement, A., Zádorová, T., & Jakšík, O. 2013. Uncertainty propagation in VNIR reflectance spectroscopy soil organic carbon mapping. Geoderma, 199, 54–63.

Brown, D. J. 2007. Using a global VNIR soil-spectral library for local soil characterization and landscape modeling in a 2nd-order Uganda watershed. Geoderma, 140, 444–453.

Brown, D. J., Shepherd, K. D., Walsh, M. G., Mays, M. D., & Reinsch, T. G. 2006. Global soil characterization with VNIR diffuse reflectance spectroscopy. Geoderma, 132, 273–290.

Browning, D. M., & Duniway, M. C. 2011. Digital soil mapping in the absence of field training data: A case study using terrain attributes and semiautomated soil signature derivation to distinguish ecological potential. Applied and Environmental Soil Science, ID 421904, 12 p.

Bruinsma, J. 2009. The resource outlook to 2050: By how much do land, water use and crop yields need to increase by 2050? Expert Meeting on How to Feed the World in 2050. Rome: FAO and ESDD.

Brunet, D., Barthes, B. G., Chotte, J.-L., & Feller, C. 2007. Determination of carbon and nitrogen contents in Alfisols, Oxisols and Ultisols from Africa and Brazil using NIRS analysis: Effects of sample grinding and set heterogeneity. Geoderma, 139(1–2), 106–117.

Brunet, D., Woignier, T., Lesueur-Jannoyer, M., Achard, R., Rangon, L., & Barthe's, B. G. 2009. Determination of soil content in chlordecone (organochlorine pesticide) using near infrared reflectance spectroscopy (NIRS). Environmental Pollution, 157, 3120–3125.

Buol, S. W., Southard, R. J., Graham, R. C., & McDaniel, P. A. 2011. Soil Genesis and Classification. 6th edition. Wiley-Blackwell, 560 p.

Buringh, P. 1960. The application of aerial photography in soil surveys. Manual of Photographic Interpretation, 631–666.

Cambule, A. H., Rossiter, D. G., Stoorvogel, J. J., & Smaling, E. M. A. 2012. Building a near infrared spectral library for soil organic carbon estimation in the Limpopo National Park, Mozambique. Geoderma, 183–184, 41–48.

Campbell, J. B., & Wynne, R. H. 2011. Introduction to Remote Sensing. 5th edition. New York: The Guilford Press, 667 p.

Campos, R. C., & Demattê, J. A. M. 2004. Cor do solo: uma abordagem da forma convensional de obtenção em oposição à automatização do método para fins de classificação de solos [Soil color: Approach to a conventional assesment method in comparison to na authomatization process for soil classification]. Brazilian Journal of Soil Science, 28, 853–863.

Canadell, J. G., Quéré, C. L., Raupach, M. R., Field, C. B., Buitenhuis, E. T., Ciais, P., Conway, T. J., Gillett, N. P., Houghton, R. A., & Marland, G. 2007. Contributions to accelerating atmospheric CO2 growth from economic activity, carbon intensity and efficiency of natural sinks. Proceedings of the National Academy of Sciences, 104, 18866–18870.

Cantarella, H., Quaggio, J. A., van Raij, B., & de Abreu, M. F. 2006. Variability of soil analysis in commercial laboratories: Implications for lime and fertilizer recommendations. Communications in Soil Science and Plant Analysis, 37, 15–20.

Casa, R., Castaldi, F., Pascucci, S., Basso, B., & Pignatti, S. 2013a. Geophysical and hyperspectral data fusion techniques for in-field estimation of soil properties. Vadose Zone Journal, 10 pg.

Casa, R., Castaldi, F., Pascucci, S., Palombo, A., & Pignatti, S. 2013b. A comparison of sensor resolution and calibration strategies for soil texture estimation from hyperspectral remote sensing. Geoderma, 197, 17–26.

Causse, E., & Sénéchal, P. 2006. Model-based automatic dense velocity analysis of GPR field data for the estimating of soil properties. Journal of Geophysics and Engineering, 3, 169–176.

Center for Food Safety. 2008. Agricultural Pesticide Use in U.S. Agriculture. Washington: Center for Food Safety, p. 20003.

Chabrillat, S., Ben-Dor, E., Viscarra Rossel, R. A., & Demattê, J. A. M. 2013. Quantitative Soil Spectroscopy (Editorial). Applied and Environmental Soil Science, Article ID 616578.

Chabrillat, S., Eisele, A., Guillaso, S., Rogaß, C., Ben-Dor, E., & Kaufmann, H. 2011. HYSOMA: An easy-to-use software interface for soil mapping applications of hyperspectral imagery. Proceedings of the 7th EARSeL SIG Imaging Spectroscopy Workshop, Edinburgh, Scotland, UK, 11–13, On CD-Rom, 7 p.

Chabrillat, S., Foerster, S., Steinberg, A., & Segl, K. 2014. Comparison of methods for the prediction of common surface soil properties based on airborne and satellite simulated EnMAP hyperspectral images, Proceedings 2014 IEEE International Geoscience and Remote Sensing Symposium, IGARSS 2014 & 35th Canadian Symposium on Remote Sensing, Québec, Canada, 13–18.

Chabrillat, S., Gholizadeh, A., Neumann, C., Berger, D., Milewski, R., Ogen, Y., & Ben-Dor, E. 2019. Preparing a soil spectral library using the Internal Soil Standard (ISS) method: Influence of extreme different humidity laboratory conditions. Geoderma, 355, 113855. https://doi.org/10.1016/j.geoderma.2019.07.013

Chabrillat, S., Goetz, A. F. H., Krosley, L., & Olsen, H. W. 2002. Use of hyperspectral images in the identification and mapping of expansive clay soils and the role of spatial resolution. Remote Sensing of Environment, 82, 431–445.

Chabrillat, S., Whiting, M. L., Guillaso, S., Eisele, A., Haubrock, S. N., & Kaufmann, H. 2012. Quantitative mapping of surface soil moisture with hyperspectral imagery using the hysoma interface. IEEE International Geoscience and Remote Sensing Symposium, IGARSS 2012, München, Germany, 22–27.

Chakraborty, S., Weindorf, D. C., Morgan, C. L. S., Ge, Y., Galbraith, J. M., Li, B., & Kahlon, C. S. 2010. Rapid identification of oil-contaminated soils using visible near-infrared diffuse reflectance spectroscopy. Journal of Environmental Quality, 39, 1378–1387.

Chakraborty, S., Weindorf, D. C., Zhu, Y., Li, B., Morgan, C. L. S., Ge, Y., & Galbraith, J. 2012a. Spectral reflectance variability from soil physicochemical properties in oil contaminated soils. Geoderma, 177–178, 80–89.

Chakraborty, S., Weindorf, D. C., Zhu, Y., Li, B., Morgan, C. L. S., Ge, Y., & Galbraith, J. 2012b. Assessing spatial variability of soil petroleum contamination using visible near-infrared diffuse reflectance spectroscopy. Journal of Environmental Monitoring, 14, 2886–2892.

Chang, C. W., & Laird, D. A. 2002. Near-infrared reflectance spectroscopic analysis of soil C and N. Soil Science, 167(2), 110–116.

Chang, C. W., Laird, D. A., Mausbach, M. J., & Hurburgh Jr., C. R. 2001. Near-infrared reflectance spectroscopy principal components regression analysis of soil properties. Soil Science Society of America Journal, 65, 480–490.

Chang, G. W., Laird, D. A., & Hurburgh, G. R. 2005. Influence of soil moisture on near-infrared reflectance spectroscopic measurement of soil properties. Soil Science, 170(4), 244–255.

Chen, F., Kissel, D. E., West, L. T., Adkins, W., Rickman, D., & Luvall, J. C. 2008. Mapping soil organic carbon concentration for multiple fields with image similarity analysis. Soil Science Society of America Journal, 72, 186–193.

Chen, H., Pan, T., Chen, J., & Lu, Q. 2011. Waveband selection for NIR spectroscopy analysis of soil organic matter based on SG smoothing and MWPLS methods. Chemometrics and Intelligent Laboratory Systems, 107(1), 139–146.

Chen, Y., & Inbar, Y. 1994. Chemical and spectrscopical analysis of organic matter transformation during composting in relation to compost maturity. In: Hoitink, H. A. J., & Keener, H. M. (Eds.), Science and Engineering of Composting: Design, Environmental, Microbiology and Utilization Aspects. Worthington, OH: Renaissance Publications, pp. 551–600.

Choe, E., Meer, F., Ruitenbeek, F., Werff, H., De Smeth, B., & Kim, K. 2008. Mapping of heavy metal pollution in stream sediments using combined geochemistry, field spectroscopy, and hyperspectral remote sensing: A case study of the Rodalquilar mining area, SE Spain. Remote Sensing of Environment, 112, 3222–3233.

Christy, C. D. 2008. Real-time measurement of soil attributes using on-the-go near infrared reflectance spectroscopy. Computers and Electronics in Agriculture, 61, 10–19.

Clark, R. N. 1999a. Project Corona. (www.geog.ucsb.edu/~kclarke/Corona/Corona.html), accessed in 14 Jan., 2014.

Clark, R. N. 1999b. Spectroscopy of rocks and minerals and principles of spectroscopy. In: Rencz, A. N. (Ed.), Manual of Remote Sensing, Vol. 3, Remote Sensing for the Earth Sciences. Toronto: John Wiley and Sons, chap. 1, pp. 3–58.

Clark, R. N., Gallagher, A. J., & Swayze, G. A. 1990. Material absorption band bepth mapping of imaging spectrometer data using a complete band shape least-squares fit with library reference spectra, Proceedings of the Second Airborne Visible/Infrared Imaging Spectrometer (AVIRIS) Workshop. JPL Publication, 90–54, 176–186.

Clark, R. N., Swayze, G. A., Wise, R., Livo, E., Hoefen, T., Kokaly, R., & Sutley, S. J. 2007. USGS digital spectral library splib06a: U.S Geological Survey. Denver: Digital Data Series. 231 p.

Coleman, T. L., Agbu, P. A., & Montgomery, O. L. 1993. Spectral differentiation of surface soils and soil properties: Is it possible from space platforms? Soil Science, 155, 283–293.

Coleman, T. L., Agbu, P. A., Montgomery, O. L., Gao, T., & Prasad, S. 1991. Spectral band selection for quantifying selected properties in highly weathered soils. Soil Science, 151, 355–361.

Collins, M. E., Puckett, W. E., Schellentrager, W. W., & Yust, N. A. 1990. Using GPR for micro-analyses of soil and karst features on the Chiefland Limestone Plain in Florida. Geoderma, 47, 159–170.

Colwell, R. N. 1997. History and place of photographic interpretation, manual of photographic interpretation. In: Philipson, W. K. (Ed.), American Society for Photogrammetry & Remote Sensing. 2nd edition. Bethesda, MD, pp. 33–48.

Condit, H. R. 1970. The spectral reflectance of American soils. Photogrammetric Engineering, 36, 955–966.

Conforti, M., Buttafuoco, G., Leone, A. P., Aucelli, P. P. C., Robustelli, G., & Scarciglia, F., 2013. Studying the relationship between water-induced soil erosion and soil organic matter using Vis-NIR spectroscopy and geomorphological analysis: A case study in southern Italy. Catena, 110, 44–58.

Cook, S. E., Corner, R. J., Groves, P. R., & Grealish, G. J. 1996. Use of airborne gamma radiometric data for soil mapping. Australian Journal of Soil Research, 34, 183–194.

Cozzolino, D., Cynkar, W. U., Dambergs, R. G., Shah, N., & Smith, P., 2013. In Situ measurement of soil chemical composition by near-infrared spectroscopy: A tool toward sustainable vineyard management. Communications in Soil Science and Plant Analysis, 44(10), 1610–1619.

Curcio, D., Ciraolo, G., D'Asaro, F., & Minacapilli, M., 2013. Prediction of soil texture distributions using VNIR-SWIR reflectance spectroscopy. In: Romano, N., Durso, G., Severino, G., Chirico, G. B., & Palladino, M. (Eds.), Four Decades of Progress in Monitoring and Modeling of Processes in the Soil-Plant-Atmosphere System: Applications and Challenges. Procedia Environmental Sciences, pp. 494–503.

D'Acqui, L. P., Pucci, A., & Janik, L. J. 2010. Soil properties prediction of western Mediterranean islands with similar climatic environments by means of mid-infrared diffuse reflectance spectroscopy. European Journal of Soil Science, 61(6), 865–876.

D'oleire-Oltmanns, S., Marzolff, I., Peter, K. D., & Ries, J. B. 2012. Unmanned aerial vehicle (UAV) for monitoring soil erosion in Morocco. Remote Sensing, 4, 3390–3416.

Dalal, R. C., & Henry, R. J. 1986. Simultaneous determination of moisture, organic-carbon, and total nitrogenby near-infrared reflectance spectrophotometry. Soil Science Society of America Journal, 50(1), 120–123.

Daniel, K. W., Tripathi, X. K., & Honda, K. 2003. Artificial neural network analysis of laboratory and in situ spectra for the estimation of macronutrients in soils of Lop Buri (Thailand). Australian Journal of Soil Research, 41(1), 47–59.

Daniels, D. J. 2004. Ground Penetrating Radar. 2nd edition. London: The Institute of Electrical Engineers.

De Jong, S. M., & van der Meer, F. D. 2004. Remote Sensing Image Analysis: Including the Spatial Domain. Berlin: Springer, 359 p.

De Tar, W. R., Chesson, J. H., Penner, J. V., & Ojala, J. C. 2008. Detection of soil properties with airborne hyperspectral measurements of bare fields. Transactions of the ASABE, 51, 463–470.

Debaene, G. 2019. Visible and near-infrared spectroscopy in Poland: From the beginning to the Polish Soil Spectral Library. Polish Journal of Agronomy, 37(cze. 2019), 3–10. https://doi.org/10.26114/pja.iung.382.2019.37.01

Debaene, G., Niedzwiecki, J., Pecio, A., & Zurek, A., 2014. Effect of the number of calibration samples on the prediction of several soil properties at the farm-scale. Geoderma, 214, 114–125.

Demattê, J. A. M. 2002. Characterization and discrimination of soils by their reflected eletromagnetic energy. Pesquisa Agropecuária Brasileira, 37, 1445–1458.

Demattê, J. A. M., Bellinaso, H., Dalmolin, R. S. D., Paiva, A. F., da, S., Quiñonez Silvero, N. E., Poppiel, R. R., Di Raimo, L. A. D. L., Souza, A. B. e., Campos, L. R., & Resende, M. E. B. de. 2020. Avaliação dos solos do estado do Maranhão por meio de sensoriamento próximo (VIS-NIR-SWIR, MIR e XRF): um enfoque pedológico. In: Silva, M. B. e., Lumbreras, J. F., Coelho, M. R., & Oliveira, V. A. de, editores. Guia de campo da XIII Reunião Brasileira de Classificação e Correlação de Solos: RCC do Maranhão. EMBRAPA. www.infoteca.cnptia.embrapa.br/handle/doc/1127220

Demattê, J. A. M., Bellinaso, H., Romero, D. J., & Fongaro, C. T. 2014. Morphological Interpretation of Reflectance Spectrum (MIRS) using libraries looking towards soil classification. Scientia Agricola, 71(6), 509–520. https://doi.org/10.1590/0103-9016-2013-0365

Demattê, J. A. M., Campos, R. C., Alves, M. C., Fiorio, P. R., & Nanni, M. R. 2004a. Visible-NIR reflectance: A new approach on soil evaluation. Geoderma, 121, 95–112.

Demattê, J. A. M., Demattê, J. L., Camargo, W., Fiorio, P., & Nanni, M. 2001. Remote sensing in the recognition and mapping of tropical soils developed on topographic sequences. Mapping Sciences and Remote Sensing, 38, 79–102.

Demattê, J. A. M., Dotto, A. C., Paiva, A. F. S., Sato, M. V., Dalmolin, R. S. D., de Araújo, M. do S. B., . . . Noronha, N. C. 2019. The Brazilian Soil Spectral Library (BSSL): A general view, application and challenges. Geoderma, 113793. https://doi.org/10.1016/j.geoderma.2019.05.043

Demattê, J. A. M., Epiphanio, J. C. N., & Formaggio, A. R. 2003b. Influência da matéria orgânica e de formas de ferro na reflectância de solos tropicais [Organic matter and iron forms influence on the reflectance of tropical soils]. Bragantia, 62, 451–464.

Demattê, J. A. M., Fiorio, P. R., & Araújo, S. R. 2010. Variation of routine soil analysis when compared with hyperspectral narrow band sensing method. Remote Sensing, 2, 1998–2016.

Demattê, J. A. M., Fiorio, P. R., & Ben-Dor, E. 2009b. Estimation of soil properties by orbital and laboratory reflectance means and its relation with soil classification. The Open Remote Sensing Journal, 2, 12–23.

Demattê, J. A. M., Galdos, M. V., Guimarães, R. V., Genú, A. M., Nanni, M. R., & Zulu, J. 2007a. Quantification of tropical soil attributes from ETM+/Landsat-7. International Journal of Remote Sensing, 24, 257–275.

Demattê, J. A. M., Gama, M. P., Cooper, M., Araújo, J. C., Nanni, M. R., & Fiorio, P. R. 2004b. Effect of fermentation residue on the spectral reflectance properties of soils. Geoderma, 120, 187–200.

Demattê, J. A. M., & Garcia, G. J. 1999. Alteration of soil properties through a weathering sequence as evaluated by spectral reflectance. Soil Science Society of America Journal, 63, 327–342.

Demattê, J. A. M., Huete, A. R., Ferreira Jr., L., Alves, M. C., Nanni, M. R., & Fiorio, P. R. 2009a. Methodology for bare soil detection and discrimination by Landsat-TM image. The Open Remote Sensing Journal, 2, 24–35.

Demattê, J. A. M., Marcondes, A., & Simões, M. S. 2004c. Metodologia para reconhecimento de três solos por sensores: laboratorial e orbital [Laboratory and orbital data: Methods on the recognation of three soils]. Brazilian Journal of Soil Science, 28, 877–889.

Demattê, J. A. M., Moretti, D., Vasconcelos, A. C. F., & Genú, A. M. 2005. Uso de imagens de satélite na discriminação de solos desenvolvidos de basalto e arenito na região de Paraguaçú Paulista [Satelite images on the discrimination of soils developed by basalt and arenit material from the region of Paraguaçu Paulista]. Pesquisa Agropecuária Brasileira, 40, 697–706.

Demattê, J. A. M., & Nanni, M. R. 2003. Weathering sequence of soils developed from basalt as evaluated by laboratory (IRIS), airborne (AVIRIS) and orbital (TM) sensors. International Journal of Remote Sensing, 24, 4715–4738.

Demattê, J. A. M., & Nanni, M. R. 2006. Comportamento spectral da linha do solo obtida por espectrorradiometia laboratorial para diferentes classes de solo. Revista Brasileira de Ciência do Solo, 30, 1031–1038.

Demattê, J. A. M., Nanni, M. R., Formaggio, A. R., & Epiphanio, J. C. N. 2007b. Spectral Reflectance for the mineralogical evaluation of Brazilian low clay activity soils. International Journal of Remote Sensing, 28, 4537–4559.

Demattê, J. A. M., Paiva, A. F. d. S., Poppiel, R. R., Rosin, N. A., Ruiz, L. F. C., Mello, F. A. d. O., Minasny, B., Grunwald, S., Ge, Y., Ben Dor, E., et al. 2022. The Brazilian Soil Spectral Service (BraSpecS): A user-friendly system for global soil spectra communication. Remote Sensing, 14(3), 740. https://doi.org/10.3390/rs14030740

Demattê, J. A. M., Pereira, H. S., Nanni, M. R., Cooper, M., & Fiorio, P. R. 2003a. Soil chemical alterations promoted by fertilizer application assessed by spectral reflectance. Soil Science, 168, 730–747.

Demattê, J. A. M., Sousa, A. A., Alves, M., Nanni, M. R., Fiorio, P. R., & Campos, R. C. 2006. Determing soil water status and other soil characteristics by spectral proximal sensing. Geoderma, 135, 179–195.

Demattê, J. A. M., & Terra, F. S. 2014. Spectral Pedology: A new perspective on evaluation of soils along pedogenetic alterations. Geoderma, 217–218, 190–200.

Demattê, J. A. M., Terra, F. S., & Quartaroli, C. F. 2012a. Spectral behavior of some modal soil profiles from São Paulo State, Brazil. Bragantia, 71, 413–423.

Demattê, J. A. M., Vasques, G. M., Corrêa, E. A., & Arruda, G. P. 2012b. Fotopedologia, espectroscopia e sistema de informação geográfica na caracterização de solos desenvolvidos do Grupo Barreiras no Amapá [Photopedology, spectroscopy and geographic information sistem in the characterization of soils developed by Barreiras Group from Amapa]. Bragantia, 71, 438–446.

Deng, F., Minasny, B., Knadel, M., McBratney, A., Heckrath, G., & Greve, M. H. 2013. Using Vis-NIR spectroscopy for monitoring temporal changes in soil organic carbon. Soil Science, 178(8), 389–399.

Denis, A., Stevens, A., Wesemael, B. V., Udelhoven, T., & Tychon, B. 2014. Soil organic carbon assessment by field and airborne spectrometry in bare croplands: Accounting for soil surface roughness. Geoderma, 226/227, 94–102.

Dent, D., MacMillan, R. A., Mayr, T. L., Chapman, W. K., & Berch, S. M. 2013. Use of Airborne Gamma radiometrics to infer soil properties for a forested area in British Columbia, Canada. Journal of Ecosystems and Management [Online], 14(1).

Derenne, S., Rouzaud, J. N., Maquet, J., Bonhomme, C., Florian, P., & Robert, F. 2003. Abundance, size and organization of aromatic moieties in insoluble organic matter of Orgueil and Murchison meteorites. In: 34th Lunar and Planetary Science Conference. Lunar and Planetary Institute, Houston.

DeTar, W. R., Chesson, J. H., Penner, J. V., & Ojala, J. C. 2008. Detection of soil properties with airborne hyperspectral measurements of bare fields. Transactions of the ASABE, 51, 463–470.

Dierke, C., & Werban, U. 2013. Relationships between gamma-ray data and soil properties at an agricultural test site. Geoderma, 199, 90–98.

Dlugokencky, E., & Pieter, T. NOAA/ESRL (www.esrl.noaa.gov/gmd/ccgg/trends/), accessed in 25 jun, 2014.

Doetterl, S., Stevens, A., Van Oost, K., & van Wesemael, B. 2013. Soil organic carbon assessment at high vertical resolution using closed-tube sampling and vis-NIR spectroscopy. Soil Science Society of America Journal, 77(4), 1430–1435.

Dogan, H. M., & Kiliç, O. M. 2013. Modelling and mapping some soil surface properties of central Kelkit basin in Turkey by using Landsat-7 ETM+ images. International Journal of Remote Sensing, 34, 5623–5640.

Doolittle, J. A., & Butnor, J. R. 2009. Chapter 6—Soils, Peatlands, and Biomonitoring Ground Penetrating Radar Theory and Applications, p. 177, 179–202.

Doolittle, J. A., Jenkinson, B., Hopkins, P., Ulmer, M., & Tuttle, J. W. 2006. Hydropedological investigation with ground-penetrating radar (GPR): Estimating water table depths and local ground-water flow pattern in areas of coarse-textured soils. Geoderma, 131, 317–329.

Doolittle, J. A., Minzenmayer, F. E., Waltman, S. W., Benham, E. C., Tuttle, J. W., & Peaslee, S. D. 2007. Ground-penetrating radar soil suitability map of the conterminous United States. Geoderma, 141, 416–421.

Doolittle, J. A., Rebertus, R. A., Jordan, G. B., Swenson, E. I., & Taylor, W. H. 1988. Improving soil-landscape models by systematic sampling with ground-penetrating radar: Implications for soil surveys. Soil Survey Horizon, 29, 46–54.

Doolittle, J. A., Zhu, Q., Zhang, J., Guo, L., & Lin, H. 2012. Chapter 13—geophysical investigations of soil—landscape architecture and its impacts on subsurface flow. Hydropedology, pp. 413–447.

Dunn, B. W., Beecher, H. G., Batten, G. D., & Ciavarella, S. 2002. The potential of near-infrared reflectance spectroscopy for soil analysis—a case study from the Riverine Plain of south-eastern Australia. Australian Journal of Experimental Agriculture, 42(5), 607–614.

Epiphanio, J. C. N., Formaggio, A. R., Valeriano, M. M., & Oliveira, J. B. 1992. Comportamento espectral de solos do Estado de São Paulo [Spectral Behavior of Soil from São Paulo state]. São José dos Campos: INPE, 131p.

FAO. 2011. The State of the World's Land and Water Resources for Food and Agriculture—Managing Systems at Risk. Rome: FAO, 47 p.

FAO. 2013a. Statistical Yearbook 2013 World food and agriculture. Rome: FAO, 289 p.

FAO. 2013b. Global Soil Partnership Technical report—State of the Art Report on Global and Regional Soil Information: Where are we? Where to go? Rome: FAO, 69 p.

Fatras, C., Frappart, F., Mougin, E., Grippa, M., & Hiernaux, P. 2012. Estimating surface soil moisture over Sahel using ENVISAT radar altimetry. Remote Sensing of Environment, 123, 496–507.

Fernandes, R. B. A., Barrón, V., Torrent, J., & Fontes, M. P. F. 2004. Quantificação de óxidos de ferro de latossolos brasileiros por espectroscopia de reflectância difusa. Revista Brasileira de Ciência do Solo, Viçosa, 28, 245–257.

Fidêncio, P. H., Poppi, R. J., & de Andrade, J. C. 2002. Determination of organic matter in soils using radial basis function networks and near infrared spectroscopy. Analytica Chimica Acta, 453(1), 125–134.

Fiorio, P. R., & Demattê, J. A. M. 2009. Orbital and laboratory spectral data to optimize soil analysis. Scientia Agricola, 66, 250–257.

Fiorio, P. R., Demattê, J. A. M., Nanni, M. R., & Formaggio, A. R. 2010. Soil spectral diferentiation using laboratory and orbital sensor. Bragantia, 69, 249–252.

Fiorio, P. R., Demattê, J. A. M., Nanni, M. R., Genú, A. M., & Martins, J. A. 2014. *In situ* separation of soil types along transects employing Vis-NIR sensors: A new view of soil evaluation. Revista Ciência Agronômica, 45, 433–442.

Fontan, J. M., Lopez-Bellido, L., Garcia-Olmo, J., & Lopez-Bellido, R. J. 2011. Soil carbon determination in a Mediterranean vertisol by visible and near infrared reflectance spectroscopy. Journal of Near Infrared Spectroscopy, 19(4), 253–263.

Formaggio, A. R., Epihanio, J. C. E., Valeriano, M. M., & Oliveira, J. B. 1996. Comportamento espectral (450–2450 nm) de solos tropicais de São Paulo. Brazilian Journal of Soil Science, 20, 467–474.

Forrester, S. T., Janik, L. J., McLaughlin, M. J., Soriano-Disla, J. M., Stewart, R., & Dearman, B. 2013. Total petroleum hydrocarbon concentration prediction in soils using diffuse reflectance infrared spectroscopy. Soil Science Society of America Journal, 77, 450–460.

Franceschini, M. H. D., Demattê, J. A. M., Sato, M. V., Vicente, L. E., & Grego, C. R. 2013. Abordagens semi-quantitativa e quantitativa na avaliação da textura do solo por espectroscopia de reflectância bidirecional no VIS-NIR-SWIR. Pesquisa Agropecuária Brasileira, 48, 1569–1582.

Francos, N., & Ben-Dor, E. 2022. A transfer function to predict soil surface reflectance from laboratory soil spectral libraries. Geoderma, 405, 115432, https://doi.org/10.1016/j.geoderma.2021.115432.

Francos, N., Heller-Pearlshtien, D., Demattê, J. A. M., Van Wesemael, B., Milewski, R., Chabrillat, S., et al. 2023. A spectral transfer function to harmonize existing soil spectral libraries generated by different protocols. Applied and Environmental Soil Science, 1–17. https://doi.org/10.1155/2023/4155390

Fystro, G. 2002. The prediction of C and N content and their potential mineralisation in heterogeneous soil samples using Vis-NIR spectroscopy and comparative methods. Plant and Soil, 246(2), 139–149.

Galvão, L. S., Formaggio, A. R., Couto, E. G., & Roberts, D. A. 2008. Relationships between the mineralogical and chemical composition of tropical soils and topography from hyperspectral remote sensing data. ISPRS Journal of Photogrammetry & Remote Sensing, 63, 259–271.

Galvão, L. S., Ícaro, V., & Formaggio, A. R. 1997. Relationships of spectral reflectance and color among surface and subsurface horizons of tropical soil profiles. Remote Sensing of Environment, 61, 24–33.

Galvão, L. S., Pizarro, M. A., & Epiphanio, J. C. N. 2001. Variations in reflectance of tropical soils: Spectral-chemical composition relationships from AVIRIS data. Remote Sensing of Environment, 75, 245–255.

Garfagnoli, F., Ciampalini, A., Moretti, S., Chiarantini, L., & Vettori, S. 2013. Quantitative mapping of clay minerals using airborne imaging spectroscopy: New data on mugello (Italy) from SIM-GA prototypal sensor. European Journal of Remote Sensing, 46, 1–17.

Ge, Y., Morgan, C. L. S., & Ackerson, J. P. 2014a. VisNIR spectra of dried ground soils predict properties of soils scanned moist and intact. Geoderma, 221, 61–69.

Ge, Y., Morgan, C. L. S., Grunwald, S., Brown, D. J., & Sarkhot, D. V. 2011. Comparison of soil reflectance spectra and calibration models obtained using multiple spectrometers. Geoderma, 161, 202–211.

Ge, Y., Morgan, C. L. S., Thomasson, J. A., & Waiser, T. 2007. A new perspective to near infrared reflectance spectroscopy: A wavelet approach. Transactions of ASABE, 50, 303–311.

Ge, Y., Thomasson, J. A., & Morgan, C. L. S. 2014b. Mid-infrared attenuated total reflectance spectroscopy for soil carbon and particle size determination. Geoderma, 213, 57–63.

Geladi, P., MacDougall, D., & Martens, H. 1985. Linerization and scatter-correction for near-infrared reflectance spectra of meat. Applied Spectroscopy, 39, 491–500.

Genot, V., Colinet, G., Bock, L., Vanvyve, D., Reusen, Y., & Dardenne, P. 2011. Near infrared reflectance spectroscopy for estimating soil characteristics valuable in the diagnosis of soil fertility. Journal of Near Infrared Spectroscopy, 19(2), 117–138.

Genú, A. M., Demattê, J. A. M., & Nanni, M. R. 2013b. Characterization and comparison of soil spectral response obtained from orbital (ASTER e TM) and terrestrial (IRIS). Ambiência, 9, 279–288.

Genú, A. M., Roberts, D., & Demattê, J. A. M. 2013a. The use of multiple endmember spectral mixture analysis for the mapping of soil attributes using Aster imagery. Acta Scientiarum Agronomy, 35, 377–386.

Gerighausen, H., Menz, G., & Kaufmann, H. 2012. Spatially explicit estimation of clay and organic carbon content in agricultural soils using multi-annual imaging spectroscopy data. Applied and Environmental Soil Science, 10 p.

Gernsheim, H., & Gernsheim, A. 1952. Re-discovery of the world's first photograph. The Photograph Journal, 1, 118–121.

Ghaemi, M., Astaraei, A. R., Sanaeinejad, S. H., & Zare, H. 2013. Using satellite data for soil cation exchange capacity studies. International Agrophysics, 27, 409–417.

Ghosh, G., Kumar, S., & Saha, S. K. 2012. Hyperspectral satellite data in mapping salt-affected soils using linear spectral unmixing analysis. Journal of the Indian Society of Remote Sensing, 40, 129–136.

Glaser, D. R., Werkema, D. D., Versteeg, R. J., Henderson, R. D., & Rucker, D. F. 2012. Temporal GPR imaging of an ethanol release within a laboratory-scaled sand tank. Journal of Applied Geophysics, 86, 133–145.

Gmur, S., Vogt, D., Zabowski, D., & Moskal, L. M. 2012. Hyperspectral analysis of soil Nitrogen, Carbon, Carbonate, and Organic matter using regression trees. Sensors, 12, 10639–10658.

Gobrecht, A., Roger, J. M., & Bellon-Maurel, V. 2014. Major issues of diffuse reflectance NIR spectroscopy in the specific context of soil carbon content estimation: A review. Advances in Agronomy, 123, 145–175.

Godfray, H. C. J., Beddington, J. R., Crute, I. R., Haddad, L., Lawrence, D., Muir, J. F., Pretty, J., Robinson, S., Thomas, S. M., & Toulmin, C. 2010. Food security: The challenge of feeding 9 billion people. Science, 327, 812–818.

Goetz, A. H. F., & Davis, C. O. 1991. The High Resolution Imaging Spectrometer (HIRIS): Science and instrument. Journal of Imaging Systems and Technology, 3, 131–143.

Gogé, F., Gomez, C., Jolivet, C., & Joffre, R. 2014. Which strategy is best to predict soil properties of a local site from a national Vis-NIR database? Geoderma, 213, 1–9.

Gogé, F., Joffre, R., Jolivet, C., Ross, I., & Ranjard, L., 2012. Optimization criteria in sample selection step of local regression for quantitative analysis of large soil NIRS database. Chemometrics and Intelligent Laboratory Systems, 110(1), 168–176.

Gomez, C., Chevallier, T., Moulin, P., Bouferra, I., Hmaidi, K., Arrouays, D., Jolivet, C., & Barthès, B. G. 2020. Prediction of soil organic and inorganic carbon concentrations in Tunisian samples by mid-infrared reflectance spectroscopy using a French national library. Geoderma, 375, 114469. https://doi.org/10.1016/j.geoderma.2020.114469

Gomez, C., Lagacherie, P., & Coulouma, G. 2008a. Continuum removal versus PLSR method for clay and calcium carbonate content estimation from laboratory and airborne hyperspectral measurements. Geoderma, 148, 141–148.

Gomez, C., Lagacherie, P., & Coulouma, G. 2012. Regional predictions of eight common soil properties and their spatial structures from hyperspectral Vis—NIR data. Geoderma, 189–190, 176–185.
Gomez, C., Viscara Rossel, R. A., & McBratney, A. B. 2008b. Soil organic carbon prediction by hyperspectral remote sensing and field vis-NIR spectroscopy: An Australian case study. Geoderma, 146(3–4), 403–411.
Gooley, L., Huang, J., Pagé, D., & Triantafilis, J. 2014. Digital soil mapping of available water content using proximal and remotely sensed data. Soil Use and Management, 30, 139–151.
Goosen, D. 1967. Aerial Photo-interpretation in Soil Survey. FAO Soils Bulletin, 6, 1–55.
Graham, D. F., & Bonham-Carter, G. F. 1993. Aiborne radiometric data: A tool for reconnaissance geological mapping using GIS. Photogrammetric Engineering & Remote Sensing, 59, 1243–1249.
Gras, J.-P., Barthes, B. G., Mahaut, B., & Trupin, S. 2014. Best practices for obtaining and processing field visible and near infrared (VNIR) spectra of topsoils. Geoderma, 214, 126–134.
Grasty, R. L. 1979. Gamma-ray spectrometric methods in uranium exploration: Theory and operational procedures. In: Geophysics and Geochemistry in Search of Metallic Ores (ed. P. J. Hood), pp. 147–161. Geological Survey of Canada Economic Report, No. 31. Geological Survey of Canada, Ottawa, Canada.
Gregory, A. F., & Horwood, J. L. 1961. A laboratory study of gamma-ray spectra at the surface of rocks. Mines Branch Research Report R.85, Ottawa: Department of Mines and Technical Surveys.
Greschuk, L. T., Araújo, M. G. da S., Albarracín, H. S. R., Bellinaso, H., Silvero, N. E. Q., Paiva, A. F. da S., et al. 2022. Combining spectral ranges for soil discrimination: A case study in the State of Maranhão—Brazil. Geoderma Regional, 29, 1–17. https://doi.org/10.1016/j.geodrs.2022.e00507
Grinand, C., Arrouays, D., Laroche, B., & Martin, M. P. 2008. Extrapolating regional soil landscapes from an existing soil map: Sampling intensity, validation procedures, and integration of spatial context. Geoderma, 143, 180–190.
Grinand, C., Barthes, B. G., Brunet, D., Kouakoua, E., Arrouays, D., Jolivet, C., Caria, G., & Bernoux, M., 2012. Prediction of soil organic and inorganic carbon contents at a national scale (France) using mid-infrared reflectance spectroscopy (MIRS). European Journal of Soil Science, 63(2), 141–151.
Guerrero, C., Stenberg, B., Wetterlind, J., Rossel, R. A. V., Maestre, F. T., Mouazen, A. M., Zornoza, R., Ruiz-Sinoga, J. D., & Kuang, B. 2014. Assessment of soil organic carbon at local scale with spiked NIR calibrations: Effects of selection and extra-weighting on the spiking subset. European Journal of Soil Science, 65(2), 248–263.
Guerschman, J. P., Hill, M. J., Renzullo, L. J., Barret, D. J., Marks, A. L., & Botha, E. J. 2009. Estimating fractional cover of photosynthetic vegetation, non-photosynthetic vegetation and bare soil in the Australian tropical savanna region upscalling the EO-1 Hyperion and MODIS sensors. Remote Sensing of Environment, 113, 928–944.
Hansena, M. K., Brown, D. J., Dennisona, P. E., Gravesa, S. A., & Bricklemyerb, R. S. 2009. Inductively mapping expert-derived soil-landscape units within dambo wetland catenae using multispectral and topographic data. Geoderma, 150, 72–84.
Hartemink, A. E., & McBratney, A. 2008. A soil science renaissance. Geoderma, 148, 123–129.
Hartemink, A. E., & Minasny, B. 2014. Towards digital soil morphometrics. Geoderma, 230–231, 305–317.
Hasan, S., Montzka, C., Rüdiger, C., Ali, M., Bogena, H. R., & Vereecken, H. 2014. Soil moisture retrieval from airborne L-band passive microwave using high resolution multispectral data. ISPRS Journal of Photogrammetry and Remote Sensing, 91, 59–71.
Haubrock, S. N., Chabrillat, S., Kuhnert, M., Hostert, P., & Kaufmann, H. 2008a. Surface soil moisture quantification and validation based on hyperspectral data and field measurements. Journal of Applied Remote Sensing, 2, 023552.
Haubrock, S. N., Chabrillat, S., Lemmnitz, C., & Kaufmann, H. 2008b. Surface soil moisture quantification models from reflectance data under field conditions. International Journal of Remote Sensing, 29, 3–29.
Hbirkou, C., Pätzold, S., Mahlein, A. K., & Welp, G. 2012. Airborne hyperspectral imaging of spatial soil organic carbon heterogeneity at the field scale. Geoderma, 175–176, 21–28.
He, Y., Huang, M., Garcia, A., Hernandez, A., & Song, H. 2007. Prediction of soil macronutrients content using near-infrared spectroscopy. Computers and Electronics in Agriculture, 58(2), 144–153.
Heinze, S., Vohland, M., Joergensen, R. G., & Ludwig, B. 2013. Usefulness of near-infrared spectroscopy for the prediction of chemical and biological soil properties in different long-term experiments. Journal of Plant Nutrition and Soil Science, 176(4), 520–528.
Hengl, T., Toomanianb, N., Reutera, H. I., & Malakoutic, M. J. 2007. Methods to interpolate soil categorical variables from profile observations: Lessons from Iran. Geoderma, 140, 417–427.

Hergarten, C., Nazarmavloev, F., & Wolfgramm, B. 2013. Building a soil spectral library for Tajikistan comparing local and global modeling approaches. In: 3rd Global Workshop on Proximal Soil Sensing. Potsdam Germany: Leibniz-Institute for Agricultural EngineeringPotsdam-Bornim, pp. 265–269.

Hershel, W. 1800. Experiments on the refrangibility of the invisible rays of the sun. Philosophical Transactions of the Royal Society, 90, 225–283.

Hewson, R. D., Cudahy, T. J., Jones, M., & Thomas, M. 2012. Investigations into soil composition and texture using infrared spectroscopy (2–14 μm). Applied and Environmental Soil Science, 12, 535646.

Hilwig, F. W., Goosen, D., & Katsieris, D. 1974. Preliminary results of the interpretation of ERTS-1 imagery for a soil survey. ITC Journal, 3, 289–312.

Hively, W. D., McCarty, G. W., Reeves III, J. B., Lang, M. W., Oesterling, R. A., & Delwiche, S. R. 2011. Use of airborne hyperspectral imagery to map soil properties in tilled agricultural fields. Applied and Environmental Soil Science, 2011, 1–13.

Hollas, J. M. 2004. Electromagnetic radiation and its interaction with atoms and molecules. In: Hollas, J. M. (Eds.), Modern Spectroscopy. Chichester: John Wiley. chap. 2, pp. 27–39.

Horwitz, H. M., Nalepka, R. F., Hyde, P. D., & Morganstern, J. P. 1971. Estimating the proportion of objects within a single resolution element of a multispectral scanner. Proceedings of the 7th International Symposium on RS of Environment, Environmental Research Institute of Michigan, Michigan, pp. 1307–20.

Hou, S., Wang, T., & Tang, J. 2011. Soil types extraction based on MODIS image. Procedia Environmental Sciences, 10, 2207–2212.

Huang, X. W., Senthilkurnar, S., Kravchenko, A., Thelen, K., & Qi, J. G. 2007. Total carbon mapping in glacial till soils using near-infrared spectroscopy, Landsat imagery and topographical information. Geoderma, 141(1–2), 34–42.

Huete, A. R. 1989. Soil influences in remotely sensed vegetation-canopy spectra. In: Ghassem, A. (Ed.), Theory and Application of Optical Remote Sensing, New York: John Wiley and Sons, pp. 107–141.

Hunt, G. R. 1982. Spectroscopic properties of rocks and minerals. In: Carmichael, R. S. (Eds.), Handbook of Physical Properties of Rocks, Volume I. Boca Raton: CRC Press, pp. 295–385.

Hunt, G. R., & Salisbury, J. W. 1970. Visible and near infrared spectra of minerals and rocks: I Silicate minerals. Modern Geology, 1, 283–300.

Hunt, G. R., Salisbury, J. W., & Lenhoff, A. 1971. Visible and near-infrared spectra of minerals and rocks: III Oxides and hydroxides. Modern Geology, 2, 195–205.

Igne, B., Reeves, J. B., III, McCarty, G., Hively, W. D., Lund, E., & Hurburgh, C. R., Jr., 2010. Evaluation of spectral pretreatments, partial least squares, least squares support vector machines and locally weighted regression for quantitative spectroscopic analysis of soils. Journal of near Infrared Spectroscopy, 18(3), 167–176.

Inman, D. J., Freeland, R. S., Ammons, J. T., & Yoder, R. E. 2002. Soil investigation using electromagnetic induction and ground-penetrating radar in southwest Tennessee. Soil Science Society of American Journal, 66, 206–211.

International Atomic Energy Agency, IAEA. 2003. Guidelines for Radioelement Mapping Using Gamma Ray Spectrometry Data. IAEATECDOC-1363, Vienna: IAEA.

Ishida, T., Ando, H., & Fukuhara, M. 1991. Estimation of complex refractive index of soil particles and its dependence on soil chemical properties. Remote Sensing of Environment, 38, 173–182.

Islam, K., Singh, B., & McBratney, A. 2003. Simultaneous estimation of several soil properties by ultra-violet, visible, and near-infrared reflectance spectroscopy. Australian Journal of Soil Research, 41(6), 1101–1114.

IUSS Working Group WRB. 2014. World Reference Base for Soil Resources 2014. International soil classification system for naming soils and creating legends for soil maps. World Soil Resources Reports No. 106. FAO, Rome.

Janik, L. J., Forrester, S. T., & Rawson, A. 2009. The prediction of soil chemical and physical properties from mid-infrared spectroscopy and combined partial least-squares regression and neural networks (PLS-NN) analysis. Chemometrics and Intelligent Laboratory Systems, 97, 179–188.

Janik, L. J., Merry, R. H., & Skjemstad, J. O. 1998. Can mid infrared diffuse reflectance analysis replace soil extractions? Australian Journal of Experimental Agriculture, 38, 681–696.

Janik, L. J., & Skjemstad, J. O. 1995. Characterization and analysis of soils using MidInfrared partial leas-squares 2 correlations with some laboratory data. Australian Journal of Soil Research, 33, 637–650.

Jean-Philippe, S. R., Labbé, N., Franklin, J. A., & Johnson, A. 2012. Detection of mercury and other metals in mercury contaminated soils using mid-infrared spectroscopy. Proceedings of the International Academy of Ecology and Environmental Sciences, 2, 139–149.

Jenny, H. 1941. Factors of Soil Formation: A System of Quantitative Pedology. New York: McGraw-Hill, p. 190.

Jensen, J. R. 2005. Introductory Digital Image Processing: A Remote Sensing Perspective, 3rd edition. Upper Saddle River, NJ: Prentice-Hall.

Jensen, J. R. 2006. Electromagnetic radiation principles. In: Jensen, J. R. (Eds.), Remote Sensing of the Environment: An Earth Resource Perspective. New Jersey: Prentice Hall, chap. 2, pp. 29–51.

Ji, W., Li, S., Chen, S., Shi, Z., Viscarra Rossel, R. A., & Mouazen, A. M. 2016. Prediction of soil attributes using the Chinese soil spectral library and standardized spectra recorded at field conditions. Soil Tillage Res. 155, 492–500.

Johnson, R. W., Glaccum, R., & Wojatasinscki, R. 1979. Application of ground-penetrating radar to soil survey. Soil and Crop Science Society of Florida Proceedings, 39, 68–72.

Jol, H. M. 2009. Ground Penetrating Radar: Theory and Applications. Amsterdam: Elsevier Science.

Kamau-Rewe, M., Rasche, F., Cobo, J. G., Dercon, G., Shepherd, K. D., & Cadisch, G. 2011. Generic prediction of soil organic carbon in Alfisols using diffuse reflectance fourier-transform mid-infrared spectroscopy. Soil Science Society of America Journal, 75(6), 2358–2360.

Karyotis, K., Tsakiridis, N. L., Tziolas, N., Samarinas, N., Kalopesa, E., Chatzimisios, P., & Zalidis, G. 2023. On-site soil monitoring using photonics-based sensors and historical soil spectral libraries. Remote Sensing, 15(6), 1624. https://doi.org/10.3390/rs15061624

Kaufmann, H., Segl, K., Chabrillat, S., Hofer, S., Stuffler, T., Mueller, A., Richter, R., Schreier, G., Haydn, R., & Bach, H. 2006. EnMAP—a hyperspectral sensor for environmental mapping and analysis (Invited paper). Proceedings of the 2006 IEEE International Geoscience and Remote Sensing Symposium (IGARSS 2006) & 27th Canadian Symposium on Remote Sensing, Denver, USA, On CD-ROM, 0-7803-9510-7/06, 2006 IEEE.

Keen, B. A., & Haines, W. B. 1925. Studies in soil cultivation. I. The evolution of a reliable dynamometer technique for use in soil cultivation experiments. Journal of Agricultural Science, 15, 375–386.

Kemper, T., & Stefan, S. 2002. Estimate of heavy metal contamination in soils after a mining accident using reflectance spectroscopy. Environment Science Technology, 36, 2742–2747.

Keshava, N., & Mustard, J. F. 2002. Spectral unmixing. IEEE Signal Processing, 19, 44–57.

King, C., Baghdadi, N., Lecomte, V., & Cerdan, O. 2005. The application of remote-sensing data to monitoring and modelling of soil erosion. Catena, 62, 79–93.

Kinoshita, R., Moebius-Clune, B. N., van Es, H. M., Hively, W. D., & Bilgili, A. V. 2012. Strategies for soil quality assessment using visible and near-infrared reflectance spectroscopy in a Western Kenya chronosequence. Soil Science Society of America Journal, 76(5), 1776–1788.

Knadel, M., Castaldi, F., Barbetti, R., Ben-Dor, E., Gholizadeh, A., & Lorenzetti, R. 2023. Mathematical techniques to remove moisture effects from visible—near-infrared—shortwave-infrared soil spectra—review. Applied Spectroscopy Reviews, 58(9), 629–662, https://doi.org/10.1080/05704928.2022.2128365

Knadel, M., Deng, F., Thomsen, A., & Greve, M. H. 2012. Development of a Danish national Vis—NIR soil spectral library for soil organic carbon determination. In: Minasny, B., Malone, B. P., & McBratney, A. B. (Eds.), Digital Soil Assessments and Beyond: Proceedings of the 5th Global Workshop on Digital Soil Mapping. Sydney, Australia. London: Taylor & Francis Group, chap. 69, pp. 403–408.

Knadel, M., Stenberg, B., Deng, F., Thomsen, A., & Greve, M. H. 2013. Comparing predictive abilities of three visible-near infrared spectrophotometers for soil organic carbon and clay determination. Journal of near Infrared Spectroscopy, 21(1), 67–80.

Knadel, M., Thomsen, A., & Greve, M. H. 2011. Multisensor on-the-go mapping of soil organic carbon content. Soil Science Society of America Journal, 75(5), 1799–1806.

Kodaira, M., & Shibusawa, S. 2013. Using a mobile real-time soil visible-near infrared sensor for high resolution soil property mapping. Geoderma, 199, 64–79.

Kokaly, R. F., Asner, G. P., Ollinger, S. V., Martin, M. E., & Wessman, C. A. 2009. Characterizing canopy biochemistry from imaging spectroscopy and its application to ecosystem studies. Remote Sensing of Environment, 113, S78–S91.

Kooistra, L., Leuven, R. S. E. W., Wehrens, R., Nienhuis, P. H., & Buydens, L. M. C. 2003. A comparison of methods to relate grass reflectance to soil metal contamination. International Journal of Remote Sensing, 24, 4995–5010.

Kooistra, L., Salas, E. A. L., Clevers, J. G. P. W., Wehrens, R., Leuven, R. S. E. W., Nienhuis, P. H., & Buydens, L. M. C. 2004. Exploring field vegetation reflectance as an indicator of soil contamination in river floodplains. Environmental Pollution, 127, 281–290.

Kopačková, V., & Ben-Dor, E. 2016. Normalizing reflectance from different spectrometers and protocols with an internal soil standard. Internationl Journal of Remote Sensing, 37, 1276–1290. https://doi.org/10.1080/01431161.2016.1148291

Kornelsen, K. C., & Coulibaly, P. 2013. Advances in soil moisture retrieval from synthetic aperture radar and hydrological applications. Journal of Hydrology, 476, 460–489.

Krinov, E. L. 1947. Spectral reflectance of natural formations. Akad. Nauk, USSR Laboratorica Aerometodov, Moscow, USSR (transl. by G. Beikov, Natl. Res. Council of Canada, Tech. Transl. TT-439).

Kuang, B., Mahmood, H. S., Quraishi, M. Z., Hoogmoed, W. B., Mouazen, A. M., & van Henten, E. J. 2012. Sensing soil properties in the laboratory, in situ, and on-line: A review. Advances in Agronomy, 114, 155–223

Kuang, B., & Mouazen, A. M. 2011. Calibration of visible and near infrared spectroscopy for soil analysis at the field scale on three European farms. European Journal of Soil Science, 62(4), 629–636.

Kuang, B., & Mouazen, A. M. 2012. Influence of the number of samples on prediction error of visible and near infrared spectroscopy of selected soil properties at the farm scale. European Journal of Soil Science, 63(3), 421–429.

Kuang, B., & Mouazen, A. M. 2013a. Effect of spiking strategy and ratio on calibration of on-line visible and near infrared soil sensor for measurement in European farms. Soil & Tillage Research, 128, 125–136.

Kuang, B., & Mouazen, A. M. 2013b. Non-biased prediction of soil organic carbon and total nitrogen with vis-NIR spectroscopy, as affected by soil moisture content and texture. Biosystems Engineering, 114(3), 249–258.

Kusumo, B. H., Hedley, C. B., Hedley, M. J., Hueni, A., Tuohy, M. P., & Arnold, G. C. 2008. The use of diffuse reflectance spectroscopy for in situ carbon and nitrogen analysis of pastoral soils. Australian Journal of Soil Research, 46(6–7), 623–635.

Kusumo, B. H., Hedley, M. J., Hedley, C. B., & Tuohy, M. P. 2011. Measuring carbon dynamics in field soils using soil spectral reflectance: Prediction of maize root density, soil organic carbon and nitrogen content. Plant and Soil, 338(1–2), 233–245.

Kweon, G., & Maxton, C. 2013. Soil organic matter sensing with an on-the-go optical sensor. Biosystems Engineering, 115(1), 66–81.

Lagacherie, P., Baret, F., Feret, J. B., Madeira Netto, J., & Robbez-Masson, J. M. 2008. Estimation of soil clay and calcium carbonate using laboratory, field and airborne hyperspectral measurements. Remote Sensing of Environment, 112, 825–835.

Lagacherie, P., Sneep, A. R., Gomez, C., Bacha, S., Coulouma, G., Hamrounid, M. H., & Mekki, I. 2013. Combining Vis—NIR hyperspectral imagery and legacy measured soil profiles to map subsurface soil properties in a Mediterranean area (Cap-Bon, Tunisia). Geoderma, 209–210, 168–176.

Lal, R. 1999. Soil management and restoration for C sequestration to mitigate the accelerated greenhouse effect. Programming in Environmental Science, 1, 307–326.

Lal, R. 2003. Soil erosion and the global carbon budget. Environment International, 29, 437–450.

Lang, S. 2008. Object-based imge analysis for remote sensing aplications: Modeling reality-dealing with complexity. In: Blaschke, T., Lang, S., & Hay, G. J. (Eds.), Object-Based Image Analysis. New York: Springer, pp. 1–25.

Legros, J. P. 2006. Mapping of the Soil. Science Pub Inc., 411 p.

Liang, S. 2004. Quantitative Remote Sensing of Land Surfaces. Hoboken: Wiley, 534 p.

Liberti, M., Simoniello, T., Carone, M. T., Coppola, R., D'Emilio, M., & Macchiato, M. 2009. Mapping badland areas using LANDSAT TM/ETM satellite imagery and morphological data. Geomorphology, 106, 333–343.

Licciardi, G. A., & Del Frate, F. 2011. Pixel unmixing in hyperspectral data by means of neural networks. IEEE Transactions on Geoscience and Remote Sensing, 49, 4163–4172.

Lillesand, T., Kiefer, R. W., & Chipman, J. 2007. Remote Sensing and Image Interpretation, 6th edition. Hoboken: Wiley, 756 p.

Liu, X., & Liu, J., 2013. Measurement of soil properties using visible and short wave-near infrared spectroscopy and multivariate calibration. Measurement, 46(10), 3808–3814.

Liu, Y. Y., Dorigo, W. A., Parinussa, R. M., de Jeu, R. A. M., Wagner, W., McCabe, M. F., Evans, J. P., & van Dijk, A. I. J. M. 2012. Trend-preserving blending of passive and active microwave soil moisture retrievals. Remote Sensing of Environment, 123, 280–297.

Liu, Y., Shi, Z., Zhang, G., Chen, Y., Li, S., Hong, Y., Shi, T., Wang, J., & Yaolin, L. 2018. Application of spectrally derived soil type as ancillary data to improve the estimation of soil organic carbon by using the Chinese soil Vis-NIR spectral library. Remote Sensing, 10. https://doi.org/10.3390/rs10111747

Lobel, D. B., & Asner, G. P. 2002. Moisture effects on soil reflectance. Soil Science Society of America Journal, 66, 722–727.

Löhrer, R., Bertrams, M., Eckmeier, E., Protze, J., & Lehmkuhl, F. 2013. Mapping the distribution of weathered Pleistocene wadi deposits in Southern Jordan using ASTER, SPOT-5 data and laboratory spectroscopic analysis. Catena, 107, 57–70.

Lu, P., Wang, L., Niu, Z., Li, L., & Zhang, W. 2013. Prediction of soil properties using laboratory VIS-NIR spectroscopy and Hyperion imagery. Journal of Geochemical Exploration, 132, 26–33.

Ludwig, B., Nitschke, R., Terhoeven-Urselmans, T., Michel, K., & Flessa, H. 2008. Use of mid-infrared spectroscopy in the diffuse-reflectance mode for the prediction of the composition of organic matter in soil and litter. Journal of Plant Nutrition and Soil Science-Zeitschrift Fur Pflanzenernahrung Und Bodenkunde, 171(3), 384–391.

Lugassi, R., Ben-Dor, E., & Eshel, G. 2014. Reflectance spectroscopy of soils post-heating—Assessing thermal alterations in soil minerals. Geoderma, 213, 268–279.

Ma, Y., Roudier, P., Kumar, K., Palmada, T., Grealish, G., Carrick, S., Lilburne, L., & Triantafilis, J. 2023. A soil spectral library of New Zealand. Geoderma Regional, 35, e00726, https://doi.org/10.1016/j.geodrs.2023.e00726.

Madeira Netto, J. 2001. Comportamento espectral dos solos. In: Meneses, P. R., & Madeira Netto, J. S. (Eds.), Sensoriamento remoto: reflectância dos alvos naturais. Brasília: UnB, Planaltina: Embrapa Cerrados, cap. 4, 127–154.

Madeira Netto, J. S. R., Robbez-Masson, J. M., & Martins, E. 2007. Visible—NIR hyperspectral imagery for discriminating soil types in the La Peyne watershed (France). In: Lagacherie, P., McBratney, A. B., & Voltz, M. (Eds.), Digital Soil Mapping: An Introductory Perspective. Amsterdam: Elsevier.

Madejová, J., & Komadel, P. 2001. Baseline studies of the clay minerals society source clays: Infrared methods. Clay and Clay Minerals, Chantilly, 49(5), 410–432.

Mahmood, H. S., Hoogmoed, W. B., & van Henten, E. J. 2012. Sensor data fusion to predict multiple soil properties. Precision Agriculture, 13(6), 628–645.

Mahmood, H. S., Hoogmoed, W. B., & van Henten, E. J. 2013. Proximal gamma-ray spectroscopy to predict soil properties using windows and full-spectrum analysis methods. Sensors, 13(12), 16263–16280.

Mahmood, K. 1987. Reservoir sedimentation—impact, extent and mitigation. Washington: World Bank Tech. Paper, No. 71.

Mahmoudzadeh, M. R., Francés, A. P., Lubczynski, M., & Lambot, S. 2012. Using ground penetrating radar to investigate the water table depth in weathered granites—Sardon case study, Spain. Journal of Applied Geophysics, 79, 17–26.

Malley, D. F., Martin, P. D., McClintock, L. M., Yesmin, L., Eilers, R. G., & Haluschak, P. 2000. Feasibility of analysing archived Canadian prairie agricultural soils by near infrared reflectance spectroscopy. In: Davies, A. M. C., & Giangiacomo, R. (Eds.), Near Infrared Spectroscopy: Proceedings of the 9th International Conference. Chichester, UK: NIR Publications, pp. 579–585.

Malley, D. F., Martin, P., & Ben-Dor, E. 2004. Application in analysis of soils. In: Craig, R., Windham, R., & Workman, J. (Eds.), Near Infrared Spectroscopy in Agriculture. A Three Society Monograph (ASA, SSSA, CSSA). Madison, WI, pp. 729–784.

Mann, K. K., Schumann, A. W., & Obreza, T. A. 2011. Delineating productivity zones in a citrus grove using citrus production, tree growth and temporally stable soil data. Precision Agriculture, 12, 457–472.

Marques, K. P., Rizzo, R., Dotto, A. C., Souza, A. B., Mello, F. A. O., Neto, L. G. M., dos Anjos, L. H. C., & Demattê, J. A. M. 2019. How qualitative spectral information can improve soil profile classification? Journal of Near Infrared Spectroscopy, 27, 156–174. https://doi.org/10.1177/0967033518821965

Martelet, G., Drufin, S., Tourliere, B., Saby, N. P. A., Perrin, J., Deparis, J., Prognon, F., Jolivet, C., Ratié, C., & Arrouays, D. 2013. Regional regolith parameter prediction using the proxy of airborne gamma ray spectrometry. Vadose Zone Journal, 12(4).

Martínez-Casasnovas, J. A. 2003. A spatial information technology approach for the mapping and quantification of gully erosion. Catena, 50, 293–308.

Masoud, A. A. 2014. Predicting salt abundance in slightly saline soils from Landsat ETM+ imagery using Spectral Mixture Analysis and soil spectrometry. Geoderma, 217–218, 45–56.

Masserschmidt, I., Cuelbas, C. J., Poppi, R. J., De Andrade, J. C., De Abreu, C. A., & Davanzo, C. U. 1999. Determination of organic matter in soils by FTIR/diffuse reflectance and multivariate calibration. Journal of Chemometrics, 13(3–4), 265–273.

McBratney, A. B., Mendonça Santos, M. L., & Minasny, B. 2003. On digital soil mapping. Geoderma, 117, 3–52.

McCarty, G. W., Hively, W. D., Reeves, J. B., III, Lang, M., Lund, E., & Weatherbee, O. 2010. Infrared Sensors to Map Soil Carbon in Agricultural Ecosystems. Proximal Soil Sensing.

McCarty, G. W., & Reeves, J. B. 2006. Comparison of near infrared and mid infrared diffuse reflectance spectroscopy for field-scale measurement of soil fertility parameters. Soil Science, 171(2), 94–102.

McCarty, G. W., Reeves, J. B., Reeves, V. B., Follett, R. F., & Kimble, J. M. 2002. Mid-infrared and near-infrared diffuse reflectance spectroscopy for soil carbon measurement. Soil Science Society of America Journal, 66(2), 640–646.

McDonald, R. A. 1997. CORONA: Beteween the Sun and the Earth: The first NRO Reconnaissance Eye in Space. Bethesda: ASP&RS, 400p.

McDowell, M. L., Bruland, G. L., Deenik, J. L., Grunwald, S., & Knox, N. M. 2012. Soil total carbon analysis in Hawaiian soils with visible, near-infrared and mid-infrared diffuse reflectance spectroscopy. Geoderma, 189, 312–320.

McGlashan, M. A., Tsoflias, G. P., Schillig, P. C., Devlin, J. F., & Roberts, J. A. 2012. Field GPR monitoring of biostimulation in saturated porous media. Journal of Applied Geophysics, 78, 102–112.

Medhat, M. E. 2012. Application of gamma-ray transmission method for study the properties of cultivated soil. Annals of Nuclear Energy, 40, 53–59.

Mendes, W. S., Boechat, C. L., Gualberto, A. V. S., Barbosa, R. S., da Silva, Y. J. A. B., Saraiva, P. C., de Sena, A. F. S., & de Duarte, L. S. L. 2021. Soil spectral library of Piauí State using machine learning for laboratory analysis in Northeastern Brazil. Revista Brasileira de Ciência do Solo, 45, e0200115.

Mendes, W. S., Demattê, J. A. M., Rosin, N. A., Terra, F. S., Poppiel, R. R., Urbina-Salazar, D. U., Boechat, C. L., Silva, E. B., Curi, N., Silva, S. H. G., et al. The Brazilian soil mid-infrared spectral library: The power of the fundamental range. Geoderma, 415, 115776.

Metternicht, G., & Zinck, J. A. 2003. Remote sensing of soil salinity: Potentials and constraints. Remote Sensing of Enviroment, 85, 1–20.

Middleton, E. M., Campbell, P. E., Huemmrich, K. F., Ong, L., Mandl, D., Frye, S., & Landis, D. R. 2013. The hyperion imaging spectrometer on the Earth Observing One (EO-1) satellite: Over a dozen years in space. Proceedings of the International Geoscience and Remote Sensing Symposium (IAGARSS'13), Melbourne, Australia, 4 pages, 2013.

Miltz, J., & Don, A. 2012. Optimising sample preparation and near infrared spectra measurements of soil samples to calibrate organic carbon and total nitrogen content. Journal of near Infrared Spectroscopy, 20(6), 695–706.

Minasny, B., & McBratney, A. B. 2006. A conditioned Latin hypercube method for sampling in the presence of ancillary information. Computers & Geosciences, 32, 1378–1388.

Minasny, B., & McBratney, A. B. 2008. Regression rules as a tool for predicting soil properties from infrared reflectance spectroscopy. Chemometrics and Intelligent Laboratory Systems, 94(1), 72–79.

Minasny, B., McBratney, A. B., Bellon-Maurel, V., Roger, J. M., Gobrecht, A., Ferrand, L., & Joalland, S. 2011. Removing the effect of soil moisture from NIR diffuse reflectance spectra for the prediction of soil organic carbon. Geoderma, 167–168, 118–124.

Minasny, B., McBratney, A. B., Tranter, G., & Murphy, B. W. 2008. Using soil knowledge for the evaluation of mid-infrared diffuse reflectance spectroscopy for predicting soil physical and mechanical properties. European Journal of Soil Science, 59(5), 960–971.

Minasny, B., McBratney, A. B., & Walvoort, D. J. J. 2007. The variance quadtree algorithm: Use for spatial sampling design. Computers & Geosciences, 33, 383–392.

Minasny, B., Tranter, G., McBratney, A. B., Brough, D. M., & Murphy, B. W. 2009. Regional transferability of mid-infrared diffuse reflectance spectroscopic prediction for soil chemical properties. Geoderma, 153(1–2), 155–162.

Minet, J., Bogaert, P., Vanclooster, M., & Lambot, S. 2012. Validation of ground penetrating radar full-waveform inversion for field scale soil moisture mapping. Journal of Hydrology, 424–425, 112–123.

Minty, B. R. S. 1997. Fundamentals of airborne gamma-ray spectrometry. AGSO Journal of Australian Geology and Geophysics, 17, 39–50.

Mladenova, I. E., Jackson, T. J., Njoku, E., Bindlish, R., Chan, S., Cosh, M. H., Holmes, T. R. H., de Jeu, R. A. M., Jones, L., Kimball, J., Paloscia, S., & Santi, E. 2014. Remote monitoring of soil moisture using passive microwave-based techniques—Theoretical basis and overview of selected algorithms for AMSR-E original research article. Remote Sensing of Environment, 144, 197–213.

Mohammedzein, M. A., Csorba, A., Rotich, B., Justin, P. N., Melenya, C., Andrei, Y., & Micheli, E. 2023. Development of Hungarian spectral library: Prediction of soil properties and applications. Eurasian Journal of Soil Science, 12(3), 244–256. https://doi.org/10.18393/ejss.1275149

Molan, Y. E., Refahi, D., & Tarashti, A. H. 2014. Mineral mapping in the Maherabad area, eastern Iran, using the HyMap remote sensing data. International Journal of Applied Earth Observation and Geoinformation, 27, 117–127.

Morgan, C. L. S., Waiser, T., Brown, D. J., & Hallmark, C. T. 2009. Simulated in situ characterization of soil organic and inorganic carbon with visible near-infrared diffuse reflectance spectroscopy. Geoderma, 151, 249–256.

Moron, A., & Cozzolino, D. 2002. Application of near infrared reflectance spectroscopy for the analysis of organic C, total N and pH in soils of Uruguay. Journal of Near Infrared Spectroscopy, 10(3), 215–221.

Mouazen, A. M., Kuang, B., De Baerdemaeker, J., & Ramon, H. 2010. Comparison among principal component, partial least squares and back propagation neural network analyses for accuracy of measurement of selected soil properties with visible and near infrared spectroscopy. Geoderma, 158(1–2), 23–31.

Mouazen, A. M., Maleki, M. R., De Baerdemaeker, J., & Ramon, H. 2007. On-line measurement of some selected soil properties using a VIS-NIR sensor. Soil and Tillage Research, 93, 13–27.

Mulder, V. L., de Bruin, S., & Schaepman, M. E. 2013. Representing major soil variability at regional scale by constrained Latin Hypercube Sampling of remote sensing data. International Journal of Applied Earth Observation and Geoinformation, 21, 301–310.

Mulder, V. L., de Bruin, S., Schaepman, M. E., & Mayr, T. R. 2011. The use of RS in soil and terrain mapping—A review. Geoderma, 162, 1–19.

Munoz, J. D., & Kravchenko, A. 2011. Soil carbon mapping using on-the-go near infrared spectroscopy, topography and aerial photographs. Geoderma, 166(1), 102–110.

Murchie, S. L. 2009. Compact reconnaissance imaging spectrometer for Mars investigation and data set from the Mars reconnaissance orbiter's primary science phase. Journal of Geophysical Research, 114, E00D07, https://doi.org/10.1029/2009JE003344.

Nadal-Romero, E., Vicente-Serrano, S. M., & Jiménez, I. 2012. Assessment of badland dynamics using multi-temporal Landsat imagery: An example from the Spanish Pre-Pyrenees. Catena, 96, 1–11.

Nagler, P. L., Inoue, Y., Glenn, E. P., Russ, A. L., & Daughtry, C. S. T. 2003. Cellulose absorption index (CAI) to quantify mixed soil plant litter scenes. Remote Sensing of Environment, 87, 310–325.

Nanni, M. R., & Demattê, J. A. M. 2006a. Spectral reflectance methodology in comparison to traditional soil analysis. Soil Science Society of America Journal, 2, 393–407.

Nanni, M. R., & Demattê, J. A. M. 2006b. Comportamento da linha do solo obtida por espectrorradiometria laboratorial para diferentes classes de solos [Soil line bevaior obtained by laboratory data for different classes of soils]. Brazilian Journal of Soil Science, 30, 1031–1038.

Nanni, M. R., Demattê, J. A. M., Chicati, M. L., Fiorio, P. R., Cézar, E., & de Oliveira, R. B. 2012. Soil surface spectral data from Landsat imagery for soil class discrimination. Acta Scientiarum, 34, 103–112.

Nanni, M. R., Demattê, J. A. M., Chicati, M. L., Oliveira, R. B., & Cezar, E. 2011. Spectroradiometric data as support to soil classification. International Research Journal of Agricultural Science, 1, 100–115.

Nanni, M. R., Demattê, J. A. M., Silva Junior, C. A., Romagnoli, F., Silva, A. A., Cezar, E., & Gasparotto, A. C. 2014. Soil mapping by laboratory and orbital spectral sensing compared with a traditional method in a detailed level. Journal of Agronomy, in press.

National Aeronautics and Space Administration. MODIS NASA, disponível em: http://modis.gsfc.nasa.gov/. Acesso em: 09 mar. 2014.

Naumann, D., Helm, D., & Labischinski, H. 1991. Microbiological characterizations byFT-IR spectroscopy. Nature, 351, 81–82.

NBSS & LUP, Reflectance libraries for development of soil sensor for periodic assessment of state of soil resources. NATP Project Report (NBSS No. 835), National Bureau of Soil Survey & Land Use Planning, Nagpur, 2005.

Nduwamungu, C., Ziadi, N., Tremblay, G. F., & Parent, L. É. 2009. Near-infrared reflectance spectroscopy prediction of soil properties: Effects of sample cups and preparation. Soil Science Society of America Journal, 7, 1896–1903.

Ng, W., Husnain, A. L., Siregar, A. F., Hartatik, W., Sulaeman, Y., Jones, E., & Minasny, B. 2020. Developing a soil spectral library using a low-cost NIR spectrometer for precision fertilization in Indonesia. Geoderma Regional, 22, e00319, https://doi.org/10.1016/j.geodrs.2020.e00319.

Nguyen, T. T., Janik, L. J., & Raupach, M. 1991. Diffuse reflectance infrared fourier treansform (DRIFT) spectroscopy in soil studies. Australian Journal of Soil Research, Collingwood, 29, pp. 49–67.

Nocita, M., Kooistra, L., Bachmann, M., Mueller, A., Powell, M., & Weel, S. 2011. Predictions of soil surface and topsoil organic carbon content through the use of laboratory and field spectroscopy in the Albany Thicket Biome of Eastern Cape Province of South Africa. Geoderma, 167–168, 295–302.

Nocita, M., Stevens, A., Noon, C., & van Wesemael, B. 2013. Prediction of soil organic carbon for different levels of soil moisture using Vis-NIR spectroscopy. Geoderma, 199, 37–42.

Nocita, M., Stevens, A., Toth, G., Panagos, P., van Wesemael, B., & Montanarella, L. 2014. Prediction of soil organic carbon content by diffuse reflectance spectroscopy using a local partial least square regression approach. Soil Biology & Biochemistry, 68, 337–347.

O'Rourke, S. M., & Holden, N. M. 2011. Optical sensing and chemometric analysis of soil organic carbon—a cost effective alternative to conventional laboratory methods? Soil Use and Management, 27, 143–155.

Obade, V. de P., & Lal, R. 2013. Assessing land cover and soil quality by RS and geographical information systems (GIS). Catena, 104, 77–92.

Obukhov, A. I., & Orlov, D. S. 1964. Spectral reflectivity of the major soils group and possibility of using diffuse reflection in soil investigation. Society Soil Science, 1, 174–184.

Ogen, Y., Faigenbaum-Golovin, S., Granot, A., Shkolnisky, Y., Goldshleger, N., & Ben-Dor, E. 2019. Removing moisture effect on soil reflectance properties: A case study of clay content prediction. Pedosphere, 29, 421–431. https://doi.org/10.1016/S1002-0160(19)60811-8

Oinuma, K., & Hayashi, H. 1965. Infrared study of mixed-layer clay minerals. American Mineralogist, Chantilly, 50, 1213–1227.

Okparanma, R. N., & Mouazen, A. M. 2013. Combined effects of oil concentration, clay and moisture contents on diffuse Reflectance Spectra of diesel-contamined soils. Water Air Soil Pollut, 224, 1539.

Orgiazzi, A., Ballabio, C., Panagos, P., Jones, A., & Fernández-Ugalde, O. 2018. LUCAS Soil, the largest expandable soil dataset for Europe: A review. European Journal of Soil Science, 69(1), 140–153, https://doi.org/10.1111/ejss.12499

Panciera, R., Walker, J. P., Jackson, T. J., Gray, D., Tanase, M. A., Ryu, D., Monerris, A., Yardley, H., Rüdiger, C., Wu, X., Gao, Y., & Hacker, J. 2014. The Soil Moisture Active Passive Experiments (SMAPEx): Towards soil moisture retrieval from the SMAP mission. IEEE Transaction on Geoscience and Remote Sensing, 52, 490–507.

Parrens, M., Calvet, J. C., de Rosnay, P., & Decharme, B. 2014. Benchmarking of L-band soil microwave emission models Original Research Article. Remote Sensing of Environment, 140, 407–419.

Peng, Y., Knadel, M., Gislum, R., Deng, F., Norgaard, T., de Jonge, L. W., Moldrup, P., & Greve, M. H. 2013. Predicting soil organic carbon at field scale using a national soil spectral library. Journal of Near Infrared Spectrosc, 21, 213–222. https://doi.org/10.1255/jnirs.1053

Perret, J., Villalobos-Leandro, J. E., Abdalla-Bolaños, K., Fuentes-Fallas, C. L., Cuarezma-Espinoza, K. M., Macas-Amaya, E. N., López-Maietta, M. T., & Drewry, D. 2020. Desarrollo de métodos de análisis de espectroscopia y algoritmos de aprendizaje automático para la evaluación de algunas propiedades del suelo en Costa Rica. Agronomía Costarricense, 44(2), 139–154. https://doi.org/10.15517/rac.v44i2.43108

Peter, K. D., d'Oleire-Oltmanns, S., Ries, J. B., Marzolff, I., & Hssaine, A. A. 2014. Soil erosion in gully catchments affected by land-levelling measures in the Souss Basin, Morocco, analysed by rainfall simulation and UAV remote sensing data. Catena, 113, 24–40.

Petersen, H., Wunderlich, T., al Hagrey, S. A., & Rabbel, W. 2012. Characterization of some middle European soil textures by gamma-spectrometry. Journal of Plant Nutrition and Soil Science, 175(5), 651–660.

Petersen, S. O., Frohne, P., & Kennedy, A. C. 2002. Dynamics of a soil microbial community under spring wheat. Soil Science Society of American Journal, 66, 826–833.

Piikki, K., Söderström, M., & Stenberg, B., 2013. Sensor data fusion for topsoil clay mapping. Geoderma, 199, 106–116.

Pirie, A., Singh, B., & Islam, K. 2005. Ultra-violet, visible, near-infrared, and mid-infrared diffuse reflectance spectroscopic techniques to predict several soil properties. Australian Journal of Soil Research, 43(6), 713–721.

Pozdnyakova, L. 1999. Electrical soil properties. Dissertation (Ph.D. in Renewable Resources)—University of Wyoming, Laramie, p. 138.

Pracilio, G., Adams, M. L., & Smettem, K. R. J. 2003. Use of airborne gamma radiometric data for soil property and crop biomass assessment. In: Stafford, J., & Werner, A. (Ed.), Precision agriculture: Papers from the 4th European Conference on Precision Agriculture, Berlin, Germany, 15–19, pp. 551–557.

Pracilio, G., Adams, M. L., Smettem, K. R. J., & Harper, R. J. 2006. Determination of spatial distribution patterns of clay and plant available potassium contents in surface soils at the farm scale using high resolution gamma ray spectrometry. Plant and Soil, 282, 67–82.

Priori, S., Bianconi, N., & Costantini, E. A. C. 2014. Can gamma-radiometrics predict soil textural data and stoniness in different parent materials? A comparison of two machine-learning methods. Geoderma, 226/227, 354–364.

Rabenarivo, M., Chapuis-Lardy, L., Brunet, D., Chotte, J.-L., Rabeharisoa, L., & Barthes, B. G., 2013. Comparing near and mid-infrared reflectance spectroscopy for determining properties of Malagasy soils, using global or LOCAL calibration. Journal of near Infrared Spectroscopy, 21(6), 495–509.

Radu, T., Gallagher, S., Byrne, B., Harris, P., Coveney, S., McCarron, S., McCarthy, T., & Diamond, D. 2013. Portable X-Ray fluorescence as a rapid technique for surveying elemental distributions in soil. Spectroscopy Letters, 46, 516–526.

Raij, B. V., Andrade, J. C., Cantarella, H., & Quaggio, J. A. 2001. Análise química para avaliação da fertilidade de solos tropicais. Campinas Instituto Agronômico, 285p.

Ramirez-Lopez, L., Behrens, T., Schmidt, K., Stevens, A., Dematte, J. A. M., & Scholten, T. 2013. The spectrum-based learner: A new local approach for modeling soil vis-NIR spectra of complex datasets. Geoderma, 195, 268–279.

Ramirez-Lopez, L., Schmidt, K., Behrens, T., Wesemael, B. V., Dematte, J. A. M., & Scholten, T. 2014. Sampling optimal calibration sets in soil infrared spectroscopy. Geoderma, 226/227, 140–150.

Ratcliff, A. W., Busse, M. D., & Shestak, C. J. 2006. Changes in microbial community structure following herbicide (glyphosate) additions to forest soils. Applied Soil Ecology, 34, 114–124.

Reeves III, J. B. 2010. Near-versus mid-infrared diffuse reflectance spectroscopy for soil analysis emphasizing carbon and laboratory versus on-site analysis: Where are we and what needs to be done? Geoderma, 158, 3–14.

Reeves III, J. B., Follett, R. F., McCarty, G. W., & Kimble, J. M. 2006. Can near or mid-infrared diffuse reflectance spectroscopy be used to determine soil carbon pools? Communications in Soil Science and Plant Analysis, 37(15–20), 2307–2325.

Reeves III, J. B., & McCarty, G. W. 2001. Quantitative analysis of agricultural soils using near infrared reflectance spectroscopy and a fibre-optic probe. Journal of near Infrared Spectroscopy, 9(1), 25–34.

Reeves III, J. B., McCarty, G. W., & Hively, W. D. 2010. Mid-Versus Near-Infrared Spectroscopy for On-Site Analysis of Soil. Proximal Soil Sensing.

Reschke, J., & Hüttich, C. 2014. Continuous field mapping of Mediterranean wetlands using sub-pixel spectral signatures and multi-temporal Landsat data. International Journal of Applied Earth Observation and Geoinformation, 28, 220–229.

Reuben, N. O., & Mouazen, A. M. 2013. Visible and near-infrared spectroscopy analysis of a polycyclic aromatic hydrocarbon in soils. The Scientific World Journal, Article ID 160360, 9.

Richards, J. A. 2013. Remote Sensing Digital Image Analysis, 5th edition. Berlin: Springer, p. 485.

Richter, N., Chabrillat, S., & Kaufmann, H. 2007. Enhanced quantification of soil variables linked with soil degradation using imaging spectroscopy. Proceedings of the 5th EARSeL Workshop "Imaging Spectroscopy: Innovation in Environmental Research", Bruges, Belgium, 23–25, Reusen, I., & Cools, J. (eds.), On CD-ROM, 7 p.

Richter, N., Jarmer, T., Chabrillat, S., Oyonart, C., Hostert, P., & Kaufmann, H. 2009. Free iron oxide determination in Mediterranean soils using diffuse reflectance spectroscopy. Soil Science Society of America Journal, Madison, 73, 72–81.

Ries, J. B., & Marzolff, I. 2003. Monitoring of gully erosion in the Central Ebro Basin by large-scale aerial photography taken from a remotely controlled blimp, Original Research Article. Catena, 50, 309–328.

Rivero, R. G. Grunwald, S., & Bruland, G. L. 2007. Incorporation of spectral data into multivariate geostatistical models to map soil phosphorus variability in a Florida wetland. Geoderma, 140, 428–443.

Roberts, D. A., Gardner, M., Church, R., Ustin, S., Scheer, G., & Green, R. O. 1998. Mapping chaparral in the Santa Monica mountains using multiple endmember spectral mixture models. Remote Sensing of Environment, 65, 267–279.

Roberts, E., Huntington, J., & Denize, R. 1997. The Australian Resource Information and Environment Satellite (ARIES), Phase A Study. Proceedings of the 11thAIAA/USU Small Satellite Conference, SSC97-III-2, Logan, Utah. 1997.

Roger, J. M., Chauchard, F., & Bellon-Maurel, V. 2003. EPO—PLS external parameter orthogonalisation of PLS application to temperature-independent measurement of sugar content of intact fruits. Chemometrics and Intelligent Laboratory Systems, 66, 191–204.

Romero, D. J., Ben-Dor, E., Demattê, J. A. M., e Souza, A. B., Vicente, L. E., Tavares, T. R., Martello, M., Strabeli, T. F., da Silva Barros, P. P., Fiorio, P. R., Gallo, B. C., Sato, M. V., & Eitelwein, M. T. 2018. Internal soil standard method for the Brazilian soil spectral library: Performance and proximate analysis. Geoderma, 312, 95–103. https://doi.org/10.1016/j.geoderma.2017.09.014

Rongjiang, Y., Jingsong, Y., Xiufang, Z., Xiaobing, C., Jianjun, H., Xiaoming, L., Meixian, L., & Hongbo, S. 2012. A new soil sampling design in coastal saline region using EM38 and VQT method. CLEAN—Soil, Air, Water, 40, 972–979.

Salisbury, J. W., & D'Aria, D. M. 1992a. Infrared (8–14 μm) remote sensing of soil particle size. Remote Sensing of Environment, 42, 157–165.

Salisbury, J. W., & D'ária, D. M. 1992b. Emissivity of terrestrial materials in the 8–14 μm atmospheric window. Remote Sensing of Environment, New York, 42, 83–106.

Salisbury, J. W., & D'ária, D. M. 1994. Emissivity of terrestrial materials in the 3–5 μm atmospheric window. Remote Sensing of Environment, New York, 47, 345–361.

Sandén, T., Lippl, M., Reiter, E., Dersch, G., Spiegel, H., & Baumgarten, A. 2022. Austrian Soil Spectral Library for future soil fertility assessments, EGU General Assembly 2022, Vienna, Austria, 23–27 May 2022, EGU22–9803, https://doi.org/10.5194/egusphere-egu22-9803, 2022.

Sankey, J. B., Brown, D. J., Bernard, M. L., & Lawrence, R. L. 2008. Comparing local vs. global visible and near-infrared (VisNIR) diffuse reflectance spectroscopy (DRS) calibrations for the prediction of soil clay, organic C and inorganic C. Geoderma, 148(2), 149–158.

Sannino, F., & Gianfreda, L. 2001. Pesticide influence on soil enzymatic activities. Chemosphere, 45, 417–425.

Santana, F. B. de, de Souza, A. M., & Poppi, R. J. 2019. Green methodology for soil organic matter analysis using a national near infrared spectral library in tandem with learning machine. Sci Total Environ. 658, 895–900. https://doi.org/10.1016/j.scitotenv.2018.12.263

Savitzky, A., & Golay, M. J. E. 1964. Smoothing and differentiation of data by simplified least squares procedures. Analytical Chemistry, 36, 1627–1639.

Schirrmann, M., Gebbers, R., & Kramer, E. 2012. Field scale mapping of soil fertility parameters by combination of proximal soil sensors. In: International Conference of Agricultural Engineering—CIGR-Ageng 2012. D Valencia, Spain, 8–12 July 2012. Diazotec, S. L., Valencia, pp. 1–6 (http://cigr.ageng2012.org/images/fotosg/tabla_2137_C0411.pdf), accessed in 2010/2010/2013.

Schmidtlein, S. 2004. Coarse-scale substrate mapping using plant functional response types. Erdkunde, 58, 137–151.

Schreier, H. 1977. Quantitative predictions of chemical soil conditions from multispectral airborne, ground and laboratory measurements. In: Proc. Of 4th Canadian Symposium on Remote Sensing, 107–112. Ottawa, Canada, 1977. CASI, Ottawa, ON, Canada.

Schuler, U., Erbe, P., Zarel, M., Rangubpit, W., Surinkum, A., & Stahr, K. 2011. A gamma-ray spectrometry approach to field separation of illuviation-type WRB reference soil groups in northern Thailand. Journal of Plant Nutrition and Soil Science, 174, 536–544.

Schulthess, C. P. 2011. Historical perspective on the tools that helped shape soil chemistry. Soil Science Society of America Journal, 75, 2009–2036.

Schwartz, G., Ben-Dor, E., & Eshel, G. 2012. Quantitative analysis of total petroleum hydrocarbons in soils: Comparison between reflectance spectroscopy and solvent extraction by 3 certified laboratories. Applied and Environmental Soil Science, 11 p.

Schwartz, G., Ben-Dor, E., & Esher, G. 2013. Quantitative assessment of hydrocarbon contamination in soil using reflectance spectroscopy: A "multipath" approach. Applied Spectroscopy, 67, 1323–1331.

Seid, N. M., Yitaferu, B., Kibret, K., & Ziadat, F. 2013. Soil-landscape modeling and RS to provide spatial representation of soil attributes for an Ethiopian watershed. Applied and Environmental Soil Science, 2013, 11p.

Selige, T., Böhner, J., & Schmidhalter, U. 2006. High resolution topsoil mapping using hyperspectral image and field data in multivariate regression modeling procedures. Geoderma, 136, 235–244.

Sellitto, V. M., Fernandes, R. B. A., Barrón, V., & Colombo, C. 2009. Comparing two different spectroscopic techniques for the characterization of soil iron oxides: Diffuse versus bi-directional reflectance. Geoderma, 149, 2–9.

Serbin, G., Daughtry, C. S. T., Hunt Jr., E. R., Reeves III, J. B., & Brown, D. J. 2009. Effects of soil composition and mineralogy on remote sensing of crop residue cover. Remote Sensing of Environment, 113, 224–238.

Serra, O. 1984. Fundamentals of Well-Log Interpretation (Vol. 1): The Acquisition of Logging Data: Dev. Pet. Sci., 15A. Amsterdam: Elsevier.
Shepherd, K. D., Ferguson, R., Hoover, D., van Egmond, F., Sanderman, J., & Ge, Y. 2022. A global soil spectral calibration library and estimation service. Soil Security, 7, 100061, https://doi.org/10.1016/j.soisec.2022.100061.
Shepherd, K. D., & Walsh, M. G. 2002. Development of reflectance spectral libraries for characterization of soil properties. Soil Science Society of America Journal, 66, 988–998.
Sherman, D. M., & Waite, T. D. 1985. Electronic spectra of Fe^{+3} oxides and oxide hydroxides in the near IR to near UV. American Mineralogist, Chantilly, 70, 1296–1269.
Shi, T., Chen, Y., Liu, Y., & Wu, G. 2014a. Visible and near-infrared reflectance spectroscopy—An alternative for monitoring soil contamination by heavy metals. Journal of Hazardous Materials, 265, 166–176.
Shi, T. Z., Liu, H. Z., Wang, J. J., Chen, Y. Y., Fei, T., & Wu, G. 2014b. Monitoring arsenic contamination in agricultural soils with reflectance spectroscopy of rice plants. Environmental Science & Technology, 48(11), 6264–6272.
Shi, Z., Wang, Q. L., Peng, J., Ji, W. J., Liu, H. J., Li, X., & Viscarra Rossel, R. A. 2014c. Development of a national VNIR soil-spectral library for soil classification and prediction of organic matter concentrations. Science China Earth Sciences, 57, 1671–1680. https://doi.org/10.1007/s11430-013-4808-x
Shih, S. F., & Jordan, D. J. 1992. Landsat mid-infrared data and gis in regional surface soil-moisture assessment. Journal of the American Water Resources Association, 28, 713–719.
Shruthi, R. B. V., Kerle, N., Jetten, V., Abdellah, L., & Machmach, I. 2014. Quantifying temporal changes in gully erosion areas with object oriented analysis. Catena. https://doi.org/10.1016/j.catena.2014.01.010.
Silva, E. B., Giasson, E., Dotto, A. C., Ten Caten, A., Demattê, J. A. M., Bacic, I. L. Z., & Veiga, M. da. 2019. A regional legacy soil dataset for prediction of sand and clay content with Vis-Nir-Swir, in Southern Brazil. Revista Brasileira de Ciência do Solo, 43, e0180174. https://doi.org/10.1590/18069657rbcs20180174
Simbahan, G. C., & Dobermann, A. 2006. Sampling optimization based on secondary information and its utilization in soil carbon mapping. Geoderma, 133, 345–362.
Singh, D., & Kathpalia, A. 2007. An efficient modeling with GA approach to retrieve soil texture, moisture and roughness from ERS-2 SAR data. Progress Electromagnetics Research, 77, 121–136.
Singh, P., & Ghoshal, N. 2010. Variation in total biological productivity and soil microbial biomass in rainfed agroecosystems: Impact of application of herbicide and soil amendments. Agriculture Ecosystems Environment, 137, 241–250.
Sobrino, J. A., Franch, B., Mattar, C., Jiménez-Muñoz, J. C., & Corbari, C. A. 2012. A method to estimate soil moisture from Airborne Hyperspectral Scanner (AHS) and ASTER data: Application to SEN2FLEX and SEN3EXP campaigns. Remote Sensing of Environment, 117, 415–428.
Soderstrom, M., & Eriksson, J. 2013. Gamma-ray spectrometry and geological maps as tools for cadmium risk assessment in arable soils. Geoderma, 192, 323–334.
Soil Survey Staff. 2013. Rapid Assessment of U.S. Soil Carbon (RaCA) project. United States Department of Agriculture, Natural Resources Conservation Service. Available online. June 1. (FY2013 official release).
Soil Survey Staff. 2014. Keys to Soil Taxonomy, 12th edition. Washington, DC: USDA-Natural Resources Conservation Service, 362 p.
Soliman, A., Brown, R., & Heck, R. J. 2011. Separating near surface thermal inertia signals from a thermal time series by standardized principal component analysis. International Journal of Applied Earth Observation and Geoinformation, 13, 607–615.
Somers, B., Asner, G. P., Tits, L., & Coppin, P. 2011. Endmember variability in Spectral Mixture Analysis: A review. Remote Sensing of Environment, 115, 1603–1616.
Sommer, M., Wehrhan, M., Zipprich, M., Weller, U., zu Castell, W., Ehrich, S., Tandler, B., & Selige, T. 2003. Hierarchical data fusion for mapping soil units at field scale. Geoderma, 112, 179–196.
Song, Y., Li, F., Yang, Z., Ayoko, G. A., Frost, R. L., & Ji, J. 2012. Diffuse reflectance spectroscopy for monitoring potentially toxic elements in theagricultural soils of Changjiang River Delta, China. Applied Clay Science, 64, 75–83.
Sørensen, L. K., & Dalsgaard, S. 2005. Determination of clay and other soil properties by near infrared spectroscopy. Soil Science Society of America Journal, 69(1), 159–167.
Soriano-Disla, J. M., Janik, L. J., Viscarra Rossel, R. A., MacDonald, L. M., & McLaughlin, M. J. 2014. The performance of visible, near-, and mid-infrared reflectance spectroscopy for prediction of soil physical, chemical and biological properties. Applied Spectroscopy Reviews, 49, 139–186.

Sousa Junior, J. G., Demattê, J. A. M., & Araújo, S. R. 2011. Modelos espectrais terrestres e orbitais na determinação de teores de atributos dos solos: potencial e custos [Terrestrial and orbital spectral models for the determination of soil attributes: Potential and costs]. Bragantia, 70, 610–621.

Spera, S. A., Cohn, A. S., VanWey, L. K., Mustard, J. F., Rudorff, B. F., Risso, J., & Adami, M. 2014. Recent cropping frequency, expansion, and abandonment in Mato Grosso, Brazil had selective land characteristic. Environmental Research Letters, 9, 064010, 12 pp.

Srasra, E., Bergaya, F., & Fripiat, J. J. 1994. Infrared spectroscopy study of tetrahedral and octahedral substitutions in an interstratified illite-smectite clay. Clay and Clay Minerals, Chantilly, 42, 237–241.

Staenz, K., Mueller, A., & Heiden, U. 2013. Overview of terrestrial imaging spectroscopy missions. Proceedings of the International Geoscience and Remote Sensing Symposium (IAGARSS'13), Melbourne, Australia, 4 pages.

Stark, E., Kes, K. L., & Margoshes, M. 1986. Near-Infrared Analysis (NIRA): A technology for quantitative and qualitative analysis. Applied Spectroscopy Reviews, 22, 335–399.

Steelman, C. M., & Endres, A. L. 2012. Assessing vertical soil moisture dynamics using multi-frequency GPR common-midpoint soundings Original Research Article. Journal of Hydrology, 436–437, 51–66.

Stenberg, B. 2010. Effects of soil sample pretreatments and standardised rewetting as interacted with sand classes on Vis-NIR predictions of clay and soil organic carbon. Geoderma, 158(1–2), 15–22.

Stenberg, B., Jonsson, A., & Börjesson, T. 2002. Near infrared technology for soil analysis with implications for precision agriculture. In: Davies, A., & Cho, R. (Eds.), Near Infrared Spectroscopy: Proceedings of the 10th International Conference. NIR Publications, Chichester, UK, Kyongju S. Korea, pp. 279–284.

Stenberg, B., & Viscarra Rossel, R. A. 2010. Diffuse reflectance spectroscopy for high-resolution soil sensing. In: Viscarra Rossel, R. A., McBratney, A. B., & Minasny, B. (Ed.), Proximal Soil Sensing. Progress in Soil Science. Dordrecht: Springer, 446 p.

Stenberg, B., Viscarra Rossel, R. A., Mouazen, A. M., & Wetterlind, J. 2010. Visible and near infrared spectroscopy in soil science. Advances in Agronomy, 107, 164–206.

Stevens, A., Miralles, I., & van Wesemael, B. 2012. Soil organic carbon predictions by airborne imaging spectroscopy: Comparing cross-validation and validation. Soil Science Society of America Journal, 76, 2174–2183.

Stevens, A., Nocita, M., Toth, G., Montanarella, L., & van Wesemael, B., 2013. Prediction of Soil Organic Carbon at the European Scale by Visible and Near InfraRed Reflectance Spectroscopy. Plos One, 8(6).

Stevens, A., Udelhoven, T., Denis, A., Tychon, B., Lioy, R., Hoffmann, L., & Wesemael, B. van. 2010. Measuring soil organic carbon in croplands at regional scale using airborne imaging spectroscopy. Geoderma, 158, 32–45.

Stevens, A., Wesemael, B. V., Bartholomeus, H., Rosillon, D., Tychon, B., & Ben-Dor, E. 2008. Laboratory, field and airborne spectroscopy for monitoring organic carbon content in agricultural soils. Geoderma, Amsterdam, 144, 395–404.

Stevens, A., Wesemael, B. van, Vandenschrick, G., Touré, S., & Tychon, B. 2006. Detection of carbon stock change in agricultural soils using spectroscopic techniques. Soil Science Society of America Journal, 70, 844–850.

Stoner, E. R., & Baumgardner, M. F. 1981. Characteristic variations in reflectance of surface soils. Soil Science Society of America Journal, Madison, 45, 1161–1165.

Stoner, E. R., Baumgardner, M. F., Biehl, L. L., & Robinson, B. F. 1980. Atlas of soil reflectance properties. Agricultural Experiment Station, Indiana Research Purdue University, West Lafayette, IN, p. 75.

Stroh, J. C., Archer, S., Doolittle, J. A., & Wilding, L. 2001. Detection of edaphic discontinuities with radar and electromagnetic induction. Landscape Ecology, 16, 377–390.

Subramanian, A., Ahn, J., Balasubramanian, V. M., & Rodriguez-Saona, L. E. 2007. Monitoring biochemical changes in bacterial spore during thermal andpressure-assisted thermal processing using FT-IR spectroscopy. Journal of Agricultural and Food Chemistry, 55, 9311–9317.

Sudduth, K. A., & Hummel, J. W. 1993. Soil organic matter, CEC, and moisture sensing with a portable NIR spectrophotometer. Transactions of the Asae, 36(6), 1571–1582.

Sudduth, K. A., Kitchen, N. R., Sadler, E. J., Drummond, S. T., & Myers, D. B. 2010. VNIR Spectroscopy Estimates of Within-Field Variability in Soil Properties. Proximal Soil Sensing.

Sullivan, D. G., Shaw, J. N., & Rickman, D. 2005. IKONOS imagery to estimate surface soil property variability in two Alabama physiographies. Soil Science Society of America Journal, 69, 1789–1798.

Sumfleth, K., & Duttmann, R. 2008. Prediction of soil property distribution in paddy soil landscapes using terrain data and satellite information as indicators. Ecological Indicators, 8, 485–501.

Summerauer, L., Baumann, P., Ramirez-Lopez, L., Barthel, M., Bauters, M., Bukombe, B., Reichenbach, M., Boeckx, P., Kearsley, E., Van Oost, K., Vanlauwe, B., Chiragaga, D., Heri-Kazi, A. B., Moonen, P., Sila, A., Shepherd, K., Mujinya, B. B., Van Ranst, E., Baert, G., Doetterl, S., & Six, J. 2021. Filling a key gap: A soil infrared library for central Africa, SOIL Discuss. [preprint], https://doi.org/10.5194/soil-2020-99, in review.

Summers, D., Lewis, M., Ostendorf, B., & Chittleborough, D. 2011. Visible near-infrared reflectance spectroscopy as a predictive indicator of soil properties. Ecological Indicators, 11(1), 123–131.

Tavares, O. C., Tavares, T. R., Pinheiro Junior, C. R., da Silva, L. M., Wadt, P. G., & Pereira, M. G. 2022. Pedometric tools for classification of southwestern Amazonian soils: A quali-quantitative interpretation incorporating visible-near infrared spectroscopy. Journal of Near Infrared Spectroscopy, 30(1), 18–30. https://doi.org/10.1177/09670335211061854

Taylor, J. A., Short, M., McBratney, A. B., & Wilson, J. 2010. Comparing the Ability of Multiple Soil Sensors to Predict Soil Properties in a Scottish Potato Production System. Proximal Soil Sensing.

Taylor, M. J., Smettem, K., Pracilio, G., & Verboom, W. 2002. Relationships between soil properties and high-resolution radiometrics, central eastern Wheatbelt, western Australia. Exploration Geophysics, 33, 95–102.

Tekin, Y., Kuang, B., & Mouazen, A. M. 2013. Potential of on-line visible and near infrared spectroscopy for measurement of pH for deriving variable rate lime recommendations. Sensors, 13, 10177–10190.

Tekin, Y., Tumsavas, Z., & Mouazen, A. M. 2012. Effect of moisture content on prediction of organic carbon and pH using visible and near-infrared spectroscopy. Soil Science Society of America Journal, 76, 188–198.

Terhoeven-Urselmans, T., Schmidt, H., Joergensen, R. G., & Ludwig, B. 2008. Usefulness of near-infrared spectroscopy to determine biological and chemical soil properties: Importance of sample pre-treatment. Soil Biology & Biochemistry, 40(5), 1178–1188.

Thomsen, I. K., Bruun, S., Jensen, L. S., & Christensen, B. T. 2009. Assessing soil carbon lability by near infrared spectroscopy and NaOCl oxidation. Soil Biology & Biochemistry, 41(10), 2170–2177.

Tian, Y., Zhang, J., Yao, X., Cao, W., & Zhu, Y. 2013. Laboratory assessment of three quantitative methods for estimating the organic matter content of soils in China based on visible/near-infrared reflectance spectra. Geoderma, 202, 161–170.

Tobehn, C., Kassebom, M., Schmälter, E., Fuchs, J., Del Bello, U., Bianco, P., & Battistelli, E. 2002. SPECTRA, Surface Process and Ecosystem Changes Through Response Analysis. Proceedings of the German Aerospace Congress DGLR-2002–185, Bonn, Germany, 4 pages.

Todorova, M., Atanassova, S., & Ilieva, R. 2009. Determination of soil organic carbon using near-infrared spectroscopy. Agricultural Science and Technology, 1(2), 45–50.

Tosti, F., Patriarca, C., Slob, E., Benedetto, A., & Lambot, S. 2013. Clay content evaluation in soils through GPR signal processing. Journal of Applied Geophysics, 97, 69–80.

Treitz, P. M., & Howarth, P. J. 1999. Hyperspectral remote sensing for estimating biophysical parameters of forests ecosystems. Progress in Physical Geography, 23, 359–390.

Tucker, C. J. 1979. Red and photographic infrared linear combinations for monitoring vegetation. Remote Sensing of Environment, 8, 127–150.

Tziolas, N., Tsakiridis, N., Ben-Dor, E., Theocharis, J., & Zalidis, G. 2019. A memory-based learning approach utilizing combined spectral sources and geographical proximity for improved VIS-NIR-SWIR soil properties estimation. Geoderma, 340, 11–24. https://doi.org/10.1016/j.geoderma.2018.12.044

Ucha, J. M., Botelho, M., Vilas Boas, G. S., Ribeiro, L. P., & Santana, P. S. 2002. Uso do radar penetrante no solo (GPR) na investigação dos solos dos Tabuleiros Costeiros do litoral norte do Estado da Bahia [Experimental use of groud-penetrating radar (GPR) to investigate tablelands in the Noethern Coast of Bahia, Brazil]. Brazilian Journal of Soil Science, 26, 373–480.

Udelhoven, T., Emmerling, C., & Jarmer, T. 2003a. Quantitative analysis of soil chemical properties with diffuse reflectance spectrometry and partial-least-square regression: A feasibility study. Plant and Soil, 251, 319–329.

Udelhoven, T., Hostert, P., Jarmer, T., & Hill, J. 2003c. Modulare Klassifikation von Getreideflächen mit mono-temporalen hyperspektralen Bilddaten des HyMap-Sensors. Photogrammetrie, Fernerkundung, Geoinformation, 1, 35–42.

Udelhoven, T., Novozhilov, M., & Schmitt, J. 2003b. The NeuroDeveloper: A tool for modular neural classification of spectroscopic data. Chemometrics and Intelligent Laboratory Systems, 66, 219–226.

Uno, Y., Prasher, S. O., Patel, R. M., Strachan, I. B., Pattey, E., & Karimi, Y. 2005. Development of field-scale soil organic matter content estimation models in Eastern Canada using airborne hyperspectral imagery. Canadian Biosystems Engineering, 47, 14 p.

USGS. United States Geological Survey. Earth Observing 1 (EO-1). USGS. Disponível em: http://eo1.usgs.gov/. Acessado em: 09 mar. 2014.

Ustin, S. L., & Gamon, J. A. 2010. Remote sensing of plant functional types. New Phytologist, 186, 795–816.

Ustin, S. L., Gitelson, A. A., Jacquemoud, S., Schaepman, M., Asner, G. P., Gamon, J. A., & Zarco-Tejada, P. 2009. Retrieval of foliar information about plant pigment systems from high resolution spectroscopy. Remote Sensing of Environment, 113, S67–S77.

Ustin, S. L., Roberts, D. A., Gamon, J. A., Asner, G. P., & Green, R. O. 2004. Using imaging spectroscopy to study ecosystem processes and properties. BioScience, 54, 523–534.

Vågen, T. G., Winowiecki, L. A., Abegaz, A., & Hadgu, K. M. 2013. Landsat-based approaches for mapping of land degradation prevalence and soil functional properties in Ethiopia. Remote Sensing of Environment, 134, 266–275.

van der Klooster, E., van Egmond, F. M., & Sonneveld, M. P. W. 2011. Mapping soil clay contents in Dutch marine districts using gamma-ray spectrometry. European Journal of Soil Science, 62(5), 743–753.

van der Marel, H. W., & Beutelspacher, H. 1975. Atlas of Infrared Spectroscopy of Clay Minerals and Their Admixtures. Amsterdam: Elsevier Scientific Publishing Company, 396 p.

van der Meer, F. D., & de Jong, S. 2003. Spectral mapping methods: Many problems, some solutions. In: 3rd EARSeL Workshop on Imaging Spectroscopy, Hersching, 13–16 May 2003, pp. 146–162.

van der Meer, F. D., van der Werff, H. M. A., van Ruitenbeek, F. J. A., Hecker, C. A., Bakker, W. H., Noomen, M. F., van der Meijde, M., Carranza, J. M., Smeth, J. B., & Woldai, T. 2012. Multi- and hyperspectral geologic remote sensing: A review. International Journal of Applied Earth Observation and Geoinformation, 14, 112–128.

van Egmond, F. M., Loonstra, E. H., & Limburg, J. 2010. Gamma Ray Sensor for Topsoil Mapping: The Mole. Proximal Soil Sensing.

van Groenigen, J. W., Mutters, C. S., Horwath, W. R., & van Kessel, C. 2003. NIR and DRIFT-MIR spectrometry of soils for predicting soil and crop parameters in a flooded field. Plant and Soil, 250(1), 155–165.

van Lier, Q. 2010. Física do solo. 298p. Viçosa: SBCS.

Vane, G., Green, R. O., Chrien, T. G., Enmark, H. T., Hansen, E. G., & Porter, W. M. 1993. The airborne visible/infrared imaging spectrometer (AVIRIS). Remote Sensing of Environment, 44, 127–143.

Varmuza, K., & Filzmoser, P. 2009. Introduction to Multivariate Statistical Analysis in Chemometrics. Boca Raton: CRC Press, 321 p.

Varshney, P. K., & Arora, M. K. (Eds.). 2004. Advanced Image Processing Techniques for Remotely Sensed Hyperspectral Data. Germany, 322 p.

Vasques, G. M., Demattê, J. A. M., Viscarra Rossel, R., Ramirez Lopez, L., & Terra, F. S. 2014. Soil classification using visible/near-infrared diffuse reflectance spectra from multiple depths. Geoderma, 223–225, 73–78.

Vaudour, E., Bel, L., Gilliot, J. M., Coquet, Y., Hadjar, D., Cambier, P., Michelin, J., & Houot, S. 2013. Potential of SPOT multispectral satellite images for mapping Topsoil Organic Carbon content over Peri-Urban Croplands Soil Science Society of American Journal, 77, 2122–2139.

Vendrame, P. R. S., Marchao, R. L., Brunet, D., & Becquer, T. 2012. The potential of NIR spectroscopy to predict soil texture and mineralogy in Cerrado Latosols. European Journal of Soil Science, 63(5), 743–753.

Vicente, L. E., & Souza Filho, C. R. 2011. Identification of mineral components in tropical soils using reflectance spectroscopy and advanced spaceborne thermal emission and reflection radiometer (ASTER) data. Remote Sensing of Environment, 115, 1824–1836.

Vicente-Serrano, S. M., Pons-Fernández, X., & Cuadrat-Pratsa, J. M. 2004. Mapping soil moisture in the central Ebro river valley (northeast Spain) with Landsat and NOAA satellite imagery: A comparison with meteorological data. International Journal of Remote Sensing, 25, 4325–4350.

Vink, A. P. A. 1964. Aerial photographs and the soil sciences. Proc. of the Toulouse Conf. on Aerial Surveys and Integrated Studies, pp. 81–141.

Viscarra Rossel, R. A. 2009. The soil spectroscopy group and the development of a global spectral library. In: 3rd Global Workshop on Digital Soil Mapping, 30 September—3 October 2008. Utah State University, Logan, UT. http://groups.google.com/group/soil-spectroscopy.

Viscarra Rossel, R. A., Adamchuk, V. I., Sudduth, K. A., & McBrown, N. J. 2007b. Using a global VNIR soil-spectral library for local soil characterization and landscape modeling in a 2nd-order Ugandan watershed. Geoderma, 140, 444–453.

Viscarra Rossel, R. A., Adamchuk, V. I., Sudduth, K. A., Mckenzie, N. J., & Lobsey, C. 2011a. Proximal soil sensing: An effective approach for soil measurements in space and time. Advances in Agronomy, 113, 237–282.

Viscarra Rossel, R. A., & Behrens, T. 2010. Using data mining to model and interpret soil diffuse reflectance spectra. Geoderma, Amsterdam, 158, 46–54.

Viscarra Rossel, R. A., Behrens, T., Ben-Dor, E., Brown, D. J., Demattê, J. A. M., Shepherd, K. D., Shi, Z., Stenberg, B., Stevens, A., Adamchuk, V., Aïchi, H., Barthès, B. G., Bartholomeus, H. M., Bayer, A. D., Bernoux, M., Böttcher, K., Brodský, L., Du, C. W., Chappell, A., Fouad, Y., Genot, V., Gomez, C., Grunwald, S., Gubler, A., Guerrero, C., Hedley, C. B., Knadel, M., Morrás, H. J. M., Nocita, M., Ramirez-Lopez, L., Roudier, P., Campos, E. M. R., Sanborn, P., Sellitto, V. M., Sudduth, K. A., Rawlins, B. G., Walter, C., Winowiecki, L. A., Hong, S. Y., & Ji, W. A global spectral library to characterize the world's soil. Earth-Science Review, 155, 198–230. https://doi.org/10.1016/j.earscirev.2016.01.012

Viscarra Rossel, R. A., Bui, E. N., Caritat, P., & Mckenzie, N. J. 2010. Mapping iron oxides and the color of Australian soil using visible—near-infrared reflectance spectra. Journal of Geophysics Research, 115, F04031.

Viscarra Rossel, R. A., Cattle, S. R., Ortega, A., & Fouad, Y. 2009. In situ measurements of soil colour, mineral composition and clay content by vis-NIR spectroscopy. Geoderma, 150(3/4), 253–266.

Viscarra Rossel, R. A., Chappell, A., De Caritat, P., & Mckenzie, N. J. 2011b. On the soil information content of visible—near infrared reflectance spectra. European of Journal Soil Science, 62, 442–453.

Viscarra Rossel, R. A., Jeon, Y. S., Odeh, I. O. A., & Mcbratney, A. B. 2008. Using a legacy soil sample to develop a mid-IR spectral library. Australian Journal of Soil Research, Collingwood, 46, pp. 1–16.

Viscarra Rossel, R. A., & Lark, R. M. 2009. Improved analysis and modelling of soil diffuse reflectance spectra using wavelets. European Journal of Soil Science, 60(3), 453–464.

Viscarra Rossel, R. A., Mcglynn, R. N., & Mcbratney, A. B. 2006b tabela. Determining the composition of mineral-organic mixes using UV-vis-NIR diffuse reflectance spectroscopy. Geoderma, Amsterdam, 137, pp. 70–82.

Viscarra Rossel, R. A., Taylor, H. J., & McBratney, A. B. 2007a. Multivariate calibration of hyperspectral γ-ray energy spectra for proximal soil sensing. European Journal of Soil Science, 58, 343–353.

Viscarra Rossel, R. A., & Walter, C. 2004. Rapid, quantitative and spatial field measurements of soil pH using an ion sensitive field effect transistor. Geoderma, 119, 9–20.

Viscarra Rossel, R. A., Walwort, D. J. J., McBratney, A. B., Janik, L. J., & Skjesmstad, J. O. 2006a texto. Visible near infrared or combined diffuse reflectance spectroscopy for simultaneous assessment of various soil properties. Geoderma, 131, 59–75.

Viscarra Rossel, R. A., & Webster, R. 2011. Discrimination of Australian soil horizons and classes from their visible-near infrared spectra. European Journal of Soil Science, 62, 637–647.

Viscarra Rossel, R. A., & Webster, R. 2012. Predicting soil properties from the Australian soil visible-near infrared spectroscopic database. European Journal of Soil Science, 63(6), 848–860.

Vohland, M., Bossung, C., & Fründ, H. C. 2009. A spectroscopic approach to assess trace—heavy metal contents in contaminated floodplain soils via spectrally active soil components. Journal of Plant Nutrition and Soil Science, 172, 201–209.

Vohland, M., & Emmerling, C., 2011. Determination of total soil organic C and hot water-extractable C from VIS-NIR soil reflectance with partial least squares regression and spectral feature selection techniques. European Journal of Soil Science, 62(4), 598–606.

Vohland, M., Ludwig, M., Thiele-Bruhn, S., & Ludwig, B., 2014. Determination of soil properties with visible to near- and mid-infrared spectroscopy: Effects of spectral variable selection. Geoderma, 223–225, 88–96.

Vrieling, A. 2006. Satellite remote sensing for water erosion assessment: A review. Catena, 65, 2–18.

Vulfson, L., Genis, A., Blumberg, D. G., Kotlyar, A., Freilikher, V., & Ben-Asher, J. 2013. Remote sensing in microwave and gamma ranges for the monitoring of soil water content of the root zone. International Journal of Remote Sensing, 34, 6182–6201.

Waiser, T. H., Morgan, C. L. S., Brown, D. J., & Hallmark, C. T. 2007. In situ characterization of soil clay content with visible near-infrared diffuse reflectance spectroscopy. Soil Science Society of America Journal, 71(2), 389–396.

Walkley, A., & Black, I. A. 1934. An examination of the Degtjareff method for determining organic carbon in soils: Effect of variations in digestion conditions and of inorganic soil constituents. Soil Science, 63, 251–263.

Wang, C.-K., Pan, X.-Z., Wang, M., Liu, Y., Li, Y.-L., Xie, X.-L., Zhou, R., & Shi, R.-J. 2013a. Prediction of soil organic matter content under Moist conditions using VIS-NIR diffuse reflectance spectroscopy. Soil Science, 178(4), 189–193.

Wang, S.-Q., Li, W.-D., Li, J., & Liu, X.-S. 2013b. Prediction of soil texture using FT-NIR spectroscopy and PXRF spectrometry with data fusion. Soil Science, 178(11), 626–638.

Waruru, B. K., Shepherd, K. D., Ndegwa, G. M., Kamoni, P. T., & Sila, A. M. 2014. Rapid estimation of soil engineering properties using diffuse reflectance near infrared spectroscopy. Biosystems Engineering, 121, 177–185.

Wetterlind, J., & Stenberg, B. 2010. Near-infrared spectroscopy for within-field soil characterization: Small local calibrations compared with national libraries spiked with local samples. European Journal of Soil Science, 61(6), 823–843.

Wetterlind, J., Stenberg, B., & Jonsson, A. 2008. Near infrared reflectance spectroscopy compared with soil clay and organic matter content for estimating within-field variation in N uptake in cereals. Plant and Soil, 302, 317–327.

Wetterlind, J., Stenberg, B., & Soderstrom, M. 2010. Increased sample point density in farm soil mapping by local calibration of visible and near infrared prediction models. Geoderma, 156(3–4), 152–160.

White, J. L., & Roth, C. B. 1986. Infrared spectrometry. In: Klute, A. (Ed.), Methods of soil analysis: Part 1—Physical and mineralogical methods. Madison: Soil Science Society of America, chap. 11, pp. 291–330.

White, W. B. 1971. Infrared characterization of water and hidroxyl ion in the basic magnesium carbonate minerals. American Mineralogist, Chantilly, 56, 46–53.

Whiting, M. L., Li, L., & Ustin, S. L. 2004. Predicting water content using Gaussian model on soil spectra. Remote Sensing of Environment, 89, 535–552.

Wijaya, I. A. S., Shibusawa, S., Sasao, A., & Hirako, S. 2001. Soil parameters maps in paddy field using the real time soil spectrophotometer. Journal of the Japanese Society of Agricultural Machinery, 63(3), 51–58.

Wijewardane, N. K., Ge, Y., Wills, S., & Libohova, Z. 2018. Predicting physical and chemical properties of US soils with a mid-infrared reflectance spectral library. Soil Science Society of America Journal, 82, 722–731. https://doi.org/10.2136/sssaj2017.10.0361

Wilford, J. 2011. A weathering intensity index for the Australian continent using airborne gamma-ray spectrometry and digital terrain analysis. Geoderma, 183–184, 124–142.

Wong, M. T. F., & Harper, R. J. 1999. Use of on-ground gamma-ray spectrometry to measure plant-available potassium and other topsoil attributes. Australian Journal of Soil Research, 37(2), 267–278.

Wong, M. T. F., Oliver, Y. M., & Robertson, M. J. 2009. Gamma-radiometric assessment of soil depth across a landscape not measurable using electromagnetic surveys. Soil Science Society of America Journal, 73, 1261–1267.

Woodhouse, I. H. 2005. Introduction to Microwave Remote Sensing, p. 400.

Wu, C., Zheng, Y., Yang, H., Yang, Y., & Wu, Z. 2021. Effects of different particle sizes on the spectral prediction of soil organic matter. Catena, 196, 104933, ISSN 0341-8162.

Xie, H. T., Yang, X. M., Drury, C. F., Yang, J. Y., & Zhang, X. D. 2011. Predicting soil organic carbon and total nitrogen using mid- and near-infrared spectra for Brookston clay loam soil in Southwestern Ontario, Canada. Canadian Journal of Soil Science, 91(1), 53–63.

Yang, H., Kuang, B., & Mouazen, A. M. 2012. Quantitative analysis of soil nitrogen and carbon at a farm scale using visible and near infrared spectroscopy coupled with wavelength reduction. European Journal of Soil Science, 63(3), 410–420.

Yochim, A., Zytner, R. G., McBean, E. A., & Endres, A. L. 2013. Estimating water content in an active landfill with the aid of GPR. Waste Management, 33, 2015–2028.

Yoder, R. E., Freeland, R. S., Ammons, J. T., & Leonard, L. L. 2001. Mapping agricultural fields with GPR and EMI to identify offsite movement of agrochemicals. Journal of Applied Geophysics, 47, 251–259.

Zhan, W., Zhou, J., Ju, W., Li, M., Sandholt, I., Voogt, J., & Yu, C. 2014. Remotely sensed soil temperatures beneath snow-free skin-surface using thermal observations from tandem polar-orbiting satellites: An analytical three-time-scale model. Remote Sensing of Environment, 143, 1–14.

Zhang, C., & Kovacs, J. M. 2012. The application of small unmanned aerial systems for precision agriculture: A review. Precision Agriculture, 13, 693–712.

Zhang, J., Lin, H., & Doolittle, J. A. 2014. Soil layering and preferential flow impacts on seasonal changes of GPR signals in two contrasting soils Original Research Article. Geoderma, 213, 560–569.

Zhang, J., Pu, R., Huang, W., Yuan, L., Luo, J., & Wang, J., 2012. Using in-situ hyperspectral data for detecting and discriminating yellow rust disease from nutrient stresses. Field Crops Research, 134(12), 165–174.

Zhang, T., Li, L., & Zheng, B. 2013. Estimation of agricultural soil properties with imaging and laboratory spectroscopy. Journal of Applied Remote Sensing, 7, 24 p.

Zhao, M. S., Rossiter, D. G., Li, D. C., Zhao, Y. G., Liu, F., & Zhang, G. L. Z. 2014. Mapping soil organic matter in low-relief areas based on land surface diurnal temperature difference and a vegetation index. Ecological Indicators, 39, 120–133.

Zhu, J., Morgan, C. L. S., Norman, J. M., Yue, W., & Lowery, B. 2004. Combined mapping of soil properties using a multi-scale spatial model. Geoderma, 118, 321–334.

Zhu, Y., Weindorf, D. C., Chakraborty, S., Haggard, B., Johnson, S., & Bakr, N. 2010. Characterizing surface soil water with field portable diffuse reflectance spectroscopy. Journal of Hydrology, 391, 133–140.

Zinck, J. A., López, J., Metternicht, G. I., Shrestha, D. P., & Vázquez-Selem, L. 2001. Mapping and modelling mass movements and gullies in mountainous areas using remote sensing and GIS techniques. International Journal of Applied Earth Observation and Geoinformation, 3, 43–53.

Zornoza, R., Guerrero, C., Mataix-Solera, J., Scow, K. M., Arcenegui, V., & Mataix-Beneyto, J. 2008. Near infrared spectroscopy for determination of various physical, chemical and biochemical properties in Mediterranean soils. Soil Biology & Biochemistry, 40(7), 1923–1930.

16 Remote Sensing of Soil in the Optical Domains

Eyal Ben-Dor and José Alexandre M. Demattê

ACRONYMS AND DEFINITIONS

ASTER	Advanced Spaceborne Thermal Emission and Reflection Radiometer
CHRIS	Compact High Resolution Imaging Spectrometer
CNES	National Centre for Space Studies
DLR	German Aerospace Center
EnMAP	Environmental Mapping and Analysis Programme
ERTS	Earth Resources Technology Satellite
GIS	Geographic Information System
GPS	Global Positioning System
IKONOS	A commercial Earth observation satellite, typically, collecting sub-meter to 5 m data
LIDAR	Light detection and ranging
MIRS	Multiple Interpretations of Reflectance Spectra
MODIS	Moderate Resolution Imaging Spectroradiometer
MSS	Multispectral Scanner
NASA	National Aeronautics and Space Administration
NIR	Near-Infrared
NOAA	National Oceanic and Atmospheric Administration
PROBA	Project for On Board Autonomy
PV	Photosynthetic Vegetation
SAR	Synthetic Aperture Radar
SPOT	Satellite Pour l'Observation de la Terre, French Earth Observing Satellites
SWIR	Shortwave Infrared
TIR	Thermal Infrared
UAV	Unmanned Aerial Vehicles
USDA	United States Department of Agriculture
VIS	Visible

16.1 INTRODUCTION

In 1987, Mulder published a book entitled *Remote Sensing in Soil Science* (Mulder, 1987) which provided a comprehensive summary and background of all of the soil remote-sensing activities known at the time. Mulder's excellent review covered theory, sensors and applications for soil using remote-sensing means. Since 1987, remarkable progress has been made in the soil remote-sensing arena, including electro-optic and space technologies, computing power, applied mathematics and artificial intelligence (AI) for data analyses, and protocols to measure, archive and harmonize soil spectral databases. After almost three decades, these significant advances in soil remote sensing have attracted many young as well as experienced users from the scientific community and from the industry. Many new users have entered the soil remote-sensing field and use the technology in different ways, making up specific scientific working groups that have created a unique subcommunity. We are

DOI: 10.1201/9781003541165-21

currently entering a new era, in which passive sensors with high spectral spatial and temporal resolutions are available in orbit, along with new portable ground sensors and platforms (e.g., drones). With better accessibility to this infrastructure, soil spectroscopy has become a very basic and powerful tool from both point and imaging spectral viewpoints. This chapter is thus aimed at covering some key historical stages of this promising technology and reviewing most of the advances in this arena to date. Based on past and present activities, this chapter also highlights the leading directions in the field, and provides some thoughts on future possibilities for the remote sensing of soils.

16.2 SOIL

16.2.1 The Soil System

Soil has been defined as "the upper layer of the earth which may be dug, plowed, specifically, the loose surface material of the earth in which plants grow" (Thompson, 1957). Soil, as an anchoring medium for roots and a supplier of nutrient for crops, is a complex system that is extremely variable in its physical and chemical composition. Soil plays an important role in both food production and environmental preservation. This is stated by the European Commission (EC), which recently published a document entitled "Caring for Soil Is Caring for Life" (Veerman et al., 2020). This important document states that "soils are fragile, and they can take thousands of years to form but can be destroyed in hours! This means that we need to take care of soils now so that they can be regenerated and safeguarded for future generations." As soil remote sensing is a promising technology to achieve the EC's goals, we first need to understand the soil arena. Soil is formed from exposed masses of partially weathered rocks and minerals of the Earth's crust. Soil formation, or genesis, is strongly dependent on the environmental conditions in both the atmosphere and lithosphere.

Soils are the product of five factors: climate, vegetation, organic matter (OM), topography and parent materials. The great variability in soils is the result of myriad interactions among these factors and their influence on the formation of different soil profiles (Buol et al., 1973). The general equation describing the final soil body is:

$$S = f\left(P, C, T, O, t\right) \tag{16.1}$$

where S represents the soil, P is the parent material, C is climate, T is topography, O is OM, and t is related to time (relative age of the soil); more recently, an anthropogenic factor has also been added to this equation. The high variation of soils makes it impossible to solve Equation 16.1 numerically or empirically. Soil serves as an important resource for food production for mankind and carries out other key environmental functions that are essential for human subsistence, such as water storage and redistribution, pollutant-filtering and carbon storage. The soil-forming factors segregate the weathered parent material into diagnostic horizons within the soil profile. In general, the profile, composed of several horizons, typically refers to the upper horizon A (termed the alluvial horizon), the intermediate horizon B (termed illuvial horizon), and the bottom horizon C (the transition to the parent material) (Figure 16.1). The number, nature, and development of the horizons are products of aforementioned five soil-forming factors, and their relationships play a major role in soil classification and mapping processes that require a description of the entire soil profile (Soil Survey Staff, 1975). Pedology is one of the most important and ancient branches of soil science that is strongly related to soil genesis, formation and mapping. Although soil mapping requires a description of the entire soil profile, observing the soil surface from close or far domains is the ultimate tool before and during any comprehensive field study aimed at generating a soil map (Simonson, 1987). In this respect, remote sensing, which sees mostly the upper part of the Earth from afar, plays a major role in soil mapping, mostly for observations of the topmost (Ao) horizon where recently some effort is being taken toward enlarging this capability into the subsurface soil domains as will be further illustrated later in this chapter.

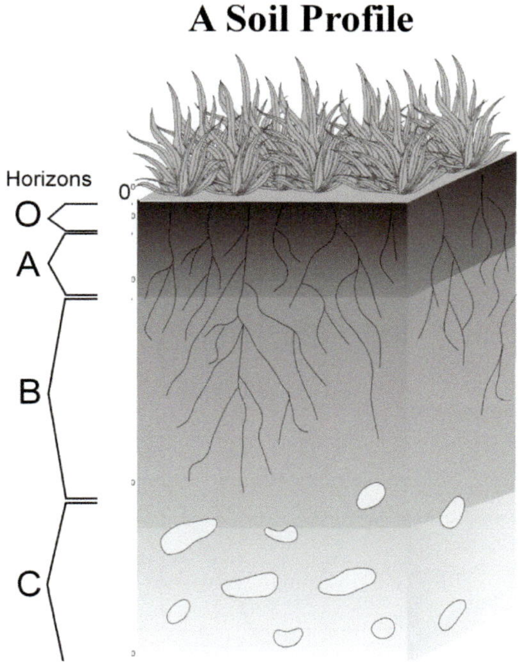

FIGURE 16.1 Illustration of the soil body along the soil profile generated during the soil formation.

16.2.2 Soil Composition

The soil body is a mixture of three phases: solid, liquid and gas. A typical soil volume may consist of about 50% pore space with temporally varying proportions of gas and liquid. The solid phase contains organic and inorganic matter in a complicated and generic mixture of primary and secondary minerals, organic components (fresh and decomposed), and salts. The solid phase consists of three main particle-size fractions: sand (2–0.2 mm), silt (0.2–0.002 mm) and clay (<0.002 mm), whose mixture is responsible for soil texture and structure. Soil texture is a function of the proportion of particle-size fractions (Figure 16.2), defining the soil as sandy, silty-loam or clayey. Soil structure is a function of the adhesive forces between the soil particles and describes the aggregation status of the solid particles (block, prism, grains and others). These two properties play a major role in the soil's behavior and govern important soil characteristics such as drainage, porosity, fertility and moisture, which have importance for plant growth and erosion process. The inorganic portion of the solid phase consists of minerals, which are generally categorized as either primary or secondary. Primary minerals are derived directly from the weathering of parent materials, and are formed under much higher temperatures and pressures than those found at the Earth's surface. Secondary minerals are formed by geochemical weathering of the primary minerals. An extensive description of minerals in the soil environment is given by Dixon and Weed (1989), and readers who wish to expand their knowledge in this area are referred to this classic comprehensive text. In general, the dominant primary minerals are quartz, feldspar, orthoclase and plagioclase. Some layer silicate minerals are mica and chlorite, and ferromagnesian silicate minerals include amphibole, peroxide and olivine. The secondary minerals in the soil body (often termed clay minerals) are aluminosilicates with a layer structure, such as smectite, illite, vermiculite, sepiolite, kaolinite, and gibbsite. The type of clay mineral present in the solid phase of the soil is strongly dependent on the weathering stage of the parent material and can be a significant indicator of the environmental conditions under which the soil was formed (Singer, 2007). Other secondary minerals in soils are

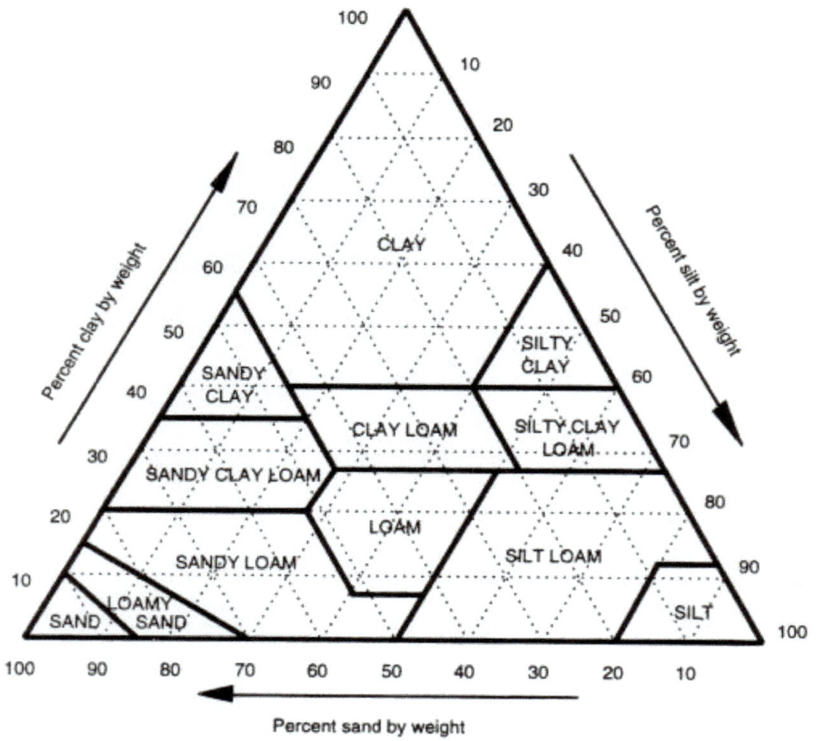

FIGURE 16.2 Soil texture, showing the 12 major textural classes and particle size scales as defined by the USDA.

aluminum and iron (Fe) oxides and hydroxides, carbonates (calcite and dolomite), sulfates (gypsum) and phosphates (apatite). Most of these minerals are relatively insoluble in water and maintain an equilibrium with the water solution. Soluble salts such as halite may also be found in soils, but they are mobile in water and are sometimes transported to the soil matrix by external forces (e.g., wind or artificial irrigation). Clay minerals are most likely found in the fine-sized soil particles (<2 μm) and are characterized by relatively high specific surface areas (50–800 m^2/g). The primary minerals and other non-clay minerals are usually found in both the sand and silt fractions, and consist of relatively low specific surface areas (<1 m^2/g). In addition to the inorganic components in the solid phase, organic components are also present. Although the OM content in mineral soils does not exceed 15% (and is usually less), it plays a major role in the soil's chemical and physical behavior (Schnitzer and Khan, 1978). OM is composed of decaying tissues from vegetation and the bodies of micro- and macrofauna. Soil OM (SOM) can be found in various stages of degradation, from coarse dead to complex fine components called humus (Stevenson, 1982). Its content is naturally higher in the upper soil horizon, making considerations of OM essential for remote-sensing applications, where only the upper thin layer is detected. SOM plays an important role in carbon sequestration processes and is a significant factor in the carbon stock in soils which is a solid indicator for tracking changes in carbon dioxide emission (Dahl et al., 2023).

The liquid and gas phases in soils are complementary to the solid phase and occupy about 50% of the soil's total volume. The liquid consists of components of water and dissolved anions and cations in various amounts and positions. The water molecules either fill the entire pore volume in the soil (saturated), occupy a portion of its pore volume (wet), or are adsorbed on the surfaces particles (air dry). Water status can be determined by the pressures needed to extract the water from the soil matrix (often call "matrix tension"). These range from 15 Atm (the dry condition; wilting point for vegetation) to 0.3 Atm (gravimetric water draining out) and 0 Atm (the saturated stage). The composition of the soil's gaseous phase is normally very similar to that of the atmosphere, except for the concentrations of oxygen and carbon dioxide which vary according to the biochemical activity in the root zone due to biogenic respiration processes. The association of all of these phases combined with plant growth status is termed "Soil Health" (Lehmann, 2020).

16.3 REMOTE SENSING

16.3.1 General

Remote sensing is the acquisition of information about an object or phenomenon without physical contact, using electromagnetic radiation (Elachi and Van Zyl, 2006). The term, "remote sensing," was first introduced in 1960 by Evelyn L. Pruitt of the U.S. Office of Naval Research (Pruitt, 1979). In general, remote sensing can be performed in two ways: passively, where the radiation is not controlled by the sensing system (e.g., the Sun's radiation in photography), and actively, where the radiation is part of the sensing system (e.g., microwaves for radar). The remote-sensing discipline mostly uses airborne and spaceborne sensors, but also portable instruments for close-range measurements in the field. As several types of electromagnetic radiation are available for both active and passive remote sensing (e.g., shortwave and longwave infrared, milli- and microwaves), we will limit our discussion to the passive radiation of the Sun, which is actually the main source of radiation for remote sensing. This decision is backed up by Mulder et al. (2011) who reviewed remote-sensing means and demonstrated that across this region, all of the soil properties that can be remotely sensed use the solar spectral region, termed "optical" region (as the foreoptics are made of glass), see Table 16.1. This region can be separated into three parts: visible (VIS) 0.4–0.7 μm, near infrared (NIR) 0.7–1.0 μm, and shortwave infrared (SWIR) 1.0–2.5 μm. Midwave infrared (MWIR) 2.5–5 μm and longwave infrared (LWIR) (8–14 μm) are other regions used in passive remote sensing of the Earth and are products of the radiation emitted by the Earth due to its black-body characteristics. Separation of the LWIR into two regions is based on the atmospheric windows that fingering where emitted photons are blocked (5–8 μm). Nonetheless, in the proximal (point) sensing discipline, which carries out measurements in the laboratory over the entire thermal region (3–25 μm), high performance in assessing many soil chromophores is demonstrated that cannot be obtained in the optical domain (e.g., quartz). In the MWIR, there is some conjunction between solar radiation and Earth radiation due the characteristics of the Planck black-body function of the two bodies (Sun and Earth). Remote sensing can be performed with one or more instruments covering part or all of the aforementioned spectral regions, the choice being based mainly on the question at hand. Remote-sensing sensors collect the radiation, disperse it into selected frequencies, measure the intensity at each frequency analogically, and then convert it to digital values for archiving and processing. The remote-sensing assembly consists of optics (e.g., lens and slit), a disperser element (e.g., prism or grating), and a detector (e.g., a charge-coupled device—CCD). For every spectral region, there are specific materials that are sensitive to that radiation's frequencies, lenses, and dispersive elements. For solar radiation, the lens and prism are made of quartz and the detectors of Si for the VIS—NIR (VNIR) region, InGaAs for the NIR region, or HgCdTe (MCT) for the SWIR region (Levinstein and Mudar, 1975). The final product of any remote sensor is governed by the quality of these components. There are several resolutions in the remote-sensing field related to the available technology as follows: spatial resolution, which refers to the size of the pixel (often termed ground sampling distance—GSD), sensor swath and dimensionality (2D—flat or 3D—elevation), spectral resolution (number of bands, band widths (that measures by the full width at

TABLE 16.1
A list of soil remote sensing applications that can be remotely sense from orbit and the systems requires for each applications (Taken from Mulders 2011)

	Radar			Optical	
Soil and terrain attributes	**Passive**	**Active**	**Lidar**	**Multispectral**	**Spectroscopy**
Terrain attributes					
Elevation	-	High	High	Medium	-
Slope	-	High	High	Medium	-
Aspect	-	High	High	Medium	-
Dissection	-	Medium-high	Medium-high	Low-medium	-
Landform unit	-	Medium-high	Medium-high	Medium-high	Low-medium
Digital soil mapping	-	High	Medium-high	Medium-high	Medium
Soil type	-	-	-	Medium	High
Soil attributes – proximal sensing					
Mineralogy	-	-	-	-	High
Soil texture	-	High	-	-	High
Iron content	-	-	-	-	Medium-high
Soil organic carbon	-	-	-	-	High
Soil moisture	High	High	-	-	High
Soil salinity	-	-	-	-	Medium
Carbonate content	-	-	-	-	Medium
Nitrogen content	-	-	-	-	High
Lichen	-	-	-	-	Medium-high
Photosynthetic vegetation	-	-	-	-	Medium-high
Nonphotosynthetic	-	-	-	-	Medium-high
Soil attributes – remote sensing					
Mineralogy	-	-	-	Medium	Medium-high
Soil texture	-	Medium	-	Medium	Medium
Iron content	-	-	-	Low	Medium
Soil organic carbon	-	-	-	Low	High
Soil moisture	Medium-high	Medium-high	-	Medium	Low-medium
Soil salinity		Medium-high	-	Low-medium	Medium
Carbonate content	-	-	-	Low-medium	Low-medium
Nitrogen content	-	-	-	-	Medium
Lichen	-	-	-	Low-medium	Medium
Photosynthetic vegetation	-	-	-	Medium	Medium-high
Nonphotosynthetic vegetation	-	-	-	Medium	Medium-high
Ellenberg indicator values	-	-	-	-	Low
Plant functional type	-	-	-	Low-medium	Low
Vegetation indices	-	-	-	High	Medium
Land cover	-	Low-medium	-	Medium-high	High
Land degradation	-	Low-medium	-	High	Low-medium

[a] Feasibility (1-5) = weighted average of scores for the number of studies reported, dataset quality, obtained result and applicability to field surveys. Low = 1, low-medium = 2, medium = 3, medium-high = 4 and high = 5.

half maximum of the band: FWHM) and sampling intervals, electronic sampling resolution (the number of bits for the stored digital data), radiometric resolution (the accuracy of conversion of the digital data into physical units), temporal resolution (time elapsed between acquisitions for the same area), and viewing resolution (the number of viewing angles capturing the same GSD).

These resolutions are interrelated: for example, high spatial resolution requires low spectral resolution to ensure a high signal-to-noise ratio (SNR) (Figure 16.3). Optical remote-sensing means are categorized according to the sensor's spectral performance as follows: monospectral (the sensor carries a monochromatic band that is often termed panchromatic), multispectral (the sensor carries between 3 and around 7 semi-broad spectral bands), superspectral (the sensor carries around 7 to 20 semi-broad spectral bands) and hyperspectral (the sensor carries over 20 narrow spectral bands). The spectral data can be acquired in image or point domains. In the image domain, the CCD works to capture a response from every pixel on the ground that corresponds to every CCD pitch, or divides onto one of the CCD axis into every ground pixel along one axis and its spectral information on the other axis. In the point domain, only one pixel is captured by the sensor, and the detector, which is formed in line-array architecture, uses it to measure the spectral information. Data acquisition in the image domain requires moving the sensor to form the image via line-by-line accumulation, known as "push broom" technology. The systems that collect pixel-by-pixel information and gather it into a final image are termed "whisk broom" sensors. The point (pixel) domain is mostly used on the ground to measure an object's response by integrating the photons of the point GSD, which is based on the field of view (FOV) of the foreoptic characteristics.

For soil applications, it is most important to retain high spectral resolution and high SNR in order to extract quantitative information on the soil object (see Section 16.8). High spatial resolution is also important, but this is strongly related to the question being asked. GSD resolutions of 1 to 30 m are very reasonable values for remote sensing of soils ranging from a selected plot to larger field coverage, respectively. Fortunately, most of the remote-sensing means today, from either air or space domains, are characterized by these GSD characteristics. If the high spatial resolution is accompanied by high spectral resolution, then the quality of the remote-sensing capability increases, if the SNR is also kept high. This capability provides a better detection limit or interpretation of the data at hand and is sometimes crucial. Unfortunately, high spectral and spatial resolution is not yet possible from orbit; however, when it does become available, the exploitation of remote-sensing means will foster new applications. New technology that steers the sensor's foreoptic to the same pixel during its overpass and provides twofold to fourfold better integration time will soon be available, enabling high spectral, spatial and temporal resolution from orbit. This has already been achieved for the air domain, based on the relatively slow sensor speed that enables collecting a sufficient

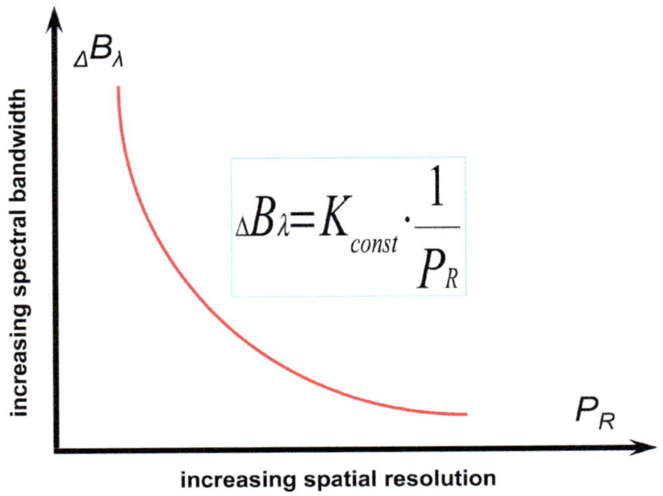

FIGURE 16.3 The relationship between geometric resolution (Pg = GSD) and the spectral resolution as measured by the bandwidth (ΔBλ).

number of photons, even from small pixels. It should be remembered, however, that sometimes high resolution is simply overkill, and a comprehensive selection of sensor performance for the question at hand is critical.

16.3.2 Historical Notes

16.3.2.1 Satellite Sensors

The orbital remote-sensing era started with the first Russian mission to space, Sputnik 1, on October 4, 1957. A handheld manual camera enabled the first large view of the Earth's surface, showing its curvature (which later became a major challenge for the remote-sensing operator who had to correct for it). However at that time, space missions were restricted to the military/defense sectors, and the civilian community was not able to gain access to the data. There was no available space-observation technology for civilian applications, and several years passed before the first civilian sensor, ERTS 1 (also known as Landsat 1) from NASA was made available to all (1972). Today, many missions, sensors, platforms and data are available to the public from several space agencies, as well as from the private sector. The distribution policy ranges from free of charge (NASA) to full-price data availability. The International Institute for Geo-Information Science and Earth Observation (ITC) has published a list of 309 orbital sensors that shows how remote sensing has advanced since ERTS 1 (www.itc.nl/research/products/sensordb/allsensors.aspx). Table 16.2(a, b) provides a list of the main sensors in orbit covering the VNIR—SWIR regions that were, are and will be available to the public and can be used for soil applications. The sensors are divided into multi-, super-, and hyperspectral sensors consisting of different spatial, temporal and spectral resolutions.

In contrast to the few, highly expensive satellites that were available 25 years ago, providing remote-sensing technology for scientists only at the demonstration and research levels, today, remote sensing from space (and the aerial domain) has become a commercial endeavor; it is less expensive, and the data are easier to process and can be used by all. Google Earth is a good example of how satellite data have become freely available, providing a new dimension in our understanding and exploration of the Earth's surface from afar. A general list of all NASA satellites (along with available data archives) is provided at http://nssdc.gsfc.nasa.gov/nmc/spacecraftSearch.do. As most of the sensors are still characterized by a broadband and low spectral resolution, effort is being devoted to placing hyperspectral—high spatial resolution sensors in space, as they can provide quantitative rather than qualitative information of small areas (see more details in this chapter). To that end, four space agencies have recently mounted hyperspectral sensors in orbit. These are: the Hyperspectral Imager Suite (HISUI) from the Japan Aerospace Exploration Agency (JAXA), the Earth Mineral Dust source InvesTigation (EMIT) instrument from the National Aeronautics and Space Administration (NASA), the Environmental Mapping and Analysis Programme (EnMAP) and DLR Earth Sensing Imaging Spectrometer (DESIS) from the German Aerospace Centre (DLR), the Geofen 5 from the China National Sapce Administration (CNSA), and PRISMA (Hyperspectral Precursor of the Application Mission) from the Italian Space Agency (ASI). These sensors have proven that hyperspectral remote sensing (HSR) from orbit is possible with added value compared to the multi- and super spectral sensors available today. It is important to mention that based on this success, many commercial enterprises are planning to mount HSR sensors to provide data and services. Future plans for HSR sensors such as NASA's Surface Biology and Geology (SBG) mission and the European Space Agency's (ESA) Copernicus Hyperspectral Imaging Mission for the Environment (CHIME) promise that the high spectral resolution from space will be maintained and improved. An important list in this regard is given by Maksimenka for ESA, providing a summary of super- and hyperspectral orbital sensors in use today and planned for the near future (http://ubuntuone.com/2QsaOhOLPL7602cCOO3IZ5). The high spectral—spatial resolution data can serve as a significant management tool for precision-farming activities covering, for example, soil cultivation and formation, contamination, degradation, fertilization and more (see further on in this chapter). Aside from spectral and spatial resolution, for soil applications, temporal resolution

TABLE 16.2a

A list of past present and future Orbital sensors for soil remote sensing and their basic characteristics

Sensor	Year	Mission	Sensor	Channels in VIS-NIR-SWIR	Spectral Range (mm)	Spatial Resolution (m)	Revisit (day)	Swath (km)	Soil Application (spectral base)	Soil Applications (spatial base)	Sector
Multi Spectral	1972, 1975,1978	EART (Landsat 1-3)	MSS	4	0.5-1.1	60	18	170×185	General, Color	Large Area General View	NASA
	1995, 1997	IRS-1C-1D-	LISS-3	4	0.52-1.70	23.5/70.5	24	141/148	General/ Semi quantitative	Medium to large coverage Land use, General View	ISRO
	1988/1991	IRS-1A-1B	LISS-1,2	4	0.45-0.86	36/72.5	22	140	General/ Semi quantitative	Medium to large coverage Land use, General View	ISRO
	1986, 1990, 1993	SPOT 1-3	HRV	4	0.50-0.89	10/20	2-3	60	General/ Color	Medium to large coverage Land use (field to landscape)	CNES
	1998	SPOT-4	HRVIR	5	0.50-1.73	10/20	2-3	60	General/ Color + non visible color	Medium to large coverage Land use (field to landscape)	CNES
	1999, 2000	IKONOS 1-2		5	0.45-0.90	¼	<3	11	General/ Color	Small coverage, Field scale	Digital-Globe
	2000, 2001	QUICK BIRD 1-2		5	0.45-0.90	0.6/2.4	<1-5	20/40	General/ Color	Small coverage, Field scale	Digital-Globe
	2002	SPOT-5		5	0.48-0.71	2.5/10	2-3	60			CNES
	2003	Orb View 3		5	0.45-0.90	¼	<3	8	General/ Color	Small coverage, Field scale	GeoEye
	2012, 2014	SPOT 6-7		5	0.45-0.89	1.5/8	2-3	60	General/ Color +	Medium to large coverage Land use (field to landscape)	CNES
	1982, 1984	Landsat 4-5	TM	6	0.45-2.35	30	18	170	General, Semi quantitative	Medium to large coverage Land use	NASA
	1999	Landsat 7	ETM+	7	0.45-2.35	30/15	18	170	General/ Semi Quantitative	Medium to large coverage Land use	NASA

TABLE 16.2b

A list of past present and future Orbital sensors for soil remote sensing and their basic characteristics

Sensor	Year	Mission	Sensor	Channels in VIS-NIR-SWIR	Spectral Range (mm)	Spatial Resolution (m)	Revisit (day)	Swath (km)	Soil Application (spectral base)	Soil Applications (spatial base)	Sector
Multi Spectral	1997	Orb View 2		8	0.4-0.86	1100	1	2800	General	Large coverage	GeoEye
	2009	World View -2		8	0.40-900	0.46/2.08	1.1	16.4	Spectral base	Small coverage, Field scale	Digital-Globe
	1997	Orb View 2		8	0.4-0.86	1100					GeoEye
	2001	EO-1	ALI	9	0.43-2.40	10/30	16	37			NASA
	2013	Landsat 8	OLI	9	0.43-2.29	30/15		170×183	Semi Quantitative	Medium to large coverage Land use (field to landscape)	NASA
Super Spectral	1999	Terra	ASTER	10	0.52-2.43	15/30	16	60	Semi quantitative/ quantitative	Medium to large coverage Land use (field to landscape)	NASA / METI
	2017	Venus		12	0.4-1.0	5.3	2	27.5	Spectral based (limited)	Small coverage, Field Scale	ISA-CNES
	2018	Copernicus	Sentinel-2	13	0.4-2.4	10/60	2/3	250	General/Spectral Base (limited)	Medium to small coverage.	ESA
	2015	World View -3	CAVIS	16	0.45-2.365	0.34/4.1	1	13.1	Spectral base	Small coverage, Field scale	Digital-Globe
	1999	Terra	MODIS	20	0.459-2.155	250/1000	1-2	2,330	General , Semi Quantitative	Large coverage General large view	NASA
Hyperspectral	2001	Chris-Proba		60	0.4-0.9	18	7	14	Spectral base (limited)	Medium Coverage	ESA
	2019	ALOS-3	HISUI	185	0.4-2.5	30		90	Spectral base	Medium Coverage	JAXA
	2017	En Map		228	0.42-2.45	30	4	30	Spectral base	Medium coverage	DLR
	2017	PRISMA		238	0.4-2.5	30	29	30	Spectral base	Medium coverage	ASI
	2001	EO-1	Hyperion	244	0.4-2.45	30	16	7.7	Spectral base	Medium coverage	NASA

is also important. A high capability to provide information in a short time domain is important for properties such as soil moisture, OM and clay content variability. The temporal resolution may vary from daily to yearly coverage. In soils, the yearly temporal coverage is important for applications such as land use and coverage, change detection (CD) in large areas and soil mapping. Figure 16.4 provides a scheme showing the temporal versus spatial resolutions required for soil applications (in comparison to climate and weather forecast applications) along with the currently operational satellites as adopted from Jensen (2011). In a new era in which microsatellites (e.g., CubeSats) will be available to all, a new initiative to mount satellite flocks into orbit will enable high temporal resolution, as done today by the company Planet Scope (Frazier and Hemingway, 2021).

16.3.2.2 Airborne Sensors

In addition to the satellite sensors that provide large coverage and high temporal resolution, there are also sensors onboard air platforms which provide better spatial (and recently also better spectral) resolution but with small coverage of a given overpass. Airborne remote sensing began many years ago with aerial photography. The history of aerial photography and remote sensing is provided with some impressive facts and illustrations by Baumann (2010) at www.oneonta.edu/faculty/baumanpr/geosat2/RS%20History%20I/RS-History-Part-1.htm; the following text reparses the main points from his site. The first aerial photograph was taken in 1858, and its first practical use emerged 50 years later during World War I. The military on both sides of the conflict saw the value of using the airplane for reconnaissance work. Aerial observers, flying in two-seater airplanes with the pilots, performed aerial reconnaissance by sketching maps and verbally conveying the conditions on the ground. Toward the end of that war, the Germans and the British were monitoring the entire front at least twice a day with a total of half a million photographs from the English side and many more from the German side. The war brought major improvements in camera and product quality. As a result, in late 1920s, the first books on aerial-photograph interpretation were published (a full list is given on Baumann's web page). World War II brought tremendous growth and recognition to the field of aerial photography

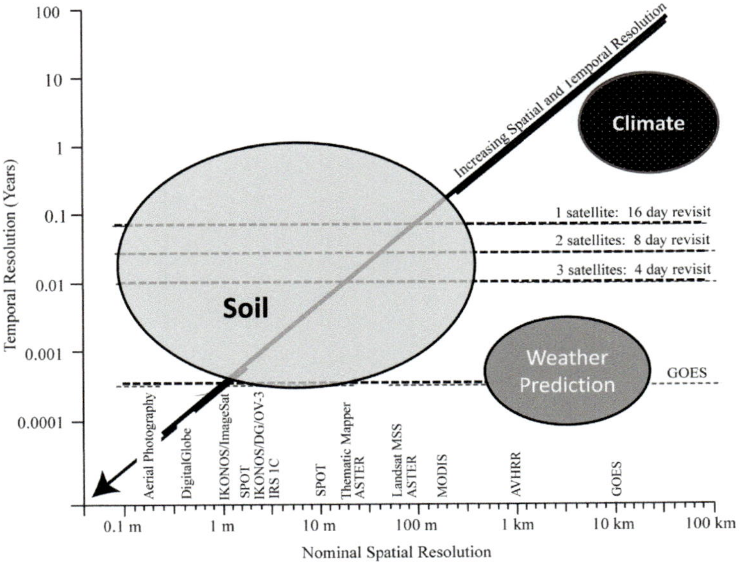

FIGURE 16.4 The spatial and temporal resolutions that are needed for soil applications and the orbital sensors available for that. Also given for comparison are the climate applications (The figure is modified from Jensen R.J. Remote Sensing of Environment: An Earth Resources Perspective. Pearson Prentice Hall press 2–11. p. 592.

which continues to this day. In 1938, the chief of the German General Staff, General Werner von Fritsch, stated "the nation with the best photoreconnaissance will win the war." Admiral J. F. Turner, Commander of the American Amphibious Forces in the Pacific, stated that the importance of "photographic reconnaissance cannot be overemphasized." In parallel to its military development, aerial photography was used for soil applications. The first report on this use of aerial photography was published in around 1927 for soil surveys of the United States (Bushnell, 1932), Australia (Prescott and Taylor, 1930) and the USSR (Levenhangst, 1930). Vink (1964) generated a checklist and bullet points on how to use aerial photographs for soil applications, and highlighted the necessary methods to interpret the images and translate them into a soil map. Figure 16.5 provides an example of a final product conducted using stereo aerial photographs to interpret "possible soil boundaries" according to Vink's (1964) instructions. It can be seen that this map is still not informative and field work is strongly needed. On the other hand, Vink (1963a, 1963b) it was the first work to prove that the use of aerial photographs diminishing the number of observation on the development of a soil map, in comparison with a work for the same area and same scale, although without the use of this product. Its invaluable need for soil survey application was later well documented by Goosen 1967. This demonstrates the impact of aerial photography at that time for soil science, that is, in providing possible boundaries based only on the gray tones of the aerial photograph and indicating less need of field observations. Aerial photography is still a common way of mapping the surface, and many mapping agencies have instrumentation to interpret the analog data which today, has mostly been converted to digital format.

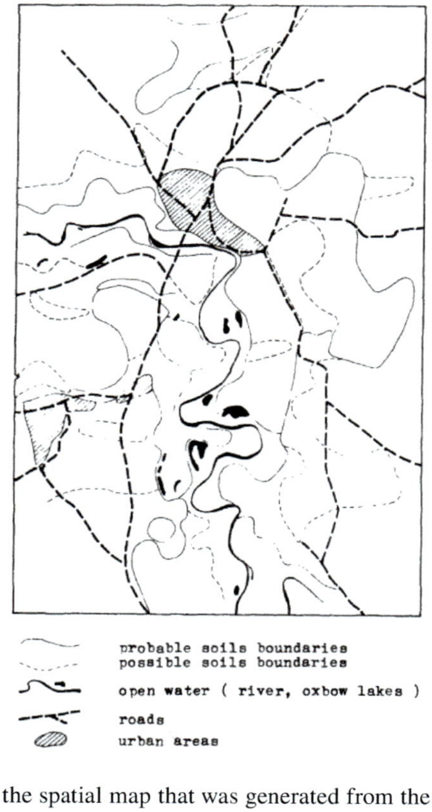

FIGURE 16.5 An example of the spatial map that was generated from the soil entity by opening trenchers and describing the entire soil profile (Taken from Vink A.P.A. j. Soil Sci., 14, 88 163a; Vink, A.P.A Planning of soil survey in land development, International Institute for Land Reclamation and Improvement, publication 10, U/D/C/632.47: 528:77, The Netherlands, 163b, 53p.)

Other sensors, such as multispectral and hyperspectral sensors, are also available from the aerial domain, and over the last decade, they have been widely used for studying particular areas and zooming in on particular fields. Unmanned platforms (unmanned aerial vehicles—UAVs) are also entering the field of aerophotography based on high-quality light cameras and GPS availability with an easy operational scheme. A recent book by Manfreda and Ben-Dor (2023) summarizes UAV capabilities and provides several case studies in precision-agriculture applications and in particular, soil mapping. The payload capacity of UAVs is now very impressive, enabling the mounting of many sensors—including a hyperspectral camera and light detection and ranging (LIDAR) sensors—on the same platform. Another example is given by d'Oleire et al. (2012), who monitored soil erosion in Morocco using an UAV. Ben-Dor et al. (2013) provide a comprehensive overview of HSR technology along with a full description of the main available airborne hyperspectral sensors worldwide; a detailed list can be found at www.tau.ac.il~rsl/rgomez. Those authors concluded that soon, HSR technology would move from the realm of scientific demonstration to become a practical commercial tool for remote sensing of the Earth's surface, including soils. Thus, there is no doubt that today (2023), HSR technology has become common for many applications, and in particular for soils (Inoue, 2020). As it is possible to obtain high spectral and spatial resolutions using this technology onboard manned and unmanned platforms, its contribution for soil applications is straightforward. Herein we present the reasons why HSR technology is promising for soil monitoring.

16.3.2.3 Sensor per Mission

Selection of a sensor that best suits a particular mission, and more specifically a particular soil application, must be tailored to the question at hand. Such questions might involve large or small areas, quantitative or qualitative information, CD or current-state position of the soil, as well as cost. Many of the satellite data available today are free of charge (mostly from governmental bodies such as NASA, The National Oceanic and Atmospheric Administration [NOAA] and ESA), whereas others can be costly, such as from DigitalGlobe ((www.digitalglobe.com/) and ImagSat (www.imagesatintl.com). As space agencies maintain open source policies, new commercial enterprises are entering the remote sensing of the Earth from orbit, reflecting the high demand for this technology. Airborne data are also costly but may provide more information on a small scale based on their high spatial resolution domain, and solve local problems by capturing high spectral/spatial information. The UAV platforms may overcome the cost issue but are still limited with respect to flying time (and hence coverage area) and with respect to carrying heavy payloads. Tactical considerations for a specific mission are important, and taking into account the processing time and spectral sensitivity of the sensor is crucial.

As across the VNIR region, both radiation and detectors are at their maximum sensitivity, this region is well covered by most of the current sensors. In the SWIR regions, however, detector sensitivity and produced radiation intensity are poor, making this region more problematic and hence less available. The evolution of remote sensing shows that the VIS sensors were the first to be used. Instead of a digital detector, a sensitive emulsion (film) for VIS radiation was used to capture visual information, allowing only limited interpretation of the remote-sensing data. Today, all sensors are equipped with a CCD assembly and the raw product is delivered in digital format. The computing power and digital visualization provide an innovative way of visualizing and analyzing the data using algorithms that can be shared with a wide spectrum of users. This is a real breakthrough for soil applications, where most of the information is hidden beneath the spectral responses of the solid and water phases. A similar problem can emerge for the MWIR and LWIR sensors. Aside from a broadband sensor, the HSR sensors in this region are limited and only a few workers are operating them in general, and for soil in particular. Nonetheless as these regions are very important for detecting minerology and organic components (Vaughan et al., 2003), we anticipate that this region will gain more attention in the future and the remote sensing of soil will advance. The forthcoming missions from space, such as Thermal infraRed Imaging Satellite for High-resolution Natural resource Assessment (TRISHNA) from the French National Centre for Space Studies (CNES) and the Indian

Space Research Organisation (ISRO) (Buffet et al., 2021), Land Surface Temperature Monitoring (LSTM) from ESA (Bernard, 2023), SBG from NASA (Cawse-Nicholson et al., 2021) and the current sensor at the International Space Station (ISS) ECOSTRESS (Fisher et al., 2020) promises that this area is still in its infancy and its future is bright.

In summary, there are different remote-sensor options related to the source of energy (passive or active sensors), the type of platform (ground, air or space), the spectral region (optical, IR, microwave), the platform trajectory, the number and width of the spectral bands (e.g., panchromatic, multispectral, superspectral, hyperspectral), the spatial resolution (high, medium, low), the spatial coverage (point or image view), the temporal resolution (e.g., hourly, daily, or weekly revisiting frequencies), the radiometric resolution (e.g., 8, 12, 16 bits) and the collection system (push broom or whisk broom). The modern sensors provide information on spatial and spectral domains from aboveground elevations of a few meters (field sensors) to 800 km (orbital sensors), with spatial resolutions varying from a few centimeters to tens of meters, and temporal coverage of minutes to days.

16.4 REMOTE SENSING OF SOILS

Remote sensing of soils refers to the product that can be obtained on soils from remote sensors. It should cover the information extracted and interpreted on a soil entity from afar. Although problems still exist in correctly remotely sensing the soil body (see further on), the advances made in both the technology and the know-how over the years is remarkable. In 1987, Mulder wrote that "*it is still impossible to extrapolate the remote sensing information for the entire profile.*" Although for the most part, optical remote-sensing means cannot detect the entire soil body that extends from the surface down to the parent material, today some innovative ideas can be used to probe below the soil surface. Most of the remote-sensing soil products in the optical domain characterize the surface, because the Sun's radiation cannot penetrate more than 50 μm of the soil's surface (Ben-Dor et al., 1999). Accordingly, only the upper (mostly Ao) horizon can be sensed, with limited ability to extrapolate the information to the soil's deeper horizons. If the parent material is known (mostly from metadata such as geological maps), the Ao information can provide inferences about the soil body if the soil in question is a direct product of the lengthy rock-weathering process. However, as the soil surface is frequently characterized by a short time process, the upper soil horizon (Ao) can provide valuable information on the environmental (soil) processes such as: soil degradation, dust accumulation, contamination, salinity, OM, crust formation, soil surface moisture, soil runoff, water infiltration, and others (see further on). This information can be valuable for the farmer, and thus the remote-sensing product can assist decision makers in selecting appropriate action. For these applications, high-spectral-resolution sensors are needed. Obtaining spectral information from afar is not simple and requires high performance from both operational and infrastructural standpoints. Nonetheless, if the spectral-reflectance information can be captured from air and space domains in a precise manner, then information on the status of the soil surface becomes invaluable and promising. This makes reflectance spectroscopy a significant factor in the remote sensing of soils. Satellites can provide overviews of larger areas than airborne sensors and have a better temporal resolution. However, high spectral and spatial resolution from orbit is still problematic, whereas data storage and computing power are no longer limiting, thanks to cloud technology and computing power that can be shared by many computers. Despite some limitations, remote sensing of soils can assist the development of digital soil mapping by providing a good basis for soil surveyors and enabling the monitoring of surface processes in a unique way. New computing power, user-friendly software and open-source algorithms combined with other sensors in the air and on the ground will enable (and in fact are enabling) new remote-sensing dimensions. It is obvious that soil, as one of the main covers of the Earth's surface, is an ideal study target for remote-sensing technology, especially from satellites. Ge et al. (2011) provide a brief description of the history of remote sensing of soils from satellite sensors. They pointed out that in the late 1960s and early 1970s, soil scientists began to understand the capabilities of Landsat 1 data, from which differences in surface soils could be delineated (Kristof, 1971). Soon after, Kristof and Zachary (1974) reported partial success in delineating soil series in an Alfisol–Mollisol region through digital analysis of aerial multispectral data. It was only in 1972, when Landsat 1 data became available to the public, that the era of remote sensing of the soil began. This was mainly based on studies that realized that soil spectral information is a key factor in exploiting soil (optical and chemical) data obtained from afar. The first investigations of soil spectroscopy were published in around 1970 (for example, Condit, 1970, 1972). Those studies showed that the spectral information can be used as a tool to discriminate between soil families. Along these lines, it was shown that a dataset consisting of 160 different soil spectra could be categorized spectrally into three groups. This actually opened up the era of soil spectroscopy as the basis for soil remote-sensing disciplines. This topic is further elaborated upon later in this chapter.

16.5 SOIL REFLECTANCE SPECTROSCOPY

16.5.1 Definition

The soil reflectance spectrum (ρ) is a collection of values obtained at every spectral band (λ) from the ratio of radiance (E) and irradiance (L) fluxes across most of the spectral region of the solar emittance function.

$$P(\lambda) = E(\lambda)/L(\lambda) \tag{2}$$

The reflectance values are traditionally described, from a practical standpoint, by a relative ratio against a perfect reflector spectrum measured at the same geometry and position as the soils (Palmer, 1982; Baumgardner et al., 1985; Jackson et al., 1987).

To illuminate the value of the soil spectrum, this section will provide some historical notes and then a comprehensive theoretical background on the interaction of electromagnetic radiation with the soil matrix described by ρ.

16.5.2 Historical Notes

When spectrometers became available in around 1960, studies were conducted to elucidate the spectral responses of soils and pure minerals. From 1970–1982, Hunt (1982), Hunt and Salisbury (1970, 1971, 1976), and Hunt et al. (1971a, 1971b, 1971c) conducted a comprehensive study on the spectral characteristics of pure minerals, which was published over a decade and provided the foundation for spectroscopy of minerals. In 1965, a first attempt to demonstrate the quantitative capabilities of soil spectral information showed a correlation between moisture content and spectral response at several wavebands (Bowers and Hanks, 1965). Later, a systematic investigation of the relationship between soil spectral information and soil properties was conducted by Condit (1970), and then by Montgomery and Baumgardner (1974), who later systematically studied the soil reflectance of American soils (Stoner and Baumgardner, 1981). These latter authors also published the first "soil reflectance atlas" (Stoner et al., 1980); they demonstrated the importance of soil spectroscopy and for the first time, the spectral grouping of mostly USA soils into five major soil types (One of them, *type 5*, was from Brazil). Their soil spectral library was initially constructed with complete soil profiles (horizons) and its classification, which quickly became a classical tool for soil scientists and a fundamental reference source for future studies. The emerging activity in soil spectroscopy over the past decade, resulting in a vast accumulation of knowledge, led workers to complete Baumgardner's work on a global basis by establishing more libraries worldwide (e.g., Viscarra Rossel, 2009). In 1991, the first portable spectrometer hit the market (ASD, www.portableas.com/index.php/manufacturers/asd/), ringing in the era of portable and facile spectral sensing of soils (as well as other Earth materials) in both field and laboratory domains (Goetz, 2009). This drove many scientists to the field of soil spectroscopy and today, there is a strong scientific community in this field of interest from many aspects, and contributing their know-how to the practical utilization of this promising tool. The significant contribution of soil reflectance lies in the possibility of extracting quantitative soil information from the spectrum by establishing a proximal sensing approach which historically, started (in soils) in around 1986; at that time, Dalal and Henry (1986) were the first to adopt the spectral data-mining technique based on Ben Gera and Norris's (1968) approach developed a decade prior for wheat grains. From 1994 on, after Ben-Dor and Banin (1994) showed the potential of the proxy technique to extract several soil properties (among them even "featureless"), soil spectral studies advanced rapidly. The spectroscopy of soils has gained a lot of attention as a practical proxy for wet chemistry evaluation. Many physical collections of soils with their spectral and chemical information have evolved into soil spectral libraries (SSLs), and the first merging of SSLs was executed by Viscarra Rossel to form the first global SSL (Viscarra Rossel et al., 2016). Since then, many attempts to build SSLs in local regional and global domains have been made (e.g., LUCAS; Orgiazzi et al., 2018, ISRIC; Shepherd et al., 2022, BSSL; Demattê et al., 2019). One of the problems with using SSLs from different sources was the different protocols in each laboratory, preventing data harmonization. To that end, in 2020, a new initiative emerged calling for a standard and protocol to harmonize diverse SSLs under the umbrella of the Institute of Electrical and Electronics Engineers (IEEE) Standards Association (SA) within the P4005 working group (WG) entitled *Standards and Protocols for Soil Spectroscopy*. The group's primary objective is to explore how the diverse range of protocols, sensors, and measurement methods can be effectively unified into a standardized protocol, enabling the harmonization of SSLs and providing more confidence for the construction of prediction models of soil attributes that can be used with remote-sensing means under multispectral and hyperspectral domains.

In 1983, the first hyperspectral airborne sensor arrived at NASA (Goetz, 2009), but it took years until this technology was adopted for a proximal-sensing approach in soils (Ben-Dor et al., 2002). Today, with some of the obstacles to extracting reflectance information from air and space domains having been partially overcome, this seems to be the direction that will enable exploitation of soil spectral information for the needs of mankind, such as soil monitoring, mapping and cultivation. In 2013, there were several hyperspectral sensors in orbit, and others are planned to be mounted in the near future (see Section 25.3.2.1). This progress will lead to a close linkage between the SSLs and orbital information and to better utilization of the new spatial domain provided by the imaging sensors.

The following sections provide a theoretical background on soil reflectance to understand its capacity in the remote-sensing field, and then its applications in soil science will be demonstrated.

16.5.3 Radiation Interactions with a Volume of Soil

The process of radiation scattering by soils results from a multitude of quantum-mechanical interactions between the enormous number and variety of atoms, molecules, and crystals in a macroscopic volume of soil. In contrast to certain absorption features, most characteristics of the scattered radiation are not attributable to a specific quantum-mechanical interaction. The effects of a particular mechanism often become obscured by the composite effect of all of the interactions. The difficulty in accounting for the effects of a large number of complex quantum-mechanical interactions often leads to the use of non-quantum-mechanical models of electromagnetic radiation. Physicists frequently resort to the classical wave theory or even to geometrical optics to elucidate the effects of a macroscopic volume of matter on radiation.

16.5.3.1 Refractive Indices

When light passes through a medium, some part of it will always be absorbed. This can be conveniently taken into account by defining a complex index of refraction,

$$\tilde{n} = n + i\kappa \tag{3}$$

Here, the real part of the refractive index n indicates the phase speed, while the imaginary part κ indicates the amount of absorption loss when the electromagnetic wave propagates through the material (i is the square root of -1). That k corresponds to absorption can be seen by inserting this refractive index into the expression for the electrical field of a plane electromagnetic wave traveling in the z-direction. The wave number is related to the refractive index by

$$k = \frac{2\pi n}{\lambda 0} \tag{4}$$

where λ is the vacuum wavelength. With complex wave number k and refractive index $n + i\kappa$, this can be inserted into the plane wave expression as:

$$\mathbf{E}(z,t) = \text{Re}(\mathbf{E}_0 e^{i(\tilde{k}z-\omega t)}) = \text{Re}(\mathbf{E}_0 e^{i(2\pi(n+i\kappa)z/\lambda_0-\omega t)})$$
$$= e^{-2\pi\kappa z/\lambda_0}\text{Re}(\mathbf{E}_0 e^{i(\kappa z-\omega t)}). \tag{5}$$

Here we see that λ gives an exponential decay, as expected from the Beer—Lambert law. Since intensity is proportional to the square of the electrical field, the absorption coefficient becomes $\frac{4\pi\kappa}{\lambda 0}$. κ is often k=0 called the extinction coefficient in physics, although this has a different definition in chemistry. Both n and κ are dependent on the frequency. Under most circumstances, k > 0 (light is absorbed) or/(light travels forever without loss). In special situations, especially in the gain medium of lasers, it is also possible for $\kappa < 0$, corresponding to an amplification of the light. In the soil matrix, the radiation travels through a thin layer of particles, is reflected back to the sensor and provides a spectrum whose shape and nature are affected by the aforementioned process (consisting of both the real and imaginary part of the complex refractive index). Any substance in the soil matrix that affects the aforementioned indices is termed a "chromophore." Knowing the chromophores behavior can shed light on the physical and chemical constituents of the soil matrix under study. This information can either be derived by the naked eye's "color vision" (across the VIS region) or by careful analysis of the spectral responses (across the VNIR—SWIR regions) according to the preceding theory. In general, due to its complexity, a given soil sample consists of a variety of chromophores, which vary with environmental conditions. In many cases, the spectral signals related to a given chromophore can overlap with other chromophores' signals, thereby hindering the assessment of the direct effect of the chromophore in question. Accordingly, it is important to understand the chromophores' physical processes as well as their origin and nature. Another point to mention is that in soil, there are many cases in which relationships between chromophoric and non-chromophoric properties exist.

We define the factors affecting soil spectra as "physical" if the real part of the refractive index is associated, and "chemical" if the spectral changes are associated with the imaginary part of the refractive index. This terminology is adopted from the weathering processes in soil where "physical" weathering refers to "size" changes in the soil matrix with no chemical alteration, and "chemical" weathering refers to chemical "alteration" of the soil materials. Figure 16.6 provides the possible light interactions within the thin layer of the soil surface.

16.5.3.2 Chemical Chromophores

16.5.3.2.1 Physical Mechanism

Chemical chromophores are those materials that absorb incident radiation at discrete energy levels. They are related to the imaginary part of the complex refractive index. The absorption process usually appears on a reflectance spectrum at positions attributed to specific chemical groups in various

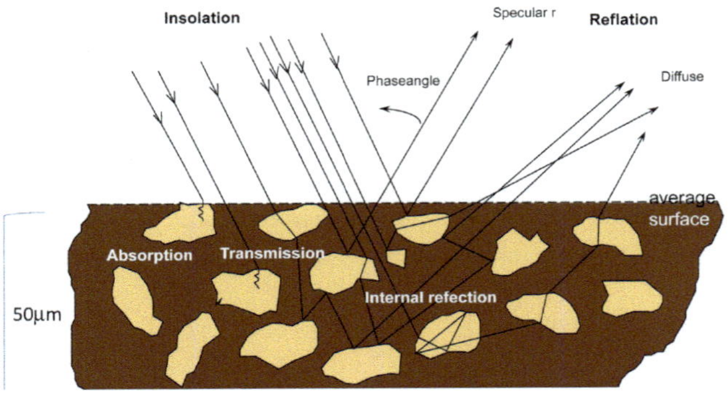

FIGURE 16.6 An illustration showing the interaction of light with the soil particles at the soil surface.

structural configurations. The interaction between radiation and matter occurs at the atomic and molecular levels. Electromagnetic radiation can be emitted or absorbed when an atom or molecule transitions between energy states. The energy of an emitted or absorbed photon equals the difference between the energy levels. Furthermore, the energy-level transitions must be accompanied by either a redistribution of the electric charge carried by electrons and nucleic protons, or a reorientation of nuclear or electronic spins before a photon is emitted or absorbed (Hunt, 1982). A comprehensive description of the physical mechanisms describing the interactions of electromagnetic radiation with diverse minerals and rocks is provided by Clark (1999). This section focuses on the most common chromophores in the soil environment and their relationship with electromagnetic radiation across the VNIR—SWIR spectral region.

16.5.3.2.2 Vibration Processes in the SWIR Region

The absorption or emission of shortwave radiation usually results from energy-level transitions accompanied by charge redistributions involving either the motion of atomic nuclei or the configuration of electrons in atomic and molecular structures. A molecule possesses several modes of vibration depending on the number and arrangement of its atoms. A molecule with N atoms may have $3N$-5 vibrational modes if the atoms are arranged linearly or $3N$-6 vibrational modes if the bonding is nonlinear (Castellan, 1983). The absorbed (or emitted) frequencies are called overtone bands when a vibrational mode transitions from one state to another that is more than one energy level above (or below) the original state. Combination bands refer to frequencies associated with transitions of more than one vibrational mode. These combined transitions occur when the energy of an absorbed photon is split between more than one mode (Castellan, 1983). Vibrational transitions corresponding to the fundamental bands are generally more likely to occur than transitions corresponding to combination and overtone bands, and are usually stronger than the overtone transition. The fundamental transition of soil chromophores occurs mostly in the IR region (>2.5 μm), whereas the overtones occur in the SWIR region, mostly above 1 μm. The two basic vibrations that correspond to the atoms' motions in the chemical bonds' molecular processes in soil minerals are stretching (ν) and bending (δ).

Aside from the overtone transitions, combination modes are also excited when the quantum mechanism enables combining different vibrational processes of the same bonds, such as stretching and bending. Combination and overtone bands associated with hydroxyl group (OH) vibrations are, for example, often apparent in soil reflectance spectra. OH groups are found on many soil minerals and the exact wavelength location of the associated bands depends upon which OH-bearing minerals are present in the soil. The one fundamental OH band due to oxygen—hydrogen stretching is found near 2.8 μm and the first overtone band due to this stretch is located near 1.4 μm. Absorption at this overtone band is the most common feature in the SWIR spectra of terrestrial materials (Hunt, 1982). The hydrogen—oxygen stretch can also be coupled with other vibrations in the molecular structure of the soil minerals to create combination band features. Bending at magnesium—OH bonds coupled with stretching results in a combination band near 2.300 μm, and bending at aluminum-OH bonds coupled with stretching produces a combination band near 2.200 μm (Hunt, 1982).

16.5.3.2.3 Electron Processes in the VNIR Region

At higher electromagnetic radiation energy (UV and VNIR), the spectral response is associated with electron transitions. The locations of these bands are due to the relatively large gaps between electron energy states. The principles of quantum mechanics dictate that each electron of an atom, ion, or molecule can exist in only certain states corresponding to discrete energy levels. There are four possible electronic possibilities, termed crystal field, charge transfer, color center and semiconductor. In soil and soil minerals, the first two are dominant. **A crystal field** can often be inferred from reflectance spectra of minerals containing transition metals. The allowable energy states for the unpaired transition-metal electrons and the gaps between states are determined primarily by the valence state of the ion and the coordination number and symmetry of the crystal site in which the ion occurs. The energy states are also influenced by the type of ligand surrounding the ion, by the interatomic metal—ligand distance, and by site distortion (Hunt, 1982). This latter author displayed the reflectance spectra of six minerals containing ferrous iron (Fe^{2+}) to demonstrate the effects of coordination number, site symmetry and site distortion. **Charge transfer** is a mechanism involving the electrons of a specific ionic bond between adjacent ions. Charge transfer occurs when an electron migrates between the adjacent ions with a corresponding change in energy state. The usually prominent decrease in reflectance in the blue region of soil reflectance spectra is due to the charge transfer between iron and oxygen in Fe-oxides (Hunt, 1982). The dark color of some minerals, such as magnetite (Fe_3O_4), which contain both ferrous (Fe^{3+}) iron, is due to charge transfer between these two ions (Nassau, 1980). **Color centers** refer to unpaired electrons, and paired valence electrons that play a role in the interaction of shortwave radiation with soils. Molecular-orbital theory describes the distribution of paired-electron charges among the atoms of a molecule or crystal. In some cases, an electron pair may remain associated with a specific bond between two adjacent atoms or ions. In other cases, the charge may be distributed over several atoms or even throughout a crystal structure. **Semiconductors** interact like solids in which the allowable energy states of the bonding electrons are divided into two broadbands. The lower energy band is called the valence band and the higher energy band, which contains all of the excited states, is called the conduction band. Between these two bands is a region called the forbidden gap within which no electrons are allowed. To be absorbed by a semiconductor, a photon must carry at least enough energy to elevate an electron from the upper level of the valence band to the lower level of the conduction band (Nassau, 1980). The reflectance spectrum of a semiconductor is distinguished by an intense absorption edge which marks the width of the forbidden gap (Hunt, 1982).

16.5.3.3 Physical Chromophores

In addition to chemical chromophores, the reflectance of light from the soil surface is dependent upon numerous physical processes, most of them related to the real part of the refractive index. Reflection, or scattering, is clearly described by Fresnel's equation and depends upon the angle of incidence radiation and upon the index of refraction of the material in question. In general, physical factors are those parameters which affect soil spectra in terms of Fresnel's equation (the real part of the refractive index), but which do not cause changes in the position of the specific chemical absorption. These parameters include particle size, particle's geometry, hydration stage, viewing angle, radiation intensity, incident angle, and azimuth angle of the source. Changes in these parameters are most likely to affect the shape of the spectral curve through changes in baseline height and absorption-feature intensities. In the laboratory, measurement conditions can be held constant; in the field, several of these parameters are unknown, which may complicate accurate assessments of their effect on soil spectra. Many studies, covering a wide range of materials, have shown that differences in particle size alter the shape of soil spectra (e.g., powdered material) (Hunt and Salisbury,

1970; Pieters, 1983; Baumgardner et al., 1985). Specifically, Hunt and Salisbury (1970) quantified particle-size difference effects of about 5% in absolute reflectance and noted that these changes occurred without altering the position of diagnostic spectral features. Under field conditions, aggregate-size rather than particle-size distribution may be more important in altering soil spectra (Orlov, 1966; and Baumgardner et al., 1985). In the field, aggregate size may change over short periods due to tillage, soil erosion, aeolian accumulation or physical-crust formation (e.g., Jackson et al., 1990). Basically, the aggregate size, or more likely roughness, plays a major role in the shape of field and airborne soil spectra (e.g., Cierniewski, 1987, 1989). Escadafal and Hute (1991) showed strong anisotropic reflectance properties in five soils with rough surfaces. Cierniewski (1987) developed a model to account for soil roughness based on the soil-reflectance parameter, illumination properties and viewing geometry for both forward and backward slopes. The model showed that the shading coefficient of the soil surface decreases with decreasing soil roughness. For soils on forward slopes of more than 20°, the shadowing coefficient also decreased with increasing solar altitude for the full interval of Sun altitudes ranging from 0° to 90°. The model indicated that the relationship for soil slopes with surface roughness lower than 0.5 might be reversed for a specified range of solar altitudes. Using empirical observations of smoothed soil surfaces, Cierniewski (1987) showed that the model closely agrees with field observations. An excellent brief summary on multiple- and single-scattering models for soil particles with respect to the roughness effect is given by Irons et al. (1989).

16.5.3.4 Chromophores in Soils

This section focuses on the most common chromophores in the soil environment and their relationship with electromagnetic radiation across the VNIR-SWIR spectral region. All features in the VNIR—SWIR spectral regions have a clearly identifiable physical basis. In soils, three major chemical chromophores can be roughly categorized as follows: minerals (mostly clay and Fe-oxides), OM (living and decomposing), and water (solid, liquid and gas phases). Physical chromophores in soils are related to particle size and measurement geometry.

16.5.3.4.1 Soil Minerals

As already discussed previously, Hunt and Salisbury (1970–1980) have studied the details of the spectral behavior of many minerals on Earth. Some minerals reported in their comprehensive review are encountered in the soil environment and will be discussed here, with additional information related to the soil medium.

Clay Minerals: Of all clay mineral elements, only the OH group is spectrally active in the VNIR—SWIR region. This group can be found as part of either the mineral structure (mostly in the octahedral position which is termed lattice water) or the thin water molecule that is directly or indirectly attached to the mineral surface (termed adsorbed water). Three major spectral regions are active for clay minerals in general and for smectite minerals in particular: around 1.3–1.4 μm, 1.8–1.9 μm and 2.2–2.5 μm. For Ca—montmorillonite (SCa-2)—a common clay mineral in the soil environment, the lattice OH features are found at 1.410 μm (assigned 2νOH, where νOH symbolizes the stretching vibration around 3630 cm^{-1}) and at 2.206 μm (assigned νOH + δOH where δOH symbolizes the bending vibration at around 915 cm^{-1}), whereas OH features of free water are found at 1.456 μm (assigned νW + 2δW, where νW symbolizes the stretching vibration at around 3420 cm^{-1}, and δW the bending vibration at around 1635 cm^{-1}), 1.910 μm (assigned νW + δW where νW symbolizes the high-frequency stretching vibration at around 3630 cm^{-1}) and 1.978 μm (assigned for νW + δW). Note that these assigned positions can change slightly from one smectite to the next, depending upon their chemical composition and surface activity. The spectra of three smectite endmembers are given in Figure 16.7, as follows: montmorillonite (dioctahedral, aluminous), nontronite (dioctahedral, ferruginous) and hectorite (trioctahedral, manganese). The OH absorption feature of the νOH + δOH in combination mode at around 2.2 μm is slightly but significantly shifted for each endmember. In highly enriched Al smectite (montmorillonite), the Al—OH bond is spectrally active at 2.16–2.17 μm. In highly enriched iron smectite (nontronite), the Fe—OH bond

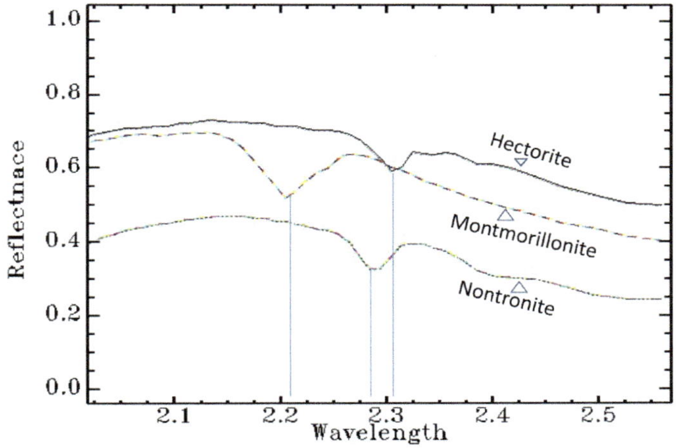

FIGURE 16.7 Reflectance spectra of three pure smectite endmembers in the SWIR region (nontronite= Fe-smectite; hectorite-Mg smectite; montmorillonite—Al-smectite). Note the different positions of combination modes of (υOH+δOH around 2.2 and 2.4 μm.

is active at 2.21–2.24 μm, and in highly enriched magnesium smectite (hectorite), the Mg—OH bond is spectrally active at 2.3 μm. Based on these wavelengths, Ben-Dor and Banin (1990a) were able to find a significant correlation between the absorbance values derived from the reflectance spectra and the total content of Al_2O_3, MgO and Fe_2O_3. Except for a significant lattice OH absorption feature at around 2.2 μm in smectite, invaluable information on OH in free water molecules can be measured μ at around 1.4 μm and 1.9 μm. Because smectite minerals contribute to the soil's relatively high specific surface area, which is covered by free and hydrated water molecules, these absorption features can be significant indicators of soil water content. Kaolinite and illite minerals are also spectrally active in the SWIR region as they both consist of octahedral OH sheets. In the case of kaolinite, a 1:1 mineral (one octahedral and one tetrahedral), the fraction of the OH group is higher than in 1:2 minerals (one octahedral and two tetrahedral), and therefore the lattice OH signals at around 1.4 μm and 2.2 μm are relatively strong, whereas the signal at 1.9 μm is very weak (because of relatively low surface area and adsorbed water molecules). In the case of gibbsite, an octahedral aluminum structure (1:0), the signal at 1.4 μm is even stronger, but that at 2.2 μm is shifted significantly to the IR region relative to kaolinite and presents an important diagnostic band at 2.265 μm. It should be noted that under relatively high SNR conditions, a second overtone feature of the structural OH (3υOH) can be observed at around 0.95 mm in OH-layer-bearing minerals as well (Goetz et al., 1991). The affinity of water molecules to clay mineral surfaces under the same atmospheric conditions is correlated to the minerals' specific surface area. For the aforementioned minerals, the specific surface area follows the order: smectite > vermiculite > illite > kaolinite > chlorite > gibbsite, and these usually provide a similar spectral sequence for the water-absorption feature near 1.8 μm (area and intensity). As smectite and kaolinite are often found in soils, they can also appear in a mixed-layer formation with spectral overlap (Kruse et al., 1991).

Carbonates: Carbonates, particularly calcite and dolomite, are found in soils that are formed from carbonic parent materials, or in a chemical environment that permits calcite and dolomite precipitation. Carbonates, and especially those of fine particle size, play a major role in many of the soil chemical processes that are most likely to occur in the root zone. The C–O bond, part of the CO_3 radical in carbonate, is the spectrally active chromophore. Hunt and Salisbury (1970, 1971) pointed out five major overtones and combination modes that describe the C–O bond in the SWIR region. Table 16.3 provides the band positions (calculated and observed from Gaffey, 1986) and their spectral assignments. In this table, υ1 accounts for the symmetric C–O stretching mode, υ2

TABLE 16.3
Band positions and assignments for calcite mineral (Taken from Gaffey 1988)

Band Position	Band Assignment
2.55 μm	$v_1 + 2v_3$
2.35 μm	$3v_3$
2.16 μm	$v_1 + 2v_3 + v_4$ or $3v_1 + 2v_4$
2.00 μm	$2v_1 + 2v_3$
1.90 um	$v_1 + 3v_3$
2.55 um	$2v_3$ + 270 + 2x416
2.37 μm	$2v_3$ + 270 + 3x416
2.54 μm	
2.533 μm	$2v_3 + v_1$
2.500 μm	"
2.330 μm	$3v_3$
2.300 μm	"

for the out-of-plane bending mode, υ_3 for the antisymmetric stretching mode, and υ_4 for the in-plane bending mode in the IR region. Gaffey (1986) added two additional significant bands centered at 2.23–2.27 μm (moderate) and at 1.75–1.80 μm (very weak), and Van der Meer (1995) summarized the seven possible calcite and dolomite absorption features with their spectral widths. It is evident that significant differences occur between the two minerals. This enabled Kruse et al. (1990), Ben-Dor and Kruse (1995) and others to differentiate between calcite and dolomite formations using airborne spectrometer data with 10-nm band widths. In addition to the seven major C–O bands, Gaffey and Reed (1987) were able to detect copper impurities in calcite minerals, as indicated by the broadband between 0.903 and 0.979 μm. However, such impurities are difficult to detect in soils because overlap with other strong chromophores may occur in this region. Gaffey (1985) showed that iron impurities in dolomite shift the carbonate's absorption bands toward longer wavelengths, whereas magnesium in calcite shifts the bands toward shorter wavelengths. As carbonates in soils are likely to be impure, it is only reasonable to expect that the carbonates' absorption-feature positions will differ slightly from one soil to the other.

Organic Matter: The wide spectral range found by different workers to assess OM content suggests that OM is an important chromophore across the entire spectral region. Figure 16.8 shows reflectance spectra of coarse OM (in the NIR—SWIR region) isolated from an Alfisol, and the humus compounds extracted from it. There are numerous absorption features that relate to the high number of functional groups in OM. These can all be spectrally explained by combination and vibration modes of organic functional groups (Chen and Inbar, 1994). Ben-Dor et al. (1997) referred the absorption peaks across the SWIR region to overtone and combination modes of nitrogen and carbon groups, and across the VIS regions to charge transfer in the remaining chlorophyll, as well as to other smeared electronic processes that affect the entire VIS spectral region, flatting the reflectance spectrum accordingly. Dry litter and fresh OM show many absorption features across the SWIR region related to many chemical chromophores, such as starch, cellulose, lignin and water.

Water: The various forms of water in soils are all active in the VNIR-SWIR region (based on the vibration activity of the OH group) and can be classified into three major categories: (1) *hydration water* which is incorporated into the lattice of the mineral (e.g., limonite ($Fe_2O_3.3H_2O$) and gypsum ($CaSO_4.4H_2O$), (2) *hygroscopic water* which is adsorbed on soil surface areas as a thin layer, and (3) *free water* which occupies the soil pores. Each of these categories influences the soil spectra differently, enabling identification of the water condition of the soil, and each is treated separately

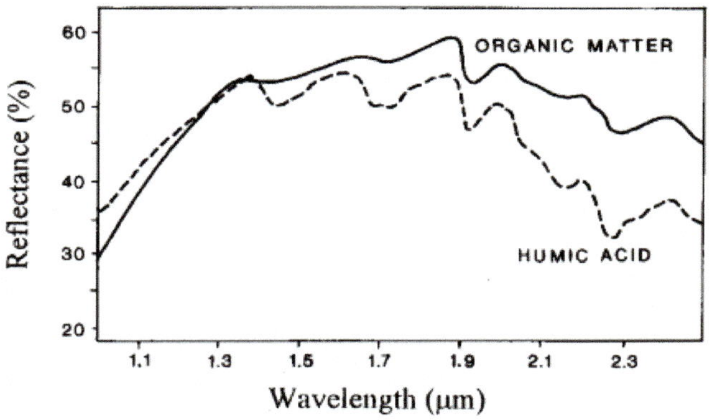

FIGURE 16.8 The reflectance spectra of soil organic matter and the humic acid extracted from it. (Taken from Ben-Dor E., et al., Soil Spectroscopy in Manual of Remote Sensing 3rd ed., A. Renz (ed.), John Wiley & Sons Inc. New York, 1999, pp 11–1989.)

in the sections that follow. Three basic fundamentals in the IR regions exist for water molecules, particularly the OH group: υwl-asymmetric stretching, δw bending and υw3-symmetric stretching vibrations. Theoretically, in a mixed system of water and minerals, combination modes of these vibrations can yield OH absorption features at around 0.95 μm (very weak), 1.2 μm (weak), 1.4 μm (strong) and 1.9 μm (very strong) related to 2υwl + υw3, υwl + υw3 + δw, υw3 + 2δw and υw3 + δw, respectively. The hydration water can be seen in minerals such as gypsum as strong OH absorption features at around 1.4 μm and 1.9 μm (Hunt et al., 1971b). However, free water changes the soil spectrum significantly as the real part of the refractive index is very dominant, causing a decrease in reflectance through the entire spectral range and consequently masking other possible features. This causes a problem in soil remote sensing when the soil is saturated or very wet.

Iron: This is the most abundant element on the Earth's surface and the fourth most abundant element in the Earth's crust (Dixon, 1989). Changes in its oxidation state, and consequently in its mobility, tend to occur under different soil conditions. The major Fe-bearing minerals in the Earth's crust are the mafic silicates, Fe-sulfides, carbonates, oxides, and smectite clay minerals. All Fe^{3+}-oxides have striking colors, ranging from red and yellow to brown, due to selective light absorption in the VIS range caused by transitions in the electron shell. It is well known that even a small amount of Fe-oxide can change the soil's color significantly. The red, brown and yellow "hue" values, all caused by iron, are widely used in soil-classification systems in almost all countries. The iron feature assignments in the VNIR region result from the electronic transition of iron cations (3 +, 2 +), either as the main constituent (as in Fe-oxides) or as impurities (as in iron smectite). Hunt et al. (1971a) summarized the physical mechanisms responsible for Fe^{2+} (ferrous) and Fe^{3+} (ferric) spectral activity in the VNIR region as follows: the ferrous ion typically produces a common band at around 1 μm due to the spin allowed during the transition between the E_g and T_{2g} quintet levels into which the D ground state splits into an octahedral crystal field. Other ferrous bands are produced by transitions from the $5T_{2g}$ to $3T_{1g}$ states at 0.55 μm, to $1A_{1g}$ at around 0.51 μm, to $3T_{2g}$ at 0.45 μm and to $3T_{1g}$ at 0.43 μm. For the ferric ion, the major bands produced in the spectrum are the result of the transition from the $6A_{1g}$ ground state to $4T_{1g}$ at 0.87 μm, $4T_{2g}$ at 0.7 μm and either $4A_{1g}$ or $4E_g$ at 0.4 μm.

Salts: Soil salts are reported to be Na_2CO_3, $NaHCO_3$ and NaCl, which are very soluble and mobile in the soil environment. In most cases, the spectra of these salts are featureless. However, indirect relationships with other chromophores can indicate their existence (e.g., OM, particle-size distribution). Hunt and Salisbury (1971c) reported an almost featureless spectrum of halite (NaCl 433B from

Kansas) whereas later, Farifteh et al. (2007) reported some features of the salt, mostly related to the adsorbed water as it is a hygroscopic material, and confirmed Hick and Russell's (1990) hypothesis that there are certain wavelengths across the VNIR—SWIR region that correlate to water features (see section 16.8.2.4). Mougenot et al. (1993) noted that in addition to an increase in reflectance with salt content, high salt content may mask ferric ion absorption in the VIS region. They concluded that salts are not easily identified in proportions below 10 or 15%. Salt is also visible in the VIS region due to its light tone, which reflects back radiation from the soil surface under dry conditions. This occurs mainly because the soluble salt migrates to the soil surface via capillary forces controlled by the evaporation process transporting water molecules from the soil body to the atmosphere.

To provide an overview of chemical chromophore activity in soils and to summarize this section, Figure 16.9 provides a spectrum from a Haploxeralf soil from Israel with the positions of all possible chromophores. Figure 16.10 provides six spectra of different soils from Israel consisting of different chromophores content as illustrates in Figure 16.9 across each spectral region segments. Figure 16.11 summarizes the chemical chromophores associated with soil and geological matter as collected from the literature and summarized by Ben-Dor et al. (1999). It also lists the intensities

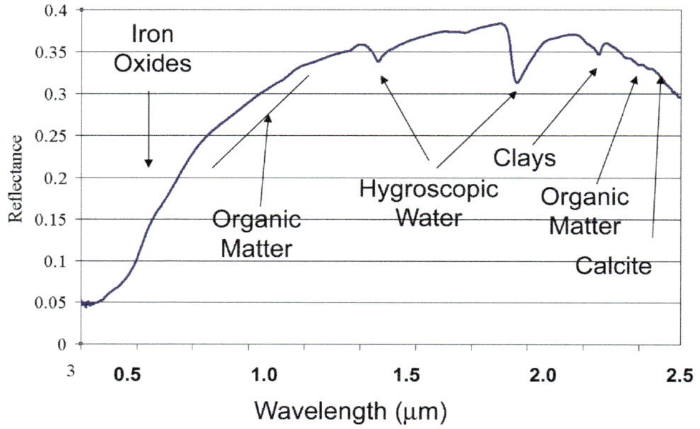

FIGURE 16.9 A representative spectrum of semiarid soil from Israel (Haploxeralf) with all possible direct chromophores associated with soil.

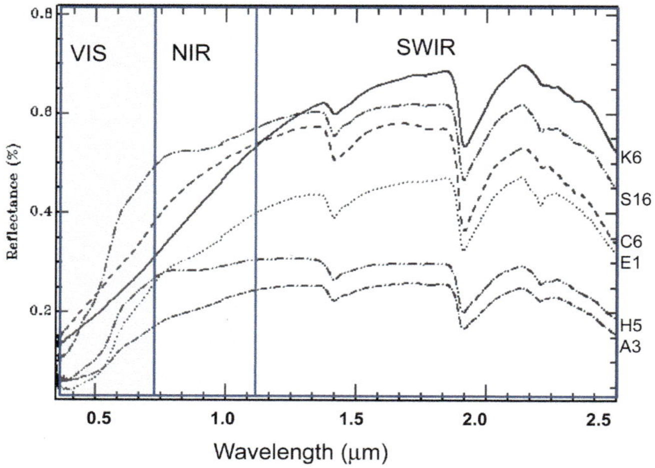

FIGURE 16.10 Reflectance spectra of six representative soil from Reflectance spectral if six representative soils from arid and semiarid areas from Israel (E1 and A3 Rhodoxeral, H5-Xerent, C6-Xerothent, S16-Thorriothent, K6-Caliorthid).

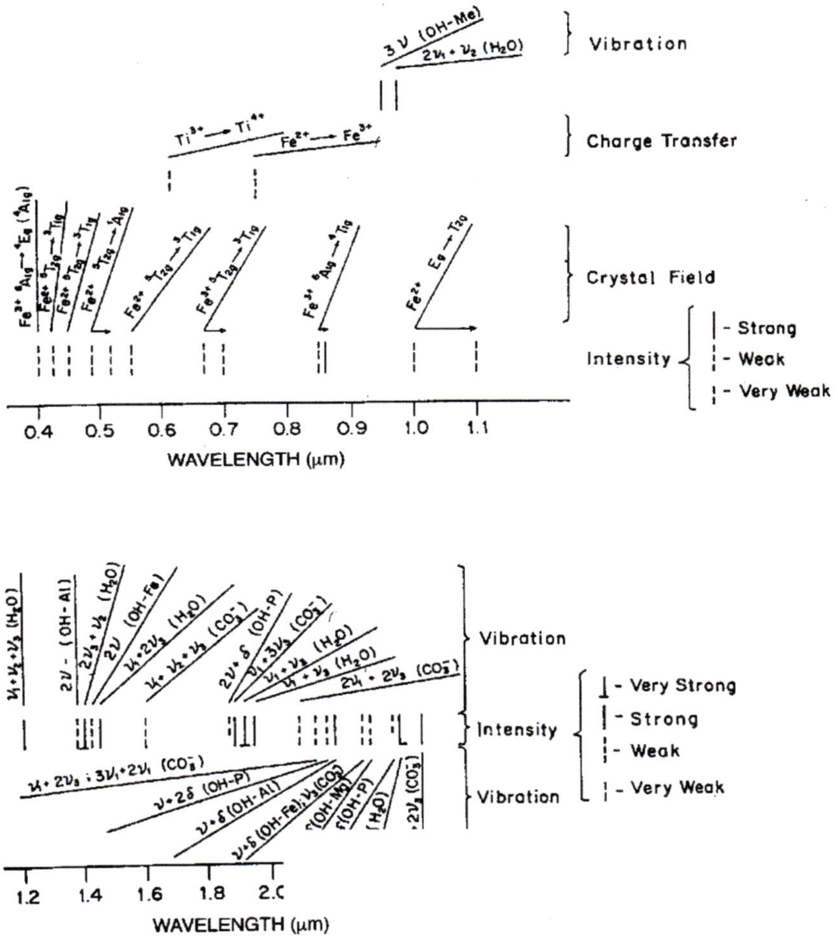

FIGURE 16.11 The active groups of the soil chromophores spectrum. For each possible mechanism, possible range and absorption feature intensity is given. The spectrum was generated using information presented in the literature. (Taken from Ben-Dor E. et al., Soil Spectroscopy, in Manual of Remote Sensing, 3rd ed., A. Rencz (ed.), John Wiley & Sons Inc., New York 1999, pp. 111–189.)

of each chromophore in the VNIR—SWIR spectral regions as they appear in those studies. The current review demonstrates that high-resolution spectral data can provide additional, sometimes quantitative information on soil properties that are strongly correlated with the chromophores, that is, primary and secondary minerals, OM, Fe-oxides, water and salt. It also demonstrates the importance of soil spectroscopy in designing a sensor for a soil mission and selecting the proper tools to interpret the results acquired by remote-sensing means using solar radiation.

16.6 RADIATION SOURCE AND ATMOSPHERIC WINDOWS

16.6.1 General

To acquire the chromophores by remote-sensing means, the radiation source and the medium through which it travels (the atmosphere) must be investigated. The Planck function is a physical expression describing the energy emitted from a black body. The Sun, as an ideal black body, is the main radiation source for remote sensing of the Earth across the VNIR-SWIR region. If a sensor is located far from the soil (air or space), radiation must travel from the source to the object and back to the sensor, thus crossing the atmosphere twice. The gasses and aerosols in the atmosphere interact with the radiation across this path and hinder soil reflectance at certain frequencies. Thus, the components in the atmosphere are spectrally active. This interaction has to be minimized as much as possible to obtain a signal from the soil. This can be done in two ways: 1) allocating the sensor bands across the high-transition spectral region of the atmosphere (known as atmospheric windows) or 2) determining the physical interaction of the radiation with the known atmospheric component using a physical calculation (known as "radiative transfer model"). Whereas in multi- and superspectral sensors, the spectral bands are usually located across atmospheric windows, in hyperspectral sensors this is not possible, as the bands cover the entire spectral region, and thus use of the radiative transfer model is called for to extract the soil reflectance. Masking the atmospheric attenuation from the sensor's radiance is termed "atmospheric correction." Figure 16.12 illustrates the spectral regions under which atmospheric attenuation can affect the soil spectrum. This figure shows the reflectance spectrum of an E-7 soil from Israel (Haploxeralf, taken from Tel Aviv University's spectral library) overlain on its simulated (soil) radiance as calculated by MODTRAN. The latter is normalized to the Planck Sun function at the top of the atmosphere to illustrate only the atmosphere transmittance. As the atmospheric attenuation remains, the most affected spectral regions can be clearly seen. The VIS region is affected by aerosol scattering (monotonous decay from 0.4 to 0.8 μm) and absorption of ozone (around 0.6 μm), water vapor (0.73 and 0.82 μm) and oxygen (0.76 μm). The NIR—SWIR regions are affected by absorption of water vapor (0.94 μm, 1.14 μm, 1.38 μm, 1.88 μm), oxygen (at around 1.3 μm), carbon dioxide (at around 1.56 μm, 2.01 μm, 2.08 μm) and methane (2.35 μm). Also seen are the absorption peaks of the soil chromophores at 2.33 μm (carbonates), 2.2 μm (clay), 1.9 and 1.4 μm (hygroscopic water), and 0.5, 0.6, and 0.9 μm (Fe-oxides) that overlap with

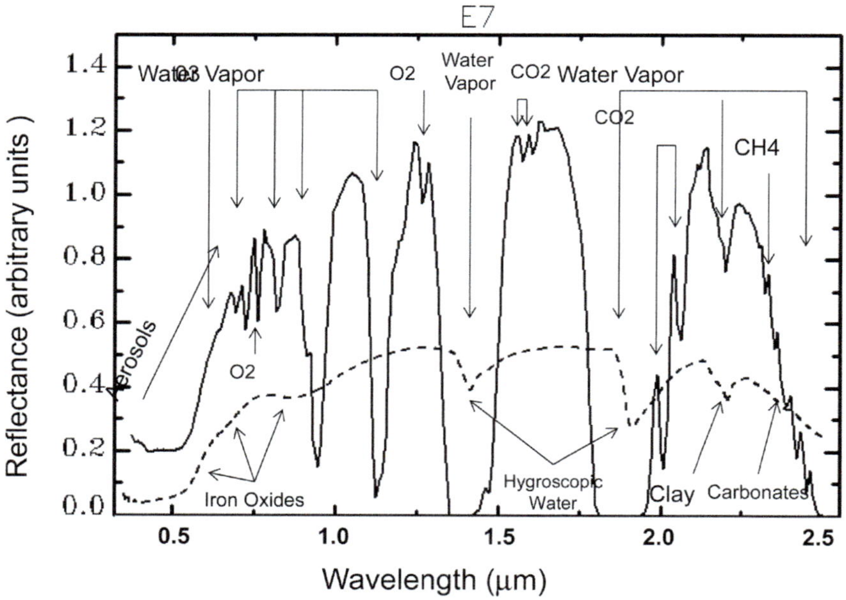

FIGURE 16.12 The reflectance spectrum of a typical soil from a semiarid environment (dotted line) and its modeled radiance using MODTRAN (solid line). The atmospheric attenuation of the soil chromophores is specified.

the above atmospheric chromophores. It can be seen that the most informative region for the soil has some overlap with the atmosphere (water vapor at 1.4 and 1.9 μm, oxygen at 0.76 μm) and if we do not allocate our spectral bands in this area, information about some soil attributes may be lost (moisture, Fe-oxides and OM).

Figure 16.12 provides also the radiation observed from a given pixel composed of the source (Sun), the atmospheric transition and the soil reflectance. The mixed radiation maintains a high response to the solar radiation source, followed by the atmospheric attenuation and lastly, the soil response. Also seen is that the energy at the end of the SWIR region (2–2.5μm) is low and may affect the quality of the information from this region based on the low SNR in this region due to the low radiation flux. It is hence demonstrated that the intensity of the soil target within the radiance observed by a sensor is quite small (>5%) and hence needs to be carefully isolated from the total radiance information at the sensor level. The atmospheric components are: water vapor (at 0.68, 0.94, 1.12, 1.4, 1.9 μm), oxygen (0.76 and 1.3 μm) and CO_2 (2.105 and 2.015 μm). Ozone also sometimes plays a role in the VIS region at around 0.45—(0.50 μm. Aerosol is also part of the atmosphere, and its scattering effect on the radiation is mostly in the VIS region (0.4–0.8 μm). The Sun's radiation interacts with the atmospheric molecules (known as *Rayleigh* absorbance and scattering) and with particles (known as Mie scattering). As previously discussed, the preceding components can be physically described by the radiative transfer equations that aim to remove the atmospheric effects, leaving the reflected information from the soil only. If the soil reflectance information can be extracted without artifacts from the sensor radiance, those soil chromophores may be useful for either qualitative or quantitative spectral utilization. A comprehensive description of methods to remove the atmospheric attenuations is given by Ben-Dor et al. (2002) and Gao et al. (2006, 2009). As already mentioned, even weak spectral features in the soil spectrum can contain very useful information. Therefore, great caution must be taken before applying any quantitative models to soil reflectance spectra derived from airborne or spaceborne hyperchannel sensors. Validation of the (atmospherically) corrected data is an essential step in ensuring that the reflectance spectrum contains reliable soil information. This section shows that atmospheric attenuation plays a major role in the final soil spectral products as the soil contributes less than 5% to the overall energy acquired by the sensor's detector.

16.6.2 Factors Affecting Soil Reflectance in Remote Sensing

Many factors can hinder the spectral information from a given soil sample or pixel. Residual atmospheric attenuation, mainly across a highly active atmospheric spectral region, may be the first to strongly affect the soil spectrum. If not effectively removed, these residuals can contribute to spectral noise signals that are not related to soil. In many studies, some spectral ranges are ignored based on strong absorption of atmospheric constituents (e.g., 1.4 and 1.9 μm due to water vapor). In the multispectral sensor, this interference is minimal as most of the bands are allocated across atmospheric windows. In this case, however, a more continuous spectral effect, such as that obtain from aerosol scattering, may affect the soil reflectance, mostly in the VIS region. Lagacherie et al. (2008) presented a comparison of field, air, and space reflectance of soils, showing differences in the spectral information mostly due to different SNR values (see discussion that follows) and the atmospheric residual in the corrected data. The soil surface is exposed to many natural effects, such as rain, erosion and deposition, as well as fire and cultivation. These effects also play an important role in the accurate identification of a soil entity. This is mainly because such effects can hinder the real soil chromophores' interactions with solar radiation. In this case, even a thin layer on the soil surface can make a difference. In this respect moisture, fire, dust, crust and roughness play a major role in the sensor's final response to the soil at any spectral resolution. The sensor quality also has an important effect on the soil spectrum. Airborne and spaceborne sensors vary in their SNR values, which are mostly lower than those used in the field and laboratory. A noisy spectral signature cannot be further analyzed as in many cases, it cannot be used for quantitative approaches. In the laboratory, soil-reflectance measurements are performed under controlled conditions and thus it is possible to standardize all spectral measurements and compare spectral libraries between users to allow a robust analysis (Piemstein et al., 2010). In the field, however, this standardization is not yet achievable and spectra may vary from one measurement to the next, mostly due to physical factors (e.g., viewing and illumination angles, particle-size and roughness distributions). Although correlations have been found between spectral responses of soils in the laboratory and field (e.g., Stevens et al., 2008), this is valid only for a select database and cannot be applied to sensor data from other areas. Some analytical manipulations (e.g., partial least squares regression [PLSR]) can decrease the variation between the two measurement domains, but upscaling the laboratory models to airborne data is still difficult. This problem calls for a common protocol and well agreed-upon measurement scheme in the field, since it is the most relevant condition for remote sensors, and quantitative models are generated for that domain. In the field, reflectance measurements are fraught not only with variations in viewing angle and changes in illumination, but also with variations in soil roughness, soil moisture and soil sealing (Ben-Dor et al., 1997). Knadel et al. (2014) studied the spectral changes in four representative soils from Denmark with water retention ranging from wet to dry. Soil reflectance was found to decrease systematically, albeit not proportionally, with decreasing matric potential and increasing molecular layers. The changes in molecular layers were best captured by the soil reflectance of clay-rich soils, demonstrating the importance of an inter-correlative approach to the quantitative analysis of soil water content under dry and semi-dry conditions, as with other soil attributes. A recent paper by Knadel et al. (2023) summarizes the physical effect of water molecules on the soil spectrum and provides the pros and cons of the recent methods to deal with water attenuation of the soil reflectance. These studies demonstrated the problem of the water's influence on soil spectra as discussed in Section 25.5.3.4.1

Acquiring soil-reflectance data from air and space involves additional difficulties, such as homogeneity of the area being observed and problems on how to represent the sensor's pixel by ground-truth measurements, as the pixel size of the sensor is a major factor affecting the soil's spectral response. Hengl (2006) provided analytical and empirical rules for selecting the proper pixel size for a particular mission. He concluded that there is no ideal grid resolution; it has to be determined according to the property in question. For soil, this is a crucial element. If salinity is important, then the ground truth needs to be taken at high spatial distance using a high-spatial-resolution sensor. Using sensors that are not adequate for the mission is a waste of resources and time. In this regard, the grid selected for the question being asked must maximize the predictive capabilities or the informational content of the final processed map. Sometimes, the pixel size plays a more dominant role than the SNR values. Asner and Heidebrecht (2003) demonstrated lower accuracy in sensing bare soil areas from the Hyperion (30 m, low SNR) than from the Airborne Visible/Infrared Imaging Spectrometer (AVIRIS; 4.5 m, high SNR), but convolving the AVIRIS spectra into Hyperion's 30-m grid gave similar accuracy. Lagacherie et al. (2008) examined the performance of clay and calcium-carbonate ($CaCO_3$) estimations using laboratory (ASD) vs. airborne (HyMap) spectral scales. A significant decrease was observed going from laboratory to field and then to airborne domains. They indicated that the main factors inducing the uncertainties were the radiometric- and wavelength-calibration uncertainties of the HyMap sensor and possible residual atmospheric effects. In general, to represent a pixel area of a given sensor on the ground, it is necessary to set a ground truth of the given pixel at 4 pixels. This means that the reflectance property of this grid has to be measured in several locations along the 4x4 GDS area and then averaged to yield the "pixel spectral response." If the pixel size is large (such as in Landsat 8, 30 m), then a large (120 × 120 m) area is the minimal area required for the ground-validation mission. As soil surface conditions may also affect the field spectrum based on the previously mentioned factors, a representative "pixel" area must be fully covered by both ground-reflectance measurements and sampling. Soil sampling must be planned according to a statistical (or other logical) framework (McKenzie and Ryan, 1999), and soil sampling should cover the 0- to 1-cm layer as much as possible (this is a compromise between what the sensor sees and what can be sampled under real field conditions). As this sampling is not easy, mis-sampling of soil can occur and hinder the classification results and accuracy based on laboratory analysis of the soil samples. As previously discussed, laboratory-based measurements provide an understanding of the chemical and physical properties of the soil reflectance. If the soil sample is not well represented, the spectral-based models or indices from the laboratory will not work for the field. This makes the models non-robust and the transformation from one sensor to another almost impossible. As the standardization process for field reflectance measurements is under construction by the P4005 WG and demonstrates a promising direction for the harmonization of field SSLs and between these and laboratory SSLs (Ben-Dor et al., 2023), the ground-truth measurement problem can be solved if one selected protocol is used by all. Another problem is vegetation (or litter), termed biospheric interference, which covers many soil surfaces worldwide and hinders direct sensing of the soil. Within the non-vegetated area, only a portion of the soils are characterized by an unaltered surface layer (e.g., as a consequence of soil tillage), and partial sensing of the natural soil surface can be interpolated into the vegetated area (Ben-Dor et al., 2002). A partial solution for this is "inferring soil properties through vegetation" (Huete, 2005) or the connection of vegetation to soil properties (e.g., Maestre and Cortina, 2002) where a study by Kopačková et al. (2014) demonstrated how this method can shed light on the information obtained from the root zone, mainly for heavy-metal content. However, it seems that this method cannot (yet) be used to obtain information on other important soil parameters, such as clay, silt and clay, OM and $CaCO_3$, and thus it is still limited.

Where spectral unmixing is often used to account for the biogenic fraction (Asner and Heidebrecht, 2002; Robichaud et al., 2007), a new and interesting approach has been carried out by Demattê et al. (2020) and independently by (Heiden et al., 2021) These authors developed a time-series method that uses a long-term database (e.g., Landsat or Sentinel 2) to select pure soil pixels using photosynthetic vegetation (PV) and non PV (NPV) thresholds over time. The synthetic image consists of the maximal number of pure soil pixels that can be used to implement the proxy models. The problem with such a technique is that the soil-surface properties may change from one season to the next, and the model may therefore not represent the exact situation on the date required by the end user.

Radiation intensity on the soil pixel can also change the reflectance properties from a topographic view, but can be corrected geometrically. A more difficult problem is the bidirectional reflectance distribution function (BRDF). This function assumes that the radiation source, the target and the sensor are all points in the measurement space, and that the ratio calculated between absolute values of radiance and irradiance is strongly dependent upon the geometry of their positions. Theories and models explaining the BRDF phenomenon in relation to soil components have been widely discussed and covered in the literature (Hapke 1981a, 1981b, 1984, 1993; Pinty et al., 1989; Jacquemoud et al., 1992; Liang and Townshed, 1996). A number of models have been developed which express soil bidirectional reflectance as a function of illumination and viewing direction. They cannot, however, be inverted to directly estimate soil properties on the basis of bidirectional reflectance observations, nor can the equations be used to predict reflectance distributions on the basis of soil-property measurements in the field. Thus the BRDF effects, although well studied, still play a major role in the final output of the soil spectral response (from all domains), and more research into this phenomenon is warranted. The soil spectral response can therefore be affected by the aforementioned factors in two ways: in its physical behavior, that is, spectral baseline position (e.g., in the case of the BRDF effect), and in its chemical behavior, that is, new or absence of spectral features (e.g., in the case of atmospheric attenuation).

TABLE 16.4
Factors attenuate soil reflectance quality from remote sensing domains in different systems and spectral regions

Spectral Range and sensing systems	Factors affecting Reflectance Retrieval from RS sensors						Effect on Soil surface chromophores (e.g. dust, water, crust)
	Atmosphere	SNR	Sensor stability	Radiometric calibration	BRDF	Spatial resolution	
VNIR-multi	+	++	++++	++	+++	++	+++
VNIR-super	++	+++	++++	+++	+++	+++	+++
VNIR-Hyper	++++	++++	++++	+++++	++++	++++	++++++
SWIR- multi	++	+++	+++++	+++	++++	+++	++++
SWIR-super	+++	+++++	+++++	++++	++++	++++	++++
SWIR-Hyper	++++	++++++	+++++++	++++++	++++++	++++++	+++++++

+ low
++ moderate
+++ average
++++ high
+++++ very high

In summary, the spectral information from a given soil pixel (from either remote-sensing means or ground-truth measurements) may be affected by a variety of factors that can change the soil spectral response significantly. This calls for user caution in upscaling data from the laboratory to the field, in analyzing the data, and in exchanging data. This issue was recently studied by Francos and Ben-Dor (2022), who used a SoilPRO® apparatus to extract the soil surface reflectance in the field, matching that seen by remote-sensing means. Ben-Dor et al. (2023) showed that a transition factor between laboratory and field SSLs can be found. Table 16.4 summarizes the factors that can affect retrieval of accurate soil reflectance from above on a qualitative scale. The SWIR region is seen to be more sensitive to these factors, along with the hyperspectral systems. This illustrates that the hyperspectral sensors are the most sensitive systems for soil remote sensing. In other words, it can be said that if all of the preceding factors remain small, then HSR can be the most sensitive way to remote sense the soil system.

16.7 QUANTITATIVE ASPECTS OF SOIL SPECTROSCOPY

16.7.1 Proximal Sensing

Proximal sensing refers to the quantitative information on soil attributes that is mined from the soil-reflectance data. Today, quantitative soil spectroscopy is a mature discipline which has come quite a long way since the mid-1960s, when Bowers and Hanks (1965) published their paper on the correlation between soil reflectance and soil moisture content. That pioneering study, followed by a series of papers by Hunt and Salisbury (1970–1980) and Hunt (1982), proved that water and minerals in the soil environment have unique spectral fingerprints that can be further used for specific recognition. Learning from several sectors' successes (e.g., food science, tobacco, textile), Dalal and Henry (1986) applied the proximal-sensing (proxy) approach to soils. In 1990, Ben-Dor and Banin demonstrated the power of reflectance spectroscopy in accounting for $CaCO_3$ content in the soil (Ben-Dor and Banin, 1990a) and later, in monitoring the structural composition of smectite soil minerals in the laboratory (Ben-Dor and Banin, 1990b). Later still, when portable field spectrometers were introduced to the market (in around 1993), more scientists realized the potential of soil spectroscopy and consequently, more spectral libraries were assembled from local and regional scales (e.g., Shepherd and Walsh, 2002; Bellinaso et al., 2010, LUCAS, BZZL). A comprehensive summary of the quantitative applications of soil reflectance spectroscopy was provided first by Ben-Dor et al. (2002), Malley et al. (2004), Viscarra Rossel (2009) and Nocita et al. (2015), and later by Poppiel et al. (2022).

In April 2009, a world soil spectroscopy group was established by Viscarra Rossel (http://groups.google.com/group/soil-spectroscopy), who gathered soil spectra and corresponding attributes from more than 80 countries worldwide to generate a global soil spectral/attribute database, providing proxy capability to all. This initiative was based on the idea that since the quantitative approach in soil sciences had become well-established and applicable, it should be more collaborative. This was the obvious step after understanding that only sharing information would help advance quantitative soil spectroscopy (e.g., Condit, 1970; Shepherd and Walsh, 2002; Brown et al., 2006). This pioneering step promoted more activity in this direction, where many local and regional SSLs are gathered in one laboratory at the University of Sao Paolo, providing data, but also models to users based on this comprehensive world SSL (Demattê, 2023, Personal Communication).

Comprehensive reviews on proxy applications for soils can be found in Malley et al. (2004) and Viscarra Rossel et al. (2006), and other important reviews focusing on soil-reflectance theory and applications can be found in Clark and Roush (1984), Irons et al. (1989), Ben-Dor et al. (1999), and Ben-Dor (2002). This activity is profiting from progress in the data sciences and AI which is increasing the number of applications and their accuracy. The number of national and international SSLs is growing, constituting a database for spectral-based model generation for the proximal-sensing strategy.

The merging or comparison of SSLs is impeded by various factors, primarily associated with instrumentation and measuring protocols. This complicates the accurate modeling of local, regional and global soil properties, thereby undermining the technology's potential for robustness and reliability. This issue has become even more critical with the use of soil spectroscopy from satellite platforms, necessitating the consolidation and harmonization of many SSLs worldwide. In response to these challenges, dedicated scientists have united their efforts in the IEEE SA P4005 WG entitled "Standards and Protocols for Soil Spectroscopy." (https://sagroups.ieee.org/4005/). The group's primary objective is to explore how the diverse range of protocols, sensors and measurement methods can be effectively unified into a standardized protocol, enabling the harmonization of SSLs. One of the main goals of this initiative is to produce a protocol to measure spectral information in the field in order to align laboratory SSLs with remote-sensing data. Ben-Dor et al. (2009) summarized the quantitative utilization of soil spectroscopy from airborne domains. They pointed out that although two communities are utilizing soil spectroscopy—soil scientists and the remote-sensing communities—there is a lack of consistency between the terminologies used by these two groups, leading to potential misunderstandings. The soil scientists define soil proximal sensing across the 400–2500 nm region as the VNIR region (as adopted from other disciplines, such as the food sciences), whereas the remote-sensing community refers to this exact spectral region as the VNIR—SWIR region (see Section 25.5.3.2). As no common agreement has yet been reached, we suggest using the terms VIS/NIR/SWIR/thermal infrared (TIR) for all soil proxy analyses to emphasize the spectral regions (and accordingly, the chromophores) in which the analysis has been done for better performance. As more and more medium- and high-spectral-resolution sensors become available in orbit, with yet others planned for the near future, a new strategy to compose SSLs is in effect. The spectral signature is extracted from the sensor directly from a pixel (after removing atmospheric attenuation) that is sampled and chemically analyzed in the laboratory. The models are then executed on the original satellite spectral data for better analytical accuracy.

16.7.2 Application Notes

In general, soil reflectance spectra are directly affected by chemical and physical chromophores, as already discussed. The spectral response is also a product of the interaction between these parameters, calling for a precise understanding of all chemical and physical reactions in soils. For example, even in a simple mixture of Fe-oxides, clay and OM, the spectral response cannot be judged simply by linear mixing models of the three endmembers. Strong chemical interactions between these components are, in most cases, nonlinear and rather complex. For instance, organic components, mostly humus, affect soil clay minerals in chemical and physical ways. Similarly, free Fe-oxides may coat soil particles and mask photons that interact with the real mineral components or the Fe-oxides themselves (and OM as well) (Heller Pearlshtien and Ben-Dor, 2020). In addition, the coating material may cohere fine particles into coarse aggregates that may physically change the soil's spectral behavior from a physical standpoint. In a recent study, Ben-Dor et al. (2022) demonstrated how micro aggregation of soils is governed by the contents of OM, Fe-oxides and clay, species which can all be captured by spectroscopy and proximal sensing. McBratney et al. (2003, 2006) have also shed light on this technology through their pioneering work over the years. Brown et al. (2006) concluded that use of the spectral proxy technique for soils has the potential to replace or enhance standard soil-characterization techniques, basing their conclusion on 3,768 soil samples from the USA; this was later confirmed by Nocita et al. (2015). As noted earlier, in view of the growing soil spectral community, Viscarra Rossel (2009) generated an initiative (Soil World Spectral Group, http://groups.google.com/group/soil-spectroscopy) in which all members of the soil spectral community were asked to contribute their local spectral library to generate a worldwide library that would be accessible to all. This pioneering world SSL and its applications were published by Viscarra Rossel et al. (2016). Besides being the first attempt to gather spectral information on the world's soils, this initiative was an important step toward more such activity as realized by The Brazilian Soil Spectral Service (Braspecs), Demattê et al. (2022). The authors gathered to create a cloud based service where the user uploads a spectra and receives the soil analysis. Despite this, the data was disposed of in a repository (Demattê et al., 2023) and published the history of all systems (Novais et al., 2024). Based on that, a new on-line global domain service gathered attention (e.g., a world spectral service by Demattê et al., (in peroration). As already discussed, this has led to the need for standard protocols and quality indicators that are accepted by all members of the growing soil-spectral community, as now being established by the IEEE SA P4005 WG.

The quantitative option and the availability of field spectrometers enabled users to show that the soil-spectral-based technology can be used in the field. Some key works (among many that have been published) are mentioned here. Genú and Demattê (2006) evaluated 3,300 samples using the multiple method with spectral proxy analysis and reached an R^2 of 0.74 and 0.53 for clay and OM, respectively. Demattê et al. (2009b) evaluated 1,000 samples and determined $R^2 = 0.85$ for clay using a laboratory spectrometer. Nanni and Demattê (2006) suggested a reflectance inflection difference (RID) index, representing the difference between reflectance values at the highest and lowest points of inflection (or amplitude of spectral data in this range—demonstrating the height of the curve between the peak and the valley). This approach led them to use only some of the spectral information from the overall 400–2500 nm range. Using multiple stepwise statistics, they obtained $R^2 = 0.91$ and 0.89 for clay and OM, respectively. Fiorio and Demattê (2009) obtained $R^2 = 0.83$ and 0.30 for clay and OM, respectively, also with ground spectra analyzing 450 samples. Viscarra Rossel and Webster (2011) mapped the distribution of the spectral proxy approach's information from Australian soils. They concluded that the technique can provide integrative measures of soil properties and can act as an alternative to the conventional analytical method that can be effectively applied for both soil classification and environmental monitoring. As an example from Israel, Schwartz et al. (2012) evaluated the total petroleum hydrocarbon (TPH) content in the soil in a field. The technique has also reached the precision agriculture discipline, as demonstrated by several workers who used it to assess important soil attributes in the field (e.g., recent research by Debaene et al. (2014) on clay and carbon, Araújo et al. (2013) with cation-exchange capacity and pH, among others, and Barnes et al. (2003) with OM and electrical conductivity). A new assembly to measure reflectance spectra in the field was recently developed by Ben-Dor et al. (2017). This assembly, termed SoilPRO®, enables extracting a soil's surface reflectance without disturbing the soil surface and thus can provide information about the soil—atmosphere contact film. Francos and Ben-Dor (2022) demonstrated that water-infiltration rate into the soil profile, which is governed by soil-surface conditions, can be estimated spectrally with the SoilPRO® assembly. Ben-Dor et al. (2023) also demonstrated that soil-surface hydrophobicity can be estimated in the field using this assembly. As the SoilPRO® assembly is simple to use and can provide the soil-surface condition at any given timepoint, it may be used as an ultimate means to spectrally describe the undisturbed soil-surface condition as seen by remote sensors.

16.7.3 Constraints and Cautions in Using Proximal Remote Sensing for Soils

In reality, the number of chromophores in soils are quite limited relative to the soil's attributes. The factors affecting the soil spectrum (see Sections 16.7 and 16.8) also hinder the proxy analysis. In addition, there is no simple correlation between the spectroscopy and the chromophore content, as it requires a sophisticated data-mining approach with significant validation tests. For example, using the spectral information for quantitative analysis, Karmanova (1981) selectively removed Fe-oxides from soil samples and concluded that the effects of various iron compounds on the spectral reflectance and color of soils were not proportional to their relative contents. Another aspect is that the proxy models are not always robust and may be related to the soil population in the analysis. In fact Araújo et al. (2014) clustered 7,125 soil samples in relation with their mineralogy, and gathered much stronger models than the global one provides. As pointed out by Bedidi et al. (1990, 1991), the normally accepted view of decreasing soil-baseline height with increasing moisture content (VIS region) does not hold for lateritic (highly leached, low pH) soils. They concluded that the spectral behavior of such soils under various moisture conditions is more complex than originally thought. In this context, Galvão et al. (1995) showed spectra from laterite soils (VNIR region) consisting of complex spectral features that appeared to deviate from those of other soils. Al-Abbas et al. (1972) found a correlation between clay content and reflectance data in the VNIR—SWIR region and suggested that this is not a direct but rather an indirect relationship, strongly controlled by the OM chromophore. Another anomaly related to the interactions between soil chromophores was identified by Gerbermann and Neher (1979). They carefully measured the reflectance properties in the VIS region of a clay—sand mixture extracted from the upper horizon of a montmorillonite soil and found that "adding of sand to a clay soil decreases the percent of soil reflectance." This observation is in contrast with the traditional expectation from adding coarse (sand) to fine (clay) particles in a mixture (soil), that is, that this will tend to increase soil reflectance. Likewise, Ben-Dor and Banin (1994, 1995a, 1995b, 1995c) concluded that inter-correlations between feature and featureless properties play a major role in assessing unexpected information about soil solely from their reflectance spectra in either the VNIR or SWIR regions. Ben-Dor and Banin (1995b) examined arid and semiarid soils from Israel and showed that "featureless" soil properties (i.e., properties without direct chromophores such as K_20, total SiO_2, and Al_2O_3) can be predicted from the reflectance curves due to their strong correlation with "feature" soil properties (i.e., properties with direct chromophores). Csilage et al. (1993) best described the effect of multiple factors indirectly affecting soil spectra in their discussion on soil salinity, which can be considered a featureless property. They stated that "salinity is a complex phenomenon and therefore variation in the [soil] reflectance spectra cannot be attributed to a single [chromophoric] soil property." To get the most out of soil spectra, they examined the chromophoric properties of OM and clay content, among others, and ran a principal component analysis to fully account for the salinity status culled from the soil reflectance spectra. With the significant progress in the data sciences and in statistical empirical approaches to mining data, soil spectral data can provide more information than ever before. New data-modeling approaches other than multiple linear regression (MLR) or PLSR are valid, with a huge number of samples and big data approaches also progressing and being gathered into a new discipline termed machine learning (ML) or Deep Learning (DL). These are specialized technologies that fall under the umbrella of AI, which is the driver behind many modern technologies.

The difference between data acquired from orbit and those from the laboratory may hinder the transfer of proxy models from the laboratory to orbital domains. This problem can be solved by generating models based on field conditions and measurements and using a better method to upscale laboratory data to the orbital domain (e.g., Francos and Ben-Dor, 2022). Another solution is to extract the spectrum directly from the satellite sensor and build the model based on these data along with chemical and physical data from the ground. The problem with this method is to find a pure soil pixel and then describe it, assuming that the surrounding pixels may affect the extracted spectrum.

16.8 SOIL REFLECTANCE AND REMOTE SENSING

16.8.1 General

A short review of remote sensing of soils from an optical perspective was published by Ge et al. (2011). A comprehensive description of soil spectral remote sensing can be found in Ben-Dor et al. (2008) and later, in Wolf et al. (2015) and Dwivedi (2017). Many studies have been conducted to classify soils and their properties using multispectral optical sensors onboard orbital satellites, starting with the original sensors such as Landsat MSS and TM, SPOT and NOAA AVHRR (e.g., Cipra et al., 1980; Frazier and Cheng, 1989; Kierein-Young and Kruse, 1989; Agdu et al., 1990; Dobos et al., 2001) and ending with the recent Sentinel 2 (e.g., Li and Chen (2014) and Landsat 8. Qualitative classification approaches have traditionally been applied to multichannel data in cases of limited spectral information, but new analysis techniques are enabling a more quantitative view (Vizzari, 2022). The quantitative approach has progressed with the use of hyperspectral data (Vane, 1993), first from airborne sensors and more recently from satellites. Important qualitative, as well as quantitative information has been obtained on SOM, soil degradation and soil conditions (Price, 1990; Ben-Dor and Banin, 1995a, Metternicht et al., 2010, Dvoakova et al., 2021). Huete (2004) summarized some remote-sensing applications of soils (using different sensors and resolutions) and discussed how "properties controlled soil reflectance." His excellent overview of soil remote sensing applications was strongly tied to the relationship between green vegetation, litter and soils as based on the fact that most of the soils are altered by these substances. Later, Huete (2005) has provided more ideas on how to use the spectral information from hyperspectral domains to measure bare soils, mixed soil-vegetation-litter, and overlaying vegetation pixels in order to investigate the soil properties. As the remote-sensing technology progresses, soil spectra are becoming an important vehicle for the remote sensing of soils, and spectral libraries are being established to cover vast geographical areas worldwide (e.g., Latz et al., 1981; Price, 1995; Viscarra Rossel, 2009; Montanarella et al., 2011; Orgiazzi et al., 2018; Aikenhead and Black, 2018; Dematte et al., 2019). Although use of these libraries has many limitations, it is understood that the spectral domain is very important for soil mapping and that effort has to be invested in super- and hyperspectral sensors. Over the past 40 years, HSR has developed to a point where it is now in high demand by many users (Ben-Dor et al., 2013; Dematte et al., 2015; Yu et al., 2020). HSR technology provides high-spectral-resolution data with the aim of giving near-laboratory-quality reflectance or emittance information for each individual picture element (pixel) from far or near distances (Vane et al., 1984). This information enables the identification of objects based on the spectral absorption features of their chromophores and has found many uses in terrestrial and marine applications (Clark and Roush, 1984; Vane et al., 1984; Dekker et al., 2001). Figure 16.13 illustrates this concept, where the spectral information of a given pixel shows a new dimension that cannot be obtained by traditional point spectroscopy, air photography or other multiband images. HSR can thus be described as an "expert" geographical information system (GIS) in which layers are built on a pixel-by-pixel basis, rather than with a selected group of points (McBratney et al., 2003). This enables spatial recognition of the phenomenon in question with a precise spatial view and use of the traditional GIS-interpolation technique in precise thematic images. Since the spatial—spectral-based view may provide better information than viewing either the spatial or spectral views separately, imaging spectroscopy serves as

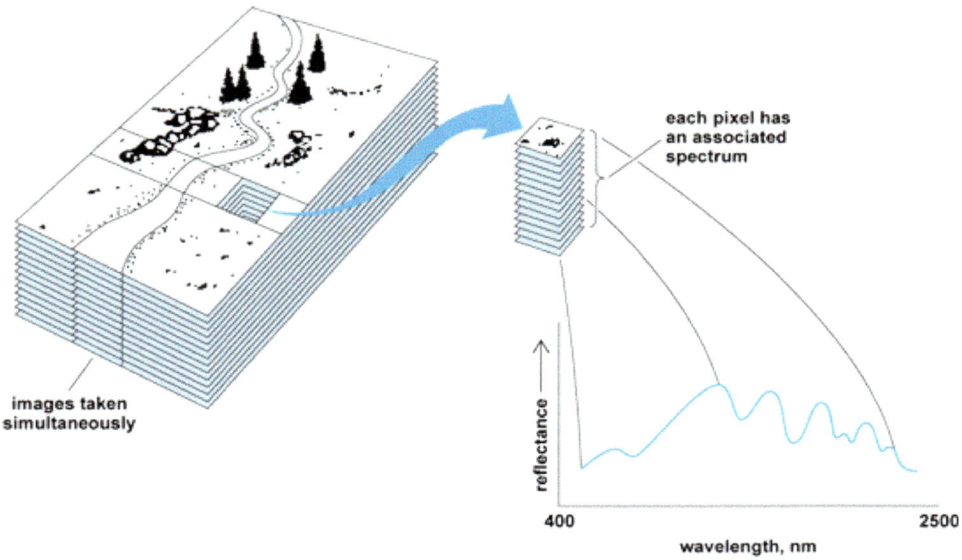

FIGURE 16.13 The spectral imaging concept where each pixel is described by the detailed spectrum constructing of many narrow continuous bands. (Taken from NASA.)

a powerful and promising tool in the modern remote-sensing arena. Since 1983, when the first airborne imaging-spectroscopy sensor (AIS, Vane et al., 1984) ushered in the HSR era, this technique has been used mostly for geology, water, and vegetation applications. Today, HSR technology is widely operated by many users, from the laboratory, field, airborne (manned and unmanned) vehicles and satellites. Although soil was not the main object of study with this technology, today it is, with the advent of better and lower-cost HSR sensors, and comprehensive studies by many scientists developing a wealth of innovative soil-spectral applications. Some unique examples of how soils are studied by HSR sensors in all domains are given here: in the laboratory, Adar et al. (2014b) studied how CD in soils can be analyzed from a HySPEX camera; in the field, Sabat-Tomala et al. (2018); in the air, Ben-Dor et al. (2002) published a pioneering paper on salinity mapping using the airborne DAIS 1759 sensor and in 2020, the Cubert UHD-185® camera on a drone; and recently, an entire space mission using JPL's EMIT has been dedicated to soil and its contribution to the world's dust. As HSR technology is based on spectral information, it is not surprising that many more unique examples can be found in the literature. Despite of the difference between laboratory, air, and spaceborne sensors and the resultant difficulties, this technology has a bright future in soil remote-sensing applications. A comprehensive overview of the pros and cons of hyperspectral technology for soils can be found in Ben-Dor et al. (2009).

16.8.1.1 The New HSR Era from Space

On March 22, 2019, when the PRISMA HSR sensor was placed in orbit, a new era of remote sensing from space began. Orbital HSR programs from different space agencies are now operational, while others are in advanced stages of development: the Earth-sensing imaging spectrometer DESIS from the DLR (Alonso et al., 2019) with 235 bands, range 400–1000 nm, 30-m GSD, the JAXA HISUI (Tanii, 2022) with 185 bands, range 400–500 nm, 20-m GSD, and NASA's EMIT with 285 bands, range 380–2500 nm, 60-m GSD, are working from the ISS (Christensen and the EMIT Team, 2022); China's Advanced Hyperspectral Imager AHSI (Liu et al., 2020) with 330 spectral bands, 30-m GSD, 60-km swath width, spectral range 400–2500 nm onboard China's GaoFen-5

(GF-5) satellite; India's Hyper-Spectral Imaging Satellite (HySIS) (Ghosh and Kumar, 2020) with 256 bands, range 400–2400 nm, 30-m GSD; the ASI's PRISMA sensor which consists of HSR and panchromatic (PAN) sensors with VNIR and SWIR detectors which can acquire a continuum of 234 spectral bands ranging from 400 to 2500 nm, with 10-nm spectral resolution, 30-m GSD, and a swath of 30 km (the PAN sensor can acquire the same area with a spatial resolution of 5-m GSD) (Cogliati et al., 2021); and the DLR's EnMAP (Bachmann et al., 2021) with 242 bands, range 420–2450 nm, 30-m GSD. In the coming years, several new programs will attempt to mount HSR sensors in space, such as the ESA's CHIME and Fluorescence Explorer (FLEX) (Nieke and Rast, 2021), and NASA's SBG mission (Cawse-Nicholson et al., 2021). As HSR technology progresses, the private sector is realizing its potential and new initiatives have been developed, such as GHOSt (Global Hyperspectral Observation Satellite, www.eoportal.org/satellite-missions/ghost#eop-quick-facts-section) launched in April 2023 that covers the VNIR—SWIR region with 512 spectral bands and 8.3 m Ground Digital Distance (GSD) and the high temporal coverage of the Pixxel HSR constellation that already has two sensors in orbit since November 2022. The sensors have 150 to 300 spectral bands with 5–10 m spatial resolution. The company plans to provide 24-hour data using 24 satellites in 2025 (www.eoportal.org/satellite-missions/pixxel#mission-capabilities). The availability of high spatial and temporal spectral data, as suggested by the private sector, may boost the use of hyperspectral technology for soil monitoring on the field scale and for monitoring frequent changes, such as in soil moisture or cultivation.

16.8.2 Application of Soil Remote Sensing: Examples

16.8.2.1 Multi- to Hyperspectral Concept in Soil

Based on the theory presented earlier, this section provides case studies exemplifying the application of remote sensing to soils taking into account the pros and cons of this technology. We will focus on the multi- to hyperspectral domains, with special emphasis on the latter. In addition, we will demonstrate that low spectral resolution can be sufficient for some soil applications (e.g., Palacios-Orueta and Ustin, 1998; Chen et al., 2000; Fox et al., 2002; Demattê et al., 2007). A good example of this is the fact that some major soil properties are correlated with color, and soil surveyors describe soil colors that can be seen with the naked eye. From a remote-sensing perspective, Chen et al. (2000) were able to map SOM content from color aerial photographs by measuring the color tones and correlating them with the SOM content. Demattê et al. (2000, 2009b) summarized the capabilities of the limited channel of the Landsat TM sensor from orbit to provide quantitative information on soils. Using the 6 TM bands in the VNIR—SWIR region, they were able to allocate a reliable quantitative model for the Al_2O_3/Fe_2O_3 ratio using band 7. In another study, Nanni et al. (2006) showed that TM band 7 is mandatory for SiO_2 and TiO_2, whereas TM band 4 was not selected. The authors speculated that band 7 might be associated with the influence of younger, more clayey soils containing kaolinite minerals that are spectrally active around band 7. The quantitative potential of Landsat TM was shown by Ben-Dor and Banin (1995c), who convolved a soil spectral library to TM channels and were able to predict $CaCO_3$, SiO_2, LOI (loss on ignition) and specific surface area solely from the reflectance data. Landsat TM has also been used to establish a strong relationship between iron and reflectance (Frazier and Cheng, 1989; White et al., 1997; Tangestani and Moore, 2000; Jarmer, 2012), where recently Kumar et al. (2013) showed that from the classified image, they can point to the best fertilization areas using soil color. Better results were obtained later with the complete spectral information, as already discussed, using either point or image spectroscopy. The technology of spectral imaging has advanced soil remote sensing, with spectral know-how in the laboratory providing an innovative starting point. In general, imaging spectroscopy (HSR) is capable of generating qualitative and quantitative spatial indicators for ecologists, land managers, pedologists, and engineers. For soils, this technology has been used since 2006 to combine spatial information with the spectral one thus providing farmers with a spatially explicit quantitative overview of the soil properties and phenomena in question. This allows them to control their resources, such as irrigation, nutrients, and cultivation, and obtain better yields per hectare. Since the HSR product is a geopositioned mosaic comprised of many spectral points, traditional (quantitative) approaches that work successfully for point-spectrometry measurements in minerals and soils (e.g., Clark and Roush, 1984) will no doubt be suitable for the imaging domain. Despite its drawbacks, most of the applications developed for point spectrometry can be immediately adapted to the imaging spectroscopy domain. The following sections provide examples in which both multi- and hyperspectral technologies have been used for various applications in soils, revealing this technology's potential for soil science.

16.8.2.2 Soil Organic Matter

SOM, or soil organic carbon (SOC) (SOM ≈ SOC × 1.72), plays a major role in many chemical and physical processes in the soil environment and therefore has a strong influence on soil-reflectance characteristics. Consequently, and as described earlier, this can be seen in the tone of the soil's color (Chen et al., 2000). SOM is a mixture of decomposing tissues of plants, animals and secreted substances. The sequence of OM decomposition in soils is strongly determined by the activity of soil microorganisms. The nature of SOM is responsible for many soil properties, such as compaction, fertility, water retention and soil-structure stability, and it constitutes one of the major resources in the global carbon cycle (Stevens et al., 2008). The mature stage refers to the final stage of microorganism activity, when new, complex compounds, often called humus, are formed. The most important factors affecting the amount of SOM are those involved with soil formation, that is, topography, climate, time, type of vegetation and oxidation state. OM, and especially humus, plays an important role in many of the soil's properties, such as aggregation, fertility, water retention, ion transformation and color. SOM is part of the upper soil horizon that serves as the interface between the body of the soil, the biosphere and the atmosphere. Since OM is mainly concentrated in the top Ao horizon that is exposed to the Sun's radiation, it is a perfect property for remote-sensing assessments. This notion is strengthened by the fact that pure OM has unique spectral fingerprints (Ben-Dor et al., 1999) that can be correlated to content, composition, and maturity (Ben-Dor et al., 1997). Much attention has been devoted to OM from many perspectives. As OM has spectral activity throughout the entire VNIR—SWIR region, especially the VIS portion, workers have extensively studied OM via remote sensing (e.g., Kristof et al., 1973). In general, in remote sensing, Baumgardner et al. (1970) noted that if the SOM drops below 2%, it has only a minimal effect on soil reflectance. Montgomery (1976) indicated that OM content as high as 9% does not appear to mask the contribution of other soil parameters to soil reflectance. Galvão and Vitorello (1998) showed how OM is affected by the Fe-oxides' influence on the spectral reflectance and color of Brazilian tropical soils and later, Heller Pearlshtien and Ben-Dor (2020) showed quantitatively how the assessment of OM is affected by the free Fe-oxide content. In another study, Schreier (1977) indicated that OM content is related to soil reflectance by a curvilinear exponential function. Mathews et al. (1973) found that OM correlates with reflectance values in the 0.5–1.2 mm range, whereas Beck et al. (1976) suggested that the 0.90–1.22 μm region is best suited for mapping OM in soils. Krishnan et al. (1980) used a slope parameter at around 0.8 μm to predict OM content, and Da-Costa (1979) found that simulated Landsat channels (# 4, 5 and 6) yield reflectance readings that are significantly correlated with SOC. The power of spectral information also led Ben-Dor et al. (1997) to exploit reflectance for the detection of SOM decomposition status and control of the soil biogenic activity aside from total SOM content. Vinogradov (1981) developed an exponential model to predict the humus content in the upper horizon of plowed forest soils by using reflectance parameters between 0.6 and 0.7 mm for two extreme endmembers (humus-free parent material and humus-enriched soil). Schreier (1977) found an exponential function to account for SOM content from reflectance spectra. Al-Abbas et al. (1972) used a multispectral scanner, with 12 spectral bands covering the 0.4–2.6 μm range, from an altitude of 1,200 m and showed that a polynomial equation will predict the OM content from only five channels. They implemented the equation on a pixel-by-pixel basis to generate an organic content map of a 25-ha field. Dalal and Henry (1986) were able to predict the OM and total organic nitrogen content in Australian soils using wavelengths in the SWIR region (1.702–2.052 μm), combined with chemical parameters derived from the soils. Using similar methodology, Morra et al. (1991) showed that the SWIR region is suitable for identification of OM composition between 1.726 μm and 2.426 μm. Evidence that OM assessment from soil-reflectance properties is related to soil texture, and most likely to soil clay, was provided by Leger et al. (1979) and Al-Abbas et al. (1972). Aber et al. (1990) noted that OM, including its stage of decomposition, affects the reflectance properties of mineral soil. Baumgardner et al. (1985) demonstrated that three organic soils with different decomposition levels yield different spectral patterns. Hill and Schütt (2000) successfully used the coefficients of a polynomial approximation of a spectral continuum between 0.4 and 1.6 μm to set up a statistical model to map organic carbon concentrations with multi- and hyperspectral imagery. As already discussed, based on the strong spectral relationship, SOM can also be estimated from soil color. Fox et al. (2002) presented a method in which the soil line Euclidean distance (SLED) could be used to estimate OM from aerial color images. They reported coefficients of determination of 0.70 and 0.78 between observed and predicted SOM contents for two study sites. Nonetheless, these results were site-dependent and did not work in another area, suggesting that the stage of SOM decomposition was not similar (see Ben-Dor et al., 1997). Ray et al. (2004) estimated SOC and nitrogen from an IKONOS multispectral image, but with only limited accuracy. This improved a little with Demattê et al.'s (2007) attempt to use Landsat TM images crossing a large geographical area (43,000 ha), which again showed "scene (local)-dependent" SOM determination from multispectral domains. In fact, research varies with OM quantification, since some works demonstrates high quantitative values and others does not. In this aspect it is interesting to note that several works performed in the tropics where OM is usual lower than in temperate areas, occur poor results. This can be probably due to the low values and variability in the tropics, which implies in the statistical models, as they are less detected. Despite this, it is likely that models for OM are more related with local situations due to the innumerous factors (climate, microorganisms, soil management). Jarmer et al. (2005) used a combination of CIE color coordinates (e.g., Escadafal, 1993) and specific spectral absorption features to parameterize statistical models to obtain maps of organic and inorganic carbon contents, as well as total iron content, on a regional scale. Chen et al. (2006, 2008) proposed using Euclidean distance for statistical clustering and the world neural network system to select fields with similar image properties, thereby ensuring the success of mapping SOC for a group of fields. Ben-Dor et al. (2002) were the first to use HSR technology from the air (DAIS 7915), adapting the spectral information modeling to quantitatively map SOM in Vertisol soils from Israel. Later, Stevens (Stevens et al., 2006; Stevens et al., 2008; Stevens et al., 2010) enlarged the HSR envelope and used CASI and SASI sensors over cultivated soils in Belgium, demonstrating the potential of HSR for mapping SOM, even with relatively low content. The assessed values of SOC ranged from a mean of 3.0% (5.8% max) to 1.7% (0.8% min). Assuming that the SOC-to-SOM ratio is about 0.58, the SOM content in these soils was rather low (6.8%—0.99%), but in some cases, still higher than the 2% threshold set by Baumgardner et al. (1985) for spectral determination. Over these low SOM areas, Stevens et al. (2008) were able to map SOC on a pixel-by-pixel basis only if the VNIR—SWIR regions were combined. Although the root mean square error (RMSE) of prediction was 0.17% SOC, which is double the value of the laboratory's accuracy, the processed SOC image was reliable, and gave the first spatial overview of SOM distribution within a given field. Additional studies, such as those of Ben-Dor et al. (2002) and Toure and Tychon (2004), also showed the ability to derive SOM, but their accuracy was rather low. Zheng (2008) summarized all spectral attempts to obtain OM and related components (such as total nitrogen and phosphorus) with point and imaging sensors from the air domain using several airborne HSR sensors. A comprehensive review of SOM estimation from reflectance data is given by Ladoni et al. (2010), showing varying results and means. In general, coefficients of determination from remote sensing of SOM vary from low (0.4 from Landsat TM; Demattê et al., 2007) to very high (0.96 from laboratory measurements; McCarty et al., 2002). The HSR sensors provided moderate (0.56, CASI, Uno et al., 2005) to high (0.89, HyMap, Selige et al., 2006) coefficients. This again demonstrates that finding a good model for predicting SOM is a challenging task with high potential. In mineral soils, the SOM content is rather low (less than 15% and about 0.1–2% in arid soils) where, as stated earlier, less than 2% is almost undetectable spectrally. As the spectral fingerprints are strongly related to the decomposition stage of the OM, different locations may have different SOM spectral features. Other problems are soil moisture, which can hinder small SOM features, BRDF effects, and sensor SNR. It is concluded that SOM is an interesting chromophore that can be problematic and challenging. There are many studies on determining SOM by remote-sensing means, with varying results. As SOM is a very important issue in soil, more such studies are warranted, as the technology progresses. Figure 16.14 provides an example of a SOM map that was generated based on spectral information and HSR data from the Hyperion satellite sensor (taken from Zheng, 2008; Zhang et al., 2013) and in Figure 16.15, OM content (with other soil attributes) are mapped in a pioneer study to demonstrate the HSR capability for soil using the DAIS 7915 airborne sensor (Ben-Dor et al., 2002). More quantitative images of SOM that were generated from airborne HSR sensors can be seen in Stevens et al. (2006). One important characteristic of SOM is its strong relation to the organic stock in the soil which, with the bulk density, can provide the quantity of organic component attached to the 30-cm plow layer. This value is used today to quantitate carbon-sequestration processes and in practice, to compensate for CO_2 emission (Poeplau and Don, 2015). Several recent papers have addressed the capability of soil spectroscopy to assess soil organic stock (e.g., Cambou et al., 2016; Allory et al., 2019; Francos et al., 2024), and others have demonstrated its applicability using HSR and multispectral sensors (e.g., Guo et al., 2019; Padarian et al., 2022); some spinoff companies are dealing with this issue from the spectral domain as well, for example, Soil Capital Farming (www.soilcapitalfarming.ag/), AgroCares (www.agrocares.com/), and TierraSpec (www.tierraspec.com/).

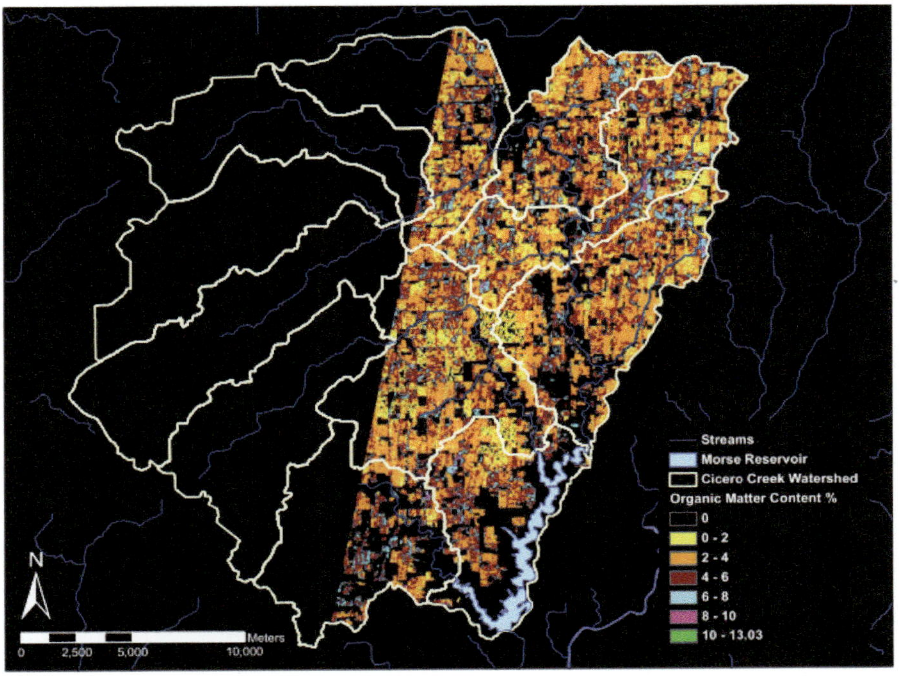

FIGURE 16.14 Prediction of soil organic matter distribution map in Cicero Creek Watershaded as generated from Hyperspectral data. (After Zebg, B., Using satellite hyperspectral imagery to map soil organic matter, total nitrogen, and total phosphorus. A master thesis submitted to the Department of Earth Science, University of Indiana, Indianapolis IN, 2008. P81.)

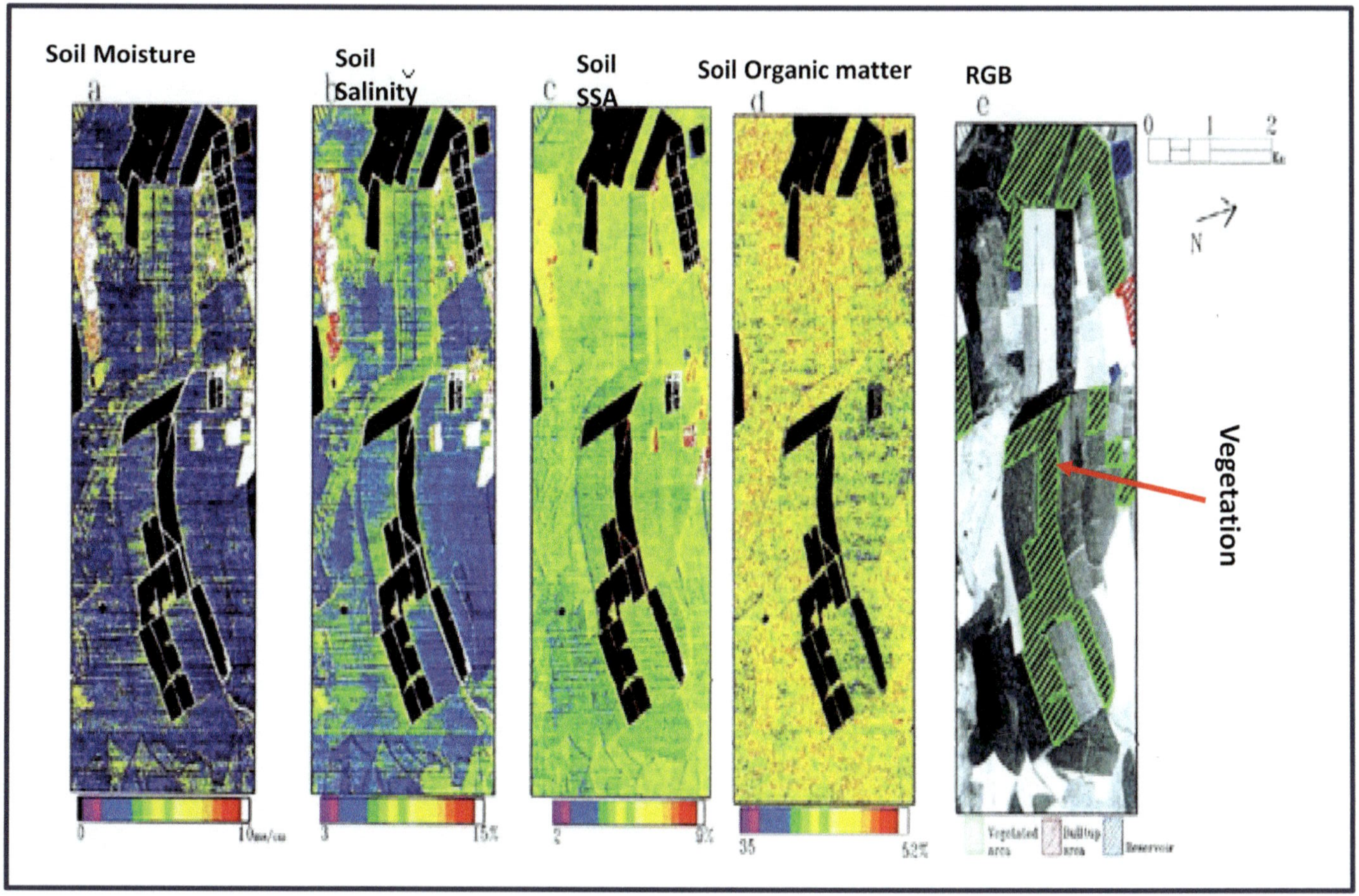

FIGURE 16.15 Classified hyperspectral images (DAIS1519) using proximal sensing model to predict soil organic matter, soil specific surface area (SSA), soil Salinity (EC), and soil moisture content over Beqat Zvaim, Israel. The black polygons represent masks of photosynthetic vegetation.

16.8.2.3 Change Detection (CD) in Soils

CD and multitemporal analysis of remote-sensing data are aimed at detecting various types of changes between two or more images taken at different times (Singh, 1989). The temporal information adds new data for the interpreter and "temporal resolution" (added to spectral and spatial resolution) to sensor-performance characterization. Adding a spectral dimension to the soil remote-sensing data provides better mapping capabilities than obtained from using limited spectral channels—even when the spatial resolution is high (e.g., Ben-Dor, 2002; Ben-Dor et al., 2005; Casa et al., 2013; Goidts and van Wesemael, 2007; Hbirkou et al., 2012; Hill et al., 2010; Stevens et al., 2010). Adding a temporal dimension to the spectral resolution provides even better capabilities. In practice, spectral resolution has the dominant role in the evolving HSR arena, and temporal resolution has been left behind. However, temporal spectral analysis can cull information on interactions between the soil surface and the surrounding environment and accordingly, can provide a better view of the factors affecting soil formation (Jenny, 1941). Rapid changes on the soil surface can occur from erosion, deposition, physical arrangement, self-segregation and man-made activity (Lemos and Lutz, 1957). More specifically, the thin upper soil layer (that is ultimately sensed by optical sensors) may be altered by dust accumulation (Offer and Goossens, 2001), rust formation (Ona-Nguema et al., 2002), plowing activity (Fu et al., 2000), changes in particle-size distribution (Sertsu and Sánchez, 1978), vegetation coverage (Zhou et al., 2006), litter (Frey et al., 2003), and the formation of physical and biogenic crusts (Bresson and Boiffin, 1990; Karnieli et al., 1999; Valentin and Bresson, 1992). Until recently, applications for high spectral and temporal resolution data were scarce, mainly due to the high cost of data acquisition by airborne HSR. However, this situation is expected to change significantly as many satellite HSR sensors are in the pipeline with high spectral, spatial and temporal resolution capabilities (Ben-Dor et al., 2013). A comprehensive overview of forthcoming HSR sensors is provided by Staenz et al. (2012). As a result, the scientific community is starting to perform controlled experiments (Buddenbaum et al., 2012) to prepare the remote-sensing community for the multiresolution (spatial, spectral, temporal) approach. Methods to account for CD between areas that are well identified either spectrally or spatially (e.g., soil to vegetation) are well known and frequently used (Adar, 2014a). Methods to account for CD between the same land-cover category (e.g., soil to soil) are more complex. In this case, the more information provided on the area in question, the better the discrimination capability will be. Spectral reflectance, as achieved from HSR sensors, can provide added information. Nonetheless, due to the aforementioned problems with factors affecting soil spectroscopy, this mission is very challenging, as a small spectral change might occur due to factors other than real change on the soil surface. Recently, Adar et al. (2014a) developed an approach in which the "factors affecting soil reflectance" can be estimated from HSR images. This method enables better CD analysis based on real spectral changes. Those authors demonstrated the approach using HyMap data acquired over an open-mining area in the Czech Republic at a 1-year interval. More recently, Adar et al. (2014b) conducted a controlled study to understand the capability of high spectral information for spatial discrimination between soil entities under optimal conditions, where "factors affecting soil reflectance" are minimized. An artificial soil matrix (made of 50 different soil samples, each in a 3 × 3 cm dish) which was measured by an image spectrometer (HySpex) under laboratory conditions provided the database for this study. Several changes were made in the soil matrix between each data acquisition along with relocating some of the soils' original position. Using the VNIR, SWIR and VNIR—SWIR spectral segments separately, with several known methods to detect possible (spectral) changes in a given pixel, it was found that the wider the spectral coverage, the better the discrimination capability. Figure 16.16 shows an example of the results obtained by this analysis using the VNIR—SWIR regions (alone and together). As seen, only when the complete spectral region is used with a specific method to assess the spectral changes, the spatial changes could be obtained (compare image d to a). The authors revealed limitations in identifying changes between different soils in three cases: 1) when the soils were within the same larger group of soil classes, very small changes could not be detected; 2) when there were opposing effects on the spectral signature,

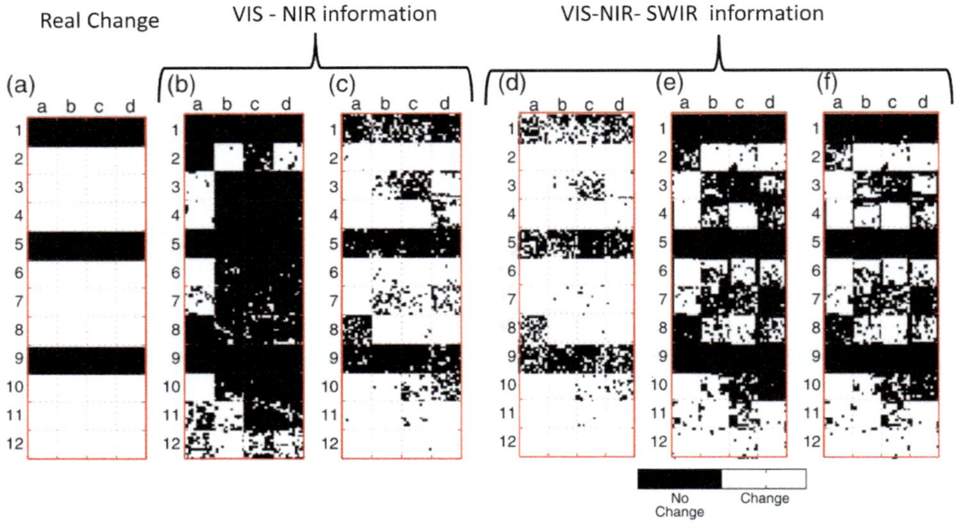

FIGURE 16.16 A demonstration of spectral change of soils as done under laboratory (ideal) conditions. Every cell represents different soil that configured a soil field which were then scanned by HySpex image spectrometer at the laboratory. Black cells in the image (a) represent soils without any change. White cells in (a) represent areas where soil was replaced by other sols. Images (b–f) provide the results of different change detection analytical methods to point out possible changes within the cells. As seen image (d) provides the best results with minimum black pixels on the white cells and minimum white pixels on the black cells. (From Adar S. et al. *Int. J. Remote Sens,* 35, 1563, 2014a, Adar S. et al., *Geoderma* 218, 19, 2014b.)

such as twice the Fe-oxides and at the same time, twice the OM, their effects might cancel each other out, resulting in very similar spectral signatures, and 3) when differences in some of the spectral absorption features are reduced as a result of similar and high average particle-fraction size, which reduces the albedo of the spectral signatures to a very similar level. Although Adar et al.'s (2014b) study demonstrated limitations for CD in soils, and indicated that different algorithms can produce better results, they also concluded that CD in soils from both spectral and spatial domains is a difficult task and calls for caution in drawing any conclusions. They did, however, suggest overcoming these limitations by fusing chemometric capabilities with CD techniques and not relying purely on the spectral information of the image. To summarize, CD in soils can shed light on some quick processes on the soil surface but at the same time, presents difficulties in significantly distinguishing between the two (or more) soils.

16.8.2.4 Soil Salinity

Soil salinity is a dynamic property that emerges on the soil's surface mostly under arid and semiarid conditions and under secondary water utilization. It can therefore be effectively monitored by remote sensing as light spots of NaCl obtained under a high-salinity regime that can be monitored by aerial photography (Rao and Venkataratnam, 1991). At the lowest saline concentrations, the unaided eye cannot detect salinity effects and a better analytical tool is required. Airborne digital multispectral cameras and videography, usually with three to four channels in the VIS and IR regions, and color IR photographs were used as tools to identify and assess problem salinity areas in US agriculture in the 1980s and 1990s. Everitt et al. (1988) used narrowband videography to detect and estimate the extent of salt-affected soils in Texas, USA, while Wiegand et al. (1991, 1992, 1994) analyzed and mapped the response of cotton to soil salinity using color IR photographs and videography with three bands (0.84–0.85 μm, 0.64–0.65 μm and 0.54–0.55 μm), and a spatial resolution of 3.4 m. By relating video and field data such as soil electrical conductivity, plant height and percent bare area, they determined the interrelations between plant, soil salinity and spectral observations. These studies found that color–IR composites and red narrowband images were better than green and NIR narrowband images (Escobar et al., 1998; Wiegand et al., 1992, 1994). Extensive research on the application of panchromatic and multispectral satellite imagery to map salt-affected areas has been conducted over the last four decades, mostly using panchromatic and multispectral (VNIR–SWIR and/or thermal) sensor. Works by Csillag et al. (1993), Epema (1990), Metternicht and Zinck (1997), Rao et al. (1995), Evans and Caccetta (2000) provide some examples of applications on different continents and in different environmental settings. All of these works were generally successful in mapping saline vs. non-saline surfaces. Some researchers have attempted to map salinity types (e.g., saline, alkaline) and degrees (e.g., low, moderate, high) (Metternicht and Zinck, 1996; Kalra and Joshi, 1996), with varying degrees of success. From the year 2000 onward, experimental hyperspectral satellite data from sensors like the CHRIS onboard the ESA mission PROBA-1, or Hyperion on EO-1, were assessed for their ability to identify and map salt-affected areas (Dutkiewicz, 2006; Schmid et al., 2007). A comprehensive description of all attempts to map soil salinity from multispectral satellite sensors, starting from old sensors such as Landsat MSS to newer ones such as IKONOS, is given in Metternicht et al. (2008). This book also covers other remote-sensing means of detecting soil salinity, such as active and passive sensors in the microwave and thermal domains.

While the preceding means were being used mostly to locate saline soil areas, especially those that are visible to the naked eye, research was being directed to assessing soil salinity that is low in content or in its first stages of development, as such soil can be agrotechnically treated to obtain optimal cultivation under extreme conditions. In this case, full spectral information is needed and sophisticated analytical approaches to mining spectral information related to salt were developed (Farifteh et al., 2004, 2006; Huang and Foo, 2002). The added value was the identification of salt-affected areas before they become visible to the naked eye, as done by Ben-Dor et al. (2002) and later also by Howari et al. (2002) and Dehaan and Taylor (2002) using hyperspectral sensors. These studies were based on Taylor et al. (1994), who were the first to show that it is possible to use airborne superspectral data to map salinity by using the 24-band airborne Geoscan, and VNIR/SWIR data, at Pyramid Hill, in Victoria, Australia.

Whereas the salinity is most important in the root zone, it is interesting to note that two recent innovative studies were able to correlate salinity level at 30-cm depth to surface reflectance as acquired by an airborne hyperspectral sensor (Figure 16.17), and form an indirect correlation between leaf reflectance of tomato and the electrical conductivity measured in the root zone. These data were then projected on a cartographic domain to generate soil salinity-affected areas for the farmer (Goldshleger et al., 2013).

Vegetation is an indirect factor that facilitates detection of salt in soils from reflectance measurements (Hardisky et al., 1983; Wiegand et al., 1994). Gausman et al. (1970), for example, pointed out that cotton leaves grown in saline soils have a higher chlorophyll content than leaves grown in low-salt soil. Hardisky et al. (1983) used the spectral reflectance of a *Spartina alterniflora* canopy to show a negative correlation between soil salinity and spectral vegetation indices. In the absence of vegetation, salt's major influence is on the structure of the upper soil surface. Figure 16.18 shows saline and non-saline spectra, taken from Everitt et al. (1988), in the VNIR region. The saline soils had relatively higher albedo than the non-saline ones. Furthermore, the saline soils had crusted surfaces that tended to be smoother than the generally rough surfaces of the non-saline soils. Although Gausman et al. (1977) and Rao et al. (1995) reported similar trends in other soils, it should be noted that in soils with relatively high salt content, the opposite behavior can also be reasonably expected. This is because salt is a very hygroscopic material, which tends to decrease the soil albedo as water content rises. It is apparent in Figure 16.19 that in artificial mixtures of soil and salt with varying concentration, a different water absorption intensity obtained due to the hygroscopic nature of the salt (Faritfreh et al., 2006). Because no direct significant spectral features are found in the VNIR—SWIR region to identify sodic soil, indirect techniques are thought to be more appropriate for classifying salt-affected areas (Verma et al., 1994; Sharma and Bhargava, 1988). Salt in water is most

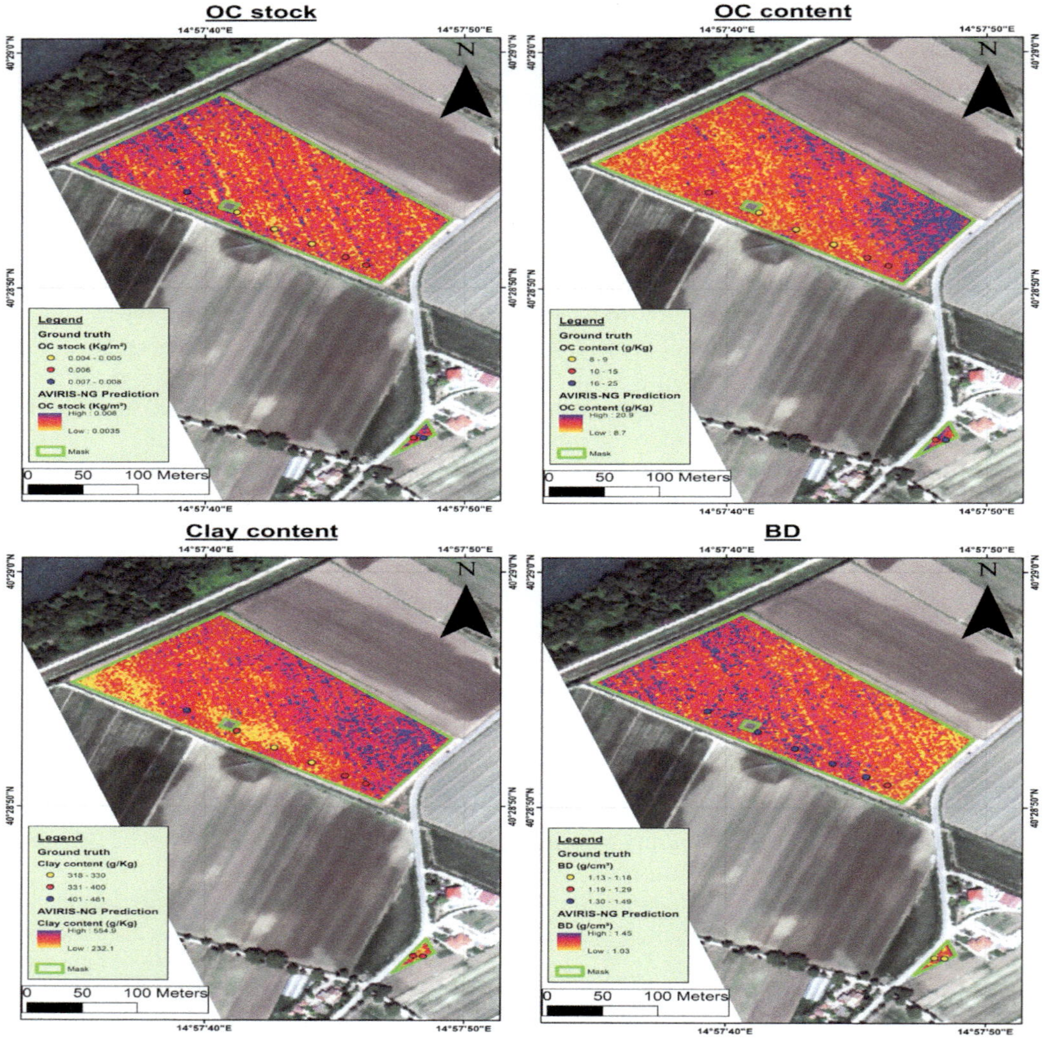

FIGURE 16.17 Classified hyperspectral images (AVIRIS NG) using proximal sensing to predict image of soil organic carbon stock (OC stock), organic carbon (OC content) content, clay (Clay content) content, and bulk density (BD) (after Frnacos et al., 2024).

likely to affect the hydrogen bond in water molecules, causing suitable spectral changes. Based on this, Hirschfeld (1985) suggested that high spectral resolution data are required. Support for this idea was given by Szilagyi and Baumgardner (1991), who reported that characterizing salinity status in soils is feasible with high-resolution laboratory spectra. A relatively high number of spectral channels are also important for identifying an indirect relationship between salinity and other soil properties that appear to consist of chromophores in the VNIR—SWIR regions. Csillag et al. (1993) analyzed high-resolution spectra taken from about 90 soils in the USA and Hungary for chemical parameters, including clay and OM content, pH and salt. They stated that because salinity is such a complex phenomenon, it cannot be attributed to a single soil property. While studying the capability of commercially available Earth-observing optical sensors, they indicated that six broadbands in the VNIR–SWIR region best discriminate soil salinity. These six channels were selected solely on the basis of their overall spectral distribution, which provided complete information about salinity status. Thus, it can be concluded that it is necessary to look at the entire spectral region to evaluate salinity levels

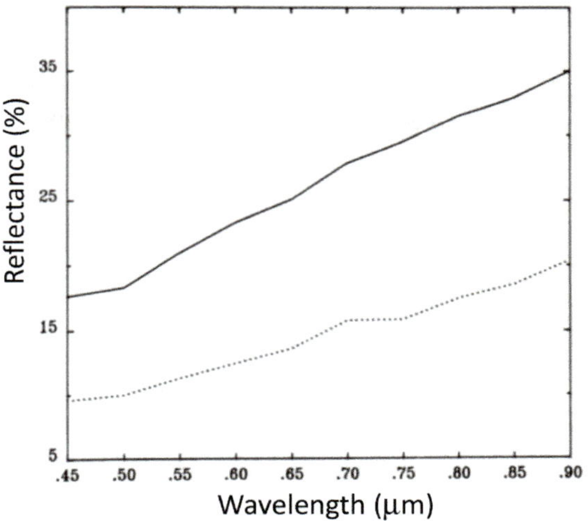

FIGURE 16.18 The field spectra across the 0.4–1.0 μm of saline and non-saline soils (After Everitt et al., 1988.)

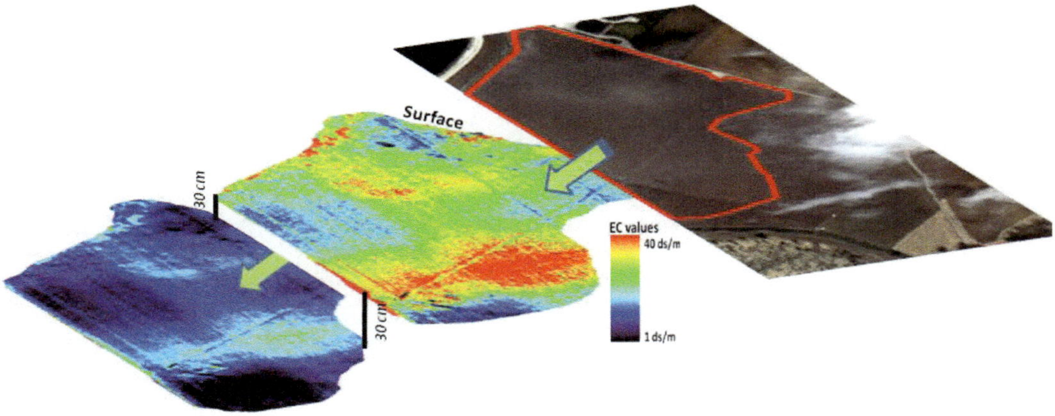

FIGURE 16.19 Soil salinity image as generated from an airborne AISA sensors (Eagle & Hawk covering the VIS-NIR-SWIR region) used a spectral model generated from samples taken at 0 and 30 cm. As seen a favorable map was obtained for the salinity at 30 cm and not only on the soil surface.

in different environments and unknown soil systems. In summary, soil salinity is a property that can be monitored by remote-sensing means, but requires high temporal, spectral and spatial resolution.

16.8.2.5 Soil Moisture

Soil moisture is an important property, not only for assessing the available water content needed for plant utilization, but also for assessing the direct exchange of soil water with the atmosphere (e.g., evaporation) and quantifying moisture effects on other chromophores. In fact, it is considered to be one of the most significant parameters in the soil system. It can vary from hygroscopic moisture (water left on the surface after equilibrium has been achieved with the atmosphere) to a saturated stage (water fills 50% of the soil pores). The effect of the water molecules on soil reflectance is strong and significant. Whereas hygroscopic water most likely shows the absorption features of OH molecules at 1.4 and 1.9 μm with a strong SWIR shoulder at 2.62 μm (demonstrating the imaginary part of the refractive index), in the saturated condition, the real part of the refractive index is dominant and hence the entire spectrum is affected such that the overall spectral-baseline height ("albedo") is lowered (visible to the naked eye as darker soil). In between these two moisture contents, the spectral signatures are affected by both mechanisms (real and imaginary parts), complicating the assessment of water content from soil reflectance. In most cases, the impact of soil moisture on the reflectance is unknown and therefore ignored. Muller and Decamps (2001) modeled reflectance changes due to soil moisture in a real field situation using multiband airborne Spot data. They showed that the impact of soil moisture on reflectance could be higher than the differences in reflectance due to the soil categories, and hence calls for caution in applying soil remote sensing under wet conditions. On the other hand, several attempts have been made to map soil water content using soil reflectance information, some under laboratory conditions, and others in the field and in air and space domains. Nevertheless, assessing soil water content from reflectance measurements is still a challenge, in particular correcting for the water-masking effect that hinders the capture of other soil chromophores' activities. Under dry soil conditions (mostly represented by hygroscopic water), the absorption features at 1.4 and 1.9 μm, and others across the SWIR region, can be correlated with water, as shown by Bowers and Hanks (1965), Dalal and Henry (1986), and Ben-Dor et al. (1999). A study by Demattê et al. (2006) assessed the 1900-nm water combination band feature and others for practical use, and found that the best interpretation of water content occurs when both dry and wet soil samples are spectrally measured.

A novel approach to reconstructing the soil's spectral signature was through the use of various water film depths related to moisture content. Thus, Bach and Mauser (1994) simulated the reflectance change in the soil spectra from dry to moist. They combined Lekner and Dorf's (1988) model for internal reflectance with the absorption coefficients from Palmer and Williams (1974) into Beer's Law. Bach and Mauser (1994) simulated dry through wet soil and applied the process for predicting water content to an AVIRIS image of a partially irrigated field and a field with dark organic soil at the Freiburg test site in Germany. Today, we have a better understanding of the causes to change the soil spectral and we are improving the methods for modeling water content in soils.

The challenge in determining soil moisture content across the VNIR—SWIR region lies in the fact that the water molecules significantly affect all other spectral chromophores and thus may hinder the quantitative spectral approach to determining chromophores, such as OM, Fe-oxide, clay and carbonates. Accordingly, the water—radiation interaction is a very important issue in the soil proximal-sensing discipline which is attracting more and more users and accumulating experience in the laboratory, field, and recently, from remote-sensing domains as well. Based on the strong effect of the real part of the refractive index, gray-level values in the VIS region enable estimating water content under certain amounts of water (Mouazen et al., 2005). Zhu et al. (2011) recently showed a good correlation between soil moisture and digital gray level under very moist soil conditions (25%—60%). Weidong et al. (2002) have shown that a better soil moisture prediction using soil spectroscopy can be determine by adjusting the soil types. Lobell and Asner (2002) demonstrated that the SWIR region is much more sensitive than the VIS region when assessing soil moisture, and described an exponential relationship between the water content and soil reflectance values. Mouazen et al. (2005) showed that the soil moisture content can be estimated using the VNIR and only part of the SWIR regions (306.5 to 1710.9 nm) where Whiting et al. (2004) suggested using the far SWIR region (2200–2500 nm) to estimate water content by fitting an inverted Gaussian function centered on the assigned fundamental water-absorption region at 2800 nm. As the far SWIR region is strongly affected by the left shoulder of the above fundamental absorption, a logarithmic soil spectrum continuum with convex hull boundary points was found to be correlated to water content (Figure 16.20). Based on the preceding method, they were also able to present a processed AVIRIS hyperspectral image that provides the soil surface moisture content (Figure 16.21, Whiting et al., 2004). The spectral approach also attracted Haubrock et al. (2008), who successfully validated a new model for predicting gravimetric soil moisture. The method was

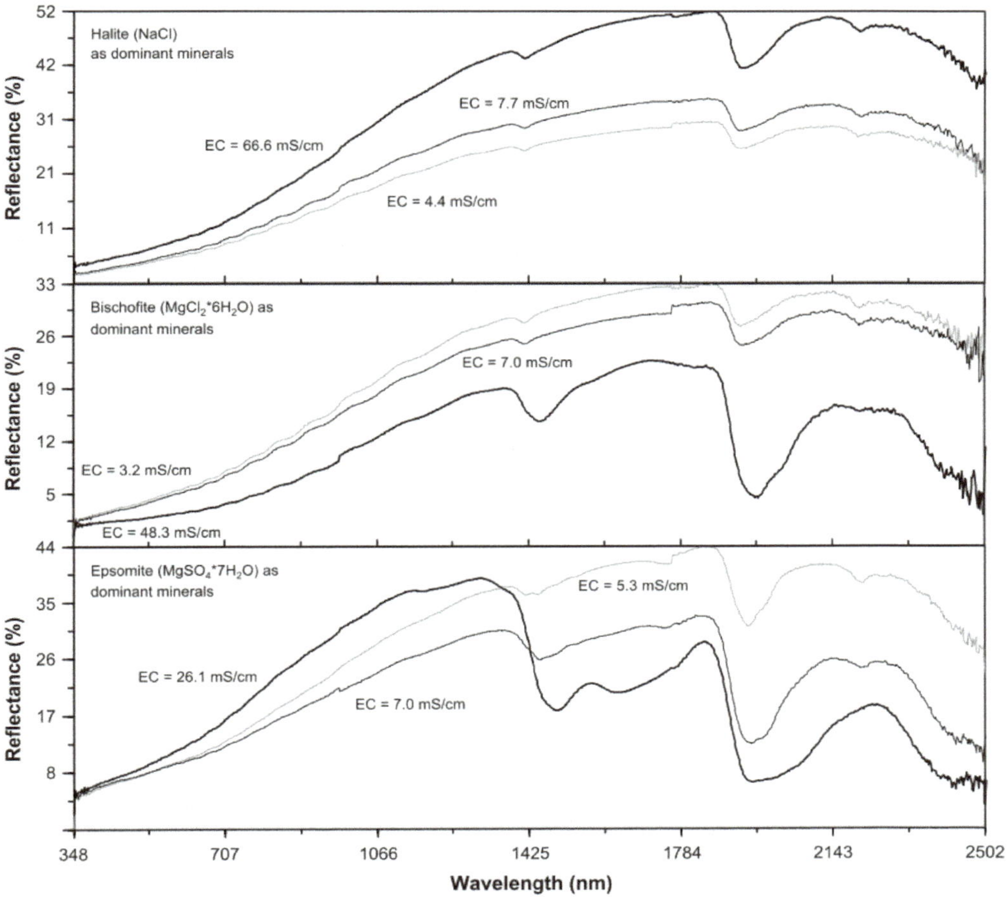

FIGURE 16.20 The reflectance spectra of a selected soil that was artificially contaminated with different evaporates in different content. (After Faritfreh et al., 2006.)

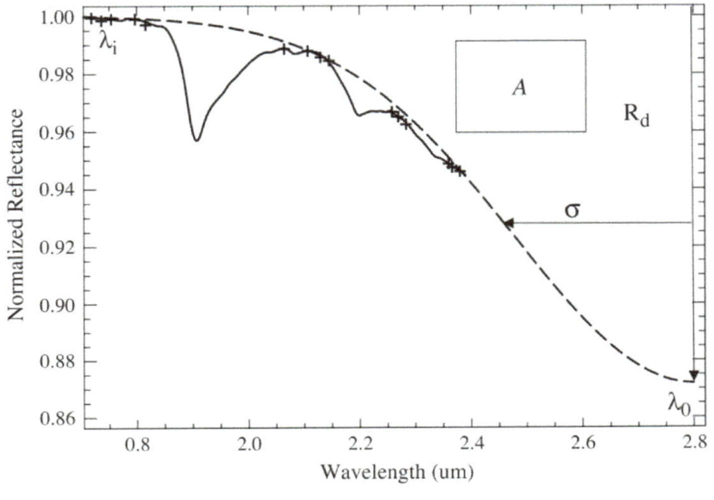

FIGURE 16.21 The fit of a logarithmic spectrum continuum with convex hull boundary points to estimate moisture in the soil (after Whiting et al. (2004)

termed Normalized Soil Moisture Index (NSMI) and combines reflectance values at 1800 nm and 2119 nm around the 1900-nm water combination bands. This index was applied to remotely sensed images and enabled the production of soil surface moisture maps, generated from HyMap airborne images, which were found to be highly correlated with the field moisture content measured at the time of the overflight (Figure 16.22; Haubrock et al., 2008). Surprisingly, neither Whiting's nor Haubrock's methods have been implemented in practice, probably because they are relevant to certain conditions in which moisture is not high. As the real part of the refractive index is dominant at high moisture levels, modeling the spectral features of the water becomes difficult in certain soil moisture ranges and the reflectance should be treated differently, as done by Zhu (2011). This author used the brightness of the water film as an indicator to determine the soil moisture from the VIS region. Ogen et al. (2019) developed a proxy model to remove the moisture effect from the soil spectrum in a method termed external parameter orthogonalization (EPO). They used a set of 830 soil samples with different soil moisture contents and converted the wet soil spectrum to an air-dried one, which enabled using other SSLs to predict clay content. A comprehensive review and mathematical methods on removing and assessing moisture signals and content from soils in the optical domain were recently provided by Knadel et al. (2023).

It should be mentioned here that in general, other remote methods exist to estimate soil water content, such as using thermal bands in the LWIR region (estimating the latent and sensible heat fluxes; Eltahir, 1998) or active microwave and millimetric wave (Eliran et al., 2012) spectral domains that are based on sensing the dielectric constant of the soil—water mixture. A comprehensive review of soil moisture assessment from orbital sensors is given by Serrano's (2010). As the water reflectance properties are important not only for determining the water content but also to sharpen and fine tune other soil properties, it is essential that this field continue to be explored. Further work in reconstructing the spectra that will combine the spectral relationships of water content and soil components based on the physical nature of the materials and photon interactions is strongly needed. Other challenges include examining the models obtained under select conditions using satellite platforms, and defining a robust way to assess water content at all levels, in all orders of soil. An excellent review on monitoring soil moisture content from all orbital remote sensing

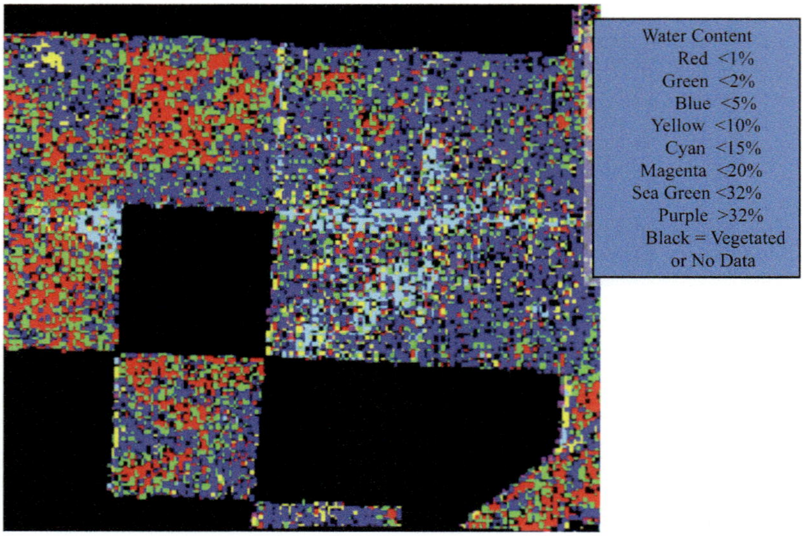

FIGURE 16.22 Surface water (gravimetric) from AVIRIS data (May 3, 2003, near Lemoore, California) as estimated with the DMGM. (Generated by Whiting M.L. et al., *Remote Sens. Environ*, 89, 535, 2004.)

means is given by Barrett and Petropoulos (2013). The optical means is partially covered by them mainly because HSR is not yet widely operational from space whereas other limitations are also reported which are identical to what was discussed in Section 16.8.1. Nevertheless, they concluded that the high spectral and spatial resolutions of the HSR technology may open up another channel to better estimate soil moisture content from orbit.

In summary, it can be concluded that the soil surface water content can be cautiously estimated using reflectance measurements, but due to the effect on other soil components, their spectral absorption requires proper attention. This approach has not yet been fully studied or developed in this innovative direction, that is, for use in HSR, although it seems to hold great promise. Other spectral regions, such as thermal, millimetric and microwaves, may also be used for this mission.

16.8.2.6 Soil Carbonates

Carbonates, particularly calcite and dolomite, are found in soils that are formed from carbonic parent materials, or in a chemical environment that permits calcite and dolomite precipitation. Carbonates, and especially those of fine particle size, play a major role in many of the soil chemical processes that are most likely to occur in the root zone. A relatively high concentration of fine carbonate particles may cause fixation of iron ions in the soil and consequently, inhibition of chlorophyll production (chlorosis-driven carbonates). On the other hand, an absence of carbonate may affect the soil's buffering capacity and thus negatively affect the biochemical and physicochemical processes. Remote sensing allows distinguishing among the common carbonate minerals on the basis of unique spectral features found in the SWIR (as well as TIR) regions. The C–O bond, part of the $-CO_3$ radical in carbonate, is the spectrally active chromophore. Hunt (Hunt et al., 1970; Hunt et al., 1971a, 1971b, 1971c, 1971d) pointed out that five major overtones and combination modes are available to describe the C–O bond in the SWIR region. Gaffey (1986) added two additional significant bands centered at 2.23–2.27 μm (moderate) and 1.75–1.80 μm (very weak), whereas Van der Meer (1995) summarized the seven possible calcite and dolomite absorption features with their spectral widths. It is evident that significant differences occur between the two minerals. This enabled Kruse et al. (1990), Ben-Dor and Kruse (1995) and others to differentiate between calcite and dolomite formations using airborne spectrometer data with band widths of 10 nm. In addition to the seven major C–O bands, Gaffey and Reed (1987) were able to detect copper impurities in calcite minerals, as indicated by the broadband between 0.903 and 0.979 μm. However, such impurities are difficult to detect in soils, because overlap with other strong chromophores may occur in this region. Gaffey (1985) showed that iron impurities in dolomite shift the carbonate's absorption bands toward longer wavelengths, whereas magnesium in calcite shifts the band toward shorter wavelengths. As soil carbonates are most likely to be impure, it is only reasonable to expect that the carbonates' absorption-feature positions will differ slightly from one soil to the next. A correlation between reflectance spectra and soil carbonate concentration was first demonstrated by Ben-Dor and Banin (1990b). They used a calibration set of soil spectra from Israel and $CaCO_3$ content to find three wavelengths that best predict the calcite content in the soil samples (1.8 μm, 2.35 μm and 2.36 μm). They concluded that the strong and sharp absorption features of the C—O bands in the examined soils provide an ideal tool for studying the soil carbonate content solely from their reflectance spectra. The best obtained performance for quantifying soil carbonate content ranged between 10 and 60%. Since that pioneering work, several proxy models to assess soil carbonate content have been published (e.g., Balsam and Deaton, 1996; Thomasson et al., 2001). The use of the SWIR region to map carbonates from airborne HSR has been shown by many users in arable lands (e.g., Lagacherie et al., 2008; Gomez et al., 2013). However, in agricultural soil, where the plowing layer is mixed, estimating $CaCO_3$ content is still a significant challenge. This is especially true in heavily leached environments (low pH) where lime is required to improve soil function. Demattê et al. (2004a,b, Hamilton) already determined alterations between controlled and limed soils. Along these lines, Viscarra Rossel et al. (2005) and later Araújo et al. (2013) developed a spectrally based concept to account, in the field, for the "lime requirement" content and demonstrated its applicability to precision agriculture. Mapping carbonate rocks from airborne domains using HSR technology is well documented (e.g., Kruse et al., 1990; Ben-Dor and Banin, 1995), as is mapping from satellites (e.g., Ninomiya et al., 2005; Gersman et al., 2008), but in soil, where the carbonate is mixed with other minerals, it is more difficult. Gmur et al. (2012) showed the efficiency of hyper-spectral analysis associated with a regression tree to increase the prediction accuracy of carbonate in the soil ($R^2 = 0.95$). Lagacherie et al. (2008) examined how reflectance spectrometry can be used to estimate clay and $CaCO_3$ contents simultaneously in soil using both field and airborne measurements. They showed nine intermediate stages from the laboratory to HyMap sensor measurements crossing spatial and sensor characteristics such as: radiometric quality, spectral resolution, spatial variability, illumination conditions, and surface status including roughness, soil moisture and presence and nature of pebbles. They found significant relationships between clay and $CaCO_3$ contents from the spectral continuum removal values computed, respectively, at 2206 nm and 2341 nm, which persisted from an ASD spectrophotometer to the HyMap spectral imaging sensor. Decreasing performance was obtained going from the laboratory to the hyperspectral domains, indicating the factors affecting reflectance spectra as discussed in Section 16.7. In summary, it can be concluded that soil calcite content has significant potential for quantitative monitoring using spectral information across the end of the SWIR spectral region, but at the same time, has some constraints related to its mixture with other soil materials (minerals, SOM and water), low solar energy and accuracy deterioration in airborne relative to field and laboratory results.

16.8.2.7 Soil Contamination

Soil contamination refers to a process in which non-pedogenic constituents enter the soil volume with no relation to the soil's natural formation or generation. This refers mostly to short-term processes in the soil. Soils can be contaminated by various sources, either anthropogenic (e.g., hydrocarbon) or natural (e.g., dust accumulation). As various contaminants may change the soil's chemistry as well as its physical behavior, one would expect to be able to monitor such processes by spectral sensing means. Demattê et al. (2004a) examined the industrial byproduct of sugarcane that was dumped into a nearby soil area and found it to significantly alter the soil's chemical properties. Accordingly, they found that this alteration was noticeable in the spectral reflectance of the soils via the magnitude of the signal, without much change in the general spectrum's shape. This is probably due to physical effects of the sugarcane byproduct that may cause different aggregation stages in the natural and contaminated soils. Chemical contamination is also an important issue in the soil environment. Heavy-metal contamination of alluvial soils on river banks has been addressed in experimental studies that used the soil proxy approach (Kooistra et al., 2004; Wu et al., 2007; Xia et al., 2007). However, probably most operational applications of airborne HSR missions for monitoring soil contamination have been performed in the context of chronic or accidental pollution resulting from metal mining. For example, Chevrel (2005) investigated six mining areas in the MINEO project, five in Europe (Portugal, UK, Germany, Austria, and Finland) and one in Greenland, using HyMap airborne imaging-spectrometry data. HSR was used for mapping the extent and type of chronic contamination with heavy metals using primarily trace minerals of pyrite oxidation as an indirect indicator of potential contamination, forming an indispensable basis for environmental impact assessment, environmental monitoring of historical mining sites, and remediation planning. Within the framework of the EO-MINERS project (2010–2013), Chevrel (2013) used two HSR sensors (HyMap and AHS) to monitor the coal-ash contamination from open mines in the nearby urban and soil environments. Ren et al. (2009) has also estimated the soil contamination by As and Cu using reflectance spectroscopy of areas near mining activities. In April 1998, the dam of a mine tailings pond in Aznalcollar, Spain, collapsed and flooded a soil area of more than 4,000 ha with pyrite-bearing sludge containing high concentrations of heavy metals. An emergency airborne remote-sensing mission, with the objective of assessing the extent of residual heavy-metal contamination after the first clean-up operations that lasted until 1999, was flown with HyMap covering the affected area in 1999 and 2000 (Kemper and Sommer, 2002, 2003; Garcia-Haro et al., 2005). As a first step, the possibility of adapting chemometric approaches to a quantitative estimation of heavy metals in the soils polluted by the mining accident was explored (Kemper and Sommer, 2002). Six months after the end of the first remediation campaign in early 1999, soil samples were collected for chemical analysis and VIS to SWIR reflectance (0.35–2.4 μm) was measured. Concentrations of As, Cd, Cu, Fe, Hg, Pb, S, Sb, and Zn in the samples were well above background values. Prediction of heavy metals was achieved by stepwise multiple linear regression analysis and by using an artificial neural network approach. This enabled the prediction of six out of the nine elements with high accuracy. The best R^2 values between the predicted and chemically analyzed concentrations were As, 0.84; Fe, 0.72; Hg, 0.96; Pb, 0.95; S, 0.87, and Sb, 0.93. Results for Cd (0.51), Cu (0.43), and Zn (0.24) were not significant.

In the second step of the study, variable multiple endmember spectral mixture analysis (VMESMA, Garcia-Haro et al., 2005) was used to analyze the HyMap data acquired in 1999 and 2000. A spectrally based zonal partition of the area was introduced to allow the application of different submodels to the selected areas. Based on an iterative feedback process, the unmixing performance could be improved in each stage until an optimum level was reached. The sludge quantities obtained by unmixing the hyperspectral data were confirmed by field observations and chemical measurements of samples taken in the area. Figure 16.23 shows the sludge-abundance map derived from the 1999 HyMap data using this iterative VMESMA approach. The semi-quantitative estimate of sludge from residual pyrite-bearing material could be transformed into quantitative information to assess acidification risk and the distribution of residual heavy-metal contaminants based on an artificial mixture experiment and the derivation of simple stoichiometric relationships. As a result, the sludge-abundance map could be rescaled to quantities of residual pyrite sludge, associated heavy metals, and acidification potential due to the need to counteract calcite buffering. Wu et al. (2005), who used reflectance spectroscopy to study the mercury contamination in suburban agricultural soils in the Nansing region, China, revealed interesting results. They found correlations between mercury concentration and goethite and clay absorbance features at 496 nm and 2210 nm, respectively. They concluded that an intercorrelation between mercury and the aforementioned constituents is the key factor for obtaining a prediction of mercury, as it has no spectral fingerprints in the VNIR—SWIR region. Although they have not yet been applied, the authors strongly recommended the use of operational remote-sensing techniques to fully implement this interesting finding

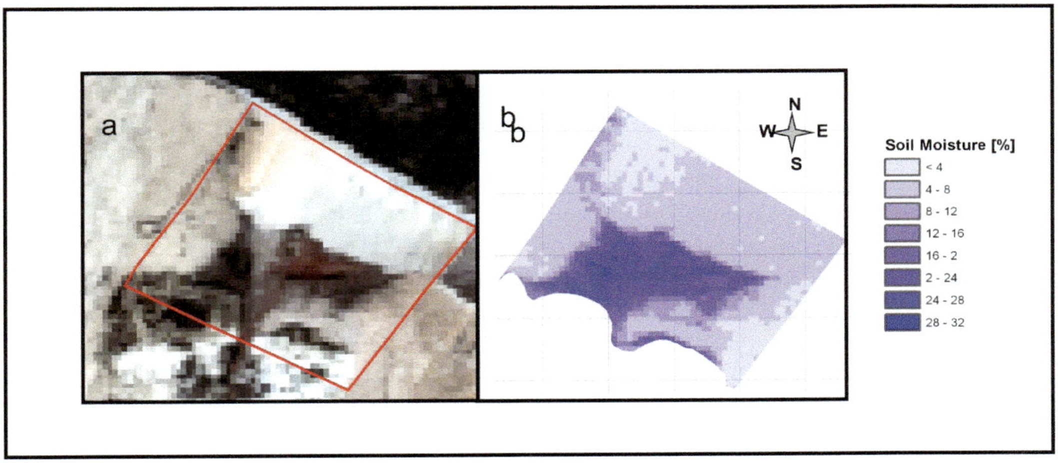

FIGURE 16.23 (a) Soil moisture content as estimated from HyMap images acquired on July 20, 2004, over Welzow, Gemmary, and (b) RGB image of the area encoded 0.619, 0.528, and 0.452 mm. (After Haubrock et al., 2008.)

for mapping of soil contamination with mercury. Another possible intercorrelation is with SOM. Malley and Williams (1997) showed that reflectance properties of sediments are associated with the content of OM, which acts as a chelating substance. As SOM has spectral fingerprints as previously discussed, the intercorrelation of the heavy metal bound to the SOM may enable extraction from the reflectance characteristics of the soil.

Several other studies have been published on the capacity of soil spectral information to detect heavy-metal content in soils. Wu et al. (2011) demonstrated that the intercorrelation of the non-spectral active constituents (Ni, Cr Co and Cd) with spectrally active soil components (Fe-oxides, SOM, and clay) is the major predictive mechanism. They showed that a correlation with total iron (including active and residual iron) is the major mechanism by which cadmium can be spectrally active in the soil environment. Looking towards showing a mechanism of public control of the environment, Araújo et al. (2014) indicated that VNIR, as combined with MWIR was able to detect Cr variation on soils caused by chemical products discarded from leather industries with a 0.93 R^2. The authors observed strong alteration on spectral features (i.e., in the 500 and 570 nm and 2,400 cm^{-1} bands) and lower intensities in all spectra when high application of the product was performed. Other workers have demonstrated similar capability to assess heavy metals spectrally, such as Pandit et al. (2010) who modeled the concentrations of lead and other heavy metals using soil reflectance, and concluded that reflectance spectroscopy has a promising potential to map the spatial distribution of lead abundance in soils. Another indirect way to study heavy-metal content in the root zone via remote sensing was demonstrated by Kooistra et al. (2004), Liu et al. (2010) and recently by Kopačková et al. (2014), who studied the relationship between the leaves' spectral response to the heavy-metal content in the root zone. They found a good correlation between leaf spectra and both aluminum and basic cations in the soil solution of the 0- to 20-cm soil (organic) horizon. Associating the results with HyMap HSR data on a small scale could indirectly result in mapping the organic horizon status of the soil underneath the vegetation. This work shows that an innovative idea can play an important role in exploiting reflectance properties in all domains to understand the soil condition in general, and pollution in particular. Recent papers reporting the use of high-spectral-resolution field and remote sensors indicate their high potential to detect heavy metals (e.g., Shi et al., 2018; Liu et al., 2017; Ghoziladeh et al., 2018). An interesting paper by Francos et al. (2022) demonstrated the spatial mapping of lead contamination over a contaminated area in the Czech Republic using a local spectral library.

Hydrocarbon contamination of soil is another important constituent that can be detected by reflectance spectroscopy (Douglas et al., 2018; Rabon and De Souza Filho, 2019). The spectral detection of hydrocarbons can be divided into two categories: direct sensing of hydrocarbons, and indirect sensing of minerals altered by the hydrocarbons. Whereas sensing of the first category relies on the spectral fingerprints of the hydrocarbon materials, the second category examines the soil matrix (minerals) that might be affected by hydrocarbon contamination. Direct detection using spectral means was reported by Malley et al. (1999), who concluded that reflectance spectroscopy has good potential to work in the laboratory. Later, Hörig et al. (2001), using a HyMap sensor, showed that HSR technology can detect hydrocarbon signals in an artificially contaminated soil environment. Another comprehensive work in this area is Winkelmann's thesis dissertation (2005), in which she systematically studied the applicability of HSR for detecting contaminated sites (including soils). Souza Filho (2013) and later, Rabon and De Souza Filho (2019) also demonstrated the ability of airborne HSR to quantitatively detect hydrocarbons from both SWIR and TIR spectral ranges. Detection via indirect sensing has been reported by several authors, such as Yang et al. (2000) and van der Meer et al. (2002), and also by Bihong et al. (2007) who used ASTER data to detect seepage from hydrocarbon reservoirs. A recent attempt to map an oil spill from Landsat data (combined with radar data) was performed by Espinosa-Hernandez (2013): Landsat images identified polluted areas over the bare soil. Several studies based on reflectance spectroscopy have demonstrated its ability to detect hydrocarbon species and their concentrations (Schwartz et al., 2011). Schwarz et al. (2012) also demonstrated that measurement of total petroleum hydrocarbon from reflectance spectroscopy is as good as measurements performed by three certified laboratories using wet chemistry analyses, and concluded that this approach has commercial applicability. Accordingly,

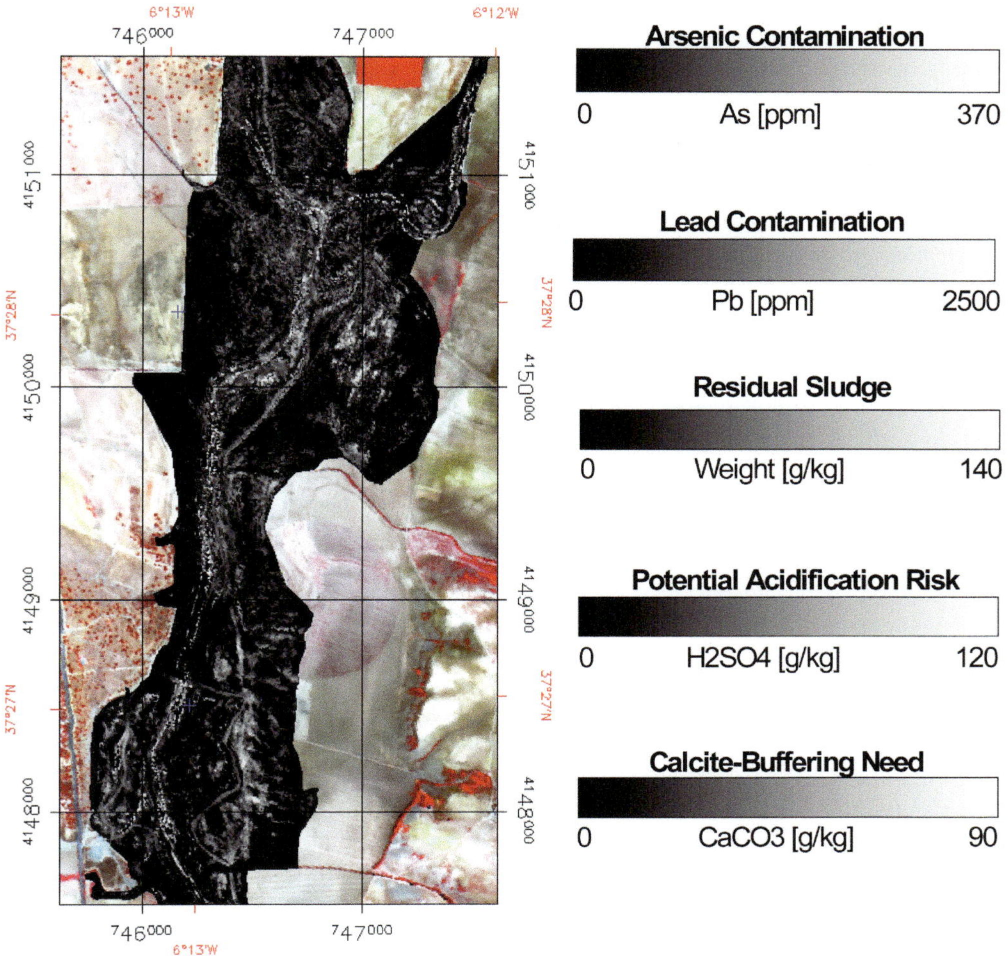

FIGURE 16.24 Contamination parameters derived from the sludge abundance image for June 1999, Arsenic and lead contaminations ppm, residual sludge in g/kg, maximum potential acidification after full oxidation of pyrite in g/kg and neutralization requirements for full oxidation of pyrite in g/kg. (Taken from Kemper T. and Sommers, *S. Environ. Sci. Technol.*, 36, 2742, 2003.)

a patent has been assigned for this application (US 20140012504 A1). Interpolating spectral measurements of total petroleum hydrocarbon in the field on a GIS-based map enabled demonstrating the spatial distribution of the main source of the oil contamination from within a gas station area. Figure 16.24 shows a case study of a contaminated gas station where point spectral measurements and GIS interpolation provides similar results as laboratory measurements of the same points.

In summary, it can be concluded that the reflectance properties of soils enable the assessment of various contaminants in their environment and that HSR technology is proving to be a promising tool for this purpose. Many more ideas and research directions are still open and workers are encouraged to further explore and study this promising field.

16.8.2.8 Soil Aggregation and Roughness

Many studies have shown that particle-size differences alter the shape of soil spectra (in powdered shape) (Hunt, 1982; Pieters, 1983; Baumgardner et al., 1985). Specifically, Hunt and Salisbury (1970) quantified an effect of about 5% in absolute reflectance due to particle-size differences, and showed that these changes occurred without altering the position of diagnostic spectral features. The physical process of the particle-size effect was related to changes in the real part of the refractive index resulting in attenuations in scattering and shading. Under field conditions, aggregate size rather than particle size may be more dominant in altering soil spectra (Orlov, 1966; Baumgardner et al., 1985). Soil aggregation is related to cementing agents in the soil such as clay, Fe-oxides and OM. Ben-Dor et al. (2008) demonstrated that rubification (the coating of quartz particles with Fe-oxides) is responsible for sand-dune stabilization due to the cementing effect of Fe-oxide. In the field, aggregate size may change over a short time frame due to tillage, soil erosion, artificial contamination (e.g., sugarcane byproduct), aeolian accumulation or physical crust formation (e.g., Dahms, 1993). Basically, the aggregate size, or more likely roughness, plays a major role in the shape of the soil spectra acquired in the field and air domains (e.g., Cierniewski, 1987, 1989; Feingrish et al., 2007). Escadafal and Hute (1991) showed strong anisotropic reflectance properties of five soils with rough surfaces. Cierniewski (1987) developed a model to account for soil roughness based on the soil reflectance parameter, illumination properties and viewing geometry for both forward and backward slopes. The model showed that the shading coefficient of the soil surface decreases with decreasing soil roughness. For soils on forward slopes of more than 20°, the shadowing coefficient also decreased when the solar altitude increased throughout the range of 0° to 90°. The model indicated that the opposite relationship might hold for soil slopes with surface roughness lower than 0.5 for a specified range of solar altitudes. Using empirical observations of smoothed soil surfaces, Cierniewski (1987) showed that the model agrees closely with field observations. A brief and excellent summary on the multiple- and single-scattering models of soil particles with respect to the roughness effect is given by Irons et al. (1989). Soil aggregation is strongly correlated to the content of cementing agents in the soil such as OM, Fe-oxides, carbonates and clay. Selige et al. (2006) studied the variability of field topsoil texture and SOM using HyMap airborne hyperspectral imagery. They found a good correlation between reflectance and two cementing agents in the soil: OM and soil texture, suggesting that SOM is the direct chromophore for soil texture. A recent study by Ben-Dor et al. (2022) demonstrated how the microaggregation stage can be quantitatively assessed by proximal sensing using these agents, and also found that the smectite clay type plays a major role.

Cierniewski et al. (2014) evaluated the fit between HSR bidirectional reflectance data of soil surfaces formed by a cultivator, a pulverizing harrow and a smoothing harrow, collected under field conditions (as illuminated by direct and diffuse solar radiation), to their bidirectional reflectance equivalents measured in the laboratory with only a direct radiation component and soil roughnesses similar to those in the field. They found that the fit increased from 400 to 450 nm, and decreased notably for wavelengths between 1950 and 2300 nm. A less significant decrease in fit was revealed at around of 700, 940 and 1140 nm. This again shows the constraint encountered when upscaling from the laboratory to field domains as a results of the soil's roughness characteristic. Piekarczyk et al. (2015) performed a controlled study using broadband (VIS) albedo characteristics to study the optimal conditions for remote sensing of soils (in terms of radiation and general geometry). Using five different roughnesses of the same soil with the same solar zenith angle, the difference between the optimal time and the available times from current Sun-synchronous satellites was examined (using NOAA-15 and Moderate Resolution Imaging Spectroradiometer [MODIS]). It was found that MODIS, crossing the equator at 1030 h in an orbit that is far from optimal for the albedo approximation, is much less useful than the NOAA-15, crossing the equator at 0730 h. Another conclusion drawn from that research was that the relationship between the broadband (blue sky) albedo of a bare cultivated soil surface and the angle of the solar zenith clearly depends on the soil roughness. A recent paper by Shi et al. (2018) investigated the relationship of spectroscopy with SOM and micro- and macroaggregates in 700 fields in Belgium using the airborne APEX HSR sensor. They were able to map the aggregation status of the soil from the spectral response of the sensor with a reliable accuracy of $R^2 > 0.5$, RPD > 1.4.

In summary, it can be concluded that soil aggregation size may affect the reflectance properties of a soil surface, with a significant impact on the overall spectral signal. The spectral features that are associated with aggregation are related to chromophore soil properties such as clay, Fe-oxides and OM. As the roughness may be associated with soil degradation, deposition or other pedogenic effects, it can serve as an indicator for the remote-sensing of these processes, but needs special attention to generate a quantitative model for that purpose.

16.8.2.9 Soil Sealing (Cover, Dust and Crust)

Soil crusts and covers can be formed by different processes. The biogenic crust is one example of such interference, and mineral alteration by fire is another. Aeolian material and desert varnish are also good examples of surface crusts. Biogenic activity can be captured by remote-sensing means via what is called a biogenic crust. This crust is made up of cyanobacteria and strongly affects the general albedo of the soil reflectance as well as the SWIR region absorption features (Karnieli, 1999). The biogenic crust shows photosynthetic activity when the soil water content is on the rise, and is less active under dry conditions. Demattê et al. (1998) observed that the spectral behavior of the biological aggregates changes according to their chemical composition, and is associated with the micro and macro fauna activity in the soil pedon. They speculated that the animals bring soil particles from below the surface to the surface, which can be distinguished using field spectroscopy. The crust is active in arable semiarid to arid soils, where cultivation is rare and vegetation even more so. Karnieli and Tsoar (1995) showed that overgrazing destroys the biogenic crust to the point that significant albedo differences can be obtained from satellite views. Similarly, they showed that these differences are significant across the Egypt—Israel border based on overgrazing activity on the Egyptian side and relatively low grazing on the Israeli side. Other studies have demonstrated similar effects across the globe (Lucht et al., 2000).

Another type of crust on the soil surface is termed "physical crust." This crust is formed by raindrops' energy (Morin et al., 1981), which causes segregation of fine particle sizes at the surface of the soil and hence affects its reflectance characteristics. This crust reduces water infiltration and accordingly, increases runoff, resulting in soil erosion. The crusting effect is more pronounced in saline soils and has been well studied in relation to the mineralogical and chemical changes in the soil surface (Sheinberg, 1992; Agassi et al., 1981) that also affect soil roughness (Cierniewski et al., 2013). The immediate observation after a rainstorm is enhanced "hue" and "value" of the soil color because of an increase in the fine fraction on the surface. One can assume that the reflectance spectrum of the "physical crust" will be totally different from that of the original soil, because it contains a greater clay fraction with a different textural component. In a study conducted by Goldshleger et al. (2009), using soils from Israel and the USA under artificial rain conditions, the spectral feature of clay minerals enabled detecting the mechanism of crust formation and distortion (under heavy rainstorm energy). The spectral changes observed at the soil surface were caused by changes in the soil's texture (fine-fraction enrichment), structure (from loose to compact), roughness and mineralogy (clay minerals rather than primary minerals). Several innovative studies have shown a significant relationship between spectral information and the infiltration rate of water into the soil profile as measured in the laboratory (e.g., Ben-Dor et al., 2003; Goldshleger et al., 2005). The next requisite step in the rain-simulation studies was to test the use of an airborne HSR sensor to characterize a structural crust in the field (Ben-Dor et al., 2004). Using a spectral-based index (the normalized spectral area; NSA), they were able to generate a possible erosion-hazard map of the soil area (Figure 16.25). An important question based on that finding was whether a generic spectral model could describe the crust's status, rather than the kinetics of the formation process. A study by Goldshleger et al. (2009) showed that the spectral model used to predict crust status might be more robust than originally thought. By using four soils from Israel and three soils from the USA subjected to rain events in a rain simulator, promising results were obtained using a combined prediction equation for infiltration rate, with a cross-validation RMSE of 15.2% and a prediction-to-deviation ratio of 1.98.

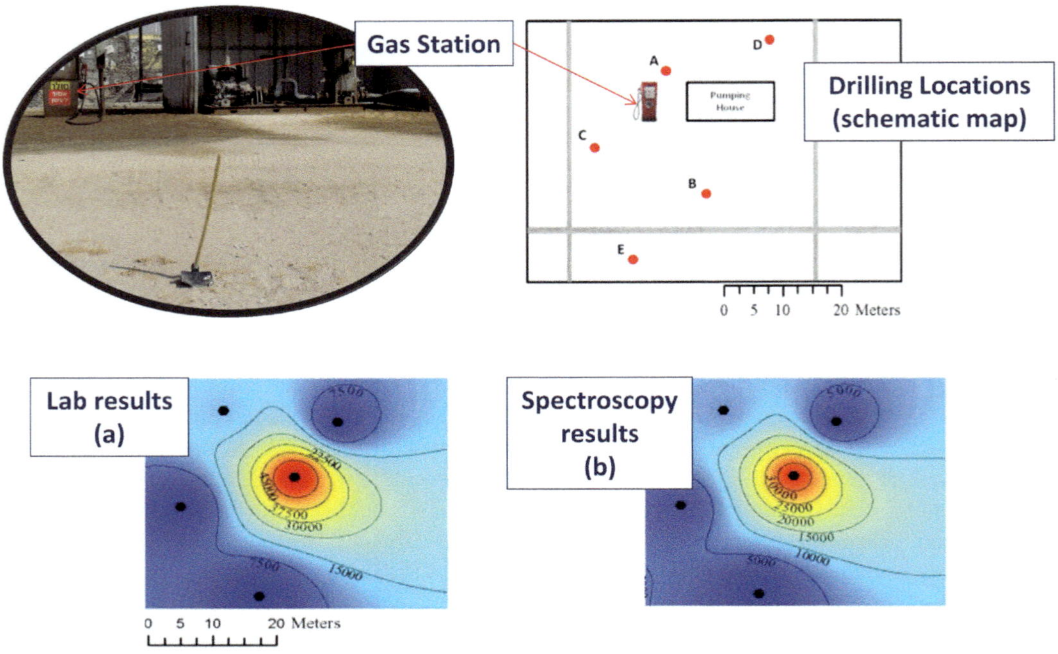

FIGURE 16.25 A contamination soil site with petroleum and the interpolation map generated from five samples using TPH analysis at the laboratory (a) and spectral based model in the field (b).

A recent study by Francos et al. (2023) demonstrated the possibility of assessing the soil crust in the field using a Cubert® HSR sensor onboard a drone platform. Using the SoilPRO® device—which can capture the surface reflectance in the field of an undisturbed soil surface—and measurements of the water-infiltration rate into the soil using microinfiltrometers, they were able to model the water-infiltration rate solely from spectroscopy in both sandy and clay soils (with different functions), and generate a water-infiltration rate map of a selected field with high spatial resolution.

Dust is another factor that can hinder remote sensing of the soil surface. The wind blowing the dust also has abrasive effects on the soil surface. Chappell et al. (2005) investigated the effect of soil-structure changes due to rain and wind-tunnel events. Their results showed that the spectral information can shed more light on the soil composition and structure generated by these two factors (rain crust and aeolian abrasion). Dust has two major effects: 1) it contaminates the atmosphere with a spectral signature that seems to interfere with soil surface sensing, and 2) it can accumulate on the soil surface as a thin layer that masks the true soil characteristics. One of the first observations of the effect of dust contamination on soil spectral information was reported by Montegomery and Baumgardner (1974). In a more recent study, Chudnovsky et al. (2009) demonstrated that an aeolian plume of Sahara dust has a significant effect on the clay mineral signals obtained from Hyperion data. They pointed out that the mineralogy of the dust plume can hinder the surface's spectral fingerprint. Dust accumulation on the soil surface has not yet been comprehensively studied. This is mainly because the spectral features of dust and the soil background may have similar features, and it is difficult to separate their contributions. This is unlike what can be seen over a vegetated background, especially if the dust is bright (Ben-Dor and Levin, 1999). Desert dust is most visible on a dark background (whereas anthropogenic dust is most visible on a light background) and hence is strongly affected by the underlying soil albedo. Li et al. (2013) demonstrated the impact of dust from a soil surface on nearby snow-covered areas as observed from MODIS, and found significant indications of the dust-source area's characteristics. Musavi et al. (2013) demonstrated the influence of dust on four different soils using MODIS thermal data. They concluded that different soil types have different effects on surface-dust detection due to their spectral mixing process, and demonstrated the connection between the real soil and the overlying accumulated dust. Okin and Paimter (2004) studied the effect of soil surface texture on spectral reflectance from the Mojave Desert. Sand plumes, eroded from the fields by wind, transported by saltation, and deposited downwind of the fields were studied based on reflectance and on its correlation with grain-size distribution in the direction of wind transport. Analysis of AVIRIS-derived apparent surface reflectance demonstrated the expected negative correlation between effective grain size of the sand in the plume and reflectance, with the most significant correlations being in the SWIR region. The change in reflectance per mm change in particle diameter was −0.06 at λ~1.7 µm and −0.08 at λ~2.2 µm with $R^2 = 0.89$ and 0.93, respectively.

NASA's recent space mission EMIT was designed to study the contribution of dust in the atmosphere based on soil minerology and particle size (Green and Thompson, 2020). The EMIT HSR sensor has been mounted on the ISS since 2022 and has acquired thousands of images worldwide. The sensor is based on the AVIRIS heritage and provides a 60-m spatial resolution with very high spectral quality (Green, 2023).

Another type of soil seal is the crust formed on the soil surface during fire events. This crust reduces infiltration and advances erosion. Postfire soil-seal effects have been described by Larsen et al. (2009) and Mataix-Solera et al. (2011). Studies by Lugassi et al. (2010, 2014) showed that the high fire temperatures can alter the surface soil mineralogy, thereby hindering the sensing of the real natural surface of soils that have been subjected to fire. They showed that Fe-oxide species formed during the heating process can be good indicators of the temperature of the fire long after it has gone out. Other reflectance changes in burned soils have been described by Kokaly et al. (2007).

In summary, it can be concluded that the effects on top of the soil surface ("sealing") are important and can be clearly observed. Although they hinder correct sensing of the real soil body, the spectral effects can provide added information on processes in the "sealed" soils, such as water infiltration (in the case of a physical crust), water runoff (in the case of burned soil) and soil aggregation.

16.8.2.10 Soil Iron

Just as OM acts as an important indicator for soils, Fe-oxides provide significant information on the soil's formation and conditions (Schwertmann, 1988). Fe-oxide content and species are strongly correlated with short- and long-term soil processes. An example of a short-term process is fire, whereas weathering exemplifies a long-term process. Fe-oxide transformation often occurs under natural soil conditions. Hematite and goethite are common Fe-oxides in soils and their relative content is strongly controlled by soil temperature, water regime, OM and annual precipitation. Fe-oxide is the major chromophore in the VNIR region, contributing a "red-brown" color to the soil. It was thus obvious that remote-sensing imagery would prove successful in assessing Fe-oxide coverage using soil color (Escadafal, 1993), and if spectral information was available, by modeling the absorption features of the active chromophore. Fe-oxides (and Fe-hydroxides) have specific absorption features that are located across the VNIR region (based on electronic processes) and can be estimated from multispectral or imaging spectrometer images (Abrams and Hook, 1995). However, due to the occurrence of spectral oversampling, it is still problematic to quantitatively assess the soil's exact Fe-oxide content (Deller, 2006). Different Fe-oxide species have different colors. Hematitic soils are reddish and goethitic soils are yellowish-brown to yellowish. Hematite (α-Fe_2O_3) has Fe^{3+} ions in octahedral coordination with oxygen. Goethite (α-FeOOH) also has Fe^{3+} in octahedral coordination, but different site distortions along with the oxygen ligand (OH) provide the main absorption features that appear near 0.9 µm. Lepidocrocite (γ-FeOOH), which is associated with goethite but rarely with hematite, is another common unstable Fe-oxide found in soils. It appears mostly in subtropical regions and is often found in the upper subsoil position (Schwertmann, 1988). Maghemite (γ-Fe_2O_3), is also found in soils, mostly in subtropical and tropical regions, but also occasionally in humid temperate areas. Ferrihydrite is a highly disordered Fe3+ oxide mineral found in soils in cool or temperate moist climates, characterized by young Fe-oxide formations and soil environments that are relatively rich in other compounds (e.g., OM, silica, etc.). Fe-oxides are secondary minerals that are sensitive indicators of pH, Eh, relative humidity, and other environmental conditions. This enabled monitoring mine-waste remediation activities over contaminated soils (Crowley et al., 2003; Zabric et al., 2005). Iron-bearing minerals in the soil precipitate, ordered according to pH, are: jarosite, pH < 3; schwertmannite, pH 2.8–4.5; mixtures of ferrihydrite and schwertmannite, pH 4.5–6.5; ferrihydrite or a mixture of ferrihydrite and goethite, pH > 6.5. Based on the Fe-oxides' spectral features, accordingly Zabcic et al. (2009), Kopačhková and others were able to map surface pH, as shown in the Figure 16.26 example.

Iron associated with clay mineral structures is also an active chromophore in both the VNIR and SWIR spectral regions. This can be seen in the nontronite-type mineral presented in Figure 16.27. Based on the structural OH—Fe features of smectite in the SWIR region, Ben-Dor and Banin (1990a) generated a predictive equation to account for the total iron content in a series of smectite minerals. The wavelengths that were selected automatically by their method were 2.2949 μm, 2.2598 μm, 2.2914 μm, and 1.2661 μm. Stoner (1979) also observed a higher correlation between reflectance in the 1.55 to 2.32 mm region and iron content in soils, whereas Coyne et al. (1989) found a linear relationship between total iron content in montmorillonite and absorbance measured in the 0.6 to 1.1 μm spectral region. Ben-Dor and Banin (1995a) used spectra of 91 arid soils to show that their total

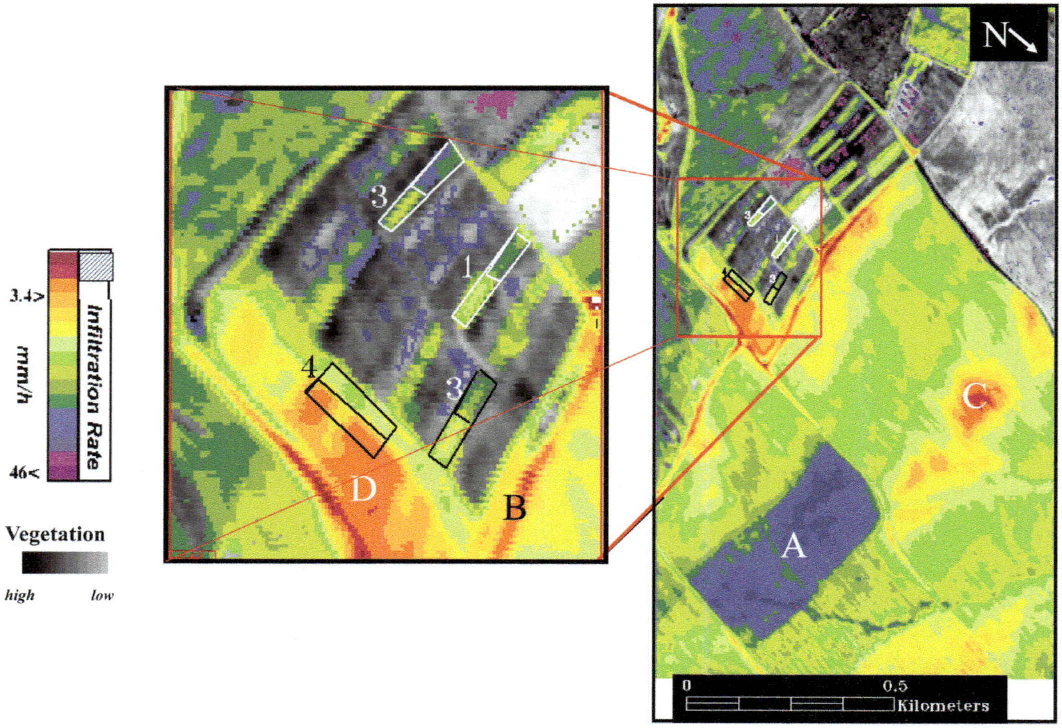

FIGURE 16.26 A false colored scaled image to account for the water infiltration rate of Loess soil as estimated from the hyperspectral data acquired by the airborne AISA Eagle sensor. The spectral information was modeled to describe the physical soil crust evolved from the rain drop energy. Given also four controlled plots where the crust was broken by a gentle cultivation ("cold" colors) and those left with the natural crust ("warm" colors). Area A represents a favorable infiltration capability and area C a hazardous area with high heavy runoff potential. (From Ben-Dor et al., 2004.)

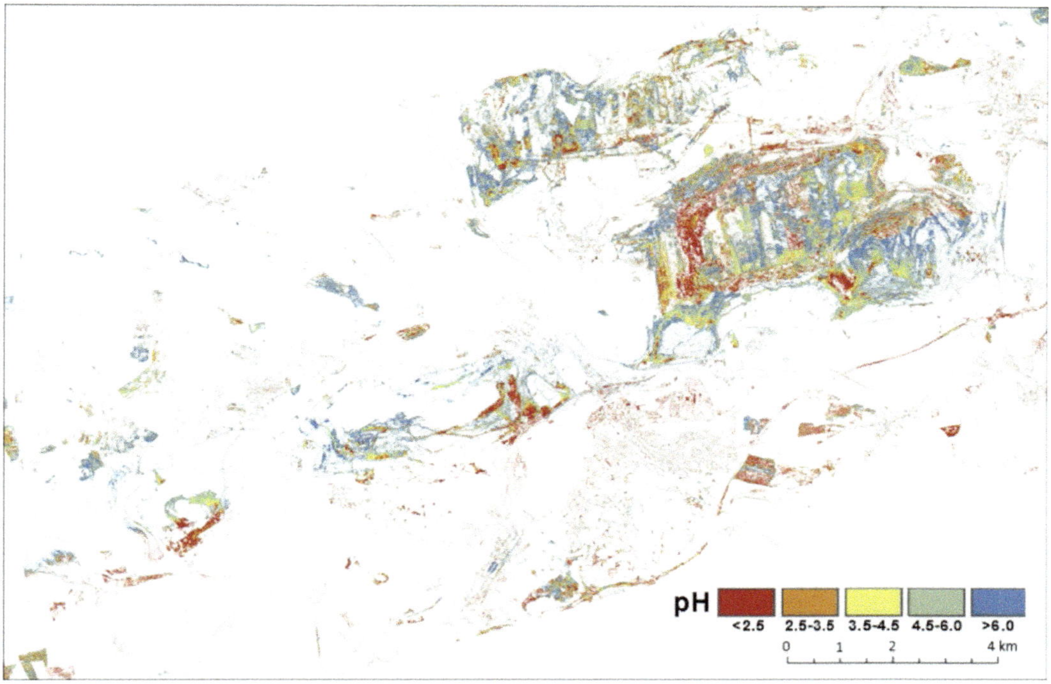

FIGURE 16.27 A pH h map of Sokolov lignite basin Czech Republic as obtained from a spectral model based on iron oxides minerals. (After Kopačková V. *Int. J. Earh Observ. Geoinform.*, 28, 28, 2014.)

iron content (both free and structural iron) can be predicted by multiple linear regression analysis, and wavelengths 1.075 μm, 1.025 mm and 0.425 μm. Obukhov and Orlov (1964) generated a linear relationship between reflectance values at 0.64 μm and the total percentage of Fe_2O_3 in other soils. Taranik and Kruse (1989) showed that a binary encoding technique for the spectral-slope values across the VNIR spectral region is capable of differentiating a hematite mineral from a mixture of hematite—goethite—jarosite. It is important to mention that iron can often have an indirect influence on the overall spectral characteristics of soils. In the case of free Fe-oxides, it is well known that soil particle size is strongly related to absolute Fe-oxide content (Soileau and McCraken, 1967; Stoner and Baumgardner, 1981; Ben-Dor and Singer, 1987): as the Fe-oxide content increases, the size fraction of the soil particles increases as well, because of the cementing effect of the free Fe-oxides. As a result, problems resulting from different scattering effects are introduced into the soil analysis. Moreover, free Fe-oxides, mostly in their amorphous state, can coat the soil particles with a film that prevents natural interaction between the soil particle (clay or non-clay minerals) and the Sun's photons. Fe-oxide minerals can be indicators for soil-stabilization processes (Ben-Dor et al., 2005). Karamanova (1981) found that well-crystallized iron compounds have the strongest effect on the spectral reflectance of soil, and that removal of non-silicate iron (mostly Fe-oxides) helps enhance other chromophores in the soil. In this respect, Kosmas et al. (1984) demonstrated a second-derivative technique in the VIS region as a feasible approach for differentiating even small features of synthetic goethite from clays, and they suggested that such a method be adopted to assess quantities of Fe-oxide in mixtures. Based on these spectral characteristics, Scott and Pain (2008) showed the possibility of spectrally assessing the alteration of regolith Gerbermann and Neher (1979) showed that soil mixtures of clay and sand can be predicted from reflectance spectra. The first to map Fe-oxides from the HSR domain were Ben-Dor et al. (2005), who modeled very low Fe-oxide absorption features in a sand dune in Ashdod, Israel, from CASI data. Using this approach,

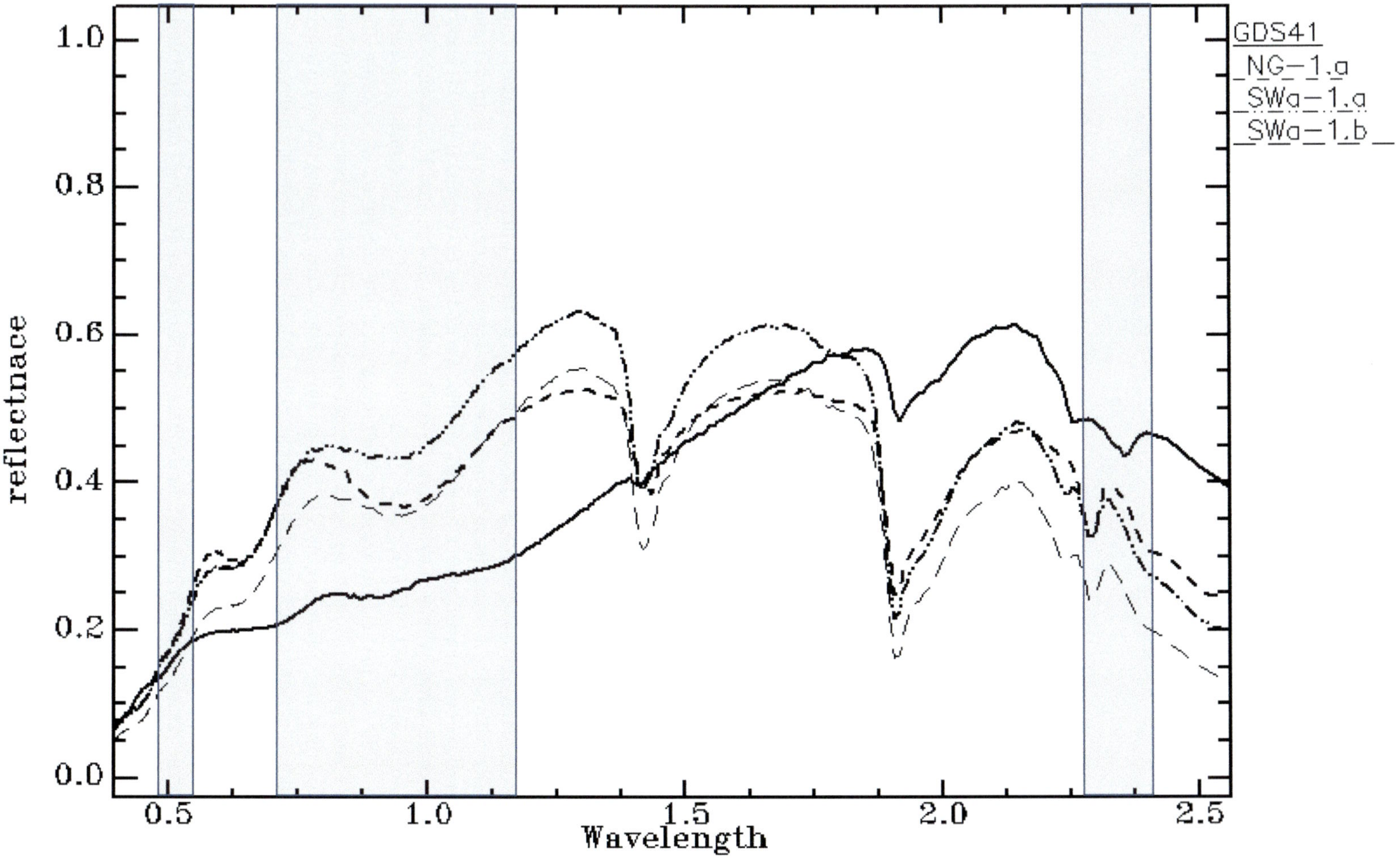

FIGURE 16.28 Reflectance spectra of three iron smectite minerals ("Nontronite") with emphasizing the SWIR spectral region where the structural iron combination mode of $\upsilon_{OH\text{-}Fe} + \delta_{OH\text{-}Fe}$ is active. Also given are possible electron transitions in the VIS-NIR region.

they were able to account for the rubification process in the sand dune and for dune stabilization and soil formation over the dune. In Figure 16.28, the area as seen from Landsat in natural RGB band combination and from CASI data which were modeled to extract the free Fe-oxide content is shown. Later, Bartholomeus et al. (2007) also applied HSR data from the ROSIS airborne image spectrometer and showed that quantification of soil iron content is possible over Mediterranean soils in Southern Spain. Fe_2O_3 was also well quantified and mapped by Landsat data by Demattê et al. (2009a). The authors also observed a high correlation between laboratory Fe_2O_3 spectral readings, with R^2 0.82 and orbital R^2 0.67 in a 500 ha area with 500 samples. Lugassi et al. (2010) showed that alteration of goethite to hematite in soils subjected to fire can be depicted from the soil spectra and further used as an indicator to assess the temperature of the fire long after it has gone out. Another recent study by Lugassi et al. (2014) showed that minerals other than iron are altered during fire events and they can be used to assess the fire characteristics long after the actual fire event. While crystalline iron, such as hematite occur in high weathered tropical soils amorphous occur in low ones. Demattê and Garcia (1999) observed that amorphous iron presents absence of spectral shape between 800–1000 nm, while crystalline irons are responsible for the convex shape. The same occur between 400–600 where Sherman and Waite (1985) indicate a narrow convex shape for goethite, and was corroborated by Demattê and Garcia (1999) which observed the absence of this convexity when extracted goethite. These findings corroborate the importance of descriptive evaluation of spectra and not only quantitative information. As the Fe-oxides' spectral features overlap with the OM spectral information, workers should be aware that proximal spectral-based modeling may be biased. Heller Pearlshtien and Ben-Dor (2020) demonstrated the effect of a mixture of Fe-oxide and OM on the overall prediction of each component. Based on their findings, Francos et al. (2023) demonstrated that the Fe-oxide content can be better analyzed spectrally if the OM is removed by heating the soil to >400°C.

It can thus be concluded that iron is a very strong chromophore in soil, and that its relationship with many of the soil's physical and chemical processes can be exploited by using its spectrum (or color) to track those processes. Based on the complexity of the iron component in the soil environment, as well as on the inter-correlation between iron and other soil components, sophisticated

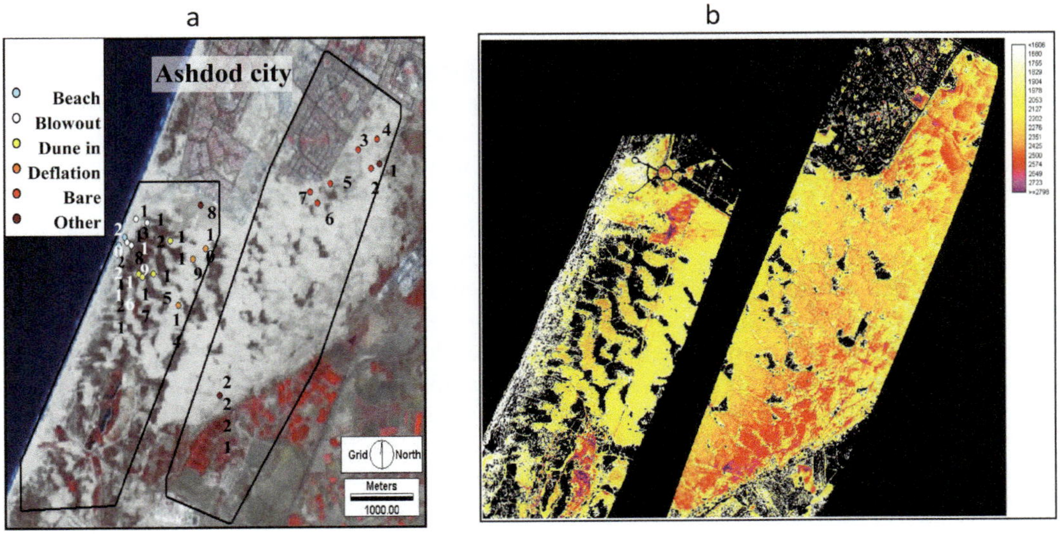

FIGURE 16.29 A color composites of Landsat 5 image (4,2,1) (a) and two strips of the hyperspectral airborne CASI sensors processed to map the iron oxide content, and (b) and sand dune under stabilization process near Ashdod city Israel. (Taken from Ben-Dor et al., *Geoderma* 131, 1, 2005.)

methods and data with relatively high spectral resolution are absolutely required to determine iron content from reflectance spectra.

16.8.2.11 Soil Classification and Taxonomy

As soil mapping requires grouping soil entities according to their chemical and physical characteristics, classification processes are a major issue. Soil classification is related to a taxonomic system that aims to give a common and agreed-upon definition and name to a given soil entity (Simonson, 1987). There are several such systems, the most common being the International Union if Soil Sciences (IUSS) classification protocol (FAO, 1998). These systems present the state and nature of several soils attributes, as well as their horizontal characteristics across the soil profile, which are then combined to define the soil name (class). Spectrally based soil classification was introduced by Stoner et al. (1980) as an Atlas which was published afterwards by Stoner and Baumgardner (1981), who were able to group American soils into five spectral categories This was the first method introduced to indicate soil entities based on soil spectra and was well studied. The limitation of this first soil's spectral Atlas was that it used only surface soil samples that could not yet represent the soil profile as needed for soil classification. Other methods projected the spectral signatures into quantitative domain and determine several soil properties solely from the reflectance information (Dalal and Henry 1986; Ben-Dor and Banin, 1995a,b,c). Later, Demattê et al. (2002) proposed a complementary method to assist in descriptive spectral evaluations, with soil classification as the major goal. The authors indicated three major spectral aspects: complete shape, reflectance intensity and absorption features. They also attempt for the necessity to evaluate spectra from each horizon, extract soil diagnostic information, compare spectra between them and merge all data to reach soil classification. In fact, they observed that Ultisols have different spectra between A and B horizons mostly related to the clay gradient. An important descriptive difference was observed at about 2,000 nm where A and B horizon spectra change because A has less clay (with more quartz) and B more clay (with less quartz). In fact, quartz raises reflectance intensity in SWIR as observed by White et al. (1997). The opposite observation was found in in oxisols. Usually oxisol description in the tropics indicates that there are very few differences between horizons, for example, in clay content and color. In fact, Demattê et al (2001, 2004b) conducted the first work to indicate a practical approach on how to perform soil classification mission using reflectance spectra. Afterwards, several works indicated the importance of spectra interpretation from all horizons on a same profile (Vasques et al., 2014; Rizzo et al., 2014). Adopting a descriptive method, Bellinaso et al. (2010) created a database using spectra of all horizons from a given profile, indicating the importance of this characterization for soil classification. In this work they revealed 75% efficiency in classifying 236 tropical soil profiles from 5 Brazilian states. Soil spectroscopy can also be used to describe variations in soil attributes along the profile, and for soil classification (Ben-Dor et al., 2008). Odgers et al. (2011) developed a system for the continuous classification of soils of 262 profiles from Australia using the MWIR (2.5–5 μm) spectral region. Using the fuzzy K-means clustering algorithm, they showed a low level of confusion. Steffens and Buddenbaum (2013) studied diagnostic horizons of Luvisols through laboratory spectroscopic-imaging technique, indicating great potential for elucidating the processes and mechanisms of soil formation. Nanni et al. (2011) evaluated 18 types of soils and indicated that in many cases, there were differences between profiles that could only be depicted by a detailed spectral shape evaluation of samples taken from the soil-diagnostic horizons. In fact, the problem was how to evaluate spectra from different horizons at the same time using an automated system. To this end, Ben-Dor et al. (2008) were the first to introduce the 3HED assembly that is inserted into a small borehole and describes the profile's chemical variations by spectral-based analysis, without having to sample the profile. They demonstrated that this profile description was much better and faster than the traditional pedagogical method. This was the first goal on gaining information from below the soil surface, since opening 2×2 m^2 trenches is a difficult, highly costly and time-consuming task in traditional work. A recent study by Murad et al. (2022) demonstrated that the 3HED idea can be further used to profile organic carbon under varying soil moisture contents with a penetrometer and spectroscopy. Elaborating on more tools for these descriptive evaluations, Demattê et al. (2014b) described the Multiple Interpretations of Reflectance Spectra method (MIRS) which relies on a detailed evaluation of spectral shapes, intensities, features, angles and complete behavior across the VNIR-SWIR region, crossing all of the diagnostic horizons along the soil's profile. Automated spectral-based systems are a more objective tool, as demonstrated by Vasques et al. (2014), which almost completely avoids interpreter subjectivity. These authors proposed the insertion of spectra from each horizon from the same profile in the same system, observing spectra across the entire VNIR-SWIR region from three sequential horizons (A, B1, B2), thus analyzing the spectra of the whole profile as a continuum, and reported 90% efficiency for soil-classification processes. Rizzo et al. (2014) ratified these findings and revealed a high correlation when classifying tropical soil profiles with the OSACA program (Carré and Jacobson, 2009). Other researchers also succeeded to classify soils based on spectroscopy, such as Nanni et al. (2004) who used stepwise multivariable analysis with 91% accuracy, Du et al. (2008) who used IR photoacoustic spectroscopy (FTIR-PAS) with 96% accuracy, and Linker (2008) who used MWIR spectroscopy with 95% accuracy. Using VNIR-SWIR descriptive analysis, Demattê et al. (2012) distinguished 17 profiles of tropical soils derived from different parent materials. This was ratified by Cezar et al. (2013), who obtained 70% agreement with traditional field classification. Both stated that statistical methods are important, but cannot detect all of the particularities "inside" the spectral information, and thus human expertise and interpretation are still required. Rizzo et al. (2014) used satellite images with different spatial, spectral and temporal resolutions and concluded that slightly different soil property maps can be evaluated from different sensors; they called for caution when using multiple sensors to classify and manage soils. Soil color is a significant parameter for soil classification. Although by remote sensing, only the surface is exposed to the sensor eyes, the surface color still holds important information regarding soil type. Rizzo et al. (2023) used temporal Landsat images (1985–2020) to show that around 39% of the bare Earth surface's color can be estimated, providing a vast understanding of soil type and condition.

Whereas soil spectroscopy can generate soil attributes objectively it cannot provide the morphological view of the soil which plays an important role in the soil classification process. Another important property that the soil spectroscopy can provide is the objectively determination of the boarders between soil horizons. This allow an accurate profile spectral pattern description which can then projected into "profile pattern" library accompanied with the soil classification information as obtained by expert. This library can be used then to classify unknown soil profile by applying multispectral classifier approaches. This technique was demonstrated by Demattê et al. (2004b), who determined soils classification by the descriptive correlation of known spectral patterns. To assist soil classification, image data can be used to support the traditional field work by indicating homogeneity and topography of the area in question (both are playing a major role in the soil classification process). Demattê et al. (2009b) showed an error of 0.028% using discriminant analysis of Landsat images for soil classification. Each (non-vegetated) pixel along the study area was classified in the field using below-surface information and traditional method. The results a indicated high performance between the traditional soil classification and the pixel (spectral) information. The spectra of the surface of each classified soil was different in intensities and tendencies on shape. Accordingly the authors, although image captures only the surface information, it has a significant relationship with the below-surface soil properties that enabling discrimination between soil classes solely from the satellite image. Nonetheless, readers need to realize that this is a very delicate procedure, and must be taken and evaluated with care, because in some cases different soil classes can have a same surface information. This strength the need of expert to evaluate and validate the remote sensing results before transfer them to the end users for operation.

Classification processes therefore represent an important application that has progressed significantly in the last few years due to the use of soil spectroscopy with field spectrometers, and airborne (manned and unmanned) and orbital HSR sensors, along with multi- and superspectral sensors such as the Landsat TM and Sentinel 2. Spectral information is particularly powerful for drawing the soil's boundaries more accurately than traditional aerial photography, and for deriving information on the soil's horizon across the soil profile, by either measuring cores or using fore optics that penetrate the soil profile. Nonetheless, for soil-mapping purposes, we strongly feel that merging remote-sensing technologies will advance soil-mapping missions. HSR with LIDAR or HSR with field profile spectroscopy are good examples of this.

A few success stories of soil classification using Landsat images strongly suggest that the HSR technology from orbit has promising potential and warrants further research. Now that this technology is available, we anticipate advances in the field. Nevertheless, the problems involved in measuring the spectral information of the soil horizons conveniently, along with the limitations in assessing soil morphology from spectroscopy are issues that need to be further studied as well.

To summarize, soil classification process as an accurate information process is an important application that has progressed significantly in the last few years due to the use of soil spectroscopy mostly in the field and laboratory. The good results from Landsat for classifying soil pixels indicate the very promising potential of the emerging HSR technology from orbit. On the other hand, soil classification by remote optical means still suffers from the inconvenience involved in determining all of the horizons in a given profile and to a lesser extent, from the fact that spectroscopy cannot provide the soil morphology.

16.8.2.12 Soil Mapping

Soil mapping involves gathering the soil's components into logical groups based on characteristics of the soil surface and profile according to taxonomical rules. Soil mapping is a basic stage in assessing a soil's agricultural potential for present and future activity. It is a difficult, time-consuming and expensive task that requires difficult field and laboratory work along with professional manpower and a complex infrastructure. Soil mapping is carried out by expert soil surveyors who have a broad background in soil formation and genesis and are often termed pedologists. An important reference to understand soil mapping is Legros (2006) who provided an excellent perspective on how and why remote sensing is an important tool for soil mapping. Along with the spatial information required for soil-mapping processes, soil-field observations such as with auger (boreholes) are used to assist detection of soil boundaries. On a next step, soil profile information is needed to classify the delineated polygon. Moreover, for the spatial domain, the soil surface's chemical and physical complexity also poses problems, significantly constraining the mapping process. Whereas the profile information is difficult to obtain with optical sensors, spatial variation can be obtained by remote sensing and, as has been demonstrated, with better accuracy if the spectral resolution is high. Airborne photographs have been intensively used in the past by soil surveyors to assist in soil-mapping procedures. The spatial variations of the soil surface were determined by gray-scale analysis or color interpretation of "soil" boundaries. Vink (1963a) presented the information that can be culled from aerial photographs for soil pedological mapping, and demonstrated the added value of this tool. Demattê and Garcia (1999) showed a significant relationship between spectral information and aerial photographs (drainage patterns) which could then be utilized in the field for soil pedological mapping. In another study, aerial photographs were used to map relief variations and were s exploited to pinpoint areas for spectral sampling and measurement reaching a soil map (Demattê et al., 2001). This approach reached 90% accuracy with traditional field systems and demonstrated the strength of soil spectroscopy. Use of multispectral sensors from orbit enables locating pure soil pixels for soil sampling and false-alarm detection of non-pure (partially vegetated or covered with litter) soil pixels. Accordingly, Demattê et al. (2009b) developed a sequence-based technique to indicate an area of bare soil using remote-sensing means. This method used atmospheric-correction data, then evaluated pixels based on the soil-line concept, use of color compositions of 5,4,3 and 3,2,1 (red, green blue of Landsat), calculating soil-adjusted and normalized-difference vegetation indices, and mostly, judging the spectral surface shape based on a soil databank from both ground and space (Landsat) domains. Based on this technique, several workers succeeded to quantify soil-surface properties of tropical Brazilian soils (Demattê and Nanni, 2003; Nanni and Demattê, 2006) using Landsat data which was then projected to obtain better soil-classification results (Demattê et al., 2007, 2009a; Nanni et al., 2012). These later authors observed that merging geological maps with spectral profiles increases mapping accuracy. Gaining experience with both field, aerial and orbital data, Demattê et al. (2001, 2004b) obtained 85% agreement between soil maps traditionally generated by soil surveyors and those using their remote-sensing method. As indicated by Legros (2006) one of the most important step in soil mapping is determine soil delineation and afterwards make their classification. But where is the limit of soils? Looking towards this task, Demattê et al. (2004b) and recently Demattê and Terra (2014) demonstrated that soils can be differentiated along toposequences by a *Spectral Pedology* analysis and thus, can detect their limits. In fact these authors observed the close relationship between the pedogenetic alterations along the toposequence which goes towards the detection of their limits. Thus, if we replicate several toposequences in a certain area, we will have several limits which can be linked using relief information (by aerial photographs for example, as indicated by Vink, 1963b) and reach the soils poligons. This method does not implicate necessarily with soil classification, instead only perform the "figure" of the map. In a subsequent step, inside the polygons, we would indicate the best spot to account for a profile; go inside with a sensor (see Viscarra Rossel et al., 2009) or make a borehole with a fiber optic method (see Ben-Dor et al., 2008). These spectral measurements can assists the classification of the soil entity and then further use with the image to generate the soil map.

In a different approach, Löhrer et al. (2013) identified and mapped the mineralogical composition of Pleistocene sediments using ASTER and SPOT-5 images associated with spectral laboratory work and also obtained good results. Although the orbital sensors see only the top 50 μm of the soil surface, the production of good soil-classification results indicates that the surface has, in some cases, a connection with below-surface dynamics of soil processes. Galvão et al. (2001) used AVIRIS data to show that surface-reflectance values and constituents (total iron, OM, TiO_2, Al_2O_3, and SiO_2) represent three important soil types from central Brazil [Terra Roxa Estruturada (S_{TE}), Latossolo Vermelho-Escuro (S_{LE}), and Areia Quartzosa (S_{AQ})]. Nonetheless, Demattê et al. (2009a) stated that some soils cannot be differentiated, mainly those that require morphological interpretation of the soil profiles. With the emergence of the new HSR technology (along with other multi- and superspectral data from orbit), the spectral approach was further extended to the notion of digital soil mapping (McBratney et al., 2003). Up until now, digital soil mappers have mainly used remote-sensing images as spatial-data inputs representing the landscape variables that are related to soil, such as vegetation, topography and parent material (soil covariates). Boettinger et al. (2008) reviewed the main indicators that could be retrieved from multispectral images for estimating these soil covariates. Along these lines, a modern remote-sensing approach can serve as a tool to assist soil scientists in obtaining up-to-date and accurate information on the soil surface in question. The recent implementation of HSR technology for the digital soil-mapping approach (Lagacherie et al., 2013) makes a significant contribution to soil-surface characterization, adding more information to both aerial photography and multispectral information. As most of the Earth's soil areas are still unmapped, and those that have been mapped need updating, new methods (such as HSR technology) to map and classify soils are crucial. This issue is most relevant today with the rapidly growing world population, as it could lead to better utilization of soil resources, a critical issue for feeding mankind.

Despite the constraints, reported in Section 16.7, to utilizing HSR technology, it has the proven capability to extract both quantitative and qualitative information on the soil surface, thereby providing a better way to account for the surface's complexity and variations.

To summarize, spectral information is particularly powerful for detecting soil limits and spots with punctual soil information with increasing observation density. Observation density is the most important rule we have to attempt on a soil mapping and is directly related with scale of the final product. This can assist on drawing the soil's boundaries, and will mostly show better use if we achieve bare soil in continuous areas as indicated by several methodologies stated before.

If combined these observations with traditional aerial photography, and for deriving information on the soil's horizon across the soil profile, by either measuring cores or using fore optics that penetrate the soil profile, we will certainly reach the best on soil mapping. Nonetheless, for soil-mapping purposes, we strongly feel that merging remote-sensing technologies will advance soil-mapping missions. HSR in both optical and thermal regions along with an active system such as LIDAR or synthetic aperture radar (SAR) accompanied by spectral information on the field surface and soil profile are good examples of this. Dealing with such a huge amount of data will also require adopting the AI approach, where machine and deep-learning technologies should be developed.

16.9 SUMMARY AND CONCLUSIONS

This chapter summarizes most of the key studies in soil remote sensing using the VNIR—SWIR spectral region, and provides the basic theory for optical remote sensing of soil. It demonstrates the significant progress that has been made in remote-sensing technology in the last 20 years. This progress has enabled the exposure and validation of many sensors from all domains for scientific and commercial use. Remote sensing of the soil is thus entering a new era, in which more data, information, infrastructure and applications can be provided more efficiently to the end user. This calls for further development of both electro-optic technology and data-mining algorithms. Mulder et al. (2011) partially summarized the soil applications for remote sensing by all means, emphasizing mainly the optical region. Although their summary ignored some of the fundamental work done with optical remote sensing of soil at that time, their conclusion is important in terms of understanding the need for remote-sensing data with high temporal and spectral resolution. The new understanding of several worldwide decision makers on the importance of soils as a primary factor in food production and with a role in environmental preservation places the remote sensing of soils at the forefront in mankind's survival.

The future will call for merging sensors, databases, data science and know-how, developing more accurate methods for data exchange, and generating more robust models to be executed by many users. Another aspect that will need further attention is the establishment of standards and protocols for the acquisition of soil spectra in both the laboratory and field domains, as well as the development of quality indicators and assurance for the new remote-sensing data. Today's space programs that are placing HSR sensors in orbit are enabling better temporal resolution and wide coverage of all soils worldwide. This calls for more attention to this direction and adoption of the comprehensive know-how that has already been achieved in soil spectroscopy. Upscaling reflectance data from the ground to the air and space domains is still a bottleneck that needs to be overcome. Mulder (1987) stated that every work exploiting remote-sensing data requires field work, to either verify or expand the information on the soil profile. This has led us to realize that new ideas on how to extract soil-profile characteristics are important but still lacking. The pioneering idea of Ben-Dor et al. (2008), later adopted by Wijewardane et al., 2020, to sense the profile "endoscopically" provides a proof of concept for remote-sensing measurements of soil profiles but needs to be further developed. Combining such a method with the spatial view obtained by remote-sensing means (mainly HSR) may pave the way for future activity in the remote sensing of soils from afar. Other directions are also welcome, and more ideas on how to improve the remote sensing of soils are required. Data science that employs an AI approach should be adopted and further developed. Nonetheless, the current achievements in remote sensing of soils are remarkable considering that not long ago, analog gray-scale aerial photos were the basic, and indeed only tool to remotely sense soils from afar. As new satellite missions covering all resolutions (spatial, spectral and temporal) are being planned, and generally available to all, it is strongly anticipated that the remote sensing of soil will undergo further development, utilization and exploitation to monitor soil status, benefiting soil productivity for all of mankind.

REFERENCES

Aber, J., Wessman, C. A., Peterson, D. L., Mellilo, J. M., & Fownes, J. H. 1990. Remote sensing of litter and soil organic matter decomposition in forest ecosystems. In: Remote Sensing of Biosphere Functioning, R. J. Hobbs and (eds.), 87–101.

Abrams, M., & Hook, S. J. 1995. Simulated ASTER data for geologic studies. Geoscience and Remote Sensing, IEEE Transactions on, 33, 692–699.

Adar, S., Shkolnisky, Y., & Ben Dor, E. 2014a. A new approach for thresholding spectral change detection using multispectral and hyperspectral image data, a case study over Sokolov, Czech republic. International Journal of Remote Sensing, 35, 1563–1584.

Adar, S., Shkolnisky, Y., & Ben-Dor, E. 2014b. Change detection of soils under small-scale laboratory conditions using imaging spectroscopy sensors. Geoderma, 216, 19–29.

Agassi, M., Shainberg, I., & Morrin, J. 1981. Effects of electrolyte concentration and soil sodicity on infiltration and crust formation. Soil Science Society of America Journal, 45, 848–851.

Agdu, P. A., Fehrenbacher, J. D., & Jansen, I. 1990. Soil property relationships with SPOT satellite digital data in the East central Illinois. Soil Science Society of American Journal, 54, 807–812.

Aikenhead, J., & Black, R. 2018. Spectral Library of Soil and Vegetation Reflectance. CRC Press, Boca Raton, FL.

Al-Abbas, H. H., Swain, H. H., & Baumgardner, M. F. 1972. Relating organic matter and clay content to multispectral radiance of soils. Soils Science, 114, 477–485.

Allory, V., Cambou, A., Moulin, P., Schwartz, C., Cannavo, P., Vidal-Beaudet, L., & Barthès, B. G. 2019. Quantification of soil organic carbon stock in urban soils using visible and near infrared reflectance spectroscopy (VNIRS) in situ or in laboratory conditions. Science of the Total Environment, 686, 764–773.

Alonso, K., Bachmann, M., Burch, K., Carmona, E., Cerra, D., de los Reyes, R., Dietrich, D., Heiden, U., Hölderlin, A., Ickes, J. 2019. Data Products, Quality and Validation of the DLR Earth Sensing Imaging Spectrometer (DESIS). Sensors, 19, 4471.

Araújo, S. R., Demattê, J. A. M., & Bellinaso, H. 2013. Analyzing the effects of applying agricultural lime to soils by VNIR spectral sensing: A quantitative and quick method. International Journal of Remote Sensing, 34, 4570–4584.

Araújo, S. R., Demattê, J. A. M., & Vicente, S. 2014. Soil contaminated with chromium by tannery sludge and identified by Vis-Nir-Mid spectroscopy techniques. International Journal of Remote Sensing.

Asner, G. P., & Heidebrecht, K. B. 2002. Spectral unmixing of vegetation, soil and dry carbon cover in arid regions: Comparing multispectral and hyperspectral observations. International Journal of Remote Sensing, 23, 3939–3958.

Asner, G. P., & Heidebrecht, K. B. 2003. Imaging spectroscopy for desertification studies: Comparing AVIRIS and EO-1 Hyperion in Argentina drylands. Geoscience and Remote Sensing, IEEE Transactions on, 41, 1283–1296.

Bach, H., & Mauser, W. 1994. Modelling and model verification of the spectral reflectance of soils under varying moisture conditions. Geoscience and Remote Sensing Symposium, 1994. IGARSS'94: Surface and Atmospheric Remote Sensing: Technologies, Data Analysis and Interpretation International, 4, 2354–2356.

Bachmann, M., et al. 2021. The EnMAP spaceborne imaging spectroscopy mission: Overview and first results. *Remote Sensing*, 13, no. 8, 1486.

Balsam, W. L., & Deaton, B. C. 1996. Determining the composition of late Quaternary marine sediments from NUV, VIS, and NIR diffuse reflectance spectra. Marine Geology, 134, 31–55.

Barnes, E., M., Sudduth, K. A., Hummel, J. W., Lesch, S. M., Corwin, D. L., Yang, G., Doughtry, C. S. T., & Bausch, W. C. 2003. Remote-and ground-based sensor techniques to map soil properties. Photogrammetric Engineering and Remote Sensing, 69, no. 6, 619–630.

Barrett, B. W., & Petropoulos, G. P. 2013. Satellite remote sensing of surface soil moisture. Remote Sensing of Energy Fluxes and Soil Moisture Content, 85.

Bartholomeus, H., Epema, G., & Schaepman, M. 2007. Determining iron content in Mediterranean soils in partly vegetated areas, using spectral reflectance and imaging spectroscopy. International Journal of Applied Earth Observation and Geoinformation, 9, 194–203.

Baumann, P. R. 2010. History of Remote Sensing, Aerial Photography. Available at: www.oneonta.edu/faculty/baumanpr/geosat2/RS%20History%20I/RS-History-Part-1.htm (Accessed: 17 June 2014).

Baumgardner, M. F., Kristof, S. J., Johannsen, C. J., & Zachary, A. L. 1970. Effects of organic matter on multispectral properties of soils. Proceedings of the Indian Academy of Science, 79, 413–422.

Baumgardner, M. F., Silva, L. F., Biehl, L. L., & Stoner, E. R. 1985. Reflectance properties of soils. Advances in Agronomy, 38, 1–44.

Beck, R. H., Robinson, B. F., McFee, W. H., & Peterson, J. B. 1976. Information Note 081176. Laboratory Application of Remote Sensing, Purdue University, West Lafayette, Indiana.

Bedidi, A., Cervelle, B., & Madeira, J. 1991. Moisture effects on spectral signatures and CIE-color of lateritic soils. Proceedings of the 5th International Colloquium, Physical Measurements and Signatures in Remote Sensing, Courchevel, France I, 209–212.

Bedidi, A., Cervelle, B., Madeira, J., & Pouget, M. 1990. Moisture effects on spectral characteristics (visible) of lateritic soils. Soil Science, 153, 129–141.

Bellinaso, H., Demattê, J. A. M., & Araújo, S. R. 2010. Spectral library and its use in soil classification. Revista Brasileira de Ciência do Solo, 34, 861–870.

Ben-Dor, E. 2002. Quantitative remote sensing of soil properties. Advances in Agronomy, 75, 173–243.

Ben-Dor, E., & Singer, A. 1987. Optical density of vertisol clays suspensions in relation to sediment volume and dithionite-citrate-bicarbonate extractable iron. Clays and Clay Minerals, 35, 311–317.

Ben-Dor, E., & Banin, A. 1990a. Diffuse reflectance Spectra of smectite minerals in the near infrared and their relation to chemical composition. Sciences Geologiques Bull, 43, no. 24, 117–128.

Ben-Dor, E., & Banin, A. 1990b. Near infrared reflectance analysis of carbonate concentration in soils. Applied Spectroscopy, 44, 1064–1069.

Ben-Dor, E., & Banin, A. 1994. Visible and near infrared (0.4–1.1mm) analysis of arid and semiarid soils. Remote Sensing of Environment, 48, 261–274.

Ben-Dor, E., & Banin, A. 1995a. Near infrared analysis (NIRA) as a rapid method to simultaneously evaluate, several soil properties. Soil Science Society of American Journal, 59, 364–372.

Ben-Dor, E., & Banin, A. 1995b. Near infrared analysis (NIRA) as a simultaneously method to evaluate spectral featureless constituents in soils. Soil Science, 159, 259–269.

Ben-Dor, E., & Banin, A. 1995c. Quantitative analysis of convolved TM spectra of soils in the visible, near infrared and short-wave infrared spectral regions (0.4–2.5mm). International Journal of Remote Sensing, 18, 3509–3528.

Ben-Dor, E., Chabrillat, S., Demattê, J. A. M., Taylor, G. R., Hill, J., Whiting, M. L., & Sommer, S. 2009. Using imaging spectroscopy to study soil properties. Remote Sensing of Environment, 113, S38–S55.

Ben-Dor, E., Francos, N., Ogen, Y., & Banin, A. 2022. Aggregate size distribution of arid and semiarid laboratory soils (< 2 mm) as predicted by VIS-NIR-SWIR spectroscopy. Geoderma, 416, 115819.

Ben-Dor, E., Goldshlager, N., Benyamini, M., & Blumberg, D. G. 2003. The spectral reflectance properties of soil's structural crust in the SWIR spectral region (1.2–2.5μm). Soil Science Society of American Journal, 67, 289–299.

Ben-Dor, E., Goldshleger, N., Braun, O., Kindel, B., Goetz, A. F. H., Bonfil, D., Margalit, N., Binaymini, Y., Karnieli, A., & Agassi, M. 2004. Monitoring infiltration rates in semiarid soils using airborne hyperspectral technology. International Journal of Remote Sensing, 25, no. 13, 2607–2624.

Ben-Dor, E., Granot, A., & Notesco, G. 2017. A simple apparatus to measure soil spectral information in the field under stable conditions. Geoderma, 306, 73–80.

Ben-Dor, E., Granot, A., Wallach, R., Francos, N., Pearlstein, D. H., Efrati, B., . . . & Schmid, T. 2023. Exploitation of the SoilPRO®(SP) apparatus to measure soil surface reflectance in the field: Five case studies. Geoderma, 438, 116636.

Ben-Dor, E., Inbar, Y., & Chen, Y. 1997. The reflectance spectra of organic matter in the visible near infrared and short wave infrared region (400–2,500nm) during a control decomposition process. Remote Sensing of Environment.

Ben-Dor, E., Irons, J. A., & Epema, A. 1999. Soil spectroscopy. In: Manual of Remote Sensing, 3rd edition, A. Rencz (ed.), J. Wiley & Sons Inc, New-York, Chichester, Weinheim, Brisbane, Singapore, Toronto, 111–189.

Ben-Dor, E., & Kruse, F. A. 1995. Surface mineral mapping of Makhtesh Ramon Negev, Israel using GER 63 channel scanner data. International Journal of Remote Sensing, 18, 3529–3553.

Ben-Dor, E., & Levin, N. 1999. Determination of surface reflectance from raw hyperspectral data without simultaneous ground truth measurements: A case study of the GER 63-channel sensor data acquired over Naan Israel. International Journal of Remote Sensing of Environment, 21, 2053–2074.

Ben-Dor, E., Levin, N., Singer, A., Karnieli, A., Braun, O., & Kidron, G. J. 2005. Quantitative mapping of the soil rubification process on sand dunes using an airborne hyperspectral sensor. Geoderma, 131, 1–21.

Ben-Dor, E., Ong, C., & Lau, I. C. 2015. Reflectance measurements of soils in the laboratory: Standards and protocols. Geoderma, 245, 112–124.

Ben-Dor, E., Patkin, K., Banin, A., & Karnieli, A. 2002. Mapping of several soil properties using DAIS-7915 hyperspectral scanner data: A case study over clayey soils in Israel. International Journal of Remote Sensing, 23, 1043–1062.

Ben-Dor, E., Schläpfer, D., Plaza, A. J., & Malthus, T. 2013. Hyperspectral remote sensing. In: Airborne Measurements for Environmental Research: Methods and Instruments, 413–456.

Ben-Dor, E., Taylor, R. G., Hill, J., Demattê, J. A. M., Whiting, M. L., Chabrillat, S., & Sommer, S. 2008. Imaging spectrometry for soil applications. Advances in Agronomy, 97, 321–392.

Ben-Gera, I., & Norris, K. H. 1968. Determination of moisture content in soybeans by direct spectrophotometry. Israeli Journal of Agriculture Research, 18, 124–132.

Bernard, F., Bourgeois, G., Manolis, I., Barat, I., Alamanac, A. B., Such-Taboada, M., . . . & Vega, I. C. 2023, July. The LSTM instrument: Design, technology and performance. International Conference on Space Optics—ICSO 2022, 12777, 1712–1729. SPIE.

Bihong, F., Zheng, G., Nonomiya, Y., Wang, C., & Sum, G. 2007. Mapping hydrocarbon-induced mineralogical alteration in the northen Tian Shan using ASTER multispectral data. Terra Nova, 19: 225–231.

Boettinger, J. L., Ramsey, R. D., Bodily, J. M., Cole, N. J., Kienast-Brown, S., Nield, S. J., & Stum, A. K. 2008. Landsat spectral data for digital soil mapping. In Digital Soil Mapping with Limited Data, Springer, The Netherlands, 193–202.

Bowers, S., & Hanks, R. J. 1965. Reflectance of radiant energy from soils. Soil Science, 100, 130–138.

Bresson, L. M., & Boiffin, J. 1990. Morphological characterization of soil crust development stages on an experimental field. Geoderma, 47, 301–325.

Brown, D. J., Shepherd, K. D., Walsh, M. G., Dewayne Mays, M., & Reinsch, T. G. 2006. Global soil characterization with VNIR diffuse reflectance spectroscopy. Geoderma, 132, 273–290.

Buddenbaum, H., Stern, O., Stellmes, M., Stoffels, J., Pueschel, P., Hill, J., & Werner, W. 2012. Field imaging spectroscopy of beech seedlings under dryness stress. Remote Sensing, 4, 3721–3740.

Buffet, L., Gamet, P., Maisongrande, P., Salcedo, C., & Crebassol, P. 2021, June. The TIR instrument on TRISHNA satellite: A precursor of high resolution observation missions in the thermal infrared domain. International Conference on Space Optics—ICSO 2020, 11852, SPIE, 300–310.

Buol, S. W., Hole, F. D., & McCracken, R. J. 1973. Soil Genesis and Classification. Iowa State University Press, Ames.

Bushnell, T. M. 1932. A new technique in soil mapping. American Soil Survey Association Bulletin, 13, 74–81.

Cambou, A., Cardinael, R., Kouakoua, E., Villeneuve, M., Durand, C., & Barthès, B. G. 2016. Prediction of soil organic carbon stock using visible and near infrared reflectance spectroscopy (VNIRS) in the field. *Geoderma*, 261, 151–159.

Carré, F., & Jacobson, V. 2009. Numerical classification of soil profile data using distance metrics. Geoderma, 148, 336–345.

Casa, R., Castaldi, F., Pascucci, S., Palombo, A., & Pignatti, S. 2013. A comparison of sensor resolution and calibration strategies for soil texture estimation from hyperspectral remote sensing. Geoderma, 197, 17–25. https://doi.org/10.1016/j.geoderma.2012.12.016.

Castellan, G. W. 1983. Physical Chemistry, 3rd edition. Addison-Wesley Publishing Co, Reading, MA, 943.

Cawse-Nicholson, K., Townsend, P. A., Schimel, D., Assiri, A. M., Blake, P. L., Buongiorno, M. F., . . . & SBG Algorithms Working Group. 2021. NASA's surface biology and geology designated observable: A perspective on surface imaging algorithms. Remote Sensing of Environment, 257, 112349.

Cezar, E., Nanni, M. R., Chicati, M. L., Oliveira, R. B., & Demattê, J. A. M. 2013. Discriminação entre solos formados em região transicional por meio de resposta alinize. Bioscience Uberlândia, 29, no. 3, 644–654.

Chappell, A., Zobeck, T., & Brunner, G. 2005. Induced soil surface change detected using on-nadir spectral reflectance to characteristics soil erodibility. Earth Surface Processes and Landforms, 30, 489–511.

Chen, F., Kissel, D. E., West, L. T., & Adkins, W. 2000. Field-scale mapping of surface soil organic carbon using remotely sensed imagery. Soil Science Society of America Journal, 64, 746.

Chen, F., Kissel, D. E., West, L. T., Adkins, W., Rickman, D., & Luvall, J. C. 2006. Feature selection and similarity analysis of crop fields for mapping organic carbon concentration in soil. Computers and Electronics in Agriculture, 54, 8–21.

Chen, F., Kissel, D. E., West, L. T., Adkins, W., Rickman, D., & Luvall, J. C. 2008. Mapping soil organic carbon concentration for multiple fields with image similarity analysis. Soil Science Society of America Journal, 72, 186–193.

Chen, Y., & Inbar, Y. 1994. Chemical and spectroscopical analysis of organic matter transformation during composting in relation to compost maturity. In: Science and Engineering of Composting: Design, Environmental, Microbiology and Utilization Aspects, H. A. J. Hoitink and H. M. Keener (eds.), Renaissance Publications, Worthington, OH, USA, 551–600.

Chevrel, S. 2005. MINEO two years later-did the project impulse a new era in imaging spectroscopy applied to mining environments. Proceedings of the 4th Workshop on Imaging Spectroscopy, Imaging Spectroscopy. New Quality in Environmental Studies, Warsaw, 26–29.

Chevrel, S. 2013. Earth observation in support of mineral industry sustainable development-the EO-MINERS project. 23rd World Mining Congress 2013 Proceedings, Montréal.

Christensen, P. R., & the EMIT Team. 2022. The Earth surface mineral dust source investigation (EMIT) mission: Objectives and capabilities. *Remote Sensing*, 14, no. 20.

Chudnovsky, A., Ben-Dor, E., Kostinski, A. B., & Koren, I. 2009. Mineral content analysis of atmospheric dust using hyperspectral information from space. Geophysical Research Letters, 36, L15811.

Cierniewski, J. 1987. A model for soil surface roughness influence on the spectral response of bare soils in the visible and near infrared range. Remote Sensing of Environment, 23, 98–115.

Cierniewski, J. 1989. The influence of the viewing geometry of bare soil surfaces on their spectral response in the visible and near infrared. Remote Sensing of Environment, 27, 135–142.

Cierniewski, J., Karnieli, A., Kuśnierek, K., & Herrmann, I. 2013. Proximating the average daily surface albedo with respect to soil roughness and latitude. International Journal of Remote Sensing, 34, 9–10, 3416–3424.

Cierniewski, J., Kaźmierowski, C., Królewicz, S., Piekarczyk, J., Wróbel, M., & Zagajewski, B. 2014. Effects of different illumination and observation techniques of cultivated soils on their hyperspectral bidirectional measurements under field and laboratory conditions. IEEE Journal of Selected Topics in Applied Earth Observations and Remote Sensing, 7, no. 6, 2525–2530.

Cipra, J. E., Franzmeir, D. P., Bauer, M. E., & Boyd, R. K. 1980. Comparison of multispectral measurements from some nonvegetated soils using Landsat digital data and a spectroradiometer. Soil Science Society of American Journal, 44, 80–84.

Clark, R. N. 1999. Spectroscopy of rocks and minerals, and principles of spectroscopy. Manual of Remote Sensing, 3, 3–58.

Clark, R. N., & Roush, T. L. 1984. Reflectance spectroscopy: Quantitative analysis techniques for remote sensing applications. Journal of Geophysical Research, 89, 6329–6340.

Cogliati, S., et al. 2021. PRISMA mission overview and first results: A new era of hyperspectral Earth observation. *Remote Sensing*, 13, no. 4, 612.

Condit, H. R. 1970. The spectral reflectance of American soils. Photogrammetric Engineering, 36, 955–966.

Condit, H. R. 1972. Application of characteristic vector analysis to the spectral energy distribution of daylight and the spectral reflectance of American soils. Applied Optics, 11, 74–86.

Coyne, L. M., Bishop, J. L., Sacttergood, T., Banin, A., Carle, G., & Orenberg, J. 1989. Near-infrared correlation spectroscopy: Quantifying iron and surface water in series of variably cation-exchanged montmorillonite clays. In: Spectorscopic Characterization of Mineral and Their Surfaces, L. M. Coyne, S. W. S. McKeever and D. F. Blake (eds.), ACS Publication, San Jose State University, Oklahoma State University, NASA—Ames Research Center, 407–429.

Crowley, J. K., Williams, D. E., Hammarstrom, J. M., Piatak, N., Chou, I. M., & Mars, J. C. 2003. Spectral reflectance properties (0.4–2.5 μm) of secondary Fe-oxide, Fe-hydroxide, and Fe-sulphate-hydrate minerals associated with sulphide-bearing mine wastes. Geochemistry: Exploration, Environment, Analysis, 3, 219–228.

Csillag, F., Pasztor, L., & Biehl, L. L. 1993. Spectral band selection for the characterization of salinity status of soils. Remote Sensing of Environment, 43, 231–242.

Da-Costa, L. M. 1979. Surface soil color and reflectance as related to physicochemical and mineralogical soil properties. Ph.D. Dissertation, University of Missouri, Columbia, 154.

Dahl, M., McMahon, K., Lavery, P. S., Hamilton, S. H., Lovelock, C. E., & Serrano, O. 2023. Ranking the risk of CO2 emissions from seagrass soil carbon stocks under global change threats. Global Environmental Change, 78, 102632.

Dahms, D. E. 1993.Mineralogical evidence for eolian contribution to soils of late Quaternary moraines, Wind River Mountains, Wyoming, USA. Geoderma, 59,175–196.

Dalal, R. C., & Henry, R. J. 1986. Simultaneous determination of moisture, organic carbon and total nitrogen by near infrared reflectance spectroscopy. Soil Science Society of America Journal, 50, 120–123.

Debaene, G., Niedźwiecki, J., Pecio, A., & Żurek, A. 2014. Effect of the number of calibration samples on the prediction of several soil properties at the farm-scale. Geoderma, 214–215, 114–125.

Dehaan, R. L., & Taylor, G. R. 2002. Field-derived spectra of salinized soils and vegetation as indicators ofirrigation-induced soil salinization. Remote Sensing of Environment, 80, 406–417.

Dekker, A. G., Brando, V. E., Anstee, J. M., Pinnel, N., Kutser, T., Hoogenboom, H. J., Pasterkamp, R., Peters, S. W. M., Vos, R. J., Olbert, C., & Malthus, T. J. 2001. Imaging spectrometry of water, Ch. 11. In: Imssaging Spectrometry: Basic Principles and Prospective Applications: Remote Sensing and Digital Image Processing, Vol. 4, Dordrecht, Kluwer Academic Publishers, 307–335.

Deller, A. M. E. 2006. Facies discrimination in laterites using Landsat Thematic Mapper, ASTER and ALI data: Examples from Eritrea and Arabia. International Journal of Remote Sensing, 27, 2389–2409.

Demattê, J. A. M. 2002. Characterization and discrimination of soils by their reflected electromagnetic energy. Pesquisa Agropecuaria Brasileira, 37, no. 10, 1445–1458.

Demattê, J. A., Alves, M. R., Gallo, B. C., Fongaro, C. T., Romero, D. J., & Sato, M. V. 2015. Hyperspectral remote sensing as an alternative to estimate soil attributes. Revista Ciência Agronômica, 46, 223–232.

Demattê, J. A. M., Bellinaso, H., & Romero, D. J. 2014a. Morfological interpretation of reflectance of spectrum (MIRS) for soil evaluation. Scientia Agricola, 71, no. 6, 509–520.

Demattê, J. A. M., Bellinaso, H., Romero, D. J., & Fongaro, C. T. 2014b. Morphological Interpretation of Reflectance Spectrum (MIRS) using libraries looking towards soil classification. Scientia Agricola, 71, 509–520.

Demattê, J. A. M., Campos, R. C., Alves, M. C., Fiorio, P. R., & Nanni, M. R. 2004b. Visible-NIR reflectance: A new approach on soil evaluation. Geoderma, 121, 95–112.

Demattê, J. A. M., Demattê, J. L. I.; Camargo, W. P.; Fiorio, P. R.; & Nanni, M. R. 2001. Remote sensing in the recognition and mapping of tropical soils developed on topographic sequences. Mapping Sciences and Remote Sensing, 38, no. 2, 79–102.

Demattê, J. A., Dotto, A. C., Paiva, A. F., Sato, M. V., Dalmolin, R. S., Maria do Socorro, B., . . . & do Couto, H. T. Z. 2019. The Brazilian soil spectral library (BSSL): A general view, application and challenges. Geoderma, 354, 113793.

Demattê, J. A. M., Fiorio, P. R., & Ben-Dor, E. 2009a. Estimation of soil properties by orbital and laboratory reflectance means and its relation with soil classification. Open Remote Sensing Journal, 2, 12–23.

Demattê, J. A. M., Galdos, M. V., Guimarães, R. V., Genú, A. M., Nanni, M. R., & Zullo Jr, J. 2007. Quantification of tropical soil attributes from ETM+/LANDSAT-7 data. International Journal of Remote Sensing, 28, 3813–3829.

Demattê, J. A. M., Gama, M. A. P., Cooper, M., Araújo, J. C., Nanni, M. R., & Fiorio, P. R. 2004a. Effect of fermentation residue on the spectral reflectance properties of soils. Geoderma, 120, 187–200.

Demattê, J. A. M., & Garcia, G. J. 1999. Alteration of soil properties through a weathering sequence as evaluated by spectral reflectance. Soil Science Society of America Journal, 63, 327–342.

Demattê, J. A. M., Huete, A. R., Ferreira Jr, L. G., Alves, M. C., Nanni, M. R., & Cerri, C. E. 2000. Evaluation of tropical soils through ground and orbital sensors. In International Conference Geospatial Information in Agriculture and Forestry, 2, 35–42.

Demattê, J. A. M., Huete, A. R., Ferreira Jr., L. G., Nanni, M. R., Alves, M. C., & Fiorio, P. R. 2009b. Methodology for bare soil detection and discrimination by Landsat TM image. Open Remote Sensing Journal, 2, 24–35.

Demattê, J. A. M., Mafra, L., & Bernardes, F. F. 1998. Comportamento alinize de materiais de solos e de estruturas biogênicas associadas [Soil spectral behavior associated with biogenic material]. Revista Brasileira de Ciência do Solo [Brazilian Journal of Soil Science], 22, 621–630.

Demattê, J. A. M., & Nanni, M. R. 2003. Weathering sequence of soils developed from basalt as evaluated by laboratory (IRIS), airborne (AVIRIS) and orbital € sensors. International Journal of Remote Sensing, 24, 4715–4738.

Demattê, J. A., Safanelli, J. L., Poppiel, R. R., Rizzo, R., Silvero, N. E. Q., Mendes, W. D. S., . . . & Lisboa, C. J. D. S. 2020. Bare earth's surface spectra as a proxy for soil resource monitoring. Scientific Reports, 10, no. 1, 4461.

Demattê, J. A. M., Sousa, A. A., Alves, M., Nanni, M. R., Fiorio, P. R., & Campos, R. C. 2006. Determing soil water status and other soil characteristics by spectral proximal sensing. Geoderma, 135, 179–195.

Demattê, J. A. M., & Terra, F. S. 2014. Spectral Pedology: A new perspective on evaluation of soils along pedogenetic alterations. Geoderma, 217–218, 190–200.

Demattê, J. A. M., Terra, F. S., & Quartarolli, C. F. 2012. Spectral behavior of some modal soil profiles from São Paulo State, Brazil. Bragantia, 71, 413–423.

Dixon, J. B., & Weed, S. B. 1989. Minerals in Soil Environments. Soil Science Society of Soil Science Society of America Publishing, Madison WI.

Dobos, E., Montanarella, L., Nègre, T., & Micheli, E. 2001. A regional scale soil mapping approach using integrated AVHRR and DEM data. International Journal of Applied Earth Observation and Geoinformation, 3, 30–42.

d'Oleire-Oltmanns, S., Marzolff, I., Peter, K. D., & Ries, J. B. 2012. Unmanned Aerial Vehicle (UAV) for monitoring soil erosion in Morocco. Remote Sensing, 4, 3390–3416.

Douglas, R. K., Nawar, S., Alamar, M. C., Mouazen, A. M., & Coulon, F. 2018. Rapid prediction of total petroleum hydrocarbons concentration in contaminated soil using vis-NIR spectroscopy and regression techniques. Science of the Total Environment, 616, 147–155.

Du, C., Linker, R., & Shaviv, A. 2008. Identification of agricultural Mediterranean soils using mid-infrared photoacoustic spectroscopy. Geoderma, 143, 85–90.

Dutkiewicz, A. 2006. Evaluating hyperspectral imagery for mapping the surface symptoms of dryland salinity. PhD dissertation, University of Adelaide, Australia.

Dvorakova, K., Heiden, U., & van Wesemael, B. 2021. Sentinel-2 exposed soil composite for soil organic carbon prediction. Remote Sensing, 13, no. 9, 1791.

Dwivedi, R. S. 2017. Hyperspectral Remote Sensing of Vegetation. CRC Press, Boca Raton, FL.

Elachi, C., & Van Zyl, J. J. 2006. Introduction to the Physics and Techniques of Remote Sensing, Vol. 28. John Wiley & Sons, Hoboken, NJ.

Eltahir, E. A. 1998. A soil moisture—rainfall feedback mechanism: 1. Theory and observations.Water Resources Research, 34, 765–776.

Eliran A., Goldshalger N, Yahalom A., Agassi M and **E. Ben Dor.** 2012. First results from a mill metric-wave measurement of soil moisture-content, Remote Sensing Letter 3, 639–645.

Epema, G. F. 1990. Effect of moisture content on spectral reflectance in a playa area in southern Tunisia. Proceedings International Symposium, Remote Sensing and Water Resources, Enschede, the Netherlands, 301–308.

Escadafal, R. 1993. Remote sensing of soil color: Principles and applications. Remote Sensing Reviews, 7, 261–279.

Escadafal, R., & Hute, A. R. 1991. Influence of the viewing geometry on the spectral properties (high resolution visible and NIR) of selected soils from Arizona. Proceedings of the 5th International Colloquium, Physical Measurements and Signatures in Remote Sensing, Courchevel, France I, 401–404.

Escobar, D. E., Everitt, J. H., Noriega, J. R., Cavazos, I., & Davis, M. R. 1998. A twelve-band airborne digital video imaging system (ADVIS). Remote sensing of environment, 66, 122–128.

Espinosa-Hernandez, A., Galvan-Pineda, J., Monsivais-Huertero, A., Jimenez-Escalona, J. C., & Ramos-Rodriguez, J. M. 2013. Delineation of hydrocarbon contaminated soils using optical and radar images in a costal region. International Geoscience and Remote Sensing Symposium (IGARSS), art. no. 6721247, 676–679.

Evans, F., & Caccetta, P. 2000. Broad-scale spatial prediction of areas at risk from dryland salinity. Cartography, 29, 33–40.

Everitt, J. D., Escobar, D. E., Gerbermann, A. H., & Alaniz, M. A. 1988. Detecting salin soils with video imagery. Photogrammetry Engineering and Remote Sensing, 54, 1283–1287.

FAO. 1998. World Reference Base for Soil Resources. Food and Agriculture Organization of the United Nations, Rome.

Farifteh, J., Bouma, A., & van der Meijde, M. 2004. A new approach in the detection of salt affected soils: Integrating surface and subsurface measurements. Proceedings of 10th European Meeting of Environmental and Engineering Geophysics, P059, Utrecht, the Netherlands.

Farifteh, J., Farshad, A., & George, R. 2006. Assessing salt-affected soils using remote sensing, solute modeling and geophysics. Geoderma, 130, 191–206.

Farifteh, J., van der Meer, F. M., Atzberger, C., & Carranza, E. J. M. 2007. Quantitative analysis of salt-affected soil reflectance spectra: A comparison of two adaptive methods (PLSR and ANN). Remote Sensing of Environment, 110, 59–78.

Feingersh, T., Ben-Dor, E., & Portugali, J. 2007. Construction of synthetic spectral reflectance imagery for monitoring of urban sprawl. Environmental Modeling and Software, 22, 335–348.

Fiorio, P. R., & Demattê, J. A. M. 2009. Orbital and laboratory spectral data to optimize soil analysis. Scientia Agricola, 66, 250–257.

Fisher, J. B., Lee, B., Purdy, A. J., Halverson, G. H., Dohlen, M. B., Cawse-Nicholson, K., . . . & Hook, S. 2020. ECOSTRESS: NASA's next generation mission to measure evapotranspiration from the international space station. Water Resources Research, 56, no. 4, e2019WR026058.

Fox, G. A., & Sabbagh, G. J. 2002. Estimation of soil organic matter from red and near-infrared remotely sensed data using a soil line Euclidean distance technique. Soil Science Society of America Journal, 66, 1922–1929.

Francos, M., Gohziladeh, S., Ben Dor, E., & Martens, J. 2022. Mapping heavy metals in soils using hyperspectral remote sensing: A case study. Remote Sensing, 14, no. 15, 3685.

Francos, N., & Ben-Dor, E. 2022. A transfer function to predict soil surface reflectance from laboratory soil spectral libraries. Geoderma, 405, 115432.

Francos, N., Chabrillat, S., Tziolas, N., Milewski, R., Brell, M., Samarinas, N., . . . & Ben-Dor, E. 2023. Estimation of water-infiltration rate in Mediterranean sandy soils using airborne hyperspectral sensors. Catena, 233, 107476.

Francos N., Nasta P, Allocca C., Sica B., Mazzitelli C., Lazzaro U., D'Urso G., Belfiore O.R., Crimaldi M., Sarghini F., Ben-Dor E., N. Romano . 2024. Mapping Soil Organic Carbon Stock Using Hyperspectral Remote Sensing: A Case Study in the Sele River Plain in Southern Italy, Remote Sens., 16(5), 897.

Frazier, A. E., & Hemingway, B. L. 2021. A technical review of planet smallsat data: Practical considerations for processing and using planetscope imagery. Remote Sensing, 13, no. 19, 3930.

Frazier, B. E., & Cheng, Y. 1989. Remote sensing of soils in the Eastern Palouse region with Landsat Thematic Mapper. Remote Sensing of Environment, 28, 317–325.

Frey, S. D., Six, J., & Elliott, E. T. 2003. Reciprocal transfer of carbon and nitrogen by decomposer fungi at the soil—litter interface. Soil Biology and Biochemistry, 35, 1001–1004.

Fu, S., Cabrera, M. L., Coleman, D. C., Kisselle, K. W., Garrett, C. J., Hendrix, P. F., & Crossley, D. A. 2000. Soil carbon dynamics of conventional tillage and no-till agroecosystems at Georgia Piedmont—HSB-C models. Ecological Modelling, 131, 229–248.

Gaffey, S. J. 1985. Reflectance spectroscopy in the visible and near infrared (0.35–2.55mm): Applications in carbonate petrology. Geology, 13, 270–273.

Gaffey, S. J. 1986. Spectral reflectance of carbonate minerals in the visible and near infrared (0.35–2.55µm): Calcite, aragonite and dolomite. American Mineralogist, 71, 151–162.

Gaffey, S. J., & Reed, K. L. 1987. Copper in calcite: Detection by visible and near infra-red reflectance. Economic Geology, 82, 195–200.

Galvão, L. S., Pizarro, M. A., & Epiphanio, J. C. N. 2001. Variations in reflectance of tropical soils: Spectral-chemical composition relationships from AVIRIS data. Remote Sensing of Environment, 75, 245–255.

Galvão, L. S., & Vitorello, I. 1998. Role of organic matter in obliterating the effects of iron on spectral reflectance and colour of Brazilian tropical soils. International Journal of Remote Sensing, 19, 1969–1979.

Galvão, L. S., Vitorello, I., & Paradella, W. L. 1995. Spectroradiometric discrimination of laterites with principle components analysis and additive modeling. Remote Sensing of Environment, 53, 70–75.

Gao, Bo-C., Davis, C. O., & Goetz, A. F. H. 2006. A review of atmospheric correction techniques for hyperspectral remote sensing of land surfaces and ocean color. 2006 IEEE International Symposium on Geoscience and Remote Sensing, 1979–1981.

Gao, Bo-C., Montes, M. J., Davis, C. O., & Goetz, A. F. H. 2009. Atmospheric correction algorithms for hyperspectral remote sensing data of land and ocean. Remote Sensing of Environment, 113, no. 1, 17–24.

García-Haro, F. J., Sommer, S., & Kemper, T. 2005. A new tool for variable multiple endmember spectral mixture analysis (VMESMA). International Journal of Remote Sensing, 26, 2135–2162.

Gausman, H. W., Allen, W. A., Cardenas, R., & Bowen, R. L. 1970. Color photos, cotton leaves and soil salinity. Photogrammetric Engineering, 36, no. 5, 454–459.

Gausman, H. W., Menges, R. M., Escobar, D. E., Everitt, J. H., & Bowen, R. L. 1977. Pubescence affects spectra and imagery of silverleaf sunflower (Helianthus argophyllus). Weed Science, 437–440.

Ge, Y., Thomasson, J. A., & Sui, R. 2011. Remote sensing of soil properties in precision agriculture: A review. Frontiers of Earth Science, 5, 229–238.

Genú, A. M., & Demattê, J. A. M. 2006. Determination of soil attribute contents by means of reflected electromagnetic energy. International Journal of Remote Sensing, 27, 4807–4818.

Gerbermann, A. H., & Neher, D. D. 1979. Reflectance of varying mixtures of a clay soil and sand. Photgrammertic Engineering and Remote Sensing, 45, 1145–1151.

Gersman, R., Ben-Dor, E., Beyth, M., Avigad, D., Abraha, M., & Kibreab, A. 2008. Mapping of hydrothermally altered rocks by the EO-1Hyperion sensor, Northern Danakil Depression, Eritrea. International Journal of Remote Sensing, 29, 3911–3936.

Ghosh, S., & Kumar, R. 2020. HySIS: An Indian hyperspectral imaging satellite for Earth observation. *Current Science*, 118, no. 8, 1215–1224.

Ghoziladeh, M., Li, X., & Zhan, L. 2018. Mapping heavy metals in soils using airborne hyperspectral imagery and machine learning techniques. Journal of Environmental Management, 226, 135–143.

Gmur, S., Vogt, D., Zabowski, D., & Moskal, L. M. 2012. Hyperspectral analysis of soil nitrogen, carbon, carbonate, and organic matter using regression trees. Sensors, 12, 10639–10658.

Goetz, A. F. H. 2009. Three decades of hyperspectral remote sensing of the Earth: A personal view. Remote Sensing of Environment, 113, S5–S16.

Goetz, F. A. H., Hauff, P., Shippert, M., & Maecher, G. A. 1991. Rapid detection and identification of OH-bearing minerals in the 0.9–1.0mm region using new portable field spectrometer. Proceeding of the 8th Thematic Conference on Geologic Remote Sensing, Denver, Colorado. I, 1–11.

Goidts, E., & van Wesemael, B. 2007. Regional assessment of soil organic carbon changes under agriculture in Southern Belgium (1955–2005). Geoderma, 141, 341–354.

Goldshleger, N., Ben-Dor, E., Benyamini., Y., & Agassi, M. 2005. Soil reflectance as a tool for assessing physical crust arrangement of four typical soils in Israel. Soil Science, 169, 677–687.

Goldshleger, N., Ben-Dor, E., Chudnovsky, A., & Agassi, M. 2009. Soil reflectance as a generic tool for assessing infiltration rate induced by structural crust for heterogeneous soils. European Journal of Soil Science, 60, 1038–1951.

Goldshleger, N., Chudnovsky, A., & Ben-Binyamin, R. 2013. Predicting salinity in tomato using soil reflectance spectra. International Journal of Remote Sensing, 34, 6079–6093.

Gomez, C., Lagacherie, P., & Coulouma, G. 2008. Continuum removal versus PLSR method for clay and calcium carbonate content estimation from laboratory and airborne hyperspectral measurements. Geoderma, 148, 141–148.

Gomez, C., Wylie, B., & Zhang, X. 2013. Soil carbonate mapping using airborne hyperspectral imagery. *Remote Sensing of Environment, 135*, 233–244.

Goosen, D. 1967. Aerial Photo Interpretation in Soil Survey. Food & Agriculture Organization, Rome.

Green, R. 2023. Imaging Spectroscopy Observations from NASA's Earth Surface Mineral Dust Source Investigation Launched in 2022 and Connections to Imaging Spectrometers for Greenhouse Gas Measurement, Europa, the Moon (No. EGU23–4510). Copernicus Meetings, Vienna.

Green, R. O., & Thompson, D. R. 2020, September. An earth science imaging spectroscopy mission: The Earth surface mineral dust source investigation (EMIT). IGARSS 2020–2020 IEEE International Geoscience and Remote Sensing Symposium, IEEE, New York, 6262–6265.

Guo, L., Zhang, H., Shi, T., Chen, Y., Jiang, Q., & Linderman, M. 2019. Prediction of soil organic carbon stock by laboratory spectral data and airborne hyperspectral images. Geoderma, 337, 32–41.

Hapke, B. W. 1981a. Bidirectional reflectance spectroscopy I. Theory. Journal of Geophysical Research, 86, 3039–3054.

Hapke, B. W. 1981b. Bidirectional reflectance spectroscopy: 2 Experiments and observation. Journal of Geophysical Research, 86, 3055–3060.

Hapke, B. W. 1984. Bidirectional reflectance spectroscopy: Correction for macroscopic roughens. Icarus, 59, 41–59.

Hapke, B. W. 1993. Theory of Reflectance and Emittance Spectroscopy. Cambridge University Press, New-York.

Hardisky, M. A., Klemas, V., & Smart, M. 1983. The influence of soil salinity, growth form, and leaf moisture on the spectral radiance of *Spartina alterniflora*. Engineering and Remote Sensing, 49, 77–83.

Haubrock, S., Chabrillat, S., Lemmnitz, C., & Kaufmann, H. 2008. Surface soil moisture quantification models from reflectance data under field conditions. International Journal of Remote Sensing, 29, 3–29.

Hbirkou, C., Pätzold, S., Mahlein, A. K., & Welp, G. 2012. Airborne hyperspectral imaging of spatial soil organic carbon heterogeneity at the field-scale. Geoderma, 175, 21–28.

Heiden, U., d'Angelo, P., Schwind, P., De Los Reyes, R., & Müller, R. 2021, July. Evaluating soil reflectance composites generated by SCMaP using different Sentinel-2 reflectance data inputs. In: 2021 IEEE International Geoscience and Remote Sensing Symposium IGARSS, IEEE, 495–498.

Heller Pearlshtien, D., & Ben-Dor, E. 2020. Effect of organic matter content on the spectral signature of iron oxides across the VIS-NIR spectral region in artificial mixtures: An example from a red soil from Israel. Remote Sensing, 12, no. 12, 1960.

Hengl, T. 2006. Finding the right pixel size. Computers & Geosciences, 32, 1283–1298.

Hick, R. T., & Russell, W. G. R. 1990. Some spectral considerations for remote sensing of soil salinity. Australian Journal of Soil Research, 28, 417–431.

Hill, J., & Schütt, B. 2000. Mapping complex patterns of erosion and stability in dry Mediterranean ecosystems. Remote Sensing of Environment, 74, 557–569.

Hill, J., Udelhoven, T., Vohland, M., & Stevens, A. 2010. The use of laboratory spectroscopy and optical remote sensing for estimating soil properties. In: Precision Crop Protection-the Challenge and Use of Heterogeneity, Springer, the Netherlands, 67–85.

Hirschfeld, T. 1985. Salinity determination using NIRA. Applied Spectroscopy, 39, 740–741.

Hörig, B., Kühn, F., Oschutz, F., & Lehaman, F. 2001. HyMap hyperspectral remote sensing to detect hydrocarbons. International Journal of Remote Sensing, 22, 1413–1422.

Howari, F. M., Goodell, P. C., & Miyamoto, S. 2002. Spectral properties of salt crusts formed on saline soils. Journal of Environmental Quality, 31, 1453–1461.

Huang, W., & Foo, S. 2002. Neural network modeling of salinity variation in Apalachicola River. Water Research, 36, 356–362.

Huete, A. R. 2004. Remote sensing of soils and soil processes. In: Remote Sensing for Natural Resource Management and Environmental Monitoring, Vol. 4, S. L. Ustin (ed.), John Wiley & Sons, Hoboken, 3–52.

Huete, A. R. 2005. Estimation of soil properties using hyperspectral VIS/IR sensors. Encyclopedia of Hydrological Sciences, 52, no. 5.

Hunt, G. R. 1982. Spectoscopic properties of rock and minerals. In: Handbook of Physical Properties Rocks, C. R. Stewart (ed.), CRC Press, Boca Raton, FL, 295.

Hunt, G. R., & Salisbury, J. W. 1970. Visible and near infrared spectra of minerals and rocks: I: Silicate minerals. Modern Geology, 1, 283–300.

Hunt, G. R., & Salisbury, J. W. 1971. Visible and near infrared spectra of minerals and rocks: II. Carbonates. Modern Geology, 2, 23–30.

Hunt, G. R., & Salisbury, J. W. 1976. Visible and near infrared spectra of minerals and rocks: XI sedimentary rocks. Modern Geology, 5, 211–217.

Hunt, G. R., Salisbury, J. W., & Lenhoff, C. J. 1971a. Visible and near-infrared spectra of minerals and rocks: III oxides and hydroxides. Modern Geology, 2, 195–205.

Hunt, G. R., Salisbury, J. W., & Lenhoff, C. J. 1971b. Visible and near-infrared spectra of minerals and rocks: IV sulfides and sulfates. Modern Geology, 3, 1–14.

Hunt, G. R., Salisbury, J. W., & Lenhoff, C. J. 1971c. Visible and near-infrared spectra of minerals and rocks: Halides, phosphates, arsenates, vandates and borates. Modern Geology, 3, 121–132.

Inoue, Y. 2020. Satellite-and drone-based remote sensing of crops and soils for smart farming: A review. Soil Science and Plant Nutrition, 66, no. 6, 798–810.

Irons, J. R., Weismiller, R. A., & Petersen, G. W. 1989. Soil reflectance. In: Theory and Application of Optical Remote Sensing, G. Asrar (ed.), Willey Series in Remote Sensing, John Wiley & Sons, New-York, 66–106.

Jackson, R. D., Moran, S., Slater, P. N., & Biggar, S. F. 1987. Field calibration of reflectance panels. Remote Sensing of Environment, 22, 145–158.

Jackson, R. D., Teillet, P. M., Slater, P. N., Fedosjsvs, G., Jasinski, M. F., Aase, J. K., & Moran, M. S. 1990. Bidirectional measurements of surface reflectance for view angle corrections of oblique imagery. Remote Sensing of Environment, 32, 189–202.

Jacquemoud, S., Baret, F., & Hanocq, J. F. 1992. Modeling spectral and bidirectional soil reflectance. Remote Sensing of Environment, 41, 123–132.

Jarmer, T. 2012. Using spectroscopy and satellite imagery to assess the total iron content of soils in the Judean Desert (Israel). In: SPIE Remote Sensing, International Society for Optics and Photonics, Washington, 853129.

Jarmer, T., Lavée, H., Sarah, P., & Hill, J. 2005. The use of remote sensing for the assessment of soil inorganic carbon in the Judean Desert (Israel). Proceedings of the 1st International Conference on Remote Sensing and Geoinformation Processing in the Assessment of Land Degradation and Desertification (RGLDD), University of Trier, Trier, Germany, 68–75.

Jenny, H. 1941. Factors of Soil Formation. McGraw-Hill Book Company, New York.

Jensen, R. J. 2011. Remote Sensing of the Environment: An Earth Resources Perspective. Pearson Prentice Hall Press, NJ, 592.

Kalra, N. K., & Joshi, D. C. 1996. Potentiality of Landsat, SPOT and IRS satellite imagery for recognition of salt affected soils in Indian arid zone. International Journal of Remote Sensing, 17, 3001–3014.

Karmanova, L. A. 1981. Effect of various iron compounds on the spectral reflectance and color of soils. Soviet Soil Science, 13, 63–60.

Karnieli, A., Kidron, G., Ghassler, C., & Ben-Dor, E. 1999. Spectral characteristics of cynobacteria soil crust in the visible near infrared and short wave infrared (400–2,500nm) in semiarid environment. International Journal of Remote Sensing, 69, 67–77.

Karnieli, A., & Tsoar, H. 1995. Spectral reflectance of biogenic crust developed on desert dune sand along the Israel-Egypt border. Remote Sensing, 16, 369–374.

Kemper, T., & Sommer, S. 2002. Estimate of heavy metal contamination in soils after a mining using reflectance spectroscopy. Environmental Science & Technology, 36, 2742–2747.

Kemper, T., & Sommer, S. 2003. Mapping and monitoring of residual heavy metal contamination and acidification risk after the Aznalcóllar mining accident (Andalusia, Spain) using field and airborne hyperspectral data. Proceedings, 3rd EARSeL Workshop on Imaging Spectroscopy EARSeL Secretariat, Paris, 333–343.

Kierein-Young, K., & Kruse, F. A. 1989. Comparison of Landsat Thematic Mapper images and geophysical and environmental reassert imaging spectrometer data for alteration mapping. Proceedings of the 7th Thematic Conference on Remote Sensing for Exploration Geology, Vol. 1, Calgary, Alberta, Canada, 349–359.

Knadel, M., Castaldi, F., Barbetti, R., Ben-Dor, E., Gholizadeh, A., & Lorenzetti, R. 2023. Mathematical techniques to remove moisture effects from visible—near-infrared—shortwave-infrared soil spectra. Applied Spectroscopy Reviews, 58, no. 9, 629–662.

Knadel, M., Deng, F., Alinejadian, A., Wollesen de Jonge, L., Moldrup, P., & Greve, M. H. 2014. The effects of moisture conditions—from wet to hyper dry—on visible near-infrared spectra of Danish reference soils. Soil Science Society of America Journal, 78, no. 4, 404–414.

Kokaly, R. F., Rockwell, B. W., Haire, S. L., & King, T. V. 2007. Characterization of post-fire surface cover, soils, and burn severity at the Cerro Grande Fire, New Mexico, using hyperspectral and multispectral remote sensing. Remote Sensing of Environment, 106, 305–325.

Kooistra, L., Salas, E. A. L., Clevers, J. G. P. W., Wehrens, R., Leuven, R. S. E. W., Nienhuis, P. H., & Buydens, L. M. C. 2004. Exploring field vegetation reflectance as an indicator of soil contamination in river floodplains. Environmental Pollution, 127, 281–282.

Kooistra, L., Wanders, J., Epema, G. E., Leuven, R. S. E. W., Wehrens, R., & Buydens, L. M. C. 2003.The potential of field spectroscopy for the assessment of sediment properties in river floodplains, Analytica Chimica Acta, 484, 198–200.

Kopačková, V. 2014. Using multiple spectral feature analysis for quantitative pH mapping in a mining environment. International Journal of Applied Earth Observation and Geoinformation, 28, 28–42, 90.

Kopačková, V., Mišurec, J., Lhotáková, Z., Oulehle, F., & Albrechtová, J. 2014. Using multi-date high spectral resolution data to assess the physiological status of macroscopically undamaged foliage on a regional scale. International Journal of Applied Earth Observation and Geoinformation, 27, 169–186.

Kosmas, C. S., Curi, N., Bryant, R. B., & Franzmeier, D. P. 1984. Characterization of iron oxide minerals by second derivative visible spectroscopy. Soil Science Society of American Journal, 48, 401–405.

Krishnan, P., Alexander, J. D., Bulter, B. J., & Hummel, J. W. 1980. Reflectance technique for predicting soil organic matter. Soil Science Society of American Journal, 44, 1282.

Kristof, S. J. 1971. Preliminary multispectral studies of soils. Journal of Soil and Water Conservation, 26, 15–18.

Kristof, S. J., Baumgardner, M. F., & Johannsen, C. J. 1973. Spectral mapping of soil organic matter. LARS Technical Reports, 26.

Kristof, S. J., & Zachary, A. L. 1974. Mapping soil features from multispectral scanner data. Photogrammetric Engineering & Remote Sensing, 40, 1427–1434.

Kruse, F. A., Kierein-Young, K., & Boardman, J. W. 1990. Mineral mapping of Cuprite, Nevada with a 63-channel imaging spectrometer. Photogrammetric Engineering and Remote Sensing, 56, 83–92.

Kruse, F. A., Thiry, M., & Hauff, P. L. 1991. Spectral identification (1.2–2.5mm) and characterization of Paris basin kaolinite/smectite clays using a field spectrometer. Proceedings of the 5th International Colloquium, Physical Measurements and Signatures in Remote Sensing, Courchevel, France I, 181–184.

Kumar, N. S., Anouncia, S. M., & Prabu, M. 2013. Application of satellite remote sensing to find soil fertilization by using soil colour. International Journal of Online Engineering, 9.

Ladoni, M., Bahrami, H. A., Alavipanah, S. K., & Norouzi, A. A. 2010. Estimating soil organic carbon from soil reflectance: A review. Precision Agriculture, 11, 82–99.

Lagacherie, P., Baret, F., Feret, J. B., Madeira Netto, J., & Robbez-Masson, J. M. 2008. Estimation of soil clay and calcium carbonate using laboratory, field and airborne hyperspectral measurements. Remote Sensing of Environment, 112, 825–835.

Lagacherie, P., Sneep, A. R., Gomez, C., Bacha, S., Coulouma, G., Hamrouni, M. H., & Mekki, I. 2013. Combining Vis—NIR hyperspectral imagery and legacy measured soil profiles to map subsurface soil properties in a Mediterranean area (Cap-Bon, Tunisia). Geoderma, 209, 168–176.

Larsen, I. J., MacDonald, L. H., Brown, E., Rough, D., Welsh, M. J., Pietraszek, J. H., . . . & Schaffrath, K. 2009. Causes of post-fire runoff and erosion: Water repellency, cover, or soil sealing? Soil Science Society of America Journal, 73, no. 4, 1393–1407.

Latz, K., Weismiller, R. A., & Van Scoyoc, G. E. 1981. A study of the spectral reflectance of selected eroded soils of Indiana in relationship to their chemical and physical properties. LARS Technical Report 082181.

Leger, R. G., Millette, G. J. F., & Chomchan, S. 1979. The effects of organic matter, iron oxides and moisture on the color of' two agricultural soils of Quebec Can. Journal of Soil Science, 59, 191–202.

Legros, J. P. 2006. Mapping of the Soil. Science Publishers, NH, 411.

Lehmann, J. 2020. Soil health and the sustainability of agricultural systems. In: Soil Health and Management, Springer, Berlin, Germany, 87–101.

Lekner, J., & Dorf, M. C. 1988. Why some things are darker when wet. Applied Optics, 27, 1278–1280.

Lemos, P., & Lutz, J. F. 1957. Soil crusting and some factors affecting it. Soil Science Society of America Journal, 21, 485–491.

Levenhangst, A. I. 1930. The use of aerial photographs in soil survey, Ponhvoedenie, 4, 116–122.

Levinstein, H., & Mudar, J. 1975. Infrared detectors in remote sensing. Proceedings of the IEEE, 63, 1.

Li, J., Okin, G. S., Skiles, S. M., & Painter, T. H. 2013. Relating variation of dust on snow to bare soil dynamics in the western United States. Environmental Research Letters, 8, 044054.

Li, S., & Chen, X. 2014. A new bare-soil index for rapid mapping developing areas using landsat 8 data. The international archives of the photogrammetry, remote sensing and spatial information sciences, 40, 139–144.

Liang, S., & Townshend, R. G. 1996. A modified Hapke model for soil biderctional reflectance. Remote Sensing of Environment, 55, 1–10.

Linker, R. 2008. Soil classification mid-infrared spectroscopy. In: IFIP International Federation for Information Processing, Vol. 259. Computer and Haifa, Israel. Computing Technologies in Agriculture, Vol. 2, Daoliang Li. Springer, Boston, 1137–1146.

Liu, H., Li, X., & Huang, H. 2020. Airborne hyperspectral sensor imaging: Advances and applications in remote sensing. *Remote Sensing*, 12, no. 23, 3911.

Liu, L., Zhang, Q., Zhang, X., & Wang, X. 2017. Spectral detection and mapping of heavy metals in soils using airborne hyperspectral imagery. International Journal of Applied Earth Observation and Geoinformation, 57, 1–10.

Liu, M., Liu, X., Li, M., Fang, M., & Chi, W. 2010. Neural-network model for estimating leaf chlorophyll concentration in rice under stress from heavy metals using four spectral indices. Biosystems Engineering, 106, 223–233.

Lobell, D. B., & Asner, G. P. 2002. Moisture effects on soil reflectance. Soil Science Society America Journal, 66, 722–727.

Löhrer, R., Bertrams, M., Eckmeier, E., Protze, J., & Lehmkuhl, F. 2013. Mapping the distribution of weathered Pleistocene wadi deposits in Southern Jordan using ASTER, SPOT-5 data and laboratory spectroscopic analysis. Catena, 100, 101–114.

Lucht, W., Hyman, A. H., Strahler, A. H., Barnsley, M. J., Hobson, P., & Muller, J. P. 2000. A comparison of satellite-derived spectral albedos to ground-based broadband albedo measurements modeled to satellite spatial scale for a semidesert landscape. Remote Sensing of Environment, 74, 85–98.

Lugassi, R., Ben-Dor, E., & Eshel, G. 2010. A spectral-based method for reconstructing spatial distributions of soil surface temperature during simulated fire events. Remote Sensing of Environment, 114, 322–331.

Lugassi, R., Ben-Dor, E., & Eshel, G. 2014. Reflectance spectroscopy of soils post-heating: Assessing thermal alterations in soil minerals. Geoderma, 213, 268–279.

Maestre, F. T., & Cortina, J. 2002. Spatial patterns of surface soil properties and vegetation in a Mediterranean semi-arid steppe. Plant and Soil, 241, 279–291.

Malley, D. F., Hunter, K. N., & Webster, G. R. 1999. Analysis of diesel fuel contamination in soils by near-infrared reflectance spectrometry and solid phase microextraction—gas chromatography. Soil and Sediment Contamination, 8, 481–489.

Malley, D. F., Martin, P., & Ben-Dor, E. 2004. Application in analysis of soils, Chapter 26. In: Near Infrared Spectroscopy in Agriculture, R. Craig, R. Windham and J. Workman (eds.), A three Societies Monograph (ASA, SSSA, CSSA), 44, 729–784.
Malley, D. F., & Williams, P. C. 1997. Use of near-infrared reflectance spectroscopy in prediction of heavy metals in freshwater sediment by their association with organic matter. Environmental Science & Technology, 31, 3461–3467.
Manfreda, S., & Eyal, B. D. (Eds.). 2023. Unmanned Aerial Systems for Monitoring Soil, Vegetation, and Riverine Environments. Elsevier, Cambridge, MA, 317.
Mapping Soil Organic Carbon Stock Using Hyperspectral Remote Sensing: A Case Study in the Sele River Plain in Southern Italy Nicolas Francos,*, Paolo Nasta, Carolina Allocca, Guido D'Urso, Oscar Rosario Belfiore, Benedetto Sica, Caterina Mazzitelli, Ugo Lazzaro, Mariano Crimaldi and Nunzio Romano, Fabrizio Sarghini, Eyal Ben-Dor, Remote Sensing Remote Sens. 2024, 16, 897. https://doi.org/10.3390/rs16050897
Mataix-Solera, J., Cerdà, A., Arcenegui, V., Jordán, A., & Zavala, L. M. 2011. Fire effects on soil aggregation: A review. Earth-Science Reviews, 109, no. 1–2, 44–60.
Mathews, H. L., Cunningham, R. L., & Peterson, G. W. 1973. Spectral reflectance of selected Pennsylvania soils. Proceedings of the Soil Science Society of America Journal, 37, 421–424.
McBratney, A. B., Minasny, B., & Rossel, R. V. 2006. Spectral soil analysis and inference systems: A powerful combination for solving the soil data crisis. Geoderma, 136, 272–278.
McBratney, A. B., Santos, M. D. L. M., & Minasny, B. 2003. On digital soil mapping. Geoderma, 117, 3–52.
McCarty, G. W., Reeves, J. B., Reeves, V. B., Follett, R. F., & Kimble, J. M. 2002. Mid-infrared and near-infrared diffuse reflectance spectroscopy for soil carbon measurement. Soil Science Society of America Journal, 66, 640–646.
McKenzie, N. J., & Ryan, P. J. 1999. Spatial prediction of soil properties using environmental correlation. Geoderma, 89, 67–94.
Metternicht, G. I., & Zinck, J. A. 1996. Modelling salinity-alkalinity classes for mapping salt-affected topsoil-sin the semi-arid valleys of Cochabamba (Bolivia). ITC-Journal, 2, 125–135.
Metternicht, G. I., & Zinck, J. A. 1997. Spatial discrimination of salt- and sodium-affected soil surfaces. International Journal of Remote Sensing, 18, 2571–2586.
Metternicht, G. I., & Zinck, J. A. 2008. Remote Sensing of Soil Salinization: Impact on Land Management. CRC Press, Boca Raton, FL, 374.
Metternicht, G., Zinck, J. A., Blanco, P. D., & Del Valle, H. F. 2010. Remote sensing of land degradation: Experiences from Latin America and the Caribbean. Journal of Environmental Quality, 39, 42–61.
Montgomery, O. L. 1976. An investigation of the relationship between spectral reflectance and the chemical, physical and genetic characteristics of soils. Ph.D. Thesis, Purdue University, West Lafayette, Indiana (Library of Congress no 79-32236).
Montgomery, O. L., & Baumgardner, M. F. 1974. The effects of the physical and chemical properties of soil and the spectral reflectance of soils. In: Information Note 1125 Laboratory for Applications of Remote Sensing, Purdue University, West Lafayette, Indiana.
Montanarella, L., Tóth, G., & Jones, A. 2011. Soil component in the 2009 LUCAS survey. In: Land Quality and Land Use Information in the European Union, G. Tóth and T. Németh (eds.), Publication Office of the European Union, Luxembourg, 209–219.
Morin, Y., Benyamini, Y., & Michaeli, A. 1981. The dynamics of soil crusting by rainfall impact and the water movement in the soil profile. Journal of Hydrology, 52, 321–335.
Morra, M. J., Hall, M. H., & Freeborn, L. L. 1991. Carbon and nitrogen analysis of soil fractions using near-infrared reflectance spectroscopy. Soil Science Society of American Journal, 55, 288–291.
Mouazen, A. M., De Baerdemaeker, J., & Ramon, H. 2005. Towards development of on-line soil moisture content sensor using a fibre-type NIR spectrophotometer. Soil and Tillage Research, 80, 171–183.
Mougenot, B. 1993. Effect of salts on reflectance on reflectance and remote sensing of salt affected soils. Cahiers Orstom Serie Pedologie XXVIII, 1, 45–54.
Mulder, M. A. 1987. Remote Sensing of Soils. Development in Soil Science 15, Elsevier Science Publishing, The Netherland, USA, 379.
Mulder, V. L., De Bruin, S., Schaepman, M. E., & Mayr, T. R. 2011. The use of remote sensing in soil and terrain mapping: A review. Geoderma, 162, 1–19.
Muller, E., & Decamps, H. 2001. Modeling soil moisture—reflectance. Remote Sensing of Environment, 76, 173–180.

Murad, M. O. F., Jones, E. J., Minasny, B., McBratney, A. B., Wijewardane, N., & Ge, Y. 2022. Assessing a VisNIR penetrometer system for in-situ estimation of soil organic carbon under variable soil moisture conditions. Biosystems Engineering, 224, 197–212.

Musavi, K. M. S., Bahrami, H. A., Effati, M., & Darvishi, B. A. 2013. Investigation of soil type effects on dust storms detection using day and night time multi-spectral MODIS images. International Journal of Agriculture: Research and Review, 3, 529–542.

Nanni, M. R., & Demattê, J. A. M. 2006. Spectral reflectance methodology in comparison to traditional soil analysis. Soil Science Society of America Journal, 70, 393–407.

Nanni, M. R., Demattê, J. A. M., Chicati, M. L., Fiorio, P. R., Cézar, E., & Oliveira, R. B. D. 2012. Soil surface spectral data from Landsat imagery for soil class discrimination. Acta Scientiarum: Agronomy, 34, 103–112.

Nanni, M. R., Demattê, J. A. M., Chicati, M. L., Oliveira, R. B., & Cezar, E. 2011. Spectroradiometric data as support to soil classification. International Research Journal of Agricultural Science And Soil Science, 1, 109–117.

Nanni, M. R., Demattê, J. A. M., & Fiorio, P. R. 2004. Soil discrimination analysis by spectral response in the ground level. Pesquisa Agropecuaria Brasileira, 39, no. 10, 995–1006.

Nassau, K. 1980. The causes of color. Scientific American, 243, 106–124.

Nieke, J., & Rast, M. 2021. The FLEX mission: Exploring global vegetation fluorescence from space. *Remote Sensing*, 13, no. 3, 400.

Nocita, M., Stevens, A., van Wesemael, B., Aitkenhead, M., Bachmann, M., Barthès, B., . . . & Wetterlind, J. 2015. Soil spectroscopy: An alternative to wet chemistry for soil monitoring. Advances in Agronomy, 132, 139–159.

Obukhov, A. I., & Orlov, D. C. 1964. Spectral reflectance of the major soil groups and the possibility of using diffuse reflection in soil investigations. Soviet Soil Science, 2, 174–184.

Odgers, N. P., McBratney, A. B., & Minasny, B. 2011. Bottom-up digital soil mapping. I. Soil layer classes. Geoderma, 163, 38–44.

Offer, Z. Y., & Goossens, D. 2001. Ten years of aeolian dust dynamics in a desert region (Negev desert, Israel): Analysis of airborne dust concentration, dust accumulation and the high-magnitude dust events. Journal of Arid Environments, 47, 211–249.

Ogen Y. N., Faigenbaum-Golovin, S., Granot, A., Shkolnisky, Y., Goldshleger, N., & Eyal, B. D. 2019. Removing moisture effect on soil reflectance properties: A case study of clay content prediction. Pedosphere, 29, no. 4, 421–431.

Okin, G. S., & Painter, T. H. 2004. Effect of grain size on remotely sensed spectral reflectance of sandy desert surfaces. Remote Sensing of Environment, 89, 272–280.

Ona-Nguema, G., Abdelmoula, M., Jorand, F., Benali, O., Gehin, A., Block, J. C., & Génin, J. M. R. 2002. Iron (II, III) hydroxycarbonate green rust formation and stabilization from lepidocrocite bioreduction. Environmental Science & Technology, 36, 16–20.

Orgiazzi, A., Ballabio, C., Panagos, P., Jones, A., & Fernández-Ugalde, O. 2018. LUCAS soil, the largest expandable soil dataset for Europe: A review. European Journal of Soil Science, 69, no. 1, 140–153.

Orlov, D. C. 1966. Quantitative patterns of light reflectance on soils I: Influence of particles (aggregate) size on reflectivity. Soviet Soil Science, 13, 1495–1498.

Padarian, J., Stockmann, U., Minasny, B., & McBratney, A. B. 2022. Monitoring changes in global soil organic carbon stocks from space. Remote Sensing of Environment, 281, 113260.

Palacios-Orueta, A., & Ustin, S. L. 1998. Remote sensing of soil properties in the Santa Monica Mountains I. Spectral analysis. Remote Sensing of Environment, 65, 170–183.

Palmer, J. M. 1982. Field standards of reflectance. Photogrametirc Engineerings and Remote Sensing, 48, 1623–1625.

Palmer, K. F., & Williams, D. 1974. Optical properties of water in the near infrared. Journal of the Optical Society of America A, 64, 1107–1110.

Pandit, C. M., Filippelli, G. M., & Li, L. 2010. Estimation of heavy-metal contamination in soil using reflectance spectroscopy and partial least-squares regression. International Journal of Remote Sensing, 31, 4111–4123.

Piekarczyk, J., Kaźmierowski, C., Królewicz, S., & Cierniewski, J. 2015. Effects of soil surface roughness on soil reflectance measured in laboratory and outdoor conditions. IEEE Journal of Selected Topics in Applied Earth Observations and Remote Sensing, 9, no. 2, 827–834.

Piemstein, A., Ben-Dor, E., & Notesko, G. 2010. Performance of three identical spectrometers in retrieving soil reflectance under laboratory conditions. Soil Science Society of America Journal, 75, 110–174.
Pieters, C. M. 1983. Strength of mineral absorption features in the transmitted component of near-infrared reflected light: First results from RELAB. Journal of Geophysical Research, 88, 9534–9544.
Pinty, B., Verstraete, M. M., & Dickson, R. E. 1989. A physical model for prediction bidirectional reflectance over bare soil. Remote Sensing of Environment, 27, 273–288.
Poeplau, C., & Don, A. 2015. Carbon sequestration in agricultural soils via cultivation of cover crops: A meta-analysis. Agriculture, Ecosystems & Environment, 200, 33–41.
Poppiel, R. R., da Silveira Paiva, A. F., & Demattê, J. A. M. 2022. Bridging the gap between soil spectroscopy and traditional laboratory: Insights for routine implementation. Geoderma, 425, 116029.
Prescott, J. A., & Taylor, J. K. 1930. The value of aerial photography in relation to soil surveys and classification. Journal CSIRO Australia, 229–230.
Price, J. C. 1990. On the information content of soil reflectance spectra. Remote Sensing of Environment, 33, 113–121.
Price, J. C. 1995. Examples of high resolution visible to near-infrared reflectance spectra and a standardized collection for remote sensing studies. International Journal of Remote Sensing, 16, 993–1000.
Pruitt, E. L. 1979. The office of naval research and geography. Annals of the Association of American Geographers, 69, 103–108.
Rao, B., Sankar, T., Dwivedi, R., Thammappa, S., Venkataratnam, L., Sharma, R., & Das, S. 1995. Spectral behaviour of salt-affected soils. International Journal of Remote Sensing, 16, 2125–2136.
Rao, B. R. M., & Venkataratnam, L. 1991. Monitoring of salt affected soils: A case study using aerial photographs, Salyut-7 space photographs, and Landsat TM data. Geocarto International, 6, 5–11.
Rabon, L. D., & De Souza Filho, C. R. 2019. Hydrocarbon contamination mapping using remote sensing: A case study. Environmental Monitoring and Assessment, 191, no. 6, 381.
Ray, S. S., Singh, J. P., Das, G., & Panigrahy, S. 2004. Use of High Resolution Remote Sensing Data for Generating Site-specific Soil Mangement Plan. XX ISPRS Congress, Commission 7. Istanbul, Turkey The International Archives of the Photogrammetry, Remote Sensing and Spatial Information Sciences: 127–131.
Ren, H. Y., Zhuang, D. F., Singh, A. N., Pan, J. J., Qiu, D. S., & Shi, R. H. 2009. Estimation of As and Cu contamination in agricultural soils around a mining area by reflectance spectroscopy: A case study. Pedosphere, 19, 719–725.
Rizzo, R., Demattê, J. A. M., & Terra, F. S. 2014. Using numerical classification of profiles on VIS-NIR spectra to distinguish soils from the Piracicaba region. Brazil: Brazilian Journal of Soil Science, 023-13.
Rizzo, R., Wadoux, A. M. C., Demattê, J. A., Minasny, B., Barrón, V., Ben-Dor, E., . . . & Salama, E. S. M. 2023. Remote sensing of the Earth's soil color in space and time. Remote Sensing of Environment, 299, 113845.
Robichaud, P. R., Lewis, S. A., Laes, D. Y., Hudak, A. T., Kokaly, R. F., & Zamudio, J. A. 2007. Postfire soil burn severity mapping with hyperspectral image unmixing. Remote Sensing of Environment, 108, 467–480.
Sabat-Tomala, A., Jarocińska, A. M., Zagajewski, B., Magnuszewski, A. S., Sławik, Ł. M., Ochtyra, A., . . . & Lechnio, J. R. 2018. Application of HySpex hyperspectral images for verification of a two-dimensional hydrodynamic model. European Journal of Remote Sensing, 51, no. 1, 637–649.
Schmid, T., Gumuzzio, J., Koch, M., Mather, P., & Solana, J. 2007. Characterizing and monitoring semi-aridwetlands using multi-angle hyperspectral and multispectral data. Proceedings Envisat Symposium, ESA SP-636, Montreux, Switzerland, 1–6.
Schnitzer, M., & Khan, S. U. 1978. Soil Organic Matter, Elsvier Publication, Amsterdam, 320.
Schreier, H. 1977. Quantitative predictions of chemical soil conditions from multispectral airborne ground and laboratory measurements. Proceedings of the 4th Canadian Symposium on Remote Sensing, Quebec, Canada, May 17–18, 1977, Canadian Aeronautics and Space Institute, Ottawa, 106–111.
Schwartz, G., Ben-Dor, E., & Eshel, G. 2012. Quantitative analysis of total petroleum hydrocarbons in soils: Comparison between reflectance spectroscopy and solvent extraction by 3 certified laboratories. Applied and Environmental Soil Science, 2012, 1–13.
Schwartz, G., Eshel, G., & Ben-Dor, E. 2011. Reflectance spectroscopy as a tool for monitoring contaminated soils. Soil Contamination, 67–90.
Schwertmann, U. 1988. Occurrence and formation of iron oxides in various pedoenvironment. In Iron in Soils and Clay Minerals, J. W. Stucki, B. A. Goodman and U. Schwertmann (eds.), NATO ASI Series, Reidel Publishing Company, Dordrecht, Boston, Lancaster, Tokyo, 267–308.

Scott, K. M., & Pain, F. P. 2008. Regolith Science. CSIRO publishing, Springer, Australia, 461.
Selige, T., Bohner, J., & Schmidhalter, U. 2006. High resolution topsoil mapping using hyperspectral image and field data in multivariate regression modeling procedures. Geoderma, 136, 235–244.
Serrano, M. H. R. L. 2010. Satellite Remote Sensing of Soil Moisture an MSc Dissertation Submitted to University of Reading Department of Meteorology, 66.
Sertsu, S. M., & Sánchez, P. A. 1978. Effects of heating on some changes in soil properties in relation to an Ethiopian land management practice. Soil Science Society of America Journal, 42, 940–944.
Sharma, R. C., & Bhargava, G. P. 1988. Landsat imagery for mapping saline soils and wet lands in north-west India. International Journal of Remote Sensing, 9, 39–53.
Sheinberg, I. 1992. Chemical and mineralogical components of crust. In: M. E. Sumner and B. A. Stewart (eds.), Soil Crusting: Advances in Soil, Science, Lewis Publishers, London, UK, 39–53.
Shepherd, K. D., Ferguson, R., Hoover, D., van Egmond, F., Sanderman, J., & Ge, Y. 2022. A global soil spectral calibration library and estimation service. Soil Security, 7, 100061.
Shepherd, K. D., & Walsh, M. G. 2002. Development of reflectance spectral libraries for characterization of soil properties. Soil Science Society of America Journal, 66, 988–998.
Sherman, D. M., & Waite, T. D. 1985. Electronic spectra of Fe3+ oxides and oxide hydroxides in the near-IR to near-UV. American Mineralogist, 70, 1262–1269.
Shi, T., Guo, L., Chen, Y., Wang, W., Shi, Z., Li, Q., & Wu, G. 2018. Proximal and remote sensing techniques for mapping of soil contamination with heavy metals. Applied Spectroscopy Reviews, 53, no. 10, 783–805.
Simonson, R. W. 1987. Historical aspects of soil survey and soil classification. Historical Aspects of Soil Survey and Soil Classification (Historicalaspec), 15–19.
Singer, A. 2007. The Soils of Israel. Springer-Verlag, Berlin and Heidelberg, 306.
Singh, A. 1989. Review article digital change detection techniques using remotely-sensed data. International Journal of Remote Sensing, 10, 989–1003.
Soil Survey Staff. 1975. Soil taxonomy. A Basic System of Soil Classification for Making and Interpreting Soil Survey, Soil Conservation Service, U.S. Department of Agriculture Handbook No 436, Washington D.C.
Soileau, J. M., & McCraken, R. J. 1967. Free iron and coloration in certain well-drained Costal Plain soils in relation to their other properties and classification. Soil Science Society of American Proceedings, 31, 248–255.
Souza Filho, C. 2013. Mapping of Geologic substrates impregnated with liquid hydrocarbons using proximal and airborne hyper spectral remote sensing: Potential applications for onshore exploration and leakage monitoring. Abstract un. IGARSS, July 25, 2013, Melbourne.
Staenz, K., & Held, A. 2012. Summary of current and future terrestrial civilian hyperspectral spaceborne systems. Geoscience and Remote Sensing Symposium (IGARSS), 2012 IEEE International, IEEE, New Jersey, USA, 123–125.
Steffens, M., & Buddenbaum, H. 2013. Laboratory imaging spectroscopy of a stagnic Luvisol profile: High resolution soil characterisation, classification and mapping of elemental concentrations. Geoderma, 195–196, 122–132.
Stevens, A., van Wesemael, B., Bartholomeus, H., Rosillon, D., Tychon, B., & Ben-Dor, E. 2008. Laboratory, field and airborne spectroscopy for monitoring organic carbon content in agricultural soils. Geoderma, 144, 395–404.
Stevens, A., Van Wesemael, B., Vandenschrick, G., Touré, S., & Tychon, B. 2006. Detection of carbon stock change in agricultural soils using spectroscopic techniques. Soil Science Society of America Journal, 70, 844–850.
Stevens, A., Udelhoven, T., Denis, A., Tychon, B., Lioy, R., Hoffmann, L., & Van Wesemael, B. 2010. Measuring soil organic carbon in croplands at regional scale using airborne imaging spectroscopy. Geoderma, 158, 32–45.
Stevenson, F. J. 1982. Humus Chemistry, John Wiley & Sons Inc., New-York, 198.
Stoner, E. R. 1979. Physicochemical, site and bidirectional reflectance factor characteristics of uniformly-moist soils. Ph.D. Thesis, Purdue University, West Lafayette, Indiana, USA.
Stoner, E. R., & Baumgardner, M. F. 1981. Characteristic variations in reflectance of surface soils. Soil Science Society of American Journal, 45, 1161–1165.
Stoner, E. R., Baumgardner, M. F., Weismiller, R. A., Biehl, L. L., & Robinson, B. F. 1980. Extension of laboratory-measured soil spectra to field conditions. Soil Science Society of America Proceedings, Madison, 4.
Szilagyi, A., & Baumgardner, M. F. 1991. Salinity and spectral reflectance of soils. Proceedings of ASPRS Annual Convention, Baltimore, 430–438.

Tangestani, M. H., & Moore, F. 2000. Iron oxide and hydroxyl enhancement using the Crosta Method: A case study from the Zagros Belt, Fars Province, Iran. International Journal of Applied Earth Observation and Geoinformation, 2, 140–146.

Tanii, K. 2022. HISUI: Hyperspectral imager suite for space-based Earth observation. *Journal of Remote Sensing*, 14, no. 15, 3087.

Taranik, D. L., & Kruse, F. A. 1989. Iron minerals reflectance in geophysical and environmental research imaging spectrometer (GERIS) data. Proceedings of the 7th Thematic Conference on Remote Sensing for Exploration Geology, Calgary, Alberta I, 445–458.

Taylor, G. R., Bennett, B. A., Mah, A. H., & Hewson, R. 1994. Spectral propertiesalinizedised land andimplications for interpretation of 24 channel imaging spectrometry. In Proceedings of the First International Remote Sensing Conference and Exhibition, Strasbourg, France, 3, 504–513.

Thomasson, J. A., Sui, R., Cox, M. S., & Al-Rajehy, A. 2001. Soil reflectance sensing for determining soil properties in precision agriculture. ASAE, 44, 1445–1453.

Thompson, L. M. 1957. Soils and Soil Fertility. McGreaw-Hill Book Company Inc, New-York.

Touré, S., & Tychon, B. 2004. Airborne hyperspectral measurements and superficial soil organic matter. Proceedings of the Airborne Imaging Spectroscopy Workshop, Noordwijk, The Netherlands.

Uno, Y., Prasher, S. O., Patel, R. M., Strachan, I. B., Pattey, E., & Karimi, Y. 2005. Development of field-scale soil organic matter content estimation models in Eastern Canada using airborne hyperspectral imagery. Canadian Biosystems Engineering, 45, 1.9–1.14.

Valentin, C., & Bresson, L. M. 1992. Morphology, genesis and classification of surface crusts in loamy and sandy soils. Geoderma, 55, 225–245.

Van der Meer, F. 1995. Spectral reflectance of carbonate mineral mixture and bidirectional reflectance theory: Quantitative analysis techniques for application in remote sensing. Remote Sensing Reviews, 13, 67–94.

Van der Meer, F. D., van Dijk, P. M., van der Werff, H. M. A., & Yang Hong. 2002. Remote sensing and petroleum seepage: A review and case study. Terra Nova, 14, 1–17.

Vane, G. 1993. Airborne imaging spectrometry, special issue. Remote Sensing of Environment, 44, 117–356.

Vane, G., Goetz, A. F., & Wellman, J. B. 1984. Airborne imaging spectrometer: A new tool for remote sensing. Geoscience and Remote Sensing, IEEE Transactions on, 546–549.

Vasques, G. M., Demattê, J. A. M., Rossel, R. V., Lopez, L. R., & Terra, F. S. 2014. Soil classification using visible/near-infrared diffuse reflectance spectra from multiple depths. Geoderma, 223–225, 73–78.

Vaughan, R. G., Calvin, W. M., & Taranik, J. V. 2003. SEBASS hyperspectral thermal infrared data: surface emissivity measurement and mineral mapping. Remote Sensing of Environment, 85(1), 48–63.

Veerman, C., Pinto Correia, T., Bastioli, C., European Commission, & Directorate-General for Research and Innovation, et al. 2020. Caring for Soil Is Caring for Life: Ensure 75% of Soils Are Healthy by 2030 for Food, People, Nature and Climate: Report of the Mission Board for Soil Health and Food, Publications Office, 2020. Available at: https://data.europa.eu/doi/10.2777/821504

Verma, K. S., Saxena, R. K., Barthwal, A. K., & Deshmukh, S. K. 1994. Remote sensing technique for mapping salt affected soils. International Journal of Remote Sensing, 15, 1901–1914.

Vink, A. P. A. 1963a. Soil survey as related to agricultural productivity. Journal of Soil Science, 14, 88–101.

Vink, A. P. A. 1963b. Planning of soil surveys in land development. International Inatitute for Land Reclamation and Improvement, Publication 10, U.D.C. 631.47: 528.77, The Netherlands, 53.

Vink, A. P. A. 1964. Aerial photographs and the soil sciences. Proceedings of the Toulouse Conference: On Aerial Surveys and Integrated Studies, 81–141.

Vinogradov, B. V. 1981. Remote sensing of the humus content of soils. Soviet Soil Science, 11, 114–123.

Viscarra Rossel, R. A. 2009. The soil spectroscopy group and the development of a global soil spectral library. EGU General Assembly Conference Abstracts, 11, 14021.

Viscarra Rossel, R.A., Behrens, T., Ben-Dor, E., Brown, D. J., Demattê, J. A. M., Shepherd, K. D., . . . & Ji, W. 2016. A global spectral library to characterize the world's soil. Earth-Science Reviews, 155, 198–230.

Viscarra Rossel, R. A., Gilbertson, M., Thylen, L., Hansen, O., McVey, S., McBratney, A. B., & Stafford, J. V. 2005. Field measurements of soil pH and lime requirement using an on-the-go soil pH and lime requirement measurement system. Precision Agriculture'05. Papers presented at the 5th European Conference on Precision Agriculture, Uppsala, Sweden. 511–520. Wageningen Academic Publishers.

Viscarra Rossel, R. A., Walvoort, D. J. J., McBratney, A. B., Janik, L. J., & Skjemstad, J. O. 2006. Visible, near infrared, mid infrared or combined diffuse reflectance spectroscopy for simultaneous assessment of various soil properties. Geoderma, 131, 59–75.

Viscarra Rossel, R. A., & Webster, R. 2011. Discrimination of Australian soil horizons and classes from their visible: Near infrared spectra. European Journal of Soil Science, 62, 637–647.

Vizzari, M. 2022. PlanetScope, Sentinel-2, and Sentinel-1 data integration for object-based land cover classification in Google Earth Engine. Remote Sensing, 14, no. 11, 2628.

Weidong, L., Baret, F., Xingfa, Qingxi, T., Lanfen, Z., & Bing, Z. 2002. Relating soil surface moisture to reflectance. Remote Sensing of Environment, 81, 238–246.

White, K., Walden, J., Drake, N., Eckardt, F., & Settlell, J. 1997. Mapping the iron oxide content of dune sands, Namib Sand Sea, Namibia, using Landsat Thematic Mapper data. Remote Sensing of Environment, 62, 30–39.

Whiting, M. L., Li, L., & Ustin, S. L. 2004. Predicting water content using Gaussian model on soil spectra. Remote Sensing of Environment, 89, 535–552.

Wiegand, C. L., Everitt, J. H., & Richardson, A. J. 1992. Comparison of multispectral video and SPOT-1 HRV observations for cotton affected by soil salinity. International Journal of Remote Sensing, 13, 1511–1525.

Wiegand, C., Richardson, A., Escobar, D., & Gerbermann, A. 1991. Vegetation indices in crop assessments. Remote Sensing of Environment, 35, 105–119.

Wiegand, C. L., Rhoades, J. D., Escobar, D. E., & Everitt, J. H. 1994. Photographic and videographic observations for determining and mapping the response of cotton to soil salinity. Remote Sensing of Environment, 49, 212–223.

Wijewardane, N. K., Hetrick, S., Ackerson, J., Morgan, C. L., & Ge, Y. 2020. VisNIR integrated multi-sensing penetrometer for in situ high-resolution vertical soil sensing. Soil and Tillage Research, 199, 104604.

Winkelmann, K. H. 2005. On the applicability of imaging spectrometry for the detection and investigation of contaminated sites with particular consideration given to the detection of fuel hydrocarbon contaminants in soil. Doctoral Dissertation, Universitätsbibliothek, Leipzig, Germany.

Wolf, B., Senesi, G. A., Lo, A. A., Gehl, R. A., & Anderson, M. H. 2015. Soil spectral remote sensing: A review of the latest advances and applications. *Remote Sensing*, 7, no. 5, 581.

Wu, Y., Chen, J., Ji, J., Gong, P., Liao, Q., Tian, Q., & Ma, H. 2007. A mechanism study of reflectance spectroscopy for investigating heavy metals in soils. Soil Science Society of America Journal, 71, 918–925.

Wu, Y. Z., Chen, J., Ji, J. F., Tian, Q. J., & Wu, X. M. 2005. Feasibility of reflectance spectroscopy for the assessment of soil mercury contamination. Environmental Science & Technology, 39, 873–878.

Wu, Y., Zhang, X., Liao, Q., & Ji, J. 2011. Can contaminant elements in soils be assessed by remote sensing technology: A case study with simulated data. Soil Science, 176, 196–205.

Xia, X. Q., Mao, Y. Q., Ji, J. F., Ma, H. R., Chen, J., & Liao, Q. L. 2007. Reflectance spectroscopy study of Cd contamination in the sediments of the Changjiang River, China. Environmental Science & Technology, 41, 3449–3454.

Yang, H., Zhang, J., van der Meer, F. D., & Kroonenberg, S. B. 2000. Imaging spectrometry data correlated to hydrocarbon microseepage. International Journal of Remote Sensing, 21, 197–202.

Yu, H., Kong, B., Wang, Q., Liu, X., & Liu, X. 2020. Hyperspectral remote sensing applications in soil: A review. Hyperspectral Remote Sensing, 269–291.

Zabcic, N., Rivard, B., & Ong, C. 2005. Mapping surface pH using airborne hyperspectral imagery at the Sotiel—Migollas Spain. Warsaw Proceedings of 4th EARSeL Workshop on Imaging Spectroscopy, EARSeL and Warsaw University, Warsaw, Poland.

Zabcic, N., Rivard, B., Ong, C., & Mueller, A. 2009. Using airborne hyperspectral data to characterize the surface pH of pyrite mine tailings. IEEE Workshop on Hyperspectral Image and Signal Processing: Evolution of Remote Sensing, August, 2009, Whispers'09, Grenoble, France, 26–28.

Zhang, T., Li, L., & Zheng, B. 2013. Estimation of agricultural soil properties with imaging and laboratory spectroscopy. Journal of Applied Remote Sensing, V.

Zheng, B. 2008. Using satellite hyperspectral imagery to map soil organic matter, total nitrogen and total phosphorus. A Master Thesis Submitted to the Department of Earth Science University of Indiana, 81.

Zhou, Z. C., Shangguan, Z. P., & Zhao, D. 2006. Modeling vegetation coverage and soil erosion in the Loess Plateau Area of China. Ecological Modeling, 198, 263–268.

Zhu, Y., Wang, Y., Shao, M., & Horton, R. 2011. Estimating soil water content from surface digital image gray level measurements under visible spectrum. Canadian Journal of Soil Science, 91, 69–76.

Part VI

Summary and Synthesis for Volume III

17 Remote Sensing Handbook, Volume III

Agriculture, Food Security, Rangelands, Vegetation, Phenology, and Soils

Prasad S. Thenkabail

ACRONYMS AND DEFINITIONS

APAR	Absorbed Photosynthetically Absorbed Radiation
ASS	Aerial Spectral Sensing
ATSR	Along-Track Scanning Radiometer
AVHRR	Advanced Very-High-Resolution Radiometer
CAI	Cellulose Absorption Indices
CDL	Cropland Data Layer
DLR	German Aerospace Center
EnMAP	Environmental Mapping and Analysis Program
EO	Earth Observation
EVI	Enhanced Vegetation Index
FAO	Food and Agriculture Organization of the United Nations
FLUXNET	A network of micrometeorological tower sites to measure carbon dioxide, water, and energy balance between terrestrial systems and the atmosphere
GEO	Group on Earth Observation
GIEWS	Global Information and Early Warning System
GIMMS	Global Inventory Modeling and Mapping Studies
GIS	Geographic Information System
GNSS	Global Navigation Satellite Systems
GLAI	Green Leaf Area Index
GLAM	Global Agricultural Monitoring
GPP	Gross Primary Production
GPS	Global Positioning System
HNB	Hyperspectral Narrow Bands
HVI	Hyperspectral Vegetation Indices
IKONOS	A commercial Earth observation satellite, typically, collecting sub-meter to 5 m data
ISS	International Space Station
JRC	Joint Research Center
LACIE	Large Area Crop Inventory Experiment
LAI	Leaf Area Index
LiDAR	Light Detection and Ranging
LSP	Land Surface Phenology

DOI: 10.1201/9781003541165-23

LSS	Laboratory Spectral Sensing
LST	Land Surface Temperature
LUE	Light Use Efficiency
LULC	Land Use, Land Cover
MARS	Monitoring Agricultural Resources action of the European Commission
MODIS	Moderate-Resolution Imaging Spectroradiometer
NASA	National Aeronautics and Space Administration
NDTI	Normalized Difference Tillage Index
NDVI	Normalized Difference Vegetation Index
NPP	Net Primary Productivity
PAR	Photosynthetically Active Radiation
PRI	Photochemical Reflectance Index
PROSAIL	Combination of PROSPECT and SAIL, the two nondestructive physically based models to measure biophysical and biochemical properties
PROSPECT	Radiative transfer model to measure leaf optical properties spectra
PV	Photosynthetic Vegetation
RANDRI	Residue Adjust Normalized Difference Residue Index
RS	Remote Sensing
SAIL	Scattering by Arbitrary Inclined Leaves
SANDRI	Soil Adjust Normalized Difference Residue Index
SAR	Synthetic Aperture Radar
SAVI	Soil Adjusted Vegetation Index
SINDRI	Shortwave Infrared Normalized Difference Residue Index
SMTs	Spectral Matching Techniques
SOC	Soil Organic Carbon
SPOT	Satellite Pour l'Observation de la Terre, French Earth Observing Satellites
SVM	Support Vector Machines
SWIR	Shortwave Infrared
TROPOMI	TROPOspheric Monitoring Instrument
UAV	Unmanned Aerial Vehicles
USAID	United States Agency for International Development
USDA	United States Department of Agriculture
VH	Vegetation Health
VHRI	Very High Resolution Imagery
VIIRS	Visible Infrared Imaging Radiometer Suite

This chapter provides a summary of each of the 16 chapters in Volume III of the six-volume *Remote Sensing Handbook* (Second Edition). The topics covered in the chapters of Volume III include (Figure 17.0) (1) vegetation and biomass, (2) agricultural croplands, (3) rangelands, (4) phenology and food security, and (5) soils. Under each of these broad topics, there are one or more chapters. For example, there are six chapters under agricultural croplands. In a nutshell, these chapters provide a complete and comprehensive overview of these critical topics, capture the advances over last 60+ years, and provide a vision for further development in the years ahead. By reading this summary chapter, a reader can have a quick understanding of what is in each of the chapters of Volume III, see how the chapters interconnect and intermingle, and get an overview on the importance of various chapters in developing complete and comprehensive knowledge of remote sensing for land resources. These chapters, together, not only capture the advances of last 60+ years and but also provide a vision for the future.

17.1 MEASURING TERRESTRIAL PRODUCTIVITY FROM SPACE

Biological activity on Earth depends ultimately on solar radiation and its conversion into biochemical energy through photosynthesis. Quantifying global terrestrial photosynthesis is essential to

Chapter 17: Summary Chapter for
Remote Sensing Handbook (Second Edition, Six Volumes): Volume III

Volume III: Agriculture, Food Security, Rangelands, Vegetation, Phenology, and Soils

Chapter 1: Measuring Terrestrial Productivity from Space

Chapter 2: Solar Induced Chlorophyll Florescence **Chapter 3: Vegetation Characterization by Physically-based Models**

Chapters 4 : Crop Biophysical and Biochemical Retrievals using Radiative transfer Models

Chapters 5 to 7: Remote Sensing of Agriculture, Mapping, Modeling, Monitoring at Local and Global Scales

Chapters 8 and 9: Precision Farming, Tillage, Crop Residues

Chapters 10: Hyperspectral Remote Sensing of Agriculture

Chapter 11 to 13: Rangeland Studies including Rangeland Mapping, Quantification, and Rangeland Biodiversity

Chapter 14: Land Surface Phenology and Food Security Analysis

Chapter 15 and 16: Remote Sensing of Soils

FIGURE 17.0 Overview of the chapters in Volume III of *Remote Sensing Handbook* (Second Edition).

understanding the global carbon cycle and the climate system and remote sensing has played a pivotal role in advancing our understanding of photosynthesis from leaf to global scale; albeit, substantial uncertainties still exist (Ryu et al., 2019). Terrestrial biological activity is fundamental to production of food, fiber, and fuel and is often considered the most important measure of global change (Running et al., 2004; Thenkabail et al., 2021b; Alves et al., 2013). The fundamental paradigm measuring photosynthesis in terrestrial vegetation was first proposed by Monteith (1972) who showed us that stress-free annual crop productivity was linearly related to vegetation absorbed Photosynthetically absorbed radiation (APAR; e.g., Figure 17.1a).

Heading here and at the start of each chapter summary?

Chapter 1 by Dr. Alfredo Huete et al. traces the development of various methods and approaches that have been applied in measuring, modeling, and mapping photosynthesis, accurately and routinely using remote sensing data. In Chapter 1, they review the integration of remote sensing with traditional in situ methods and the more recent eddy covariance tower approach for estimating gross primary productivity (GPP) and net primary productivity (NPP) at global scales. Solar-induced chlorophyll fluorescence (SIF) has an apparent near-linear relationship with gross primary production (GPP) (Yang et al., 2021). LiDAR integration with field inventory plots now provide calibrated estimates of aboveground carbon stocks, which can be scaled-up using satellite data of vegetation cover, topography, and rainfall to model carbon stocks. A series of 6 productivity models are presented and discussed, based on the light-use-efficiency (LUE) concept and primarily dependent on satellite data inputs. These include: (a) NPP derived from the integral of growing season normalized difference vegetation index (NDVI), as surrogate of vegetation absorbed PAR radiation and, more recently, integral of enhanced vegetation index (iEVI); (b) BIOME-BGC (BioGeochemical Cycles) model that calculates daily GPP as a function of incoming solar radiation, light use conversion coefficients, and environmental stresses; (c) Vegetation Index-tower GPP relationships where spectral vegetation indices (VIs) are directly related to eddy covariance tower carbon flux measurements; (d) Temperature and Greenness (T-G) model which combines land surface temperature and EVI products from MOD€ (e) Greenness and Radiation (G-R) model where chlorophyll-related spectral indices are coupled with measures of light energy, Absorbed Photosynthetic Active Radiation (FAPAR) (e.g., Figure 17.1a) and PAR, to provide robust estimates of GPP; and (f) satellite-based Vegetation Photosynthesis Model (VPM) that estimates GPP using satellite inputs of EVI and the land surface

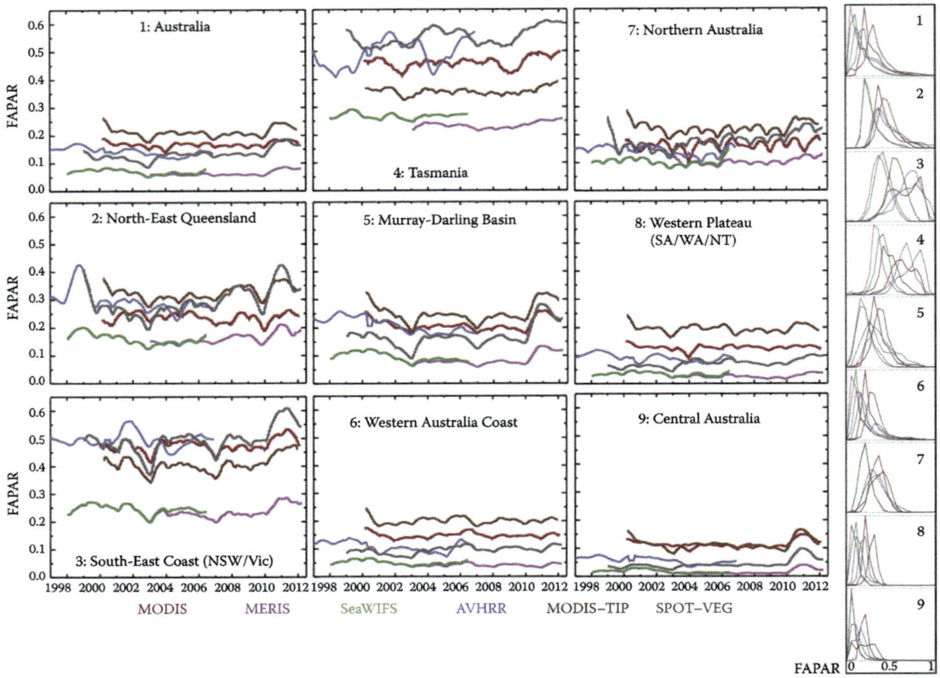

FIGURE 17.1A Time-series of the different Fraction of Absorbed Photosynthetic Active Radiation (FAPAR) satellite products (persistent vegetation component) for Australia and eight Australian drainage divisions. The plots on the right present the frequency histograms of each product (full FAPAR signal) for each region. (Source: Pickett-Heaps et al., 2014.)

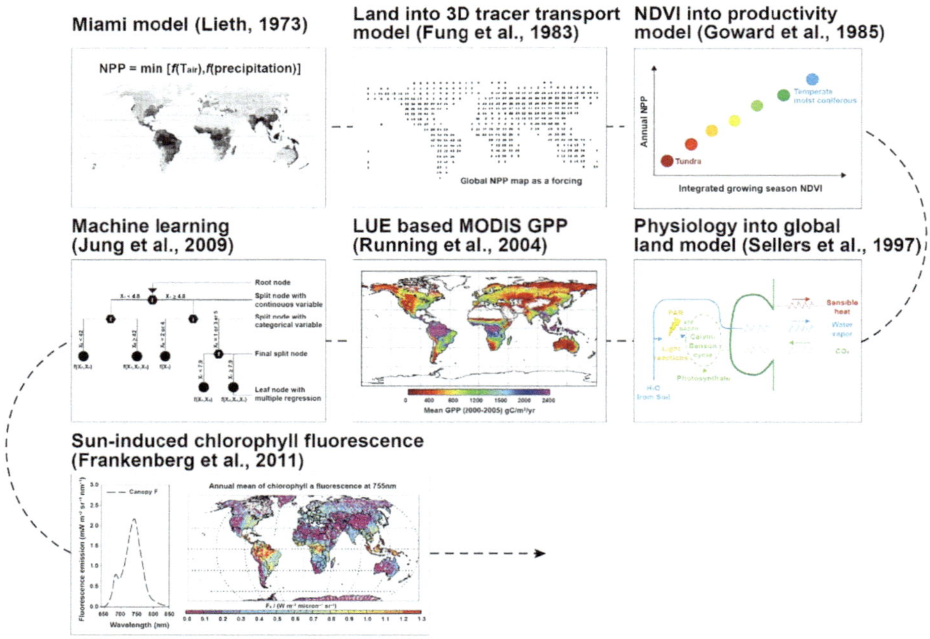

FIGURE 17.1B Milestone works in global photosynthesis estimates. All figures were redrawn or reprocessed. (Source: Ryu et al., 2019.)

water index (LSWI), along with phenology and temperature scalars. Many of the limitations in productivity assessments concern the difficulty in deriving independent estimates of light-use efficiency, and the hyperspectral-based Photochemical Reflectance Index (PRI), a scaled LUE measure based on light absorption processes by carotenoids, is discussed as a way to advance the accuracies of remote sensing retrievals of productivity. Significant and promising advances in direct estimates of GPP, even under stress conditions, have been demonstrated with new spaceborne measures of solar-induced chlorophyll fluorescence (SIF) based on near-infrared light re-emitted from illuminated plants, as a by-product of photosynthesis and thereby strongly correlated with GPP. Milestone works in global photosynthesis estimates are illustrated in Figure 17.1b by Ryu et al. (2019).

Remote sensing offers the ability to model and map productivity for the entire planet due to its synoptic and temporal coverage, and thus complements other conventional approaches that are costly, time consuming, and very limited in coverage. Indeed despite the rapid growth of FLUXNET, there is not even a single tower present in many parts of the world. FLUXNET has been shown invaluable for validating remotely sensed measurements and process based productivity models (Williams et al., 2009; Zhu et al., 2020). However, the combination of in situ with satellite data offers much more than calibration opportunities. It will be through a better understanding of why satellite and in situ relationships hold, or don't hold, that will greatly advance and contribute to our comprehension of the carbon cycle mechanisms and scaling factors at play (Williams et al., 2009; Zhu et al., 2020). Recent advances in satellite remote sensing are helping reduce uncertainties in measurement of photosynthesis. For example, the Orbiting Carbon Observatory-2 (OCO-2) represents a major advance in satellite SIF remote sensing. The Ecosystem Spaceborne Thermal Radiometer Experiment on Space Station (ECOSTRESS), launched in June 2018, observes the land surface temperature (LST) at different times of day with high spatial resolution (70 m × 70 m) from the International Space Station (ISS), overcomes lack the capability to examine the diurnal cycles of GPP with the polar-orbiting satellites (e.g., Landsat, Sentinel, Terra, Aqua, Suomi NPP, JPSS, OCO-2) because they observe the Earth's surface at the same time of day (Li et al., 2021).

17.2 REMOTE SENSING OF SOLAR-INDUCED CHLOROPHYLL FLORESCENCE

Solar-induced chlorophyll fluorescence (SIF) is an excellent measure of plant photosynthesis and is critical to understand, model and map plant parameters such as chlorophyll, plant health/stress, and gross primary productivity (GPP). Plants absorb solar radiation and emit it as fluorescence, which in turn is detected as SIF. The SIF is an emission of light in the 650–850 nm spectral range from the excited state of the chlorophyll-a pigment after absorption of photosynthetically active radiation (PAR). Mohammed et al. (2019) conduct an excellent review of solar-induced chlorophyll fluorescence (SIF) over the last 50 years and highlight SIF heritage and complementarity within the broader field of fluorescence science, the maturation of physiological and radiative transfer modeling, SIF signal retrieval strategies, techniques for field and airborne sensing, advances in satellite-based systems, and applications of these capabilities in evaluation of photosynthesis and stress effects. Du et al. (2019) used a spectroradiometer with the spectral range of 649–805 nm and peak signal-to-noise ratio (SNR) of over 1000 to make it possible to retrieve the full-band-width SIF spectrum with two peaks at 685 and 740 nm that were found to be ideal for measuring SIF (Du et al., 2019). Terrestrial SIF is emitted throughout the red and near-infrared spectrum and is characterized by two peaks centered around 685 nm and 740 nm, respectively. Zhao et al. (2024) reconstruct the terrestrial SIF spectrum from measurements by TROPOspheric Monitoring Instrument (TROPOMI) on board the Sentinel-5 precursor mission (Figure 17.2).

In Chapter 2, Juan Quiros-Vargas et al. provide clear definitions and understanding of SIF measurements from uncrewed aerial vehicles (UAVs) (Mishra et al., 2023), other airborne measurements, and from satellites. They review several state-of-the-art UAV-based systems including the most interesting hyperspectral sensors for retrieving SIF including measurements from 760 nm

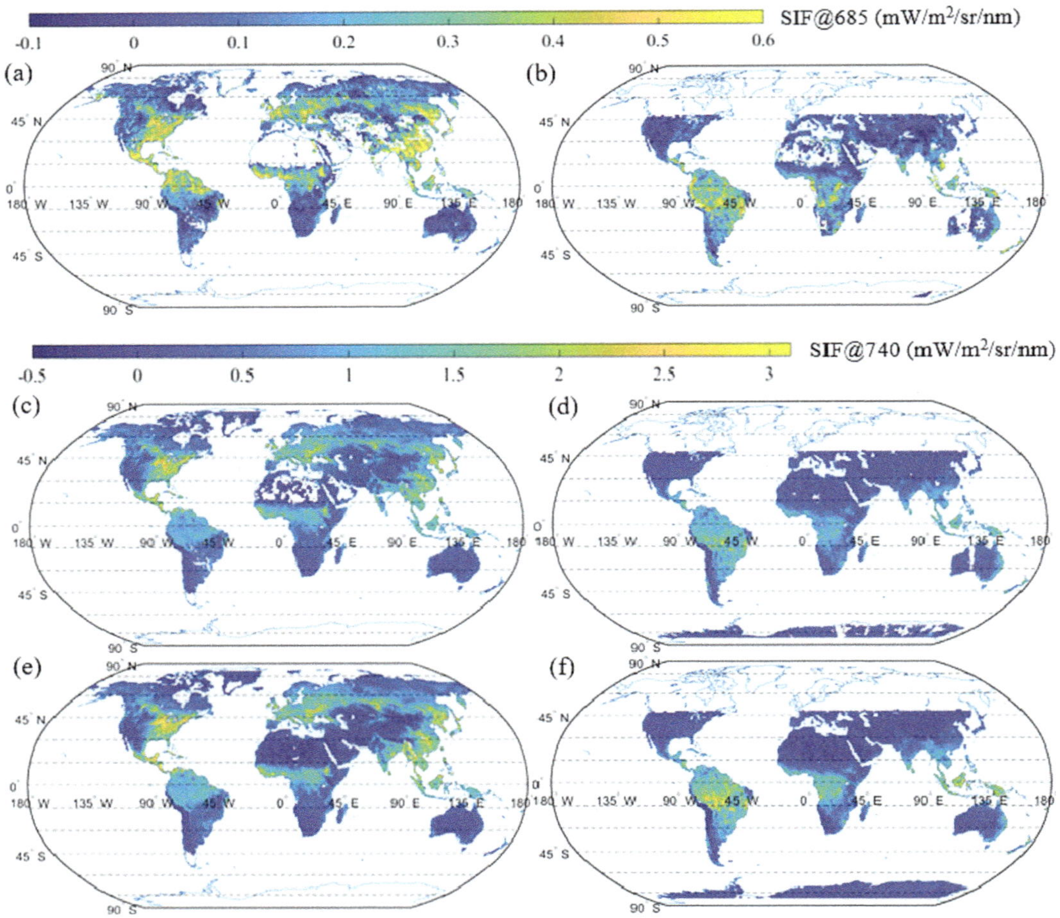

FIGURE 17.2 Global maps for July 1–7 (left panels) and December 1–7 (right panels), 2021, of red SIF retrievals (a and b), far-red SIF retrievals (c and d), and Caltech far-red SIF (e and f). The horizontal bar at the top indicates the scale of red SIF values in the first row, while middle and bottom rows share the color scale above them. (Source: Zhao et al., 2024.)

band. The UAV based SIF measurements are best for field-scale measurements. Nevertheless, for large-scale understanding other airborne and spaceborne SIF measurements that cover larger areal extents, yet at high spatial resolution, are valuable. Early satellite-based SIF data was used for atmospheric chemistry. But they are increasingly used for wide array crop studies such as plant stress/health and plant physiological parameters such as gross primary productivity (GPP), plant nitrogen and photosynthetic activity such as chlorophyll a. Chapter 2 further discusses SIF measurements and advances achieved from orbiting carbon observatory 2 (OCO-2) and tropospheric monitoring instrument (TROPOMI) onboard the Sentinel-5 precursor satellite. A critical part of the chapter is discussing the scale issues of SIF measurements from different platforms.

17.3 VEGETATION CHARACTERIZATION USING PHYSICALLY BASED MODELS SUCH AS SAIL, PROSPECT

Canopy biophysical variables (e.g., LAI, FAPAR; see Table 17.1) are retrieved through methods such as: (a) direct destructive measurements, (b) indirect non-destructive statistical modeling by relating biophysical quantities to spectral reflectivity in various wavebands or to indices derived from these wavebands, and (c) non-destructive physically based models such as the canopy bidirectional reflectance model called Scattering by Arbitrary Inclined Leaves (SAIL), leaf optical properties model based on the radiative transfer model called PROSPECT, and a combination of these two called PROSAIL (Han et al., 2023; Sun et al., 2023). Direct measurements of biophysical quantities (e.g., LAI, FAPAR, biomass, plant height, and canopy cover) are most accurate, but require destructive sampling that is resource (e.g., time, money) intensive. Statistical approaches of estimating biophysical quantities by relating them to reflectivity of spectral wavebands or various vegetation indices, is non-destructive, less resource intensive compared to in situ methods, and often explain over 80% variability in data (e.g., Figure 17.3). Physically based methods rely on inverting surface reflectance properties in various wavebands to determine biophysical quantities. Non-destructive methods are not as accurate as destructive methods, but often explain over 80% variability in each of the biophysical quantity, are less resource intensive, and avoid destructive sampling. Darvishzadeh et al. (2008) showed a carefully selected spectral subset (Table 17.1) contains sufficient information for a successful model inversion.

Chapter 3, written by Dr. Frédéric Baret, focuses on retrieving canopy biophysical quantities based on various physically based models, through model inversion. They demonstrate how green leaf area index (GLAI; m^2/m^2), LAI, and fraction of the photosynthetically active radiation (FAPAR) are retrieved through physically based radiative transfer model inversion methods. They address this through two approaches: 1. Radiometric driven approach which minimizes the

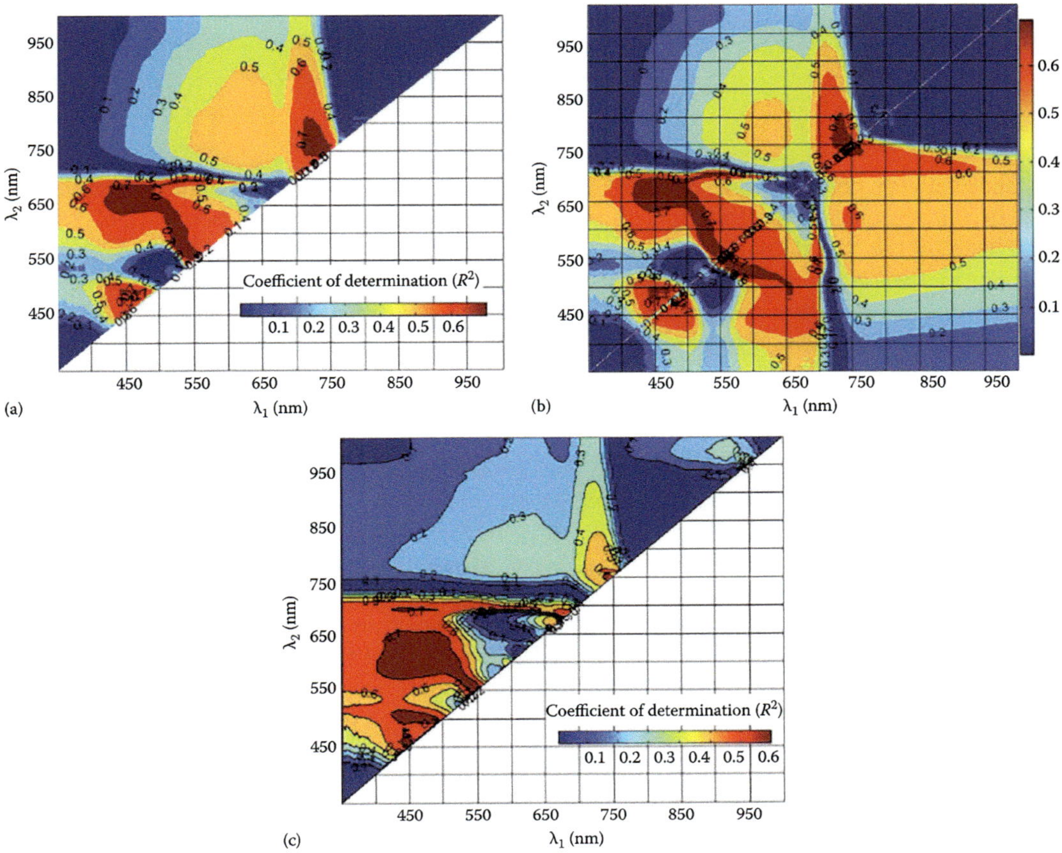

FIGURE 17.3 Contour maps for coefficients of determination (R2) between rice leaf nitrogen concentrations and normalized (a), ratio (b), and difference (c) indices using two spectral bands (λ1 and λ2) (n = 312). (Source: Tian et al., 2011.)

TABLE 17.1
A Spectral Subset Used for Successful Model Inversion in Physically Based Models Like PROSAIL, SAIL, PROSPECT

Wavelength (nm)	Vegetation Parameters	Reference
466	Chlorophyll b	Curran (1989)
695	Total chlorophyll	Gitelson and Merzlyak (1997); Carter (1994)
725	Total chlorophyll, leaf mass	Horler et al. (1983)
740	Leaf mass, total	Horler et al. (1983)
786	Leaf mass	Guyot and Baret (1988)
845	Leaf mass, total, chlorophyll	Thenkabail et al. (2004a)
895	Leaf mass, LAI	Schlerf et al. (2005); Thenkabail et al. (2004a)
1114	Leaf mass, LAI	Thenkabail et al. (2004a)
1215	Plant moisture, cellulose, starch	Curran (1989); Thenkabail et al. (2004a)
1659	Lignin, leaf mass, starch	Thenkabail et al. (2004a)
2173	Protein, nitrogen	Curran (1989)
2359	Cellulose, protein, nitrogen	Curran (1989)

(Source: Darvishzadeh et al., 2008).

distance between observed and simulated reflectance and emphasizes on the outputs (radiometric data driven approach) of the radiative transfer model; and 2. Canopy biophysical variables which are based on the vegetation indices approaches or are based on machine learning approaches and emphasizes on the inputs (the canopy biophysical variables driven approach) of the radiative transfer model. The chapter then goes on to discuss the pros and cons of retrieval approaches. The chapter then provides limitations of these models and enumerate on the strategies of reducing their uncertainties.

Later in Chapter 10, some of the statistical approaches and methods of determining biophysical quantities from remote sensing are discussed. Readers can also look into recent works of Marshall and Thenkabail (2016) on in situ and statistical methods as well as other novel work on statistical models for biophysical retrieval in Chapter 10 and by Thenkabail et al. (2000, 2014, 2013, 2011, 2004b, 2004c, 2004d). Suitability of coupled PROSPECT and SAIL radiative transfer models (PROSAIL), for the retrieval of biophysical and biochemical variables in the context of agricultural crop monitoring using advanced hyperspectral sensors are reviewed by Berger et al. (2018). The most important recommendations of the study when hyperspectral data are to be elaborated with the PROSAIL model were to use (Berger et al., 2018) (1) physically based PROSAIL models instead of simple empirical (parametric) models that lack transferability, (2) machine learning regression algorithms to process hyperspectral data, and (3) suitable approaches to estimate plant pigments from hyperspectral data, such as carotenoids and anthocyanins, which have been implemented in the recent PROSPECT versions. Lu et al. (2021) investigated three PROSPECT models (i.e., PROSPECT-5, PROSPECT-D, and PROSPECT-5M) and established that the three models achieved similar accuracies for simulating spectra of photosynthetic leaves, but PROSPECT-5M performed better than the other two models for non-photosynthetic leaves,

17.4 AGRICULTURAL CROP BIOPHYSICAL AND BIOCHEMICAL QUANTITY RETRIEVALS FROM REMOTE SENSING, MACHINE LEARNING, AND RADIATIVE TRANSFER MODELS

Agricultural biophysical and biochemical quantities have been retrieved using remote sensing techniques for a while (Mariotto et al., 2013; Marshall and Thenkabail, 2014, 2015a, 2015b, 2016; Enclona et al., 2004; Thenkabail et al., 2003; Thenkabail, 2003) with robust techniques and methods. Nevertheless, recent advances in remote sensing in data acquisition on multiple-platforms, more frequently covering the planet, large archives from multiple satellites, and acquiring data in hyperspectral, and hyperspatial necessitates novel approaches to modeling, mapping, and classifying massively large big-data. As a result, researchers are exploring biophysical and biochemical retrievals using machine learning, deep learning, artificial intelligence and through cloud computing (e.g., Aneece and Thenkabail, 2018, 2021, 2022). For example, Danner et al. (2021) evaluated efficient radiative-transfer model-based (RTM-based) training of four machine learning algorithms (MLA's), namely artificial neural networks (ANN), random forest regression (RFR), support vector machine regression (SVR), and Gaussian process regression (GPR), to estimate biophysical and biochemical variables of unseen targets with high performance (relative error scores < 10%). They also showed that the artificial neural networks (ANNs) excelled in terms of accuracy, model size and execution time when the 242 hyperspectral bands were transformed into 15 principal components, the signals of which were scaled by a z-transformation. Wolanin et al. (2019) highlight MLA RTM models to map crop productivity from new satellite sensors at a global scale with the help of current Earth Observation cloud computing platforms (Figure 17.4).

Chapter 4 provides most comprehensive coverage of agricultural crop biophysical and biochemical quantity modeling and mapping using wide array of remote sensing big data including new generation of hyperspectral sensors such as the PRISMA and EnMAP. They discuss various radiative transfer modeling (e.g., PROSAIL, PROSPECT, SAIL, ACRM, SCOPE, WCM) and the computing challenges they offer when working over large areas and large datasets and process of overcoming the same using emulators. This is followed by a discussion of various cloud computing environments such as the Google Earth Engine (GEE), ArcGIS, the Sentinel Application Platform (SNAP) (Zhou et al., 2020; Xie et al., 2019), which all offer substantial scalability but lack flexibility. This flexibility is offered by cloud platforms such as ARTO and EOLDAS (Zhou et al., 2020; Xie et al., 2019), but they lack support. These issues are overcome by the MULTIPLY cloud framework (Badri et al., 2023). They highlight the need for assimilating remote sensing data into crop growth models (CGMs) to enable better decision making by reducing uncertainties. A unique feature of the chapter

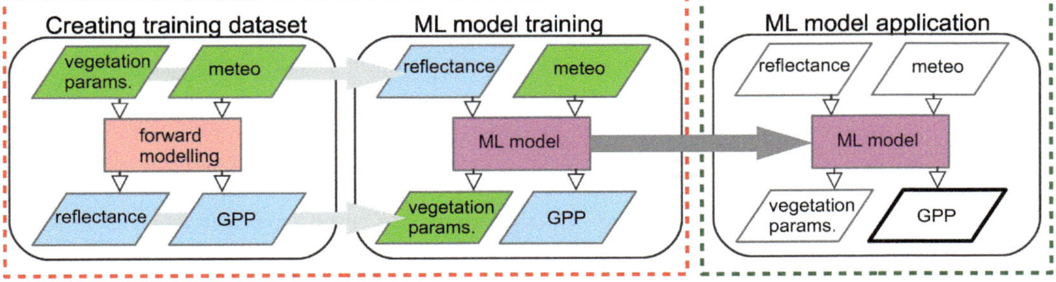

FIGURE 17.4 Flow chart of the processing chain applied in this work. The ML model is trained on the dataset created from soil-canopy energy balance radiative transfer model (SCOPE) simulations, afterwards the ML model is applied to the satellite and meteorological data. (Source: Wolanin et al., 2019.)

is the studies presented and discussed using data from optical, thermal, SAR, LiDAR, and SIF data in crop biophysical and biochemical modeling Considerable emphasis is provided in discussing modeling as well as their uncertainties, errors, and accuracies pertaining to specific parameters such as leaf area index (LAI), chlorophyll, nitrogen, canopy temperature, and water stress.

17.5 REMOTE SENSING OF AGRICULTURE

One of the main applications of remote sensing has always been agriculture. When the first Landsat was launched in 1972, one of the main uses of its data was for agriculture through such programs as the Large Area Crop Inventory Experiment (LACIE), and the agriculture and resources inventory surveys through aerospace remote sensing (AgRISTARS). Pioneering work of Compton Tucker (1979) showed the use of red and near-infrared bands for computing the widely used, normalized difference vegetation index (NDVI) from any sensor that carries these wavebands. Remote sensing of agriculture is now common from any platform: ground-based, airborne, spaceborne, or UAVs using sensors gathering data in wavelengths including optical, thermal, radar, and LiDAR.

In Chapter 5 by Dr. Clement Atzberger et al., the importance of remote sensing of agriculture, type of its applications, and its evolution over last 50 years is well documented. They identify remote sensing application of agriculture into following main area:

A. Qualitative crop monitoring involving changes that take place from within and across seasons. Changes such as deviation from normal conditions, and changes from croplands to cropland fallows (also see Chapter 6);
B. Cropland classification and mapping including crop type identification leading to acreage estimation, phenological studies, identifying shifts in cultivation (also see Chapter 6);
C. Regression modeling involving spectral indices, fAPAR (fraction of absorbed photosynthetically active radiation) and/or wavebands to predict crop growth and yield variables such as grain yield and biomass.
D. Physical modeling of crop growth through remote sensing data assimilation in dynamic (process-driven) crop growth models; and
E. Data mining approaches in cropland studies.

Chapter 5 makes an assessment of various remote sensing data used for different types of agricultural applications such as coarse resolution high revisit data for crop yield and biomass estimation, hyperspectral data in quantifying biophysical and biochemical quantities (also see Chapter 9), and multispectral broadband data in cropland classification.

Over the years, a cropland study from remote sensing still provides a number of challenges. The issues involved include imagery resolution, time of acquisition, and number of images available during the season and their frequency, pre-processing and atmospheric correction and classification methods and approaches (e.g., Figure 17.5). In Figure 17.5, multispectral Quickbird 2.44 m imagery is used for classification and mapping of crops and other land use using a number of different classification algorithms. When object-oriented segmentation approaches are used with different classification algorithms, they provide much better results. Similarly crop biophysical and biochemical modeling faces a number of challenges. Nevertheless, modern remote sensing, is increasingly overcoming these challenges. For example, availability of imagery from multiple sensors (e.g., Landsat, Sentinel, IRS, and SPOT) provides more frequent coverage of the same area that will overcome cloud cover issues and advance our ability to more precisely monitor phenology. Temporal images will also increase the classification accuracies.

Chapter 5 also provides an overview of existing global and regional remote sensing data based agricultural monitoring systems such as Group on Earth Observation's (GEO) Global Agricultural Monitoring (GLAM), USAID's Famine Early Warning (FEWS-NET), United Nations' Food and

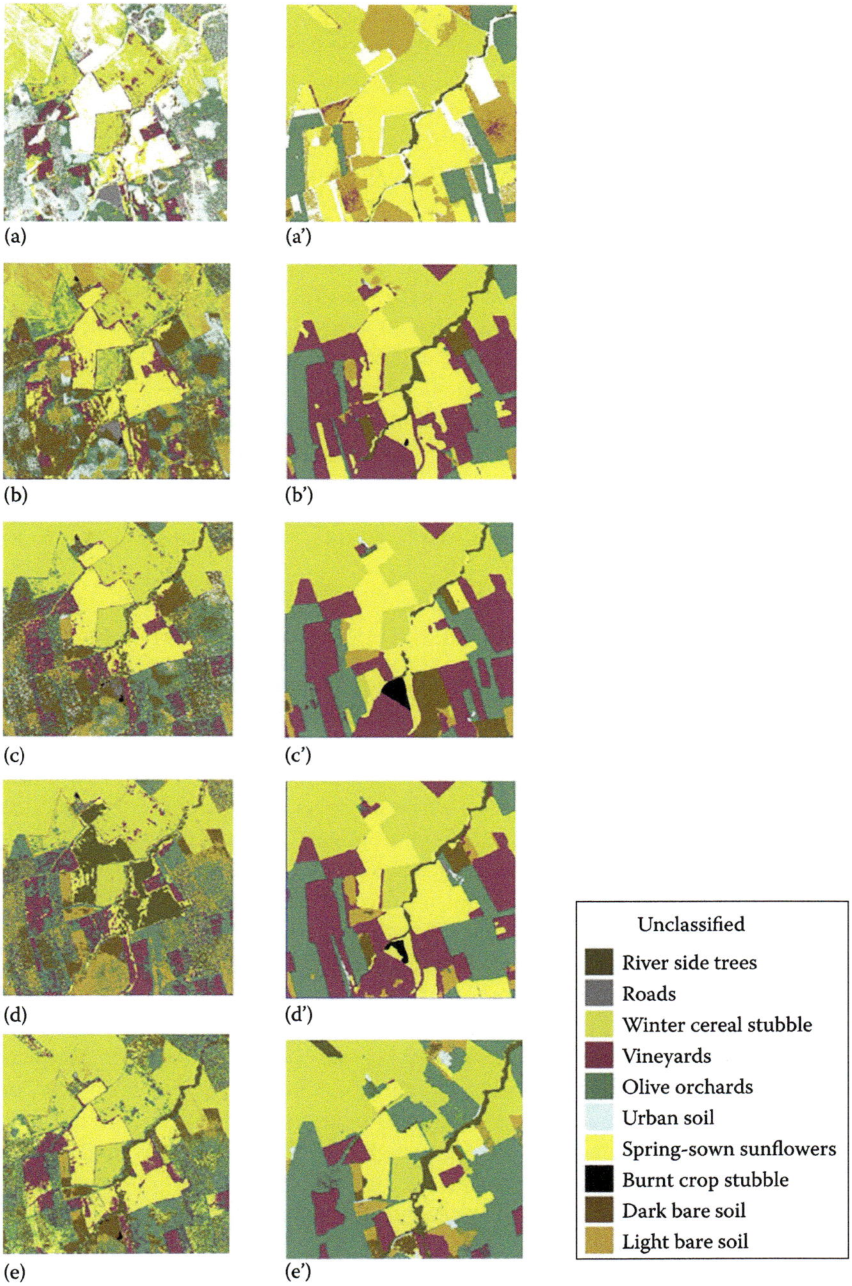

FIGURE 17.5 Result of the least (a, b, c, d, e) and most accurate (a′, b′, c′, d′, e′) land use classifications based on Quickbird 2.44 m imagery with and without segmentation based on number of well-known algorithms; (a) P*, (b) MD*, (c) MC*, (d) SAM*, (e) ML* for pan-sharpened image and pixel as Minimum Information Unit (MIU); (a′) P, (b′) MD, (c′) MC, (d′) SAM for multispectral image and pixel + object as MIU; (e′) ML for pan-sharpened image and pixel + object as MIU. Note: P*: Parallelepiped; MD*: Minimum Distance; MC*: Mahalanobis Classifier Distance; SAM*: Spectral Angle Mapper; ML*: Maximum Likelihood. (Source: Castillejo-González et al., 2009.)

Agriculture Organization (FAO) Global Information and Early Warning System (GIEWS), JRC's Monitoring Agricultural Resources action of the European Commission (MARS), European Union Global Monitoring of Food Security program (GMFS), and Crop Watch Program at the Institute of Remote Sensing Applications of the Chinese Academy of Sciences (CAS) (CropWatch). Evolution of these various crop monitoring systems, from various agencies of the world is the testament of the progress and maturity of remote sensing data in agricultural studies.

Great advances are taking place in quantifying and modeling cropland characteristics because of several factors that include improved time-series remote sensing data availability, easy access to data, ability to code and compute on the cloud, advances in methods and techniques involving machine learning, cloud computing. For example, the phenology metrics obtained from MODIS-NDVI accurately reflect the corn characteristics and can be used for large-scale yield prediction (Ji et al., 2021). Duncan et al. (2015) demonstrate the potential of satellite-observed crop phenology to enhance yield gap assessments in smallholder landscapes. Artificial neural networks (ANNs) provide information about the yield that can be obtained even a few months before harvest, which is extremely valuable for adopting an appropriate strategy in the import and export of agricultural products (Hara et al., 2021). Convolutional Neural Networks (CNNs), a deep learning methodology, was able to reduce crop yield prediction uncertainty considerably (Zhou et al., 2020; Xie et al., 2019). Such and many other advances are well enumerated and discussed in Chapter 5 by Atzberger et al.

17.6 AGRICULTURAL SYSTEMS STUDIES THROUGH REMOTE SENSING

Humans started domesticating the agriculture some 10–12 thousand years ago. Early civilizations were primarily agrarian (Ellis et al., 2021). Agricultural sysems of the world have evolved over millennia and with these evolutions have also seen major changes in how agriculture is farmed and managed by humans. The present-day agriculture systems include both croplands and rangelands and occupy nearly one-third of the global terrestrial area. So, an understanding and study of agricultural systems would involve every component of croplands and rangelands and their associations (e.g., Figure 17.6). Since croplands and rangelands are spread across the world and are dynamic by nature remote sensing provides an ideal platform to characterize, model, and manage agricultural systems of the world routinely. Remote sensing has the capacity to assist the adaptive evolution of agricultural practices in order to face this major challenge, by providing repetitive information on crop status throughout the season at different scales and for different actors (Weiss et al., 2020).

Chapter 6 by Dr. Agnès Bégué et al. provides approaches on applying remote sensing to study agricultural systems. They show us how remote sensing data are used to derive land cover, land use, and land systems which are then tied to croplands, cropping systems, and cropland systems, respectively. Crop type is the first criterion used to characterize agricultural systems at the regional scale, followed by the cropping pattern, the water supply, and the cropland extent and fragmentation. Image resolutions (spatial, spectral, radiometric, and temporal) are all important in discerning particular features of agricultural systems. Especially, when the sensor spatial resolution is smaller than the objects of interest (fields, trees, etc.) the agricultural landscape can be described as a mosaic of patches and thus agricultural systems can be characterized and mapped directly using object-based analysis and landscape metrics (landscape agronomy). When the sensor spatial resolution is larger than the object of interest, the agricultural systems should be characterized indirectly by computing a large variety of satellite-derived indices, environmental and socio-economic variables further processed with data mining techniques in order to stratify the agricultural lands. Dr. Agnès Bégué et al. in Chapter 6 provides concrete illustrations of mapping agroforestry in Bali using sub-meter to 2-m Quickbird imagery and double cropping in Brazil using high temporal MODIS data (direct approach), and rainfed cropping in Mali also using MODIS time series data (indirect approach). Overall, they show us the potential of remote sensing to study agricultural systems at various levels such as region, landscape, field, and plant using appropriate imagery.

The relevance of agriculture to food, water, climate, economy, and peace are enormous (Thenkabail et al., 2021a, 2021b). A scientometric analysis revealed that the research on remote sensing of irrigation

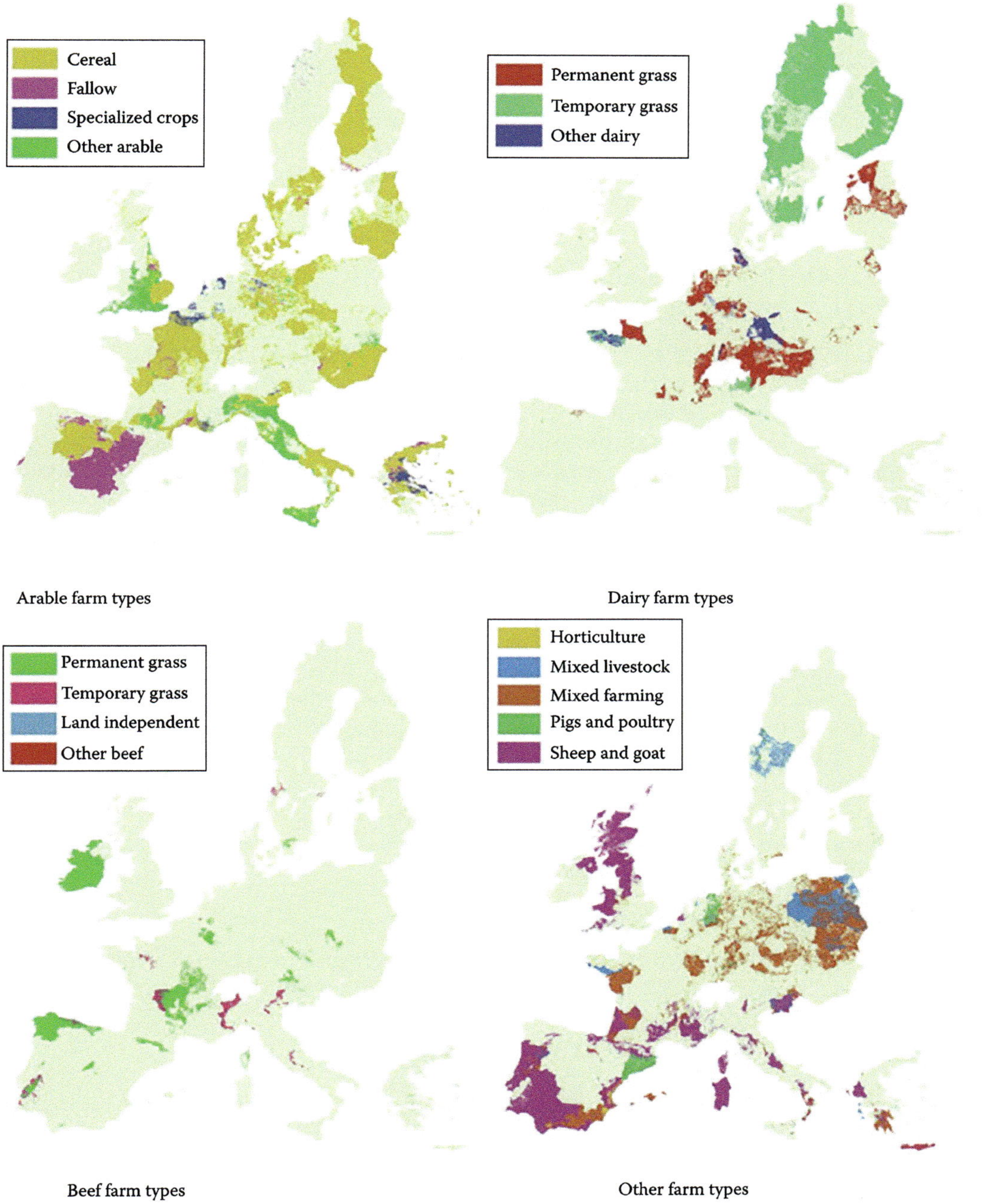

FIGURE 17.6 The distribution of arable farm types in agri-environmental zones dominated by arable farm types, Dairy farm types in agri-environmental zones dominated by arable farm types, beef farm types in agri-environmental zones dominated by beef farm types and other farm types in agri-environmental zones dominated by other farm types. (Source: Kempen et al., 2011.)

had grown exponentially over the last two decades, indicating a growing interest in this field and facilitated through data from various satellite missions, specifically open-access very high spatial-resolution imagery (Manivasagam, 2024). Recent advances in remote sensing and Artificial Intelligence (AI), has enabled us to quantify field scale phenotypic information accurately and integrate the big data into predictive and prescriptive management tools (Jung et al., 2021). A big leap in understanding, modeling, and mapping agricultural cropland systems is made possible by novel technologies such as Internet of Things (IoTs). For example, reference data required to train, test, and validate models can now be gathered using Mobile Apps (Thenkabail et al., 2021a, 2021b; Phalke and Özdoğan, 2018; Fritz et al., 2010). Availability of modern technologies and tools will help climate-smart, economic-smart agriculture to ensure best crop productivity and crop water productivity (Foley et al., 2023). Going forward, the goal will be to produce more food from less land and less water by using the power the advances in IoT suite technologies (Thenkabail et al., 2012, 2021a, 2021b).

17.7 GLOBAL CROPLAND AREA DATABASE FROM EARTH OBSERVATION SATELLITES

Globally, there is between 1.5 to 1.7 billion hectares of croplands of which around 400 million hectares are irrigated fully or partially when you consider cropping intensity (i.e., how many crops are grown over same area in a calendar year; so one may account an area twice when there are two crops a year over the same land) (Thenkabail et al., 2012, 2021a, 2021b). The overwhelming proportion of rainfed croplands are cropped only once a year, during the rainy season. An overwhelming proportion of irrigated areas is cropped more than once a year. Cropland area estimates include continuous plantation crops. The importance of understanding croplands is essential for managing its crop productivity (productivity per unit of land; kg/m^2) and water productivity (productivity per unit of water; m^3/m^2). All of this has huge implications on managing food security (Foley et al., 2020, 2023).

Chapter 7 by Dr. Pardhasaradhi Teluguntla et al. provides a comprehensive overview of the state-of-art of global agricultural croplands. A common application of remote sensing over the last 5 decades has been agricultural cropland mapping and separating them from other land cover (e.g., Figure 17.7a). However, most of these applications were limited to smaller areas. Few large area applications exist, such as the USDA cropland data layer (CDL). Global cropland products are limited to few studies, but they provide large uncertainties due to factors such as coarse resolution of the imagery used, and lack of field data and understanding needed to develop and test algorithms, so, uncertainties in existing cropland products are substantial. Generally, it is agreed that crop products derived from remote sensing should include:

- Cropland extent and areas;
- Watering method: irrigated or rainfed (e.g., separating irrigated from rainfed: Figure 17.7b);
- Cropping intensity: single, double, triple, or continuous cropping;
- Crop type: major global crops;
- Change: how crops change over a given location over space and time;

Accurate production of such products will also lead to far greater accuracies in cropland water use. Since, 70–80% of all human water use goes towards agriculture to produce food (Thenkabail et al., 2021a, 2021b; Foley et al., 2020), it is of utmost importance. Accurate production of the aforementioned products will lead to better assessment, planning, and management of crop productivity (productivity per unit of land; kg/m^2) and water productivity (productivity per unit of water; m^3/m^2), which are very essential for increasing food production from population that is currently at little over 7.2 billion, but is expected to reach 9–10 billion by the year 2050 (Thenkabail et al., 2012). Chapter 7 shows us the remote sensing data, approaches, methods, and synthesis involved in use of remote sensing Earth Observation (EO) data for global cropland studies at various resolutions from 30 m (e.g., Landsat), 250 m (e.g., MODIS), to 1,000 m (e.g., AVHRR). Strengths and limitations of various remote sensing EO data, methods and algorithms required to routinely and repeatedly producing accurate cropland products, and approaches of using these data in food security analysis have been presented and highlights.

Great advances in global and regional cropland mapping using Landsat, and Sentinel, and other remote sensing big data along with machine learning, deep learning, and cloud computing have been achieved, as documented in several well cited publications (Thenkabail et al., 2021a, 2021b; Xiong et al., 2017a, 2017b; Teluguntla et al., 2017, 2018; Gumma et al., 2020; Oliphant et al., 2019; Phalke et al., 2020; Massey et al., 2018, 2017).

17.8 PRECISION FARMING AND REMOTE SENSING

The concept and the idea of precision farming were established in the early quarter of the last century. Precision farming is variously referred to as site specific farming, variable rate technology, or prescription farming. Concisely, precision farming can be understood as a customized sub-field agricultural management decision system relying on information from advanced technologies such as remote sensing, GIS, and GPS. Raj et al. (2021) and Jung et al. (2021) define precision farming as the application of a combination of advanced technologies to: (a) improve agricultural crop productivity and (b) reduce environmental pollution through quantitative and qualitative information of within field variability (or site specific variability) due to natural and human causes. Adoption

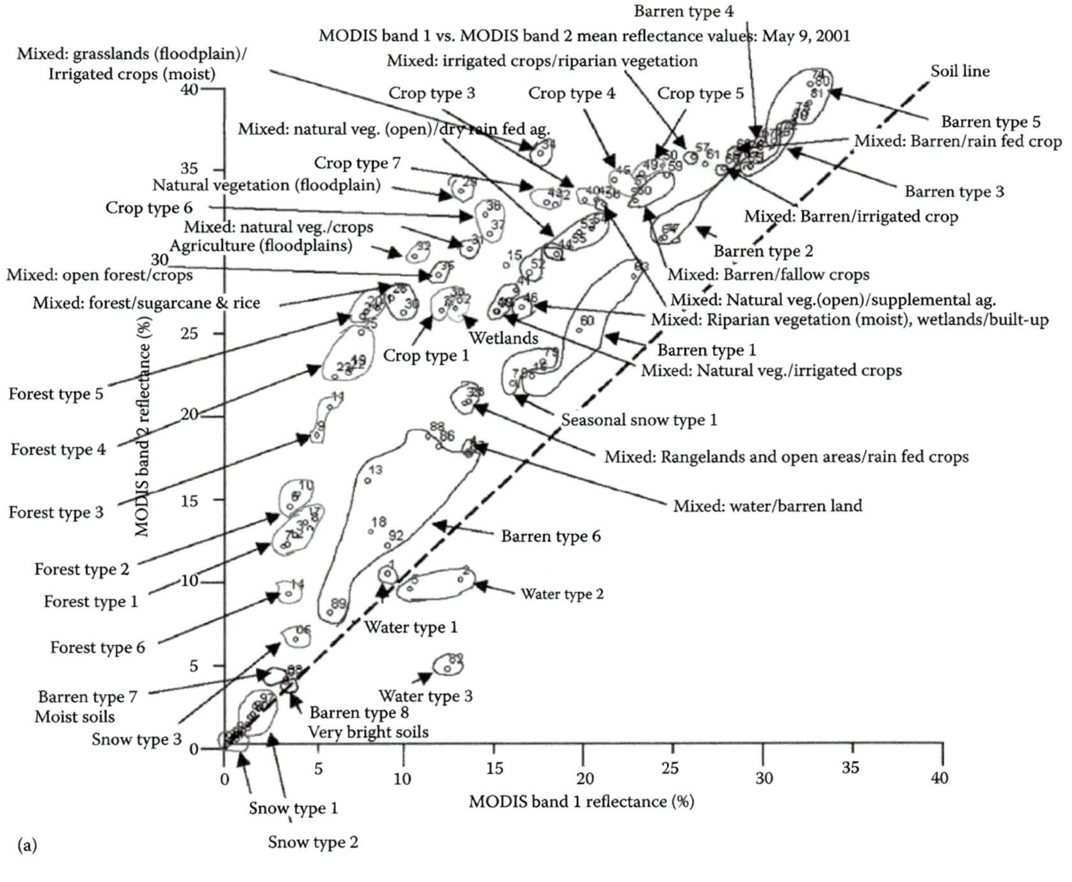

FIGURE 17.7A RED-NIR single dates (RN-SDs) plot of 100 unsupervised classes. The 100 unsupervised classes are plotted taking mean class reflectance in MODIS band 1 (red) and band 2 (NIR). The classes are shown in brightness—greenness—wetness (BGW) feature space and their preliminary class names identified for further investigations during ground truthing. Similar to figure shown above RN-SDs were plotted for each of the 42 dates. (Source: Thenkabail et al., 2005.)

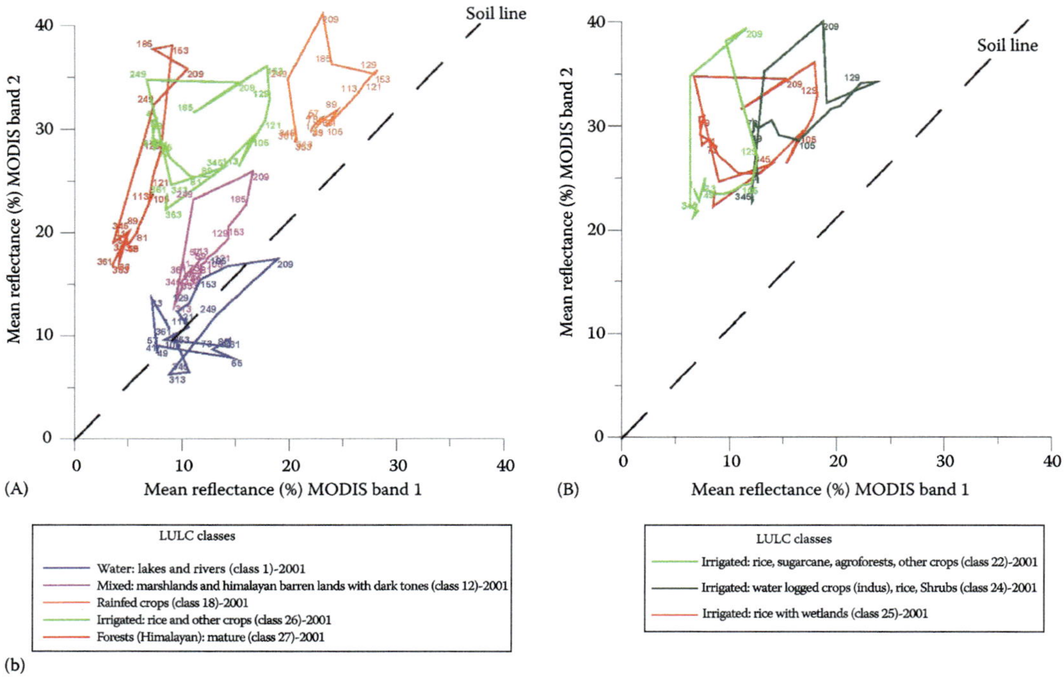

FIGURE 17.7B Space-time spiral curves (ST-SCs) to study subtle and not-so-subtle changes in LULC spectral separability. The ST-SCs are a unique and powerful representation of observing subtle and not so subtle changes over time mapped in two-dimensional feature space. MODIS band reflectance in band 1 (red) and band 2 (NIR) are used to plot ST-SCs for: (a) five spectrally distinct LULC classes and (b) six spectrally similar irrigated area classes. As the spectral properties of classes change over time, we can observe dates on which two or more classes spectral intersect (no spectral separability) or stay spectrally separate highlighting the near-continuous interval multi-temporal data in LULC studies. (Thenkabail et al., 2005.)

of precision farming accelerated as sophisticated technologies such as remote sensing, GPS, GIS, and yield sensors improved and decreased in cost. An integrated approach for the use of these technologies has made precision farming realistic. The target of precision farming is to identify, map, quantify, and assess farm by farm (but spread across entire regions; e.g., Figure 17.8)) spatial and temporal variability for maximizing profits, sustainable production, and protecting the environment. Information requirements of precision farming (Zheng et al., 2021; Raj et al., 2021; Jung et al., 2021) include:

- Seasonally stable factors
  Soil properties, topography, prior management history, etc. (e.g., Figure 17.8a)
- Seasonally variable factors
  Crop biophysical parameters, crop growth, crop yield (e.g., Figure 17.8b), phenology, crop disease, weed infestation, nutrient deficiencies, soil characteristics, and evapotranspiration rates

Based on the preceding, precision farming seeks to diagnose the cause of crop growth and/or yield variability and develop management strategies.

Remote sensing technology is ideally suited to answer questions such as: how much are we producing (e.g., grain or biomass, or leaf area index) per unit of land (kg/m$^{2;}$ Figure 17.8b)?, and what is the variability in these production quantities within and between agricultural fields? (e.g., Figure 17.8). Providing such information at high-spatial resolutions (typically <10 m; but preferably sub-meter to a few meters) becomes invaluable for increasing crop productivity by targeting management to areas that produce less or are more responsive. Remote sensing is a generally well established, powerful, and accurate technology for quantifying crop biophysical quantities such as biomass, leaf area index, plant density, crop vigor, grain yield, and even plant height and canopy cover. As a result, remote sensing has been widely applied for better management of crops over last 25 years. Especially, when remote sensing is combined with other spatial technologies such as GIS, GPS, and spatial modeling, they become a powerful tool to understand and manage agricultural crops leading to improved productivity. However, precision farming requires an intimate understanding of crop stresses and their causes (e.g., nutrient, pest, disease, water deficiency, weeds, and even macro and micro nutrients). Whereas, remote sensing can detect stress accurately, its ability to detect the nature of stress (e.g., whether from pests or nutrients) is still uncertain. Hyperspectral data acquired at very high spatial resolution and often combined with other type of data (e.g., thermal) have been helpful in making advances. For example, hyperspectral narrow bands (HNBs) and hyperspectral vegetation indices (HVIs) from specific wavelengths have enabled advances.

Dr. David J. Mulla, in Chapter 8 identifies specific HVIs for detecting N, P, or K, disease, insects, and weeds. For example, Dr. Mulla shows us that NDVI, SAVI, and red-edge are good for detecting N, P, or K; fluorescence and PRI are good for detecting disease, Aphid index (AI) damage index, leaf hopper index are good for insects, and RVI good for weeds. Of course, there are indices like NDVI or SAVI that can detect multiple types of crop stress. Yet, the uncertainty in understanding and modeling specific stress causes using remote sensing is very high. Some of these uncertainties can be reduced substantially if we have sufficient spatial resolution (e.g., centimeters) along with hyperspectral and

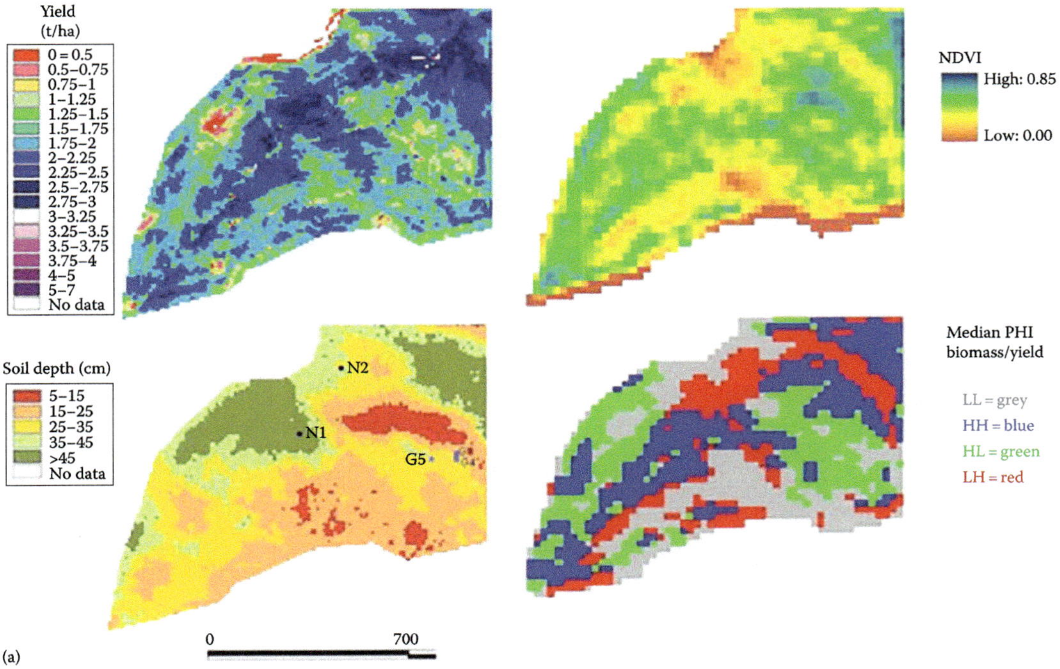

FIGURE 17.8A Yield, NDVI, soil depth, and NDVI (biomass)/yield classification on a single field at Buntine, Western Australia. For biomass/yield classification LL = low biomass and low yield; HH = high biomass and high yield; HL = high biomass and low yield; LH = low biomass and high yield. (Source: Robertson et al., 2007).

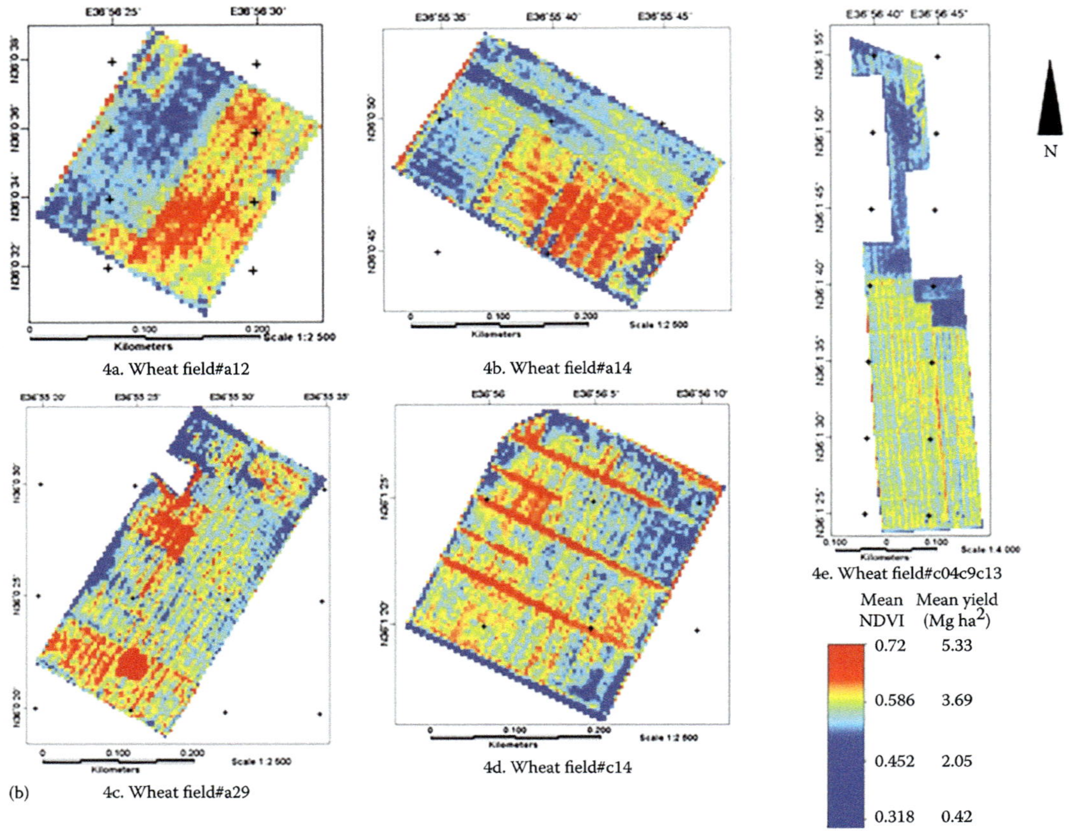

FIGURE 17.8B Understanding, modeling, and mapping crop yields to improve crop productivity (productivity per unit of land; kg/m^2) and water productivity (productivity per unit of water; m^3/m^2). Spatial distribution of wheat yield based on Landsat TM data. These are the typical maps used in precision farming to understand and improve crop and water productivity of crops.

other supplementary data (e.g., thermal). Chapter 8 by Dr. Mulla provides an overview of remote sensing technology use over last 25 years and establishes the current state-of-art. There is great scope for further research and development of precision farming applications using remote sensing to reduce uncertainties and increase accuracies of prediction. UAVs offer a new platform with much higher spatial resolution (e.g., few centimeters) as well as ability to fly multiple sensors (e.g., hyperspectral, thermal), and offer new opportunities for precision farming applications of remote sensing. But, these need to go hand in hand with methodological improvements. UAVs will, however, come with their own limitations such as the ability to get permission to fly, inability to carry heavier sensors, errors due to stability issues, and inability to cover large areas. Nevertheless, a combination of multi-sensor approach, with greater efforts in ideal sensor design (e.g., hyperspectral sensor that is of few centimeter resolution), and improved methodological efforts will lead to wider application of remote sensing technology in precision farming leading to improved productivity in crop grain and biomass.

Remote sensing can be used to collect quantitative crop information on sub-areas within a farm, including:

- Crop biophysical quantities-LAI, biomass;
- Within field variations over time (temporal changes);

- Spatial variations in growth stage, phenology;
- Weed and pest infestations;
- Crop stress, disease; and
- Quantitative and qualitative changes within and between seasons.

The preceding information can be linked to decision support systems such as pest or drought early warning systems, leading to county or regional scale agricultural decision support systems, etc. The present advances in precision farming are in the role the Artificial Intelligence (AI) and Internet of Things (IoT) that play a vital role in our modern day-to-day applications making more precise and profitable (Raj et al., 2021). The five key technologies of agricultural IoT are (1) sensor perception technology involving a wide array of sensors such as moisture sensors, chlorophyll sensors, plant stress sensors, and other sensors like spectroradiometers that measure multiple traits; (2) node location technology involving wireless sensor networks (WSN) such as WiFi, Bluetooth; (3) information processing technologies such as cloud computing, machine learning, deep learning, data mining, AI; (4) radio-frequency identification; and (5) 3S technologies a collective term for remote sensing (RS), global navigation satellite system (GNSS), and geographic information system (GIS). Precision farming these days includes drones carrying a wide array of sensors (e.g., hyperspatial, hyperspectral, thermal), drones and aircrafts for spraying, GPS technology (e.g., for spraying, combine harvesting, gathering reference data), satellite remote sensing data from multitude of platforms and sensors, web access to data and analytics, cloud computing, machine learning, AI, and robotics all working in tandem to make farming smarter, cheaper, environmentally friendly, sustainable, profitable, and climate friendly. For example, the AI is now helping us quantify field scale phenotypic information accurately and integrate the big data into predictive and prescriptive management tools (Jung et al., 2021). The high-throughput phenotyping technology has benefitted immensely from both remote sensing big-data and machine learning, especially when simultaneous use of multiple sensors (e.g., high-resolution RGB, multispectral, hyperspectral, chlorophyll fluorescence, and light detection and ranging (LiDAR)) are used to allow for a range of spatial and spectral resolutions depending on the trait in question (Zheng et al., 2021). A review of 239 articles on precision farming showed that the precision farming IoT technologies such as robotic mechanical weeding and precision harrowing allowed the weed management in a more sustainable manner by removing up to 86 % of the weeds and help reduce pesticide use by up to 97% (Anastasiou et al., 2023).

17.9 MAPPING TILLAGE VERSUS NON-TILLED LANDS AND ESTABLISHING CROP RESIDUE STATUS OF AGRICULTURAL CROPLANDS USING REMOTE SENSING

The importance of mapping tilled versus no-tilled agricultural lands is discussed in Chapter 9 by Dr. Baojuan Zheng et al. Conservation tillage practices improve soil and cropland management. Conventional tillage practices often have detrimental impacts, including soil erosion, loss of organic matter, and leaching of nutrients. Conservation tillage practices leave crop residues (e.g., Figure 17.9) on the soil surface to retain moisture and resist soil and wind erosion. So, the ability to reliably monitor tilled versus no-till lands on a field-by-field basis becomes an important component of soil and landscape management. Dr. Baojuan Zheng et al. survey state-of-the-art strategies to map tillage practices by estimating crop residue cover. Further, they survey applications of remote sensing tillage assessment using:

1. Optical remote sensing
2. SAR remote sensing

For optical remote sensing, SWIR bands in range of 2,000 to 2,300 nm are critical to distinguish live crops versus crop residue versus soil. Cellulose Absorption Bands centered at 2,100 nm are

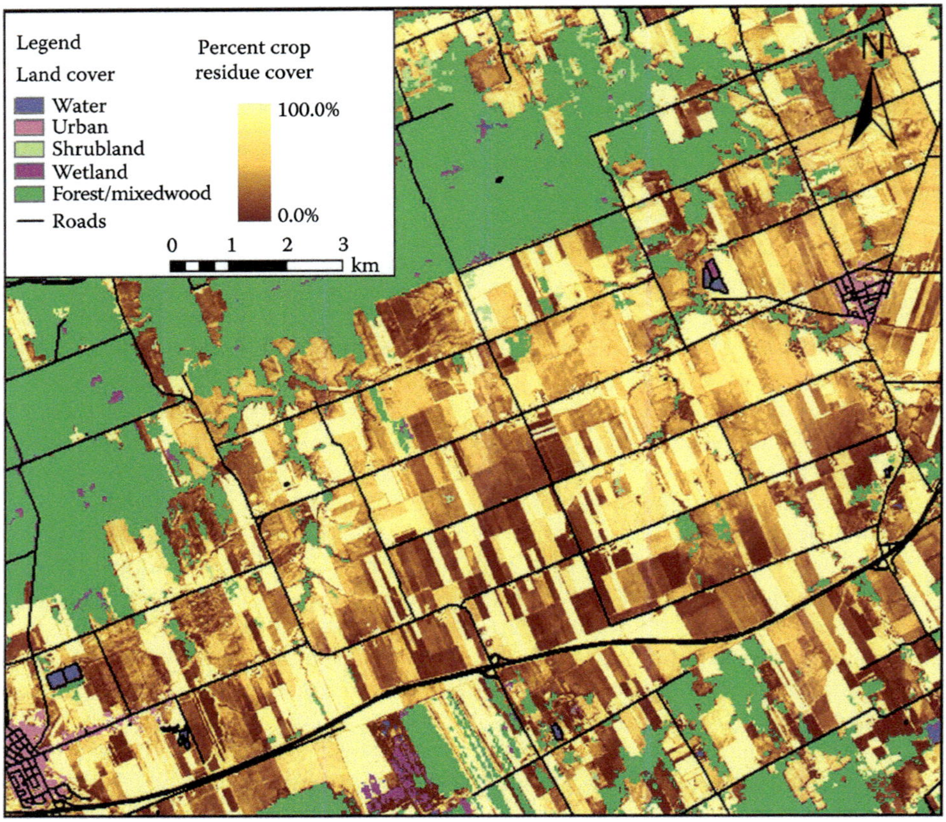

FIGURE 17.9 Percent crop residue cover map over the Casselman/St. Isidore study site derived from spectral mixture analysis on a SPOT image acquired on November 9, 2007. A land cover mask (nonagricultural areas) is overlaid on the residue cover map. (Source: Pacheco and McNaim, 2010.)

especially valuable. Landsat bands 5 and 7 (for calculation of the Normalized Difference Tillage Index (NDTI)), are widely used in detecting crop residue, thereby distinguishing tilled and no-till conditions. Classification methods such as Maximum Likelihood, spectral angle mapping, and data mining (e.g., the Random Forests classifier and support vector machines) are widely used. However, narrow bands centered at 2,030 nm, 2,100 nm, and 2,210 nm with bandwidths of $\sim$ 10 nm or less are likely to provide the best results. When hyperspectral data is available (e.g., Hyperion, EnMAP, and HyspIRI), they can be used to compute Cellulose Absorption Indices (CAI). Accurate tillage assessment depends upon observing crop residue in fields within a short interval at the start of the growing season, just as farmers begin to plant crops. So, timeliness of imagery is key to successful mapping of tillage status using optical images, which are subject to effects of cloud cover. Further, Zheng et al. explore SAR images acquired from multiple platforms at X, C, and L bands, showing that SAR backscatter coefficient thresholds are key to distinguishing tilled from no-till conditions. SAR images perform best when more than one frequency and polarization are used. Cloud penetration of SAR images offers important advantages. Research to investigate the role of polarimetric images will likely form an important topic supporting SAR applications for tillage assessment.

Tillage indices along with Sentinel-2A textural feature indicators helped increase accuracies of maize residue cover estimations (Xiang et al., 2022). Gao et al. (2022) suggested a Sentinel-2 data-based Residue Adjust Normalized Difference Residue Index (RANDRI) and combined with Soil Adjust Normalized Difference Residue Index (SANDRI) to effectively predict the proportion of conventional tillage, low residue tillage, conservation tillage, and high residue tillage, thus providing the data support for improving soil quality, agricultural product quality, and farmland system management. What is exciting is for crop tillage and residue mapping is the data from new generation of sensors such as the hyperspectral PRISMA, EnMAP, SBG along with the Landsat Next instrument suite. All of these will have specific 2.5 to 12.5 nanometer (nm) wide narrowbands spread across the electromagnetic spectrum that will offer specific narrowbands to study specific quantities. For example, a recent study reports the findings of a Landsat Next expert review panel that evaluated the use of narrow shortwave infrared (SWIR) reflectance bands to measure ligno-cellulose absorption features centered near 2100 and 2300 nm, with the objective of measuring and mapping non-photosynthetic vegetation (NPV), crop residue cover, and the adoption of conservation tillage practices within agricultural landscapes (Hively et al., 2021), They report a Shortwave Infrared Normalized Difference Residue Index (SINDRI) which relies on two narrowbands centered at 2,260 nm and 2,210 nm to study non-photosynthetic vegetation (NPV).

17.10 HYPERSPECTRAL REMOTE SENSING FOR TERRESTRIAL APPLICATIONS

Traditional remote sensing from sensors such as Landsat, SPOT, and IRS gathered data in broad spectral wavebands across the electromagnetic spectrum, typically in spatial resolution of greater than 20 m. The new generation hyperspatial sensors (e.g., IKONOS, Quickbird) acquired data in very high spatial resolution (e.g., sub-meter to 5 m), but also in broad spectral wavebands across the electromagnetic spectrum. However, there is great scientific interest and need to gather data near-continuously across the electromagnetic spectrum. This need is fulfilled by hyperspectral or imaging spectroscopy data.

Chapter 10 by Dr. Prasad S. Thenkabail defined hyperspectral data as follows:

> *Remote sensing data is called hyperspectral when the data is collected contiguously over a spectral range, preferably in narrow bandwidths and in reasonably high number of bands.*

So, a typical hyperspectral (Imaging Spectroscopy) data is gathered in very narrow bands (~1 nm to 10 nm bandwidths), contiguously across electromagnetic spectrum (e.g., 400–2,500 nm), resulting in several 10s, or 100s, or 1000s of narrowbands of data.

Chapter 10 begins by enumerating key characteristics of ground based, airborne, and spaceborne hyperspectral sensors. The spatial, spectral, radiometric, and temporal characteristics of the key spaceborne hyperspectral sensors are presented and discussed. One of the first steps in hyperspectral data analysis is data mining. The data mining is extremely important from point of view of reducing the data volumes to eliminate redundant bands, especially in age of "big data." The value, importance, and approach of overcoming Hughes' phenomenon is discussed. Greater number of hyperspectral narrowbands (HNBs) are very important to increase classification accuracies as well as to develop unique and more powerful hyperspectral vegetation indices (HVIs). Chapter 10 shows that it is feasible to achieve increased accuracies of 30% or higher when ~20 HNBs are involved relative to 7 Landsat broadbands. The chapter also shows that there are several unique two-band HVIs (TBHVIs; e.g., Table 17.2) targeted to study specific biophysical and biochemical quantities. Further multi-band HVIs (MBHVIs) involving 3 to 10 HNBs often provide much higher R-square values relative to commonly known two-band indices like NDVI.

The key highlights of Chapter 10 include summary of:

1. 28 optimal HNBs for studying agricultural crops and vegetation;
2. .6 categories of important HVIs for modeling crop and vegetation biophysical and biochemical quantities; and
3. 4, 6, 8, 10, 12, 16, and 20 HNB combinations to best classify croplands and vegetation.

TABLE 17.2
Some of the Two-Narrowband Hyperspectral Vegetation Indices (HVIs) as per Early Research by Thenkabail et al. (2000). Band centers (λ1 and λ2) and band widths (Δλ1 and Δλ2) in nanometers. For example, an HVI involving two bands, one centered at 682 nm (band width = 28 nm) and another centered at 918 nm (band width 20 nm), provide the best index for modeling wet biomass (WBM) of cotton crop. Ranking of indices: Index 1 always has highest R-square value and hence ranked higher, and index 7 has the lowest R-square of the seven indices listed for each variable of each crop. Note: these were linear models. However, often, two band nonlinear and/or multi-band linear models involving more than two narrowbands provide significantly improved results. (Source: Thenkabail et al., 2000.)

Crop (Sample Size)	Crop Variable	Band Center and Width (nm)	Band Centers (λ1 and λ2) and Band Widths (Δλ1 and Δλ2) for Two-Band Vegetation Indices						
			Index 1 (nm)	Index 2 (nm)	Index 3 (nm)	Index 4 (nm)	Index 5 (nm)	Index 6 (nm)	Index 7 (nm)
1. Cotton (73) except for yield that has a sample size of 50	WBM (kg/m²) (see Figure 17.4)	λ_1	682	682	568	555	615	525	982
		$\Delta\lambda_1$	28	28	10	20	175	4	10
		λ_2	918	845	918	666	925	540	940
		$\Delta\lambda_2$	20	250	10	5	20	7	10
	LAI (m²/m²)	λ_1	682	550	678	550	568	525	940
		$\Delta\lambda_1$	15	30	15	50	4	60	10
		λ_2	940	682	865	675	915	540	980
		$\Delta\lambda_2$	60	20	275	50	15	60	10
	yield: lint ha)	λ_1	540	696	678	510	690	678	670
		$\Delta\lambda_1$	30	4	30	40	10	30	50
		λ_2	678	940	940	684	720	860	970
		$\Delta\lambda_2$	20	20	50	20	20	290	20
2. Potato (25)	WBM (kg/m²)	λ_1	550	550	678	682	720	682	615
		$\Delta\lambda_{1.}$	20	30	10	20	30	10	70
		λ_2	682	682	920	940	790	710	935
		$\Delta\lambda_2$	4	26	20	40	20	20	50
	LAI (m²m²) (see Figure 17.5)	λ_1	682	472	682	678	550	682	625
		$\Delta\lambda_{1:}$	28	15	28	28	20	7	50
		λ_2	982	790	738	860	688	940	940
		$\Delta\lambda_2$	36	20	45	100	28	30	8
3. Soybean (27)	WBM (kg./m²) (see Figure 17.4)	λ_1	725	732	696	565	550	635	490
		$\Delta\lambda_1$	25	10	28	50	20	30	75
		λ_2	845	758	791	875	755	682	682
		$\Delta\lambda_2$	70	10	10	130	10	10	28
	LAI (m²/m²)	λ_1	625	495	495	730	682	600	418
		$\Delta\lambda_1$	30	30	30	20	10	40	510
		λ_2	688	685	670	840	790	860	855
		$\Delta\lambda_2$	15	20	40	60	20	100	90
4. Corn (17)	WBM (kg/m²)	λ_1	720	505	715	620	620	620	490
		$\Delta\lambda_1$	18	10	4	30	30	30	50
		λ_2	820	645	990	830	1.000	940	825
		$\Delta\lambda_2$	160	6	20	160	30	70	110
	LAI (m²/m²)	λ_1	620	550	635	470	495	495	650
		$\Delta\lambda_{1:}$	20	30	20	8	50	50	4

(Continued)

TABLE 17.2 (*Continued*)
Some of the Two-Narrowband Hyperspectral Vegetation Indices (HVIs) as per Early Research by Thenkabail et al. (2000). Band centers (λ1 and λ2) and band widths (Δλ1 and Δλ2) in nanometers. For example, an HVI involving two bands, one centered at 682 nm (band width = 28 nm) and another centered at 918 nm (band width 20 nm), provide the best index for modeling wet biomass (WBM) of cotton crop. Ranking of indices: Index 1 always has highest R-square value and hence ranked higher, and index 7 has the lowest R-square of the seven indices listed for each variable of each crop. Note: these were linear models. However, often, two band nonlinear and/or multi-band linear models involving more than two narrowbands provide significantly improved results. (Source: Thenkabail et al., 2000.)

			Band Centers (λ1 and λ2) and Band Widths (Δλ1 and Δλ2) for Two-Band Vegetation Indices						
Crop (Sample Size)	Crop Variable	Band Center and Width (nm)	Index 1 (nm)	Index 2 (nm)	Index 3 (nm)	Index 4 (nm)	Index 5 (nm)	Index 6 (nm)	Index 7 (nm)
	(see Figure 17.5)	λ_2	590	684	720	740	760	1.000	800
		$\Delta\lambda_2$	10	28	10	2	40	25	165
5. All crops	WBM	λ_1	682	655	525	660	525	640	510
(151)-	(kg/(m^2)	$\Delta\lambda_1$	30	90	20	30	50	120	100
		λ_2	910	920	682	875	675	880	670
		$\Delta\lambda_2$	20	20	25	250	30	280	60
	LAI (m^2/m^2)	λ_1	540	682	550	682	682	670	490
		$\Delta\lambda_1$	20	10	40	28	35	40	30
		λ_2	682	756	682	910	754	910	965
		$\Delta\lambda_2$	10	20	30	200	40	200	20

Further, Chapter 10, presents and illustrates various hyperspectral data analysis methods that are broadly grouped under two categories: Feature extraction methods; and Information extraction methods. The various hyperspectral classification methods discussed and illustrated include spectral matching techniques (SMT), spectral mixture analysis (SMA), support vector machines (SVM), and tree-based ensemble classifiers (e.g., Random Forest and Adaboost).

Currently, we are in a new era of hyperspectral remote sensing with data acquisition from several new generation spaceborne sensors. These include data from, already in orbit sensors such as the (1) German Deutsches Zentrum fur Luftund Raumfahrt (DLR's) Earth Sensing Imaging Spectrometer (DESIS) sensor onboard the International Space Station (ISS), (2) Italian Space Agency (ASI) PRISMA (Hyperspectral Precursor of the Application Mission), (3) German DLR's The Environmental Mapping and Analysis Program (EnMAP, and 4. two hyperspectral sensors called Tanager in 202 from Planet Labs PBC. Further, the NASA is planning hyperspectral sensor Surface Biology and Geology (SBG) mission. In addition, we already have over 83,000 hyperspectral images of the planet acquired from NASA's Earth Observing-1 (EO-1) Hyperion that are freely available to anyone from U. S. Geological Survey's data archives (Thenkabail et al., 2021a). These suites of sensors acquire data in 200 plus hyperspectral narrowbands (HNBs) in 2.55 to 12 nm bandwidth, either in 400–1000 or 400–2500 nm spectral range with SBG also acquiring data in thermal range. HNBs provide data as "spectral signatures" in stark contrast to "a few data points along the spectrum" provided by multispectral broadbands (MBBs) such as the Landsat satellite series. Advent of these new generation of hyperspectral sensors in tandem with major developments in machine learning, deep learning, AI and cloud computing has made it possible to model and map products with significantly greater accuracies as demonstrated by Aneece and Thenkabail (2018, 2021, 2022) and Thenkabail et al. (2021a).

17.11 RANGELANDS: A GLOBAL VIEW

Rangelands represent relatively arid (and semi-arid) sites where potential vegetation is predominantly comprised of grasses, forbs and shrubs (e.g., Figure 17.10). Examples include savannas, shrub- and grasslands, tundra, open woodlands, and chaparral. Rangelands are globally important as sources of forage for both domesticated and wild animals. Additionally, rangelands support unique flora and fauna, and provide numerous ecosystem services. Roughly, 24% of the terrestrial area or about 3 billion hectares (double the area of croplands) can be considered rangeland. Omitting deserts from the global terrestrial area, however, increases rangeland proportion to as much as 52% (Wolf et al., 2021), also as mentioned in Chapter 11 of this Volume. So, definition is key to how rangelands are assessed and accounted for.

Given the vastness and inaccessibility of rangelands, remote sensing offers the best opportunity to study rangelands. The key factors of rangeland studies using remote sensing are presented in detail in Chapter 11 by Dr. Matthew Reeves et al. These factors are:

1. Rangeland degradation studies that involve quantifying and modeling degradation of land and soil and ensuing desertification
2. Biomass, net primary productivity (NPP) and forage quantification and modeling
3. Assessment of rangeland fires that include: (a) burned area determination, (b) burned area impact on land and atmosphere, and (c) fire progression and intensity monitoring
4. Land cover studies in rangelands
5. Rangeland extent mapping and monitoring

Chapter 11 also provides a history of rangeland studies and its evolution over time, starting from early days of Landsat in 1970s to the current period. A wide array of satellites and sensors are routinely used in rangeland studies. For example wide area fire studies are best conducted using thermal imagery from sensors such as Along-Track Scanning Radiometer (ATSR-2), Advanced ATSR

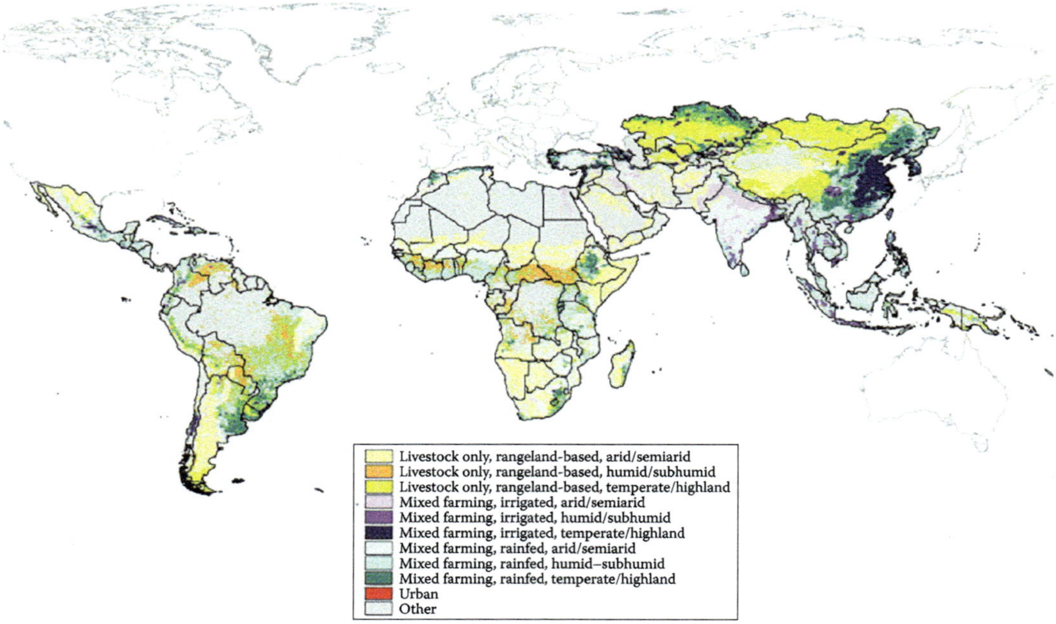

FIGURE 17.10 Global livestock production systems. (Source: Kruska et al., 2003.)

(AATSR), and MODIS. Degradation studies typically use indices such as NDVI, tasseled cap or classification approaches. Given the large areal extent of rangelands, routine monitoring of large landscapes can be done using coarse spatial but high temporal resolution, imagery such as AVHRR, MODIS, VIIRS, and Sentinel. But, more detailed assessments of rangelands are often conducted using either hyperspatial or hyperspectral imagery.

The land surface phenology (LSP) spectral characteristics (e.g., normalized difference vegetation index or NDVI) derived from the global archives of the early satellite sensors such as Landsat, AVHRR, and MODIS played a critical role in tracking and monitoring the seasonality of rangeland ecosystems (Matongera et al., 2021). Innovations in machine learning using random forest model and cloud-based computing were merged with historical remote sensing and field data to provide the first moderate resolution, annual, percent cover maps of plant functional types across the Western United States rangeland ecosystems to predict per-pixel percent cover of annual forbs and grasses, perennial forbs and grasses, shrubs, and bare ground over from 1984 to 2017 (Jones et al., 2018). The power of machine learning and the Google Earth Engine (GEE) cloud computing in processing very large volumes of Landsat remote sensing data for rangeland monitoring is demonstrated for the Western United States (Zhou et al., 2020) and Australia (Xie et al., 2019).

17.12 RANGELAND BIODIVERSITY STUDIES

Rangelands are often the ideal landscapes for study using remote sensing (e.g., Figure 17.11). They are vast and contiguous, relatively uninhabited, have a clear phenology based on precipitation, and have characteristic vegetation comprising of grasslands, shrublands, or some mixture of these. Maintenance of biodiversity is the goal for managing rangelands sustainably; rangeland health is monitored to prevent an irreversible loss of biodiversity. Dr. Raymond Hunt et al. in Chapter 12 identify 19 indicators of rangeland health of which 3 are the most important: (a) bare ground, (b) vegetation composition by plant functional type, and (c) presence of invasive plants. Spatial, spectral, temporal, and radiometric properties of sensors play key roles in determining what rangeland indicators can be studied and at what accuracy. Spectral unmixing of either medium-resolution or hyperspectral data are important for estimating the cover of bare ground and the amount of soil erosion. Land characterizations need to have sufficient temporal resolution; for example, functional types of the rangeland grasses (e.g., tallgrass C_3, tallgrass C_4, shortgrass C_3, and shortgrass C_3 study reported in Chapter 12) are best distinguished using temporal data. Better characterization of NPP, biomass, and rangeland plant functional types can be made using new satellite sensors that have high temporal coverage with medium spatial resolution (between 30m and 100m pixels). Accurate characterization of rangeland species types and invasive species within these rangelands and study of their health will require either hyperspectral or hyperspatial data (< 10 cm pixels). Invasive species are often a significant problem in most rangelands and hence Hunt et al. provide substantial discussion on remote sensing of invasive plants in Chapter 12. They found spectral matching algorithms like spectral angle mapper (SAM) and spectral diversity of classified objects have potential to estimate rangeland biodiversity directly, but the spectral similarity of green leaves will be a limit to the number of species individually characterized. For managing biodiversity, hyperspatial imagery acquired from low-flying crewed or uncrewed aircraft can be used to estimate most of the rangeland health indicators, especially bare ground, vegetation composition, and invasive species that are not spectrally unique. New ways of analyzing hyperspatial data include statistical analysis of transects over the landscape and methods based on computer vision research.

Invasive species such as cheatgrass (Bromus tectorum L.) reduce biodiversity and increase fire hazard. Dahal et al. (2022) used fully automated and multi-task deep-learning framework to simultaneously predict and generate weekly, near-seamless composites of Harmonized Landsat Sentinel-2 (HLS) spectral data to map invasive species exotic annual grass (e.g., cheatgrass or Bromus tectorum) and medusahead (Taeniatherum caput-medusae), in rangeland ecosystems of the western United States. These invasive species affects wildlife habitats, increases wildfire frequency, and

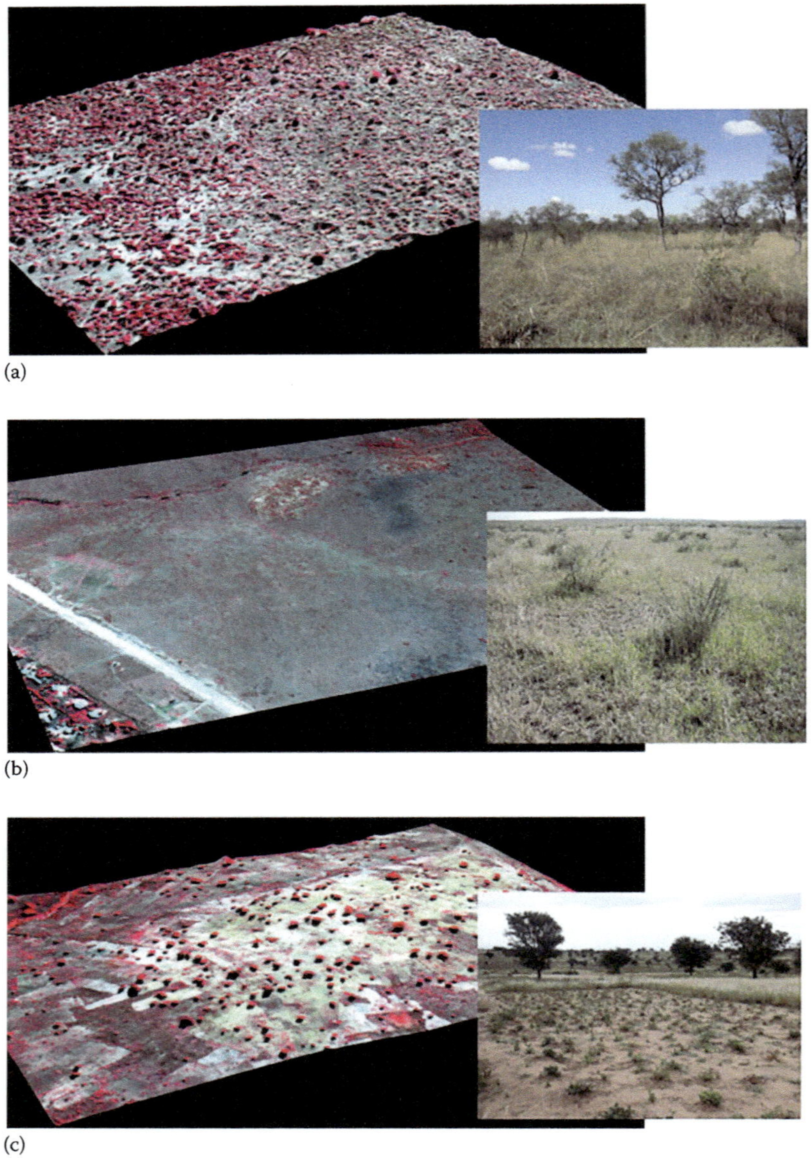

FIGURE 17.11 False color images draped over the surface height model derived from Lidar data for selected sites. Data recorded by Carnegie Airborne Observatory. (a) Dense woodland in private reserve, L3-granite-SabiSand, (b) highly impacted rangeland in communal rangeland with very low woody vegetation cover, L6-gabbro-rangeland; and (c) cultivated and fallow fields in communal areas with large trees, L6-granite-fields. (Source: Wessels et al., 2011.)

adds to land management costs (Dahal et al., 2022). Pastick et al. (2020) developed fully automated and scalable workflow that integrates field observations, HLS data, maps of landscape conditions (i.e., soils, topography), and machine-learning techniques to construct a time series of fractional estimates of exotic annual grass cover and associated uncertainties across dryland ecoregions in the western United States. Grassland degradation and pasture conditions impacting biodiversity under varying climate and human activities in Oklahoma were evaluated using the seasonal dynamics of leaf area index (LAI) and aboveground biomass (AGB) using Sentinel-1, Landsat-8, and Sentinel-2 data, both individually and integrally, by applying three widely used algorithms: Multiple Linear Regression (MLR), Support Vector Machine (SVM), and Random Forest (RF) (Wang et al., 2019).

17.13 METHODS OF CHARACTERIZING, MAPPING, AND MONITORING RANGELANDS

Remote sensing methods, approaches, and techniques of rangeland characterization, mapping, and monitoring are many and depend on the parameter studied. Broadly, these are grouped as follows as per Chapter 13 by Lalit Kumar et al:

1. Rangeland phenology and productivity
   Natural as well as human induced (e.g., grazing) changes in rangeland dynamics are studied using vegetation indices such as the Normalized Difference Vegetation Index (NDVI), Enhanced Vegetation Index (EVI), and a host of other vegetation indices. Rangeland productivity parameters modeled using NDVI, EVI and other variables include biomass, leaf area index (LAI), percent cover, fractional cover, species dominance, and species type. Phenology is characterized by taking, for example, a wide range of NDVI characteristics: cumulative NDVI over a season, NDVI at peak, NDVI at the start of season and/or end of season, NDVI amplitude, and so on (e.g., Figure 17.12).
2. Rangeland ecological characteristics
   Distinct ecologies of rangelands are distinguished using differences in NDVI magnitude and timing.
3. Rangeland biological diversity, fuel analysis, and change detection
   The biological diversity, fuel loadings, and change detection are determined using vegetation indices, well known methods of classification such as Decision Trees (DTs), various supervised and unsupervised classification methods as well as newer methods such as neural networks, random forest, and object oriented classifications.
4. Rangeland change detection
   Apart from classification approaches, image differencing methods are of great importance for rangeland change detection.
5. Vegetation continuous fields
   Vegetation continuous fields are mapped through the use of digital remote sensing data coupled with field-based measurements and advanced statistical modeling techniques.

A persistent monitoring gap between plot-level inventories and the scale at which rangeland assessments are conducted has required decision makers to fill data gaps with statistical extrapolations or assumptions of homogeneity and equilibrium (Jones et al., 2020). This gap is now being bridged with spatially comprehensive, annual, rangeland monitoring data across all western US rangelands to assess vegetation conditions at a resolution appropriate to inform cross-scale assessments and decisions (Jones et al., 2020). Rigge et al. (2020) mapped nine rangeland ecosystem components (percent shrub, sagebrush (Artemisia), big sagebrush, herbaceous, annual herbaceous, litter, bare ground cover, sagebrush, and shrub heights), were quantified at Landsat 30 m resolution. They used very high spatial-resolution imagery (VHRI) from WorldView-2, WorldView-3, QuickBird, or Pleaides resampled to 2 m nominal resolution for the first scale for ground level interaction and measurement, followed by Landsat 8 imagery acquired between 2013 and 2017 for the second scale for landscape modeling. Deep learning is increasing in popularity and may improve taxonomic classifications (Retallack et al., 2023).

17.14 LAND SURFACE PHENOLOGY IN FOOD SECURITY ANALYSIS

Dr. Molly E. Brown et al. in Chapter 14 identify the two basic approaches to measure food insecurity directly: anthropometrics measuring body weight, height, and age of the population; and determining the individual consumption of food per day compared to average requirements. Both approaches require extensive, time consuming, and costly data to be collected in households and communities. Early warning organizations require much more rapid and timely information about the probability of food insecurity in response to droughts, floods and other environmental shocks. Modern day remote sensing can provide a proxy for possible food insecurity through the measurement of land surface phenology (LSP). The NDVI time-series provides a good proxy for LSP. Sensors such as the MODIS 250m (6.25 hectares per pixel) are ideal in time-series coverage of LSP. The idea here is to look at LSP for normal, food secure year and compare it with food insecure years where food production has been affected. The ability to make these observations for every pixel enables us to study very small areas in regions with food insecure communities (e.g., Figure 17.13). Chapter 14 provides illustration of this in highly food insecure country like Niger. Developing countries that have food insecure populations are also characterized by small, irregularly shaped fields where a variety of crops are cultivated (Paracchini et al., 2020). Ideally, LSP studied using pixel

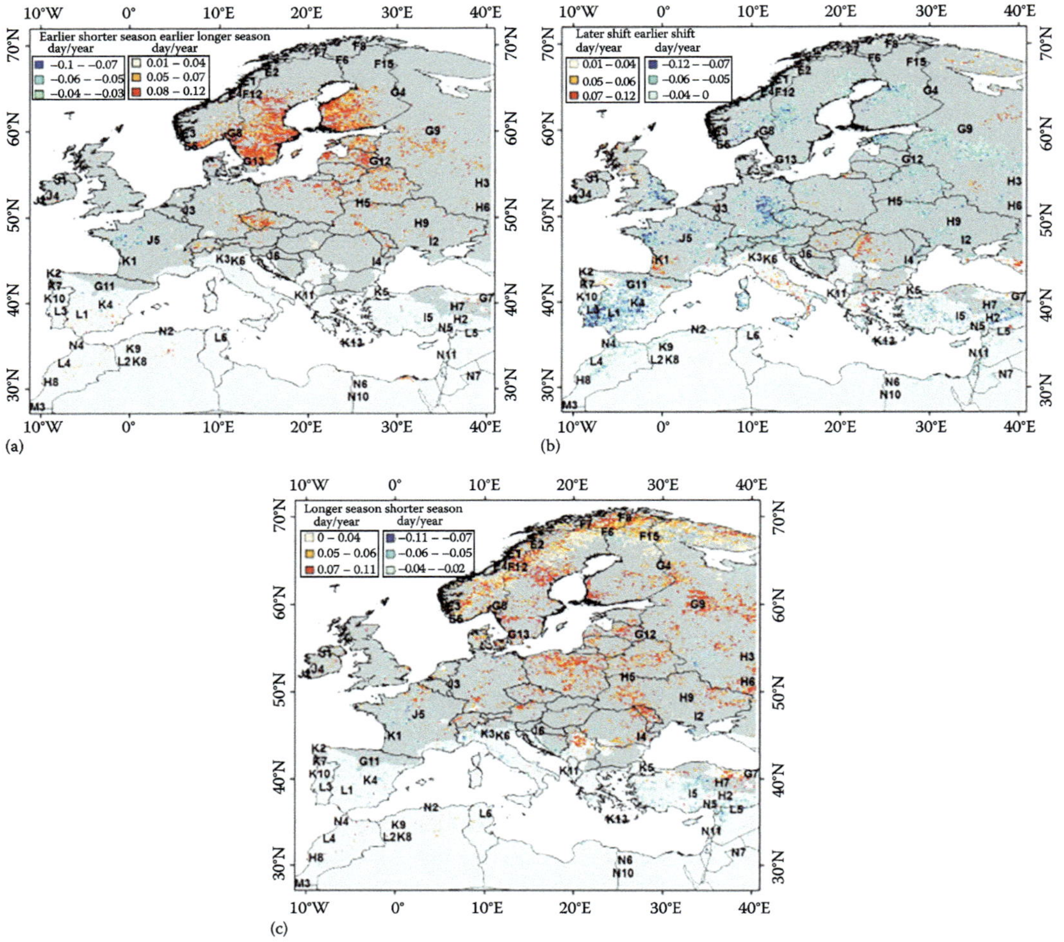

FIGURE 17.12 Spatial distribution of the phenological trend combinations using Global Inventory Modeling and Mapping Studies (GIMMS) data over the period 1982–2006 over Europe. In the background the primarily precipitation driven areas (Mediterranean, light gray) and the primarily temperature driven areas (central-northern Europe, dark gray) for phenological development are shown delineated after the Köppen-Geiger climate classification. (Source: Ivtis et al., 2012).

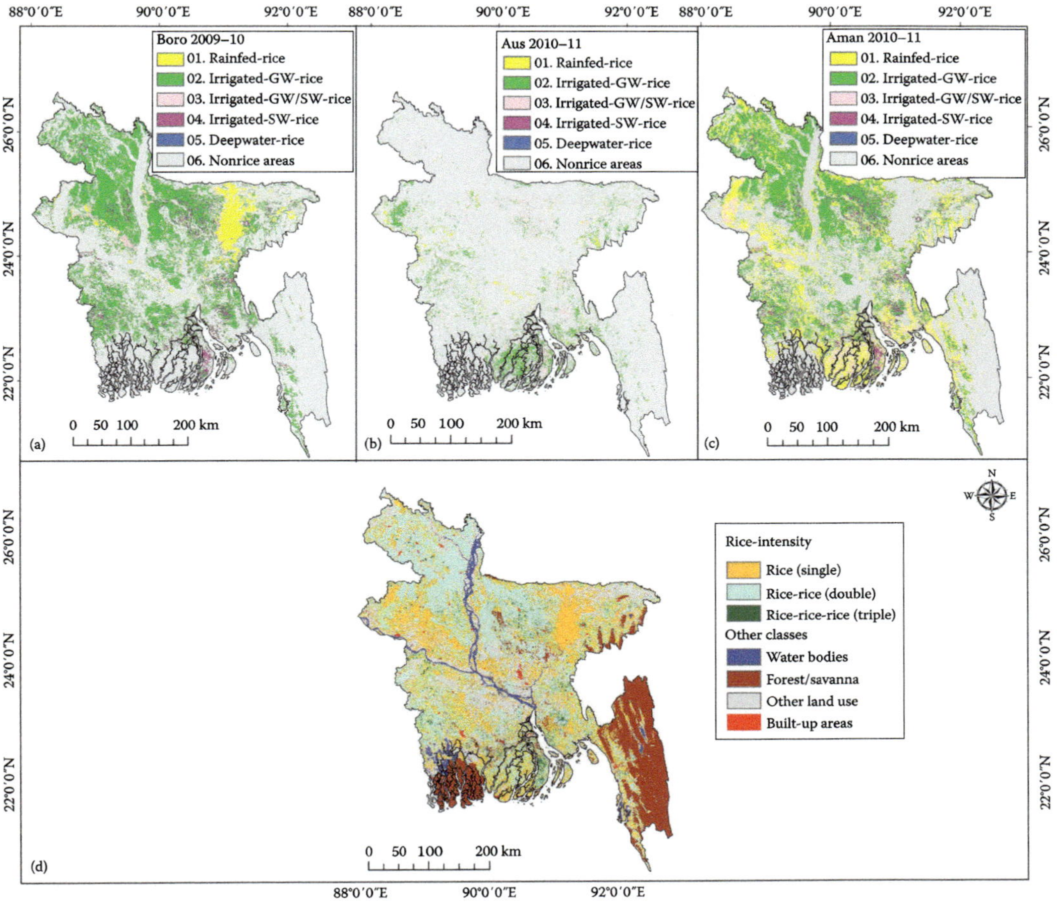

FIGURE 17.13 Spatial distribution of rice cultivation with season and irrigation source. (a) Spatial distribution of boro, (b) aus, (c) aman rice, and (d) net rice with other land use/land cover. (Source: Gumma et al., 2010.)

resolutions that can capture crop types would be ideal. Remote sensing data from high resolution sensors like Landsat have sufficient resolution for identifying crop types in small fields, but often do not have sufficient cloud free images during the growing season to use the sensor for agricultural monitoring. MODIS 250 m data is ideal in terms of temporal coverage but often the pixel will have more than one crop due to the small field size. Even then, LSP studies from a pixel with multiple crops will suffice in food security analysis. However, the change from one year to next may just be due to change in the distribution of crop types within a pixel area and not a change in agricultural productivity. The use of LSP in global food security studies can have powerful impact in understanding and monitoring food security. The ideal remote sensing platforms for LSP studies will be to acquire frequent data of good quality and sufficient resolution (e.g., 30 m or less) throughout the growing season. However, the current MODIS 250 m data that has excellent daily coverage is of great value.

Remote sensing plays a critical role in food security analysis. Kogan (2019) provides a comprehensive remote sensing derived vegetation health (VH) methods for monitoring the environment and improving global food security. The VIIRS-based VH will detect drought early, monitor accurately at 0.5 km^2 resolution, provide drought intensity, duration and predict agricultural loss two months ahead of crop harvest, assisting in global food security assessments (Kogan et al., 2019). Such early estimates will predict food security situation. Examples in this chapter prove high accuracy of vegetation health assessment, drought-triggered crop stress and the resulting grain production loss. These applications provide two to four months of advanced predictions of global food insecurity and early assessments of food assistance for the countries in need. Information on crop growth (e.g., yield and leaf area index), irrigation (e.g., irrigated areas, crop water use), and crop losses (e.g., pasts, diseases, weeds) (Karthikeyan et al., 2020) assist in food security analysis. Rembold et al (2019) develop two-step new early warning decision support system ASAP (Anomaly hot Spots of Agricultural Production): 1. A 10-day rainfall and remote sensing-derived vegetation index per-pixel anomaly, 2. A monthly analysis that takes a number of factors including Landsat and Sentinel imagery analysis on the Google Earth Engine (GEE) (Xiong et al., 2017a, 2017b).

17.15 SPECTRAL SENSING OF SOILS

Spectral sensing implies gathering near-continuous or non-continuous spectral data of targets as images or spectral behaviors from different platforms (i.e., remote and proximal sensing). Depending on the level of acquisition, José A.M. Demattê et al., the authors of Chapter 15, classified spectral sensing into: laboratory spectral sensing (LSS), field spectral sensing (FSS), ground spectral sensing (GSS), aerial spectral sensing (ASS), and space spectral sensing (SSS). Chapter 15 by José A.M. Demattê dwells deep into how soils can be evaluated by spectral sensing using platforms from ground to space. The chapter shows a new way to "see" soils and study their characteristics by spectral sensing point of view. In this way, the chapter indicates the study of soil properties of all types such as soil organic matter and carbon, pH, plant nutrients (e.g., N, Ca, Mg, K, P, and Na), soil particle size (clay, sand, and silt) content, moisture, and color. A large portion of the chapter includes summaries on how the spectral bands and indices are correlated with soil properties using linear and nonlinear modeling. Linear modeling includes statistical methods involving linear and multi-linear regressions, principal components analyses, and partial least squares regression. Nonlinear modeling methods include support vector machine (SVM), boosted regression trees (BT), and artificial neural networks (ANN). An extensive discussion of literature shows how various wavebands (absorption features) in spectral sensing help decipher soil information. What is important to note is that one to multiple wavebands or indices can be used to obtain important correlations with soil properties (typically, R^2 of 0.80 or above). For example, the most important wavebands for predicting soil water are 1,350 to 1,450 nm, 1,890 to 1,990 nm, and 2,220 to 2,280 nm. Minerals like goethite and hematite are predicted with R^2 values as high as 0.8 using a simple spectral band depth calculation in the visible wavelengths.

As also pointed out in the chapter, no single problem has plagued soil scientists more than the identification of the spatial boundaries of an individual soil body on the landscape (e.g., Figure 17.14). Chapter 15 by José A.M. Demattê demonstrates several strategies on how to use spectral data to assessment soils indicating advantages and limitations of each platform. Further demonstrates the ability of spectral sensing to assist and produce accurate (approximately 80%) pedological maps comparable to traditional approaches. The aspect of studying soils from vegetated areas involves their use and coverage, photosynthetic vegetation (PV), and non-photosynthetic vegetation (NPV), and has been discussed in this chapter. Use of soil mapping for precision farming needs detailed spatial information of soil physical and chemical properties like nutrient status and water holding capacity where field spectral sensing has been applied. Soil conservation requires a large scale understanding of relief and slope, erosion susceptibility, drainage systems, and vegetation cover which aerial and space spectral sensing can be very useful. Study of soil profiles at depths up to 2 m can employ new ground penetrating spectral sensing equipment that will help establish soils information. Gamma ray spectroscopy helps produce fast and cost effective soil maps for soil properties associated with parent material. Radar helps penetrate soils to few centimeters to study soil moisture, salinity, and other properties. Further, Chapter 15 deals with variations about study of soil properties from different platforms as well as building spectral libraries from these data. The

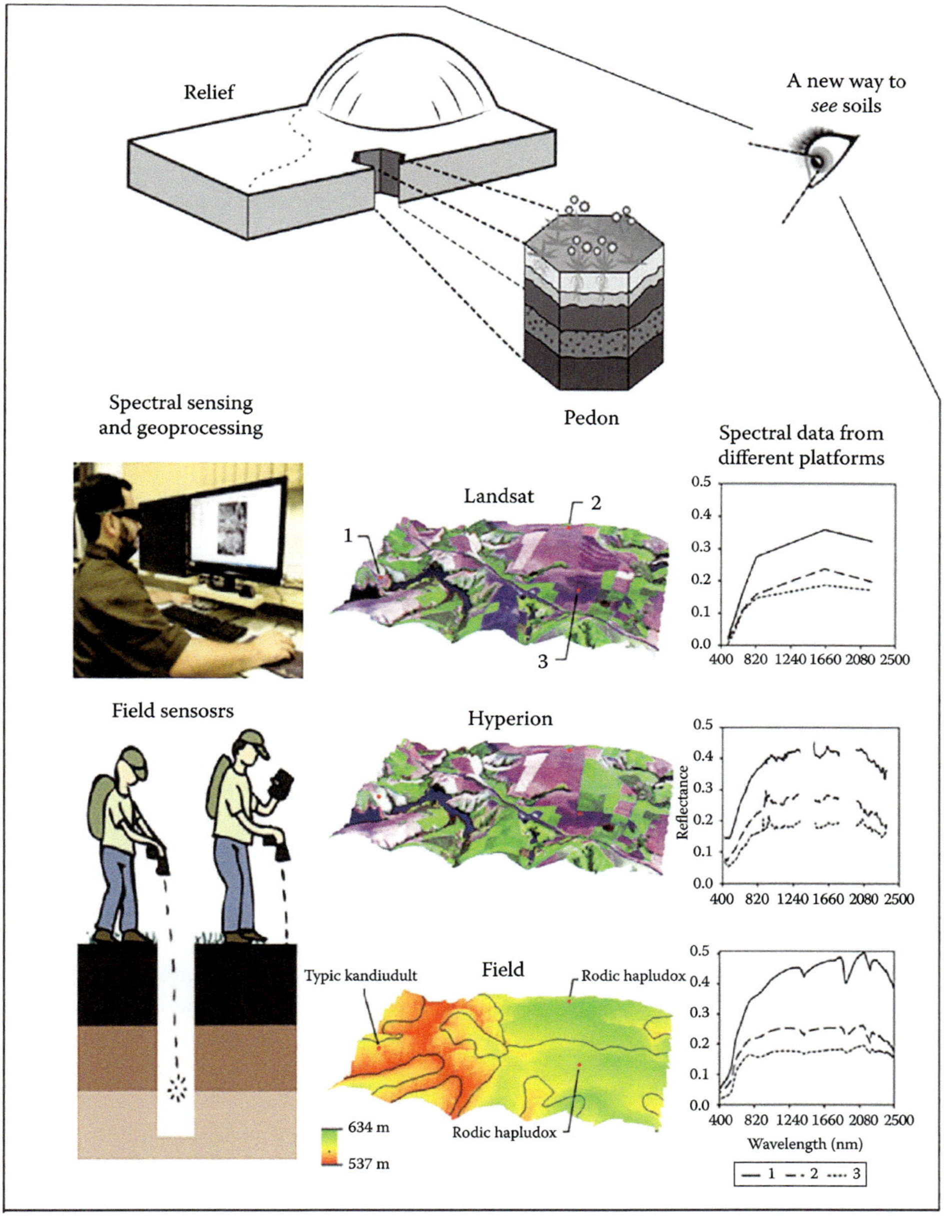

FIGURE 17.14 A new integrated perspective of soil assessment by all spectral sensing (remote and proximal) platforms. (Courtesy: Dr. Alexandre Dematte; Lead Author of Chapter 15).

chapter also indicates on how to use spectral sensing associated with pedotransference system to build pedological maps.

An exciting option for soil spectroscopy, going forward, comes from the new generation of hyperspectral sensors. These sensors include EnMAP (Environmental Mapping and Analysis Program) of Germany, HISUI (Hyperspectral Imager Suite) of Japan, HyspIRI (USA), HypXIM (France), PRISMA (PRecursore IperSpettrale della Missione Applicativa) of Italy, SHALOM (Spaceborne Hyperspectral Applicative Land and Ocean Mission) of Israel-Italy, NASA-SBG (Surface Biology and Geology), and the Sentinel-10/CHIME satellite proposed as ESA candidate mission (Copernicus Hyperspectral Imaging Mission for the Environment). The idea of spectral sensing from space (SSS) and building spectral libraries for soil applications is an exciting one, given the uniformity of such a data collection on a routine temporal basis. Soil spectral libraries (SSLs) to predict soil properties such as soil organic carbon in support of the United Nations Sustainable Developmental Goals (SDGs) is well established (Tziolas et al., 2020). Wang et al., (2022) used USDA public soil spectral library with 37,540 records of soil 350–2500 nm reflectance to assess the performance of the state-of-the-art machine learning and spectral preprocessing algorithms on SOC concentration predictions. These machine learning algorithms include Partial-Least Squares Regression (PLSR), Random Forest (RF), K-Nearest Neighbors (KNN), Ridge, Artificial Neural Networks (ANN), Convolutional Neural Networks (CNN), and Long Short-Term Memory (LSTM) (Wang et al., 2022). Of these LSTM provided the best prediction fr soil organic carbon (SOC). Overall, development of robust soil spectral libraries from various sensors such as hyperspectral and multispectral are crucial for soil characterization and study of various soil quantities.

17.16 SOIL STUDIES FROM REMOTE SENSING

Soils are foundation of agriculture and all vegetation on the planet and have any number of other uses in preserving our environments and sustaining our livelihoods. Ideal soils for agriculture are balanced in contributions from mineral components (sand: 0.05–2 mm, silt: 0.002–0.05 mm, clay: <0.002 mm), soil organic matter (SOM), air, and water (Parikh and James, 2012). Soils are also places house many living beings such as microbes, fungi, earthworms, and mites. So, study of basic soil properties, understanding their fertility, and soil degradation is of utmost importance for agriculture, carbon storage, biomass sustainability, and livelihood of plants, animals, and humans. Soil formation is a result of 5 factors (climate, parent material, time, organic matter and topography) that leads the world soils to vary widely, from location to location, even within a small area. Soil formations have occurred over thousands of years and are heavily influenced by climate as enumerated by the International Soil Reference and Information Center (ISRIC) (Yu et al., 2018). The ISRIC defines the soils of the world into these broad climate driven themes: (1) Tropical soils: strongly weathered and leached with low nutrient with only lush vegetation to replenish soils, (2) arid soils: low precipitation and high evapotranspiration leading easily soluble components like calcium carbonate and gypsum left behind after evaporation of water, (3) temperate climate soils: soil formation restricted to warmer part of the season and hence less deep, but less weathered, (4) subarctic and northern temperate soils: melting of large glaciers from last ice removed most of the soils and hence new soils have formed after the ice retreat and hence are relatively young and immature, and (5) arctic climate soils: soil formation is highly restricted and is permanently frozen (permafrost). So, a global or a local study of soils using wide array of satellite sensors is considered both cost effective and powerful. The World Reference Base for Soil Resources (WRB; Figure 17.15).

In Chapter 16, Dr. Eyal Ben-Dor and Dr. José Demattê provide a comprehensive assessment of soil studies using optical remote sensing. They show us how remote sensing is used widely, and successfully, to map soil's properties: (1) organic matter, (2) salinity, (3) degradation and change, (4) moisture, (5) carbonates, (6) contamination, (7) aggregation and roughness, (8) sealing, (9) classification and taxonomy, and (10) pedo-mapping. It is clear from their synthesis that much of success in is achieved in characterizing, and/or quantifying, and and/or mapping surface soil moisture, organic matter, texture, and color. Organic Matter has spectral activity throughout the entire VIS—NIR—SWIR region. Researchers have shown wavelengths such as 425 to 695 nm, 500–1,200 nm, 900 to 1,220 nm, 1,926–2,032 nm, 1,726 to 2,426 nm as effective is soil organic matter studies. Saline vs. non-saline soils as well as salinity types (e.g., saline, alkaline) and salinity degrees (e.g., low, moderate, high) are successfully delineated using optical remote sensing using data from VIS—NIR—SWIR spectrum. However, often the uses of multiple bands across the 400–2500 nm when used to classify and determine soil properties provide far better results. This may involve, for example, use

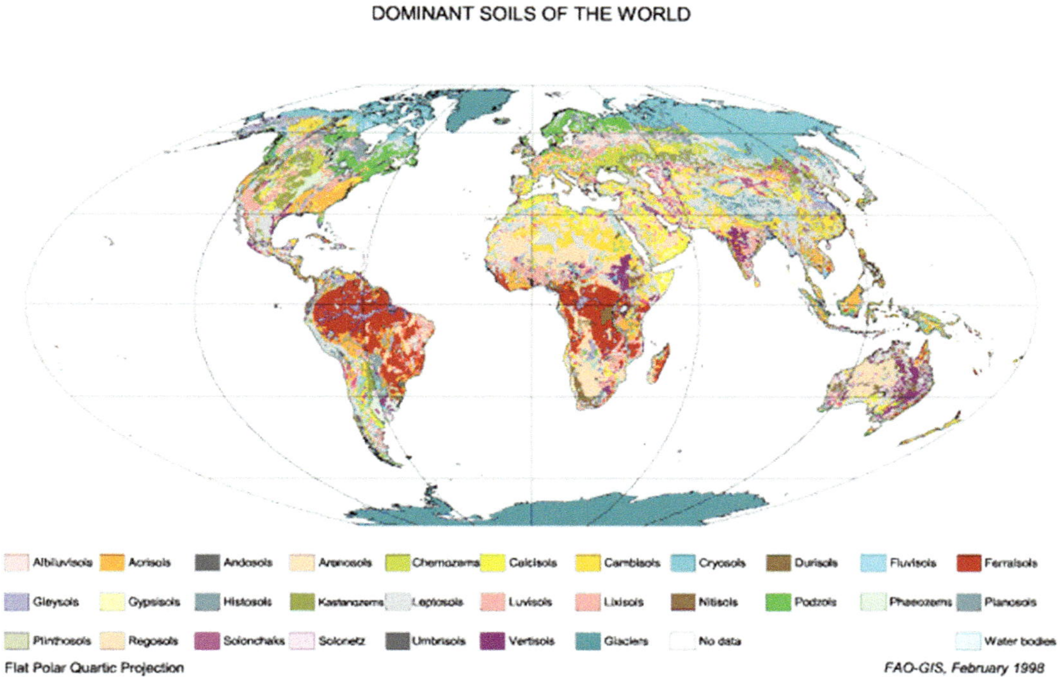

FIGURE 17.15 Dominant reference soils groups of the world.

of 10 or 20 hyperspectral narrow bands used to classify an area and determine soil characteristics like organic matter, salinity, and moisture. However, optical data can only penetrate soils to some degree, resulting in measuring, to most extent, only surface properties. In a summary of all soil applications obtained from remote sensing they concluded that the optical region is the most widely used. Nonetheless, a soil study conducted using non optical remote sensing has great value but accordingly is not part of Chapter 16. For example, thermal data is, often, used in determining salt affects and soil moisture. Microwave (both active and passive) data is widely used to quantify and map surface soil moisture as well as near-surface (<20 cm) soil moisture.

Hyperspectral data help advance the study of various soil characteristics such as soil texture, moisture, minerals, organic matter, nutrients, and salinity. Yu et al. (2018) showed that the stepwise regression models based on the satellite hyperspectral image enhanced spectral variables produced reasonable spatial distributions of the soil properties and the relative RMSE values of 68.9, 46.3, 31.4 and 45.5% for soil organic carbon, total nitrogen, total phosphorus and total potassium, respectively. Through this they demonstrate the ability of hyperspectral data to predict soil properties. The total nitrogen, available phosphorus, and available potassium of soil were predicted using 115 hyperspectral narrowband (HNB) data in a study area based on auxiliary variables after dimensionality reduction, along with stepwise linear regression (SLR), support vector machine (SVM), random forest (RF) and back-propagation neural network (BPNN) models (Song et al., 2018). Xu et al. (2021) used hyperspectral data and a variety of nonlinear machine learning techniques, such as artificial neural networks (ANN), cubist regression tree (Cubist), k-nearest neighbor (KNN), support vector machine regression (SVMR) and extreme gradient boosting (XGBoost), to compare with a partial least squares regression (PLSR) to determine the most suitable model for the prediction of the various soil Nitrogen (N) fractions. Overall, the results showed that nonlinear techniques performed better than PLSR in most cases, with a high coefficient of determination (R2) and low root mean square error (RMSE) (Yu et al., 2018). From Chapters 15 and 16, it is now clearly established that most of the soil properties can be accurately and efficiently quantified and classified using: (a) well established solid spectral library, (b) hyperspectral data from new generation of sensors on their own or in addition to multispectral data, and (c) machine learning and AI on the cloud to process large data volumes.

ACKNOWLEDGMENTS

I would like to thank the lead authors and co-authors of each of the chapters for providing their insights and edits of my chapter summaries. Any use of trade, firm, or product names is for descriptive purposes only and does not imply endorsement by the U.S. government.

REFERENCES

Alves, M.C., Carvalho, L.G., and Oliveira, M.S. 2013. Terrestrial Earth couple climate: Carbon spatial variability and uncertainty, Global and Planetary Change, Volume 111, December, Pages 9–30, ISSN 0921-8181, https://doi.org/10.1016/j.gloplacha.2013.08.009

Anastasiou, E., Fountas, S., Voulgaraki, M., Psiroukis, V., Koutsiaras, M., Kriezi, O., Lazarou, E., Vatsanidou, A., Fu, L., Di Bartolo, F., Barreiro-Hurle, J., and Gómez-Barbero, M. 2023. Precision farming technologies for crop protection: A meta-analysis, Smart Agricultural Technology, Volume 5, Page 100323, ISSN 2772-3755, https://doi.org/10.1016/j.atech.2023.100323. (www.sciencedirect.com/science/article/pii/S2772375523001521)

Aneece, I., and Thenkabail, P.S. 2018. Accuracies achieved in classifying five leading world crop types and their growth stages using optimal earth observing-1 Hyperion hyperspectral narrowbands on Google Earth Engine, Remote Sensing, Volume 10, Page 2027, https://doi.org/10.3390/rs10122027, IP-097093

Aneece, I., and Thenkabail, P.S. 2021. Classifying crop types using two generations of hyperspectral sensors (Hyperion and DESIS) with machine learning on the cloud, Remote Sensing, Volume 13, Page 4704, https://doi.org/10.3390/rs13224704, IP-128072

Aneece, I., and Thenkabail, P. 2022. New generation hyperspectral sensor (DESIS and PRISMA) performances in agriculture, Photogrammetric Engineering and Remote Sensing, Volume 88, Issue 11, Pages 715–729, https://doi.org/10.14358/PERS.22-00039R2

Badri, S., Alghazzawi, D.M., Hasan, S.H., Alfayez, F., Hasan, S.H., Rahman, M., and Bhatia, S. 2023. An efficient and secure model using adaptive optimal deep learning for task scheduling in cloud computing, Electronics, Volume 12, Page 1441, https://doi.org/10.3390/electronics12061441

Berger, K., Atzberger, C., Danner, M., D'Urso, G., Mauser, W., Vuolo, F., and Hank, T. 2018. Evaluation of the PROSAIL model capabilities for future hyperspectral model environments: A review study, Remote Sensing, Volume 10, Issue 1, Page 85, https://doi.org/10.3390/rs10010085

Carter, G.A. 1994. Ratios of leaf reflectances in narrow wavebands as indicators of plant stress, International Journal of Remote Sensing, Volume 15, Pages 697–703.

Castillejo-González, I.L., López-Granados, F., García-Ferrer, A., Peña-Barragán, J.M., Jurado-Expósito, M., Sánchez de la Orden, M., and González-Audicana, M. 2009. Object- and pixel-based analysis for mapping crops and their agro-environmental associated measures using QuickBird imagery, Computers and Electronics in Agriculture, Volume 68, Issue 2, October, Pages 207–215, ISSN 0168-1699, https://doi.org/10.1016/j.compag.2009.06.004

Curran, P.J. 1989. Remote sensing of foliar chemistry, Remote Sensing of Environment, Volume 30, Issue 3, Pages 271–278.

Dahal, D., Pastick, N.J., Boyte, S.P., Parajuli, S., Oimoen, M.J., and Megard, L.J. 2022. Multi-species inference of exotic annual and native perennial grasses in rangelands of the western United States using Harmonized Landsat and Sentinel-2 data, Remote Sensing, Volume 14, Issue 4, Page 807, https://doi.org/10.3390/rs14040807

Danner, M., Berger, K., Wocher, M., Mauser, W., and Hank, T. 2021. Efficient RTM-based training of machine learning regression algorithms to quantify biophysical & biochemical traits of agricultural crops, ISPRS Journal of Photogrammetry and Remote Sensing, Volume 173, Pages 278–296, ISSN 0924-2716, https://doi.org/10.1016/j.isprsjprs.2021.01.017. (www.sciencedirect.com/science/article/pii/S092427162100023X)

Darvishzadeh, R., Skidmore, A., Schlerf, M., and Atzberger, C. 2008. Inversion of a radiative transfer model for estimating vegetation LAI and chlorophyll in a heterogeneous grassland, Remote Sensing of Environment, Volume 112, Issue 5, 15 May, Pages 2592–2604, ISSN 0034-4257, https://doi.org/10.1016/j.rse.2007.12.003

Du, S., Liu, L., Liu, X., Guo, J., Hu, J., Wang, S., and Zhang, Y. 2019. SIFSpec: Measuring solar-induced chlorophyll fluorescence observations for remote sensing of photosynthesis, Sensors, Volume 19, Issue 13, Page 3009, https://doi.org/10.3390/s19133009

Duncan, J.M.A., Dash, J., and Atkinson, P.M. 2015. The potential of satellite-observed crop phenology to enhance yield gap assessments in smallholder landscapes: Review article, Frontiers in Environmental Science, Volume 3, https://doi.org/10.3389/fenvs.2015.00056

Ellis, E.C., Gauthier, N., Goldewijk, K.K., Bird, R.B., Boivin, N., Díaz, S., Fuller, D.Q., Gill, J.L., Kaplan, J.O., Kingston, N., Locke, H., McMichael, C.N.H., Ranco, D., Rick, T.C., Rebecca Shaw, M., Stephens, L., Svenning, J.C., and Watson, J.E.M. 2021. People have shaped most of terrestrial nature for at least 12,000 years, Proceedings of the National Academy of Sciences of the United States of America, Volume 118, Issue 17, Article e2023483118, https://doi.org/10.1073/pnas.2023483118

Enclona, E.A., Thenkabail, P.S., Celis, D., Diekmann, J. 2004. Within-field wheat yield prediction from IKONOS data: A new matrix approach, International Journal of Remote Sensing, Volume 25, Issue 2, Pages 377–388.

Foley, D.J., Thenkabail, P.S., Aneece, I.P., Teluguntla, P.G., and Oliphant, A.J. 2020. A meta-analysis of global crop water productivity of three leading world crops (wheat, corn, and rice) in the irrigated areas over three decades, *International Journal of Digital Earth*, Volume 13, Issue 8, Pages 939–975, https://doi.org/10.1080/17538947.2019.1651912

Foley, D., Thenkabail, P., Oliphant, A., Aneece, I., and Teluguntla, P. 2023. Crop water productivity from cloud-based Landsat helps assess California's water savings, Remote Sensing, Volume 15, Issue 19, Page 4894, https://doi.org/10.3390/rs15194894

Fritz, S., See, L., Bayas, J.C.L., Waldner, F., Jacques, D., Becker-Reshef, I., Whitcraft, A., Baruth, B., Bonifacio, R., Crutchfield, J., Rembold, F., Rojas, O., Schucknecht, A., Van der Velde, M., Verdin, J., Wu, B., Yan, N., Gallaun, H., Zanchi, G., Nabuurs, G.-J., Hengeveld, G., Schardt, M., and Verkerk, P.J. 2010. EU-wide maps of growing stock and above-ground biomass in forests based on remote sensing and field measurements, Forest Ecology and Management, Volume 260, Issue 3, 30 June, Pages 252–261, ISSN 0378-1127, https://doi.org/10.1016/j.foreco.2009.10.011

Gao, L., Zhang, C., Yun, W., Ji, W., Ma, J., Wang, H., Li, C., and Zhu, D. 2022. Mapping crop residue cover using adjust normalized difference residue index based on Sentinel-2 MSI data, Soil and Tillage Research, Volume 220, Page 105374, ISSN 0167-1987, https://doi.org/10.1016/j.still.2022.105374. (www.sciencedirect.com/science/article/pii/S0167198722000605)

Gitelson, A.A., and Merzlyak, M.N. 1997. Remote estimation of chlorophyll content in higher plant leaves, International Journal of Remote Sensing, Volume 18, Issue 12, Pages 2691–2697.

Gumma, M.K., Thenkabail, P.S., Maunahan, A., Islam, S., and Nelson, A. 2010. Mapping seasonal rice cropland extent and area in the high cropping intensity environment of Bangladesh using MODIS 500 m data for the year 2010, ISPRS Journal of Photogrammetry and Remote Sensing, Volume 91, May 2014, Pages 98–113, ISSN 0924-2716, https://doi.org/10.1016/j.isprsjprs.2014.02.007

Gumma, M.K., Thenkabail, P.S., Teluguntla, P., Oliphant, A., Xiong, J., Giri, C., Pyla, V., Dixit, S., and Whitbread, A.M. 2020. Agricultural cropland extent and areas of South Asia derived using Landsat satellite 30m time-series big-data using random forest machine learning algorithms on the Google Earth Engine cloud, GIScience & Remote Sensing, Volume 57, Issue 3, Pages 302–322, https://doi.org/10.1080/15481603.2019.1690780, IP-111091.

Guyot, G., and Baret, F. 1988. Utilisation de la haute resolution spectrale pour suivre l'état des couverts végétaux, Proceedings 4th International Colloquium on Spectral Signatures of Objects in Remote Sensing, ESA SP-287, Aussois, France, pp. 279–286.

Han, D., Liu, J., Zhang, R., Liu, Z., Guo, T., Jiang, H., Wang, J., Zhao, H., Ren, S., and Yang, P. 2023. Evaluation of the SAIL radiative transfer model for simulating canopy reflectance of row crop canopies, Remote Sensing, Volume 15, Page 5433. https://doi.org/10.3390/rs15235433

Hara, P., Piekutowska, M., and Niedbała, G. 2021. Selection of independent variables for crop yield prediction using artificial neural network models with remote sensing data, Land, Volume 10, Issue 6, Page 609. https://doi.org/10.3390/land10060609

Ji, Z., Pan, Y., Zhu, X., Wang, J., and Li, Q. 2021. Prediction of crop yield using phenological information extracted from remote sensing vegetation index, Sensors, Volume 21, Issue 4, Page 1406, https://doi.org/10.3390/s21041406

Jones, M.O., Allred, B.W., Naugle, D.E., Maestas, J.D., Donnelly, P., Metz, L.J., Karl, J., Smith, R., Bestelmeyer, B., Boyd, C., Kerby, J.D., and McIver, J.D. 2018. Innovation in rangeland monitoring: Annual, 30 m, plant functional typepercent cover maps for U.S. rangelands, 1984–2017, Ecosphere, Volume 9, Issue 9, Page e02430, 10.1002/ecs2.2430

Hively, W.D., Lamb, B.T., Daughtry, C.S.T., Serbin, G., Dennison, P., Kokaly, R.F., Wu, Z., and Masek, J.G. 2021. Evaluation of SWIR crop residue bands for the Landsat next mission, Remote Sensing, Volume 13, Issue 18, Page 3718, https://doi.org/10.3390/rs13183718

Horler, D.N.H., Dockray, M., and Barber, J. 1983. The red edge of plant leaf reflectance, International Journal of Remote Sensing, Volume 4, Issue 2, Pages 273–288.

Ivits, E., Cherlet, M., Tóth, G., Sommer, S., Mehl, W., Vogt, J., and Micale, F. 2012. Combining satellite derived phenology with climate data for climate change impact assessment, global and planetary change, Volumes 88–89, May 2012, Pages 85–97, ISSN 0921-8181, https://doi.org/10.1016/j.gloplacha.2012.03.010

Jones, M.O., Naugle, D.E., Twidwell, D., Uden, D.R., Maestas, J.D., and Allred, B.W. 2020. Beyond inventories: Emergence of a new era in rangeland monitoring, Rangeland Ecology & Management, Volume 73, Issue 5, Pages 577–583, ISSN 1550-7424, https://doi.org/10.1016/j.rama.2020.06.009. (www.sciencedirect.com/science/article/pii/S1550742420300750)

Jung, J., Maeda, M., Chang, A., Bhandari, M., Ashapure, A., and Landivar-Bowles, J. 2021. The potential of remote sensing and artificial intelligence as tools to improve the resilience of agriculture production systems, Current Opinion in Biotechnology, Volume 70, Pages 15–22, ISSN 0958-1669, https://doi.org/10.1016/j.copbio.2020.09.003. (www.sciencedirect.com/science/article/pii/S0958166920301257)

Karthikeyan, L., Chawla, I., and Mishra, A.K. 2020. A review of remote sensing applications in agriculture for food security: Crop growth and yield, irrigation, and crop losses, Journal of Hydrology, Volume 586, Page 124905, ISSN 0022-1694, https://doi.org/10.1016/j.jhydrol.2020.124905. (www.sciencedirect.com/science/article/pii/S0022169420303656)

Kempen, M., Elbersen, B.S., Staritsky, I., Andersen, E., and Heckelei, T. 2011. Spatial allocation of farming systems and farming indicators in Europe, Agriculture, Ecosystems & Environment, Volume 142, Issues 1–2, July, Pages 51–62, ISSN 0167-8809, https://doi.org/10.1016/j.agee.2010.08.001

Kogan, F. 2019. Remote Sensing for Food Security. Springer, Cham, p. 255, ISBN: 978-3-319-96255-9, https://doi.org/10.1007/978-3-319-96256-6

Kogan, F., Guo, W., and Yang, W. 2019. Drought and food security prediction from NOAA new generation of operational satellites, Geomatics, Natural Hazards and Risk, Volume 10, Issue 1, Pages 651–666, https://doi.org/10.1080/19475705.2018.1541257

Kruska, R.L., Reid, R.S., Thornton, P.K., Henninger, N., and Kristjanson, P.M. 2003. Mapping livestock-oriented agricultural production systems for the developing world, Agricultural Systems, Volume 77, Issue 1, July, Pages 39–63, ISSN 0308-521X, https://doi.org/10.1016/S0308-521X(02)00085-9

Li, X., Xiao, J., Fisher, J.B., and Baldocchi, D.D. 2021. ECOSTRESS estimates gross primary production with fine spatial resolution for different times of day from the International Space Station, Remote Sensing of Environment, Volume 258, Page 112360, ISSN 0034-4257, https://doi.org/10.1016/j.rse.2021.112360. (www.sciencedirect.com/science/article/pii/S003442572100078X)

Lu, B., Proctor, C., and He, Y. 2021. Investigating different versions of PROSPECT and PROSAIL for estimating spectral and biophysical properties of photosynthetic and non-photosynthetic vegetation in mixed grasslands, GIScience & Remote Sensing, Volume 58, Issue 3, Pages 354–371, https://doi.org/10.1080/15481603.2021.1877435

Manivasagam, V.S. 2024. Remote sensing of irrigation: Research trends and the direction to next-generation agriculture through data-driven scientometric analysis, Water Security, Volume 21, Page 100161, ISSN 2468-3124, https://doi.org/10.1016/j.wasec.2023.100161. (www.sciencedirect.com/science/article/pii/S2468312423000299)

Mariotto, I., Thenkabail, P.S., Huete, H., Slonecker, T., and Platonov, A. 2013. Hyperspectral versus multispectral crop-biophysical modeling and type discrimination for the HyspIRI mission, Remote Sensing of Environment, Volume 139, Pages 291–305, IP-049224.

Marshall, M., and Thenkabail, P. 2016. Developing in situ non-destructive estimates of crop biomass to address issues of scale in remote sensing. *Remote Sensing,* Volume 2015, Issue 7, Pages 808–835, https://doi.org/10.3390/rs70100808

Marshall, M.T., and Thenkabail, P.S. 2014. Biomass modeling of four leading world crops using hyperspectral narrowbands in support of HyspIRI mission, Photogrammetric Engineering and Remote Sensing, Volume 80, Issue 4, Pages 757–772, IP-052043.

Marshall, M.T., and Thenkabail, P.S. 2015a. Developing in situ non-destructive estimates of crop biomass to address issues of scale in remote sensing, Remote Sensing, Volume 7, Issue 1, Pages 808–835, https://doi.org/10.3390/rs70100808, IP-060652.

Marshall, M.T., and Thenkabail, P.S. 2015b. Advantage of hyperspectral EO-1 Hyperion over multispectral IKONOS, GeoEye-1, WorldView-2, Landsat ETM+, and MODIS vegetation indices in crop biomass estimation, International Society of Photogrammetry and Remote Sensing (ISPRS) Journal of Photogrammetry and Remote Sensing (ISPRS P&RS), Volume 108, Pages 205–218, https://doi.org/10.1016/j.isprsjprs.2015.08.001, IP-060745.

Massey, R., Sankey, T.T., Congalton, R.G., Yadav, K., Thenkabail, P.S., Ozdogan, M., and Sánchez Meador, A.J. 2017. MODIS phenology-derived, multi-year distribution of conterminous U.S. crop types, Remote Sensing of Environment, Volume 198, Pages 490–503, ISSN 0034-4257, https://doi.org/10.1016/j.rse.2017.06.033, IP-081309.

Massey, R., Sankey, T.T., Yadav, K., Congalton, R.G., and Tilton, J.C. 2018. Integrating cloud-based workflows in continental-scale cropland extent classification. Remote Sensing of Environment, Volume 219, pp. 162–179. https://doi.org/10.1016/j.rse.2018.10.013

Matongera, T.N., Mutanga, O., Sibanda, M., and Odindi, J. 2021. Estimating and monitoring land surface phenology in Rangelands: A review of progress and challenges, Remote Sensing, Volume 13, Issue 11, Page 2060, https://doi.org/10.3390/rs13112060

Mishra, M., Avtar, R., Prathiba, A.P., Mishra, P.K., Tiwari, A., Sharma, S.K., Singh, C.H., Yadav, B.C., and Jain, K. 2023. Uncrewed aerial systems in water resource management and monitoring: A review of sensors, applications, software, and issues, *Advances in Civil Engineering*, Volume 2023, Article ID 3544724, Page 28, https://doi.org/10.1155/2023/3544724

Mohammed, G.H., Colombo, R., Middleton, E.M., Rascher, U., van der Tol, C., Nedbal, L., Goulas, Y., Pérez-Priego, O., Damm, A., Meroni, M., Joiner, J., Cogliati, S., Verhoef, W., Malenovský, Z., Gastellu-Etchegorry, J.-P., Miller, J.R., Guanter, L., Moreno, J., Moya, I., Berry, J.A., Frankenberg, C., and Zarco-Tejada, P.J. 2019. Remote sensing of solar-induced chlorophyll fluorescence (SIF) in vegetation: 50 years of progress, Remote Sensing of Environment, Volume 231, Page 111177, ISSN 0034-4257, https://doi.org/10.1016/j.rse.2019.04.030. (www.sciencedirect.com/science/article/pii/S0034425719301816)

Monteith, J.L. 1972. Solar radiation and productivity in tropical ecosystems, Journal of Applied Ecology, Volume 9, Issue 3, Page 747.

Oliphant, A., Thenkabail, P.S., Teluguntla, P., Xiong, J., Gumma, M.K., Congalton, R., and Yadav, K. 2019. Mapping cropland extent of Southeast and Northeast Asia using multi-year time-series Landsat 30m data using random forest classifier on Google Earth Engine, International Journal of Applied Earth Observation and Geoinformation, Volume 81, Pages 110–124, https://doi.org/10.1016/j.jag.2018.11.014

Pacheco, A., and McNairn, H. 2010. Evaluating multispectral remote sensing and spectral unmixing analysis for crop residue mapping, Remote Sensing of Environment, Volume 114, Issue 10, 15 October, Pages 2219–2228, ISSN 0034-4257, https://doi.org/10.1016/j.rse.2010.04.024

Paracchini, M.L., Justes, E., Wezel, A., Zingari, P.C., Kahane, R., Madsen, S., Scopel, E., Héraut, A., Bhérer-Breton, P., Buckley, R., Colbert, E., Kapalla, D., Sorge, M., Adu Asieduwaa, G., Bezner Kerr, R., Maes, O., and Negre, T. 2020. Agroecological practices supporting food production and reducing food insecurity in developing countries: A study on scientific literature in 17 countries, EUR 30329 EN, Publications Office of the European Union, Ispra, 2020, ISBN 978-92-76-21077-1, https://doi.org/10.2760/82475, JRC121570

Parikh, S.J., and James, B.R. 2012. Soil: The foundation of agriculture, Nature Education Knowledge, Volume 3, Issue 10, Page 2.

Pastick, N.J., Dahal, D., Wylie, B.K., Parajuli, S., Boyte, S.P., and Wu, Z. 2020. Characterizing land surface phenology and exotic annual grasses in dryland ecosystems using Landsat and Sentinel-2 data in harmony, Remote Sensing, Volume 12, Issue 4, Page 725, https://doi.org/10.3390/rs12040725

Phalke, A.R., and Özdoğan, M. 2018. Large area cropland extent mapping with Landsat data and a generalized classifier, Remote Sensing of Environment, Volume 219, Pages 180–195.

Phalke, A.R., Özdoğan, M., Thenkabail, P.S., Erickson, T., Gorelick, N., Yadav, K., and Congalton, R.G. 2020. Mapping croplands of Europe, Middle East, Russia, and Central Asia using Landsat, random forest, and Google Earth Engine, ISPRS Journal of Photogrammetry and Remote Sensing, Volume 167, Pages 104–122, ISSN 0924-2716, https://doi.org/10.1016/j.isprsjprs.2020.06.022, IP-116983. (www.sciencedirect.com/science/article/pii/S0924271620301805)

Pickett-Heaps, C.A., Canadell, J.G., Briggs, P.R., Gobron, N., Haverd, V., Paget, M.J., Pinty, B., and Raupach, M.R. 2014. Evaluation of six satellite-derived Fraction of Absorbed Photosynthetic Active Radiation (FAPAR) products across the Australian continent, Remote Sensing of Environment, Volume 140, January, Pages 241–256, ISSN 0034-4257, https://doi.org/10.1016/j.rse.2013.08.037

Raj, E.F.I., Appadurai, M., and Athiappan, K. 2021. Precision farming in modern agriculture. In: Choudhury, A., Biswas, A., Singh, T.P., and Ghosh, S.K. (eds.), Smart Agriculture Automation Using Advanced Technologies: Transactions on Computer Systems and Networks. Springer, Singapore. https://doi.org/10.1007/978-981-16-6124-2_4

Rembold, F., Meroni, M., Urbano, F., Csak, G., Kerdiles, H., Perez-Hoyos, A., Lemoine, G., Leo, O., and Negre, T. 2019. ASAP: A new global early warning system to detect anomaly hot spots of agricultural production for food security analysis, Agricultural Systems, Volume 168, Pages 247–257, ISSN 0308-521X, https://doi.org/10.1016/j.agsy.2018.07.002. (www.sciencedirect.com/science/article/pii/S0308521X17309095)

Retallack, A., Finlayson, G., Ostendorf, B., Clarke, K., and Lewis, K. 2023. Remote sensing for monitoring rangeland condition: Current status and development of methods, Environmental and Sustainability Indicators, Volume 19, Page 100285, ISSN 2665-9727, https://doi.org/10.1016/j.indic.2023.100285. (www.sciencedirect.com/science/article/pii/S2665972723000624)

Rigge, M., Homer, C., Cleeves, L., Meyer, D.K., Bunde, B., Shi, H., Xian, G., Schell, S., and Bobo, M. 2020. Quantifying Western U.S. rangelands as fractional components with multi-resolution remote sensing and in situ data, Remote Sensing, Volume 12, Issue 3, Page 412. https://doi.org/10.3390/rs12030412

Robertson, M., Isbister, B., Maling, I., Oliver, Y., Wong, M., Adams, M., Bowden, B., and Tozer, P. 2007. Opportunities and constraints for managing within-field spatial variability in Western Australian grain production, Field Crops Research, Volume 104, Issues 1–3, October–December, Pages 60–67, ISSN 0378-4290, https://doi.org/10.1016/j.fcr.2006.12.013

Running, S.W., Heinsch, F.A., Zhao, M., Reeves, M., Hashimoto, H., and Nemani, R.R. 2004. A continuous satellite-derived measure of global terrestrial primary production, BioScience, Volume 54, Issue 6, Pages 547–560.

Ryu, Y., Berry, J.A., and Baldocchi, D.D. 2019. What is global photosynthesis? History, uncertainties and opportunities, Remote Sensing of Environment, Volume 223, Pages 95–114, ISSN 0034-4257, https://doi.org/10.1016/j.rse.2019.01.016. (www.sciencedirect.com/science/article/pii/S0034425719300161)

Schlerf, M., Atzberger, C., and Hill, J. 2005. Remote sensing of forest biophysical variables using HyMap imaging spectrometer data, Remote Sensing of Environment, 95(2), 177–194.

Song, Y.-Q., Zhao, X., Su, H.-Y., Li, B., Hu, Y.-M., and Cui, X.-S. 2018. Predicting spatial variations in soil nutrients with hyperspectral remote sensing at regional scale, Sensors, Volume 18, Issue 9, Page 3086, https://doi.org/10.3390/s18093086

Sun, Q., Jiao, Q., Chen, X., Xing, H., Huang, W., and Zhang, B. 2023. Machine learning algorithms for the retrieval of canopy chlorophyll content and leaf area index of crops using the PROSAIL-D model with the adjusted average leaf angle, Remote Sensing, Volume 15, Page 2264, https://doi.org/10.3390/rs15092264

Teluguntla, P., Thenkabail, P.S., Oliphant, A., Xiong, J., and Gumma, M.K. 2018. A 30m Landsat-derived cropland extent product of Australia and China using random forest machine learning algorithm on Google Earth Engine cloud computing platform, ISPRS Journal of Photogrammetry and Remote Sensing, Volume 144, Pages 325–340, ISSN 0924-2716, https://doi.org/10.1016/j.isprsjprs.2018.07.017

Teluguntla, P., Thenkabail, P.S., Xiong, J., Gumma, M.K., Congalton, R.G., Oliphant, A., Poehnelt, J., Yadav, K., Rao, M., and Massey, R. 2017. Spectral matching techniques (SMTs) and automated cropland classification algorithms (ACCAs) for mapping croplands of Australia using MODIS 250-m time-series (2000–2015) data, International Journal of Digital Earth, Volume 10, Issue 9, https://doi.org/10.1080/17538947.2016.1267269, IP-074181

Thenkabail, P.S. 2003. Biophysical and yield information for precision farming from near-real time and historical Landsat TM images, International Journal of Remote Sensing, Volume 24, Issue 14, Pages 2879–2904.

Thenkabail, P.S., Aneece, I., Teluguntla, P., and Oliphant, A. 2021a. Hyperspectral narrowband data propel gigantic leap in the earth remote sensing, highlight article, Photogrammetric Engineering and Remote Sensing, Volume 87, Issue 7, Pages 461–467, https://doi.org/10.14358/PERS.87.7.461, IP-127022. (www.asprs.org/a/publications/pers/2021journals/07-21_July_Flipping_Public.pdf).

Thenkabail, P.S., Enclona, E.A., Ashton, M.S., Legg, C., and De Dieu, M.J. 2004a. Hyperion, IKONOS, ALI, and ETM+ sensors in the study of African rainforests, Remote Sensing of Environment, Volume 90, Issue 1, Pages 23–43.

Thenkabail, P.S., Enclona, E.A., Ashton, M.S., Legg, C., and Jean De Dieu, M. 2004b. Hyperion, IKONOS, ALI, and ETM+ sensors in the study of African rainforests, Remote Sensing of Environment, Volume 90, Pages 23–43.

Thenkabail, P.S., Enclona, E.A., Ashton, M.S., and Van Der Meer, V. 2004c. Accuracy assessments of hyperspectral waveband performance for vegetation analysis applications, Remote Sensing of Environment, Volume 91, Issues 2–3, Pages 354–376.

Thenkabail, P.S., Gumma, M.K., Teluguntla, P., and Mohammed, I.A. 2014. Hyperspectral remote sensing of vegetation and agricultural crops: Highlight article, Photogrammetric Engineering and Remote Sensing, Volume 80, Issue 4, Pages 697–709.

Thenkabail, P.S., Hall, J., Lin, T., Ashton, M.S., Harris, D., Enclona, E.A. 2003. Detecting floristic structure and pattern across topographic and moisture gradients in a mixed species Central African forest using IKONOS and Landsat-7 ETM+ images, International Journal of Applied Earth Observation and Geoinformation, Volume 4, Pages 255–270.

Thenkabail, P.S., Knox, J.W., Ozdogan, M., Gumma, M.K., Congalton, R.G., Wu, Z., Milesi, C., Finkral, A., Marshall, M., and Mariotto, I. 2012. Assessing future risks to agricultural productivity, water resources and food security: How can remote sensing help? Photogrammetric Engineering and Remote Sensing, Volume 78, Issue 8, Pages 773–782.

Thenkabail, P.S., Lyon, G.J., and Huete, A. 2011. Book Entitled: "Hyperspectral Remote Sensing of Vegetation." CRC Press-Taylor and Francis group, Boca Raton, London, New York, pp. 781 (80+ pages in color). (www.crcpress.com/product/isbn/9781439845370)

Thenkabail, P.S., Mariotto, I., Gumma, M.K., Middleton, E.M., Landis, D.R., and Huemmrich, F.K. 2013. Selection of hyperspectral narrowbands (HNBs) and composition of hyperspectral twoband vegetation indices (HVIs) for biophysical characterization and discrimination of crop types using field reflectance and Hyperion/EO-1 data. IEEE Journal of Selected Topics in Applied Earth Observations and Remote Sensing, Volume 6, Issue 2, April, Pages 427–439, https://doi.org/10.1109/JSTARS.2013.2252601

Thenkabail, P.S., Schull, M., and Turral, H. 2005. Ganges and indus river basin land use/land cover (LULC) and irrigated area mapping using continuous streams of MODIS data, Remote Sensing of Environment, Volume 95, Issue 3, 15 April, Pages 317–341, ISSN 0034-4257, https://doi.org/10.1016/j.rse.2004.12.018

Thenkabail, P.S., Smith, R.B., and De-Pauw, E. 2000. Hyperspectral vegetation indices for determining agricultural crop characteristics, Remote Sensing of Environment, Volume 71, Pages 158–182.

Thenkabail, P.S., Stucky, N., Griscom, B.W., Ashton, M.S., Diels, J., Van Der Meer, B., and Enclona, E. 2004d. Biomass estimations and carbon stock calculations in the oil palm plantations of African derived savannas using IKONOS data, International Journal of Remote Sensing, Volume 25, Issue 23, Pages 5447–5472.

Thenkabail, P.S., Teluguntla, P.G., Xiong, J., Oliphant, A., Congalton, R.G., Ozdogan, M., Gumma, M.K., Tilton, J.C., Giri, C., Milesi, C., Phalke, A., Massey, R., Yadav, K., Sankey, T., Zhong, Y., Aneece, I., and Foley, D. 2021b. Global cropland-extent product at 30-m resolution (GCEP30) derived from Landsat satellite time-series data for the year 2015 using multiple machine-learning algorithms on Google Earth Engine cloud: U.S, Geological Survey Professional Paper 1868, p. 63, https://doi.org/10.3133/pp1868

Tian, Y.C., Yao, X., Yang, J., Cao, W.X., Hannaway, D.B., and Zhu, Y. 2011. Assessing newly developed and published vegetation indices for estimating rice leaf nitrogen concentration with ground- and space-based hyperspectral reflectance, Field Crops Research, Volume 120, Issue 2, 31 January, Pages 299–310, ISSN 0378-4290, https://doi.org/10.1016/j.fcr.2010.11.002

Tucker, C.J. 1979. Red and photographic infrared linear combinations for monitoring vegetation, Remote Sensing of Environment, Volume 8, Pages 127–150.

Tziolas, I., Tsakiridis, N., Ogen, Y., Kalopesa, E., Ben-Dor, E., Theocharis, J., and Zalidis, G. 2020. An integrated methodology using open soil spectral libraries and earth observation data for soil organic carbon estimations in support of soil-related SDGs, Remote Sensing of Environment, Volume 244, 2020, Page 111793, ISSN 0034-4257, https://doi.org/10.1016/j.rse.2020.111793. (www.sciencedirect.com/science/article/pii/S0034425720301632)

Wang, S., Guan, K., Zhang, C., Lee, D., Margenot, A.J., Ge, Y., Peng, J., Zhou, W., Zhou, Q., and Huang, Y. 2022. Using soil library hyperspectral reflectance and machine learning to predict soil organic carbon: Assessing potential of airborne and spaceborne optical soil sensing, Remote Sensing of Environment, Volume 271, Page 112914, ISSN 0034-4257, https://doi.org/10.1016/j.rse.2022.112914. (www.sciencedirect.com/science/article/pii/S0034425722000281)

Wang, J., Xiao, X., Bajgain, R., Starks, P., Steiner, J., Doughty, R.B., and Chang, Q. 2019. Estimating leaf area index and aboveground biomass of grazing pastures using Sentinel-1, Sentinel-2 and Landsat images, ISPRS Journal of Photogrammetry and Remote Sensing, Volume 154, Pages 189–201, ISSN 0924-2716, https://doi.org/10.1016/j.isprsjprs.2019.06.007. (www.sciencedirect.com/science/article/pii/S0924271619301480)

Weiss, M., Jacob, J., and Duveiller, G. 2020. Remote sensing for agricultural applications: A meta-review, Remote Sensing of Environment, Volume 236, Page 111402, ISSN 0034-4257, https://doi.org/10.1016/j.rse.2019.111402. (www.sciencedirect.com/science/article/pii/S0034425719304213)

Wessels, K.J., Mathieu, R., Erasmus, B.F.N., Asner, G.P., Smit, I.P.J., van Aardt, J.A.N., Main, R., Fisher, J., Marais, W., Kennedy-Bowdoin, T., Knapp, D.E., Emerson, R., and Jacobson, J. 2011. Impact of communal land use and conservation on woody vegetation structure in the Lowveld savannas of South Africa, Forest Ecology and Management, Volume 261, Issue 1, 1 January, Pages 19–29, ISSN 0378-1127, https://doi.org/10.1016/j.foreco.2010.09.012

Williams, M., Richardson, A.D., Reichstein, M., Stoy, P.C., Peylin, P., Verbeeck, H., Carvalhais, N., Jung, M., Hollinger, D.Y., Kattge, J., Leuning, R., Luo, Y., Tomelleri, E., Trudinger, C.M., and Wang, Y.-P. 2009. Improving land surface models with FLUXNET data, Biogeosciences, Volume 6, Pages 1341–1359, https://doi.org/10.5194/bg-6-1341-2009

Wolanin, A., Camps-Valls, G., Gómez-Chova, L., Mateo-García, G., van der Tol, C., Zhang, Y., and Guanter, L. 2019. Estimating crop primary productivity with Sentinel-2 and Landsat 8 using machine learning methods trained with radiative transfer simulations, Remote Sensing of Environment, Volume 225, Pages 441–457, ISSN 0034-4257, https://doi.org/10.1016/j.rse.2019.03.002. (www.sciencedirect.com/science/article/pii/S0034425719300938)

Wolf, J., Chen, M., and Asrar, G.R. 2021. Global rangeland primary production and its consumption by livestock in 2000–2010, Remote Sensing, Volume 13, Page 3430, https://doi.org/10.3390/rs13173430

Xiang, X., Du, J., Jacinthe, P., Zhao, B., Zhou, H., Liu, H., and Song, K. 2022. Integration of tillage indices and textural features of Sentinel-2A multispectral images for maize residue cover estimation, Soil and Tillage Research, Volume 221, Page 105405, ISSN 0167-1987, https://doi.org/10.1016/j.still.2022.105405. (www.sciencedirect.com/science/article/pii/S0167198722000915)

Xie, Z., Phinn, S.R., Game, E.T., Pannell, D.J., Hobbs, R.J., Briggs, P.R., and McDonald-Madden, E. 2019. Using Landsat observations (1988–2017) and Google Earth Engine to detect vegetation cover changes in rangelands: A first step towards identifying degraded lands for conservation, Remote Sensing of Environment, Volume 232, Page 111317, ISSN 0034-4257, https://doi.org/10.1016/j.rse.2019.111317. (www.sciencedirect.com/science/article/pii/S0034425719303360)

Xiong, J., Thenkabail, P.S., Gumma, M., Teluguntla, P., Poehnelt, J., Congalton, R., and Yadav, K. 2017a. Automated cropland mapping of continental Africa using Google Earth Engine cloud computing. ISPRS Journal of Photogrammetry and Remote Sensing, Volume 126, Pages 225–244, https://doi.org/10.1016/j.isprsjprs.2017.01.019

Xiong, J., Thenkabail, P.S., Tilton, J.C., Gumma, M.K., Teluguntla, P., Oliphant, A., Congalton, R.G., Yadav, K., and Gorelick, N. 2017b. Nominal 30m cropland extent map of continental Africa by integrating pixel-based and object-based algorithms using Sentinel-2 and Landsat-8 data on Google Earth Engine, Remote Sensing, Volume 9, Issue 10, Page 1065, https://doi.org/10.3390/rs9101065, IP-088538. (www.mdpi.com/2072-4292/9/10/1065).

Xu, S., Wang, M., Shi, X., Yu, Q., Zhang, Z. 2021. Integrating hyperspectral imaging with machine learning techniques for the high-resolution mapping of soil nitrogen fractions in soil profiles, Science of The Total Environment, 754, 2021, 142135, ISSN 0048-9697, https://doi.org/10.1016/j.scitotenv.2020.142135.

Yang, P., van der Tol, C., Campbell, P.K.E., and Middleton, E.M. 2021. Unraveling the physical and physiological basis for the solar-induced chlorophyll fluorescence and photosynthesis relationship using continuous leaf and canopy measurements of a corn crop, Biogeosciences, Volume 18, Pages 441–465, https://doi.org/10.5194/bg-18-441-2021

You, L., Gilliams, S., Mücher, S., Tetrault, R., Moorthy, I., and McCallum, I. 2019. A comparison of global agricultural monitoring systems and current gaps, Agricultural Systems, Volume 168, Pages 258–272, ISSN 0308-521X, https://doi.org/10.1016/j.agsy.2018.05.010. (www.sciencedirect.com/science/article/pii/S0308521X17312027)

Yu, H., Kong, B., Wang, G., Du, R., and Qie, G. 2018. Prediction of soil properties using a hyperspectral remote sensing method, Archives of Agronomy and Soil Science, Volume 64, Issue 4, Pages 546–559, https://doi.org/10.1080/03650340.2017.1359416

Zhao, F., Ma, W., Zhao, J., Guo, Y., Tariq, M., and Li, J. 2024. Global retrieval of the spectrum of terrestrial chlorophyll fluorescence: First results with TROPOMI, Remote Sensing of Environment, Volume 300, Page 113903, ISSN 0034-4257, https://doi.org/10.1016/j.rse.2023.113903. (www.sciencedirect.com/science/article/pii/S0034425723004546)

Zheng, C., Abd-Elrahman, A., and Whitaker, V. 2021. Remote sensing and machine learning in crop phenotyping and management, with an emphasis on applications in strawberry farming, Remote Sensing, Volume 13, Issue 3, Page 531. https://doi.org/10.3390/rs13030531

Zhou, B., Okin, G.S., and Zhang, J. 2020. Leveraging Google Earth Engine (GEE) and machine learning algorithms to incorporate in situ measurement from different times for rangelands monitoring, Remote Sensing of Environment, Volume 236, Page 111521, ISSN 0034-4257, https://doi.org/10.1016/j.rse.2019.111521. (www.sciencedirect.com/science/article/pii/S0034425719305401)

Zhu, N., Liu, C., Laine, A.F., and Guo, J. 2020. Understanding and modeling climate impacts on photosynthetic dynamics with FLUXNET data and neural networks, Energies, Volume 13, Page 1322. https://doi.org/10.3390/en13061322

Index

Note: Page numbers in *italics* indicate figures, and page numbers in **bold** indicate tables on the corresponding page.

A

B

C

D

E

F

H

M

N

O